Methods in Enzymology

Volume 75

CUMULATIVE SUBJECT INDEX

Volumes XXXI, XXXII, XXXIV–LX

METHODS IN ENZYMOLOGY

EDITORS-IN-CHIEF

Sidney P. Colowick Nathan O. Kaplan

Methods in Enzymology

Volume 75

Cumulative Subject Index

Volumes XXXI, XXXII, XXXIV–LX

EDITED BY

Martha G. Dennis, Ph.D.

COMPUTER SCIENTIST
LA JOLLA, CALIFORNIA

Edward A. Dennis, Ph.D.

DEPARTMENT OF CHEMISTRY
UNIVERSITY OF CALIFORNIA AT SAN DIEGO
LA JOLLA, CALIFORNIA

1982

ACADEMIC PRESS

A Subsidiary of Harcourt Brace Jovanovich, Publishers

New York London
Paris San Diego San Francisco São Paulo Sydney Tokyo Toronto

ACADEMIC PRESS, INC.
111 Fifth Avenue, New York, New York 10003

United Kingdom Edition published by
ACADEMIC PRESS, INC. (LONDON) LTD.
24/28 Oval Road, London NW1 7DX

LIBRARY OF CONGRESS CATALOG CARD NUMBER: 54-9110

ISBN: 0-12-181975-2

PRINTED IN THE UNITED STATES OF AMERICA

82 83 84 85 9 8 7 6 5 4 3 2 1

Preface

This is the second cumulative index we have prepared for "Methods in Enzymology." It employs computer programs and procedures developed for the first cumulative index (Volume XXXIII). These were developed in response to a challenging question: Could current computer technology be employed to generate a cumulative index for Volumes I through XXX of "Methods in Enzymology"? For this series, each of the indexes had been prepared without any need for compatibility with the index of any other volume. To prepare a cumulative index manually from the approximately 100,000 lines involved would have been a monumental undertaking.

This cumulative index was prepared using a Burroughs 7800 computer. It was printed by photo-offset methods from pages composed automatically from the computer-generated index. The indexes of the individual volumes were used as the source for entries.

It was felt that if the nomenclature and usage in the original articles were retained wherever possible, this index would be most useful to the reader in locating information. Thus, rather than using a standard format and prescribed nomenclature, as is usually done, we used the entries in their original form. Similar entries were merged into one entry only when they differed at most in capitalization, typeface, punctuation, or number (singular or plural). As a result, some inconsistency of nomenclature occurs. Entries with similar meaning but different phraseology were generally not modified or combined. It is an old adage in computer science that the computer can do only that for which humans can define rules. Rules for dealing with the semantics of language are very difficult to define and this led us to be cautious in computer manipulation of different phraseologies.

There are two well-known pitfalls in computerized indexing: information loss due to oversimplification and information loss due to the presence of too much information. We have attempted to reach a balance between these two extremes by using the computer for simplification and combination wherever possible, but only when it was clear that no information would be lost in the process. In order to use the resulting index most effectively, it is suggested that the reader become familiar with the "Notes on the Use of the Index" (see p. vii) in which the occurrence of entries for a given subject under a variety of names and phraseologies and the order of entries in the index are discussed.

We wish to thank Dr. Nathan O. Kaplan and Dr. Sidney P. Colowick, Editors-in-Chief of "Methods in Enzymology," for their help and encouragement in the preparation of the index. Also, we wish to thank the staff of Academic Press for their assistance and confidence in the success of this project. We were aided appreciably in this undertaking by Mr. Raymond A. Deems who carried out the complicated and laborious computer processing required and Mrs. Dolores Wright who provided valuable editorial and proofreading assistance. We also wish to thank Mr. Henry Fischer and his excellent staff of the University of California at San Diego Computer Center for their cooperation and help.

MARTHA G. DENNIS
EDWARD A. DENNIS

Notes on the Use of the Index

Multiple Entries

Because a given enzyme or compound may have several different names, including both trivial and systematic ones, all possible names should be checked when looking for a particular entry. Cross-references are often included to aid in this process, but, in general, only if they were supplied in the index of one of the individual volumes. Although different phraseologies were used to express a single idea, entries of similar content in different volumes were not necessarily combined by the computer into a single entry, so it is essential that adjacent entries and differing phraseologies for a given subject be checked.

Order of Entries

Although most entries are listed in simple alphabetical order, the index contains a number of complex entries for which this is not adequate. An explanation of the method used in these cases is presented below.

The index is comprised of main entries and subentries. Subentries appear indented on separate lines below the main entry, or on the same line separated from the main entry by a comma. An entry should be considered to consist of the following four components:

1. the basic subject
2. subscripts and superscripts applying to the basic subject
3. phrases modifying the basic subject
4. prepositions (and conjunctions) preceding the basic subject

For example, in the entry

with cytochrome c_1

the basic subject is "cytochrome c," "with" is a preceding preposition, and "$_1$" is a subscript. In the entry

D-Glucose 6-phosphate

the basic subject is "Glucose phosphate," and "D" and "6" are both modifying phrases. There is always a basic subject; the other components may or may not appear.

The actual order in which entries appear may be determined by considering each of these components in turn, if present, until an order is obtained. For example, with the two entries

Glucose
Glucose 6-phosphate

the order is determined by the basic subject component. With the entries

Glucose 1-phosphate
Glucose 6-phosphate

the basic subject component does not determine an order, hence, the modifying phrase component is considered, which does determine an order.

For those readers interested, a somewhat more detailed definition of each component and the manner in which it should be used to determine order is given below.

1. Entries are ordered first according to the basic subject. The basic subject consists of those characters remaining in an entry once subscripts, superscripts, modifying phrases, and preceding prepositions have been removed. The basic subject is ordered first according to Greek letters and alphabetics, with Greek letters ordered before alphabetics. Then, if an order has not been determined, numbers and primes appearing in the basic subject are considered, and "full alphanumeric ordering" applied. In full alphanumeric ordering, Greek letters are ordered before alphabetics which are ordered before numbers; if necessary, primes are then considered and are ordered last.

2. If an order still has not been achieved, the subscript and superscript sequences applying to the basic subject should be considered as they occur from left to right. If a subscript or superscript occurs in one entry at a given position but does not occur at that position in an otherwise similar entry, the entry lacking the subscript or superscript appears first. The ordering within subscripts and superscripts follows full alphanumeric order discussed in Step 1.

3. Modifying phrases should be considered next. These phrases are sequences of characters set off from the basic entry by dashes or arrows, and fall into one of the following categories:

(a) Greek letters
(b) the single alphabetics A, C, D, F, H, M, N, O, P, R, S, T, d, i, m, n, o, p, s, t, D, L
(c) signs +, −
(d) the Roman numerals I, II, III
(e) the symbols or words γA, γG, all, allo, anti, arabino, cis, dl, erythro, exo, im, iso, meso, meta, myo, neo, ortho, para, scyllo, sec, sym, syn, tert, threo, trans
(f) sequences of numbers (optionally followed by Greek letters or by the letters a, b, R, S, H)

Modifying phrases in entries should be examined phrase by phrase from left to right. If a modifying phrase occurs in one entry at a given position but does not occur at that position in an otherwise similar entry, the entry lacking the modifying phrase appears first. The ordering within modifying phrases follows full alphanumeric order discussed in Step 1. Subscripts and superscripts applying to the modifying phrase should be considered along with the phrase after the alphanumerics of that phrase have been considered.

4. If an order still has not been obtained, preceding prepositions (or conjunctions), which are ordered alphabetically, should be considered. A preceding preposition is one of the following words appearing at the beginning of an entry:

and, as, at, by, for, from, in, of, on, to, with

Italic characters should be treated in the same manner as normal characters in locating an entry.

Outline of Volumes XXXI, XXXII, XXXIV–LX

VOLUME XXXI

Biomembranes (Part A)

Section I. Multiple Fractions from a Single Tissue
Section II. Isolation of Purified Subcellular Fractions and Derived Membranes (from Mammalian Tissue Excluding Nerve)
A. Plasma Membrane
B. Golgi Complex
C. Rough and Smooth Microsomes
D. Nuclei
E. Mitochondria
F. Lysosomes
G. Peroxisomes
H. Secretory Granules
I. Specialized Cell Fractions
Section III. Subcellular Fractions Derived from Nerve Tissue
Section IV. Cell Fractions Derived from Plant Tissue
A. General Aspects
B. Specific Organelles and Derived Components
Section V. Preparations Derived from Unicellular Organisms
Section VI. General Methodology

VOLUME XXXII

Biomembranes (Part B)

Section I. Characterization of Membranes and Membrane Components
A. Electron Microscopy
B. Protein Components of Membranes
C. Phospholipid Components of Membranes
D. Biophysical Techniques
Section II. Isolation of Selected Membrane Components
A. From Animals
B. From Plants
C. From Unicellular Organisms
Section III. Reconstitution of Membranes or Membrane Complexes
Section IV. Model Membranes

Section V. Isolation and Culture of Cells
- A. Basic Methodology for Cell Culture
- B. Isolation of Specific Cell Types
- C. Isolation and Culture of Specific Cell Types
- D. Membranes and Membrane Variants from Microorganisms

VOLUME XXXIV

Affinity Techniques

Section I. Introduction

Section II. Coupling Reactions and General Methodology

Section III. Attachment of Specific Ligands and Methods for Specific Proteins
- A. Coenzymes and Cofactors
- B. Sugars and Derivatives
- C. Amino Acids and Peptides
- D. Nucleic Acids, Nucleotides, and Derivatives
- E. Other Systems

Section IV. Purification of Synthetic Macromolecules

Section V. Purification of Receptors

Section VI. Immunological Approaches

VOLUME XXXV

Lipids (Part B)

Section I. Fatty Acid Synthesis

Section II. Fatty Acid Activation and Oxidation

Section III. HMG-CoA Enzymes

Section IV. Hydrolases

Section V. Miscellaneous Enzymes

Section VI. General Analytical Methods

Section VII. Synthesis or Preparation of Substrates

Section VIII. Preparation of Single Cells

Section IX. *In Vivo* and Perfusion Techniques

VOLUME XXXVI

Hormone Action (Part A: Steroid Hormones)

Section I. Hormone-Binding Proteins and Assays for Steroid Hormones

Section II. Serum-Binding Proteins for Steroid and Thyroid Hormones

Section III. Cytoplasmic Receptors for Steroid Hormones

Section IV. Nuclear Receptors for Steroid Hormones
Section V. Purification of Receptor for Steroid Hormones
Section VI. Steroid Hormone Effects on Biochemical Processes
Section VII. Isolation of Biologically Active Metabolites of Steroid and Thyroid Hormones

VOLUME XXXVII

HORMONE ACTION (PART B: PEPTIDE HORMONES)

Section I. Hormone Assays
Section II. Hormone Receptors
Section III. Evaluation of Biological Effects of Hormones
Section IV. Purification and Synthesis of Hormones

VOLUME XXXVIII

HORMONE ACTION (PART C: CYCLIC NUCLEOTIDES)

Section I. Extraction and Purification of Cyclic Nucleotides
Section II. Assay of Cyclic Nucleotides
Section III. Biosynthesis of Cyclic Nucleotides
Section IV. Degradation of Cyclic Nucleotides
Section V. Cyclic Nucleotide-Dependent Protein Kinases and Binding Proteins
Section VI. Synthetic Derivatives of Cyclic Nucleotides and Their Precursors

VOLUME XXXIX

HORMONE ACTION (PART D: ISOLATED CELLS, TISSUES, AND ORGAN SYSTEMS)

Section I. Kidney
Section II. Liver
Section III. Heart
Section IV. Skeletal and Smooth Muscle
Section V. Endocrine and Reproductive Tissue
Section VI. Neural Tissue
Section VII. Exocrine Tissue
Section VIII. Vascular Tissue
Section IX. Unicellular Organisms
Section X. Substrate and Ion Fluxes

VOLUME XL

Hormone Action (Part E: Nuclear Structure and Function)

Section I. The Cell Nucleus and Cell Division
Section II. The Cell Nucleus and Chromatin Proteins
Section III. General Methods for Evaluating Hormone Effects

VOLUME XLI

Carbohydrate Metabolism (Part B)

Section I. Analytical Methods
Section II. Enzyme Assay Procedures
Section III. Preparation of Substrates
Section IV. Oxidation-Reduction Enzymes
Section V. Epimerases and Isomerases
Section VI. Miscellaneous Enzymes

VOLUME XLII

Carbohydrate Metabolism (Part C)

Section I. Kinases
Section II. Aldolases
Section III. Dehydratases
Section IV. Phosphatases
Section V. Mutases
Section VI. Carboxylases and Decarboxylases
Section VII. Glycosidases

VOLUME XLIII

Antibiotics

Section I. Methods for the Study of Antibiotics
Section II. Antibiotic Biosynthesis
- A. Aminoglycoside Antibiotics
- B. β-Lactam Antibiotics
- C. Erythromycin
- D. Indolmycin
- E. Novobiocin
- F. Nucleoside Antibiotics
- G. Patulin
- H. Peptide Antibiotics
- I. Tetracyclines

Section III. Antibiotic Inactivation and Modification
A. Aminoglycoside Antibiotics
B. β-Lactam Antibiotics
C. Chloramphenicol
D. Clindamycin
E. Nucleoside Antibiotics
F. Peptide Antibiotics

VOLUME XLIV

Immobilized Enzymes

Section I. General Comments
Section II. Immobilization Techniques
A. Covalent Coupling
B. Adsorption
C. Entrapment and Related Techniques
D. Aggregation
E. Miscellaneous
F. Other Techniques and General Classification
Section III. Assay Procedures
Section IV. Characterization by Physical Methods
Section V. Characterization by Chemical Methods
Section VI. Kinetics of Immobilized Enzymes
Section VII. Multistep Enzyme Systems
Section VIII. The Application of Immobilized Enzymes to Fundamental Studies in Biochemistry
Section IX. Application of Immobilized Enzymes
A. Analytical Area
B. Medical Area
C. Enzyme Engineering (Enzyme Technology)
D. Application of Immobilized Enzymes in Organic Chemistry
Section X. Immobilized Coenzymes
Section XI. Miscellaneous

VOLUME XLV

Proteolytic Enzymes (Part B)

Section I. General Aspects
Section II. Blood Clotting Enzymes
A. Enzymes of Blood Coagulation
B. Snake Venom Enzymes with Coagulating Activities
Section III. Enzymes of Clot Lysis

Section IV. Kallikreins
Section V. Proteases from Gametes and Developing Embryos
Section VI. Dipeptidases
Section VII. Endopeptidases
Section VIII. Exopeptidases
A. Aminopeptidases
B. Carboxypeptidases
Section IX. Naturally Occurring Protease Inhibitors
A. Specific Inhibitors of Clotting and Lysis in Blood
B. Inhibitors from Bacteria
C. Inhibitors from Plants
D. Protease Inhibitors from Various Sources

VOLUME XLVI

Affinity Labeling

Section I. General Methodology
Section II. Specific Procedures for Enzymes, Antibodies, and Other Proteins
A. Proteolytic Systems
B. Nucleotide and Nucleic Acid Systems
C. Carbohydrate Systems
D. Amino Acid Systems
E. Steroid Systems
F. Antibodies
G. Other Proteins
H. Receptors and Transport Systems
Section III. Specific Procedures for Nucleic Acids and Ribosomal Systems

VOLUME XLVII

Enzyme Structure (Part E)

Section I. Amino Acid Analysis
Section II. End Group Methods
Section III. Chain Separation
Section IV. Cleavage of Disulfide Bonds
Section V. Selective Cleavage by Chemical Methods
Section VI. Selective Cleavage with Enzymes
Section VII. Separation of Peptides
Section VIII. Sequence Analysis
Section IX. Chemical Modification
Section X. Methods of Peptide Synthesis

VOLUME XLVIII

Enzyme Structure (Part F)

Section I. Molecular Weight Determinations and Related Procedures
Section II. Interactions

VOLUME XLIX

Enzyme Structure (Part G)

Section I. Conformation and Transitions
Section II. Conformation: Optical Spectroscopy
Section III. Resonance Techniques

VOLUME L

Complex Carbohydrates (Part C)

Section I. Analytical Methods
Section II. Preparations
Section III. Carbohydrate-Binding Proteins
Section IV. Biosynthesis
Section V. Degradation

VOLUME LI

Purine and Pyrimidine Nucleotide Metabolism

BIOSYNTHETIC ENZYMES
Section I. Activation of Ribose Phosphate
Section II. *De Novo* Pyrimidine Biosynthesis
A. Single Enzymes
B. Enzyme Complexes
Section III. *De Novo* Purine Biosynthesis
Section IV. Deoxynucleotide Synthesis
DEGRADATIVE AND SALVAGE ENZYMES
Section V. Nucleotidases and Nucleosidases
Section VI. Pyrimidine Metabolizing Enzymes
A. Kinases
B. Deaminases
C. Phosphorylases and *trans*-Deoxyribosylase
Section VII. Purine Metabolizing Enzymes
A. Kinases
B. Deaminases
C. Phosphorylases
D. Phosphoribosyltransferases
E. General Methods

VOLUME LII

BIOMEMBRANES (PART C: BIOLOGICAL OXIDATIONS)

Section I. General Overview
Section II. Microsomal Electron Transport and Cytochrome *P*-450 Systems
- A. Complex and Resolved Systems
- B. Reconstitution
- C. General Methods
- D. Specific Assay Methods

Section III. Other Hemoprotein Systems
- A. General
- B. Hemoglobin and Myoglobin
- C. Catalases and Peroxidases

VOLUME LIII

BIOMEMBRANES (PART D: BIOLOGICAL OXIDATIONS)

Section I. Electron Transfer Complexes
Section II. Cytochromes
Section III. Nonheme Metalloproteins
Section IV. Flavoproteins
Section V. Quinones
Section VI. Other Microbial Electron Transport Systems

VOLUME LIV

BIOMEMBRANES (PART E: BIOLOGICAL OXIDATIONS)

VOLUME LV

BIOMEMBRANES (PART F: BIOENERGETICS)

Section I. Organelles and Membranes
Section II. ATP Synthesis and Regulation
- A. Phosphorylation Reactions and Regulation
- B. ATPase Complexes
- C. Coupling Factors
- D. Specialized Reagents
- E. Model Systems

Section III. Membrane Potential and Intramitochondrial pH
Section IV. Reconstitution

VOLUME LVI

Biomembranes (Part G: Bioenergetics)

Section I. Biogenesis of Mitochondria
Section II. Genetic Approaches
Section III. Membrane Organization
A. Compartmentalization
B. Sidedness Reagents
Section IV. Transport
A. Methods for Measuring Transport
B. Isolation of Carriers and Reconstitution
Section V. Specialized Techniques
A. Function
B. Structure

VOLUME LVII

Bioluminescence and Chemiluminescence

Section I. Firefly Luciferase
Section II. Bacterial Luciferase
Section III. *Renilla reniformis* Luciferase
Section IV. Aequorin
Section V. *Cypridina*
Section VI. Earthworm Bioluminescence
Section VII. *Pholas dactylus*
Section VIII. Chemiluminescent Techniques
Section IX. Instrumentation and Methods

VOLUME LVIII

Cell Culture

Section I. Basic Methods
A. Laboratory Requirements and Media
B. General Cell Culture Techniques
Section II. Specialized Techniques
A. Metabolism
B. Genetics, Hybridization, and Transformation
C. Virus Preparation
Section III. Specific Cell Lines

VOLUME LIX

Nucleic Acids and Protein Synthesis (Part G)

Section I. Transfer RNA and Aminoacyl-tRNA Synthetases
Section II. Ribosomes

VOLUME LX

Nucleic Acids and Protein Synthesis (Part H)

Section I. Initiation of Protein Synthesis
Section II. Peptide Chain Elongation

METHODS IN ENZYMOLOGY

EDITED BY

Sidney P. Colowick and Nathan O. Kaplan

VANDERBILT UNIVERSITY
SCHOOL OF MEDICINE
NASHVILLE, TENNESSEE

DEPARTMENT OF CHEMISTRY
UNIVERSITY OF CALIFORNIA
AT SAN DIEGO
LA JOLLA, CALIFORNIA

I. Preparation and Assay of Enzymes
II. Preparation and Assay of Enzymes
III. Preparation and Assay of Substrates
IV. Special Techniques for the Enzymologist
V. Preparation and Assay of Enzymes
VI. Preparation and Assay of Enzymes (*Continued*)
Preparation and Assay of Substrates
Special Techniques
VII. Cumulative Subject Index

METHODS IN ENZYMOLOGY

EDITORS-IN-CHIEF

Sidney P. Colowick   Nathan O. Kaplan

VOLUME VIII. Complex Carbohydrates
Edited by ELIZABETH F. NEUFELD AND VICTOR GINSBURG

VOLUME IX. Carbohydrate Metabolism
Edited by WILLIS A. WOOD

VOLUME X. Oxidation and Phosphorylation
Edited by RONALD W. ESTABROOK AND MAYNARD E. PULLMAN

VOLUME XI. Enzyme Structure
Edited by C. H. W. HIRS

VOLUME XII. Nucleic Acids (Parts A and B)
Edited by LAWRENCE GROSSMAN AND KIVIE MOLDAVE

VOLUME XIII. Citric Acid Cycle
Edited by J. M. LOWENSTEIN

VOLUME XIV. Lipids
Edited by J. M. LOWENSTEIN

VOLUME XV. Steroids and Terpenoids
Edited by RAYMOND B. CLAYTON

VOLUME XVI. Fast Reactions
Edited by KENNETH KUSTIN

VOLUME XVII. Metabolism of Amino Acids and Amines (Parts A and B)
Edited by HERBERT TABOR AND CELIA WHITE TABOR

VOLUME XVIII. Vitamins and Coenzymes (Parts A, B, and C)
Edited by DONALD B. MCCORMICK AND LEMUEL D. WRIGHT

VOLUME XIX. Proteolytic Enzymes
Edited by GERTRUDE E. PERLMANN AND LASZLO LORAND

VOLUME XX. Nucleic Acids and Protein Synthesis (Part C)
Edited by KIVIE MOLDAVE AND LAWRENCE GROSSMAN

VOLUME XXI. Nucleic Acids (Part D)
Edited by LAWRENCE GROSSMAN AND KIVIE MOLDAVE

VOLUME XXII. Enzyme Purification and Related Techniques
Edited by WILLIAM B. JAKOBY

VOLUME XXIII. Photosynthesis (Part A)
Edited by ANTHONY SAN PIETRO

VOLUME XXIV. Photosynthesis and Nitrogen Fixation (Part B)
Edited by ANTHONY SAN PIETRO

VOLUME XXV. Enzyme Structure (Part B)
Edited by C. H. W. HIRS AND SERGE N. TIMASHEFF

VOLUME XXVI. Enzyme Structure (Part C)
Edited by C. H. W. HIRS AND SERGE N. TIMASHEFF

VOLUME XXVII. Enzyme Structure (Part D)
Edited by C. H. W. HIRS AND SERGE N. TIMASHEFF

VOLUME XXVIII. Complex Carbohydrates (Part B)
Edited by VICTOR GINSBURG

VOLUME XXIX. Nucleic Acids and Protein Synthesis (Part E)
Edited by LAWRENCE GROSSMAN AND KIVIE MOLDAVE

VOLUME XXX. Nucleic Acids and Protein Synthesis (Part F)
Edited by KIVIE MOLDAVE AND LAWRENCE GROSSMAN

VOLUME XXXI. Biomembranes (Part A)
Edited by SIDNEY FLEISCHER AND LESTER PACKER

VOLUME XXXII. Biomembranes (Part B)
Edited by SIDNEY FLEISCHER AND LESTER PACKER

VOLUME XXXIII. Cumulative Subject Index Volumes I-XXX
Edited by MARTHA G. DENNIS AND EDWARD A. DENNIS

VOLUME XXXIV. Affinity Techniques (Enzyme Purification: Part B)
Edited by WILLIAM B. JAKOBY AND MEIR WILCHEK

VOLUME XXXV. Lipids (Part B)
Edited by JOHN M. LOWENSTEIN

VOLUME XXXVI. Hormone Action (Part A: Steroid Hormones)
Edited by BERT W. O'MALLEY AND JOEL G. HARDMAN

VOLUME XXXVII. Hormone Action (Part B: Peptide Hormones)
Edited by BERT W. O'MALLEY AND JOEL G. HARDMAN

VOLUME XXXVIII. Hormone Action (Part C: Cyclic Nucleotides)
Edited by JOEL G. HARDMAN AND BERT W. O'MALLEY

VOLUME XXXIX. Hormone Action (Part D: Isolated Cells, Tissues, and Organ Systems)
Edited by JOEL G. HARDMAN AND BERT W. O'MALLEY

VOLUME XL. Hormone Action (Part E: Nuclear Structure and Function)
Edited by BERT W. O'MALLEY AND JOEL G. HARDMAN

VOLUME XLI. Carbohydrate Metabolism (Part B)
Edited by W. A. WOOD

VOLUME XLII. Carbohydrate Metabolism (Part C)
Edited by W. A. WOOD

VOLUME XLIII. Antibiotics
Edited by JOHN H. HASH

VOLUME XLIV. Immobilized Enzymes
Edited by KLAUS MOSBACH

VOLUME XLV. Proteolytic Enzymes (Part B)
Edited by LASZLO LORAND

VOLUME XLVI. Affinity Labeling
Edited by WILLIAM B. JAKOBY AND MEIR WILCHEK

VOLUME XLVII. Enzyme Structure (Part E)
Edited by C. H. W. HIRS AND SERGE N. TIMASHEFF

VOLUME XLVIII. Enzyme Structure (Part F)
Edited by C. H. W. HIRS AND SERGE N. TIMASHEFF

VOLUME XLIX. Enzyme Structure (Part G)
Edited by C. H. W. HIRS AND SERGE N. TIMASHEFF

VOLUME L. Complex Carbohydrates (Part C)
Edited by VICTOR GINSBURG

VOLUME LI. Purine and Pyrimidine Nucleotide Metabolism
Edited by PATRICIA A. HOFFEE AND MARY ELLEN JONES

VOLUME LII. Biomembranes (Part C: Biological Oxidations)
Edited by SIDNEY FLEISCHER AND LESTER PACKER

VOLUME LIII. Biomembranes (Part D: Biological Oxidations)
Edited by SIDNEY FLEISCHER AND LESTER PACKER

VOLUME LIV. Biomembranes (Part E: Biological Oxidations)
Edited by SIDNEY FLEISCHER AND LESTER PACKER

VOLUME LV. Biomembranes (Part F: Bioenergetics)
Edited by SIDNEY FLEISCHER AND LESTER PACKER

VOLUME LVI. Biomembranes (Part G: Bioenergetics)
Edited by SIDNEY FLEISCHER AND LESTER PACKER

VOLUME LVII. Bioluminescence and Chemiluminescence
Edited by MARLENE A. DELUCA

VOLUME LVIII. Cell Culture
Edited by WILLIAM B. JAKOBY AND IRA H. PASTAN

VOLUME LIX. Nucleic Acids and Protein Synthesis (Part G)
Edited by KIVIE MOLDAVE AND LAWRENCE GROSSMAN

VOLUME LX. Nucleic Acids and Protein Synthesis (Part H)
Edited by KIVIE MOLDAVE AND LAWRENCE GROSSMAN

VOLUME 61. Enzyme Structure (Part H)
Edited by C. H. W. HIRS AND SERGE N. TIMASHEFF

VOLUME 62. Vitamins and Coenzymes (Part D)
Edited by DONALD B. MCCORMICK AND LEMUEL D. WRIGHT

VOLUME 63. Enzyme Kinetics and Mechanism (Part A: Initial Rate and Inhibitor Methods)
Edited by DANIEL L. PURICH

VOLUME 64. Enzyme Kinetics and Mechanism (Part B: Isotopic Probes and Complex Enzyme Systems)
Edited by DANIEL L. PURICH

VOLUME 65. Nucleic Acids (Part I)
Edited by LAWRENCE GROSSMAN AND KIVIE MOLDAVE

VOLUME 66. Vitamins and Coenzymes (Part E)
Edited by DONALD B. MCCORMICK AND LEMUEL D. WRIGHT

VOLUME 67. Vitamins and Coenzymes (Part F)
Edited by DONALD B. MCCORMICK AND LEMUEL D. WRIGHT

VOLUME 68. Recombinant DNA
Edited by RAY WU

VOLUME 69. Photosynthesis and Nitrogen Fixation (Part C)
Edited by ANTHONY SAN PIETRO

VOLUME 70. Immunochemical Techniques (Part A)
Edited by HELEN VAN VUNAKIS AND JOHN J. LANGONE

VOLUME 71. Lipids (Part C)
Edited by JOHN M. LOWENSTEIN

VOLUME 72. Lipids (Part D)
Edited by JOHN M. LOWENSTEIN

VOLUME 73. Immunochemical Techniques (Part B)
Edited by JOHN J. LANGONE AND HELEN VAN VUNAKIS

VOLUME 74. Immunochemical Techniques (Part C)
Edited by JOHN J. LANGONE AND HELEN VAN VUNAKIS

VOLUME 75. Cumulative Subject Index Volumes XXXI, XXXII, and XXXIV-LX
Edited by EDWARD A. DENNIS AND MARTHA G. DENNIS

VOLUME 76. Hemoglobins
Edited by ERALDO ANTONINI, LUIGI ROSSI-BERNARDI, AND EMILIA CHIANCONE

VOLUME 77. Detoxication and Drug Metabolism
Edited by WILLIAM B. JAKOBY

VOLUME 78. Interferons (Part A)
Edited by SIDNEY PESTKA

VOLUME 79. Interferons (Part B)
Edited by SIDNEY PESTKA

VOLUME 80. Proteolytic Enzymes (Part C)
Edited by LASZLO LORAND

VOLUME 81. Biomembranes (Part H: Visual Pigments and Purple Membranes, I)
Edited by LESTER PACKER

VOLUME 82. Structural and Contractile Proteins (Part A: Extracellular Matrix)
Edited by LEON W. CUNNINGHAM AND DIXIE W. FREDERIKSEN

VOLUME 83. Complex Carbohydrates (Part D)
Edited by VICTOR GINSBURG

VOLUME 84. Immunochemical Techniques (Part D: Selected Immunoassays)
Edited by JOHN J. LANGONE AND HELEN VAN VUNAKIS

VOLUME 85. Structural and Contractile Proteins (Part B: The Contractile Apparatus and the Cytoskeleton)
Edited by DIXIE W. FREDERIKSEN AND LEON W. CUNNINGHAM

VOLUME 86. Prostaglandins and Arachidonate Metabolites
Edited by WILLIAM E. M. LANDS AND WILLIAM L. SMITH

VOLUME 87. Enzyme Kinetics and Mechanism (Part C: Intermediates, Stereochemistry, and Rate Studies)
Edited by DANIEL L. PURICH

VOLUME 88. Biomembranes (Part I: Visual Pigments and Purple Membranes, II)
Edited by LESTER PACKER

VOLUME 89. Carbohydrate Metabolism (Part D)
Edited by WILLIS A. WOOD

VOLUME 90. Carbohydrate Metabolism (Part E) (in preparation)
Edited by WILLIS A. WOOD

VOLUME 91. Enzyme Structure (Part I) (in preparation)
Edited by C. H. W. HIRS AND SERGE N. TIMASHEFF

VOLUME 92. Immunochemical Techniques (Part E: Monoclonal Antibodies and General Immunoassay Methods) (in preparation)
Edited by JOHN J. LANGONE AND HELEN VAN VUNAKIS

VOLUME 93. Immunochemical Techniques (Part F: Conventional Antibodies, Fc Receptors, and Cytotoxicity) (in preparation)
Edited by JOHN J. LANGONE AND HELEN VAN VUNAKIS

VOLUME 94. Polyamines (in preparation)
Edited by HERBERT TABOR AND CELIA WHITE TABOR

Methods in Enzymology

Volume 75

CUMULATIVE SUBJECT INDEX

Volumes XXXI, XXXII, XXXIV–LX

Subject Index

Roman numerals indicate volume number

A

Aabomycin, XLIII, 137
AAG, *see* α_1-Acid glycoprotein
A–204 antibiotic
as ionophore, LV, 443
solutions, LV, 448
A–218 antibiotic, LV, 445
A3823 antibiotic, LV, 445
A23187 antibiotic, LV, 445
calcium flux, LV, 453, 454
as ionophore, LV, 443, 447
solutions, LV, 448
structure, LV, 437, 445
A28695A antibiotic, LV, 445
A28086A (B or C) antibiotic, LV, 445
A2 + APG medium, LVIII, 57
Abbe 31 refractometer, XXXII, 120
Abortive-complex affinity, XXXIV, 115, 601
Abortive infection, *see* Infection, abortive
Abrin
assay, L, 324, 325
immunization, L, 328, 329
purification, L, 326–328
toxic effects, L, 323
Abrus, agglutinin
carbohydrate specificity, L, 323
mitogenic activity, L, 324
purification, L, 326–328
Abrus precatorius
lectin, XXXIV, 333; L, 323–330, *see also* Abrin; *Abrus* agglutinin
properties, L, 329
purification of antibodies against, L, 329
Absorbance
of atomized sample, definition, LIV, 463
change, computation, LIV, 54, 55
high-pressure studies, XLIX, 16
nonspecific changes, correction, LVI, 308, 309
optical cell, design, XLIX, 23
Absorbance indicator
metal ions, LVI, 303, 304
penetration, LVI, 305, 306
practical considerations, LVI, 307, 308
calibration, LVI, 309, 310
instrumentation, LVI, 309
measurements, LVI, 308, 309
purification, LVI, 308
response time, LVI, 307
selection of wavelength, LVI, 308
selectivity, LVI, 306
side effects, LVI, 306
solution, LVI, 308
specificity, LVI, 306
Absorbance optic, sedimentation equilibrium, XLVIII, 171–173
Absorbance spectroscopy, *see* specific types
Absorption anisotropy
for bacteriorhodopsin, LIV, 60
computation, LIV, 55–57
Absorption anisotropy factor, XLIX, 180, 181, 185
Absorption coefficient, oscillatory part
definition, LIV, 340
Fourier transforms, LIV, 341
Absorption spectrometry, *see* specific type
Absorption tube, long-path, LIV, 462
7-ACA, *see* 7-Aminocephalosporanic acid
Acanthamoeba castellanii
disruption, LV, 136
membrane isolation, XXXI, 686
Acanthoptilum gracile
luciferase-GFP interaction, specificity, LVII, 264
source, LVII, 248
Acatalasemia, XLIV, 684–688
Acceleration gradient, in plant-cell fractionation, XXXI, 509–511
Acceptor control ratio, of hepatoma mitochondria, LV, 87
ACD reagent, *see* Acid-citrate-dextrose reagent
Acer, cell cultures, XXXII, 727, 728

Acetaldehyde, XLIV, 39, 40, 128–131, 133, 134, 863
permeability data, XLIV, 297, 302
in preparation of deuterated NAD^+, LIV, 226, 228
product of ethoxycoumarin *O*-dealkylation, LII, 373
substrate, of transaldolase, XLVII, 495
Acetal phosphatidalcholine, intermediate in plasmalogen synthesis, XXXV, 484
*P*¹-2-Acetamido-4-*O*-(2-acetamido-2-deoxy-β-D-glucopyranosyl)-2-deoxy-α-D-glucopyranosyl *P*²-dolichyl pyrophosphate, synthesis, L, 134–137
*P*¹-2-Acetamido-4-*O*-(2-acetamido-3,4,6-tri-*O*-acetyl-2-deoxy-β-D-glucopyranosyl)-3,6-di-*O*-acetyl-2-deoxy-α-D-glucopyranosyl *P*²-dolichyl pyrophosphate, preparation, L, 136
2-Acetamido-*N*-(ε-aminocaproyl)-2-deoxy-β-D-glucopyranosylamine, XXXIV, 344
9-(*p*-Acetamidobenzyl)adenine, synthesis, XLVI, 331
3′-Acetamido-3′-deoxyadenosine, XLIII, 154
2-Acetamido-2-deoxy-3-*O*-α-L-fucopyranosyl-D-glucose, synthesis, L, 114–116
2-Acetamido-2-deoxy-α-D-galactopyranoside, XXXIV, 363
2-Acetamido-2-deoxy-*O*-β-D-galactopyranosyl-(1→4)-*O*-β-D-galactopyranosyl-(1→4)-D-glucose, preparation, L, 116–118
*P*¹-2-Acetamido-2-deoxy-α-D-glucopyranosyl *P*²-dolichyl pyrophosphate, synthesis, L, 131–132
3-*O*-(2-Acetamido-2-deoxy-β-D-glucopyranosyl)-D-galactitol, L, 17
2-Acetamido-2-deoxy-α-mannopyranose, ^{13}C nuclear magnetic resonance spectrum, L, 43
2-Acetamido-2-deoxy-β-mannopyranose, ^{13}C nuclear magnetic resonance spectrum, L, 43
O-(2-Acetamido-3,4,6-tri-*O*-acetyl-2-deoxy-β-D-glucopyranosyl)-(1→4)- 2-acetamido-3,6-di-*O*-acetyl-2-deoxy-α-D-glucopyranosyl chloride, synthesis, XLVI, 406
O-(2-Acetamido-3,4,6-tri-*O*-acetyl-2-deoxy-β-D-glucopyranosyl)-(1→4)- 2-[^3H]acetamido-1,3,6-tri-*O*-acetyl-2-deoxy-α-D-glucopyranose, synthesis, XLVI, 408, 409
O-(2-Acetamido-3,4,6′-tri-*O*-acetyl-2-deoxy-β-D-glucopyranosyl)-(1→4)-2-acetamido-1,3,6-tri-*O*-acetyl-2- deoxy-α-D-glucopyranose, synthesis, XLVI, 405, 406
O-(2-Acetamido-3,4,6-tri-*O*-acetyl-2-deoxy-β-D-glucopyranosyl)-(1→4)-2-amino-1,3,6-tri-*O*-acetyl-2-deoxy-α-D-glucopyranose hydrochloride, synthesis, XLVI, 408
2-Acetamido-3,4,6-tri-*O*-acetyl-2-deoxy-α-D-glucopyranosyl chloride, XXXIV, 343
synthesis, XLVI, 409
*P*¹-2-Acetamido-3,4,6-tri-*O*-acetyl-2-deoxy-α-D-glucopyranosyl *P*²-dolichyl pyrophosphate, preparation, L, 133, 134
2-Acetamido-3,4,6-tri-*O*-acetyl-2-deoxy-α-D-glucopyranosyl phosphate, preparation, L, 131, 132
Acetaminophen, toxicity, mechanism, LII, 70, 71
p-Acetaminophenylethoxy methacrylate, XLIV, 71–72
Acetanilide, *N*-hydroxylation, Ah^b allele, LII, 231
Acetanilide 4-hydroxylase, Ah^b allele, LII, 231
Acetate, XLIV, 129, 133, *see also* Acetic acid
ΔpH determination, LV, 229, 562, 563
assay
with acetyl-CoA synthetase, XXXV, 302–307
with CoA-transferase, XXXV, 298–301
by gas-liquid chromatography, XXXV, 307–311
calcium uptake, LVI, 339
^{14}C-labeled, fatty acid synthesis precursor, XXXV, 279–282
covalent binding to fatty acid synthase, XXXV, 83

flow dialysis experiments, LV, 683, 684, 685
calculation of concentration gradient, LV, 687, 688
of free acetate in upper chamber, LV, 686
of uptake, LV, 686, 687
hemin requirement, LVI, 172
in media, LVIII, 66
mitochondrial matrix, LVI, 253
oxidation, XLIV, 315, 316
product, of glycine reductase, LIII, 373
substrate
acetyl-CoA synthetase, XXXV, 302, 307
CoA-transferase, XXXV, 291
succinyl-CoA:propionate CoA-transferase, XXXV, 235, 242
tissue content, XXXV, 307
Acetate-ammonium hydroxide buffer, purification of cytochrome *c*-551, LIII, 657, 658
Acetate buffer, *see also* Potassium acetate; Sodium acetate
assay of cytochrome *c*, LIII, 160
of glucose, LVII, 453
preparation of luciferyl adenylate, LVII, 27
purification of green-fluorescent protein, LVII, 309, 310
of *Pholas* luciferase, LVII, 387
of *Pholas* luciferin, LVII, 388
Acetate kinase
activation, XLIV, 889
activity assay, XLIV, 893, 894
cysteine residues, XLIV, 887–889
immobilization, XLIV, 891, 892
phosphorylation potential, LV, 237
Acetate membrane, electrochemical methods, LVI, 455
Acetazolamide, XLIX, 168, 169
autoradiography, XXXVI, 150
Acetic acid, XLIV, 23, 77, 132, 392, 762, 804, 847, *see also* Acetate
acid precipitation procedure, LI, 7–9, 43, 324, 334
assay of aspartate transcarbamylase, LI, 123
of carbamoyl-phosphate synthetase, LI, 122
for destaining, LIX, 507
determination of radioactivity in slab gels, LIX, 521
dialytic permeability data, XLIV, 297
electrophoretic analysis of ribosomal proteins, LIX, 506, 544, 545
extraction of ribosomal proteins, LIX, 431, 446, 447, 453, 470, 506, 517–519, 538, 541, 542, 836
fingerprinting procedure, LIX, 66
for fixing RNA in gels, LIX, 567
formaldehyde assay, LII, 299
in gel staining solution, LIX, 540, 546
glacial
in 6-amino-1-hexanol phosphate preparation, LI, 253
in diphenylamine reagent, LI, 248
synthesis of 7-hydroxynaphthalene-1,2-dicarboxylic anhydride, LVII, 436
of 4-[4-methoxyphenyl]butyric acid, LVII, 434
hydrogen bromide cleavage, XLVII, 546, 547, 603
metal-free, preparation, LIV, 476
in ninhydrin solution, LIX, 777
peptide hydrolysis, XLVII, 42, 147
perdeuterated, preparation of deuterated bacteria, LIX, 640
in postcoupling wash, XLVII, 326
preparation of ribosomal RNA, LIX, 557, 558, 586
purification of cytochrome *c*-551, LIII, 657
of 4-[*N*-ethyl-*N*-4-(*N*-phthalimido)butyl]amino-*N*-methylphthalimide, LVII, 433
of flavin peptides, LIII, 453
of succinate dehydrogenase, LIII, 485
redistillation, XLVII, 359
removal, LIII, 458
of trichloroacetic acid from gels, LIX, 546
reverse salt gradient chromatography, LIX, 215, 217
siroheme demetallation, LII, 443
solvent, for hydrazides, LIX, 178
synthesis of adenosine derivative, LVII, 116
of coenzyme Q analogs, LIII, 595

thin-layer chromatography, XXXVIII, 29, 30, 34, 36
tryptophan modification, XLVII, 447
washing of DEAE-cellulose, LIX, 194
wash solution, for gel electrophoresis, XXXII, 96
Acetic acid-chloroform, in iodometric assay of lipid hydroperoxides, LII, 307
Acetic acid-formic acid, in electrophoretic separation of MS2 peptides, LIX, 301
Acetic acid-methanol, for gel destaining, LIX, 313
Acetic acid-pyridine
in electrophoretic separation of MS2 peptides, LIX, 301
purification of 8α-flavins, LIII, 458, 459
Acetic acid–sodium acetate
synthesis of 8-bromoadenosine 5′-monophosphate, LVII, 117
of *N*-(ethyl-2-diazomalonyl)streptomycyl hydrazone, LIX, 815
Acetic acid-sodium chloride, in hemin purification, LII, 453
Acetic acid-trifluoroacetic acid, identification of cysteinyl flavins, LIII, 464
Acetic anhydride, XLVII, 529, 531
acetylation of flavin peptides, LIII, 458, 459
modification of lysine, LIII, 139
of tyrosine, LIII, 172
redistillation, XLVII, 359
removal, LIII, 458
synthesis of *N*-methyl-4-nitrophthalimide, LVII, 427
of 4-nitrophthalic anhydride, LVII, 427
Acetic anhydride-acetic acid, coupling solvent, XLVII, 359, 362
Acetimidic acid, ^{14}C-labeled, use, XLVII, 203
Acetoacetate
assay, LV, 207
inhibitor of succinate dehydrogenase, LIII, 33
synthesis, control studies, XXXVII, 292
S-Acetoacetyl-*N*-acetylcysteamine, assay of reductase activity of fatty acid synthase, XXXV, 50
Acetoacetyl-CoA reductase, XXXV, 122
Acetoacetyl-CoA thiolase, *see* Acetyl-CoA acetyltransferase
Acetoacetyl coenzyme A
absorption, at 300 nm (of Mg^{2+} complex), XXXV, 156, 162, 168, 169, 174
in assay for L-3-hydroxyacyl-CoA dehydrogenase, XXXV, 123
inhibitor of crotonase, XXXV, 150
inhibitor of mitochondrial acetoacetyl-CoA thiolase, XXXV, 135
substrate
for acetoacetyl-CoA thiolase, XXXV, 134, 167
for 3-hydroxy-3-methylglutaryl-CoA synthase, XXXV, 155, 160, 173
for 3-ketoacyl-CoA thiolase, XXXV, 136
synthesis, XXXV, 129
Acetoacetylpantetheine
in assay for L-3-hydroxyacyl-CoA dehydrogenase, XXXV, 123
substrate for acetoacetyl-CoA thiolases, XXXV, 134
Acetobacter, oxidase, LII, 15, 19
Acetobacter melanogenum, 5-keto-D-fructose reductase, XLI, 128, 137
Acetobacter pasteurianum, cytochrome a_1, LIII, 206, 207
Acetobacter suboxydans
aldohextose dehydrogenase, XLI, 147
5-keto-D-fructose produced, XLI, 85
o-type cytochrome, LIII, 208
Acetobacter xylinum
pyruvate orthophosphate dikinase, XLII, 187, 200
assay, purification, and properties, XLII, 192–199
Acetobromo sugar, XXXIV, 321
Acetocarmine, as plant nuclei stain, XXXI, 560
α-Acetochloromannose, XLVI, 367
Acetoin dehydrogenase
assay, XLI, 529, 530
chromatography, XLI, 531
properties, XLI, 532, 533

purification, XLI, 530–532
Acetolactate decarboxylase, from *Aerobacter aerogenes*, XLI, 526–529
assay, XLI, 526, 527
chromatography, XLI, 527
electrophoresis, XLI, 528
properties, XLI, 528, 529
purification, XLI, 527–529
Acetolactate-forming enzyme, from *Aerobacter aerogenes*, XLI, 519–526
assay, XLI, 519, 520
chromatography, XLI, 521, 522
crystallization, XLI, 522–524
inhibitors, XLI, 526
properties, XLI, 525, 526
purification, XLI, 520–522, 525
Acetone, XLIV, 33, 34, 41, 42, 50, 99, 103, 112, 113, 117, 140, 187, 230, 384, 480, 506, 632, 762, 852, 867; XLIX, 438
analysis of ribosomal proteins, LIX, 507
assay of amidophosphoribosyltransferase, LI, 173
chemiluminescent reaction, LVII, 523
crystal morphology, XLIX, 470
dehydration, XLV, 38
electron spin resonance spectra, XLIX, 383
endogenous substrate depletion, LII, 271
extraction of heme A, LII, 423
of quinone intermediates, LIII, 605, 608
of ubiquinone, LIII, 573, 574, 581, 639
in fatty acid unsaturation position determination, XXXV, 347
fluorescamine assay, LIX, 501
lipid extraction of microsomal fraction, LII, 98
phosphodiesterase precipitation, XXXVIII, 252
precipitation of liver proteins, LII, 507
preparation of disuccinimidyl succinate, LIX, 159
of firefly luciferase, LVII, 6
of luciferyl adenylate, LVII, 27
of ribosomal proteins, LIX, 431, 447, 470, 518, 520, 777
purification of adenine phosphoribosyltransferase, LI, 563
of *Cypridina* luciferase, LVII, 370
of cytochrome oxidase subunits, LIII, 76
of dihydroorotate dehydrogenase, LI, 65
of flavocytochrome b_2, LIII, 241, 242, 252
of guanine phosphoribosyltransferase, LI, 554
of guanylate kinase, LI, 478
of high-potential iron-sulfur protein, LIII, 336
of hydroxanthine phosphoribosyltransferase, LI, 554
of nitrate reductase, LIII, 642
of superoxide dismutase, LIII, 388
redistillation, XLVII, 339
reductive alkylation, XLVII, 471, 472
for removal of ATPase and DPNH oxidase activities from crude extracts, LI, 475
solvent
of benzopyrene, LII, 409, 414
for fluorescamine, XLVII, 17
of 2-hydroxy-5-nitrobenzylbromide, LIII, 170
of *n*-octane, LIII, 356
synthesis of adenosine derivative, LVII, 118
of oleylpolymethacrylic acid resin, LIII, 214
Acetone L-dihydroxypropylphosphonic acid, diethyl ester, intermediate in synthesis of phosphatidic acid analogues, XXXV, 503, 504
D-Acetoneglycerol, *see* Isopropylidene-D-glycerol
Acetone-hexane, extraction of phenolic metabolites, LII, 409
Acetone-hydrogen chloride
globin isolation, LII, 489
preparation, LII, 440
of apomyoglobin, LII, 477

in siroheme extraction and characterization, LII, 440
Acetone-methanol, preparation of 3′-amino-3′-deoxy-ATP, LIX, 136
Acetone powder
heptane-butanol extraction, XLV, 38
of mitochondria, LV, 386
preparation from chicken liver, LI, 198
from Ehrlich ascites cells, LI, 221
from *Pholas* luminescent organs, LVII, 387
from *Photinus pyralis* lanterns, LVII, 6, 7
from rabbit tissue, LI, 208
from rat liver, LI, 14
Acetone-water
heme extraction, LII, 424, 426
synthesis of *N*-(3-carbethoxy-3-diazo)acetonylblasticidin S, LIX, 814
of *N*-(3-carbethoxy-3-diazo)acetonylgougerotin, LIX, 814
of *N*-(ethyl-2-diazomalonyl)blasticidin S, LIX, 814
Acetone-water-formic acid, chromatographic identification of siroheme, LII, 442
Acetonitrile, XXXIV, 546; XLIX, 383, 438, 470
assay reagent, XLIV, 572
commercial source, XLVII, 46
drying, XXXVIII, 412
electron-transfer chemiluminescence, LVII, 499
for elution of polycyclic aromatic hydrocarbons, LII, 284
high-sensitivity sequence analysis, XLVII, 326, 327
for removal of contaminants from sytrene oxide, LII, 416
solvent
in carboxyl group activation, XLVII, 556
for cyanogen bromide, XLIV, 28
in peptide synthesis, XLVII, 550, 558
in radical ion chemiluminescent reactions, LVII, 498, 507, 511, 525
for styrene oxide, LII, 194
synthesis of *Diplocardia* luciferin, LVII, 377
trypsin conjugate active site titration, XLIV, 391
Acetoorcein, autoradiography prestain, LVIII, 286
Acetophenylhydrazide, for bone marrow induction, LI, 238
Acetothiokinase, in glyoxysomes, XXXI, 569
[^{14}C]Acetoxy-chloramphenicol derivative, XLIII, 743
(±)-6β-Acetoxy-3α-hydroxytropane, XLIV, 836
4-Acetoxymercuriestradiol, XXXVI, 422, 423
Acetoxyprogesterone, reduction by 20β-hydroxysteroid dehydrogenase, XXXVI, 398
Acetylacetone
in formaldehyde assay, LII, 298, 299, 345
in hydrogen peroxide determination, LII, 345
N-Acetyl-4-*O*-acetylneuraminic acid, bacterial neuraminidases, L, 374
N-Acetyl-7-*O*-acetylneuraminic acid, *N*-acylneuraminate-9(7)-*O*-acetyltransferase, L, 381
N-Acetyl-9-*O*-acetylneuraminic acid
N-acylneuraminic-9(7)-*O*-acetyltransferase, L, 381
occurrence, L, 385
Acetyl-DL-alanine, maximal reactor space velocity, XLIV, 753
N-Acetylalanylalanylalanylalanylchloromethyl ketone, XLVI, 206
Acetylalanylalanylazoalanine-*o*-nitrophenol, XLVI, 211
reaction with elastase, XLVI, 214
N-Acetylalanylalanylprolylalanylchloromethyl ketone, XLVI, 206
Acetylalanylalanylprolyl hydroxide, XLVI, 202
N-Acetylalanylalanylprolylvaline chloromethyl ketone, XLVI, 201, 202

Acetylalanylazophenylalanine-*o*-nitrophenol, XLVI, 210
reaction with chymotrypsin A, XLVI, 213, 214
N-Acetylalanylchloromethyl ketone, XLVI, 206
Acetyl-D-alanyl-D-glutamic acid, XLIII, 697
Acetylalanyl hydrazine, XLVI, 210
N-Acetyl-*p*-aminobenzylsuccinic acid, XLVI, 226
2-Acetylaminofluorene, Ah^b allele, LII, 231
N-Acetyl-D-arginine, XLVI, 226
Acetylarsanilic acid, XLVI, 500
synthesis, XLVI, 498
Acetylarylamine *N*-hydroxylase, Ah^b allele, LII, 231
Acetylation, in regulation of protein synthesis, LX, 534, 535
N-Acetyl-(*p*-azobenzenearsonic acid)-L-tyrosine, XLVI, 494, 500
m-Acetylbenzene sulfonamide, inhibitor, of bovine carbonic anhydrase, XLIX, 246
4-Acetylbenzoylpentagastrin, XLVI, 89
N^α-Acetyl-*N*′-carbethoxyhistidine, difference extinction coefficient, XLVII, 437
Acetylcarnitine, XXXV, 273, 274
brown adipose tissue mitochondria, LV, 76
tissue content, LV, 221
transport, LVI, 248, 249
Acetylcarnitine transferase, *see* Carnitine acetyltransferase
[^{14}C]Acetyl-chloramphenicol, XLIII, 743–746, *see also* Chloramphenicol acetyltransferase, radioactive assay
Acetyl chloride, XLVII, 295
Acetylcholine, XLVI, 578–580; XLIX, 494
adenylate cyclase, XXXVIII, 173
binding site, XLVI, 85
biosynthesis, XXXII, 766
cyclic nucleotide levels, XXXVIII, 94
effect on melanin, XXXVII, 129
identification, XXXII, 782, 783, 785
in nerve function, XXXII, 309
photochemically active derivatives, LVI, 660
receptors, XLVI, 75, 573, 582–590
photoaffinity labeling, LVI, 672
reduction, XLVI, 586
Reinickate salt, XLIV, 594
substrate, of cholinesterase, XLIV, 593, 594, 653
Acetylcholinesterase, XXXIV, 571–591; XLVI, 22, 25, 515; LIV, 52
assay, XXXII, 775–777; XXXIV, 575
brain mitochondria, LV, 58, 59
column mechanism and capacity, XXXIV, 584, 585
in cultured cells, XXXII, 766, 768, 772, 773
in erythrocyte ghost sidedness assay, XXXI, 176, 177, 179
gel electrophoresis, XXXII, 88
in nerve impulse, XXXII, 310
in neurobiology, XXXII, 765
purification, XXXIV, 575–580
covalent affinity technique, XXXIV, 581–591, *see also* Covalent affinity chromatography
11 S, XXXIV, 575–578
14 S, XXXIV, 578–580
18 S, XXXIV, 578–580
separation of 18 S and 14 S AChE, XXXIV, 580
spacers, XXXIV, 588–591
spin-label studies, XLIX, 445, 446
Acetyl-CoA acetyltransferase
assay, XXXV, 129, 130, 167–169
cytoplasmic, XXXV, 128–134, 167–173
inhibitor, XXXV, 134, 173
specificity, XXXV, 129, 134
in glyoxysomes, LII, 502
mitochondrial, XXXV, 128–135
activation by K^+ and NH_4^+, XXXV, 129, 130, 135
inhibitor, XXXV, 135
specificity, XXXV, 129, 134
Acetyl-CoA acyltransferase
assay, XXXV, 129, 130
stability, XXXV, 132
tissue contents of individual thiolases, XXXV, 133

Acetyl-CoA:amine acetyltransferase, XXXV, 247–253, *see also* D-Amino acid acetyltransferase
assay, XXXV, 247
inhibitor, XXXV, 253
pH optimum, XXXV, 252
specificity, XXXV, 253
stability, XXXV, 252
Acetyl-CoA carboxylase
carboxylation of acetyl–fatty acid synthase, XXXV, 83
from *Escherichia coli*, XXXV, 17–37
biotin carboxylase, XXXV, 25–31
biotin carboxyl carrier protein, XXXV, 17–25
carboxyltransferase, XXXV, 32–37
from rat liver, XXXV, 3–11
activation of crude and pure enzyme, XXXV, 8, 9
activator, XXXV, 8, 9
assay, XXXV, 3–5
biotin content, XXXV, 10
gel filtration of crude enzyme, XXXV, 4
inhibitor, XXXV, 8, 9
phosphate content, XXXV, 10
stability, XXXV, 8, 9
from rat mammary gland, XXXV, 11–17
activator, XXXV, 17
assay, XXXV, 11–13
inhibitor, XXXV, 17
stability, XXXV, 16
Acetyl-CoA:carnitine *O*-acetyltransferase, *see* Carnitine acetyltransferase
Acetyl-CoA hydrolase, brown adipose tissue mitochondria, LV, 76
Acetyl-CoA:long chain base acetyltransferase, XXXV, 242–247
assay, XXXV, 242–245
inhibitors, XXXV, 247
specificity, XXXV, 246, 247
stability, XXXV, 246
Acetyl-CoA synthetase
in acetate assay, XXXV, 302–307
brain mitochondria, LV, 58
Acetyl coenzyme A, XLIII, 521, 540; XLIV, 458, 474
N-acylneuraminate-9(7)-*O*-acetyltransferase, L, 381–384
assay, LV, 221
of acetyl-CoA-pantetheine transacylase, XXXV, 49
of covalent binding sites of acetyl groups to fatty acid synthase, XXXV, 51, 52
of fatty acid synthase, XXXV, 37, 60, 66, 75, 85
in carbohydrate metabolism, XXXVII, 292, 294, 295
carnitine assay, LV, 211
citrate synthase *in situ*, LVI, 550
coenzyme A assay, LV, 210
content
of hepatocytes, LV, 214
of rat heart, LV, 216, 217
of rat liver, LV, 213
in glyoxylate cycle, XXXI, 565, 566
hydrolysis, phosphorylation potential, LV, 237
inhibitor of fatty acid synthase, XXXV, 80
oxaloacetate assay, LV, 208
purification, XXXV, 74
stability, XXXV, 75
substrate
for acetoacetyl-CoA thiolase, XXXV, 134, 167
for acetyl-CoA:amine acetyltransferase, XXXV, 253
for citrate synthase, XXXV, 273, 274, 299, 302, 303
for CoA-transferase, XXXV, 235, 242, 299
for fatty acid synthase, XXXV, 312
for 3-hydroxy-3-methylglutaryl-CoA synthase, XXXV, 155, 160, 173
synthesis, XXXV, 48, 66, 74
N-Acetylcysteine
assay of creatine kinase isoenzymes, LVII, 60
Nbf-Cl, LV, 490
N-Acetylcytidine, substrate, of uridine-cytidine kinase, LI, 313

N-Acetyl-2-deoxyglucosamine derivative, XXXIV, 342–345
N-Acetyl-4,9-di-*O*-acetylneuraminic acid, occurrence, L, 386
N-Acetyl-7,9-di-*O*-acetylneuraminic acid, *N*-acylneuraminate-9(7)-*O*-acetyltransferase, L, 381
N-Acetyldioxindolylalanine spirolactone, absorption spectra, XLVII, 463
Acetylene, XLVI, 29
  inhibitor of nitrous oxide reductase, LIII, 646
  reduction, assay of nitrogenase, LIII, 329
β,γ-Acetylenic amine, XLVI, 161
Acetylenic inhibitor, XLVI, 158–164
  isotope effect, XLVI, 159
Acetylenic steroid, XLVI, 461–468
  conversion to allenic, XLVI, 466–468
Acetylenic thioester, XLVI, 158
Acetylesterase
  activity assay, XLVII, 78
  cephalosporin, *see* Cephalosporin acetylesterase
  gel electrophoresis, XXXII, 88
N-Acetyl-D-galactosamine
  Dictyostelium discoideum agglutinin inhibition, L, 311
  pallidin inhibition, L, 315
  4-*O*-sulfated, L, 451
N-Acetylgalactosamine-agarose derivative, XXXIV, 347
α-*N*-Acetylgalactosaminidase, for glycolipid structure studies, XXXII, 363
endo-α-*N*-Acetylgalactosaminidase
  from *Diplococcus pneumoniae*
    assay, L, 561–563
    properties, L, 566, 567
    purification, L, 563–565
N-Acetylgalactosaminitol, electrophoresis, L, 51
N-Acetylgalactosaminyl galactosylglucosylceramide, Tay-Sachs ganglioside, preparation of specifically labeled ganglioside, XXXV, 541–548
N-Acetyl-*O*-α-D-galactosaminylpolyacrylamide, XXXIV, 367
N-Acetylgalactosaminyltransferase
  A1 protein as acceptor, XXXII, 324
    assay, XXXII, 327–329
N-Acetylglucosamine, XXXIV, 318, 360, 639
  adsorbents, XXXIV, 360
  potato lectin inhibition, L, 344
  saccharide peptide structure, L, 113
  substrate, of lysozyme, XLIV, 290, 445
  in sugar transfer, XXXI, 23
N-Acetyl-β-glucosamine, XXXIV, 347
N-Acetyl-D-glucosamine, XXXIV, 334, 335
  as eluent, XXXIV, 346
  2′,3′-epoxypropyl β-glycoside, synthesis, XLVI, 404–410
  lectin purification, XXXIV, 333, 336
  wheat germ agglutinin column, XXXIV, 669, 670
N-Acetylglucosamine-agarose, XXXIV, 327
N-Acetylglucosamine binding protein
  affinity chromatography, L, 289–291
  binding assay, L, 289, 290
  isolation from chicken liver, L, 289–291
N-Acetyl-α-D-glucosamine 1,6-diphosphate, preparation, XLI, 83
N-Acetylglucosamine 2-epimerase, from hog kidney, XLI, 407–411
  assay, XLI, 408, 409
  properties, XLI, 411
  purification, XLI, 409–411
N-Acetylglucosamine galactosyltransferase, as diagnostic enzyme, XXXI, 20, 24
N-Acetylglucosamine kinase, from hog spleen, XLII, 58–62
  assay, XLII, 58–60
  chromatography, XLII, 61
  inhibitors, XLII, 62
  properties, XLII, 62
  purification, XLII, 60–62
N-Acetylglucosamine-6-phosphate deacetylase
  in *Bacillus subtilis*, XLI, 497
  in *Bifidobacterium bifidum*, XLI, 497
  from *Escherichia coli*, XLI, 497–502
    assay, XLI, 498

chromatography, XLI, 499
inhibitors, XLI, 501
properties, XLI, 500, 501
purification, XLI, 498–500
N-Acetylglucosamine tetrasaccharide, XXXIV, 639, 640
α-*N*-Acetylglucosaminidase
assay, L, 449, 450
distribution, L, 449
Sanfilippo B syndrome, L, 450
β-Acetylglucosaminidase
assay, XXXI, 348
in liver, XXXI, 97
in plasma membrane, XXXI, 92, 100
in PMN granules, XXXI, 349–351
β-*N*-Acetylglucosaminidase
from hen oviduct
assay, L, 580, 581
properties, L, 583, 584
purification, L, 581–583
from *Streptomyces plicatus*
assay, L, 575, 576
properties, L, 578–580
purification, L, 576–578
endo-β-*N*-Acetylglucosaminidase C_I, from *Clostridium perfringens*
assay, L, 568, 569
properties, L, 571–574
purification, L, 569–571
reaction, L, 567
endo-β-*N*-Acetylglucosaminidase C_{II}, from *Clostridium perfringens*
assay, L, 568–569
properties, L, 571–574
purification, L, 569–571
reaction, L, 567
endo-β-*N*-Acetylglucosaminidase D, from *Diplococcus pneumoniae*
assay, L, 556
properties, L, 557–559
purification, L, 556, 557
β-*N*-Acetylglucosaminide, XXXIV, 334
N-Acetylglucosaminitol, electrophoresis, L, 51
N-Acetylglucosaminyl-*N*-acetylglucosamine, *see* *N,N*′-Diacetylchitobiose
N^{α}-β-1,4-*N*-Acetylglucosaminyl-*N*-acetylmuramyl-L-alanyl-D-isoglutaminyl-N^{ε}-(pentaglycyl)-L-lysyl-D-alanyl-D-alanine, XLIII, 697
N-Acetyl-D-glucosaminyl-agarose, XXXIV, 328
N-Acetylglucosaminyltransferase, in Golgi apparatus, XXXI, 191
N-Acetyl-β-glucuronidase, in macrophage fractions, XXXI, 340
N-Acetyl-L-glutamate, requirement, carbamoyl-phosphate synthetase, LI, 21
Acetyl L-glutamic acid diamide, substrate, of papain, XLIV, 425
O-Acetylglycosyl bromide, XXXIV, 364, 365
β-*N*-Acetylhexosaminidase, XXXIV, 3
from bovine testes
assay, L, 520, 521
properties, L, 522, 523
purification, L, 521, 522
in brain fractions, XXXI, 469–471
for glycolipid structure studies, XXXII, 363
N-Acetylhexosaminyl residue, deacetylation, L, 432
N-Acetylhistidine, assay of ethoxyformic anhydride, LVII, 175
N-α-Acetyl-L-histidine, synthesis of 8α-*N*-histidylriboflavin isomers, LIII, 458
N-Acetylhomocysteine thiolacetone, XLIV, 87
N-Acetylhomocysteine thiolactone, XXXIV, 236, 682
oxidation by Br_2, LV, 537
Acetylimidazole, modification of tyrosine, LIII, 172
N-Acetylimidazole, XLIV, 532; XLVI, 64; XLVII, 601
N-Acetylindoxyl oxidase, LVI, 475
Acetyllasalocid, code number, LV, 445
N^{α}-Acetyl-L-lysyl-D-alanyl-D-alanine, XLIII, 697
N-Acetylmannosamine kinase, from *Salmonella typhimurium*, XLII, 53–58
assay, XLII, 54–56
chromatography, XLII, 57
properties, XLII, 57, 58
purification, XLII, 56, 57
Acetyl mercaptoethanol, internal standard, XLVII, 34

Acetyl-DL-methionine
maximal reactor space velocity, XLIV, 753
substrate, aminoacylase, XLIV, 747
N-Acetyl-L-methionine, substrate, of aminoacylase, XLIV, 199
Acetylmethionylpuromycin, assay for ribosome dissociation factor, LX, 296, 297
N-Acetylmorpholine, methyldithioacetate synthesis, XLVII, 290, 291
N-Acetylneomycin, XLIII, 120
N-Acetylneuramic acid, in glycoprotein hormones, XXXVII, 321–322
N-Acetylneuraminate lyase, labeling, XLVI, 139
N-Acetylneuraminate monooxygenase
assay, L, 375–377
glycoprotein synthesis, L, 380
isolation from porcine submandibular glands, L, 377–378
occurrence, L, 380
properties, L, 378, 379
reaction, L, 375
N-Acetylneuraminic acid
N-acylneuraminate-9(7)-*O*-acetyltransferase, L, 381
determination, XXXII, 361
in oligosaccharides, L, 53
^{3}H isotope effect with *N*-acetylneuraminate monooxygenase, L, 379
lectin, XXXIV, 333, 335
N-Acetylneuraminic acid hydroxylase, LII, 12
N-Acetylneuraminosylgalactosylglucosylceramide
preparation
from erythrocytes, XXXV, 546
of specifically labeled ganglioside, XXXV, 541–548
4-*O*-Acetyl-1,2,3,5,6-penta-*O*-methyl-D-galactitol, L, 24
Acetyl-DL-phenylalanine, maximal reactor space velocity, XLIV, 753
Acetyl-L-phenylalanine methyl ester, substrate, of chymotrypsin, XLIV, 409
N-Acetylphenylalanylglycylalanylleucylchloromethyl ketone, XLVI, 206
Acetylphenylalanyl-transfer ribonucleic acid, assay for initiation factor IF-3, LX, 230–239
Acetyl[^{14}C]phenylalanyl-transfer ribonucleic acid, determination of [^{3}H]puromycin specific activity, LIX, 360
Acetylphosphatase, in plasma membranes, XXXI, 88
Acetyl phosphate, XLIV, 891–894
coenzyme A assay, LV, 210
Acetylprolylalanylproline, XLVI, 224
Acetylprolylalanylprolylalaninal, XLVI, 225
Acetylprolylalanylprolylphenylalaninal, synthesis, XLVI, 224
Acetylpyridine, XLVI, 149, 251
3-Acetylpyridine adenine dinucleotide, XLVI, 250
NADH dehydrogenase, LVI, 586, 587
substrate, of NADH dehydrogenase, LIII, 17, 18
transhydrogenase assay, LV, 270, 272
3-Acetylpyridine-*n*-alkyladenine dinucleotide analog, XLVI, 250
[ω-(Acetylpyridinio)-*n*-alkyl]adenosine pyrophosphate, XLVI, 252
spectra, XLVI, 254
Acetylpyridinio-*n*-alkyl monophosphate, synthesis, XLVI, 251, 252
Acetylpyridinio-*n*-alkyl polyphosphoric acid, XLVI, 252
3-Acetylpyridinio-*n*-butyladenosine pyrophosphate, XLVI, 257
2-Acetylpyridiniopropyladenosine pyrophosphate, XLVI, 257
3-Acetylpyridiniopropyladenosine pyrophosphate, XLVI, 257
4-Acetylpyridiniopropyladenosine pyrophosphate, XLVI, 257
N-Acetylserotonin
fluorometric assay, XXXIX, 384
gas chromatography-mass spectrometry, XXXIX, 386, 387
Acetylthiocholine, as esterase substrate, XXXII, 88, 89
Acetylthiocholine iodide, substrate, of cholinesterase, XLIV, 658
Acetyltransferase, *see also* specific acetyltransferase; specific type
assay, LX, 535

catalytic effects, LX, 537, 539, 540
isolation, LX, 536–538
varieties, LX, 537, 539
N^2-Acetyl-N^{10}-trifluoroacetylpteroic acid, XXXIV, 283
Acetyl-DL-tryptophan, maximal reactor space velocity, XLIV, 753
N-Acetyl-D-tryptophan, inhibitor, of α-chymotrypsin, XLIV, 565
N-Acetyl-L-tryptophan, inhibitor of tryptophanyl-tRNA synthetase, LIX, 252
N-Acetyltryptophanamide, magnetic ellipticity, XLIX, 157, 162
Acetyl-L-tyrosinamide, substrate, of chymotrypsin, XLIV, 410
Acetyltyrosine, XLIV, 75
N-Acetyl-L-tyrosine ethyl ester
substrate, of chymotrypsin, XLIV, 31, 34, 36, 73–76, 266, 410, 411, 530, 536, 561, 562
Acetyl-DL-valine, maximal reactor space velocity, XLIV, 753
Acholeplasma, cultivation, XXXII, 459, 460
Acholeplasma laidlawii, membranes
differential scanning calorimetry, XXXII, 265
NMR studies, XXXII, 202, 203
reconstitution, XXXII, 464, 465
Achromobacter
fungal penicillin acylase, XLIII, 722
oxidase, LII, 11, 17, 19–21
penicillin acylase, XLIII, 719
Achromobacter-Alcaligenes, D-gluconate dehydratase activity, XLII, 304
Achromobacter liquidum, L-histidine ammonia-lyase, XLIV, 745
Acid, *see also* specific type
concentration in histone extraction, XL, 107
extraction, of chick bone collagen, XL, 339
fluorometric assay of mixtures of organic, XLI, 56
metal-free, preparation, LIV, 476
in plant cells, XXXI, 501
production, of bone, XL, 328
Acid-acetone precipitation, in purification of NAD^+, LIV, 227
Acid-ammonium sulfate, dissociation of flavoproteins, LIII, 430–432
Acid chloride, coupling to inorganic supports, XLIV, 145
Acid-citrate-dextrose formula A, LII, 467
Acid-citrate-dextrose reagent, XXXII, 115
for affinity chromatography, preparation, XXXVI, 106, 107
Acid cleaning, LVIII, 5
Acid-copper reagent, for colorimetric assay of reducing sugars and hexuronic acids, XLI, 29, 30
Acid deoxyribonuclease
in brain homogenates, XXXI, 464, 466, 467
in chromaffin granule, XXXI, 381
in lysosomes, XXXI, 330
in macrophage fractions, XXXI, 340
Acid esterase, in brain fractions, XXXI, 469–471
Acid gel, for polyacrylamide electrophoresis, XXXII, 79
Acid glycerophosphatase
in brain homogenates, XXXI, 464, 466
in macrophage fractions, XXXI, 340
α_1-Acid glycoprotein, XXXVI, 91
molecular weight, XXXVI, 99
steroid bonding, XXXVI, 93, 95, 97
Acid hydrolase
in brain fractions, XXXI, 464–471
in lysosomes, XXXI, 23, 24
in macrophage fractions, XXXI, 340
Acid hydrolysis
determination, covalent coupling group, XLIV, 16
identification of amino acid bound to flavin ring, LIII, 463, 464
Acid *p*-nitrophenolphosphatase, in brain homogenates, XXXI, 464, 466, 467
Acid nucleotidase, LI, 271–275
activators, LI, 274
assay, LI, 272
inhibitors, LI, 274
kinetic properties, LI, 274
molecular weight, LI, 275
pH optimum, LI, 274
purification, LI, 273, 274
purity, LI, 274

from rat liver lysosomes, LI, 271–275
substrate specificity, LI, 274
temperature optimum, LI, 274
Acidosis, metabolite content of kidney cortex, LV, 220
Acid phosphatase, XLVI, 3
in aleurone grains, XXXI, 576
in amoebic phagosomes, XXXI, 697
assay, XXXI, 329, 361, 362
by Fast Analyzer, XXXI, 816
associated with cell fractions, XL, 86
in brain fractions, XXXI, 469–471
in brush borders, XXXI, 121
cytochemistry, XXXIX, 150, 151
as diagnostic enzyme, XXXI, 20, 186, 187, 326, 327, 330, 743, 744
for lysosomes, XXXI, 330, 334, 337, 338, 344, 735
in liver-cell fractions, XXXI, 209
in lung lamellar bodies, XXXI, 423, 424
lysosomes, LV, 101, 103, 104
assay, LV, 102
as marker enzyme, XXXII, 15
partial purification, LI, 273
in peroxisomes, XXXI, 361, 362
viscosity barrier centrifugation, LV, 133, 134
Acid precipitation
of liver proteins, LII, 507
purification of adenylate kinase, LI, 461
of aspartate carbamyltransferase, LI, 43
of cytochrome *c*-551, LIII, 657
of hypoxanthine phosphoribosyltransferase, LI, 545
of OPRTase-OMPdecase complex, LI, 160
of phosphoribosylpyrophosphate synthetase, LI, 7–9
of pyrimidine nucleoside monophosphate kinase, LI, 324
of pyrimidine nucleoside phosphorylase, LI, 433, 434
of ribosephosphate pyrophosphokinase, LI, 14, 15
of UMP-CMP kinase, LI, 334
Acid protease, in aleurone grains, XXXI, 576
Acid proteinase, as marker enzyme, XXXI, 745
Acid ribonuclease
in brain homogenates, XXXI, 464, 466, 467
in lysosomes, XXXI, 330
in macrophage fractions, XXXI, 340
Acid silica gel plate, isolation of quinone intermediates, LIII, 607
Aconitase, *see also* Aconitate hydratase
citrate assay, LV, 211
in glyoxysomes, XXXI, 565, 569
cis-Aconitate, activator, of adenylate kinase, LI, 465
Aconitate hydratase
latency, intact mitochondria, LV, 143
in microbodies, LII, 496
toluene-treated mitochondria, LVI, 547
ACP, *see* Acyl carrier protein
AcPyAD, *see* 3-Acetylpyridine adenine dinucleotide
Acquisition time, choice, LIV, 158, 159
Acridine, XLVII, 350
as mutagen, XLIII, 35
as optical probes, LV, 573
Acridine carboxylic acid, chemiluminescence, LVII, 423
Acridine dye, LVI, 31
fluorescence quenching, assay in mutants, LVI, 114
Acridine orange, XLVI, 563
cell viability, LVIII, 152
as curing agent, XLIII, 51
DNA fixation, XL, 33
as fluorescent probes, XXXII, 235, 239, 240
as lysosome stain, XXXI, 472
use in assessing viability of fat cells, XXXV, 560
Acridone anion, chemiluminescence, LVII, 417, 423
Acriflavin
in electron microscope cytochemistry of nucleic acids, XL, 24, 32
photochemically active derivative, LVI, 660
sulfate incorporation into mitochondrial protein, LVI, 60

Acrolein, XLVI, 33
synthesis of *Cypridina* luciferin, LVII, 369
Acrolein fixative, XXXIX, 138
Acrosin, XXXIV, 3; XLVI, 206
sperm
assay, XLV, 331
active-site titration, with *p*-nitrophenyl-*p*′-guanidinobenzoate, XLV, 334
amidase activity, XLV, 334
combined N^{α}-benzoyl-L-arginine ethyl ester/alcohol dehydrogenase, XLV, 333
comparison and evaluation, XLV, 335
direct N^{α}-benzoyl-L-arginine ethyl ester, XLV, 333
esterase activity, XLV, 332
proteolytic activity estimation, XLV, 331
composition, XLV, 341, 342
function, XLV, 325
inhibitors, XLV, 337
kinetics, XLV, 340, 341
molecular weight, XLV, 340
properties, XLV, 330, 338
purification, XLV, 336, 338
specificity, XLV, 342
stability, XLV, 339
Acrylamide, XLIV, 57, 171–177, 185, 194, 198, 273, 741, 750, 889, 891; LIX, 65
as carrier, XXXIV, 3, 4, 7
coentrapment procedure, XLIV, 475, 476
copolymer, with vinylated proteins, XLIV, 195–201
electrophoretic analysis of oligonucleotides, LIX, 68, 69
of protein-RNA complexes, LIX, 567
of ribosomal proteins, LIX, 431, 506, 507, 539, 544
of ribosomal RNA, LIX, 458
of RNA fragments, LIX, 574
entrapment for enzyme electrode, XLIV, 600
estradiol adsorbents, XXXIV, 687, 688
fluorescence quenching agent, XLIX, 225, 231, 233, 235
for gel electrophoresis, XXXII, 74, 76, 85, 94, 95
as neurotoxin, XXXII, 73
for gel preparation, LI, 342; LII, 325
inhibitor, of phagocyte chemiluminescence, LVII, 492
photochemical polymerization, XLIV, 571, 572
SDS-acrylamide gels, LIX, 509, 539, 544
separation of bacterial ribosomal subunits, LIX, 398
support
for trypsin, XLIV, 662
for trypsin/chymotrypsin, XLIV, 571, 572
Acrylamide/acrylic acid copolymer, XLIV, 180
1-ethyl-3-(3-dimethylaminopropyl) carbodiimide activation, XLIV, 62
multistep enzyme system, XLIV, 454–456
negative charge, advantage, XLIV, 63
Acrylamide/1-acryloylamino-2-(4-nitrobenzoylamino)ethane copolymer
characterization, XLIV, 92
mechanical stability, XLIV, 92
reduction of aryl nitro group, XLIV, 92
synthesis, XLIV, 91, 92
thermolysin immobilization, XLIV, 93, 94
Acrylamide/*N*-acryloyl-*N*′-*t*-butyloxycarbonyl hydrazine copolymer, *see* Enzacryl AH
Acrylamide/*N*-acryloyl-4-carboxymethyl-2,2-dimethylthiazolidine copolymer, XLIV, 87
Acrylamide gel electrophoresis, *see* Polyacrylamide gel electrophoresis
Acrylamide/2-hydroxyethylmethacrylate ester copolymer, XLIV, 85
Acrylamide/methylacrylate copolymer, XLIV, 87
hydrophobicity studies, XLIV, 63, 64
Acrylamide/4-nitroacrylanilide copolymer, *see* Enzacryl AA

Acrylic acid, XLIV, 57
anionic copolymers, XLIV, 108
Acrylic acid-2,3-epoxypropyl ester, structural formula, XLIV, 197
Acrylic acid/3-isothiocyanatostyrene copolymer
papain immobilization, XLIV, 114
polymerization, XLIV, 113, 114
swellability, XLIV, 113, 114
Acrylic acid-*O*-succinimide ester, structural formula, XLIV, 197
Acrylic acid-2,3-thioglycidyl ester, structural formula, XLIV, 197
Acrylic plastic rod, for light guide, LVII, 320
Acryloylaminoacetaldehyde dimethylacetal, XLIV, 88
Acryloylaminoacetaldehyde dimethylacetal/*N,N*′-methylene*di* acrylamide copolymer, XLIV, 105
1-Acryloylamino-2-(4-nitrobenzoylamino)-ethane, XLIV, 91
Acryloyl-*N,N*-bis(2,2-dimethoxyethyl)amine, XLIV, 88, 90
preparation, XLIV, 102
Acryloyl-*N,N*-bis(2,2-dimethoxyethyl)amine/*N,N*′-methylene*dia* crylamide copolymer
β-D-glucosidase immobilization, XLIV, 103
preparation, XLIV, 102
N-Acryloyl-*N*′-*t*-butoxycarbonyl hydrazide, XLIV, 87, 92, 93
Acryloyl morpholine, XLIV, 102
Acryloyl morpholine/acryloyl-*N,N*-bis(2,2-dimethoxyethyl)amine copolymer, XLIV, 88, 90
preparation, XLIV, 102, 103
characterization, XLIV, 103
β-D-glucosidase immobilization, XLIV, 103
storage, XLIV, 103
N-Acryloyloxysuccinimide, XLIV, 889–891
optimum concentration, XLIV, 896, 897
synthesis, XLIV, 894
Acrylyl chloride, XLIV, 101, 102, 196, 198, 894
structural formula, XLIV, 197
ACTH, *see* Adrenocorticotropic hormone; Adrenocorticotropin
Actidione, *see* Cycloheximide
Actin, XLVI, 294
antibody binding, L, 55
effect on bilayers, XXXII, 500, 535
polymerization, XLVI, 293, 294
saturation-transfer studies, XLIX, 495
Actinamine, XLIII, 216, 217
Actinoidin, XLIII, 127
Actinomycete, *see also* specific type
antibiotic fermentation, XLIII, 18–21, *see also* specific antibiotic
media for maintaining, XLIII, 5–7, *see also* Media, for actinomycetes
soil culture, XLIII, 8
Actinomycin
chiroptical studies, XLIII, 352
circular thin-layer chromatography, XLIII, 180
countercurrent distribution, XLIII, 327
high-pressure liquid chromatography, XLIII, 313–319
paper chromatography, XLIII, 150, 151
in promoter site mapping studies, LIX, 847
as protein synthesis inhibitor, XXXII, 870
solvent systems for countercurrent distribution, XLIII, 334
synthesis, XLIII, 763–767, *see also* Actinomycin lactonase
thin-layer chromatography, XLIII, 187, 202
Actinomycin D, LVI, 31
^{3}H-labeled, binding, XL, 12
RNA inhibitor, XL, 278, 280
Actinomycin lactonase, XLIII, 763–767
Actinoplanes, XLIII, 765, 766
ammonium sulfate fractionation, XLIII, 765
assay, XLIII, 764
calcium phosphate gel chromatography, XLIII, 765, 766
crude extract preparation, XLIII, 765
culture preparation, XLIII, 764, 765
DEAE-cellulose column chromatography, XLIII, 766

inactivation product, XLIII, 767
properties, XLIII, 767
purification, XLIII, 764–766
reaction scheme, XLIII, 763
Sephadex G–200 column chromatography, XLIII, 766
Actinoplanes missouriensis
actinomycin, XLIII, 763, *see also* Actinomycin lactonase
peptide antibiotic lactonase, XLIII, 767, *see also* Peptide antibiotic lactonase
Actinospectacin
paper chromatographic data, XLIII, 121
solvent system, XLIII, 119, 120
Actinospectinoic acid, XLIII, 216, 217
Action potential, in culture, LVIII, 307
Actithiazic acid, XLIII, 334
Activating Factor, purification, LIII, 321
Activation energy, XLIX, 27, 35
Activation enthalpy, LIV, 508
distribution curves, for heme proteins, LIV, 523
Activation entropy, LIV, 508
Active site
effect on residue reactivity, XLIV, 15
geometry, electron spin resonance spectra, XLIX, 420, 421
involved in adsorption, XLIV, 153
protection during coupling, XLIV, 15
residue identification, electrophoretic mobility, XLVII, 66, *see also* Affinity labeling
titration, XLIV, 390–393, 436
Active transport
across artificial enzyme membranes, XLIV, 921–924
carbodiimide-resistant mutant, LVI, 171
model system, XLIV, 434, 435
in substrate entry, XXXIX, 498–502
Activity coefficient, equation, LVI, 360, 365
Acumycin, XLIII, 134
Acute myelocytic leukemia blast cell, *see* Leukemia
Ac–X–537A antibiotic, LV, 445
S-Acyl-*N*-acetylcysteamine derivative, synthesis, XXXV, 48, 49
p-Acylamidobenzoic acid
esterification, LII, 517
synthesis, LII, 516, 517
Acylamino acid-releasing enzyme
activators, XLV, 559
assay, by ninhydrin, XLV, 552
composition, XLV, 558, 560
distribution, XLV, 559
function, XLV, 552
for peptide sequence determination, XLV, 561
inhibitors, XLV, 559
isoelectric point, XLV, 557
kinetics, XLV, 557, 558
molecular weight, XLV, 557
pH optimum, XLV, 557
properties, XLV, 557
purification, XLV, 556, 557
residues, terminal, XLV, 558–560
specificity, XLV, 557
stability, XLV, 557
structure, subunit, XLV, 557
synthetic substrate, XLV, 553
N-Acylamino sugar, XXXIV, 324–327, *see also* specific sugars
Acylase, XXXIV, 105
Acylation, monomers used, XLIV, 197
Acylatractyloside, spin-labeled, preparation, LV, 527
Acyl azide, XLVI, 78
activation of collagen films, LV, 744, 745
activation mechanism, XLIV, 19
thermolysin binding, XLIV, 95
Acyl azide derivative, XXXIV, 46–48, 52
Acyl carnitine
in assay of long-chain fatty acyl-CoA synthase, XXXV, 117
hydrolysis, LV, 202, 203
Acyl carrier protein, XXXV, 84, 90, 95–101, 110–114
from *Euglena gracilis*, XXXV, 110–114
dimer, XXXV, 113
phosphopantetheine content, XXXV, 114
stability, XXXV, 113, 114
Acyl carrier protein hydrolase, XXXV, 101

Acyl carrier protein synthetase, XXXV, 95–101
activators, XXXV, 100
apo-peptide acceptors, XXXV, 100, 101
assay, XXXV, 95–97, 99, 100
inhibitors, XXXV, 100, 101
stability, XXXV, 100
Acylcholine acyl-hydrolase, XLIV, 647
Acyl-CoA:6-aminopenicillanic acid acyltransferase, XLIII, 474–476
activators, XLIII, 476
ammonium sulfate precipitation, XLIII, 475
assay, XLIII, 474, 475
hydroxyapatite gel adsorption and elution, XLIII, 475
Penicillium chrysogenum preparation, XLIII, 475
pH effect, XLIII, 476
purification, XLIII, 475, 476
Sephadex G–200 column chromatography, XLIII, 475
specific activity, XLIII, 474
specificity, XLIII, 476
unit definition, XLIII, 474
Acyl-CoA dehydrogenase, LIII, 502–518
activity, LIII, 502
assay, LIII, 503, 504
from beef heart, LIII, 506, 507, 510–512, 518
catalytic properties, LIII, 506, 507
deficiency, in *E. coli* mutants, XXXII, 859
electrophoretic properties, LIII, 508, 509, 512
flavin content, LIII, 508
optical properties, LIII, 504, 505, 508, 510, 511
from pig liver, LIII, 506–518
stability, LIII, 518
substrate specificities, LIII, 502, 503
Acyl-CoA hydrolase, in assay of long-chain fatty acyl-CoA, XXXV, 273
Acyl-CoA oxidase, in glyoxysomes, XXXI, 369
Acyl-CoA synthetase, localization, LV, 9
Acyl coenzyme A
AIP synthesis, LV, 77
assay, XXXV, 66
derivative, assay of acyl-CoA dehydrogenases, LIII, 502, 503, 506
long chain, hydrolysis, LV, 202, 203
transport, LVI, 248
1-Acylglycerol-3-phosphate acyltransferase, in *E. coli* auxotrophs, XXXII, 864
Acyl-*N*-glycoside, XXXIV, 323, 324, *see also* specific glycosides
Acyl hydrazide, XLVI, 209
Acyl hydrazide derivative, XXXIV, 34
Acyl hydrazide group, assay, by titration, XLIV, 93
Acyl hydrolase, lipolytic
enzymatic activity, XXXI, 521, 522, 525
inhibition, XXXI, 524
in plant extracts, XXXI, 527
Acyl hydroxamate, XLIII, 477
chromatography, L, 81
1-Acyl-lysolecithin, *see* DL-Stearoyl-3-lysolecithin
2-Acyl-lysophosphatidylethanolamine, *see* 2-*O*-Palmitoyl-*sn*-glycerol-3-(2′-aminoethyl hydrogen phosphate)
Acylmannosamine
from acylneuraminic acids, L, 81, 82
thin layer chromatography, L, 81, 82
N-Acylneuraminate-4-*O*-acetyltransferase, from equine submandibular gland, reaction, L, 386
N-Acylneuraminate-7-*O*-acetyltransferase
N-acylneuraminate-9(7)-*O*-acetyltransferase, L, 381
reaction, L, 381
N-Acylneuraminate-9(7)-*O*-acetyltransferase
assay, L, 381–384
glycoprotein synthesis, L, 386
isolation from bovine submandibular gland, L, 384
occurrence, L, 385, 386
properties, L, 384, 385
reaction, L, 381
Acylneuraminate pyruvate-lyase
acylneuraminic acid cleavage, L, 81, 82
sialic acid determination, L, 75, 76

Acylneuraminic acid
fractionation, L, 69
isolation, L, 67, 68
purification, L, 68, 69
erythro-*N*-Acyl-D-sphingosine 2-aminoethylphosphonate, intermediate in synthesis of sphingomyelin analogues, XXXV, 526
O-Acylthiamin hydroxylase, LII, 16
Acyltransferase, *see* Acyl-CoA:6-aminopenicillanic acid acyltransferase
N-Acylurea, formation, XLVII, 280, 281
AD, *see* Acyl-CoA dehydrogenase
ADA, *see* Azodianiline
Adair expression, XLVIII, 286, 288
positive cooperativity, XLVIII, 293
Adams Suction Apparatus, XXXII, 98, 99
ADC, *see* Analog-to-digital converter
Adenine, XLVI, 19
cAMP incorporation, XXXIX, 266–268
chromatographic separation, LI, 559, 560
in cyclic AMP assay, XXXI, 105–109
determination, LI, 263
electrophoretic mobility, LI, 570
electrophoretic separation, LI, 568–570
inhibitor of adenosine deaminase, LI, 506
of nucleoside deoxyribosyltransferase, LI, 455
of orotate phosphoribosyltransferase, LI, 166
in media, LVIII, 54, 66
phosphodiesterase, XXXVIII, 244
product of adenosine monophosphate nucleosidase, LI, 263
radiolabeled
incorporation into cAMP, XXXII, 679, 680
purification, LI, 575
substrate of adenine phosphoribosyltransferase, LI, 558
of tryptophanyl-tRNA synthetase, LIX, 251, 252
Adenine deaminase, in bacteria, LI, 263
Adenine nucleotide, *see also* specific compounds
analogs, XLVI, 259, 302–307
assay, LV, 668
binding by ATPase, LVI, 527
binding sites of ATPases, LV, 300
calcium influx, LVI, 340
cell fractionation by digitonin method, LVI, 212, 213
distribution in mitochondria and cytosol using digitonin fractionation, LVI, 213–215
cavitation procedure, LVI, 219
extraction methods, LVII, 81
in mitochondria, LV, 239
production in isolated renal tubules, XXXIX, 13, 16, 18
ratio, constancy, LV, 230–232
spin-labeled, XLVI, 260
stock solutions, LVI, 531, 532
transport, XLVI, 83
transporter, reconstituted, LV, 706, 710
activity, LV, 705
Adenine phosphoribosyltransferase, LI, 558–580
activity, LI, 558, 559
amino acid analysis, LI, 580
in 5′-amino-1-ribosyl-4-imidazolecarboxamide 5′-phosphate synthesis, LI, 189
assay, LI, 559–561, 568–570, 574, 575
cation requirement, LI, 556, 573
cellular localization, LI, 559, 568
from *E. coli*, LI, 558–567
effectors, LI, 566, 573
from human erythrocytes, LI, 568–574
isoelectric point, LI, 573, 579, 580
kinetic properties, LI, 565, 573, 580
molecular weight, LI, 565, 573, 579
from monkey liver, LI, 574
mouse cells deficient in, in cell hybridization studies, XXXII, 578

pH optimum, LI, 566, 574, 580
in phosphoribosylpyrophosphate synthetase assay, LI, 5
physiological function, LI, 566, 567
purification, LI, 554, 561–564, 570–572, 575–579
purity, LI, 565, 579
from rat liver, LI, 574–580
reaction stoichiometry, LI, 565
regulation, LI, 558, 559
sedimentation coefficient, LI, 573
in situ properties, LI, 567
specific activity, LI, 565
stability, LI, 564, 565, 579
Stokes radius, LI, 573
substrate specificity, LI, 565, 572, 573
sulfhydryl content, LI, 580
tissue distribution, LI, 568
turnover number, LI, 565
Adenine-pseudouridine base pairing, LIX, 47, 48
Adenocarcinoma
cervical, monolayer culture, LVIII, 133
fucose glycolipid, XXXII, 365
renal, LVIII, 376
Adenosine, XXXIV, 3; XLVI, 19
adenylate cyclase, XXXVIII, 169
analogs, inhibitors, of GMP synthetase, LI, 224
chromatographic separation, LI, 508; LIX, 188
derivatives, XLVI, 240–249
diethylcarbonate treatment, XLVII, 432, 442
in DNA synthesis, XL, 42
electrophoretic mobility, LI, 570
inhibitor of GMP synthetase, LI, 224
of orotate phosphoribosyltransferase, LI, 166
optical spectra, XLIX, 164, 165
phosphodiesterase, XXXVIII, 244
photochemically active derivative, LVI, 660
residue, cleavage site, LIX, 105
resistant strains, LVIII, 314
substrate of adenosine deaminase, LI, 502, 506, 507, 508, 511
of nucleoside phosphotransferase, LI, 391, 392
of purine nucleoside phosphorylase, LI, 521
of tryptophanyl-tRNA synthetase, LIX, 251, 252
Adenosine-2′,5′-bisphosphate, XLIV, 864
Adenosine-5′-(β-bromoethane phosphonate), inhibitor of tryptophanyl-tRNA synthetase, LIX, 255
Adenosine-5′-(β-bromoethane pyrophosphonate), inhibitor of tryptophanyl-tRNA synthetase, LIX, 255
Adenosine-5′-(β-chloroethyl phosphate), inhibitor of tryptophanyl-tRNA synthetase, LIX, 255
Adenosine-5′-chloromethane pyrophosphonate, inhibitor of tryptophanyl-tRNA synthetase, LIX, 255
Adenosine-5′-chloromethyl phosphonate, inhibitor of tryptophanyl-tRNA synthetase, LIX, 251
Adenosine 3′,5′-cyclic monophosphate, *see* Cyclic adenosine monophosphate
Adenosine-2′,3′-cyclic phosphate hydrolase, *see* Cyclic adenosine phosphodiesterase
Adenosine deaminase, XXXIV, 3; XLVI, 19, 25, 327–335; LI, 502–512
activity, LI, 502, 508
assay, LI, 502, 503, 508, 509
congenital immunodeficiency, LI, 506
difference spectrum with inhibitors, XLVI, 333
from *E. coli*, LI, 508–512
friction ratio, LI, 505
6-halopurine ribosides as substrates, XLVI, 328
from human erythrocytes, LI, 502–507
inactivation kinetics, XLVI, 334
inhibitors, LI, 506, 507, 511
kinetic properties, LI, 506, 511
molecular weight, LI, 505, 512
in nucleotidase assay, XXXII, 127, 369
partial specific volume, LI, 505

properties, LI, 505, 506, 511, 512
purification, LI, 503–505, 509–511
purity, LI, 511
reaction, XLVI, 332, 333
sedimentation coefficient, LI, 505
stability, LI, 505, 512
Stokes radius, LI, 505
substrate specificity, LI, 505–507, 511
subunit structure, LI, 505
Adenosinediphosphatase, in plasma membranes, XXXI, 88
Adenosine 5′-diphosphate, XXXIV, 486; XLIV, 507; XLVI, 64, 65
acceptor control ratios, LVI, 687
activator, of adenylate deaminase, LI, 496
allosteric effector, of glutamate dehydrogenase, XLIV, 507, 512, 513
analogs, XLVI, 241
aqueous stock solution, stability, LVII, 38
assay, LV, 221; LVII, 74
of creatine kinase isoenzymes, LVII, 60, 63
of mitochondrial ATP production, LVII, 41–44
of photophosphorylation, LVII, 55
of succino-AICAR synthetase, LI, 187
binding proteins, XLVI, 248, 249
bound, phosphorylation, LVI, 496
cGMP assay, XXXVIII, 112
chloroplast ATPase, LV, 300
chromatographic separation, LI, 459, 460
CoA-binding protein, XXXIV, 270
Co(III) complex, XLVI, 313
content
of brain, LV, 218
of hepatocytes, LV, 214
of kidney cortex, LV, 220
of rat heart, LV, 216, 217
of rat liver, LV, 213
coupling factor, LV, 189, 190, 192
E. coli ATPase, LV, 795, 796
enzymic conversion to ATP, LVII, 78
equivalence to ATP, LV, 231
estimation, XXXVII, 248
exchange, LVI, 253
exchange reactions, LV, 253, 254
F_1 ATPase, LV, 324, 328
in firefly luciferase reagents, LVII, 8, 88
hepatoma mitochondria, LV, 87
inhibitor of amidophosphoribosyltransferase, LI, 178
of inorganic pyrophosphatase, LVII, 29
of orotate phosphoribosyltransferase, LI, 163
of orotidylate decarboxylase, LI, 163
of phosphoribosylpyrophosphate synthetase, LI, 11
of rat liver carbamoyl-phosphate synthase, LI, 119
of ribosephosphate pyrophosphokinase, LI, 17
interference, in luciferase assay, LVII, 11
as iron chelator, LII, 309
in isolated renal tubules, XXXIX, 16, 18
as ligands, XXXIV, 482
microorganismal mitochondria, LV, 143
mitochondrial protein synthesis, LVI, 21, 22
onset of labeling during synthesis by chloroplasts, LV, 246, 247
oxygen ratio, of adrenal mitochondrial studies, XXXVII, 300, 301
oxygen uptake, LV, 226
Pasteur effect, LV, 290, 291
pet9 mitochondria, LVI, 127, 128
phosphorylation potential, LV, 238, 239
photochemically active derivative, LVI, 656
in polarographic studies of membrane phase changes, XXXII, 260
product of adenylate kinase, LI, 459
of carbamoyl-phosphate synthetase, LI, 21, 29
of deoxycytidine kinase, LI, 337
of deoxythymidine kinase, LI, 360
of FGAM synthetase, LI, 193

of guanylate kinase, LI, 473, 483
of *Lactobacillus* deoxynucleoside kinases, LI, 346
of phosphoribosylglycinamide synthetase, LI, 179
of pyrimidine nucleoside monophosphate kinase, LI, 321
of succino-AICAR synthetase, LI, 186
of uridine-cytidine kinase, LI, 299, 331
proton-motive force, LV, 242
recycling enzyme system, XLIV, 216
removal of ATP, LVII, 61
respiratory control ratio, LV, 8
ribosylation
accompanied by dansylation, LX, 718, 719
assay for EF-2, LX, 677, 678, 706, 710, 711
separation, LV, 285, 288
procedure, LV, 251, 252
reagents and materials, LV, 250, 251
spin-labeled, XLVI, 285, 286
substrate
of adenylate kinase, XLIV, 892
of creatine kinase, LVII, 58
of nucleoside diphosphokinase, LI, 375
of pyruvate kinase, LVII, 74, 82
of ribonucleoside diphosphate reductase, LI, 235, 236, 245
translocator, binding properties of specific inhibitors, LV, 531
transport, LVI, 247–250
rate, LVI, 246
Adenosine diphosphate-adenosine triphosphate, external/internal ratios, LVI, 254
Adenosine diphosphate-adenosine triphosphate carrier
genetic modification, LVI, 125–130
inhibitors, LVI, 251
isolation, LVI, 251, 252, 407–409
preparation of BKA-protein complex, LVI, 412–414
principle, LVI, 409, 410
purification of CAT-protein complex, LVI, 410–412
Adenosine 5′-diphosphate-agarose, XXXIV, 477
Adenosine kinase, preparation of 3′-amino-3′-deoxy-ATP, LIX, 135, 136
Adenosinemonophosphatase, in microsomal fractions, LII, 83
Adenosine 2′-monophosphate
substrate, of acid nucleotidase, LI, 274
transhydrogenase, LV, 270
Adenosine 2′(3′)-monophosphate, electrophoretic separation, LIX, 177, 180
Adenosine 3′-monophosphate
phosphate donor, of nucleoside phosphotransferase, LI, 391, 392
substrate, of acid nucleotidase, LI, 274
Adenosine 5′-monophosphate, XXXIV, 242–253; XLVI, 296, 297
aminoalkyl esters, XLVI, 26
analogs, XXXIV, 125, 126; XLVI, 299
isozymes, XXXIV, 597
aqueous stock solution, characteristics, LVII, 38
stability, LVII, 38
assay, LV, 208, 209, 221
of *p*-hydroxybenzoate hydroxylase, LIII, 545
of melilotate hydroxylase, LIII, 553
of mitochondrial ATP production, LVII, 41, 42
of nitric oxide reductase, LIII, 645
bromination, LVI, 647
in chemotaxis studies, XXXIX, 491
chromatographic mobility, LIX, 73
chromatographic separation, LI, 459, 460, 559, 560
chromatography, XLVI, 281
Co(III) complex, XLVI, 313
in complex II, LIII, 24
content
of brain, LV, 218
kidney cortex, LV, 220
rat heart, LV, 216, 217
rat liver, LV, 213
conversion to FMN, LIII, 423
cyclic, photochemically active derivatives, LVI, 656

derivatives, TLC chromatography, XXXIV, 246
determination by enzymic analysis, LIII, 422, 423
by fluorometric analysis, LIII, 423, 425–429
electrophoretic mobility, LI, 570; LVII, 117; LIX, 64
electrophoretic separation, LI, 568–570
elution, XXXIV, 236
enzymic conversion to ATP, LVII, 78, 84, 95, 96, 99, 101
estimation, XXXVII, 248
extinction coefficient, LI, 220; LIII, 421
formation, during aminoacylation assay, LIX, 287, 288
as general ligand, XXXIV, 229–253, *see also* specific ligand
immobilization, XLIV, 887
inhibitor of adenine phosphoribosyltransferase, LI, 566
of adenylosuccinate synthetase, LI, 212
of amidophosphoribosyltransferase, LI, 178
of orotate phosphoribosyltransferase, LI, 163
of orotidylate decarboxylase, LI, 163
of ribosephosphate pyrophosphokinase, LI, 17
in isolated renal tubules, XXXIX, 16, 18
labeled
chemical synthesis, XXXVIII, 414–416
cyclization, XXXVIII, 416–418
determination of specific activity, XXXVIII, 419
8α-linkage, LIII, 449–452
in media, LVIII, 66
molar fluorescence, LIII, 428
M value range, LIX, 89
phosphate donor, of nucleoside phosphotransferase, LI, 391, 392
phosphodiesterase, XXXVIII, 256
photochemically active derivatives, LVI, 656
product of adenine phosphoribosyltransferase, LI, 558, 568
of adenylosuccinate AMP-lyase, LI, 202
of GMP sythetase, LI, 213, 219
of luciferase, LVII, 3, 37, 58
synthesis of 8-bromoadenosine 5′-monophosphate, LVII, 117
of phosphoribosylpyrophosphate synthetase reaction, LI, 3, 12
in reconstitution of melilotate hydroxylase, LIII, 555
redox potential, LIII, 400
reduced, product, of D-lactate dehydrogenase, LIII, 519
in salicylate hydroxylase, LIII, 528, 534
separation, LV, 285
sites, XLVI, 241
spectrophotometric determination, LI, 219, 220
spin-labeled, XLVI, 285, 286
structural formula, LI, 203
substrate of acid nucleotidase, LI, 274
of adenosine monophosphate nucleosidase, LI, 263, 264
of adenylate deaminase, LI, 490, 497
of adenylate kinase, LI, 459
of glycine reductase, LIII, 373
of D-lactate dehydrogenase, LIII, 519
in succinate dehydrogenase, LIII, 30
tritium labeling rate, LIX, 336, 342
ubiquinone redox state, LIII, 583
Adenosine 5′-monophosphate-agarose, XXXIV, 229, 262, 263, 477
Adenosine monophosphate-dialdehyde, stability, LIX, 174–177, 180
Adenosine monophosphate nucleosidase, *see* AMP nucleosidase
Adenosine nucleotide analog, XLVI, 289
Adenosine 5′-phosphate, *see* Adenosine 5′-monophosphate
Adenosine phosphate-agarose, XXXIV, 490

Adenosine 5′-phosphate aminohydrolase, *see* Adenosine (phosphate) deaminase
Adenosine (phosphate) deaminase, XLVI, 305
Adenosine-3′-phosphate 5′-sulfatophosphate, LII, 69
Adenosine 5′-phosphoromorpholidate, XLVI, 252
Adenosinetriphosphatase, XXXI, 591; XXXIV, 135; XLVI, 62, 75, 87, 89, 260, 295, 318
  activity, of carbamoyl-phosphate synthetase subunits, LI, 26
  assay, XXXII, 293, 429; LV, 320, 329, 788, 789, 801; LVI, 168
    in carbodiimide-resistant mutants, LVI, 168
    complex V, LV, 313, 314
    factors affecting, LVI, 168, 169
    inhibitors, LVI, 585, 586
    using firefly luciferase, LVII, 51, 53, 54
  assembly, inhibition, LVI, 55
  aurovertin-resistant, assay, LVI, 180, 181
  azide, LV, 516
  of beef heart mitochondria, LV, 317, 318
    definition of unit and specific activity, LV, 304
    ligand binding, LVI, 527–530
    measurement of protein concentration, LV, 304, 305
    properties, LV, 319
    purification procedure, LV, 305–308, 318, 319
  of brown adipose tissue mitochondria, LV, 77, 78
  Ca^{2+}-activated
    assay, XXXII, 303
    isolation, from kidney cells, XXXII, 303–306
    properties, XXXII, 305, 306
  calcium-binding assay, XXXII, 293, 294, 295–297
  (Ca^{2+},Mg^{2+})-activated
    aerobic photosensitivity, LIV, 52
    probe binding, LIV, 51
    rotation studies, LIV, 58
  carbodiimide-resistant mutants, LVI, 163, 164
    biochemical analysis, LVI, 167–170
    selection, LVI, 164–167
    strain RF-7, LVI, 170–173
  4-chloro-7-nitrobenzofurazan, LV, 489, 490
  chromatography, XXXII, 431, 432
  complex V, LVI, 584
  continuous measurement with firefly luciferase, LVI, 530, 531
    materials and instrumentation, LVI, 531–533
    precautions, LVI, 514–544
    qualitative aspects, LVI, 533
    quantitation, LVI, 534–537
    rapid kinetics, LVI, 540, 541
    rate of change of ATP concentration, LVI, 537–540
  continuous sucrose gradient resolution, XXXII, 398–404
  cross-linking, LVI, 629
  cyclase preparation, XXXVIII, 143, 151
  DBCT, LV, 511
  DCCD, LV, 499
  as diagnostic enzyme, XXXI, 20, 36–38
  dicyclohexylcarbodiimide-sensitive
    reconstitution
      detergents, LV, 702–704
      purple membrane vesicles, LV, 778
    of $TF_0{\cdot}F_1$, LV, 369, 370
      reconstitution of vesicles, LV, 370–372
    of thermophilic bacterium
      assay methods, LV, 364–368
      preparation of $TF_0{\cdot}F_1$, LV, 368, 369
      properties of reconstituted vesicles, LV, 372
  dissociation and reconstitution from *E. coli*, LV, 800, 801
    assay methods, LV, 801, 802
    properties, LV, 805
      of reconstituted ATPase, LV, 809, 810

purification of wild type and mutant ATPases, LV, 802–804
reversible dissociation, LV, 805–808
in vitro complementation of mutant and wild type, LV, 810
in *E. coli* cell envelope, XXXII, 91
effect of DABs or PMPs labeling, LVI, 621
efrapeptin, LV, 494, 495
electron microscopy, XXXII, 31
energy-transducing
binding sites, LV, 299–302
distribution, LV, 297
physical properties, LV, 298, 299
F_1, XLVI, 83, 277, 278, 287
in fat cells, XXXI, 67
gel electrophoresis, XXXII, 297
hepatoma mitochondria, LV, 81, 84, 88
inhibition by arylazido-β-alanine ATP, XLVI, 276, 277
Keilin-Hartree preparation, LV, 121, 124, 125
latent
assay, activity, LV, 191, 192
DCCD-sensitive, solubilization and purification, LV, 193–195
properties, LV, 192
unmasking, LV, 188
in lung lamellar bodies, XXXI, 423, 424
membrane-bound in plants
isolation, XXXII, 392–406
properties, XXXII, 403
membrane depletion, LV, 110–112
Mg^{2+}-activated, XXXII, 289
assay in mutants, LVI, 113
everted vesicles, LVI, 235, 237, 238
Mg^{2+}-Na^+-K^+-dependent, in microsomal fractions, LII, 88
in microvillous membrane, XXXI, 130
miscellaneous inhibitors, LV, 515–518
mit^- mutants, LVI, 16
mitochondrial, assay, LVI, 101
mutant, complementation, LV, 810
in myosin measurement, XXXII, 744
Na^+,K^+-activated, XLVI, 523–531
active site studies, XXXVI, 489
assay, XXXII, 285–289
8-azido-ATP, LVI, 646
Ca^{2+}-activated ATPase compared, XXXII, 305, 306
circular dichroism studies, XXXII, 230–233
gel electrophoresis, XXXII, 288, 289
isolation, XXXII, 277–290
incubation, XXXII, 281, 282
tissue preparation, XXXII, 280, 281
molecular activity, XXXII, 286
ouabain-binding capacity, XXXII, 287, 288
properties, XXXII, 279; XXXVI, 438, 439
purification, XXXII, 282–285
steroid effects, XXXVI, 434–439
zonal centrifugation, XXXII, 283, 284, 289
NCCD, LV, 500, 501
of neurosecretory granules, XXXI, 403
in nuclear membrane, XXXI, 290
oligomycin, LV, 503, 504
oligomycin-sensitive
2-azido-4-nitrophenol labeling, LVI, 683
components, LVI, 12, 40
cytoplasmic petite mutants, LVI, 155
mitochondria, LVI, 11
mutants, LVI, 105
polypeptides, LVI, 597–600
organotins, LV, 509, 510
in oxyntic cells, XXXII, 716, 717
perchloric acid, LV, 201
phosphorylating vesicles, LV, 168
in plasma membranes, XXXI, 67, 88, 148, 171, 185
preparation
by chloroform technique, LV, 337, 338
from beef heart mitochondria, LV, 338

concentration and further purification, LV, 338–341
energy-conserving membranes, LV, 341–343
properties, LV, 343
from *E. coli*
comments, LV, 792–795
preparation of membrane particles, LV, 790
properties, LV, 795, 796
purification, LV, 791, 792
release, LV, 791
protein kinase assay, XXXVIII, 289
quercitin, LV, 491, 492
reconstituted vesicles, optical probes, LV, 573
removal, by acetone, LI, 475
in rough microsomal subfractions, LII, 75, 76
rutamycin-sensitive
assay method, LV, 315, 316
preparation procedure, LV, 317
in sarcoplasmic reticulum, XXXI, 241; XXXII, 79–81
assay, XXXI, 245
isolation, XXXII, 291, 292
from *S. faecalis* membrane, XXXII, 428–439
properties, XXXII, 433, 434
soluble, inhibitors interacting, LV, 478, 495
of spermatozoan mitochondria, LV, 22
spleen mitochondria, LV, 21
stability, LV, 795
subunits, LV, 298, 299, 324, 347–351, 353
synthesis, LV, 350
in synaptosome plasma membrane, XXXI, 450, 452
tentoxin, LV, 493
$TF_0{\cdot}F_1$ activity, assay, LV, 365
uncouplers, LV, 463
venturicidin, LV, 505, 506
in yeast, measurement, XXXII, 842, 843
Adenosinetriphosphatase F_1
antibodies, LVI, 227
2-azido-4-nitrophenol labeled, LVI, 677–680
mutants, LVI, 115
immunological assay, LV, 802
subunits, LVI, 596–602
assay, LV, 320, 324, 325
functions, LV, 297
from *N. crassa*, LV, 148
nucleotide depleted, LV, 377, 378
preparation, LV, 378, 379
reaction with adenine nucleotides and analogs, LV, 379, 380
preparation, LV, 720–722, 742–744
properties, LV, 381
proteoliposomes, LV, 761, 762, 765, 771
purification from rat liver mitochondria
method of Catterall and Pedersen, LV, 320, 324
of Lambeth and Lardy, LV, 324, 327
properties, LV, 324, 327, 328
of rat liver mitochondria
characterization of crystals, LV, 334–337
purification and crystallization, LV, 333, 334
reassociation with membrane factor covalently bound to collagen membrane, LV, 744–746
submitochondrial particles, LV, 105, 108
from thermophilic bacterium
assay methods, LV, 781, 782
properties of TF_1, LV, 783, 784
purification of TF_1, LV, 782, 783
of TF_1 subunits, LV, 784, 785
reconstitution of $TF_0{\cdot}F_1$ from subunits, LV, 786, 787
of TF_1 from subunits, LV, 786
subunit properties, LV, 785
Adenosinetriphosphatase inhibitor
from *Candida utilis* mitochondria
assay methods, LV, 421–423
general comments, LV, 426
properties, LV, 425
purification, LV, 423–425
cross reactivity, LV, 405
mitochondrial
applications, LV, 407
properties, LV, 399–407
release, LV, 400

properties, LV, 383
of rat liver mitochondria
analytical procedures, LV, 408, 409
final note, LV, 413, 414
purification, LV, 409–411
general properties, LV, 412, 413
Adenosinetriphosphatase inhibitor-agarose, XXXIV, 566–571
Adenosinetriphosphatase inhibitor protein, XXXIV, 566–571
column, XXXIV, 566–571
Adenosinetriphosphatase TF_1
stability, LV, 783
subunits
properties, LV, 785
purification, LV, 784, 785
reconstitution of TF_1, LV, 786
of $TF_0 \cdot F_1$, LV, 786, 787
from thermophilic bacterium, crystallization, LV, 372–377
Adenosinetriphosphatase $TF_0 \cdot F_1$, reconstitution, LVI, 600, 601
Adenosine 5′-triphosphate, XXXIV, 135, 258, 259, 486
activator of aspartate transcarbamoylase, LI, 35
of ribonucleoside diphosphate reductase, LI, 238, 245
active transport, XLIV, 921–924
ADP exchange reaction, XXXV, 31
amidophosphoribosyltransferase assay, LI, 172
aminoacylation of tRNA, LIX, 126–128, 130, 132, 133, 183, 216, 221, 231, 288, 298, 818
p-aminophenyl ester, XXXIV, 255, 259, 260
analog, XXXIV, 4; XLVI, 241, 259, 295, 302
characterization, XLIV, 868, 869
coenzymic activity, XLIV, 873–876
immobilization, XLIV, 869–873
inhibitors of tryptophanyl-tRNA synthetase, LIX, 254–256
synthesis, XLIV, 866–868
aqueous stock solution, stability, LVII, 38
assay, LV, 221
of adenosine monophosphate nucleosidase, LI, 264
of bacterial count, LVII, 70
comparative sensitivity of commercial enzyme preparations, LVII, 554, 555
continuous, LVII, 36–50
of creatine kinase isoenzymes, LVII, 61
of cyclic nucleotide phosphodiesterases, LVII, 97
effect of arsenate, LVII, 13, 14
with firefly luciferase, procedure, LVII, 11, 77, 556
internal light-color control, LVII, 3
of luciferase, LVII, 30
in mitochondria, LVII, 37–49
principle, LVII, 37
procedure, LVII, 39–49
of pyrimidine nucleosidase monophosphate kinase, LI, 322, 323
of ribonucleoside triphosphate reductase, LI, 247–250
sensitivity, LVII, 83, 84
sources of error, LVII, 5, 29, 83, 93
of tryptophanyl-tRNA synthetase, LIX, 237, 238
brown adipose tissue mitochondria, LV, 74, 77, 78
calcium transport, LVI, 238, 239, 338
cAMP assay, XXXVIII, 56
cell-free protein synthesis system, LIX, 300, 448, 459, 850
cellular level, drug metabolism, LII, 66, 69
chromatographic mobility, LIX, 73
chromatographic separation, LI, 459, 460
CoA-binding protein, XXXIV, 270
Co(III) complex, XLVI, 313, 314, 318
codon-anticodon recognition studies, LIX, 296
concentration
rapid kinetics, LVI, 540, 541
rate of change, LVI, 537–540
relationship to biomass, LVII, 84, 85

concentration curves, comparative, for evaluation of photometers, LVII, 558
content
per average bacterium, LVII, 66
of brain, LV, 218
of hepatocytes, LV, 214
of kidney cortex, LV, 220
of rat heart, LV, 216, 217
of rat liver, LV, 213
Co(III)-phen complex, XLVI, 314, 315, 318–321
coupling factor, LV, 192
cyanine fluorescence, LV, 579
cyanylation, XLVII, 131
cyclase assay, XXXVIII, 131, 141, 146
cyclic nucleotide immunoassay, XXXVIII, 102, 103
determination of aminoacylation isomeric specificity, LIX, 272
disodium salt
self-association, XLVIII, 116
storage, LI, 338
effect of vacuum filtration, LVII, 80
efflux kinetic studies, LVI, 259
enzymic conversion to ADP, LVII, 89
estimation, XXXVII, 248
exchange, LVI, 253
exchange reactions, LV, 253, 254
measurement, LV, 258
extraction from bacteria, LVII, 68
in fat cells, proteolytic enzyme effects, XXXVII, 213
fluorescence, LVI, 499, 500
formation from cAMP, XXXVIII, 62
in hexokinase assay, XLIV, 342, 455, 457, 476
hydrogen ion extrusion, LV, 227
hydrolysis during work-up, LV, 238
immobilization, on Sepharose, XLIV, 882
inhibitor of adenylate deaminase, LI, 496
of amidophosphoribosyltransferase, LI, 178
of aspartate carbamyltransferase, LI, 57
of orotate phosphoribosyltransferase, LI, 163
of orotidylate decarboxylase, LI, 163
of tryptophanyl-tRNA synthetase, LIX, 252
initial rate of synthesis
artificial electrochemical proton gradient, LVI, 494–495
respiration, LVI, 496
internal standard solution, LVII, 55, 61, 70, 75
application, LVII, 64, 77, 78, 83
intramitochondrial, phosphorylation potential, LV, 241, 242
in isolated renal tubules, XXXIX, 16, 18
labeled
chromatography, XXXVIII, 67, 68
determination of activity, LX, 513–516
impurities, XXXVIII, 45, 46
preparation, XXXVIII, 107, 108; LX, 497, 512, 513
use in lysate phosphoprotein synthesis, LX, 511–516
in protein kinase assay, LX, 499
level, in kidney perfusion, XXXIX, 10
as ligand, XXXIV, 482
luciferase assay, LIII, 559
in media, LVIII, 66
Mg^{2+}-activated, in preparation of chromophore labeled sarcoplasmic reticulum vesicles, LIV, 51
misactivation studies, LIX, 286
mitochondrial, XXXIV, 566–571
mitochondrial protein synthesis, LVI, 21, 22
Na^+,K^+-activated, XXXIV, 4
^{18}O content, reactions contributing, LV, 256
onset of labeling during synthesis by chloroplasts
procedure, LV, 247
reagents, LV, 246, 247
Pasteur effect, LV, 291
perchlorate, LV, 202
in perfused hemicorpus, XXXIX, 80–82
P_1 exchange reaction, XXXV, 31

phosphate donor, of adenylate kinase, LI, 459
of deoxycytidine kinase, LI, 337
of deoxythymidine kinase, LI, 360, 361, 364
of guanylate kinase, LI, 473, 483, 489
of *Lactobacillus* deoxynucleoside kinases, LI, 346
of nucleoside diphosphokinase, LI, 377
of pyrimidine nucleoside monophosphate kinase, LI, 321
of thymidine kinase, LI, 355, 358, 365, 370
of uridylate-cytidylate kinase, LI, 331, 332, 336
phosphodiesterase, XXXVIII, 154, 155, 233, 248
phosphofructokinase, XXXVIII, 397, 398
photochemically active derivatives, LVI, 656
potassium salt, dissociation constant, LV, 237
preparation of 3′-amino-3′-deoxy-ATP, LIX, 135
of *E. coli* crude extract, LIX, 297
product, adenylate kinase, XLIV, 892
production, determination, LI, 468
in protein synthesis, XXXVII, 247–248
purification, XXXVIII, 79
radiolabeled
electrophoretic mobility, LI, 355
synthesis, LI, 3, 12
rate of labeling with submitochondrial particles, LV, 249–250
regenerating system, in adenylate cyclase assay, XXXI, 112
regeneration, XLIV, 887, 888
removal, XXXVIII, 25
requirement
messenger ribonucleic acid binding to 40 S ribosomal subunits, LX, 409
polyphenylalanine synthesis, LX, 683, 684
protein kinase inhibition of protein synthesis, LX, 287–290, 503
reverse salt gradient chromatography, LIX, 216
ribosomal protein synthesis, LVI, 27
separation, XXXI, 107, 108; LV, 285, 287, 288
procedure, LV, 251, 252
reagents and materials, LV, 250, 251
spin-labeled, XLVI, 285, 286
stabilizer of ribonucleoside diphosphate reductase, LI, 228, 238, 241
submitochondrial particles, LV, 111, 112
substrate
for acetyl-CoA carboxylase, XXXV, 3, 16, 17, 25
of carbamoyl-phosphate synthetase, LI, 21, 29, 33, 106, 111, 112, 122
of cytidine triphosphate synthetase, LI, 79, 84
of FGAM synthetase, LI, 193, 195, 196, 197
of firefly luciferase, LVII, 3, 37, 58, 74
of GMP synthetase, LI, 213, 219
for palmityl-CoA synthase, XXXV, 117, 122
of phosphoribosylglycinamide synthetase, LI, 179, 180
of phosphoribosylpyrophosphate synthetase, LI, 3, 10, 12
of succino-AICAR synthetase, LI, 186
of tryptophanyl-tRNA synthetase, LIX, 247, 248, 249
of uridine-cytidine kinase, LI, 299, 300, 305, 315, 320
synthesis
artificially imposed electrochemical potential of protons, LV, 367, 368
by carbamoyl-phosphate synthetase subunits, LI, 26, 27
by cyanide-inhibited mitochondria, LVII, 44, 45
energetics, LIII, 374

implications of electrochemical proton gradient for mechanism, LVI, 496
in intact mitochondria, LVII, 42, 43
ionic composition, LVII, 49
by light-stimulated purple membrane vesicles
assay of ATP synthesis, LV, 778, 779
of membrane potential and proton gradient, LV, 779, 780
principle, LV, 777
properties, LV, 780
reconstitution of vesicles, LV, 777, 778
measurement, LVI, 494
model reaction, LV, 538, 539
in muscle, substrate entry, XXXIX, 495
in plant, inhibition, XXXI, 531
rapid kinetics, LVII, 45–50
rate, LV, 118
reconstitution under imposed pH gradient, LV, 746–748
resolution and reconstitution, LV, 736, 737
methods, LV, 737–741
by sonicated inner-membrane vesicles, LVII, 43, 44
steroid hydroxylation, LV, 10
using K^+ concentration gradients, LV, 666, 667
cation fluxes mediated by carboxylic ionophores, LV, 674, 675
dependence of energy transduction on gradient, LV, 672, 673
experimental procedure, LV, 667, 668
general features, LV, 669, 670
properties, LV, 670–672
summary, LV, 675
yield and reproducibility, LV, 673, 674
transport, LVI, 247–250
uptake, mitochondrial composition and, LVI, 565, 566
utilization by *in vitro* placenta, XXXIX, 249
in vitro labeling of ribosomal proteins, LIX, 517, 527
[α-^{32}P]Adenosine triphosphate, in tRNA end-group labeling, LIX, 103
[γ-^{32}P]Adenosine triphosphate
inosamine kinase assay, XLIII, 451
preparation, LIX, 61
of labeled oligonucleotides, LIX, 58, 75
in tRNA end-group labeling, LIX, 71, 103
[^{14}C]Adenosine triphosphate
assay of AMP formation during aminoacylation, LIX, 287
of nucleotide incorporation, LIX, 182
[^{3}H]Adenosine triphosphate, assay of CTP(ATP):tRNA nucleotidyltransferase, LIX, 123
[^{32}P]Adenosine triphosphate
assay of tryptophanyl-tRNA synthetase, LIX, 238
removal, LIX, 238
Adenosine triphosphate-adenosine diphosphate exchange, assay, LV, 285, 286, 288
Adenosine 5′-triphosphate-agarose, XXXIV, 253, 261, 477
meromyosin, XXXIV, 479
Adenosine triphosphate-P_i exchange complex, components, LVI, 580, 584, 585, 600
Adenosine triphosphate nitrene, XLVI, 264
Adenosine triphosphate-phosphate exchange
assay, LV, 286, 287, 288, 310–313
ATPase inhibitor, LV, 404, 412
DCCD, LV, 499
OSCP assay, LV, 392
preparation from *E. coli*
assay, LV, 358, 359
characteristics, LV, 360–363
preparation of membrane particles, LV, 359, 360
reconstituted, LV, 704
activity, LV, 703, 705, 708–710
$TF_0 \cdot F_1$ activity, assay, LV, 365, 366

Adenosine triphosphate-pyrophosphate exchange, tRNA misactivation, LIX, 283, 285
Adenosine triphosphate-Sepharose, binding capacity, LI, 238
Adenosine triphosphate synthetase
DCCD, LV, 497, 499
inhibition data
interpretation, LV, 473, 474
requirements for publishing, LV, 474
inhibitors
availability, LV, 474, 475
physical properties, LV, 475–477
types, LV, 472, 473
oligomycin, LV, 503
organotins, LV, 508
preparation of components, LV, 718–722
ATPase F_1, LV, 720–722
membrane sector (F_0), LV, 718–720
other coupling factors, LV, 722
proteolipid ionophore, LV, 414–421
reconstitution from heart mitochondria
assay methods, LV, 711, 712
preparation of hydrophobic proteins, LV, 713, 714
reconstitution of vesicles, LV, 714, 715
venturicidin, LV, 506
Adenosine triphosphate triethylammonium salt, XLVI, 263
Adenosine triphosphate-Tris, assay of polypeptide synthesis, LIX, 367
L-Adenosylcobalamin, effector, of ribonucleoside triphosphate reductase, LI, 258
S-Adenosylhomocysteine, inhibitor of tRNA methyltransferase, LIX, 202
S-Adenosylmethionine, XLVI, 560
puromycin assay, XLIII, 509
in studies of A1 protein, XXXII, 329
S-Adenosyl-L-methionine
stability, LIX, 203
substrate of tRNA methyltransferases, LIX, 202
S-[Me-^{14}C]-Adenosyl-L-methionine, in studies of ubiquinone biosynthesis, LIII, 608
S-Adenosylmethionine:dedimethylamino-4-aminoanhydrotetracycline *N*-methyltransferase, XLIII, 603–606
assay, XLIII, 603–605
properties, XLIII, 605, 606
reaction scheme, XLIII, 603
Streptomyces aureofaciens preparation, XLIII, 605
Streptomyces rimosus preparation, XLIII, 604
S-Adenosylmethionine:erythromycin C *O*-methyltransferase, XLIII, 487–498
S-adenosylmethionine, XLIII, 490, 491
ammonium sulfate precipitation, XLIII, 492, 494
assay, XLIII, 488–491
cell-free extract, XLIII, 492
cellular location, XLIII, 495, 496
L-cladinose moiety, radioactivity, XLIII, 489
erythromycin A
derivatization, XLIII, 494
measurement of formation from erythromycin C, XLIII, 489
recrystallization, XLIII, 490
erythromycin C and spiroketal (6 → 9:12 → 9), XLIII, 490
incubation, XLIII, 488, 489
inhibitors, XLIII, 496
kinetic properties, XLIII, 496, 497
partition chromatography, XLIII, 492, 493
pH effect, XLIII, 496
purification, XLIII, 491, 492
radioactivity measurement, XLIII, 495
scintillation system, XLIII, 495
specificity, XLIII, 497
Streptomyces erythreus preparation, XLIII, 491
temperature dependence, XLIII, 496
thin-layer chromatography, XLIII, 493, 494
transmethylase reaction product, XLIII, 495
transmethylase reaction reversibility, XLIII, 495
S-Adenosylmethionine:indolepyruvate 3-methyltransferase, *see* Indolepyruvate methyltransferase

S-Adenosyl-L-[*methyl*-^{3}H]methionine, assay of tRNA methyltransferase, LIX, 190, 191
Adenovirus
availability, LVIII, 426
cell lines for study, XXXII, 584, 585
cell transformation, XXXII, 590, 591
human, LVIII, 425–435
isolation of DNA, LVIII, 426
large-scale growth, LVIII, 431–433
media, LVIII, 427–431
monolayer culture, LVIII, 426, 430
purification, LVIII, 431–433
serotypes, LVIII, 425
suspension culture, LVIII, 426, 430
transformation, LVIII, 370, 426
tumorigenicity, LVIII, 425
Adenovirus DNA, LVIII, 433, 434
Adenylate, *see also* Adenylic acid
energy charge, *see* Energy charge
pool, size, LV, 233
Adenylate cyclase, XLVI, 591, 598
activation, XXXIX, 253, 256
adenylyliminodiphosphate, XXXVIII, 420, 421
assay, XXXVIII, 42–46, 72; XXXIX, 480
general principles
determination of residual substrate, XXXVIII, 124, 125
divalent cations, XXXVIII, 120
enzyme concentration, XXXVIII, 117
labeled nucleoside triphosphates, XXXVIII, 118, 119
linearity of cyclic nucleotide production, XXXVIII, 122
nucleoside triphosphate concentration and volume, XXXVIII, 115–117
protection of cyclase activity, XXXVIII, 121, 122
purification, proof of identity, determination of product, XXXVIII, 123, 124
reduction of product degradation, XXXVIII, 120, 121
of substrate degradation, XXXVIII, 119, 120
termination of cyclase reaction, XXXVIII, 122, 123
methods, XXXI, 103–114
procedure, XXXI, 111, 112
stopping solution, XXXI, 112
automated chromatographic assay, XXXVIII, 125, 126, 134, 135
cyclase preparation, XXXVIII, 128, 129
identification of product, XXXVIII, 132
materials, XXXVIII, 127, 128
nucleotide separation, XXXVIII, 126, 127
procedure, XXXVIII, 129, 130
sensitivity and reproducibility, XXXVIII, 131
bovine brain, preparation, XXXVIII, 128, 129
Brevibacterium liquefaciens, XXXVIII, 160
forward reaction assay, XXXVIII, 161
properties, XXXVIII, 167–169
purification, XXXVIII, 163–167
reverse reaction assay, XXXVIII, 162, 163
in brush borders, XXXI, 122
cAMP inhibitor formation, XXXVIII, 274, 275
cytochemical localization, XXXIX, 480–482
erythrocyte
assay, XXXVIII, 170, 171
properties, XXXVIII, 173
purification, XXXVIII, 171, 172
Escherichia coli, XXXVIII, 155
assay, XXXVIII, 156–158
properties, XXXVIII, 159, 160
purification, XXXVIII, 159
in fat-cell plasma membrane, XXXI, 65, 67
hormonal regulation, XXXI, 68, 70
follicle stimulating hormone effects, XXXIX, 253
gonadotropin regulation, XXXIX, 237

heart and skeletal muscle
assay, XXXVIII, 143, 144
preparation, XXXVIII, 144–146
properties, XXXVIII, 146–149
solubilization, XXXVIII, 149
in hormone antagonist assay, XXXVII, 435
hormone-sensitive
assay, XXXVIII, 190
binding to isoproterenol on glass beads, XXXVIII, 189
preparation, XXXVIII, 188
inhibition, XLVI, 600
kidney, preparation, XXXVIII, 150–152
myocardial, solubilization and role of phospholipids, XXXVIII, 174–180
neonatal bone, XXXVIII, 152, 153
norepinephrine activation, XXXIX, 376, 380
particulate from brain, XXXVIII, 135, 136
assay, XXXVIII, 136
characteristics, XXXVIII, 140
dispersion, XXXVIII, 138–140
preparation, XXXVIII, 136–138
in pineal, XXXIX, 396
in plasma membranes, XXXI, 88, 90, 143, 145–148
rod outer segments, XXXVIII, 154, 155
solubilized from brain, characteristics, XXXVIII, 140–143
testis mitochondria, LV, 11
water flow, XXXVII, 255, 256
Adenylate cyclase-cAMP system, TSH effects, XXXVII, 262
Adenylate deaminase, *see* AMP deaminase
Adenylate energy charge
adenylate deaminase activity, LI, 496–497
definition, LVII, 73; LX, 578, 584
determination, procedure, LVII, 78, 79
establishment, LX, 585
regulation
elongation complex formation, LX, 587–589
ternary initiation complex formation, LX, 585–588
Adenylate kinase, XLVI, 21, 25, 298, 299; LI, 459–467
activation, XLIV, 889
activators, LI, 464–466
activity, LI, 459, 473; LVII, 74, 90, 108
activity assay, XLIV, 892, 893
affinity for calcium phosphate gel, LI, 584
amino acid analysis, LI, 463, 464
AMP assay, LV, 208
assay, LI, 459, 460, 468, 469; LV, 285, 286, 288; LVII, 30
of cyclic nucleotide phosphodiesterase, LVII, 95–98
of modulator protein, LVII, 108–109
associated with cell fractions, XL, 86
cAMP assay, XXXVIII, 62, 63, 65
cellular function, LI, 467
contaminant
in glycine reductase preparations, LIII, 381
in luciferase preparation, LVII, 7, 29, 35, 38, 56, 62
cysteine residues, XLIV, 887–889
distribution, LI, 459
effect of prostaglandins, LI, 466
energy charge, LV, 231–233
exchange reactions, LV, 258
in GMP synthesis assay, LI, 219
hepatoma mitochondria, LV, 88
from human erythrocytes, LI, 467–473
immobilization, in polyacrylamide gel, XLIV, 890, 891
immunological properties, LI, 473
insect muscle mitochondria, LV, 23
interference, in GTP determinations, LVII, 93, 94
in luciferase assay, LVII, 11
isoelectric point, LI, 464, 472
kinetic properties, LI, 473
luminescent determination of ATP, LVI, 536, 537, 543
as marker enzymes, LVI, 210–211
mitochondrial subfractions, LV, 94
molecular weight, LI, 463, 464, 472

multiple forms, LI, 467
electrophoretic patterns, LI, 469, 470
partial specific volume, LI, 464
perchloric acid, LV, 201
purification, LI, 460–462
purity, LI, 463, 472
from rat liver, LI, 459–467
sedimentation coefficients, LI, 464
stability, LI, 463, 472; LVII, 109
substrate specificity, LI, 464, 473; LVII, 82
subunit structure, LI, 463, 464, 473
Adenyl cyclase, *see* Adenylate cyclase
Adenylic acid, *see also* Adenylate
in preparation of luciferyl adenylate, LVII, 27
2′-Adenylic acid, XLVII, 441
Adenyl imidodiphosphate, LV, 377
attachment to ATPase F_1, LV, 379, 380
purification, LV, 379
Adenylosuccinase, *see* Adenylosuccinate lyase
Adenylosuccinate
product, of adenylosuccinate synthetase, LI, 207
structural formula, LI, 203
substrate, of adenylosuccinate AMP-lyase, LI, 202
Adenylosuccinate adenylate-lyase, *see* Adenylosuccinate lyase
Adenylosuccinate AMP-lyase, *see* Adenylosuccinate lyase
Adenylosuccinate lyase, XLVI, 302, 305, 306; LI, 202–207
activity, LI, 202, 203
aggregated forms, LI, 206
assay, LI, 202, 203
intragenic complementation, LI, 206, 207
molecular weight, LI, 206
from *Neurospora crassa*, LI, 202–207
pH optimum, LI, 206
properties, LI, 206, 207
in succino-AICAR synthesis, LI, 189
in succino-AICAR synthetase assay, LI, 186, 187
from yeast, preparation, LI, 190
Adenylosuccinate synthetase, LI, 207–213
assay, LI, 207, 208, 212, 213
buffers, LI, 212
crystallization, LI, 210, 211
inhibitors, LI, 212
molecular weight, LI, 211
pH optimum, LI, 211
properties, LI, 211–213
purification, LI, 208–211
purity, LI, 211
from rabbit tissues, LI, 207–213
sources, LI, 212
stability, LI, 211
Adenylstreptomycin, XLIII, 121
Adenyluridylylguanylate, initiation factor complexes, LX, 4–6, 18, 19, 37, 49, 256–265, 431, 432, 453, 454
Adenylyl cyclase, *see* Adenylate cyclase
Adenylylimidodiphosphate, XXXIV, 253
adenylate cyclase, XXXVIII, 169
DNA transcription, XXXVIII, 375
preparation, XXXVIII, 420, 421
principle, XXXVIII, 422, 423
procedure, XXXVIII, 423–425
tetrasodium iminodiphosphate, XXXVIII, 421, 422
5-Adenylylimidodiphosphate, in studies on adenyl cyclase, XXXIX, 480, 482
5′-Adenylylimidodiphosphate, in adenylate cyclase assay, XXXI, 114
Adenylyl imidophosphate
crystallization of ATPase TF_1, LV, 373
Adenylylsulfate kinase, activity, LVII, 245
Adherence, nonspecific, XXXIV, 754
Adhesion
of cells, XXXII, 597–600
cellular, LVIII, 369
to vessels, LVIII, 137
Adhesion protein, LVIII, 267
preparation, LVIII, 273, 274
Adiabaticity, calorimeter design, XLIX, 5
Adipic acid, XXXIV, 477; XLVI, 506
Adipic acid dihydrazide, XXXIV, 45, 74, 477, 491; XLIV, 121, 127, 640
preparation, LIX, 179
Sepharose activation, XLIV, 882

use in affinity chromatography of flavoproteins, LII, 90, 92
Adipic acid dihydrazide-agarose derivative, XXXIV, 74
Adipic hydrazide, nucleotide derivative, XXXIV, 478
Adipocyte, *see also* Fat cells
differentiation, LVIII, 105
hormone antagonist from, XXXVII, 431–438
prostaglandin binding, XXXII, 109
subcellular fractionation of, XXXI, 60–71
Adipose cell, XXXIV, 6
isolation, XXXIX, 466
in mammary glands, XXXII, 697–701
Adipose tissue, *see also* Fat cell
brown
carnitine, LV, 221
mitochondria, LV, 16, 17, 65–78
dorsal interscapular, rat, source of brown fat cells, XXXV, 560
(Na^+,K^+)-ATPase isolation, XXXII, 280
protein kinase, hormonal regulation, XXXVIII, 363–366
white, mitochondria, LV, 15, 16
N^7-Adipyl-L-arginine, XLVII, 157, 158
ADP, *see* Adenosine 5′-diphosphate
ADPase, *see* Adenosinediphosphatase
ADP crystal, LIV, 8
Adrenal cell
2′,3′-cyclic nucleotide-3′-phosphohydrolase, XXXII, 128
hormone production, XXXIX, 302–328
superfusion studies, XXXIX, 306–317
intact, isolation, XXXII, 632
isolation, XXXII, 673–693
techniques compared, XXXII, 692
separation, XXXIX, 229
Adrenal cell Y1, *see* Y1 adrenal cell line
Adrenal cortex
bovine, isolation of adrenodoxin reductase, LII, 133, 134
carcinoma, LVIII, 376
cell
adrenocorticotropin analog studies, XXXIX, 347–359
electron microscopy, XXXIX, 160–165
fixation and embedding, XXXIX, 161
preparation, XXXIX, 348, 349
cow, glucose-6-phosphate dehydrogenase, assay, purification, and properties, XLI, 188–196
mitochondria, LV, 9, 10
properties, LV, 10, 11
oxidase, LII, 13, 14
Adrenal gland
bovine, purification of adrenodoxin, LII, 135, 136
cAMP receptor, XXXVIII, 380, 381
chromaffin granule isolation, XXXI, 379–389
cortical cell culture, XXXIX, 159
ferredoxin, purification, LIII, 267
mammalian, oxidases, LII, 9
perfused, XXXIX, 328–336
in situ studies, XXXIX, 336–347
Adrenal gland mitochondria, ACTH effects, XXXVII, 295–304
Adrenal medulla
optical probes, LV, 573
oxidase, LII, 12
Adrenal tissue, superfusion, XXXIX, 303–306
Adrenal tumor, plating, LVIII, 573
Adrenergic neuron, LVIII, 575
Adrenergic receptor, XXXIV, 695–700, *see also* specific type
agarose derivatives, XXXIV, 696, 697
in parotid tissue enzyme secretion and K^+ release, XXXIX, 461–466
quantitation of ligand, XXXIV, 698
α-Adrenergic receptor, in parotid slices, XXXIX, 461–466
β-Adrenergic receptor, XXXIV, 695; XLVI, 131, 578, 579, 591–601
action of propranolol, XLVI, 600
inhibition, XLVI, 599, 600
in parotid slices, XXXIX, 461–466

β-Adrenergic receptor-adenylate cyclase complex
glass bead-immobilized proterenol, XXXVIII, 187, 188
adenylate cyclase preparation, XXXVIII, 188
binding, XXXVIII, 189
cyclase assay, XXXVIII, 190
epinephrine-sensitive membrane fraction, XXXVIII, 188, 189
erythrocyte preparation, XXXVIII, 188
properties, XXXVIII, 190, 191
Adrenocortical hormone, biorhythms in secretion, XXXVI, 480
Adrenocortical steroid, effects on water flow, XXXVII, 256
Adrenocortical Y1 cell, *see* Y1 adrenal cell line
Adrenocorticotropic hormone, *see* Adenocorticotropin
Adrenocorticotropin, XXXV, 181, 187
adrenal mitochondria, LV, 10, 11
analogs, structure-activity studies, in adrenal cortex cells, XXXIX, 347–359
bioassay, XXXVII, 121–130
cAMP and corticosterone in response to, XXXII, 680–686
effect on adrenal cell morphology, XXXIX, 161–163
on adrenal cortex cells, XXXII, 674, 680–688
on adrenal mitochondria, XXXVII, 295–304
on corticosterone output, XXXIX, 317–322, 326–328
on fat-cell hormones, XXXI, 68
on fat cells, modification, XXXVII, 213
on perfused adrenal gland, XXXIX, 333–335
on Y_1 cell line, LVIII, 571
heterogeneity, XXXVII, 40–41
hydroxylamine cleavage, XLVII, 140
iodinated, properties, XXXVII, 230
models for action, XXXIX, 353–358
pituitary tumor cell, LVIII, 527
protein kinase, XXXVIII, 364
radioimmunoassay, XXXVII, 24, 25, 27, 32, 34–38
standards, XXXVII, 30
release in GH cells, LVIII, 534
standard, XXXII, 677
stock solutions, XXXIX, 341
synthesis, XLVII, 584
tritium labeling, XXXVII, 315, 316, 321
in tumor extracts, XXXVII, 35
ultrastructural localization, XXXIX, 153, 154
in vivo decay hypothesis, XXXIX, 314, 315
Adrenodoxin, XLIX, 137, 175
in adrenal gland, LII, 124
antibody against
effects, LII, 244
inhibitor of mitochondrial enzymes, LII, 248
preparation, LII, 242
characterization, LIII, 274
coupling to Sepharose-4B-200, LII, 133
crystallization, LII, 137
in cytochrome P–450 $_{scc}$ assay, LII, 125, 131, 139
in cytochrome P–450 $_{11\beta}$ assay, LII, 126
kinetics, LII, 141
mitochondria, LV, 10, 11
molecular weight, LII, 141
optical properties, LII, 141
prosthetic group, LII, 141
purification procedure, LII, 135, 136
reconstitution of iron-sulfur chromophore, LII, 136, 137
reference spectrum, LIII, 271
requirement by mitochondrial P_{450}, XXXVII, 304
solubilization, LII, 140, 141
stability, LII, 136
Adrenodoxin reductase, *see* Ferredoxin-$NADP^+$ reductase
Adsorbent, *see also* specific substances
for affinity chromatography, XL, 267
dilution, XXXIV, 676, 677
types, XLIV, 43
Adsorbosil 3 silica gel, XLVII, 348

Adsorption, XLIV, 148–169
applications, XLIV, 165, 166
bonding mechanisms, XLIV, 149–153
carrier choice, XLIV, 154–160
to DEAE-Sephadex, of aminoacylase, XLIV, 748–751
definition, XLIV, 149
effect, XXXIV, 16–18, 108–125
active-site ligands, XXXIV, 116
association constant, XXXIV, 111
compound affinity, XXXIV, 115
control and elimination of, XXXIV, 116–118
deforming buffers, XXXIV, 112
dissociation constant, XXXIV, 111
on enzyme activity, XLIV, 44, 45
on enzyme concentration, XLIV, 153, 154
β-galactosidase, XXXIV, 118, 119
of hydrogen ion concentration, XLIV, 153
hydrophilic spacer, XXXIV, 117, 118
ionic groups, XXXIV, 113
ionic strength, XXXIV, 116; XLIV, 153
lactate dehydrogenase, XXXIV, 120–212
nonbiospecific, XXXIV, 108, 114
nonspecific, XXXIV, 663, 664
polar spacer, XXXIV, 117, 118
ribonuclease A, XXXIV, 119, 120
spacers, XXXIV, 113, 114, 117, 118, 123
temperature, XXXIV, 116, 117; XLIV, 45, 154
of time, XLIV, 154
enzyme cleavage, XLIV, 153
hydrophobic, XLIV, 43–45
loading method, XLIV, 160
in-column, XLIV, 160
shaking bath, XLIV, 160
methods, XLIV, 43–45, 160–163
regeneration, of enzyme and carrier, XLIV, 164, 165
Adsorption chromatography, XLIII, 173, 174
calculation, XLIII, 174
of D-β-hydroxybutyrate apodehydrogenase, XXXII, 389, 390
nonionic, XLIII, 275–280
destomycin, XLIII, 277
kanamycin, XLIII, 276, 277
neomycin, XLIII, 277
1,5-AEDANS, *see N*-(Iodoacetylaminoethyl)-5-naphthylamine-1-sulfonic acid
AEDANS derivative
fluorescent donor, XLVIII, 362, 363
spectral properties, XLVIII, 362
structural formulas, XLVIII, 362
Aedes aegypti
media, LVIII, 457, 462
suspension culture, LVIII, 454
Aedes albopictus
media, LVIII, 462
suspension culture, LVIII, 454
Aedes ω-albus, media, LVIII, 462
Aedes novalbopictus, media, LVIII, 462
Aedes taeniorhynchus, media, LVIII, 462
Aedes vexans, media, LVIII, 457
AEFS spectrum, *see* Absorption edge fine structure spectrum
Aequorea, luciferase, LIII, 560
Aequorea aequorea
bioluminescence, stability, LVII, 275, 276
coelenteramide, LVII, 341, 344
emission of wavelength, LVII, 341
collection, LVII, 303
fluorescent accessory protein, LVII, 259
harvesting of photogenic organs, LVII, 277, 278, 303–305
nomenclature, LVII, 272
photoprotein system, LVII, 355
Aequorea forskalea, *see Aequorea aequorea*
Aequorin, LVII, 269–328
absorption spectrum, LVII, 289
aggregation, LVII, 283
assay, LVII, 279, 293–301
apparatus, LVII, 293–297
of calcium, advantages and disadvantages, LVII, 292, 293
calibration, LVII, 300, 301

effect of ammonium sulfate, LVII, 298
of pH, LVII, 298
of temperature, LVII, 298
linearity, LVII, 299, 323, 324
principle, LVII, 293
reproducibility, LVII, 298
sensitivity, LVII, 327
bioluminescent reactions, LVII, 284, 285
calcium binding affinity, LVII, 326–328
calcium measurement, LVI, 303, 445
commercial source, LVII, 302
copurification with GFP, LVII, 308, 309
crude extract preparation, LVII, 277–280, 304–307
isoelectric point, LVII, 283, 308, 310
isomeric forms, LVII, 310
kinetics of luminescence, LVII, 289, 290
light-emitter mechanism, LVII, 285–291
luminescence spectrum, LVII, 289
microinjection into cells, LVII, 312–318
molecular weight, LVII, 282, 283
purification, LVII, 280, 281, 307–311
quantum yield, LVII, 279, 284, 289
regeneration, LVII, 289–291
solubility, LVII, 283
spectral properties, LVII, 283
storage, LVII, 281, 311, 312
Aeration, mitochondrial protein synthesis, LVI, 23
Aerobacter, oxidase, LII, 21
Aerobacter aerogenes
acetolactate decarboxylase, assay, purification, and properties, XLI, 527–529
acetolactate-forming enzymes, XLI, 519
D-arabinose isomerase, assay, purification, and properties, XLI, 462–465
as bacterial associate, XXXIX, 485
2,3-butanediol biosynthetic system, XLI, 518–533
carbamoyl-phosphate synthetase, LI, 29
Dictyostelium discoideum growth, L, 308
D-fructose-1-phosphate kinase, assay, purification, and properties, XLII, 63–66
β-glucoside kinase in metabolism of cellobiose and gentiobiose, XLII, 3
glycerol dehydratase, assay, purification, and properties, XLII, 315–323
growth, LI, 182
5-keto-D-fructose reductase, XLI, 137
β-lactamase, XLIII, 673
D(−)-lactate dehydrogenase, XLI, 294
nitrate reductase, LIII, 642
phosphocellobiase, XLII, 494, 497
phosphoribosylglycinamide synthetase, LI, 179–185
Polysphondylium pallidum growth, L, 313
pullulanase, XLIV, 62, 63
L-ribulose-5-phosphate 4-epimerase, assay, purification, and properties, XLI, 412–419, 422, 423
in D-ribulose preparation from ribitol, XLI, 104–106
Aerobacter cloacae, 5-keto-D-fructose reductase, XLI, 137
Aerococcus, oxidase, LII, 11
Aeromonas
aminopeptidase
assay
endopeptidase, XLV, 534
by leucyl-β-naphthylamide, XLV, 542
by leucyl-*p*-nitroanilide, XLV, 542
characteristics, XLV, 530
hydrolysis, qualitative, XLV, 540
inhibition, XLV, 538, 541
isolation, XLV, 406, 407, 533
kinetics, XLV, 539
production, XLV, 531
properties, XLV, 536
physical, XLV, 537
purification, XLV, 531, 536
specificity, XLV, 537
stability, XLV, 537

neutral protease
assay, XLV, 404
furylacryloyl peptides, XLV, 405
hemoglobin, XLV, 406
composition, XLV, 414
isolation, XLV, 409, 413
production of, XLV, 406
properties, XLV, 412, 414
purity, XLV, 412
stability, XLV, 415
Aeromonas proteolytica, XLV, 406, 408
Aerosol, biohazard, LVIII, 39, 40
AF, *see* Activating Factor
AF–400, aequorin reaction product, LVII, 284, 285
Affi–Gel 501, purification of glycine reductase, LIII, 379
Affi–Gel 703, XLIV, 311, 312
Affinity, metal ion indicators, LVI, 304, 305
Affinity chromatography, XLIV, 38; XLVI, 216
adrenodoxin reductase purification, LII, 133
adsorbents, XL, 267
association constant, XXXIV, 141–147
BCF_0–BCF_1, LV, 194
binding
constants, XXXIV, 141–147
effectiveness, XXXIV, 143
function of K_1, XXXIV, 146–148
model, XXXIV, 146, 155
biospecific, adsorbents, XLIV, 32
column preparation, XL, 270
competitive inhibitor, XXXIV, 155–158
contaminant recovery, XXXIV, 150
of corticosteroid-binding globulin, XXXVI, 104–109
coupling factor, LV, 189, 190
covalent, XXXIV, 531–544, 581–591
for acetylcholinesterase, XXXIV, 581–591
arm attachment to matrix, XXXIV, 588, 589
capacity, XXXIV, 586
ligand attachment, XXXIV, 589
ligand synthesis, XXXIV, 586–588
mechanism, XXXIV, 584
penicillin-binding components, XXXIV, 404
rational, XXXIV, 581–591
spacers, XXXIV, 591
specificity, XXXIV, 586
by thiol-disulfide interchange, XXXIV, 531–544, *see also* Papain; Thiol-disulfide interchange
of cytoplasmic progesterone receptors, XXXVI, 207, 208
elution
calculation of conditions, XXXIV, 147, 148, 153–155
by competitive inhibitor, XXXIV, 155–158
with inhibitor, XXXIV, 147–148
multiple, XXXIV, 153–155
multiple washes, XXXIV, 151–155
without washing, XXXIV, 148–151
enrichment of polysomes synthesizing specific protein, XL, 266, 271
flavoprotein purification, LII, 90
gel preparation, LI, 366, 367
gel storage buffer, LI, 367
of glutamate carrier, LVI, 425, 426
plant peroxidase purification, LII, 514–521
with polycytidylate-cellulose, LX, 457
protein-RNA interactions, LIX, 568, 569
purification of adenosine deaminase, LI, 504, 505
of adenosine monophosphate nucleosidase, LI, 269
of adenylate deaminase, LI, 493, 494, 500
of blue-fluorescent protein, LVII, 232
of carbamoyl-phosphate synthetase, LI, 31–33
of cytochrome bc_1 complex, LIII, 99–106
of deoxythymidine kinase, LI, 363, 364
of *Lactobacillus* deoxynucleoside kinase, LI, 348–350
of phosphoribosylpyrophosphate synthetase, LI, 8, 9

of ribonucleoside diphosphate reductase, LI, 230, 231, 241, 242, 246
of ribonucleoside triphosphate reductase, LI, 256
of RNase III, LIX, 829
of thymidine kinase, LI, 366, 367, 368, 369
of uridine-cytidine kinase, LI, 312, 313
recovery
calculation, XXXIV, 151–153
prediction, XXXIV, 151–153
theory, XXXIV, 140–162
resin preparation, LII, 92, 515–517
Rhizobium cytochrome *P*-450 purification, LII, 164
ribonucleic acid-cellulose, LX, 247–255
separation
on aggregated enzyme, XLIV, 277
of nucleoside deoxyribosyltransferase activities, LI, 449
of right-side-out from inside-out particles, LV, 114
Sepharose-heparin, LX, 124–135
of TeBG, XXXVI, 109–120
Affinity constant, *see* Association constant
Affinity density perturbation, XXXIV, 171–177
method, XXXIV, 176
Affinity electrophoresis, XXXIV, 178–181
disc gel apparatus, XXXIV, 180
gel preparation, XXXIV, 179, 180
phytohemagglutinins, XXXIV, 181
Affinity elution, XXXIV, 163–171, *see also* Affinity method, theory; Aminoacyl-tRNA synthetase, affinity elution
enzyme-ligand interaction, quantitation, XXXIV, 165
quantitation, XXXIV, 164, 165
Affinity elution chromatography
purification of CTP(ATP):tRNA nucleotidyltransferase, LIX, 184, 185
of tRNA methyltransferases, LIX, 195, 196
Affinity gel, XXXIV, 178
Affinity labeling, XLVI, 3–14; LX, 720, 724–728, 733–745, *see also* Photoaffinity
acetylenic inhibitors, XLVI, 158–164
controls, XLVI, 110, 111
covalent bond formation, XLVI, 10, 11
criteria, XLVI, 5, 6, 29, 55, 482, 483, 575
DBCT, LV, 511
definition, XLVI, 3–5
diradicals, XLVI, 82–91
free radicals, XLVI, 82–91
historical, XLVII, 479
inactivation, half-time, XLVI, 6
instability, XLVI, 186
kinetics, XLVI, 5–7, 28, 203–205, 211–214, 481, 482, 576, 577, 581, 582
macromolecules, XLVI, 183–186
multicomponent systems, XLVI, 180–194
multisubstrate analogs, XLVI, 17
negative results, LVI, 654
polyfunctional labels, XLVI, 12
reagents, XLVII, 479–498, *see also* specific types
antibody, XLVI, 486
choice, XLVI, 483–486
concentration, XLVI, 107, 108
k_{cat}, XLVI, 9, 10
K_s, XLVI, 8, 9
K_s‡, XLVI, 9, 16
triplet-state, XLVI, 11
of receptors, XLVI, 572–582
reversal by thiolysis, XLVI, 315
of ribosomes, XLVI, 187
specificity, XLVII, 479
of steroid, for binding site studies, XXXVI, 374–410
preparation, XXXVI, 374–388
for steroid receptor proteins, XXXVI, 411–426
stoichiometry, XLVI, 316
sulfhydryl groups, XLVII, 409, 422
time course, XLVI, 315
in vitro applications, XLVI, 13, 14
in vivo applications, XLVI, 14
Affinity model system, XXXIV, 140–162, *see also* Affinity method, theory

African green monkey kidney cell
microcarrier culture, LVIII, 190
SV40, LVIII, 404
Agalacto-orosomucoid, *N*-acetylglucosamine binding protein, L, 289
Agallia constricta, media, LVIII, 462
Agar
amphipathic gels, XLIV, 43–45
Bennett's, XLIII, 6
biological resistance, XLIV, 23
chemical resistance, XLIV, 23
Czapek's solution, XLIII, 4
derivatization, XLIV, 23, 24
dilution assay, XLIII, 62
economic considerations, XLIV, 24, 25
effect on enzymatic activity, XLIV, 24
epichlorohydrin cross-linking, procedure, XLIV, 23
malt extract, XLIII, 5
matrices, activity, XLIV, 24
mechanical stability, XLIV, 21, 22
nutrient, XLIII, 5
N–Z amine-starch-glucose, XLIII, 7
oatmeal, XLIII, 6
potato dextrose, XLIII, 4
space-fitting properties, XLIV, 22, 23
support properties, XLIV, 20
TGY, *see* Tryptone glucose yeast extract
tomato paste oatmeal, XLIII, 6
trypticase-yeast extract, XLIII, 6
uses, XLIV, 20, 21
vicinal hydroxyl, XLIV, 27
yeast extract, XLIII, 5
Agar assay, soft, LVIII, 297, 299, 300
Agar gel electrophoresis, XLIII, 287, 289
relative mobilities of antibiotics, XLIII, 288
Agaricus, oxidase, LII, 20
Agarose, XXXIV, 19, *see also* specific type
activated
commercial source, XLIV, 29
handling and storage, XLIV, 29, 33
activation methods, XXXIV, 77–102
cyanogen bromide, XXXIV, 77–80
hazards, XXXIV, 80
leakage, XXXIV, 96–102
periodate oxidation, XXXIV, 80–85
spacers, XXXIV, 93–95
stable derivatives, XXXIV, 85–93, *see also* specific derivatives
activation procedure with bisoxirane, XLIV, 33, 34
with cyanogen bromide, XLIV, 28, 29
advantages, XLIV, 45
aminoalkyl, and photochemical immobilization, XLIV, 284, 285
azide derivatives, XXXIV, 84
bead
formation mechanism, XLIV, 25
lectin coupling, XXXIV, 338
chemical resistance, XLIV, 23
column, stability, XXXIV, 494
copolymer with polyacrylamide, XLIV, 273
coupling
pH effect, XXXIV, 88
selectivity, XXXIV, 89
cross-linking, XXXIV, 338–341
cyanogen bromide activation, XXXIV, 77–80, *see also* Cyanogen bromide
definition, XLIV, 21
derivatives
macromolecular, XXXIV, 95
stable activated, XXXIV, 85–93, *see also* specific derivative
derivatization, XLIV, 38–42, 44
with 5-fluoro-2′-deoxyuridylate, LI, 100
to hexyl agarose, XLIV, 44
to mercapto-agarose, XLIV, 41
to oxo-agarose, XLIV, 40
to *p*-phenylenediamine agarose, XLIV, 39, 40
in electrophoretic analysis of protein-RNA complexes, LIX, 567
hydrocarbon-coated, XXXIV, 128–131
alkyl-agarose, XXXIV, 130, 131
ω-aminoalkyl-agarose, XXXIV, 131
homologous series, XXXIV, 129
mechanism, XXXIV, 137–139
N-hydroxysuccinimide esters, XXXIV, 86–90
immunodiffusion, XLIV, 716, 717

ligands, stability, XXXIV, 96–102, *see also* Ligand
mechanical stability, XLIV, 21, 22
periodate activation, XXXIV, 80–85
periodate oxidation, XLIV, 40
p-phenylenediamine derivative, preparation, XLIV, 39, 40
purification of ribosome protein S1, LX, 450, 451
reductive amination of oxidized, XXXIV, 82, 83
in sample wells, in slab gel electrophoresis, LIX, 458
in separation of bacterial ribosomal subunits, LIX, 398
stable derivatives, XXXIV, 73, 85–93, *see also* specific derivatives
uses, XLIV, 20, 21
Agarose/acrylamide electrophoresis, inter-α-trypsin, XLV, 767
Agarose-anthranilate derivative, XXXIV, 377–385
Agarose gel filtration chromatography
of cytoplasmic progesterone receptors, XXXVI, 200, 201
DNA binding assay, XXXVI, 311–313
in purification of CPSase-ATCase complex, LI, 109
of dihydroorotate dehydrogenase, LI, 61
of phosphoribosylpyrophosphate synthetase, LI, 15
Agarose hydrazide, XXXIV, 477, 478
Agarose-poly(I):poly(C), purification of RNase III, LIX, 829
AG 1–X2 chloride column, preparation of 3′-amino-3′-deoxy-ATP, LIX, 136
AGDA, *see* Acetyl L-glutamic acid diamide
Age pigment, *see* Lipofuscin
Agglutinability, transformed cells, LVIII, 368, 369
Agglutination
assay, using lectins, XXXII, 615–621
measurement, XXXII, 618–620
Agglutinin, *see also* Lectin
wheat germ
isolation, XXXII, 611–615
reductive methylation, XLVII, 475
Aggregate, polyoxyethylene detergents, LVI, 742, 744
Aggregation, XLIV, 263–280
applications, XLIV, 276–279
optimization, XLIV, 274–276
procedure, XLIV, 265–274
in frozen state, XLIV, 269, 270, 277, 278
with isolated enzyme, XLIV, 265–267
using inert proteic feeder, XLIV, 268–270
storage, XLIV, 266
Aging
cell, LVIII, 30
cell culture studies, XXXII, 562
Aging Cultured Cell Repository, LVIII, 442
Agkistrodon halys blomhoffi, venom proteinases, XLV, 459, *see also* Proteinase
Agkistrodon piscivorus leucostoma, venom, chromatography, XLV, 399, *see also* Proteinase
Aglycone, of olivomycin, XLIII, 334
Agmatine, XLV, 210
β-Agonist, XLVI, 599
AG1–8X resin, XLIV, 845
Agrobacterium
hexopyranoside:cytochrome *c* oxidoreductase, assay, purification, and properties, XLI, 153–158
oxidase, LII, 19
in 3-ulose determination, XLI, 25, 26
Agrobacterium tumefaciens
chloramphenicol acetyltransferase, XLIII, 738
as crown gall pathogen, XXXI, 554
D-gluconate dehydratase, XLII, 304
Agroclavine
optical-difference spectrum produced, LII, 259, 262, 266
type of binding reaction with cytochrome *P*-450, LII, 264
Ah^b allele
fluorometric macroassay, LII, 234–236
phenotypic test, LII, 232, 233
AHH, *see* Aryl hydrocarbon hydroxylase
Ah locus, LII, 226–240
alleles, LII, 228
genetic crosses involving, LII, 229
hemoprotein spin states, LII, 228

monooxygenase induction associated, LII, 231, 232
in mouse, LII, 229, 230
phenotypic testing, LII, 232, 233
in rat, LII, 230
AH-Sepharose, *see* Aminohexamethyl-Sepharose
A. H. Thomas homogenizer, HeLa cell disruption, LVI, 68
AIB, *see* α-Aminoisobutyric acid
AIBN, *see* 2,2′-Azobisisobutyronitrile
AICA, *see* 5-Amino-4-imidazole carboxamide
AICAR, *see* 5-Amino-1-ribosyl-4-imidazolecarboxyamide 5′-phosphate
A23187 ionophore, calcium transport, LVI, 322, 323
AIR, *see* 5-Amino-1-ribosylimidazole 5′-phosphate
Air/acetylene flame, LIV, 464
Airborne contamination, *see* Contamination, airborne
Air circulation
laminar flow system, LVIII, 10, 40, 154
safety factor, LVIII, 40, 43
Air monitoring, XLIV, 647–658
constant current supply, XLIV, 655
electrochemical cell, XLIV, 654
mechanical components, XLIV, 656
platinum electrodes, XLIV, 654, 655
starch pad, XLIV, 652–654
strip chart recorder, XLIV, 656
Air Product Model LTD–3–110, LIII, 488
Ajmaline, LV, 516
Alamethicin
cation-transporting conformation, LV, 437, 438
chemical cleavage, XLVII, 146
effect on bilayers, XXXII, 500, 552
as ionophore, LV, 441, 442
Alanine, XLIV, 13
aminoacylation isomeric specificity, LIX, 274, 275, 279, 280
assay, LV, 221
conformational state, XLIX, 103
content of rat heart, LV, 216
effect on rat liver carbamoyl-phosphate synthase, LI, 119
gluconeogenesis, XXXVII, 294
in media, LVIII, 63
misactivation, LIX, 289
production in enzyme reactor, XLIV, 880–882
in Sepharose, XLIV, 497
solid-phase peptide synthesis, XLVII, 610
thermolysin hydrolysis, XLVII, 175
transport, by reconstituted vesicles, LVI, 434, 435
uptake to perfused liver, XXXIX, 34
β-Alanine
activator, of aspartate carbamyltransferase, LI, 49
derivative, XLIV, 284
L-Alanine
activator, of aspartate carbamyltransferase, LI, 49
isolation, XLIV, 756, 757
Alanine aminotransferase, assay by Fast Analyzer, XXXI, 816
β-Alanine buffer, in pyrophosphate determination, LI, 275
DL-Alanine-*N*-carboxyanhydride, XXXIV, 75, 76
D-Alanine carboxypeptidase, XXXIV, 398–401, 405
Alanine carrier, solubilization and purification, LVI, 430, 431
alanine transport by reconstituted vesicles, LVI, 434, 435
general principle, LVI, 431
procedure, LVI, 431, 432
reconstitution of vesicles, LVI, 433, 434
specific activity and purity of preparation, LVI, 432, 433
Alanine dehydrogenase, in enzyme reactor, XLIV, 880–882
L-Alanine:1D-1-guanidino-1-deoxy-3-keto-*scyllo*-inositol aminotransferase, XLIII, 462–465
assay, XLIII, 462–464
biological distribution, XLIII, 464
inosamine transaminase separation, XLIII, 463, 464
specificity, XLIII, 464
stability, XLIII, 465
α-Alanine *p*-nitrophenyl ester, diazotization, XLVI, 263

Alanosine, in cell hybridization studies, XXXII, 578
D-Alanyl-D-arginine, XXXIV, 413
D-Alanyl-D-arginine-agarose, XXXIV, 412
β-Alanylhexamethylenediamine-polydimethylacrylamide resin, XLVII, 265
  capacity, XLVII, 270
  degradative efficiency, XLVII, 276
  DITC-activation, XLVII, 271
  stability, XLVII, 273
  synthesis, XLVII, 267
Alanylphenyllysylmethyl chloride, XLVI, 206
Alanyl-tRNA synthetase, XXXIV, 170
  subcellular distribution, LIX, 233, 234
Albery equation, electrodes, LVI, 457
L-Albizziin, XLVI, 415, 418
  inhibitor, of FGAM synthetase, LI, 201
Albomycetin, XLIII, 131
Albomycin, effects on membranes, XXXII, 882
Alborixin, LV, 446
Albumin, XXXIV, 6, 592, 707, 717; XLVI, 89, 90, *see also* Bovine serum albumin
  bovine serum, *see* Bovine serum albumin
  cyclase assays, XXXVIII, 122
  gel electrophoresis, XXXII, 101
  hydrazido derivative, XXXIV, 92
  immobilization
    in liposomes, XLIV, 227
    by microencapsulation, XLIV, 214
  in kidney perfusion studies, XXXIX, 8
  labeling with fluorescein, XLVI, 565, 566
  in media, LVIII, 88, 500
  for pancreas perfusate, XXXIX, 371
  slide coating, XXXVII, 136–137
  as spacer, XXXIV, 94, 611
  sulfenylation, XXXIV, 193
  synthesis, in perfused liver, XXXIX, 35
  testosterone antiserum preparation, XXXIX, 262, 263
Albumin A, sulfenylation, XXXIV, 193
Albumin-agarose, XXXIV, 95, 660
  denatured, XXXIV, 95
Albumin gland inhibitor, snail
  assay, XLV, 786
  carbohydrate composition, XLV, 791
  characteristics, XLV, 785
  composition, XLV, 791
  distribution, XLV, 790
  isoinhibitor separation, XLV, 787, 788
  kinetics, XLV, 790
  properties, XLV, 789, 790
  purification, XLV, 786, 789
  reactive site, XLV, 790
  specificity, XLV, 790
  stability, XLV, 789
Albumin medium, for fat-cell separation, XXXI, 104
Alcaligenes
  D-gluconate dehydratase, assay, purification, and properties, XLII, 301–304
  hydrogenase reaction, LIII, 287
  oxidase, LII, 21
Alcaligenes eutrophus H16, hydrogenase, LIII, 290
Alcaligenes faecalis
  aureovertin, LV, 485
  cytochrome *c*, LIII, 210
  penicillin acylase, XLIII, 718
Alcohol, *see also* specific type
  effect on optical-difference spectroscopy measurements, LII, 264
    on radical ion chemiluminescent reaction, LVII, 497, 498
  enzyme electrode, XLIV, 585, 617
  inhibitors, of bacterial luciferase, LVII, 151
  oxidation, in microbodies, LII, 498
  solvent, of organic hydroperoxide, LII, 510
Alcohol-chloroform extraction, of hypothalamic peptides, XXXVII, 402–407
Alcohol dehydrogenase, XXXIV, 3, 122, 238, 246, 251; XLVI, 21, 33, 83, 145, 147, 148, 249–258, 260
  active site, XLVII, 66
  alcohol oxidation, XLIV, 838–840

assay, LII, 361; LVII, 210, 214
by Fast Analyzer, XXXI, 816
of glucose, LVII, 212
associated with cell fractions, XL, 86
atomic emission studies, LIV, 458
atomic fluorescence studies, LIV, 469
binding site, XLVI, 75
coenzyme oxidation, XLIV, 863
conjugate
activity, XLIV, 271
kinetics, XLIV, 406, 407
stability, XLIV, 841
from *Drosophila melanogaster*, XLI, 374–379
assay, XLI, 376
electrophoresis, XLI, 379
molecular weight, XLI, 378
properties, XLI, 378, 379
purification, XLI, 376–378
enzymic activity, XLVI, 254
in ethanol metabolism, LII, 355
from human liver, XLI, 369–374
assay, XLI, 369
chromatography, XLI, 371
inhibitors, XLI, 373
properties, XLI, 372–374
purification, XLI, 370–373
hydrophobicity studies, XLIV, 63, 64
immobilization, XLIV, 836, 837; LVII, 212, 214
in hollow-fiber membrane-device, XLIV, 307
on inert protein, XLIV, 903, 908
in multistep enzyme system, XLIV, 315, 316
on nylon tube, XLIV, 120
on water-insoluble carriers, XLIV, 108
from liver
histidine content, XLVII, 431
reductive alkylation, XLVII, 474
as marker enzyme, XXXII, 733
membrane osmometry, XLVIII, 75
oxaloacetate assay, LV, 207
polarography, LVI, 478
in preparation of deuterated NAD^+, LIV, 226, 228
pyridine nucleotide assay, LV, 264, 265, 267–270
reaction with arylazido-β-alanine, XLVI, 282, 284
with arylazido-β-alanine NAD^+, XLVI, 282, 284
reduction of NAD^+ analogs, XLIV, 873, 874
of $NADP^+$ analogs, XLIV, 874–876
removal from liver microsomal components, LII, 359–361
from *Rhodotorula*, XLI, 361–364
assay, XLI, 361
chromatography, XLI, 363
electrofocusing, XLI, 363
inhibitors, XLI, 364
properties, XLI, 364
purification, XLI, 361–363
stereospecific reduction of ketone, XLIV, 837, 838
from yeast
affinity labeling, XLVII, 423, 424, 428
histidine modification, XLVII, 433
peptide separation, XLVII, 205
Alcohol nitroxide, *see* 1-Oxyl-2,2,6,6-tetramethyl-4-piperidinal
Alcohol oxidase, LII, 17, 18
from basidiomycetes, XLI, 364–369
assay, XLI, 365
production, XLI, 366, 367
properties, XLI, 368
purification, XLI, 366–368
in enzyme electrode, XLIV, 590
immobilization, LVI, 483, 489
sources and stability, LVI, 467
substrates, XLIV, 615
Aldehyde, XLVI, 209, *see also* specific type
assay
with bacterial luciferases, LIII, 564
of luciferase, LIII, 565
fluorescent labeling, XLVIII, 359
inhibitor, of bacterial luciferase, LVII, 138, 139, 152
long-chain, assay, LVII, 189–194
solution preparation, LVII, 196
substrates, of bacterial luciferase, LVII, 151
sulfite reaction, XXXI, 539

Aldehyde dehydrogenase
from baker's yeast, XLI, 354–360
assay, XLI, 354, 355
inhibition, XLI, 360
properties, XLI, 359, 360
purification, XLI, 355–359
immobilization, in enzyme system, XLIV, 315, 316
from *Pseudomonas aeruginosa*, XLI, 348–354
amino acids, XLI, 353
assay, XLI, 348
chromatography, XLI, 351
properties, XLI, 352–354
purification, XLI, 349–352
from *Pseudomonas oleovorans*, XLII, 313–315
assay, XLII, 313, 314
chromatography, XLII, 315
properties, XLII, 315
purification, XLII, 314315
Aldehyde-glass linkage, XXXIV, 66, 69
Aldehyde mutant, bioassays of myristic acid and long-chain aldehydes, LVII, 189–194
Aldehyde oxidase, XXXIV, 3; LII, 17, 18; LIII, 401
reversible dissociation, LIII, 435
Aldehyde reductase, *see* Alcohol dehydrogenase
Aldehydic residue, total, measurement, XL, 362
Aldehydrol
content, assay, XLIV, 105
coupling, of urease, XLIV, 105
D-Aldohexopyranoside dehydrogenase, *see* D-Glucoside 3-dehydrogenase
Aldohexose, qualitative analysis, XLI, 18
Aldohexose dehydrogenase
from *Acetobacter suboxydans*, XLI, 147
from *Gluconobacter cerinus*, XLI, 142–147
assay, XLI, 142
inhibitors, XLI, 145, 146
occurrence, XLI, 146
properties, XLI, 145
purification, XLI, 143, 144
from pseudomonad MSU-1, XLI, 147–150
assay, XLI, 147, 148
chromatography, XLI, 149
properties, XLI, 150
purification, XLI, 148
Aldolase, XXXIV, 165; XLII, 223–297; XLVI, 23, 25, 49, 132, 383, *see also* specific types
activity determination, XLVII, 495
assay by Fast Analyzer, XXXI, 816
conjugate
assay, XLIV, 336, 337
of monomeric derivatives, XLIV, 495
tetramer dissociation, XLIV, 491
denaturation studies, XLIV, 498–500
detection, by tetrazolium dye reduction, XLI, 66, 67
electrophoresis, XLI, 67–73
enzymic detection, XLI, 68, 70
epoxide synthesis, XLVII, 490, 491
site-specific reagent, XLVII, 483
gluconeogenic catalytic activity of, XXXVII, 281, 286
hollow-fiber retention data, XLIV, 296
immobilization
in polyacrylamide gel, XLIV, 902, 909
on Sepharose 4B, XLIV, 493–501
membrane osmometry, XLVIII, 75
phosphate assay, LV, 212
in sedoheptulose, 1,7-biphosphate preparation, XLI, 77–79
tryptophan content, XLIX, 162
Aldopentose, qualitative analysis, XLI, 18
L-*arabino*-Aldose dehydrogenase, *see* L-Arabinose dehydrogenase
Aldose 1,6-diphosphate, preparation, XLI, 79–84
Aldose 1-epimerase
from bovine kidney, substrate specificities, XLI, 487
from *Capsicum frutescens*, XLI, 484–487
inhibitors, XLI, 487
purification, XLI, 485, 486
substrate specificities, XLI, 487
distribution and abundance, XLI, 478

from *Escherichia coli,* substrate specificities, XLI, 487
from higher plants, XLI, 484–487
assay, XLI, 484
chromatography, XLI, 485
inhibitors, XLI, 486
properties, XLI, 486, 487
purification, XLI, 485, 486
from kidney cortex, XLI, 471–484
assay, XLI, 473–478
catalytic coefficients, XLI, 484
chromatography, XLI, 479
inhibitors, XLI, 481, 482
molecular weights, XLI, 483
properties, XLI, 480–484
purification, XLI, 478–480
from *Penicillium notatum,* substrate specificities, XLI, 487
sources, XLI, 471–473
Aldose reductase
from mammalian tissues, XLI, 159–165
activators and inhibitors, XLI, 164
assay, XLI, 159–161
chromatography, XLI, 162, 163
immunological data, XLI, 165
properties, XLI, 164, 165
purification, XLI, 161–164
sources, XLI, 160, 165
from seminal vesicle and placenta of ruminants, XLI, 165–170
activators and inhibitors, XLI, 168
assay, XLI, 165, 166
chromatography, XLI, 167
properties, XLI, 168
purification, XLI, 166, 167, 169
Aldosterone
of adrenal tissue capsules, XXXII, 686–688
adsorption to various materials, XXXVI, 93, 94
effects on ($Na^+ + K^+$)-ATPase, XXXVI, 435
on sodium transport, XXXVI, 439–444
metabolites, synthesis and isolation, XXXVI, 503–512
nuclear receptors, differential extraction, XXXVI, 286–292
pineal effects, XXXIX, 382
radioimmunoassay, XXXVI, 21
receptor-protein assay, XXXVI, 49
in superfused adrenal tissue; XXXIX, 307–314
Aldosterone-binding system, XXXIV, 6
Aldulosonic acid
estimation with diphenylamine, XLI, 33
with thiobarbituric acid, XLI, 32
Aleurone vacuole, isolation of, XXXI, 575, 576
Alfalfa
enzyme induction, LII, 235
ferredoxin
proton magnetic resonance studies, LIV, 204, 206
purification, LIII, 267
Algae, *see also* specific types
blue-green, *see* Blue-green algae
chloroplasts, XXXI, 733
countercurrent distribution, XXXI, 768
disruption, LV, 135, 138
free cells, XXXII, 723
fructose-diphosphate aldolase
from blue-green, assay, purification and properties, XLII, 228–234
from green, XLII, 234
hydrolyzate, deuterated, preparation of deuterated bacteria, LIX, 641
organelles, XXXI, 509
3-phosphoglycerate phosphatase, XLII, 409
Alginase, XXXIV, 3
Alginic acid, XXXIV, 3
Alkali, ATPase inhibitor extraction, LV, 409, 423, 424
Alkaline bismuth stain, XXXIX, 144, 145
Alkaline copper reagent, in quantitative histone determination, XL, 110
Alkaline phosphatase, XXXIV, 105; XLVI, 670
assay
by Fast Analyzer, XXXI, 813, 816, 831–833
fluorometric, XLIV, 626, 627
associated with cell fractions, XL, 86
atomic emission studies, LIV, 458
atomic fluorescence studies, LIV, 469

bacterial, in tRNA sequence analysis, LIX, 61, 75, 99, 102
as brush-border marker, XXXI, 120–122, 125
calf intestinal, in tRNA sequence analysis, LIX, 61
characterization of tRNA digestion products, LIX, 171
cobalt-substituted, magnetic circular dichroism, XLIX, 168
conjugate
kinetics, XLIV, 426
optimal effective pH, XLIV, 429
conjugation, with antibody, XLIV, 711
cytochemistry, XXXIX, 151
determination of deuterium content of RNA, LIX, 646
digestion of tRNA 3′-terminus, LIX, 175, 186, 343
fluorine nuclear magnetic resonance, XLIX, 273, 274
chemical shift, XLIX, 292, 293
spectra, XLIX, 276, 279, 280, 286, 287
gel electrophoresis, XXXII, 83, 90
immobilization
in collodion membrane, XLIV, 904, 908
in hollow-fiber membrane, XLIV, 308, 309
by microencapsulation, XLIV, 214
on water-insoluble supports, XLIV, 108
in intestinal mucosa, XXXII, 667
in lung lamellar bodies, XXXI, 423, 424
in microvillous membrane, XXXI, 130, 132
phosphodiesterase assay, XXXVIII, 260
in phosphohydrolase assay, XXXII, 131
in PMN granules, XXXI, 345
reductive denaturation studies, XLIV, 517
SDS complex, properties, XLVIII, 6
site-specific reagent, XLVII, 484
stain for detection, XLIV, 713
Alkaline phosphodiesterase I, in plasma membrane, XXXI, 24
Alkaline phosphomonoesterase, *see* Alkaline phosphatase
Alkaloid, enzymatic oxidation, LII, 151
Alkane, oxidation, in microbodies, LII, 496
α,ω-Alkanediamine, XXXIV, 243
Alkane hydroxylase, *see* Alkane 1-monooxygenase
Alkane 1-hydroxylase, *see* Alkane 1-monooxygenase
Alkane 1-monooxygenase, LII, 12, 14, 15, 318–324
composition, LIII, 360
inhibitor, LIII, 360
molecular weight, LIII, 359
prosthetic group, LIII, 360
purification, LIII, 357–359
purity, LIII, 359
stability, LIII, 359
substrate specificity, LIII, 360
2-Alkanol, steric analysis of 2-D-phenylpropionate derivatives by gas-liquid chromatography, XXXV, 329
3-Alkanol, steric analysis of 2-D-phenylpropionate derivatives by gas-liquid chromatography, XXXV, 329
Alkenyl *O*-glycoside, XXXIV, 178, 361–367, *see also* *O*-Glycosyl polyacrylamide
1-Alkyl 2-acyl glycerophosphatide desaturase, LII, 13
Alkylagarose, XXXIV, 130, 131
Alkyl alcohol oxidase, LII, 17
Alkyl alkanethiosulfonate, XLVII, 425–430
reaction with protein sulfhydryls, XLVII, 426
Alkylamine
carrier, XXXIV, 67–69
coupling, XLIV, 139, 140
by aqueous silanization, XLIV, 139
by organic silanization, XLIV, 140
glass bead, XXXIV, 289
Alkylamine glass
long-chain, synthesis, XXXVIII, 183, 184
synthesis, XXXVIII, 181, 182

Alkylammonium tetrahaloferrate, osmometric studies, XLVIII, 79
Alkyl arylamine oxygenase, LII, 15
Alkylating agent
mutation induction, XLIII, 32–35, *see also* specific agent
reactions with tributylphosphine, XLVII, 113
Alkylation, XXXIV, 49, 50
covalent bonding, XLVI, 28
monomers used, XLIV, 197
reductive
of protein amino groups, XLVII, 465–478
radioactive labeling, XLVII, 476
of thiol groups, XLVII, 113–116
Alkyl azide, XLVI, 80
Alkylbetain, as solubilizers, LVI, 747
Alkyl β-D-glucopyranoside tetraacetate, XLVII, 492
Alkylguanidine, activator of NADH dehydrogenase, LIII, 20
Alkyl halide derivative
limitation, XLIV, 49
preparation, XLIV, 49
Alkylhydrazine oxidase, LII, 21
2-Alkyl-3-hydroxy-1,4-naphthoquinone
inhibitor,of complex I–III, LIII, 9
of complex III, LIII, 38
2-Alkyl-4-hydroxyquinoline *N*-oxide
complex III, LVI, 586
inhibitor, of complex I–III, LIII, 9
Alkyl imidate, cleavable, bifunctional, XLIV, 555, 556
Alkyl phenylpolyoxyethylene, structure, LVI, 736
Alkyl polyoxyethylene, structure, LVI, 736
Alkyl sulfate, partial specific volume, XLVIII, 18
Alkyl sulfonate, partial specific volume, XLVIII, 18
Alkyltrichlorosilane, immobilized mitochondria, LVI, 550, 551
Allantoin, XLIV, 591
Allantoinase, in microbodies, LII, 496, 500
Allen formula, in hormone spectrophotometry, XXXIX, 242
Allenic steroid, XLVI, 461–468
Allenic thioester, XLVI, 158
Allergen-cellulose, XXXIV, 707
Allohydroxy-D-proline oxidase, LII, 21
Allopurinol, XXXIV, 6
inhibitor, of orotate phosphoribosyltransferase, LI, 166
1-Allopurinol 5′-phosphate, inhibitor of orotidylate decarboxylase, LI, 153
Allosteric constant, XLVIII, 290
Allosteric interaction, XLVIII, 305
Allosteric regulation, definition, XLIV, 504
Allosteric system, relaxation kinetics, LIV, 81, 82
Allouridylate
inhibitor of orotate phosphoribosyltransferase, LI, 163
of orotidylate decarboxylase, LI, 163
Alloxan, in cytochemical localization of adenyl cyclase, XXXIX, 480, 481
Allyl *O*-(2-acetamido-3,4,6-tri-*O*-acetyl-2-deoxy-β-D-glucopyranosyl)- (1→4)-2-acetamido-3,6-di-*O*-acetyl-2-deoxy-β-D-glucopyranoside, synthesis, XLVI, 406, 407
N^6-Allyladenosine, inhibitor, of GMP synthetase, LI, 224
Allyl alcohol, XLIV, 84, 108; XLVI, 33
alcohol oxidase, LVI, 467
Allyl amine, XLVI, 33
Allyl chloride, as inhibitors, XLVI, 36–38
Allyldichlorophosphine, intermediate in synthesis of phosphatidylcholine analogues, XXXV, 521, 522
Allyl glucoside, XLVI, 365
Allyl glycoside, XXXIV, 178, 362–367
O-acetylated, XXXIV, 365, 366
Allyl β-D-glycoside, XXXIV, 364, 365
Allyl octadecyl ether, intermediate in synthesis of phosphatidic acid analogues, XXXV, 508
1-Allyloxy-3-(*N*-aziridine)-2-propanol, XLIV, 198, 199
Allyloxy-2,3-epoxypropane, XLIV, 196
1-Allyloxy-3-(*N*-ethyleneimine)-2-propanol
structural formula, XLIV, 197
synthesis, XLIV, 196

1-(2-Allylphenoxy)-3-isopropylaminopropanol, *see* Alprenolol
Allyl sulfate, as inhibitors, XLVI, 36, 37
Alpha-MEM medium, LVIII, 56
Alprenolol, XLVI, 591
metabolism, LII, 65–69
Alternaria tenuis, tentoxin, LV, 492
Alternariol, XLIII, 163
Alumina
controlled-pore, glucose isomerase immobilization, XLIV, 161–164
efrapeptin purification, LV, 494
hydrogen ion concentration operating range, XLIV, 163
preparation, LII, 429, 430
of *Bacillus* crude extract, LIX, 439
of *E. coli* crude extract, LIX, 314, 363, 451, 554, 827, 835, 839
purification of cyclic GMP, LVII, 99
of luciferin sulfokinase, LVII, 249
for separation of porphyrin ester, LII, 432
Alumina A–301, in bacterial cell extract preparation, LI, 519
Alumino A–305, preparation of *E. coli* crude extract, LIX, 445
Alumina Cγ gel
in cytochrome *P*–450 purification, LII, 129
protein kinase, XXXVIII, 303, 311
purification of biotin carboxyl carrier protein, XXXV, 21
Alumina column chromatography
cyclic nucleotide separation, XXXVIII, 41, 42, 45, 46, 88
column preparation, XXXVIII, 43, 44
cyclase assays, XXXVIII, 42, 43
interfering compounds, XXXVIII, 44
material and methods, XXXVIII, 42
procedure, XXXVIII, 44
Alumina gel, *see* Alumina; specific type
Alumina oxide
purification of cytochrome c_3, LIII, 622, 623
of desulfovibrione electron carriers, LIII, 634
of rubredoxin:NAD oxidoreductase, LIII, 619
washing procedure, LIII, 615
Aluminum
atomic emission studies, LIV, 455
sample cell, in neutron scattering experiments, LIX, 657
Aluminum-27, lock nuclei, XLIX, 348; LIV, 189
Aluminum chlorhydroxide complex, XLIV, 652, 653
Aluminum chloride
anhydrous, XLVII, 360
synthesis of 4-(4-methoxyphenyl)-4-oxobutyric acid, LVII, 434
Aluminum foil, in flash photolysis system, LIV, 99
Aluminum oxide, XLIV, 158
durability, XLIV, 135
pore properties, XLIV, 137
rhein purification, LV, 459
Aluminum sulfate
osmometric studies, XLVIII, 78
purification of cytochrome *c*, LIII, 128
Aluminum tubing, in freeze-quench technique, LIV, 89
Alundum, LIX, 439
Alveolar macrophage, *see* Macrophage, alveolar
α-Amanitin, XXXVI, 319, 320; LVI, 31
resistance, LVIII, 309, 314
RNA inhibitor, XL, 282
Amaromycin, XLIII, 131
AMBD 647/3 medium, LVIII, 59
Amberlite
chromatographic determination of histones, XL, 119
in plant-tissue fractionation, XXXI, 535, 536
Amberlite CG–50, *see also* Amberlite IRC–50
purification, of cytochrome b_5 reductase, LII, 468, 469
of radiolabeled edeine, LX, 559
Amberlite IR-120, L-amino acid purification, XLIV, 757
Amberlite IRA 400–Cl^-, purification of high-potential iron-sulfur protein, LIII, 338
Amberlite IRC–50, *see also* Polymethacrylic acid resin
alternate names, LIII, 133

purification of carbon monoxide-binding heme protein P–460, LIII, 638
of cytochrome *c*, LIII, 129, 130, 131, 133, 148
of cytochrome *c* derivatives, LIII, 172, 174, 177
Amberlite MB–1, XXXI, 4
for deionizing urea solutions, LVII, 172
Amberlite XAD–2, for detergent removal, LII, 113, 203
5A medium, *see* McCoy's medium 5A
Amelanotic mutant, *see* Mutant, amelanotic
American Type Culture Collection, LVIII, 119, 440, 441
Ames test, LVIII, 302
Amethopterin
in bacterial growth medium, LI, 553
DNA inhibitor, XL, 276
in DNA synthesis, XL, 42
Amicetin
paper chromatography, XLIII, 153
solvent system for countercurrent distribution, XLIII, 334
Amicon Diaflo membrane filter, LIX, 500
Amicon ultrafilter device, XXXI, 17
Amidase, XLIII, 208; XLVI, 22
Amide
fluorescent, continuous rate assay, XLV, 180, 185
osmometric studies, XLVIII, 78
Amide bond, in polypeptide backbone, chromophore classification, LIV, 254
Amide-glass linkage, XXXIV, 70, 71
Amide group, XXXIV, 34
planar trans, vibrational modes, XLIX, 100–102
Amidination, pH, LVI, 623
Amidine, formation, XLIV, 321
Amidinomycin, XLIII, 145, 146
Amidinotransferase, XLIII, 451–458, *see also* L-Arginine:inosamine-P amidinotransferase
Amido Black
for gel electrophoresis, XXXII, 74–77, 80, 81
in myelin basic protein studies, XXXII, 342
precipitin lines, LVI, 225
for protein, XXXII, 867
Amido Black 10B, LIX, 496, 503
staining procedure, LIX, 507, 540
Amidophosphoribosyltransferase
assay, LI, 171–173
from chicken liver, LI, 171
inhibitors, LI, 178
kinetic properties, LI, 178
molecular weight, LI, 177
properties, LI, 177, 178
purification, LI, 173–177
low ionic strength precipitation, LI, 174
purity, LI, 177
Amido Schwarz
as lysosome stain, XXXI, 473
as protein gel dye, XXXVI, 450
Amidotransferase, XLVI, 414–427
Aminco-Bowman spectrofluorometer, XXXII, 678; LVII, 227
Aminco-Chance dual-wavelength spectrophotometer, LII, 223
Aminco DW–2 dual wavelength/split beam spectrophotometer, LII, 214, 265, 361
Aminco-French press, LII, 160
Aminco French pressure cell, LIX, 435
Aminco-Morrow stopped-flow apparatus, LII, 223
Aminco photometer, *see* Chem-Glow photometer
Amine
acetylenic, XLVI, 161
aromatic, XXXIV, 371
labeled, filter paper assay, factor XIII, XLV, 182
organic, lasalocid, LV, 446
primary, XXXIV, 33
secondary, conversion to primary, XLVII, 8
spot test, XXXIV, 482
tertiary
activation catalyst, XLIV, 37
production of amine oxides, XLIV, 850, 854, 855
uptake, LV, 564
Amine drug, enzymatic oxidation, LII, 151

Amine oxidase, XLVI, 160; LII, 10, 18
flavin-containing
absorption spectra, LVI, 696
antibody
affinity chromatography, LVI, 700
conjugation to ferritin, LVI, 711, 712
specificity, LVI, 690
assay, LIII, 496; LV, 101, 102
from beef liver mitochondria, LIII, 495–501
coupling to Sepharose 4B, LVI, 698, 699, 700
digitonin, LVI, 692
distribution
on inner surface of outer membrane, LVI, 686
on mitochondrial outer membrane, LVI, 684, 691
on outer surface of outer membrane vesicles, LVI, 688
flavin content, LIII, 500, 501
flavin linkage, LIII, 450
hepatoma mitochondria, LV, 81, 82, 88
inactivation, LIII, 441
inhibitors, LIII, 501
isolation of flavin peptide, LIII, 454
molecular weight, LIII, 500
outer membrane, LV, 101, 103, 104
pH optimum, LIII, 501
preparation and properties, LVI, 694, 695
purification, LIII, 495, 496, 498–500
purity, LIII, 500
release from outer membrane, LV, 91, 93, 94
storage, LIII, 500
formaldehyde generation, LII, 297
2-Amine-4-pentynoic acid, *see* Propargylglycine
Aminex A–6, resolution, XLVII, 220
Amino acid, *see also* specific type
activation, in protein turnover, XXXVII, 247
active ester
definition, XLVII, 557, 558
peptide bond formation, XLVII, 560–562
physical properties, XLVII, 559
stability, XLVII, 558
synthesis, XLVII, 559, 560
of aldehyde dehydrogenase from *Pseudomonas aeruginosa*, XLI, 353
analysis, XL, 359; XLVII, 1–69
for bound protein determination, XLIV, 73, 388, 389
column preparation, XLVII, 11
derivatives, XLVII, 45–51
eluent linear flow velocity, XLVII, 20, 21
high-pressure, XLVII, 23, 24
hydrolysis, XLVII, 17, 18
intermediate pressure, XLVII, 24–27
internal standards, XLVII, 31–40
ion-exchange chromatography, XLVII, 3–7, 19–34
improved resins, XLVII, 19–31
internal standards, XLVII, 32–34
of nonhistone chromosomal proteins, XL, 158
peptide hydrolysis, XLVII, 40–45
peptide synthesis monitoring, XLVII, 598, 599
preparation of human biopsy tissue, XL, 360
resin cross-linkage, XLVII, 22, 23
sample preparation, XLVII, 17, 18
in subnanomole range, XLVII, 3–18
time
control, factors affecting, XLVII, 23–27
particle size, XLVII, 21, 22
two-column system, advantages, XLVII, 9
anilinothiazolinone derivative
automated conversion to phenylthiohydantoin derivative, XLVII, 385–391
enhanced solubility, XLVII, 339
extraction, solvents, XLVII, 345
hydrolysis, XLVII, 369–373

identification, XLVII, 374–385
N^{α}-phenylthiocarbamylmethylamide derivative
identification, XLVII, 379–381
preparation, XLVII, 375–377, 391
quantitation, XLVII, 381, 382
assay for transport with intact *M. phlei* cells, LV, 179, 180
azide derivative
infrared absorption, XLVII, 565
synthesis, XLVII, 566–568
basic, analysis, XLVII, 9–11
benzyl ester, preparation, XLVII, 524
benzyloxycarbonyl derivative
commercial availability, XLVII, 592, 593
properties, XLVII, 512
salts, XLVII, 593
support attachment, XLVII, 591
synthesis, XLVII, 511, 512
2-(*p*-biphenyl)-2-propyloxycarbonyl derivatives, XLVII, 594
bromoprogesterone reactions with, XXXVI, 387, 388
tert-butyl ester, preparation, XLVII, 525
tert-butyloxycarbonyl derivative
properties, XLVII, 513–515
synthesis, XLVII, 513–516
α-carboxyl group blocking by esterification, XLVII, 522–526
carboxymethylated, elution profile, XXXVI, 403
catabolism, substrate entry, XXXIX, 495
cesium salt, peptide synthesis, XLVII, 590
chemical modification, XLVI, 4; XLVII, 405–498, 609–616
composition of ATPase inhibitors, LV, 400, 404
of ATPase proteins, LV, 350, 351
coupling factor B, LV, 389, 390
dansyl derivative, analysis, XLVII, 47, 49
derivatives
high-performance liquid chromatography, XLVII, 7, 45–51
solid-phase peptide synthesis, XLVII, 609–617
detection, radioactive phenylisothiocyanate, XLVII, 247, 248, 254, 255
determination of aminoacylation isomeric specificity, LIX, 272
2,4-dinitrophenyl derivatives, analysis, XLVII, 47, 48
of 1,3-diphosphoglycerate phosphatase, XLII, 422–424
of enolase from *Escherichia coli*, XLII, 327
essential, in media, LVIII, 53, 54, 62
esters, properties, XLVII, 523
ethyl ester, preparation, XLVII, 523
of fructose-diphosphate aldolase from lobster muscle, XLII, 226, 227
from *Peptococcus aerogenes*, XLII, 256, 257
of fructose-1,6-diphosphate phosphohydrolase from spinach chloroplasts, XLII, 402
of β-galactosidases from *Neurospora crassa*, XLII, 501
in glucose-6-phosphate dehydrogenase from human erythrocytes, XLI, 214
in glyceraldehyde-3-phosphate dehydrogenase from *Bacillus stearothermophilus*, XLI, 272
of glycerol kinase, XLII, 154
in glycerol-3-phosphate dehydrogenase from honey bees, XLI, 245
HeLa cell labeling, LVI, 68, 75
of hexokinase of rat brain, XLII, 24
in histone fraction from calf thymus, XL, 105
hydrazide derivative, synthesis, XLVII, 565, 566
hydrophobic nature, differences, XLIV, 13, 14
1-hydroxybenzotriazole ester, fragment condensation, XLVII, 568–571
N-hydroxysuccinimidyl ester
fragment condensation, XLVII, 568, 569
synthesis, XLVII, 558–560
in insect media, LVIII, 455

isocyanate derivative, infrared absorption, XLVII, 565
of 2-keto-3-deoxy-6-phosphogluconic aldolase from *Pseudomonas putida*, XLII, 262
magnetic circular dichroism, XLIX, 159–164
metabolism, mitochondrial transport, LVI, 248, 249
methyl ester, preparation, XLVII, 523
mitochondrial protein synthesis, LVI, 21
mixture, for embryo studies, XXXIX, 299
p-nitrobenzyl ester, synthesis, XLVII, 524
p-nitrophenyl ester
 solid-phase peptide synthesis, XLVII, 597
 synthesis, XLVII, 559
o-nitrophenylsulfenyl derivative
 storage, XLVII, 518
 synthesis, XLVII, 518
nonessential, in media, LVIII, 63
number of available modification reactions, XLIV, 14
OSCP, LV, 397
partial specific compressibility, XLIX, 17
2-phenyl-2-propyloxycarbonyl derivatives, XLVII, 594
phenylthiohydantoin derivative, XLVII, 153, 247, 336
 analysis, XLVII, 46–48, 248, 249, 348–351
 choice of label, XLVII, 326
 ^{14}C-labeled, analysis, XLVII, 248, 249
 hydrolysis, XLVII, 351, 371
 labeled, identification, XLVII, 327
 recovery, XLVII, 347, 348
 separation, gradient elution system, XLVII, 349, 350
phosphodiesterase activator composition, XXXVIII, 269–271
of phosphofructokinase from yeast, XLII, 85
of phospho-β-galactosidase, XLII, 493
in 6-phosphogluconate dehydrogenase from *Candida utilis*, XLI, 240
 from human erythrocytes, XLI, 226
of 3-phosphoglycerate kinase from baker's yeast, XLII, 138
 from human erythrocytes, XLII, 148
of phosphoglycerate mutase from liver, XLII, 445
phosphorescence, XLIX, 242–244
polymers, conformation studies, using vacuum ultraviolet circular dichroism, XLIX, 219–221
in polypeptide hormones, tritium labeling, XXXVII, 313–321
procedures for incorporation
 chloramphenicol, LVI, 57, 58
 cycloheximide, LVI, 55–57
production systems, XLIV, 184
racemization during peptide synthesis, XLVII, 509
radiolabeled
 incorporation procedure, XLVII, 249, 250, 260
 in quantitation of L-amino acids, LIX, 315, 316
  tRNA aminoacylation, LIX, 288
regeneration
 from anilinothiazolinones, XLVII, 369–373
 from thiohydantoins, XLVII, 368
residue, identity of bonding site in flavin compounds, LIII, 463, 464
of ribulose-1,5-diphosphate carboxylase, XLII, 471
secondary functional group
 blocking, XLVII, 526–544, 595, 609–616
 deprotection, XLVII, 602–604
side chain, chromophore classification, LIV, 254
specific radioactivity, measurement, LIX, 316, 317
thiazolinone derivative, thermal conversion, XLVII, 390, 391
thiohydantoin derivative
 hydrolysis, XLVII, 368
 identification, XLVII, 363–368
 mass spectra, XLVII, 367
 preparation, XLVII, 360

retention times, XLVII, 366
R_f values, XLVII, 365
transport, XLVI, 151, 607–613
comparison of ghosts and ETP, LV, 186, 187
K^+ gradient, LV, 679
in proteoliposomes
preparation of proteoliposomes, LV, 197
reconstitution of proline uptake, LV, 197–199
transport
in bacteria, XXXII, 843–849
by *Halobacterium* vesicles, LVI, 404–407
2,4,5-trichlorophenyl ester, XLVII, 558, 559
synthesis, XLVII, 559
trityl derivatives, XLVII, 520
β,γ-unsaturated, as inhibitors, XLVI, 31–34
D-Amino acid
enzyme electrode, XLIV, 585, 589, 590
response time, XLIV, 608
inhibitor, of thermolysin, XLVII, 177
preparation, XLIV, 758
L-Amino acid
enzyme electrode, XLIV, 585, 589, 590
quantitation by aminoacylation method, LIX, 314
Amino acid activating enzyme, XLVI, 24, 26
Amino acid analyzer
commercial, XLVII, 6, 7
conversion to high-sensitivity, XLVII, 7–9
diagram, XLVII, 4
preparation of enzyme digestion reaction mixture, XLVII, 81
Amino acid antibiotic, paper chromatography, XLIII, 144–152, *see also* specific antibiotic
Amino acid chloromethyl ketone, XLVI, 200
purification, XLVI, 611
synthesis, XLVI, 609, 610
L-Amino acid decarboxylase
assay, XXXIX, 388
in melatonin biosynthesis, XXXIX, 379
Amino acid derivative, in media, LVIII, 63
Amino acid oxidase, XLVI, 37; LII, 18, *see also* specific types
immobilization, LVI, 484, 485
in microbodies, LII, 496, 498
mixed function, in liver microsomes, LII, 142–151
D-Amino acid oxidase, LIII, 402; LVI, 474
apoenzyme, assay of FAD, LIII, 422
dissociation kinetics, LIII, 430
in enzyme electrode, XLIV, 589
holoenzyme reconstitution, LIII, 436
immobilization, in gel, XLIV, 908
inactivation, LIII, 442, 446, 448
in peroxisomes, XXXI, 364, 367
redox potential, LIII, 400
reversible dissociation, LIII, 429–431, 433
substrates, XLIV, 615
L-Amino acid oxidase, LVI, 474
conjugate, activity, XLIV, 271
in enzyme electrode, XLIV, 589
immobilization
in gel, XLIV, 908
on inert protein, XLIV, 908
inactivation, LIII, 443, 445
substrates, XLIV, 615
9-Aminoacridine, measurement of ΔpH, LV, 229, 563–567, 779–781, 787
Aminoacylase
assay, XLIV, 199–201, 747, 748
entrapment, in polyacrylamide gel, XLIV, 750
fiber entrapment, industrial application, XLIV, 241
immobilization on DEAE-Sephadex, XLIV, 748, 749
on iodoacetylcellulose, XLIV, 749, 750
by protein copolymerization, XLIV, 198–200
regeneration of activity during, XLIV, 164, 165
immobilized, kinetic properties, XLIV, 750, 751
industrial application, XLIV, 746–759
native, preparation, XLIV, 748

source, XLIV, 196
supports, XLIV, 759
thermal stability, XLIV, 751, 752
vinylation, effect of pH, XLIV, 199
L-Aminoacylase, commercial use, XLIV, 165
Aminoacylation, LIX, 216, 217, 221, 298
assay of hydrolyzed tRNA, LIX, 183
determination of *in vivo* levels, LIX, 268–271
2′-hydroxyl specific, mechanism, LIX, 275–277
3′-hydroxyl specific, mechanism, LIX, 274–276
in intact cells, LIX, 270, 271
isomeric specificity, LIX, 272–282
in *E. coli*, LIX, 272–276, 280, 281
in wheat germ, LIX, 279–281
mechanisms, LIX, 274–279, 281
reaction parameters, LIX, 231
using nonradioactive amino acids, LIX, 288, 289
using radioactive amino acids, LIX, 288
α-Aminoacylpeptide hydrolase, *see* specific types
Aminoacyltransferase I, zone sedimentation patterns, XLVIII, 252
Aminoacyl-transfer ribonucleic acid, XXXIV, 4; XLVI, 103, 104, 624, 626, 629–631, 707; *see also* Methionyl-transfer ribonucleic acid; Phenylalanyl-transfer ribonucleic acid
acetylation, LIX, 818
aromatic ketone derivatives, XLVI, 676–683
binding
as assay of ribosomal activity, LIX, 459, 460
initiation factor IF-2, LX, 219
70 S ribosomes, LX, 625, 626
complex with elongation factors and guanosine nucleotides, LX, 615, 616, 622, 624, 625, 663, 698
deacylation, LIX, 312
isoaccepting species, amino acid transfer among, LIX, 293–295
isolation, LIX, 310–312
preparation, LX, 92, 93, 140, 598, 620–623
reaction with amino acid hydrazides, LIX, 173
steady-state, radiolabeling, LIX, 271
use in protein synthesis, LX, 143–145, 161–163
Aminoacyl-tRNA synthetase, XXXIV, 7; XLVI, 170, 683
adenylyliminodiphosphate synthesis, XXXVIII, 423
affinity elution, XXXIV, 163–171, *see also* specific enzyme
alanine, XXXIV, 170
aspartic acid, XXXIV, 170
assay system, XXXIV, 166, 167
KC concentration, XXXIV, 168
phosphocellulose capacity, XXXIV, 167, 168
assay, LX, 620
from Chinese hamster ovary cells
assay, LIX, 231, 232
distribution in subcellular fractions, LIX, 233, 234
classes, LIX, 234
crude preparation, LIX, 220
misactivation, LIX, 282–291
mutants, LVIII, 319
polynucleotide removal, XXXIV, 166–168
preparation, LX, 92, 110–112, 277, 619–621
of aminoacylated tRNA, LIX, 130, 221
purification from *E. coli,* LIX, 314, 315
from yeast, LIX, 257–267
RNA complex, XLVI, 170
tRNA separation, XXXIV, 168, 169
use in aminoacylation of transfer ribonucleic acids, LX, 620, 622, 623
in protein synthesis, LX, 113, 121, 122
zone sedimentation patterns, XLVIII, 254
DL-α-Aminoadipic acid ethyl amide, XLIII, 411, 412
γ-(α-Aminoadipyl)cysteinylvaline, XLIII, 471
γ-(α-Aminoadipyl)cysteinylvaline synthetase, XLIII, 471–473
assay, XLIII, 471

Cephalosporium acremonium
preparation, XLIII, 473
specificity, XLIII, 473
α-Amino alcohol, XLVI, 221
α-Amino aldehyde, XLVI, 221
ω-Aminoalkylagarose, XXXIV, 106, 107, 126, 131, 679
Aminoalkyl nucleotide, XXXIV, 229
m-Aminobenzamidine, XXXIV, 441, 442
thrombin, XXXIV, 445
p-Aminobenzamidine, XXXIV, 445
thrombin, XXXIV, 445
yeast cell disruption, LV, 416
m-Aminobenzamidine-agarose, XXXIV, 443, 736, 737
thrombin, XXXIV, 446
m-Aminobenzamidine dihydrochloride, XLVI, 123
m-Aminobenzamidine-polyacrylamide, XXXIV, 737
p-Aminobenzamidoalkylagarose, XXXIV, 107
p-Aminobenzamidoethylagarose, XXXIV, 626
2-Amino-3-(3-benzamidopropyl)-5-(3-indolyl)pyrazine, synthesis of *Cypridina* luciferin, LVII, 369
2-Amino-5-benzamidovaleramidine dihydrobromide, synthesis of *Cypridina* luciferin, LVII, 369
p-Aminobenzoic acid, XLVI, 103, 515
in media, LVIII, 64
as spacer, XXXIV, 72
substrate, of *p*-hydroxybenzoate hydroxylase, LIII, 549, 550
synthesis of affinity matrix, LII, 515, 517
9-(*p*-Aminobenzyl)adenine, XLVI, 328, 329
synthesis, XLVI, 330
6-(*p*-Aminobenzylamino)purine, XXXIV, 106
(*p*-Aminobenzyl)cellulose, XLIV, 46
2-Amino-3-benzyl-5-(*p*-methoxyphenyl)pyrazine, *see* Etioluciferin
p-Aminobenzylpenicillin-agarose, XXXIV, 399
3-Amino-1-bromo-4-methyl-2-hexanone, XLVI, 152
4-Aminobutanal dehydrogenase, stereospecificity determination, LIV, 229, 231
DL-2-Aminobutanoic acid, XLVI, 40
2-Amino-3-butenoate, *see* Vinylglycine
6-*N*-(4-Aminobutyl)amino-2,3-dihydrophthalazine-1,4-dione
chemiluminescence of ligand conjugates, LVII, 440
structural formula, LVII, 428
synthesis, LVII, 432, 433
6-[*N*-(4-Aminobutyl)-*N*-ethyl]amino-2,3-dihydrophthalazine-1,4-dione
chemiluminescence of ligand conjugates, LVII, 440
structural formula, LVII, 428
synthesis, LVII, 433
of 6-[*N*-ethyl-*N*-(4-thyroxinylamido)butyl]amino-2,3-dihydrophthalazine, LVII, 434
7-[*N*-(4-Aminobutyl)-*N*-ethyl]aminonaphthalene-1,2-dicarboxylic acid hydrazide
chemiluminescence of ligand conjugates, LVII, 440
structural formula, LVII, 429
synthesis, LVII, 434–437
4-Amino-2-butynoic acid, XLVI, 163
L-α-Aminobutyrate, activator, of aspartate carbamyltransferase, LI, 49
4-Aminobutyrate aminotransferase, brain mitochondria, LV, 58
γ-Aminobutyrate-α-ketoglutarate aminotransferase, XLVI, 33, 37
γ-Aminobutyrate-α-ketoglutarate transaminase, GABA transaminase, *see* γ-Aminobutyrate-α-ketoglutarate aminotransferase
4-Aminobutyrate transaminase, *see* 4-Aminobutyrate aminotransferase
α-Aminobutyric acid
identification, XXXII, 780
internal standard, XLVII, 34, 39
in media, LVIII, 63
recovery, XLVII, 371, 372
γ-Aminobutyric acid, XLVI, 33, 163
activator, of aspartate carbamyltransferase, LI, 49
content of brain, LV, 219

γ-Amino-*n*-butyric acid methyl ester hydrochloride, XLVII, 361

6-Aminocaproic acid, XXXIV, 426, 591; XLIV, 70, 76, 77
  internal standard, XLVII, 34, 39

ε-Aminocaproic acid-agarose, as spacer, XXXIV, 72

N[6]-Aminocaproyladenosine 5′-triphosphate, XXXIV, 253, 254

ε-Aminocaproyl-L-Ala-L-Ala-L-Ala, XXXIV, 4

N-ε-Aminocaproyl-*N*-ε-aminocaproyl-α-D--galactopyranosylamine, XXXIV, 348

N-(ε-Aminocaproyl)-2-amino-2-deoxy-D-glucopyranose, XXXIV, 326

N-(ε-Aminocaproyl)-2-amino-2-deoxy-D-glucopyranose-agarose, XXXIV, 328

[*N*-(ε-Aminocaproyl)-*p*-aminophenyl]-trimethylammonium bromide, XXXIV, 572

[*N*-(ε-Aminocaproyl)-*p*-aminophenyl]-trimethylammonium bromide hydrobromide, XXXIV, 575, 576

N-(ε-Aminocaproyl-*p*-aminophenyl)trimethylammonium bromide hydrobromide-agarose, XXXIV, 576

ε-Amino-*n*-caproyl-D-arginine, XXXIV, 413

ε-Amino-*n*-caproyl-D-arginine-agarose, XXXIV, 412

ε-Amino-*n*-caproyl-D-arginine-CM-Sephadex, XXXIV, 412

ε-Amino-*n*-caproyl-D-arginine-polyacrylamide, XXXIV, 412

N[6]-Aminocaproyl-3′,5′-cyclic adenosine monophosphate, XXXIV, 5

N[6]-ε-Aminocaproyl-cyclic adenosine monophosphate, preparation, XXXVIII, 386

N-ε-Aminocaproyl-α-D-galactopyranosylamine, XXXIV, 348

N-(ε-Aminocaproyl)-β-D-galactopyranosylamine, XXXIV, 324

ε-Aminocaproyl-glycyl-leucine, XXXIV, 5

ε-Amino-*n*-caproyl-D-phenylalanine, XXXIV, 413

ε-Amino-*n*-caproyl-D-phenylalanine-CM-Sephadex, XXXIV, 412

6-Aminocaproyl-L-thyroxine-agarose, XXXIV, 386

ε-Amino-*n*-caproyl-D-tryptophan, XXXIV, 413

ε-Amino-*n*-caproyl-D-tryptophan-agarose, XXXIV, 412

ε-Aminocaproyl-D-tryptophan methyl ester, XXXIV, 4

ε-Amino-*n*-caproyl-D-tryptophan methyl ester, XXXIV, 415, 416

ε-Amino-*n*-caproyl-D-tryptophan methyl ester-agarose, XXXIV, 415
  preparation, XXXIV, 419

8-Aminocaprylic acid, XXXIV, 591

L-3-Amino-3-carboxypropanesulfonamide, XLVI, 415, 419

N[3]-(3-L-Amino-3-carboxypropyl)uridine
  chromatographic behavior, LIX, 169
  tRNA modification, LIX, 165–171

3-(3-Amino-3-carboxypropyl)uridine, XLVI, 632, 684, 686

7-Aminocephalosporanic acid, XLIII, 699
  bioautography, XLIII, 117
  CMR spectroscopy, XLIII, 411, 412
  β-lactamase hydrolysis, XLIII, 690
  paper chromatography, XLIII, 116–118
    solvent systems, XLIII, 116

p-Aminochloramphenicol hydrochloride, XLVI, 705

3-Amino-4-chlorobenzenesulfonylfluoride, XLVI, 118

L-(2*S*,5*S*)-2-Amino-3-chloro-4,5-dihydro-5--isoxazoleacetic acid, *see* Isoxazole

2-Amino-5-chlorolevulinic acid, XLVI, 152

9-Amino-6-chloro-2-methoxyacridine, ATPase assay, LV, 801

3-Amino-1-chloro-4-methyl-2-pentanone, XLVI, 152

3-Amino-1-chloro-4-phenyl-2-butanone, XLVI, 151

1-Amino-3-chloro-2-propanone, XLVI, 151

2-Aminocinnamic acid, substrate, of melilotate hydroxylase, LIII, 557

2-Amino-4-cyanobutyric acid, XLVII, 598

Aminocyclitol antibiotic, XLIII, 215–217, *see also* Aminoglycoside-modifying antibiotic

7-Aminodeacetylcephalosporanic acid, XLIII, 730

2′-Amino-2′-deoxyadenosine, isomers, LIX, 138, 139
3′-Amino-3′-deoxyadenosine
electrophoretic properties, LIX, 135, 137
preparation of 3′-amino-3′-deoxy-ATP, LIX, 135
substrate, of adenosine deaminase, LI, 507
2′-Amino-2′-deoxyadenosinc monophosphate-transfer ribonucleic acid, in determination of aminoacylation isomeric specificity, LIX, 272, 273
3′-Amino-3′-deoxyadenosine monophosphate-transfer ribonucleic acid, in determination of aminoacylation isomeric specificity, LIX, 272, 273
2′-Amino-2′-deoxyadenosine triphosphate
incorporation into tRNA, LIX, 141
preparation, LIX, 138–140
3′-Amino-3′-deoxyadenosine triphosphate
incorporation into tRNA, LIX, 141
preparation, LIX, 135–138
purity, LIX, 137, 138
3-Amino-3-deoxy-D-glucose, XLIII, 272
Aminodeoxy-*scyllo*-inositol, XLIII, 432
ATP:inosamine phosphotransferase assay, XLIII, 445–448
L-glutamine:keto-*scyllo*-inositol aminotransferase assay, XLIII, 440
myo-inositol:NAD$^+$ 2-oxidoreductase assay, XLIII, 434
1-Amino-1-deoxy-*scyllo*-inositol 4-phosphate, XLIII, 432
1,D-1-Amino-1-deoxy-*scyllo*-inositol 6-phosphate, XLIII, 432
2-Amino-2-deoxy-*neo*-inositol 5-phosphate, XLIII, 457
9-(3′-Amino-3′-deoxy-β-D-ribofuranosyl)-6-dimethylamino-9*H*-purine, XLIII, 211
5′-Amino-5′-deoxythymidine, XXXIV, 5
5-Amino-4-deoxy-α,α-trehalose, XLIII, 269, 270, 272
5′-Amino-5′-deoxyuridine
coupling to Sepharose 4B, LI, 309–311
inhibitor of uridine-cytidine kinase, LI, 309, 313
2-Amino-4,6-dichloro-*s*-triazine, XXXIV, 716
preparation, XLIV, 50
5-Amino-3,5-dideoxy-D-glycero-D-galactononulosonic acid, *see* Neuraminic acid
5-Amino-2,3-dihydrophthalazine-1,4-dione, *see* Luminol
6-Amino-2,3-dihydrophthalazine-1,4-dione, *see* Isoluminol
3-Amino-4,7-dihydroxy-8-methylcoumarin, XLIII, 506
p-Amino-α-dimethylaminotoluene, XLVI, 583
p-Aminodimethylaniline, *see* *N,N*-Dimethyl-*p*-phenylenediamine dihydrochloride
ω-Aminododecylagarose, XXXIV, 133
3-Amino-enolpyruvate phosphate, preparation, and complex formation with enolase, XLI, 121–124
4-Aminoesterone methyl ether, XLVI, 455, 456
Aminoethanesulfonic acid, XLVI, 37
S-(β-Aminoethanethiol)-*N*-acetyl-L-cysteine, disulfide derivatives, XLVII, 430
9-[*p*-(β-Aminoethoxy)phenyl]guanine, XXXIV, 525, 527
N^6-(2-Aminoethyl)adenosine 5′-monophosphate
electrophoretic mobility, LVII, 117
millimolar extinction coefficient, LVII, 116
structure, LVII, 116
synthesis, LVII, 115–117
C^8-(2-Aminoethylamino)adenosine 5′-monophosphate
electrophoretic mobility, LVII, 118
millimolar extinction coefficient, LVII, 118
structure, LVII, 116
synthesis, LVII, 117, 118
N-(2-Aminoethyl)-3-aminopropyl glass, XLVII, 264
large proteins, XLVII, 333
synthesis, XLVII, 266
N-(2-Aminoethyl)-3-aminopropyl triethoxysilane, XLVII, 267
S-Aminoethylation, XLVII, 115, 116, 125
Aminoethyl Bio-Gel P, XLIV, 56

(Aminoethyl)cellulose, XXXIV, 713; XLIV, 46, 48
glutaraldehyde binding, XLIV, 51
matrix for carboxypeptidase B, XXXIV, 411
methotrexate attachment, XXXIV, 272
S-2-Aminoethylcysteine, hydrolysis, XLVII, 172–174
Aminoethyl derivative, XXXIV, 35, 37, 43, 44
Aminoethylglutaramyl-hydrazide derivative, XXXIV, 42, 43
2-Amino-1-ethylglycoside, of glucose, XXXIV, 318
Aminoethyl group, titration, XLIV, 445
3-(2-Aminoethyl)indole, XLVII, 446
6-(2-Aminoethyl)-9-(2′,3′-*O*-isopropylidene-β-ribofuranosyl)adenine
electrophoretic mobility, LVII, 118
structure, LVII, 116
synthesis, LVII, 118
2-Aminoethyl *p*-nitrophenyl methylphosphonate, XXXIV, 588
DL-2-Aminoethyl 2-octadecoxy-3-octadecoxypropylphosphonate, intermediate in synthesis of phosphatidylethanolamine analogues, XXXV, 512, 513
N-Aminoethyl polyacrylamide resin, XLVII, 265
capacity, XLVII, 270
preparation, XLVII, 293
stability, XLVII, 273
synthesis, XLVII, 267
2-Aminoethylpolystyrene-divinyl benzene resin, sequence analysis, XLVII, 297
Aminoethyl-Sepharose, XXXIV, 304
2-Aminofluorene, XLVI, 646
Amino gel, dinitrophenylation and trinitrophenylation, XXXIV, 48, 49
1-Aminoglucose, XLVI, 75
Aminoglutethimide, cytochrome *P*–450 difference spectra, LII, 268
Aminoglycolipid, isolation, XXXII, 345
1-Aminoglycoside, XLVI, 21
Aminoglycoside antibiotic, *see also* Aminoglycoside-modifying antibiotics
gas-liquid chromatography, XLIII, 217–218
high-pressure liquid chromatography, XLIII, 278
ion-exchange chromatography, XLIII, 263–278, *see also* specific antibiotic
Amberlite CG–50 column, XLIII, 266
ammonia gradient, XLIII, 270
carboxylic acid resins, XLIII, 268
commercial cation exchangers, XLIII, 266
Dowex 1–X2, XLIII, 267
phosphonic resins, XLIII, 272, 273
resin columns preparation, XLIII, 267, 268
sulfonic acid resins, XLIII, 272, 273
nonionic adsorption chromatography, XLIII, 275–280
paper chromatography, XLIII, 119–122, *see also* specific antibiotic
Aminoglycoside-modifying enzyme, XLIII, 611–632, *see also* specific enzyme
assay, XLIII, 612–619
cofactors, XLIII, 617
crude extract preparation, XLIII, 620
DEAE-cellulose chromatography, XLIII, 622, 623
enzymic lysis by lysozyme *Escherichia coli*, XLIII, 621
by lysostaphin, XLIII, 621, 622
French pressure cell, XLIII, 621
molecular weight, XLIII, 623
osmotic shock, XLIII, 620
polymyxin B, XLIII, 622
properties, XLIII, 623–628
purification, XLIII, 622, 623
sonication, XLIII, 620
streptomycete, growth medium preparation, XLIII, 619
Amino group
aliphatic, XXXIV, 57
assay, XXXIV, 57, 107
blockage, XLVII, 149–155, 279, 305, 510–520, 616, 617
concentration determination, XLIV, 384, 533
deprotection, XLVII, 592–595

determination, XLVII, 302, 303, 305
using ^{36}Cl, XLIV, 384, 385
introduction, procedure, XLIV, 115
neutralization, XLVII, 595
primary, reactions, XLIV, 263
reductive alkylation, XLVII, 469–478
source, XLIV, 115
N-terminal, XLIV, 12
coupling with activated Sepharose, XLIV, 17
ε-Amino group
and covalent coupling, XLIV, 12
fluorescent labeling, XLVIII, 358, 359
Aminoguanidine, XLVI, 549, 550
incorporation in tRNA, LIX, 118
L-2-Amino-4-guanidinobutyric acid, internal standard, XLVII, 34, 39
1,D-1-Amino-3-guanidino-1,3-deoxy-*scyllo*-inositol 6-phosphate, XLIII, 432
L-2-Amino-3-guanidinopropionic acid, internal standard, XLVII, 34, 36, 37
Aminohexamethyliminoagarose, XXXIV, 378, 384
Aminohexamethyl-Sepharose, derivatized
preparation, LIII, 246
purification of flavocytochrome b_2, LIII, 248
2-Aminohexanoate, as spacer, XXXIV, 444
6-Amino-1-hexanol, in 6-amino-1-hexanol phosphate preparation, LI, 253
6-Amino-1-hexanol phosphate, XXXIV, 482
preparation, LI, 253
of *N*-trifluoroacetyl 6-amino-1-hexanol phosphate, LI, 253
6-Aminohexanoylglutamine-Sepharose, preparation, LI, 31–32
ε-Aminohexanoyl-NAD^+, XXXIV, 229
ε-Aminohexanoyl-NAD^+-agarose, XXXIV, 232, 233
P^1-Aminohexanoyl-NAD^+-agarose, XXXIV, 246
P^1-Aminohexanoyl-$NADP^+$-agarose, XXXIV, 246
6-Amino-1-hexyl-2-acetamido-2-deoxy-β-D-glycopyranoside, XXXIV, 318–320
N^6-(6-Aminohexyl)adenosine 5′-monophosphate, XXXIV, 229–232, 242, 243, 486, 490
molar absorption coefficient, XLIV, 867
N^6-(6-Aminohexyl)adenosine 5′-monophosphate-agarose, XXXIV, 229–232, 246, 397
P^1-(6-Aminohexyl)-P^2-(adenosine)-pyrophosphate-agarose, XXXIV, 246
N^6-(6-Aminohexyl)adenosine triphosphate, Sepharose-bound, for affinity chromatography, LI, 9
ω-Aminohexylagarose, XXXIV, 274, 275
N^6-(6-Aminohexyl)-8-aminoadenosine 5′-phosphate, XXXIV, 487
N^8-(6-Aminohexyl)-8-aminoadenosine 5′-phosphate, XXXIV, 490
N^8-(6-Aminohexyl)-8-aminoadenosine 5′-triphosphate, XXXIV, 490
8-(6-Aminohexyl)amino-cyclic adenosine monophosphate-agarose, XXXIV, 262–264
N-(ω-Aminohexyl)-L-aspartate, XXXIV, 406
N^6-[(6-Aminohexyl)carbamoylmethyl]-adenosine 5′-diphosphate, synthesis, XLIV, 868
N^6-[(6-Aminohexyl)carbamoylmethyl]-adenosine 5′-monophosphate, synthesis, XLIV, 868
N^6-[(6-Aminohexyl)carbamoylmethyl]-adenosine 5′-triphosphate, synthesis, XLIV, 867, 868
N^6-[(6-Aminohexyl)carbamoylmethyl]-NAD^+
molar absorption coefficient, XLIV, 864
synthesis, XLIV, 863–864
N^6-[(6-Aminohexyl)carbamoylmethyl]-$NADP^+$
molar absorption coefficient, XLIV, 866
synthesis, XLIV, 866
P^3-(6-Aminohex-1-yl)deoxyguanosine triphosphate, preparation, LI, 255, 256
6-Amino-1-hexylglycoside, of glucose, XXXIV, 318
N^6-(6-Aminohexyl)-NAD, XXXIV, 243, 244

N^6-(6-Aminohexyl)-NAD-agarose, XXXIV, 246
6-Amino-1-hexyl nucleoside ester, XXXIV, 484, 485
6-Amino-1-hexyl nucleoside ester-agarose, XXXIV, 484, 485
6-Amino-1-hexyl nucleoside phosphoester, XXXIV, 481
Aminohexyl-Sepharose, in purification of bacterial luciferase, LVII, 147
4-Amino-5-hexynoic acid, XLVI, 163
2-Amino-4-hydroxy-6,7-dimethyltetrahydropteridine, substrate of phenylalanine hydroxylase, LIII, 278
2-Amino-4-hydroxy-6,7-dimethyl-5,6,7,8-tetrahydropteridine, inhibitor of brain α-hydroxylase, LII, 317
2-Amino-4-hydroxy-6-lactyl-7,8-dihydropteridine, substrate of phenylalanine hydroxylase, LIII, 285
2-Amino-2-(hydroxymethyl)-1,3-propanedial, XLIV, 684
2-Amino-4-hydroxy-6-methyltetrahydropteridine, substrate of phenylalanine hydroxylase, LIII, 278
6-*N*-(3-Amino-2-hydroxypropyl)amino-2,3-dihydrophthalazine-1,4-dione
  chemiluminescence of ligand conjugates, LVII, 440
  structural formula, LVII, 428
  synthesis, LVII, 427, 430
    of 6-*N*-[(3-biotinylamido)-2-hydroxypropyl]amino-2,3-dihydrophthalazine-1,4-dione, LVII, 430
6-[*N*-(3-Amino-2-hydroxypropyl)-*N*-ethyl]-amino-2,3-dihydrophthalazine-1,4-dione
  chemiluminescence of ligand conjugates, LVII, 440
  structural formula, LVII, 428
  synthesis, LVII, 431
    of 6-(*N*-ethyl-*N*-[2-hydroxy-3-(thyroxinylamido)propyl])-amino-2,3-dihydrophthalazine-1,4-dione, LVII, 432
2-Amino-4-hydroxy-6-[1,2,3-trihydroxypropyl(6-*erythro*-)]tetrahydropteridine, *see* Tetrahydroenopterin
4-Aminoimidazole-5-carboxamide, substrate, of adenine phosphoribosyltransferase, LI, 572
5-Aminoimidazole-4-carboxamide
  inhibitor, of guanine deaminase, LI, 516
  in succino-AICAR synthesis, LI, 189
5-Aminoimidazole-4-carboxamide ribonucleoside, inhibitor, of guanine deaminase, LI, 516
5-Aminoimidazole ribonucleotide carboxylase, in succino-AICAR synthetase assay, LI, 187
5-Aminoimidazole ribonucleotide synthetase, preparation, LI, 194
5-Amino-4-imidazole-*N*-succinocarboxamide ribonucleotide lyase, *see* Adenylosuccinate lyase
α-Aminoisobutyric acid
  mammary accumulation, XXXIX, 454
  in *in vitro* ovary, XXXIX, 237
δ-Aminolevulinate synthase
  assay, LII, 351–353
  deficiency, LVI, 118, 119
  tissue levels, problems in determination, LII, 353, 354
δ-Aminolevulinic acid
  heme-deficient mutants, LVI, 559–562
  heme protein labeling, LVI, 45
  mutants requiring, LVI, 173, 567
  growth media, LVI, 123, 124
  nutritional requirement, LVI, 122
  strain variation in cytochrome spectra, LVI, 122, 123
  structural formula, XLIV, 845
  substrate, δ-aminolevulinic acid dehydratase, XLIV, 844, 845
  synergistic and temperature effects, LVI, 121, 122
  types, LVI, 121
5-Amino[^{3}H]levulinic acid, for labeling of cytochrome c_1, LIII, 108, 109
δ-Aminolevulinic acid dehydratase, *see* Porphobilinogen synthase
δ-Aminolevulinic acid synthetase, *see* δ-Aminolevulinate synthase
6′-Aminoluciferin, structural formula, LVII, 27

4-Amino-N-methylphthalimide
synthesis, LVII, 427
of 4-N-(3-chloro-2-hydroxypropyl)amino-N-methylphthalimide, LVII, 427
of 4-N-ethylamino-N-methylphthalimide, LVII, 431
of N-methyl-4N-[4-(N-phthalimido)butyl]aminophthalimide, LVII, 432
2-Amino-2-methyl-1,3-propanediol, as inhibitor for D-and L-arabinose isomerase, XLI, 465
2-Amino-2-methylpropanol-HCl buffer, *see* 2-Amino-2-methylpropanol-hydrochloric acid buffer
2-Amino-2-methylpropanol-hydrochloric acid buffer, alkaline phosphatase assay, XLIV, 627
4-Amino-10-methylpteroylglutamic acid, *see* Methotrexate
7-Aminonaphthalene-1,2-dicarboxylic acid
synthesis, LVII, 436
of dimethyl 7-aminonaphthalene-1,2-dicarboxylate, LVII, 436
7-Aminonaphthalene-1,2-dicarboxylic acid hydrazide, derivatives, as labeling compounds, LVII, 424
2-Aminonaphthalene-6-sulfonate, XLVII, 418
1-Amino-2-naphthol-4-sulfonic acid
in deoxythymidylate phosphohydrolase assay, LI, 286
reagant, XXXI, 92
reducing agent, phosphatase assay, XXXII, 396
6-Aminonicotinic acid, substrate of p-hydroxybenzoate hydroxylase, LIII, 548
2-Amino-4-nitrophenol, tritiated, synthesis, LVI, 659, 661, 662
Aminonucleoside, XLVI, 634
L-2-Amino-4-oxo-5-chloroketone, activator, of glutamine-dependent carbamoyl-phosphate synthetase, LI, 28
L-2-Amino-4-oxo-5-chloropentanoic acid, XLVI, 415–417, *see also* Chloroketone
Aminooxyacetate
aspartate loading, LVI, 260, 272
as enzyme inhibitor, XXXVII, 294
6-Aminopenicillanic acid, XXXIV, 532; XLIV, 241; XLVI, 536
chromatographic assay, XLIII, 700
differential pulse polarography, XLIII, 387
hydroxylamine assay, XLIII, 699–705, *see also* Penicillin acylase
α-lactamase hydrolysis, XLIII, 690
paper chromatography, XLIII, 110–116, *see also* Penicillin, paper chromatography
penicillins separated, XLIII, 115, 116
production methods, XLIV, 759, 760, 765–768
batch process, XLIV, 766
recirculation process, XLIV, 745, 767, 768
quantitative estimation, XLIII, 115, 116
structural formula, XLIV, 760
6-Aminopenicillanic acid acyltransferase, XLIII, 474–476, *see also* Acyl-CoA:aminopenicillanic acid acyltransferase
Aminopeptidase, XXXIV, 4
clostridial
assay, XLV, 544
inhibitors, XLV, 549
kinetics, XLV, 550, 551
metal ion requirements, XLV, 549
properties, XLV, 547
purification, XLV, 545, 547
specificity, XLV, 550
stability, XLV, 547
cytosol, XLVII, 542, 577
action, XLV, 504
activity, XLV, 513
assay, XLV, 518
esterase activity, XLV, 517
isolation, XLV, 505
kinetics, XLV, 515–517
macrocrystals, XLV, 505
from porcine kidney specificity, LIX, 600
preparation, XLV, 506
properties, XLV, 507
purity, XLV, 507
specificity, XLV, 515
stability, XLV, 509

structure, XLV, 511
human liver
action, XLV, 495
assay
with aminoacyl-p-nitroanilide substrate, XLV, 498
with amino-β-naphthylamide substrate, XLV, 496, 498
composition, XLV, 503
properties, XLV, 501
physical, XLV, 502
purification, XLV, 500
specificity, XLV, 495
in plasma membranes, XXXI, 86, 88, 89
spot test, XXXIV, 449
Aminopeptidase I, thermophilic
action, XLV, 522, 525
assay, XLV, 523
cell disintegration, XLV, 525
kinetics, XLV, 530
metal chelators, XLV, 528
properties, XLV, 527
purification, XLV, 524
purity, XLV, 527
specific activity, XLV, 524
specificity, XLV, 530
stability, XLV, 528
structure, XLV, 528
subunits, separation, XLV, 529
units, XLV, 524
Aminopeptidase M, XLVII, 542, 577
activity assay, XLVII, 77
for amino acid analysis of flavin peptides, LIII, 456
complete hydrolysis, XLVII, 41, 43
immobilization, XLVII, 44
o-Aminophenol, as mixed-function oxidase substrate, XXXI, 233
p-Aminophenol, colorimetric determination, in aniline hydroxylase assay, LII, 409
2-Aminophenol oxidase, LVI, 475
m-(m-Aminophenoxybutoxy)benzamidine-agarose, XXXIV, 736
m-(o-Aminophenoxybutoxy)benzamidine-agarose, XXXIV, 736
m-[4-(o-Aminophenoxy)butoxy]benzamidine p-toluenesulfonic acid, XLVI, 118
m-(o-Aminophenoxyethoxy)benzamidine, XLVI, 118
p'-Aminophenoxypropoxybenzamidine, XXXIV, 441
4-(4'-Aminophenoxypropoxy)benzamidine, XXXIV, 5
4-(p-Aminophenylazo)phenylarsonic acid, XXXIV, 190
4-[4-(4-Aminophenyl)butanamido]-phenyl-β-D-fucopyranoside, XXXIV, 370
2-(m-Aminophenyl)-1,3-dioxolane, XLIV, 115
p-Aminophenyl ester
of ATP, XXXIV, 255, 259, 260
of dATP, XXXIV, 255
β-(p-Aminophenyl)ethylamine, sugar reaction, L, 172
p-Aminophenylethylenediaminoagarose, XXXIV, 682
4-Aminophenyl-β-D-fucopyranoside-agarose, XXXIV, 373
p-Aminophenyl-β-p-D-galactoside, XXXIV, 351
p-Aminophenylglycidyl ether, XLIV, 38
Aminophenyl-β-glycoside, XLVI, 518
p-Aminophenylglycoside, XXXIV, 49
p-Aminophenylguanidine, XXXIV, 441
p-Aminophenylguanidine hydrochloride, XXXIV, 442
3'-Amino-9-phenylguanine, XXXIV, 524
4'-Amino-9-phenylguanine, XXXIV, 524
p-Aminophenyl-β-lactoside, XLVI, 519
4-Aminophenyl-α-D-mannoside, XLVI, 75
p-Aminophenylmercuric acetate, XXXIV, 245
N-(4-Aminophenyl)oxamic acid, XXXIV, 5, 7, 106
3'-(4-Aminophenylphosphoryl)deoxythymidine 5'-phosphate, XXXIV, 492
5'-(4-Aminophenylphosphoryl)guanosine 2'(3')-phosphate, XXXIV, 492
3'-(4-Aminophenylphosphoryl)thymidine 5'-phosphate, XXXIV, 496
5'-(4-Aminophenylphosphoryl)uridine (2')3'-phosphate, XXXIV, 492, 493
p-Aminophenyl-β-D-thiogalactopyranoside, XXXIV, 353–355
p-Aminophenyl-β-D-thiogalactopyranoside polyacrylamide, XXXIV, 352
p-Aminophenyl-α-thiogalactoside, XXXIV, 6

p-Aminophenyl-β-D-thiogalactoside, XXXIV, 7, 351
p-Aminophenylthiogalactoside
  polybovine γ-globulin, XXXIV, 368
4-Aminophenyl-tri-O-acetyl-β-D-fucopyranoside, XXXIV, 369
3-Aminophthalic acid
  fluorescence emission spectra, LVII, 413, 415
  product, of luminol oxidation, LVII, 412, 563
Aminopolystyrene, XLVII, 264, 301
  capacity, XLVII, 269
  coupling reaction, choice, XLVII, 271
  dilution with glass beads, XLVII, 272
  stability, XLVII, 265, 273
  synthesis, XLVII, 265
Aminopropanol, amino group determination, XLVII, 302
3-Aminopropanol, synthesis of *Diplocardia* luciferin, LVII, 377
N-(3-Aminopropyl)diethanolamine, XLIV, 50
Aminopropyl glass, XLVII, 264
  capacity, XLVII, 302, 303
  carboxyl-terminal coupling, XLVII, 271, 272
  derivatized, storage, XLVII, 303
  peptide attachment, XLVII, 303–306
  p-phenylene diisothiocyanate activation, XLVII, 303
  pretreatment, XLVII, 301, 302
  stability, XLVII, 273, 274
  synthesis, XLVII, 266, 267
α-Aminopropyltriethoxysilane, XXXIV, 290; XLIV, 480
  catecholamine glass beads, XXXVIII, 181
3-Aminopropyltriethoxysilane, XXXIV, 64; XLVII, 267, 302
  aqueous silanization, XXXIV, 64
  glass bead activation, XLIV, 325, 506
  glass sequenator cup derivatization, XLVII, 308
  organic silanization, XXXIV, 64
  preparation of alkylamine glass, XLIV, 139, 140, 670, 777, 931
  support of immobilized enzymes, LVI, 490
Amino protecting group, XXXIV, 507, 508
6-Amino-9-D-psicofuranosylpurine, *see* Psicofuranine
Aminopterin, XXXIV, 288; XLVI, 68
  in bacterial growth medium, LI, 553
  DNA inhibitor, XL, 276
Aminopyrine
  inhibitor of horseradish peroxidase, LII, 349
  as mixed-function oxidase substrate, XXXI, 233
  type of binding reaction with cytochrome P–450, LII, 267
Aminopyrine N-demethylase
  Ah locus, LII, 232
  assay, LII, 362
  in isolated MEOS fraction, LII, 362, 363
  in microsomal fractions, LII, 88
5-Amino-1-ribosyl-4-imidazolecarboxamide 5′-phosphate, LI, 186
  preparation, LI, 189, 190
  product, of adenylosuccinate AMP-lyase, LI, 202
  structural formula, LI, 203
5-Amino-1-ribosyl-4-imidazolecarboxylic acid 5′-phosphate
  substrate, of succino-AICAR synthetase, LI, 186
5-Amino-1-ribosylimidazole 5′-phosphate, LI, 186
N-(5-Amino-1-ribosyl-4-imidazolylcarbonyl)-L-aspartic acid 5′-phosphate, LI, 186
  preparation, LI, 189
  product, of succino-AICAR synthetase, LI, 186
  structural formula, LI, 203
  substrate, of adenylosuccinate AMP-lyase, LI, 202
  in succino-AICAR synthetase assay, LI, 187
N-(5-Amino-1-ribosyl-4-imidazolylcarbonyl) L-aspartic acid 5′-phosphate synthetase
  activity, LI, 186
  assay, LI, 187, 188
  bacterial sources, LI, 190
  from chicken liver, LI, 190
  properties, LI, 193
  purification, LI, 190–192
  stability, LI, 192

4-Aminosalicylic acid, XLIV, 88
substrate of salicylate hydroxylase, LIII, 536, 538, 539
5-Aminosalicylic acid, XLIV, 88
Amino-Spheron, XLIV, 76–80
p-Aminostyrene, glucose oxidase immobilization, XLIV, 320
Amino sugar, *see also* specific types
cyclic, 1,2-derivatives, L, 106
microdetermination, L, 52
in plasma membranes, XXXI, 165, 167
synthesis, L, 105–108
2-amino-2-methoxyethyl thioglycosides, L, 113
glucose tetrasaccharide formation, L, 231–233
4-Amino-2,2,6,6-tetramethylpiperidine, CAT_n spin probe synthesis, LVI, 518
Aminothiol cysteamine, substrate of dimethylaniline monooxygenase, LII, 151
Aminotransferase, XXXIV, 295–299; XLVI, 31, 163, 164, *see also* L-Alanine:1D-1-guanidino-1-deoxy-3-keto-*scyllo*-inositol aminotransferase; L-Glutamine:keto-*scyllo*-inositol aminotransferase; specific enzymes
assay by Fast Analyzer, XXXI, 816
in microbodies, LII, 499, 500
Aminotriazole, in studies on plant catalase, XXXI, 497, 500
3-Amino-1,2,4-triazole, glucose oxidase, LVI, 490
3-Aminotyrosine, internal standard, XLVII, 34, 39
5-Aminouridine, substrate, of uridine-cytidine kinase, LI, 313
Aminovinyl, as optical probes, LV, 573
Ammonia, XLIV, 50, 80, 92, 104, 117, 231, 762
activator, of glutamine-dependent carbamoyl-phosphate synthetase, LI, 22, 27
assay, LV, 209
blood level, XLIV, 681–684
determination by microdiffusion technique, LI, 491, 497, 503
by NADH oxidation, LI, 503
EDTA particle
coupling factor B assay, LV, 384
nature, LV, 380
preparation, LV, 385, 386
extraction of OSCP, LV, 395
infiltration, in plant-cell fractionation, XXXI, 505
for nucleotide elution, LI, 419
oxidation, by cell free extracts, LIII, 636, 637
peptide cleavage from support, XLVII, 604
product of adenosine deaminase, LI, 502
of adenylate deaminase, LI, 490, 497
of cytidine deaminase, LI, 401
of dCTP deaminase, LI, 418
of deoxycytidylate deaminase, LI, 412
of glycine reductase, LIII, 373
of guanine deaminase, LI, 512
submitochondrial particles, LV, 111, 112
substrate of carbamoyl-phosphate synthetase, LI, 21
of CPSase-ATCase-dihydroorotase complex, LI, 119
of cytidine triphosphate synthetase, LI, 79
of GMP synthetase, LI, 213, 217, 219
Ammonia buffer
chemiluminescent assay of glucose, LVII, 452
for TEAE-cellulose chromatography, XLVII, 208
Ammonia lyase, in glyoxysomes, LII, 500
Ammonia oxidase
in *Nitrosococcus oceanus*, LIII, 636
in *Nitrosomonas europaea*, LIII, 636
Ammonium acetate
activator, of dihydroorotate dehydrogenase, LI, 62
analysis of oligonucleotides, LIX, 343, 344
assay of adenine phosphoribosyltransferase, LI, 559
of cyclic AMP phosphodiesterase, LVII, 96
in chromatographic separation of bases and nucleotides, LI, 551

in 5-fluoro-2′-deoxyuridine 5′-(p-nitrophenylphosphate) synthesis, LI, 99
of 5-methyluridine methyltransferase, LIX, 191
codon-anticodon recognition studies, LIX, 296
cyclic nucleotide separation, XXXVIII, 30, 31
desalting procedure of purified aequorin, LVII, 311, 312
extraction of nucleic acids, LIX, 312
in formaldehyde assay, LII, 299, 345
oligonucleotide digestion, LIX, 79, 81, 82
preparation of aminoacylated tRNA isoacceptors, LIX, 298
of crystals, LIX, 5
of mitochondrial enzyme complexes, LIII, 3, 6
of protein-depleted 30 S subunits, LIX, 621
of 30 S subunits, LIX, 618
purification of complex III, LIII, 86, 87
of cytochrome *c* derivatives, LIII, 176
reconstitution of ribosomal subunits, LIX, 447
tRNA labeling procedure, LIX, 71
separation of amino acids and tRNA, LIX, 312
Ammonium acetate buffer
gel filtration eluent, XLVII, 105
staphylococcal protease specificity and, XLVII, 190
Ammonium bicarbonate
determination of deuterium content of RNA, LIX, 646
digestion of MS2 peptides, LIX, 301
of tRNA, LIX, 343
in modification of cytochrome *c* residues, LIII, 171, 172
purification of adenosine derivatives, LVII, 117, 118
Ammonium bicarbonate buffer, staphylococcal protease specificity, XLVII, 190
Ammonium carbonate
analysis of modified tRNA, LIX, 178
in preparation of deuterated NAD^+, LIV, 226
of deuterated NADH, LIV, 226
Ammonium carbonate-ethyl acetate, for drug isolation, LII, 333
Ammonium chloride, XLIV, 508, 881
in AICAR synthesis, LI, 189
aminoacylation of tRNA, LIX, 818
assay of carbamoyl-phosphate synthetase, LI, 105, 106
of GTP hydrolysis, LIX, 361
of phosphoribosylglycinamide synthetase, LI, 180
of polypeptide chain elongation, LIX, 357
of radiolabeled ribosomal particles, LIX, 779
of RNase III, LIX, 825
for translocation, LIX, 359
cell-free protein synthesis system, LIX, 300, 367, 448, 459, 850
crystallization of transfer RNA, LIX, 6
filter binding assay, LIX, 848
gradient purification
elongation factors, LX, 652–655, 669
initiation factors, LX, 8, 206, 207, 228, 338, 339
ribosome protein S1, LX, 449, 450
leukocyte isolation, LVII, 467, 470
preparation of *Bacillus* ribosomes, LIX, 373, 438
of chloroplastic ribosomes, LIX, 434, 435
of cross-linked ribosomal subunits, LIX, 535
of *E. coli* endogenous polysomes, LIX, 353–356, 363
of *E. coli* ribosomes, LIX, 451, 554, 647, 752, 817, 818, 838
of postmitochondrial supernatant, LIX, 515
of radiolabeled ribosomal protein, LIX, 777, 783
of ribosomal subunits, LIX, 403, 404, 444, 554, 752
of ribosome-DNA complexes, LIX, 841
of RNase III, LIX, 826–829
of S100 extract, LIX, 357

purification of aminoacyl-tRNA synthetase, LIX, 258
of tryptophanyl-tRNA synthetase, LIX, 241
reconstitution of ribosomal subunits, LIX, 447, 451, 654, 864
release of ribosomal proteins, LIX, 818
removal, LIX, 830
RNase III digestion buffer, LIX, 833, 834
separation of bacterial ribosomal subunits, LIX, 399
synthesis of adenosine derivative, LVII, 116
Ammonium citrate buffer, in amidophosphoribosyltransferase purification, LI, 173, 175
Ammonium formate
assay of deoxythymidine kinase, LI, 362
of phosphoribosylglycinamide synthetase, LI, 181
chromatographic separation of 3′ end nucleoside, LIX, 188
cyclic nucleotide separation, XXXVIII, 31
separation of nucleosides and deoxynucleotides, LI, 338, 347
Ammonium fumarate, substrate, aspartase, XLIV, 740–743
Ammonium hydroxide
in Cu^{2+} Chelex preparation and regeneration, LI, 539
in deacylation, LIX, 312
for elution of radiolabeled nucleotides, LIX, 192
metal-free, preparation, LIV, 476
in synthesis of dimethyl 7-aminonaphthalene-1,2-dicarboxylate, LVII, 436
Ammonium ion
activator of guanylate kinase, LI, 474, 481, 482, 489
of ribonucleoside triphosphate reductase, LI, 258
content of brain, LV, 219
effect on *Cypridina* luminescence, LVII, 351
mitochondrial matrix, LVI, 253
M. phlei ETP, LV, 186
uptake, pH change, LV, 564
Ammonium molybdate
assay of acid nucleotidase, LI, 272
of deoxythymidylate phosphohydrolase, LI, 286
assay reagent, XLIV, 664
in negative stains, XXXII, 22, 24, 29, 30, 34
osmolality, XXXII, 30
phosphatase assay, XXXII, 396
Ammonium molybdate tetrahydrate, assay of amidophosphoribosyltransferase, LI, 173
Ammonium peroxodisulfate, LIX, 506
Ammonium peroxydisulfate, XLIV, 198
Ammonium persulfate, XLIV, 57, 58, 102, 177, 185, 187, 188, 193, 196, 455, 458, 476, 662; XLVII, 267; LIX, 65, 431, 509, 539, 544, 545
in catalyst solution, XXXII, 74
electrophoretic removal, LI, 343
in gel preparation, LI, 342, 343; LII, 329
Ammonium phosphate, activator, of dihydroorotate dehydrogenase, LI, 62
Ammonium-PIPES buffer, aminoacylation of tRNA, LIX, 126, 127, 130
Ammonium purpurate, *see* Murexide
Ammonium sulfamate
assay of FGAM synthetase, LI, 195
of phosphoribosylglycinamide synthetase, LI, 180
Ammonium sulfate, XLIV, 162, 374, 748, 749, 889–891
activator, of dihydroorotate dehydrogenase, LI, 62
assay of GMP synthetase, LI, 214
chromatographic separation of nucleotides, LI, 419
computation of saturation level, 184n, 185n, LIII, 533
crystallization of adenosine monophosphate nucleosidase, LI, 267
of azurin, LIII, 659
of cytochrome *c*-551, LIII, 658
of *p*-hydroxybenzoate hydroxylase, LIII, 547

of *Pseudomonas* cytochrome oxidase, LIII, 652
of purine nucleoside phosphorylase, LI, 535
of rubredoxin, LIII, 626
of transfer RNA, LIX, 7
cyclic nucleotide separation, XXXVIII, 32–35
in cytosol preparation, XXXVI, 353–355
dissociation of flavoproteins, LIII, 430–432
in elution procedures, for purification of erythrocyte enzymes, LI, 584, 593
fractionation
antisera to *Pseudomonas* toxin A, LX, 785
elongation factors, LX, 639, 641, 652, 656, 659, 666, 675, 681
heme-regulated translational inhibitors, LX, 471, 485, 486, 495
initiation factors
eIF, LX, 20, 22, 127, 146, 148–150, 197–199, 242–244, 310
EIF-2 and Co-EIF-1, LX, 40, 54, 56
IF, LX, 6, 7, 9, 206, 227, 228
ribosome protein SI, LX, 449
guanylate cyclase, XXXVIII, 193
interference in glutaraldehyde cross-linking, XLIV, 551, 641
in 5′-nucleotidase purification, XXXII, 370–372
pH adjustment, LIII, 56
preparation of aequorin crude extract, LVII, 278, 280, 307
of coliphage R17 RNA, LIX, 364
of crystals, LIX, 5
of *E. coli* ribosomal subunits, LIX, 752
of hydrogenase/thiosulfate-sulfite reductases extract, LIII, 619
of IF fraction, LIX, 364
of mitochondrial enzyme complexes, LIII, 3, 11
of RNase III, LIX, 828
in progesterone receptor precipitation, XXXVI, 193, 194, 198–200
purification of acyl-CoA dehydrogenases, LIII, 512, 514, 517, 518
of adenine phosphoribosyltransferase, LI, 562, 563, 571, 576
of adenosine deaminase, LI, 504
of adenosine monophosphate nucleosidase, LI, 266, 268
of adenylate deaminase, LI, 499, 500
of adenylosuccinate adenylate-lyase, LI, 204, 205
of adenylsuccinate synthetase, LI, 209, 210
of aequorin, LVII, 281, 307, 308
of AIR synthetase, LI, 194
of amidophosphoribosyltransferase, LI, 174, 175, 176
of aminoacyl-tRNA synthetase, LIX, 259, 261, 263
of aspartate carbamyltransferase, LI, 44, 45, 53
of bacterial luciferase, LIII, 567; LVII, 146, 147
of blue-fluorescent protein, LVII, 228
of carbamoyl-phosphate synthetase, LI, 32
of carbon monoxide-binding heme protein P–460, LIII, 638, 639
of complex II, LIII, 22, 23
of complex III, LIII, 36, 86, 87, 94
of complex III peptides, LIII, 83, 84
of complex IV, LIII, 43
of CPSase-ATCase complex, LI, 109
of CPSase-ATCase-dihydroorotase complex, LI, 115, 124, 125
of CTP(ATP):tRNA nucleotidyltransferase, LIX, 123, 184
of *Cypridina* luciferase, LVII, 370
cytidine deaminase, LI, 397, 398, 402, 403, 406, 410
of cytidine triphosphate synthetase, LI, 80, 81, 86–88
of cytosine deaminase, LI, 396
of cytochrome a_1, LIII, 639, 640
of cytochrome b_1, LIII, 234, 235

of cytochrome bc_1, complex, LIII, 114, 115
of cytochrome *c*, LIII, 129
of cytochrome c_1, LIII, 183–186, 223
of cytochrome oxidase, LIII, 57–63, 75
of deoxycytidine kinase, LI, 340
of deoxycytidylate deaminase, LI, 415, 416, 420
of deoxythymidine kinase, LI, 363
of dihydroorotate dehydrogenase, LI, 61, 65
of electron-transferring flavoprotein, LIII, 512, 514, 516
of ferredoxin, LIII, 623
of FGAM synthetase, LI, 199
of firefly luciferase, LVII, 7–10, 35, 62
of 8α-flavins, LIII, 458
of flavocytochrome b_2, LIII, 242, 247, 252, 253
of flavodoxin, LIII, 624, 625
of formate dehydrogenase, LIII, 367
of glycine reductase, LIII, 377, 381
of guanine deaminase, LI, 513–515
of guanine phosphoribosyltransferase, LI, 554, 555
of guanylate kinase, LI, 478, 485
of guanylate synthetase, LI, 215, 222
of high molecular weight cytochrome c_3, LIII, 625
of high-potential iron-sulfur protein, LIII, 333, 336, 338
of hydrogenase, LIII, 311
of hypoxanthine phosphoribosyltransferase, LI, 546, 547, 554, 555
of *p*-hydroxybenzoate hydroxylase, LIII, 546, 547
of ω-hydroxylase, LIII, 358
D-lactate dehydrogenase, LIII, 521
of *Lactobacillus* deoxynucleoside kinases, LI, 348
of melilotate hydroxylase, LIII, 554
of NADH dehydrogenase, LIII, 15
of nitrate reductase, LIII, 642
of nitric oxide reductase, LIII, 644
of nitrite reductase, LIII, 643
of nucleoside deoxyribosyltransferase, LI, 449
of nucleoside diphosphokinase, LI, 374, 382
of nucleoside phosphotransferase, LI, 388, 389, 390
of nucleoside triphosphate pyrophosphohydrolase, LI, 276, 277, 280, 281, 282
of OPRTase-OMPdecase complex, LI, 137, 138, 146, 149, 160, 162
of orotate phosphoribosyltransferase, LI, 71
of orotidylate decarboxylase, LI, 75
of oxidoreductase, LVII, 203
of phenylalanine hydroxylase, LIII, 281, 282
of *Pholas* luciferase, LVII, 387, 390
of phosphoribosylglycinamide synthetase, LI, 183, 184
of phosphoribosylpyrophosphate synthetase, LI, 7–9
of *Pseudomonas* cytochrome oxidase, LIII, 650
of purine nucleoside phosphorylase, LI, 519, 520, 522, 523, 527, 528, 533
of pyrimidine nucleoside monophosphate kinase, LI, 324, 325
of pyrimidine nucleoside phosphorylase, LI, 434
of ribonucleoside diphosphate reductase, LI, 230, 240
of ribonucleoside triphosphate reductase, LI, 252
of ribosephosphate pyrophosphokinase, LI, 15
of rubredoxin, LIII, 343
of salicylate hydroxylase, LIII, 533
of succinate dehydrogenase, LIII, 28, 475, 486
of succino-AICAR synthetase, LI, 191, 192

of superoxide dismutase, LIII, 391
of thymidine kinase, LI, 357, 367, 368
of thymidine phosphorylase, LI, 439, 441, 444
of thymidylate synthetase, LI, 92, 94
of tRNA methyltransferase, LIX, 195, 201
of tryptophanyl-tRNA synthetase, LIX, 240
of uridine-cytidine kinase, LI, 302, 303, 318
uridine nucleosidase, LI, 293, 294
of uridine phosphorylase, LI, 427, 428
of uridylate-cytidylate kinase, LI, 334, 335
separation of tRNAs by reverse salt gradient chromatography, LIX, 217
solution, purification with amyl alcohol, LIII, 56
in storage buffer of aequorin, LVII, 281
transhydrogenase, LV, 270
types, for crystallization procedures, LIII, 615
Ammonium sulfate fractionation
chloroperoxidase purification, LII, 527
cytochrome b_5 reductase purification, LII, 468
cytochrome *m* purification, LII, 181, 182, 183
cytochrome P–450_{cam} purification, LII, 153, 154
epoxide hydrase purification, LII, 195, 196
hemoglobin preparation, LII, 488
liver microsomal protein, LII, 118–120, 123, 146
mitochondrial protein, LII, 128, 129
myoglobin isolation, LII, 474
putidaredoxin purification, LII, 179
putidaredoxin reductase purification, LII, 176
Rhizobium protein, LII, 160
serum proteins, LII, 243
Ammonium thiocyanate, XLIV, 132
coupling agent, XLVII, 359, 362
recrystallization, XLVII, 359
Ammonolysis, reversal of amidination, LVI, 628
Amniotic fluid
drug isolation, LII, 333
prolactin in, XXXVII, 390
purification, XXXVII, 399–400
Amoeba
axenic growth medium, XXXI, 687
growth and harvesting, XXXIX, 486, 487
liquid culture, with bacteria, XXXIX, 487, 488
phagosome membrane isolation, XXXI, 686–698
plasma membrane isolation, XXXI, 161, 686–698
Amoxicillin, in media, LVIII, 112, 114
AMP, *see* Adenosine 5′-monophosphate
AMP aminohydrolase, *see* AMP deaminase
AMP deaminase, LI, 490–502
activity, LI, 490, 491, 497
amino acid analysis, LI, 495
assay, LI, 491, 492, 497, 498
distribution, LI, 490, 491
from human erythrocytes, LI, 497–502
inhibitors, LI, 496
kinetic properties, LI, 496, 501, 502
metal requirement, LI, 495, 496
molecular weight, LI, 495, 501
pH optimum, LI, 496
purification, LI, 492–494, 498–501
purity, LI, 493, 500
from rat skeletal muscle, LI, 490–497
regulation, LI, 496, 497
relationship to erythrocyte membrane, LI, 502
sedimentation coefficient, LI, 501
stability, LI, 495, 501
substrate specificity, LI, 501
subunit structure, LI, 495
Amphetamine, salivary secretion effects, XXXIX, 475, 476
d-Amphetamine, type of binding reaction with cytochrome P–450, LII, 264, 266
Amphibia, *see also* specific genera
balanced salt solution, osmolarity adjustment, LVIII, 469

incubation temperature, LVIII, 471
media, LVIII, 469
skin culture, LVIII, 473
Amphiphilic gel, XLIV, 43
Amphiphilic spin label, membrane surface potential measurement, LVI, 515–517
experimental, LVI, 518–526
principle, LVI, 517
sample calculation, LVI, 526
Ampholine, purification of luciferase, LVII, 62
Amphotericin A, XLIII, 135
Amphotericin B, XLIII, 136, 200
effect on bilayers, XXXII, 500, 535
inhibitor, of phagocyte chemiluminescence, LVII, 492
in media, LVIII, 112–114, 116
phospholipid bilayers, LV, 766
in tissue culture media, XXXIX, 111, 112
Amphotericin B methyl ester, in media, LVIII, 112
Ampicillin, XXXIV, 4; XLIV, 615; XLVI, 536
differential pulse polarography, XLIII, 382
high-pressure liquid chromatography, XLIII, 309–311, 319
in media, LVIII, 112, 114, 116
penicillin acylase assay, XLIII, 701
AMP kinase, *see* Adenylate kinase
Amplifier
electrophysiological techniques, LV, 661
output, in detector system for nanosecond absorbance spectroscopy, LIV, 41–43
Amplifier current transducer, circuit diagram, LVII, 531
AMP nucleosidase, LI, 263–271
activity, LI, 263
assay, LI, 263, 264
from *Azotobacter vinelandii*, LI, 263
crystallization, LI, 267
from *E. coli*, LI, 263
kinetic properties, LI, 270, 271
molecular weight, LI, 270
purification, LI, 265–271
from *Azotobacter vinelandii*, LI, 265–267
from *Escherichia coli*, LI, 267–271
purity, LI, 267, 271
stability, LI, 267, 270
subunit structure, LI, 270
AMP–S, *see* Adenylosuccinate
Ampule, shipment, LVIII, 34
Amyl alcohol, purification of ammonium sulfate solution, LIII, 56
Amylase, XXXIV, 592
identification of luminous bacteria, LVII, 161, 162
in pancreas, XXXI, 53
secretion by parotid slices, XXXIX, 463, 464
tissue culture, LVIII, 125
α-Amylase
assay, spectrophotometric, XLIV, 98
in cellular DNA isolation, XXXVI, 296
conjugate
activity, XLIV, 100, 271
stability, XLIV, 100, 101
storage, XLIV, 98
diazo binding, XLIV, 98
fluorine nuclear magnetic resonance, XLIX, 274
immobilization
by adsorption, effects on activity, XLIV, 44
on Enzacryl AA, XLIV, 98, 99
on Enzacryl AH, XLIV, 99, 100
in polyacrylamide gel, XLIV, 902, 909
on silk, XLIV, 909
isothiocyanato coupling, XLIV, 99
purification with starch, XXXIV, 164
source, XLIV, 98
starch hydrolysis, XLIV, 784, 788, 789
β-Amylase, XLIV, 63
assay, spectrophotometric, XLIV, 98
conjugate
activity, XLIV, 100
stability, XLIV, 100, 101
storage, XLIV, 99
conjugation with oxo-agarose, procedure, XLIV, 40
diazo binding, XLIV, 98

entrapment, in *N,N*′-methylenebisacrylamide, XLIV, 171
immobilization on Enzacryl AA, XLIV, 98, 99
on Enzacryl AH, XLIV, 99, 100
on hexyl agarose, XLIV, 44
in polyacrylamide gel, XLIV, 902
immobilized, relative activity, XLIV, 40
isothiocyanate coupling, XLIV, 99
in purification of mycobacterial polysaccharide MMP, XXXV, 91–93
source, XLIV, 98
γ-Amylase, in brush border, XXXI, 130
Amyloglucosidase, *see* Glucoamylase
Amylo-α-1,4-α-1,6-glucosidase
conjugate, pH optimum shift, XLIV, 468, 469
immobilization multistep enzyme system, XLIV, 460, 461
Amyloid A protein, sequence determination, XLVII, 91
Amyloplast, in plant cells, XXXI, 493, 495
Amylose, support, for penicillin acylase, XLIV, 765
N^{α}-*tert*-Amyloxycarbonyl arginine, XLVII, 612
Amytal
as anesthetic, contraindication in kidney removal, XXXIX, 13
inhibitor of complex I, LIII, 13
of complex I–III, LIII, 8
ubiquinone redox state, LIII, 583
Anabaena, lysis, XXXI, 681
Anabaena cylindrica, nitrogenase, LIII, 328
Anabaena variabilis, aldolase, XLII, 233
Anabaenopsis, aldolase, XLII, 233
Anacystis nidulans, fructose-diphosphate aldolase, assay, purification, and properties, XLII, 228–234
Anaerobic technique
assay for formate dehydrogenase, LIII, 364
column chromatography, LIII, 367–369
continuous titration, LIV, 124–130
cylindrical glove box, LIII, 367
demountable glass apparatus, LIV, 120, 121, 122
for EPR sample preparation, LIV, 114–131
in freeze-quench experiments, LIV, 89
gas analysis, LIV, 119, 120
gas choice, LIV, 114, 115
gas delivery, LIV, 115–117
gas manifold, LIV, 118, 119
for NMR sample preparations, LIV, 199
oxygen removal, LIV, 117, 118
preparation for crude extracts, LIII, 366
preparative acrylamide gel electrophoresis, LIII, 320
purification of hydrogenase, LIII, 287
of nitrogenase, LIII, 315
in redox potentiometry, LIV, 421
solid reactant addition, LIV, 122
Anaerobiosis
nutritional requirements, LVI, 172, 174, 175
sedimentation, LVI, 283
Analog Device, 311K, ion selective electrodes, LVI, 364
Analog peak-hold circuit, LVII, 539
Analog-to-digital converter
dynamic range, 182n, 183n
rate, LIV, 157, 158
word length, aqueous solutions, LIV, 183
Analtech silica gel GF 250 TLC plate, LIX, 815
Anasil S, in glycolipid isolation, XXXII, 350–352, 354
Anchorage dependence, LVIII, 44, 45, 48, 81, 133, 137, 152
Ancrod, XLV, 205
assay, XLV, 205
coagulant activity, XLV, 205
esterolytic activity, XLV, 206
composition, XLV, 213, 232
inhibitors, XLV, 210, 211
properties, XLV, 210
physical, XLV, 212
purification, XLV, 207, 208
purity, XLV, 212
specificity, XLV, 210
stability, XLV, 212

Anderson's critical point method, XXXIX, 155
Andosterone, metabolites, gas chromatographic and mass spectral properties, LII, 386
Androgen, *see also* specific types
  chromatographic separation, XXXVI, 488
  mercurated, XXXVI, 424, 425
  metabolism, in tfm mice and rats, XXXIX, 454–460
  receptor binding and nuclear retention, XXXVI, 313–319
  receptor-protein assay, XXXVI, 49
  testicular, in interstitial cells, XXXIX, 256
  in testicular lymph, XXXIX, 276
  in vivo production and interconversion, XXXVI, 67–75
Androgen receptor
  in mouse kidneys, XXXIX, 459, 460
    physical properties, XXXIX, 460
  purification, XXXVI, 366–374
    cytoplasmic 8 S type, XXXVI, 369–372
Androstanediol, in oxidoreductase studies, XXXVI, 468, 473
5α-Androstane-3α,17β-diol
  chromatographic separation, LII, 379
  hydroxylated metabolites
    derivatization, LII, 380, 381
    identification, LII, 381–388
    TLC, LII, 380
  hydroxylation, LII, 379, 380
  metabolites, gas chromatographic and mass spectral properties, LII, 386
  protein-binding assay, XXXVI, 37
  tissue production and interconversion rates, XXXVI, 74
5α-[4-^{14}C]Androstane-3α,17β-diol, preparation, LII, 378, 379
5α-Androstane-3β,17β-diol
  protein-binding assay of, XXXVI, 37
  tissue production and interconversion rates of, XXXVI, 74
Androstanediol 3-hemisuccinate, preparation, XXXVI, 111
Androstanediol-Sepharose, preparation, XXXVI, 112
5α-Androstane-2β,3α,17β-triol, identification, LII, 381
5α-Androstane-3α,7α,17β-triol, identification, LII, 381
5α-Androstane-3α,7β,17β-triol, identification, LII, 381
5α-Androstane-3α,17β,18-triol, identification, LII, 381
5α-Androstane-3β,17β,18-triol, identification, LII, 381
5α-Androstan-17β-ol-3-one
  diazo derivatives, XXXVI, 415–417
  oximino derivatives, XXXVI, 415, 416
Δ^4-Androstenediol, thin-layer chromatography, XXXIX, 262
Δ^5-Androstenediol, thin-layer chromatography, XXXIX, 262
4-Androstene-3β,17β-diol, protein binding assay, XXXVI, 37
5-Androstene-3β,17β-diol, protein binding assay, XXXVI, 37
Androstenedione
  chromatographic separation, XXXVI, 487, 488
  derivatives, XXXVI, 19
  electron-capture detection, XXXVI, 60–63, 67
  nuclear receptor binding, XXXVI, 317–319
  as 5α-oxidoreductase substrate, XXXVI, 473
  in vivo production and interconversion, XXXVI, 68
Δ^4-Androstenedione
  in testicular lymph, XXXIX, 276
  testicular production, XXXIX, 282
  as testosterone precursor, XXXIX, 256
  thin-layer chromatography, XXXIX, 262
Δ^5-Androstene-3,17-dione, XLVI, 461
4-Androstene-3,17-dione
  competitive inhibitor of cytochrome P-$450_{11\beta}$, LII, 132
  metabolites, gas chromatographic and mass spectral properties, LII, 385
4[4-^{14}C]Androstene-3,17-dione, in 5α-[1-^{14}C]androstane-3α,17β-diol synthesis, LII, 378
5-Androstene-17β-thiol-3β-ol, synthesis, XXXVI, 421
4-Androstene-17β-thiol-3-one, synthesis, XXXVI, 419–422

5-Androsten-3β-ol-17-one dibenzyl thioketal, synthesis, XXXVI, 420, 421
5-Androsten-3β-ol-17-thione, synthesis, XXXVI, 421
4-Androsten-3-one-17β-carboxylate, diazoprogesterone, XXXVI, 413
Anemia, *see* specific type
Anemonia sulcata, XLV, 881–888, *see also* Sea anemone
Anesthetic
  for kidney removal, XXXIX, 13
  luminescence, LVI, 543
  for rodent brain surgery, XXXIX, 166, 167
  spin-label studies, XXXII, 177
Angiokeratoma corporis diffusum, *see* Fabry's disease
Angiosperm cell, culture, XXXII, 723–732
Angiotensin, XLVI, 235
  antisera, XXXVI, 17
  free peptide hydrogen exchange, XLIX, 30
  iodinated, properties, XXXVII, 230
  logit-log assay, XXXVII, 9
Angle rotor, for centrifugation, XXXI, 717
Angolamycin, XLIII, 134
Angustmycin, XLIII, 154
Anhydrid-Acrylharzperlen, support, for penicillin acylase, XLIV, 764
Anhydride, mixed, fluorescent labeling and, XLVIII, 359
Anhydride analog, XLVI, 302–307
2,2′-Anhydro-1-β-D-arabinofuranosylcytosine, *see* Cyclocytidine
Anhydrochlorotetracycline, XLIII, 199
Anhydrochymotrypsin, XXXIV, 4
Anhydrocycloheximide, XLIII, 158
2,6-Anhydro-1-diazo-1-deoxy-D-glycero-L-*manno*-heptitol, XLVI, 36
Anhydroerythromycin, XLIII, 309
1,6-Anhydro-β-D-glucosamine
  peptidoglycan synthesis, L, 109
  Tay–Sachs trisaccharide, L, 109
1,6-Anhydroglucosan, from dolichol monophosphate glucose, L, 431
1,2-Anhydro-L-gulitol 6-phosphate, XLVI, 383; XLVII, 493, 494
  synthesis, XLVII, 490, 491
1,2-Anhydrohexitol 6-phosphate, XLVI, 381–387
1,2-Anhydro-D-mannitol 6-phosphate, XLVI, 383; XLVII, 493, 494
  affinity labeling, XLVII, 482
  structure, XLVII, 482
  synthesis, XLVII, 490
Anhydrooxytetracycline, XLIII, 199
2,5-Anhydro-D-talose, L, 18
Anhydrotetracycline, XLIII, 199, 317
Anileridine, XLVI, 604
Aniline
  blocking reagent, of activated support, XLIV, 494
  inhibitor, β-fructofuranosidase, XLIV, 408
  as mixed-function oxidase substrate, XXXI, 233
  N-substituted, XLVI, 95
  optical-difference spectrum produced, LII, 259
  preparation of aniline hydrochloride, LIX, 62
  for Sepharose 4B binding, LII, 127
  type of binding reaction with cytochrome *P*-450, LII, 264
Aniline hydrochloride
  in aniline hydroxylase assay, LII, 409
  chemical cleavage of tRNA, LIX, 100, 102
  preparation, LIX, 62
Aniline hydroxylase
  Ah locus, LII, 232
  assay, LII, 408, 409
  in isolated MEOS fraction, LII, 363
  in microsomal preparations, LII, 87, 88
Anilinonaphthalenesulfonic acid
  fluorescent acceptor, XLVIII, 362, 363
  spectral properties, XLVIII, 362
  structural formula, XLVIII, 362
1-Anilinonaphthalene-8-sulfonic acid, XLIV, 363
  characteristics, LVI, 497
  fluorescence changes, energy, LVI, 497
1,5-Anilinonaphthalenesulfonic acid, luminescence, LVI, 543
1,8-Anilinonaphthalenesulfonic acid, XXXIV, 388

8-Anilino-1-naphthalenesulfonic acid, LV, 574
counterstain, in protein gel electrophoresis, XXXII, 97, 99, 100, 102
fluorescent probes, XXXII, 234, 235
fluorescence emission spectra, XXXII, 243–245
in histone assay, XL, 115
inhibitor of bacterial luciferase, LVII, 152
of bioluminescence, LVII, 49
luciferase binding studies, LVII, 133
membrane potential, LV, 559, 571, 573, 585, 694, 779, 780
membranes studied, XXXII, 234, 235, 237, 238
polarization measurements, XXXII, 245, 246
properties, XXXII, 238
quantum yield measurements, XXXII, 240–243
submitochondrial particles, LV, 112
transhydrogenase assay, LV, 270
X-ray studies using, XXXII, 219
N-(1-Anilinonaphthyl-4-)maleimide, XLVII, 418
Animal, *see also* specific type
administration of inhibitors
orally by gastric intubation or in drinking water or diet, LVI, 35
parenterally, LVI, 38, 39
laboratory, pre-experimental care, XXXVI, 480–481
mitochondrial DNA, LVI, 4
Animal tissue, *see also* specific animal
D-2-hydroxy acid dehydrogenase, assay, preparation, and properties, XLI, 323–329
Anion
ATPases, LV, 301, 302
concentration determination in perfusion experiments, LII, 58
exchange, organotins, LV, 509
gradients, LVI, 254, 255
Hofmeister lyotropic, inhibitors of adenylosuccinate synthetase, LI, 212
inhibitors, of firefly luciferase, LVII, 15, 49
lipophilic
ionophores, LV, 439, 448
uptake, transhydrogenase assay, LV, 262, 272–274
mitochondrial transport, LV, 452, 453; LVI, 247–249
Anion-exchange paper, for separation of nucleosides and deoxynucleotides, LI, 338, 347
Anion-exchange resin, in nonionic adsorption chromatography, XLIII, 275–280
Anion transport, XLIX, 14
energetics, LVI, 255
Anisaldehyde reagent, XXXVI, 467; L, 124
p-Anisidine
preparation of luciferin, LVII, 16, 19
structural formula, LVII, 18
Anisole
cation scavenger in peptide synthesis, XLVII, 538, 542, 547, 575, 602, 615
synthesis of 4-(4-methoxyphenyl)-4-oxobutyric acid, LVII, 434
Anisomycin, LVI, 31
protein inhibitor, XL, 287, 288
Anisotropy
antisymmetric, LIV, 238
average depolarization factor, XLVIII, 370, 372, 373
emission
measurement, XLVIII, 369, 370
steady-state, XLVIII, 352
optical, definition, XLVIII, 440, 441
symmetric, LIV, 238
Anisotropy factor, definition, LIV, 253
Anomerase activity, of glucosephosphate isomerase from baker's yeast and other sources, XLI, 57–61
ANOVA, variance analysis, in radioligand assay data analysis, XXXVII, 17, 19
Anoxia, rat heart metabolites, LV, 217
ANP, *see* 2-Nitro-4-azidophenyl group

ANS, *see* Anilinonaphthalenesulfonic acid; 8-Anilino-1-naphthalenesulfonic acid
ANSA, *see* 8-Anilino-1-naphthalenesulfonic acid
Antagonist, to hormones, from adipocytes, XXXVII, 431–438
Antamanide, XLVI, 84
Antenna complex, reaction center proteoliposomes, LV, 760, 765, 771
Anterior pituitary cell, *see* Pituitary cell
Anterior pituitary gland, *see* Pituitary gland, anterior
Anterior pituitary hormone, purification of, XXXVII, 360–380
Anthelvencin, XLIII, 145, 146
Antheraea eucalypti, LVIII, 451
  media, LVIII, 457
  storage, LVIII, 453
Anthocidaris crassispina, hatching enzyme, XLV, 371
Anthracene
  acceptor, in chemiluminescent reactions, LVII, 516, 563
  structural formula, LVII, 498
Anthracene compound, chemiluminescent reactions, LVII, 507
Anthracycline, XLIII, 141–144
Anthracyclinone, XLIII, 141–144
Anthranilate-agarose derivative, XXXIV, 378–385
  charge effects, XXXIV, 384
  prepared by diazotization, XXXIV, 384
  spacers, XXXIV, 380, 381, 384
Anthranilate dihydroxylase, LII, 12
Anthranilate 5-phosphoribosylpyrophosphate phosphoribosyltransferase, XXXIV, 377, 389–393
  inactivation on column, XXXIV, 392
Anthranilate phosphoribosyltransferase, XXXIV, 377–385, *see also* Anthranilate-agarose derivative
Anthranilate-PRtransferase, *see* Anthranilate-5-phosphoribosylpyrophosphate phosphoribosyltransferase
Anthranilate synthase, XXXIV, 377, 389–394; XLVI, 420
Anthranilate synthetase, *see* Anthranilate synthase; Anthranilate synthetase complex
Anthranilate synthetase complex, XXXIV, 377, 389–394
  feedback-inhibitor as ligand, XXXIV, 390
  inactivation on column, XXXIV, 392
  tryptophan-agarose, XXXIV, 390–392
Anthraquinone 1,5-disulfonate, as mediator-titrant, LIV, 408
Anthraquinone 2,6-disulfonate, as mediator-titrant, LIV, 409, 423, 433
Anthraquinone sulfonate, for alkali removal from gases, LII, 222
Anthraquinone 2-sulfonate, as mediator-titrant, LIV, 409, 423, 433
Anti–A hemagglutinin, XXXIV, 367
Antiandrogen, XL, 292
Anti-*p*-azobenzoate, from rabbit, hapten binding studies, XLVIII, 280–282
Anti–B hemagglutinin, XXXIV, 367
Antibinitrophenyl antibody, XXXIV, 7
Antibiotic, XLVI, 633–636; LIX, 371, *see also* specific compound
  alicyclic, XLIII, 156–159
  antitumor activity, screening, XLIII, 203
  aromatic
    gas-liquid chromatography, XLIII, 228–234
    high-pressure liquid chromatography, XLIII, 301–305
    paper chromatography, XLIII, 159–163
    solvent systems, XLIII, 163
  binding to ribosomal particles, measurement, LIX, 862–866
  biosynthesis, ^{13}C-labeling, XLIII, 404–425, *see also* Nuclear magnetic resonance spectroscopy, carbon-13
  in cell culture, LVIII, 95, 110–116
    antimicrobial spectrum, LVIII, 112, 113
    criteria for usefulness, LVIII, 110
  commercial source, LVIII, 113–115
  cytotoxicity, LVIII, 111
  DNA determination, LVIII, 147

effect on membranes, XXXII, 881–893
on protein synthesis, LIX, 851–862
excipients, LVIII, 111
for frog haploid cell culture, XXXII, 798, 799
for granulosa cell culture, XXXIX, 208
heterocyclic, XLIII, 157
nitrogen-containing, XLIII, 152–154
oxygen-containing, XLIII, 154–157
inhibitory coefficient, XLIII, 66–69
interaction with biopolymer, XLIII, 367–369
introduction into biomembranes, XXXII, 545
as ionophores, LV, 442, 443
β-lactam, and DD-carboxypeptidases-transpeptidases, interactions, XLV, 623, 625, 635, 636
macrocyclic, XLIII, 139
as uncouplers, LV, 471, 472
production in liquid culture, XLIII, 11–14
fermentation, XLIII, 12
inoculum, XLIII, 11, 12
medium A–4, XLIII, 13
medium A–9, XLIII, 14
medium A–12, XLIII, 14
reaction with biomembranes, XXXII, 500, 535, 544, 552
relative inhibitory coefficient, XLIII, 66–69
resistance, LVIII, 19, 115, 116
mutants, LVI, 14
petite mutants, LVI, 159
temperature-sensitive mutants, LVI, 134, 135
ribosomes, LVI, 10, 11
screening for antitumor activity, XLIII, 203
solubility, LVIII, 113
stability, LVIII, 113
sterility, LVIII, 111
in storage procedure, XLIV, 270
strategy of use, LVIII, 115, 116
use in preparing hepatocytes, XXXV, 582, 592
water-insoluble, solvents, LIX, 853
Antibiotic-producing microorganism, XLIII, 3–21, *see also* Antibiotic, production in liquid culture; Culture; Media
Antibody, *see also* Radioimmunoassay; specific type
affinity labeling reagents, XLVI, 486
affinity purification, L, 171–175
antigen binding, L, 54–64
antiviral, titration, LVIII, 419, 420
bacterial-luciferase coupled, assay, LVII, 404
carrier protein, LVI, 252
on cell surface, XXXIV, 197, 220
SEM, XXXII, 58
columns, XXXIV, 188, 189, *see also* Nitrotyrosine peptide
conformation studies, XLIX, 185, 186
coupling, XXXIV, 725, 726
cross linking, *see* Cross linking
cytochrome *c*, LVI, 695–697
affinity chromatography, LVI, 699, 700
for cytochrome oxidase, LVI, 693, 694, 697, 702
conjugation to ferritin, LVI, 710, 711
specificity, LVI, 690
effects on enzyme activities, LII, 243–250
enzyme inhibition, LVI, 227, 228
enzyme labeling, XLIV, 709–717
procedure, XLIV, 710, 711
ferritin-conjugated, preparation, XXXII, 61–63
glucose-oxidase coupled, assay, LVII, 403, 404
hybridized, XXXII, 61
preparation, XXXII, 63, 64
immunoadsorbents, XXXIV, 703, 706, 707
iodination, L, 58
Bolton–Hunter reagent, L, 59
lactoperoxidase, L, 58
latent ATPase, LV, 192
as ligand, XXXIV, 6, 7
measurement, XXXIV, 708–711
in membrane component localization, XXXII, 60–70
method of labeling, XLVI, 491, 492

modification of protein, XXXIV, 183, 184
for monoamine oxidase, LVI, 697
affinity chromatography, LVI, 700
conjugation to ferritin, LVI, 711, 712
specificity, LVI, 690
monoclonal, LVIII, 350
monospecific
for membrane enzymes
assay, LVI, 701–708
conjugation to ferritin, LVI, 708–715
labeling by ferritin-antibody, LVI, 715–717
preparation of, LVI, 698–701
for rabbit IgG, LVI, 700, 701
for studying sidedness of membrane components, LVI, 223–228
to morphine, XXXIV, 621, 622
M. phlei membrane systems, LV, 186
peroxidase coupled, binding studies, LVII, 406
polyinosinic:polycytidylic acid, for detecting double-stranded ribonucleic acid, LX, 553, 554
preparation
in protein determination, XL, 243
purification, LII, 241–244
production, injection schedule, LII, 242
to prostaglandin, XXXV, 287–298
to prostaglandin metabolites, XXXV, 296, 297
Pseudomonas toxin A
coupling with cyanogen-activated Sepharose 4B, LX, 786
production, LX, 785
use in purifying *Pseudomonas* toxin A, LX, 786, 787
purification, *see* Immunoglobulin purification
quantitation, XLIV, 714–716
reaginic, XXXIV, 709
SI IgG, LX, 427, 429, 430, 434, 436, 440–444
in steroid quantification, XXXVI, 16–34
subcellular localization, LII, 250, 251
$TF_0 \cdot F_1$, LV, 370
to TSH, XXXIV, 694
use, LII, 240, 244–251
to determine immunochemical similarities of enzymes, LII, 245, 246, 247–249
as diagnostic biochemical probes, LII, 240, 250, 251
Antibody-agarose, XXXIV, 186, 747, 748, 751
column, XXXIV, 191
Antibody binding, XLVIII, 280–282
Antibody complex, antigen-insoluble, XXXIV, 727
Antibody-hapten binding, XLVIII, 276
Antibody labeling, fluorescent, LVIII, 175, 176
Antibody-producing cell, XXXIV, 720–722
Antibody-secreting cell, XXXIV, 219
Antichromosomal protein, XL, 191
Anticodon loop, XLVI, 90
Anticooperativity, *see* Cooperativity, negative
Anti-dinitrophenyl agarose, XXXIV, 186
Anti-dinitrophenyl antibody, XXXIV, 6, 191; XLVI, 76, 83, 479–492
labeled chain, XLVI, 487, 488
labeled residues, XLVI, 488, 489
peptides, XXXIV, 183, 185
Anti-dinitrophenyl antibody-agarose, XXXIV, 186, 709
Anti-dinitrophenyl immunoglobin, XLVI, 531
Antidiuretic hormone, LVIII, 559
effect on water flow, XXXVII, 251–256
Antiestrogen, XL, 292
Antiferromagnetic exchange coupling, LIV, 205–207
Antifoam agent, LI, 43, 182; LII, 152–170; LIII, 649; LVII, 228; LIX, 204
Antifreeze glycoprotein, peanut agglutinin, L, 367
Antigen, *see also* specific type
carbohydrate-protein coupling
antibody specificity, L, 160
carbodiimide, L, 160–162
cyanoborohydride, L, 155–160
carcinoembryonic, LVIII, 184

on cell membranes, antibodies in study, XXXII, 60–70
cell surface type, spin-label studies, XXXII, 184
cross-linking to γ-globulins, LVI, 225, 226
identification in SDS-polyacrylamide gels, L, 54–64
for immunization of protein, XL, 242
localization, XLIV, 712–714
nitrotyrosine peptides, XXXIV, 188, 189
P, L, 247
p^k, L, 247
production of antibodies, L, 158
P_1, L, 247
in plasma membranes, XXXI, 90
preparation, cyclic nucleotide assay, XXXVIII, 98
quantitation, XLIV, 714–716
tumor-specific cell surface, LVIII, 371
Antigen-agarose, XXXIV, 718
Antigen–antibody complex, XLVI, 505–508
disrupted, electrophoresis, in SDS-acrylamide gels, XL, 248
Antigen-antibody reaction
cell monitoring, LVIII, 174–178
mixed association, XLVIII, 122
Antigen-binding cell
affinity fractionation, XXXIV, 197
characterization, XXXIV, 215
fractionation, XXXIV, 214, 215
isolation, XXXIV, 212, 213
Antigen-coated particle, XXXIV, 717
Antigenicity, measurement by quantitative microcomplement fixation, XL, 195
Anti-γ-globulin, in steroid radioimmunoassay, XXXVI, 29, 30
Antiglucocorticoid, XL, 292
Antihapten antibody, XXXIV, 704, 717
Antihapten lymphocyte, XXXIV, 719
Antihemophilic factor, *see* Factor VIII
Anti–H hemagglutinin, XXXIV, 329–331, 367
Antihistamine, XLIV, 855
Antihormone, XL, 291, 292
Anti-human M subunit-inhibiting antibody, assay of creatine kinase isoenzymes, LVII, 61
Anti-immunoglobulin serum, XXXIV, 216
Anti-insulin-agarose, XXXIV, 747
Anti-lactose antibody, XLVI, 516–523
Antimannotetraose, inhibition, L, 168
Antimineralocorticoid, XL, 292
Antimony, volatilization losses, in trace metal analysis, LIV, 481
Antimony-121, lock nuclei, XLIX, 349; LIV, 189
Anti-mouse immunoglobulin, XXXIV, 219
Antimycin, LVI, 33
chiroptical studies, XLIII, 352
Crabtree effect, LV, 297
cyanine dyes, LV, 693
cytochrome *b-c*, region, LV, 461, 462
ferricyanide reduction, LVI, 231–233
fluorescence, LVI, 499, 500
glutamate transport, LVI, 421
immobilized mitochondria, LVI, 557
M. phlei membrane systems, LV, 185
N. crassa mitochondria, LV, 147
paper chromatography, XLIII, 137
resistance, LVI, 140
solvent systems for countercurrent distribution, XLIII, 334
sources, LV, 462
Antimycin A, XLIII, 135, 253
activation of succinate dehydrogenase, LIII, 472, 473
assay of succinate dehydrogenase, LIII, 468, 470
carnitine-acyl carnitine translocase, LVI, 375
complex III, LVI, 586
efflux kinetic studies, LVI, 259
inhibitor of bacterial cytochromes, LIII, 206
of brain α-hydroxylase, LII, 317
of complex III, LIII, 13, 38
of complex I–III, LIII, 9
of duroquinol-cytochrome *c* reductase, LIII, 91
of reconstituted electron-transport system, LIII, 50
of succinate-cytochrome *c* reductase, LIII, 50
of succinoxidase activity, LIII, 50

for loading of mitochondria in purification of complex III, LIII, 93
photochemically active derivative, LVI, 660
purification of complex III peptides, LIII, 84
resistance, LVI, 118
of small cytochrome bc_1 complex, LIII, 105
Antimycin-binding protein, of complex III, characterization, LIII, 82
Anti-*p*-nitrophenyl antibody, XLVI, 505–508
polymer formation, XLVI, 507, 508
Antinucleoside
reaction with human metaphase chromosomes, XL, 304
in study of nucleic acid in cells, XL, 302
Antinucleotide, XL, 302, 304, *see also* Antinucleoside
Antioxidant
as enzyme inhibitors, XXXI, 524
for plant-cell extraction, XXXI, 538–540
function, LII, 304, 305
inactivators, of cytochrome *P*-450 peroxidase activity, LII, 411
Antipain, *see* [(*S*)-1-Carboxy-2-phenylethyl]carbamoyl-L-arginyl-L-valylargininal
Antipyrine-acetyl monoxime, assay of aspartate transcarbamoylase, LI, 46
Antipyrine reagent, assay of dihydroorotase, LI, 123
Antipyrine-sulfuric acid, assay of aspartate carbamyltransferase, LI, 51
Antipyrylazo III
absorbance spectra, LVI, 326–328
calcium measurement, LVI, 326, 327
comparison to other calcium indicators, LVI, 327, 329–332
properties, LVI, 330
Antiserum, *see also* Antibody
to ATPase, preparation, LV, 802
cell monitoring, protocol, LVIII, 176–178
coupling factor B, LV, 390
cyclic nucleotides, sensitivity and selectivity, XXXVIII, 104
cytochrome oxidase polypeptides, LVI, 596
fluorescent, LVIII, 175, 176
microencapsulation, XLIV, 215
to *N. crassa* ATPase, preparation, LV, 345, 346
potency, determination, LVI, 225
preparation, LVIII, 174, 175, 276, 277
Anti-θ serum, XXXIV, 216
Antistaphylococcal penicillinase, XLIII, 87
Anti-Stokes Raman scattering, XLIX, 69, 73
frequency shift, XLIX, 83
Anti-Stokes Raman spectroscopy, coherent, XLIX, 94
Antithrombin, assay, XLV, 657
Antithrombin-heparin cofactor
assay, XLV, 654, 656
characteristics, XLV, 653
properties, XLV, 666
purification, XLV, 658
affinity chromatography, XLV, 664
classic procedure, XLV, 659
purity, XLV, 666
reactive sites, XLV, 667
specificity, XLV, 667
stability, XLV, 666
Anti-thymocyte agarose, XXXIV, 753
Antitrypsin, XXXIV, 6
α-Antitrypsin, XXXIV, 338, 707
Aorta
oxidase, LII, 13
preparation for collagen studies, XL, 311
6-APA, *see* 6-Aminopenicillanic acid
Apamin, histidine modification, XLVII, 433, 438, 440
A particle, OSCP assay, LV, 392, 393
APEMA, *see p*-Acetaminophenylethoxy-methacrylate
Aperature foci, LIV, 37, 38
Apiezon N grease, LIV, 121
removal, LIV, 122
Aplanospore, free cell culture, XXXII, 723

Apo-acyl carrier protein
analogues, XXXV, 101
in assay of acyl carrier protein synthetase, XXXV, 95, 96
substrate inhibition with ACP synthetase, XXXV, 101
synthesis, XXXV, 96
Apoadrenodoxin
identification of iron-sulfur centers, LIII, 274
preparation, LII, 136
Apoadrenodoxin reductase, preparation, LII, 134, 135
Apoenzyme, XLVI, 444, 445
Apoferredoxin, identification of iron-sulfur centers, LIII, 274
Apoferritin
in ferritin samples, XXXII, 61
molecular weight, LIII, 350, 351
removal, LVI, 709, 710
Apoflavodoxin
assay of FMN, LIII, 422
purification of flavin peptides, LIII, 455
Apoflavoprotein, preparation, LIII, 429–437
Apogon ellioti
luciferase, species cross-reactivity, LVII, 371
luciferin, LVII, 342
Apomyoglobin
preparation, LII, 477, 478
from sperm whale, sequence analysis, XLVII, 373
Apoprotein, tissue factor, *see* Tissue factor apoprotein
Apoxytetracycline, XLIII, 199
Apparatus constant, determination, XLVIII, 89
Appeal soap, for glassware cleaning, XXXII, 818
Apple
organelle studies, XXXI, 541
protein, XXXI, 529, 543
A–protein, of lactose synthetase, XXXIV, 359, 360
A1 protein
assay, XXXII, 324, 325
biological role, XXXII, 339, 340
of CNS myelin, XXXII, 323, 324
encephalitogenic activity, XXXII, 325–327, 340
in experimental allergic encephalomyelitis, XXXII, 324
immunological properties, XXXII, 340, 341
physical properties, XXXII, 336–339
purification, XXXII, 330–335
purity, XXXII, 335, 336
APRT, *see* Adenine phosphoribosyltransferase
APS-kinase, *see* Adenylylsulfate kinase
p-*ApUpG, XLVI, 623
ApUpGpU*, XLVI, 623
Apyrase, hydrolysis of ATP, LVII, 69, 70, 71, 72
Aquacide I, purification of cytidine deaminase, LI, 410
Aquacide II, LIII, 231
Aquasol scintillation fluid, LI, 13; LIX, 192, 217, 222, 328, 344
Arabidopsis, cell cultures, XXXII, 729
L-Arabinofuranose, tricyclic orthoester derivative, L, 100
Arabinose
chromatographic constants, XLI, 17
potato lectin, L, 340
L-Arabinose, XXXIV, 333, 368
metabolism in pseudomonad MSU-1, XLII, 269
D-fuconate dehydratase, XLII, 305
L-Arabinose dehydrogenase, from pseudomonad MSU-1, XLI, 150–153
assay, XLI, 151
chromatography, XLI, 152
properties, XLI, 153
purification, XLI, 151, 152
D-Arabinose isomerase, from *Aerobacter aerogenes*, XLI, 462–465
assay, XLI, 462, 463
inhibitors, XLI, 465
molecular weight, XLI, 465
properties, XLI, 465
purification, XLI, 463–465
L-Arabinose isomerase
from *Escherichia coli*, XLI, 453–458
assay, XLI, 453, 454
chromatography, XLI, 455
inhibition, XLI, 457
properties, XLI, 456–458

purification, XLI, 454–456
from *Lactobacillus gayonii*, XLI, 458–461
assay, XLI, 458
chromatography, XLI, 459
inhibition constants of pentitols, XLI, 461
molecular weight, XLI, 460
properties, XLI, 460, 461
purification, XLI, 458–460
L-Arabinose operon, XXXIV, 368
Arabinosyladenine, substrate of adenosine deaminase, LI, 507, 511
β-D-Arabinosyladenosine triphosphate, interaction with GMP synthetase, LI, 224
Arabinosyl cytidylate, substrate of pyrimidine nucleoside monophosphate kinase, LI, 329
α-D-Arabinosylcytosine, substrate of deoxycytidine kinase, LI, 345
Arabinosyl-6-thiopurine, inhibitor of adenosine deaminase, LI, 507
β-D-Arabinosyl xanthylate, substrate of GMP synthetase, LI, 224
Arachidonate cyclooxygenase, LII, 19
Arachidonic acid, in biomembrane phospholipids, XXXII, 542
Arachis hypogea, lectin, XXXIV, 334
Arachis hypogea var. *varginia*, source of phospholipase D, XXXV, 228
araC protein, XXXIV, 368–373
adsorbent, XXXIV, 369–371
affinity chromatography procedure, XXXIV, 372, 373
assay, XXXIV, 371, 372
ligand, XXXIV, 368, 369
spacer requirement, XXXIV, 373
Araldite, as embedding plastic, XXXIX, 139
Arbacia lixula, hatching enzymes, XLV, 371
Arbacia punctulata, hatching enzymes, XLV, 371
Arbor tissue press, XXXI, 95
Arbovirus, interferon producing, LVIII, 293
Archibald technique, sedimentation equilibrium, XLVIII, 168
Arc source, LIV, 456–457
Arene oxide, bond definition, LII, 228
Arginase, renal mitochondria, LV, 12
Arginine
activator of NADH dehydrogenase, LIII, 20
aminoacylation isomeric specificity, LIX, 274, 275, 279, 280
analysis, for bound protein determination, XLIV, 389
chemical properties, XLIV, 12–18
cleavage, by clostripain, XLVII, 165–170
by trypsin, at alkaline pH, XLVII, 173, 174
dipeptides containing, identification, XLVII, 396
effect on bioluminescence, LVII, 129, 166
enzyme aggregation, XLIV, 263
δ-guanidino group blocking, XLVII, 527, 532–534, 611, 612
in isoelectric focusing of cytochrome *c* derivatives, LIII, 149
in media, LVIII, 53, 62
methylated, in A1 protein, XXXII, 324
modified, identification with nickel ion, XLVII, 160
N-*o*-nitrophenylsulfenylation, XLVII, 519
oxidative decarboxylase, LII, 16
peptide, hydrazinolysis, XLVII, 284, 285
in protein precipitation, LII, 93
repressor, of glutamine-dependent carbamoyl-phosphate synthetase synthesis, LI, 21, 30
residue, modification, LIII, 143
reversible blocking, by cyclohexanedione, XLVII, 156–161
synthesis
in *E. coli*, LI, 21
in *Neurospora*, LI, 105
in *Salmonella typhimurium*, LI, 29, 30
in ureotelic vertebrates, LI, 21
N^{ω}-tosylation, XLVII, 533, 534
Arginine-agarose, XXXIV, 432–435
Arginine chloromethyl ketone, XLVI, 201, 229–235

Arginine decarboxylase, conjugate, activity, XLIV, 271
Arginine deiminase, effect on immune response, XLIV, 705
L-Arginine iminohydrolase, *see* Arginine deiminase
L-Arginine:inosamine-phosphate amidinotransferase, XLIII, 451–458
assay
canavanine, XLIII, 454, 455
L-[guanidino-^{14}C]arginine, XLIII, 452–454
hydroxylamine, XLIII, 455, 456
ATP:inosamine phosphotransferase assay, XLIII, 447
biological distribution, XLIII, 457
canavanine:ammonium hydroxide transamination, XLIII, 458
chemically phosphorylated inosamine derivative, XLIII, 453, 454
inhibitors, XLIII, 458
natural acceptors preparation, XLIII, 453
properties, XLIII, 457, 458
purification, XLIII, 453, 456, 457
specific activity unit, XLIII, 456
specificity, XLIII, 457
unit definition, XLIII, 456
Arginine kinase, XLVI, 21
Arginine methylase, assay, XXXII, 329–334
Arginine:ornithine exchange reaction, XLIII, 452
Arginine vasopressin
adenylate cyclase, XXXVIII, 151, 152
radioimmunoassay, XXXVII, 34
Arginyl insulin
formation, XXXVII, 335
isolation, XXXVII, 337
Arginyl methyl ester, activator of NADH dehydrogenase, LIII, 20
L-Arginyl-β-naphthylamide, as enzyme substrate, XXXII, 89
Arginyl-tRNA synthetase, XXXIV, 170
subcellular distribution, LIX, 233, 234
Argon, XLIV, 222, 223, 889–892
in anaerobic procedures, LIV, 52, 114, 115, 421, 430, 468, 501
deoxygenation, LVI, 384; LVII, 110, 251, 366–368
as inert atmosphere, XXXI, 540
Argon gas laser, wavelength, XLIX, 85, 86
Argon ion laser, use in resonance Raman spectroscopy, LIV, 235, 246
Argon-krypton gas laser, wavelength, XLIX, 85
Argon-krypton-xenon flowing gas laser, wavelength, XLIX, 85
Arg-tRNA, XXXIV, 7
Arkopal–9 detergent, partial specific volume, XLVIII, 19, 20
Arkopol–13 detergent, LVI, 741
partial specific volume, XLVIII, 19, 20
Arm, *see* Spacer
Armigeries subalbatus, media, LVIII, 462
L-Aromatic amino acid decarboxylase, in neurobiological studies, XXXII, 787, 788
Arrhenius graph, LIV, 106
in polarographic data analysis, XXXII, 261, 262
Arrhenius relation, ligand binding, LIV, 508–530
Arrhenius transition, LIV, 527
p-Arsanilazo-*N*-succinylcarboxypeptidase A, XXXIV, 190
Arsanilazotyrosyl peptide, XXXIV, 190
isolation, XXXIV, 184
Arsenate, *see also* Arsenic acid
activator of phosphoribosylpyrophosphate synthetase, LI, 16
complex V, LVI, 586
Crabtree effect, LV, 296
substrate of purine nucleoside phosphorylase, LI, 529
Arsenazo III, LVII, 292
calcium, LVI, 305, 317–326
chromaffin vesicle calcium transport, LVI, 322, 323
comparison to other calcium indicators, LVI, 327, 329–332
mitochondrial calcium transport, LVI, 320, 321
properties, LVI, 318–320, 330
purification, LVI, 308, 317, 318
relaxation time, LVI, 307

sarcoplasmic reticulum calcium transport, LVI, 321, 322
single cell calcium transport, LVI, 323–326
spectral characteristics, LVI, 318–320
Arsenic, volatilization losses, in trace metal analysis, LIV, 466, 480, 481
Arsenic-75, lock nuclei, XLIX, 349; LIV, 189
Arsenic acid, *see also* Arsenate
assay of luciferase, LVII, 30
effect on luciferase bioluminescence, LVII, 13, 14
and glucose isomerization, XLIV, 811
luciferase reagent, LVII, 76, 88
purification of luciferase, LVII, 31, 32, 35
stabilizer, of luciferase, LVII, 199
uncoupler, of substrate level phosphorylation, LVII, 38
Arsenite
as enzyme inhibitor, XXXVII, 294
as uncoupler, LV, 471
Arsenomolybdate reagent, in colorimetric assay of hexuronic acids, XLI, 29, 30
Artemia salina
cysts
for developmental studies, LX, 298
disruption, LX, 304, 305
incubation, LX, 302, 696
initiation factor eIF-2 preparation, LX, 310, 311
purification, LX, 300–302, 696
ribosomal subunits, LX, 308, 309, 560, 561
ribosome preparation, LX, 307, 308, 660, 661, 688
transfer ribonucleic acid preparation, LX, 305, 306
use in polyamine studies, LX, 555, 559
viability, LX, 304
nauplii, LX, 304, 305
prenauplii, LX, 303, 304
Artery, isolation of mitochondria, LV, 60, 61
Arthrobacter
oxidase, LII, 13, 16–18
protease, amino acid composition, XLV, 423
Arthrobacter oxidans, D-6-hydroxynicotine oxidase from, LIII, 450
Arthrobacter simplex, immobilization, by polyacrylamide entrapment, XLIV, 184–190
Arthropod, oxidase, LII, 10
Arthus reaction, XLIV, 709
Artichoke tuber, enzyme studies, XXXI, 521
Aryl-alcohol oxidase, LII, 18
P-Arylamine/CPG-550 A glass bead, for enzyme immobilization, LVII, 204
Arylamine derivative
glass, XXXIV, 71, 72
with protein, XXXIV, 53, 54
Arylamine glass, XXXIV, 65, 620
production, XXXVIII, 182, 183
Arylamine ligand, alkylation, XXXIV, 50
Arylamine support, diazotized, for immobilized enzymes, LVI, 491
Aryl amino group, assay, by radioestimation with ^{36}Cl, XLIV, 92
Aryl azide, XLVI, 78, 80
displacement of halide ion, XLVI, 99
groups, XLVI, 99–104, 627
quantitation, XLVI, 99, 105
stability, XLVI, 98, 105
synthesis, XLVI, 98–105
Arylazidoadenine nucleotide
synthesis, XLVI, 260–279
solvents, XLVI, 260, 261
Arylazidoadenosine diphosphate, assay of binding protein, LVI, 417, 418
Arylazido-β-alanine, *see* *N*-(4-Azido-2-nitrophenyl)alanine
Arylazido-β-alanine ATP, *see* 3′-*O*-(3-[*N*-(4-Azido-2-nitrophenyl)amino]propionyl)adenosine triphosphate
Arylazido-β-alanine NAD^+, *see* 3′-*O*-(3-[*N*-(4-Azido-2-nitrophenyl)amino]propionyl)nicotinamide adenine dinucleotide
Arylazido-β-alanine nicotinamide adenine dinucleotide, XLVI, 279–282; LVI, 656
Arylazido-β-alanine nucleotide, XLVI, 270

Arylazido-β-alanine pyridine nucleotide analogs, XLVI, 279–285
chromatography, XLVI, 281
Arylazido-4-aminobutyric adenosine triphosphate, reactions, XLVI, 276
Arylazido-6-aminocaproic adenosine triphosphate, reactions, XLVI, 276
Arylazidoatractyloside, assay of binding protein, LVI, 417, 418
Arylazido coenzyme A, analogs, XLVI, 260
Arylazido nucleotide, analogs, XLVI, 259–288
Aryldiazirine, XLVI, 73, 74
Aryldiazomethane, XLVI, 73, 74
Arylesterase, gel electrophoresis, XXXII, 88
Aryl-β-hexosidase, from bovine liver
assay, L, 524, 525
properties, L, 527, 528
purification, L, 525–527
Aryl hydrocarbon hydroxylase, *see also* Flavoprotein-linked monooxygenase
hormone induction, XXXIX, 39
Aryl hydrocarbon monooxygenase, *see* Flavoprotein-linked monooxygenase
Aryl hydroxylase, in nuclear membrane, XXXI, 290–291
Aryl keto group, XLVI, 627
Aryl nitrene diradical, XLIV, 282
Aryl sulfatase
bioluminescence assay, LVII, 257
in brain fractions, XXXI, 469–471
as diagnostic enzyme, XXXI, 20, 31, 186, 187
in macrophage fractions, XXXI, 340
synthesis of benzyl luciferyl sulfate, LVII, 252
Arylsulfatase A, XXXIV, 4
assay, L, 537–543
barium ion inactivation, L, 450
metachromatic leukodystrophy, L, 471–474
mucolipidoses II and III, L, 455
multiple sulfatase deficiency, L, 453, 454, 474
properties, L, 546, 547
purification, L, 543, 544
specificity, L, 537
substrates, L, 472, 473
Arylsulfatase B
assay, L, 450, 451, 474, 475, 537–543
distribution, L, 450
Maroteaux–Lamy syndrome, L, 451
multiple sulfatase deficiency, L, 453, 454, 474
properties, L, 546, 547
purification, L, 544, 545
specificity, L, 537
Arylsulfatase C, multiple sulfatase deficiency, L, 454, 474
Aryl sulfotransferase, assay of PAP, LVII, 254, 256, 257
AS, *see* Anthranilate synthase
Asbestos, platinized, fibrous, LVII, 137, 138
Ascites cell, *see also* Dipeptidase, ascites tumor
Ehrlich, messenger ribonucleic acid from, LX, 394
homogenization, XLV, 388
Krebs
elongation factors, LX, 649, 651
initiation factors, LX, 88, 90–92
propagation, LX, 651
optical probes, LV, 573
Ascites hepatoma, dispersion, LVIII, 125
Ascites tumor, LVIII, 376
suspension culture, LVIII, 203
Ascorbate 2,3-dioxygenase, LII, 11
Ascorbate oxidase, LII, 9
prosthetic group, LII, 4
Raman frequencies and assignments, XLIX, 144
reaction mechanism, LII, 39
resonance Raman spectrum, XLIX, 142
Ascorbate-phenazine methosulfate
as electron donor, LVI, 386
fluorescence changes, LVI, 501
Ascorbate-TMPD oxidase, effect of DABS or PMPS labeling, LVI, 621
Ascorbate-TPD, amino acid transport, LV, 186, 187
Ascorbic acid, XLIV, 256, 257
assay of acid nucleotidase, LI, 272
of *Pholas* luciferase, LVII, 386
cytochrome *c* reduction, LV, 110
effects on fluorescence probes, XXXII, 242

flow dialysis experiments, LV, 683, 685, 686
inhibitor of *Pholas* luciferin oxidation, LVII, 399, 400
level in plasma, LVI, 449
in media, LVIII, 54, 64
membrane autoxidation, LII, 305
membrane vesicle isolation, LVI, 382, 383
in plant-cell fractionation, XXXI, 505
polarography, LVI, 453, 457, 463, 483
proline uptake by proteoliposomes, LV, 197
as redox titrant, LIV, 409
reductant of azurin, LIII, 661
of cytochrome *c*, LIII, 159, 160
of cytochrome *c*-551, LIII, 661
of cytochrome c_1, LIII, 119, 226; LIV, 137
spin probe measurements, LVI, 522, 523
stimulator of *Pholas* light emission, LVII, 397
substrate of *Pholas* luciferase, LVII, 398
transport studies, LVI, 257, 260
L-(+)-Ascorbic acid, purification of complex III, LIII, 103
Ascorbic acid-2-sulfate, arylsulfatase A, L, 474
Ascosin, XLIII, 136
AI–SDS complex, human serum high-density lipoprotein, properties, XLVIII, 6
AII–SDS complex, human serum high-density lipoprotein, properties, XLVIII, 6
Ashing procedure
in iron analysis, LIV, 439–441
in trace metal analysis, LIV, 480–482
Asialofetuin
electrolectin, L, 294
Polysphondylium pallidum agglutinin, L, 316
Asialo-G_{M1}-ganglioside, β-galactosidase, L, 482
Asialoglycophorin
isolation, L, 367
peanut agglutinin, L, 367
Asialoglycoprotein, XXXIV, 688–691, *see also* Asialoglycoprotein receptor
liver membrane binding, L, 113
Asialoglycoprotein receptor, XXXIV, 688, 689
assay, XXXIV, 690
purification, XXXIV, 690, 691
solubilization, XXXIV, 689, 690
Asialo-orosomucoid, XXXIV, 690
^{125}I-labeled, XXXIV, 690
Asialo-Tay–Sachs ganglioside, *see* 2-Acetamido-2-deoxy-*O*-β-D-galactopyranosyl-(1→4)-*O*-β-D-galactopyranosyl-(1→4)-D-glucose
Asolectin
assay of complex I, LIII, 14
of cytochrome oxidase, LIII, 74
of NADH-cytochrome *c* reductase, LIII, 9
in lipid bilayers, XXXII, 516, 517
preparation of suspension, LV, 311, 316, 364
reconstituted vesicles, LV, 371
solution, LV, 759
treatment of filters, LV, 589, 599
Asparaginase, XLVI, 22, 35
active site, XLVI, 432–435
assay, XXXIV, 407
peptide active site, XLVI, 434
purification, XXXIV, 409, 410
reaction with DONV, XLVI, 434, 435
L-Asparaginase
activity
assay procedure, native protein, XLIV, 690, 691
relative, conjugate, XLIV, 271
specific conjugate, XLIV, 255
effect on immune response, XLIV, 705
encapsulated, exposed antigenic sites, XLIV, 692
entrapment in polyacrylamide gel, XLIV, 692
in poly-2-hydroxyethylmethacrylate, XLIV, 176, 692
fiber entrapment
biomedical application, XLIV, 242
in cellulose triacetate fiber, XLIV, 693

immobilization on collagen, XLIV, 255, 260, 693
on Dacron, XLIV, 693
on glass plates, XLIV, 693
on inert protein, XLIV, 909
on nylon tube, XLIV, 120, 693
on polymethylmethacrylate, XLIV, 693
microencapsulation, XLIV, 212–214
storage in solution, XLIV, 212, 213
therapeutic application, XLIV, 679, 689–694
Asparagine, XLIV, 13, 242
aminoacylation isomeric specificity, LIX, 274, 275, 279, 280
assay, LV, 209
carbodiimide activation, XLVII, 553
β-carboxamido group blocking, XLVII, 612
dipeptides containing, identification, XLVII, 396
enzyme electrode, XLIV, 585
esterification, XLVII, 525
identification, XLVII, 83, 84
isoxazolium salt activation, XLVII, 556
in media, LVIII, 63
phenylthiohydantoin derivative, identification, XLVII, 347
residue, effect on Raman spectra, XLIX, 116
hydrogen exchange, XLIX, 25
saccharide linkage, L, 113
side chain cyclization, chain cleavage, XLVII, 132–145
thermolysin hydrolysis, XLVII, 175
Asparagine-glycine bond
probability of occurrence, XLVII, 142
specific cleavage, XLVII, 132–145
mechanism, XLVII, 143, 144
Asparagine synthetase, XLVI, 420
Asparaginyl-tRNA synthetase, subcellular distribution, LIX, 233, 234
Asparagus
cell cultures, XXXII, 729
from protoplasts, LVIII, 367
Aspartase, XXXIV, 407, 410, 411
L-Aspartase, *see* L-Aspartate acetyltransferase
Aspartate, *see also* Aspartic acid
assay of carbamoyl-phosphate synthetase, LI, 105
substrate of aspartase carbamyltransferase, LI, 51
of succino-AICAR synthetase, LI, 186
D-Aspartate, activator of aspartate carbamyltransferase, LI, 49
L-Aspartate, *see also* L-Aspartic acid
assay of carbamoyl-phosphate synthetase, LI, 122
substrate of adenylosuccinate synthetase, LI, 207
of aspartate carbamyltransferase, LI, 35, 41, 51, 111, 123
Aspartate acetyltransferase
pH activity profile, XLIV, 741, 742
thermal stability, XLIV, 741
Aspartate aminotransferase, XXXIV, 4; XLVI, 31, 32, 37, 52, 163, 432, 445
in L-asparaginase assay, XLIV, 690
assay by Fast Analyzer, XXXI, 816
brain mitochondria, LV, 58
coimmobilization, XLIV, 472
cyanylation, XLVII, 131
elimination from mitochondria, LVI, 423
in glyoxysomes, XXXI, 569
immobilization, on collagen film, XLIV, 908
inactivation, XLVI, 45
α-ketoglutarate assay, LV, 206
oxaloacetate assay, LV, 207
in plasma membrane, XXXI, 89
purification, LV, 208
syncatalytic modification, XLVI, 42–44
Aspartate carbamoyltransferase, XLVI, 21, 90; LI, 35–58, 105–134
activators, LI, 48, 50
active site titration, with tritiated N-phosphonacetyl-L-aspartate, LI, 126
activity, LI, 35
allosteric activator site, LI, 50
allosteric interaction, XLVIII, 305, 306
assay, LI, 41, 42, 106, 107, 122, 123
of carbamoyl-phosphate synthetase, LI, 105

binding studies, XLIV, 557
carbamoyl-phosphate synthetase activity, LI, 57
cyanylation, XLVII, 131
from *E. coli*, LI, 35–41
energy transfer studies, XLVIII, 378
in enzyme complexes, LI, 105–134
inhibitors, LI, 48
kinetic properties, LI, 48–50, 57, 58
molecular weight, LI, 46, 55, 121
properties, LI, 46–50, 120
from *Pseudomonas fluorescens*, LI, 51–58
purification, LI, 43–46, 52–56
from rat cells, LI, 111–121
regulation, LI, 35
renaturation, LI, 55, 56
stability, LI, 46, 47
from *Streptococcus faecalis*, LI, 41–50
subunit dissociation, LI, 36
identification, LI, 36–38
isolation, LI, 37–39
purity, LI, 39, 40
structure, LI, 55
Aspartate β-decarboxylase, XXXIV, 407
L-Aspartate β-decarboxylase, XLVI, 37, 427–432
labeling, XLVI, 429–432
Aspartate-α-ketoglutarate aminotransferase, in spinach, XXXI, 741
D-Aspartate oxidase, LII, 18
Aspartate-tRNA synthetase, XXXIV, 170
Aspartate transaminase, gluconeogenic catalytic activity, XXXVII, 281
Aspartate transcarbamylase, *see* Aspartate carbamoyltransferase
Aspartic acid, *see also* Aspartate
aminoacylation isomeric specificity, LIX, 274, 275, 279, 280
assay, LV, 207, 221
β-carboxyl group blocking, XLVII, 526, 528–530, 612
cleavage rates, XLVII, 342–344
content
of brain, LV, 218
of hepatocytes, LV, 214
of kidney cortex, LV, 214
of rat heart, LV, 216
of rat liver, LV, 213
cyclic imide formation, XLVII, 280
detection by fluorescence, XLVII, 242
efflux, kinetic analysis, LVI, 271–276
exchange, LVI, 254
hydrogen bromide cleavage, XLVII, 604
loading for efflux studies, LVI, 260
mobility reference, XLVII, 56
phenylthiohydantoin derivative, identification, XLVII, 348
problems in peptide synthesis, XLVII, 606, 607
reaction with diethylpyrocarbonate, XLVII, 440
transport, LVI, 248, 249
L-Aspartic acid, XXXIV, 405–411; *see also* L-Aspartate
assay, XXXIV, 407
bioassay, XLIV, 739
chemical properties, XLIV, 12–18
factors affecting production, XLIV, 741–744
group-specific adsorption, XXXIV, 407–409
industrial production, XLIV, 744
uses, XLIV, 739
Aspartic acid-agarose, XXXIV, 406–411
Aspartylglucosamine
human urine, L, 229
structure, L, 229
Aspartylglucosaminuria, urinary oligosaccharides, L, 229–233
β-Aspartylglycine dipeptide, in urine, XLVII, 144
Aspartyl-prolyl bond, chemical cleavage, XLVII, 145–149
Aspartyl-tRNA synthetase, subcellular distribution, LIX, 233, 234
Aspergillopeptidase B, amino acid composition, XLV, 423
Aspergillopeptidase C
action, XLV, 428, 429
amino acid composition, XLV, 423
Aspergillus, proteinase, barley
composition, XLV, 727, 728
inhibitor, assay, by casein, XLV, 724
kinetics, XLV, 727
properties, XLV, 726
purification, XLV, 725
purity, XLV, 726

specificity, XLV, 726
units, XLV, 724
Aspergillus niger
catalase, XLIV, 481
disruption, LV, 139
β-galactosidase, XLIV, 793
D-gluconate dehydratase activity, XLII, 304
glucose oxidase, XLIV, 196, 268, 481
oxidase, LII, 18
Aspergillus ochraceus, XLIII, 721
Aspergillus oryzae
carboxylpeptidase, XLIV, 517
culture, XLIV, 748
disruption, LV, 139
Aspergillus oxidase, LII, 12, 16, 20
Aspergillus ustus, XLIII, 323
Asperlin, XLIII, 409
Assay, *see also* Microassay; specific types
antibiotic, XLIII, 55–69
diffusion, XLIII, 60, 61
dilution, XLIII, 62–65, *see also* Dilution assay
of diverse biological samples, XLIII, 65, 66
microbiological, XLIII, 55–69
organisms used, XLIII, 57, 58
photometric method, XLIII, 63–65
quantitative aspects, XLIII, 58, 59
relative inhibitory coefficient, XLIII, 66–69
screening methods, XLIII, 58
theoretical equations, XLIII, 66, 67
colorimetric, for hexuronic acids and keto sugars, XLI, 29–31
colorimetric ultramicro, for reducing sugars, XLI, 27–29
Assembly, study, singlet-singlet energy transfer, XLVIII, 347–379
Association constant, XXXIV, 111, 141–147, *see also* Adsorption effect
evaluation from Hill plot, XLVIII, 288
from Scatchard plot, XLVIII, 386
lower limit, XXXIV, 145
Association equilibrium, ligand-mediated, XLVIII, 239
Association reaction, *see also* Binding; Cooperativity; Dimerization; Dissociation; Interaction; Ligand binding; Self-association
determination of rate constants, XLVIII, 388, 389
equilibrium equations, XLVIII, 385–387
kinetic equations, XLVIII, 387–392
mixed, XLVIII, 121–133
analysis by Gibbs-Duhem equation, XLVIII, 124
by Nichol and Winzor procedure, XLVIII, 125
by Steiner method, XLVIII, 128
ideal, XLVIII, 75, 125–128
nonideal, XLVIII, 75, 128–131
simulated example, XLVIII, 131–133
reaction order, determination of, XLVIII, 387–388
Astasia longa, disruption, LV, 138
Asterias forbesi, N-glycolyl-8-*O*-methylneuraminic acid, L, 65
Asterina pectinifera, N-acetyl-8-*O*-methylneuraminic acid, L, 65
Astrocyte
chemical composition, XXXV, 578
2′,3′-cyclic nucleotide-3′-phosphohydrolase, XXXII, 128
fixation for electron microscopy, XXXV, 557
isolation, XXXV, 567–570
morphology, XXXV, 576, 577
ASW, *see* Seawater, artificial
Asymmetry parameter, LIV, 348, 351, 352, 354
ATCase, *see* Aspartate carbamoyltransferase
ATCC, *see* American Type Culture Collection
Atebrin
fluorescence quenching, LVI, 181
assay in mutants, LVI, 114
as fluorescent probe, XXXII, 235, 239, 240
M. phlei membrane systems, LV, 185
ATEE, *see N*-Acetyl-L-tyrosine ethyl ester
Atlas G 2127 detergent, LVI, 741

Atomic absorption spectroscopy
atomization cells, LIV, 464–467
burner system, LIV, 451
calcium distribution, LVI, 302
principles, LIV, 462–464
spectrometer demography, LIV, 462
Atomic emission spectroscopy, LIV, 447–459
atomization/excitation process, LIV, 451, 452
burner systems, LIV, 450, 451
compound formation, LIV, 452, 453
detection limit, LIV, 453–455
detectors, LIV, 459–462
excitation sources, LIV, 449–459
principles, LIV, 448, 449
sensitivity, LIV, 453
Atomic fluorescence spectroscopy
atomization cells, LIV, 468–470
principles, LIV, 467, 468
Atomic spectroscopy, LIV, 446–484, *see also* specific technique
experimental parameters, LIV, 471–483
recovery studies, LIV, 481, 482
sample preparation, LIV, 479–482
standards, LIV, 482, 483
Atomization cell, LIV, 464–467
Atomization/excitation process, LIV, 451–453
ATP, *see* Adenosine 5′-triphosphate
ATPase, *see* Adenosinetriphosphatase
ATP(CTP):tRNA nucleotidyltransferase, *see* CTP(ATP):tRNA nucleotidyltransferase
ATP:dihydrostreptomycin-6-phosphate 3′,α-phosphotransferase, *see* Dihydrostreptomycin-6-phosphate 3′α-kinase
ATP hydrolase, *see* Adenosinetriphosphatase, Ca^{+}-activated
ATP:inosamine phosphotransferase, *see* Inosamine kinase
ATP-phosphoribosyltransferase, protein-tRNA interaction studies, LIX, 323, 324, 328, 329
ATP photometer
JRB Model 2000, LVII, 75, 92, 550
operational characteristics, LVII, 551, 553–555
JRB Model 3000, LVII, 217, 218
ATP:streptomycin 3′-phosphotransferase, *see* Streptomycin 6-kinase
ATP:streptomycin 3″-phosphotransferase, *see* Streptomycin 3″-kinase
ATP-sulfurylase, *see* Sulfate adenylyltransferase
ATR, *see* Attenuated total reflectance technique
Atractylic acid
ADP, ATP carrier, LVI, 407, 408
transport, LVI, 248
Atractylis gummifera, labeling with [^{35}S]sulfate, LV, 520
Atractyloside
arylazido derivatives, LV, 526, 527
brown adipose tissue mitochondria, LV, 71, 72
cell fractionation, LVI, 217
chemically labeled
method, LV, 523
principle, LV, 522
remarks, LV, 522, 523
yield, LV, 523
cytosol ATP/ADP ratio, LV, 244
derivatives
criteria of purity, LV, 530–532
preparation, LV, 524–527
electron flow, LV, 242, 243
labeled
extraction, LV, 521
purification, LV, 521, 522
yield, LV, 522
mitochondrial protein synthesis, LVI, 20, 22
pet9 mitochondria, LVI, 128
photochemically active derivative, LVI, 660
transport, LVI, 292
Atractyloside-binding protein, purification from yeast mitochondria, LVI, 414, 415, 417
characterization, LVI, 417, 418
procedure, LVI, 415, 417
Atropa, from protoplasts, LVIII, 367
Atropa × petunia, LVIII, 365
Atropine, cyclic nucleotide levels, XXXVIII, 94

Attenuated total reflectance technique, LIV, 323
AUG, *see* Adenyluridylylguanylate
Auramine O
  as fluorescent probe, XXXII, 235
    properties, XXXII, 239
  as optical probe, LV, 573
Aureofungin, XLIII, 136
Aureolic acid, XLIII, 127
Aureomycin, XLIII, 334
Aurin tricarboxylic acid
  effect on protein synthesis, LIX, 855, 858
  inhibitor of polypeptide synthesis initiation, LIX, 359
  protein inhibitor, XL, 287, 288
Aurovertin
  ATPase, LV, 300–302, 338, 343, 401
  binding by ATPase, LVI, 527
  chemical properties, LV, 481
  comparative potencies, LV, 483, 484
  as fluorescent probes, LV, 485
  inhibitory effects
    on chloroplasts and bacteria, LV, 483, 485
    on mitochondrial systems, LV, 481–483
  preparation, LVI, 179
    analysis of extracts, LV, 479
    growth conditions, LV, 478, 479
    isolation and purification, LV, 480
    large-scale cultures, LV, 479
    liquid chromatography, LV, 480
    radioactive, LV, 480, 481
    small-scale cultures, LV, 479
  structure, LV, 478
  toxicology, LV, 481
Aurovertin B, physical properties, LV, 476
Aurovertin D, physical properties, LV, 476
Australian emperor gum moth, media, LVIII, 457
Autoanalyzer cup
  assay of malate dehydrogenase, LVII, 183, 187
  polystyrene, for bioluminescent assays, LVII, 218, 221
Autoclaving, LVIII, 6, 7
Autocorrelation function, *see also* Power spectrum
  amplitude
    for non-spherical scatterers, XLVIII, 436–437
    for spherical scatterers, XLVIII, 427
  field, XLVIII, 423
    for depolarized scattering from anisotropic molecules, XLVIII, 440–442
    for dilute solutions, XLVIII, 427
    for isomerization, XLVIII, 447–449
    for moving charged particles, XLVIII, 431
    for particle undergoing translation, XLVIII, 480
    for rodlike molecules, XLVIII, 437, 438
    rotational contributions, XLVIII, 437, 440–442
    for spherical scatterers undergoing translational diffusion, XLVIII, 427
    for two-component biopolymer solution, XLVIII, 428
    for uniform translational motion, XLVIII, 430
  normalized
    first order, XLVIII, 423, 424
      for rodlike molecules, XLVIII, 438
      for several relaxation processes, XLVIII, 474
      for uniform translational motion, XLVIII, 430
    second order, XLVIII, 422
      for dilute solution of identical, spherical independent particles, XLVIII, 451
      for number fluctuation studies, XLVIII, 451
      for rodlike molecules, XLVIII, 438
      for uniform translational motion, XLVIII, 431
  optical field, for two relaxation process system, XLVIII, 479
  orientational, XLVIII, 440

photocurrent, XLVIII, 423, 424
dc component, XLVIII, 423, 424, 467
heterodyne
for moving charged particles, XLVIII, 432
for spherical scatterers, XLVIII, 427
for two-component biopolymer solution, XLVIII, 428, 429
homodyne
for bacteriophage T4 assembly, XLVIII, 444, 445
data analysis, XLVIII, 473–479
dependence on coherence areas, XLVIII, 461
distortion, XLVIII, 478, 479
for rodlike molecules, XLVIII, 439
for spherical scatterers, XLVIII, 427
for two-component biopolymer solution, XLVIII, 428, 429
for uniform translational motion, XLVIII, 431
for two relaxation process system, XLVIII, 479
Auto Densiflow apparatus, XXXI, 652
Autoinducer, LVII, 127, 128, 164
Autoinduction, in luminescent bacteria, LVII, 127, 128, 164
Automatic Diluting Station apparatus, XXXVII, 33
Automation, of analysis, *see also* Continous flow analyzer
using immobilized enzymes, XLIV, 633–646
Autoradiography, XLVI, 113; LVIII, 279–292
assay of nucleoside diphosphokinase, LI, 372, 373
artifacts, LVIII, 290–292
cell culture, LVIII, 281, 282
cell synchrony, LVIII, 249–251
chemography, LVIII, 291
darkroom, LVIII, 280, 281
safelight, LVIII, 181
developer, LVIII, 285
diffusible substances, LVIII, 289
double isotope, LVIII, 287, 288
electron microscope
of chromosomes, XL, 71, 72
coating, XL, 6
dipping method, XL, 7
loop method, XL, 6
cytological procedures, XL, 5
exposure, XL, 9
grain density, results, XL, 18
interpretation of data, XL, 5
localization of proteins and nucleoproteins, XL, 5
quantitative analysis of results, XL, 15
sectioning, XL, 6
of steroid hormones, XXXVI, 138, 139
vessels, washing, XL, 3
environmental radiation, LVIII, 292
exposure, LVIII, 285
film, XLVII, 328; LVIII, 287
fixation, LVIII, 282
of gels, LIX, 833, 834
of gel slabs, LVI, 64–66, 71, 606
high-speed scintillation, LVIII, 288, 289
of hormone receptors, XXXVII, 145–167
light exposure, LVIII, 291
photographic processing, LVIII, 290, 291
poststaining, LVIII, 286, 287
pressure effects, LVIII, 291
prestaining, LVIII, 286
procedure, LVIII, 240, 281–286
protocol, LVIII, 250, 251
in tRNA sequence analysis, LIX, 68, 70, 104
self-absorption, LVIII, 289, 290
staining, LVIII, 285, 286
static discharges, LVIII, 291
of steroid hormones, XXXVI, 135–156
Autoranging instrument amplifier, LVII, 537, 538
circuit diagram, LVII, 535
AUTOTURB system, XLIII, 56, 69
Autoxidation, polyoxyethylene chains, LVI, 742
Auxin
in media, LVIII, 478

as plant growth substances, XXXII, 731
Auxotroph, XLIII, 38
selective media, LVIII, 352
Auxotrophy, as genetic marker, XLIII, 44
Avenaciolide, glutamate transport, LVI, 421, 427
Avian myeloblastosis virus, atomic emission studies, LIV, 458
Avian myeloblastosis virus p19 polypeptide, peptide separation, XLVII, 229, 230, 232
Avian pineal gland, cell dissociation, LVIII, 130
Avian retrovirus, LVIII, 379
Avidin, XXXIV, 265–267
gel electrophoresis, XXXII, 101
homogeneous competitive protein-binding assay, LVII, 441, 442
inhibition of acetyl-CoA carboxylase, XXXV, 17
progesterone-induced, XXXVI, 187, 293
tRNA mapping, LVI, 9
tryptophan content, XLIX, 162
Avilamicin, XLIII, 128
Axenic growth, of bacteria, XXXIX, 488, 489
Axon
fluorescent probe studies, XXXII, 234
squid, optical probes, LV, 573, 580, 581
8-Azaadenine, resistant strains, LVIII, 314
8-Azaadenosine, substrate of adenosine deaminase, LI, 506, 507
Aza-amino acid, XLVI, 208–216
kinetics, XLVI, 211–214
reaction with enzymes, XLVI, 214–216
reaction mechanisms, XLVI, 209
synthesis, XLVI, 209–211
Azacolutin, XLIII, 136
5-Azacytidine
RNA inhibitor, XL, 278, 282
substrate of cytidine deaminase, LI, 407
of uridine-cytidine kinase, LI, 313, 319
6-Azacytidine, substrate of cytidine deaminase, LI, 407
6-Azadeoxythymidine, substrate of deoxythymidine kinase, LI, 364
8-Azaguanine
enzyme-deficient mutants using, XXXIX, 123
hybrid selection, LVIII, 352
substrate of guanine deaminase, LI, 516
of guanine phosphoribosyltransferase, LI, 557
8-Azaguanine-resistant cell line, LVIII, 376
8-Azaguanylate, substrate of guanylate kinase, LI, 488, 489
Azalomycin, XLIII, 137
Azapeptide, XLVI, 208–216
design, XLVI, 208
kinetics, XLVI, 211–214
reaction with enzymes, XLVI, 214–216
synthesis, XLVI, 209–211
Azapeptide *p*-nitrophenol ester, hydrolysis rate, XLVI, 209
Azaphenylalanine phenyl ester, XLVI, 216
Azaquinone, reaction intermediate in chemiluminescence, LVII, 419
Azaserine, XLVI, 35, 36, 415
^{14}C-labeled, XLVI, 424, 425
synthesis, XLVI, 416
L-Azaserine
of FGAM synthetase, LI, 193, 201
inhibitor of amidophosphoribosyltransferase, LI, 178
6-Azathymine, substrate of orotate phosphoribosyltransferase, LI, 152
7-Azatryptophan, substrate of tryptophanyl-tRNA synthetase, LIX, 249
Azauracil, inhibitor of orotate phosphoribosyltransferase, LI, 166
6-Azauracil, substrate of orotate phosphoribosyltransferase, LI, 152
Azauridine, inhibitor of orotate phosphoribosyltransferase, LI, 166
6-Azauridine, substrate of uridine-cytidine kinase, LI, 313, 319

8-Azauridine diphosphate, substrate of nucleoside diphosphokinase, LI, 385
6-Azauridine 5′-phosphate, inhibitor of ortidylate decarboxylase, LI, 153
6-Azauridylate
inhibitor of orotate phosphoribosyltransferase, LI, 163
of ototidylate decarboxylase, LI, 163, 164
8-Azaxanthylate, substrate of GMP synthetase, LI, 224
Azelaic acid dihydrazide, preparation, LIX, 179
Azetidinecarboxylic acid, resistant strains, LVIII, 314
Azide, *see also* Sodium azide
adenylate cyclase assays, XXXVIII, 119
aliphatic, absorption maxima, LVI, 664
in antigen studies, XXXII, 67
aromatic
photolysis, LVI, 665
spectra, LVI, 664
assay of modulator protein, LVII, 109
ATPase, LVI, 181
binding, to ferricytochrome *c*, LIV, 174, 177, 181
in coelomic cell homogenization buffer, LVII, 376
competition with 2-azido-4-nitrophenol for binding, LVI, 669
complex IV, LVI, 586
crystallization of ATPase TF_1, LV, 373
guanylate cyclase assay, XXXVIII, 196
hydrogen peroxide stabilization, LVI, 459
immobilized mitochondria, LVI, 556
inhibitor of complex IV, LIII, 45
of *Cypridina* luciferin, LVII, 339
of formate dehydrogenase, LIII, 371
of phagocyte chemiluminescence, LVII, 491, 492
of *Pseudomonas* cytochrome oxidase, LIII, 657
of reconstituted electron-transport system, LIII, 50, 51
inhibitory effects, LV, 516
interaction with cytochrome *c* oxidase, LIII, 193, 194, 196, 197
in IR spectroscopy, LIV, 304
purification of *Diplocardia* luciferase, LVII, 376
stretching frequency, LIV, 311
synthesis, potential explosion hazards, LVI, 652, 653
Azide method, of carboxyl group activation, XLVII, 564–568
Azide reagent, aromatic, XLVI, 486
4-Azidoacetanilido-tRNA, XLVI, 85
Azidoacetylchloramphenicol, XLVI, 84
3-Azidoacetylstrophanthidin, XLVI, 87
3-Azidoacridine, XLVI, 645, 647
9-Azidoacridine, XLVI, 645, 647
8-Azidoadenosine, LVI, 660
8-Azidoadenosine 3′,5′-cyclic monophosphate, *see* 8-Azido cyclic adenosine monophosphate
8-Azidoadenosine 5′-diphosphate, XLVI, 83; LVI, 656
photoaffinity labeling, LVI, 671
8-Azidoadenosine 3′,5′-monophosphate, XLVI, 99, 241; LVI, 656
8-Azidoadenosine 5′-monophosphate, LVI, 656
synthesis, LVI, 647–649
8-Azidoadenosine triphosphate, XLVI, 83, 99; LVI, 656
analogs, LVI, 645
synthesis, LVI, 650, 651
5′,8-Azidoadenylyl-β,γ-imidodiphosphate, synthesis, LVI, 652
5′,8-Azidoadenylyl-β,γ-methylenediphosphonate, synthesis, LVI, 652
Azidoalkyl fatty acid, XLVI, 84
Azido-4-aminobutyric adenosine triphosphate, XLVI, 269
Azido-6-aminocaproic adenosine triphosphate, XLVI, 269
γ-(4-Azidoanilide), of ATP, XLVI, 85
γ-(*n*-Azidoanilide)-adenosine triphosphate, inhibitor of tryptophanyl-tRNA synthetase, LIX, 252, 256
Azidoaryl fatty acid, XLVI, 84
Azidoaryl reagent
intermediates, XLVI, 100
synthesis, XLVI, 100–104

p-Azidobenzaldehyde
modification of elongation factor EF–G, LX, 720, 721
preparation, LX, 721
4-Azidobenzenesulfonamide, XLVI, 87
p-Azidobenzoic acid, XLVI, 264, 266
4-Azidobenzoic acid hydrazide, preparation, LIX, 179
N-4-Azidobenzoyl-3,4-dichloroaniline, XLVI, 106
N-(4-Azidobenzoyl)guanylylimidodiphosphate, LVI, 656
N-(4-Azidobenzoyl)pentagastrin, XLVI, 85
4-Azidobenzoyl pentagastrin, LVI, 660
4-Azidobenzylamine, XLVI, 658, 659
2-Azidobiphenyl, XLVI, 78
4-Azido-α-bromoacetophenone, XLVI, 87
p-Azido-N-t-butyloxycarbonyl-Phe-$(Gly)_n$-COOH, XLVI, 709
p-Azido-N-t-butyloxycarbonyl-Phe-$(Gly)_n$-Phe-tRNA, XLVI, 709, 710
p-Azido-N-t-butyloxycarbonyl-Phe-tRNA-, XLVI, 708, 709
8-Azido-cAMP, see 8-Azido cyclic adenosine monophosphate
p-Azidochloramphenicol, XLVI, 706
synthesis, XLVI, 705, 706
4-Azido-2-chlorophenoxyacetic acid, XLVI, 84
4-Azido-7-chloroquinoline, synthesis, XLVI, 646, 647
4-Azidocinnamoyl-α-chymotrypsin, XLVI, 84
p-Azidocinnamoyl methyl ester, XLVI, 108
8-Azido cyclic adenosine monophosphate, XLVI, 83, 339–346
2′-Azido-2′-deoxycytidine, synthesis, XLVI, 324
2′-Azido-2′-deoxy-3′,5′-diacetyluridine, XLVI, 324
2′-Azido-2′-deoxyuridine, synthesis, XLVI, 323, 324
5′-Azido-5′-deoxyuridine, inhibitor of uridine-cytidine kinase, LI, 313
2-Azido-3,5-dinitrobiphenyl, XLVI, 80
3-Azido-4,6-dinitrophenyl cytochrome c, XLVI, 86
4-Azido-3,5-dinitrophenyl cytochrome c, LVI, 656
Azidoestradiol, XLVI, 84
Azidoestrone, XLVI, 84
2-Azidofluorene, synthesis, XLVI, 646
8-Azidoguanosine 5′-monophosphate, synthesis, LVI, 649, 650
8-Azidoguanosine 5′-triphosphate, synthesis, LVI, 651, 652
3-Azidohexestrol, XLVI, 58, 84
2-Azidoinosine 5′-triphosphate, LVI, 656
1-Azido-4-iodobenzene, LVI, 660
Azidoisophthalic acid, XLVI, 87
Azidomorphine, LVI, 660
β-Azido-NAD^+, see β-Azidonicotinamide adenine dinucleotide
1-Azidonaphthalene, XLVI, 86; LVI, 660
β-Azidonicotinamide adenine dinucleotide, XLVI, 83
Azidonitrobenzoic acid hydrazide, preparation, LIX, 179
5-Azido-2-nitrobenzoic acid N-hydroxysuccinimide ester, modification of tRNA, LIX, 171
N-(5-Azido-2-nitrobenzoyl)pentagastrin, XLVI, 85
N-(4-Azido-2-nitrobenzoyl)-Phe-tRNA, XLVI, 84
(4-Azido-2-nitrobenzyl)triethylammonium, XLVI, 85
4-Azido-2-nitrobenzyltriethylammonium fluoroborate, LVI, 660
acetylcholine receptor, LVI, 672
4-Azido-2-nitrobenzyltrimethylammonium, photolysis, LVI, 658
4-Azido-2-nitrobenzyltrimethylammonium fluoroborate, XLVI, 580, 583
4-Azido-2-nitrofluorobenzene, XLVI, 98, 101, 105
2-Azido-4-nitrophenol, XLVI, 83; LVI, 660
absorption spectra, LVI, 664
binding by mitochondria, LVI, 666–668, 673–683
inhibition, LVI, 669, 670
complex V, LV, 315
triplet nitrene, LVI, 658
tritiated, synthesis, LVI, 663
N-(4-Azido-2-nitrophenoxy-4′-phenylacetyl)-Phe-$tRNA^{Phe}$, XLVI, 85
N-(4-Azido-2-nitrophenyl)-α-alanine, XLVI, 268
^{14}C-labeled, XLVI, 269

N-(4-Azido-2-nitrophenyl)-β-alanine, XLVI, 267, 268, 270, 279
N-(4-Azido-2-nitrophenyl)-ω-amino alkyl ester, of ATP, XLVI, 86
N-(4-Azido-2-nitrophenyl)-4-aminobutyric acid, XLVI, 268
N-(4-Azido-2-nitrophenyl)-4-aminobutyrylatractyloside, LVI, 660
4-Azido-2-nitrophenylaminobutyrylatractyloside
  preparation, LV, 526, 527
  structure, LV, 526
N-(4-Azido-2-nitrophenyl)-6-aminocaproic acid, XLVI, 268
N-(4-Azido-2-nitrophenyl)aminocarboxylic acid, XLVI, 264, 265, 267, 268
N-(4-Azido-2-nitrophenyl)-12-aminododecanoic acid, XLVI, 268
N-(4-Azido-2-nitrophenyl)aminoethyl-1-thio-β-D-galactopyranoside, XLVI, 86
N-(4-Azido-2-nitrophenyl)-*p*-aminophenyl-β-lactoside, XLVI, 520
3[*N*-(4-Azido-2-nitrophenyl)]aminopropionic acid, XLVI, 270
3-(4-Azido-2-nitrophenylamino)propionyladenosine 5′-triphosphate, LVI, 656
3′-*O*-(3-[*N*-(4-Azido-2-nitrophenyl)amino]-propionyl)adenosine triphosphate, XLVI, 265
  hydrolysis, XLVI, 271, 272
  inhibition of ATPase, XLVI, 276, 277
  irradiation, XLVI, 272, 273
  NMR data, XLVI, 271
  radioactive preparations, XLVI, 270, 271
  reactions, XLVI, 273–279
  synthesis, XLVI, 268, 269
3′-*O*-(3-[*N*-(4-Azido-2-nitrophenyl)amino]-propionyl)nicotinamide adenine dinucleotide, LVI, 656
  characterization, XLVI, 280
  structure, XLVI, 280–282
  synthesis, XLVI, 279, 280
N-(4-Azido-2-nitrophenyl)-5-aminovaleric acid, XLVI, 268
4-Azido-2-nitrophenyl antibody, XLVI, 580
4-Azido-2-nitrophenyl conalbumin, XLVI, 86
4-Azido-2-nitrophenyl cytochrome *c*, LVI, 656
N-(4-Azido-2-nitrophenyl)erythromycylamine, XLVI, 85
N-(4-Azido-2-nitrophenyl)glycine, XLVI, 268, 696
  N-hydroxysuccinimide ester, XLVI, 686, 691, 692
N-(4-Azido-2-nitrophenyl)glycyl-Phe-$tRNA^{Phe}$, XLVI, 85
N-(4-Azido-2-nitrophenyl)insulin, XLVI, 85
ε-(4-Azido-2-nitrophenyl)-L-lysine, XLVI, 580
N-ε-(4-Azido-2-nitrophenyl)lysine, XLVI, 86
N-ε-(5-Azido-2-nitrophenyl)lysine, XLVI, 86
4-Azido-2-nitrophenyl-α-D-mannopyranoside, synthesis, XLVI, 367, 368
N-(4-Azido-2-nitrophenyl)taurine, XLVI, 86, 87
4-Azido-2-nitrophenyl-1-thio-β-D-galactopyranoside, XLVI, 86
N-(4-Azido-2-nitrophenyl)-11-undecanoic acid, XLVI, 268
Azido nucleotide phosphate, XLVI, 99
12-Azidooleic acid, extinction coefficient, LVI, 664
p-Azidophenacyl bromide, XLVI, 684, 685
  coupling to tRNA, XLVI, 692–696
  modification of cytochrome *c*, LIII, 175
  reaction conditions, XLVII, 420
  synthesis, XLVI, 686–688
p-Azidophenacyl bromoacetic acid, XLVI, 690, 691
  synthesis, XLVI, 688, 689
p-Azidophenacyl iodoacetic acid, XLVI, 688–691
4-(4′-Azidophenacylmercapto)butyrimidyl cytochrome *c*, LVI, 656
p-Azidophenacylpseudouridine, XLVI, 694
S-(*p*-Azidophenacyl)-4-thiouridine, XLVI, 694
4-Azidophenacyl-Val-$tRNA_1^{Val}$, XLVI, 85
4-Azidophenol, XLVI, 651, 652
Azidophenylalanine, XLVI, 84
4-Azidophenylalanine, LVI, 660
β-(4-Azidophenyl ester), of GDP, XLVI, 85

5-(3-Azidophenyl)-5-ethylbarbituric acid, LVI, 660
N-(2-*p*-Azidophenylethyl)norlevorphanol, XLVI, 604
  synthesis, XLVI, 605
N-(4-Azidophenylethyl)norlevorphanol, XLVI, 85
p-(4-Azidophenyl)guanosine 5′-diphosphate, LVI, 656
4-Azidophenyl guanosine diphosphate, XLVI, 649–656
  photolysis, XLVI, 653–656
  properties, XLVI, 649, 650
  synthesis, XLVI, 650–653
1-(4-Azidophenyl)-2-(5′-guanyl)pyrophosphate, XLVI, 633
1-(4-Azidophenyl)imidazole, LVI, 660
4-Azidophenyl-α-D-mannopyranoside, XLVI, 85
4-Azidophenylphosphoric acid, XLVI, 652
Azidopurine, XLVI, 79
p-Azidopuromycin, *see* 6-Dimethylamino-9-[3′-deoxy-3′(*p*-azido-L-phenylalanylamino)-β-D-ribofuranosyl]purine
3-Azidopyridine adenine dinucleotide, LVI, 656
Azidopyrimidine, XLVI, 79
Azidoquinone, XLVI, 99
4-Azido-β-D-xylopyranoside, XLVI, 86
Azimuthal angle, LIV, 196
Aziridinum ion, XLVI, 578
p-Azobenzene arsonate, modification of carboxypeptidase A, XXXIV, 184
p-Azobenzene arsonate antibody, XLVI, 492–501
p-Azobenzoate hapten, XLVI, 62
2,2′-Azobisisobutyronitrile, XLIV, 113
Azocasein, substrate, bdellin assay, XLV, 798
Azo coupling, to inorganic supports, XLIV, 143, 144
Azo derivative, XXXIV, 49, 104, 105
Azodianiline
  in affinity chromatography, XXXVI, 110, 111
  structure, XXXVI, 111
Azo dye, as optical probe, LV, 573
Azofer, *see* Nitrogenase
Azofermo, *see* Nitrogenase
Azoferredoxin, *see also* Nitrogenase
  purification, LIII, 317, 318, 320–322, 324, 326, 328
  specific activity, LIII, 319
Azohistidyl linkage, XXXIV, 658
Azo linkage, XXXIV, 20, 48, 49, 53–55, 71, 102–108
  ω-aminoalkylagarose, XXXIV, 106, 107
  p-aminobenzamidoalkylagarose, XXXIV, 107, 108
  glass, XXXIV, 71, 72
  spacer, XXXIV, 103, 104
p-Azophenyl 2-acetamido-2-deoxy-β-D-glucopyranoside-bovine serum albumin, *Bandeiraea simplicifolia* lectin II activity, L, 350, 351
Azophenyl-β-lactoside, XXXIV, 719
p-Azophenyl-β-lactoside, XLVI, 516
Azotobacter
  hydrogenase reaction, LIII, 287
  oxidase, LII, 19
  transhydrogenase, LV, 272
Azotobacter agilis, cell-wall isolation, XXXI, 655
Azotobacter chroococcum
  crude extract preparation, LIII, 327
  growth, LIII, 326
  nitrogenase, LIII, 319, 326–328
Azotobacter vinelandii
  adenosine monophosphate nucleosidase, LI, 263, 265–267, 270
  crude extract preparation, LIII, 325
  cytochrome c_4, CD spectrum, LIV, 277
  growth, LI, 264, 265; LIII, 324
  iron-sulfur protein I, nomenclature, LIII, 265
  membrane isolation, XXXI, 701
  nitrogenase, LIII, 319, 324–326
    Mössbauer studies, LIV, 371
  transport systems, XXXI, 704
  ubiquinone, LIII, 578
Azotyrosyl linkage, XXXIV, 658
Azoxyketone, XLIII, 352
Azurin, *see also* Cytochrome c_{551}-azurin
  activity, LIII, 657
  assay of *Pseudomonas* cytochrome oxidase, LIII, 648

circularly polarized luminescence, XLIX, 186
crystallization, LIII, 659
electron-transfer reactions, LIII, 661
extinction coefficient, LIII, 648
molecular weight, LIII, 660, 661
from *Pseudomonas aeruginosa*, LIII, 646–648
from *Pseudomonas fluorescens*, thermolysin hydrolysis products, XLVII, 176
purification, LIII, 657–659
purity, LIII, 659
reaction with *Pseudomonas* cytochrome oxidase, LIII, 656
redox potential, LIII, 661
spectral properties, LIII, 659, 660
stability, LIII, 659
Azurophile granule, isolation, XXXI, 345–353

B

B_{12}, *see* Vitamin B_{12}
BAA, *see* Benzoyl-L-arginine amide
BA–6903 antibiotic, solvent system, XLIII, 334
BA–90912 antibiotic, solvent system, XLIII, 334
BA–181314 antibiotic, solvent system, XLIII, 334
BACA, *see* 6-Bromoacetamidocaproic acid
Bacillus
cell-wall isolation, XXXI, 661
oxidase, LII, 14
Bacillus brevis
edeines, XLIII, 560, 564
gramicidin S, XLIII, 567
tyrocidine synthetase, XLIII, 587, 588
Bacillus brevis 9999, 5-keto-D-fructose reductase, XLI, 137
Bacillus Calmette-Guérin bacteria, for stimulation of alveolar macrophage production, LVII, 471
Bacillus cereus
cell-wall isolation, XXXI, 655, 656
β-lactamase, XLIII, 98, 640–652, *see also* β-Lactamase, from *Bacillus cereus*
methyl group source, LIX, 202
phospholipase C, XXXII, 131, 139, 161
purification, XXXII, 154–161
ribonuclease, in tRNA sequence analysis, LIX, 106, 107
Bacillus circulans
aminoglycoside antibiotics, XLIII, 265
cell-wall hydrolyzing enzyme, XXXI, 610
circulin, XLIII, 579
peptidoglutaminase
activators, XLV, 492
assay, XLV, 485
inhibitors, XLV, 492
kinetics, XLV, 489
pH, XLV, 492
properties, XLV, 489
purification, XLV, 486
purity, XLV, 489
specificity, XLV, 489, 491
temperature, XLV, 492
units, XLV, 486
Bacillus coagulans, XLIV, 811, 815
Bacillus licheniformis
bacitracin, XLIII, 548
growth, LIX, 438
β-lactamase, XLIII, 93, 94, 653–664, *see also* β-Lactamase, from *Bacillus licheniformis*
pyruvate kinase, assay, purification, and properties, XLII, 157–166
ribosomes
preparation, LIX, 438–441
reconstitution, LIX, 442, 443
transport systems, XXXI, 705
Bacillus megaterium
cell-wall isolation, XXXI, 654, 663
5-keto-D-fructose reductase, XLI, 137
oxidase, LII, 14
penicillin acylase, XLIII, 711–721, *see also* Penicillin acylase, from *Bacillus megaterium*
steroid hydroxylase system, LII, 378
transport systems, XXXI, 704
Bacillus mycoides, hydrolase, XLIII, 735
Bacillus polymyxa
crude extract preparation, LIII, 315
ferredoxin, LIII, 263, 273, 274, 628
proton magnetic resonance studies, LIV, 204, 208

growth, LIII, 315
nitrogenase, LIII, 315–319
polymyxin, XLIII, 17, 18, 579–581
Bacillus stearothermophilus
ATPase, LV, 299
ferredoxin, LIII, 629
Mössbauer studies, LIV, 371
glucose-6-phosphate isomerase, assay, purification, and properties, XLI, 383–387
glycerol-3-phosphate dehydrogenase, assay, purification, and properties, XLI, 268–273
growth, LI, 303
pyrimidine nucleoside phosphorylase, LI, 437
uridine-cytidine kinase, LI, 299
purification procedure, LI, 303–305
Bacillus subtilis
N-acetylglucosamine-6-phosphate deacetylase, XLI, 497
adenylosuccinate synthetase, LI, 212
aminoglycosidic antibiotic, XLIII, 120
α-amylase, XLIV, 98
cephalosporin, XLIII, 117
cephalosporin acetylesterase, XLIII, 731–734, *see also* Cephalosporin acetylesterase, from *Bacillus subtilis*
cytochrome a_3, LIII, 207
deoxythymidylate phosphohydrolase, LI, 285–290
efrapeptin, LV, 495
enzyme A, XLIII, 735
everninomicin group, XLIII, 128
growth, LIX, 438
phage infection, LI, 286, 287
5-keto-D-fructose reductase, XLI, 137
β-lactamase, XLIII, 247
L-lactate dehydrogenase, assay, purification, and properties, XLI, 304–309
lincomycin group, XLIII, 128
lysis, LIX, 439
media, XLIII, 16, 17
methyl group source, LIX, 202
oxidase, LII, 15
penicillin, XLIII, 111
ribosomes
preparation, LIX, 439–441
reconstitution, LIX, 441, 442
S30 extract, preparation, LIX, 439
tetracycline, XLIII, 140
transport systems, XXXI, 704, 709
Bacillus subtilis W23, membrane studies, XXXII, 890
Bacillus subtilis W168
crude extract preparation, LIX, 374
growth, LIX, 373, 374
lysis, LIX, 374
Bacillus thermoproteolyticus, XLIV, 89
Bacitracin
biosynthesis, XLIII, 548–559, *see also* Bacitracin synthetase
countercurrent distribution, XLIII, 326, 330
solvent system, XLIII, 334
for structural studies, XLIII, 330
effects on membranes, XXXII, 882
high-pressure liquid chromatography, XLIII, 314, 319
hollow-fiber dialytic permeability, XLIV, 297
hollow-fiber retention data, XLIV, 296
in media, LVIII, 565
paper chromatography, XLIII, 146
structure, XLIII, 548
Bacitracin synthetase, XLIII, 548–559
ammonium sulfate precipitation, XLIII, 558
assay, XLIII, 551–556
ATP-$^{32}PP_i$ exchange measurement, XLIII, 551, 553, 555, 558, 559
Bacillus licheniformis preparation, XLIII, 198
cell, harvesting and storage, XLIII, 556
cell lysis, XLIII, 556
DEAE-cellulose chromatography, XLIII, 557
diafiltration and concentration, XLIII, 557
disc gel electrophoresis, XLIII, 554, 558
fermentation, XLIII, 554–556
Micrococcus flavus preparation, XLIII, 551
Millipore filter test, XLIII, 552, 553, 558
molecular weight, XLIII, 555

protein determination, XLIII, 554
purification, XLIII, 556–558
radio thin-layer chromatography, XLIII, 552–554
reaction scheme, XLIII, 549
SDS disc gel electrophoresis, XLIII, 554
Sephadex G–50 chromatography, XLIII, 556
Sephadex G–200 chromatography, XLIII, 557
sucrose density gradient centrifugation, XLIII, 554
unit definition, XLIII, 549
Back-scattering amplitude function, LIV, 340, 341, 342
Back-titration, LIX, 284, 285, 288, 289
Bacteria, *see also* specific type
amino acid transport, assay methods, XXXII, 843–849
assay of iron transport
construction and use of filtration apparatus, LVI, 392–394
materials and methods, LVI, 389–391
principle, LVI, 388, 389
procedure, LVI, 391, 392
ATPase, isolation by chloroform method, LV, 341
cell envelopes, enzyme complex reconstitution, XXXII, 449–459
countercurrent distribution, XXXI, 761, 768
cytochromes, LIII, 202–212
energy-conserving systems, aureovertins, LV, 483, 485
ferredoxins, LIII, 262, 263
gram-negative, cytoplasmic membrane isolation, XXXI, 642–653
isolation of membrane vesicles, LVI, 379–383
luciferase, LIII, 560
luminous
bioassay, LVII, 127, 130
cell death, effect on intracellular ATP concentration, LVII, 66–68
cell extract, adenylate energy charge, determination, LVII, 73–85
effect of arginine on, LVII, 166
of salt on, LVII, 166
fermentation products, LVII, 165
growth media, LVII, 141, 154–156
identification, LVII, 160–163
by carbon source use, LVII, 160–162
on solid media, LVII, 156
by testing for extracellular enzymes, LVII, 161, 162
using luciferase kinetics, LVII, 161–163
inhibitors, in complete medium, LVII, 164
isolation, LVII, 158–160
mutagenesis, LVII, 167
mutants, LVII, 131, 166–171, 189–194
aldehyde, LVII, 131, 168
in bioassays, LVII, 130, 131, 189–194
classes, LVII, 130, 166
conditional, LVII, 131
dark, LVII, 130, 131
isolation technique, LVII, 168
in luciferase structure, LVII, 130, 168–171
minimal bright, LVII, 168
resistant to arginine stimulation, LVII, 129
to catabolite repression, LVII, 128
UY-437, LVII, 128
numbers, in fluids, determination, LVII, 65–72
as oxygen indicator, LVII, 223–226
storage media, LVII, 155, 156
taxonomy, LVII, 153
membrane
ATPase, XXXII, 428–439
nectin, XXXII, 428–439
mutants, for membrane study, XXXII, 883
negative staining, XXXII, 33
oxidases, LII, 9–21
as plant fractionation contaminant, XXXI, 549, 553
preparation of mitochondria, LVI, 18
preservation, for transmission electron microscopy, XXXIX, 155

rubredoxins, LIII, 262
in sucrose gradients, XXXI, 720
thermophilic
crystallization of ATPase (TF_1), LV, 372–377
DCCD-sensitive ATPase, LV, 364–372
isolation of alanine carrier, LVI, 430–435
phospholipid molecular species, LV, 371
transhydrogenase assay, LV, 263, 266, 269, 270
transport in membrane vesicles, XXXI, 698–709
transport mutants, XXXII, 849–856
Bacterial contamination, *see* Contamination, bacterial
Bacterial luminescence, *see* Luminescence, bacterial
Bacteriochlorophyll, LIV, 23
ATP ratio to, during photophosphorylation, LVII, 52
bleaching kinetics, LIV, 25–32
reaction center complex, proteoliposomes, LV, 760, 762, 764, 765, 768, 770, 771
Bacteriophage, *see also* specific type
DNA, preparation, XXXVIII, 374
E. coli
growth, LX, 315, 631, 632
source of labeled messenger ribonucleic acid, LX, 316
of Qβ ribonucleic acid replicase, LX, 632
Bacteriophage φX174
DNA-ribosome binding site studies, LIX, 849, 850
fMet-dipeptides from cell-free system, LIX, 850
Bacteriophage MS2 peptide
electrophoretic separation, LIX, 301–309
growth and harvest, LIX, 298, 299
RNA extraction, LIX, 298–300
strain source, LIX, 297
Bacteriophage MS2-specific protein, in codon-anticodon interaction M studies, LIX, 296–309
Bacteriophage P1
in *Escherichia coli*, XLIII, 45–47
mutation MA1 mapping, LVI, 181, 182
Bacteriophage P22, in *Salmonella typhimurium*, XLIII, 47
Bacteriophage PBS2, enzyme induction in *Bacillus subtilis*, LI, 285
Bacteriophage R17, quasi-elastic laser light scattering, XLVIII, 493
Bacteriophage R17 RNA
in antibiotic action studies, LIX, 853, 856
assay of initiation factor activity, LIX, 367, 368
preparation, as messenger RNA, LIX, 364
in studies of protein-RNA interactions, LIX, 581
in vitro preparation of polysomes, LIX, 368
Bacteriophage T2, deoxycytidylate deaminase induction, LI, 412
Bacteriophage T4
gene 32 protein, LIX, 840–842, 847
head-tail attachment, kinetics, XLVIII, 445–447
Bacteriorhodopsin, LIV, 57
envelope vesicles, LVI, 402
free peptide hydrogen exchange, XLIX, 30
light-dependent potential changes, LV, 774, 775
liposomes containing, assay of photopotentials, LV, 608–611
membrane potential reconstitution, LV, 587, 588, 592, 601
photopotentials, LV, 420
planar membranes containing
device, LV, 752, 753
membrane-forming mixture, LV, 752
problem of asymmetry, LV, 753–755
in protein rotation studies, LIV, 58–60
proteoliposomes, LV, 758–760, 762, 764, 765, 768–771
sheet, association with planar black membranes, LV, 772
transport, LVI, 399

vesicle, reconstitution, LV, 704, 707, 709, 710
detergents, LV, 702, 704
Bacterioruberin, in *Halobacterium* red membrane, XXXI, 677
Bacteroid, isolation, XXXI, 518
Bacteroides symbiosis
D-fructose-1-phosphate kinase, XLII, 63
pyruvate orthophosphate dikinase, assay, purification, and properties, XLII, 187–197, 199–209
BADE, *see* 1-Bromoacetyl-1′-dinitrophenylethylenediamine
BADL, *see* N^{α}-Bromoacetyl-N^{ε}-dinitrophenyl-L-lysine
BAEE, *see* N-α-Benzoyl-L-arginine ethyl ester; N^{α}-Benzoyl-L-arginine ethyl ester
Bakers' yeast, *see Saccharomyces cerevisiae;* Yeast
Balanced salt solution, *see* Salt solution, balanced
BALB/c cell, *see* BALB/c 3T3 cell line
BALB/c 3T3 cell line, LVIII, 96
effect of hormone, LVIII, 101, 102
properties and uses, XXXII, 584
BALB/c 3T12 cell line, properties and uses, XXXII, 584
Ball mill, for cell rupture, XXXI, 660
Ballou mixer, LIV, 90
Balzers' freeze-etch machine, XXXII, 37, 40, 41, 44–46
Banana, cell organelle studies, XXXI, 532, 541
Bandeiraea simplicifolia, *see also* Isolectin
isolectins, L, 345–350
lectin, XXXIV, 344; L, 345–354
lectin II, L, 350–354
Band Emission Average technique, LVII, 585–587
Band frequency, in IR spectroscopy, LIV, 307–309
Band height, in IR spectroscopy, LIV, 312
Banding, *see also* specific types
chromosomal, LVIII, 324
techniques, problems, LVIII, 337
isopycnic viral, LVIII, 415
reverse, LVIII, 325
Band width
fourth derivative analysis of spectra, LVI, 506, 507
half, notation, LIV, 311
in IR spectroscopy, LIV, 311, 312
Bandwidth variation, of amplifier, in nanosecond absorbance spectroscopy, LIV, 43
BAPA, *see* Benzoyl-DL-arginine p-nitroanilide
Barbital buffer
chemiluminescence and, LVII, 438
diluent, for microperoxidase, LVII, 425
in heterogeneous competitive protein-binding assay, LVII, 442
Barbiturate, *see also* Barbituric acid
complex I, LVI, 586
inhibitor of reconstituted electron-transport systems, LIII, 50
Barbituric acid, *see also* Barbiturate
inhibitor of orotate phosphoribosyltransferase, LI, 166
Bard-Parker Tru-Touch surgical glove, XXXII, 94
Barium
atomic emission studies, LIV, 455
lasalocid, LV, 446
Barium bromide, XLVII, 488
Barium carbonate, cAMP isolation, XXXVIII, 40, 41
Barium chloride, test for sulfate ion, LIX, 195
Barium citrate, chromatography
factor X, bovine, XLV, 102
prothrombin purification, XLV, 128
Barium fluoride, infrared spectral properties, LIV, 316
Barium hydroxide
5-fluoro-2′-deoxyuridine 5′-(p-nitrophenylphosphate) synthesis, LI, 99
in formaldehyde assay, LII, 299
in purification of cyclic AMP, LVII, 98
Barium ion
antipyrylazo III, LVI, 326, 328
eriochrome blue, LVI, 315

inhibitor, of *Cypridinium* luminescence, LVII, 349–351
of deoxythymidine kinase, LI, 365
of guanine phosphoribosyltransferase, LI, 557
of hypoxanthine phosphoribosyltransferase, LI, 557
of phosphoribosylpyrophosphate synthetase, LI, 11
microsome aggregation, LII, 87
Barium sulfate
cAMP isolation, XXXVIII, 40, 41
in preparation of diffusing plate, LVII, 580, 598
Barium thiocyanate-acetone precipitation, in purification of NADH, LIV, 227
Barker biopsy needle, in testicular transplantation, XXXIX, 288
Barley
aleurone grain, XXXI, 576
protoplast isolation, XXXI, 580
Barley mesophyll × soybean culture, LVIII, 365
Barley root, ATPase isolation, XXXII, 395, 405
Barley seed, oxidase, LII, 19
Barnase, XXXIV, 4
Barrier, energy
multiple, data evaluation, LIV, 512–518
theoretical, LIV, 507, 510, 511
Barrier width, LIV, 528
Barr-Stroud BC6 polarizing beam splitter, LIV, 54
Barstar, XXXIV, 4
Base, metal-free, preparation, LIV, 476, 477
Base analog, as mutagen, XLIII, 35
Base-base interaction, in transfer RNA, LIX, 17, 27–30
Basement membrane, LVIII, 265
carbohydrate composition, XL, 353
collagen, LVIII, 266
components, in media, LVIII, 269
globular solubilization, XXXI, 788
in malignancy, LVIII, 266
protein, cyanylation, XLVII, 132
reconstituted rafts, LVIII, 263–278
Base pair, proton NMR studies, LIX, 22, 27–30
Base stacking, osmometric studies, XLVIII, 78
BASIC computer program, for prostaglandin assay analysis, XXXII, 110, 121–123
Basidiomycete
alcohol oxidase, assay, purification, and properties, XLI, 364–369
5-keto-D-fructose production, XLI, 84
oxidase, LII, 17, 21
Batch culture method, limitations, LVI, 571–573
Batch elution, XXXIV, 684
Batch growth, cost, LVIII, 206
Bathocuproine, in copper analysis, LIV, 444
Bathophenanthroline
in iron analysis, LIV, 439
solution preparation, LIV, 437
Bathophenanthroline-iron(II) chelate, LV, 517
ATPase, LV, 401
Bathophenanthrcline-nickel(II) chelate, LV, 517
Bathophenanthroline sulfonic acid, LV, 517
Bathorhodopsin, *see* Prelumirhodopsin
Batrachotoxin, neurotoxic action, XXXII, 310
Batroxobin
activators, XLV, 222
assay, XLV, 214
on benzoylphenylalanylvalylarginine *p*-nitroanilide, XLV, 216
on bovine fibrinogen, XLV, 215
on human plasma, XLV, 214
immunopherograms, qualitative determination, XLV, 217
inactivators, XLV, 222
molecular weight, XLV, 220
properties, XLV, 220
chemical, XLV, 221
immunological, XLV, 221
physical, XLV, 220
purification, XLV, 218
purity, XLV, 220
serum, quantitative determination, XLV, 215

specificity, XLV, 221
stability, XLV, 221
Batten's disease, *see* Neuronal ceroid lipofuscinosis
Bauhinia purpurea, agglutinin, *see* Lectin, from *Bauhinia purpurea*
Bauhinia variegata, lectin, XXXIV, 334
Bausch and Lomb Tri-simplex viewer, XXXIX, 201
BB–K8 antibiotic, *see also* Amilcacin
structure, XLIII, 612
BBOT, *see* 2,5-Bis(5-*tert*-butylbenzoxazol-2-yl)thiophene
BBS–3 Biosolvent, XLVIII, 336
BC, *see* Benzocoumarin
BCCP, *see* Biotin carboxyl carrier protein
B cell, XXXIV, 216, 716, 719; LVIII, 178, 493
antigen-specific, XXXIV, 219
B_{16} cell line, LVIII, 564
BD-cellulose, *see* (Diethylaminoethyl)cellulose, benzoylated
Bdellin, leech
assay, XLV, 798
characteristics, XLV, 797
composition, XLV, 804
homology, XLV, 805
inhibition specificity, XLV, 803
molecular weight, XLV, 804
properties, XLV, 803
purification, XLV, 799
reactive sites, XLV, 804
separation, XLV, 800
from hirudin, XLV, 799
structure, XLV, 805
units, XLV, 798
Bead
for column sieving fractionation, XXXI, 549–551
polymerization
advantages, XLIV, 64, 65, 68
column procedure, XLIV, 191–195
procedure, XLIV, 57–60, 176, 177, 662, 663
sizing, XLIV, 195
Beam profile, XLVIII, 457, 458
Beam splitter, LIV, 5, 10, *see also* specific type
in flash photolysis apparatus, LIV, 54
Bean
agglutinin isolation, XXXII, 615
aleurone grain, XXXI, 576
chloroplast isolation, XXXI, 602
enzyme studies, XXXI, 522, 523
organelles, XXXI, 491, 493, 533, 542
BEA technique, *see* Band Emission Average technique
Beaufay's rotor, isopycnic equilibration, XXXI, 349–351, 357, 367
Beauvericin
as ionophore, LV, 442, 444
structure, LV, 436
Beckman AA–15 resin, XLVII, 231–233
particle size, XLVII, 227
Beckman AA–20 resin, particle size, XLVII, 227
Beckman Acta C III recording spectrophotometer, LII, 523
Beckman automatic peptide synthesizer, XXXVII, 417, 418
Beckman dynograph recorder, XXXIX, 56
Beckman Fraction Recovery System, XXXVI, 162
Beckman M71 cation-exchange resin, for separation of 3′ end nucleoside, LIX, 188
Beckman microfuge test tube, XXXII, 111
Beckman Model 141 gradient pump, LIX, 451
Beckman Readi-solve VI, LVII, 556
Beckman-Spinco centrifuge, liver homogenization in, XXXI, 196, 197
Beckman W–1 resin, XLVII, 231
particle size, XLVII, 227
Bee, phospholipase A_2 in venom, XXXII, 147, 148
Beef
heart
acyl-CoA dehydrogenases, LIII, 506, 507, 510–512, 518
electron-transferring flavoprotein, LIII, 506, 507, 510–512, 518
mitochondria
ATPase inhibitor, LV, 402, 404, 405

aureovertin, LV, 482
DCCD-binding protein, LV, 428, 429
improved method for ATPase isolation, LV, 317–319
isolation, LV, 385
preparation of ATPase by chloroform method, LV, 338–341, 342
of ATPase, LV, 304–308
in high yield, LV, 46–50
of rutamycin-sensitive ATPase, LV, 315–317
purification of transhydrogenase, LV, 276, 283
reconstitution
of ATP synthetase, LV, 711–715
of pyridine nucleotide transhydrogenase, LV, 811–816
submitochondrial particles, aureovertin, LV, 482
liver
L-α-glycerophosphate dehydrogenase, assay, purification, and properties, XLI, 259–264
mitochondria preparation, LIII, 498
monoamine oxidase, LIII, 495–501
pancreas, crude extract preparation, LIX, 240
Beeman slit camera, XXXII, 212
Beer–Lambert law, XL, 217
Beer's law, XXXI, 818; LVI, 463
Beetle, flight muscles, mitochondria, LV, 22–24
Bellow, stainless-steel, flexible, LIV, 119
Beneckea, characteristics, LVII, 153, 157
Beneckea harveyi
aldehyde mutant, LVII, 189
growth, LVII, 190
alternative names, LVII, 157
base ratio, LVII, 161
catabolite repression, LVII, 164
culture, LIII, 566, 567
diagnostic taxonomic characters, LVII, 161
growth and harvesting, LVII, 141–142, 202, 203
growth range, temperature, LVII, 157
isolation, LVII, 158
luciferase, LIII, 560, 566–570; LVII, 132, 133, 135–152, 172, 179, 180, 197, 200, 201
assay, LVII, 138, 140
specific activity, LVII, 150
luminescence
characteristics, LVII, 125, 128, 129
mutants, LVII, 130, 131, 166–171
metabolic studies, LVII, 546, 547
natural habitats, LVII, 157, 160
osmotic lysis, LVII, 145
oxidase, LII, 17
storage temperature, LVII, 156, 157
thermogram, LVII, 546
Beneckea natriegens, cytochrome *c*, LIII, 207
Beneckea splendida, LVII, 157
Bentonite, XXXIV, 219
in plant-cell fractionation, XXXI, 504, 505
preparation of total ribosomal RNA, LIX, 452
separation of ribosomal subunits, LIX, 451
surface area, XLIV, 155
Bentonite-SF, preparation of ribosomal RNA, LIX, 446
Benz[*a*]anthracene 5,6-oxide, substrate of epoxide hydrase, LII, 199, 200
Benzaldehyde, XLVI, 127, 210
molar extinction value, LIII, 496
reductive alkylation, XLVII, 472, 474
Benzalkonium chloride, preservative, XLIV, 791
Benzamidine, XXXIV, 3, 5, 733; XLIV, 212, 529, 532; XLVI, 117
coupling, XLVII, 44
in hamster cell harvesting, LI, 124
inhibitor
of clostripain, XLVII, 168
of trypsin, XLIV, 428, 476; XLIX, 447–449, 456, 457
storage buffer of ribosomal protein S2, LIX, 498
m-Benzamidine-agarose, XXXIV, 293, 739

Benzamidine-agarose reagent, factor XI, XLV, 66
Benzamidine hydrochloride, preparation of ribosomal proteins, LIX, 484, 490
m-Benzamidine-polyacrylamide, XXXIV, 735
Benzamidine sulfonyl fluoride, XLVI, 115, 116
  synthesis, XLVI, 117–119
Benzene, XLIV, 131, 141; XLVII, 346; LIV, 19, 21, 23
  benzo[*a*]pyrene recrystallization, LII, 237
  for cell path length determination, LIV, 318
  chromatographic separation of fatty acids, LII, 321
  cytochrome P–450_{scc} assay, LII, 125
  disadvantage, XLVII, 338
  effect on Spheron swelling, XLIV, 69
  hemin derivative crystallization, LII, 429
  α-hydroxylase assay, LII, 311
  nuclear magnetic resonance standard, XLIX, 260
  preparation of 2-cyano-6-methoxybenzothiazole, LVII, 20, 21
  purification of *Cypridina* luciferin, LVII, 366
  pyridine purification, LII, 423
  solvent
    for dimethyldichlorosilane, LIX, 4
    in osmometric studies, XLVIII, 78, 79, 118, 120
  synthesis of chemiluminescent aminophthalhydrazides, LVII, 436
    of adenosine derivative, LVII, 118
    of *Diplocardia* luciferin, LVII, 377
    of ethyl-4-chloro-2-diazoacetoacetate, LIX, 811
Benzene-acetone, for chromatogram development, LII, 126
Benzene dihydroxylase, LII, 12
Benzene dioxygenase, LII, 11
Benzene-ethanol, for chromatographic separation of steroids, LII, 379
Benzene ether
  for cerebronic acid extraction, LII, 314
  for chromatographic separation of fatty acids, LII, 321
Benzene-ethyl acetate-methanol-butanol, for chromatographic analysis of sirohydrochlorin esters, LII, 445
Benzene–methanol
  purification of dimethyl 7-[*N*-ethyl 4-(*N*-phthalimido)butyl]aminonaphthalene-1,2-dicarboxylate, LVII, 437
    of dimethyl 7-*N*-(4-phthalimidobutyl)aminonaphthalene-1,2-dicarboxylate, LVII, 436
Benzene-pyridine-water, in hemin reduction, LII, 433
Benzenesulfonanilide hydroxylase, *Ah* locus, LII, 232
Benzenesulfonic acid, XLVII, 524
Benzenethiol, identification of iron-sulfur centers, LIII, 269, 271
Benzhydrylamine resin, in TRF synthesis, XXXVII, 412, 413
Benzidine, XLIV, 120
  staining, LVIII, 510
Benzidine dihydrochloride, for identification of cytochrome P–450 in gels, LII, 325, 327
  sensitivity, LII, 331
Benzidine-water-pyridine, for identification of peroxidatic activity, LII, 442
Benzil
  calibration standard, XLVIII, 89, 118
  solute, in vapor pressure osmometer, XLVIII, 153
Benzimidazole, as uncoupler, LV, 468
N-[*p*-(2-Benzimidazolyl)phenyl]maleimide, XLVII, 408
Benzoate dihydroxylase, LII, 12
Benzoate 4-hydroxylase, LII, 12
Benzocoumarin
  chemiluminescent reaction, LVII, 520–523
  structural formula, LVII, 520
5,6-Benzoflavone
  associated form of cytochrome b_5, LII, 109
  cytochrome binding, LII, 279
  inducer of cytochrome P–450_{LM_4}, LII, 203

injection, for induction of liver enzymes, LII, 114
Benzohydroxamic acid-peroxidase complex, apparent dissociation constants, LII, 516
Benzoic acid
care of enzyme-activated electrodes, LVI, 472, 473
inhibitor of phagocyte chemiluminescence, LVII, 492
pseudosubstrate of salicylate hydroxylase, LIII, 528, 537, 539
substrate of *p*-hydroxybenzoate hydroxylase, LIII, 548
Benzomorphan, XLVI, 603
Benzonitrile, solvent, in radical ion chemiluminescent reactions, LVII, 498
Benzopyrene, as mixed-function oxidase substrate, XXXI, 233
Benzo[*a*]pyrene
carcinogen, LII, 279, 291
chemiluminescent reaction, LVII, 505, 507
metabolism, LII, 66–69, 234, 283–287
recrystallization procedure, LII, 237
removal of contaminants, LII, 414
structural formula, LVII, 498
substrate of benzopyrene hydroxylase, LII, 409
of cytochromes *P*–450 and *P*–448, LII, 117, 122
Benzo[*a*]pyrene dihydrodiol, chromatographic properties, LII, 286, 294–296
Benzo[*a*]pyrene 4,5-dihydrodiol
chromatographic separation, LII, 285, 286
enantiomers, chromatographic separation, LII, 289–291
Benzo[*a*]pyrene 7,8-dihydrodiol
chromatographic separation, LII, 285, 286
enantiomers, absolute stereochemistry, LII, 293
Benzo[*a*]pyrene 9,10-dihydrodiol, chromatographic separation, LII, 285, 286
Benzo [*a*]pyrene 11,12-dihydrodiol, chromatographic separation, LII, 285, 286
Benzo[*a*]pyrene-7,8-diol, chemiluminescence, LVII, 568
Benzo[*a*]pyrene 7,8-diol 9,10-epoxide
absolute stereochemistry, LII, 293
chromatographic separation, LII, 287–289
metabolism, LII, 287–289
nucleoside adducts, chromatographic separation, LII, 291, 292
Benzo[*a*]pyrene epoxide hydrase, contaminant of cytochrome *P*-450LM preparations, LII, 116, 121
Benzo[*a*]pyrene hydroxylase, *see* Flavoprotein-linked monooxygenase
Benzo[*a*]pyrene monooxygenase, *see* Flavoprotein-linked monooxygenase
Benzo[*a*]pyrene 4,5-oxide
metabolite of benzo[*a*]pyrene, LII, 284
substrate of epoxide hydrase, LII, 199, 200, 284
Benzo[*a*]pyrene 7,8-oxide
metabolite of benzo[*a*]pyrene, LII, 284
substrate of epoxide hydrase, LII, 199, 200, 284
Benzo[*a*]pyrene 9,10-oxide
metabolite of benzo[*a*]pyrene, LII, 284
substrate of epoxide hydrase, LII, 199, 200, 284
Benzo[*a*]pyrene 11,12-oxide, substrate of epoxide hydrase, LII, 199, 200
Benzopyrene-4,5-oxide hydrase, specificity, alternate activity, LII, 416
Benzo[*a*]pyrene phenol, chromatographic separation, LII, 286, 287, 293
Benzo[*a*]pyrene 1,6-quinone, chromatographic separation, LII, 285, 286
Benzo[*a*]pyrene 3,6-quinone, chromatographic separation, LII, 285, 286
Benzo[*a*]pyrene 6,12-quinone, chromatographic separation, LII, 285, 286
Benzoquinone
model redox reactions, LV, 543
modification of lysine, LIII, 140
proline uptake by proteoliposomes, LV, 197, 199

support activation procedure, XLIV, 35

p-Benzoquinone
electron-transfer chemiluminescence, LVII, 497, 501
structural formula, LVII, 498

2-Benzothiazolesulfonyl chloride, XLVI, 540

Benzoxycarbonyl dipeptide, carboxypeptidase C activity, XLV, 562, 567

N-α-Benzoyl-L-arginine amide
partitioning effect, XLIV, 402, 403
substrate of papain, XLIV, 110, 111, 425

N-Benzoylarginine ethyl ester, substrate, of clostripain, XLVII, 165, 166

N-α-Benzoyl-L-arginine ethyl ester
substrate of bromelain, XLIV, 406–408
of ficin, XLIV, 338, 339
of papain, XLIV, 111
of trypsin, XLIV, 63, 346, 428, 476, 519, 561, 664

N^{α}-Benzoyl-DL-arginine ethyl ester, substrate
acrosin assay, XLV, 333
bdellin assay, XLV, 798
kallikrein assay, XLV, 300, 313
Leucostoma peptidase A assay, XLV, 398
proacrosin assay, XLV, 326
sea urchin egg protease assay, XLV, 346

DL-Benzoylarginine *p*-nitroanilide, substrate, of trypsin, XLIV, 312

N-Benzoyl-DL-arginine-*p*-nitroanilide, substrate, of trypsinlike enzyme, XLVII, 78

N-Benzoyl-L-arginine-*p*-nitroanilide, substrate, of trypsin, XLIV, 198, 428

N^{α}-Benzoyl-DL-arginine *p*-nitroanilide, substrate
acrosin assay, XLV, 335, 338
albumin gland inhibitor assay, XLV, 786
bdellin assay, XLV, 798
bronchial mucus assay, XLV, 870
sea anemone inhibitor assay, XLV, 882
seminal plasma inhibitors, XLV, 835
submandibular gland inhibitors, XLV, 860
trypsin inhibitor assay, XLV, 807, 826
trypsin-kallikrein inhibitor, XLV, 774, 793

4-Benzoylbenzoylpentagastrin, XLVI, 89

3-*O*-Benzoyl-D-ceramide, intermediate in sphingomyelin synthesis, XXXV, 502

Benzoyl chloride, XLVI, 124
purification of luciferin, LVII, 338

N-Benzoylcytdine, substrate of uridine-cytidine kinase, LI, 313

N^6-Benzoyl-cyclic adenosine monophosphate, preparation, XXXVIII, 406

(Benzoyl diethylaminoethyl)cellulose, XLIV, 46

Benzoylglycine ethyl ester, substrate of papain, XLIV, 430

N^{α}-Benzoyl-DL-lysine *p*-nitroanilide, substrate, acrosin, XLV, 336

(Benzoylnaphthoyl diethylaminoethyl)cellulose, XLIV, 46

Benzoyl peroxide, XLIV, 289

3-(4-Benzoylphenyl)propionic acid, synthesis, XLVI, 678, 679

3-(4-Benzoylphenyl)propionic acid *N*-hydroxysuccinimide ester, XLVI, 679

3-(4-Benzoylphenyl)propionyl-Phe-tRNA, XLVI, 676, 677

3-(4-Benzoylphenyl)propionyl-Phe-$tRNA^{Phe}$, XLVI, 89, 682
synthesis, XLVI, 679, 680

3-Benzoylpropionic acid, synthesis, XLVI, 678, 679

3-Benzoylpropionic acid *N*-hydroxysuccinimide ester, XLVI, 679

3-Benzoylpropionyl-Phe-tRNA, XLVI, 676, 677

3-Benzoylpropionyl-Phe-$tRNA^{Phe}$, XLVI, 680

3-*O*-Benzoylsphingosine, intermediate in sphingomyelin synthesis, XXXV, 500

3-*O*-Benzoyl-DL-sphingosine sulfate, intermediate in sphingomyelin synthesis, XXXV, 499, 500
Benzoyl-L-tyrosine ethyl ester, substrate of α-chymotrypsin, XLIV, 446, 523
  chymotrypsin inhibitor, potato, XLV, 729
  trypsin inhibitor, garden bean, XLV, 710
Benzphetamine
  metabolite binding to reduced cytochrome, LII, 273–275
  in studies of multiple forms of cytochrome *P*-450, LII, 275–279
  substrate of cytochrome *P*-450, LII, 117, 122, 205, 215
    of NADPH-cytochrome *P*-450 reductase, LII, 91, 96
    of reconstituted MEOS system, LII, 367
Benzphetamine *N*-demethylase
  assay, LII, 362
  in isolated MEOS fraction, LII, 362, 363
d-Benzphetamine *N*-demethylase, *Ah* locus, LII, 232
N^6-Benzyladenine, as cytokinin, LVIII, 478
Benzyl alcohol, XLVII, 424, 524, 528, 544
  identification of substituted flavins, LIII, 464
N-Benzyl-3-(*m*-amidinophenoxy)propylamine *p*-toluenesulfonic acid, XLVI, 124
Benzylamine, XLVI, 33
  assay of monoamine oxidase, LIII, 496; LIV, 492
  oxidation, LVI, 474
Benzylamine oxidase, LII, 10
8-Benzylamino-cyclic guanosine monophosphate, protein kinase, XXXVIII, 337–339
β-Benzyl-L-aspartic acid, esterification, XLVII, 528, 529
Benzyl bromide, XLVII, 543
Benzyl carbonium ion, XLVI, 217
Benzyl chloride, XLVII, 539, 541
N-Benzyl-3-(3-cyanophenoxy)propylamine, XLVI, 124
S-Benzyl-L-cysteine, synthesis, XLVII, 539
γ-Benzyl-L-glutamic acid, esterification, XLVII, 528, 529
2-*O*-Benzylglycerol, intermediate in lysoglycerophospholipid synthesis, XXXV, 490, 491
Benzylglyoxal, diethyl acetal, synthesis of luciferyl sulfate analogs, LVII, 251
Benzyl group, *see also* specific types
  removal, XLVII, 521, 546–548, 612–614
  stability, XLVII, 521
N^{im}-Benzyl group, stability, XLVII, 540
O-Benzyl group
  removal, XLVII, 542
  stability, XLVII, 542
S-Benzyl group, stability, XLVII, 537, 538
N^{im}-Benzyl-L-histidine, synthesis, XLVII, 541
Benzylhydroxamic acid, determination, XLVII, 138
Benzylidinemalononitrile, as uncoupler, LV, 468, 469
Benzyl luciferin
  assay of *in vitro* energy transfer, LVII, 260, 261
  methyl ether
    stability, LVII, 250, 251
    structural formula, LVII, 264
    synthesis, LVII, 251
  stability, LVII, 250, 251
  structural formula, LVII, 238, 246
  synthesis, LVII, 251
Benzyl luciferyl disulfate
  stability, LVII, 251
  structural formula, LVII, 246
  synthesis, LVII, 252
Benzyl luciferyl sulfate
  extinction coefficients, LVII, 252, 253
  stability, LVII, 251
  structural formula, LVII, 238, 246
  substrate of luciferin sulfokinase, LVII, 240
  synthesis, LVII, 252, 253
Benzylmalonate, transport, LVI, 292
N-Benzyl-6-methyl nicotinamide, XXXIV, 3

Benzyloxycarbonyl, as blocking group, XLVI, 199
Benzyloxycarbonyl-amino acid, source, XLVII, 382
Benzyloxycarbonyl-amino acid methylamine
conversion to PTMA-amino acid, XLVII, 379
synthesis, XLVII, 384, 385
N-(*N*-Benzyloxycarbonyl-ε-aminocaproyl-*N*-ε-aminocaproyl)-α-D-galactopyranosylamine, XXXIV, 348
[*N*-(*N*-Benzyloxycarbonyl-ε-aminocaproyl)-*p*-aminophenyl]trimethylammonium iodide, XXXIV, 576
N-(*N*-Benzyloxycarbonyl-ε-aminocaproyl)-*N*′,*N*′-dimethyl-*p*-phenylenediamine, XXXIV, 575
N-(*N*-Benzyloxycarbonyl-ε-aminocaproyl)-α-D-galactopyranosylamine, XXXIV, 348
4-(Benzyloxycarbonylamino)-2-nitrophenol, XLVI, 367
4-(Benzyloxycarbonylamino)-2-nitrophenyl-α-D-mannopyranoside tetraacetic acid, synthesis, XLVI, 367
N-Benzyloxycarbonyl-L-aspartic acid anhydride, synthesis, XLVII, 529
N-Benzyloxycarbonyl-L-aspartic acid α-*p*-nitrobenzyl ester, synthesis, XLVII, 529, 530
N-Benzyloxycarbonyl-β-*tert*-butyl-L-aspartic acid dicyclohexylamine salt, synthesis, XLVII, 530
N-Benzyloxycarbonyl-β-*tert*-butyl-L-aspartic acid α-*p*-nitrobenzyl ester, synthesis, XLVII, 530
N-Benzyloxycarbonyl-γ-*tert*-butyl-L-glutamic acid, synthesis, XLVII, 532
N-Benzyloxycarbonyl-γ-*tert*-butyl-L-glutamic acid α-methyl ester, synthesis, XLVII, 532
Benzyloxycarbonyl chloride, XLVII, 511, 535
Benzyloxycarbonyl dipeptide, substrate, of carboxypeptidases, XLVII, 75–77, 85, 86
N-Benzyloxycarbonyl-L-glutamic acid, XLVII, 531
N-Benzyloxycarbonyl-L-glutamic acid anhydride, synthesis, XLVII, 531
N-Benzyloxycarbonyl-L-glutamic acid α-methyl ester, dicyclohexylamine salt, synthesis, XLVII, 531
N-Benzyloxycarbonylglycine-*p*-nitrophenol ester, substrate, thermomyocolin assay, XLV, 416
N-Benzyloxycarbonylglycylleucylphenylalanine chloromethyl ketone, XLVI, 202
Benzyloxycarbonyl group
α-amino group blocking, XLVII, 510–512
misnomer, XLVII, 610
removal, XLVII, 511, 535, 546–548
stability, XLVII, 511
N^{ε}-Benzyloxycarbonyl-L-lysine, synthesis, XLVII, 535
Benzyloxycarbonyl-L-phenylalanyl-L-leucine, substrate, carboxypeptidase Y assay, XLV, 569
N^{α}-Benzyloxycarbonyl-N^{ω}-tosyl-L-arginine, synthesis, XLVII, 533, 534
Benzylpenicillin, XLIII, 476; XLIV, 241, *see also* Penicillin G
6-aminopenicillanic acid conversion, XLIII, 111
instrumentation for conversion, XLIV, 766, 767
K_m and V_{max} values, XLIII, 689
β-lactamase assay, XLIII, 71
in lactose reduction reactor, XLIV, 826
penicillin acylase assay, XLIII, 699
Streptomyces R39 and R61 enzymes, titration, XLV, 628, 629, 636
structural formula, XLIV, 760
substrate, penicillin acylase, XLIV, 760, 761
Benzylpenicillin-enzyme complex, XLIII, 688
Benzylpenicillin hydrolysis, XLIV, 195
Benzylpenicillin isocyanate, XLVI, 532, 534, 535, 537
O-Benzyl-L-serine, XLVII, 543
DL-1-*O*-Benzyl-2-stearoylglycerol-3-phosphorylcholine monobenzyl ester; intermediate in lysoglycerophospholipid synthesis, XXXV, 491
Benzyl succinate, XLVI, 21
α-Benzylsulfonyl-*p*-aminophenylalanine, XXXIV, 152

α-Benzylsulfonyl-*p*-nitrophenylalanine, XXXIV, 453
α-Benzylsulfonyl-*p*-nitrophenylalanine-agarose, XXXIV, 453
2-Benzylthiohypoxanthine, XXXIV, 525
S-(Benzylthiol)-*N*-acetyl-L-cysteine disulfide derivatives, XLVII, 430
O-Benzyl-L-threonine, synthesis, XLVII, 544
O-Benzyl-L-threonine benzyl ester hemioxalate, synthesis, XLVII, 544
Benzyl tricarboxylate
  carrier sites, LVI, 251
  cell fractionation, LVI, 217
  transport, LVI, 248, 292
Benzyl 2-tritylaminoethyl phosphate, intermediate in lysoglycerophospholipid synthesis, XXXV, 492, 493
N-Benzyl-type protector
  removal by catalytic hydrogenation, XLVII, 547, 548
    by hydrogen bromide in acetic acid, XLVII, 546, 547
O-Benzyl-L-tyrosine, snythesis, XLVII, 542
Benzyl viologen
  assay of formate dehydrogenase, LIII, 363
  as mediator-titrant, LIV, 409, 410, 423
  in redox titration, LIII, 491
  substrate of hydrogenase, LIII, 314
BEPTI, *see* Trypsin inhibitor, black-eyed pea
Beryllium, atomic emission studies, LIV, 455
Beryllium-9, lock nuclei, XLIX, 348
Beryllium ion, inhibitor of deoxythymidine kinase, LI, 365
Beta, *see also* Damköhler number
  definition, XLVIII, 123
  optical anisotropy, XLVIII, 440, 441
Beta function, protein shape, XLVIII, 155
Bethanechol, cyclic nucleotide levels, XXXVIII, 94
Bethune and Kegeles technique, XLVIII, 213, 237, 238
Better Built infrared drier, XXXII, 818
BF-5 freezing plug, XXXII, 795
BFP, *see* Blue-fluorescent protein
BGEE, *see* Benzoylglycine ethyl ester
BGJb medium, modified, composition, XXXIX, 402
BGM medium
  formulation, LVII, 154, 155
  species supported, LVII, 165
BHK21/C_{13} cell, plasma membrane isolation from, XXXI, 161
BHK cell line
  cell adhesion studies, XXXII, 606, 607, 609–611
  trypsinization, LVIII, 298
BHK 21 cell line, LVIII, 297, 377
  properties and uses, XXXII, 584, 587, 588
BHT, *see* Hydroxytoluene, butylated
Biacetyl, interceptor, in chemiluminescent reaction, LVII, 520
1,1′-Bianthracene-2,2′-dicarboxylic acid, circularly polarized luminescence, XLIX, 184, 185
Bicarbonate
  as buffer, LVIII, 199, 200
  F_1 ATPase, LV, 328
  inhibitor of succinate dehydrogenase, LIII, 33
  measurement of ΔpH, LV, 562
  in media, LVIII, 70
Bicarbonate assay
  for acetyl-CoA carboxylase, XXXV, 3, 11
  for acyl carrier protein, XXXV, 110
  of acyl carrier protein synthetase, XXXV, 96, 97
  for biotin carboxylase, XXXV, 26–28
  carbon dioxide exchange component reaction of fatty acid synthase, XXXV, 49
Bicarbonate buffer solution, XL, 327
Bicarbonate ion, substrate of carbamoylphosphate synthetase, LI, 21, 29, 33, 111
Bicarbonate reagent, for plasma membrane isolation, XXXI, 77
Bicine buffer, *see N,N*-Bis(2-hydroxyethyl)glycine buffer
Bifidobacterium bifidum, N-acetylglucosamine-6-phosphate deacetylase, XLI, 497

Bifunctional reagent, XLIV, 124–128, *see also* specific compounds
Biimidate, as cross-linking reagents, LVI, 622, 623
Bilayer, XXXII, 489–501
  component incorporation, XXXII, 500, 501
  lipid
    asymmetric, XXXII, 554
    planar, XXXII, 513–539
  osmotic measurements, XXXII, 500
  permeability measurements, XXXII, 496–500
  principle of formation, XXXII, 546
  tracer measurements, XXXII, 499, 500
Bile acid, lipophilic Sephadex chromatography, XXXV, 395
Bile acid sulfate, lipophilic Sephadex chromatography, XXXV, 393
Bile salt
  ionic strength, LVI, 746
  in membrane isolation, XXXI, 789
  osmometric studies, XLVIII, 79
  partial specific volumes, XLVIII, 21; LVI, 737
  properties and uses, LVI, 748, 749
Bilirubin, XXXIV, 6
  conjugation, LVIII, 184
  formation, LII, 368, 369
  as mixed-function oxidase substrate, XXXI, 233
Biliverdin, formation, LII, 45
Biliverdin IXα, product of heme oxygenase, LII, 367
Biliverdin reductase, in heme oxygenase assay, LII, 368–370
Bimolecular reaction, time range, LIV, 1, 2
Binding, *see also* Association reaction; Cooperativity; Dimerization; Dissociation; Interaction; Ligand binding; Self-association; specific types
  absorbance indicators, LVI, 305, 306
  cooperative, for nonassociating macromolecules, XLVIII, 302, 305
  data, interpretation guidelines, XLVIII, 306, 307
  equations, XLVIII, 385–387
  at equilibrium, XXXIV, 143
  mechanism, for collagen supports, XLIV, 261–263
  nonspecific, XXXIV, 727
Binding affinity, XLVIII, 284, 285
Binding competition, XLVI, 188
Binding constant, XXXIV, 141–147
  intrinsic, average, evaluation, XLVIII, 300–303, 307
Binding protein, for 1,25-dihydroxyvitamin D_3 preparation, LII, 391
Binding site
  discrete, ligand binding, XLVIII, 270–295
  linear lattice, ligand binding to, XLVIII, 295–299
Binding study, XLVI, 109, 110, 181, 182, 312–321
Binocular operational microscope, use in rodent surgery, XXXIX, 170, 171
Bioaffinity chromatography, XXXIV, 13–30, *see also* specific listings
  adsorbate-adsorbent association complex, XXXIV, 29
  adsorbents, group-specific, XXXIV, 30
  adsorption effects, XXXIV, 108–126
  agarose, XXXIV, 19
  cellulose, XXXIV, 18
  chaotropic ions, XXXIV, 16
  compound affinity, XXXIV, 115–118
  desorption, XXXIV, 16
  displacement, XXXIV, 16–18
    nonspecific, one-step elution, XXXIV, 16, 17
    specific, XXXIV, 17
  divinyl sulfone-agarose, XXXIV, 19
  for β-galactosidase, XXXIV, 118, 119
  glass, XXXIV, 18
  gradient elution, XXXIV, 17, 18
  immunosorption, XXXIV, 16
  interference, XXXIV, 111–115
    elimination of, XXXIV, 116–118
  matrix selection, XXXIV, 18, 19
  polyacrylamide, XXXIV, 18
  quantitation, XXXIV, 14, 15
  for ribonuclease A, XXXIV, 119, 120
  Sephadex, XXXIV, 19
  solvents, XXXIV, 29
  spacers, XXXIV, 29

ternary complex formation, XXXIV, 17
Bioassay, of peptide hormones, radioreceptor assay comparison, XXXVII, 68, 69
Bioautography, XLIII, 105–110, *see also* specific antibiotic
antibiotics
antibacterial, XLIII, 106, 107
antifungal, XLIII, 107, 108
antileptospiral, XLIII, 109
antiprotozoal, XLIII, 109
antiviral, XLIII, 109
cytotoxic, XLIII, 108, 109
correlative assays, XLIII, 110
paper electrophoresis, XLIII, 282
penicillin, XLIII, 111
phage-inducing antibiotics, XLIII, 108
thin-layer chromatography, XLIII, 179–186
Bio-Beads SM–2, in nicotinic receptor purification, XXXII, 319
Bio–Beads S–X1, XLVII, 265, 266
Biocytin, column, for purification of avidin, XXXIV, 265
Biocytin-agarose, XXXIV, 265
Biocytin-Sepharose, for purification of avidin, XXXIV, 267
Bioelution, XXXIV, 109
Bioenergetic system, ionophores, LV, 450–454
Biofiber 20, XLIV, 293, 296–298
hydraulic permeability, XLIV, 298
Biofiber 50, XLIV, 293, 296–298, 308, 312
hydraulic permeability, XLIV, 298
Biofiber 80, XLIV, 293, 296–298, 307
hydraulic permeability, XLIV, 298
Biofiber 50 hollow-fiber beaker device, LII, 176
Biofiber 50 hollow-fiber miniplant, LII, 177, 180, 181
BioGel, *see also* Polyacrylamide
microbore column packing, XLVII, 213
BioGel A, XLVII, 393
purification of cytidine triphosphate synthetase, LI, 81
BioGel A 0.5
purification of guanine phosphoribosyltransferase, LI, 555
of hypoxanthine phosphoribosyltransferase, LI, 555
BioGel A 1.5, *see also* Agarose gel filtration
purification of initiation factors, LX, 97, 98, 253, 254
BioGel Affi–Gel 10
derivatized in affinity chromatography, LII, 515–517
solubilization, LII, 517
BioGel A–0.5m
ATPase, LV, 345, 353
purification of cytochrome oxidase subunits, LIII, 79
of superoxide dimutase, LIII, 391
studies of protein-RNA interactions, LIX, 567
zone gel filtration, LIX, 328
BioGel A–5M, purification of CPSase-ATCase-dihydroorotase complex, LI, 125
BioGel A–15m, purification of nitrate reductase, LIII, 349
BioGel A–50m, purification of ω-hydroxylase, LIII, 358
BioGel CM–2, XLIV, 56
BioGel CM–100, XLIV, 61–63
BioGel HPA, subunit purification, LV, 355
BioGel HTP powder, LI, 209
BioGel P, XLIV, 56, 85–86
BioGel P–2, XLIV, 445, 446; XLVII, 83, 137
analysis of modified tRNA, LIX, 178, 179
desalting procedure, XLVII, 208
preparation of cross-linked ribosomal subunits, LIX, 536
purification of cytochrome *c* derivatives, LIII, 152
of deflavoenzyme, LIII, 435
of shortened tRNA species, LIX, 186, 188
BioGel P–4, purification of cytochrome *c* derivatives, LIII, 172

BioGel P–6
purification of cytochrome bc_1 complex, LIII, 115
replacement of guanidine by urea, LV, 784
in retrieval of stored aequorin, LVII, 281
BioGel P–10, XLVII, 216, 219
in epidermal growth factor isolation, XXXVII, 427, 428
BioGel P–30, in insulin isolation, XXXVII, 335, 338
BioGel P–60
cytochrome b_5 isolation, LII, 469
purification of adenosine deaminase, LI, 504, 505
of cytochrome b_5 reductase, LII, 468
BioGel P–100
purification of cytochrome, LII, 181
of guanylate kinase, LI, 486
of luciferin sulfokinase, LVII, 249
of putidaredoxin, LII, 179
of putidaredoxin reductase, LII, 177
BioGel P–150
purification of adenine phosphoribosyltransferase, LI, 563
of guanine phosphoribosyltransferase, LI, 554
of hypoxanthine phosphoribosyltransferase, LI, 554
studies of protein-RNA interactions, LIX, 565
BioGel P–300
protein kinase, XXXVIII, 312
for sheep antibody, XLIV, 715, 716
in spectrin purification, XXXII, 277
Biohazard, LVIII, 36–43
aerosols, LVIII, 39, 40
infectious virus, LVIII, 37, 42, 43
lymphoid cells, LVIII, 37, 42
Biological clock
central, XXXVI, 475
evidence, XXXVI, 476–479
function, evolution, XXXVI, 475, 476
homeostatic, XXXVI, 475
peripheral, XXXVI, 475
types, XXXVI, 475
Bioluminescence
bacterial
bioassays, LVII, 127
energetics, LVII, 129, 130
principles, LVII, 125–127
quantum yield, LVII, 151
reactions involved, LVII, 194
stoichiometry, LVII, 152
general mechanism, LVII, 563, 564
nonradiative energy transfer, LVII, 240–242, 257–267
assay, LVII, 260–262
mechanism, LVII, 262, 263
species specificity, LVII, 263, 264
photon emissions, absolute measurement, LVII, 583–589
quantum yield
computation, LIII, 565
relative measurement, LVII, 587–589
reaction, LIII, 558
of firefly, LIII, 559
mechanisms, LIII, 559, 561
relation to reaction rate, LVII, 361–364
wavelength variation, LVII, 562
Bioluminescent system
autoinduction, LVII, 127, 128
enzyme-substrate, LVII, 277
photoprotein, LVII, 272, 276, 277
transcriptional control, LVII, 128, 129
Biomass
determination, principles, LVII, 73
relationship to ATP concentrations, LVII, 84, 85
Biomembrane, *see also* Membrane
infrared absorption bands, XXXII, 249
from lipid monolayers, XXXII, 545–554
negative staining, XXXII, 21, 29–31
optical rotation data, XXXII, 220–233
photoaffinity labeling, LVI, 666–668
spin-label measurements, XXXII, 161, 162
Biomphalaria glabrata, media, LVIII, 466
Biopsy
for human cells, XXXII, 806–808

human tissue, preparation for amino acid analysis, XL, 360
sampling methods, XXXVIII, 4, 5
skin, *see* Skin biopsy
Biopterin, prosthetic group, LII, 4
Bioquest hood, XXXII, 815
Bio-Rad AG–11A8 resin, for removal of hydrochloric acid, LIX, 313
Bio-Rad AG 1–X2
cyclic nucleotide purification, XXXVIII, 25
for enzyme purification, XXXII, 125
Bio-Rad AG–1–X8, cGMP separation, XXXVIII, 76, 77, 91
Bio-Rad Dowex 1–X8, for removal of sodium dodecyl sulfate, LIX, 545
Biorad Econo column, LI, 229
Bio-Rad G–11A8 resin, for removal of acidity, LIX, 317
Bio–Rad hollow-fiber device, LII, 174
Bio–Rex 70, XLVII, 267, *see also* Amberlite IRC–50
Biosil A, in glycolipid analysis, XXXII, 361, 362
BioSil HA, phospholipid preparation, LV, 717, 718
Bio–Solv BBS 2, XXXIX, 302
in scintillation cocktail, LI, 91
Bio–Solv BBS 3, in scintillation cocktail, LI, 419
Biosonik II sonic oscillator, LI, 374
Biosonik IV sonic oscillator, LI, 414, 415
Biosonik sonicator, LII, 134, 139
Biospecific adsorption, *see* Bioaffinity chromatography
Biotin
activity of analogues with carboxyltransferase, XXXV, 37
in assay for biotin carboxyl carrier protein, XXXV, 18
homogeneous competitive protein-binding assay, LVII, 441, 442
in media, LVIII, 54, 64
protection against inhibition by avidin, XXXV, 17
for purification of avidin, XXXIV, 265–267
tRNA mapping, LVI, 9
substrate for biotin carboxylase, XXXV, 26, 31
synthesis of 6-*N*-[(3-biotinylamido)-2-hydroxypropyl]amino-2,3-dihydrophthalazine-1,4-dione, LVII, 430
transport, XLVI, 608, 613–617
assay, XLVI, 616
Biotin carboxylase, XXXV, 25–31
activation, XXXV, 31
assay, XXXV, 26–28
inhibition, XXXV, 31
reactions catalyzed, XXXV, 25, 31
Biotin carboxyl carrier protein, XXXV, 17–25
assay of carbon dioxide acceptor-donor activity, XXXV, 18, 19
biotin content, XXXV, 24
molecular species, XXXV, 18, 24, 25
reaction with biotin carboxylase and transcarboxylase, XXXV, 17, 18
stability, XXXV, 20, 23
Biotin-cellulose, XXXIV, 265
Biotin-Sephadex, XXXIV, 265
Biotinyl-β-alanyl-*p*-nitrophenyl ester, XLVI, 615
6-*N*-[(3-Biotinylamido)-2-hydroxypropyl]-amino-2,3-dihydrophthalazine-1,4-dione
chemiluminescence of ligand conjugates, LVII, 444
homogeneous competitive protein-binding assay, LVII, 441
structural formula, LVII, 428
synthesis, LVII, 430
Biotinyl-γ-aminobutyryl-*p*-nitrophenyl ester, XLVI, 615
Biotinyl-ε-aminocaproyl-*p*-nitrophenyl ester, XLVI, 615
Biotinyl-δ-aminovaleryl-*p*-nitrophenyl ester, XLVI, 615
Biotinylglycyl-*p*-nitrophenyl ester, XLVI, 615
Biotinyl-*N*-hydroxysuccinimide ester, XXXIV, 266, 267; XLVI, 615, 616
ε-*N*-Biotinyl-L-lysine, XXXIV, 266
Biotinyl-*p*-nitrophenyl ester, XLVI, 613, 615
synthesis, XLVI, 614, 615
Biotinyl-polylysyl-agarose, XXXIV, 265
Biphenyl
chromatographic properties, LII, 405
for HPLC column efficiency determination, LII, 322

hydroxylation
fluorometric assay, LII, 400–403
tissue activity levels, LII, 404
metabolites of, chromatographic separation, LII, 403–407
polychlorinated, sand, LV, 125, 126
substrate of cytochrome *P*–450LM forms, LII, 117
of microsomal aryl hydrocarbon hydroxylase, LII, 399
Biphenyl 2-hydroxylase
Ah[b] allele, LII, 231
in *in vitro* test for chemical carcinogens, LII, 400
Biphenyl 4-hydroxylase
Ah[b] allele, LII, 231
in microsomal fraction, LII, 88, 89
2-(*p*-Biphenyl)isopropyloxycarbonyl, XXXIV, 508
2-(*p*-Biphenyl)isopropyloxycarbonyl-L-phenylalanine, XXXIV, 515
2-(*p*-Biphenyl)-2-propyloxycarbonyl group
peptide synthesis, XLVII, 519
removal, XLVII, 593, 594
2-(4-Biphenylyl)-2-propyloxycarbonyl group, XXXIV, 514
2,3-Biphospho-D-glycerate:2-phospho-D-glycerate phosphotransferase, *see* Phosphoglyceromutase
2,2′-Bipyridine, XLIII, 502
2,2′-Bipyridine-ruthenium complex, electron-transfer chemiluminescence, LVII, 499
Bipyridyl-cobalt(III) complex, XLVI, 315
Birch and Pirt's medium, LVIII, 58
Bis, *see N,N*′-Methylenebisacrylamide
BIS, *see N,N*′-Methylenebisacrylamide
Bisacrylamide, LIX, 431, 458, 506, 507, 520, *see also N,N*-Methylenebisacrylamide
N,N′-Bisacryloylethylenediamine, XLVII, 267
Bis(6-aminocaproic acid), XXXIV, 591
Bis(4-antipyrylazo)-4,5-dihydroxy-2,7-naphthalenedisulfonic acid, *see* Antipyrylazo III
Bisaziridine, coupling methods, XXXIV, 27
2,4-Bis(bromomethyl)estradiol-17β 3-methyl ether, XLVI, 452–454
2,4-Bis(bromomethyl)estrone methyl ether, XLVI, 452, 453
2,5-Bis(5-*tert*-butylbenzoxazol-2-yl)thiophene, XLVII, 364
N,N′-Bis(2-carboximidoethyl)tartaramidedimethyl ester, cross-linking, LVI, 631, 632
N,N′-Bis(2-carboxyimidoethyl)tartarimide, LIX, 550
1,3-Bischloroacetylglycerol 2-phosphate
affinity labeling, XLVII, 484
structure, XLVII, 484
synthesis, XLVII, 490
Bis(cyclohexanone)oxalyl dihydrazone, for removal of copper from ammonium sulfate solutions, LIII, 56
Bisdiazobenzidine, XLIV, 263
Bisdiazobenzidine-2,2-disulfonic acid, XLIV, 904
N,N-Bis(2,2-dimethoxyethyl)amine, XLIV, 101, 102
4,4′-Bisdimethylaminodiphenylcarbinol, reaction conditions, XLVII, 413
Bisepoxide, coupling methods, XXXIV, 24, 25
Bis(ethyldiazomalonyl)adenosine 3′,5′-phosphate, LVI, 656
2′,3′-Bis-*p*-fluorosulfonylbenzoyladenosine, XLVI, 243, 244
Bis(hexafluoroacetonyl)acetone
complex V, LV, 315
synthesis, LV, 470
uncoupling activity, LV, 465
Bishydroxyethylaniline, XLIV, 83
1,1′-Bis(hydroxyethyl)-4,4′-bipyridyl dihalide, as mediator-titrant, LIV, 409, 410
N,N-Bis(2-hydroxyethyl)glycine buffer, XXXI, 20; XLVII, 489, 493, 496
effect on chemiluminescence, LVII, 439
in α-hydroxylase assay, LII, 311, 312
mitochondrial protein synthesis, LVI, 21
preparation of aminoacylated tRNA, LIX, 132, 133
storage of ribosomal protein S2, LIX, 498
1,1′-Bishydroxymethylferrocene, as mediator-titrant, LIV, 408

Bisimidate, XLIV, 125
Bisiodoacetamide, XLVII, 408
Bismaleimide, XLVII, 408
4,4′-Bis(2-methoxybenzene diazonium) chloride, XLIV, 147
Bis(4-methoxyphenyl)methyl group, carboxamido group blocking, XLVII, 612, 614
1,4-Bis-2-(4-methyl-5-phenyloxazolyl)benzene
as scintillator, LVIII, 288, 289
in scintillation cocktail, LII, 396
Bismuth-209, lock nuclei, XLIX, 349
Bismuth stain, XXXIX, 144, 145
Bis-*p*-nitrophenyl ester, of dicarboxylic acid, XLVI, 505–508
synthesis, XLVI, 506
Bis(*p*-nitrophenyl)methyl phosphonate, XXXIV, 587
Bis(*P*-nitrophenyl) phosphate, titration of carboxylesterase, XXXV, 214
Bisoxirane, XLIV, 26
coupling methods, XXXIV, 24, 25
1,4-butanediol-diglycidyl ether, XXXIV, 25
stability, XXXIV, 26
Sepharose activation, XLIV, 32, 33
4,4′-Bisphenyl diazonium disulfide, fluoroborate salt, cross-linking, LVI, 631, 632
1,4-Bis[2-(5-phenyloxazolyl)]benzene, *see* 2,2′-(1,4-Phenylene)bis(5-phenyloxazole)
2,3-Bisphosphoglycerate, substrate, phosphoglycerate mutase, XLVII, 496
Bisphosphoglycerate phosphatase
amino acids, XLII, 422–424
assay, XLII, 410–412
from bovine brain, preparation and purification, XLII, 415, 417
from heart muscle, preparation and purification, XLII, 418, 419
from human erythrocyte, preparation and purification, XLII, 419, 420
inhibition, XLII, 424, 425
molecular weight, XLII, 421, 422
preparation, from horse liver, XLII, 413, 415, 416
properties, XLII, 420–426
purification, from horse muscle, XLII, 412–414
sources, XLII, 409
Bis(pyridine)-adenosine triphosphate, derivatization, of transfer RNA, LIX, 14
Bis(pyridine) osmate, derivatization of transfer RNA, LIX, 14
Bis(pyridine) osmate-adenosine monophosphate, derivatization of transfer RNA, LIX, 14
Bis(pyridine) osmate-cytidine triphosphate, derivatization of transfer RNA, LIX, 14
Bis(pyridine) osmate-guanosine triphosphate, derivatization of transfer RNA, LIX, 14
Bis(pyridine) osmate-uridine triphosphate, derivatization of transfer RNA, LIX, 14
1,3-Bis(1-stearoyl-2-oleyl-*sn*-glycerol-3-phosphoryl)glycerol, intermediate in diphosphatidylglycerol synthesis, XXXV, 470, 471
N,N′-Bis(3-succinimidyloxycarbonylpropyl)tartaramide, cross-linking, LVI, 631, 632, 639
Bis-2,4,6-trichlorophenyloxalate, in chemiluminescence system, LVII, 459, 460
N,O-Bis(trimethylsilyl)trifluoroacetamide, XLVII, 395
Bistrimethysilylacetamide, silylating reagent, LII, 332, 405
Bistris, in electrophoretic separation of ribosomal proteins, LIX, 431, 542, 544
Bisulfite
interaction with cytochrome *c* oxidase, LIII, 195
wash solution, for gel electrophoresis, XXXII, 96
Biuret method, LVIII, 176
for plant protein determination, XXXI, 542–544
Black lipid film, *see* Lipid, black film
Bladder, toad, in water flow studies, XXXVII, 251–256
Blast cell
crude extract preparation, LI, 367
thymidine kinase, LI, 365–371

Blasticidin S, XLIII, 47 fractionation, XXXII, 637-647 by countercurrent distribution, for photoaffinity labeling, LIX, 798, XXXII, 642, 643 guanylate cyclase, XXXVIII, 200 in synthesis of N-(3-carbethoxy-3-Blood clot, scanning electron microscopy, diazo)acetonylblasticidin S, LIX, XXXII, 51 Blood clotting enzyme of N-(ethyl-2diazomalonyl)blasticidin S, cascade, XLV, 33 LIX, 814 coagulation factors, XLV, 32, 34 Bleaching, XLIX, 94 components interrelationship, XLV, 31 Bleomycin group, XLIII, 148, 150 protein, XLV, 32 Bligh-Dyer extraction of lipid, XXXII, pathways, XLV, 31 504 Blood group Blister formation, MDCK cells, LVIII, ABH-active glycosphingolipid, isolation, L, 207-211 Block averaging, LIV, 183 A glycosphingolipid, structure, L, 273 Block staining, for electron microscopy of B antigen, XXXIV, 347 liver cells, XXXI, 22, 23 determinant, neutral Blood glycosphingolipid, L, 247 androgen production and glycoprotein, XXXIV, 337 interconversion, XXXVI, 68-71 structure, L, 279 cholesterol measurement, LVI, oligosaccharides from human urine, 470 - 472L, 226, 227 cuvet reagent, LVI, 472 substance, XXXIV, 712-715 for erythrocyte ghost isolation, XXXI, A, XXXIV, 338 171 lectin purification, XXXIV, 336 estrogen assay, XXXVI, 51 Blood serum, oxidases, LII, 9, 10 galactose measurement, LVI, 466 Blood vessel, in rat brain surgery, glucose electrodes, LVI, 462-464 XXXIX, 173, 174 human Blowfly, salivary glands, hormone studies, XXXIX, 467-469 culture of cells, XXXII, 799, 800 Blue copper protein, XLIX, 140–143 screening for NTPH activity, LI, Blue dextran in nicotinic receptor assay, XXXII, lymphocyte isolation, XXXII, 633-636 monocyte isolation, XXXII, 759, 760 Sepharose-bound, in affinity perfusion, XLIV, 679, 680 chromatography, LI, 8, 9, 349 for red cell isolation, XXXIX, 27 void volume marker, LI, 24 testosterone and androstenedione in, Blue-fluorescent protein, LVII, 226-234, interconversion rates, XXXVI, 71 - 73copurification, with luciferase, LVII, triglyceride, analysis by gas 227, 229, 230 chromatography-mass fluorescence emission spectra, LVII, spectrometry, XXXV, 359 urate measurement, LVI, 468 molecular weight, LVII, 227 Blood bank, as platelet source, XXXI, from Photobacterium phosphoreum, 151 LVII, 226-234 product, of aequorin reaction, LVII, Blood cell countercurrent distribution, XXXI, properties, LVII, 285 761

purity, LVII, 232–234
relative, determination, LVII, 229
role in *Aequorea* bioluminescence, LVII, 285, 287
spectral properties, LVII, 226, 227
Blue-green algae, gas-vesicle isolation, XXXI, 678–686
Bluensidine, XLIII, 429
moiety, equation, XLIII, 430, 444
Bluensomycin
biosynthesis
of guanidinated inositol moeties, XLIII, 429–433
inosamine kinase, XLIII, 444–451, *see also* Inosamine kinase
bluensidine moiety, XLIII, 430, 444
chiroptical studies, XLIII, 352
solvent system, XLIII, 119, 120
structure, XLIII, 429
Blue Sepharose CL-6B, purification of *Lactobacillus* deoxynucleoside kinases, LI, 348–350
B lymphocyte, XXXIV, 223, 716, 720
separation, XXXII, 636
BME, *see* Eagle's basal medium
B_{16} melanoma M2R cell, LVIII, 96
BNPS-skatole, *see* 2-(2-Nitrophenylsulfenyl)-3-methyl-3-bromoindolenine
Boar, seminal plasma inhibitor, XLV, 834–844
Boc, *see tert*-Butyloxycarbonyl
BOC-ON, *see* 2-*tert*-Butyloxycarbonyloxyimino-2-phenylacetonitrile
Bodine motor, for perfusion apparatus, XXXIX, 25
Bohr magneton number, LIV, 382
Boltzmann distribution, spin probes, LVI, 523, 525
Bombyx mori, media, LVIII, 457
Bonding
covalent, in affinity labeling, XLVI, 10, 11
ionic, in adsorption, XLIV, 149–153
Bone
cancellous, preparation for amino acid analysis, XL, 360
cell isolation, mechanical disruption, XL, 317
collagen
extraction, chick, XL, 339
metabolic studies, XL, 326
methods of organ culture, XL, 331, 333, 334
specific, XL, 328
defatting, drying, and demineralization, XL, 312
enzymatic digestion, XL, 316
isolation of cells, XL, 316, 317
neonatal, adenylate cyclase, XXXVIII, 152, 153
oxygen consumption, XL, 328
preparation for collagen studies, XL, 311
procollagenase, mouse, XL, 352
Bone marrow, short-term cultures, in erythropoietin studies, XXXVII, 115, 116
Bone marrow cell
cloning, LVIII, 163
human, growth factor, LVIII, 163
mouse, growth factor, LVIII, 163
Bongkrekic acid
ADP, ATP carrier, LVI, 408, 409, 412–414
biosynthetic labeling, LV, 529
carrier protein, LVI, 252
carrier sites, LVI, 251
chemical labeling
method, LV, 530
principle, LV, 529, 530
preparation
chemical criteria of purity, LV, 529, 531, 532
extraction, LV, 528
ion-exchange chromatography, LV, 529
thin-layer chromatography, LV, 529
resistance, LVI, 125
transport, LVI, 248
Bonse and Hart collimation system, XXXII, 212
Borate, *see also* Boric acid
as antioxidant, XXXI, 539–540
Borate buffer, XLVII, 11, 42
advantages, as peroxidase eluent, LII, 518

in coelomic cell homogenization buffer, LVII, 376
coupling, XLVII, 44
preparation, XLVII, 17
for protein labeling, XLVIII, 332
purification of *Diplocardia* luciferin, LVII, 376
for short analysis time, XLVII, 25, 27
for TEAE-cellulose chromatography, XLVII, 208
Borate complex, of saccharides, automated assay by ion-exchange chromatography, XLI, 10–21
Borate-phosphate buffer, preparation of cytochrome *c* reductase complexes, LIII, 58
Borate-potassium hydroxide buffer, assay of glucose, LVII, 453
Borax, XLIV, 765
Bordetella pertussis, in studies of A1 encephalitogenic activity, XXXII, 327
Boric acid, *see also* Borate
electrophoretic analysis of ribosomal proteins, LIX, 506
preparation of water-soluble luminol salt, LVII, 477
reductive methylation reaction, LIX, 784
tRNA sequence analysis, LIX, 67, 69
separation of bacterial ribosomal subunits, LIX, 398
2-Bornanone, *see d*-Camphor
Borohydrazide reduction, XXXIV, 178
Borohydride, *see also* Cyanoborohydride
flavoprotein inactivation, LIII, 447
inactivator of protein B, LIII, 381
modification of lysine, LIII, 141
preparation of duroquinol, LIII, 90
Boron-10, lock nuclei, XLIX, 348
Boron-11
lock nuclei, XLIX, 348, 349; LIV, 189
resonance frequency, LIV, 189
Boron carbide, beam block, in neutron scattering experiments, LIX, 666, 667
Boronic acid, XLVI, 19, 22, 27
Boronic acid polyacrylamide, XXXIV, 47, 48
Boron trifluoride diethyl etherate, XLIV, 126
Boron trifluoride etherate, XLIV, 117
cleavage agent, XLVII, 341, 342
as lock nuclei source, LIV, 189
redistillation, XLVII, 341
Boron trifluoride-methanol
in fatty acid methylation, LII, 321
in synthesis of methyl 4-(4-methoxyphenyl)butyrate, LVII, 435
Borotritide, XLVII, 476
Bostrycoidin, XLIII, 334
Bothrops, venom, *see* Batroxobin
Bothrops atrox, venom, phosphodiesterase, XXXIV, 608, 609
Botrycidin, nature, LV, 502, 505
Bottle cap, for embedding, XXXII, 4, 7
Bottromycin, inhibitor of polypeptide chain elongation, LIX, 359
Botulinum, XXXIV, 610
Bouin's fixative, XXXVII, 158
in endocrine tissue fixation, XXXVII, 136
Boundary centroid, definition, XLVIII, 196
Boundary pattern, simulation, XLVIII, 196–201
Boundary profile
asymptotic, XLVIII, 195–212
flux equations, XLVIII, 196–201
hypersharp, XLVIII, 210, 211
Bovine brain, 1,3-diphosphoglycerate, preparation and purification, XLII, 415, 417
Bovine heart, phosphofructokinase, purification, XLII, 107, 110
Bovine herpesvirus, LVIII, 28
Bovine kidney
mutarotase, substrate specificities, XLI, 487
phosphofructokinase, purification, XLII, 107, 110
Bovine liver
cytidine triphosphate synthetase, LI, 84–90
fructose-1,6-diphosphatase, assay, purification, and properties, XLII, 363–369
homogenization, LI, 86
2-keto-4-hydroxyglutarate aldolase, assay, purification, and properties, XLII, 280–285

phosphofructokinase, purification, XLII, 107, 110
Bovine pancreas, fractionation, XXXI, 49–59
Bovine rhinotracheitis virus, infectious, LVIII, 28
Bovine serum
fetal
freezing, LVIII, 31
preservation, LVIII, 31
growth promotion, LVIII, 20, 21
mycoplasmal testing, LVIII, 22–27
sterility, LVIII, 19–21
tests, LVIII, 20, 21
toxicity, LVIII, 20
viral contamination, LVIII, 28, 29
virus testing, LVIII, 28, 29
Bovine serum albumin, XXXI, 720; XXXIV, 252
activator of dihydroorotate dehydrogenase, LI, 62
adenylate cyclase, XXXVIII, 141, 168, 177
aminoacylation of tRNA, LIX, 216
antibody binding, L, 55
assay of adenine phosphoribosyltransferase, LI, 570
of L-amino acids, LIX, 315
of ATP, LVII, 98
of complex, III, LIII, 39
of creatine kinase isoenzymes, LVII, 60
of cyclic nucleotide phosphodiesterases, LVII, 97
of luciferase, LIII, 565; LVII, 137, 139
of modulator protein, LVII, 109
of penicillopepsin, XLV, 443
of polypeptide chain elongation, LIX, 358
of protease, LVII, 200
of RNase III, LIX, 826
of rubredoxin, LIII, 341
of succinate dehydrogenase, LIII, 26, 27, 34
of translocation, LIX, 360
brown adipose tissue mitochondria, LV, 70, 76
carrier, in phosphoribosylpyrophosphate synthetase assay, LI, 4, 12
citrate impurity, XXXV, 598
complex V, LV, 315
crystalline, in assay, of deoxycytidine kinase, LI, 338
for culture medium, XXXIX, 205
defatted, preparation, LV, 712, 713
for depletion of endogenous substrate, LII, 272
disadvantage, in tryptophanyl-tRNA synthetase assay, LIX, 237
effect on electrophoretic pattern of cytochrome *P*–450, LII, 328
electron spin resonance spectra, XLIX, 411, 425
electrophoresis reference, XLIV, 474
electrophoretic light scattering properties, XLVIII, 432, 433, 435
enhancer, aromatic hydrocarbon sensitivity, LII, 236
enzyme immobilization buffer, LVII, 204
in estrogen receptor studies, XXXVI, 175
fatty acids, LV, 472
synthesis rate, XXXV, 598
fluorescent peptide map, XLVII, 30
fragments, molecular weights, XLVIII, 68
free fatty acid-poor, preparation, XXXVII, 431
glutaraldehyde insolubilization, XLIV, 553
for hepatocyte isolation, LII, 62
hollow-fiber dialytic permeability, XLIV, 297
hollow-fiber retention data, XLIV, 296
in immunoassay, XLIV, 716
for inert proteic matrix, XLIV, 264, 265, 268, 269, 271, 903, 904, 912, 925
inhibitor, in fatty acid bioassay, LVII, 193
of phospholipase D, XXXV, 231
intact cell oxidative phosphorylation, LV, 175

isolation of chloroplasts, LIX, 434
of mitochondria, LV, 4, 15, 16, 17, 80, 81, 86, 116
of synthetic peptides, LIX, 388
ligand binding assays, LVII, 114
lipolysis in perifused fat cells, XXXV, 609, 611, 612
in liver perfusion, XXXV, 598
for liver perfusion medium, XXXIX, 27
as liver sample preservative, XXXI, 19
luciferase reagent, LVII, 55, 60, 218, 219, 221
advantage, LVII, 122
luciferase renaturation buffer, LVII, 178
lysine content, XXXV, 339
methylated, LX, 553
complex with neutral glycolipid acid, L, 140
microorganismal mitochondria, LV, 142
in mitochondria isolation procedure, XXXI, 306, 310, 531
mitochondrial incubation, LV, 7
molecular weight marker, LIII, 351, 353
NADPH-cytochrome *P*–450 reductase assay, LII, 91
osmometric studies, XLVIII, 67, 68, 76
perfusion fluid, LII, 49, 50, 51
phosphorescence spectra, XLIX, 245
phosphorylated, coupling of histones, XL, 192
piericidin A, LV, 457
in plant-cell extraction, XXXI, 524, 525, 536, 537, 592, 599, 603
preparation of bacterial luciferase subunits, LVII, 173
pressure-jump studies, XLVIII, 310
protection against inhibition by long-chain fatty acyl-CoA, XXXV, 9, 17, 88, 105
purification of mitochondrial tRNA, LIX, 204, 205
pyrene-maleimide labeling, XLVII, 416, 417
rat hepatocyte isolation, LII, 358
refolding studies, LII, 249, 250
reverse salt gradient chromatography, LIX, 216
rhein, LV, 459
ribonuclease-free
in tRNA sequence analysis, LIX, 71
storage, LIX, 61
tRNA end-group labeling, LIX, 103
tRNA reconstitution, LIX, 140, 141
rotenone removal, LV, 455
SDS complex, properties, XLVIII, 6
solution density, XLVIII, 27–29
in spin label introduction into membranes, XXXII, 193, 194
stabilizer of apolipoamide dehydrogenase, LIII, 433
of hydrogenase, LIII, 291, 300
of phosphoribosylpyrophosphate synthetase, LI, 16
of thymidine kinase, LI, 355, 358, 369
of uridylate-cytidylate kinase, LI, 336
steroid conjunction, XXXVI, 21
storage solution of purified luciferase, LVII, 62
studies of protein-RNA interactions, LIX, 563
submitochondrial particles, LV, 107
surface-caused enzyme inactivation, XLIV, 307
temperature-dependent transitions, XLIX, 4
in ultrafiltration studies, XLIV, 300
use in preparing isolated fat cells, XXXV, 555, 557
viscosity, XLVIII, 53, 55
volume change during fragment recombination, XLVIII, 33
Bovine spinal cord, source of sphingosine, dihydrosphingosine, XXXV, 529
Bovine submaxillary mucin, structure, L, 279
Bowen quantum counter, LVII, 591–593, 598
Bowman-Birk soybean inhibitor
amino acid sequence, XLV, 697, 698, 704
biological activity, XLV, 707
chemical modifications, XLV, 705

enzyme treatment, effect, XLV, 699
kinetics, XLV, 704
properties, XLV, 704
Bowman's capsule, in glomerular isolation, XXXII, 653
Bowman Infusion Pump, XXXVI, 72
BP, *see* Benzo[*a*]pyrene
B particle, radiation properties, XXXVII, 146, 147
BP-LH_2, *see* Luciferin-binding protein, calcium-triggered
BPOC group, *see* 2-(*p*-Biphenyl)isopropyloxycarbonyl
BPS, *see* Sodium bathophenanthrolinesulfonate
BQ, *see* *p*-Benzoquinone
Bradykinin
spin-spin distance, XLIX, 424
substrate, kallikrein, XLV, 312
Brain, *see also* specific types
adenylate cyclase
particulate, XXXVIII, 135, 136
assay, XXXVIII, 136
characteristics, XXXVIII, 140
dispersion, XXXVIII, 138–140
preparation, XXXVIII, 136–138
preparation, XXXVIII, 128, 129
solubilized, characteristics, XXXVIII, 140–143
bovine, preparation
delipidated tissue powders, XLV, 38
ethylenediamine tetraacetate-washed, XLV, 46
cAMP receptor, XXXVIII, 380
cell
cloning, LVIII, 163
dispersion, LVIII, 123
enzymes, XXXII, 773
2′,3′-cyclic nucleotide-3′-phosphohydrolase, XXXII, 128, 131
enzymes, rapid inactivation, XXXVIII, 5, 6
heat-stable factor, preparation, LII, 312, 313
homogenization procedure, LII, 312
human, endothelial cell isolation, XXXII, 717–722
lipids, in bilayer formation, XXXII, 491, 492, 515
long chain fatty acid α-hydroxylase, LII, 310–318
lysosome isolation, XXXI, 457–477
metabolite content, LV, 218, 219
of mice and rats, basic proteins, XXXII, 335
microsomes, ATPase, XXXII, 305, 306
mitochondria
isolation, materials and methods, LV, 51, 52
of nonsynaptic origin, LV, 17, 18
synaptosomal, LV, 18
modulator protein, bovine, XLVII, 29
peptide separation, XLVII, 235
nuclei isolation from cells, XXXI, 452–457
oxidase, LII, 12, 13, 16, 18, 20
phosphodiesterase, XXXVIII, 223, 224
assay, XXXVIII, 132–134, 224, 225
properties, XXXVIII, 232–239
purification, XXXVIII, 225–232
regional and subcellular distribution, XXXVIII, 237, 238
phosphodiesterase activator, XXXVIII, 262, 263
assay, XXXVIII, 263, 264
properties, XXXVIII, 267–273
purification, XXXVIII, 264–267
protein kinases, XXXVIII, 298, 365, 366
source of neurons and astrocytes, XXXV, 566, 575
tumor cell, growth factor, LVIII, 163
Branson Sonifier, LI, 303, 311, 439
HeLa cell subcellular fraction, LVI, 69
Model W185, LIX, 373
Branton-de Silva extraction apparatus, XXXI, 541, 570
Bratton-Marshall method, LI, 187, 194
Braun homogenizer, XXXI, 658; LIII, 251
Model MSK Mechanical Cell homogenizer, LI, 43
yeast cell breakage, LVI, 47, 61
properties, XXXI, 662
Braunitzer reagents, peptide methodology and, XLVII, 260–263
Bray device, LIV, 86

Bray scintillation fluid, LII, 415; LIX, 222, 315, 361
Breast milk, drug isolation, LII, 333
Breast tissue, estrogen receptors, XXXVI, 168, 175
Breathing model
in nucleic acids, XLIX, 32
for protein hydrogen exchange, XLIX, 35
quantitative aspects, XLIX, 36, 37
Brendler homogenizing apparatus, XXXI, 43
Brevibacterium liquefaciens
adenylate cyclase, XXXVIII, 160
forward reaction assay, XXXVIII, 161
properties, XXXVIII, 167–169
purification, XXXVIII, 163–167
reverse reaction assay, XXXVIII, 162, 163
Brevibacterium sterolicum, oxidase, LII, 18
Brevicid, XLIII, 202
Brice-Phoenix light-scattering photometer, LVII, 39
Bridge formation, inorganic, XLIV, 166–169
Briggs and King nuclear transfer technique, XXXII, 789
Brigl's anhydride, *see* Tri-*O*-acetyl-1,2-epoxy-α-D-glucopyranose
Brij detergent, LVI, 741
extraction of mitochondria, LVI, 411, 413
partial specific volume, XLVIII, 19, 20
Brij 35 detergent, XLVII, 17
activator of dihydroorotate dehydrogenase, LI, 62
in *E. coli* lysis solution, LIX, 354
for microsome isolation, XXXI, 216, 224
M. phlei membrane systems, LV, 185
of organic hydroperoxides, LII, 510
protein assay in lipid extracts, XXXV, 336–338
Brij-58 surfactant
partial specific volume, XLVIII, 19, 20
preparation of ribosomal subunits, LIX, 412
Brij-96 surfactant, partial specific volume, XLVIII, 19, 20
Brilliant Cresyl Blue, LIX, 301
Brillouin function, LIV, 381, 382
Brine shrimp, *see Artemia salina*
Brinkmann silica gel GF 250 TLC plate, LIX, 813
Brinster culture dish, XXXIX, 298
Britton-Robinson buffer, XLIV, 81
Bromamphenicol, XLVI, 706
synthesis, XLVI, 703, 704
Bromelain, XXXIV, 4, 418, 532
conjugate, kinetics, XLIV, 406–408, 437
enzyme, nomenclature, XLV, 475
fruit
amino acid composition, XLV, 480
sequence, XLV, 484
properties, XLV, 484
purification, XLV, 483
immobilization
on (carboxymethyl)cellulose, XLIV, 52
property changes, XLIV, 53
stem
activators, XLV, 482
amino acid composition, XLV, 480
sequence, XLV, 481
assay, by *p*-nitrophenyl-N^{α}-benzyloxycarbonyl-L-lysine, XLV, 740
chemical modifications, XLV, 482
enzymatic mechanism, XLV, 482
inhibitors, XLV, 482
kinetics, XLV, 482
properties, XLV, 479
purification, XLV, 476, 743
purity, XLV, 478
specificity, XLV, 482
Bromelain A, titration, XLV, 9
burst, XLV, 10
sulfhydryl group, XLV, 10
Bromelain inhibitor, pineapple stem
amino acid composition, XLV, 747
sequence, XLV, 749
assay, by *p*-nitrophenyl-N^{α}-benzyloxycarbonyl-L-lysine, XLV, 742
properties, XLV, 748
protein chemistry, XLV, 748

purification, XLV, 745
Bromide
activator of succinate dehydrogenase, LIII, 33, 472, 474, 475
inhibitor, of firefly luciferase, LVII, 15
phase shift studies, LIV, 344
Bromine
in preparation of eosin isothiocyanate, LIV, 49
synthesis of 8-bromoadenosine 5′-monophosphate, LVII, 117
Bromine-79, lock nuclei, XLIX, 349; LIV, 189
Bromine-81, lock nuclei, XLIX, 349; LIV, 189
9-(*m*-Bromoacetamidobenzyl)adenine, XLVI, 334
9-(*o*-Bromoacetamidobenzyl)adenine, XLVI, 334
9-(*p*-Bromoacetamidobenzyl)adenine, XLVI, 328, 329, 332–334
synthesis, XLVI, 329–332
Bromoacetamidobutylguanidine, XLVI, 226, 228
6-Bromoacetamidocaproic acid, XXXIV, 39, 40
Bromoacetamidocaproyl-hydrazide derivative, XXXIV, 40, 41
β-Bromoacetamido-*trans*-cinnamoyl coenzyme A, reaction conditions, XLVII, 422
4-Bromoacetamidoestrone methyl ether, XLVI, 455–457
Bromoacetamidoethylagarose, XXXIV, 353
Bromoacetamidoethyl side chain, XXXIV, 356
Bromoacetamidophenyl derivative, XLVI, 669
5′-[4-(Bromoacetamido)phenylphospho]adenylyl-(3′-5′)-uridylyl-(3′-5′)-guanosine, XLVI, 669, 670, 673
2,4-Bromoacetamidopropranolol, synthesis, XLVI, 594
Bromoacetic acid, XLVI, 56, 157, 400, 568; XLVII, 488; LIX, 157
p-azidophenacyl ester, XLVI, 684, 685, 696
N-hydroxysuccinimide ester, *see* Succinimidyl bromoacetate
modification of histidine, LIII, 137
of methionine, LIII, 138, 150
preparation of bromoacetamidoethyl-Sepharose, LI, 310
Bromoacetic anhydride, XLVI, 154, 363; XLVII, 489
synthesis, XLVI, 157, 363, 364, 400
Bromoacetol phosphate, synthesis, XLVI, 141
16α-Bromoacetoxyestradiol 3-methyl ether, XLVI, 449–452
16α-Bromoacetoxy-1,3,5(10)-estratriene-3,17β-diol 3-methyl ether, XLVI, 449–452
3-Bromoacetoxyestrone, XLVI, 56, 58
12β-Bromoacetoxy-19-nor-4-androstene-3,17-dione, XLVI, 457–460
Bromoacetoxyprogesterone derivative
in enzyme active site studies, XXXVI, 390
synthesis, XXXVI, 378–383
Bromoacetylagarose derivative, XXXIV, 658
Bromoacetylalk-3-ynoyl coenzyme A compound, inhibitor of mitochondrial acetoacetyl-CoA thiolase, XXXV, 135
3′-(4-Bromoacetylamidophenylphosphoryl)-deoxythymidine 5′-phosphate, XXXIV, 185, 187
N-Bromoacetyl-*p*-aminobenzyl-L-succinic acid, XLVI, 226, 227, 229
p-(*N*-Bromoacetyl-*p*-aminophenylazo)phenyl-β-lactoside, XLVI, 519, 522
N-Bromoacetyl-*p*-aminophenyl-β-lactoside, XLVI, 519, 522
α-*N*-Bromoacetyl-D-arginine, XLVI, 226, 227
Bromoacetylarsanilic acid, XLVI, 493, 497, 498, 500
Bromoacetylation, XLVI, 363–365, 457
alkylation of ligand, XXXIV, 49, 50
Bromoacetyl bromide, XLIV, 49, 749; XLVI, 154, 227; XLVII, 487
3-Bromoacetyl-1-carboxyethyl pyridinium, XLVI, 146
(Bromoacetyl)cellulose, XLIV, 36
preparation, XLIV, 749
(Bromoacetyl)cellulose derivative, XXXIV, 704, 705
Bromoacetylcholine, XLVI, 578

Bromoacetylcholine bromide, XLVI, 582
Bromoacetyl coenzyme A, inhibitor of mitochondrial acetoacetyl-CoA thiolase, XXXV, 135
N-Bromoacetyldinitrophenylaniline, synthesis, XLVI, 488, 489
α-*N*-Bromoacetyl-γ-*N*-dinitrophenyl-α,γ-diamino-L-butyric acid, XLVI, 501
N-Bromoacetyl-*N*′-dinitrophenylethylenediamine, XLVI, 489, 490
1-Bromoacetyl-1′-dinitrophenylethylenediamine, XXXIV, 185
N^{α}-Bromoacetyl-N^{ε}-dinitrophenyl-L-lysine, XXXIV, 186, 191; XLVI, 501
  labeled protein-315, XXXIV, 188
N-(N^{α}-Bromoacetyl-N^{ε}-dinitrophenyl-L-lysyl)-*p*-aminophenyl-β-lactoside, XLVI, 519, 520, 522
N-Bromoacetylethanolamine phosphate, XLVII, 423
  affinity labeling, XLVII, 483, 494–497
  chromatographic behavior, XLVII, 487
  radioactive labeling, XLVII, 488
  structure, XLVII, 483
  synthesis, XLVII, 487, 488
N-Bromoacetyl-L-fucosylamine, XLVI, 364
N-Bromoacetyl-D-galactopyranosylamine
  structure, XLVII, 482
  synthesis, XLVII, 489, 490
N-Bromoacetyl-β-D-galactosylamine, XLVI, 364
  inactivation, XLVI, 401, 402
  synthesis, XLVI, 400, 401
N-Bromoacetylglucosylamine, XLVI, 399
N-Bromoacetyl-D-glucosylamine, XLVI, 364
N-Bromoacetyl-L-glucosylamine, XLVI, 364
Bromoacetylglycosylamine, XLVI, 363–365
Bromoacetyl group, XLVI, 185, 484, 486, 578
Bromoacetyl hydroxysuccinimide, synthesis, XLVI, 360
Bromoacetyl-*N*-hydroxysuccinimide, synthesis, XLVI, 568, 569
O-Bromoacetyl-*N*-hydroxysuccinimide, XXXIV, 49
Bromoacetyl-*N*-hydroxysuccinimide ester, XLVI, 154
  synthesis, XLVI, 157
N-Bromoacetyl-α-lactosylamine, XLVI, 364
Bromoacetyllysine, XLVI, 185
Bromoacetyllysine-tRNA, XLVI, 193, 194
Bromoacetylmethionine, XLVI, 185
N-Bromoacetyl-*N*-methyl-L-phenylalanine, XLVI, 226–228
Bromoacetylmono(*p*-azobenzenearsonic acid)-L-tyrosine, XLVI, 493, 494, 500
  synthesis, XLVI, 495, 496
Bromoacetylpeptide-tRNA, XLVI, 193
Bromoacetylphenylalanine, XLVI, 185
Bromoacetylphenylalanine-tRNA, XLVI, 193
3-Bromoacetylpyridine, XLVI, 145, 146, 149
  reactions with cysteine, XLVI, 150
  synthesis, XLVI, 149
3-Bromoacetylpyridinio-*n*-butyladenosine pyrophosphate, XLVI, 257
[4-(3-Bromoacetylpyridinio)-*n*-butyl]adenosine pyrophosphate, XLVI, 255
(Bromoacetylpyridiniopropyl)adenosine pyrophosphate, hydrolysis rates, XLVI, 254, 255
2-Bromoacetylpyridiniopropyladenosine pyrophosphate, XLVI, 257
3-Bromoacetylpyridiniopropyladenosine pyrophosphate, XLVI, 257; XLVII, 423
4-Bromoacetylpyridiniopropyladenosine pyrophosphate, XLVI, 257
P^1-[3-(3-Bromoacetylpyridinium)propyl]-P^2-5′-adenos-5′-yl diphosphate, XLVI, 153
Bromoacetyl reagent, XLVI, 153, 154, 358–362
  homologous series, XLVI, 156
  identification, XLVI, 156
  properties, XLVI, 154
  specificity, XLVI, 154
N-Bromoacetyl-L-thyroxine, XLVI, 435–441
  ^{14}C-labeled, XLVI, 439–441
  synthesis, XLVI, 436–439

8-Bromoadenosine 3′,5′-cyclic monophosphate, *see* 8-Bromo cyclic adenosine monophosphate
8-Bromoadenosine monophosphate, XLVI, 99
conversion to 8-azidoAMP, LVI, 647, 648
purification, LVI, 647
8-Bromoadenosine 5′-monophosphate, XXXIV, 486
electrophoretic mobility, LVII, 117
millimolar extinction coefficient, LVII, 117
synthesis, LVII, 117
of adenosine derivatives, LVII, 117
4-(β-Bromoaminoethyl)-3-nitrophenyl azide, XLVII, 421
Bromobenzene
assay of deuterium oxide, in water samples, LIX, 644, 645
density gradient, LIX, 5
toxicity mechanism, LII, 70, 71
o-Bromobenzyl alcohol, XLVII, 536
3-Bromobenzyl group, tyrosine blocking, XLVII, 615
o-Bromobenzyl-*p*-nitrophenyl carbonate, synthesis, XLVII, 536
2-Bromobenzyloxycarbonyl group, XLVII, 610
N^{ε}-*o*-Bromobenzyloxycarbonyl-L-lysine, synthesis, XLVII, 536
3-Bromo-2-butanone-1,4-diol diethyl ketal, synthesis, XLVI, 393, 394
N-(4-Bromobutyl)phthalimide
synthesis of dimethyl 7-*N*-(4-phthalimidobutyl)aminonaphthalene-1,2-dicarboxylate, LVII, 436
of *N*-methyl-4*N*-[4-(*N*-phthalimido)butyl]aminophthalimide, LVII, 432
Bromocolchicine, XLVI, 567–571
binding to brain extracts, XLVI, 570, 571
synthesis, XLVI, 569
Bromoconduritol, XLVI, 369, 370
inhibition of glucosidase, XLVI, 381
synthesis, XLVI, 380, 381
Bromocresol purple, transport, LVI, 292
α-Bromo-ω-(3-cyanophenyl)alkane, XLVI, 122
8-Bromo cyclic adenosine monophosphate, XXXIV, 261; XLVI, 341–346
8-Bromo-cyclic guanosine monophosphate, protein kinase, XXXVIII, 337, 339
6-Bromo-6-deoxyconduritol B epoxide, active site labeling, XLVI, 377–380
5-Bromodeoxycytidine
substrate of cytidine deaminase, LI, 411
of thymidine kinase, LI, 370
5-Bromodeoxycytosine triphosphate, activator of thymidine kinase, LI, 358
Bromodeoxyuridine
effect on proliferating cell culture, XXXIX, 115
enzyme-deficient mutants using, XXXIX, 123
resistant cell line, LVIII, 376
5-Bromodeoxyuridine
as mutagen, LVIII, 316
resistant strains, LVIII, 314
substrate of deoxythymidine kinase, LI, 364
of pyrimidine nucleoside phosphorylase, LI, 436, 437
of thymidine kinase, LI, 358, 370
of thymidine phosphorylase, LI, 441
5-Bromo-2′-deoxyuridine, DNA inhibitor, XL, 278, 280
3-Bromo-1,4-dihydroxy-2-butanone 1,4-bisphosphate, XLVI, 142, 392, 396, 397
synthesis, XLVI, 392–395
3-Bromo-1,4-dihydroxy-3-butanone 1,4-bisphosphate, XLVI, 152
3-Bromo-1,4-dihydroxy-2-butanone 1,4-bisphosphate diethyl ketal, synthesis, XLVI, 394
2-Bromoethyl-*N*,*N*-dimethylamine hydrobromide, intermediate in lysoglycerophospholipid synthesis, XXXV, 491
2-Bromoethylphosphonic acid
intermediate in synthesis of phosphatidylcholine analogues, XXXV, 515

monoanilinium salt, intermediate in synthesis of phosphatidylcholine analogues, XXXV, 514, 515
β-Bromoethylphosphoryl dichloride, intermediate in phosphatidylcholine synthesis, XXXV, 453
(2-Bromoethyl)trimethylammonium bromide, reaction conditions, XLVII, 410
8-Bromoguanosine monophosphate
conversion to 8-azido-GMP, LVI, 649, 650
synthesis and purification, LVI, 649
Bromohydroxyacetone phosphate, XLVII, 66
6α-Bromo-17β-hydroxy-17α-methyl-4-oxa-5α-androstan-3-one, in androgen receptor studies, XXXVI, 371, 372
α-Bromo-4-hydroxy-3-nitroacetophenone
preparation, XLV, 8
titration, active site, XLV, 4
2-Bromo-4′-hydroxy-3-nitroacetophenone, XLVI, 151
(*R,S*)-2-Bromo-3-hydroxypropionate 3-phosphate
affinity labeling, XLVII, 484, 487
structure, XLVII, 484
synthesis, XLVII, 492
α-Bromoisocaproic acid, XLVI, 235
D-α-Bromoisocaproic acid, XLVI, 236, 237
L-α-Bromoisocaproic acid, XLVI, 236, 237
α-Bromoisocaproyl peptide, XLVI, 236, 237
Bromoisocolchicine, XLVI, 568
Bromoketone, XLVI, 131, 198, 199
Bromolasalocid, code number, LV, 445
Bromomethyl-2-hydroxy-3-naphthoxypropylaminomethyl ketone, XLVI, 597, 598
Bromopentol B, XLVI, 375, 376
p-Bromophenol, XLIV, 133, 386
Bromopractolol, synthesis, XLVI, 592–594
Bromoprogesterone
in enzyme active site studies, XXXVI, 390, 392–399
synthesis, XXXVI, 385–388
3-Bromopropene, structural formula, XLIV, 197
3-Bromopropionic acid, alkylation of protein A, LIII, 379
Bromopyruvate, XLVI, 45–47, 67, 132–139
labeling, XLVI, 139
residues, modified, XLVI, 134
as substrate, XLVI, 137
Bromostyrylpyridinium salt, XXXIV, 4
N-Bromosuccinimide, XLVII, 409, 456
extraction of adenine nucleotides, LVII, 81
of guanine nucleotides, LVII, 88
in hemoprotein conformation studies, LIV, 270, 279
methionine oxidation, XLVII, 456
modification of methionine, LIII, 139, 171
of tryptophan, XXXV, 128; XLVII, 443, 452, 459; LIII, 171
in studies of luciferase active site, LVII, 180
8α-Bromotetraacetylriboflavin
synthesis, LIII, 458
of 8α-*S*-cysteinylriboflavin, LIII, 460
of 8α-*N*-histidylriboflavin isomers, LIII, 458
6-Bromo-3,4,5-trihydroxycyclohex-1-ene, *see* Bromoconduritol
Bromotrimethylcolchicinic acid, XLVI, 569–571
5-Bromouracil
determination, spectrophotometric, LI, 433
extinction coefficient, LI, 433
inhibitor of orotate phosphoribosyltransferase, LI, 166
substrate of pyrimidine nucleoside phosphorylase, LI, 436
5-Bromouridylate
inhibitor of orotate phosphoribosyltransferase, LI, 163
of orotidylate decarboxylase, LI, 163
Bromphenol blue, XLIV, 385; XLVI, 563
marker dye, LIX, 67, 69, 429, 431, 509
percentage in gel, LIX, 567, 574
in tRNA digestion mixture, LIX, 107

as tracking dye, XXXII, 86, 96
Bromthymol blue
indicator for lipids, XXXV, 406
internal pH, LV, 565
Bronchial mucus inhibitor, human
amino acid composition, XLV, 874
assay
chymotrypsin inhibition, XLV, 870
elastase inhibition, XLV, 870
pronase inhibition, XLV, 870
trypsin inhibition, XLV, 870
characteristics, XLV, 869
properties, XLV, 873
purification, XLV, 871
purity, XLV, 873
Bronson sonic oscillator, LIII, 311, 554
Bronwill Biosonik II sonicator, LI, 215
Brownian diffusion, rotational, isotropic, model, XLIX, 489
Browning, of plant tissue, XXXI, 532
Bruker HX270 spectrometer, LIX, 646
Bruker 360 MHz NMR spectrometer, LIV, 201
Bruker WHX–90 spectrometer, LIV, 190
Brunswick motor, for perfusion apparatus, XXXIX, 5
Brush border
disruption, XXXI, 126–129
electron microscopy, XXXI, 119
electrophoretic purification, XXXI, 756–757
human, preparation, XXXI, 131, 132
from intestinal mucosa, isolation, XXXII, 672, 673
isolation from heart muscle, XXXI, 134–144
from intestine, XXXI, 123–134
from kidney, XXXI, 115–123
from skeletal muscle, XXXI, 134–144
from smooth muscle, XXXI, 134–144
negative staining, XXXII, 31, 33
purity criteria, XXXI, 120, 121
storage, XXXI, 122, 123
subfractionation, XXXI, 126
Br–X–537A antibiotic, LV, 445
Bryan high-titer strain, Rous sarcoma virus, LVIII, 380
BSA, *see* Bovine serum albumin
BSC-1 cell line, properties and uses, XXXII, 585
BTEE, *see* Benzoyl-L-tyrosine ethyl ester
BTS R3–111 catalyst, LIV, 117, 404
Bucher medium, LII, 379, 380
Buchler Densi-Flow apparatus, XXXVI, 162, 163
Buchler Mixing Device, XXXVI, 160
Buchler Polystaltic pump, LIX, 373
Buchler Universal Piercing Unit, XXXVI, 162
Buffer, *see also* specific type
for adrenal tissue incubation, XXXVII, 298
for affinity chromatography, XXXVI, 113
alumina column chromatography, XXXVIII, 46
boiling
for extraction of adenine nucleotides, LVII, 81
of guanine nucleotides, LVII, 88
calcium, LVI, 351, 352
changes, sequence system, XLVII, 12
choice
for microsomal protein solubilization, LII, 202, 203
for use with cyclohexanedione, XLVII, 158
in chromosome isolation, XL, 77, 87
coupling, XLVII, 42
coupling yields, XLVII, 270–272
for cytoplasmic receptor extraction, XXXVI, 189
degradation, nonaqueous, XLVII, 272, 273
deoxygenation, LIII, 315
diethylpyrocarbonate stability, XLVII, 436
digestion, LVIII, 542, 543
effect on activity of entrapped enzymes, XLIV, 187
on chemiluminescence, LVII, 438, 439
on culture, LVIII, 70, 72, 73
for estrogen-receptor purification, XXXVI, 352

for ferrocytochrome *c* oxidation determination
for gel electrophoresis, XXXII, 95–96
for gel filtration eluent, XLVII, 105
heat inactivation, of thymidylate synthetase, LI, 96
heat of protonization, XLIV, 664
for hepatocyte isolation, LII, 61, 62
hydrocarbon/water interface, LV, 773
inorganic, for peptide separation, XLVII, 204, 205, 207, 208
interference
in amino acid analysis, XLVII, 9–11, 13–15
in nuclear magnetic resonance studies, XLIX, 255
in OPRTase-OMPdecase assay and storage, LI, 167
in quantitative histone determination, XL, 112
for iron transport studies, LVI, 389
isolation of mitochondria, LV, 4
in media, LVIII, 199, 200
metal-free, preparation, LIV, 475, 476
for monolayer formation, XXXII, 553
nonvolatile, disadvantages, XLVII, 222
for optical-difference spectroscopy techniques, LII, 263
preparation of ribosomal fragments, LIX, 462, 471, 472
for quinone determination, LIII, 586
for quinone extraction, LIII, 580
for short analysis time, XLVII, 24–27
staphylococcal protease specificity, XLVII, 190, 191
for thermolysin hydrolysis, XLVII, 183
for thin-layer electrophoresis, XLVII, 196, 200
for transmission electron microscopy, XXXIX, 135, 136
ultraviolet absorption monitoring, XLVII, 204–206, 214
volatile, XLVII, 214
for microbore column chromatography, XLVII, 214, 222, 233
use in sequence analysis, XLVII, 260
Buffer system, discontinuous, SDS-polyacrylamide gel electrophoresis, XLVIII, 8
Bumble bee, glycerol-3-phosphate dehydrogenase, XLI, 244
α-Bungarotoxin, XXXII, 309–323; XLVI, 573
effect on acetylcholine, XXXII, 310
preparation, XXXII, 311–313
properties, XXXII, 311–313
receptors, XXXII, 314–317
Bungarus multicinctus, α-toxin, XXXII, 310
Bunte salt group, *see* *S*-sulfo group
Buoy, magnetic suspension
construction, XLVIII, 48–50
ferromagnetic core, XLVIII, 48, 49
glass-jacketed, XLVIII, 49
for osmometer, XLVIII, 62–63
polypropylene-jacketed, XLVIII, 49
rotatable, XLVIII, 59–61
Burner system, for atomic emission spectroscopy, LIV, 450, 451
Busycon canaliculatum, hemocyanin, XLIX, 137, 140
2,3-Butanediol
biosynthetic system, in *Aerobacter aerogenes,* XLI, 518–533
synthesis in *Aerobacter aerogenes,* XLI, 519
Butanediol bisphosphate, XLVII, 496
2,3-Butanediol/borate buffer
effectiveness in minimizing alkaline rearrangements or degradative reactions, XLI, 21
in ion-exchange chromatography of saccharides, XLI, 11
1,4-Butanediol-diglycidyl ether, XXXIV, 25
1,4-*n*-Butanediol diglycidyl ether, XLIV, 33
Butanediol divinyl ether-malate copolymer, XXXIV, 503
2,3-Butanedione, LV, 518
modification of arginine, LIII, 143
2,3-Butanedione monoxime, in dihydroorotase assay, LI, 123
Butanol
in adenine determination, LI, 263

chromatographic separation of bases and nucleotides, LI, 551
of nucleosides, LI, 509
of pyrimidines, LI, 409
in dihydroorotate dehydrogenase purification, LI, 65
for endogenous substrate depletion, LII, 271
for extraction of adenine, LI, 575
substrate, of reconstituted MEOS system, LII, 356, 367
type of binding reaction with cytochrome *P*-450, LII, 264
n-Butanol, XLIV, 58, 94, 539, 540
for extraction of adenine nucleotides, LVII, 81
of guanine nucleotides, LVII, 88
in purification of *Cypridina* luciferin, LVII, 367
substrate, of alcohol dehydrogenase, XLIV, 64, 407
thioacetylation, XLVII, 292
tert-Butanol, XLVII, 341
solvent, in peroxyoxalate chemiluminescence reactions, LVII, 458
stopped-flow nuclear magnetic resonance, XLIX, 321
1-Butanol
purification of flavocytochrome b_2, LIII, 241, 251, 253
synthesis of 8α-flavins, LIII, 458
1-Butanol-acetic acid-water, purification of flavin peptides, LIII, 456, 459
Butanol-hydrochloric acid-water, in tRNA sequence analysis, LIX, 62
Butanol-pyridine-acetic acid-water, chromatographic separation of MS2 peptides, LIX, 301
2-Butanone, *see also* Methyl ethyl ketone method
electron spin resonance spectra, XLIX, 383
siroheme solubility, LII, 437
synthesis of ethyl-4-iodo-2-diazoacetoacetate, LIX, 811
Butesin-agarose, XXXIV, 428–430
Butirosin
solvent system, XLIII, 119, 120
structure, XLIII, 613
n-Butyl acetate, XLIV, 230
tert-Butyl acetate, XLVII, 525
n-Butyl alcohol, *see n*-Butanol
tert-Butyl alcohol, *see tert*-Butanol
n-Butylamine, electron spin resonance, XLIX, 383
Butyl-*p*-aminobenzoate-agarose, XXXIV, 428
β-*tert*-Butyl-L-aspartic acid
N^{α}-benzyloxycarbonyl derivatives, XLVII, 529
synthesis, XLVII, 530
tert-Butylazidoformate
lysine derivatization, XLVII, 537
stability, XLVII, 516
toxicity, XLVII, 513, 516
sec-Butylbenzene, solvent, in saturation-transfer spectroscopy, XLIX, 501, 509
n-Butyl benzoate, XLIV, 204–206
effects on membrane permeability, XLIV, 206
Butyl biguanide, XLVI, 549
tert-Butyl-(*N*-*tert*-butyloxycarbonyl-2-aminoethyl) phosphate, intermediate in phosphatidylethanolamine synthesis, XXXV, 457, 458
γ-*tert*-Butyl-L-glutamic acid
N^{α}-benxyloxycarbonyl derivatives, XLVII, 531, 532
synthesis, XLVII, 532
β-*tert*-Butyl group, XLVII, 522, 542
removal, XLVII, 545, 546, 607, 614
Butylguanidine, inhibitor, of clostripain, XLVII, 168
tert-Butyl hypochlorite, XLVII, 456
n-Butylisocyanate, XLVI, 534
Butyl malonate
cell fractionation, LVI, 217
transport, LVI, 248, 292, 355
tert-Butyl nitrate, XLVII, 384
source, XLVII, 382
tert-Butyloxycarbonyl, XXXIV, 508
blocking group, XXXVII, 418; XLVI, 199, 200
N-t-Butyloxycarbonyl-α-alanine, XLVI, 270
esterification, XLVI, 265, 266
N-t-Butyloxycarbonyl-α-alanylimidazole, XLVI, 262

N-tert-Butyloxycarbonyl-β-alanyl-*N'*-acryloylhexamethylenediamine, XLVII, 267
N-t-Butyloxycarbonyl amino acid, XLVI, 610
tert-Butyloxycarbonylamino acid methylamide
conversion to PTMA-amino acid, XLVII, 379
synthesis, XLVII, 384
N-t-Butyloxycarbonyl-*p*-aminophenylalanine, XLVI, 98
tert-Butyloxycarbonylaspartic acid-β-*tert*-butyl ester, dicyclohexylammonium salt, conversion to free amino acid, XLVII, 382
tert-Butyloxycarbonyl azide, amino group blocking, XLVII, 279
N-tert-Butyloxycarbonyl-*p*-azidophenylalanine, synthesis of puromycin derivative, LIX, 808
Butyloxycarbonyl-*O*-benzyl-L-tyrosine-4-nitrophenyl ester, XXXIV, 505
N-tert-Butyloxycarbonylethanolamine, intermediate in phosphatidylethanolamine synthesis, XXXV, 457
tert-Butyloxycarbonyl-glycine resin, in LH-RH synthesis, XXXVII, 418, 419
t-Butyloxycarbonylglycylleucylphenylalanylchloromethyl ketone, XLVI, 206
tert-Butyloxycarbonyl group
α-amino group blocking, XLVII, 512–516
removal, XLVII, 267, 513, 535, 545, 546, 593, 609
stability, XLVII, 513
6'-*N-tert*-Butyloxycarbonylkanamycin, XLIII, 289, 290
N^{ϵ}-*tert*-Butyloxycarbonyl-L-lysine, synthesis, XLVII, 537
2-*tert*-Butyloxycarbonyloxyimino-2-phenylacetonitrile, XLVII, 516, 517
N^{α}-*tert*-Butyloxycarbonyl-N^{im}-tosyl-L-histidine, synthesis, XLVII, 541
tert-Butyloxycarbonyl-L-tryptophan, source, LIX, 236
N-tert-Butyloxycarbonyl-L-tyrosine, XLVI, 494
tert-Butyloxycarbonylvaline diazomethyl ketone, XLVI, 202
tert-Butyloxycarbonylvalylhydroxide, XLVI, 202
Butyl–PBD
in scintillation fluid, LI, 91; LIX, 358
tert-Butyl phosphate, intermediate in phosphatidylethanolamine synthesis, XXXV, 461
Butyl rubber tubing, in low-temperature EPR spectroscopy, LIV, 119
n-Butyl stearate, internal standard, XLVII, 34
S-n-Butyl thiuronium iodide, XLVII, 358
tert-Butyl-type protector
removal by formic acid method, XLVII, 545, 546
by trifluoroacetic acid method, XLVII, 545
Butyraldehyde, reductive alkylation, XLVII, 472
Butyribacterium rettgeri, D(−)-lactate dehydrogenase, assay, purification, and properties, XLI, 299–303
Butyric acid
activator of aspartate carbamyltransferase, LI, 49
ethyl ester
inhibitor of carboxylesterase, XXXV, 191, 209
substrate for carboxylesterase, XXXV, 191, 209
measurement of ΔpH, LV, 562
synthesis by fatty acid synthase, XXXV, 73, 81, 83
γ-Butyrobetaine hydroxylase, LII, 12
n-Butyryl-cAMP, *see n*-Butyryl cyclic adenosine monophosphate
Butyrylcholinesterase, *see also* Cholinesterase
turnover number, XLIV, 647
Butyryl-CoA dehydrogenase, LIII, 402
Butyryl coenzyme A
inhibitor of acetyl-CoA:amine acetyltransferase, XXXV, 253
primer for mammalian fatty acid synthase, XXXV, 38
product of fatty acid synthase, XXXV, 81
substrate for fatty acid synthase, XXXV, 58, 65, 73

Butyryl-cyclic adenosine monophosphate, in studies on parotid slices, XXXIX, 461–464
n-Butyryl cyclic adenosine monophosphate, XLVI, 90
Butyrylthiocholine, as cholinesterase substrate, XXXII, 88, 89
Butyrylthiocholine iodide, substrate, cholinesterase, XLIV, 637, 648, 655, 658
Bühler homogenizer, freeze-stop tissue, LVI, 203

C

C–55, inhibitor, in fatty acid bioassay, LVII, 194
Ca^{2+}-activated ATPase, *see* Adenosinetriphosphatase, Ca^{+}-activated
Cabbage
  enzyme studies, XXXI, 521, 525, 527
  spherosome isolation, XXXI, 578
Cabbage looper, media, LVIII, 457
Cacodylate buffer, *see* Cacodylic acid buffer
Cacodylate-Tris buffer, *see* Cacodylic acid-tris(hydroxymethyl)aminomethane buffer
Cacodylic acid
  crystallization of transfer RNA, LIX, 6
  NMR studies of transfer RNA, LIX, 25
Cacodylic acid buffer, XXXI, 20, 21, 185; XXXIX, 135; LI, 100
  assay of cytochrome *c*, LIII, 160
  determination of ferrocytochrome *c* oxidation, LIII, 45
  for Raman spectroscopy, XXXII, 255
Cacodylic acid-tris(hydroxymethyl)aminomethane buffer, purification of nitrogenase, LIII, 315
$C\bar{1}$ activator, XLV, 752
Cadmium
  atomic emission studies, LIV, 455
  in ferritin purification, XXXII, 61
  volatilization losses, in trace metal analysis, LIV, 480
Cadmium-111, lock nuclei, LIV, 189
Cadmium-113, lock nuclei, LIV, 189
Cadmium acetate, iodometric assay of lipid hydroperoxides, LII, 307
Cadmium gas laser, wavelength, XLIX, 85
Cadmium ion
  activator of uridine-cytidine kinase, LI, 319
  inhibitor of
    phosphoribosylpyrophosphate synthetase, LI, 11
    of uridine nucleosidase, LI, 294
  as uncoupler, LV, 471
Cadoxen, *see* Triethylenediaminecadmium hydroxide
Caeruloplasmin, partial specific volume, XLVIII, 29
Caffeate 3,4-dioxygenase, LII, 19
Caffeine
  cAMP inhibitor, XXXVIII, 274, 275
  cyclase assays, XXXVIII, 121
  in melatonin bioassay, XXXIX, 384
  as mutagen, XLIII, 36
  phosphodiesterase, XXXVIII, 244
  protein kinase, XXXVIII, 363
Calcarisporium arbuscula, aureovertin preparation, LV, 478–481
Calciferol, *see* Vitamin D
Calcimycin, LV, 445
Calcined alumina, *see* Alumina oxide
Calcite polarizer, LIV, 5
Calcitonin
  adenylate cyclase, XXXVIII, 151–153
  effect on collagens, and administration, XL, 319
Calcium
  atomic emission studies, LIV, 455
  in cell, measurement, XXXIX, 515–520
  cellular uptake, XXXIX, 520–528
  deficiency, XL, 321
  deprivation, synchrony, LVIII, 261, 262
  effect on cytoplasmic estrogen receptors, XXXVI, 178, 180
    on endocrine cell medium, XXXIX, 119
  efflux
    kinetic analysis, XXXIX, 547–555

methods for measurement, LVI, 349, 350
factor V, bovine, XLV, 117
flux, hormone effects, *in vitro* studies, XXXIX, 513–573
influx
kinetic analysis, XXXIX, 528–546
measurement
direct methods
inhibitor-stop technique, LVI, 343–346
labeled calcium uptake, LVI, 342, 343
use of calcium-selective electrodes, LVI, 347–349
indirect methods
activation of respiratory chain, LVI, 339–341
ejection of protons, LVI, 341, 342
redox shift of respiratory chain carriers, LVI, 341
in media, LVIII, 68
in mitochondrial matrix, LVI, 254
phospholipase C requirement, XXXII, 161
prothrombin, XLV, 151, 153
role in isolation of subcellular membrane, LII, 86, 87
salivary gland uptake, XXXIX, 473–476
sarcoplasmic reticulum uptake, XXXII, 475, 476, 480
steady state fluxes, perturbations, XXXIX, 555–563
transport
in chromaffin vesicles, LVI, 322, 323
hydrogen ion transport, LVI, 336–338
measurement, everted membrane vesicles, LVI, 233–241
by mitochondria, LVI, 320, 321
parathyroid hormone effects, XXXIX, 20
in sarcoplasmic reticulum, LVI, 321, 322
in single cells *in situ*, LVI, 323–326
Calcium-45
in calcium flux studies, XXXIX, 514, 515
in cellular uptake studies, XXXIX, 520–528
desaturation curves, XXXIX, 571
in efflux studies, XXXIX, 547–555
in influx studies, XXXIX, 529
in membrane studies, XXXII, 884–893
properties, XXXII, 884
Calcium acetate
assay of aequorin bioluminescence, LVII, 278, 279
dissociation of flavoproteins, LIII, 435
Calcium-binding protein, lanthanide ions, XLV, 198
Calcium carbonate, in seawater complete medium, LVII, 155
Calcium chloride, XLIV, 89–91, 94, 121, 130, 324, 446, 447, 523, 529, 531, 863, 865, 866
assay of aequorin, LVII, 298
of cyclic nucleotide phosphodiesterases, LVII, 97
of modulator protein, LVII, 109
of PAP, LVII, 253
of succinate dehydrogenase, LIII, 468
calibration of calcium-aequorin assay, LVII, 325
in coupling buffer, XLVII, 44
drying firefly lanterns, LVII, 6
isolation of mitochondria, LIII, 22
Keilin-Hartree preparation, LV, 125
for microsomal protein precipitation, LII, 83–89, 191
preparation of calcium phosphate gel, LIII, 182; LVII, 8
of chromophore-labeled sarcoplasmic reticulum vesicles, LIV, 51
of mitochondrial ribosomes, LIX, 425
purification of adenosine deaminase, LI, 509, 510, 512
of CPSase-ATCase-dihydroorotase complex, LI, 124
of luciferin sulfokinase, LVII, 249
of monoamine oxidase, LIII, 497, 498

reversible dissociation of flavoproteins, LIII, 434
Calcium-EGTA buffers, *see* Calcium-ethyleneglycol-bis(β-aminoethyl ether)-*N,N*′-tetraacetic acid buffer
Calcium-ethyleneglycol-bis(β-aminoethyl ether)-*N,N*′-tetraacetic acid buffer, for renal tubule studies, XXXIX, 14, 15
Calcium fluoride, infrared spectral properties, LIV, 316
Calcium fluoride cell, in IR spectroscopy, LIV, 303, 315, 316
Calcium hydroxide, XLIV, 284, 863
Calcium ion
A23187, LV, 449
accumulation, respiration, LV, 8, 9
activator of adenine phosphoribosyltransferase, LI, 573
of deoxycytidine kinase, LI, 345
of deoxycytidylate deaminase, LI, 417
of nucleoside diphosphokinase, LI, 385
of phospholipase D, XXXV, 230
of pyrimidine nucleoside monophosphate kinase, LI, 329
of thymidine kinase, LI, 370
adrenal mitochondria, LV, 10, 11
assay of aequorin, LVII, 293, 297, 298, 301
binding affinity to aequorin, LVII, 326–328
binding resins, LVII, 312
biochemical systems, LVI, 445
brown adipose tissue mitochondria, LV, 76
buffers for mitochondrial studies, LVI, 351, 352
coarctation, XLIV, 179
coelenterate luciferases, LIII, 560, 562
cofactor, for apyrase, LVII, 70
of modulator protein, LVII, 108
as contaminant, control, LVII, 301, 302
effect on liver fractionation, XXXI, 8, 9
energy-independent binding to mitochondria, LVI, 341, 345, 351
enzyme activation, XXXI, 591, 592
eriochrome blue, LVI, 315
in glassware, LVII, 282, 301
glucose isomerase inhibitor, XLIV, 816
heart mitochondria, LV, 14
hemocyanin association studies, XLVIII, 260–265
indicators
antipyrylazo III, LVI, 326–328
arsenazo III, LVI, 317–326
comparison, LVI, 327–332
inhibitor of deoxythymidine kinase, LI, 365
of guanine phosphoribosyltransferase, LI, 557
of hypoxanthine phosphoribosyltransferase, LI, 557
of mitochondrial nuclease activity, LIX, 423–425
of nucleoside phosphotransferase, LI, 393
of phosphoribosylpyrophosphate synthetase, LI, 11
intramitochondrial, exchange, LVI, 350
lung mitochondria, LV, 20
measurement
of H^+/site ratios, LV, 642, 643, 648, 649
methods, LVI, 301, 302
atomic absorption spectroscopy, LVI, 302
isotype distribution, LVI, 302
photoluminescent, fluorescence and absorbance indicators, LVI, 302. 303
specfic electrodes, LVI, 302
of proton translocation, LV, 635
membrane potential, LV, 229
opsonization, LVII, 487
oxidative phosphorylation, LVI, 533, 534
phosphodiesterase activator, XXXVIII, 272, 273
phosphorylation potential, LV, 237
reconstitution studies, LV, 709

renal mitochondria, LV, 12
role in *Aequorea* bioluminescence, LVII, 276, 287–291
in *Cypridina* luminescence, LVII, 349–351, 371
in *Renilla* bioluminescence, LVII, 243
using aequorin, LVII, 282, 292–328
calibration, LVII, 300, 301, 325, 326
data analysis, LVII, 321–326
instrumentation, LVII, 293–297
light recording, LVII, 318–321
microinjection techniques, LVII, 312–318
null method, LVII, 324, 326
procedure, LVII, 297, 298
sensitivity, LVII, 298, 299, 322
in sarcoplasmic reticulum, XXXI, 241
site ratio, measurement in respiring mitochondria, LV, 649, 650
isotopic Ca^{++}-jump procedure, LV, 650–652
apparatus, LV, 650, 651
calculations, LV, 652
components of system, LV, 651
procedure, LV, 651, 652
simple steady-state rate method, LV, 652–655
simultaneous determination of Ca^{++}/site, H^+/site and H^+/Ca^{++} ratios, LV, 655, 656
smooth muscle mitochondria, LV, 15, 62, 64, 65
thermolysin hydrolysis, XLVII, 183, 185, 188
transhydrogenase, LV, 270
translocation, electrogenicity, LV, 639, 640
uptake, K^+ diffusion potential, LV, 673
Calcium ion pump, reconstitution, LV, 701, 702, 704
activity, LV, 703, 705, 708, 709
phospholipids, LV, 701, 706
Calcium-magnesium-free phosphate-buffered saline, composition, XXXII, 618
Calcium phosphate gel
cGMP-dependent protein kinase, XXXVIII, 332, 334, 335
chromatographic, preparation, LIII, 182, 183
column preparation, LI, 582
cytochrome purification, LII, 120
for detergent removal, LII, 113, 115, 203
fractionation of elongation factors, LX, 652, 656
in hydroxylapatite column preparation, LI, 114
for partial purification of NADPH-cytochrome *c* reductase, LII, 364
phosphodiesterase purification, XXXVIII, 221, 227
preparation, LVII, 8, 9
protein kinase regulatory subunit, XXXVIII, 327
purification of adenine phosphoribosyltransferase, LI, 563
of adenosine deaminase, LI, 503
of CPSase-ATCase complex, LI, 109
of cytidine deaminase, LI, 410
of cytidine triphosphate synthetase, LI, 86, 87
of deoxycytidine kinase, LI, 340, 341
of deoxythymidine kinase, LI, 363
of erythrocyte enzymes, LI, 583, 584
of FGAM synthetase, LI, 200
of firefly luciferase, LVII, 8, 9
of guanylate kinase, LI, 485
of melilotate hydroxylase, LIII, 554
of nitrate reductase, LIII, 642
of nucleoside diphosphokinase, LI, 374, 378, 379
of nucleoside triphosphate pyrophosphohydrolase, LI, 282
of OPRTase-OMPdecase complex, LI, 148, 164
of phenylalanine hydroxylase, LIII, 281
of purine nucleoside phosphorylase, LI, 520, 533

of pyrimidine nucleoside monophosphate kinase, LI, 324, 325
of pyrimidine nucleoside phosphorylase, LI, 434
of succinate dehydrogenase, LIII, 477, 478, 485
of succino-AICAR synthetase, LI, 191
of uridine-cytidine kinase, LI, 302–304
relative enzyme affinities, LI, 584
Calcium phosphate gel:cellulose
column preparation, LI, 582
preparation, LIII, 342
purification of cytochrome b_1, LIII, 234, 235
of cytochrome c_1, LIII, 185
of erythrocyte enzymes, LI, 584, 585
of purine nucleoside phosphorylase, LI, 534
of rubredoxin, LIII, 342
Calcium-selective electrode, use, LVI, 347–349
Calcium-triggered luciferin-binding protein, *see* Luciferin-binding protein, calcium triggered
Caldariomyces fumago
chloroperoxidase, LII, 521; LIV, 378
growth conditions, LII, 524, 525
Calex bead, in storage of aequorin, LVII, 312
Calf
anterior pituitary cell, suspension culture, LVIII, 207, 208
corpus callosum and centrum semiovale, source of oligodendroglia, XXXV, 567
pancreas, fractionation, XXXI, 49–59
serum, LVIII, 382
fetal, LVIII, 95, 213
thymus
crude extract preparation, LI, 339
deoxycytidine kinase, LI, 339
Calibration
absorbance indicators, LVI, 309, 310
solution, for ion selective electrodes, LVI, 365
Calliphora erythrocephala, salivary gland studies, XXXIX, 467–469
Callus culture, LVIII, 478
establishment, LVIII, 479, 480
Calmagite, magnesium ions, LVI, 316
Calomel electrode, LIV, 422
saturated, in cyclic voltammetry, LVII, 511
standard potentials, LIV, 423
Calonectria, XLIII, 721
Calorimeter
commercial models available, XLIX, 5
design, XLIX, 5–7
sensitivity, XLIX, 7
Calorimeter-light detector unit, calibration, LVII, 543–545
Calorimetry
differential heat capacity, applications, XLIX, 3–14
differential scanning type, XXXII, 262–272
principles, LVII, 541–543
sources of error, XLIX, 9
Calsequestrin, preparation, XXXII, 299–301
Camera, *see also* Streak camera
in nanosecond absorbance spectroscopy, LIV, 40
for neutron scattering
D11, LIX, 754, 755
D17, LIX, 755
for small-angle X-diffraction, XXXII, 212, 213
(Ca^{2+}, Mg^{2+})ATPase, *see* Adenosinetriphosphatase, (Ca^{2+}, Mg^{2+})-activated
cAMP, *see* Cyclic adenosine monophosphate
d-Camphor
in *Pseudomonas putida* growth medium, LII, 151
removal from cytochrome P–450_{cam}, LII, 155, 156
as stabilizing factor *Rhizobium* P-450, LII, 160
type of binding reaction with cytochrome P–450, LII, 264, 276
Camphor hydroxylase, system, LII, 151–157
Camphor 5-methylene hydroxylase, LII, 14
Camphor 5-monooxygenase, system, LII, 167–169

Camphor-10-sulfonic acid, magnetic ellipticity, XLIX, 156
(+)-Camphor-10-sulfonic acid, circular dichroism standard, LIV, 252, 290
Camptothecin, LVI, 31
  inhibitor of DNA and RNA, XL, 278, 280
Canalco electrophoresis apparatus, XXXII, 72
Canavalia ensiformis, lectin, XXXIV, 334
Canavanine, XLIII, 454, 455
  ammonium hydroxide transamination, XLIII, 458
Cancer
  cultured cells, XXXII, 800, 801
  glycosphingolipid change, XXXII, 347
Cancer cell membrane, X-ray studies, XXXII, 220
Cancer magister, hemocyanin, XLIX, 137–140
Candicidin, XLIII, 136
Candida albicans, stendomycin assays, XLIII, 771
Candida krusei, cytochrome *c*, LIII, 143, 151
  CD spectrum, LIV, 276, 278
Candida tropicalis, cytochrome *P*–450 enzyme system, LII, 202
Candida utilis
  disruption, LV, 140, 141
  fructose-1,6-diphosphatase, assay, purification, and properties, XLII, 347–353
  glucose-6-phosphate dehydrogenase, assay, purification, properties, XLI, 205–208
  growth conditions, LV, 423
  mitochondria, ATPase inhibitor, LV, 402, 404, 405, 421–426
  NADH dehydrogenases
    piericidin A, LV, 457
    rotenone, LV, 456
  oxidase, LII, 19
  6-phosphogluconate dehydrogenase, assay, purification, and properties, XLI, 237–240
  D-ribose-5-phosphate isomerase, assay, purification, and properties, XLI, 427–429
  sedoheptulose-1,7-diphosphatase, assay, purification, and properties, XLII, 347–353
  spheroplasts, XXXI, 613, 614
  transaldolase, assay, purification, and properties, XLII, 290–297
  vacuole isolation, XXXI, 574
Canine thymus cell, LVIII, 414
Cann and Goad method, XLVIII, 213, 238, 239
Cannula
  glass, for thyroid gland perfusion, XXXIX, 361
  for liver studies, XXXIX, 26, 30
  for pituitary stalk, XXXIX, 177
Cannula holder, XXXIX, 182
  diagram, XXXIX, 181
Cannulation
  aortic, in hemicorpus perfusion, XXXIX, 77–79
  of heart, XXXIX, 45, 46
  in kidney perfusion, XXXIX, 6, 7
  of pulmonary artery, XXXIX, 46, 47
Cap, glass, commercial source, LIV, 126
Capacitor
  high-voltage, safety measures, LIV, 95, 96
  preferred type in luminescence assays, LVII, 539
Capacity
  agarose columns, XXXIV, 494
  cysteine-containing proteins, XXXIV, 549
  for papain, XXXIV, 545, 546
  for thiol compounds, XXXIV, 545, 546
Capillary, artificial, cell culture, LVIII, 178–184
Capillary culture unit, LVIII, 178–184
  construction, LVIII, 179–181
  histology, LVIII, 183
  occlusion, LVIII, 182
  oxygenation, LVIII, 183
  perfusion, LVIII, 181–183
  seeding, LVIII, 182
  sterilization, LVIII, 182
  use, LVIII, 183, 184
Capillary tube, cleaning procedure, LIX, 8
Capreomycin, XLIII, 146
Caprine, *see* Goat

Caprolactam, XLIV, 118, 126
Caproyl coenzyme A
in assay of carbon dioxide exchange component reaction of fatty acid synthase, XXXV, 49
synthesis, XXXV, 48
Caproylpantetheine
in assay for acyl carrier protein, XXXV, 110
for acyl carrier protein synthetase, XXXV, 96
synthesis, XXXV, 96
Caprylic acid, purification of adenylate kinase, LI, 470
Capsicum frutescens, mutarotase, purification, inhibitors, and properties, XLI, 485–487
Caragana arborescens, lectin, XXXIV, 334
Carbamate pesticide, detection, XLIV, 648, 657, 658
covalent binding, XLIV, 61, 62, 87, 118–120, 141, 142, 144–145
mechanism, XLIV, 17, 18
Carbamic acid, formation, XLIV, 27
Carbamoylcholine, XLVI, 587
Carbamoyl-β-alanine, activator, of aspartate carbamyltransferase, LI, 49
Carbamoyl-L-aspartate
colorimetric determination, LI, 123
product, of aspartate transcarbamcylase, LI, 35, 41, 51, 111
S-Carbamoyl-L-cysteine, XLVI, 418
S-Carbamoyl cysteine *O*-carbamoyl serine, XLVI, 415
Carbamoyl group hydrolysis, XXXIV, 46
2-Carbamoyl-6-methoxybenzothiazole
chromatographic properties, LVII, 26
preparation, LVII, 19
spectral properties, LVII, 26
structural formula, LVII, 19
Carbamoyl phosphate
product, of glutamine-dependent carbamyl-phosphate synthase, LI, 21, 29, 111
reaction with biotin carboxylase, XXXV, 31
recrystallization, LI, 52
substrate of aspartate transcarbamylase, LI, 35, 41, 51, 111
Carbamoyl-phosphate synthetase, XLVI, 152, 420, 425, 426; LI, 21–35, 105–134
activity, LI, 21–23, 29, 30
assay, LI, 30, 31, 105–107, 122
dissociation, LI, 34, 35
from *Escherichia coli*, LI, 21–29
kinetics, LI, 33, 34
molecular weight, LI, 23, 33, 121
from *Neurospora*, LI, 105
preparation, LI, 23–25
properties, LI, 25–27, 33, 34
purification, LI, 31–33
from rat cells, LI, 111–121
reconstituted, enzymic activities, LI, 26
regulation, LI, 21–23, 30, 32
from *Salmonella typhimurium*, LI, 29–35
stability, LI, 118
subunit interactions, LI, 27–29
subunit properties, enzymic, LI, 35
Carbamoyl-phosphate synthetase (glutamine):aspartate carbamoyltransferase:dihydroorotase enzyme complex, LI, 111–134
aggregation studies, LI, 126–133
assay, LI, 112, 113, 122, 123
dissociation, LI, 120, 121
from hamster cells, LI, 121–134
isoelectric point, LI, 118
kinetic properties, LI, 118, 119
molecular weight, LI, 118, 126, 133
primary structure, LI, 133, 134
properties, LI, 118–120, 126–134
purification from mutant hamster cells, LI, 123–126
from rat ascites hepatoma cells, LI, 114–116
from rat liver, LI, 116–118
purity, LI, 126
sedimentation coefficient, LI, 118
sources, LI, 119, 120, 121
stability, LI, 134
N-terminus, LI, 133
tissue distribution, LI, 119, 120

Carbamoyl-phosphate synthetase (glutamine):aspartate carbamyltransferase complex, LI, 105–111
assay, LI, 105–107
inhibitors, LI, 111
kinetic properties, LI, 111
molecular weight, LI, 111
from *Neurospora*, LI, 105–111
purification, LI, 107–110
purity, LI, 110
stability, LI, 110
O-Carbamoyl-L-serine, XLVI, 418
Carbamoylthiocarbonylthioacetic acid
preparation, LVII, 16
structural formula, LVII, 18
Carbamyl, *see* Carbamoyl
Carbamyl phosphate synthetase, *see* Carbamoyl-phosphate synthetase
Carbanion
generation, XLVI, 164
intermediates, XLVI, 164
paracatalytic oxidation, XLVI, 48–50
role in biological oxidations, LIII, 401
Carbazole, XLVI, 78
N-tosylated, chemiluminescent reaction, LVII, 524–526
O-Carbazyl L-serine, XLVI, 415, 419
Carbene, XLVI, 10, 579, 640
in photoaffinity labeling, XLVI, 72–77, 645
Carbenicillin, in media, LVIII, 112, 114
Carbenicillin isocyanate, XLVI, 533–535
O-Carbethoxy-*N*-acetyltyrosine ethyl ester, XLVII, 438
2-Carbethoxy-2-diazoacethydrazide
structural formula, LIX, 812
synthesis, LIX, 811
of *N*-(ethyl-2-diazomalonyl)streptomycyl hydrazone, LIX, 815
2-[^{14}C]Carbethoxy-2-diazoacethydrazide, microscale synthesis, LIX, 814
N-(3-Carbethoxy-3-diazo)acetonylblasticidin S
structural formula, LIX, 812
synthesis, LIX, 814
N-(3-Carbethoxy-3-diazo)acetonylgougerotin
structural formula, LIX, 812
synthesis, LIX, 814
S-(3-Carbethoxy-3-diazo)acetonyl-7-thiolincomycin
structural formula, LIX, 813
synthesis, LIX, 814
N-Carbethoxyhistidine, difference extinction coefficient, XLVII, 434
N-Carbethoxyimidazole
decarbethoxylation, XLVII, 438
difference extinction coefficient, XLVII, 434
stability, XLVII, 441
Carbethoxylation, XLVII, 434–438
Carbobenzoxyalanylglycylphenylalanylchloromethyl ketone, XLVI, 206
Carbobenzoxyamino acid imidazolide, reduction, XLVI, 222
Carbobenzoxy-ε-aminocaproic acid anhydride, preparation, XXXVIII, 386
N^6-Carbobenzoxy-ε-aminocaproyl-cyclic adenosine monophosphate, preparation, XXXVIII, 386
N-Carbobenzoxy-3-bromopropyl ether, XLVI, 124
N-Carbobenzoxy-3-(3-cyanophenoxy)propylamine, XLVI, 124
Carbobenzoxy-L-glutamine, peptidoglutaminase assay, XLV, 485
Carbobenzoxyglycylleucylphenylalanylchloromethyl ketone, XLVI, 201, 202, 206
Carbobenzoxylysine ester, kinetics, XLV, 19
Carbobenzoxylysylchloromethyl ketone, XLVI, 206
Carbobenzoxy-L-phenylalanine, XLVII, 441
Carbobenzoxyphenylalanylbromomethane, XLVI, 206
Carbobenzoxyphenylalanylchloromethyl ketone, XLVI, 206
N-Carbobenzoxyserine benzyl ester, intermediate in phosphatidylserine synthesis, XXXV, 465
N-Carbobenzoxy-L-tyrosine-*p*-nitrophenyl ester, XLVI, 116, 117

Carbobenzoyl-L-glutamyl-L-tyrosine, substrate for penicillocarboxypeptidases S–1 and S–2, XLV, 591
Carbodiimide, *see also* specific compounds
activation, XLVII, 280–282, 305, 552–554
amide group modification, XXXIV, 34
condensation, in steroid condensation, XXXVI, 22
coupling to glass, XXXIV, 67, 68
physical properties, LV, 477
for plastic, XXXIV, 223
with proteins, XXXIV, 52, 53
spin-labeled
availability and preparation, LV, 500
biological properties, LV, 500, 501
structure, LV, 500
Carbogen gas, LII, 61, 65, 66
Carbohydrase, immobilization, choice of pore diameter, XLIV, 160
Carbohydrate, *see also* specific type
analogs, XLVI, 362
binding sites, XLVI, 362–368
complex, *see also* specific types of animals
carbohydrate-amino acid linkages, L, 280
disaccharide units, L, 272–284
monosaccharide components
of HeLa cells, L, 177
of human colonic mucosa, L, 177
of L-cells, L, 177
of 3T3 cells, L, 177
quantitation of biosynthesis, L, 191
radioactive monosaccharide precursor labeling, L, 175–204
inhibitors of protein synthesis, L, 187–189
pH effects, L, 189
cytochrome oxidase, LIII, 76
effect on factor V activity, XLV, 118
in ω-hydroxylase, LIII, 360
hydroxylysine-linked, XL, 353
labeled, determination of internal water volume, LV, 549, 550
in media, LVIII, 66
metabolism
hormone effects, metabolic crossover plots, XXXVII, 277–295
in perfused liver, XXXIX, 34
regulatory enzymes, XXXVII, 280–282
in plasma membranes, XXXI, 87
whole cell entrapment, XLIV, 184
Carbohydrate antibiotic, XLIII, 119–129, *see also* specific antibiotic
Carbohydrate exchange chromatography, nucleic acid purification, XXXIV, 499–502
Carbohydrate exchange resin, XXXIV, 500
Carbohydrazide, modification of tRNA, LIX, 147
Carbohydrazide-HCl, *see* Carbohydrazide-hydrochloride
Carbohydrazide-hydrochloride, in determination of tRNA ester modification, LIX, 161
Carbomycin, XLIII, 131, 198; LVI, 31
mitochondria, LVI, 32
Carbon-12, nucleus
coherent scattering length, LIX, 672
scattering cross section, LIX, 672
Carbon-13
chemical shift range, XLIX, 271, 334
lock nuclei, XLIX, 348, 349; LIV, 189
Carbon-14, in steroid autoradiography, XXXVI, 153
Carbon arc, relative brightness, LIV, 95
Carbon-carbon electrode, for freeze-fracture, XXXII, 43, 44
Carbon coating, for scanning electron microscopy, XXXIX, 155
Carbon dioxide
contamination, LVIII, 202
effect on culture, LVIII, 70, 72, 135
on virus yield, LVIII, 401
evolved, XLIV, 348
high pressure ion-exchange chromatography, XXXVIII, 21, 22
as inert atmosphere, XXXI, 540
inhibitor, XLIV, 346
in media, LVIII, 201, 202
Pasteur effect, LV, 294, 295

preparation of triethylammonium bicarbonate, LIX, 61
product of firefly luciferase, LVII, 3, 58, 74
of orotidylate decarboxylase, LI, 74
radiolabeled, determination, LI, 156
reduction to acetate pathway, LIII, 361
solid, for *E. coli* lysis, LIX, 365
substrate of formate dehydrogenase, LIII, 360, 370
transport studies, XLIV, 920, 921
Carbon dioxide-free media, LVIII, 86, 87
Carbon disulfide, XLVII, 291
Carbon-hydrogen group, Raman spectra, XLIX, 98
Carbonic acid medium, growth of *Chromatium vinosum*, LIII, 309, 310
Carbonic anhydrase, XXXIV, 4; XLVI, 87, 155
atomic emission studies, LIV, 458
atomic fluorescence spectroscopy, LIV, 469
bovine, phosphorescence studies, XLIX, 246
carbon dioxide transport, XLIV, 920, 921
cobalt-substituted, magnetic circular dichroism, XLIX, 168–170
conjugate, activity, XLIV, 271
gel electrophoresis, XXXII, 101
immobilization
on membrane, XLIV, 905, 909, 920, 921
by microencapsulation, XLIV, 215
labeling, XLVI, 139
in oxyntic cells, XXXII, 716, 717
progestin assay, XXXVI, 457
spin-label studies, XLIX, 463–466
therapeutic use, XLIV, 697
Carbonic anhydrase isozyme, XXXIV, 595
Carbonium ion
formation, XLVI, 39
inhibition, XLVI, 216
intermediates, XLVI, 26
peptide deblocking, XLVII, 602, 603, 614, 615
production, XLVI, 216
Carbon membrane
advantages, LIX, 843
preparation, LIX, 615
Carbon molecular resonance spectroscopy, *see* Nuclear magnetic resonance spectroscopy, carbon-13
Carbon monoxide
in anaerobic procedures, LIV, 114, 115
in anaerobic reductase assay, LII, 91
binding to cytochrome oxidase, LIV, 531, 532
to hemoglobin, LIV, 82–84, 94, 95, 99, 100, 515, 523, 530, 531
to myoglobin, LIV, 514, 515, 523, 529, 530
to protoheme, LIV, 507–511, 513
complex IV, LVI, 586
as contaminant, LVIII, 202
in cytochrome b_5 assay, LII, 110
cytochrome oxidase, LVI, 693
dissociation from heme proteins, effect of temperature, LIV, 107
inhibitor of bacterial cytochromes, LIII, 206, 207
of complex IV, LIII, 45
of cytochrome *P*–450, mechanism, LII, 272
of hydrogenase, LIII, 296
of *Pseudomonas* cytochrome oxidase, LIII, 657
interaction with cytochrome *c* oxidase, LIII, 193, 194
in IR studies, LIV, 304, 308–310
stretching frequency, LIV, 311
Carbon monoxide-binding heme protein P–460
ammonia oxidation, LIII, 638
molecular weight, LIII, 639
from *Nitrosomonas*, LIII, 638, 639
purification, LIII, 638, 639
spectral properties, LIII, 639
Carbonmonoxyhemoglobin
preparation, LII, 458
spectral properties, LII, 459–463
Carbon paste, in optically transparent electrodes, LIV, 406
Carbon source test, identification of luminous bacterial species, LVII, 160–162

Carbon tetrachloride, XLIV, 99
IR bandwidth, LIV, 311
Carbonyl, unsaturated, as plant extraction interferents, XXXI, 531
Carbonyl cyanide *m*-chlorophenyl hydrazone, XXXII, 883
complex V, LV, 315
cyanine dye, LV, 579, 693
glutamate transport,
hepatoma mitochondria, LV, 85
inhibition of binding of 2-azido-4-nitrophenol, LVI, 670
inhibitor of luciferase assay, LVII, 49
luminescence, LVI, 543
model redox reactions, LV, 542, 543
phosphate esterification, LV, 175
proteolipid ionophores, LV, 419
uncoupling activity, LV, 465, 469, 470
Carbonyl cyanide phenylhydrazone
M. phlei membrane systems, LV, 186
photochemically active derivative, LVI, 660
solubility, LV, 463
sources and synthesis, LV, 469, 470
Carbonyl cyanide *p*-trifluoromethoxyphenylhydrazone
ATPase inhibitor, LV, 407
ATP-P_i exchange, LV, 360, 363
ATP synthesis, LV, 118
in fluorescence probe studies, XXXII, 240
measurement of proton translocation, LV, 637
transhydrogenase assay, LV, 273
uncoupling activity, LV, 465
Carbonyldiimidazole
activation, XLVII, 280, 591
synthesis of 6-[*N*-ethyl-*N*-(4-thyroxinylamido)butyl]-amino-2,3-dihydrophthalazine-1,4-dione, LVII, 433
in *N*-trifluoroacetyl-6-aminohexanol 1-pyrophosphate preparation, LI, 254
N,N'-Carbonyldiimidazole, XLIV, 288–289
guanylyliminodiphosphate preparation, XXXVIII, 425, 427
Carbonyl group, concentration determination, XLIV, 385
Carbowax, *see also* Polyethylene glycol
as plant-cell protectant, XXXI, 599
Carboxin
complex II, LVI, 586
inhibitor of succinate-ubiquinone reductase, LIII, 26
succinate oxidation, LV, 460, 461
synthesis, LV, 461
Carboxy-AIR, *see* 5-Amino-1-ribosyl-4-imidazolecarboxylic acid 5'-phosphate
Carboxyatractylic acid
ADP, ATP carrier, LVI, 407–412
carrier proteins, LVI, 251, 252
carrier sites, LVI, 251
transport, LVI, 248, 292
Carboxyatractyloside
decarboxylation, LV, 523
labeled
extraction, LV, 521
purification, LV, 521, 522
yield, LV, 522
Carboxyatractyloside-binding protein, from *N. crassa*, LV, 148
Carboxybenzoyl-L-tyrosine-*p*-nitrophenyl ester, substrate, of chymotrypsin, XLIV, 523
Carboxybiotin, assay for biotin carboxylase, XXXV, 26–28
4-Carboxy-2,6-dinitrophenyllysine, extinction coefficient, LIII, 152
Carboxyethyldisulfide monosulfoxide, reaction conditions, XLVII, 412
Carboxyhemoglobin, labeling procedure, XLVIII, 332, 333
Carboxyhemoglobin A_0^3, dissociation, XLVIII, 331–342
2-Carboxy-3-ketoribitol 1,5-bisphosphate, XLVI, 390
N-Carboxyl anhydride, XLIV, 18
Carboxylase, XLII, 457–487, *see also* specific types
Carboxylate carrier, glass, XXXIV, 70, 71
Carboxyl derivative
by deamidation, XXXIV, 35, 36
formed during formation of hydrazide, XXXIV, 36, 37
of glass, XXXIV, 66

Carboxylesterase
absorbance at 280 nm, XXXV, 224
active site titration, XXXV, 191
amino acid composition, XXXV, 209
assay, XXXV, 191
chemical properties, XXXV, 199, 207
from chicken liver, XXXV, 209–213
crystallization, XXXV, 196, 211
gel electrophoresis, XXXII, 88
from horse liver, XXXV, 218, 219
from ox liver, XXXV, 202, 204
phosphorylation of enzyme, XXXV, 202
physical properties, XXXV, 197, 198, 206, 213, 217, 220
from pig liver, XXXV, 192–197
purity, XXXV, 197, 204, 213, 216, 219
rate constant and K_m values for various substrates, XXXV, 200, 207
from sheep liver, XXXV, 214–216
stability, XXXV, 197, 205, 206, 213
stereospecificity, XXXV, 201
N-Carboxy-L-leucine anhydride, XXXIV, 712
Carboxyl group
activation, XLIV, 17, 18, 542
by active esters method, XLVII, 557–562, 568–571
by azide method, XLVII, 564–568
by carbodiimide method, XLVII, 552–554
by isoxazolium salts method, XLVII, 556, 557
by mixed anhydride method, XLVII, 554–556
blocking, XLVII, 520–526, 607, 612
concentration determination, XLIV, 131, 384
determination, XXXIV, 57; XLVII, 360, 361
ligands, XXXIV, 48
Carboxylic acid
activated, fluorescent labeling, XLVIII, 359
assay, by titration, XLIV, 92, 93
in plants, pH effects, XXXI, 534
Carboxylic ester hydrolase, conjugate, microfluorometric assay, XLIV, 374–376
Carboxylic-phosphoric anhydride, XLVI, 302–307
synthesis of AMP analog, XLVI, 303, 304
of ATP analog, XLVI, 304, 305
Carboxylmethylcysteine, XLIV, 392
Carboxyl-Spheron, XLIV, 76–80, *see also* 6-Aminocaproic acid
Carboxyltransferase, XXXV, 32–37, *see also* specific types
assay, XXXV, 32–34
reaction, XXXV, 32–34
Carboxymethylamino acid, XLVI, 423, 492
N^6-Carboxymethyladenosine 5′-triphosphate
molar absorption coefficient, XLIV, 867
synthesis, XLIV, 866, 867
Carboxymethylated amino acid, in enzyme active site studies, XXXVI, 403–405
Carboxymethylation, XLVII, 115
with iodoacetic acid, micromethod, XLVII, 199
of protein sulfhydryls, XXXII, 72
(Carboxymethyl)cellulose, XLIV, 46, 51
activation mechanism, XLIV, 18, 19
adenylate cyclase, XXXVIII, 172
alanine carrier isolation, LVI, 432
chromatography, of prolactin, XXXVII, 395, 396
factor B, LV, 387, 388
hydrazide
preparation, LX, 747, 751
reaction with periodate-oxidized polyuridylate, LX, 751–753
paper disk, assay of tryptophanyl-tRNA synthetase, LIX, 238, 239
penicillin acylase purification, XLIV, 762, 763
peptide separation, XLVII, 207
preparation of ribosomal proteins, LIX, 518, 519, 528, 557, 649–651
purification of adenine phosphoribosyltransferase, LI, 571, 572
of adenylate kinase, LI, 471, 472
of cytochrome *c*, LIII, 130, 131, 133
of cytochrome *c*-551, LIII, 657, 658

of cytochrome *c* derivatives, LIII, 149, 150, 153, 167, 170
of *Diplocardia* luciferin, LVII, 376
of electron-transferring flavoprotein, LIII, 516
of high-potential iron-sulfur protein, LIII, 337–339
of initiation factor Ef-1, LX, 643, 644
of *Pseudomonas* cytochrome oxidase, LIII, 653
of subunit 4, LV, 355
(Carboxymethyl)cellulose azide, preparation, LIX, 385
(Carboxymethyl)cellulose C–25
OSCP preparation, LV, 395
TF_1 subunits, LV, 784
Carboxymethylcysteine, XLVI, 396, 423
S-Carboxymethylcysteine sulfone, XLVI, 423
Carboxymethylcysteine thiohydantoin, marker, XLVII, 364
Carboxymethyl cytochrome *c*, temperature jump relaxation kinetics, LIV, 80, 81
Carboxymethyl disulfide, inactivation of fatty acid synthase, XXXV, 55
Carboxymethylhistidine, XLVI, 153, 423
S-Carboxymethylhomocysteine, XLVI, 423
S-Carboxymethylhomocysteine sulfone, XLVI, 423
Carboxymethylhomoserine, XLVI, 156
O-(Carboxymethyl)hydroxylamine for oxime preparation, XXXVI, 17, 18
N^{ε}-Carboxymethyl-L-lysine, synthesis, XLVI, 498, 499
3-Carboxymethylmorphine-albumin, XXXIV, 621
N^6-Carboxymethyl-NAD^+, XXXIV, 233, 234
synthesis, XLIV, 862, 863
N^6-Carboxymethyl-$NADP^+$
molar absorption coefficient, XLIV, 866
synthesis, XLIV, 864–866
6-Carboxymethyloxime-17β-estradiol, XXXIV, 675
(Carboxymethyl)Sephadex, *see also* Sephadex C–50
matrix for carboxypeptidase B, XXXIV, 411
protein kinase, XXXVIII, 302, 303, 305, 306
catalytic subunit, XXXVIII, 306, 307
purification of purine nucleoside phosphorylase, LI, 522
of thymidylate synthetase, LI, 94
(Carboxymethyl)Sephadex C–25
purification of 30 S ribosomal proteins, LIX, 486–489
of 50 S ribosomal proteins, LIX, 491–495
(Carboxymethyl)Sephadex C–50
colicin K, LV, 535
for myoglobin purification, LII, 479
in cytochrome *P*–450 purification, LII, 121, 123
for hemoglobin isolation, LII, 488
for hemoglobin reconstitution, LII, 454
purification of CTP(ATP):tRNA nucleotidyltransferase, LIX, 184, 185
of OPRTase-OMPdecase complex, LI, 151
of *Pseudomonas* cytochrome oxidase, LIII, 651
(Carboxymethyl)Sephadex C–50–120, purification of nucleoside triphosphate pyrophosphohydrolase, LI, 283
(Carboxymethyl)Sepharose, chromatography of elongation factors, LX, 679–682
O-Carboxymethylserine, XLVI, 423
O-Carboxymethyl-L-tyrosine, synthesis, XLVI, 498, 499
Carboxypeptidase, XLVI, 155, 156, 206
acid, XLVI, 206
active site, XLVI, 228, 229
diisopropyl fluorophosphate-inhibited, isolation, sources, XLV, 599
nitrotyrosine, XXXIV, 189
DD-Carboxypeptidase
alanine, chemical estimation of free, XLIII, 690, 691

D-alanine, enzymic estimation of, XLIII, 691
assay, XLIII, 690–692
crude extract preparation, XLIII, 692
fluorodinitrobenzene assay, XLIII, 690
incubation conditions, XLIII, 690
radioactive [^{14}C]Ac_2-L-Lys-D-Ala dipeptide estimation, XLIII, 691, 692
Streptomyces preparation, XLIII, 692
substrates, XLIII, 690
unit definition, XLIII, 690
Carboxypeptidase A, XXXIV, 4, 418; XLVI, 22, 225–229
arsanilazotyrosyl peptide, XXXIV, 190
atomic emission studies, LIV, 458
atomic fluorescence studies, LIV, 469
cobalt-substituted, magnetic circular dichroism, XLIX, 168
conjugate, thermal stability, XLIV, 537
crystal
immobilization, XLIV, 548, 551
substrate binding studies, XLIV, 557
X–ray analysis, XLIV, 555, 556
in digestion of oxyhemoglobin, LIV, 488
immobilization, on Sephadex G–200, XLIV, 529, 533
immobilized-peptide cleavage, XLVII, 277
infrared spectroscopy, XXXII, 252
labeling, XLVI, 228
modified, XXXIV, 184
phosphorescence studies, XLIX, 248
thermolysin hydrolysis products, XLVII, 176
unit cell dimensions, XLIV, 550
Carboxypeptidase B, XXXIV, 411–414, 418; XLVI, 225–229
activity resembling, in beta granules, XXXI, 378
alkylated, XXXIV, 414
inhibition, by amino group alkylation, XLVII, 474
modified proteins, XXXIV, 414
phosphorescence studies, XLIX, 239, 248
separation of fragments, XLVII, 99
Carboxypeptidase C, XLVI, 206; XLVII, 73–84
assay, XLVII, 76, 77
by Cbo-Leu-Phe or Cbo-Glu-Tyr, XLV, 562, 563
colorimetric, XLVII, 77
spectrophotometric, XLVII, 76, 77
commercial preparations of, contaminants, XLVII, 74–79
function, XLV, 561
preparation, XLVII, 75, 76
pretreatment with DFP, XLVII, 78, 79
properties, XLV, 566, 567
purification, XLV, 563, 565
reaction conditions, XLVII, 79, 80
reaction mixture analysis, XLVII, 80, 81
sample preparation, XLVII, 79
specificity, XLV, 566–568; XLVII, 74, 75
stability, XLVII, 78
units, XLV, 563
Carboxypeptidase inhibitor
from potatoes, XLV, 736
amino acid composition and sequence, XLV, 739
assay, XLV, 736
chemical cleavage, XLVII, 147, 148
properties, XLV, 738
purification, XLV, 736
D D-Carboxypeptidase-transpeptidase, from *Streptomyces* , XLV, 610, 611, 614, 615, 619
assay, XLV, 611
for D,D-carboxylase activity, XLV, 611
for β-lactamase, XLV, 613
electrophoresis, polyacrylamide gel, XLV, 620
hydrolysis reaction, XLV, 631
transfer reaction, XLV, 632, 635
by β-lactam antibiotics, inhibition, XLV, 635
reaction, XLV, 623, 625, 636
titration, XLV, 627
molecular weight, XLV, 619, 620
properties, XLV, 619
purification, XLV, 615, 617, 618

Carboxypeptidase Y, XLVI, 206; XLVII, 84–93
activators, XLV, 579
activity, potential, XLV, 575
amidase activity, XLV, 584, 585
assay, XLV, 568
anilidase activity, XLV, 571
esterase activity, XLV, 570, 584; XLVII, 86, 87
peptidase activity, XLV, 569; XLVII, 86
characteristics, XLV, 568
complete hydrolysis, XLVII, 41, 43
composition, XLV, 578
DFP sensitivity, XLVII, 89
distribution, XLV, 586
esterase activity, XLV, 584
hydrolysis, limited, XLV, 585
inhibitors, XLV, 579, 580; XLVII, 89, 90
kinetics, XLV, 581–583, 585
metal ion sensitivity, XLVII, 89
nomenclature, XLV, 586
peptidase activity, XLV, 581
properties, XLV, 576, 586; XLVII, 84, 85, 88, 90
chemical, XLV, 578
physical, XLV, 577
purification, XLV, 572; XLVII, 87, 88
purity, XLV, 576
reaction conditions, XLVII, 91, 92
sequence analysis, XLV, 585
specificity, XLV, 581; XLVII, 90
stability, XLV, 576; XLVII, 88
from yeast, specificity, LIX, 600
[(*S*)-1-Carboxy-2-phenylethyl]carbamoyl-L-arginyl-L-valylargininal
assay, XLV, 683
cathepsins A and B, XLV, 683
papain, XLV, 683
properties
biological, XLV, 685
physicochemical, XLV, 684
purification, XLV, 684
trypsin, XLV, 683
Carboxypropionylphenylalanine-*p*-nitroanilide, substrate, of chymotrypsin-like enzyme, XLVII, 77
2-Carboxy-D-ribitol 1,5-bisphosphate, XLVI, 390
N-Carboxy-L-tyrosine anhydride, XLIV, 327
N-Carboxytyrosyl anhydride reagent, XXXVII, 344
Carcinoembryonic antigen, LVIII, 184
Carcinogen, *see also* specific substances
cell transformation, XXXII, 586
chemical, transformation, LVIII, 370
solution, LVIII, 298, 299
Carcinogenesis, XLVI, 644
postulated mechanisms, LII, 228, 279, 412
Carcinogenicity
chemical, LVIII, 296–302
testing, LVIII, 301
survival curve, LVIII, 301
transformant induction, LVIII, 301
Carcinoma, *see also* specific types
adrenal cortex, LVIII, 376
cervical, LVIII, 376
colon, LVIII, 184
embryonal
effect of hormone, LVIII, 105
mouse, LVIII, 97
Lewis lung, LVIII, 376
lung, LVIII, 376
mammary, estrogen receptor, XXXVI, 248–254
nasopharyngeal, LVIII, 376
transplantable, LVIII, 276, 277
Cardiac glycoside, XLVI, 523
analog, XLVI, 524
photoaffinity labeling, XLVI, 527–530
Cardiac muscle, *see* Heart, muscle
Cardiolipid
freeze-fracture, XXXII, 46
phospholipase A_2 effects, XXXII, 148
as phospholipase C substrate, XXXII, 160
Cardiolipin, *see also* Diphosphatidylglycerol
adenylate cyclase, XXXVIII, 177, 180
ATPase
depletion of particles, LV, 111, 112
inhibitor, LV, 407
column chromatography, XXXV, 417
in complex I–III, LIII, 8
in *E. coli* 0111a membrane, XXXI, 642
in mitochondria, XXXI, 302, 303

plasma membrane purity, XXXI, 84, 86, 87
polyribitol phosphate polymerase, L, 389, 391, 392
of thermophilic bacterium PS3, LV, 370
molecular species, LV, 371
thin-layer chromatography, XXXV, 400
Carnitine
assay, LV, 211, 212
brown adipose tissue mitochondria, LV, 74, 76
incubation of mitochondria, LV, 8
tissue content, LV, 221
transport, LVI, 248, 249
Carnitine acetyltransferase
in assay of long chain fatty acyl-CoA, XXXV, 273
carnitine assay, LV, 211, 212
in microbodies, LII, 496, 500, 501
Carnitine-acylcarnitine translocase
assay
calculation, LVI, 377
method, LVI, 376
preparation and loading of mitochondria, LVI, 375, 376
principle, LVI, 374
reagents, LVI, 375
remarks, LVI, 377
properties, LVI, 377, 378
Carnitine derivative, partial specific volume, XLVIII, 22
Carnitine octanyltransferase, in microbodies, LII, 496, 500, 501
Carnitine palmitoyltransferase
assay, LVI, 368, 369
of CoASH released, LVI, 371, 372
hydroxamate, LVI, 372
isotope exchange, LVI, 370, 371
radioactive forward method, LVI, 372
spectophotometric, LVI, 369, 370
brown adipose tissue mitochondria, LV, 76
distribution, methods of evaluation, LVI, 372–378
properties, LVI, 374
purification, LVI, 373, 374
Carnoy's fixative, XXXVII, 158
Carotene
in chromoplasts, XXXI, 495
electrochromism, LV, 557, 558
β-Carotene 15,15′-dioxygenase, LII, 11
γ-Carotene hydroxylase, LII, 20
Carotenoid
measurement, XXXII, 866
as optical probe, LV, 573, 585
Carotenoid pigment, binding to orange peel protein, XXXI, 531
Carrier, *see also* Translocator
mitochondrial, LVI, 248
molecular approach, LVI, 250, 251
Carrier transport
constant, interpretation, XXXIX, 513
passive, in substrate entry, XXXIX, 498–502
Carrot
crude extract preparation, LI, 388
disk inoculation method, for crown gall, XXXI, 555, 556
enzyme studies, XXXI, 521, 525
nucleoside phosphotransferase, LI, 387–394
from protoplast, LVIII, 367
Carrot × petunia, LVIII, 365
Carrot root, sucrose phosphatase, purification, XLII, 344, 345
Carr-Purcell pulsed Fourier-transform method, for T_2 evaluation, XLIX, 344, 345
CARS, *see* Anti-Stokes Raman spectroscopy, coherent; Coherent anti-Stokes Raman scattering
Cartesian diver
applications, XLVIII, 18–22
as basis for microgasometric measurements, XXXIX, 403–407
calibration, XLVIII, 25, 26
Cartilage, LVIII, 560, *see also* specific types
isolation of cells, XL, 314
method of isolation, XL, 315
preparation for collagen studies, XL, 311
in somatomedin bioassay, XXXVII, 93–109
Cary 61 CD spectrophotometer, LIV, 251
Cary Model 14 spectrophotometer, LIV, 404; LVII, 205, 228

Cary Model 15 spectrophotometer, LVII, 386
Cary-Tolbert ionization chamber, LIII, 304
Casamino acid
in determination of aminoacylation isomeric specificity, LIX, 273
low-phosphate, preparation, LIX, 831
Casassa and Eisenberg method, for sedimentation studies, XLVIII, 13, 14
Casein
cAMP-binding assay, XXXVIII, 368
labeled
assay for phosphatase, LX, 526
preparation by phosphorylation, LX, 524
mammary epithelial cell synthesis, XXXIX, 450
assay, XXXIX, 452, 453
protein kinase, XXXVIII, 328, 340, 348, 351, 357, 358
purification and dephosphorylation, XXXVIII, 67
substrate
of chymotrypsin, XLIV, 31, 523
of thermolysin, XLIV, 89, 95–97
α-Casein, boundary complexes, XLVIII, 212
β-Casein, boundary patterns, XLVIII, 212
Casein hydrolyzate, as growth promoting substance, XXXII, 730
Casein kinase, *see also* Protein kinase
catalytic properties, LX, 503, 504, 524, 525
criteria of purity, LX, 503
purification, LX, 499–502, 507
Caseinolytic activity
determination, XLV, 26
assay procedure, XLV, 27
unit, definition, XLIV, 96
Casitone growth medium, composition, XXXI, 635, 636
Castor bean
glyoxysome isolation, XXXI, 565–571
spherosome isolation, XXXI, 578
Castor oil, oxidase, LII, 20
Cat, kidney cell, media, LVIII, 58
Catabolite repression, in bacterial luminescent system, LVII, 128, 164
Catalase, XXXI, 40
activity studies with salicylate hydroxylase, LIII, 539
affinity for calcium phosphate gel, LI, 584
in aggregation procedure, XLIV, 268
amino acid oxidase, LVI, 484
assay, XXXI, 361, 743
with luminol, LVII, 404
perborate, XLIV, 685
of phenylalanine hydroxylase, LIII, 279
of superoxide dismutase, LIII, 385
associated with cell fractions, XL, 86
atebrin fluorescence quenching, LVI, 114
catalytic reaction, XLIV, 478, 479
circular dichroism studies, LIV, 283
in coelomic cell homogenization buffer, LVII, 376
commercial source, XLIV, 481
conformational composition, LIV, 269, 271
conjugate, activity, XLIV, 156, 200, 271, 272, 482, 483
as electron microscopic tracer, XXXIX, 147
in enzyme thermistor, XLIV, 674, 675
formaldehyde generation, LII, 297, 344
gel electrophoresis, XXXII, 101
glucose electrode, LVI, 482, 483
in glucose oxidase assay, XLIV, 348, 349
in glyoxysomes, XXXI, 565, 569
as heme-containing protein, in chemiluminescent oxidation system, LVII, 445
heme-deficient mutants, LVI, 559
in hydrogen peroxide assay, LII, 344
immobilization on collagen, XLIV, 153, 255
on controlled-pore ceramics, XLIV, 152, 153, 156–160
in enzyme system, XLIV, 315, 316
on glass beads, XLIV, 671
with glucose oxidase, XLIV, 181, 472, 473, 478, 488
with inert protein, XLIV, 903, 904, 908

by microencapsulation, XLIV, 207, 212–214, 684, 685
in porous matrix, XLIV, 270–272
immobilized, effect of inhibitor, XLIV, 916, 917
inhibitor
Dextran 500, LV, 131
of phagocyte chemiluminescence, LVII, 492
interference
in aromatic hydroxylation assays, LII, 412
in hydrogen peroxide assay, LII, 344, 349, 350
as iodination interferent, XXXII, 108, 109
kinetics, XLIV, 483–488, 686
in liver-cell fractions, XXXI, 209
luciferase-mediated oxygen assay, LVII, 225
as marker enzyme, XXXI, 20, 46, 496, 735, 741–743; XXXII, 15, 29, 35
in microbodies, LII, 495–497
molecular weight, LIII, 350, 351
nicotinic acid receptors, XXXII, 320
in oxygen consumption studies, LIV, 487, 492, 493
in oxygen determination, LIV, 501
in oxygen scavenging system, LII, 223
in peroxisomes, XXXI, 361, 364, 367
in plant microbodies, XXXI, 489–501
polarographic measurement, LVI, 459
preparation of cross-linked ribosomal subunits, LIX, 537
properties, XLIV, 479
removal from liver microsomal components, LII, 359–361, 364
role in ethanol metabolism, LII, 355
SDS complex, properties, XLVIII, 6
soluble, assay, XLIV, 479, 482
specific activity, XLIV, 255
spin diameter, XLIV, 160
storage in solution, XLIV, 212, 213
test, LII, 361
therapeutic application, XLIV, 679, 681, 684–688
thymol-free, lipid peroxidation, LII, 304
unit cell dimensions, XLIV, 159
viscosity barrier centrifugation, LV, 133, 134, 135
Catalysis, mechanism, XLI, 121
Catalyst, *see also* specific substance
pellicular particles, XLIV, 179
Catalytic competence, XLVI, 11, 12, 54–58
definition, XLVI, 55
Cataract, risk from X-rays, XXXII, 215
CAT averaging, *see* Computer, of average transient
Catechol
biosynthesis, XXXII, 766
inhibitor of phenylalanine hydroxylase, LIII, 285
of *Pholas* luciferin oxidation, LVII, 399
stimulator of *Pholas* light emission, LVII, 397
Catecholamine, XXXIV, 105, 695; XXXV, 181
adenylate cyclases, XXXVIII, 147–149, 170, 173, 176
binding sites, XLVI, 336
cAMP inhibitor, XXXVIII, 275
effect on melanin, XXXVII, 129
on perfused liver, XXXIX, 34
on renin production, XXXIX, 22
glass bead-immobilized, XXXVIII, 180, 181
alkylamine glass synthesis, XXXVIII, 181, 182
catecholamine glass production, XXXVIII, 184
arylamine glass production, XXXVIII, 182, 183
succinyl glass and long-chain alkylamine glass synthesis, XXXVIII, 183, 184
washing procedure, XXXVIII, 185, 186
instability, XXXIV, 700
lipolytic response, XXXVII, 213
metabolism, in pineal, XXXIX, 395, 396
Catecholamine-agarose, XXXIV, 696–697
Catechol 1,2-dioxygenase, LII, 11
Catechol 1,6-dioxygenase, LII, 19
Catechol 2,3-dioxygenase, LII, 11

Catechol *O*-methyltransferase, XXXIV, 700; XLVI, 554–561
affinity labeling, XLVI, 559–561
assay, XXXII, 777–779
isolation, XLVI, 559
in neurobiology, XXXII, 765
in pineal, XXXIX, 396
Catechol oxidase, LII, 19
inhibitors, XXXI, 539, 540
Cathepsin, in lysosomes, XXXI, 330
Cathepsin A, inhibited by antipain, XLV, 683
Cathepsin B, XLVI, 206
inhibited by antipain, XLV, 683
by leupeptin, XLV, 679
Cathepsin D
in brain homogenates, XXXI, 464, 466, 467, 469–471
effect on purified NADH-cytochrome b_5 reductase, LII, 108
as lysosome marker, XXXI, 406, 408
in macrophage fractions, XXXI, 340, 344, 355
Cathepsin G, XLVI, 201, 206
Cathodoluminescence, in scanning electron microscopy, XXXII, 50, 53, 58, 59
Cation
antibiotics, membranes, XXXII, 881–893
ATPase activity, LV, 795, 796
ATP-P_i exchange, LV, 360, 361
chelation, adenylate kinase, LI, 466, 467
concentration determination, in perfusion experiments, LII, 57, 58
divalent
adenylate cyclase, XXXVIII, 120, 141, 142, 146, 160, 169, 173
guanylate cyclase, XXXVIII, 195, 198, 199, 201
inhibitor of phosphoribosylpyrophosphate synthetase, LI, 5, 11
phosphodiesterase, XXXVIII, 223, 233, 234, 243, 248, 256
polyoxyethylene detergents, LVI, 743
protein kinases, XXXVIII, 314, 322, 343, 344
fluxes mediated by carboxylic ionophores, LV, 674, 675
ionophores, LV, 436–438
of mitochondrial matrix, LVI, 254
monovalent, activators of ribonucleoside triphosphate reductase, LI, 258
polyvalent, lasalocid, LV, 446
radioisotopes for membrane studies, XXXII, 884
transport, simultaneous measurement with other absorbance changes, LVI, 332–338
cytochrome oxidation-reduction, LVI, 334–336
hydrogen and calcium ion transport, LVI, 336–338
swelling-shrinkage of cells and cell fractions, LVI, 334
Cavernularia
coelenteramide, LVII, 344
luciferin sulfokinase, LVII, 248
Cavitation method, separation of particulate and cytosolic cell fractions, LVI, 214, 215
application to studies of metabolic regulation, LVI, 220, 221
calculation of result, LVI, 218–220
combination with digitonin method, LVI, 221–223
criteria for satisfactory separation, LVI, 217, 218
experimental procedure, LVI, 215–217
C–banding, *see* Centromere banding
CBG, *see* Corticosteroid-binding globulin
4CC, *see* Condensation reaction, four-component
CCCP, *see* Carbonylcyanide *m*-chlorophenylhydrazone
C_6 cell, *see* Rat, glial C_6 cell
c–C–PR antibiotic
as ionophore, LV, 443, 448
structure, LV, 441
CD, *see* Circular dichroism
CDAK, XLIV, 293
CDNB, *see* 4-Chloro-3,5-dinitrobenzoic acid
Cecropia silkmoth, tissues, protein kinases, XXXVIII, 335
Cefazolin, XLIII, 41, 73

Celery, enzyme studies, XXXI, 521 Celesticetin, XLIII, 336 Celestosaminide, XLIII, 253 Celite assay of L-amino acids, LIX, 315 estrogen purification, XXXVI, 51, 52 in heme isolation, LII, 424 in α -hydroxylase assay, LII, 311, 313 packing procedure, LII, 424, 425 pituitary gland chromatography on, XXXI, 411-412 preparation, LII, 423 of aequorin crude extract, LVII, 280, 306 recrystallization of bile salts, LIII, 4, Celite 505, XLIV, 762 Celite chromatography, in nucleoside triphosphate pyrophosphohydrolase purification, LI, 280, 281 Cell, XXXII, 555-893, see also specific types affinity fractionation of, see Cell, fractionation anchorage dependence, LVIII, 44, 45, 48, 81, 133, 137, 152 antibody-secreting, XXXIV, 219, 720 - 722antigen-binding, see Antigen-binding artificial, XLIV, 206, 209, 217, 218, see also Microencapsulation binding to fibers, XXXIV, 205–220 calcium measurements, XXXIX, 515 - 520characteristics, monitoring, LVIII, 164 - 178in circular dichroism, XL, 229 differentiated, LVIII, 80, 263-278 culture, LVIII, 264 media, LVIII, 58 DNA content, LVIII, 148 effective mating rate, XXXII, 582 enlargement in plants, XXXI, 494 enucleated, LVIII, 359 flow dialysis, LV, 682, 683 fractionation, XXXIV, 195-225, 716–720; LVIII, 222, see also Cell, surface antibody methods, XXXIV, 197

column procedures, XXXIV, 197

fiber, XXXIV, 198, 225, see also Fiber fractionation glycosphingolipids, XXXII, 345–367 hard-to-grow, media, LVIII, 56 human media, LVIII, 56, 57 tumorigenic, LVIII, 376 hybridization, XXXII, 575-583 terms describing, XXXII, 582 in incubated muscle geometry in relation to incubation, XXXIX, intact, phosphorylation potential, LV, 243-245 intended, LVIII, 81 internal volumes, LV, 551 isolated, suspensions, XL, 312 lifespan, LVIII, 134 metaphase, see Metaphase cell microinjection technique, LVII, 312-318 mitotic, see Mitotic cell nontransformed, LVIII, 45 media, LVIII, 56, 58 normal, LVIII, 45 media, LVIII, 85, 90 population studies, XXXIV, 220 preparation, LVIII, 12 protein content, LVIII, 145 pulse-labeled, LVIII, 244 purity, XXXIV, 196 for rapid separation from medium, LV, 203-205 receptors, XXXIV, 197 relative culture age, XXXII, 563, 564 removal from fiber, XXXIV, 208-210 plucking, XXXIV, 203 respiratory-deficient preparation, LVI, 175 separation, XXXIV, 195-225, 714 applications, XXXIV, 211–225 conditions, XXXIV, 205 quantitation, XXXIV, 208 variables, XXXIV, 205 in situ, calcium transport, LVI, 323-326 steroid dynamics in, tracer superfusion method, XXXVI, 75 - 88

steroid receptors, assay, XXXVI, 52–58
substrate entry, XXXIX, 495–513
surface, XXXIV, 195–225
antibody, XXXIV, 197, 220
binding sites, XXXIV, 197
markers, XXXIV, 195–225
receptors, XXXIV, 197
surface-nuclear interactions, XXXIV, 221
suspensions, for substrate entry studies, XXXIX, 505, 506
swelling-shrinkage, metal ion transport, LVI, 334
transformed, LVIII, 80, 81, 134, 296
effect of lectins, LVIII, 368
properties, LVIII, 368
virus-infected
cloning, LVIII, 155
microcarrier culture, LVIII, 191, 192
whole
immobilization with collagen, XLIV, 244
with polyacrylamide gel, XLIV, 185, 740, 741, 745
negative staining, XXXII, 20, 31
in reactor, for glucose isomerization, XLIV, 768–776
thin-sectioning procedure, XXXII, 872
β-cell, energy metabolism, XXXIX, 424
Cell adhesion, XXXII, 579, 600, *see also* Adhesion, cellular
assay, XXXII, 600–602
cell-layer assay, XXXII, 603–609
nonspecific, inhibition, XXXII, 609–611
Cell-cell interaction, XXXIV, 221, 224
Cell classification, criteria, LVIII, 44
Cell counting, XXXII, 566; LVIII, 143
data expression, LVIII, 150
direct, LVIII, 143
electronic counter, LVIII, 149
of granulosa cells, XXXIX, 192–194
liver cells, LVIII, 543, 544
microcarrier, LVIII, 188, 189
Cell culture, XXXII, 555–819, *see also* Media; specific cell type
basic methodology, XXXII, 583–592
cell-cell interactions, XXXII, 597–611
cell count, XXXII, 566
cell lines used, XXXII, 584, 585
cell synchronization studies, XXXII, 592–597
characterization, LVIII, 164–178
continuous, LVIII, 216
cost, LVIII, 195, 204–206
defined, LVIII, 263
effect of buffer, LVIII, 72, 73
of carbon dioxide, LVIII, 72, 135
of humidity, LVIII, 73, 74, 136
of light, LVIII, 136
of osmolality, LVIII, 73, 136, 137
of oxygen, LVIII, 74, 136
of pH, LVIII, 72, 135
of temperature, LVIII, 55, 135, 450
of endocrine cells, XXXIX, 165
equipment, *see* Laboratory equipment
exponential, execution point, XL, 46
facility, *see* Laboratory facility
of haploid frog cells, XXXII, 789–799
of heart cells, XXXII, 740–745
of higher plant cells, XXXII, 723–732
of hormone-dependent lines, XXXII, 557–561
of human mutant cells, XXXII, 799–819
large-scale, LVIII, 211–221
adenoviruses, LVIII, 431–433
chondrocytes, LVIII, 560–564
cost, LVIII, 205
equipment, LVIII, 221–229
growth methods, LVIII, 194–210
microcarrier, LVIII, 193
Rous sarcoma virus, LVIII, 395–403
of liver cells, XXXII, 733–740
of macrophages, XXXII, 762–765
management, LVIII, 116–119
materials, sources, LVIII, 196
mixed
cytopathogenicity, LVIII, 421–423
neurons and nonneuronal cells, LVIII, 581
monolayer, LVIII, 46
for neurobiological studies, XXXII, 765–788
packaging, LVIII, 36

of phagocytes, XXXII, 758–765
population doubling, XXXII, 562–564
preparative operations, LVIII, 195–202
roller bottles, XXXII, 566, 567
shipment, LVIII, 236
regulations, LVIII, 34–36
sources, LVIII, 119
sterility testing, LVIII, 21
synchronized, execution point, XL, 48
synchrony of cell cycles, XXXII, 567, 568
of testis, XXXIX, 288–296
thyroid cells, XXXII, 745–758
Cell cycle
analysis, LVIII, 232–247
DNA content, LVIII, 243–246
parameters, LVIII, 241
phases, XL, 58; LVIII, 248–262
quantitation, LVIII, 241
in vivo analysis
colcemid, XL, 61
colchicine, XL, 61
cytochemistry, quantitative, XL, 62
by direct observation, XL, 58
isotopic labeling, XL, 59
metabolic inhibitors, XL, 61
percent labeled mitosis curve, XL, 60
Cell density, LVIII, 47, 218, 219
analysis, in exponential cultures, XL, 46
mating efficiency, LVI, 141
measurement, LVIII, 225
Cell dispersion, LVIII, 239
fluorescence, LVIII, 246, 247
mechanical, LVIII, 130, 131
protocol, LVIII, 246, 247
shearing, LVIII, 130, 131
Cell disruption, LVIII, 223
Cell disruption bomb, XXXI, 293, 294
Cell dissociation
after culture, LVIII, 129, 130
with collagenase, LVIII, 128, 129
epithelial cell, LVIII, 129
of explants, LVIII, 129, 130
muscle, LVIII, 513–516
by organ perfusion, LVIII, 128, 129
with pronase, LVIII, 129
sorting, LVIII, 239
Cell division
action of chemical agents on cycle, XL, 44
cell nucleus, XL, 1–89
doubling time, XL, 55
execution point, XL, 45, 46, 48
inhibitors, XL, 279, 284
in absence, XL, 56
termination point, XL, 45
measurement in absence of steady-state cell cycle conditions, XL, 49
under steady-state cell cycle conditions, XL, 52
Cell extract, guanosine nucleotide determinations, LVII, 85–94
Cell filter, LVIII, 514
Cell fixation, LVIII, 140
Cell fusion, LVIII, 324, 345–359
polyethylene glycol, LVIII, 353, 356–359
protocol, LVIII, 353, 354
Sendai virus, LVIII, 353–356
suppression of function, LVIII, 347
virus detection, LVIII, 348, 349
Cell growth
analysis, LVIII, 223
curves, LVIII, 75, 76, 215
data evaluation, LVIII, 150
interactions, LVIII, 78, 79
kinetics, LVIII, 133, 214–217, 241, 245
large-scale, LVIII, 13
measurements, LVIII, 142–150
capillary culture unit, LVIII, 183
chemical, LVIII, 144–149
electronic, LVIII, 149
hematocrit, LVIII, 149
microscopy, LVIII, 149
visual, LVIII, 142–144
on microcarrier, LVIII, 187–190
parameters, LVIII, 214–217
plateau region, LVIII, 75, 76
population-dependent, LVIII, 47
quantitation, LVIII, 141–152
requirements, LVIII, 51–79
synchronous, LVIII, 218
Cell harvest, LVIII, 221–229
Cell hybrid, *see* Hybrid cell

Cell identification, LVIII, 164–178
  karyotyping, LVIII, 323, 324
Cell layer assay, of cell adhesion, XXXII, 603–609
Cell lectin receptor, energy transfer studies, XLVIII, 378
Cell line, *see also* specific types
  cloning, LVIII, 155
  heteroploid, LVIII, 323
    mammalian, LVIII, 371
  invertebrates, LVIII, 450–466
  mixed, LVIII, 117
  nontumorigenic, LVIII, 378
  permanent, LVIII, 45
  quality control, LVIII, 439
  sources, LVIII, 132, 439–444
  stable, LVIII, 439–444
  suspension culture, LVIII, 202–206
  tumorigenic, LVIII, 371, 376–378
    human, LVIII, 376
    mouse, LVIII, 377
    various species, LVIII, 377
Cell membrane, countercurrent distribution, XXXI, 761, 762
Cell monitoring, *see* Monitoring
Cell movement, inhibitors, XL, 279, 284
Cell number, LVIII, 218
Cellobiose, XXXIV, 334; XLVI, 365
  chromatographic constants, XLI, 17
  lectin, XXXIV, 334
  metabolism
    β-glucoside kinase, XLII, 3
    phosphocellobiase, XLII, 494
Cellobiose–albumin conjugate
  antibody formation, L, 158
  production, L, 155–157
    antibodies, L, 158
  quantitation, L, 157
  serological tests, L, 159
Cellogel electrophoresis, in quantitative histone assay, XL, 122
Cellogel strip, LIX, 65
Celloidin, for negative staining grids, XXXII, 24, 25
Cellophane
  commercial source, XLIV, 270, 271
  disk, preparation for growth of *N. crassa*, LV, 659
  electrode current, LVI, 458
  glucose oxidase immobilization, LVI, 487
Cellophane square test, XXXIX, 490
Cell preservation, LVIII, 29–36
Cell radius, equation, XXXIX, 325, 326
Cell receptor
  description and activity, XXXVII, 67
  in peptide hormone assay, XXXVII, 66–81
Cell recovery, LVIII, 212
Cell repository material, LVIII, 12, 440–442
Cell rupture, for liver fractionation, XXXI, 78
Cell shearing, LVIII, 223
Cell size, LVIII, 238
Cell sorting, LVIII, 239
  efficiency, LVIII, 240
Cell staining, *see* Staining
Cell storage, LVIII, 12, 29–36, 220, 221
Cell surface protein, LVIII, 267
  purification by immunoprecipitation, XLVII, 250
    by molecular weight sieving, XLVII, 250
  radiolabeling procedure, XLVII, 249, 250
  sequence analysis, XLVII, 256, 257
Cell synchronization, XXXII, 592–597, *see also* Synchrony
  basic methods, XXXII, 592, 593
  in chromosome isolation, XL, 77
Celltester model 1030, LVII, 218, 219
Cell transfer, precautions, LVIII, 50
Cell transformation, LVIII, 368–370, *see also* Cell, transformed
  adenovirus, LVIII, 370
    human, LVIII, 426
  basic methodology, XXXII, 583–592
  chemical carcinogen, LVIII, 370
  criteria, LVIII, 368
  defined, LVIII, 368
  DNA tumor virus, LVIII, 370
  helper virus, LVIII, 369
  malignant, LVIII, 296, 368
  murine lymphoma virus, LVIII, 369
  murine sarcoma virus, LVIII, 369
  polyoma virus, LVIII, 370
  RNA tumor virus, LVIII, 369
  Rous sarcoma virus, LVIII, 369, 370

SV40, LVIII, 370
high, LVIII, 369, 370
low, LVIII, 369, 370
temperature-sensitive mutant, LVIII, 370
Cellulase, XLIV, 108, 165, 166
in protoplast isolation, XXXI, 582, 583
Cellulose, XXXIV, 704
activation with halogens, XLIV, 36
aminoethyl, and photochemical immobilization, XLIV, 284, 285
bonding to phosphatase, LX, 749
to ribonucleic acids, LX, 248–250
bridge formation, XLIV, 166
bromoacetyl derivative, preparation, XLIV, 749
carboxymethyl, *see* (Carboxymethyl)cellulose
as carrier, XXXIV, 3–6
chemical properties, XLIV, 46
chromatography, polysomal ribonucleic acid, LX, 403, 404
composite formation, with iron oxide, XLIV, 326
derivatives commercially available, XLIV, 46
diethylaminoethyl, *see* (Diethylaminoethyl)cellulose
enzyme attachment sites, XLIV, 47–49
fiber
elastic support, XLIV, 560
preparation, XLIV, 230, 231
hollow-fiber device, XLIV, 293
ion-exchange resin, peptide separation, XLVII, 204–210
as matrix, XXXIV, 18
modification with polycytidylate, LX, 457
nucleotide elution, LIX, 63
osmotic pressure studies, XLVIII, 74
periodate methods for activation, XXXIV, 82
physical properties, XLIV, 46
purification of *Cypridina* luciferin, LVII, 367, 368
of double-stranded ribonucleic acid, LX, 552
rehydration procedure, XLIV, 46, 47
substrate, of cellulase, XLIV, 166
sulfoethyl, XLIV, 46
support properties, XLIV, 20
thin-layer plate
identification of ribosomal proteins, LIX, 777, 778
tRNA sequence analysis, LIX, 62
separation of modified nucleosides, LIX, 181
of mononucleotides, LIX, 192
thin-layer sheet, separation of bases and nucleotides, LI, 551, 559
Cellulose acetate
electrophoresis, LX, 328–331
hollow-fiber device, XLIV, 293
strip, LIX, 65
Cellulose acetate butyrate, XLVII, 267
Cellulose MN 300, for phosphohydrolase assay, XXXII, 129
Cellulose nitrate
microencapsulation with, XLIV, 204–207, 212
tube, LIX, 373
steroid adsorption, XXXVI, 161, 207
Cellulose-phosphate
purification of adenylate deaminase, LI, 492, 493
of pyrimidine nucleoside monophosphate kinase, LI, 325, 326, 328
Cellulose polyacetate strip, LI, 37
Cellulose triacetate, fibers, preparation, XLIV, 230
Cellulose tube, XXXII, 111
Cellulose wall, of plant tissues, XXXI, 501
Cell viability, LVIII, 131, 141–152, 218
assay, LVIII, 150
measurement, LVIII, 150–152
Cell wall
bacterial, neural oligosaccharide composition, L, 256–260
disintegrators, XXXI, 662, 663
of gram-negative bacteria, isolation, XXXI, 642–653
of gram-positive bacteria, isolation, XXXI, 653–667
hydrolyzing enzymes, XXXI, 610, 611
in plants, formation, XXXI, 493
Cell yield, LVIII, 131
C3 endopeptidase, XXXIV, 732

Centaurea, nuclei isolation, XXXI, 563
Central nervous system, hormone effects, XXXIX, 429–440
CentrifiChem Fast Analyzer, XXXI, 800
Centrifugal force, calculation, XXXI, 9
Centrifugation, XXXI, 713–733; LVIII, 222, *see also* specific type
ancillary equipment, XXXI, 718, 719
of cells and nuclei, XXXI, 721–723
for concentration, LII, 105
convex-concave gradient, preparation, LII, 75
density-dependent banding, XXXI, 714, 715
in detergent-containing gradient, LII, 82, 83
differential, isolation of mitochondria, LV, 546
dye-density equilibrium, LVIII, 408, 410
effect on nucleotide concentrations, LVII, 88
equipment, XXXI, 716–721
gradient
isolation of mitochondria, LV, 6
preparation, XXXI, 735
shape, XXXI, 720–721
unloading, XXXI, 739, 740
grinding of material, XXXI, 737
isolation of mitochondria, LV, 36–38, 43
isopynic density gradient, of microsomes, LII, 76–79
large-scale preparations, XXXI, 741, 742
of liver homogenate, XXXI, 78, 79; LII, 72, 73
low-speed, rat liver homogenate, LV, 35
maximum rotor speed, determination, LIX, 419
media, XXXI, 719, 720
of membrane-bound elements, XXXI, 726–734
nomenclature, XXXI, 714–716
purity and recentrifugation, XXXI, 740, 741
rapid
measurement of glutamate binding, LVI, 427, 428
for plant-cell extraction, XXXI, 541, 542
rate-differential
for rough microsome subfractionation, LII, 74–76, 81
for smooth microsome subfractionation, LII, 78
of RNA and RNP particles, XXXI, 723–726
rotors, XXXI, 716–718
of subcellular fractions, XXXII, 3–20
with three-layered Cs^+-containing discontinuous sucrose gradient, LII, 72, 73
types, XXXI, 713–714
zonal, *see* Zonal centrifugation
Centrifugation speed, sedimentation equilibrium, XLVIII, 171
Centrifuge
for liver homogenates, XXXI, 7
safety equipment, LVIII, 40
separation of cells and organelles, LV, 552
Centrifuge bucket evacuation device, LIX, 657, 659
Centrifuge tube
cleaning, LV, 41
conical, XXXII, 3
conical-tipped, LII, 314
Corex, LII, 316
graduated, LII, 332
Centromere banding, LVIII, 325
Cephalexin, XLIII, 73, 411, 412
in media, LVIII, 116
Cephalin, *see also* Phosphatidylethanolamine
IR spectroscopy, XXXII, 253
Cephaloglycin, XLIII, 117, 387
Cephaloridine, XLIII, 73
hydrolysis by *Enterobacter cloacae* P99 cephalosporinase, XLIII, 687
interaction with staphylococcal penicillinase, XLIII, 671
β-lactamase assay, XLIII, 71
thin-layer chromatography in pharmacokinetics studies, XLIII, 211
Cephalosporin, XLVI, 531
bioautography, XLIII, 117
β-lactamase assays, XLIII, 71

in media, LVIII, 116
paper chromatography, XLIII, 116–118, 159
semisynthetic, XLIII, 199
solvent systems for countercurrent distribution, XLIII, 334
thin-layer chromatography, XLIII, 207, 208
titration of *Streptomyces* R39 enzymes, XLV, 629
Cephalosporin acetylesterase
from *Bacillus subtilis*, XLIII, 731–734
assay, XLIII, 732, 733
culture preparation, XLIII, 733
inhibitors, XLIII, 734
kinetic properties, XLIII, 734
molecular weight, XLIII, 734
pH optimum, XLIII, 734
purification, XLIII, 733, 734
specific activity, XLIII, 733
stability, XLIII, 734
substrates, XLIII, 734
unit definition, XLIII, 733
citrus, XLIII, 728–731
activators, XLIII, 730
assay, XLIII, 728–730
manometric, XLIII, 728, 729
potentiometric, XLIII, 729
crude extract preparation, XLIII, 729
kinetic properties, XLIII, 731
pH optimum, XLIII, 731
properties, XLIII, 730, 731
reaction scheme, XLIII, 728
specific activity, XLIII, 729
specificity, XLIII, 730
stability, XLIII, 730, 731
unit definition, XLIII, 729
Cephalosporinase, *see* Penicillinase
Cephalosporin C, XXXIV, 5, *see also* Cephalosporin
bioautography, XLIII, 117
CMR spectroscopy, XLIII, 408, 410–425, *see also* Nuclear magnetic resonance spectroscopy, carbon-13
high-pressure liquid chromatography, XLIII, 311, 312, 319
macroreticular resin chromatography, XLIII, 299
paper chromatography, XLIII, 117, 118
Cephalosporin C–agarose, XXXIV, 404
Cephalosporin P series, *see* Antibiotic, alicyclic
Cephalosporium acremonium
δ-(L-α-aminoadipyl)-L-cysteinyl-D-valine, XLIII, 471
carbon molecular resonance spectroscopy, XLIII, 411, 414
fungal penicillin acylase, XLIII, 721
media, XLIII, 14
Cephalosporoic acid, XLIII, 71
Cephalothin, XLIII, 72, 73
bioautography, XLIII, 117
interaction with staphylococcal penicillinase, XLIII, 671
in media, LVIII, 112, 114, 116
Cephamandole, XLIII, 73
Cephamycin, XLIII, 71, 110
bioautography, XLIII, 117
paper chromatography, XLIII, 116–118
Cephem antibiotic, XLIII, 410
Ceramic support, *see also* Glass, controlled pore; Glass bead; Support, inorganic
comparative half-life data, XLIV, 779–783
controlled-pore, surface area, XLIV, 155–158
physical parameters, XLIV, 777
porous, physical properties, XLIV, 135–137
relative cost, XLIV, 776, 777
types, XLIV, 777
Ceramidase, assay, L, 463–465
Ceramidase deficiency disease, *see* Farber's lipogranulomatosis
Ceramide, *see also* Glycosphingolipid
^{14}C–labeled, preparation, L, 463
column chromatography, XXXV, 421
Farber's lipogranulomatosis, L, 463
glycosphingolipids, L, 236
stain, XXXII, 357
thin-layer chromatography, XXXII, 356, 358; XXXV, 396, 400, 402
Ceramide oligohexoside, isolation, XXXII, 346, 352–355
Ceramide pentasaccharide, isolation, XXXII, 349, 350, 355

Ceramide-1-phosphoryl-*N,N*-dimethylethanolamine, intermediates in synthesis of choline-labeled sphingomyelins, XXXV, 533
Ceramide polyhexoside, isolation, XXXII, 346, 352–355
Ceramide trihexosidase, L, 494
from human placenta
assay, L, 533, 534
properties, L, 535, 536
purification, L, 534, 535
treatment of Fabry's disease, L, 536, 537
Ceramide trihexoside, NMR spectroscopy, XXXII, 367
Ceratotherium simum, iron protein, LIV, 375
Cereal
enzyme studies, XXXI, 522
leaves, protoplast isolation, XXXI, 579–581
Cereal grain, proteinase inhibitors, XLV, 723
Cerebron, optical rotation, XXXII, 365
Cerebronic acid
in α-hydroxylase assays, LII, 311–315
in saponification mixture, LII, 312
synthesis, LII, 311
(*R*)-Cerebronic acid, inhibitor of brain α-hydroxylase, LII, 317
Cerebroside, *see also* Glycosylceramide
column chromatography, XXXV, 417
thin layer chromatography, XXXII, 356; XXXV, 425
Ceroid
isolation from brain, XXXI, 478–485
lipofuscins, XXXI, 425
Ceroid storage disease, XXXI, 478
Cerulenin, inhibitor, in lipase bioassay, LVII, 194
Ceruloplasmin, *see also* Ferroxidase
electron paramagnetic resonance studies, XLIX, 514
Raman frequencies and assignments, XLIX, 144
resonance Raman spectrum, XLIX, 142
structure, L, 273
Cervix
adenocarcinoma, monolayer, culture, LVIII, 133
carcinoma, LVIII, 376
Cesium, fluorescence quencher, XLIX, 225, 233
Cesium-133, lock nuclei, XLIX, 349
Cesium-137
in membrane studies, XXXII, 884–893
properties, XXXII, 884
Cesium chloride
DNA extraction, LVIII, 408, 409
equilibrium density gradient centrifugation, analysis of native ribosomal subunits, LIX, 416–421
in purification of *Pholas* luciferase, LVII, 387
separation into alkali, LV, 205
solution
gravimetric density determination, LIX, 418
as reference density solutions, LIX, 5
refractive index, LIX, 5
relationship to density, LIX, 10
solution centrifugation
assay of binding of initiation factor complex to ribosome subunits, LX, 275–280
preparation of double-stranded ribonucleic acid, LX, 386
of viruslike particles from yeast, LX, 550, 551
Cesium chloride gradient
in myelin isolation, XXXI, 441–442
properties, XXXI, 508, 509
Cesium ion, activator of ribonucleoside triphosphate reductase, LI, 258
Cesium salt, of amino acid, peptide synthesis, XLVII, 590
C1 esterase, XXXIV, 731–746, *see also* Complement
inhibitors, XLVI, 115, 116, 120
Cetavlon, *see* Cetyltrimethylammonium bromide
Cetyltrimethylammonium bromide
assay of CTP(ATP):tRNA nucleotidyltransferase, LIX, 123
crystallization of transfer RNA, LIX, 7
determination of aminoacylation isomeric specificity, LIX, 273

hydrocarbon/water interface, LV, 773
precipitation of initiation factor complexes, LX, 73, 76–78, 86
for rhodopsin isolation, XXXII, 307
tRNA precipitation, LIX, 126–128, 132, 133, 140
solubilizer, of organic hydroperoxides, LII, 510
structure, LVI, 735
CF–3 cell protein content, LVIII, 145
C–gamma gel
purification of adenine phosphoribosyltransferase, LI, 563
of guanine phosphoribosyltransferase, LI, 554
of hypoxanthine phosphoribosyltransferase, LI, 554
of OPRTase-OMPdecase complex, LI, 146–149
of orotate phosphoribosyltransferase, LI, 71
CGD, *see* Chronic granulomatous disease
C_6 glial cell, *see* Rat, glial C_6 cell
cGMP, *see* Cyclic guanosine monophosphate
CGN, *see* *p*-Nitrophenyl *N*-benzyloxycarbonyl glycinate
Chaetopterus, photoprotein system, LVII, 355
Chain elongation
action of antibiotics, LIX, 852–856
assay, LIX, 356–359
effect of magnesium ion, LIX, 370
of temperature, LIX, 371
Chain length, heterogeneity, effect on vibrational frequency of helical structures, XLIX, 172
Chain separation, by gel filtration, XLVII, 97–107
Chamber method, in water flow studies, XXXVII, 253–255
Champamycin, XLIII, 136
Chaney adapter, XXXVI, 32
Chang cell, protein content, LVIII, 145
Chaotrophic agent, LII, 305
Chaotropic anion, LIII, 27
activator of succinate dehydrogenase, LIII, 33
complex II resolution, LIII, 52
Chaotropic ion, membrane destabilization, XXXI, 770–790
antichaotropes, XXXI, 784–787
applications, XXXI, 771, 772
kinetics and thermodynamics, XXXI, 777, 784
in membrane resolution, XXXI, 772–777
weak chaotropes, XXXI, 787–789
Chara, free cells, XXXII, 723
Charcoal
activated
assay of GTP hydrolysis, LIX, 361
glutaraldehyde treatment, XLIV, 550, 551
microencapsulation, XLIV, 216
cyclic nucleotide preparation, XXXVIII, 10, 20
purification, XXXVIII, 79
dextran-coated
in estrogen receptor studies, XXXVI, 249–254
for steroid adsorption, XXXVI, 40
nucleotide separation, LV, 251
plasma-coated, as hormone adsorbent, XXXVII, 34
in serum or plasma treatment, XXXVI, 91–97
treatment with AMP, LV, 251
washing, LV, 250
Charcoal disk, LIX, 236
purification of tryptophanyl-tRNA synthetase, LIX, 242
Charge fractionation, XXXIV, 722
Chase solution, mitochondrial labeling, LVI, 60
CHD, *see* Cyclohexanedione
Chelating agent, *see also* specific type
adenylate cyclase, XXXVIII, 142
broad-spectrum, for preparation of metal-free buffers, LIV, 475
as enzyme inhibitor, XXXI, 524, 525
in media, LVIII, 70
Chelex 100 resin, XLIX, 336, 460
affinity for calcium ion, LVII, 302, 307

arsenazo III purification, LVI, 318
assay of nucleoside phosphorylase, LI, 539
for preparation of metal-free buffers, LIV, 476
removal of calcium ions, LVII, 301, 302
of paramagnetic contaminants, LIV, 154
Chelocardin, XLIII, 367
Chem-Glow photometer, LVII, 75, 205, 218, 477, 478, 550
advantages, LVII, 479, 480
operational characteristics, LVII, 551, 553–555
Chemical bond, vibration, time range, LIV, 2
Chemical carcinogenicity, *see* Carcinogenicity, chemical
Chemical exchange
determination, using spin-lattice relaxation time, LIV, 181
in saturation transfer experiments, LIV, 172–177, 180, 181
Chemical potential
definition, XLVIII, 82, 83
Hill plot, XLVIII, 286
sedimentation equilibrium, XLVIII, 164, 165
Chemical proofreading, LIX, 283
Chemical quench technique, LIV, 2, 93
Chemical shift, LIV, 348, 351
accuracy of measurement, LIV, 200
components, LIV, 194, 196, 197
principle, LIV, 193
sign convention, LIV, 200
units, LIV, 200
Chemical shift anisotropy, XLIX, 278
Chemiluminescence, *see also* Electrochemiluminescence; Luminol
assay systems using, LVII, 424–462
data reproducibility, LVII, 441
procedure, LVII, 426, 427
sensitivity, LVII, 438–441
chemical perturbation, LVII, 568
definition, LVII, 409
for detection of enzymically generated peroxide, LVII, 445–462
of dimers, LVII, 568
direct, LVII, 563
effect of buffers, LVII, 438, 439
of proteins, LVII, 439, 440
of solvent, LVII, 497, 498, 563
efficiency, instantaneous, LVII, 514
electrogenerated, LVII, 510
electron-transfer, LVII, 494–526
energy balance considerations, LVII, 500–502
prototype reactions, LVII, 496–500
yields, theoretical expectations, LVII, 508–510
elimination with hydrogen peroxide, LI, 122
general mechanism, LVII, 410, 563
intensity, LVII, 570
lipid peroxidation, LII, 304
of luminol, LVII, 409–423
in phagocytic cells, LVII, 462–494
general principles, LVII, 462–466, 468, 469
instrumentation, LVII, 473–476
photon emissions, absolute measurement, LVII, 583–589
protein-chromophore interaction, LVII, 567
quantum yield, LVII, 570
absolute measurement, LVII, 583–589
relative measurement, LVII, 587–589
quenching, LVII, 569, 570
reaction subclasses, LVII, 563
sensitized, LVII, 563
solute-solvent interaction, LVII, 566, 567
S route, LVII, 502, 519
tertiary emission standards, LVII, 598
T route, LVII, 502, 518, 519
Chemiosmotic hypothesis, ionophores, LV, 450, 451
Chemography, LVIII, 291
Chemostat, mass culture, LVIII, 205
Chemostat culture, glucose-limited
culturing procedure, LVI, 575
general considerations, LVI, 573–575
manipulation of unsaturated fatty acid composition of mitochondria, LVI, 575–577
Chemotaxis, assay, in amoebae, XXXIX, 490, 491

Chenodeoxycholate, sodium salt, partial specific volume, XLVIII, 21
Chick
  bioassay, of vitamin D_3 metabolites, XXXVI, 533, 534
  embryo, oxidase, LII, 12
Chicken
  breast muscle
    glycerol-3-phosphate dehydrogenase, assay, purification, and properties, XLI, 245–249
    phosphoglycerate mutase, preparation, purification, and properties, XLII, 435, 440–442, 446–449
  embryo cell
    cloning medium, LVIII, 384
    media, LVIII, 56, 58
    primary culture, LVIII, 385, 386
    secondary culture, LVIII, 388–390
    tissue dispersion, LVIII, 126
  embryo extract, preparation, LVIII, 526, 527
  embryo fibroblast, LVIII, 59
    freezing, LVIII, 392, 393
    heart, media, LVIII, 90
    preparation, LVIII, 381–384
    storage, LVIII, 392, 393
  fibroblast, microcarrier culture, LVIII, 189
  liver
    acetone powder, preparation, LI, 190, 198
    amidophosphoribosyltransferase, LI, 173–177
    FGAM synthetase, LI, 197–200
    phosphofructokinase, purification, XLII, 110
    succino-AICAR synthetase, LI, 190
  nontransformed cell, media, LVIII, 56
  red cell × tobacco mesophyll, LVIII, 365
  serum, LVIII, 382
    in media, LVIII, 139
Chick helper factor, LVIII, 387, 388
Chikugunya virus, interferon-producing, LVIII, 293, 294
Chinese hamster brain
  crude extract preparation, LI, 545
  hypoxanthine phosphoribosyltransferase, LI, 543–549
Chinese hamster cell line, tumorigenicity testing, LVIII, 374, 375
Chinese hamster karyotype, LVIII, 340–342
Chinese hamster ovary cell, LVIII, 53–55, 86, 377
  cell cycle, LVIII, 242
  Colcemid treatment, LVIII, 255, 333
  culture conditions, LIX, 219, 232
  in determination of *in vivo* tRNA aminoacylation, LIX, 269–271
  growth, for mutagenesis, LVIII, 310
  media, LVIII, 58, 89, 310
  microcarrier culture, LVIII, 189–191
  mutagenesis, LVIII, 309
  plasma membrane isolation, XXXI, 161
  tRNA isoacceptors, LIX, 218–229
  strains, LVIII, 310
  subcellular fractions
    aminoacyl-tRNA synthetases distribution, LIX, 233, 234
    preparation, LIX, 232, 233
Chinese hamster V79 tissue culture cell
  crude extract preparation, LI, 541
  purine nucleoside phosphorylase, LI, 538–543
Chiral molecule, optical properties, XLIX, 179, 180
Chironomus thummi thummi, hemoglobin, LIV, 73, 275
Chiroptical method, *see* Spectropolarimetry
Chitin, XXXIV, 5, 639
  affinity chromatography, L, 352
  carboxymethylated, XXXIV, 639
  deaminated, XXXIV, 4
  purification of lysozyme, XXXIV, 163
Chitin-cellulose, XXXIV, 639
Chitin-starch, XXXIV, 330
Chitobiose, XLVI, 365, 366
Chitodextrin, *N*-acetylated, lectin, XXXIV, 337
Chitotriose, XLVI, 365, 366
  interaction with egg white lysozyme, XLVIII, 272, 273

Chlamydomonas
aldolases, XLII, 228–234
free cells, XXXII, 723
Chlamydomonas reinhardi
membrane properties, XXXII, 865–871
wild and mutant strains, chloroplast membranes, XXXII, 871–880
Chlorambucyl group, XLVI, 627
Chloramine-T, XXXVII, 49; XLVII, 456
procedure, of peptide hormone labeling, XXXVII, 29, 147, 225, 226, 325
Chloramphenicol, XLVI, 90, 336, 630, 634, 635; LVI, 31, 33
O-acetoxy derivatives, XLIII, 738
administration to animals, LVI, 35
amino acid incorporation, LVI, 57, 58
analogs, XLVI, 702–707
assay of radiolabeled ribosomal particles, LIX, 779, 780
binding to ribosomal particles, LIX, 865, 866
biotransformation, XLIII, 208
cell starvation, LV, 169
chiroptical studies, XLIII, 352
cyanide-insensitive respiration, LV, 26
detoxification, LVI, 38
differential pulse polarography, XLIII, 375–377, 380, 382, 383
effect on protein synthesis, LIX, 853, 854, 862
electron transport, LV, 516
enantiomeric *p*-phenyl analog, XLIII, 366
gas liquid chromatography, XLIII, 228–232
calculation, XLIII, 232
HeLa cell mitochondrial proteins, LVI, 78
hydrolysis at amide bond, XLIII, 734–737
inhibitor of polypeptide chain elongation, LIX, 359, 370
in media, LVIII, 112, 114, 116
mitochondria, LVI, 32
paper chromatography, XLIII, 159, 160, 162
for photoaffinity labeling, LIX, 798, 802, 803, 805
preparation of polysomes, LIX, 365
preparative thin-layer chromatography, XLIII, 185
as protein synthesis inhibitor, XXXII, 869
resistance, XLIII, 737; LVI, 140
gene mapping, LVI, 185
resistant bacteria, XLIII, 737–754
solution spectrophotometry with thin-layer chromatography, XLIII, 186
solvent system for countercurrent distribution, XLIII, 336
spectropolarimetry, XLIII, 364–366
Streptomyces venezuelae, media, XLIII, 18
structure of compounds related, XLIII, 737
sulfate incorporation into mitochondrial protein, LVI, 60
in vivo administration, LVI, 36–39
in *in vivo* preparation of radiolabeled RNA, LIX, 831, 834
Chloramphenicol acetyltransferase, XLIII, 739–754
acetylation techniques, XLIII, 739
[^{14}C]acetyl-chloramphenicol direct measurement, XLIII, 743–746
activators, XLIII, 755
acyl acceptor, specificity, XLIII, 753
acyl donor, specificity, XLIII, 753
affinity chromatography, XLIII, 750
alumina gel procedure, XLIII, 750
ammonium sulfate precipitation, XLIII, 748
assay, XLIII, 740–746, *see also* specific assay
reaction sequence, XLIII, 740
bacterial growth, XLIII, 747, 748
catabolite repression of synthesis, XLIII, 738, 739
chromatographic detection of acetylation, XLIII, 740, 741
coenzyme A sulfhydryl group, XLIII, 742
crude extract preparation, XLIII, 748
DEAE-cellulose chromatography, XLIII, 749
gel filtration, XLIII, 749
genetic mode of synthesis, XLIII, 754
heat stability, XLIII, 754

hybrid formation, XLIII, 755
inhibitors, XLIII, 755
K_m values, XLIII, 754
molecular weight, XLIII, 752
native tetrameric, XLIII, 755
phages carrying gene, XLIII, 739
pH optimum, XLIII, 752
properties, XLIII, 752–754
purification, XLIII, 746–752
from *Escherichia coli*, XLIII, 747–751
from other bacteria, XLIII, 751, 752
pyruvate-kinase system, XLIII, 744
radioactive assay, XLIII, 739, 743–746
[^{14}C]acetyl-CoA, XLIII, 744
culture preparation, XLIII, 745, 746
specificity, XLIII, 753
spectrophotometric assay, XLIII, 739, 742, 743
stability, XLIII, 755
from *Staphylococcus*, XLIII, 751, 752
streptomycin sulfate precipitation, XLIII, 748
structural gene, XLIII, 738
tetrameric, XLIII, 755
thin-layer chromatography, XLIII, 739–741
Chloramphenicol hemisuccinate, LVI, 34
Chloramphenicol hydrolase, *Streptomyces*, XLIII, 734–737
acetone powder, XLIII, 736
assay, XLIII, 735
culture preparation, XLIII, 736
properties, XLIII, 736
purification, XLIII, 736
reaction scheme, XLIII, 734
Chloramphenicol palmitate, XLIII, 59
Chloramphenicol stearate, XLIII, 159
Chloramphenicol succinate, XLIII, 159
Chlorate, as electron acceptor, LVI, 386
Chlorcylizine *N*-demethylase, *Ah* locus, LII, 232
Chlorella
countercurrent distribution, XXXI, 768, 769
free cells, XXXII, 723
membrane mutants, XXXII, 865
Chlorella pyrenoidosa, aldolase, XLII, 234
Chlorhexidine, XLVII, 107
Chlorhydrol, *see* Aluminum chlorhydroxide complex
Chloride
activator of succinate dehydrogenase, LIII, 33
inhibitor of firefly luciferase, LVII, 15
of *p*-hydroxybenzoate hydroxylase, LIII, 551
of salicylate hydroxylase, LIII, 541
in media, LVIII, 68
stabilizer of luciferase, LVII, 199
Chloride ion
E. coli ATPase, LV, 796
effect on aspartate carbamyltransferase, LI, 50
permeation, oligomycin, LV, 504
Chlorin, in bacterial cytochromes, LIII, 203, 212
Chlorine-35, lock nuclei, XLIX, 348
Chlorine-37, lock nuclei, XLIX, 348
Chlorine gas, optical filter, XLIX, 192
Chloroacetaldehyde, 1,N^6-ethenoadenosine 3′,5′-monophosphate preparation, XXXVIII, 429
Chloroacetamide, for sulfhydryl group alkylation, LI, 247
p-Chloroacetanilide, *N*-hydroxylation, Ah^b allele, LII, 231
Chloroacetic acid, XLVII, 490
reaction with tributylphosphine, XLVII, 113
testosterone derivative, EC detection, XXXVI, 59
Chloroacetic anhydride, XLVII, 490
Chloroacetol, XLVII, 488
Chloroacetol phosphate, XLVI, 141
synthesis, XLVI, 141
3-Chloroacetol phosphate, XLVII, 494
Chloroacetol sulfate, XLVI, 145
3-Chloroacetol sulfate, XLVII, 498
affinity labeling, XLVII, 483
structure, XLVII, 483
synthesis, XLVII, 488
Chloroacetol sulfate dimethyl ketal, chromatographic behavior, XLVII, 489

Chloroacetoxyprogesterone, reduction by 20β-hydroxysteroid dehydrogenase, XXXVI, 398
Chloroacetyl chloride, synthesis of ethyl-4-chloro-2-diazoacetoacetate, LIX, 811
1(3)-Chloroacetylglycerol 2-phosphate, synthesis, XLVII, 490
N-[α'-(Chloroacetyl)phenethyl]-*p*-toluenesulfonamide, XLVI, 151
Chloroacetyl-D-phenylalanine, XXXIV, 435, 438
3-Chloroacetylpyridine adenine dinucleotide, XLVI, 150
  synthesis, XLVI, 149
3-Chloroacetylpyridinium adenine dinucleotide, XLVI, 148
β-Chloroalanine, XLVI, 37, 427–432
  synthesis of ^{14}C-labeled, XLVI, 428, 429
cis-3-Chloroallylamine, XLVI, 37
2-Chloroallylamine, XLVI, 38
N-Chloroambucilyl-tryptophanyl-tRNA^{Trp}, inhibitor of tryptophanyl-tRNA synthetase, LIX, 256, 257
o-Chloroaniline, as mixed-function oxidase substrate, XXXI, 233
1-Chlorobenzotriazole, XLVII, 457
4-Chlorobenzylamine, XXXIV, 5
2-Chlorobenzyloxycarbonyl group, XLVII, 610
Chlorobium, cytochrome *c*-533, CD spectrum, LIV, 277
Chlorobium thiosulfatophilum, cytochrome c_{553}, LIII, 450, 454
Chlorobutane, extraction, XLVII, 345
5-Chloro-3-*tert*-butyl-2′-chloro-4′-nitrosalicylanilide
  complex V, LV, 315
  hepatoma mitochondria, LV, 85
  submitochondrial particles, LV, 108
  uncoupling activity, LV, 465
Chlorocarbonate derivative, of steroids, XXXVI, 19
Chlorocruorin, LII, 4, 5, 16
7-Chloro-6-demethyl-5a,6-anhydrotetracycline, XLIII, 140
7-Chloro-6-demethyl-4-didemethylaminotetracycline, XLIII, 140
2′-Chloro-2′-deoxycytidine, XLVI, 324
  synthesis, XLVI, 322, 323
2′-Chloro-2′-deoxycytidine 5′-monophosphate, XLVI, 324, 325
2′-Chloro-2′-deoxycytidine 5′-phosphate, XLVI, 324, 325
2′-Chloro-2′-deoxy-4-thiouridine, XLVI, 322
5-Chlorodeoxyuridine
  substrate of deoxythymidine kinase, LI, 364
    of thymidine kinase, LI, 358
Chlorodifluoromethane, *see* Freon 22
4-Chloro-3,5-dinitrobenzoic acid, modification of lysine, LIII, 151, 152
1-Chloro-2,3-epoxypropane, XLIV, 126
  synthesis of 4-*N*-(3-chloro-2-hydroxypropyl)amino-*N*-methylphthalimide, LVII, 430
    of 4-*N*-[(3-chloro-2-hydroxypropyl)-*N*-ethyl]amino-*N*-methylphthalimide, LVII, 431
2-Chloroethanol, XLVII, 89
  decomposition, XXXII, 469
2-Chloroethyldiazoacetate, XLVI, 40
N-2-Chloroethyl-*N*-methyl-2-acetoxyethylamine, XLVI, 578
β-Chloroethylphosphate of 3-*O*-benzoylceramide, intermediate in sphingomyelin synthesis, XXXV, 500, 501
2-Chloroethylphosphoryl dichloride, intermediate in sphingomyelin synthesis, XXXV, 502
N-2-Chloroethyl-*N*-propyl-2-benzilylethylamine, XLVI, 578
N-2-Chloro-5-fluorosulfonyl-*O*-(4-nitrophenyl) carbamate, XLVI, 118
m-[*o*-(2-Chloro-5-fluorosulfonylphenylureido)methoxy]benzene, XLVI, 120
m-(4-[*o*-(2-Chloro-5-fluorosulfonylphenylureido)phenoxy]butoxy)benzamidine, XLVI, 118, 120
Chloroform, XLIV, 58, 141, 145, 177, 207, 222, 480, 662, 839, 855, 932
  in copper analysis, LIV, 444
  cytochrome *P*–450 assay, LII, 125, 126
  deuterated, XLIII, 393
  determination of flavins, LIII, 424
    of *in vivo* levels of aminoacylated tRNA, LIX, 271
  drug isolation, LII, 334

electron spin resonance spectra, XLIX, 401, 402
7-ethoxyresorufin crystallization, LII, 374
extraction of adenine nucleotides, LVII, 81
of guanine nucleotides, LVII, 88
fatty acid extraction, LII, 313
heme extraction, LII, 423, 429
identification of cysteinyl flavins, LIII, 464
isolation of ATPase, LV, 338, 345
of 2-octaprenyl-6-methoxy-1,4-benzoquinone, LIII, 607
of quinone intermediates, LIII, 607
magnetic circular dichroism, XLIX, 159
porphyrin ester synthesis, LII, 432
preparation of chloroplastic tRNA, LIX, 210
of metal-free buffers, LIV, 475
of MS2 RNA, LIX, 299, 300
of RPC resin, LIX, 221
purification, LII, 423
of 2-carbethoxy-2-diazoacethydrazide, LIX, 811
as red blood cell hemolysing agent, LII, 448
for removal of low molecular weight dimethylsiloxanes, XLVII, 396
resin swelling, XLVII, 307
solvent
for spectral analysis of porphyrins, LII, 445
for thin-layer chromatography, XLIII, 205
of vacuum grease, LIV, 122
in vapor pressure osmometer, XLVIII, 153
sterilizing agent, XLIV, 783–785
synthesis of *Diplocardia* luciferin, LVII, 377
of ethyl-4-iodo-2-diazoacetoacetate, LIX, 811
of 8α-flavins, LIII, 458
TF_1 extraction, LV, 782
Chloroform-acetone powder, preparation from liver, XXXV, 193, 209, 214
Chloroform-ethanol treatment, purification of superoxide dismutase, LIII, 388
Chloroform-ethyl acetate, solvent system, for 2-cyano-6-hydroxybenzothiazole, LVII, 21
Chloroform-heptane-ethanol, for chromatographic separation of steroids, LII, 379
Chloroform-hexane
for chromatographic separation of vitamin D metabolites, LII, 398
purification of coenzyme Q_2 analog, LIII, 595
Chloroform-hexane-ether, purification of coenzyme Q_2 analog, LIII, 595
Chloroform-*n*-hexane-methanol, for chromatographic separation of porphyrins, LII, 452
Chloroform-isoamyl alcohol, in tRNA labeling procedure, LIX, 71
Chloroform-isopentyl alcohol, preparation of mitochondrial tRNA, LIX, 207
Chloroform-light petroleum
extraction of quinone intermediates, LIII, 602, 607
isolation of 2-octaprenyl-6-methoxy-1,4-benzoquinone, LIII, 607
Chloroform-methanol
aromatic hydrocarbon extraction, LII, 238
DCCD-binding protein, LV, 428, 430, 431
denaturation, purification of adenylate kinase, LI, 469, 470
extraction, HeLa cell mitochondrial fraction, LVI, 69, 76, 77
α-hydroxylase assay, LII, 311, 313
iodometric assay of lipid hydroperoxides, LII, 307
isolation of 3-octaprenyl-4-hydroxybenzoic acid, LIII, 609
proteolipid extraction, LV, 417
steroid hydroxylation, LII, 380
steroid synthesis, LII, 378
synthesis of *N*-(3-carbethoxy-3-diazo)acetonyl-7-thiolincomycin, LIX, 814
vitamin D metabolite extraction, LII, 394
Chloroform-pyridine, in heme extraction, LII, 424
β-Chloroglutamic acid, XLVI, 37

4-*N*-(3-Chloro-2-hydroxypropyl)amino-*N*-methylphthalimide
synthesis, LVII, 430
of *N*-methyl-4-*N*-[2-hydroxy-3-(*N*-phthalimido)propyl]-aminophthalimide, LVII, 430
4-*N*-[(3-Chloro-2-hydroxypropyl)-*N*-ethyl]-amino-*N*-methylphthalimide
synthesis, LVII, 431
of 4-(*N*-ethyl-*N*-[2-hydroxy-3-(*N*-phthalimido)propyl]amino)-*N*-methylphthalimide, LVII, 431
Chloroketone, XLVI, 29, 199, 415
^{14}C-labeled, XLVI, 425–427
synthesis, XLVI, 415–417
Chlorolactic acid phosphate, preparation, and reaction with enolase, XLI, 121
Chloromaleic acid anhydride, XLIV, 197
Chloromecodrin, *see* Neohydrin
2-Chloromercuri-1,4-androstadiene-3,17-dione, synthesis, XXXVI, 424
2-Chloromercuri-1,4,6-androstatriene-3,17-dione, synthesis, XXXVI, 424
2-Chloromercuri-5α-androst-1-ene-3,17-dione, synthesis, XXXVI, 424
p-Chloromercuribenzenesulfonic acid, proline uptake, LV, 199
p-Chloromercuribenzoate, XXXIV, 548; XLVII, 407
adenylate cyclase, XXXVIII, 169
analog, XLVII, 425
cAMP-receptor protein, XXXVIII, 381
complex V, LV, 315
effect on *Cypridina* luciferase, LVII, 353
inactivator of L-3-hydroxyacyl dehydrogenase, XXXV, 128
inhibitor, XLIV, 615
of adenosine deaminase, LI, 505, 511
of adenylate kinase, LI, 465
of brain α-hydroxylase, LII, 317
of dCTP deaminase, LI, 422
of GMP synthetase, LI, 223
of guanine deaminase, LI, 516
of guanylate kinase, LI, 489
of NADH dehydrogenase, LIII, 20
of nucleoside diphosphokinase, LI, 385
of orotidylate decarboxylase, LI, 79
of purine nucleoside phosphorylase, LI, 530, 536
of thymidylate synthetase, LI, 96
M. phlei membrane systems, LV, 185
phosphate transport, LVI, 359
phosphodiesterase, XXXVIII, 244
p-Chloromercuribenzoic acid, *see* *p*-Chloromercuribenzoate
1-(3-Chloromercuri-2-methoxypropyl)urea, *see* Neohydrin
2-Chloromercuri-17α-methyl-1,4,6-androstatrien-17β-ol-3-one, synthesis, XXXVI, 425
2-Chloromercuri-4-nitrophenol, XLVII, 408
p-Chloromercuriphenyl sulfonate, guanylate cyclase, XXXVIII, 195
p-Chloromercuriphenyl sulfonic acid
complex V, LV, 315
coupling factor B, LV, 390
inhibitor of NADH dehydrogenase, LIII, 20
preparation of aposulfite reductase, LIII, 436
2-Chloromercuri-1,4,6-pregnatriene-3,20-dione, synthesis, XXXVI, 425, 426
p-Chloro-*N*-methylaniline, in microsomal fractions, LII, 88
p-Chloro-*N*-methylaniline demethylase, in microsomal fractions, LII, 89
Chloromethylated resin, in TRF synthesis, XXXVII, 412
5-Chloro-3-methylcatechol dioxygenase, LII, 11
Chloromethyl ethyl ether, XLVII, 585
Chloromethyl ketone, XLVI, 130, 197, 234
synthesis, XLVI, 200, 609, 610
Chloromethyl methyl ether, toxicity, XLVII, 585
4-Chloro-1-naphthol, as hormone stain, XXXVII, 138, 140
4-Chloro-7-nitrobenzofurazan
availability, LV, 488, 489
biological properties, LV, 489–491
chemical properties, LV, 489
physical properties, LV, 476
structure, LV, 488

7-Chloro-4-nitrobenzofurazan, ATPase, LVI, 181
7-Chloro-4-nitrobenzo-2-oxa-1,3-diazole, XLVII, 8, 67, 408
  modification of lysine, LIII, 142
6-Chloro-9-(*p*-nitrobenzyl)purine, synthesis, XLVI, 331
Chloronitrophenol, XLIII, 168
Chloroperoxidase
  activity, LII, 521, 522
  assay, LII, 522–524
  in *Caldariomyces fumago*, LII, 521–529
  chemical properties, LII, 528
  inhibitors, LII, 529
  magnetic circular dichroism, XLIX, 170
  Mössbauer studies, LIV, 355, 378
    compound I, LIV, 379
  molecular weight, LII, 528
  purification procedure, LII, 525–527
  purity determination, LII, 524
  specificity, LII, 529
  spectral properties, LII, 528, 529
  stability, LII, 528
3-Chloroperoxybenzoic acid, XLVII, 492
4-Chlorophenylalanine
  inhibitor of phenylalanine hydroxylase, LIII, 285
  substrate of phenylalanine hydroxylase, LIII, 284
Chlorophyll, LIV, 18, 23, *see also* specific types
  in chloroplasts, XXXI, 606
    identification, XXXI, 735
  electrochromism, LV, 557
  lipid interactions with, NMR studies, XXXII, 211
  measurement, XXXII, 866
  RR spectra, LIV, 240, 245
Chlorophyll α fluorescence lifetime, LIV, 25
Chlorophyll *a*, LIV, 62, 63
  circularly polarized luminescence, XLIX, 184
Chlorophyll *b*, LIV, 62, 63
Chloroplast
  ATPase, LV, 298–300
    isolation by chloroform method, LV, 341, 342
  chlorophyll, XXXI, 606, 744
  countercurrent distribution, XXXI, 761, 764–768
  cytochrome isolation, XXXII, 406–422
  differential centrifugation, XXXI, 727, 732, 744, 745
  Dio-9, LV, 515
  efrapeptin, LV, 495
  energy-conserving systems, aureovertins, LV, 483
  fluorescent probe studies, XXXII, 234, 238, 240
  fragments, water compartment, LV, 550–552
  function, XXXI, 495, 496
  galactolipids, XXXII, 542
  internal volume, LV, 551
  ionophores, LV, 449, 452, 453
  isolation, XXXI, 501, 507–510, 519, 520, 522–524, 544–553; LIX, 209, 434, 435
    interferents, XXXI, 531, 532
    rapid method, XXXI, 600–606
  leucinostatin, LV, 514
  measurements of ΔpH, LV, 563
  membrane, XLVI, 106
    isolation, XXXII, 876–878
    optical probes, LV, 573
    in wild and mutant *C. reinhardi*, XXXII, 871–880
  Nbf-Cl, LV, 490
  onset of ATP and ADP labeling, LV, 246, 247
  organotins, LV, 509
  preservation, XXXI, 3
  quercitin, LV, 491, 492
  X-ray studies, XXXII, 219
  spinach, ATPase inhibitor, LV, 402, 404, 405
Chloropractolol, XLVI, 591
3-Chloropropionic acid *p*-nitrophenyl ester, XLVI, 261
6-Chloropurine ribonucleoside, substrate, of adenosine deaminase, LI, 507
6-Chloropurine ribonucleoside 5′-phosphate, XLVI, 299–302
6-Chloropurine riboside, synthesis of adenosine derivative, LVII, 118
6-Chloropurine riboside phosphate, XXXIV, 230, 231

6-Chloropurine riboside 5′-phosphate, synthesis of adenosine derivative, LVII, 115
6-Chloropurine 5′-triphosphate, XXXIV, 254
Chloroquine, preparation of aminoacylated tRNA, LIX, 132, 133
4-Chlororesorcinol, as antioxidant, XXXI, 538
4-Chlorostyrene, XLIV, 289
L-Chlorosuccinate, LVI, 586
N-Chlorosuccinimide, XLVII, 8, 456
Chlorothricin, XLIII, 409
Chlorotriazine, multifunctional agent, XLIV, 263
5-Chlorouracil, inhibitor of orotate phosphoribosyltransferase, LI, 166
Chlorox-benzidine indicator, for sphingolipid and gangliosides, XXXV, 406
Chlorphentermine, in studies of phagocytosis, LVII, 493
Chlortetracycline
  calcium or magnesium, LVI, 303
  electroanalytical techniques, XLIII, 385
  high-pressure liquid chromatography, XLIII, 317
  in media, LVIII, 112, 114
  paper chromatography, XLIII, 140
  Streptomyces aureofaciens, media, XLIII, 19
  thin-layer chromatography, XLIII, 199
CHO cell, *see* Chinese hamster ovary cell
Cholate, *see also* Sodium cholate
  asolectin solution, LV, 759
  calcium ion pump, LV, 702
  complex V preparation, LV, 310
  cytochrome oxidase, LV, 699, 701, 749
  extraction of F_0, LV, 719, 720
  hydrophobic protein preparation, LV, 713, 714
  M. phlei membrane fragments, LV, 195
  phospholipid solubilization, LV, 714
  preparation of F_1 and membrane factor, LV, 743, 744
    of $TF_0{\cdot}F_1$, LV, 368
  purification of BCF_0, LV, 195
  rutamycin-sensitive ATPase, LV, 317
  sodium salt, partial specific volume, XLVIII, 21
  solubilization of BCF_0-BCF_1, LV, 194
  submitochondrial particle fractionation, LV, 277
  transhydrogenase chromatography, LV, 279
  transhydrogenase reconstitution, LV, 813, 815
  treatment of ETP, LV, 193
Cholate-dialysis procedure, reconstitution studies, LV, 701–703, 722, 723
Cholera toxin, XXXIV, 613–619
  cAMP, XXXVIII, 3
  crude, affinity chromatography of, XXXIV, 616, 618, 619
  ganglioside-agarose, XXXIV, 610–619, *see also* Ganglioside-agarose
  ^{125}I-labeled, XXXIV, 613–616
    assay method, XXXIV, 616
  polyfunctional spacer arms, XXXIV, 94
  purified, affinity chromatography of, XXXIV, 616, 617
Cholera toxin receptor, gangliosides, L, 250
5α-Cholestane, standard, in gas chromatography, LII, 381
Cholestan-3β-ol, sterol estimation, LVI, 562, 563
3-Cholestanone, derivatives, in spin-label studies, XXXII, 164
Cholesterol, XLIX, 410
  assay, fluorometric, XLIV, 633
  in biomembranes, XXXII, 543
  content of cellular membranes, LVI, 208
  in cytochrome *P*–450 assay, LII, 125, 126, 139
  electron spin resonance spectra, XLIX, 407
  as endogenous substrate in adrenal mitochondria, LII, 271
  filter membranes, LV, 604
  lecithin interactions, XXXII, 208, 211
  in lipid bilayers, XXXII, 516, 517, 519, 538

lipophilic Sephadex chromatography, XXXV, 388, 389, 392
liposome formation, XLIV, 222, 633
in liposomes, XXXII, 505
in liver fractions, XXXI, 29, 35
in media, LVIII, 54, 68
metabolism, in adrenal gland, LII, 124
in microsomal fraction, LII, 88
in nuclear envelope, XXXI, 289
partial specific volume, XLVIII, 22
in plasma membranes, XXXI, 87, 101, 149, 165–167
polarography, LVI, 452, 468–472
side chain cleavage, XXXVII, 296
solubilization, XXXIX, 241
synthesis, control, XXXVII, 292
as testosterone precursor, XXXIX, 256
thin-layer chromatography, XXXV, 397, 407
transport in bloodstream, LVI, 468

Cholesterol desmolase, *see* Cytochrome *P*-450

Cholesterol ester hydrolase, XLIV, 633
in microvillous membrane, XXXI, 130
as myelin marker, XXXI, 444
polarography, LVI, 471
sources, LVI, 471

Cholesterol-lecithin bilayers, formation, XXXII, 490–492, 515

Cholesterol oxidase, LII, 18
electrode system, LVI, 485
in enzyme thermistor, XLIV, 674
in macrophage fractions, XXXI, 340, 344
sources, LVI, 471

Cholesterol sulfate sulfatase, multiple sulfatase deficiency, L, 474

Cholesteryl ester, lipophilic Sephadex chromatography, XXXV, 388, 392

7α-Cholesteryl oleate, radiolabeled, XXXII, 143

Cholesteryl sulfate, multiple sulfatase deficiency, L, 474

Cholic acid, *see also* Potassium cholate; Sodium cholate
alanine carrier isolation, LVI, 431
ATPase, XXXIV, 570
membrane solubilization, LVI, 423–425
micelles, ionic strength and pH effects, LVI, 746
potassium salt, preparation, LIII, 74
preparation of mitochondrial enzyme complexes, LIII, 3, 11
purification of elongation factor EF-1$\beta\gamma$, LX, 670, 702
recrystallization procedure, LIII, 3, 4, 56, 74
structure, LVI, 735

Choline, XLVI, 75
in media, LVIII, 54, 68

Choline acetyltransferase, XXXIV, 4
affinity labeling, XLVII, 428
assay, XXXII, 781–785
in cultured cells, XXXII, 766, 772, 773
in neurobiology, XXXII, 765

Choline alkyl phosphonate, intermediate in synthesis of phosphatidylcholine analogues, XXXV, 517–520, 525

Choline dehydrogenase, LIII, 405
effect of extraction methods, LIII, 409

Choline-methyl chloride, radioactive, microsome labeling, XXXI, 218

Choline phosphotransferase
as diagnostic enzyme, XXXI, 24, 303
microsomes, LV, 101, 104
assay, LV, 102

Cholinergic receptor, XXXIV, 6

Cholinesterase
assay
by Fast Analyzer, XXXI, 816
fluorometric, XLIV, 626
gel electrophoresis, XXXII, 88
immobilized
activity assay, XLIV, 653
by entrapment, XLIV, 173, 178
in monitoring system, XLIV, 647–658
in starch pad, XLIV, 652, 653
inactivation, XLVI, 85
pesticide inhibitors, XLIV, 637
detection, XLIV, 647–658

Cholismate mutase, XXXIV, 4

Chondrocyte
from cartilage, LVIII, 560
characterization, LVIII, 563
isolation, XXXVII, 105–109

large-scale preparation, LVIII, 560–565
media, LVIII, 562
source, LVIII, 561
staining, LVIII, 563
suspension culture, LVIII, 563
Chondroitin sulfate, LVIII, 268
chondrocyte culture, LVIII, 563
repeating disaccharide structure, L, 280
somatomedin stimulation, XXXVII, 94
Chondromucoprotein, somatomedin stimulation, XXXVII, 94
Chondrus crispus
aldolase, XLII, 234
oxidase, LII, 10
Chopper, light, XLIX, 239, 240
Chopping machine, for plant material, XXXI, 548
Choriocarcinoma, LVIII, 184
Chorionic gonadotropin, *see* Gonadotropin, chorionic
Chorismic acid
intermediate in ubiquinone biosynthesis, LIII, 600, 601
structural formula, LIII, 601
Choristoneura fumiferana, media, LVIII, 457
CHP, *see* Cumene hydroperoxide
Christmas factor, *see* Factor IX
Christ Omega Ultracentrifuge, live microsome isolation, XXXI, 197, 198
Chromaffin granule
ionophores, LV, 454
isolation and disassembly, XXXI, 379–389
measurement of ΔpH, LV, 563
membrane potential, LV, 555
membrane proteins, XXXI, 387, 388
Chromaffin vesicle, calcium transport, LVI, 322, 323
Chromatin
characterization, XXXI, 278, 279
constituent components, XL, 145
separation, in analysis of chromosomal nonhistone proteins, XL, 162
cytoplasmic receptor binding, XXXVI, 188
differential centrifugation, XL, 94
fraction, properties, XL, 100
fractionation, XL, 93
by agarose gel filtration, XL, 96
on hydroxyapatite, XL, 160
magnesium chloride, XL, 98
in template active and inactive regions
evidence, XL, 99
procedure, XL, 98
hormone induction of biorhythmic activity, XXXVI, 480
isolation, XXXVI, 296
in bulk quantity, XXXI, 276, 277
characterization of nonhistone chromosomal proteins, XL, 144
preparation
in analysis of chromosomal nonhistone proteins, XL, 162
Bonner method, XL, 144
receptor binding, XXXVI, 304–306
structure and function, XL, 104
template active and inactive regions
chemical composition, XL, 100
isolation, XL, 97
thermal elution from hydroxyapatite, XL, 94
Chromatin protein, *see* Protein, chromatin
Chromatin protein kinase, XL, 198
assay of activity, XL, 201
fractionation, XL, 199
substrate specificities and effects of cyclic AMP, XL, 205
typical results, XL, 203
heterogeneity, XL, 206
lability, XL, 207
Chromatium
coenzyme Q homolog, LIII, 593
cytochrome c_{552}, LIII, 449, 450, 452, 453, 462, 465
cytochrome c', LIII, 210
high-potential iron-sulfur protein
EPR characteristics, LIV, 139, 144
Mössbauer studies, LIV, 371
proton magnetic resonance studies, LIV, 204, 206, 207
redox studies, LIV, 429
hydrogenase, LIII, 290
oxygen-stable hydrogenase, LIII, 297

ribulose-5-phosphate kinase, assay, purification, and properties, XLII, 115–119
Chromatium D
cytochrome *c*-552, CD spectrum, LIV, 276, 278
oxidase, LII, 21
Chromatium vinosum
cell extract, preparation, LIII, 311
ferredoxin
characterization, LIII, 273
nomenclature, LIII, 265
physical properties, LIII, 263
purification, LIII, 267
growth, LIII, 309, 310, 335
high-potential iron-sulfur protein from, LIII, 329, 330, 332, 333, 339
pure culture techniques, LIII, 310
Chromatoelectrophoresis
in hormone assay, XXXVII, 4
in radioimmunoassay, XXXVII, 33, 34
Chromatogram
paper, *see* Paper chromatography
pH, XLIII, 164–168
salting out, XLIII, 162–164
Chromatography, *see also* specific type
agar and agarose, XLIV, 21
of aldose reductase and L-hexonate dehydrogenase, XLI, 162
DNA-cellulose, XL, 181
gel permeation, XLVII, 97–107
ion-exchange, of saccharide borate complexes, XLI, 10–21
of 5-keto-D-fructose, XLI, 86
on microbore columns, XLVII, 210–236
analytical procedures, XLVII, 225–230
buffers, XLVII, 214, 222, 223
column elution, XLVII, 214, 215
elution profile markers, XLVII, 233, 234
instrumentation, XLVII, 211–213, 225–228
packing procedure, XLVII, 213, 227
preparative separations, XLVII, 230–234
reproducibility, XLVII, 217, 218
resolution, XLVII, 219, 220
sample application, XLVII, 213, 228, 229
sample load, XLVII, 217, 218
sample preparation, XLVII, 221, 222
pumps, XLVII, 211
in quantitative histone assay, XL, 117
R_f values, XLIII, 182, 183
R_m values, in structural analysis, XLIII, 183, 184
simulation, XXXIV, 158
in sugar analysis, XLI, 3
system for peptide purification, XXXVII, 319
of 3-uloses for detection, XLI, 158
Chromatophore
electrical potential differences, LIV, 63, 64
hydrogen ion transport, LVI, 312
measurement of ΔpH, LV, 563
membrane potentials, LV, 452
photosynthetic redox chain, planar membranes, LV, 756
Chromato-vue cabinet, *rec* A mutants, LVI, 116
Chromatronix plastic valves and tubing, XXXIX, 374
Chrome, XLIV, 248
Chromin, XLIII, 135
Chromium
atomic emission studies, LIV, 455
in media, LVIII, 69
volatilization losses, in trace metal analysis, LIV, 466, 480, 481
Chromium(III)-adenosine 5′-triphosphate, conformation studies, XLIX, 355
Chromium agent, for oxygen removal, LIV, 117
Chromium chloride, as antibody coupling agent, XXXII, 65, 66
Chromium sulfate, XLIV, 248
Chromogranin
description, XXXI, 383
isolation, XXXI, 374–387
Chromomembrin, isolation, from chromaffin granules, XXXI, 388, 389
Chromomycin, XLIII, 352

Chromomycin A_3, XLIII, 127
fluorescence spectrum, LVIII, 238
fluorescent dye, LVIII, 236–238, 240, 242
preparation, LVIII, 247
Chromophore
choice, XLIX, 96, 97
classifications, LIV, 253, 254
definition, LIV, 253
in excited state, circularly polarized luminescence, XLIX, 184
intrinsic, in flash photolysis technique, LIV, 57, 58
resonance Raman scattering, XLIX, 72
Chromoplast, function, XXXI, 495
Chromosome
analysis, of haploid frog cells, XXXII, 795–797
autoradiography, XL, 71
in cell fusion, XXXII, 575
critical point drying, XL, 68
DNase digestion, XL, 87
examination, LVIII, 170–174
identification, LVIII, 322
in interphase cells, XL, 63
isolation, XL, 64
buffers, XL, 77
preparation for electron microscopy, XL, 67
stability, XL, 87
universal medium, XL, 76
loss, LVIII, 322, 348
metaphase
composition, XL, 82
fractionation, XL, 81
isolation, XL, 65
methods and properties, XL, 78, 80
parallel procedures with mitotic apparatus and nuclei, XL, 75, 79
preparation, XL, 305
reaction with antinucleoside antibodies, XL, 304
microscopy, LVIII, 336
photography, LVIII, 336–339
slide preparation, LVIII, 326–334
studies using hybrid somatic cells, XXXIX, 123
whole, electron microscope autoradiography procedure, XL, 73
Chromosome banding, LVIII, 324, 325
Chromosome deletion, LVIII, 323
Chromosome inversion, LVIII, 323
Chromosome localization, LVIII, 348
SV40, LVIII, 349
Chromosome marker, LVIII, 174
Chromosome number, LVIII, 323
Chromosome rearrangement, LVIII, 323
Chromosome staining, LVIII, 324–326, 334–336
Chromosome translocation, LVIII, 323
Chromosome variation, LVIII, 322, 323
Chromosorb G DMCS column, LII, 323
Chromous sulfate, for oxygen removal, LIV, 117
Chromsorb W, XLVII, 395
Chronic granulomatous disease, *see* Granulomatous disease, chronic
Chrysene dihydrodiol, chromatographic properties, LII, 294–296
CHT, *see* Chymotrypsin
Chylomicron, phospholipid exchange, XXXII, 140
Chymopapain, XXXIV, 532, 542, 546
Chymostatin
activity, XLV, 687
assay, XLV, 686
chymotrypsins, inhibition, XLV, 685
properties, XLV, 687
Chymotrypsin, XXXIV, 7, 415–420, 585; XLVI, 22, 39, 75, 84, 89, 90, 96, 156, 197, 201, 206, 207, 215–220, 431, 478, 537, 538; *see also* specific type
active site, XXXVI, 390; XLIV, 410
activity, XLIV, 30, 31, 34, 36, 200, 266, 267, 271, 446, 447, 530, 561
acylation, XLIV, 447, 562
from bovine pancreas, specificity, LIX, 600
conformation studies, XLIV, 363, 365, 366
conjugate
active site titration, XLIV, 391, 392

effect of heat treatment, XLIV, 535, 536
of support stretching, XLIV, 562–568
kinetics, XLIV, 425
storage, XLIV, 75, 76
cyanogen bromide binding, XLIV, 52, 72–76
dye binding, XLIV, 535
effect on luciferase activity, LVII, 201
on purified NADH-cytochrome b_5 reductase, LII, 108
energy transfer studies, XLVIII, 378
in hormone receptor studies, XXXVII, 212
immobilization
on activated agarose, XLIV, 33–36
on activated Spheron, XLIV, 72–76, 83
by adsorption, effects on activity, XLIV, 44
by aggregation, XLIV, 265–267
on anhydride derivative of hydroxyalkylmethacrylate gel, XLIV, 83
in cellophane membrane, XLIV, 905, 909
on cellulose, XLIV, 52
in hollow-fiber membrane device, XLIV, 310, 311
on imidoester-containing polyacrylonitrile, XLIV, 323, 324
on inert protein, XLIV, 909
on polyacrylamide beads, XLIV, 446
in polyacrylamide gel, XLIV, 902
on Sephadex, XLIV, 30, 529
using acyl azide intermediate method, XLIV, 446
using glutaraldehyde, XLIV, 265–267
matrix blocking on Sephadex, XLIV, 24
maximum loading, XLIV, 727
modification, XLIV, 310
effect on activity, XLIV, 409–412, 448–450
in pancreas, XXXI, 53
pH-activity profiles, XLIV, 404
pressure-jump studies, XLVIII, 314
purification of flavin peptides, LIII, 453
resolution of organic acids, XLIV, 833–835
selective ester hydrolysis, XLIV, 835, 836
soybean, membrane osmometry, XLVIII, 75
tissue culture, LVIII, 125
s-triazine binding, XLIV, 52
α-Chymotrypsin, XXXIV, 514
active site conformation, XLIX, 440, 447–450
molecular orientation, XLIX, 470–478
acylated, stability of enantiomers, XLIX, 471–473
amino acid analog substrates, XLVII, 430
carbethoxylation of active site serine, XLVII, 439
crystalline
g value, XLIX, 473
spin-labeling procedure, XLIX, 470, 471
symmetry relationships, XLIX, 473–476
diethylpyrocarbonate treatment, XLVII, 433, 438
fluorescence quenching, XLIX, 222
immobilization, XLVII, 44
immobilized-peptide cleavage, XLVII, 277
inactivation, XLVII, 172
molecular weight determination using sedimentation equilibrium, XLVIII, 180
multiple-site binding model, XLIX, 450, 451
peptide separation, XLVII, 206
phosphorescence studies, XLIX, 246
protein digestion, concentration and incubation period, XL, 38
purification of commercial, XXXIV, 420
reductive methylation, XLVII, 475
self-association, XLVIII, 116
sequence analysis, XLVII, 180
spin-label studies, XLIX, 436–443
after phosphorylation, XLIX, 445, 446

sample preparation, XLIX, 438
stopped-flow Fourier-transform nuclear magnetic resonance spectra, XLIX, 300, 321
after sulfonylation, XLIX, 447–450
titration, XLV, 18
tosyl hole, XLIX, 447
tryptophan content, XLIX, 162
Chymotrypsin A, XXXIV, 418; XLVI, 213, 214
Chymotrypsin B, XXXIV, 418
Chymotrypsin inhibitor, XLV, 579, 580, 687, 870; LIX, 598
groundnut amino acid composition, XLV, 721
assay, XLV, 716
enzyme treatment, effect, XLV, 699
esterolysis inhibition, XLV, 717
potentiometric, XLV, 717
spectrophotometric, XLV, 718
purification, XLV, 718
purity, XLV, 721
specificity, XLV, 721
stability, XLV, 721
units, XLV, 717, 718
yield and potency, XLV, 719
soybean, XLV, 700
Chymotrypsin inhibitor I, potato
amino acid composition, XLV, 733, 735
assay
immunological, XLV, 730
spectrophotometric, XLV, 729
kinetics, XLV, 733
molecular weight, XLV, 733
properties, XLV, 732
immunological, XLV, 733
physical, XLV, 733
purification, XLV, 731
purity, XLV, 732
specificity, XLV, 733
stability, XLV, 732
units, XLV, 731
Chymotrypsin-like enzyme, activity assay, XLVII, 77
Chymotrypsinogen, XXXIV, 415, *see also* Chymotrypsinogen A
acquisition of tertiary structure, XLIV, 521–526
activation, rate determination, XLIV, 572
HABA column, XXXIV, 266
heat-capacity profile, XLIX, 10
membrane osmometry, XLVIII, 75
pressure-temperature transition
gel electrophoresis, XLIX, 19
map, XLIX, 18
SDS complex, properties, XLVIII, 6
steric exclusion of trypsin, XLIV, 667
unfolding, XLIX, 4
α-Chymotrypsinogen, hollow-fiber retention data, XLIV, 296
Chymotrypsinogen A, XXXIV, 417
conjugate, reductive denaturation studies, XLIV, 522–526
immobilization on glass beads, XLIV, 520–522
reductive denaturation studies, XLIV, 518
Chymotrypsin-Sepharose, XLVII, 42
Cibacron Blue F 3GA, XXXIV, 5
for affinity chromatography, LI, 348
Cibacron Blue-Sepharose 4B, in purification of blue-fluorescent protein, LVII, 232
CIEEL, *see* Luminescence, chemically initiated electron exchange
Ciliate, mitochondrial DNA, LVI, 5
cIMP, *see* Cyclic inosine monophosphate
$C\bar{1}$ inactivator
assay, XLV, 752
characteristics, XLV, 751
composition, XLV, 759, 760
kallikreins, XLV, 314
properties, XLV, 755, 756
purification, XLV, 753
purity, XLV, 756
specificity, XLV, 756
stability, XLV, 755
units, XLV, 753
variants, XLV, 758
Cinerubin, XLIII, 145
Cinnamate-4-hydroxylase, LII, 15
trans-Cinnamate-2-hydroxylase, LII, 20
Cinnamic acid, pseudosubstrate of melilotate hydroxylase, LIII, 557
Circadian rhythm, examples, XXXVI, 474

Circular dichroism, XL, 214, *see also* Spectropolarimetry
analysis of nucleoprotein complexes, XL, 209
of biomembranes, XXXII, 220–233
calibrations, XXXII, 222, 223
conformation studies, XLIV, 366, 367
corrections, XXXII, 220–233
pseudo reference state approach, XXXII, 225, 226
difference, XL, 225
ellipticity in analysis, XL, 216
equations, XLIII, 355
experimental parameters, XL, 227
fluorescence-detected, XLIX, 181–184, 199–214
calibration, XLIX, 210, 211
filters, XLIX, 209, 210
instrumentation, XLIX, 200, 206–210
instrument proportionality constant, XLIX, 202
momentum operator, XLIX, 204
photomultiplier and preamplifier, XLIX, 207–209
photoselection, XLIX, 183, 184
position operator, XLIX, 204
sample cell, XLIX, 206, 209
theoretical considerations
applicability, XLIX, 181, 182
for isotropic case, XLIX, 200–204
for photoselected case, XLIX, 204–206
of fluorophore excited state, *see* Circularly polarized luminescence
of fluorophore ground state, *see* Circular dichroism, fluorescence-detected
instrumental artifact, XXXII, 222
instrumentation, XL, 221
calibration, XL, 227
modifications, XL, 224
slit width of instrument, XL, 230
optical parameters, XL, 229
phototube dynode voltage, XXXII, 222, 223
presentation of data, XL, 231
principles of measurement, XL, 222
tRNA, XLIX, 212–214
sample calculation, XXXII, 231–233
spectra, relationship with structure, XL, 231
in studies of protein-RNA interactions, LIX, 579
vacuum ultraviolet, XLIX, 214–221
applications, XLIX, 218–221
instrumentation, XLIX, 215–218
wavelength scanning speed, relation with time constant, XL, 230
Circular dichroism spectroscopy, LIV, 249–284
aromatic absorption region, definition, LIV, 254
data analysis
isodichroic method, LIV, 257, 258, 268, 269
multiple-component analysis approach, LIV, 257
phenomenological approach, LIV, 257–259, 267
reference protein method, LIV, 258, 259, 268, 269
experimental parameters, LIV, 252
instrumentation, LIV, 251, 252
intrinsic absorption region, definition, LIV, 254, 265
nomenclature for spectral regions, LIV, 254, 265
normalization of raw data, LIV, 252
Soret region, definition, LIV, 265
spectral data parameters, LIV, 253
Circularly polarized luminescence, XLIX, 179–199
applicability to chromophores in excited state, XLIX, 184
to conformation studies, XLIX, 184–188
calibration procedure, XLIX, 195–198
conformation studies, XLIX, 184–188
depolarizer, XLIX, 194, 195
instrumentation, XLIX, 188–199
light modulator, XLIX, 193
light source, XLIX, 190, 191
monochromator, XLIX, 191
optical filters, XLIX, 191–193
photomultiplier, XLIX, 195
photoselection, XLIX, 182–184, 187, 188
sample cell, XLIX, 195
sensitivity, XLIX, 198, 199

theoretical applicability, XLIX, 180–182
Circulin, XLIII, 579
Circumfusion, dual-rotary, LVIII, 14
Cirramycin, XLIII, 336
Cirripede, spermatozoa of, electron microscopy, XXXII, 33
C_{55}–isoprenoid alcohol phosphokinase, *see* Isoprenoid-alcohol kinase
Citraconic anhydride
acylation, XLVII, 149
for modification of green-fluorescent protein, LVII, 265
Citrate buffer
in microsomal protein solubilization, LII, 202
preparation for amino acid analysis, XLVII, 16
purity check method, XLVII, 15
for short analysis time, XLVII, 25, 27
two-dimensional gel electrophoresis, LIX, 68
Citrate cleavage enzyme, in parenchyma cells, XXXII, 706
Citrate lyase, citrate assay, LV, 211
Citrate-phosphate buffer, for enzyme assay, XXXII, 130
Citrate synthase
in acetate assay, XXXV, 299, 302
activity assay, XLIV, 458–460, 474
in assay of carboxyltransferase, XXXV, 32–34
of long-chain fatty acyl-CoA, XXXV, 273
coenzyme A assay, LV, 210
immobilization in multistep enzyme system, XLIV, 181, 457–460, 473–475
in glyoxysomes, XXXI, 565, 569
kinetics, in multistep enzyme system, XLIV, 434, 466–468
as marker enzyme, LVI, 217, 218
in microbodies, LII, 496
oxaloacetate assay, LV, 208
toluene-treated mitochondria, LVI, 546, 547
regulatory properties, LVI, 549, 550
Citrate synthetase, *see also* Citrate synthase
Citreoviridin
ATPase, LV, 343; LVI, 181
availability, LV, 486
bacterial systems, LV, 487
fluorescence properties, LV, 488
inhibitory effects on mitochondrial systems, LV, 487, 488
physical properties, LV, 476
preparation
analysis of extracts, LV, 486, 487
extraction, LV, 486
further purification, LV, 487
structure, LV, 486
toxicity, LV, 487
Citreoviridin monoacetate, physical properties, LV, 476
Citric acid
activation of acetyl-CoA carboxylase, XXXV, 3, 8, 9, 17
activator of adenylate kinase, LI, 465
assay, LV, 210, 211, 221
brain mitochondria, LV, 57
coenzyme A assay, LV, 210
content
of brain, LV, 218
of hepatocytes, LV, 214
of kidney cortex, LV, 220
of rat heart, LV, 216, 217
of rat liver, LV, 213
exchange, LVI, 253
fep$^-$ mutant, LVI, 396
fluorescence quenching agent, XLIX, 233
impurity in bovine serum albumin, XXXV, 598
inhibitor of adenine phosphoribosyltransferase, LI, 573
of rat liver acid nucleotidase, LI, 274
intact mitochondria, LV, 143
iron solutions, LVI, 391
in isolated renal tubules, XXXIX, 18
in liver homogenization, XXXI, 77
Pasteur effect, LV, 291, 296
purification of complex III, LIII, 88
reagent for nuclei isolation, XXXI, 272
renal tubule permeability, XXXIX, 15, 16

stabilizer, of luciferase, LVII, 199
transport, LVI, 248–251, 292
yeast mitochondria, LV, 150
Citric acid cycle, intermediates
assay
methological update and correction, LV, 200–203
new, modified and alternative methods, LV, 205–215
brown adipose tissue mitochondria, LV, 77
mitochondria, LV, 7, 8
Citric acid-sodium citrate buffer, XXXIX, 136
Citrulline
colorimetric determination, LI, 105
continuous production, XLIV, 745
internal standard, XLVII, 34, 39, 40
transport, LVI, 248, 249
Citrus, cephalosporin acetylesterase, XLIII, 728–731
L-Cladinose, erythromycin A moiety, XLIII, 489
Clam, luciferase, LIII, 560
Clapeyron equation, XLVIII, 88
Clark oxygen electrode, XXXI, 19; XXXII, 258; LIII, 445; LVII, 39
assay of salicylate hydroxylase, LIII, 530
CLASS computer language, XLIX, 406
Clauberg progestin assay, XXXVI, 456, 457
Clausius-Clapeyron relation, XLIX, 19
Clay-Adams black desiccator box, XXXVI, 144, 145
Cleaning facility, LVIII, 3–7
Clegg signal-averaging method, XLVIII, 313
Cleland method, LVI, 265, 266
Cleland's reagent, *see also* Dithiothreitol
in assay of nitrate reductase, LIII, 348, 350
in plant-cell fractionation, XXXI, 505
in purification of nitrate reductase, LIII, 348
Clerodendrum viscosum, lectin, XXXIV, 334
Cleves acid, in nitrite determination, LIII, 638
Clindamycin
biosynthesis, XLIII, 755–759, *see also* Clindamycin phosphotransferase
gas-liquid chromatography, XLIII, 237–239
phosphorylation, XLIII, 755–759, *see also* Clindamycin phosphotransferase
Clindamycin palmitate, XLIII, 243–245
Clindamycin phosphate
gas-liquid chromatography, XLIII, 242, 243
high-pressure liquid chromatography, XLIII, 306–307, 319
Clindamycin 3-phosphate, XLIII, 756
Clindamycin phosphotransferase, XLIII, 755–759
ammonium sulfate fractionation, XLIII, 757, 758
assay, XLIII, 756, 757
crude extract preparation, XLIII, 757
DEAE-cellulose column chromatography, XLIII, 758
nucleoproteins removal, XLIII, 757
properties, XLIII, 758, 759
purification, XLIII, 757–759
reaction scheme, XLIII, 755
specific activity, XLIII, 757
specificity, XLIII, 758
stability, XLIII, 758
stoichiometry, XLIII, 759
unit definition, XLIII, 756
Clipping, XLVIII, 470, 471
CLN, *see* *p*-Nitrophenyl-N^{α}-benzyloxycarbonyl-L-lysinate
Cloacin DF13, LX, 215
Clock, biological, *see* Biological clock
Clofibrate, activator of microbody enzymes, LII, 501, 502, 504
Clomiphene citrate, antiestrogen, XL, 292
Clone, *see also* Cloning
culture, media, LVIII, 56, 58, 59
growth, LVIII, 46, 83
media, LVIII, 56–59, 86, 87, 89, 90
Clone 3T3–L1, effect of hormone, LVIII, 105, 106
Cloning, LVIII, 46, 152–164
anchorage-dependent cells, LVIII, 152
protocol, LVIII, 158, 159

anchorage-independent, LVIII, 153
protocol, LVIII, 159–161
of anterior pituitary cells, XXXIX, 128–132
capillary technique, LVIII, 153
efficiency, LVIII, 509
of endocrine cell lines, XXXIX, 117, 118
HTC cells, LVIII, 547–549
layer preparation, LVIII, 161–164
media, LVIII, 155, 384, *see also* specific types
preparation, LVIII, 161–164
melanoma cell, LVIII, 566
morphology, LVIII, 139
in multiwell plastic trays, LVIII, 321, 322
mutant, LVIII, 321, *see also* Mutant
neuronal cell, LVIII, 588, 589
plant cells, LVIII, 482–484
protocol, LVIII, 154, 155
Rous sarcoma virus, LVIII, 390–392
Cloning cylinder, XXXII, 583
Clostridiopeptidase B, *see* Clostripain
Clostridium
D-gluconate dehydratase activity, XLII, 304
pyruvate formate-lyase, XLI, 518
Clostridium acidiurici
ferredoxin, LIV, 205, 207, 208
purification, LIII, 267
formate dehydrogenase, LIII, 372
pyruvate-ferredoxin oxidoreductase, assay, purification, and properties, XLI, 334–337
Clostridium formicoaceticum, formate dehydrogenase, LIII, 372
Clostridium histolyticum
collagenase, XXXI, 104
proteinase, in fat cell isolation, XXXVII, 211
Clostridium lentoputrescens, *see* *Clostridium malenominatum*
Clostridium malenominatum, culture, LIII, 375
Clostridium pasteurianum
crude extract preparation, LIII, 321, 322
Dio 9, LV, 515
ferredoxin, LIV, 205, 207, 208
characterization, LIII, 273
nomenclature, LIII, 265
physical properties, LIII, 262, 263
purification, LIII, 267
reference spectrum, LIII, 271
D-gluconate dehydratase, XLII, 304
growth, LIII, 321
nitrogenase, LIII, 319, 321, 322; LIV, 338
phosphofructokinase, assay, purification, and properties, XLII, 86–91
rubredoxin, LIII, 340, 341, 344, 345; LIV, 359, 360
nomenclature, LIII, 265
physical properties, LIII, 262
purification, LIII, 267
Clostridium perfringens
acylneuraminate pyruvate-lyase, L, 75, 76
neuraminidase, XLIV, 223; L, 67
in sialylated hormone hydrolysis, XXXVII, 324
Clostridium sticklandii
crude extract preparation, LIII, 276
culture, LIII, 375
Clostridium thermoaceticum
anaerobic crude extract preparation, LIII, 366
culturing methods, LIII, 361–363
formate dehydrogenase, LIII, 360–372
Clostridium welchii, phospholipase C, XXXII, 139, 159
Clostripain, XLVI, 206, 229
active site properties, XLVII, 165–169
from *Clostridium histolyticum* specificity, LIX, 600
inhibition, XLVI, 234
inhibitors, XLVII, 167–169
protein cleavage, XLVII, 165–170
utility as sequence tool, XLVII, 169, 170
Cloth, for plating, LVIII, 320
Clot lysis
streptokinase, XLV, 245
urokinase, XLV, 239
Cloudman S91 melanoma, LVIII, 564
Cloxacillin, XLIII, 671
Clupeine, XLIII, 336

$C_{17\text{-}20}$–lyase, inhibitor, XXXIX, 282
CM, *see* (Carboxymethyl)cellulose
CMC, *see* 1-Cyclohexyl-3-(2-morpholinoethyl)carbodiimide metho-*p*-toluenesulfonate
CM–cellulose, *see* (Carboxymethyl)cellulose
7C's medium, LVIII, 58, 89
CMK, *see* Tryptophan, chloromethyl ketone analog
CMOS counter, LVII, 534, 535, 538
CMOS switch, LVII, 534, 535, 538
CMP, *see* Cytidine 5′-monophosphate
CMRL 1066 medium, LVIII, 57, 88, 155
  composition, LVIII, 62–70
CMRL 1415 medium, LVIII, 58, 87, 90, 91
  composition, LVIII, 62–70
CMRL 1415–ATM medium, LVIII, 58, 87
CMRL 1969 medium, LVIII, 59, 87, 91
  composition, LVIII, 62–70
CMR spectroscopy, *see* Nuclear magnetic resonance spectroscopy, carbon-13
CM-Sephadex, *see* (Carboxymethyl)Sephadex
CNP, *see* 2′:3′-Cyclic-nucleotide-3′-phosphohydrolase
Co α-(aden-9-yl)adenosylcobamide, effector, of ribonucleoside triphosphate reductase, LI, 258
Co α-(benzimidazoyl)-*Co* β-adenosylcobamide, effector of ribonucleoside triphosphate reductase, LI, 258
CoA, *see* Coenzyme A
ε-CoA, *see* 1,N^6-Etheno-coenzyme A
Coagulant enzyme, from *Bothrops atrox* venom, XLV, 214, *see also* Batroxobin
Coagulant protein
  Russell's viper venom, XXXIV, 592–594; XLV, 191
    assay, XLV, 192
      continuous spectrophotometric, XLV, 194
      one stage, XLV, 192
      two stage, XLV, 193
    characteristics, XLV, 202
    kinetics, XLV, 203
    properties, XLV, 202
      metal binding, XLV, 205
      physical, XLV, 204
    purification, XLV, 196, 200, 201
    specificity, XLV, 202
Coagulating solvent
  for cellulose triacetate, XLIV, 230
  of ethyl cellulose, XLIV, 231
  for nitrocellulose, XLIV, 230
Coagulation factor, XXXIV, 6
Coarctation, XLIV, 179
CoA-transferase, *see* Succinyl-CoA:propionate CoA-transferase
Coaxial cable
  as delay lines, LIV, 44
  nanosecond absorbance spectroscopy, LIV, 41
Cobalamin, electron paramagnetic resonance studies, XLIX, 514
Cobalamin analog, effectors of ribonucleoside triphosphate reductase, LI, 258
Cobalt
  atomic emission studies, LIV, 455
  in media, LVIII, 69
  in reconstituted hemoglobin, LII, 491, 492
  in tertiary bonding assignments, LIX, 45, 46
Cobalt(II)
  catalyst, in luminol chemiluminescence, LVII, 448
  complex ions, magnetic circular dichroism, XLIX, 167, 175
  correlation time evaluation, XLIX, 346
  electron paramagnetic resonance studies, XLIX, 514
  interatomic distance calculations, XLIX, 331
Cobalt(III)
  catalyst, in luminol chemiluminescence, LVII, 448
  electron paramagnetic resonance studies, XLIX, 514
Cobalt-59, lock nuclei, XLIX, 348; LIV, 189
Cobalt-60
  in membrane studies, XXXII, 884–893
  properties, XXXII, 884
Cobalt acetate, XLIV, 162

Cobalt chloride, XLIV, 198, 747, 748, 774, 793
  chemiluminescence assays, LVII, 427
Cobalt(III) complex, XLVI, 312–321
  ATP-phen, XLVI, 314, 315, 318–321
  stoichiometry, XLVI, 316, 317
  synthesis, XLVI, 313–315
    air oxidation, XLVI, 314, 315
    electrolytic, XLVI, 313, 314
    oxidation by hydrogen peroxide, XLVI, 315
  thiolysis, XLVI, 315, 316
Cobalt(III)-ethylenediaminetetraacetic acid, Cr^{3+}-doped, *g* value, XLIX, 523
Cobalt fluoride, for drying hydrogen fluoride, LIII, 176
Cobalt ion
  activator of adenine phosphoribosyltransferase, LI, 573
    of dCTP deaminase, LI, 422
    of deoxycytidine kinase, LI, 345
    of deoxythymidylate phosphohydrolase, LI, 289
    of guanylate kinase, LI, 481
    of nucleoside diphosphokinase, LI, 385
    of nucleoside phosphotransferase, LI, 393
    of succino-AICAR synthetase, LI, 193
    of uridine-cytidine kinase, LI, 319
  glucose isomerase activation, XLIV, 815
  inhibitor of *Cypridina* luminescence, LVII, 349, 350, 351
    of deoxythymidine kinase, LI, 365
    of phosphoribosylpyrophosphate synthetase, LI, 11
    of uridine nucleosidase, LI, 294
  Pholas luciferin chemiluminescence, LVII, 400
Cobalt(III)-nucleotide complex, XLVI, 313, 314, 318
Cobaltous sulfate, magnetic circular dichroism standard, LIV, 290
Cobalt-substituted cytochrome *c*
  midpoint redox potential, LIII, 181
  preparation, LIII, 177, 178
  properties, LIII, 178–181
Cobalt-substituted enzyme, magnetic circular dichroism, XLIX, 166–170
Cobalt sulfate
  extinction coefficient, XLIX, 156
  magnetic ellipticity, XLIX, 157
  for petite-negative yeast mutant enrichment, XXXII, 839, 840
Cobra
  neurotoxins, XXXII, 310
    preparation, XXXII, 312, 313
  venom, phospholipase A, XXXII, 379
Cobramine B, classification of tyrosyl residues, XLIX, 119
Cocarboxypeptidase A, XLIX, 137
Coconut charcoal, preparation of para hydrogen, LIII, 306
Coconut milk
  as growth promoting substance, XXXII, 730, 731
  in plant-cell fractionation, XXXI, 505, 518
  for plant tissue purification, XXXII, 405
Cocoonase, XLVI, 206
Codeine, as mixed-function oxidase substrate, XXXI, 233
Codling moth, media, LVIII, 457
Codon-anticodon recognition, LIX, 292–309
Coelenteramide
  emission wavelengths, LVII, 341
  product of BFP reaction, LVII, 284, 287
  structural formula, LVII, 284, 341
Coelenteramine
  product of aequorin denaturation, LVII, 284, 289
  structural formula, LVII, 284
Coelenterazine, LVII, 343, 344
  aequorin regeneration, LVII, 289–291
  structural formula, LVII, 289, 341
Coelomic cell, of *Diplocardia longa*, collection, LVII, 375, 376
Coelosphaerium, lysis, XXXI, 681
Coenzyme, *see also* specific type
  alkylated, spectral properties, XLIV, 868
  alkylation, XLIV, 859–862
  analogs, R_f values, XLIV, 868, 869
  derivative synthesis, XLIV, 859–862

enzyme electrode, XLIV, 603
^{19}F-substituted, XLIV, 366
immobilization, XXXIV, 242; XLIV, 859–887
support pretreatment, XLIV, 155
Coenzyme A, XXXIV, 267–271, *see also* specific types
activator of fatty acid synthase, XXXV, 89, 90
assay, XXXIV, 268
content of rat liver, LV, 213
dephospho, XXXIV, 270
derivatives, assay, LV, 209–211, 221
fluorometric assay, XXXV, 278
hepatocyte content, LV, 214
inhibitor of cytoplasmic thiolases, XXXV, 134
labeled, LX, 535
in media, LVIII, 64
nitroprusside reaction, XXXIV, 268
oxidized, XXXIV, 270
preparation of crude, XXXIV, 270
product and inhibitor of acetyl-CoA:amine acetyltransferase, XXXV, 253
of acetyl-CoA:long-chain base acetyltransferase, XXXV, 242, 247
protection during hydrolysis, LV, 202
purification, XXXVIII, 81
recycling, in assay of long-chain fatty acyl-CoA, XXXV, 274, 275
released from palmitoyl-CoA, assay, LVI, 371, 372
separation from long-chain fatty acyl-CoA, XXXV, 275, 276
spectrophotometric assay, XXXV, 276
substrate
for acetoacetyl-CoA thiolase, XXXV, 167
for acetyl-CoA synthetase, XXXV, 302
for acyl carrier protein synthase, XXXV, 95, 100
for carnitine acetyltransferase, XXXV, 273, 274
for 3-hydroxy-3-methylglutaryl-CoA synthase, XXXV, 155, 160, 173
for 3-ketoacyl-CoA thiolase, XXXV, 128, 130
in succinyl-CoA generating system, LII, 352
Coenzyme A analog, conformation studies, XLIX, 355
Coenzyme A-agarose, XXXIV, 267, 268
Coenzyme analog, inhibitor constants, XLVI, 257
Coenzyme A-binding protein-agarose, XXXIV, 270
ADP, XXXIV, 270
ATP, XXXIV, 270
dephospho-CoA, XXXIV, 270
oxidized CoA, XXXIV, 270
Coenzyme A thioester, *see also* specific esters
purification, XXXV, 139
Coenzyme Q, LIII, 591–599, *see also* Ubiquinone
bacteriochlorophyll proteoliposomes, LV, 760
carboxamides, LV, 461
nomenclature, LIII, 591, 592
piericidin A, LV, 456
proteoliposomes, LV, 767, 768
rotenone, LV, 455
2-theoyltrifluoroacetone, LV, 460
Coenzyme Q_1
analog, LIII, 594
structural formula, LIII, 594
synthesis, LIII, 593, 594, 596, 597
Coenzyme Q_2
analog, LIII, 594
structural formula, LIII, 594
synthesis, LIII, 594, 596, 597
Coenzyme Q_3
analog, LIII, 594
structural formula, LIII, 594
synthesis, LIII, 594, 596, 597
Coenzyme Q_6, commercial source, LIII, 599
Coenzyme Q_9, commercial source, LIII, 599
Coenzyme Q_{10}, LIII, 405, 409, 412
commercial source, LIII, 599
structural formula, LIII, 594
Coenzyme Q analog
identification by mass spectroscopy, LIII, 595, 596
by nuclear magnetic resonance spectroscopy, LIII, 596

synthesis, LIII, 594–597
Coenzyme Q-cytochrome *c* reductase, *see also* Ubiquinol-cytochrome *c* reductase
components, LVI, 580, 582, 583
mitochondria, LVI, 11
peptides, LVI, 12
mutants, LVI, 105
Coenzyme Q homolog
biochemical assay, LIII, 597, 598
biological sources, LIII, 591–593
commercial sources, LIII, 599
stability, LIII, 594
synthesis, LIII, 593, 594
Coenzyme Q reductase, XLVI, 260; LIII, 414, 415, 416, 418
reaction with arylazido-β-alanine NAD$^+$, XLVI, 284, 285
with arylazido-β-alanine NADH, XLVI, 284, 285
Cofactor, as eluent, XXXIV, 237
Coformycin, XLVI, 25
inhibitor of adenosine deaminase, LI, 506, 507, 511
Coherence area, XLVIII, 461, 462
estimation, XLVIII, 462
Coherent scattering length, *see* Neutron scattering length
Cohn fraction IV precipitate, for affinity chromatography, XXXVI, 112, 113
Coil state, polypeptide chain, XLIX, 114
Colcemid
inhibitor of chromosome movement, XL, 279, 284
mitotic cells, LVIII, 255
mitotic spindle inhibitor, LVIII, 327
slide preparation, LVIII, 329, 331, 343
synchrony, LVIII, 261
in vivo cell cycle analysis, XL, 61
Colchicine, XXXIV, 623
glycoprotein biosynthesis, L, 188, 189
inhibitor of chromosome movement, XL, 279, 284
microtubular disruptive, XL, 322
mitotic inhibitor, LVIII, 256, 327
resistance, LVIII, 309, 314
in vivo cell cycle analysis, XL, 61
Colchicine-agarose, XXXIV, 624
Colchicine analog, XLVI, 567–571
Colchium mesophyll × soybean culture, LVIII, 365
Colicin
effects, LV, 532, 533
on membranes, XXXII, 882
use
intact cells, LV, 535, 536
membrane vesicles, LV, 536
Colicin E1, preparation, LV, 533
Colicin I, purification, LV, 533
Colicin K
assay systems
spot test, LV, 533, 534
survival test, LV, 534
bacterial strain and media, LV, 533
purification, LV, 534, 535
Coliphage K29, XXXIV, 173, 174
Collagen
acid-soluble
biosynthesis in adult bone, XL, 329
purification, XL, 338
alpha chains, XXXIV, 422
antibody binding, L, 55
assay of enzyme bound
by cysteine analysis, XLIV, 256, 257
by tryptophan analysis, XLIV, 255, 256
biosynthesis
in bone, XL, 328
formation rate, XL, 330
resorption rate, XL, 330
rate, XL, 309
chemical modification, XLIV, 244
chemical properties, XLIV, 243
chick bone, extraction, XL, 339
cleavage, XLVII, 133, 139, 140, 142, 143
complexes
binding mechanism, XLIV, 261–263
kinetic behavior, XLIV, 259–261
storage stability, XLIV, 258
composition, XL, 335
cross-linking, XL, 360
as culture substrate, LVIII, 266
cysteamine-soluble, purification, XL, 337
fibrils, XXXIV, 420

flexible pore diameter, XLIV, 155
floating substrates, LVIII, 266
free peptide hydrogen exchange, XLIX, 30
glutaraldehyde insolubilization, XLIV, 553
hormone effect, XL, 307
on degradation, XL, 349
experimental procedures, XL, 319
metabolic studies, XL, 324
human cells synthesizing, XXXII, 799
immobilization procedures, XLIV, 245–250
electrocodeposition method, XLIV, 248–250
macromolecular complexation, XLIV, 247
membrane impregnation method, XLIV, 246, 247
for whole microbial cells, XLIV, 250
labeled
[^{14}C]glycine, preparation, XL, 350
in vivo, XL, 322
lyophilized, purified, reconstitution, XL, 340
membrane, XLIV, 902, 903
covalent binding of membrane factor and reassociation with F_1, LV, 744–746
loading capacity, XLIV, 254, 255
source, LV, 744
microbial degradation, XLIV, 258, 259
native, incubation with thiosemicarbazide, XL, 363
neutral-salt-soluble
extraction, XL, 336
purification
^{14}C-labeled, XL, 337
in supernatant, XL, 337
physical properties, XLIV, 244
precursor, XXXVII, 326
preparation, LVIII, 270–273
purification, XL, 335; LVIII, 270, 526
by digestion with purified collagenase, XL, 340
radioactive, assay in presence of other proteins, XL, 342
as raft, LVIII, 271
cell separation, LVIII, 275
preparation, LVIII, 270
somatomedin stimulation of synthesis, XXXVII, 94
source, XLIV, 245
subunits, chromatographic separation, XL, 358
support properties, XLIV, 243–263
treatment of culture dish, LVIII, 525, 526
Collagen-agarose, XXXIV, 421–424
Collagenase, XXXIV, 420–424
in adrenal cortex cell isolation, XXXII, 673, 674, 688, 692
cell dissociation, LVIII, 128, 129
from *Clostridium histolyticum,* use in preparing isolated fat cells, XXXV, 555, 556
crude, LVIII, 126
digestion
in measurement of [^{14}C]hydroxylysine, XL, 354
technique, for Langerhans islet isolation, XXXI, 374–376
as dispersal agent for interstitial cells, XXXIX, 257, 258
in fat cell isolation, XXXI, 60, 104; XXXVII, 211, 274, 275
for hepatocyte isolation, LII, 60–62
hormone effect, XL, 349
in media, LVIII, 139
in oxyntic cell isolation, XXXII, 712
in pancreas superfusion studies, XXXIX, 374
in parenchymal cell isolation, XXXII, 625–627, 632
purification, XL, 341
assay of radioactive collagen in presence of other proteins, XL, 342
digestion of collagen, XL, 340
partial, XL, 350
in renal cell isolation, XXXIX, 21
in renal tubule isolation, XXXIX, 12
separation from inhibitory serum antiproteases, XL, 352
tadpole, zymogen, XL, 351
in thyroid cell isolation, XXXII, 748
tissue culture, LVIII, 125
types, LVIII, 126
Collagen-polyacrylamide, XXXIV, 423
sym-Collidine buffer, XXXIX, 135

Colligative property, definition, XLVIII, 70
Collimation, in neutron scattering experiments, LIX, 662–665
Collodion
  in microcapsules, XLIV, 684, 685
  preparation, XLIV, 204
  support properties, XLIV, 904, 906
Collodion bag, LI, 229
  for concentration of ribosomal subunits, LIX, 505
  for ultrafiltration, LII, 468
Collodion film, bacteriorhodopsin incorporation, LV, 601
Colon, carcinoma, LVIII, 184
Colony stimulating activity, LVIII, 162, 163
Colorimetric assay, *see also* Colorimetry
  of 3-deoxy-2-ketoaldonic acids, XLI, 32, 33
  for hexuronic acids and keto sugars, XLI, 29–31
  of sugars, XLI, 3
Colorimetric ultramicroassay, for reducing sugars, XLI, 27–29
Colorimetry, test for enzyme leakage, XLIV, 276
Color temperature, definition, LVII, 597
Colostrum, cow, trypsin inhibitor, XLV, 806, *see also* Trypsin inhibitor
Column, *see also* specific type
  construction and materials, LII, 174
  design
    connections, XLVII, 212, 227, 228
    for gel chromatography, XLVII, 211, 396
    for ion-exchange chromatography, XLVII, 211, 225–227
  durability, XLVII, 323
  fabrication in laboratory, XLVII, 101, 102
  geometry, XXXIV, 250, 251
  large-diameter, XLVII, 50
  O ring, composition, XLVII, 230
  packing procedure, XLVII, 101–103
    for ion-exchange resins, XLVII, 227
    for microbore design, XLVII, 213
  parameters affecting, XLVII, 50
  pH artifact, correction, XLIX, 33
  reversed-phase, XLVII, 50
  for solid-phase sequence analysis, XLVII, 369
Column chromatography, *see also* specific type
  advantages of discontinuous gradient, LIII, 621
  on alumina, *see* Alumina column chromatography
  of phosphorylated nonhistone proteins, XL, 200
  steroid purification, XXXVI, 39
Column Coat, XXXII, 73, 84
Column photolysis, XXXVIII, 395
Column procedure for antibody, XXXIV, 726–728
Column sieving fractionation, of plant tissue, XXXI, 549–553
Co-methylcobamide, in acetate formation pathway, LIII, 361
Compartmentalization, as protoplasm protectant, XXXI, 532
Compartmentation, metabolite binding, XXXVII, 286–288
Compensation law, LIV, 520
Compensation temperature, LIV, 520
Complement, XXXIV, 731–746, *see also* specific components
  assay, XXXIV, 733–735
  C1, XXXIV, 733, 741–746
  C3, XXXIV, 717
  C3 endopeptidase, XXXIV, 732
  C1 esterase, XXXIV, 731–746
  C3Pase, XXXIV, 731
  C3 proactivator, XXXIV, 731
  C1q, XXXIV, 733–735, 741–746
    purification, XXXIV, 741–746
  C1s, XXXIV, 295, 735, 736, 738–741
    purification, XXXIV, 738–741
  inactivation, LVI, 224
  spacers, XXXIV, 735
Complementation
  genetic, LVIII, 347
    protoplast fusion, LVIII, 367
  temperature sensitive mutants, LVI, 138
  test, *unc* mutants, LVI, 115, 116
    new mutants, LVI, 116, 117
Complement complex, XXXIV, 217
Complement fixation, quantitative, XL, 195

Complete Growth Medium, for frog haploid cell culture, XXXII, 798
Complexation, XLIV, 149
macromolecular, XLIV, 247
selectivity, of neutral ionophores, factors affecting, LV, 440
Complex I, LIII, 11–14
activity, LIII, 11
composition, LIII, 12
inhibitors, LIII, 13
iron-sulfur centers, LIII, 12, 16
iron-sulfur protein, LIII, 15–21
NADH dehydrogenase, LIII, 15–21, 414, 415, 418
polypeptides, LVI, 587–589
preparation, LIII, 3, 4, 6, 7, 11, 12; LV, 716
properties, LIII, 12, 13
purification from complex I–III, LIII, 11, 12
substrate specificity, LIII, 12, 13
ubiquinone, LIII, 576
Complex II, LIII, 21–35
activators, LIII, 25
activity, LIII, 21, 24, 25
composition, LIII, 24
error in protein determination, LIII, 4
inhibitors, LIII, 26
polypeptides, LVI, 589, 590
preparation, LIII, 3, 4, 6
purification, LIII, 22, 23
reconstitution, LIII, 50, 52–54
spectral properties, LIII, 25
substrate specificity, LIII, 24
succinate dehydrogenase
activity studies, LIII, 410, 411
assay, LIII, 466, 474, 475
purification, LIII, 27–35, 478
ubiquinone, LIII, 576
Complex III, LIII, 35–40, 80–91, 92–112
activity, LIII, 35, 37, 90
assay, LIII, 38–40, 90, 91
from beef heart, LIII, 35–40, 80–91
components after cross-linking, LVI, 637
composition, LIII, 36, 37, 89, 96, 97
cytochrome *b*-containing peptide
characterization, LIII, 82, 83
isolation, LIII, 83, 85
cytochrome composition, LIII, 85
electron paramagnetic resonance studies, LIII, 119–121
error in protein determination, LIII, 4
inhibitors, LIII, 38
kinetic properties, LIII, 117
molar ratio of peptides, LIII, 82, 83, 100, 110, 111
molecular weight, LIII, 37
monomer, LVI, 633
of subunits, LIII, 36, 37
from *Neurospora crassa*, LIII, 98–112
peptide subunits, LIII, 80–83
molecular weights, LIII, 80
nomenclature, LIII, 82
polypeptides, LVI, 590–592, 637
preparation, LIII, 3, 4, 6
purification by affinity chromatography, LIII, 99–106
as bc_1 complex, LIII, 92–95
from complex I–III, LIII, 35, 36
by hydroxylapatite chromatography, LIII, 92–95
importance of starting material, LIII, 86
redox centers, LIII, 126, 127
resolution methods, comparison, LIII, 85–90
spectral properties, LIII, 37, 38, 117–121, 126
substrate specificity, LIII, 37
ubiquinone, LIII, 576
Complex I–II, partial purification, LIII, 57–59, 62
Complex I–III, LIII, 5–10
activity, LIII, 5
composition, LIII, 7, 8
enzymic properties, LIII, 8
inhibitors, LIII, 8, 9
reconstitution, LIII, 49
spectral properties, LIII, 8
transhydrogenation activity, LIII, 8
Complex II–III
partial purification, LIII, 57–59, 62
preparation, LIII, 6, 7
reconstitution, LIII, 49, 50
Complex IV, LIII, 40–48
activity, LIII, 40
assay, LIII, 45–47
composition, LIII, 44
error in protein determination, LIII, 4

heme *a*, LIII, 44
inhibitors, LIII, 44, 45
kinetic properties, LIII, 44
oxidase reaction, kinetic constants, determination, LIII, 46, 47
polypeptides, LVI, 592–596
preparation, LIII, 3, 4, 6, 40–43
Complex U, *see* Orotate phosphoribosyltransferase:orotidylate decarboxylase complex
Complex V
assay for ATPase reaction, LV, 313, 314
for ATP-P_i exchange reaction
principle, LV, 310, 311
procedure, LV, 312, 313
reagents, LV, 311, 312
preparation, LIII, 3, 4, 6
properties
activators and inhibitors, LV, 315
composition, LV, 314, 315
purification procedure, LV, 308, 309
starting material for preparation, LIII, 6
Composite agarose-polyacrylamide gel
analysis of ribosomal fragment components, LIX, 471
of 30 S pre-rRNA cleavage products, LIX, 833
of 16 S and 23 S RNAs, LIX, 556, 567
purification of 30 S pre-rRNA, LIX, 830, 832
separation of bacterial ribosomal subunits, LIX, 398, 399, 401
of ribosomal fragments, LIX, 468
Compound affinity, XXXIV, 115
Compressibility, isothermal, determination, XLVIII, 32
Computer
of average transient, averaging, LIV, 34, 39
curve fitting, LVI, 266–268
in EPR spectroscopy, XXXII, 196–198
experimental data
analysis methods, LVI, 264–268
graphical methods, LVI, 265, 266
in Fast Analyzer, XXXI, 801
methodology for reconstruction of mitochondria
alignment, LVI, 721–723
allocation of profile labels, LVI, 723
data entry, LVI, 721
three-dimensional reconstructions, LVI, 723
in NMR spectroscopy, of glycolipids, XXXII, 367
in protein-ligand studies, XXXVI, 9
radioimmunoassay programs, XXXVI, 31
in study of calcium fluxes, XXXIX, 556–563
program, XXXIX, 563–565, 567–570
system for reconstruction of mitochondria
computer, LVI, 727
coordinate digitizer and stylus, LVI, 723, 726
data retrieval and display, LVI, 727, 728
data storage, LVI, 727
zero filling, LIV, 159
Computer analysis
for 3–D reconstruction, XLIX, 59
of electron paramagnetic resonance spectra, XLIX, 378, 406, 528
integration, XLIX, 408–412
subtraction, XLIX, 428
titration, XLIX, 412–414
for estimating tumbling rates, XLIX, 422
filtering, XLIX, 60–63
for image reconstruction, XLIX, 40–42, 58–63
Raman spectroscopy, XLIX, 89
substrate conformation, XLIX, 354, 355
word length, XLIX, 334
Computer of average transient, LIX, 25
Concanavalin, as carrier, XXXIV, 7
Concanavalin A, XXXIV, 6, 7, 173, 199, 329; XLVI, 85, 368
in affinity chromatography, LII, 515
in agglutination studies, XXXII, 617
binding to Sephadex, XXXIV, 329
binding site, XLVI, 75

carbohydrate-exchange resin, XXXIV, 500
coupling to coliphage, XXXIV, 175
crystal immobilization, XLIV, 548
effect on hepatocytes, XXXV, 593
electron paramagnetic resonance studies, XLIX, 514
iodination, XXXIV, 175; L, 57, 58
labeled, binding assay, XXXII, 621–625
lectin, XXXIV, 338
lymphocyte reaction, XXXII, 636
as mitogen, LVIII, 487, 492
preparation, XXXII, 622
Concanavalin A-agarose, XXXIV, 500, 669, 670
Concanavalin A-nylon, XXXIV, 204, 207, 211
Concentration, error in logarithm, XLVIII, 183
Concentration-jump procedure, XLVIII, 309
Concentration polarization, definition, XLIV, 299, 300
Concentration response curve, analysis, XXXIX, 349–352
Concentration variable, in temperature jump experiment, LIV, 68
Conchoecia, luminescence, LVII, 339
Condensation reaction, four-component, enzyme immobilization on nylon, XLIV, 127–129, 133, 134
Condensing vacuole, of pancreas, XXXI, 41, 42
Conditioning, mechanism, LVIII, 47
Condon, Alter, and Eyring theory, LIV, 254
Conductivity, change, assay using, XLIV, 348, 349
Conduritol B, XLVI, 369
synthesis, XLVI, 373
Conduritol B epoxide, XLVI, 369
active site labeling, XLVI, 377–380
labeling, XLVI, 370
reaction with β-glucosidases, XLVI, 369
synthesis, XLVI, 371–376
Conduritol B tetraacetate, synthesis, XLVI, 373
Confluence, LVIII, 133
Conformation
circular dichroism studies, XXXII, 220, 221
perception, XLIV, 567, 568, 571
stretching studies, XLIV, 560, 561
techniques for studying, XLIV, 361–370
fluorescence, XLIV, 362–366
transitions
in immobilized proteases, XLIV, 528–538
kinetics, XLIV, 528, 529
Conformational state
definition, XLIX, 102
of protein backbone, XLIX, 102–104
Congo red, XLIV, 654
Conjugation, *see also* Immobilization
of antibiotic, XLIII, 210
Connective tissue
mammalian, oxidase, LII, 10
separation from homogenized freeze-stop tissue, LVI, 203
transformed, LVIII, 376
Constant, *see* specific constant
Contact feeding, LVIII, 315
Contact hyperfine shifts, LIV, 167, 194, 196, 197
Contact inhibition, LVIII, 134
Contaminant, *see* Contamination; specific type
Contamination
airborne, elimination, in trace metal analysis, LIV, 479
bacterial, LVIII, 18, 19
tests, LVIII, 20
carry-over, LVIII, 83
culture, LVIII, 201
detection, LVIII, 18–29, 116
evaluation, large-scale, LVIII, 219, 220
by extraneous cells, LVIII, 165
fungal, LVIII, 18, 19
test, LVIII, 20
microbial, LVIII, 110
microbiological, resazurin as indicator, XLI, 56
mycoplasmal, LVIII, 18, 21–28, 219, 220, 439
testing, LVIII, 21–28
toxic compound, LVIII, 5

viral, LVIII, 28, 29
Continuous flow analyzer
bis-acid hydrazide spacers, XLIV, 645
carry-over, XLIV, 644, 645
enzyme stability, XLIV, 644
kinetics, XLIV, 642, 643
method of operation, XLIV, 638
sample rate, XLIV, 644
Continuous-flow apparatus, testing and performance, LV, 622
Continuous-flow harvesting, in density-gradient centrifugation of plant cells, XXXI, 518, 519
Continuous flow method
macromolecular interaction kinetics, XLVIII, 308
pH measurement, LV, 618, 627
Continuous free-flow preparative electrophoresis, XXXI, 746–761
apparatus, XXXI, 747, 748
application, XXXI, 751–761
of cell organelles, XXXI, 752–755
of membranous components, XXXI, 755–759
operational aspects, XXXI, 748–751
Continuous wave spectroscopy, LIX, 25, 26
Contractile system, model, saturation-transfer studies, XLIX, 494, 495
Contrast, definition, LIX, 674, 675
Contrast variation
in H_2O/D_2O mixture, LIX, 681–683
by solvent exchange, LIX, 676–678
with specifically deuterated ribosomes, LIX, 682, 683
Convolution difference technique, LIV, 161
Convolution function
noise suppression, LVI, 512, 514
spectral analysis, LVI, 509
Coomassie Blue
for gel electrophoresis, XXXII, 74–77, 80, 81, 87, 289
gel staining, LVI, 64, 70, 72, 605
precipitin lines, LVI, 225
as protein stain, XXXII, 96, 102, 105, 106, 444, 867, 878
Coomassie Brilliant Blue, LVII, 176, 234; LIX, 313, 497, 545, 546, 560, 567
gel staining, *see* Polyacrylamide gel electrophoresis
protein determination, LX, 30, 231
as protein gel dye, XXXVI, 450
as protein receptor stain, XXXVI, 373
staining procedure, LIX, 509
Coon's medium F12, LVIII, 86
Cooperativity, *see also* Association reaction; Binding; Dimerization; Dissociation; Interaction; Ligand binding; Self-association
determination, using Hill plot, XLVIII, 287–293
models, XLVIII, 289–294
negative, XLVIII, 276, *see also* Site heterogeneity
ligand-mediated binding, XLVIII, 302
qualitative effects, XLVIII, 276–278
Scatchard plot, XLVIII, 283–285
positive, XLVIII, 274
protein-nucleic acid binding, XLVIII, 297
Copoliodal, XLIV, 84
Copolymer, *see also* specific types
compositional heterogeneity, XLIV, 181
Copolymer ethyl maleic anhydride, as carrier, XXXIV, 6
Copper
active site types, LII, 4
alkaline reagent, in quantitative histone determination, XL, 110
atomic emission studies, LIV, 455
in clam luciferase, LIII, 560, 562
complex IV, LIII, 44; LVI, 580, 583, 584
of cytochrome aa_3, stoichiometry, LIV, 145
in cytochrome oxidase, LIII, 64, 76
ligand binding, LIII, 191
deficiency, XL, 321
EPR signals, LIV, 134, 136
of heme a_3, oxidation kinetics, LIV, 110
inhibitor, of brain α-hydroxylase, LII, 317

Keilin-Hartree preparation, LV, 123
in luciferase of *Diplocardia*, LVII, 380, 381
of *Pholas dactylus*, LVII, 388
in media, LVIII, 52, 69
metallic, for oxygen removal, LIV, 117, 118
in oxidases, LII, 4
removal, from ammonium sulfate solutions, LIII, 56
in superoxide dismutase, LIII, 389
wire, in gel electrophoresis of phosphodiesterase I, XXXII, 90
Copper(II)
catalyst, in luminol chemiluminescence, LVII, 448
correlation time evaluation, XLIX, 346
electron paramagnetic resonance studies, XLIX, 514
in hemocyanin, XLIX, 140, 141
interatomic distance calculations, XLIX, 331
Copper-63, lock nuclei, XLIX, 349; LIV, 189
Copper-64
in membrane studies, XXXII, 884–893
properties, XXXII, 884
Copper-65, lock nuclei, XLIX, 349; LIV, 189
Copper analysis, LIV, 443–445
procedure, LIV, 444, 445
reagents, LIV, 443, 444
Copper carbonate, XLVII, 535, 536
Copper(II) Chelex, in nucleoside synthesis assay, LI, 538, 539
Copper(II)–chlorate, *g* value, XLIX, 523
Copper compound, *K*-edge transition ranges, LIV, 339, 340
Copper(II)-ethylenediaminetetraacetic acid, *g* value, XLIX, 523
Copper ion, interference, in iron analysis, LIV, 440
Copper-ligand vibrational mode, XLIX, 141, 144
Copper standard, preparation, LIV, 443, 444
Copper-substituted cytochrome *c*
preparation, LIII, 177
properties, LIII, 178–181
Copper sulfate, XLIV, 387
assay of adenosine monophosphate nucleosidase, LI, 264
inhibitor of *Cypridina* luminescence, LVII, 349, 350
modification of cysteine, LIII, 144
optical filter, XLIX, 192
Copper sulfur hydrate, XLVII, 542, 543
Copper tetraphenylporphyrin, X-ray absorption spectroscopy, LIV, 332
Copper tubing
in anaerobic procedure, LIV, 116
in purge gas connections, LIV, 404
in redox potentiometry apparatus, LIV, 421
Copra, bongkrekic acid preparation, LV, 528
Coproporphyrinogen oxidase, LII, 21
C3 opsonin, LVII, 482, 483
Coral snake, neurotoxin, XXXII, 312
Cordycepin, XLIII, 154; LVI, 31
RNA inhibitor, XL, 279, 282
substrate of adenosine deaminase, LI, 506
Core protein 1, of complex III
characterization, LIII, 82
isolation, LIII, 83
in Triton X–100-isolated complex III, LIII, 95
Core protein 2, of complex III
characterization, LIII, 82, 83
isolation, LIII, 83
in Triton X–100-isolated complex III, LIII, 95
Corn, organelle studies, XXXI, 561, 562
Cornea
cell, rabbit, LVIII, 414
tissue culture, fish, LVIII, 473
Corner-Allen progestin assay, XXXVI, 456
Corn mesophyll × soybean culture, LVIII, 365
Corn oil
cytochrome activities induced, LII, 377
for injection, LII, 114, 118, 233
Corn root, ATPase isolation, XXXII, 395, 405
Cornsteep liquor, XLIV, 190

Corn syrup solid, substrate, glucoamylase, XLIV, 778
Coronene, activator, oxidation potential, LVII, 522
Coronilla varia, lectin, XXXIV, 334
Corpus luteum
 cytochrome P_{450}, XXXVII, 304
 by follicle transformation, XXXIX, 185
 slice incubation studies, XXXIX, 238–244
Corpus luteum cell
 hormone effects, XXXIX, 424
 separation, XXXIX, 229
Correlation spectroscopy, LIX, 26, 27
Correlation time
 from associated water protons, XLIX, 347, 350
 for dipolar interaction, XLIX, 324, 345–351, 424
 electron paramagnetic resonance linewidth, XLIX, 350
 from frequency dependence of ligand nucleus, XLIX, 346, 347
 from relaxation rates ratio, XLIX, 350, 351
 rotational
  from electron spin resonance spectra, XLIX, 421, 422
  for fast tumbling region, XLIX, 466–470
  for membrane-bound protein, XLIX, 494
  for nitroxide compound VII, XLIX, 468, 469
  for saturation-transfer studies, XLIX, 489
Cortexolone
 cytochrome P_{450} in formation, XXXVII, 310
 as 5α-oxidoreductase substrate, XXXVI, 473
Cortical granule protease, *see* Sea urchin egg protease
Corticoid, lipophilic Sephadex chromatography, XXXV, 394
Corticosteroid
 enzyme induction, XXXII, 733, 740
 receptor, XLVI, 76
Corticosteroid-binding globulin, XXXIV, 6; XXXVI, 91
 affinity chromatography, XXXVI, 104–109, 113
 as contaminant in progesterone receptor preparations, XXXVI, 209
 density gradient centrifugation, XXXVI, 165
 isolation and purification, XXXVI, 36, 37, 104–109
 progesterone-binding protein compared, XXXVI, 120
 steroid binding, XXXVI, 97
 in steroid hormone assay, XXXVI, 34, 36–38, 47, 48
Corticosteroidogenesis, adrenal cortex mitochondria, LV, 10
Corticosterone, XLVI, 76
 analysis, XXXII, 677, 678
 assay, XXXIX, 345–348
 autoradiography, XXXVI, 150, 151, 155
 conversion to aldosterone, in adrenal cortex cell studies, XXXII, 686–688
 effects, (Na^+ + K^+)-ATPase sodium transport, XXXVI, 444
 extraction and measurement, XXXVII, 299
 formation, LV, 10
  in situ adrenal, XXXIX, 341–345
  in superfused adrenals, XXXIX, 307–314, 317–322
 as 5α-oxidoreductase substrate, XXXVI, 473
 PBP affinity, XXXVI, 125
 pineal effects, XXXIX, 382
 plasma levels, rhythms, XXXVI, 479
 protein-binding assay, XXXVI, 36
 radioimmunoassay, XXXVI, 21
 R_f value, LII, 126
 in superfused adrenal tissue, XXXIX, 307–314
Corticosterone-binding globulin assay, for progestins, XXXIX, 217
Cortisol
 adsorption to various materials, XXXVI, 91, 93, 94
 assay, by thin-layer chromatography, XLIV, 187

CBG binding, XXXVI, 97
cytoplasmic receptor kinetics, XXXVI, 258
density gradient centrifugation, XXXVI, 159
effects on glucose pulse, XXXVI, 431, 432
on mercaptide formation, XXXVI, 389
on sodium transport, XXXVI, 444
logit-log assay, XXXVII, 9
metabolites, isolation, XXXVI, 499–503
molar absorption coefficient, XLIV, 186
as 5α-oxidoreductase substrate, XXXVI, 473
PBP affinity, XXXVI, 125
in plasma, XXXVI, 39
protein-binding assay, XXXVI, 36
receptor-protein assay, XXXVI, 49
substrate, of 3-ketosteroid-Δ^1-dehydrogenase, XLIV, 186
Cortisol hemisuccinate, XXXIV, 6
for affinity chromatography, XXXVI, 105
Cortisol-Sepharose, for affinity chromatography, XXXVI, 107, 108
Cortisone
antibodies to derivatives, XXXVI, 54
antiglucocorticoid, XL, 292
in cytoplasmic receptor kinetic studies, XXXVI, 258
effects on $(Na^+ + K^+)$-ATPase, XXXVI, 435
iodo derivatives
in enzyme active site studies, XXXVI, 390, 392
synthesis, XXXVI, 383–385
as 5α-oxidoreductase substrate, XXXVI, 473
radioimmunoassay, XXXVI, 21
reduction by 20β-hydroxysteroid dehydrogenase, XXXVI, 398
Cortol, isolation, XXXVI, 501
Cortolone, isolation, XXXVI, 501
Cortrosyn, as ACTH standard, XXXII, 677
Corynebacterium simplex, *see Arthrobacter simplex*
Costal cartilage, for somatomedin bioassay, XXXVII, 95–104
Cotton bollworm, media, LVIII, 457
Cotton effect, XLIII, 349
of cytochrome *c*-types, LIV, 264
in cytochrome oxidase spectrum, LIV, 264
dipole-dipole coupling, LIV, 256
in heme proteins, LIV, 260–262, 272
in hemoglobin spectra, LIV, 273–275
Cotton leaf, peroxidase in, binding properties, LII, 516, 521
Cottonseed
aleurone vacuoles, XXXI, 576
organelles, XXXI, 536, 537, 540
spherosome isolation, XXXI, 578
Coulometry, titrant generation, instrumentation, LIV, 406
Coulter counter, XXXII, 598, 599; XXXIX, 488
cell-adhesion studies using, XXXII, 600–603, 620
p-Coumarate 3-hydroxylase, *see* Coumarate 3-monooxygenase
Coumarate 3-monooxygenase, LII, 9
Coumarin, sulfite reaction with, XXXI, 539
Coumarin 6G dye laser
for flash photolysis, LIV, 53
relative brightness, LIV, 95
Coumeromycin, XLIII, 336
Countercurrent distribution, XXXI, 761–769; XLIII, 320–346
apparatus, XLIII, 341–346
bacitracin, XLIII, 326
of blood cells, XXXII, 637–647
of chloroplasts, XXXI, 761, 764–768
distribution assembly, XLIII, 344
isolation from growth medium, XLIII, 320–324
of microorganisms, XXXI, 768, 769
molecular weight determination, partial substitution, XLIII, 328–330
nonideality, XLIII, 331
phase systems, XXXI, 763–765
selection, XXXII, 640–642
phenol as solvent for highly polar solutes, XLIII, 333, 334
polypeptide antibiotics, XLIII, 324–327

Post trains, XLIII, 343, 344
principles, XXXI, 762, 763
purification, XLIII, 324–327
purity, testing, XLIII, 327, 328
Raymond trains, XLIII, 343
rotatory evaporation apparatus, XLIII, 345, 346
solvent systems, XLIII, 331–341
structural studies, XLIII, 330, 331
tube design of counter-double-current distribution train, XLIII, 345
Count-off detergent, LIV, 478
Coupling
brown adipose tissue mitochondria, LV, 70
capacity, XLVII, 268–270, 303, 362
conditions, XLVII, 337
efficiency, XLVII, 304, 305
definition, XLIV, 795
low yields, XLVII, 332, 333
reaction, choice, XLVII, 271, 272
reagents, XLVII, 270, 271, 337
solid-phase sequencing, XLVII, 277–288
tightness, criteria, LV, 225
Coupling factor, *see also* specific type
assay for activity, LV, 190, 191
Keilin-Hartree preparation, LV, 126
overview, LV, 380–383
Coupling factor A, nature, LV, 382
Coupling factor B
properties, LV, 381
purification
AE particle preparation, LV, 385, 386
assay method, LV, 384
isolation procedure, LV, 386–388
mitochondrial isolation, LV, 385
other B-type factors, LV, 391
properties, LV, 388–391
Coupling factor F_3, LV, 383
Coupling factor F_6
assay, LV, 398, 399
preparation, LV, 399
properties, LV, 381, 383
Coupling factor-latent ATPase
purification by affinity chromatography, LV, 189, 190
removal from ETP, LV, 187
solubilization, LV, 189
Coupling procedure, XXXIV, 19–30, *see also* specific types
activation step, XXXIV, 20, 21
association constants, XXXIV, 14, 15
bisoxirane method, XXXIV, 20
cyanogen bromide method, XXXIV, 21–24, *see also* Cyanogen bromide coupling method
derivatized flat surface, XXXIV, 221
epoxide method, XXXIV, 20
isocyanide reaction, XXXIV, 21
quantitation, XXXIV, 14, 15
theoretical aspects, XXXIV, 13–16
Coupling system, assay in heart mitochondria, LV, 48, 49
Covalent affinity chromatography, *see* Affinity chromatography, covalent
Covalent binding
adsorption, XLIV, 149
choice of technique, XLIV, 148
determination of group involved, XLIV, 16
direct, XLIV, 26–37
general experimental conditions, XLIV, 11, 12
indirect, XLIV, 37–42
advantages, XLIV, 38
on inorganic supports, XLIV, 134–148
irreversible, XLIV, 45
to polyacrylic copolymers, XLIV, 84–107
reactive residues, XLIV, 12–15
reversible, XLIV, 45
via chelation, XLIV, 88
Cow, granulosa cell isolation, XXXIX, 199
Cox method, XLVIII, 213
CP 38295 antibiotic, LV, 446
D-CPase, *see* D-Alanine carboxypeptidase
C3Pase, XXXIV, 731
CPC, *see* Carboxypeptidase C
C–peptide
immunoassay, XXXVII, 343, 344
as insulin precursor, plasma levels, XXXVII, 340–343
CP_4F
fluorescent acceptor, XLVIII, 378
structural formula, XLVIII, 379
CPG, *see* Glass, controlled pore

CPG 10–75 glass bead, XLVII, 266, 269, 272
CPK, *see* Creatine phosphokinase
CPL, *see* Circularly polarized luminescence
C3 proactivator, XXXIV, 731
CPSase, *see* Carbamoyl-phosphate synthetase
CPY, *see* Carboxypeptidase Y
C1q, *see* Complement
Crabtree effect, expression, LV, 296
Crayfish, oxidase, LII, 21
Creatine
  content of rat heart, LV, 217
  urinary, assay, fluorometric, XLIV, 630, 631
Creatine kinase, XXXIV, 246; XLVI, 21
  activity, LVII, 58
  affinity labeling, XLVII, 427
  assay
    bioluminescent, LVII, 63–65
    of thymidine kinase, LI, 366
  associated with cell fractions, XL, 86
  heart mitochondria, LV, 14
  isolation of mitochondria, LV, 4
  in *in vitro* labeling of ribosomal proteins, LIX, 517, 527
Creatine kinase isozyme
  assay, using firefly luciferase, LVII, 56–65
    spectrophotometric, LVII, 59, 60
  separation, LVII, 56–58
    by ion-exchange chromatography, LVII, 62
Creatine phosphate
  content of brain, LV, 218
    of rat heart, LV, 216, 217
  estimation, XXXVII, 248
  inhibitor, of adenylate deaminase, LI, 496
    of luciferase, LVII, 59
  in perfused hemicorpus, XXXIX, 80–82
  radiolabeled, LX, 512, 513
  in *in vitro* labeling of ribosomal proteins, LIX, 517, 527
Creatine phosphokinase, XXXIV, 532
  assay
    cGMP, XXXVIII, 73, 78, 79
    creatine, XLIV, 630, 631
    cyclase, XXXVIII, 119, 154
    fluorometric, XLIV, 628, 629
Creatinine
  hollow-fiber dialytic permeability, XLIV, 297
  hollow-fiber retention data, XLIV, 296
  in kidney function studies, XXXIX, 7–9
Creatinine kinase, assay by Fast Analyzer, XXXI, 816
o-Cresol, glucose oxidase, LVI, 461
m-Cresol 6-hydroxylase, LII, 15
2,6-Cresotic acid, *see* 6-Methylsalicylic acid
Cristae, *see also* Mitochondria
  membrane
    electric capacitance, LV, 628
    proton conductance, LV, 629
Crithidia fasciculata, disruption, LV, 136, 137
Crithidia oncopelti, cytochrome *c*-557, LIV, 212
Critical point drying, of chromosomes, XL, 68
  apparatus, XL, 70
Cross feeding, LVIII, 315
Cross-linkage, *see also* Aggregation; Cross linking
  agents, *see also* specific compounds
    bifunctional, XLIV, 263, 458
    electron beam irradiation, XLIV, 178, 179
    gamma-irradiation, XLIV, 178
    multifunctional, XLIV, 263
    optimum concentration, XLIV, 173
    photochemical, XLIV, 280–288
    ultraviolet radiation, XLIV, 248
  effect on swelling capacity, XLIV, 108
  enzyme bonding heterogeneity, XLIV, 358–360
Cross linking, XLVI, 90, 641, 642
  antibody, XLVI, 501–504
  artifacts, XLVI, 174, 175
  available reagents and reaction conditions, LVI, 630–636
  characterization, XLVI, 172–180
  efficiency, XLVI, 170–172
  intermolecular, XLVI, 90
  irradiation, XLVI, 170–172

methods, XLVI, 170–172
nucleic acid-protein, XLVI, 168–180
of oligonucleotides, XLVI, 676
products
analysis, LVI, 636
components present, LVI, 636–639
quantitation, XLVI, 171, 178
reagents, LVI, 623, 624
regions, XLVI, 175, 176
procedures, XLVI, 176–178
reversal, LVI, 628
Crossover analysis, of substrate entry, XXXIX, 495–497
Crossover theorem, application to hormonal effects, on carbohydrate metabolism, XXXVII, 277–295
Cross reaction, immunological, LVIII, 178
Cross term scattering function, LIX, 677
Crotalaria juncea, lectin, XXXIV, 334
Crotalaria zanzibarica, lectin, XXXIV, 334
Crotalase, *see Crotalus adamanteus* serine proteinase
Crotalus adamanteus
serine proteinase
assay, XLV, 228, 230
esterolytic, XLV, 229
fibrinogen clotting, XLV, 229
composition, XLV, 231
distribution, XLV, 236
function, XLV, 223
hydrolysis
of ester substrates, XLV, 232, 233
of oligopeptide substrates, XLV, 233, 234
inactivators, XLV, 235
kinetic, XLV, 232
properties, XLV, 231
purification, XLV, 224, 228
purity, XLV, 227
specificity, XLV, 233
stability, XLV, 230
Crotalus horridus horridus, venom, XLV, 231
Crotalus terrificus terrificus venom phosphodiesterase, LI, 238
Crotonase, *see also* Enoyl-CoA hydratase
affinity labeling, XLVII, 422
Crotonic anhydride, in synthesis of crotonyl-CoA, XXXV, 139
S-Crotonyl-*N*-acetylcysteamine
assay of reductase activity of fatty acid synthase, XXXV, 51
synthesis, XXXV, 48, 49
Crotonyl coenzyme A
assay, XXXV, 140
inhibitor of crotonase, XXXV, 148
stability, XXXV, 140
synthesis, XXXV, 139
Crotyl alcohol, alcohol oxidase, LVI, 467
18-Crown-6 ether, XLVII, 590
Crown gall
inoculation methods, XXXI, 555, 556
plant cell studies, XXXI, 553–558
tumor tissue from, culture, XXXI, 556, 557
maintenance, XXXI, 558
Crown polyether, LV, 447
Cryogenics
storage, LVIII, 31
supply, LVIII, 17
Cryoprecipitate, factor VIII isolation, XLV, 86
Cryopreservation, HTC cells, LVIII, 551, 552
Cryoprotectant, LVIII, 12
effect on carbamoyl-phosphate synthetase, LI, 118
Cryoprotection, of biological material, XXXII, 36, 37
Cryo-pump, XXXVI, 142–144
Cryostat
frozen-sectioning, in autoradiography, XXXVI, 142
for low-temperature MCD studies, LIV, 295
Cryosublimation, XLVIII, 325
β-Cryptoxanthin hydroxylase, LII, 20
Crystal, *see also* specific types
density measurement, LIX, 5
heavy-atom derivatives, use, LIX, 12
mounting, LIX, 5, 8, 9, 11
preparation, LIX, 4, 5
procedure, LIX, 4, 5
second harmonic generating, LIV, 5, 7, 8, 21, 23
Crystal filtering, for elimination of fold-over, LIV, 160

Crystal holder, XLIX, 471, 472, 478–480
Crystallin, bovine, thermolysin hydrolysis products, XLVII, 176
Crystallization, adenylate cyclase, XXXVIII, 166, 167
Crystal violet, microcarrier culture, LVIII, 189
C1s, *see* Complement
CSA, *see* Chemical shift anisotropy
CTA-Br, *see* Cetyltrimethylammonium bromide
Ctenophore, bioluminescence, LVII, 274
CTPase, *see* Cytidinetriphosphatase
CTP(ATP):tRNA nucleotidyltransferase
  activity, LIX, 135
  assay, LIX, 123
    of nucleotide incoporation, LIX, 182
    of tRNA hydrolysis, LIX, 183
  from *E. coli*, LIX, 61, 135–141
  in incorporation of terminal nucleoside analogs, LIX, 188
  purification, LIX, 123–126, 183–185, 262
  in reconstitution of tRNA, LIX, 127, 128, 133, 134, 140, 141
  in tRNA end-group labeling, LIX, 103
  stability, LIX, 185
  from yeast, LIX, 123–126
CTP synthetase, XLVI, 420; LI, 79–90
  activators, LI, 82
  activity, LI, 79, 84
  assay, LI, 79, 80, 84–86
  from bovine calf liver, LI, 84–90
  from *E. coli*, LI, 79–83
  inhibitors, LI, 89
  kinetics, LI, 83, 84, 89, 90
  molecular weight, LI, 82, 90
  properties, LI, 82, 83, 89, 90
  purification, LI, 80–82, 86–89
  sedimentation coefficients, LI, 90
  self-association studies, XLVIII, 306
  storage, LI, 82
  structure, LI, 82, 83
Culex quinquefasciatus, suspension culture, LVIII, 454
Culex tritaeniorhynchus, suspension culture, LVIII, 454
Culiseta inornata, media, LVIII, 457
Culture, *see also* specific types
  collections
    American Type Culture Collection, XLIII, 3
    Centraalbureau voor Schimmelcultures, XLIII, 3
    Commonwealth Mycological Institute, XLIII, 3
    Institute for Fermentation, XLIII, 3
    National Collection of Industrial Bacteria, XLIII, 3
    Northern Utilization Research and Development Division, XLIII, 3
  liquid, XLIII, 11–14, *see also* Antibiotic, production in liquid culture
  maintenance, XLIII, 3–8, *see also* Media
  media, *see* Media
  of mutants, purification, LVI, 120
  preservation, XLIII, 8–11
    freezing, XLIII, 9
    liquid nitrogen, XLIII, 9
    lyophilization, XLIII, 9–11
    mineral oil, XLIII, 9
    periodic transfer, XLIII, 8
    soil culture, XLIII, 8
  stock, maintenance, LVI, 120
Culture chamber, disposable, XXXIX, 202
Culture flask, LVIII, 13
Culture media, *see also* Media
  for granulosa cells, XXXIX, 204–210
Culture plate, *see also* Plating
  collagen-coated, LVIII, 158
  fibrin-coated, LVIII, 158
  microtest, LVIII, 156
Culture system, LVIII, 12–15, *see also* specific system
Culture vessel, LVIII, 12–15
Cumene hydroperoxide
  in benzopyrene hydroxylase assay, LII, 408, 409
  extinction coefficient, LII, 307
  in glutathione peroxidase assay, LII, 506
  hydrogen donor, for cytochrome *P*-450 peroxidase activity, LII, 411

oxygen donor, to cytochrome P–450, LII, 407
Cup, for forming lipid bilayers, XXXII, 492, 493
Cuprammonium hydroxide, XLIV, 326
Cupric acetate
in cupric oleate synthesis, LII, 312
in diphenylamine reagent, LI, 248
Cupric chloride
binding to Chelex 100, LI, 539
inhibitor of deoxythymidylate phosphohydrolase, LI, 290
Cupric ion
activator of thymidine kinase, LI, 370
inhibitor of cytidine deaminase, LI, 411
of nucleoside phosphotransferase, LI, 393
of phosphoribosylpyrophosphate synthetase, LI, 11
in uridine nucleosidase, LI, 294–296
Cupric oleate, synthesis, LII, 312
Cupric *o*-phenanthroline, cross-linking, LVI, 631
Cuprizone, *see* Bis(cyclohexane)oxalyl dihydrazone
Cuprophane, electrode current, LVI, 458
Cuprous protein, oxygenated forms, spectra, LII, 36
Curamicin, XLIII, 128
Curare, effect on bungarotoxin, XXXII, 310
Curie law
Clostridium pasteurianum ferredoxin, LIV, 205
magnetic susceptibility, LIV, 381, 382, 384
Curie paramagnetism, LIV, 386
Curie spin contribution, LIV, 198
Curing agent, XLIII, 49–52, *see also* specific agent
Curtius azide method, XLIV, 48
Curtius rearrangement, XLVII, 564, 565
Curtius-Schmidt rearrangement, XLVI, 78
Curvularia lunata, immobilization, by polyacrylamide entrapment, XLIV, 184–190
Cut-stem inoculation method, for crown gall, XXXI, 555
Cutting technique, for plant material, XXXI, 502, 547, 548
Cuttlefish, isoinhibitors, XLV, 772, 774
amino acid composition, XLV, 796
trypsin-kallikrein, XLV, 792, *see also* Trypsin-kallikrein inhibitors
Cuvette, *see also* Sample cell
double-septum seal, LIII, 270
flat-bottom, in aequorin assay, LVII, 293, 297
CV-1 cell, plasma membrane isolation from, XXXI, 161
CV-1 cell line, properties and uses, XXXII, 585
Cyanamide gel, XLIV, 30
Cyanate, lability, XLIV, 27
Cyanate ion
inhibitor of FGAM synthetase, LI, 201
in IR spectroscopy, LIV, 304
stretching frequency, LIV, 311
Cyanein, XLIII, 137
Cyanide, *see also* Potassium cyanide
activation of succinate dehydrogenase, LIII, 472
ATPase synthesis via K^+ gradient, LV, 670, 671
complex IV, LVI, 586
cytochrome *c* reduction, LV, 110
ferricyanide reduction, LVI, 231
immobilized mitochondria, LVI, 556
inactivator, of flavoprotein enzymes, LIII, 440
inhibitor of clostridial hydrogenase, LIII, 296
of complex IV, LIII, 45
of cuprozinc superoxide dismutase, LIII, 384, 386
of cytochrome *c* oxidase, LIII, 192
of formate dehydrogenase, LIII, 371
of ω-hydroxylase, LIII, 360
of *Pseudomonas* cytochrome oxidase, LIII, 657
of reconstituted electron-transport system, LIII, 50, 51
interaction with bacterial cytochromes, LIII, 206
with cytochrome *c* oxidase, LIII, 192–194

interference, in iron analysis, LIV, 440
in IR spectroscopy, LIV, 304
M. phlei membrane systems, LV, 185
NADH oxidase, LV, 266, 270
N. crassa mitochondria, LV, 147
stretching frequency, LIV, 311
in studies of plant cells, XXXI, 500, 505
Cyanide ion
for check of silver contamination, LVII, 325
inhibitor of *Cypridina* luciferase, LVII, 339, 349, 352
of *Pholas* luciferase, LVII, 398, 399
of thymidylate synthetase, LI, 96
Cyanine (2,7-dimethyl-3,6-diazacyclohepta-1,6-diene) perchlorate, optical filter, XLIX, 192
Cyanine dye
adsorption to glass, LV, 576
membrane potentials, LV, 559, 571, 573, 585, 586
metabolic process, LV, 693
use for determination of membrane potentials, LV, 689, 690
calibration, LV, 691, 692
comments, LV, 694
instruments, LV, 690
optimization of procedure, LV, 690, 691
other classes of dyes, LV, 694, 695
precautions, LV, 692–694
reagents, LV, 690
reversibility, LV, 691
studies with single cells, LV, 695
m-Cyanobenzyltriphenylphosphonium bromide, XLVI, 125, 127
Cyanoborohydride, XXXIV, 82, 83
periodate method of activation of agarose, XXXIV, 80
α-Cyanocinnamic acid, cell fractionation, LVI, 217
Cyanocobalamine, *see* Vitamin B_{12}
1-Cyano-4-dimethylaminopyridinium salts, XLVII, 411
Cyanogen bromide, XLVI, 114
activation of acrylamide/hydroxyethylmethacrylate polymer, XLIV, 61, 62
of cellulose, XLIV, 48, 52
of Dextran T40, XLIV, 872
of hydroxyalkyl methacrylate, XLIV, 70
of polysaccharide support, XLIV, 27–30
of porous glass, XLIV, 146, 147
of Sephadex, LIX, 463
of Sephadex A–50, XLIV, 540
of Sephadex DEAE–50, XLIV, 542
of Sephadex G–50, XLIV, 457
of Sephadex G–100, XLIV, 540, 542
of Sephadex G–200, XLIV, 529, 764, 765
of Sepharose, XXXVIII, 386, 387; LII, 127, 133, 138
of Sepharose 4B, XLIV, 16, 17, 454, 457, 458, 493, 529, 930
of Spheron gels, XLIV, 72, 73
agarose activation methods, XXXIV, 77–102
buffer, XXXIV, 23
buffer method, XXXIV, 78, 79
carbohydrates, XXXIV, 24
conditions of use, XXXIV, 186
coupled ligands, stability, XXXIV, 99–101
coupling conditions variations, XXXIV, 23
coupling method, XXXIV, 21–24
hazards, XXXIV, 80
imido carbonate formation, XXXIV, 23
leakage, XXXIV, 93–95, 96–102
ligands, stability, XXXIV, 96–102
for matrix activation, LI, 100
products, XXXIV, 77
sodium periodate method, XXXIV, 80–85
soybean trypsin inhibitor, treatment, XLV, 702
spacers, XXXIV, 79, 80, 93–95
stability, XXXIV, 24, 96–102
stable activated derivatives, XXXIV, 85–93, *see also* specific derivatives

in studies of protein-RNA interactions, LIX, 576
titration methods, XXXIV, 77
Cyanogen bromide activated-agarose, XXXIV, 84
Cyanogen-oxygen flame, LIV, 450
Cyanogum 41 gelling agent, XXXI, 383; XXXVI, 449
2-Cyano-6-hydroxybenzothiazole
chromatographic properties, LVII, 26
electrophoretic mobility, LVII, 21
preparation, LVII, 20, 21
spectral properties, LVII, 26
structural formula, LVII, 18
Cyanohydroxycinnamic acid, transport, LVI, 248, 292
Cyanomethemoglobin
preparation, LII, 459
spectral properties, LII, 460–462
2-Cyano-6-methoxybenzothiazole
chromatographic properties, LVII, 26
preparation, LVII, 19, 20
removal, LVII, 21
spectral properties, LVII, 26
structural formula, LVII, 18
synthesis of luciferin, LVII, 16
m-Cyanophenol, XLVI, 122–124
m-(3-Cyanophenoxy)propylamine hydrobromide, XLVI, 124
1-(3-Cyanophenylethyl)-2-phenylethylene, XLVI, 125, 127
1-(3-Cyanophenyl)-4-phenylbutane, XLVI, 127
2-(3-Cyanophenyl)-1-phenylethane, XLVI, 127
2-(3-Cyanophenyl)styrene, XLVI, 127
Cyanuric chloride, XXXIV, 713
cellulose activator, XLIV, 19, 37, 47, 49, 50, 120
Cyanylation, XLVII, 130
Cyathin, XLIII, 201
Cyclibillin, XLIV, 615
Cyclic adenosine monophosphate, XLIV, 884; XLVI, 75, 76, 90, 95
2′-*O*-acyl derivatives, preparation and properties, XXXVIII, 402–404
in adrenal cortex cells, XXXII, 678, 691
analysis, XXXII, 678–686
agarose coupling, XXXIV, 263
8-(6-aminohexyl)amino, XXXIV, 262–264
analogs, XLVI, 339–346
assay, XXXI, 105–109; XXXIX, 222, 223, 345, 348
in adrenals, XXXIX, 344, 345
luciferin-luciferase system, XXXVIII, 62–65
of modulator protein, LVII, 109
protein kinase activation, XXXVIII, 66–73
radioimmunoassay, XXXVIII, 96–105
receptor protein binding, XXXVIII, 49–57
sample preparation, XXXVIII, 3–20, 54, 55, 100
simultaneous, with cGMP assay, XXXVIII, 60, 61
from ATP, XXXI, 103
bacterial luminescence and, LVII, 128
binding sites, XLVI, 339–346
8-bromo, XXXIV, 261
cGMP-dependent protein kinase, XXXVIII, 335–338
chemotaxis, in amoebae, XXXIX, 490, 491
cholera toxin, XXXVIII, 3
chromatography, XXXIV, 263, 264
N^6-2′-*O*-diacyl derivatives, preparation, XXXVIII, 404, 405
diazomalonyl derivatives, XXXVIII, 387, 388
applications, XXXVIII, 397, 398
Dimroth rearrangement, XXXVIII, 392, 393
procedures, XXXVIII, 393–397
syntheses, XXXVIII, 388–393
effect on adrenal gland *in situ*, XXXIX, 341–345
on calcium efflux, XXXIX, 554, 561
on calcium influx, XXXIX, 523
on corpus luteum, XXXIX, 241, 244
on corticosterone output by superfused adrenals, XXXIX, 317–322
on estrogen formation, XXXIX, 251

on insect salivary gland secretion, XXXIX, 469, 474–476
on protein kinases, XL, 205, 206
on Purkinje cell discharge, XXXIX, 439
on renin production, XXXIX, 22
on synthesis of fatty acid synthase, XXXV, 44
epinephrine-dependent, XLVI, 601
excretion, XXXVIII, 163
from perfused kidney, XXXIX, 9, 10
formation, measurement, XXXIX, 266–268
in granulosa cells, levels, XXXIX, 219–223, 230
^{3}H-labeled, XLVI, 337–339
hypophysiotropic regulation, XXXVII, 92
identification, XXXVIII, 132
inhibitor, XXXVIII, 273, 274, 282, 283
assay, XXXVIII, 278–282
formation in cell-free incubations, XXXVIII, 274–276
purification, XXXVIII, 276–278
in interstitial cell metabolism, XXXIX, 259, 260
in isolated thyroid cells, XXXII, 746, 758
isolation, XXXI, 112–114
logit-log assay, XXXVII, 9, 13
in luteinization control, XXXIX, 184–188
measurement, MDCK, LVIII, 557–559
melanoma cell cycle, LVIII, 567
metabolism, in epididymal capillary cells, XXXIX, 479–482
N^{6}-monoacyl derivatives, preparation, XXXVIII, 405, 406
norepinephrine release, XXXIX, 376, 380
in perfused liver, XXXIX, 34
in perfused muscle, XXXIX, 73
photolabeling, XLVI, 344
in pineal, XXXIX, 396
^{32}P–labeled, XXXVIII, 410–412
general experimental details, XXXVIII, 412–414
preparation of cyclic nucleotides, XXXVIII, 414–420
prostaglandins, XXXII, 109
protamine kinase, XXXIV, 261–264
protein kinase classification, XXXVIII, 290–292, 295, 296
purification, LVII, 98, 99
radioimmunoassay, XXXVII, 34; XXXIX, 267
standard, XXXIX, 210
synthesis and release, XXXIX, 267, 268
Cyclic adenosine monophosphate-agarose, XXXIV, 261
Cyclic adenosine monophosphate-binding protein
in cytoplasmic progesterone receptor preparations, XXXVI, 210, 211
preparation, XXXVIII, 49–52
Cyclic adenosine monophosphate-dependent protein kinase, *see* Protein kinase
Cyclic adenosine monophosphate phosphodiesterase, assay, XXXIX, 480
Cyclic adenosine monophosphate-receptor protein
animal cells, XXXVIII, 376, 377
application, XXXVIII, 381, 382
assay, XXXVIII, 377, 378
characteristics of interaction, XXXVIII, 378–380
detection in tissues, XXXVIII, 380, 381
properties, XXXVIII, 381
Escherichia coli, XXXVIII, 367
assay, XXXVIII, 368
bacteria, XXXVIII, 368, 369
DNA preparation, XXXVIII, 372, 373
growth of bacteria, XXXVIII, 373, 374
ion-exchange resin preparation, XXXVIII, 371
mechanism of action, XXXVIII, 375, 376
properties, XXXVIII, 376
purification, XXXVIII, 369–371
transcription assay, XXXVIII, 372
Cyclic adenosine monophosphate-Sepharose, preparation, XXXVIII, 385–387

Cyclic adenosine phosphodiesterase, hydrolysis rate, XXXII, 128
Cyclic adenylic acid, *see* Cyclic adenosine monophosphate
Cyclic AMP, *see* Cyclic adenosine monophosphate
Cyclic AMP phosphodiesterase
  activity, LVII, 96, 108
  assay, LVII, 94–106
    modulator-deficient, preparation, LVII, 109
    of modulator protein, LVII, 108, 109
Cyclic AMP phosphorylase, in fat cells, XXXI, 67, 68
Cyclic cytidine phosphate, substrate, for ribonuclease, XLIV, 36
Cyclic cytosine phosphodiesterase, hydrolysis rate, XXXII, 128
Cyclic deoxyribonucleotide, *see also* specific type
  preparation, XXXVIII, 420
Cyclic GMP, *see* Cyclic guanosine monophosphate
Cyclic GMP phosphodiesterase
  activity, LVII, 96
  assay, LVII, 94–106
Cyclic guanosine monophosphate
  assay
    elongation factor Tu, XXXVIII, 85
    enzymatic cycling, XXXVIII, 73, 74
    labeled GDP formation, XXXVIII, 106
  bacterial luminescence and, LVII, 128
  cAMP-dependent protein kinases, XXXVIII, 336–338
  ^{3}H-labeled, XLVI, 337–339
  in insect issues, XXXIX, 474, 475
  molar extinction coefficient, LVII, 99
  phosphodiesterase, XXXVIII, 235, 237, 242–244, 259
  preparation of extracts, XXXVIII, 85–87
    radioimmunoassay, XXXVIII, 13
    receptor protein binding, XXXVIII, 57–60
  protein kinases, XXXVIII, 329, 330
    characterization, XXXVIII, 335–349
    purification, XXXVIII, 330–335
    standard assay, XXXVIII, 330
  purification, XXXVIII, 76; LVII, 99
    column chromatography, XXXVIII, 76, 77
    thin-layer chromatography, XXXVIII, 77, 78
  separation, XXXVIII, 13
Cyclic guanosine monophosphate-binding protein, preparation, XXXVIII, 57
Cyclic guanosine monophosphate-dependent protein kinase, XXXVIII, 329
Cyclic guanylic acid, *see* Cyclic guanosine monophosphate
Cyclic hydrazide, chemiluminescence, LVII, 416, 417
Cyclic imide group, cleavage by hydroxylamine, XLVII, 134–137
Cyclic inosine monophosphate, reactivity in immunoassay, XXXVIII, 103
Cyclic isoadenosine monophosphate, activity, XXXVIII, 401
Cyclic nucleotide, XLVI, 335–339, *see also* specific types
  acylated derivatives, XXXVIII, 399–401
    butyryl, XXXVIII, 406–409
    cAMP, XXXVIII, 402–406
    principle of synthetic methods, XXXVIII, 401, 402
  analogs, XLVI, 293–295
    protein kinases, XXXVIII, 339
      Dowex-50, XXXVIII, 10–13
      ion-exchange resin chromatography, XXXVIII, 9, 10
      (polyethyleneimine)cellulose, XXXVIII, 16–20
      quaternary aminoethyl Sephadex, XXXVIII, 13–16
  effect on melanin, XXXVII, 129
  labeling, XLVI, 337–339
  metabolism, in pineal, XXXIX, 395, 396
  protein kinase inhibitor, XXXVIII, 357
  purification, XXXVIII, 9, 20, 27, 38, 41, 208

radioimmunoassay methods
preparation of materials, XXXVIII, 96–101
procedure, XXXVIII, 101–105
thin-layer chromatography, XXXVIII, 158
2′:3′-Cyclic-nucleotide 2′-phosphodiesterase, occurrence and activity, XXXII, 127
2′:3′-Cyclic-nucleotide 3′-phosphodiesterase
activity, XXXII, 128
assay, XXXII, 124–131
as marker enzyme, XXXII, 124
as myelin marker, XXXI, 443
occurrence, XXXII, 127
2′:3′-Cyclic-nucleotide-3′-phosphohydrolase, *see* 2′:3′-Cyclic-nucleotide 3′-phosphodiesterase
Cyclic nucleotide phosphodiesterase, *see* Phosphodiesterase
Cyclic uridine monophosphate, substrate, ribonuclease A, XLIV, 519
Cyclocytidine, LI, 407
Cyclodepsipeptide, as ionophore, LV, 441, 442, 444
1,3-Cyclohexadienyl-5-aminocarboxylic acid, XLVI, 32, 33
Cyclohexane, XLIV, 207
in diene conjugation assay, LII, 308
preparation of reduced ubiquinone-2, LIII, 39
purification of complex II, LIII, 23
solvent
in osmometric studies, XLVIII, 79
in self-association studies, XLVIII, 118
substrate, of cytochrome P–450LM$_2$, LII, 204, 205
of ω-hydroxylase, LIII, 360
in thymus nuclei isolation, XXXI, 252
type of binding reaction with cytochrome P–450, LII, 264
Cyclohexane-benzene, purification of 2-carbethoxy-2-diazoacethydrazide, LIX, 811
1,2-Cyclohexanedione
arginine blockage, XLVII, 156–161
buffer selection, XLVII, 158
^{14}C-labeled, XLVII, 161
Cyclohexane-ethyl acetate, for chromatogram development, LII, 126
Cyclohexanol
alcohol oxidase, LVI, 467
pore distribution in Spheron, XLIV, 67
Cyclohexanone
oxidative lactonase, LII, 17
reductive alkylation, XLVII, 472, 474
Cyclohex-2-en-1-ol, XLIV, 837
Cyclohexenone, XLVI, 476, 477
Cycloheximide, LVI, 31
amino acid incorporation, LVI, 55–57
in antigen studies, XXXII, 67
determination of *in vivo* levels of aminoacylated tRNA, LIX, 271
gas-liquid chromatography, XLIII, 234–237
HeLa cell labeling, LVI, 68
mitochondrial protein labeling, LVI, 44, 60, 99
mitochondrial translation products, LVI, 11, 33, 34
paper chromatography, XLIII, 156–158
preparation of ribosomal subunits, LIX, 413
protein inhibitor, XL, 287, 288
rho$^-$ mutants, LVI, 159
ribosomes, LVI, 10, 11
solvent systems for countercurrent distribution, XLIII, 336
in specific labeling of cytochrome *b* polypeptides, LIII, 108
sulfate incorporation into mitochondrial protein, LVI, 60
use in flux studies, LV, 664
Cycloheximide resistant strain, LVIII, 314
1,2-*O*-Cyclohexylidene *myo*-inositol, synthesis, XLVI, 372
1,2-*O*-Cyclohexylidene tetra-*O*-acetyl-*myo*-inositol, synthesis, XLVI, 372
Cyclohexyl isocyanide, conjugation agent, XLIV, 40
1-Cyclohexyl-3-morpholinoethyl carbodiimide, XLVI, 223

1-Cyclohexyl-3-(2-morpholinoethyl)carbodiimide metho-*p*-toluenesulfonate, XXXIV, 67, 89, 379, 464, 696; XLIV, 62, 141, 932, 933; XLVII, 42
Cyclopaldic acid, XLIII, 162
Cyclopentanone
  oxidative lactonase, LII, 17
  reductive alkylation, XLVII, 472
β-Cyclopiazonate dehydrogenase
  flavin fluorescence, LIII, 462
  flavin interactions, LIII, 465
  flavin linkage, LIII, 450
  flavin peptide
    amino acid analysis, LIII, 457
    isolation, LIII, 454
β-Cyclopiazonate oxidocyclase, *see* β-Cyclopiazonate dehydrogenase
Cyclopropylamine, as inhibitors, XLVI, 38
Cyclostome
  balanced salt solution osmolarity adjustment, LVIII, 469
  incubation temperature, LVIII, 471
Cymarin, XLVI, 75, 524–526, 529
Cymbidium, cell cultures, XXXII, 729
Cynthia moth, media, LVIII, 457
Cypridina
  bioluminescence system, LVII, 329–372
    effect of experimental parameters, LVII, 360
    kinetics, LVII, 355–358
    simplicity, LVII, 353–355
  bioluminescent reaction, LIII, 561
  collection, LVII, 365
  historical observations, LVII, 333, 334
  luciferase, LIII, 559, 560, 561; LVII, 258, 344, 345
  in luciferin-luciferase specificity studies, LVII, 336, 371
  oxidase, LII, 20
Cypridina bairdii, LVII, 336
Cypridina hilgendorfii, LVII, 334–337
  bioluminescent characteristics, LVII, 364, 365
Cypridina noctiluca, LVII, 336
Cypridina serrata, LVII, 336
Cyproterone, antihormone, XL, 291, 292
Cystamine
  effect on 1-guanidino-1-deoxy-*scyllo*-inositol-4-phosphate phosphohydrolase, XLIII, 461
  as inosamine-phosphate amidinotransferase inhibitor, XLIII, 458
  as spacer, XXXIV, 43, 44, 72
Cystathionase, *see* Cystathionine γ-lyase
L-Cystathionine, polarography, LVI, 477
Cystathionine γ-lyase, XLVI, 31, 32, 163, 445
Cysteamine
  extraction, XL, 338
  substrate of dimethylaniline monooxygenase, LII, 142
  uncoupling by CCP, LV, 469
Cysteamine dioxygenase, LII, 11
Cysteamine-soluble collagen, *see* Collagen
Cysteic acid
  elution profile marker, XLVII, 233
  internal standard, XLVII, 34, 39
  peptide electrophoretic mobility, XLVII, 59–61
Cysteine, XXXIV, 547; XLIV, 81, 88, 110, 111, 476, 926
  aminoacylation isomeric specificity, LIX, 279, 280
  assay, for bound enzyme determination, XLIV, 256, 257
  chain cleavage, XLVII, 129–132
  chemical properties, XLIV, 12–18
  in electrophoresis buffer, LI, 357
  enzyme aggregation, XLIV, 263
  in fatty acid synthase, XXXV, 45, 46, 59, 73
  inhibitor of nucleoside phosphotransferase, LI, 393
    of *Pholas* luciferin chemiluminescence, LVII, 400
  insolubilized, XXXIV, 535
  in media, LVIII, 53, 62, 87
  membrane autoxidation, LII, 305
  misactivation, LIX, 289
  modification, XLVII, 407–430, 482, 483, 494, 496–497
  oxidation, with DMSO/HCl, XLVII, 445, 448, 457
  peptide synthesis, XLVII, 510
  in plant-cell fractionation, XXXI, 505

in protein-turnover studies, XXXVII, 246, 247
in rat liver extract preparation, LI, 460
reactions with bromoacetylpyridine, XLVI, 150
in reactivation reagent, LVII, 180
residue, modification, LIII, 143
reversible dissociation of flavoproteins, LIII, 434, 435
sulfhydryl group blocking, XLVII, 407–430, 526, 537–539, 614
thiol derivatives, alkylation, XLVII, 115, 116
D-Cysteine, synthesis of luciferin, LVII, 15, 16
L-Cysteine, XXXIV, 541
activator of pyrimidine nucleoside monophosphate kinase, LI, 330
in affinity labeling of steroids, XXXVI, 386–388, 411
enzyme active site studies, XXXVI, 390, 393, 419
for cytidine triphosphate synthetase storage, LI, 87
Cysteine-containing protein, XXXIV, 547–552
binding capacity, XXXIV, 549
histone preparation, XXXIV, 549–552
organomercurial-agarose preparation, XXXIV, 548, 549
Cysteine dioxygenase, LII, 11
Cysteine glutathione, activator of formate dehydrogenase, LIII, 371
D-Cysteine hydrochloride, preparation of luciferin, LVII, 21, 22
L-Cysteine hydrochloride, synthesis of 8α-*S*-cysteinylriboflavin, LIII, 460
5-*S*-Cysteine-6-hydrouracil, XLVI, 174
Cysteine peptide, XXXIV, 191, 192
Cysteine protease, XLVI, 220–225
activation, XLV, 9
active site titration, XLV, 3
kinetics, XLV, 6
papain, XLV, 4
reaction site, XLV, 5
spectrophotometric, XLV, 7
inhibitors, acidic, XLV, 740
Cysteine thioether, XLVI, 135
Cysteinyl residue, modification with *N*-ethylmaleimide, LVII, 177
8α-*S*-Cysteinylriboflavin
properties, LIII, 465
synthesis, LIII, 460
Cysteinyl-tRNA synthetase, subcellular distribution, LIX, 233, 234
Cystine
in media, LVIII, 53, 62
quinone binding, XXXI, 537
Cystine-binding protein, from *E. coli*, isolation, XXXII, 422–427
Cystine as inosamine phosphate amidinotransferase, inhibitor, XLIII, 458
Cystine residue, vibrational frequency of side chains, XLIX, 116, 117
Cytidine
chromatographic separation, LI, 409; LIX, 188
effect on aspartate carbamyltransferase, LI, 57
inhibitor of cytosine deaminase, LI, 400
radiolabeled, chromatographic separation, LI, 300, 315, 316
residue, site of attachment of fluorescent probe, LIX, 146, 147
substrate of cytidine deaminase, LI, 395, 401
of deoxycytidine kinase, LI, 345
of nucleoside phosphotransferase, LI, 391
of uridine-cytidine kinase, LI, 299, 305, 308, 313
Cytidine aminohydrolase, *see* Cytidine deaminase
Cytidine deaminase, LI, 394–412
activity, LI, 401, 408
apparent Stokes radius, LI, 404
assay, LI, 395, 401, 402, 405, 406, 408, 409
from *E. coli*, LI, 401–405
effects of deuterium oxide, LI, 404, 407
genetic polymorphism, LI, 412
heat sensitivity, LI, 401
from human liver, LI, 405–407
inhibitors, LI, 405, 411
kinetic properties, LI, 400
from leukemic mouse spleen, LI, 408–412
molecular weight, LI, 400, 404, 411

from normal mouse spleen, LI, 411, 412
partial specific volume, LI, 404
pH effects, LI, 404, 407, 412
product specificity, LI, 411
properties, LI, 404, 405, 407, 411, 412
purification, LI, 397–399, 406, 409, 410
sedimentation coefficient, LI, 404
sources, LI, 408, 412
stability, LI, 404, 411
substrate specificity, LI, 404, 407, 411
from yeast, LI, 394–401
Cytidine 5′-diphosphate
chromatographic separation, LI, 315, 316, 323, 333
effect on orotidylate decarboxylase, LI, 79
inhibitor of aspartate carbamyltransferase, LI, 57
of cytidine deaminase, LI, 400
of cytosine deaminase, LI, 400
product of pyrimidine nucleoside, monophosphate kinase, LI, 321
of uridylate-cytidylate kinase, LI, 331
substrate of nucleoside diphosphokinase, LI, 375
of pyruvate kinase, LVII, 82
of ribonucleoside diphosphate reductase, LI, 228, 235, 236, 245
Cytidine diphosphate ribitol, polyribitol phosphate polymerase, L, 389–394
Cytidine 2′(3′)-monophosphate, electrophoretic mobility, LIX, 177
Cytidine 5′-monophosphate
chromatographic separation, LI, 315, 316, 323, 333
effect on aspartate carbamyltransferase, LI, 57
on orotidylate decarboxylase, LI, 79
inhibitor of cytidine deaminase, LI, 400
of cytosine deaminase, LI, 400
of orotate phosphoribosyltransferase, LI, 163
of orotidylate decarboxylase, LI, 163
product of uridine-cytidine kinase, LI, 299, 308
substrate of pyrimidine nucleoside monophosphate kinase, LI, 321, 329
of uridylate-cytidylate kinase, LI, 331, 336
Cytidine 5′-monophosphate-agarose, XXXIV, 477
Cytidine monophosphate deaminase, XLVI, 25
Cytidine 5′-phosphate
electrophoretic mobility, LIX, 64
M value range, LIX, 89
Cytidinetriphosphatase, as marker enzyme, for plasma membranes, XXXI, 165–167
Cytidine 5′-triphosphate
assay of CTP(ATP):tRNA nucleotidyltransferase, LIX, 123
of tRNA hydrolysis, LIX, 183
ATP-P_i exchange preparation, LV, 361, 362
^{14}C-labeled, assay of nucleotide incorporation, LIX, 182
chromatographic separation, LI, 315, 316, 323, 333
extinction coefficient, LI, 80, 85
inhibitor of adenylate deaminase, LI, 496
of aspartate carbamyltransferase, LI, 35, 57
of carbamoyl-phosphate synthetase, LI, 119
of cytidine deaminease, LI, 400
of cytosine deaminase, LI, 400
of uridine-cytidine kinase, LI, 307, 320
interaction with GMP synthetase, LI, 224
preparation of aminoacylated tRNA, LIX, 126, 127, 221, 231
product of cytidine triphosphate synthetase, LI, 79, 84
reconstitution of tRNA, LIX, 126, 127, 133, 140
regulator of pyrimidine synthesis, LI, 299
tRNA end-group labeling, LIX, 103
separation, LV, 287–289

substrate of phosphoribosylpyrophosphate synthetase, LI, 10
Cytidine 5′-triphosphate-agarose, XXXIV, 477
Cytidine triphosphate synthetase, *see* CTP synthetase
Cytidylate-deoxycytidylate kinase, *see* Cytidylate kinase
Cytidylate kinase, in bacterial extracts, LI, 332
Cytidylyl-(3′-5′)-3′-*O*-phenylalanyladenosine, in photoaffinity labeling studies, LIX, 802, 804, 805
Cytisus sesslifolius, lectin, XXXIV, 334
Cytochalasin B
glycoprotein biosynthesis, L, 188, 189
inhibitory properties, XL, 284, 286
Cytochemistry, in transmission electron microscopy, XXXIX, 149–154
Cytocholasin B, in studies of phagocyte chemiluminescence, LVII, 491
Cytochrome, *see also* specific type
absorption spectra
determination, LVI, 125
fourth derivative analysis, LVI, 503
pet9 mutants, LVI, 126, 127
strain variations, LVI, 122, 123
temperature-sensitive mutants, LVI, 137, 138
assay in bovine heart mitochondria, LV, 49, 50
bacterial
a-type, properties, LIII, 206–208
b-type, LIII, 208, 210, 211
content, determination, LIII, 205
c-type
CD spectra, LIV, 263
properties, LIII, 207, 209–212
redox titrations, LIV, 429
differential absorptivities, LIII, 205
d–type, properties, LIII, 212
pyridine hemochromes, LIII, 203, 204
spectral characterization, LIII, 202–212
brown adipose tissue mitochondria, LV, 75
complex I, LIII, 12; LVI, 581
composition, respiratory chain assembly, LVI, 567, 568
content of mitochondria
brain, LV, 59
N. crassa, LV, 26, 147
plant, LV, 25
cytoplasmic petite mutants, LVI, 155
deficiency, identification, LVI, 118
of depleted ETP, LV, 188
detergent-treated ETP, LV, 193
diffusion-limited reaction, LII, 211
heme-deficient mutants, LVI, 560
Keilin-Hartree preparation, LV, 123
lack, in promitochondria, XXXI, 627
in liver microsomes, XXXI, 85
low temperature kinetics, LIV, 102–111
of microsomes, measurement of difference spectra, LII, 212–220
microspectrophotometry, XXXIX, 415
M. phlei
ETP, LV, 184, 185
membrane fragments, LV, 196
nomenclature, LIII, 203
oxidation-reduction, metal ion transport, LVI, 334–336
plasma membrane, XXXI, 90
proposed role in membrane autoxidation, LII, 305
protoplast ghosts, LV, 182
red-green, separation, LIII, 5, 6, 42; LV, 309
spectrophotometric assay, XXXII, 416, 417
in yeast, measurement, XXXII, 842
Cytochrome *a*
in complex IV, LIII, 44
ligand binding, LIII, 191, 196, 197
midpoint redox potential, LIII, 197
of *Nitrosomonas europaea*, LIII, 207
prosthetic group, LIII, 203
redox titration, LIV, 430
resonance Raman studies, LIV, 245
Cytochrome a_1, LII, 15
from *Acetobacter pasteurianum*, LIII, 206, 207
from *Nitrosomonas*, LIII, 639, 640
prosthetic group, LII, 5
Cytochrome a_2, LII, 15
prosthetic group, LII, 5

Cytochrome a_3, LII, 15
azide binding, LIII, 197
from *Bacillus subtilis*, LIII, 207
blocked, LIV, 76
carbon monoxide binding, LIII, 197–199
in complex IV, LIII, 44
cyanide binding, LIII, 192, 193
in flash photolysis experiments, LIV, 101
fluoride binding, LIII, 200, 201
ligand binding, LIII, 191–201; LIV, 531, 532
MCD studies, LIV, 296, 301
midpoint redox potential, LIII, 197
from *Mycobacterium phlei*, LIII, 207, 208
nitrosyl binding, LIII, 199, 200
prosthetic group, LII, 5
redox titrations, LIV, 429–431
Cytochrome a_4, LII, 15
prosthetic group, LII, 5
Cytochrome aa_3, LIII, 54, 55, 155, *see also* Cytochrome *c* oxidase
complex IV, LVI, 580, 583, 584
complex V, LVI, 580
organ absorbance spectrophotometry, LII, 56
oxidation, proton release, LV, 624–626
Cytochrome ($a + a_3$), in mitochondria, XXXI, 23, 24
Cytochrome *b*
absorbance coefficient, LIII, 127
absorption maxima, LVI, 503
amino acid composition, LIII, 221
from beef heart, LIII, 213, 221
complex II, LVI, 580, 581, 590
complex III, LVI, 580, 582, 583, 590–592; LIII, 36, 37, 89, 96
in complex I–III, LIII, 7, 8
complex V, LVI, 580
concentration determination, LIII, 127
in cytochrome bc_1 complex, LIII, 116
dimeric
molecular weight, LIII, 106
purification, LIII, 105, 106
EPR signal properties, LIV, 137
from *Escherichia coli*, LIII, 208
heme center, properties, LIII, 218, 219
isolation from complex III, LIII, 84, 85
from *Locusta migratoria*, LIII, 213, 221
measurement, LV, 668
from *Micrococcus lysodeikticus*, LIII, 208
mit mutants, LVI, 16
molar absorbance coefficients, LIII, 219
molecular weight, LIII, 219, 220
M. phlei ETP, LV, 184, 185
from *Neurospora crassa*, LIII, 212–221; LV, 148
oxidation
calcium uptake, LVI, 341
proton uptake, LV, 627
prosthetic group, LIII, 203
purification, LIII, 213–218
resonance Raman studies, LIV, 245
solubility in detergents, LIII, 219
spectral properties, LIII, 126, 127, 218, 219
structural gene, genetic loci, LVI, 196
subunit composition, LIII, 219, 220
visible CD spectra, LIV, 282, 283
Cytochrome b_K, *see also* Cytochrome *b*-562
absorption maximum, LVI, 503
Cytochrome b_T, absorption maxima, LVI, 503
Cytochrome b_1
assay, LIII, 232, 233
association with nitrate reductase, LIII, 355
from *Escherichia coli*, LIII, 232–237
molar extinction coefficient, LIII, 233
molecular weight, LIII, 237
potential modifying protein, LIII, 237
spectral properties, LIII, 236
Cytochrome b_2
conformational composition, LIV, 269, 272
inactivation, LIII, 439
magnetic circular dichroism, XLIX, 170
MCD studies, LIV, 295
reaction with cytochrome *c*, LIII, 156

from yeast, LIII, 238, 439
Cytochrome b_5, LIII, 156, 238
aggregation, LII, 101
antibodies against, preparation, LII, 242, 243
assay, LII, 97, 110, 207
binding to liposomes, procedure, LII, 208–210
in column eluate, LII, 93
conformational composition, LIV, 269
contaminant
in cytochrome *P*–450LM preparations, LII, 116, 121
in *Rhizobium P*–450 preparation, LII, 159
distribution, LII, 101
in erythrocytes, LII, 464, 465
assay, LII, 466, 467
human, properties, LII, 471, 472
inactivation, by membrane lipid peroxidation, LII, 304
in isolated MEOS fraction, LII, 362, 363
magnetic circular dichroism, XLIX, 170
MCD studies, LIV, 295, 298, 299
measurement of interaction with cytochrome b_5 reductase, LII, 207
in microsomes, LII, 45, 46, 88
millimolar difference extinction coefficient, LII, 215
molecular weight, LII, 100
Mössbauer studies, LIV, 372, 373
multiple forms, electrophoretic separation, LII, 469
mutants, LVI, 122, 560
NADH-cytochrome b_5 reductase assay, LII, 103
in nuclear membrane, XXXI, 290–292
plasma membrane, XXXI, 90
purification procedure, LII, 97–100
from human erythrocytes, LII, 468–470
reactivity, LII, 101
reduction, LII, 45
requirement for regeneration of stearyl-CoA desaturase activity, XXXV, 258, 259
retention in DEAE-cellulose column, LII, 147
in smooth 11 microsome fraction, LII, 73
stability, LII, 101
in stearyl-CoA desaturase activity assay, LII, 189
storage, LII, 100
structure, LII, 100, 101
subcellular localization, LII, 251
from yearling steer liver, LII, 97–101
Cytochrome b_6, spectrophotometric assay, XXXII, 416
Cytochrome b_{559}
from chloroplasts, XXXII, 91, 406–422
chromatography, XXXII, 413, 414
properties, XXXII, 421, 422
purification, XXXII, 419–422
Cytochrome *b*–555, CD spectrum, LIV, 283
Cytochrome *b*–557.5
in complex II, LIII, 24
difference spectrum, LIII, 25
properties, LIII, 24
Cytochrome *b*–558, spectral properties, LIII, 127
Cytochrome *b*–559
mid-potential, LIII, 97
redox titrations, LIV, 429
Cytochrome *b*–561
EPR signal properties, LIV, 136, 137
mid-potential, LIII, 97
Cytochrome *b*–562
CD spectrum, LIV, 260, 283
in complex III, LIII, 85, 86
EPR characteristics, LIII, 120, 121
spectral properties, LIII, 86, 127
Cytochrome *b*–563, CD spectrum, LIV, 283
Cytochrome *b*–565
in complex III, LIII, 85, 86
EPR signal properties, LIV, 137
reduction, brown adipose tissue mitochondria, LV, 72
Cytochrome *b*–566, spectral properties, LIII, 38, 127
Cytochrome bc_1 complex, *see also* Complex III
from beef heart complex III, LIII, 92–98

binding of cytochrome *c*, LIII, 104, 105
composition, LIII, 116, 117
heme centers, properties, LIII, 110, 112
iron-sulfur center, EPR signal properties, LIV, 138, 140, 141, 143, 144, 146, 147
isolation, LIII, 92–95
kinetic properties, LIII, 117
peptide
composition, LIII, 95, 96, 106, 107
molecular weight, LIII, 117
physicochemical data, LIII, 97, 98
purification, LIII, 223
by affinity chromatography, LIII, 100–104
by ammonium sulfate fractionation, LIII, 114, 115
small
amino acid composition, LIII, 110
definition, LIII, 99
peptides
amino acid compositions, LIII, 110
molar ratios, LIII, 111
molecular weights, LIII, 111
purification, LIII, 105
spectral properties, LIII, 112
spectral properties, LIII, 112, 117–121
ubiquinone extraction, LIII, 576
from yeast, LIII, 114–121
Cytochrome *b* complex, peptide, LVI, 12, 40
Cytochrome b-c_1 region, inhibitors, LV, 461, 462
Cytochrome *b* polypeptide
amino acid compositions, LIII, 110
selective labeling, LIII, 108
Cytochrome b_5 reductase
assay, LII, 207
binding to liposomes, LII, 209, 210
partial purification, LII, 99
reactivity, LII, 101
in stearyl-CoA desaturase activity assay, LII, 189
Cytochrome *c*, *see also* Metallocytochrome *c*; Porphyrin cytochrome *c*
absorbance coefficient, LIII, 128
absorption maxima, LVI, 503
effects of pH, LIII, 166
absorption spectrum, LIV, 277
activity determinations with mitochondrial cytochrome *c* oxidase, LIII, 158–164
with redox proteins, LIII, 155–158
from *Alcaligenes faecalis*, LIII, 210
antibodies, LVI, 227, 695–697
assay of complex III, LIII, 91
of NADH-cytochrome *c* reductase, LIII, 9, 10
of rubredoxin, LIII, 618
of succinate dehydrogenase, LIII, 466–468
axial ligands, XLIX, 172, 173
p-azidophenacyl bromide derivative
preparation, LIII, 174, 175
properties, LIII, 175
bacterial, purification by isoelectric focusing, LIII, 229–231
α-band, extinction coefficient, LIII, 45
from *Beneckea natriegens*, LIII, 207
binding to purified cytochrome bc_1 complex, LIII, 104, 105
N-bromosuccinamide-tryptophanyl derivative
preparation, LIII, 171
properties, LIII, 166, 168, 169
structural formula, LIII, 171
from *Candida krusei,* solid-phase sequencing, XLVII, 269
4-carboxy-2,6-dinitrophenyllysyl derivative, preparation, LIII, 152, 153
CD spectrum, LIV, 277, 279
deoxycholate, LIV, 281, 282
charge distribution, LIII, 136, 146, 151
chemical modification, LIII, 134–153
chromatographic behavior, LIII, 147, 148
cobalt-substituted, *see* Cobalt-substituted cytochrome *c*
complex IV, LVI, 587, 594, 596
complex V, LVI, 580
conformation, LIII, 135
conformational composition, LIV, 268, 270, 272

contaminant
of high-potential iron-sulfur protein preparations, LIII, 339
in *Rhizobium P*–450 preparations, LII, 159, 163
copper-substituted, *see* Copper-substituted cytochrome *c*
coupling to Sepharose 4B, LIII, 101
cross-linking to oxidase, LVI, 637
in cytochrome b_5 assay, LII, 207
in cytochrome b_5 reductase assay, LII, 465
cytochrome oxidase proteoliposomes, LV, 761
derivatives
absorbance maxima, effects of pH, LIII, 166
chemical properties, LIII, 169
homogeneity, criteria, LIII, 153, 154
spectral properties, LIII, 168
differential absorptivity, LIII, 40
diiodotyrosyl derivative
preparation, LIII, 172, 173
properties, LIII, 166, 168, 169
structural formula, LIII, 173
dinitroazidophenyl derivative
preparation, LIII, 173, 174
properties, LIII, 174
double-resonance experiments, LIV, 213, 214
effect of freezing, LIV, 378
electron acceptor, from NADPH-cytochrome *P*–450 reductase, LII, 90, 96
as electron microscopic tracer, XXXIX, 146, 147
EPR signal properties, LIV, 136
excited-state angular momentum, XLIX, 171
extinction coefficient, LIII, 10
Fe *K*-edge derivative spectra, LIV, 338
ferricyanide reduction, LVI, 231
fold, LIII, 135
formyl-tryptophanyl derivative
preparation, LIII, 167, 168
properties, LIII, 166, 168–170
structural formula, LIII, 167
heme-deficient mutants, LVI, 559
heme modification, LIII, 175–181
heme peptide, XXXIV, 194
hemochrome structure, LIII, 149
hollow-fiber retention data, XLIV, 296
from horse heart, chemical cleavage, XLVII, 460, 464–467
2-hydroxy-5-nitrobenzylbromide-tryptophanyl derivative
preparation, LIII, 170
structural formula, LIII, 170
hyperfine resonance shifts, LIV, 167, 169, 172, 174
immobilized, electron transfer, XLIV, 545, 546
inhibitor of brain α-hydroxylase, LII, 317
intact mitochondria, LV, 143
intrinsic CD spectrum, LIV, 267, 270
iron removal, LIII, 176, 177
iso–1, preparation, LIII, 133
iso–2, preparation, LIII, 133
isolation of mitochondria, LV, 4
Keilin-Hartree preparation, LV, 120, 124
kinetic properties, LIII, 158–164
localization, LVI, 708
magnetic circular dichroism, XLIX, 170, 171
manganese-substituted, *see* Manganese-substituted cytochrome *c*
MCD studies, LIV, 295
membrane orientation, XLIV, 371
membrane preparations enriched, LV, 109, 110
metal-substituted derivatives
preparation, LIII, 177–179
properties, LIII, 178–181
midpoint redox potential, LIII, 181
modification of aromatic amino acids, LIII, 165–173
molecular weight, 229n, LIII, 351
monocarboxymethylmethionyl-80 derivative, preparation, LIII, 150, 151
monoiodotyrosyl-74 derivative
preparation, LIII, 148, 149
monomeric form, preparation, LIII, 161, 162
Mössbauer studies, LIV, 372, 373

in NADPH-cytochrome *P*–450 reductase assay, LII, 91
nickel-substituted, *see* Nickel-substituted cytochrome *c*
nitrotyrosyl derivative
preparation, LIII, 172
properties, LIII, 166, 168, 169
structural formula, LIII, 172
oxidation
by bacterial cytochromes, LIII, 207
proton release, LV, 624–626
oxidized
assay of complex III, LIII, 39
of superoxide dismutase, LIII, 383
conformation studies, XLIV, 368, 369
molar extinction coefficient, LIII, 383
product of complex IV, LIII, 40
of cytochrome oxidase, LIII, 73
reduction, spectrophotometric determination, LIII, 9
substrate of complex III, LIII, 35
of cytochrome oxidase, LIII, 54
of NADH:cytochrome *c* oxidoreductase, LIII, 5
of *Paracoccus dinitrificans*, LIII, 207, 229–231
partial deficiency, LVI, 122, 123
peptide mapping, LIII, 154
phosphorylation site, LV, 240
photoaffinity labeling, LIII, 173–175
photochemically active derivative, LVI, 656
physical properties, LIV, 209
plasma membrane, XXXI, 90
polarographic assay, LIII, 160–164
polymerization, LVI, 696, 697
prosthetic group, LIII, 203
proton magnetic resonance studies, LIV, 156, 209–214
proton translocation, LV, 751
purification, LIII, 10, 128–134, 247
purity, LIII, 133, 134
reconstitution of mitochondrial inner membrane, LVI, 117, 118
reduced
of cytochrome oxidase, LIII, 54, 73
of NADH:cytochrome *c* oxidoreductase, LIII, 5
extinction coefficient, LII, 91; LIII, 40
oxidation, spectrophotometric determination, concentration dependence, LIII, 45
product of complex III, LIII, 35
substrate of complex IV, LIII, 40
time-dependent spectrum, LIV, 18
reduction, assay of rubredoxin, LIII, 341
removal from submitochondrial particles, LV, 114
residue modifications, LIII, 137–153, 165
resonance Raman studies, LIV, 239, 245, 378
saturation transfer experiments, LIV, 174
SDS complex, properties, XLVIII, 6
Soret CD spectrum, LIV, 260, 262
spectral assay, LIII, 162–164
in spreading solution, LIX, 842, 847
steady-state kinetic analysis, LIII, 160–162
structure, XLIX, 128; LIII, 135, 147, 149
in studies of enzyme action of submitochondrial vesicles, XXXI, 298
substrate of acyl-CoA dehydrogenases, LIII, 503, 506
of NADH dehydrogenase, LIII, 18, 413
thermolysin cleavage sites, XLVII, 177
treatment to remove bound ions, LIII, 45
tryptophan peptides, XXXIV, 193
unfolding, XLIX, 4
visible CD spectra, LIV, 276–279
from yeast, LIII, 137, 143, 144, 146
isolation, LVI, 44, 45, 49, 50
zinc-substituted, *see* Zinc-substituted cytochrome *c*
Cytochrome c_1
absorption maximum, LVI, 503

complex III, LVI, 580, 582, 583, 587, 590, 591
in complex III, LIII, 36, 37, 85, 86, 89, 96
in complex I–III, LIII, 7, 8
complex V, LVI, 580
concentration determination, LIII, 127, 128
in cytochrome bc_1 complex, LIII, 116
difference spectrum, LIII, 38
EPR characteristics, LIII, 120, 121; LIV, 136, 137
extinction coefficient, LIII, 189
heme *c*, LIII, 189, 190
heme-deficient mutants, LVI, 568
kinetic properties, LIII, 190, 191
mammalian, CD spectrum, LIV, 277, 278
from mammalian heart, LIII, 181–191
midpoint redox potential, LIII, 97, 190
molecular weight, LIII, 225–227, 229
polypeptide composition, LIII, 226, 227
preparation, LIII, 183–187, 191
properties, LIII, 188–191
protease lability, LIII, 227–229
purification, LIII, 222–225
purity, LIII, 225
from *Saccharomyces cerevisiae*, LIII, 222–231
spectral properties, LIII, 126–128, 188, 189, 226
stability, LIII, 229
subunits, LIII, 188
amino acid composition, LIII, 189
yeast, LVI, 41

Cytochrome c_2
CD spectra, LIV, 276, 277, 278
membrane vesicles, LVI, 382
from *Rhodospirillum rubrum*, LIII, 209, 210

Cytochrome c_3
assay, LIII, 616
from *Desulfovibrio gigas*, LIII, 616
CD spectrum, LIV, 276, 278
from desulfovibriones, LIII, 613
from *Desulfovibrio vulgaris*, LIII, 210
high molecular weight species
activity, LIII, 628
amino acid composition, LIII, 627
presence in *Desulfovibrio* strains, LIII, 632
prosthetic groups, LIII, 627
purification, LIII, 622, 623, 625
spectral properties, LIII, 628
in hydrogenase reactions, LIII, 287
low molecular weight species
amino acid composition, LIII, 627
presence in *Desulfovibrio* strains, LIII, 632
prosthetic groups, LIII, 626, 627
purification, LIII, 621–623
redox properties, LIII, 626
role in respiratory chain, LIII, 627
proton magnetic resonance studies, LIV, 212
Soret CD spectrum, LIV, 260, 263, 264

Cytochrome *c*′
from *Chromatium*, LIII, 210
MCD studies, LIV, 298
properties, LIII, 207, 209, 210
from *Rhodospirillum rubrum*, LIII, 207

Cytochrome *c*–550, from *Paracoccus denitrificans*, LIII, 210

Cytochrome *c*–551
activity, LIII, 657
assay of *Pseudomonas* cytochrome oxidase, LIII, 647
crystallization, LIII, 658, 659
electron-transfer reactions, LIII, 661
extinction coefficient, LIII, 648
molecular weight, LIII, 660
from *Pseudomonas aeruginosa*, LIII, 210, 647
purification, LIII, 657–659
purity, LIII, 659
reaction with *Pseudomonas* cytochrome oxidase, LIII, 656
redox potential, LIII, 661
spectral properties, LIII, 659, 660
stability, LIII, 659

Cytochrome *c*–552
CD spectrum, LIV, 264, 276, 278
flavin linkage, LIII, 449, 450, 452
flavin peptide
fluorescence, LIII, 462
interactions, LIII, 465

isolation, LIII, 453, 454
in hydrogenase reactions, LIII, 287
Cytochrome *c*-553
CD spectrum, LIV, 277
in desulfovibriones, LIII, 632, 634
flavin linkage, LIII, 450
flavin peptide, isolation, LIII, 454
Cytochrome *c*-555, Soret CD spectrum, LIV, 260
Cytochrome *c*-556, from *Pseudomonas aeruginosa*, LIII, 647
Cytochrome *c*-557, from *Crithidia oncopelti*, LIV, 212
Cytochrome *c*-562, from *Escherichia coli*, LIII, 210
Cytochrome *c*-551–azurin, relaxation kinetics, LIV, 78–80
Cytochrome *c* + c_1, redox changes, determination, LV, 668
Cytochrome *c*-depleted mitochondria, *see* Mitochondria, cytochrome *c*-depleted
Cytochrome *c* oxidase, XLVI, 86; LII, 15; LIII, 54–79, *see also* Complex IV
absorbance coefficients, LIII, 127
activity, LIII, 54, 73
amino acid composition, LIII, 71, 73
assay, XXXII, 398, 400; LIII, 73, 74
α-band, extinction coefficients, LIII, 65, 66
from beef heart, LIII, 54–66
carbohydrate, LIII, 76
composition, LIII, 64, 76
concentration determination, LIII, 127
conformational composition, LIV, 269, 271
Cu *K*-edge spectra, LIV, 335, 336
Cu *K*-edge transitions, LIV, 339
determination of cytochrome *c* activity, LIII, 158–164
EPR signal properties, LIV, 135, 136
extinction coefficient, 68n, LIII, 71
in flash photolysis experiments, LIV, 100, 102, 105–110
in freeze-quench experiments, LIV, 92
heme stoichiometries, LIV, 145
interaction with azide, LIII, 194, 196, 197
with carbon monoxide, LIII, 194, 197–199
with cyanide, LIII, 192–194
with fluoride, LIII, 194, 200, 201
with formate, LIII, 201
with hydrogen sulfide, LIII, 194, 201
with hydroxylamine, LIII, 199
with isonitriles, LIII, 201
interference in superoxide dismutase assay, LIII, 385
IR studies, LIV, 309, 310, 312
ligand, LIII, 191–201
binding studies, LIV, 531, 532
inhibition mechanism, LIII, 193
from *Locusta migratoria*, LIII, 66, 72, 73
as marker enzyme, XXXI, 735, 739–742; XXXII, 393
MCD studies, LIV, 295–298, 301
mitochondrial, assay, LIV, 492, 493
from *Neurospora crassa*, LIII, 66–73
oxidation kinetics, LIV, 108–110
oxidation-reduction midpoint potentials, LIII, 65, 66
oxygen complex, LIV, 282
of *Paracoccus denitrificans*, LIII, 207
phospholipid-deficient, preparation, LIII, 60, 61
phospholipid-sufficient, preparation, LIII, 61
plasma membrane, XXXI, 90, 148, 249
from *Polytoma mirum*, LIII, 66, 72, 73
in protein rotation studies, LIV, 58
purification, LIII, 55–63, 74–76
chromatographic, LIII, 67–69, 75
purity, LIII, 76
from *Rattus norwegicus*, LIII, 66, 72, 73
reaction mechanism, LII, 39, 40
redox centers, LIII, 126, 127, 191
reductive titration, LIV, 124
resonance Raman spectroscopy, LIV, 240
from *Saccharomyces cerevisiae*, LIII, 73–79
Soret CD spectrum, LIV, 260, 262–264
spectrally invisible redox components, LIV, 398

spectral properties, LIII, 65, 66, 69, 72, 126, 127
stability, LIII, 63, 76
storage, LIII, 56
subunit composition, LIII, 54, 70–72, 76, 77
molecular weights, LIII, 70, 72, 73
properties, LIII, 79
purification, LIII, 76–79
temperature-jump relaxation kinetics, LIV, 76, 77
turnover number, LIII, 64, 71
visible CD spectra, LIV, 279–282
from *Xenopus muelleri*, LIII, 66, 72, 73
Cytochrome c_1 peptide
amino acid composition, LIII, 110
of complex III
characterization, LIII, 82, 83
isolation, LIII, 83, 85
selective labeling, LIII, 108, 109
Cytochrome *c* peroxidase, LIII, 156
interference in superoxide dismutase assay, LIII, 385
in redox potentiometry, LIV, 419
Cytochrome *c* reductase, XXXIV, 301, *see also* NADH dehydrogenase
assay, XXXI, 19
as marker enzyme, XXXI, 735, 743, 745
in plasma membranes, XXXI, 165, 166
Cytochrome *c*-type
bacterial, properties, LIII, 207, 209–212
effector, of nitric oxide reductase, LIII, 645, 646
from *Pseudomonas perfectomarinus*, LIII, 645
Cytochrome *d*, LIII, 202, 203
from *Escherichia coli*, LIII, 207
from *Paracoccus denitrificans*, LIII, 207
Cytochrome *e*, *see* Cytochrome *c*
Cytochrome *f*
from chloroplasts, XXXII, 406–422
chromatography, XXXII, 413, 414
estimation, XXXII, 867
gel electrophoresis, XXXII, 410
properties, XXXII, 419
purification, XXXII, 415–419
redox titrations, LIV, 429
Cytochrome *h*, LII, 15
prosthetic group, LII, 5
Cytochrome *m*
assay, LII, 169
crystallization, LII, 183
gel electrophoresis, LII, 175
heme, concentration determination, LII, 175
isolation and purification
flow chart, LII, 174
procedure, LII, 180–183, 278
reagents, LII, 175
optical properties, LII, 184, 185
production during *P. putida* growth cycle, LII, 171
purity criteria, LII, 184
stability, LII, 183, 184
substrate removal, LII, 183
Cytochrome *o*, LII, 15
from *Acetobacter suboxydans*, LIII, 208
from *Escherichia coli*, LIII, 207
from *Micrococcus pyogenes* var. *albus*, LIII, 207
prosthetic group, LII, 5, 15
from *Vitreoscilla*, LIII, 209
Cytochrome oxidase, XLIV, 371, 372, 545, *see also* Cytochrome *c* oxidase
abnormal composition, LVI, 565
absorption spectra, LVI, 693
antibody, LVI, 227
affinity chromatography, LVI, 690, 700
conjugation to ferritin, LVI, 710, 711
specificity, LVI, 690
assay, LVI, 101, 102
assembly, inhibition, LVI, 55
biogenesis, blockage, LVI, 41
in biomembrane reconstitution, XXXII, 554
bovine heart, cross-reaction with rat liver enzyme, LVI, 706, 707
components, LVI, 12, 40, 580, 583, 584
coupling to Sepharose 4B, LVI, 698, 699
cross-linking, LVI, 637, 641

as diagnostic enzyme, XXXI, 334, 337, 338, 361, 393
distribution on inner membrane-matrix particle, LVI, 686
on inverted inner membrane vesicles, LVI, 689
electron spin resonance spectra, XLIX, 405
free subunit, labeling kinetics, LVI, 54
HeLa cell, LVI, 78
hepatoma mitochondria, LV, 88
inner membrane, LV, 101, 103
interference in cytochrome P–450 difference spectrum measurements, LII, 213
in intestinal mucosa, XXXII, 667
Keilin-Hartree preparation, LV, 126
labeled immunoglobulin, LVI, 228
in liver-cell fractions, XXXI, 97, 208, 209, 262
localization, LVI, 707
luciferase-mediated oxygen assay, LVII, 226
in lung lamellar bodies, XXXI, 423
magnetic circular dichroism, XLIX, 170
as marker enzyme, XXXII, 15; LV, 5
membrane potential, LV, 595
mit^- mutants, LVI, 15
in mitochondria, XXXI, 24, 52, 326, 327
mitochondrial mRNA, LVI, 10, 11
mitochondrial subfractions, LV, 94, 104
monomeric complex, LVI, 635
mutants, LVI, 105
polypeptides, screening, LVI, 602, 603
of *N. crassa*, LV, 26, 148
nuclear mutations, LVI, 13
oxy form, spectrum, LII, 36
oxygen affinity, LVII, 223
planar membranes containing, LV, 755, 756
in plasma membrane, XXXI, 100
polypeptides, LVI, 592–596, 637
molecular weights, LVI, 595, 606
properties, LVI, 693, 694
proteoliposomes, LV, 761, 762, 765, 771
reconstitution, LV, 699, 703, 706–710
of smooth muscle mitochondria, LV, 62
submitochondrial particles, LV, 109
subunits
labeling kinetics, LVI, 52–54
resolution, LVI, 602–606
temperature-sensitive mutants, LVI, 139
turnover number, LVII, 223
vesicle
activity, LV, 703, 705, 709
preparation, LV, 749
viscosity barrier centrifugation, LV, 133, 134
yeast
components, LVI, 41
isolation, LVI, 44, 45, 49, 50
Cytochrome P–420, LII, 123
in cytochrome m° samples, LII, 183
denatured product, LII, 153, 215
prevention of formation, LII, 215
Cytochrome P–420_{cam}, carbon monoxide binding, IR studies, LIV, 310
Cytochrome P–448
absorption spectra, LII, 122
aryl hydrocarbon hydroxylase activity, LII, 399
assay, LII, 117
difference spectra, LII, 274
extinction coefficient, LII, 117, 122
immunologic properties, LII, 122
molecular weight, LII, 122
spin states, LII, 228
stability, LII, 121
Cytochrome P–450
absorption spectra, LII, 122
activity, LII, 46, 47, 109
from adrenocortical mitochondria, LII, 139
assay, LII, 65, 117, 122
by CO difference spectrum, LII, 138, 213, 270–272
axial ligands, XLIX, 173
from bacteria, LII, 151–166
CO binding, mechanism, LII, 272
in column eluate, LII, 93
compounds binding, LII, 264
content in hamster tissues, LII, 319
in microsomal protein, LII, 200

drug interactions, LII, 278
drug metabolism, LII, 67–69
drug toxicity, LII, 70, 71
effect of alcohols, LII, 263
of *in vivo* pretreatment, LII, 270–272
electron paramagnetic resonance, LII, 252–257
extinction coefficient, LII, 65, 110, 117, 122, 127
fatty acid hydroxylation, LII, 318–324
formaldehyde-generating reactions, LII, 297, 298
genetics, LII, 226–240
high spin, detection and quantitation, LII, 256, 257
immunologic properties, LII, 122
inactivation, by membrane lipid peroxidation, LII, 304
inhibitors, binding types, LII, 279
interaction with mediator-titrant, LIV, 426
in isolated MEOS fraction, LII, 362, 363
isolation, LII, 139, 362–364
kinetics, LII, 141
ligand binding, characteristics, LII, 262, 263
from liver microsomes, LII, 109–117
low-spin, detection and quantitation, LII, 254–256
magnetic circular dichroism, XLIX, 170, 173
MCD studies, LIV, 295, 296, 298
microsomes, LII, 46, 88, 89; LV, 101, 104
assay, LV, 102
millimolar difference extinction coefficient, LII, 215, 218
mitochondria, LV, 10, 11, 117
liver, LV, 21
renal, LV, 12
as mixed-function oxidase substrate, XXXI, 233
mixed-function oxidation systems, LII, 13
reconstitution, LII, 200–206
molecular weight, LII, 122, 141
multiple forms
detection by difference spectra, LII, 275–279
drug induction, LII, 399
effect on difference spectra, LII, 266, 267
ethanol induction, LII, 364, 365
identification with gel electrophoresis, LII, 328, 329
isolation from liver microsomes, LII, 109–117
molecular weights, LII, 275, 329
reactive intermediate ratios, LII, 228, 229
mutants, LVI, 122
NADPH-dependent reduction, kinetics, LII, 224–226
in nuclear membrane, XXXI, 290
optical properties, LII, 141
oxy form, spectrum, LII, 36
oxygen-donating compounds, LII, 407
peroxidase activity, LII, 407–412
assay, LII, 410–412
plasma membrane, XXXI, 90
prosthetic group, LII, 141
from *Pseudomonas putida*, LIII, 207, 209
from rabbit liver, LII, 109–117
from rat liver, LII, 117–123
reaction mechanism, LII, 39
reduced, types of complexes, LII, 272–275
reduction, LII, 45
for enzyme assay, LII, 91
of *Rhizobium*
assay, LII, 158
localization, LII, 157
molecular weight, LII, 166
multiple forms, LII, 158
optical properties, LII, 164–166
protein purity determination, LII, 158
subspecies separation, LII, 161
in smooth II microsome fraction, LII, 73
solubilization, LII, 141
spectral properties, XXXVII, 307–310
EPR, XXXVII, 308, 309
light-action spectrum, XXXVII, 309, 310
spin state
of free species, LII, 252
quantitation, LII, 267–270

stability, LII, 121, 140
steroid hydroxylations, catalyzed, LII, 377–388
in steroid-producing tissue, XXXVII, 295
preparation, XXXVII, 304–310
Cytochrome *P*–450$_{cam}$
assay, LII, 152, 153
bacterial, purification, LII, 152–156
carbon monoxide binding, IR studies, LIV, 310
CO difference spectrum, LII, 152
crystalline spectra, LII, 158
molecular weight, LII, 156
monooxygenase components, LII, 187
Mössbauer spectrum, LIV, 353–355, 357
prosthetic group, LII, 157
of *Pseudomonas putida*
difference spectra, LII, 259–261
electron paramagnetic spectra, LII, 252, 253
spin states, LII, 259–261, 267
specific activity determination, LII, 156
spectral properties, LII, 131, 155
Cytochrome *P*–450$_{scc}$
activity, LII, 14, 124, 131
in adrenal cortex mitochondria, LII, 14
assay, LII, 125, 126, 139
complexes, *g* value, LII, 254
immunologic properties, LII, 124, 131
inhibitors, LII, 132
kinetics, LII, 131
low-spin species
compounds causing shift, LII, 131
preparation, LII, 129
molecular weight, LII, 130
spectral properties, LII, 131
spin states
amine difference spectra, LII, 268
in bound species, LII, 253, 254
stability, LII, 131
substrate specificity, LII, 132
Cytochrome *P*–450$_{11\beta}$
activity, LII, 124, 131
assay, LII, 126, 127
complexes, *g* value, LII, 254
immunologic properties, LII, 124, 131
inhibitors, LII, 132
kinetics, LII, 131
low-spin species
compounds causing shift, LII, 131
preparation, LII, 130
molecular weight, LII, 130
spectral properties, LII, 131
stability, LII, 131
substrate specificity, LII, 132
Cytochrome *P*$_1$-450, *see* Cytochrome *P*-448
Cytochrome *P*–450-dependent ethanol oxidase, LII, 16
Cytochrome *P*–450LM
carbon monoxide binding, IR studies, LIV, 310
multiple forms, LII, 203
Cytochrome *P*–450LM$_2$
absorption spectrum, LII, 117
assay of hydroxylation activity, LII, 205
carbohydrate content, LII, 117
heme content, LII, 116
isolation and purification, LII, 111–114
molecular weight, LII, 117
in reconstituted mixed-function oxidase system, LII, 200–206
solubility, LII, 116
substrate specificity, LII, 117
Cytochrome *P*–450LM$_4$
absorption spectrum, LII, 117
carbohydrate content, LII, 117
heme content, LII, 116
isolation and purification, LII, 114, 115
molecular weight, LII, 117
solubility, LII, 116
substrate specificity, LII, 117
Cytochrome *P*–450 system, LII, 17
Cytocplasm, mitochondrial protein synthesis, LVI, 74–76
Cytogenetics, LVIII, 322
Cytokinin, in media, LVIII, 478
Cytological hybridization, *see* Hybridization, cytological
Cytopathogenicity, mixed culture, LVIII, 421–423

Cytoplasmic estrogen receptor
artifacts from extraction, XXXVI, 176–182
density gradient centrifugation, XXXVI, 166–176
Cytoplasmic membrane
from gram-negative bacteria, XXXI, 642–653
chemical composition, XXXI, 649
from yeast and fungi, XXXI, 609–626
Cytoplasmic steroid hormone receptor, XXXVI, 133–264
artifacts, XXXVI, 176–182
autoradiography, XXXVI, 135–156
binding site measurements, XXXVI, 193–198
filter technique, XXXVI, 234–239
density gradient centrifugation, XXXVI, 156–165
extraction and quantification, XXXVI, 187–211
gel electrophoresis, XXXVI, 222–228
gel filtration, XXXVI, 213–222
intracellular kinetics, XXXVI, 255–264
isoelectric focusing, XXXVI, 228–234
in mammary tumors, XXXVI, 248–254
physicochemical studies, XXXVI, 211–234
8 S androgen complex, XXXVI, 369–372
5 S receptor, purification, XXXVI, 362, 363
stability, XXXVI, 191–193
tissue sources, XXXVI, 188, 189
Cytosine
chromatographic mobility, LIX, 73
chromatographic separation, LI, 316
effect on aspartate carbamyltransferase, LI, 57
inhibitor of orotate phosphoribosyltransferase, LI, 116
substrate of cytosine deaminase, LI, 395, 399
Cytosine aminohydrolase, *see* Cytosine deaminase
Cytosine arabinoside
DNA inhibition, XL, 276
as mutagen, LVIII, 316
substrate of cytidine deaminase, LI, 407, 408, 411
Cytosine-2′,3′-cyclic-phosphate hydrolase, *see* Cyclic cytosine phosphodiesterase
Cytosine deaminase, LI, 394–401
assay, LI, 395
inhibitors, LI, 399, 400
kinetic properties, LI, 399
molecular weight, LI, 399
purification, LI, 396, 397
from yeast, LI, 394–401
Cytosine deoxyriboside, substrate of cytidine deaminase, LI, 404, 407
Cytosine nucleotide, separation by paper electrophoresis, LI, 323
Cytosol, XLVI, 336, 338, 453, 479
assay, XXXI, 410
ATP/ADP ratio, LV, 243, 244
cAMP receptor, XXXVIII, 380
enzyme marker, XXXII, 671
estrogen receptor preparation, XXXVI, 336–349
marker enzymes, XXXI, 740
of oviduct, preparation, XXXVI, 190, 293, 294
preparation, XXXVI, 177, 337
proteins of, XXXI, 45, 46
requirement, in DNA replication, XL, 43
5 S receptor, purification, XXXVI, 362, 363
Cytotoxicity assay, XXXIV, 216
Czerlinski and Eigen temperature-jump method, XLVIII, 308

D

2,4–D, *see* (2,4-Dichlorophenoxy)acetic acid
DAB, *see* 3,3′-Diaminobenzidine; 3,3′-Diaminobenzidine tetrahydrochloride
Dacron
cloth, preparation of aequorin, LVII, 280
support, for L-asparaginase, XLIV, 693
DAD, *see* Diaminodurol

DAHP synthetase, *see* 3-Deoxy-D-*arabino*-heptulosonate 7-phosphate synthetase
Dallner method, of microsome fractionation, XXXI, 228–230
Dalton, definition, XXXII, 35
Damköhler number, XLIV, 182, 183
Damping constant, XLIX, 78, 80
Damping term, LIV, 235
S-(2-Dansylaminoethyl)-protein adduct, XLVII, 419
N^{α}-Dansylarginine, mobility reference, XLVII, 56, 63, 64
Dansylation, of protein, XL, 159
Dansylaziridine, PMPS labeling, LVI, 619
N-Dansylaziridine, XLVII, 418, 419
Dansylcadaverine, factor XIII assay, XLV, 178, 180, 181, 185
Dansyl chloride, XLVI, 441
  adsorption on cellulose, LX, 715
  fluorescence labeling of initiation factor EF-2, LX, 715
  ^{3}H-labeled, use, XLVII, 203
Dansyl derivative
  fluorescent donor, XLVIII, 362, 363
  special properties, XLVIII, 362
  structural formula, XLVIII, 362
Dansylglycine, in modification of tRNA, LIX, 156
N^{ε}-Dansyllysine, elution profile marker, XLVII, 234
Dansyl sulfonic acid, mobility reference, XLVII, 57
DAPI, *see* 4′,6-Diamidino-2-phenylindole
DAPP, *see* Diadenosine pentaphosphate
Darco G, for steroid adsorption, XXXVI, 40
Darco G–60, XXXII, 88; LI, 189
  in Schiff reagent purification, XXXII, 97
Dark mutant, LVII, 130, 131
Dark reaction, photoaffinity labeling, XLVI, 107, 111
Darkroom
  for autoradiography, XXXVII, 159; LVIII, 280, 281
  safelight, LVIII, 281
Datalab DL 102A signal averager, LIV, 54
Daucus, cell cultures, XXXII, 729
Daunomycin, XLIII, 144, 318, 319
  DNA, RNA inhibitor, XL, 281
Dawe Instruments Sonitype Probe, XXXII, 207
DB, *see* 2,3-Dimethoxy-5-methyl-6-decyl-1,4-benzoquinone
DBAE-cellulose, *see* *N*-[*N*′-*m*-Dihydroxyborylphenyl succinamyl]aminoethylcellulose
DBCT, *see* Dibutylchloromethyltin chloride
DBPO, *see* Dibenzoyl peroxide
DCCD, *see* Dicyclohexylcarbodiimide
DCCI, *see* Dicyclohexylcarbodiimide
dc component, XLVIII, 423, 424, 467
  elimination, XLVIII, 474
DCIP, *see* 2,6-Dichlorophenolindophenol
DCPIP, *see* 2,6-Dichlorophenolindophenol
dc polarography, *see* Direct current polarography
DCR, *see* Decidual Cell Response, progestin assay
dCTP deaminase, LI, 418–423
  activity, LI, 418
  assay, LI, 418–420
  cation requirement, LI, 422
  inhibitors, LI, 422
  kinetic properties, LI, 423
  molecular weight, LI, 423
  pH optimum, LI, 422
  purity, LI, 422
  from *Salmonella typhimurium*, LI, 418
  stability, LI, 422
  substrate specificity, LI, 422
DDA, *see* Dimethyldibenzylammonium ion
DDL, *see* 3,5-Diacetyl-1,4-dihydrolutidine
Deacetylcephaloglycin, XLIII, 117
Deacetylcephalosporanic acid, XLIII, 730
Deacetylcephalosporin C, XLIII, 117, 730
Deacetylcephalothin, XLIII, 730
Deacetylcolchicine, XLVI, 568
  synthesis, XLVI, 569
N-Deacetyl-*N*-methylcolchicine, metaphase preparations, LVIII, 170
Deacetylrifampin, XLIII, 305
Deactivation, of enzymes, cause, XLIV, 240

Deacylation, by alkaline hydrolysis, LIX, 270, 271, 273, 312
DEAE, *see* Diethylaminoethyl
DEAE-cellulose, *see* (Diethylaminoethyl)cellulose
DEAE-11 column, in preparation of deuterated NADH, LIV, 226
Deamidation, preparation of carboxyl derivatives, XXXIV, 35, 36
3,8-Deamino-3,8-diazidoethidium bromide, LVI, 660
DE-3936 antibiotic, LV, 445
3-Deazauridine 5′-triphosphate, inhibitor of cytidine triphosphate synthetase, LI, 89
Deblocking, XLVI, 200, 201
Debye-Hückel ion atmosphere, thickness, XLVIII, 434–435
Debye length, spin probes, LVI, 524, 525
Debye magneton unit, LIV, 253
Debye-Waller factor, LIV, 341, 345, 373, 374
2*S*, 9*R*, 10*R* -*trans* -2-Decalol, XLIV, 837
9*S*, 10*S* -2-Decalone, XLIV, 837, 838
(±)-*trans* -2-Decalone, XLIV, 837
Decanal
  assay of luciferase, LIII, 565, 566, 570; LVII, 136–138, 140, 205
    of luciferase reaction intermediate, LVII, 196
    of malate, LVII, 187
    of malate dehydrogenase, LVII, 183
    of oxaloacetate, LVII, 187
  in luciferase-mediated assay, LVII, 218, 221
  quantum yield, in bioassay, LVII, 192
  solution, preparation and storage, LVII, 184, 218
Decanoic acid
  internal standard, XLVII, 34
  quantum yield in bioassay, LVII, 192
1-Decanol, electron spin resonance spectra, XLIX, 383
Decanoylcarnitine, as enzyme inhibitor, XXXVII, 294
Decaprenol, LIII, 592
Decarboxylase, XLII, 457–487, *see also* specific types
β-Decarboxylase, XLVI, 132
α-Decarboxylation, XLVI, 427
Decarboxyluciferin, structural formula, LVII, 27
Decay hypothesis, *in vivo*, of steroid output, XXXIX, 314, 315
Dechlorogriseofulvin, XLIII, 234
Decidual Cell Response progestin assay, XXXVI, 457–459
Decontamination, LVIII, 39
Decoyenine, inhibitor, of GMP synthetase, LI, 224
n -Decyl aldehyde, in assay of protease, LVII, 200
α-D_2-Decynoyl-*N* -acetylcysteamine, XLVI, 159
3-Decynoyl-*N* -acetylcysteamine, XLVI, 158
3-Decynoyl thiolester, inactivator of hydroxydecanoyl thiolester dehydrase, LIII, 438
Dedimethylamino-4-aminoanhydrotetracycline, *see* *S* -Adenosylmethionine:dedimethylamino-4-aminoanhydrotetracycline *N* -methyltransferase
de Duve plot, of enzyme activities, XXXII, 671
DeeO liquid, XLIV, 157
Defatting, of tissue, for collagen studies, XL, 312
Deformamidoazidoantimycin A, LVI, 660
Deforming buffer, XXXIV, 112, 128
Deformino LL–AC541 antibiotic, XLIII, 258–260
Degalan bead, XXXIV, 754
Dehydrase, XLVI, 158, *see also* specific enzymes
Dehydratase, XLII, 301–338, *see also* specific types
Dehydrocholate, sodium salt, partial specific volume, XLVIII, 21
7-Dehydrocholesterol, in lipid bilayers, XXXII, 517, 538
Dehydrocycloheximide, XLIII, 158
Dehydroepiandrosterone, as testosterone precursor, XXXVI, 68, 69
Δ^7-Dehydroepiandrosterone, as steroid precursor, XXXIX, 251
Dehydroepiandrosterone sulfatase, multiple sulfatase deficiency, L, 474

Dehydrogenase, XXXIV, 229, 246; XLVI, 149, 240–249, *see also* specific type
  anodic activity of NADH, LVI, 449, 477, 478
  conformation studies, XLIX, 354, 357
  fluorometric determination of activity using resorufin, XLI, 53–56
  membrane-bound, ferricyanide method for determining sidedness, LVI, 229–233
  NAD-linked, XXXIV, 125, 126
  specific ligand affinity chromatography, XXXIV, 598
$\Delta^{1\text{-}2}$-Dehydrogenase, XLIV, 471
Dehydrogriseofulvin, XLIII, 234
Dehydroluciferin
  chromatographic properties, LVII, 26
  inhibitor of firefly luciferase light reaction, LVII, 4, 12, 33
  reaction with ATP, LVII, 4, 5
  separation on Sephadex gels, LVII, 29, 31, 35, 36
  spectral properties, LVII, 26
  structural formula, LVII, 4, 16, 27
  synthesis, LVII, 25
Dehydroluciferol
  chromatographic properties, LVII, 26
  spectral properties, LVII, 26
  structural formula, LVII, 16
  synthesis, LVII, 25
Dehydroluciferyl adenylate
  chromatographic properties, LVII, 26
  spectral properties, LVII, 26
  structural formula, LVII, 27
  synthesis, LVII, 25, 27, 28
Delesse's principle, in morphometric determinations, XXXII, 10
Delipidation, detergents, LVI, 737
Delrin, for Mössbauer spectroscopy sample cells, LIV, 376
Delta, definition, XLVIII, 109
Demagnetization recovery, XLIX, 339, 340
Demerol
  complex I, LVI, 586
  inhibitor of complex I, LIII, 13
    of complex I–III, LIII, 8
N-Demethylase
  in nuclear membrane, XXXI, 290
  plasma membrane, XXXI, 90
Demethylation
  heme-deficient mutants, LVI, 558
  of phosphatidylcholine, XXXV, 535, 536
  of sphingomyelin, XXXV, 535, 536
6-Demethyl-7-chlortetracycline, in media, LVIII, 112, 114
Demethyldecarbamylnovobiocin, XLIII, 304
O-Demethylpuromycin, XLIII, 509–513, *see also* Puromycin *S*-adenosylmethionine:*O*-demethylpuromycin *O*-methyltransferase
Demetric acid, XLIII, 336
Demineralization, of tissue, for collagen studies, XL, 312
Denaturation, XL, 306; XLVI, 108
  agent, chemical cleavage, XLVII, 147, 149
  conformation studies, XLIV, 369, 370
  for covalent coupling, XLIV, 12
  detergents, LVI, 738, 739
  in diffusion controlled systems, XLIV, 428
  reductive, reversibility studies, XLIV, 516–526
  resistance, XLIV, 383
  using high-pressure techniques, XLIX, 14–24
  using temperature, XLIX, 3–14
Dendraster excenticus, hatching enzymes, XLV, 372
Denier, definition, XLIV, 235
Denitrification, principle, LIII, 641, 642
4-Denitro-4-azidochloramphenicol, XLVI, 84
Densimeter, *see also* Viscometer-densimeter
  magnetic, design of basic unit, XLVIII, 30–32, 43, 45–50
Densitometry, after gel electrophoresis, XXXII, 78
Density
  crystal, measurement, LIX, 5
  determination, XLVIII, 14, 15
    factors affecting, XLVIII, 26–29
    using Cartesian diver, XLVIII, 23–29
    using magnetic balancing, XLVIII, 29–68

in viscometer-densimeter, XLVIII, 56
partial specific volume, XLVIII, 15
protein solution
determination, XLVIII, 161, 162
estimation, XLVIII, 160, 162
Density centrifugation, freeze-stop tissue fractionation, LVI, 203, 204
Density gradient
analysis, *see* Cesium chloride solution centrifugation; Glycerol gradient centrifugation; Sucrose gradient centrifugation
preparation, LIX, 5
Density gradient centrifugation
anomalous zone broadening, XXXI, 511, 512
of BCF_0-BCF_1, LV, 194
in brain nuclei isolation, XXXI, 461–463
of *E. coli* 0111a cell extracts, XXXI, 637
of estrogen-receptor proteins, XXXVI, 166–176
fractionation and recovery, XXXVI, 161, 163
of gonadal receptors, XXXVII, 184, 185, 187, 188
gradient preparation, XXXVI, 160, 161, 171
gradient solutions, XXXVI, 157–160
M. phlei membrane fragments, LV, 195, 196
in plant-cell fractionation, XXXI, 505–519
continuous-flow harvesting, XXXI, 518, 519
gradients, XXXI, 507–514
isopycnic separations, XXXI, 515–518
rate separations, XXXI, 509–519
quasi-elastic laser light scattering, XLVIII, 492, 493
results analysis, XXXVI, 163–166
sample preparation and layering, XXXVI, 161, 167–171
of steroid hormone receptors, XXXVI, 156–165
of submitochondrial membrane fragments, LV, 108, 109
sucrose gradients, XXXVI, 157–158
Density gradient method of hydrogen-exchange, XLVIII, 325
Dental burr, in rat brain surgery, XXXIX, 174, 175
Dental carie, resazurin in diagnostic test, XLI, 56
Dental wax, in crystal mounting procedure, LIX, 9
Deo regulon, LI, 438
Deoxo Gas Purifier, XXXI, 629
Deoxyadenosine
in media, LVIII, 66
substrate of deoxycytidine kinase, LI, 345
of nucleoside phosphotransferase, LI, 391
of purine nucleoside phosphorylase, LI, 521
2′-Deoxyadenosine
DNA inhibitor, XL, 276
substrate of adenosine deaminase, LI, 507, 511
of deoxynucleoside kinase, LI, 346
of nucleoside phosphotransferase, LI, 392
3′-Deoxyadenosine
substrate of adenosine deaminase, LI, 511
of nucleoside phosphotransferase, LI, 392
Deoxyadenosine diphosphate
activator of thymidine kinase, LI, 358
product of guanylate kinase, LI, 473
substrate of nucleoside diphosphokinase, LI, 375
Deoxyadenosine triphosphate
activator of ribonucleoside triphosphate reductase, LI, 258
of thymidine kinase, LI, 358
adenylate cyclase, XXXVIII, 160, 169, 173
colorimetric determination, LI, 247, 248
effector of ribonucleoside diphosphate reductase, LI, 236, 246
phosphate donor of adenylate kinase, LI, 464
of guanylate kinase, LI, 473
of pyrimidine nucleoside monophosphate kinase, LI, 329

of thymidine kinase, LI, 358, 370
of uridine-cytidine kinase, LI, 306, 313, 320
of uridylate-cytidylate kinase, LI, 336
2′-[^{3}H]Deoxyadenosine 5′-triphosphate, preparation of 2′-deoxyadenosine-terminal tRNA, LIX, 129, 134
3′-[^{3}H]Deoxyadenosine 5′-triphosphate, preparation of 3′-deoxyadenosine-terminal tRNA, LIX, 128, 129
Deoxyadenosine triphosphate-agarose, XXXIV, 253–261
Deoxyadenosine triphosphate-Sepharose binding capacity, LI, 238
purification of ribonucleotide diphosphate reductase, LI, 229, 230, 241
2′-Deoxyadenosylcobalamin, effector of ribonucleoside triphosphate reductase, LI, 258
2′-Deoxyadenylate, radiolabeled, determination, LI, 238, 239
5′-Deoxyadenylate, product of deoxynucleoside kinases, LI, 346
Deoxyadenylate kinase, *see* Adenylate kinase
6-Deoxyaldohexose, qualitative analysis, XLI, 18
3-Deoxyaldulosonic acid, detection on chromatograms, XLI, 96
5′-Deoxy-5′-aminothymidine 3-*O*-thiophosphate 5′-*O*-thymidine ester, XXXIV, 606
2′-Deoxy-2′-azidocytidine 5′-diphosphate, inactivation, XLVI, 326
6-Deoxy-6-bromoconduritol B epoxide, synthesis, XLVI, 374–376
Deoxycarnitine, carnitine-acylcarnitine translocase, LVI, 378
2-Deoxy-2′-chlorocytidine 5′-diphosphate, inactivation, XLVI, 325, 326
Deoxycholate, *see* Deoxycholic acid
Deoxycholic acid, *see also* Potassium deoxycholate; Sodium deoxycholate
alanine carrier isolation, LVI, 431
assay of cytochrome oxidase, LIII, 74
ATPase, XXXIV, 570
ATPase isolation, XXXII, 281, 282; LV, 332
for ATPase purification, XXXII, 291, 292, 295–297
calcium ion pump, LV, 702
complex V preparation, LV, 309
cross-linking, LVI, 633
cytochrome oxidase, LV, 699
CD spectrum, LIV, 281, 282
hydrophobic protein preparation, LV, 714
inhibitor, of complex II, LIII, 25
of dihydroorotate dehydrogenase, LI, 62
isolation of polysomes, LIX, 365
of ribosomes, LIX, 505, 526
lysis of mitochondria, LVI, 24
membranes, LVI, 748, 749
micelles, ionic strength and pH effects, LVI, 746
potassium salt, preparation, LIII, 74
preparation of F_1 and membrane factor, LV, 743
of mitochondrial enzyme complexes, LIII, 3, 6
purification of cytochrome oxidase, LIII, 74, 75
reagent, XXXII, 477
recrystallization procedure, LIII, 3, 4, 56, 74
sodium salt, partial specific volume, XLVIII, 21
solubilization of lipid-depleted microsomal preparation, XXXV, 119
Deoxycoformycin, inhibitor of adenosine deaminase, LI, 506, 507
Deoxycorticosterone
adsorption to various materials, XXXVI, 93, 94
bioassay, XXXVI, 458, 465
cytochrome *P*–450 assay, LII, 125, 126, 139
cytochrome *P*–450 purification, LII, 127, 128
derivatives, XXXVI, 20
hydroxylation, LV, 10
metabolism, in adrenal gland, LII, 124
mitochondrial hydroxylation, XXXVII, 296, 301–304
by cytochrome P_{450}, XXXVII, 307
as 5α-oxidoreductase substrate, XXXVI, 473
PBP affinity, XXXVI, 125

protein-binding assay, XXXVI, 36
radioimmunoassay, XXXVI, 20, 21
R_f value, LII, 126
stabilizer of purified cytochrome P-$450_{11\beta}$, LII, 131, 132
Deoxycorticosterone ethyl diazomalonate, synthesis, XXXVI, 418, 419
Deoxycorticosterone hemisuccinate, XXXIV, 6
Deoxycorticosterone 21-hemisuccinate, in progesterone receptor chromatography, XXXVI, 207, 208
Deoxycorticosterone 11β-hydroxylase, *see* Cytochrome P–$450_{11\beta}$
Deoxycortisol
cytochrome P-$450_{11\beta}$, LII, 131, 132
protein-binding assay, XXXVI, 36
Deoxycytidine
in media, LVIII, 67
substrate of nucleoside phosphotransferase, LI, 391
of thymidine kinase, LI, 370
2′-Deoxycytidine
substrate of cytidine deaminase, LI, 395, 411
of deoxycytidine kinase, LI, 337
of deoxynucleoside kinase, LI, 346, 352
Deoxycytidine-deoxyadenosine kinase, LI, 346–354
assay, LI, 346, 347
isoelectric point, LI, 353
kinetic properties, LI, 353, 354
molecular weight, LI, 353
pH optimum, LI, 353
purification by affinity chromatography, LI, 348–350
Deoxycytidine diphosphate
activator of thymidine kinase, LI, 358
product of pyrimidine nucleoside monophosphate kinase, LI, 321
radiolabeled, determination, LI, 228, 229
substrate of nucleoside diphosphokinase, LI, 375
Deoxycytidine kinase, LI, 337–345
activators, LI, 345
activity, LI, 337
assay, LI, 337–339
from calf thymus, LI, 337–345
cation requirement, LI, 345
isoelectric point, LI, 345
kinetic properties, LI, 345
molecular weight, LI, 345
regulation, LI, 345
stability, LI, 345
substrate specificity, LI, 344, 345
Deoxycytidine triphosphate
activator, for ribonucleoside triphosphate reductase, LI, 258
of thymidine kinase, LI, 358
inhibitor of deoxycytidine kinase, LI, 345
of deoxycytidylate deaminase, LI, 418
phosphate donor of pyrimidine nucleoside monophosphate kinase, LI, 329
of thymidine kinase, LI, 358
of uridine-cytidine kinase, LI, 306, 313, 320
stimulator of ribonucleoside diphosphate reductase, LI, 236
substrate of dCTP deaminase, LI, 418
of deoxycytidylate deaminase, LI, 417
Deoxycytidine triphosphate deaminase, *see* dCTP deaminase
Deoxycytidylate
radiolabeled, determination, LI, 238
substrate of deoxycytidylate deaminase, LI, 412
of pyrimidine nucleoside monophosphate kinase, LI, 321, 329
of uridylate-cytidylate kinase, LI, 332, 336
2′-Deoxycytidylate, product of deoxynucleoside kinases, LI, 346, 352
5′-Deoxycytidylate, product of deoxycytidine kinase, LI, 337
Deoxycytidylate deaminase, LI, 412–418
amino acid analysis, LI, 417
assay, LI, 413
kinetic properties, LI, 417
molecular weight, LI, 417
purification, LI, 413–417
purity, LI, 417
regulation, LI, 412, 417, 418
stability, LI, 417
substrate specificity, LI, 417

subunit structure, LI, 417
from T2-infected *E. coli*, LI, 412–418
Deoxycytidylate kinase, from calf thymus, LI, 332
2-Deoxy-2,3-dehydro-*N*-acetylneuraminic acid crystallization, L, 70
3′-Deoxydihydrostreptomycin 6-phosphate, XLIII, 634
2-Deoxygalactose, XLVI, 26
2-Deoxyglucose, for petite-negative yeast mutant enrichment, XXXII, 839, 840
2-Deoxy-D-glucose, glycoprotein incorporation of sugars, L, 188
2-Deoxyglucose 6-phosphate, preparation, XLI, 99
Deoxyglucosylthymine, inhibitor of uridine phosphorylase, LI, 429
Deoxyguanosine
absorption maximum, LI, 525
in media, LVIII, 67
substrate, of deoxycytidine kinase, LI, 345
of nucleoside phosphotransferase, LI, 391
of purine nucleoside phosphorylase, LI, 521, 529, 530, 537
2′-Deoxyguanosine
product, of deoxynucleoside kinase, LI, 346
radiolabeled, determination, LI, 239
substrate, of deoxynucleoside kinase, LI, 346, 352
Deoxyguanosine-deoxyadenosine kinase, LI, 346–354
assay, LI, 346, 347
isoelectric point, LI, 353
molecular weight, LI, 353
pH optimum, LI, 353
purification by affinity chromatography, LI, 348–350
Deoxyguanosine diphosphate
activator of thymidine kinase, LI, 358
product of guanylate kinase, LI, 473, 483
substrate of nucleoside diphosphokinase, LI, 375
Deoxyguanosine monophosphate, guanylate cyclase, XXXVIII, 199
Deoxyguanosine triphosphate, XXXIV, 4
activator of ribonucleoside triphosphate reductase, LI, 258
of thymidine kinase, LI, 358
as ligands, XXXIV, 482
phosphate donor of adenylate kinase, LI, 464
of thymidine kinase, LI, 358
of uridine-cytidine kinase, LI, 306, 313, 320
stimulator of ribonucleoside diphosphate reductase, LI, 236
Deoxyguanosine triphosphate-Sepharose, structural formula, LI, 253
Deoxyguanylate
substrate of deoxythymidylate phosphohydrolase, LI, 289
of guanylate kinase, LI, 473, 483, 488, 489
Deoxyhemoglobin
electrode calibration, LVI, 555
preparation, LII, 458
spectral properties, LII, 459–463
3-Deoxyheptonic acid, methyl ester, preparation, XLI, 94
3-Deoxyheptonic acid 7-phosphate, preparation, XLI, 98
3-Deoxy-D-*arabino*-heptulosonate 7-phosphate synthase, labeling, XLVI, 139
3-Deoxy-D-*arabino*-heptulosonate 7-phosphate synthetase, XXXIV, 394, 397
phenylalanine-sensitive, XXXIV, 397
tyrosine-sensitive, XXXIV, 397
3-Deoxy-D-*arabino*-hept-2-ulosonic acid
ammonium salt, preparation, XLI, 94, 95
estimation with thiobarbituric acid, XLI, 32
3-Deoxy-D-*arabino*-hept-2-ulosonic acid 7-phosphate, preparation, XLI, 98
3-Deoxy-D-*erythro*-hex-2-ulosonic acid
estimation with thiobarbituric acid, XLI, 32
preparation, XLI, 96
3-Deoxy-D-*threo*-hex-2-ulosonic acid
ammonium salt, preparation, XLI, 95, 96
estimation with thiobarbituric acid, XLI, 32

3-Deoxy-D-*erythro*-hex-2-ulosonic acid 6-phosphate, preparation, XLI, 97
Deoxyinosine
substrate of nucleoside deoxyribosyltransferase, LI, 447
of purine nucleoside phosphorylase, LI, 521, 530
2′-Deoxyinosine, inhibitor of adenosine deaminase, LI, 507
3-Deoxy-2-ketoaldonic acid
estimation, XLI, 32–34
phosphorylation, XLI, 97–99
preparation, XLI, 94–97
D-4-Deoxy-5-ketoglucarate hydro-lyase
assay, XLII, 273, 274
chromatography, XLII, 275
inhibition, XLII, 276
molecular weight, XLII, 275
properties, XLII, 275, 276
in pseudomonads, XLII, 272
purification, XLII, 274, 275
Deoxynucleoside, storage, LI, 338
2′-Deoxynucleoside 5′-diphosphate, XLVI, 321
Deoxynucleoside kinase, LI, 346–354
assay, LI, 346, 347
kinetic properties, LI, 353, 354
from *Lactobacillus acidophilus*, LI, 346–354
properties, LI, 352, 353
purification, LI, 347–351
stability, LI, 352, 353
substrate specificity, LI, 352
Deoxynucleotide, separation on anion-exchange paper, LI, 338, 347
3-Deoxyoctulosonic acid, 5-*O*-substituted, estimation with thiobarbituric acid, XLI, 33
3-Deoxy-D-*manno*-octulosonic acid
estimation
with diphenylamine, XLI, 33
with thiobarbituric acid, XLI, 32
Deoxyribonuclease, XXXIV, 4, 468
adenylate cyclase purification, XXXVIII, 159, 172
cAMP-binding protein preparation, XXXVIII, 369
in cell-adhesion studies, XXXII, 601
digestion of chromosomes and nuclei, XL, 87
isolation of acidic chromatin proteins, XL, 174
membrane preparation, LV, 782, 803
membrane vesicle preparation, LVI, 381, 383, 402
pancreatic, XXXIV, 464, 468
sequence determination, XLVII, 91
phosphodiesterase purification, XXXVIII, 251
preparation of aminoacyl-tRNA synthetase, LIX, 314
of *Bacillus* ribosomes, LIX, 439
of CTP(ATP):tRNA nucleotidyltransferase, LIX, 123
of *E. coli* ribosomes, LIX, 444, 445, 451, 554
of hydrogenase/thiosulfate-sulfite reductases extract, LIII, 618
of MS2 RNA, LIX, 299
of polysomes, LIX, 355, 365
protein digestion, concentration and incubation period, XL, 38
in protein isolation, LII, 153, 160, 178
protoplast ghost preparation, LV, 181
purification of adenylate nucleosidase, LI, 268
of dihydroorotate dehydrogenase, LI, 61
of double-stranded ribonucleic acid, LX, 552
of ω-hydroxylase, LIII, 358
of *P*-hydroxybenzoate hydroxylase, LIII, 546
of nitrate reductase, LIII, 348
of nitrite reductase, LIII, 643
of nitrogenase, LIII, 316, 318, 322, 327
of *Nitrosomonas* ubiquinone, LIII, 639
of *Pseudomonas* cytochrome oxidase, LIII, 650
of rubredoxin, LIII, 342
of rubredoxin:NAD oxidoreductase, LIII, 619
of uridine-cytidine kinase, LI, 311
for purple-membrane isolation, XXXI, 670
spheroplasts, LVI, 175
tissue culture, LVIII, 125, 126

Deoxyribonuclease I, RNase-free, preparation of *E. coli* crude extract, LIX, 194
Deoxyribonucleic acid, XXXIV, 4, 5; XLVI, 90, 171, 335, 358
λ*pgal* 25*Sam* 7, preparation, XXXVIII, 372–374
alkali denaturation, procedure, LIX, 840
of animal cells, preparation, XXXVI, 296–298
assay, XXXIV, 464; XXXVI, 301–303
association with protein, in preparation of nucleoprotein complexes, XL, 227
binding of phosphorylated acidic chromatin proteins, XL, 181
comparison of procedures, XL, 190
measurement, XL, 187
bone content, XL, 328
in cell calcium studies, XXXIX, 521
cell content, LVIII, 148
in cell cycle, LVIII, 24246
centrifugal sedimentation, XXXI, 722, 723, 727
in chromatin, XXXI, 278, 279
collagen content, measurement, XL, 365
complicating binding, XXXIV, 205
cytoplasmic receptor binding, XXXVI, 188
defined sequences, XXXIV, 6
determination, protocol, LVIII, 146–149
electron microscope cytochemistry, XL, 28
extraction, LVIII, 406–409
in fat-cell nuclei, XXXI, 65
fragment maps from three strains of yeast, LVI, 183
geometric parameters, XL, 223
Hirt procedure, LVIII, 406–409
injection into cells, LVIII, 412
isolation, from human adenovirus, LVIII, 426
lambda, XXXIV, 466
linear duplexes, LVIII, 411
in liver cell fractions, XXXI, 208, 209
localization
labeling conditions, XL, 12
techniques, XL, 28
mapping of promoter sites, LIX, 846, 847
melting, thermodynamic data, XLIX, 19
messenger, in polypeptide synthesis, LIX, 850, 851
methylation, XL, 186
in mitochondria, XXXI, 24
mitochondrial
cloning, LVI, 5
DAPI staining, LVI, 729, 731–733
map for yeast, LVI, 194, 195
percent of total, LVI, 146
replication, LVI, 6
D loop, LVI, 6
size, LV, 9; LVI, 4
transcripts, LVI, 7
use of mapping in membrane research, LVI, 194–197
value of sequencing, LVI, 197
in nucleus, XXXI, 23, 24
petite mitochondrial
arrangement, LVI, 156
consolidation of map, LVI, 188–191
construction of overlap map, LVI, 188
hybridization experiments, LVI, 191, 192
in plant cells, XXXI, 495, 516, 519
isolation, XXXI, 545
in plasma membrane, XXXI, 87, 101
poly(U,G) strand separation, LIX, 840
protein binding, XLVIII, 295–299
purification, XXXIV, 499–502
receptor binding, XXXVI, 307–313
relaxed circles, LVIII, 411
removal, in SE preparation of nonhistone chromosomal protein, XL, 152
replication, from HeLa cells, nuclear system, XL, 41
replication-chromosome-condensation cycle, XL, 45
requirements for transformation, XLIII, 49
restriction fragment assay, LVIII, 434, 435
separation, LVIII, 411
single-stranded, XXXIV, 6

staining
fluorescent, LVIII, 246
procedure, LVIII, 240
in steroid receptor studies, XXXVI, 300
structure
relationship with CD spectra, XL, 232
saturation-transfer studies, XLIX, 496, 497
supercoiled, LVIII, 411
synthesis
assay, XL, 42
inhibitors, XL, 275, 280
in mammary epithelial cells, XXXIX, 451
rhythmicity, XXXVI, 477, 480
somatomedin effects, XXXVII, 94
template, LX, 323
thymidine incorporation, LVIII, 251, 252
transcription, assay, XXXVIII, 372
viral, *see* Virus, DNA; specific types
Deoxyribonucleic acid-ribosome complex
electron microscopy, LIX, 841–843
mapping of binding sites, LIX, 844–846
preparation, LIX, 841
Deoxyribonucleoside, *see also* specific type
phosphorylation, XXXVIII, 411, 412
Deoxyribonucleoside triphosphate, product of ribonucleoside triphosphate reductase, LI, 246
Deoxyribonucleotide, *see also* specific types
product of ribonucleoside diphosphate reductase, LI, 227
requirement, in DNA replication, XL, 43
thin-layer chromatography, XXXVIII, 30
Deoxyribose, in media, LVIII, 66
2-Deoxyribose alcohol, electrophoresis, L, 51
Deoxyribose 1-phosphate, substrate of uridine phosphorylase, LI, 425
2-Deoxy-D-ribose 1-phosphate, product of thymidine phosphorylase, LI, 437, 442
Deoxyribose 5-phosphate aldolase
regulation, LI, 438, 517
from *Salmonella typhimurium*, XLII, 276–279
assay, XLII, 276, 277
molecular weight, XLII, 279
properties, XLII, 279
purification, XLII, 277–279
7-Deoxyribosylxanthine, product of nucleoside deoxyribosyltransferase, LI, 453
9-Deoxyribosylxanthine, product of nucleoside deoxyribosyltransferase, LI, 452
2-Deoxystreptamine, XLIII, 444, 445, 449, 450
2-Deoxystreptamine derivative, XLIII, 124
Deoxythymidine
chromatographic separation, LI, 354, 355
substrate of deoxythymidine kinase, LI, 360, 364
of thymidine kinase, LI, 354, 358, 365, 370
Deoxythymidine 3′-*p*-aminophenylphosphate, XLVI, 358
Deoxythymidine 5′-*p*-aminophenylphosphate, XLVI, 358
Deoxythymidine 3′-*p*-aminophenylphosphate 5′-phosphate, XLVI, 358, 360
Deoxythymidine diphosphate
assay of nucleoside diphosphokinase, LI, 377
inhibitor of deoxycytidine kinase, LI, 345
substrate of nucleoside diphosphokinase, LI, 375
Deoxythymidine 3′,5′-diphosphate, XXXIV, 633; XLVI, 358
Deoxythymidine 3′,5′-diphosphate derivative, XXXIV, 183
Deoxythymidine kinase, LI, 360–365, *see also* Thymidine kinase
activity, LI, 360, 361
assay, LI, 361, 362
inhibitors, LI, 365
kinetic properties, LI, 364
molecular weight, LI, 365
pH optimum, LI, 364

purification, LI, 362–364
regulation, LI, 365
sedimentation coefficient, LI, 365
substrate specificity, LI, 364
Deoxythymidine-3′-(4-nitrophenylphosphate), in affinity gel preparation, LI, 366
Deoxythymidine 3′-*p*-nitrophenylphosphate 5′-phosphate, XLVI, 358
Deoxythymidine 3′-phosphate *p*-nitrophenyl ester, XLVI, 358, 360
Deoxythymidine 5′-phosphate *p*-nitrophenyl ester, XLVI, 358
Deoxythymidine triphosphate
activator of ribonucleoside diphosphate reductase, LI, 238, 245
of ribonucleoside triphosphate reductase, LI, 258
inhibitor of adenylate deaminase, LI, 496
of cytidine deaminase, LI, 400
of cytosine deaminase, LI, 400
of dCTP deaminase, LI, 423
of deoxycytidylate deaminase, LI, 418
of deoxythymidine kinase, LI, 365
of thymidine kinase, LI, 358, 371
phosphate donor, for uridine-cytidine kinase, LI, 306, 313
stimulator, of ribonucleoside diphosphate reductase, LI, 236
Deoxythymidylate
chromatographic separation, LI, 354, 355
determination, LI, 362
inhibitor of cytidine deaminase, LI, 400
product of deoxythymidine kinase, LI, 360
5′-Deoxythymidylate, substrate of deoxythymidylate phosphohydrolase, LI, 285, 286
Deoxythymidylate-5′-phosphatase, LI, 285–290
activity, LI, 285, 290
from *Bacillus subtilis*, LI, 285–290
inhibitors, LI, 289
kinetic properties, LI, 289
molecular weight, LI, 290
pH optimum, LI, 289
purification, LI, 286–288
salt effects, LI, 289, 290
stability, LI, 290
substrate specificity, LI, 285, 288, 289
temperature optimum, LI, 289
Deoxythymidylate phosphohydrolase, *see* Deoxythymidylate-5′-phosphatase
Deoxythymine pyrophosphate, conformational study, XLIX, 354
6-Deoxy-2,3,5-tri-*O*-methyl-D-galactitol, acetylation, L, 4
Deoxyuridine
substrate of deoxythymidine kinase, LI, 364
of nucleoside phosphotransferase, LI, 391
of pyrimidine nucleoside phosphorylase, LI, 436
of thymidine kinase, LI, 358
of thymidine phosphorylase, LI, 441
of uridine phosphorylase, LI, 424, 429
2′-Deoxyuridine
product of cytidine deaminase, LI, 395
radiolabeled, determination, LI, 239
5′-Deoxyuridine, inhibitor of uridine-cytidine kinase, LI, 313
Deoxyuridine diphosphate
inhibitor of carbamoyl-phosphate synthetase, LI, 119
substrate of nucleoside diphosphokinase, LI, 375
2′-Deoxyuridine 5′-(6-*p*-nitrobenzamido)hexylphosphate, XXXIV, 522
Deoxyuridine triphosphate
inhibitor of carbamoyl-phosphate synthetase, LI, 119
of dCTP deaminase, LI, 423
phosphate donor, for uridine-cytidine kinase, LI, 306, 313, 320
product of dCTP deaminase, LI, 418
Deoxyuridylate
inhibitor of orotate phosphoribosyltransferase, LI, 163
of orotidylate decarboxylase, LI, 163

product of deoxycytidylate deaminase, LI, 412
substrate of deoxythymidylate phosphohydrolase, LI, 289
of thymidylate synthetase, LI, 90, 91
2′-Deoxyuridylate-agarose, XXXIV, 520–523
2′-Deoxyxanthylate, substrate of GMP synthetase, LI, 224
DEP, *see* Diethylpyrocarbonate
Dephospho coenzyme A, XXXIV, 270
Dephosphorylation, of translational components, LX, 522, 527–530, 532
Depolarization factor
average, definition, XLVIII, 370
dynamic, XLVIII, 371
dynamic transfer, definition, XLVIII, 373
static transfer, definition, XLVIII, 373
Depolarization ratio, XLVIII, 441; XLIX, 75–77, 81, 96; LIV, 238
for heme proteins, XLIX, 130
Depression plate, coating, with nonwetting agent, LIX, 4
Dermatan sulfate, LVIII, 268
multiple sulfatase deficiency, L, 474
Desalanine insulin, formation, XXXVII, 335
Desalting
by dialysis, XLVII, 199
by gel filtration, XLVII, 199, 207, 208
Desaspidin, uncoupling activity, LV, 465, 472
Δ^9-Desaturase enzyme, lack of, in yeast lipid mutants, XXXII, 820, 829
Desicote, XXXIX, 307
Desmearing, of neutron scattering data, LIX, 712
Desmosome
negative staining, XXXII, 31, 32
of plasma membranes, XXXI, 85
Desorption, and ionic strength, XLIV, 153
Destomycin, XLIII, 277
Desulfovibrio, hydrogenase reaction, LIII, 287
Desulfovibrio africanus, electron-carrier proteins, LIII, 632
Desulfovibrio desulfuricans
cytochrome c_3, CD spectrum, LIV, 276
electron-transfer components, LIII, 613, 632
ferredoxin, LIII, 629
oxygen-stable hydrogenase, LIII, 297, 310
sulfite reductase, spectral characteristics, LII, 439
Desulfovibrio gigas
cell extract preparation, LIII, 621, 622
cultural purity, maintenance, LIII, 614, 615
cytochrome c_3, LIV, 212
desulfoviridin, LII, 447
spectral characteristics, LII, 439
electron-transfer components, LIII, 613–634
rubredoxin, LIII, 344
stock culture, storage, LIII, 615
Desulfovibrione
cell extract preparation, LIII, 633
electron carriers, LIII, 613–634
purification of electron-carriers, general scheme, LIII, 633, 634
Desulfovibrio salexigens
cytochrome c_3, CD spectrum, LIV, 276
electron carrier proteins, LIII, 632
Desulfovibrio vulgaris
cytochrome c_3, LIII, 210; LIV, 212
desulfoviridin, LII, 447
spectral characteristics, LII, 439
electron-transfer components, LIII, 613, 632
hydrogenase, LIII, 290
rubredoxin, LIII, 344
Desulfoviridin, LIII, 622, 623, 625
spectral characteristics, LII, 439
Desulfurization, XLVII, 112, 113
Detection interference, XXXIV, 111
Detergent, *see also* specific type
alkyl ionic, properties and uses, LVI, 738–740
in ATPase isolation, XXXII, 281, 282
catonic, hydrocarbon/water interface, LV, 773
choice, XXXIV, 569, 570

effect on phospholipase activity, XXXII, 135, 136
as eluent, XXXIV, 666, 668, 710
general properties
amount to use, LVI, 737, 738
critical micelle concentration and micelle size, LVI, 736, 737
partial specific volume, LVI, 737
immunochemical isolation of cytochrome oxidase and cytochrome *c*, LVI, 45, 49
ionic, partial specific volume, XLVIII, 18
in media, LVIII, 70
micelles, ionic strength, LVI, 738
for microsome isolation, XXXI, 215–224
mixtures, for plant-nuclei extraction, XXXI, 559
in NMR studies of membranes, XXXII, 211
nonionic
partial specific volume, XLVIII, 19
with polyoxyethylene or sugar head groups, properties and uses, LVI, 740–745
removal, XLVIII, 8, 9
for nuclei isolation, XXXI, 254
partial specific volume, LVI, 737
for receptor solubilization, XXXIV, 655, 656
reconstitution studies, LV, 707
removal, LII, 113, 121, 123
solubilization of lipid-associated proteins, XLVIII, 11
stock solutions, storage and stability, LII, 145, 146
structures, LVI, 735
systems, ATPase, XXXIV, 566–571
trade names and sources, LVI, 741
use in bead, suspension, polymerization, XLIV, 176, 177
in gel electrophoresis of enzymes, XXXII, 82–85
Detergent-dialysis procedure, LV, 701–703
Detergent-dilution procedure, reconstitution studies, LV, 703, 704
Deuteration
of exchangeable protons, procedure, LIV, 155
specific, LIX, 676
Deuterium
assay
in protein, LIX, 645, 646
in RNA, LIX, 646, 647
exchange assay, of hydrogenase, LIII, 298, 300–304
labeling of phosphoenolpyruvate and pyruvate, XLI, 110–115
of pyruvic acid, XLI, 106–110
lock nucleus, LIV, 189
nucleus
coherent scattering length, LIX, 630, 672
scattering cross section, LIX, 672
resonance frequency, LIV, 189
solvent for nuclear magnetic resonance, XLIX, 255, 256
Deuterium lamp, LIII, 590, 591
in photometer calibration, LVII, 578
Deuterium oxide
antichaotropes, XXXI, 784–787
assay of hydrogenase, LIII, 301
in water samples, LIX, 644, 645
concentration determination, XLIX, 28, 29
as deuterium-field lock, LIV, 154, 155
effect on cytidine deaminase, LI, 404, 407
on dissociation of fatty acid synthase, XXXV, 70
as gradient, properties, XXXI, 508
impurities, XLIX, 255
infrared spectrum, LIV, 303
in IR studies, LIV, 304, 318
as lock nuclei source, LIV, 189
neutron scattering studies of ribosomal subunits, LIX, 655, 666
pH measurements, XLIX, 256
in picosecond continuum generation, LIV, 12
preparation of deuterated bacteria, LIX, 641–643
of deuterated NADH, LIV, 226
in reaction mechanism studies of guanine deaminase, LI, 517
recycling, LIX, 643, 644
Developer, photographic, XL, 9

Dexamethasone
effects $(Na^+ + K^+)$-ATPase, XXXVI, 435
on sodium transport, XXXVI, 444
for steroid-receptor binding studies, XXXVI, 235–248
Dextran, XXXIV, 338, 709
barrier solution, for erythrocyte ghost isolation, XXXI, 173
centrifugal layer filtration, LVI, 285
for countercurrent distribution, XXXI, 763, 764
effect of solvents on volume, XLIV, 22
efflux kinetic studies, LVI, 259
as electron microscopic tracer, XXXIX, 149
enzyme modification, XLIV, 310
in extraction medium, XXXI, 546, 599
gradient properties, XXXI, 508, 509
as hormone adsorbent, XXXVII, 34
in leukocyte isolation, LVII, 467, 470
osmometric studies, XLVIII, 75, 76
periodate method for activation, XXXIV, 82
as plant-cell protectant, XXXI, 41, 536
purification elongation factor, LX, 665, 666, 677, 690, 693, 699
Qβ replicase, LX, 633
ribosomal protein S1, LX, 458
purification of monoamine oxidase, LIII, 498, 499
separation of mitochondria from medium, LV, 204
support for coenzymes, XLIV, 871–873
for NAD^+, XLIV, 872
for $NADP^+$, XLIV, 872, 873
for penicillin acylase, XLIV, 765
properties, XLIV, 20
Dextran 110, hollow-fiber retention data, XLIV, 296
Dextran 500, viscosity barrier, LV, 131
Dextran antibody, XXXIV, 712
Dextranase, XXXIV, 720
immobilization, by adsorption, effects on activity, XLIV, 44
source, XLIV, 30
Dextran-coated charcoal, *see* Charcoal, dextran-coated
Dextran-polyethylene glycol system, in plasma membrane isolation, XXXI, 162
Dextran T40, XLIV, 189; XLVIII, 335
Dextran T70
differential molecular weight distribution, XLVIII, 103
number average molecular weight, XLVIII, 102
osmometric studies, XLVIII, 100–103
osmotic pressure second virial coefficient, XLVIII, 102
Dextran T500, for phase systems, XXXII, 638, 639
Dextrorphan-glass, XXXIV, 623
Dextrose, *see* Glucose
DFP, *see* Diisopropyl fluorophosphate
DHCH-arginine, *see* N^7-N^8-(1,2-Dihydroxycyclohex-1,2-ylene)-L-arginine
DHOase, *see* Dihydroorotase
DHT, *see* Dihydrotestosterone
Diabetes
glucose electrodes, LVI, 451
C-peptide levels, XXXVII, 344
1,2,5,6-Diacetone-D-mannitol, intermediate in phosphatidic acid synthesis, XXXV, 431, 434
N,N′-Diacetylchitobiose, XXXIV, 334
protein, attachment, L, 425
synthesis, Koenigs–Knorr reaction, L, 108, 109
Diacetylcolchicine-agarose, XXXIV, 627
N,N′-Diacetyl-3,5-diamino-2,4,6-triiodobenzoate, *see* Urografin
3,5-Diacetyl-1,4-dihydrolutidine, in formaldehyde assay, LII, 298
N^α,N^ε-Diacetyl-L-lysyl-D-alanyl-D-alanine, XLIII, 697
substrate, DD-carboxypeptidase activity, XLV, 611, 612
Diacetylmonoxime-acetic acid, assay of aspartate carbamyltransferase, LI, 51
Diacetyl reductase, *see* Acetoin dehydrogenase
1,5-Di-*O*-acetyl-2,3,4,6-tetra-*O*-methyl-D-glucitol, L, 24
1,2-Diacyl-L-glycerol-3-iodohydrin, intermediate in phosphatidic acid synthesis, XXXV, 436, 437

1,2-Diacyl-*sn*-glycerol-3-(β,β,β-trichloroethyl) carbonate, intermediate in synthesis of phosphatidylethanolamine analogues, XXXV, 509, 510
N,N′-Diacylhydrazide, XXXIV, 38
Diadenosine pentaphosphate
 assay of creatine kinase isoenzymes, LVII, 60
 inhibitor, of adenylate kinase, LVII, 59, 65
N′,*N*²-Diadenosyl pyrophosphate, formation, LV, 538
Diaflo XM–100A membrane, LII, 121
Diaflo XM–300A membrane, LII, 120
Dialkyl arylamine *N*-oxidase, LII, 17
Dialysis
 critical micelle concentration, LVI, 737
 device for reconstitution experiments, LV, 812
 equilibrium
  for measurement of ligand binding, XLVIII, 307
  nuclear magnetic resonance studies, XLIX, 325
 hydrogen exchange, XLVIII, 326, 327, 335
 of mitochondrial complexes, LIII, 6
 purification of cytochrome *c*, LIII, 130
 rapid, preparation of ribosomal protein, LIX, 453
 reconstitution of vesicles, LV, 371, 372, 715
  alanine transport, LVI, 434
 of spin-labeled sample, XLIX, 431, 441
 stripped membrane particles, LV, 796
 technique, LX, 147, 148
Dialysis membrane
 for equilibrium dialysis, LIX, 864
 preparation, XLIV, 877, 878
Dialysis photolysis, XXXVIII, 396, 397
Dialysis rack, LIX, 453
Dialysis tubing, LIII, 521
 cleaning procedure, LIV, 478, 479
 electrode current, LVI, 458
 preparation, LI, 415; LVII, 10
 for use with ribosomal proteins, LIX, 482, 483
Diamagnetic chemical shift, LIV, 194, 197
Diamagnetic compound, Mössbauer spectroscopy, LIV, 351, 352
Diamagnetic proton sample, sampling rate, LIV, 158
4′,6-Diamidino-2-phenylindole
 source, handling and storage, LVI, 729
 staining
  microscopy, LVI, 731
  postvital, LVI, 730
  significance of results, LVI, 731–733
  vital, LVI, 730, 731
 uses, LVI, 729
Diamine oxidase, LII, 10
Diaminobenzidine
 as hormone stain, XXXVII, 138
 rho⁺ mutants, LVI, 148
3,3′-Diaminobenzidine
 for catalase determination, LII, 495
 hydrogen donor, for cytochrome *P*–450 peroxidation function, LII, 411
 incubation medium containing, XXXI, 497, 499, 500
3,3′-Diaminobenzidine tetrahydrochloride, XLIV, 713, 912
1,4-Diaminobutylphosphonic acid, XLIII, 458
1,4-Diamino-2-butyne, XLVI, 162
L-2,4-Diaminobutyrate activating enzyme, XLIII, 579, *see also* Polymyxin synthetase:L-2,4-diaminobutyrate activating enzyme
L-2,4-Diaminobutyric acid, internal standard, XLVII, 34, 39
trans-Diaminodichloroplatinate, in derivatization of transfer RNA, LIX, 14
1,D-1,3-Diamino-1,3-dideoxy-*scyllo*-inositol phosphate, XLIII, 432
p,p′-Diaminodiphenylmethane, XLIV, 129, 132
3,3′-Diaminodipropylamine, XXXIV, 611
 in affinity chromatography, XXXVI, 105
 reaction with Sepharose, LV, 525
 as spacer, XXXIV, 80

Diaminodipropylamine B_{12}-agarose, XXXIV, 307
Diaminodipropylaminoagarose, XXXIV, 682
3,3′-Diaminodipropylaminosuccinyl-agarose, XXXIV, 657, 658
Diaminodurene, model redox reactions, LV, 543
Diaminodurol, LIV, 429, 431, 432, 433
 as redox mediator, LIV, 423
1,6-Diaminohexane, XLIV, 76, 77, 85, 118, 125, 127, 130, 458, 859–868
 in agarose derivatization, LI, 100, 269
 preparation of derivatized Sepharose, LVII, 147
Diaminohexane-agarose, XXXIV, 602
1,6-Diaminohexane-Sepharose 4B, in AMP nucleosidase purification, LI, 269
1,5-Diaminopentane, XXXIV, 589
Diaminopimelic acid, internal standard, XLVII, 34, 39
1,3-Diaminopropanol derivative, XLIV, 284
N,N′-Diaminopropylamine, XXXIV, 696
2,7-Diaminopurine, substrate of adenine phosphoribosyltransferase, LI, 565, 572
2,6-Diaminopurine ribonucleoside, substrate of adenosine deaminase, LI, 507
1,D-1,3-Diamino-1,2,3-trideoxy-*scyllo*-inositol 6-phosphate, XLIII, 432
Dianemycin
 cation fluxes, LV, 675
 as ionophore, LV, 443, 446
o-Dianisidine, XLIV, 460, 474, 476, 716
 assay of superoxide dismutase, LIII, 387
 glucose oxidase, LVI, 461
Diaphorase, XXXIV, 288, *see also* Dihydrolipoamide reductase
Diaphragm
 dissection, XXXIX, 85, 86
 of rat, anatomy, XXXIX, 84, 85
 in vitro preparations, XXXIX, 82–94
Diarginyl insulin, isolation, XXXVII, 337
Diarrhea virus, LVIII, 28
Diastase, XLIV, 108
1,5-Diazabicyclo[3.4.0]non-5-ene, XLVI, 243
Diazaquinone, reaction intermediate in chemiluminescence, LVII, 420
Diazidoethidium bromide, XLVI, 84
3,6-Diazido-10-methylacridinium chloride, LVI, 660
Diazinon, detection, XLIV, 658
3-Diazoacetoxymethylpyridine adenine dinucleotide, LVI, 656
Diazoacetylamphenicol, XLVI, 706
 synthesis, XLVI, 704, 705
Diazoacetyl compound, XLVI, 73–75
 absorption maximum, LVI, 663
 synthesis, XLVI, 93–95
Diazoacetylglycine methyl ester, XLVI, 240
Diazoacetyl-DL-norleucine methyl ester, XLVI, 240
O-Diazoacetyl-L-serine, XLVI, 414, 415, *see also* Azaserine
α-Diazoamide, XLVI, 97
2-Diazo-5α-androstan-17β-ol-3-one, synthesis, XXXVI, 415–416
Di-(*p*-azobenzenearsonic acid)-*N*-*t*-butyloxycarbonyl-L-tyrosine, XLVI, 494
Diazobenzenesulfonic acid, XLIV, 445
 preparation, LVI, 615, 621
α-Diazobenzylphosphonate, XLVI, 74
1,4-Diazobicyclo[2.2.2]octane, XLVII, 457
Diazocarbonyl compound, XLVI, 73
 absorption maxima, LVI, 663, 664
α-Diazocarbonyl derivatives, synthesis, LIX, 810–815
Diazo compound
 N-acylated, XLVI, 74
 quantitation, XLVI, 97
 triazole form, XLVI, 74
Diazoester, as inhibitors, XLVI, 35, 36
Diazoethane, ethylation of drugs and drug metabolites, LII, 332
Diazo group, concentration determination, XLIV, 385, 386
Diazoketone, XLVI, 74, 96, 640
 nitrophenyl-based, XLVI, 508–516
 synthesis, XLVI, 96
Diazoketone compound, XLVI, 76, 485, 486
Diazomalonyl chloride, XLVI, 95

Diazomalonyl compound, synthesis, XLVI, 95, 96
α-Diazomalonyl compound, XLVI, 74, 75
Diazomalonyl cyclic adenosine monophosphate, XLVI, 241
Diazomethane, XLVI, 40, 95, 200, 231, 391, 433
  ethereal, preparation, LII, 444
  fatty acid methylation, LII, 321
  hazard, LV, 510
  methylation of drugs and drug metabolites, LII, 332
2-Diazo-17α-methyl-5α-androstan-17β-ol-3-one, synthesis, XXXVI, 416–417
Diazomethyl ketone, XLVI, 199
  derivatives of amino acids and peptides, XLVI, 201
  synthesis, XLVI, 200
Diazomycin, XLIII, 336
p-Diazonium benzene[^{35}S]sulfonic acid, labeling of mitochondria, LVI, 594
Diazonium coupling, XLIV, 18, 38, *see also* Azo linkage
Diazonium ion, XLVI, 217, 578
Diazonium phenyl β-glycoside, XLVI, 517, 518
Diazonium reagent, XLVI, 358–362, 483, 486
Diazonium salt, XXXIV, 53–55, *see also* Azo linkage
  from aryl amines, XXXIV, 53, 54
  reactivity, LVI, 615
Diazonium 1*H*-tetrazole, for histidine modification, LIII, 137
Diazooxonorleucine
  binding studies, to amidophosphoribosyltransferase, LI, 178
    to cytidine triphosphate synthetase, LI, 83
  half of the sites reactivity, LI, 83
  inhibitor of amidophosphoribosyltransferase, LI, 178
    of FGAM synthetase, LI, 193, 201
    of GMP synthetase, LI, 217
6-Diazo-5-oxo-L-norleucine, XLVI, 35, 36, 414–416
  ^{14}C-labeled, XLVI, 424
  synthesis, XLVI, 416
5-Diazo-4-oxo-L-norvaline, XLVI, 35, 36, 415, 432–435
  synthesis, XLVI, 433, 434
3′-(4-Diazophenylphosphoryl)deoxythymidine 5′-phosphate, XXXIV, 106
5′-(4-Diazophenylphosphoryl)uridine 2′(3′)-phosphate, XXXIV, 185
  peptide, affinity-labeled, XXXIV, 187
21-Diazoprogesterone, synthesis, XXXVI, 417
Diazo steroid, in receptor protein studies, XXXVI, 413–419
Diazosulfanilic acid, XLVI, 64
α-Diazothiol ester, XLVI, 73
Diazotization, XLVI, 94, 98
  capacity assay, XLIV, 133
  method of making azo linkage, XXXIV, 20, 48, 49, 53–55, 71, 102–108, *see also* specific compounds
    glass, XXXIV, 71, 72
  with nitrous acid, XLIV, 88, 91
  with sodium nitrite, XLIV, 93, 115–117
Diazotrifluoropropionic acid, XLVI, 73
2-Diazo-3,3,3-trifluoropropionyl *N*-acetylcysteine methyl ester, photolysis, LVI, 665
Dibenz[*a,h*]anthracene dihydrodiol, chromatographic properties, LII, 294–296
Dibenz[*a,h*]anthracene 5,6-oxide, substrate of epoxide hydrase, LII, 199, 200
5-Dibenzosuberol, reaction conditions, XLVII, 410
o-Dibenzoylbenzene, detection of singlet oxygen, LVII, 499
1,4-Di-*O*-benzoyl-3-bromo-1,2,4-butanetriol, synthesis, XLVI, 392
cis-1,4-Di-*O*-benzoyl-2-butene-1,4-diol, synthesis, XLVI, 392
Dibenzoyl peroxide, XLIV, 110, 112–114
1,1′-Dibenzyl 4,4′-bipyridylium dichloride, as mediator-titrant, LIV, 409, 410, 423
1,4-Di-*O*-benzyl-3-bromo-2-butanone, synthesis, XLVI, 393
Dibenzyline, adenyl cyclase, XXXVIII, 173

Dibenzyl phosphate, intermediate in phosphatidylglycerol synthesis, XXXV, 467
α,α'-Dibromoacetic anhydride, XLVI, 674
Dibromoacetone, XLVII, 408
9,10-Dibromoanthracene, interceptor in chemiluminescent reaction, LVII, 520, 521
(1,2,4/3,5,6)-4,6-Dibromocyclohexanetetrol, synthesis, XLVI, 374, 375
Dibromofluorescein acetate, monomercurated, properties, XLVII, 415
1,3-Dibromopropane, XLVI, 122
Dibromotetrol D, XLVI, 374, 375
Dibromotriphenoxyphosphorane, intermediate in synthesis of phosphatidylcholine analogues, XXXV, 525
Dibutylchloromethyltin chloride
  availability, LV, 510
  biological properties, LV, 511
  isotopically labeled, LV, 511
  physical properties, LV, 477
  synthesis, LV, 510
Dibutylferrocene, model redox reactions, LV, 542–544
3,5-Di-*tert*-butyl-4-hydroxybenzylidinemalononitrile
  synthesis, LV, 468, 469
  uncoupling activity, LV, 465
Di-*tert*-butyl nitroxide
  anisotropic spectra, XLIX, 377
  electron spin resonance parameters
    isotropic, XLIX, 383, 422, 423
    solvent dependence, XLIX, 384, 422, 423
  structure, XLIX, 500
  triplet quencher, XLIX, 425
Di-(*sec*-butyl)peroxydicarbonate, XLIV, 175
Di-*tert*-butyl phosphate
  intermediate in phosphatidic acid synthesis, XXXV, 437–439, 457, 458
  silver salt, intermediate in phosphatidic acid synthesis, XXXV, 438
Di-*tert*-butyl phosphite, intermediate in phosphatidic acid synthesis, XXXV, 438
Dibutyryl-cyclic adenosine monophosphate
  effect on granulosa cell culture, XXXIX, 212, 213
    on growth, LVIII, 245
    on iodide transport, XXXVII, 258–262
N^2-2′-*O*-Dibutyryl-cyclic guanosine monophosphate
  preparation, XXXVIII, 406, 407
  protein kinases, XXXVIII, 338, 339
N^6-2′-*O*-Dibutyryl-8-mercapto-cyclic adenosine monophosphate, preparation, XXXVIII, 408, 409
Dicarboxylic acid, bis-*p*-nitrophenyl esters, XLVI, 505–508
1,3-Dicarboxymethylhistidine, XLVI, 423
Dicetyl phosphate, XLIV, 222, 227
2,6-Dichlorobenzoquinone-4-chloroimine, ethanolic, LII, 403
2,6-Dichlorobenzyl bromide, XLVII, 543
O-2,6-Dichlorobenzyl group, stability, XLVII, 542
O-2,6-Dichlorobenzyl-L-tyrosine, synthesis, XLVII, 543, 615
Dichlorodifluoromethane, tissue freezing, XXXVIII, 4
Dichlorodimethylsilane, XLVII, 395
  for silanizing glass, LII, 326
1,2-Dichloroethane, XLIV, 231; XLVII, 267, 326
  extraction, XLVII, 345
  support swelling, XLVII, 272, 298
4-Di-(2-chloroethyl)aminophenylacetic acid, LV, 515
Dichlorofluorescein, indicator for fatty acids, lipids and phospholipids on thin-layer chromatography, XXXV, 319, 324, 325, 327, 328, 406
L-*erythro*-2,2-Dichloro-*N*-[β-hydroxy-α-(hydroxymethyl)-*p*-nitrophenyl] acetamide, in photoaffinity labeling studies, LIX, 803
2,6-Dichloroindophenol, *see* 2,6-Dichlorophenolindophenol
Dichloroisoproterenol, adenylate cyclases, XXXVIII, 147
Dichloromethane, XLIV, 117, 126, 187, 198, 230, 838
  in corticosterone assay, XXXIX, 346

in 1,25-dihydroxyvitamin D_3 assay, LII, 391, 392
for gas chromatography, LII, 332
IR bandwidth, LIV, 311
solvent in peptide synthesis, XLVII, 550
synthesis of 2-carbethoxy-2-diazoacethydrazide, LIX, 814
of *Diplocardia* luciferin, LVII, 377
of ethyl-2-diazomalonylchloride, LIX, 813
of oleylpolymethacrylic acid resin, LIII, 214
1,3-Dichloro-5-methoxytriazine, XLIV, 141
enzyme immobilization, LVI, 491
1,3-Dichloromethyl methyl ether, toxicity, XLVII, 585
2,6-Dichlorophenolindophenol, XLVI, 50, 51
assay of acyl-CoA dehydrogenase, LIII, 503, 504
of D-lactate hydrogenase, LIII, 519
of succinate dehydrogenase, LIII, 27, 34, 466, 467
of succinate-ubiquinone reductase, LIII, 26
carboxamides, LV, 461
complex II, LVI, 586
electron acceptor
for dihydroorotate dehydrogenase, LI, 59, 60, 64
for erythrocyte cytochrome b_5 reductase, LII, 465
from NADPH-cytochrome P–450 reductase, LII, 90, 96
extinction coefficient, LII, 92; LIII, 21, 27, 34
Keilin-Hartree preparation, LV, 126
membrane vesicle isolation, LVI, 382, 383
polarography, LVI, 478
as mediator-titrant, LIV, 408
reduction
by complex I, LIII, 13
by NADH-cytochrome b_5 reductase, LII, 107
spectrophotometric determination, LIII, 21, 26, 27
substrate
of acyl-CoA dehydrogenases, LIII, 503
of adrenodoxin reductase, LII, 134
of complex II, LIII, 24
of NADH dehydrogenase, LIII, 18, 20, 413
2,4-Dichlorophenoxyacetic acid
as auxin, LVIII, 478
effect on plant growth, XXXII, 731
4,7-Dichloroquinoline, XLVI, 646
2,2-Dichlorovinyl dimethyl phosphate, inhibitor, cholinesterase, XLIV, 651, 656, 657
Dichromate solution, interference by, in assays utilizing UV difference spectra, LI, 424
Dicloxacillin, XLIV, 615
Dicoumarol
effect on oxidative phosphorylation, XXXI, 531
M. phlei membrane systems, LV, 185, 186
uncoupling activity, LV, 465, 471
Dictyosome, *see also* Golgi complex
in plant cells, XXXI, 489, 491, 518
Dictyostelium discoideum
agglutinin, *see* Lectin, from *Dictyostelium discoideum*
N-acetyl-D-galactosamine inhibition, L, 311
affinity chromatography, L, 309, 310
biological significance, L, 311, 312
hemagglutination assay, L, 307, 308
culture conditions, L, 308, 309
as model study system, XXXIX, 485–492
phosphodiesterase, XXXVIII, 244, 245
assay, XXXVIII, 245
properties, XXXVIII, 247, 248
purification, XXXVIII, 245–247
Dicumarol, *see* Dicoumarol
Dicyclocarbodiimide, XLVI, 200
Dicyclohexylamine, XLVII, 518, 529–531
Dicyclohexyl carbodiimide, *see* *N,N'*-Dicyclohexyl carbodiimide

N,N′-Dicyclohexyl carbodiimide, XXXIV, 68, 78, 87, 89, 255, 300, 413, 510, 522, 612, 736; XLIV, 48, 146, 384; XLVI, 157, 210, 227, 437, 438, 506, 568, 614, 670; XLVII, 358, 359, 361, 488, 552, 554, 558, 559, 561, 562, 568, 570; LIX, 157
for affinity chromatography, XXXVI, 105
analysis, LV, 496
ATPase, LV, 302, 303, 329, 330, 333, 343, 350; LVI, 163–173, 181
ATP-P_1 exchange, LV, 360, 362
availability, LV, 495
BCF_0-BCF_1, LV, 194, 195
biological properties, LV, 497–500
in bromoacetamidoethyl-Sepharose preparation, LI, 311
carboxyl-terminal attachment, XLVII, 305, 306, 308, 591
chemical properties, LV, 496
complex V, LV, 315; LVI, 584, 586, 597, 600
energy-linked function, LV, 502
in 5′-fluoro-2′-deoxyuridine 5′-(*p*-nitrophenylphosphate) synthesis, LI, 98
intrinsic membrane protein of *M. phlei*, LV, 188
N. crassa mitochondria, LV, 147
peptide bond formation, XLVII, 595–598
physical properties, LV, 477
preparation of disuccinimidyl succinate, LIX, 159
of luciferin adenylate, LVII, 25, 27
proteolipid ionophore, LV, 419
stimulation of energy-linked function, LV, 502
structure, LV, 495
submitochondrial particles, LV, 106
toxicology, LV, 496
use, LV, 498, 500
Dicyclohexylcarbodiimide-binding protein
extraction and isolation
from beef heart mitochondria, LV, 428, 429
from *E. coli*, LV, 429–434
nature, LV, 426–428, 434
Dicyclohexylcarbodiimide-sensitive adenosinetriphosphatase, *see* Adenosinetriphosphatase, dicyclohexylcarbodiimide-sensitive
Dicyclohexyl-18-crown-6
as ionophore, LV, 443
structure, LV, 436
Dicyclohexylurea, preparation, LIX, 159
2,3-Dideoxyadenosine, substrate of adenosine deaminase, LI, 511
3′,4′-Dideoxykanamycin B, XLIII, 271, 272, 278
DIDS, *see* 4,4′-Diisothiocyano-2,2′-stilbenedisulfonic acid
Diencephalon, ventral, of rat, exposure, XXXIX, 166–175
Diene, conjugation assay for determination of lipid peroxidation, LII, 308, 310
Dierucoyl-L-α-phosphatidylcholine, transhydrogenase reconstitution, LV, 814, 815
Diester phosphonate analogue, intermediate in phosphatidic acid analogue synthesis, XXXV, 503–506
Diet
inhibitor administration, LVI, 35
rachitogenic, for chicks, XXXVI, 515
Diether analogue, intermediate in synthesis of phosphatidic acid analogues, XXXV, 508
Diether lecithin, intermediate in phosphatidylcholine synthesis, XXXV, 452
Diether phosphonate analogue, of phosphatidic acids, intermediate in synthesis of phosphatidic acid analogues, XXXV, 506, 507
Diethyl adipimidate, XLIV, 124, 127
Diethyl allylphosphonate, intermediate in synthesis of phosphatidic acid analogues, XXXV, 505, 507
Diethylaminoethyl-agarose
in glutathione peroxidase assay, LII, 507, 508
purification of luciferin sulfokinase, LVII, 249
Diethylaminoethyl-BioGel A
ATPase purification, LV, 339, 340
in purification of luciferin sulfokinase, LVII, 249

(Diethylaminoethyl)cellulose, XLIV, 46
acyl-cAMP derivatives, XXXVIII, 408, 409
adenylate cyclase, XXXVIII, 159, 165, 167, 176
adenylyliminodiphosphate, XXXVIII, 424, 425
alanine carrier isolation, LVI, 432
aminoacylation of tRNA isoacceptors, LIX, 298
antibody purification, LII, 243
ATPase, LV, 345, 353
ATPase subunits, LV, 806, 807
batch elution technique, LII, 154
batch variability of pH, LII, 197
benzoylated
buffers, LIX, 213
purification of chloroplast tRNA^{Phe}, LIX, 213, 214
of mitochondrial tRNA, LIX, 207
of modified tRNA, LIX, 170
of rat liver tRNA, LIX, 230
of yeast tRNA^{Phe}, LIX, 185
in separation of chloroplastic tRNAs, LIX, 211, 212
of tRNA isoacceptors, LIX, 298
source, LIX, 190
brand differences, LII, 123
cAMP binding protein, XXXVIII, 50, 51, 370
cAMP inhibitor, XXXVIII, 277
cGMP-dependent protein kinase, XXXVIII, 331, 332
chromatography, of prolactin, XXXVII, 393–395
conversion to phosphate form, LI, 582
cyclic nucleotides, XXXVIII, 416, 418, 419
diazomalonyl derivatives of cAMP, XXXVIII, 380, 390, 392
enzyme modification, XLIV, 310
in epidermal growth factor isolation, XXXVII, 428
guanylate cyclase, XXXVIII, 193, 194
labeled atractyloside, LV, 523
as microcarrier, LVIII, 186
microsomal fractionation, LII, 360
paper electrophoresis, in identification of modified nucleotides, LIX, 87–95
peptide separation, XLVII, 206, 207
phosphate form, preparation, LVII, 145
phosphodiesterase
brain, XXXVIII, 230–232, 234
Escherichia coli, XXXVIII, 254
heart, XXXVIII, 219, 220
liver, XXXVIII, 258
muscle, XXXVIII, 241, 242
slime mold, XXXVIII, 246, 247
phosphodiesterase activator, XXXVIII, 265
plate, for homochromatography of tRNA hydrolyzates, LIX, 63
preparation, XXXVIII, 371; LVIII, 186
of aminoacyl-tRNA synthetase, LIX, 314
of chloroplastic tRNA, LIX, 210
of mitochondrial tRNA, LIX, 207
pretreatment, LIX, 236; LX, 17, 188, 467
protein kinase, XXXVIII, 297, 299–302, 312, 313, 318, 324, 325
catalytic subunit, XXXVIII, 314, 315
preparation, XXXVIII, 67
regulatory subunit, XXXVIII, 327
protein kinase inhibitor, XXXVIII, 354, 355
purification
acetyltransferase, LX, 536–539
of acid nucleotidase, LI, 273
of acyl-CoA dehydrogenases, LIII, 516
of adenine phosphoribosyltransferase, LI, 571
of adenosine deaminase, LI, 504, 510, 511
of adenylate deaminase, LI, 499
of adenylate kinase, LI, 471
of adenylosuccinate adenylate-lyase, LI, 205
adrenodoxin, LII, 135, 136
of aequorin, LVII, 281
of amidophosphoribosyltransferase, LI, 174, 176
of aminoacyl-tRNA synthetases, LIX, 263; LX, 110, 111, 619
antibody, LII, 243

of aspartate carbamyltransferase, LI, 45
of bacterial luciferase, LIII, 567; LVII, 145–147
camphor 5-monooxygenase system, LII, 173, 174
of carbamoyl-phosphate synthetase, LI, 32, 33
casein kinase, LX, 499, 500, 502
chloroperoxidase, LII, 526, 527
of CPSase-ATCase complex, LI, 109
of CTP(ATP):tRNA nucleotidyltransferase, LIX, 123, 124
of *Cypridina* luciferase, LVII, 370
of cytidine deaminase, LI, 403, 410
of cytochrome *b*, LIII, 214, 217
cytochrome b_5, LII, 99, 468, 469
cytochrome b_5 reductase, LII, 467
of cytochrome c_1, LIII, 224
of cytochrome c_3, LIII, 623
of cytochrome oxidase, LIII, 67–69, 75
cytochrome *m*, LII, 178
cytochrome *P*–450, LII, 111, 153, 154, 161–163, 364
of cytosine deaminase, LI, 396, 397
of cytoplasmic progesterone receptors, XXXVI, 201–204
of deoxycytidine kinase, LI, 341
of deoxycytidylate deaminase, LI, 415
of deoxythymidylate phosphohydrolase, LI, 287
of desulfovibrione electron carriers, LIII, 622, 633
of dihydroorotate dehydrogenase, LI, 62, 66
dimethylaniline monooxygenase, LII, 147, 148
diphosphorylated polyuridylate, LX, 749
of *Diplocardia* luciferin, LVII, 376
of electron-transferring flavoprotein, LIII, 514–516
elongation factor EF–1, LX, 681
elongation factor EF–2, LX, 653, 654, 656, 707, 709
epoxide hydrase, LII, 197, 198
of erythrocyte enzymes, LI, 585
in estrogen receptor, XXXVI, 355, 356
of ferredoxin, LIII, 622, 623
of FGAM synthetase, LI, 199, 200
of flavin peptides, LIII, 454, 455
of flavocytochrome b_2, LIII, 242, 247, 251–253
of flavodoxin, LIII, 622–624
of formate dehydrogenase, LIII, 369, 370
of guanine deaminase, LI, 513–515
of guanylate kinase, LI, 479, 480, 486
heme-regulated translational inhibitor, LX, 471–473, 477, 486, 495
of high-potential iron-sulfur protein, LIII, 336–338
of hydrogenase, LIII, 288, 289, 311
of *p*-hydroxybenzoate hydroxylase, LIII, 546
of ω-hydroxylase, LIII, 359
of hypoxanthine phosphoribosyltransferase, LI, 547
initiation factor
  Co-EIF, LX, 54, 192, 193
  EF, LX, 142
  EIF, LX, 39–45
  eIF, LX, 20–22, 24–28, 104–107, 128–133, 149, 150, 152, 158, 187, 188, 242, 244, 310, 311, 469
  IF, LX, 98, 227, 228
of luciferase, LVII, 352
of luciferin sulfokinase, LVII, 249
of melilotate hydroxylase, LIII, 555
NADH-cytochrome b_5 reductase purification, LII, 104
of nitrate reductase, LIII, 349
of nitric oxide reductase, LIII, 645
of nitrite reductase, LIII, 643
of nitrogenase, LIII, 316–318, 320, 322–327
of nucleoside diphosphokinase, LI, 374
of nucleoside phosphotransferase, LI, 388, 389

of nucleoside triphosphate pyrophosphohydrolase, LI, 277–279, 281
of OPRTase-OMDdecase complex, LI, 137, 138, 146, 148–151
of orotate phosphoribosyltransferase, LI, 72
of phenylalanine hydroxylase, LIII, 282
phosphoprotein phosphatases, LX, 528, 531
of phosphoribosylglycinamide synthetase, LI, 184
of protein A, LIII, 377, 378
of protein B, LIII, 380, 381
protein kinases, LX, 509
of *Pseudomonas* cytochrome oxidase, LIII, 652, 653
Pseudomonas toxin A, LX, 784
of purine nucleoside phosphorylase, LI, 520, 527, 528, 533
putidaredoxin, LII, 178–181
putidaredoxin reductase, LII, 176–178
of pyrimidine nucleoside monophosphate kinase, LI, 325
of pyrimidine nucleoside phosphorylase, LI, 434
Qβ ribonucleic acid replicase, LX, 634
of reconstituted tRNA, LIX, 134
ribosome protein S1, LX, 420–423, 440, 441, 449, 450
of tRNA methyltransferase, LIX, 194, 197
of RNase III, LIX, 829
of rubredoxin, LIII, 342, 343, 625
of rubredoxin:NAD oxidoreductase, LIII, 619
of salicylate hydroxylase, LIII, 533
of succino-AICAR synthetase, LI, 192
of superoxide dismutase, LIII, 388, 389, 391
TF_1, LV, 782, 783
of thymidine phosphorylase, LI, 440
translational control ribonucleic acid, LX, 543
of tryptophanyl-tRNA synthetase, LIX, 241
of uridine-cytidine kinase, LI, 304, 312, 317
wheat germ initiation factors, LX, 198–201
separation of aggregated forms of OPRTase-OMPdecase complex, LI, 151, 152
of aminoacylated tRNA, LIX, 216
of aspartate transcarbamylase subunits, LI, 37–39
of mixed-function oxidase components, LII, 202
of nucleotides and bases, LI, 543, 544
of ribosome-protected messenger ribonucleic acid fragments, LX, 330, 331, 349, 350
succinyl-atractyloside, LV, 524
TF_1 subunits, LV, 784, 785
washing procedure, LI, 205; LIII, 615; LIX, 194
(Diethylaminoethyl)cellulose DE–23, ATPase, LV, 341
(Diethylaminoethyl)cellulose DE–52
ATPase, LV, 353, 791–794, 804
inhibitor, LV, 424
subunits, LV, 355
bongkrekic acid, LV, 529
coupling factor B, LV, 387
DCCD-binding protein, LV, 431, 432
preparation of $TF_0{\cdot}F_1$, LV, 368, 369
(Diethylaminoethyl)cellulose disc, for separation of nucleosides and nucleotides, LI, 13, 355, 362, 366
2-Diethylaminoethyl-2,2-diphenylvalerate
inhibitor of bacterial luciferase, LVII, 152
of blue-fluorescent protein, LVII, 230
of firefly luciferase, LVII, 35
of NADH oxidoreductase, LVII, 203
of NADPH oxidoreductase, LVII, 203
of *Pholas* luciferase, LVII, 387
of *Pholas* luciferin, LVII, 390

separation of creatine kinase isoenzymes, LVII, 62
2-Diethylaminoethyl-2,2-diphenylvalerate hydrochloride, *see* SKF-525A
N,N-Diethylaminoethyl methacrylate, XLIV, 71
(Diethylaminoethyl)Sephadex, *see also* Sephadex A-50
column photolysis, XXXVIII, 395
in cytochrome b_5,purification, LII, 469, 470
cytochrome P–450_{cam} purification, LII, 153–155
for drug metabolite isolation from urine and plasma, LII, 334
guanylyliminodiphosphate, XXXVIII, 427
labeled ATP preparation, XXXVIII, 67, 68
as microcarrier, LVIII, 186, 209
in myoglobin purification, LII, 477
NADH-cytochrome b_5 reductase purification, LII, 104
protein kinase, XXXVIII, 311, 324–326
purification of cytochrome oxidase subunits, LIII, 78
of deoxythymidylate phosphohydrolase, LI, 288
of GMP synthetase, LI, 215, 216
of hypoxanthine phosphoribosyltransferase, LI, 546, 548
of nucleoside diphosphokinase, LI, 374, 375
of purine nucleoside phosphorylase, LI, 542
of thymidine phosphorylase, LI, 444
of thymidylate synthetase, LI, 95
for reductase purification, LII, 93
support, for penicillin acylase, XLIV, 765
(Diethylaminoethyl)Sephadex A–20–150, purification of nucleoside triphosphate pyrophosphohydrolase, LI, 283, 284
(Diethylaminoethyl)Sephadex A–25
ATPase, LV, 326
preparation of [γ-^{32}P]ATP, LIX, 61
purification of CTP(ATP):tRNA nucleotidyltransferase, LIX, 185
of 50 S ribosomal proteins, LIX, 492
support for aminoacylase, XLIV, 748, 749
(Diethylaminoethyl)Sephadex A–50
preparation, LV, 306
purification of adenine phosphoribosyltransferase, LI, 576
of adenosine monophosphate nucleosidase, LI, 266, 268, 269
of aspartate carbamyltransferase, LI, 45
of bacterial luciferase, LIII, 567
of cytidine triphosphate synthetase, LI, 81
of dCTP deaminase, LI, 420, 421
of ferredoxin forms, LIII, 624
of GMP synthetase, LI, 215, 221, 222
of OPRTase-OMPdecase complex, LI, 162–164
of orotidylate decarboxylase, LI, 76
of uridine phosphorylase, LI, 427, 428
(Diethylaminoethyl)Sepharose CL–6B
purification of D-lactate dehydrogenase, LIII, 521, 522
of luciferase, LVII, 31–33
transhydrogenase, LV, 277–279
Diethylaminophthalhydrazide, structural formula, LVII, 428
N^6,$O^{2'}$-Di(ethyl-2-diazomalonyl)-cyclic adenosine monophosphate, synthesis, XXXVIII, 390, 391
N,N-Diethyl-2,4-dichloro-(6-phenylphenoxy)ethylamine, inhibitor of bacterial luciferase, LVII, 152
Diethyldithiocarbamate
inhibitor of phenylalanine hydroxylase, LIII, 285
of *Pholas* luciferase, LVII, 388, 398
purification of *Pholas* luciferin, LVII, 388
Diethylene glycol, XLIV, 117

N,N-Diethylethanolamine, XLVII, 385
source, XLVII, 382
Diethyl ether, XLIV, 117, 834–836, 838
DCCD-binding protein, LV, 431
extraction, XLVII, 345
in fluorography, LIX, 192
identification of quinone intermediates, LIII, 604, 605
isolation of quinone intermediates, LIII, 606–608
preparation of labeled tRNA hydrolyzate, LIX, 71
proteolipid preparation, LV, 417
for removal of trichloroacetic acid, LIX, 497
in tRNA sequence analysis, LIX, 99
solvent, synthesis of chemiluminescent aminophthalhydrazides, LVII, 430, 432
in synthesis of 8α-flavins, LIII, 458, 460
Diethyl ether-methanol, purification of flavin compounds, LIII, 463
Diethylisoluminol, in chemiluminescence assays, LVII, 438, 439
3,4-Di-*O*-ethyl-1-*O*-methyl-D-erythritol, L, 28
O,O-Diethylphosphofluoride, XXXIV, 586
Diethyl pyrocarbonate, XXXI, 720; XLIII, 540
concentration determination, XLVII, 434
histidyl residue modification, XLVII, 431–442
inactivator of tryptophanyl-tRNA synthetase, LIX, 250
of ribonucleases, LIX, 464, 554
labeling, XLVII, 434
in plant-cell fractionation, XXXI, 504, 505
in polyribosome isolation, XXXI, 586–589
reaction conditions, XLVII, 436, 437
ribosomal RNA isolation, LVI, 90
side reactions, XLVII, 439, 440
stability, XLVII, 434, 436
toxicity, XLVII, 434
use in nucleic acid research, XLVII, 441, 442
Diethylstilbestrol, XLVI, 57, 58
Diethyl sulfate
synthesis of dimethyl 7-[*N*-ethyl-4-(*N*-phthalimido)butyl]aminonaphthalene-1,2-dicarboxylate, LVII, 436
of 4-*N*-ethylamino-*N*-methylphthalimide, LVII, 431
of 4-[*N*-ethyl-*N*-4-(*N*-phthalimido)butyl]amino-*N*-methylphthalimide, LVII, 433
Difco growth medium, for yeast, XXXI, 628
Difference reactivity method, XLVIII, 321, 322
Difference spectra, measurement, LII, 212–220
absorbance changes due to substrate addition, LII, 216–219
during aerobic steady state, LII, 216–219
artifact sources, LII, 219, 220
base line changes, causes, LII, 213, 218, 220
base line check, LII, 213
effect of organic solvents, LII, 218
of microsomal protein concentration, LII, 213, 219, 220, 225, 226
general protocol, LII, 213
isosbestic points, LII, 219, 220
multiple forms of cytochrome *P*-450, LII, 266, 267
Differential absorptivity
for adenosine to inosine, LI, 503, 508
for adenylate to inosinate, LI, 491
for 6-azacytidine deamination, LI, 405, 406
for cytidine to uridine, LI, 395, 401, 402, 405
for cytosine to uracil, LI, 395
for dCTP to dUTP, LI, 418
for deoxyadenosine-cytosine conversion, LI, 448
for 2′-deoxycytidine to 2′-deoxyuridine, LI, 395
for deoxyguanosine-cytosine conversion, LI, 448
for deoxyinosine-adenine conversion, LI, 448

for deoxyinosine-cytosine conversion, LI, 448
for deoxyinosine to hypoxanthine, LI, 447
for deoxyuridine to uracil, LI, 425
for guanosine to guanine, LI, 532
for inosine to uric acid, LI, 518, 532
for 5-methylcytosine to thymine, LI, 395
for thymidine-cytosine conversion, LI, 448
for thymidine to thymine, LI, 425, 442
for uridine to uracil, LI, 425
Differential centrifugation
in brain nuclei isolation, XXXI, 461
definition, XXXI, 714
in glyoxysome isolation, XXXI, 568
liver homogenate, XXXI, 726, 728
plant cells, XXXI, 545
in subcellular fractionation, XXXI, 738
Differential conductivity, assay, XLIV, 348, 349
Differential interference microscopy, *see* Nomarski imaging
Differential labeling, XLVI, 4, 5, 59–69
definition, XLVI, 59, 60
dual isotopes, XLVI, 66, 67
kinetics, XLVI, 63, 64
principle, XLVI, 63, 64
procedure, XLVI, 60–62
protective ligand, XLVI, 63
variations, XLVI, 62, 63
Differential pulse polarography, XLIII, 373–388, *see also* specific antibiotic
advantages, XLIII, 375–377
analysis procedure, XLIII, 380, 381
dropping mercury electrode, XLIII, 374, 375
equations, XLIII, 380, 381
equipment, XLIII, 377–379
limitations, XLIII, 375–377
Differential refractometer, XLVII, 212, 216
disadvantages, LII, 281
Differential scanning calorimetry, XXXII, 262–272
calibration and calculations, XXXII, 271, 272
calorimeter operation, XXXII, 269–271
instruments, XXXII, 266–268
principles, XXXII, 263–266
sample preparation, XXXII, 268, 269
Differential thermal analysis, XXXII, 263
Differentiation, of ovarian cell, *in vitro*, XXXIX, 183–230
Diffraction, *see* specific type
Diffractometer
Duke design, XLIX, 47
EMBL design, XLIX, 43–46
Diffusing screen
preparation, LVII, 580
substitute, LVII, 593
Diffusion
activated, XLIV, 175
assay, XLIII, 60, 61; XLIV, 351, 352
particle size, XLIV, 136, 138
rotational
fluorescence emission polarization, XLVIII, 350–352
of protein, data analysis, LIV, 54–57
simulation, XLVIII, 224–227
of substrate
external resistances, XLIV, 413–420
internal resistances, XLIV, 414, 420–426
Diffusion cell, artificial enzyme membranes, XLIV, 907, 910, 911
Diffusion coefficient
average
calculation, XLVIII, 219–224
definition, XLVIII, 219
of membrane protein, computation, LIV, 56, 57
translational, evaluation, XLVIII, 417–420
Diffusion constraint, determination, XLIV, 269
Diffusion potential, across membrane, LIV, 63
Diffusion time, rotational, estimation, XLIX, 429
Difluorescein isothiocarbamidocystamine, properties, XLVII, 415
p,p'-Difluoro-*m,m'*-dinitrophenylsulfone, XLIV, 263

2,2-Difluorosuccinate, succinate dehydrogenase, LVI, 586
5,7-Difluorotryptophan, inhibitor of tryptophanyl-tRNA synthetase, LIX, 252
N,N'-Diformyl-1,6-diaminohexane, synthesis, XLIV, 130
Difucosyl-*p*-lacto-*N*-hexaose
  isolation from human milk, L, 220
  structure, L, 217
Difucosyllacto-*N*-hexaose I, structure, L, 217
Difucosyl-*p*-lacto-*N*-neohexaose
  isolation from human milk, L, 220
  structure, L, 217
Digalactosyl ceramide, structure, L, 273
Digestion, proteolytic, isolation of flavin peptides, LIII, 452, 453
Digestion buffer, liver cells, LVIII, 542, 543
Digital filtering, LIV, 170
  noise suppression, LVI, 512, 514
Digital integrator, LVII, 538, 539
  circuit diagram, LVII, 536
  timed integration control circuit, LVII, 539
Digitizer, LIV, 39
Digitonin
  adenylate cyclase preparation, XXXVIII, 128, 129
  ATPase preparation, LV, 331
  for breakage of mitochondrial outer membrane, LVI, 692
  cell disruption, LV, 243, 244
  isolation of mitochondria, LVI, 25, 687
  lysis, of mycoplasmas, XXXII, 461
  in membrane isolation, XXXI, 789
  method for separation of particulate and soluble cell fractions, LVI, 208, 209
    application to studies of metabolic regulation, LVI, 220, 221
    calculation of results, LVI, 212
    combination with cavitation method, LVI, 221–223
    criteria for satisfactory separation, LVI, 210–212
    experimental procedure, LVI, 209, 210
    validity, LVI, 212, 213
  mitochondrial inner membranes, LVI, 615
  partial specific volume, XLVIII, 22
  phosphodiesterase preparation, XXXVIII, 132
  in progesterone-metabolite precipitation, XXXVI, 496
  purification, LV, 116
  solution, preparation, XXXI, 314
  treatment of mitochondria, LV, 117
Digitoxin, antisera, XXXVI, 17
1,2-Diglyceride, intermediate in synthesis of phosphatidylethanolamine analogues, XXXV, 508
2,3-Diglyceride, intermediate in synthesis of phosphatidylethanolamine analogues, XXXV, 509
Diglycolamic acid 1e, structure, LV, 441
Digoxin, XXXIV, 4; XLVI, 523
  autoradiography, XXXVI, 150
1,3-Diguanidino-1,3-dideoxy-*scyllo*-inositol, XLIII, 432
1,D-1,3-Diguanidino-1,3-dideoxy-*scyllo*-inositol 6-phosphate, XLIII, 432
  paper chromatographic data, XLIII, 121
  streptomycin 6-phosphotransferase, XLIII, 629, 631
  structure, XLIII, 429, 614
Di-*n*-hexadecanoyl peroxide
  synthesis, LIII, 596
    of coenzyme Q analog, LIII, 596
1,2-Di-*O*-hexadecyl *sn*-glycerol, intermediate in phosphatidylcholine synthesis, XXXV, 452
Di-*n*-hexanoyl peroxide
  synthesis, LIII, 596
    of coenzyme Q_1 analog, LIII, 596
Dihydrazide
  periodate method of activation, XXXIV, 80, 81
  preparation, XXXIV, 476–478
7,8-Dihydro-2-amino-4-hydroxy-6-[1,2-dihydroxypropyl(L-*erythro*-)]pteridine, *see* 7,8-Dihydrobiopterin
7,8-Dihydrobiopterin, substrate of phenylalanine hydroxylase, LIII, 278, 285

erythro-DL-Dihydroceramide aminoethylphosphonate, intermediate in synthesis of sphingomyelin analogues, XXXV, 528
Dihydrocoumarin, synthesis of melilotate, LIII, 553
5,6-Dihydrocytidine, substrate of cytidine deaminase, LI, 404
1,4-Dihydro-6,7-dimethoxy-3(2*H*)-isoquinolone, XLVI, 219
Dihydroepiandrosterone, metabolites, gas chromatographic and mass spectral properties, LII, 387
Dihydrofolate, as eluent, XXXIV, 286
Dihydrofolate reductase, XXXII, 576; XXXIV, 48, *see also* Tetrahydrofolate dehydrogenase
  methotrexate-agarose purification, XXXIV, 272–281
    aminoethyl cellulose attachment, XXXIV, 272
    ω-aminohexylagarose, XXXIV, 274, 275
    assay, XXXIV, 277–281
    formylaminopterin acrylamide, XXXIV, 274
    methotrexate-AH-agarose, XXXIV, 276
    spacer effect, XXXIV, 273
  peptide separation, XLVII, 206
  pteroyllysine-agarose in purification, XXXIV, 281–288
    assay, XXXIV, 282, 283
    coupling of agarose, XXXIV, 284
    folate as eluent, XXXIV, 286
    pteroic acid, XXXIV, 283
    N^{α}-pteroyl-L-lysine, XXXIV, 283
    purification, XXXIV, 284
    stability, XXXIV, 288
7,8-Dihydrofolic acid, XLVI, 307
  in thymidylate synthetase assay, LI, 91
1,6-Dihydro-6-hydroxymethylpurine ribonucleoside, inhibitor of adenosine deaminase, LI, 507
Dihydrolipoamide, XXXIV, 294
Dihydrolipoamide reductase
  assay, LII, 244
  flavoprotein classification, LIII, 397
  holoenzyme reconstitution, LIII, 436
  NAD^+, immobilization, in enzyme system, XLIV, 315, 316
  in preparation of deuterated NADH, LIV, 226
  reversible dissociation, LIII, 433
Dihydrolipoate
  reducing substrate for ribonucleoside triphosphate reductase, LI, 247, 258
  sodium salt, preparation, LI, 248
5,6-Dihydro-2-methyl-1,4-oxathiin-3-carboxanilide, *see* Carboxin
Dihydromorphine, XXXIV, 622; XLVI, 603
3,4-Dihydronaphthalene-1,2-dicarboxylic acid anhydride
  synthesis, LVII, 435
    of 7-methoxynaphthalene-1,2-dicarboxylic anhydride, LVII, 436
Dihydroneopterin triphosphate synthetase, XXXIV, 4
Dihydroorotase, LI, 111–121, 135–142
  assay, LI, 123
  association with OPRTase:OMPdecase complex activity, LI, 135, 138, 140–142
  molecular weight, LI, 121
  properties, LI, 120
  from rat cells, LI, 111–121
Dihydroorotate
  inhibitor of orotate phosphoribosyltransferase, LI, 166
  product of dihydroorotase, LI, 111
  substrate of dihydroorotate dehydrogenase, LI, 63
Dihydroorotate dehydrogenase, LI, 58–69
  assay, LI, 58–60, 64
    as oxidase, LI, 59
  associated cofactors, LI, 68
  cytochrome, LVI, 172
  from *E. coli*, LI, 58–63
  electron transport and, LI, 68, 69
  inhibitors, LI, 68
  kinetics, LI, 62
  molecular weight, LI, 62, 67
  from *Neurospora*, LI, 63–69
  oxidase activity, LI, 69
  properties, LI, 62, 63, 67–69
  purification, LI, 60–62, 65–67

purity, LI, 67, 68
stability, LI, 62
Dihydroorotate oxidase, LII, 17, 19
L-4,5-Dihydroorotate oxidase, LVI, 474
Dihydro-L-orotic acid, in dihydroorotase assay, LI, 123
Dihydropteridine reductase
activity, LIII, 278
assay of phenylalanine hydroxylase, LIII, 278, 279
Dihydropyran, preparation of dehydroluciferol, LVII, 25
Dihydropyridine ring, stereospecificity, LIV, 223
Dihydroriboflavin, deoxygenation of gases, LVI, 384
Dihydrosphingosine, silica gel chromatography, XXXV, 529–533
Dihydrostreptomycin
binding to ribosomal particles, LIX, 865, 866
for haploid cell culture, XXXII, 798, 799
in media, LVIII, 112, 114, 116
[^{3}H]Dihydrostreptomycin
effect on protein synthesis, LIX, 860
for photoaffinity labeling, LIX, 798
Dihydrostreptomycin-3′′-phosphate, XLIII, 637
Dihydrostreptomycin-6-phosphate, XLIII, 634
[3′,α-^{3}H]Dihydrostreptomycin-6-phosphate, XLIII, 466
Dihydrostreptomycin-6-phosphate 3′,α-kinase, XLIII, 634–637
assay, XLIII, 635, 636
biological distribution, XLIII, 636, 637
properties, XLIII, 636, 637
reaction scheme, XLIII, 634
specificity, XLIII, 637
stability, XLIII, 636
Streptomyces bikiniensis, preparation, XLIII, 636
Dihydrostreptomycin-6-phosphate 3′,α-phosphotransferase, XLIII, 634–637
Dihydrotestosterone
effects, (Na^+ + K^+)-ATPase, XXXVI, 435
nuclear receptor binding, XXXVI, 313–319
as 5α-oxidoreductase substrate, XXXVI, 468
protein-binding assay, XXXVI, 37
receptor, purification, XXXVI, 366–374
receptor binding, XXXVI, 411–412
in TeBG binding studies, XXXVI, 112
in vivo production and interconversion, XXXVI, 68, 74
5α-Dihydrotestosterone
metabolites, gas chromatographic and mass spectral properties, LII, 386
in testicular lymph, XXXIX, 276
testicular production, XXXIX, 282
testosterone conjugate reaction, XXXIX, 263
thin-layer chromatography, XXXIX, 262
Dihydrotestosterone 17β-acetate, reduction, XXXVI, 73
Dihydroubiquinone
of NADH:ubiquinone oxidoreductase, LIII, 11
product, of complex II, LIII, 24
substrate of complex III, LIII, 35
Dihydroubiquinone:cytochrome *c* oxidoreductase, *see* Dihydroubiquinone:cytochrome *c* reductase
Dihydroubiquinone:cytochrome *c* reductase, assay, LIII, 90, 91, *see also* Complex III
5,6-Dihydrouridine
chromatographic mobility, LIX, 73
electrophoretic mobility, LIX, 91
identification, LIX, 79, 80, 92, 110
inhibitor of cytidine deaminase, LI, 405
loop, XLVI, 90
Dihydroxyacetone
galactose polarography, LVI, 466
renal gluconeogenesis, XXXIX, 14
Dihydroxyacetone phosphate, XLVI, 18, 49, 139, 143, 383; XLVII, 490, 498
assay, LV, 221
4,5-Dihydroxyanthraquinone-2-carboxylate, *see* Rhein
2,3-Dihydroxybenzoate 2,3-dioxygenase, LII, 9, 19

2,3-Dihydroxybenzoic acid, substrate of salicylate hydroxylase, LIII, 538, 539
2,4-Dihydroxybenzoic acid
 substrate of *p*-hydroxybenzoate hydroxylase, LIII, 549, 550
  of salicylate hydroxylase, LIII, 536, 538
2,5-Dihydroxybenzoic acid, substrate of salicylate hydroxylase, LIII, 538, 539
2,6-Dihydroxybenzoic acid, substrate of salicylate hydroxylase, LIII, 538, 539
3,4-Dihydroxybenzoic acid, substrate of salicylate hydroxylase, LIII, 538
3,5-Dihydroxybenzoic acid, substrate of salicylate hydroxylase, LIII, 538
2,2-Dihydroxybiphenyl, chromatographic properties, LII, 405
2,3-Dihydroxybiphenyl, chromatographic properties, LII, 405
2,5-Dihydroxybiphenyl, chromatographic properties, LII, 405
3,4-Dihydroxybiphenyl, chromatographic properties, LII, 405
4,4′-Dihydroxybiphenyl, chromatographic properties, LII, 405
N-[*N*′-*m*-Dihydroxyborylphenyl succinamyl]aminoethyl-cellulose, separation of tRNA, LIX, 128, 134
1,25-Dihydroxycholecalciferol, renal mitochondria, LV, 11, 12
20α,22*R*-Dihydroxycholesterol, substrate of cytochrome P–450_{scc}, LII, 132
2,3-Dihydroxycinnamic acid, pseudosubstrate of melilotate hydroxylase, LIII, 557
N^7,N^8-(1,2-Dihydroxycyclohex-1,2-ylene)-L-arginine, XLVII, 157, 158
 characterization, XLVII, 158, 159
 R_f values, XLVII, 159
2,2′-(1,8-Dihydroxy-3,6-disulfonaphthalene-2,7-bisazo)bisbenzenearsonic acid, *see* Arsenazo III
Dihydroxyfumaric acid
 inhibitor of *Pholas* luciferin oxidation, LVII, 399
 stimulator of *Pholas* light emission, LVII, 397
2,6-Dihydroxy-1,1,1,7,7,7-hexafluoro-2,6-bis(trifluoromethyl)heptan-4-one, LV, 470
2,3-Dihydroxyindole 2,3-dioxygenase, LII, 19
L-1,2-Dihydroxy-3-iodopropane, *see* L-Glycerol-α-iodohydrin
7,8-Dihydroxykynurenate 8,8α-dioxygenase, LII, 11
2,8-Dihydroxyphenoxazine, as mediator-titrant, LIV, 408
3,4-Dihydroxyphenylacetate 2,3-dioxygenase, LII, 11
3,4-Dihydroxyphenylacetate 3,4-dioxygenase, LII, 11
3,4-Dihydroxyphenylalanine, XLVII, 36
 in cultured cell systems, XXXII, 787
 identification, XXXII, 787, 788
3,5-Dihydroxyphenylalanine, paper chromatography, XXXII, 788
L-Dihydroxyphenylalanine oxidase
 assay, XXXI, 745
 as marker enzyme, XXXI, 735, 744, 745
Dihydroxyphenyl derivative, XXXIV, 6
2,3-Dihydroxyphenylpropionate 1,2-dioxygenase, LII, 11
2,3-Dihydroxyphenylpropionic acid, pseudosubstrate of melilotate hydroxylase, LIII, 557
11β,21-Dihydroxypregn-4-ene-3,20-dione, *see* Corticosterone
17α,21-Dihydroxypregn-4-ene-3,20-dione, *see* Deoxycortisol
17α,20α-Dihydroxypregn-4-en-3-one, as enzyme inhibitor, XXXIX, 282
11β,20α-Dihydroxyprogesterone synthesis, LVIII, 571
9-(2′,3′-Dihydroxypropyl)adenine, inhibitor of tryptophanyl-tRNA synthetase, LIX, 252
Dihydroxypropylphosphonic acid
 intermediate in synthesis of phosphatidic acid analogues, XXXV, 504–506
  of phosphatidylethanolamine analogues, XXXV, 512–514
L-Dihydroxypropylphosphonic acid diethyl ester, intermediate in synthesis of phosphatidic acid analogues, XXXV, 503–505

2,4-Dihydroxypteridine, substrate of orotate phosphoribosyltransferase, LI, 152
6,8-Dihydroxypurine, in media, LVIII, 67
2,5-Dihydroxypyridine 5,6-dioxygenase, LII, 11
3,4-Dihydroxypyridine 2,3-dioxygenase, LII, 19
2,6-Dihydroxypyridine 3-hydroxylase, LII, 16
Dihydroxyvitamin D_3
chemical synthesis, XXXVI, 522–529
isolation, XXXVI, 518
preparation, XXXVI, 513–528
1,25-Dihydroxyvitamin D_3
competitive binding assay, LII, 391–394
tritium-labeled, preparation, LII, 391
vitamin D_3 metabolism, LII, 388–391
24(*R*),25-Dihydroxyvitamin D_3
assay, LII, 397, 398
normal serum level, LII, 398
tritium-labeled, preparation, LII, 397
vitamin D metabolism, LII, 389
25,26-Dihydroxyvitamin D_3, vitamin D_3 metabolite, LII, 389
Diimidazole-ferrohemochrome, oxidation
by oxygen, LV, 537
phosphorylation coupled, LV, 537–539
1,4-Diisocyanatobenzene, XLIV, 116
1,6-Diisocyanohexane, synthesis, XLIV, 129, 130
Diisopropyl allylphosphonite, intermediate in synthesis of phosphatidylcholine analogues, XXXV, 521, 522
N,N-Diisopropylethylamine, XLVII, 564, 595, 600
Diisopropyl fluorophosphate, XLIV, 15; XLIX, 443; LIX, 235
inhibitor of palmityl thioesterase I, XXXV, 102, 106
in phospholipase A_2 purification, XXXII, 148
protease action, LIX, 267
purification of aminoacyl-tRNA synthetases, LIX, 263
of cytochrome c_1, LIII, 228
of tryptophanyl-tRNA synthetase, LIX, 240, 241
titration, XLIV, 524
toxicity, 228n
Diisopropyl fluorophosphate inhibitor
of blood factors, XLV, 49, 55, 96
of crotalase, XLV, 235
of kallikrein, XLV, 314
of plasmin, XLV, 263
plasminogen, rabbit, XLV, 277
in plasminogen complex, XLV, 269
of thrombin, XLV, 173
of urokinase, XLV, 242
Diisopropylidine-D-mannitol, *see* 1,2,5,6-Diacetone-D-mannitol
Diisopropylphosphofluoridate, *see* Diisopropylfluorophosphate
Diisopropyl-plasmin chromatography, urokinase, XLV, 250
Diisothiocyanate activation, XLVII, 283–285
Diisothiocyanate-carbodiimide procedure, XLVII, 286
4,4′-Diisothiocyano-2,2′-stilbenedisulfonic acid, inhibitor, of anion transport, XLIX, 14
Diketopiperazine derivative, XLIII, 144
Dilantin, *see* Diphenylhydantoin
Dilatometer, applications, XLVIII, 18–22
Dilauroylglyceryl-3-phosphorylcholine
in cytochrome *P*–450LM$_2$ assay, LII, 205
for reconstitution of mixed-function oxidase system, LII, 204
Dilauroylphosphatidylcholine, in reductase assay, LII, 91
1,2-Dilinoleyl-*rac*-glycero-3-phosphorylserine, intermediate in phosphatidylserine synthesis, XXXV, 463–465
Dilution assay, XLIII, 62–65
agar method, XLIII, 62
photometric method, XLIII, 63–65, *see also* Photometric method, for dilution assays
serial dilution in tubes, XLIII, 62, 63
Dilution factor, calculation, LVI, 261, 262
Dilution jump method, for rate constant determination, XLVIII, 390, 391
Dilution plating, techniques, LVIII, 156–158
Dilution rate, growth in chemostat culture, LVI, 574, 575

Dimedone, in monochlorodimedone synthesis, LII, 523
Dimer, chemiluminescence, LVII, 568
Dimerization, *see also* Association reaction; Binding; Cooperativity; Dissociation; Interaction; Ligand binding; Self-association
irreversible, theoretical sedimentation patterns, XLVIII, 252, 253
ligand-facilitated
definition, XLVIII, 248, 260, 303
model, XLVIII, 301, 303–305
reaction mechanism, XLVIII, 262–265
ligand-mediated
definition, XLVIII, 248, 260
model, XLVIII, 300–303, 305
reaction mechanism, XLVIII, 262, 265
theoretical sedimentation patterns, XLVIII, 259
nonmediated, theoretical sedimentation patterns, XLVIII, 257, 258
reversible, sedimentation studies, XLVIII, 260–270
sedimentation velocity, XLVIII, 242, 243
Dimerization constant, XLVIII, 302
1,6-Di-*O*-methanesulfonyl-D-mannitol, XLVI, 384
2,4-Dimethoxybenzyl group, carboxamido group blocking, XLVII, 612, 614
3,4-Dimethoxy-5-benzyloxybenzaldehyde, XLVI, 557
3,5-Dimethoxy-4-benzyloxybenzaldehyde, XLVI, 556, 557
3,4-Dimethoxy-5-benzyloxy-β-nitrostyrene, XLVI, 557
3,5-Dimethoxy-4-benzyloxy-β-nitrostyrene, XLVI, 557
1,2-Dimethoxyethane, solvent, in radical ion chemiluminescence reaction, LVII, 498
3,4-Dimethoxy-5-hydroxyphenylethylamine, XLVI, 555
3,5-Dimethoxy-4-hydroxyphenylethylamine, XLVI, 555
3,4-Dimethoxy-5-hydroxyphenylethylamine hydrochloride, XLVI, 558
3,5-Dimethoxy-4-hydroxyphenylethylamine hydrochloride, XLVI, 557, 558
6,7-Dimethoxy-3-isochromanone, XLVI, 219
2,3-Dimethoxy-5-methyl-1,4-benzoquinone, synthesis of coenzyme Q analogs, LIII, 595
2,3-Dimethoxy-5-methyl-6-decyl-1,4-benzoquinone
analog of coenzyme Q_2, LIII, 594
biological activity, LIII, 597–599
structural formula, LIII, 594
2,3-Dimethoxy-5-methyl-6-farnesylbenzoquinone, *see* Coenzyme Q_3
2,3-Dimethoxy-5-methyl-6(all-*trans*)-farnesyl farnesyl-1,4-benzoquinone, *see* Coenzyme Q_6
2,3-Dimethoxy-5-methyl-6-geranylbenzoquinone, *see* Coenzyme Q_2
2,3-Dimethoxy-5-methyl-6-(3′-methyl-2′-butenyl)benzoquinone, *see* Coenzyme Q_1
2,3-Dimethoxy-5-methyl-6-pentadecyl-1,4-benzoquinone
analog, of coenzyme Q_3, LIII, 594
biological activity, LIII, 597–599
structural formula, LIII, 594
synthesis, LIII, 595
2,3-Dimethoxy-5-methyl-6-pentyl-1,4-benzoquinone
analog of coenzyme Q_1, LIII, 594
assay of succinate dehydrogenase, LIII, 466
biological activity, LIII, 597–599
structural formula, LIII, 594
synthesis, LIII, 595
2,2-Dimethoxypropane, XLVI, 289
N-6,9-Dimethyladenine, osmometric studies, XLVIII, 79
Dimethylalkyl amine oxide, partial specific volume, XLVIII, 19
Dimethylalkylammoniopropane sulfonate, partial specific volume, XLVIII, 18
Dimethylalkyl phosphine oxide, partial specific volume, XLVIII, 19
N,N′-Dimethyl-*N*-allylamine buffer, XLVII, 262, 303, 306
properties, XLVII, 322
N,N′-Dimethylallylaminetrifluoroacetic acid buffer, XLVII, 270
Dimethylaminoazobenzene *N*-demethylase, Ah^b allele, LII, 231

p-Dimethylaminobenzaldehyde, XLIV, 256
penicillin analysis, XLIV, 760
p-Dimethylaminobenzeneazobenzene, XLVII, 350
6-Dimethylamino-9-[3′-deoxy-3′-(*p*-azido-L-phenylalanylamino)-β-D-ribofuranosyl]purine
in photoaffinity labeling studies, LIX, 808–810
structural formula, LIX, 812
Dimethylaminoethanol, XLVII, 590
Dimethyl 7-aminonaphthalene-1,2-dicarboxylate
synthesis, LVII, 436
of dimethyl 7-*N*-(4-phthalimidobutyl)aminonaphthalene-1,2-dicarboxylate, LVII, 436
1-Dimethylaminonaphthalene-5-sulfonyl chloride, *see* Dansyl chloride
5-Dimethylaminonaphthalene-1-sulfonyllysine, XLVII, 153
4-Dimethylamino-3-nitro(α-benzamido)cinnamic acid
electronic absorption spectrum, XLIX, 146
resonance Raman spectrum, XLIX, 147
structure, XLIX, 145
4-(4-Dimethylamino-3-nitro)benzylidene-2-phenyloxazolin-5-one, Raman spectrum, XLIX, 148
p-Dimethylaminophenyl isothiocyanate, excess amino group blocking and, XLVII, 286–287
β-Dimethylaminopropionitrile, XLIV, 194, 196, 198, 741, 750
N-Dimethylaminopropyl-*N*′-ethyl carbodiimide, XLVII, 282
[3-(Dimethylamino)propyl] ethyl carbodiimide, XXXIV, 454
1,1-Dimethylamino-2-propyne, inactivator of flavoprotein enzymes, LIII, 441
6-Dimethylaminopurine, phosphodiesterase, XXXVIII, 244
Dimethylaniline monooxygenase, LII, 142–151
activators and inhibitors, LII, 148, 149
activity, LII, 142
assay, LII, 142, 143
in vitro, LII, 144
conjugate
activity assay, XLIV, 852, 853
concentration determination, XLIV, 852
properties, XLIV, 853, 854
distribution, LII, 142
flavin content, LII, 148
immobilization, on glass beads, XLIV, 851, 852
molecular weight, LII, 148
oxygen sensitivity, XLIV, 853
postmortem inactivation, LII, 143
purification procedure, LII, 143–148
soluble
properties, XLIV, 849–851
substrate specificity, XLIV, 849, 850
specificity, LII, 149, 150
N,N-Dimethyl-*p*-anisidine, chemiluminescent reaction, LVII, 507
9,10-Dimethylanthracene
chemiluminescence, LVII, 504, 505
emission spectrum, temperature dependence, LVII, 506
structural formula, LVII, 498
7,12-Dimethylbenz[*a*]anthracene 5,6-oxide, nucleoside adducts, chromatographic separation, LII, 291, 292
1,6-Dimethylbenzo[*a*]pyrene
chemiluminescent reaction, LVII, 504
structural formula, LVII, 498
Dimethylbenzylamine, XLVII, 338
3,4-Dimethylbenzyl group, sulfhydryl group blocking, XLVII, 614
1,1′-Dimethyl 4,4′-bipyridylium dichloride, as mediator-titrant, LIV, 409, 410, 423
Dimethylcarbamate ester, inhibitor of carboxylesterase, XXXV, 192
1,1-Dimethyl-4-chloro-3,5-cyclohexanedione, *see* Monochlorodimedone
Dimethyldibenzylammonium chloride, transport, LVI, 407
Dimethyldibenzylammonium ion
membrane potential, LV, 678
M. phlei ETP, LV, 186
proline uptake, LV, 199

Dimethyldichlorosilane, nonwetting agent, for coating glassware, LIX, 4
α,α′-Dimethyl-3,5′-dimethoxybenzyloxycarbonyl group, peptide synthesis, XLVII, 519, 520
Dimethyldioxetanone
chemiluminescent reaction, LVII, 523
structural formula, LVII, 523
2,2-Dimethyl-5,5-dipentyl-*N*-oxyloxazolidine, as yeast spin label, XXXII, 828, 831, 833, 837
Dimethyl 3,3′-dithiobispropionimidate, LIX, 550
cleavage cross-linking, LVI, 628, 631, 632
2,2-Dimethyl-5,5-ditridecane-*N*-oxyloxazolidine, as yeast spin label, XXXII, 828, 831
Dimethyldodecanedioate, internal standard, XLVII, 34
Dimethylethylenediamine-cobalt(III) complex, XLVI, 315
Dimethyl 7-[*N*-ethyl-4-(*N*-phthalimido)butyl]aminonaphthalene-1,2-dicarboxylate
synthesis, LVII, 436, 437
of 7[*N*-(4-aminobutyl)-*N*-ethyl]aminonaphthalene-1,2-dicarboxylic acid hydrazide, LVII, 437
1,1′-Dimethylferrocene, as mediator-titrant, LIV, 408
Dimethylformamide, *see N,N*-Dimethylformamide
Dimethyl-d_7 formamide, XLIII, 393
N,N-Dimethylformamide, XLIV, 115, 116, 131, 289; XLVII, 279, 322, 489; LIX, 157
in affinity gel preparation, LI, 367
for carboxyl group titration, XLIV, 384
coupling yields, XLVII, 270
in cytochrome P–450_{scc} assay, LII, 125, 126
electron spin resonance spectra, XLIX, 383
in 5-fluoro-2′-deoxyuridine 5′-(*p*-nitrophenylphosphate) synthesis, LI, 98
ionophores, LV, 448
for luminol chemiluminescence reactions, LVII, 410
radical ion chemiluminescent reactions, LVII, 498, 503, 517
synthesis of chemiluminescent aminophthalhydrazides, LVII, 430, 432–434
in modification of tRNA, LIX, 160
in Mössbauer studies, LIV, 378
in peptide synthesis
advantages, XLVII, 550
disadvantages, XLVII, 555, 557
polystyrene derivatization, XLVII, 265
protein solubility, XLVII, 113
purification, XXXVI, 379
solvent, of *d*-camphor, LII, 170
for highly apolar substances, XLVII, 563
support swelling, XLVII, 298
synthesis of affinity matrix, LII, 517
of 8α-flavins, LIII, 458, 459
thioacetylation, XLVII, 292
in tributylammonium phosphate preparation, LI, 254
in *N*-trifluoroacetyl 6-amino-1-hexanol phosphate preparation, LI, 254
trypsin conjugate active site titration, XLIV, 391
as uncoupler solvent, LV, 464
Dimethylglutaric acid, as buffer, LV, 658
Dimethylglycine dehydrogenase, flavin linkage, LIII, 450
2-Dimethylguanosine
chromatographic mobility, LIX, 73
electrophoretic mobility, LIX, 91
2,5-Dimethylhexane, substrate of ω-hydroxylase, LIII, 360
2,2-Dimethyl-5-hexylmethylundecanoate-*N*-oxyloxazolidine, as yeast spin label, XXXII, 828, 837
1,1-Dimethylhydrazine, incorporation in tRNA, LIX, 118
5,5-Dimethylluciferin, structural formula, LVII, 27
5,5-Dimethylluciferyl adenylate, structural formula, LVII, 27
ε-Dimethyllysine, chromatographic separation, LIX, 788

N^{ε},N^{ε}-Dimethyllysine, identification, XLVII, 477
Dimethylmaleic anhydride, modification of green-fluorescent protein, LVII, 265
4,6-Di-*O*-methyl-D-mannose, L, 8
7α,17α-Dimethyl-19-nor-5α-dihydrotestosterone, nuclear receptor binding, XXXVI, 319
7α,17α-Dimethyl-19-nortestosterone, nuclear receptor binding, XXXVI, 315, 317, 319
Dimethyl oxalate, synthesis of ethyl[2-hydroxy-3-methoxycarbonyl-5-(4-methoxyphenyl)]valerate potassium enolate, LVII, 435
5,5-Dimethyloxazolidine-2,4-dione, measurement of ΔpH, LV, 229, 681, 683–685
4′,4′-Dimethyloxazolidine-*N*-oxyl derivative, XLIX, 370
5,5-Dimethyloxyluciferin, structural formula, LVII, 27
5,9-Dimethylpentadecanoate, deuterated, mass spectrum, XXXV, 345
2,9-Dimethyl-1,10-phenanthroline, assay of adenosine monophosphate nucleosidase, LI, 264
Dimethyl 2,2′-(1,4-phenylene)bis(5-phenyloxazole)
  in scintillation fluid, LI, 362
Dimethylphenylenediamine, hydrogen donor, for cytochrome *P*–450 peroxidation function, LII, 411
N,*N*-Dimethyl-*p*-phenylenediamine dihydrochloride, determination of acid-labile sulfide, LIII, 276
1,5-Dimethyl-2-phenyl-3-pyrazolone, *see* Antipyrene
Dimethyl 7-*N*-(4-phthalimidobutyl)aminonaphthalene-1,2-dicarboxylate
  synthesis, LVII, 436
    of dimethyl 7-[*N*-ethyl-4-(*N*-phthalimido)butyl]aminonaphthalene-1,2-dicarboxylate, LVII, 437
Dimethyl POPOP, *see* Dimethyl 2,2′-(1,4-phenylene)bis(5-phenyloxazole)
N,*N*-Dimethyl-1,3-propanediamine, XLIV, 120–122
6,7-Dimethylpterin, substrate of phenylalanine hydroxylase, LIII, 284
3,5-Dimethylpyrazole-1-carboxamidine, XLVI, 548–554
2,5-Dimethyl-1-pyrroline 1-oxide, XLIX, 371
5,5-Dimethyl-1,1-pyrroline 1-oxide, XLIX, 371
2,2-Dimethylsilapentane-5-sulfonate, standard, in proton NMR, LIX, 22
2,2-Dimethyl-2-silapentane-5-sulfonate, as resonance standard, LIV, 200, 206, 207, 210, 211, 212, 215
Dimethyl suberate, inner membrane enzymes, LVI, 629
Dimethyl suberimidate, XLIV, 640; LV, 518; LIX, 550
  cross-linking, LVI, 623
Dimethyl sulfate, XLIV, 124
Dimethyl sulfoxide, XXXIV, 546, 656; XLIV, 187, 284, 384, 834, 890, 891; LVIII, 12, 32; LIX, 147, 157
  assay of carbamoyl-phosphate synthetase, LI, 112, 122
  in cell freezing preservation, XXXIX, 121
  drying, XXXVIII, 412
  effect on carbamoyl-phosphate synthetase, LI, 119, 120
    on mouse erythroleukemia cells, LVIII, 506
  electron spin resonance spectra, XLIX, 383
  fluorescein labeling of tRNA, LIX, 149, 150
  in freezing, LVIII, 30, 448
  identification of iron-sulfur centers, LIII, 269, 271, 272, 274
  interaction with haloacids, XLVII, 443, 444, 454
  interference, in Mössbauer spectroscopy, LIV, 375
  ionophores, LV, 448
  as liver sample preservative, XXXI, 3, 4, 19
  long-term storage of mitochondria, LV, 29, 31, 32
  in low-temperature solvent mixture, LIV, 105, 108
  luciferin luminescence in, LVII, 348

modification reaction, of tRNA, LIX, 160
in ninhydrin reagent, XLVII, 8, 16
preparation of modified tRNA, LIX, 167, 168, 170, 171
of 30 S precursor-rRNA, LIX, 832
purification of CPSase-ATCase complex, LI, 114
of reconstituted tRNA, LIX, 134
redistillation, XLVI, 223
solvent
in EPR spectroscopy, LIV, 150
for highly apolar substances, XLVII, 563
for luminol chemiluminescence reactions, LVII, 410, 477, 563
of water-insoluble antibiotics, LIX, 853
storage of activated gel, XLIV, 33
synthesis of 4-[4-methoxyphenyl]butyric acid, LVII, 435
as uncoupler solvent, LV, 464
Dimethyl-d_6 sulfoxide, XLIII, 393
Dimethyl sulfoxide/hydrochloric acid, cleavage of tryptophanyl peptide bond, XLVII, 459–469
Dimethyl sulfoxide/hydrogen bromide, specificity, XLVII, 468
2,2-Dimethyl-5-tetradecane-5-propionic acid-*N*-oxyloxazolidine, as yeast spin label, XXXII, 828, 836
Dimethyl-3,3′-(tetramethylenedioxy)dipropionate dihydrochloride, cross-linking agent, LI, 131
3-(4,5-Dimethylthiazolyl-2)-2,5-diphenyl tetrazolium bromide, glucose oxidase staining, XLIV, 713, 905
3,4-Di-*O*-methyl-2-*O*-trideuteriomethyl-L-rhamnose, L, 33
Dimroth rearrangement, XLVI, 74
cAMP derivatives, XXXVIII, 392, 393
Dimyristoyl lecithin
desaturase activity, LII, 192
liposome preparation, LII, 208
NMR studies, XXXII, 205–211
Dimyristoylphosphatidylcholine, electron spin resonance spectrum, XLIX, 385
Dimyristoylphosphatidylcholine liposome, NADH-cytochrome b_5 reductase binding, LII, 108
Dimyristoylphosphatidylcholine vesicle, LIV, 59
Dimyristoylphosphatidylethanolamine, phase transition studies, XLIX, 498, 499
2,4-Dinitroaniline, XLVI, 488
Di-*N*-(2-nitro-4-azidophenyl)cystamine *S,S*-dioxide, XLVII, 422
2,4-Dinitrobenzenesulfonic acid, XXXIV, 49
Dinitrobenzoic acid hydrazide, preparation, LIX, 179
2,4-Dinitrobenzoic acid *N*-hydroxysuccinimide ester, preparation of modified tRNA, LIX, 170
4-(2,4-Dinitro)benzylidene-2-phenyloxazol-in-5-one, Raman spectrum, XLIX, 148
3,5-Dinitro-4-chlorobenzoate, modification of lysine, LIII, 143
Dinitrofluorobenzene, XLVII, 408
2,4-Dinitrofluorobenzene, site-specific inhibitor of nucleoside phosphotransferase, LI, 393
2,4-Dinitro-1-fluorobenzene, XLVI, 289
2,4-Dinitro-5-fluorophenylazide
millimolar extinction coefficient, LIII, 174
modification of cytochrome *c*, LIII, 173
of lysine, LIII, 141
4′-*N*-(2,4-Dinitrofluorophenyl)pyridoxamine, synthesis, XLVI, 445
4′-*N*-(2,4-Dinitro-5-fluorophenyl)pyridoxamine-5′-phosphate, XLVI, 442
reactions, XLVI, 444, 445
synthesis, XLVI, 442–444
Dinitrophenol
antibodies, XXXIV, 722
isolation of protein-315, XXXIV, 815
antigens, XXXIV, 718
derivatives, XXXIV, 711
inhibitor of luciferase assay, LVII, 49
2,4-Dinitrophenol
ATPases, LV, 301, 302
ATP synthesis, LV, 672
binding by ATPase, LVI, 527
complex V, LV, 315
cyanine dyes, LV, 579, 693

hepatoma mitochondria, LV, 84, 85, 88
immobilized mitochondria, LVI, 556
inhibitor of brain α-hydroxylase, LII, 317
M. phlei membrane systems, LV, 186
Pasteur effect, LV, 290
photochemically active derivative, LVI, 660
uncoupling activity, LV, 465
2,4-Dinitrophenol azide, XLVI, 83
2,4-Dinitrophenylacetate, hydrolysis, to measure rate constant, LVI, 494
P′-Dinitrophenyladenosine 5′-tetraphosphate
in ligand binding assays, LVII, 119–122
structure, LVII, 114
Dinitrophenylagmatine, mobility reference, XLVII, 57
Dinitrophenylalanine, XLVI, 76, 510
synthesis, XLVI, 510, 511
Dinitrophenylalanyl acid chloride, synthesis, XLVI, 511
2,4-Dinitrophenylalanyldiazoketone, XLVI, 509
synthesis, XLVI, 510–512
Dinitrophenyl-albumin, XXXIV, 191, 209
N-2,4-Dinitrophenyl-*p*-aminobenzoic acid, XLVI, 76
N-2,4-Dinitrophenyl-ε-aminocaproic acid, XLVI, 76
Dinitrophenylaminoethyl derivative, XXXIV, 49
Dinitrophenyl antibody, XLVI, 86, 479–492
Dinitrophenyl azide, XLVI, 99, 508–516
2,4-Dinitrophenyl-1-azide, XLVI, 510, 580
synthesis, XLVI, 511, 512
Dinitrophenylbacitracin, XLIII, 336
Dinitrophenyl-bovine serum albumin nylon, XXXIV, 218, 219
N^{γ}-Dinitrophenyl-N^{α}-bromoacetyl-L-diaminobutyric acid, XLVI, 490
γ-Dinitrophenyl-α-bromoacetyl-L-diaminobutyric acid bromoacetyl hydrazide, XLVI, 503
N^{ε}-Dinitrophenyl-N^{α}-bromoacetyl-D-lysine, synthesis, XLVI, 490
N^{ε}-Dinitrophenyl-N^{α}-bromoacetyl-L-lysine, synthesis, XLVI, 490
N^{δ}-Dinitrophenyl-N^{α}-bromoacetyl-L-ornithine, synthesis, XLVI, 490
γ-Dinitrophenyl-α-carbobenzoxy-L-diaminobutyric acid hydrazide, XLVI, 503
γ-Dinitrophenyl-α-carbobenzoxy-L-diaminobutyric acid methyl ester, synthesis, XLVI, 503
γ-Dinitrophenyl-L-diaminobutyric acid hydrazide dihydrobromide, XLVI, 503
γ-Dinitrophenyl-L-diaminobutyric acid hydrochloride, synthesis, XLVI, 502
γ-Dinitrophenyl-L-diaminobutyric acid methyl ester hydrochloride, XLVI, 502, 503
N-Dinitrophenyl-*N*′-*Z*-ethylenediamine, synthesis, XLVI, 489
N-Dinitrophenylethylenediamine hydrobromide, synthesis, XLVI, 489
Dinitrophenyl-gelatin, XXXIV, 209
Dinitrophenyl group, removal, XLVII, 611
Dinitrophenyl-hemocyanin, XXXIV, 219
2,4-Dinitrophenylhydrazine, XLVII, 489; LIX, 110
incorporation in tRNA, LIX, 114, 115, 118
Dinitrophenyllysine, XXXIV, 186
N^{ε}-Dinitrophenyllysine, mobility reference, XLVII, 56, 57
S-(Dinitrophenyl)-6-mercaptopurine riboside monophosphate, synthesis, XLVI, 290, 291
S-(Dinitrophenyl)-6-mercaptopurine riboside triphosphate, synthesis, XLVI, 289–291
N-2,4-Dinitrophenylmethionine, marker, of total available column volume, LI, 24
Dinitrophenyl-nylon, XXXIV, 204, 212, 219
N-Dinitrophenyl-L-ornithine hydrochloride, synthesis, XLVI, 490
Dinitrophenyl-poly(prolylglycylproline), substrate
aminopeptidase assay, XLV, 544
dipeptidyl carboxypeptidase assay, XLV, 600
Dinitrophenylprotamine sulfate, preparation, XLV, 25

4′-*N*-(2,4-Dinitrophenyl)pyridoxamine-5′-phosphate, XLVI, 442
reactions, XLVI, 444, 445
synthesis, XLVI, 443
2,4-Dinitrophenyl residue, in ligand binding studies, LVII, 120–122
Dinitrophenylstreptothricin, XLIII, 336
Dinitrosalicylate reagent, assay, of reducing sugar, XLIV, 98
2,4-Dinitrostyrene, XLIII, 72
Dinoseb, herbicide, LV, 470
Dio-9 complex
availability, LV, 514
biological properties, LV, 514, 515
nature, LV, 473
physical properties, LV, 477
structure, LV, 514
1,2-Di(9-*cis*-octadecenyloxy)-*rac*-glycerol-3-(2′-aminoethyl) phosphate, intermediate in phosphatidylethanolamine synthesis, XXXV, 459, 460
3,4-Dioctadecoxybutylphosphonic acid, intermediate in synthesis of phosphatidylcholine analogues, XXXV, 518, 519
3,4-Dioctadecoxybutylphosphonylcholine, intermediate in synthesis of phosphatidylcholine analogues, XXXV, 519, 520
L-α-Dioleoyllecithin, phospholipid, in reconstituted MEOS system, LII, 366
Dioleyl-L-α-lecithin, intermediate in phosphatidylcholine synthesis, XXXV, 443, 444
Dioleyl-L-α-phosphatidylcholine, transhydrogenase reconstitution, LV, 812, 813
Dio–9 reagent, XXXII, 843
Dioxane, XLIV, 50, 117, 126, 146, 749; XLIX, 438; LIX, 158
in CO_2 determination, LI, 156
crystallization of transfer RNA, LIX, 7
crystal morphology, XLIX, 470
impurity test, XLVII, 382
inhibitor of deoxythymidine kinase, LI, 365
peptide hydrolysis, XLVII, 41, 89
preparation of crystals, LIX, 5
of disuccinimidyl succinate, LIX, 159
in scintillation cocktail, LI, 52
in Sepharose substitution reaction, LI, 311
solvent, in peroxyoxalate chemiluminescent assays, LVII, 458
thioacetylation, XLVII, 292
p-Dioxane
in scintillation cocktail, LII, 392
synthesis of octylamine-substituted Sepharose, LII, 138
in Triton X–45 anionic derivative preparation, LII, 145
1,4-Dioxetanedione, reaction intermediate, in peroxyoxalate chemiluminescence, LVII, 456, 457
Dioxindolylalanine, XLVII, 447
absorption spectra, XLVII, 463
Dioxygenase, LII, 6, 7
Dioxygenation
definition, LII, 6
mechanism, LII, 38, 39
Dipalmitoyl-L-dihydroxypropylphosphonic acid, intermediate in synthesis of phosphatidic acid analogues, XXXV, 506
1,2-Dipalmitoyl-*sn*-glycerol-3-(2′-aminoethyl) phosphonate, intermediate in synthesis of phosphatidylethanolamine analogues, XXXV, 510, 511
Dipalmitoyl-DL-α-glycerol phosphate, intermediate in phosphatidic acid synthesis, XXXV, 440
Dipalmitoyl-*sn*-glycerol-3-phosphorylethanolamine, intermediate in phosphatidylethanolamine synthesis, XXXV, 456
1,2-Dipalmitoyl-*sn*-glycerol-3-(2′-trimethylammonium ethyl) phosphonate, intermediate in synthesis of phosphatidylcholine analogues, XXXV, 515–517
Dipalmitoyllecithin
freeze-fracture, XXXII, 48
NMR studies, XXXII, 204
phase transition temperature, XXXII, 197
Raman spectroscopy, XXXII, 257

spin-label studies, XXXII, 174
Dipalmitoylphosphatidylcholine, phase transition studies, XLIX, 498, 499
Dipalmitoylphosphatidylcholine vesicle, LIV, 59, 60
Dipeptidase
ascites tumor, XLV, 386
assay, XLV, 387
distribution, XLV, 392
inhibitors, XLV, 392
metal ions, effect, XLV, 391
molecular weight, XLV, 390
pH dependence, XLV, 392
properties, XLV, 390
purification, XLV, 387
purity, XLV, 390
specificity, XLV, 391
stability, XLV, 390
zinc content, XLV, 391
Cucurbita maxima, cotyledons, XLV, 393
E. coli B, XLV, 377
assay
absorbance difference, XLV, 379
with substrate Ala-Gly, XLV, 377
distribution, XLV, 386
inhibitors, XLV, 385
metal ions, effect, XLV, 385
pH dependence, XLV, 385
properties, XLV, 384
purification, XLV, 381
specificity, XLV, 385
stability, XLV, 384
thiols and SH reagents, effect, XLV, 385
units, XLV, 380
Dipeptide
alignment, XLVII, 399–403
analysis by gas chromatography-mass spectrometry, XLVII, 391–404
procedure, XLVII, 395–399
overlapping, preparation, XLVII, 394, 399
polarography, LVI, 477
thermolysin hydrolysis, XLVII, 178
trimethylsilylation, XLVII, 395
Dipeptidyl aminopeptidase I
assay, XLVII, 393
preparation, XLVII, 392, 393
Dipeptidyl aminopeptidase IV
preparation, XLVII, 394
stability, XLVII, 394
Dipeptidyl aminopeptidase V, XLVII, 404
Dipeptidyl carboxypeptidase
assay
benzyloxycarbonyltetraalanine, XLV, 600
ninhydrin method, XLV, 599
potentiometric method, XLV, 602
distribution, XLV, 609
immunodiffusion and immunoelectrophoresis, XLV, 607
kinetics, XLV, 608
metal ion requirement, XLV, 607
molecular weight, XLV, 607
properties, XLV, 607
purification, XLV, 603, 604
purity, XLV, 607
specificity, XLV, 608, 609
stability, XLV, 607
Diphenhydramine, adenylate cyclases, XXXVIII, 148, 178
o-Diphenol, protection against, XXXI, 539, 540
Diphenol oxidase, *see* Catechol oxidase
m-Diphenol oxidase, LII, 10
o-Diphenol oxidase, *see* Catechol oxidase
Diphenoyl peroxide
caged ion pair, LVII, 520
chemiluminescent reaction, LVII, 520, 523
structural formula, LVII, 520
Diphenylamine
assay, LI, 442, 443
in estimation of 3-deoxy-2-ketoaldonic acids, XLI, 33, 34
indicator for glycolipids, XXXV, 407
Diphenylamine reagent, preparation, LI, 248
9,10-Diphenylanthracene
acceptor, in chemiluminescent reaction, LVII, 563
activator, oxidation potential and role, LVII, 522
Diphenylcarbamyl chloride, XLVII, 360

Diphenylcarbamyl fluoride, inhibitor, kallikrein, XLV, 314
Diphenyl chlorophosphate, synthesis of 8-azido nucleotides, LVI, 650, 651
4,4′-Diphenyl-2,2′-dipyridine, LV, 517
Diphenyleneiodonium, binding site, LVI, 588
Diphenylhydantoin
  gas chromatographic determination, LII, 337–342
  metabolism, LII, 66
Diphenylhydantoin hydroxylase, *Ah* locus, LII, 232
1,3-Diphenylisobenzofuran, detection of singlet oxygen, LVII, 499
Diphenylmethanol, XLVII, 539
S-Diphenylmethyl-L-cysteine, synthesis, XLVII, 539
Diphenylmethyl ester moiety, peptide synthesis, XLVII, 522
S-Diphenylmethyl group
  removal, XLVII, 539
  stability, XLVII, 538
2,5-Diphenyl-1,3,4-oxadiazole
  chemiluminescent reaction, LVII, 511, 512, 519, 525, 526
  electrochemiluminescence step experiment, LVII, 512
  electron-transfer chemiluminescence, LVII, 497
  structural formula, LVII, 498
2,5-Diphenyloxazole, LIX, 192, 232, 236, 388, 459
  gel autoradiography, LVI, 606
  in scintillation cocktail, XXXII, 111; LI, 4, 30, 157, 292, 362, 409, 540; LII, 312, 315, 321, 392, 396, 398, 404; LVIII, 288
4,7-Diphenyl-1,10-phenanthroline, *see* Bathophenanthroline
Diphenylphosphoamide, preparation, XXXVIII, 421
Diphenyl phosphorazidate, LV, 516
Diphenylphosphoryl chloride, synthesis of adenosine derivatives, LVII, 119
α,α′-Diphenyl-β-picrylhydrazyl
  electron spin resonance spectra, XLIX, 416
  g value, XLIX, 523
  as reference, XLIX, 377
1,3-Diphenylpropane-1,3-dione, LV, 517
Diphenylthiocarbazone, *see* Dithizone
Diphosphatidyl(β-acyl)glycerol, intermediate in diphosphatidylglycerol synthesis, XXXV, 471
Diphosphatidylglycerol, *see also* specific compounds
  activator of dihydroorotate dehydrogenase, LI, 62
  amino acid transport, LV, 188
  methods of synthesis, XXXV, 469–471
3′,5′-Diphosphoadenosine
  assay, LVII, 239, 240
    calibration curve, LVII, 254, 255
    of PAPS, LVII, 245
    procedure, LVII, 253
    sensitivity, LVII, 246, 247, 253, 254
  substrate, of luciferin sulfokinase, LVII, 237
  tissue levels, LVII, 256
2,3-Diphosphoglycerate
  hemoglobin ligand binding and, LIV, 216, 217, 219
  inhibitor of adenylate deaminase, LI, 502
  nuclear magnetic resonance spectra, XLIX, 338
2,3-Diphosphoglycerate mutase, *see* Phosphoglycerate phosphomutase
1,3-Diphosphoglycerate phosphatase, *see* Bisphosphoglycerate phosphatase
1,3-Diphosphoglyceric acid, inorganic phosphate assay, LV, 212
Diphosphopyridine dinucleotide, estimation, XXXVII, 248
Diphosphopyridine nucleotide, cGMP assay, XXXVIII, 74, 78, 80–83
Diphtheria toxin
  catalysis of adenosine diphosphate ribosylation, LX, 677, 678, 706, 710
  inhibition of eukaryotic protein synthesis, LX, 780
  resistant strains, LVIII, 314
Diphtheria toxoid, antibody response, XLIV, 706
Diphytanoylphosphatidylcholine, in lipid bilayers, XXXII, 516

Diplocardia, luciferase structure, LVII, 258
Diplocardia longa
bioluminescent system, LVII, 375
collection, LVII, 375
Diplococcus pneumoniae
chloramphenicol acetyltransferase, XLIII, 738
neuraminidase, L, 560
Diploid analysis, random, LVI, 142–144
bias, of crosses, LVI, 143
Diploid strain, maintenance of recessive nuclear mutations, LVI, 120
Dipole-dipole coupling mechanism, LIV, 254–256
Dipole moment
electric dipole transition moment, definition, XLIX, 78, 79, 204
induced, XLIX, 72, 73
magnetic, of metal cation, XLIX, 424
Di-*n*-propylamine, electron spin resonance spectra, XLIX, 383
3,3′-Dipropylthiodicarbocyanine iodide, availability, LV, 690
2,2′-Dipyridine, inhibitor of phenylalanine hydroxylase, LIII, 285
2,2′-Dipyridine disulfide, activation agent, XLIV, 41
Dipyridine heme, preparation, LII, 432–435
Dipyridine heme dimethyl ester, proton paramagnetic resonance chemical shifts, LII, 434
α,α'-Dipyridyl
effect on epoxide hydrase, LII, 200
phosphodiesterase, XXXVIII, 256
2,2′-Dipyridyl disulfide, XXXIV, 533
Diquat, as mediator-titrant, LIV, 409, 410
Direct current polarography, XLIII, 374
Direct photoaffinity labeling, *see* Photoaffinity labeling, direct
Disaccharidase, in brush borders, XXXI, 125
Disaccharide
qualitative analysis, XLI, 18
synthesis, 2-nitrosoglycopyranosyl chlorides, L, 119–121
Disaggregation, with enzymes, LVIII, 125, 126
Discarine B, LV, 516
Disc gel electrophoresis
membrane protein separation, XXXII, 105, 106
with SDS, XXXII, 71
Discoidin I, *see also* Lectin, from *Dictyostelium discoideum*
properties, L, 311
separation from discoidin II, L, 310
Discoidin II, properties, L, 311
Disequilibrium, metabolic control, XXXVII, 278–280
Dishon, Weiss and Yphantis method, XLVIII, 213, 239–242
Disialoganglioside, isolation, XXXII, 354, 355
Disialomonofucosyllacto-*N*-hexaose, structures, L, 225
Disialomonofucosyllacto-*N*-neohexaose, structure, L, 225
Disialyllacto-*N*-tetraose, structure, L, 225
Disialylmonofucosyllacto-*N*-neooctaose, structure, L, 224
Disialylmonofucosyllacto-*N*-octaose, structure, L, 224
Disodium 2,6-dibromobenzenone-indo-3′-carboxyphenol, as mediator titrant, LIV, 408
Disodium *p*-nitrophenylphosphate hexahydrate, in 5-fluoro-2′-deoxyuridine 5′-(*p*-nitrophenylphosphate) synthesis, LI, 98
Disodium oxalate, preparation of derivatized aminohexaethyl-Sepharose, LIII, 246
Dispersant solution, LVIII, 124
Dispersion number, definition, XLIV, 802, 803
Dispersion theory, XLIX, 78, 79
Disse's space, XXXI, 102
Dissimilarity index, LIX, 604
Dissociation, *see also* Association reaction; Binding; Cooperativity; Dimerization; Interaction; Ligand binding; Self-association
enzymic, effect on viability, LVIII, 138
of protein, kinetics, XLVIII, 327–331

rate constant determination by dilution jump method, XLVIII, 245, 390, 391
Dissociation buffer, for sarcoplasmic reticulum studies, XXXII, 477
Dissociation constant, XXXIV, 111; XLVIII, 321, *see also* Adsorption effect
determination by tritium exchange
data analysis, XLVIII, 338–342
procedure, XLVIII, 333–336
regression algorithm, XLVIII, 339
theoretical derivation, XLVIII, 328
of enolase-inhibitor complexes, XLI, 122
metal ion-indicator, LVI, 304, 305
of redox reactions, definition, LIV, 416, 417
Dissociation media, XXXII, 741
for cell-adhesion studies, XXXII, 601
Distamycin A, XLIII, 145, 146
Distance
electron-nuclear, XLIX, 423
interatomic
C values, XLIX, 331, 353
from electron paramagnetic resonance spectra, XLIX, 352
Solomon-Bloembergen equations and, XLIX, 352, 353
sources of error, XLIX, 325, 326, 345
spin label-paramagnet, XLIX, 424
Distance distribution, LIX, 697–699
1,2-Distearoyl-*sn*-glycerol, intermediate in synthesis of phosphatidylethanolamine analogues, XXXV, 510
1,2-Distearoyl-*sn*-glycerol-3-phosphoryl-DL-serine, intermediate in phosphatidylserine synthesis, XXXV, 462
L-α-Distearoyl-lecithin, intermediate in phosphatidylcholine synthesis, XXXV, 441, 442
Distillation, azeotropic, XLVII, 524
Distolasterias nipon, *N*-acetyl-8-*O*-methylneuraminic acid, L, 65
Disuccinimidyl succinate
properties, LIX, 159
reaction with modified tRNA, LIX, 160, 163
with poly(C), LIX, 162
Disuccinimidyl tartaric acid, cross-linking, LVI, 631, 632
N^{α},N^{ε}-Disuccinyl-L-lysyl-D-alanyl-D-glutamic acid, XLIII, 697
Disulfide, in membrane proteins, reduction, XXXII, 72
Disulfide bond cleavage, XLVII, 79, 109–126
reductive, XLVII, 111–116
theory, XLVII, 111–113
with tributylphosphine, XLVII, 111–116
Disulfide-containing reagent, in histone determination, XL, 111
Disulfide exchange, yeast mitochondria preparation, LVI, 128, 129
DITC-glass, *see* Isothiocyanato glass
4,4′-Dithiobenzoic acid dihydrazide, preparation, LIX, 179
Dithiobisalkylimidate, cross-linking, LVI, 631
Dithiobischoline, XLVI, 587
5,5′-Dithiobis(2-nitrobenzoic acid), XLIII, 742, 743; XLIV, 458, 459, 474, 522; XLVI, 291, 391; XLVII, 130. 408, 428, 489, 497
in assay of α,β-unsaturated acyl-CoA derivatives, XXXV, 140, 141
carnitine assay, LV, 211
coenzyme A assay, LVI, 371, 372
cysteinyl residue measurement, XLIV, 16
inactivator of L-3-hydroxyacyl-CoA dehydrogenase, XXXV, 128
inhibitor of guanylate kinase, LI, 489
of NADPH-cytochrome *P*-450 reductase, LII, 96
of orotidylate decarboxylase, LI, 79
of purine nucleoside phosphorylase, LI, 536, 538
of thymidylate synthetase, LI, 96
of uridine phosphorylase, LI, 429
of palmityl thioesterases, XXXV, 102, 103, 107
phosphodiesterase, XXXVIII, 244
reagent for acetylcholinesterase accessibility assay, XXXI, 117

Dithiobis(succinimidyl propionate), LIX, 157
cross-linking, LVI, 631, 632
properties, LIX, 159, 160
radiolabeled, LIX, 166
reaction with modified tRNA, LIX, 160, 161
Dithiobis(succinimidyl) propionimidate, LIX, 550
Dithiocarbamic acid, XLVI, 165
Dithiodiglycolic acid, synthesis of dithiodiglycolic dihydrazide, LIX, 383
cellulose-bound, coupling with poly(U), LIX, 386, 387
preparation, LIX, 386
synthesis, LIX, 383–385
3,3′-Dithiodipropionic acid, di-*N*-hydroxysuccinimide ester, *see* Dithiobis(succinimidyl propionate)
2,2′-Dithiodipyridine, XLIV, 385
4,4′-Dithiodipyridine, XLVII, 408
Dithioerythitol, reducing substrate, of ribonucleoside diphosphate reductase, LI, 247
Dithioerythritol, LII, 489
in cytochrome b_5 purification, LII, 99
in plant-cell extraction, XXXI, 603
purification of aminoacyl-tRNA synthetases, LIX, 259
of peptide components of cytochrome bc_1 complex, LIII, 108
tRNA-protein coupling, LIX, 166
for storage of ribosomal proteins, LIX, 498
Dithioerythrol, assay of formate dehydrogenase, LIII, 364
Dithionite, *see also* Sodium dithionite
assay, LVII, 170, 175
of hydrogenase, LIII, 290, 291, 294, 295
catecholamine glass beads, XXXVIII, 186
cytochrome oxidase, LVI, 693
cytochrome spectra, LVI, 177
NADH dehydrogenase spectral properties, LIII, 19
oxidation, assay of nitrogenase, LIII, 329
preparation of arylamine glass, XLIV, 932
purification of hydrogenase, LIII, 289
reductant of cytochrome b_1, LIII, 232
of yeast cytochrome c_1, LIII, 226
Dithiooxamide, in gel electrophoresis of enzymes, XXXII, 89
Dithiothreitol, XXXIV, 538, 547; XLIV, 41, 316, 514, 846, 889–891; XLVI, 587–590; LIX, 235
activator, of cytidine deaminase, LI, 407, 411, 412
of deoxycytidine kinase, LI, 345
of formate dehydrogenase, LIII, 371
of GMP synthetase, LI, 223
of pyrimidine nucleoside monophosphate kinase, LI, 330
acyl-CoA hydrolysis, LV, 202
adenylate cyclase assays, XXXVIII, 122, 139, 140, 141, 173
adrenodoxin chromophore reconstitution, LII, 137
analysis of mitochondrial ribosomal proteins, LIX, 429, 431
assay of bacterial count, LVII, 70
of cyclic nucleotide phosphodiesterases, LVII, 97
of formate dehydrogenase, LIII, 364
of GTP hydrolysis, LIX, 361
of malate dehydrogenase, LVII, 183
of modulator protein, LVII, 109
of nucleotide incorporation, LIX, 182
for translocation, LIX, 359
bacterial luciferase reagent, LVII, 200
cell-free protein synthesis system, LIX, 358, 367, 850
coelomic cell homogenization buffer, LVII, 376
coenzyme A, LV, 211
cytochrome isolation, LII, 118–121, 127, 128, 140, 183, 195
cytochrome P–450_{scc} assay, LII, 126
in cytochrome P–450 isolation, LII, 364

for dehydrogenase isolation, XXXII, 380
dissociation of carbamoyl-phosphate synthetase, LI, 24, 25
effect on orotate phosphoribosyltransferase activity, LI, 152
on polymerization of polyacrylamide, XLIV, 896
in erythrocyte lysis, LI, 282
guanylate cyclase, XXXVIII, 202
harvesting of CHO cells, LIX, 219, 232
Hummel-Dreyer gel filtration, LIX, 326
incorporation of terminal nucleoside analogs, LIX, 188
inhibitor of brain α-hydroxylase, LII, 317
of glutaminase activity, LI, 25
as interference, in hemoprotein staining, LII, 330
with protein determination, LI, 160
isolation of translating ribosomes, LIX, 388, 391
luciferase reagent, LVII, 70, 218, 219, 221
luciferase renaturation buffer, LVII, 178
in microsomal protein solubilization, LII, 202, 360
in PED buffer, LI, 137
phosphodiesterase, XXXVIII, 246–248, 251, 255
photoaffinity labeling, LVI, 646
in plant-cell extraction, XXXI, 603
in postcoupling wash, XLVII, 326
preparation of aminoacylated tRNA, LIX, 132, 133, 818
of cross-linked ribosomal subunits, LIX, 535
of deuterated NADH, LIV, 226
of deuterated NADPH, LIV, 232
of *E. coli* crude extract, LIX, 194
of luciferase subunits, LVII, 172, 173, 178
of mitochondrial ribosomes, LIX, 425, 426
of neutron-scattering samples, LIX, 655
of polysomes, LIX, 363
of radiolabeled protein, LIX, 794, 795
of rat liver tRNA, LIX, 230
of ribosomal proteins, LIX, 485, 518, 519
of ribosomal subunits, LIX, 404, 411, 412, 752
in protein binding buffer, LIX, 322, 323, 328, 330
purification of bacterial luciferase, LIII, 567; LVII, 146, 147, 148
of creatine kinase isoenzymes, LVII, 62
of CTP(ATP):tRNA nucleotidyltransferase, LIX, 184
of glycine reductase, LIII, 376, 378, 379, 381
of luciferase, LVII, 62
of NADH dehydrogenase, LIII, 16
of nitrogenase, LIII, 323, 327
of oxidoreductase, LVII, 203
of reductase, LII, 93, 94
of tRNA methyltransferase, LIX, 194, 195, 197, 198, 200, 201
of succinate dehydrogenase, LIII, 28, 478
of tryptophanyl-tRNA synthetase, LIX, 239, 241, 242
in pyrophosphate determination, LI, 275
reconstitution of melilotate hydroxylase, LIII, 555
of tRNA, LIX, 127
reducing substrate of ribonucleoside triphosphate reductase, LI, 247, 258
reduction of methionine sulfoxide, XLVII, 458
reductive denaturing agent, XLIV, 522
removal, LVII, 204; LIX, 387
reversible dissociation of flavoproteins, LIII, 435
tRNA labeling procedure, LIX, 71, 103, 104
scavenger, of flavoprotein inactivators, LIII, 445, 447
in separation of tRNA isoacceptors, LIX, 222

stabilizer of adenosine monophosphate nucleosidase, LI, 264, 265
of adenylate kinase, LI, 468, 471
of adenylosuccinate synthetase, LI, 207, 208, 210
of carbamoyl-phosphate synthetase, LI, 34, 112, 122, 124
of cytidine deaminase, LI, 410
of deoxycytidine kinase, LI, 338
of deoxythymidylate phosphohydrolase, LI, 228
of erythrocyte enzymes, LI, 584
of FGAM synthetase, LI, 197
of GMP synthetase, LI, 215, 216
of hypoxanthine phosphoribosyltransferase, LI, 543, 545
of luciferase, LVII, 552
of nucleoside diphosphokinase, LI, 380, 381, 386
of nucleoside phosphotransferase, LI, 393
of OPRTase-OMPdecase complex, LI, 156, 162
of orotidylate decarboxylase, LI, 158
of phosphoribosylpyrophosphate synthetase, LI, 16
of protein B1, LI, 228, 230
of purine nucleoside phosphorylase, LI, 536, 538, 539, 553
of pyrimidine nucleoside monophosphate kinase, LI, 322–325, 328
of pyrimidine nucleoside phosphorylase, LI, 433, 434
of ribonucleoside diphosphate reductase, LI, 229–233, 238–240
of thymidine kinase, LI, 366–368
of uridine-cytidine kinase, LI, 302
storage buffer of ribosomal proteins, LIX, 585, 837
for storage of deoxycytidine kinase, LI, 341, 344
in storage medium for bacterial luciferase, LVII, 150
for enzyme rods, LVII, 204, 205, 207
studies of aminoacylation isomeric specificity, LIX, 279
of protein-RNA interactions, LIX, 323, 586
substrate of dimethylaniline monooxygenase, LII, 151
sucrose gradient buffers, LIX, 329
synthesis of *S*-(3-carbethoxy-3-diazo)acetonyl-7-thiolincomycin, LIX, 814
tritium labeling studies, LIX, 342
viscosity measurements, XLVIII, 54
Dithizone
in copper analysis, LIV, 444
for preparation of metal-free buffers, LIV, 475
Dithymidine 3′,5′-thiophosphate, XXXIV, 606
Diumycin, XLIII, 128
Di-*n*-undecanoyl peroxide
synthesis, LIII, 596
of coenzyme Q_2 analog, LIII, 595
Diver
for microgasometric measurements, XXXIX, 404–407
filling, XXXIX, 406, 407
1,4-Divinylbenzene, XLIV, 84, 110, 112–114
Divinyl ether, XLIV, 117
Divinyl ketone, XLIV, 35
Divinyl sulfone
activation, XLIV, 26, 34
cross-linking agent, XLIV, 22, 492
Divinylsulfone-agarose, XXXIV, 19
Divinylsulfone coupling method, XXXIV, 27–29
Divoklor soap, for glassware cleaning, XXXII, 818
DMA, *see* 9,10-Dimethylanthracene
DMB, *see* 5,5-Dimethyloxazolidine-2,4-dione
1,6-DMBP, *see* 1,6-Dimethylbenzo-[*a*]pyrene
199 D medium, for granulosa cell culture, XXXIX, 205, 209, 211
DME/F12 mixed medium, LVIII, 103, 104, 106
DMF, *see* *N,N*′-Dimethylformamide
DM medium, LVIII, 59
DM medium 120, LVIII, 57, 88
DM medium 145, LVIII, 57, 62–70

DM medium 160, LVIII, 57, 88
DMPE, *see* Dimyristoylphosphatidylethanolamine
DMSO, *see* Dimethyl sulfoxide
DM unit, *see* Debye magneton unit
DNA, *see* Deoxyribonucleic acid
DNA adenovirus, LVIII, 433, 434
DNA-agarose, XXXIV, 464–468
  entrapment, XXXIV, 469
DNA animal virus, LVIII, 404
DNA-binding protein, XXXIV, 6
DNA-cellulose, XXXIV, 469, 470
DNA-cellulose chromatography, XL, 181
  of cytoplasmic progesterone receptors, XXXVI, 204, 205
DNA-dependent RNA, *see* Ribonucleic acid, DNA-dependent
DNA-dependent RNA polymerase, *see* RNA polymerase, DNA-dependent
DNA-dextran, permissive tests, LVIII, 412
DNA inhibitor, in cell synchronization studies, XXXII, 596
DNA polymerase, XXXIV, 4, 468; XLVI, 90, 171
  associated with cell fractions, XL, 85, 86
  conformational studies, XLIX, 354
  differential centrifugation, XXXI, 722
  electron-nuclear distance, XLIX, 423
  in mammalian tissue, XXXIX, 454
  in mitochondria, XXXI, 24
  in nucleus, XXXI, 23, 24
  RNA-dependent, XXXIV, 5, 472, 473
    in virus transformed cell assay, XXXII, 591
DNA polymerase I, XXXIV, 469
  DNA-dependent, XLVI, 357
DNA polymerase II, XXXIV, 469
DNA polyoma, extraction, LVIII, 406, 407
DNase, *see* Deoxyribonuclease
DNase-agarose, XXXIV, 517–520
DNase-I inhibitor protein, XXXIV, 517–520
DNA-Sephadex, XXXIV, 463–465
  quantitation, XXXIV, 464
DNA simian virus
  assay, LVIII, 411, 412
  isolation, LVIII, 405
  preparation, LVIII, 406–410
  quantitation, LVIII, 411, 412
  radiolabeling, LVIII, 410, 411
DNA tumor virus, LVIII, 425
  transformation, LVIII, 370
DNA virus, cell transformation, XXXII, 586–587
DNBS, *see* 2,4-Dinitrobenzenesulfonic acid
Dnph, *see* 2,4-Dinitrophenylhydrazine
DNPP, *see* Dinitrophenylprotamine sulfate
Docosahexanoic acid, in mass spectrometry of triglycerides, XXXV, 355
Docosanoic acid, deuterated, mass spectrum, XXXV, 342
Dodecanal
  assay of luciferase, LIII, 565, 566, 570; LVII, 136, 137, 138, 162
  quantum yield, in bioassay, LVII, 192
Dodecanoic acid
  electron spin resonance spectra, XLIX, 403
  quantum yield, in bioassay, LVII, 192
1-*O*-(Dodecen-1′-yl)-2-stearoyl-*rac*-glycerol-3-dibenzylphosphate, intermediate in plasmalogen synthesis, XXXV, 482, 483
1-*O*-(Dodecen-1′-yl)-2-stearoyl-*rac*-glycerol-3-(*N,N*-dimethylaminoethyl) phosphate, intermediate in plasmalogen synthesis, XXXV, 483, 484
Dodecyl alcohol, and pore distribution in Spheron, XLIV, 67
Dodecyl aldehyde, assay of FMN, LIII, 422
Dodecylamine, XXXIV, 7
Dodecylammonium phosphate, self-association, XLVIII, 118–121
Dodecylammonium propionate
  osmometric studies, XLVIII, 79
  self-association, XLVIII, 118
Dodecyl benzene sulfonate, in *E. coli* lysis solution, LIX, 354, 356
Dodecyl sulfate, *see also* Sodium dodecyl sulfate
  dissociation of immunoprecipitate, LV, 347

inhibitor, of dihydroorotate dehydrogenase, LI, 62
Dodecyl sulfate gel electrophoresis, for soluble enzyme aggregate characterization, XLIV, 474, *see also* Sodium dodecyl sulfate
Dodecyl trimethylammonium bromide, micelles, ionic strength, LVI, 738
Dog
heart, mitochondria, characteristics, LV, 45
infused testis, XXXIX, 273–277
oxyntic cell isolation, XXXII, 709
Dolichol
reaction with 2,3,4,6-tetra-*O*-acetyl-β-D-glucopyranosyl phosphate, L, 129
with 2,3,4,6-tetra-*O*-acetyl-β-D-mannopyranosyl phosphate, L, 126, 127
Dolichol derivative, thin layer chromatography, L, 427
Dolichol diphosphate, test, L, 433, 434
Dolichol diphosphate-[^{14}C]-*N*-acetylglucosamine, synthesis, L, 416, 417
Dolichol diphosphate-[^{14}C]-*N*-acetylglucosaminyl-*N*-acetylglucosamine, synthesis, L, 417
Dolichol diphosphate-*N*-acetylglucosaminyl-[^{14}C]-*N*-acetylglucosamine, synthesis, L, 418
Dolichol diphosphate-[^{14}C]-*N*-acetylglucosaminyl-*N*-acetylglucosaminyl-mannose, synthesis, L, 422–424
Dolichol diphosphate-oligosaccharide-[^{14}C]glucose
alkaline treatment, L, 432
assay, L, 409
[^{14}C]glucose-oligosaccharide transfer to protein, L, 414–416
paper chromatography of [^{14}C]glucose-oligosaccharide, L, 412, 413
purification, L, 410, 411
synthesis, L, 408–410
Dolichol diphosphate-oligosaccharide-[^{14}C]-mannose
deacetylation, L, 433
preparation, L, 419–422
purification, L, 422
Dolichol intermediate
acid hydrolysis, L, 426–431
alkaline breakdown, L, 431–433
lipid-soluble product analysis, L, 426–428
methanolysis, L, 426
molecular weight determination, L, 435
phenol treatment, L, 430
protein glycosylation, L, 402–435
from uridine diphosphate glucose, L, 408–416
Dolichol monophosphate
assay, L, 403–406
extraction from pig liver, L, 406–408
test, L, 433, 434
Dolichol monophosphate-[^{14}C]glucose
assay, L, 409
glucose transfer, L, 411, 412
purification, L, 410, 411
synthesis, L, 408, 409
Dolichol monophosphate-[^{14}C]mannose, synthesis, L, 419
Dolichos biflorus, lectin, XXXIV, 333
P^1-Dolichyl P^2-diphenyl pyrophosphate, preparation from dolichyl phosphate, L, 132, 133
Dolichyl β-D-glucopyranosyl phosphate, preparation, L, 128–130
Dolichyl β-D-mannopyranosyl phosphate, purification and characterization, L, 127, 128
synthesis, L, 124–128
Dolichyl phosphate, synthesis, L, 130, 131
Dolichyl phosphate intermediate
chemical synthesis, L, 122–137
thin-layer chromatography, L, 123, 124
Domain
definition, XLIX, 4
in human erythrocyte membrane, XLIX, 14
DON, *see* 6-Diazo-5-oxonorleucine
Don-C cloned strain, Chinese hamster, XL, 76
Donnan equilibrium, XLVIII, 90, 91
Donnan potential, probe calibration, LV, 584
Donor quenching, *see* Fluorescence spectroscopy

DONV, *see* 5-Diazo-4-oxo-L-norvaline
Dopa, inhibitor of phenylalanine hydroxylase, LIII, 285
Dopa decarboxylase, *see* L-Amino acid decarboxylase
Dopamine, inhibitor of phenylalanine hydroxylase, LIII, 285
Dopamine β-hydroxylase, LII, 9
  in chromaffin granules, XXXI, 383–386
Dopa-oxidase, *see* Monophenol monooxygenase
Doppler shift spectroscopy, XLVIII, 416
Double-labeling technique, rat liver ribosomal proteins, LIX, 517, 527
Double-resonance technique, LIV, 213
Douglas fir seed, spherosome isolation from, XXXI, 577
Dounce homogenizer, XXXI, 10, 11, 17, 116, 145, 156, 161, 162, 253; XXXII, 120; LIX, 412
Dow-Corning light silicone oil DC200, XLVIII, 313, 314
Dow-Corning 30% silicone, LVII, 228
Dowex, *see also* specific type
  chromatography, elastatinal, XLV, 688
  purification of translational control ribonucleic acid, LX, 546
  resin, for cyclic AMP assay, XXXI, 106–108
Dowex 1
  cAMP chromatography, XXXVIII, 55
  cAMP inhibitor, XXXVIII, 277
  column, in preparation of deuterated NAD^+, LIV, 227
  cGMP assay, XXXVIII, 59, 88
  nucleotide separation, XXXVIII, 126–128, 130
Dowex 1–X2, XLIV, 865–867
Dowex 1–X4, XLIV, 864; LIX, 236, 254
Dowex 1–X8
  iodination of cytochrome *c*, LIII, 149
  purification of ADP, LVII, 61
  synthesis of adenosine derivatives, LVII, 116, 117
Dowex 2
  cAMP inhibitor, XXXVIII, 276
  phosphodiesterase assay, XXXVIII, 250
Dowex 50, XLIV, 878
  cAMP inhibitor, XXXVIII, 277
  column
    assay of cytidine triphosphate synthetase, LI, 85
      of phosphoribosylglycinamide synthetase, LI, 181
      of ribonucleoside diphosphate reductase, LI, 228
    preparation of derivatized Sepharose, LI, 253
  cyclic nucleotide chromatography, XXXVIII, 55, 144
    resin preparation, XXXVIII, 10, 11
    separation procedure, XXXVIII, 11–13
  dibutyryl-cGMP, XXXVIII, 407
  succinyl-cAMP purification, XXXVIII, 97
Dowex 50–X4, XLIV, 868; LIX, 236, 254
Dowex 50–X8
  purification of cyclic AMP, LVII, 98
    of cyclic nucleotides, LVII, 98, 99
Dowex 50-type resin, for microbore columns, XLVII, 227
Dowex AGI–X4, separation of nucleotides, LV, 248, 249, 251, 252
Dowex AG 50W–X8, in δ-aminolevulinic acid synthetase assay, LII, 352, 353
Dowex AG 501–X8, preparation of sucrose from cane sugar, LIX, 555
Dow polyglycol P–2000, LIX, 205
Doxycycline, XLIII, 385
  in media, LVIII, 112, 114
Doxylcyclohexane, electron spin resonance spectra, XLIX, 396
Doxyl moiety, definition, XLIX, 370
8-Doxyl palmitoylcholine, structure, XLIX, 494
16-Doxylphosphatidylcholine, XLIX, 405, 411
  electron spin resonance spectra, XLIX, 405
5-Doxylstearic acid, electron spin resonance spectra, XLIX, 407, 408, 410, 413
7-Doxylstearic acid, electron spin resonance spectra, XLIX, 407

12-Doxylstearic acid
electron spin resonance spectra, XLIX, 398
methyl ester, electron spin resonance spectra, XLIX, 399–401
14-Doxylstearic acid, electron spin resonance spectra, XLIX, 410
DP, *see* Polymerization, degree of
DPB, *see* 2,3-Dimethoxy-5-methyl-6-pentyl-1,4-benzoquinone
DPDA, *see* *N,N*-Diethyl-2,4-dichloro-(6-phenylphenoxy)ethylamine
DPEA, *see* 2,3-Dichloro-(6-phenylphenoxy)ethylamine
DPF, *see* Diisopropylfluorophosphate
DPL, *see* Dipalmitoyllecithin
DPN, *see* Nicotinamide adenine dinucleotide
DPNH, *see* Nicotinamide adenine dinucleotide, reduced
DPNH-cytochrome *c* reductase, *see* NADH-cytochrome *c* reductase
DPNH dehydrogenase, *see* NADH dehydrogenase
DPNH oxidase, removal, by acetone treatment, LI, 475
DPPC, *see* Dipalmitoylphosphatidylcholine
DPPH, *see* α,α'-Diphenyl-β-picrylhydrazyl
Drabkin's solution, hemoglobin concentration, LVI, 555
Drill, for rat skull surgery, XXXIX, 173
Drinking water, *see* Water, drinking
Driving block, continuous-flow pH measurement, LV, 619, 620
Driving-mixing device, pH measurement, LV, 619–622
Droplet sedimentation, in density gradient centrifugation, XXXI, 505, 513
Dropping mercury electrode, XLIII, 374, 375
Drosophila, raising and collection, XLI, 375
Drosophila immigrans, media, LVIII, 464
Drosophila melanogaster
alcohol dehydrogenase, assay, purification, and properties, XLI, 374–379
fructose-diphosphate aldolase, XLII, 223
karyotype, LVIII, 342
media, LVIII, 464
suspension culture, LVIII, 454, 455
Drosophila virilis, media, LVIII, 464
Drosopholin, XLIII, 336
Drug, *see also* specific substance
binding to receptor, XLVI, 574
isolation, LII, 333
metabolism
conjugation reactions, LII, 69
in hepatocytes, LII, 67–69
metabolites, analysis by gas chromatography and mass spectroscopy, LII, 331–342
microtubular disruptive, XL, 322
opiate, *see* Opiate receptor
toxicity, possible mechanism, LII, 228
Drug resistance, nuclear modifiers, LVI, 142
Drug-resistance mutant, *see* Mutant, drug-resistance
Dry-autoradiography, of steroid hormones, XXXVI, 139–146
Drying, of tissue, for collagen studies, XL, 312
Drying procedure, in trace metal analysis, LIV, 480
Dry weight method, for protein concentration determination, XLVIII, 155–162
accuracy, XLVIII, 156, 157
DSC, *see* Differential scanning calorimetry
DTA, *see* Differential thermal analysis
DTDI, *see* Dimethyl-3,3′-(tetramethylenedioxy)dipropionimidate dihydrochloride
DT-diaphorase, rhein, LV, 459
DTNB, *see* 5,5′-Dithiobis(2-nitrobenzoic acid)
dTPP, *see* Deoxythymine pyrophosphate
DTSP, *see* Dithiobis(succinimidyl propionate)
DTT, *see* Dithiothreitol
Duall tissue grinder, XXXI, 14
Dual-wavelength spectrophotometry, fast spectrophotometry, LIV, 36
Dubnoff shaking incubator, XXXIX, 240

Duchenne muscular dystrophy, urinary glucose-containing tetrasaccharide, L, 231
Duck, embryo fibroblast, media, LVIII, 59
Dulbecco buffer, LIV, 493
Dulbecco's modified Eagle's medium, XXXVII, 82; LVIII, 58, 86, 91, 95, 97, 100, 101, 213, 423, 499
  composition, LVIII, 62–70
Dulbecco's phosphate-buffered salt solution, LVIII, 120, 121, 142
  in studies of phagocyte chemiluminescence, LVII, 477
DuPont Cronex "Lightning Plus" intensifying screen, LIX, 70
DuPont Model 760 luminescence biometer, LVII, 115, 425, 550
  operational characteristics, LVII, 551, 553–555
Du Pont permaphase ODS column, LII, 285, 287
Du Pont Zorbax ODS column, LII, 285, 287
Du Pont Zorbax SIL adsorptive column, LII, 287
Duramycin, XLIII, 336
Duro–E–Pox E 5 cement, preparation of immobilized enzyme rods, LVII, 204
Duroquinol
  assay of complex III, LIII, 91
  preparation, LIII, 90, 91
  substrate of complex III, LIII, 37
Duroquinone, LIV, 429, 433
  preparation of duroquinol, LIII, 90, 91
Durrum electrophoresis apparatus, XXXVI, 130, 131
Durrum resin, XLVII, 23–31, 235
  particle sizes, XLVII, 227
Dursban pesticide, detection, XLIV, 658
Duysens absorption flattering effect, in circular dichroism, XXXII, 224
DVB, *see* 1,4-Divinylbenzene
Dye, XLVI, 90
  purity, monitoring, LV, 574
  sensitization, XLVI, 562, 563
  trinuclear heterocyclic, as optical probe, LV, 573
Dye-density equilibrium centrifugation, LVIII, 408, 410
Dye exclusion test, for cell viability, LVIII, 151
Dye-histone interaction, in direct quantitative determination, XL, 115
Dye laser
  in flash photolysis technique, LIV, 96, 97
  pulse length, LIV, 97
  in Raman spectroscopy, LIV, 246
Dye-ligand conjugate, XLVI, 563
Dye sensitization, luminol chemiluminescence, LVII, 421
Dynode chain, epoxy potting, LVII, 531, 533
Dynode feedback voltage meter, evaluation of photomultiplier overload, LII, 213
Dyno Mill, yeast cell breakage, LVI, 46
Dysprosium, magnetic circular dichroism, XLIX, 177
Dysprosium acetate, derivatization of transfer RNA, LIX, 14

E

E, *see* Redox potential
EAC$\bar{1}$,4 formation, XXXIV, 734
EAC$\bar{4}$,2 formation, XXXIV, 734
Eadie and Hofstee method, LVI, 266
Eadie plot, in enzymology, XXXVI, 4
EaDM, *see* Erythromycin C
EAE, *see* Encephalomyelitis, experimental allergic
Eagle's basal medium, XXXII, 749; LVIII, 84, 85, 91
  composition, LVIII, 62–70
  mixture with Hanks' Balanced Salt Solution, for cell culture, XXXII, 565
Eagle's medium, XXXI, 157
Eagle's minimum essential medium, LVIII, 54, 84, 85, 91, 155, 405
  composition, LVIII, 62–70
  for hepatic explants, XXXIX, 38
  HTC cells, LVIII, 545, 546
  for human cell culture, XXXII, 803
  for lymphocytes, LVIII, 486
  for macrophages, LVIII, 499
  myoblasts, LVIII, 522, 523
  for testis culture, XXXIX, 285

Earle's balanced salt solution, XXXII, 748; LVIII, 120, 121, 142
Eastman chromagram 6065, LI, 551, 559
Eastman Kodak AR–10 film, LVIII, 287
Eastman Kodak NTB nuclear emulsion, LVIII, 250, 279, 280, 283
Eastman White Reflecting barium sulfate paint, LVII, 580
Eberbach Animal Board, XXXIX, 373
EB-protein, *see* Estrogen-binding protein
EBS silicon diode image intensifiers, XXXII, 214
ECDI, *see* 1-Ethyl-3(3-dimethylaminopropyl)carbodiimide-hydrogen chloride
Ecdysone, autoradiography, XXXVI, 150
Ecdysone oxidase, LII, 21
Echelle grating, LIV, 461
Echelon
  index, matched glass, LIV, 5
  principles, LIV, 13, 14
  reflection, schematic representation, LIV, 13
  size limitations, LIV, 19
  transmission, schematic representation, LIV, 13
Echinarachnius parma, hatching enzymes, XLV, 372
Echinocardium cordatum, N-glycolyl-8-*O*-sulfoneuraminic acid, L, 65
Echinometra vanbrunti, hatching enzyme, XLV, 373
Echinomycin
  paper chromatography, XLIII, 150, 152
  peptide antibiotic lactonase, XLIII, 767–773
  Sarcina lutea, XLIII, 771
  structure, XLIII, 768
ECL, *see* Electrochemiluminescence
ECTEOLA, *see* Epichlorohydrin triethanolamine-cellulose
ECTHAM-cellulose, chromatography on, XL, 95
EDC, *see* 1-Ethyl-3(3-dimethylaminopropyl) carbodiimide
Edebo press, XXXI, 660
Edeine, *see also* Polyamine
  binding to ribosomal subunits, LX, 561
  biosynthesis, XLIII, 562–564
  countercurrent distribution, XLIII, 335, 339
  inhibition of protein synthesis, LX, 79, 80, 84, 86, 556, 575–577
  labeling with radioiodine, LX, 559, 560
  mediation of abnormal complex formation, LX, 366
  paper chromatography, XLIII, 146
  of 46 S complex formation, LX, 335, 337, 343
  of 60 S ribosomal subunit joining, LX, 73, 79, 80, 82, 84, 86, 123, 564–566
  structure, XLIII, 337
Edeine A
  biosynthesis, XLIII, 562–564
  structure, XLIII, 559, 560
Edeine B
  biosynthesis, XLIII, 562–564
  structure, XLIII, 559, 560
Edeine synthetase, XLIII, 559–567
  activation reaction, XLIII, 560
  amino acids activation, XLIII, 562
  assay, XLIII, 560–564
  Bacillus brevis preparation, XLIII, 564
  binding reaction, XLIII, 561
  biosynthesis of edeines A and B, XLIII, 562–564
  crude extract preparation, XLIII, 564
  DEAE-cellulose chromatography, XLIII, 564, 565
  inhibitors, XLIII, 566
  molecular weight, XLIII, 567
  pH optimum, XLIII, 566
  polymerization reaction, XLIII, 561
  properties, XLIII, 566, 567
  purification, XLIII, 563–566
  Sephadex G–200 filtration, XLIII, 565–567
  stability, XLIII, 566
Edestin, oligosaccharide-phenethylamine conjugates, L, 163–169
Edman degradation, XLVI, 114, 167, 380, 483
Edsall's fluid, XXXII, 305
EDTA, *see* Ethylenediaminetetraacetic acid
Edward medium, modified, for mycoplasma culture, XXXII, 459

EEDQ, *see* 1-Ethoxycarbonyl-2-ethoxy-1,2-dihydroquinoline
Egriwe's reagent, L, 78
Eel, electroplax, $(Na^+ + K^+)$-ATPase isolation, XXXII, 280
Effectiveness factor, XLIV, 412, 423–426
  definition, XLIV, 182, 412
  influence of particle radius, XLIV, 727–729
  for inhibition, definition, XLIV, 427
  limiting value, XLIV, 423
Effector, as ligand, XXXIV, 394
Efflux
  measurement, LVI, 289, 290
  studies, LVI, 257, 258
EFO, *see* Eicosafluoroundecanoic acid
Efrapeptin
  ATPase, LV, 343
  availability and preparation
    analysis, LV, 494
    growth conditions, LV, 494
    isolation and purification, LV, 494
  biological properties, LV, 494, 495
  physical properties, LV, 476
  structure, LV, 493
EGDMA, *see* Ethyleneglycol dimethacrylate
EGF, *see* Epidermal growth factor
Egg, phosphatidylcholine, XXXII, 485
Egg albumin, XXXI, 537
  steroid reaction, XXXVI, 389
Egg lecithin liposome
  desaturase activity, LII, 192
  preparation, LII, 208
  in stearyl-CoA desaturase activity assay, LII, 189
Egg yolk
  emulsion of, for phospholipase A_2 studies, XXXII, 149
  lipoprotein purification, XXXII, 155
  phosphatidylcholine, XXXII, 142
EGTA, *see* Ethylene glycol bis(β-aminoethyl ether)*N,N'*-tetraacetic acid
Ehrlich ascites cell
  acetone powder preparation, LI, 221
  adenylosuccinate synthetase, LI, 212
  Crabtree effect, LV, 296
  crude extract preparation, LIX, 412, 413
  efficiency of Pasteur effect, LV, 293
  GMP synthetase, LI, 219–224
  growth and harvest, LI, 221
  homogenate preparation, LI, 160
  native ribosomal subunits, preparation, LIX, 411–416
  OPRTase-OMPdecase complex, LI, 160–164
  plasma membranes, XXXI, 144, 161
Ehrlich-Lettré mouse ascites tumor cell, dipeptidase, XLV, 386, *see also* Dipeptidase, ascites tumor
Ehrlich reaction, indirect, glycolipid characterization, XXXII, 364, 365
Ehrlich reagent, modified, XLIV, 845
Eicosafluoroundecanoic acid, testosterone derivative, EC detection, XXXVI, 59
Ekman's reagent, arylamine group detection, L, 172
Elaioplast, in plant cells, XXXI, 495
Elasmobranch
  balanced salt solution osmolarity adjustment, LVIII, 469
  incubation temperature, LVIII, 471
Elastase, XXXIV, 4; XLVI, 20, 22, 198, 201, 206, 207, 214, 215
  amino acid composition, XLV, 423
  pancreatic, dissociation of rat liver, enzyme complex, LI, 121
  peptide separation, XLVII, 206
  spin-label studies, XLIX, 445, 446
  tissue culture, LVIII, 125
Elastase inhibitor, *see also* Bronchial mucus inhibitor; Submandibular gland inhibitor
  assay, XLV, 861, 870
  by elastatinal, XLV, 687
Elastatinal
  assay, by elastin, XLV, 687
  biological activity, XLV, 689
  inhibition of elastase, XLV, 687
  kinetics, XLV, 689
  properties, XLV, 688
  purification, XLV, 688
ELDOR, *see* Electron-electron double resonance technique
Electrical charge, transport, LVI, 248, 250
Electrical potential, assay across envelope vesicles and plasma membranes, LV, 612

Electrical potential difference, across biomembranes, LIV, 61–64
Electric charge, generation by proteoliposomes associated with planar phospholipid membranes, LV, 770, 771
Electric dipole transition moments, LIV, 235
Electric field, *see also* Autocorrelation function
  of laser light source, XLVIII, 425
  of photocathode surface, XLVIII, 424
  at photodetector, from light scattering from dilute solutions, XLVIII, 425
  of scattered light, XLVIII, 418, 424
    from dilute solution, XLVIII, 426
Electric organ, fluorescent probe studies, XXXII, 234
Electric tissue
  current output, XXXII, 310
  nicotinic receptors, XXXII, 315–317
  postsynaptic membranes, XXXII, 317
  source, XXXII, 316
  toxin-binding membranes, purification, XXXII, 317, 318
  toxin-binding protein, XXXII, 322
Electrochemical method
  general, LVI, 453
  polarography, LVI, 453, 454
Electrochemical proton gradient, artificial, initial rate of ATP synthesis, LVI, 494, 495
Electrochemiluminescence
  data analysis, LVII, 514, 515
  definition, LVII, 510
  effects of magnetic field, LVII, 518–520
  interception, LVII, 516, 517
  light generation, using rotating ring-disc electrode, LVII, 515, 516
    with step techniques, LVII, 512–516
  measurement, with cyclic voltammetry, LVII, 511, 512
Electrochromism, LIV, 62–64
Electrode, *see also* specific type
  air-gap, XLIV, 613
  amino acid, XLIV, 615, 617
  ammonia gas, XLIV, 582, 588
  ammonium-ion, XLIV, 347, 348, 582, 587, 590, 593
  ammonium selective, glutamate dehydrogenase, LVI, 483, 484
  amygdalin, XLIV, 585, 592, 617
    interferences, XLIV, 614
    response time, XLIV, 604–609
    stability, XLIV, 602
  bromide ion, XLIV, 582
  cadmium ion, XLIV, 582
  calcium ion, XLIV, 582
  calcium-sensitive, LV, 652
    calibration, LV, 653, 654
    of probe signal, LV, 581
  carbon dioxide, XLIV, 582, 588, 589
  chloride ion, XLIV, 582
  cholinesterase, XLIV, 593, 594
  cleaning, LVI, 473
  copper ion, XLIV, 582
  creatinine, XLIV, 617
  cyanide ion, XLIV, 582, 614
  divalent ion, XLIV, 582
  for electrophoretic light scattering, XLVIII, 489, 490, 492
  enzyme, *see* Enzyme electrode
  fluoride ion, XLIV, 582
  fluoroborate ion, XLIV, 582
  glass, *see* Glass electrode
  glucose, XLIV, 585–587, 602–616
    commercial, XLIV, 616
    interferences, XLIV, 614
    response time, XLIV, 607
    stability, XLIV, 602
  glutamine, response time, XLIV, 608
  hydrogen cyanide gas, XLIV, 582
  hydrogen fluoride gas, XLIV, 582
  hydrogen sulfide gas, XLIV, 582
  hydroxide gas, XLIV, 582
  iodide ion, XLIV, 582, 586, 587
  ion selective, XLIV, 347, 348
    biochemical systems, LVI, 445, 446
    definition, XLIV, 583
    description, XLIV, 581–583
    electrometers, LVI, 363, 364
    general background, LVI, 359–361
    glass, LVI, 361, 362
    liquid membrane, LVI, 362
    making measurements, LVI, 364–368

microelectrodes, LVI, 362, 363
neutral ionophores, LVI, 441–445
reference electrode, LVI, 363
solid state, LVI, 362
types, LVI, 361
commercially available, XLIV, 582
useful concentration range, XLIV, 581
ion-sensitive, measurement of proton translocation, LV, 631, 632
ion specific
measurement of membrane potential, LV, 556
pH changes, LV, 564
lead ion, XLIV, 582
for microiontophoresis, XXXIX, 430–433
monovalent ion, XLIV, 582
net K^+ and H^+ movements, LV, 668
nitrate ion, XLIV, 528
optically transparent, LIV, 405, 406
oxygen, XLIV, 347
in catalase assay, XLIV, 482
commercial source, XLIV, 582
in glucose assay, XLIV, 586
glucose oxidase assay, XLIV, 481
measurement of proton translocation, LV, 633
washing, XLIV, 611
penicillin, *see* Penicillin electrode
perchlorate ion, XLIV, 582
platinum, preparation, XLIV, 654, 655
porous cup, LIV, 456
potassium ion, XLIV, 582
for potentiometric-spectrophotometric titration, LIV, 405, 406
for redox potentiometry, LIV, 421, 422
reference, *see* Reference electrode
rotating disc, LIV, 456
routine care, LIV, 422
rubidium ion, XLIV, 582
silver ion, XLIV, 582
sodium ion, XLIV, 582
specific, calcium, LVI, 302
sulfide ion, XLIV, 582
sulfur dioxide gas, XLIV, 582
thiocyanate ion, XLIV, 582
urea
comparison of types, XLIV, 616
interferences, XLIV, 612, 613, 615
optimum enzyme concentration, XLIV, 611
response time, XLIV, 604, 608–611
urease, XLIV, 593
Electrode-monitoring, assay, XLIV, 346–348
Electrodeposition adsorption method, XLIV, 160, 248–250
Electrogenerated chemiluminescence, *see* Chemiluminescence, electrogenerated
Electrolectin, L, 291–302
affinity chromatography, L, 294, 295
assay, L, 292, 293
cellular adhesion, L, 300–302
isolation from *Electrophorus electricus,* L, 293–297
myoblast fusion, L, 292, 300, 301
properties, L, 297–299
saccharide specificity, L, 299
tissue distribution, L, 299, 301
Electrometer
ion selective electrodes, LVI, 363, 364
measurement of proton translocation, LV, 633
rapid proton-transfer reactions, LV, 615, 616, 619
vibrating-reed, LIII, 304
Electromotive force, estimation, LV, 592–595
Electron absorption, of γ-radiation, LIV, 375
Electron beam evaporator, LIX, 613–615
Electron-capture technique
acyl derivatives, XXXVI, 59, 60
for steroid analysis, XXXVI, 58–67
Electron density, computation from x-ray data, LIX, 11
Electron-density map, preparation, LIX, 12, 14
Electron donor, artificial, in bacterial transport studies, XXXI, 700, 701
Electron-electron double resonance technique, XLIX, 484, 503, 510
Electroneutrality, relation to molecuiar weight, XLVIII, 91
Electronic energy level scheme, LIV, 48

Electronic flash tube, photoaffinity labeling, LVI, 672
Electronic gyromagnetic ratio, LIV, 196
Electronik 15 potentiometer recorder, LIII, 304
Electron microscope autoradiography, *see* Autoradiography, electron microscope
Electron microscopy, *see also* Shadowing, high-resolution; specific types
  of adrenal cells, XXXIX, 160–165
  antibody use, XXXII, 60–70
  artificial enzyme membranes, XLIV, 912, 913
  ATPases, LV, 298
  ATPase (TF_1) from thermophile, LV, 373, 375–377
  of chromosomes, XL, 67
    autoradiography, XL, 72
  cytochemistry
    methods staining both nucleic acids, XL, 23
    of nucleic acids, XL, 22
    of nucleoproteins, XL, 19
      localization, XL, 19
  of endocrine tissue, XXXIX, 133–157
  ferritin-antibody labeled membranes, LVI, 716
  hepatoma mitochondria, LV, 87
  identification of mtDNA fragments, LVI, 191
  of liver organelles, XXXI, 20–23
  localization of cellular constituents, XLIV, 713
  of membranes and membrane components, XXXII, 1–70
  of membrane vesicles reacted with ferritin-conjugated immunoglobulin, LVI, 228
  mitochondrial integrity, LV, 144
  of mitochondrial ribosomes
    negative staining, LVI, 87, 88
    positive staining, LVI, 86, 87
  of pancreatic subcellular fractions, XXXI, 53, 54
  preshadowed carbon replica, preparation, LIX, 617
  of protein-depleted ribosomal 30 S subunits, LIX, 623–625
  of protein-RNA complexes, LIX, 581, 582
  of ribosomal 30 S subunits, LIX, 618–621
  scanning methods, XXXII, 50–60
  scanning type, XXXIX, 154–156
  of 16 S RNA-protein S4 complexes, LIX, 625–629
  of subcellular fractions, XXXII, 3–20
  supporting film preparation, LIX, 615, 616
  techniques
    analysis of micrographs, LVI, 721
    fixation and embedding procedures, LVI, 719, 720
    freeze-fracture, XXXII, 35–44
    negative staining, XXXII, 20–35
    photography, LVI, 721
    reagents, LVI, 719
    serial sectioning, LVI, 720, 721
  transmission type, XXXIX, 133–154
  using high voltage, XXXIX, 156, 157
Electron multiplier phototube, *see* Photomultiplier tube
Electron paramagnetic resonance
  characterization of iron-sulfur proteins, LIII, 265, 266, 268, 269, 273, 274
  cytochrome P–450 detection and quantitation, LII, 252–257
  of respiratory chain, ubiquinone extraction, LIII, 576
  techniques, studies with succinate dehydrogenase, LIII, 487–490
Electron paramagnetic resonance spectroscopy, *see also* Saturation-transfer spectroscopy
  acyl enzyme spin-labeling, XLIX, 438
  binding studies, XLIX, 324
  computer use, XXXII, 196–198
  correlation with Mössbauer spectroscopy, LIV, 363
  dialysis, XLIX, 431
  distance between spin label and fluorophore, XLIX, 425
    nucleus and, XLIX, 423, 424
    paramagnet and, XLIX, 424
  effects of isotropic and anisotropic motion, XLIX, 378–382
  field/frequency lock, XLIX, 414, 415
  on freeze-quenched samples, LIV, 85
  immunoassay, XLIX, 463

instrumentation, XLIX, 389–391, 435, 436
labeling compounds, XLIX, 369–371, 434, 443, 444
linear, definition, XLIX, 481
line shape, LIV, 2
low-temperature, LIV, 111–132
combination with optical spectroscopy, LIV, 131, 132
with Mössbauer spectroscopy, LIV, 132, 363
continuous titration, LIV, 124–130
of cytochromes, LIV, 102–111
data acquisition, LIV, 103, 104
design philosophy, LIV, 103, 104
discontinuous titration with solid titrant, LIV, 130
log time base, LIV, 104
sample preparation, LIV, 105, 120
time-sharing equipment, LIV, 104
titrant flask design, LIV, 126
titration procedure, LIV, 128–130
triple-trapping method, LIV, 105
magnetic field calibration, XLIX, 521–523
microwave cavity, XLIX, 391
microwave power saturation, XLIX, 522
polarity, XLIX, 422, 423
procedure, XLIX, 525, 526
protein applications, XLIX, 420–427, 512–528
protein concentration, LIV, 112
relaxing drunken sailor model, XLIX, 481, 482
sample handling, XLIX, 391–394, 435, 436
spectra
angular dependence, XLIX, 373–378, 386–389
anisotropic motion, XLIX, 373–378, 380–382, 430
artifacts, XLIX, 431, 432
autodigestion, XLIX, 454–457
from hydrolytic release of spin label, XLIX, 431, 450–454
macromolecular, XLIX, 457
broadening, XLIX, 424
conformational changes, XLIX, 432, 433
effect of temperature, XLIX, 519–521
field inhomogeneity, XLIX, 397, 400, 401
field modulation amplitude, XLIX, 394–396
filter time constant, XLIX, 397, 400
of free tumbling nitroxide, XLIX, 441
g value anisotropy, XLIX, 526, 527
integration, XLIX, 408–412, 524, 525, 527
kinetic information, XLIX, 441–443
microwave power, XLIX, 394, 396, 397
nitroxide-nitroxide interactions, XLIX, 385–389, 402–406, 424
optimal, parameters, XLIX, 527, 528
order parameter, definition, XLIX, 381, 382
orientation shifts, XLIX, 433
oxygen, XLIX, 401
powder, XLIX, 429, 440
proteolysis, XLIX, 431, 432
rapid rotation, XLIX, 380, 381
rapid wobble, XLIX, 381
restricted random walk, XLIX, 381, 382
rigid glass limit, XLIX, 378–380, 389, 440, 507
scaling, XLIX, 407, 408
scan time, XLIX, 397–399
screw axes, XLIX, 426
sensitivity, XLIX, 420, 429
solvent, XLIX, 382–385, 468
time averaging, XLIX, 406, 407
titration, XLIX, 412–417
tumbling rate, XLIX, 429–431
urea, XLIX, 438, 439
viscosity, XLIX, 378–380, 429, 430, 469
spectrometer, XLIX, 389–397
spin-labeling technique, XLIX, 369–418
spin parameters, XLIX, 372, 373
of transition metals in proteins, XLIX, 512–528

Electron paramagnetic resonance tube
  aerobic filling, LIV, 113
  cleaning, LIV, 121, 122
  demountable, types, LIV, 121
  in freeze-quench technique, LIV, 91
Electron spin resonance, *see also* Electron paramagnetic resonance spectroscopy
  trypsin active site titration, XLIV, 392
Electron transfer
  enzyme antibodies, LVI, 707
  in immobilized cytochrome *c*, XLIV, 545, 546
  reversed, OSCP assay, LV, 392, 393
Electron transfer component, temperature-sensitive mutants, LVI, 139
Electron-transfer process, chemiluminescence, LVII, 494–526
Electron-transferring flavoprotein, LIII, 502–518
  activity, LIII, 502
  assay, LIII, 503, 504
  catalytic properties, LIII, 506, 507
  dehydrogenase, LIII, 402, 404, 406
  electrophoretic properties, LIII, 508, 509, 512
  flavin content, LIII, 508
  from pig liver, LIII, 506–518
  role in electron-transport, LIII, 402
  spectral properties, LIII, 508, 510, 511
  stability, LIII, 518
  substrate specificities, LIII, 502, 503
Electron transfer system, anaerobic, transport energized, LVI, 383–388
Electron-transfer system component
  EPR signal properties, LIV, 133–145
  stoichiometries, LIV, 145
Electron transport
  cyanide-insensitive, plant mitochondria, LV, 25
  reconstruction with hemin, LVI, 176
  superoxide anion, LI, 69
  tentoxin, LV, 493
Electron transport complex, components, LVI, 3
Electron-transport component
  of bovine heart mitochondria, reconstitution, LIII, 48–54
  in nitrification pathways, LIII, 635
Electron transport particle
  amino acid transport, comparison to ghosts, LV, 186, 187
  depleted
    preparation, LV, 187
    properties and reconstitution, LV, 187, 188
  detergent-treated, properties and reconstitution, LV, 192, 193
  of *M. phlei*
    preparation, LV, 182, 183
    properties, LV, 183–185
  phosphorylating, of mitochondria, fluorescent probe studies, XXXII, 237
  preparation, LV, 309
Electron transport system
  cytochrome *P*-450-mediated immunological comparisons, LII, 247–249
  microsomal
    absorbance changes during aerobic steady state, LII, 216–219
    genetic differences, LII, 226–240
Electron tunneling phenomenon, LIV, 102
Electropherogram, two-dimensional, interpretation, LVI, 639–641
Electrophilic labeling, XLVI, 70
Electrophoresis, XLIII, 279–291, *see also* specific type
  agar gel, XLIII, 287–289
  of aldolase, XLI, 67–73
  amidinated mitochondria, LVI, 626–629
  ascending and descending boundaries, XLVIII, 203–207
  asymptotic boundary profiles, XLVIII, 201–208
  of ATPase, LV, 308
    beef heart mitochondria, LV, 319
    chloroform-solubilized, LV, 338
    E. coli, LV, 805, 806, 809, 810
    N. crassa, LV, 345, 347, 348
    oligomycin-sensitive, LV, 333
    prepared by chloroform method, LV, 343
    purified subunits, LV, 356
    yeast, LV, 353

of atractylose and carboxyatractyloside, LV, 521
2-azido-4-nitrophenol labeled material, LVI, 674–683
in bacterial membrane separation, XXXI, 650–653
band, quasi-elastic laser light scattering, XLVIII, 494
carrier-free continuous, XLIII, 289, 290
cell monitoring, LVIII, 167, 168
continuous free-flow type, for cell fractionation, XXXI, 746–761
of coupling factor 6, LV, 399
of coupling factor B, LV, 388, 389
cross-linked components, LVI, 636–639
cyclic nucleotides, XXXVIII, 417
cytochrome oxidase, LVI, 693, 694
of DCCD-binding protein, LV, 431, 432, 434
determination of isoenzyme content in lactate dehydrogenase, XLI, 47
disc gel, analysis of nonhistone chromosomal proteins, XL, 154
of disrupted antigen–antibody complex in SDS-acrylamide gels, XL, 248
enzyme activity, XLI, 66
exponential gradient gel, as second dimension, LVI, 610, 611
finite difference boundaries, XLVIII, 206–208
gel slabs, LVI, 64
HeLa cell mitochondrial proteins
  cylindrical gels, LVI, 69, 70
  slab gels, LVI, 70–72
high-voltage, XLIII, 281, 282
of labeled mitochondrial inner membranes, LVI, 617–619
mitochondrial RNA, LVI, 91
of monoamine oxidase, LVI, 695
of OSCP, LV, 396, 397
paper, XLIII, 280–286, *see also* Paper electrophoresis
of pyruvate kinase, XLI, 68–70
in quantitative histone assay, XL, 122
  in polyacrylamide gels, XL, 127
quasi-elastic laser light scattering, XLVIII, 420, 421
in tRNA sequence analysis, LIX, 65
slab gel
  methodology
    high concentration polyacrylamide step gels, LVI, 603, 604
    procedure, LVI, 605, 606
    solutions, LVI, 604, 605
  mitochondrial products, LVI, 104, 105
of TF_1, LV, 783, 784
of $TF_0 \cdot F_1$, LV, 370
thin-layer, XLIII, 286, 287
for thin-layer peptide mapping
  buffer, XLVII, 196, 200
  instrumentation, XLVII, 197, 198
of transhydrogenase, LV, 281, 282
two-dimensional, mitochondrial proteins, LVI, 78

Electrophoretic elution, *see* Elution, electrophoretic

Electrophoretic light scattering, XLVIII, 431–436
  advantages and disadvantages, XLVIII, 432
  bovine serum albumin and, XLVIII, 432, 433
  cell design, XLVIII, 485, 486
  electrodes, XLVIII, 489, 490, 492
  electroosmosis, XLVIII, 486, 487
  joule heating, XLVIII, 487–489
  power supply, XLVIII, 486
  reaction kinetics, XLVIII, 449, 450
  resolution, XLVIII, 432–435

Electrophoretic mobility, XLVIII, 3, 4
  accuracy, XLVII, 57
  amide assignments, XLVII, 62
  applicability to thin-layer electrophoresis, XLVII, 58
  factors affecting, XLVII, 53–57
  internal standards, XLVII, 56, 57
  at low pH, XLVII, 63
  molecular weight determination, XLVII, 61, 62
  of peptide derivatives, XLVII, 64–67
  peptide synthesis, XLVII, 68, 69
  theoretical considerations, XLVII, 52, 53

Electrophorus electricus, electric tissue, XXXII, 316
Electro-Photonics Model 43 dye laser, LIV, 53
Electrophysiological technique
in hormone-nerve studies, XXXIX, 429–440
extracellular recording, XXXIX, 434, 435
intracellular, XXXIX, 435–438
for transport linked membrane potential
growth and handling of cells, LV, 658, 659
procedure, LV, 661, 662
recording apparatus, LV, 659–661
α,β-Elimination, XLVI, 428
β-Elimination
determination of periodate-oxidized tRNA stability, LIX, 174, 175, 180, 181
in methylation analysis, L, 20
polysaccharide degradation, L, 33–38
Elliot Automation, XXXII, 212
Elliott double bent mirror, XXXII, 213
Elliott-Franks camera, XXXII, 213
Ellman's reagent, XLVII, 411, 426, 439, 466, 489
in factor XIII assay, XLV, 183
titration, of bromelain A, XLV, 9
of protein A, LIII, 380
Elongation, *see* Chain elongation
Elongation factor 2, preparation of ribosomal subunits, LIX, 404
Elongation factor eEF–1(EF–1)
affinity for guanosine nucleotides, LX, 583, 584, 589
assay by polyphenylalanine synthesis, LX, 115, 142, 640, 677, 689, 690
function, LX, 162, 163, 578, 579, 638, 657, 676, 695
nomenclature, LX, 686, 687
polyacrylamide gel electrophoresis, LX, 176, 177
preparation from *Artemia salina* cysts, LX, 690–694
from ascites cells, LX, 92
from reticulocyte supernatant, LX, 140–142, 582, 583, 638, 639, 641–644, 647, 648
from yeast, LX, 678–681
properties, LX, 658, 686, 694–696
separation from elongation factor eEF–2, LX, 640, 641
size classes, LX, 694, 695
Elongation factor eEF–2(EF–2)
adenosine diphosphate ribosyl derivative
assay, LX, 677, 678, 704, 706
dansylated, LX, 718
preparation, LX, 710–712
assay
by binding of guanosine nucleotides, LX, 704, 705
by polyphenylalanine synthesis, LX, 115, 142, 583, 650, 651, 662, 664, 704, 705
for *Pseudomonas* toxin A, LX, 788, 789
by ribosome-dependent guanosine triphosphate hydrolysis, LX, 704, 705, 708, 710
binding of guanosine nucleotides, LX, 583, 584, 708, 710, 711
contaminant of 80 S ribosomes, LX, 580
fluorescence polarization, LX, 712–714, 716–719
functional role, LX, 162, 613, 638, 649, 657, 676, 702, 703
labeling with dansyl chloride, LX, 715
molecular shape, LX, 718, 719
polyacrylamide gel electrophoresis, LX, 176, 177
preparation from Krebs ascites cells, LX, 652–654, 656
from pig liver, LX, 667, 668
from reticulocyte supernatant, LX, 140–142, 639–641, 645–647
from wheat germ, LX, 706–708
properties, LX, 658, 668, 681, 682, 708, 710
separation from other elongation factors
eEF–1, LX, 640, 641
EF–1α, LX, 666
stimulation by elongation factor EF-1β, LX, 656

Elongation factor eEF–Ts(EF–Ts), *see also* Elongation factor EF-1$\beta\gamma$
activity in protein synthesis, LX, 593, 594, 605, 606, 702, 703
assay
binding of aminoacyl-transfer ribonucleic acid to 80 S ribosomes, LX, 700
exchange of guanosine diphosphate in its complex with elongation factor EF-Tu, LX, 636, 637, 700
protein synthesis from aminoacyl-transfer ribonucleic acid, LX, 598, 599
stimulation of polyphenylalanine synthesis, LX, 597, 598, 700
complex with elongation factor EF–Tu, LX, 602–605, 616–618, 626, 627, 636–638, *see also* Elongation factor EF–$\alpha\beta\gamma$
constituent of intiation factor EF–1, LX, 686, 687, 694, 703
functional role, LX, 700
isolation, LX, 602–604, 700, 701
nomenclature, LX, 687
properties, LX, 701–703
Elongation factor eEF–Tu(EF–Tu), *see also* Elongation factor EF–1α
activity in protein synthesis, LX, 593, 594, 605, 615
assay
guanosine diphosphate binding, LX, 594, 636
protein synthesis from aminoacyl-transfer ribonucleic acid, LX, 598, 599
stimulation of polyphenylalanine synthesis, LX, 597, 598
complex with elongation factor EF–Ts, LX, 602–605, 616–618, 626, 627, 636–638, 703, *see also* Elongation factor EF–$\alpha\beta\gamma$
with guanosine diphosphate, LX, 616–618, 622
with guanosine triphosphate, LX, 627
with guanosine triphosphate and aminoacyl-transfer ribonucleic acid, LX, 615, 616
with guanylyl imidodiphosphate and aminoacyl-transfer ribonucleic acid, LX, 622, 624, 625
constituent of EF–1, LX, 194, 197, 686, 687, 703
isolation
using Sephadex A–50, LX, 602–604
using Sephadex C–50, LX, 696, 697, 699
using Sephadex G–100, LX, 599–602
nomenclature, LX, 687
properties, LX, 697, 698
regulation of binding of aminoacyl-transfer ribonucleic acid to 70 S ribosomes, LX, 615, 616
removal from elongation factor EF–G, LX, 610, 611, 613
Elongation factor EF–1_H, LX, 649, 656, 658, 686, 687
Elongation factor EF–3
assay, LX, 679, 682
functional role, LX, 676, 677
separation from elongation factor EF–2, LX, 681, 682
Elongation factor EF–1α, *see also* Elongation factor EF–Tu
assay, LX, 650, 662–665
complex with guanosine diphosphate, LX, 663
with guanosine triphosphate and aminoacyl-transfer ribonucleic acid, LX, 663
functional role, LX, 649, 658
nomenclature, LX, 686, 687
preparation, LX, 654, 656, 668, 670
properties, LX, 654, 658, 670
separation from elongation factor EF–2, LX, 666, 667
stimulation by elongation factor EF–1β, LX, 656
varieties, LX, 670
Elongation factor EF–1$\alpha\beta\gamma$, *see also* Elongation factor EF–Ts, complex with EF–Tu
composition, LX, 675, 676
high-molecular-weight form of EF–1, LX, 659

preparation from pig liver mitochondria, LX, 659
purification, LX, 674, 675
Elongation factor EF–1β
assay, LX, 651, 662–664
elution, LX, 655, 656
functional role, LX, 649, 656, 658, 673, 674
isolation, LX, 672–674
molecular weight, LX, 657, 671, 672
nomenclature, LX, 686, 687
Elongation factor EF–1βγ, *see also* Elongation factor EF–Ts
assay, LX, 662–665
gel electrophoresis, LX, 671, 672
molecular weight, LX, 671
nomenclature, LX, 686, 687
purification, LX, 670, 671
separation into EF–1β and EF–1γ, LX, 659, 672, 673
Elongation factor EF–1γ
isolation, LX, 672–674
molecular weight, LX, 671, 672
nomenclature, LX, 686, 687
Elongation factor EF–G
activity in protein synthesis, LX, 593, 594, 605, 606, 761, 762, 775–779
adjacency of sulfhydryl group and binding center, LX, 744, 745
assay
by protein synthesis from aminoacyl-transfer ribonucleic acid, LX, 598, 599
as quaternary complex, LX, 607–609
by stimulation of guanosine triphosphatase, LX, 595–597
of polyphenylalanine synthesis, LX, 597, 598
binding to ribosomal subparticles, LX, 719, 720
complex with guanosine triphosphate and ribosomes, LX, 726–738
active site of binding of guanosine triphosphate, LX, 733–737, 739–745
detachment from ribosomes as requirement for further elongation, LX, 775–779
enzymic homogeneity, LX, 614
free vs. ribosome-bound, binding of guanosine nucleotides, LX, 737–741
functional role, LX, 719, 720
isolation, LX, 599–602, 608–613
labeling with iodoacetamide, LX, 743
limited trypsinolysis, LX, 743, 744
molecular weight, LX, 614
photoactivated analogs
complex with guanosine triphosphate and ribosomes, LX, 722, 723
preparation, LX, 720–722, 742
use in proof of interaction with ribosomal subunits, LX, 723–726
primary structure, LX, 742
tests of purity, LX, 613, 614
translation in ribosomes induced by attachment, LX, 766–774
Elongation factor G
assay of GTP hydrolysis, LIX, 361
of polypeptide chain elongation, LIX, 358
for translocation, LIX, 359
radiolabeled, preparation, LIX, 790, 791
ribosome release, LIX, 862
Elongation factor T, assay of polypeptide chain elongation, LIX, 358
Elongation factor Tu, XLVI, 151
cGMP assay, XXXVIII, 85–90
in protein-tRNA interaction studies, LIX, 324
radiolabeled, preparation, LIX, 790, 791
Elon solution, assay of amidophosphoribosyltransferase, LI, 172, 173
ELS, *see* Electrophoretic light scattering
Elton Developing Agent, XXXI, 246
EL–4 tumor cell, antigen studies, XXXII, 68, 69
Eluent
concentration, XXXIV, 248
selection, XXXIV, 246
Eluent line flow velocity, definition, XLVII, 20
Eluent pump pressure, cause, XLVII, 20
Eluotropic series, XLIII, 292, 293

Elution
of antigen, XXXIV, 726–728
batchwise, estradiol receptors, XXXIV, 684
electrophoretic, in analysis of nonhistone chromosomal proteins, XL, 158
gradient, XXXIV, 237–239
linear velocity, XLVII, 214
pH jump, XXXIV, 252
profile, variability, factors causing, XLVII, 20
temperature, XXXIV, 252
ternary complex formation, XXXIV, 238, 239
Elutriation, centrifugal, LVIII, 258
EMA, *see* Ethylene/maleic anhydride copolymer
Emanuel-Chaikoff orifice-type homogenizer, LI, 301
Emasol 1130 detergent, purification of cytochrome oxidase, LIII, 56, 58–60
Emasol 4130 detergent, LVI, 741
partial specific volume, XLVIII, 19, 20
purification of cytochrome oxidase, LIII, 58
Embedding
electron microscopy, LVI, 720
of subcellular particles, XXXII, 6, 7
Embedding plastic, for transmission electron microscopy, XXXIX, 139–141
Embryo, *see also* Fetus
carcinoma, LVIII, 97, 105
from mice, preparation, XXXIX, 297–302
preimplantation, culture, XXXIX, 298, 299
Embryo chick cartilage, in somatomedin bioassay, XXXVII, 104–106
Embryonic cell, *see also* specific types
cloning, LVIII, 155
Emericellopsis minima, fungal penicillin acylase, XLIII, 723
Emericid, LV, 445
Emetine, LVI, 31
HeLa cell labeling, LVI, 68, 77
protein inhibitor, XL, 287, 288
resistant strains, LVIII, 314
Emission, sensitized, *see* Fluorescence spectroscopy
Emission anisotropy factor
behavior across emission bands, XLIX, 186
conformation studies, XLIX, 185
definition, XLIX, 180, 181
Emission spectrum, effect on units on, shape, LVII, 573, 574
EM-1 medium, LVIII, 105
EMS, *see* Ethylmethanesulfonate
Emulgen 911, LII, 120, 121, 123
absorbance, LII, 198
for eluting epoxide hydrase, LII, 197
inhibitor of epoxide hydrase, LII, 200
Emulgen 913, in cytochrome *P*–450 purification, LII, 140
Emulgophene BC–720 detergent, LVI, 741
in rhodopsin purification, XXXII, 308, 309
Emulsion
choice, LVIII, 279, 280
coating, in electron microscope autoradiography, XL, 6
interference, color, XL, 8
dipping, LVIII, 284, 285
drying, LVIII, 284, 285
nuclear, LVIII, 279, 280
photographic, XXXVII, 159, 160
preparation, LVIII, 283, 284
pressure-sensitive, LVIII, 291
Encephalomyelitis
A1 protein, XXXII, 324, 325–327, 340
experimental allergic, XXXII, 324
5′-End-group labeling, LIX, 74–76, 100, 102, 103
End-group method, XLVII, 71–107
Endoaffinity labeling, XLVI, 12
reagents, XLVI, 115, *see also* specific substances
Endocrine cell
differentiation, LVIII, 269, 270
experimental systems, used in study, XXXIX, 165
isolated, hormone effects, XXXIX, 403–425
microspectrophotometry, XXXIX, 413–422

ultrastructural studies, XXXIX, 157–165
Endocrine cell line
freezing preservation, XXXIX, 120–122
isolation, cloning, and hybridization, XXXIX, 109–128
Endocrine gland
cytochrome *P*
properties and preparation, XXXVII, 304–310
ultrastructural studies, XXXIX, 157–165
Endocrine tissue, XXXIX, 107–425
electron microscopy, XXXIX, 133–157
fixation, XXXVII, 136
hormone production, XXXIX, 302–328
6,14-Endoethenotetrahydrothebaine, XLVI, 602
Endogluconase, as cell wall hydrolyzing enzyme, XXXI, 610
Endoglycosidase, L, 272
bacterial viruses, L, 269
Endometrium, nuclei preparation, XXXVI, 321
Endomycin, XLIII, 135, 336
Endopeptidase
activity assays, XLVII, 77, 78
production by *Aeromonas proteolytica*, XLV, 406, 407
Endoplasmic reticulum, *see also* specific type
N-acetylgalactosaminyltransferase, XXXII, 328
AHH activity, LII, 238–240
characterization, XXXI, 23, 33, 35
diagnostic enzyme, XXXI, 20, 25, 165, 743
differential centrifugation, XXXI, 727
electron microscopy, XXXI, 26, 29
enzyme, XXXII, 86, 87
enzyme markers, XXXII, 393, 402
fluorescent probe studies, XXXII, 234
IR spectroscopy, XXXII, 250
in liver preparations, XXXI, 7
melanin formation, XXXI, 389
negative staining, XXXII, 31
in plant cells, XXXI, 489, 491
ribosome binding, XXXI, 227, 228
rough
mammary formation, XXXIX, 454
of pancreas, XXXI, 41, 42
rough and smooth microsomes, XXXI, 191
by centrifugation, XXXI, 724, 725
rough and smooth types, separation, XXXI, 225–237
Endothelial cell
isolation of, from brain, XXXII, 717–722
from epididymal capillaries, XXXIX, 479–482
vascular, monolayer, LVIII, 133
Enediol, XLVI, 145
Energy, requirements for mitochondrial protein synthesis, LVI, 21, 22, 30
Energy charge
measurement of responses
concentration of substrates, LV, 234
effects of metal ions and phosphate, LV, 233, 234
establishment of desired values, LV, 232, 233
size of adenylate pool, LV, 233
presentation of results, LV, 234, 235
relevance, LV, 229–232
Energy conservation
reconstitution, LV, 715, 716
assay
of oxidative phosphorylation
at site I, LV, 723–726
site III, LV, 731, 732
of transhydrogenase, LV, 726–728
conclusions, LV, 735, 736
formation of phospholipid vesicles containing complex I and ATP synthetase, LV, 716–722
of vesicles
by cholate-dialysis technique, LV, 722, 723, 729, 730, 733, 734
cholate-dilution technique, LV, 730, 731, 734, 735

containing succinate dehydrogenase, complex III, cytochrome *c*, cytochrome oxidase ATP synthetase, LV, 732, 733
oxidative phosphorylation
site I, LV, 716
site II, LV, 732
site III, LV, 728, 729
Energy coupling, by submitochondrial membrane fragments, LV, 108, 109
Energy-linked process, fluorescence, LVI, 496, 497
apparatus, LVI, 497, 498
measurements, LVI, 498–501
remarks, LVI, 498
Energy mechanism, hormone effects, XXXVI, 429–433
Energy transduction, dependence on K^+ gradient, LV, 672, 673
Energy transfer
efficiency, LVII, 589, 590
future use, XLVIII, 375–377
intramolecular, hydrazide chemiluminescence, LVII, 417
nonradiative, *see* Fluorescence
singlet-singlet, XLVIII, 347–379
ENJ–3029 emulsifier, XLIV, 211, 330
Enniatin A
as ionophore, LV, 442, 444
transport, LV, 680
Enolase, XLII, 323–338; XLVI, 23, 381, 382
conformation studies, XLIX, 357, 358
from *Escherichia coli*, XLII, 323–329
activation and inhibition, XLII, 327
amino acid composition, XLII, 327
assay, XLII, 323
chromatography, XLII, 326
molecular weight, XLII, 327
preparation and purification, XLII, 324–328
properties, XLII, 327, 329
from fish muscle, XLII, 329–334
assay, XLII, 330
molecular weight, XLII, 334
preparation and purification, XLII, 329–334
properties, XLII, 334
gluconeogenic catalytic activity, XXXVII, 281
from human muscle, XLII, 335–338
assay, XLII, 335
molecular weight, XLII, 338
preparation and purification, XLII, 335–337
properties, XLII, 337, 338
membrane osmometry, XLVIII, 75
site-specific reagent, XLVII, 483
from thermophilic bacteria, properties, XLII, 329
transition-state analogs, and active site-specific reagents, XLI, 120–124
Enolase-inhibitor complex, dissociation constants and stoichiometry, XLI, 122
Enoyl-CoA hydratase, XXXV, 136–151
assay, XXXV, 137–139
effect of chain length on activity, XXXV, 149, 150
in glyoxysomes, XXXI, 569; LII, 502
inhibition by sulfhydryl reagents, XXXV, 150
inhibitors, XXXV, 149, 150
β-oxidation, XXXV, 136–151
regulation of β-oxidation of fatty acids, XXXV, 149, 150
stability, XXXV, 138, 145
stereospecificity, XXXV, 148
Entamoeba histolytica, pyruvate orthophosphate dikinase, XLII, 187, 199, 200
Enterobacter
cephalosporinase, XLIII, 678, 679, *see also* β-Lactamase, from *Enterobacter*
β-lactamase, *see* β-Lactamase, from *Enterobacter*
Enterobacter aerogenes, *see Aerobacter aerogenes*
Enterochelin
estimation, LVI, 397
isolation, LVI, 394, 395
materials, LVI, 396–398
principle, LVI, 395, 396
properties, LVI, 398
Enterokinase, XXXIV, 449
in brush border, XXXI, 130

Enthalpy
cAMP hydrolysis, XXXVIII, 238, 239
difference, in thermal transition, calculation, XLIX, 7–10
Entrapment, *see also* Microencapsulation
aggregation, XLIV, 265
by cross-linking polymers, XLIV, 177–179
in gels, XLIV, 169–183, 902
of multistep enzyme system, XLIV, 181, 182, 475–478
parameters effecting, XLIV, 169–171
by solution polymerization, XLIV, 172–176
by suspension polymerization, XLIV, 176, 177
of whole cells, XLIV, 183–190, 740, 741, 768–776
effects of buffering, XLIV, 187
Envelope vesicle, *see also* Membrane vesicle
assay of electrical potentials, LV, 612
Environmental sample, biomass determination, LVII, 73–85
Enzacryl AA, XLIV, 56, *see also* Polyacrylamide gel
amylase immobilization, XLIV, 98–100
components, XLIV, 86, 98
enzyme coupling, XLIV, 87
Enzacryl AH, XLIV, 56, *see also* Polyacrylamide gel
acyl azide derivative, XLIV, 99
amylase immobilization, XLIV, 99, 100
characterization, XLIV, 93
components, XLIV, 87
synthesis, XLIV, 92, 93
thermolysin immobilization, XLIV, 95
Enzacryl polyacetal, XLIV, 105
structural features, XLIV, 88
Enzacryl polyaldehyde, immobilization reations, XLIV, 89
Enzacryl polyaldehyde A, XLIV, 105
Enzacryl polyaldehyde B, XLIV, 105
Enzacryl polythioacetone, XLIV, 87, 88
Enzacryl polythiol, XXXIV, 535; XLIV, 87
Enzacryl support, structural features, XLIV, 86
Enzite, *see* (Carboxymethyl)cellulose
Enzyme, *see also* specific substance
action, mechanistic studies, XLIV, 560–566
active sites, XLVI, 3, 115, 197
agar gels, XLIV, 22
aggregation, *see* Aggregation
allergic reaction, XLIV, 685–687, 689, 692, 699, 704, 708, 709
assay
using Fast Analyzer, XXXI, 826–828
of lactic acid, XLI, 41–44
in nonaqueous tissue fractions, LVI, 204
procedures, XLI, 47–73
of resazurin and resorufin, XLI, 55
associated with cell fractions, XL, 86
in bacterial membranes, XXXI, 649
binding mechanism
to collagen supports, XLIV, 261–263
covalent, XLIV, 25–27, 37, 38, 41, 47
bonding to anodes, LVI, 460
of brain mitochondria, LV, 17, 18
specific activity, LV, 58
in cloning, LVIII, 125
commercial source, XLIV, 584
conformation, changes, XLVI, 41–47
coupling by γ-radiation, XLIV, 37
decay of activity following effect of inducing agents, XL, 261
following inhibition of protein synthesis, XL, 262
deficiency, treatment, XLIV, 684–688, 700–703
degradation in animal tissue, analysis, XL, 241
diagnostic, for liver cells, XXXI, 19, 20
digestion
interpretation of results, XL, 39
procedure, XL, 36
direct polymerization onto electrode membrane, LVI, 486
drug-metabolizing, in microsomes, XXXI, 84, 85
electrode applications, kinetic considerations, LVI, 481, 482
as electron microscopic tracers, XXXIX, 145–149

entrapment in gels, XLIV, 169–183
extract preparation, LVIII, 167
fiber-entrapped, XLIV, 227–242
  assay, XLIV, 231, 232, 237
  efficiency parameters, XLIV, 235–238
  stability, XLIV, 238–240
half-life
  hormone-stimulated rate of synthesis, XL, 262
  in short-term incorporation of isotope, XL, 251
of hepatoma mitochondria, LV, 82–84, 88
hormone-stimulated rate of synthesis, XL, 262
hydrolytic, XL, 353
  in liver homogenates, XXXI, 7
immobilization, LVI, 480, 481
  on (aminochloro-*s*-triazinyl)cellulose, XLIV, 50, 51
  on cellulose-transition metal salt complex, XLIV, 48
  choice of coupling technique, XLIV, 148
    of pore diameter, XLIV, 156–160
  coupling time, XLIV, 144
  effect of concentration, XLIV, 143
    on thermal stability, XLIV, 97, 98
  in hollow-fiber membranes, XLIV, 306–310
  to inert protein, XLIV, 709–717
  on magnetic particles, XLIV, 724
immobilized
  active site titration, XLIV, 390–393
  activity, XLIV, 3, 4
    parameters effecting, XLIV, 136–138
  amino acid analysis, XLVII, 40–45
  automated analysis, XLIV, 633–646
  as biological model systems, XLIV, 63, 64
  bonding heterogeneity, XLIV, 358–360
  bound protein determination, XLIV, 386–390
  on cellulose, properties, XLIV, 46–53
  classification, XLIV, 676, 677
  as commercial products, XLIV, 584, 596
  conformation studies, XLIV, 361, 362
  definition, XLIV, 201, 202
  effect of organic solvent, XLIV, 843
  injection, XLIV, 678, 679
  kinetic behavior, XLIV, 397–443
  physical properties, XLIV, 357–372
  preparation, XLVII, 41, 42
    for microflow immobilized enzyme reactors, LVI, 489–491
  reactivation, XLIV, 164
  routes of administration, XLIV, 678–681
  stability, XLIV, 97, 98, 361, 362, 643, 644
  theoretical enzymology, XLIV, 559, 560
  therapeutic applications, XLIV, 676–699
  types, XLIV, 202
  use, as analytic reagent, XLIV, 579, 580, *see also* Enzyme electrode; Organic synthesis; Reactor
immunoassay, XLIV, 714–716
immunotitration, XL, 243
induction, in perfused liver, XXXIX, 35
inhibition
  by antibodies, LVI, 227, 228
  by phospholipid analogues, XXXV, 503
inhibitors, XLIV, 615
  in plant-cell extraction, XXXI, 524
inner membrane
  effect of cross-linking, LVI, 629
    of labeling, LVI, 619, 621
inorganic supports, LVI, 490, 491
isolation, chaotic ions used, XXXI, 771
of Keilin-Hartree preparation, LV, 124
leakage, test, XLIV, 276

level, steady state kinetics, XL, 252
lipolytic, in plants, problems, XXXI, 520–528
in liver fractions, XXXI, 31–39
loading, definition, XLIV, 170
lysosomal, hormone effect, XL, 349
marker, *see also* Marker enzyme
distribution in submitochondrial subfractions, LV, 93, 94
outer membrane, LV, 101
of membranes
gel electrophoresis, XXXII, 82–85
as markers, XXXII, 393
microassays, *see* Microassay
microencapsulation, XLIV, 201–218
theoretical considerations, XLIV, 202, 203
of microvillous membrane, XXXI, 129, 130
mitochondrial, purity, LVI, 692–696
modification, effect on activity, XLIV, 448–450
molecular size, effect on immobilization, XLIV, 27
NAD-linked, hydride transfer stereospecificity, LIV, 223–232
in neuron communication, XXXII, 765
in nuclear envelope, XXXI, 290–292
oxidation-reduction, XLI, 127–379
in oxyntic cells, XXXII, 716
pad, preparation, XLIV, 652–654
peptidylproline and peptidyllysine hydroxylase activity, XL, 343
of plant
assay, XXXI, 543, 544
inhibitors, XXXI, 538–540
for plant-cell rupture, XXXI, 503
in plasma membranes, XXXI, 88–102
polymerization, immunogenicity, LVI, 696
proteolytic, plasma membrane, XXXI, 90
rate of turnover, estimation, XL, 260
regulatory, in carbohydrate metabolism, XXXVII, 280–282
relevance of adenylate energy charge, LV, 229–232
return to steady state following irreversible inhibition, XL, 264
spin diameter, XLIV, 156
stabilization, LVIII, 126
steroid binding sites, XXXVI, 390–405
substrate conformation, XLIX, 353–359
subunits
activity, XLIV, 501–503
immobilization, XLIV, 491–503
synthesis in animal tissue
altered rate, XL, 251
analysis, XL, 241
system, *see* Multistep enzyme system
thiolation, by 3-mercaptopropioimidate, XLIV, 42
for tissue dispersion, LVIII, 125
Enzyme complex, definition, LVI, 578
Enzyme deficiency disease, L, 441
Enzyme electrode
for analysis of substrate levels, XLIV, 277, 278, 580–593, 877–880
artificial enzyme membranes, XLIV, 913
characteristics, XLIV, 585
construction, XLIV, 596–601
for air and water monitoring, XLIV, 654, 655
of dialysis membrane type, XLIV, 599, 600
by physical entrapment, XLIV, 600
effect of interferences, XLIV, 612–615
with enzyme, XLIV, 614, 615
with sensor electrode, XLIV, 612–614
for glutamate determination, XLIV, 877–880
for lysine determination, XLIV, 277, 278
preparation
direct polymerization onto electrode membrane, LVI, 486
entrapped systems, LVI, 486
gel system, LVI, 486
glucose oxidase immobilized in cellophane matrix, LVI, 487
glucose oxidase-polyacrylamide adduct, LVI, 486
soluble immobilized systems, LVI, 485
for pyruvate determination, XLIV, 877–880

range of substrate determinable, XLIV, 611, 612
response time, XLIV, 604–611
dialysis membrane thickness, XLIV, 608, 609
electrode sensor response speed, XLIV, 609, 610
enzyme concentration, XLIV, 606, 610, 611
gel-layer thickness, XLIV, 608, 609
pH, XLIV, 607, 608
stirring rate, XLIV, 604, 605
substrate concentration, XLIV, 605, 606
temperature, XLIV, 608
stability, XLIV, 601
storage, XLIV, 601, 602
theory, XLIV, 580, 581
types, XLIV, 585
utilizing dextran-bound NAD^+, XLIV, 877–880
washing, XLIV, 611
Enzyme-inhibitor complex, XLVI, 204, 468
Enzyme-ligand interaction, quantitation, XXXIV, 165
Enzyme reactor, immobilized enzyme
hollow column, LVI, 489
microflow, LVI, 487–489
kinetics, LVI, 487
packed pellicular, LVI, 488, 489
Enzyme rod
enzyme assays, LVII, 210
preparation, LVII, 204, 205
recycling, LVII, 206, 210, 212, 213
substrate assay, LVII, 210–214
Enzyme solution, for hypophysiotropic substance assay, XXXVII, 82–93
Enzyme-substrate binding
electron spin resonance studies, XLIX, 436–478
nuclear magnetic resonance studies, XLIX, 322–359
Enzyme-substrate bioluminescent system, LVII, 277
Enzyme-substrate complex, large, agar supports, XLIV, 23
Enzyme substrate-enzyme inhibitor binding, relaxation kinetics, LIV, 75
Enzyme-substrate intermediate
carbanion, XLVI, 48–50
covalent, XLVI, 7–10; XLIX, 440, 441
Enzyme thermistor, XLIV, 472, 667, 676
instrumentation, XLIV, 668–670
Enzyme turnover, time range, LIV, 2
Eosin
autoradiography poststain, LVIII, 287
bound, determination, LIV, 52
conjugation to membrane proteins, LIV, 50, 51
derivatization, LIV, 49, 50
as triplet probe, LIV, 49, 54
Eosin 5-isothiocyanate
extinction coefficient, LIV, 52
preparation, LIV, 49, 50
structural formula, LIV, 50
Eosin-NCS, *see* Eosin 5-isothiocyanate
Eosinophil, chemiluminescence, LVII, 465
EPA, *see* Ethyl ether/isopentane/ethyl alcohol
Ependymal cell, motility after treatment with trypsin, XXXV, 575
4-Epianhydrotetracycline, XLIII, 199, 317
Epichlorhydrin
in affinity chromatography, XXXVI, 110, 111
agar cross-linking, XLIV, 23, 39, 41
Epichlorohydrin triethanolamine-cellulose, XLIV, 46
chromatography, hirudin, XLV, 671
purification, of guanine phosphoribosyltransferase, LI, 555, 558
of hypoxanthine phosphoribosyltransferase, LI, 555, 558
Epidermal growth factor, LVIII, 78, 100, 101, 103, 109
bioassay of, XXXVII, 425–427
^{125}I-labeled, binding to HeLa cells, LVIII, 107–109
isolation of, XXXVII, 427–429
lacking terminal residue, XXXVII, 428–429
as mitogen, LVIII, 109
preparation of, XXXVII, 424–430
properties of, XXXVII, 430
radioimmunoassay, XXXVII, 32

yield of, XXXVII, 429–430
Epidermophyton floccosum, fungal penicillin acylase, XLIII, 721
Epididymal capillary, endothelial cell isolation, XXXIX, 479–482
Epididymal fat pad, fat cell preparation, XXXI, 60, 61
6-Epihetacillin, XLIII, 362, 363
Epimerase, *see also* specific types
calculation of activity, XLI, 64
Epinephrine, XXXIV, 105, 699
adenylate cyclases, XXXVIII, 147–149, 170, 173
autoxidation, as source of free radicals, LIII, 387
binding to adipocytes, XXXI, 70–71
cAMP inhibitor assay, XXXVIII, 279–281
in chromaffin granule, XXXI, 381
derivative, XLVI, 592, 594
effect on β-cell, XXXIX, 425
on fat-cell hormones, XXXI, 68
on hepatocytes, XXXV, 592, 593
on renal gluconeogenesis, XXXIX, 15
on renin production, XXXIX, 22
lipolysis stimulation in adipose tissue, XXXV, 187
in perifused fat cells, XXXV, 609–612
protein kinase, XXXVIII, 363–365
Epinephrine-glass, XXXIV, 699
Epitestosterone, as 5α-oxidoreductase substrate, XXXVI, 473
4-Epitetracycline, XLIII, 199, 317
Epithelial cell
culture, LVIII, 263
dissociation, LVIII, 129
electron microscopy, XXXII, 32
enrichment, LVIII, 277
freezing, LVIII, 31
intestinal, XXXI, 123, 124; XXXIV, 212
protein content, LVIII, 145
of mammary gland, developmental studies, XXXIX, 443–454
rat, media for, LVIII, 57
scanning electron microscopy, XXXII, 52, 57
suspension, LVIII, 270
Epithelial-mesenchymal interaction, LVIII, 265
in adult, LVIII, 266
Epithelium, feeder layer, LVIII, 265
Epon
as embedding plastic, XXXIX, 140
embedding reagent, for subcellular fractions, XXII, 4, 7
in tissue embedding, XXXVII, 139, 144
Epon-araldite embedding plastic, XXXIX, 140
Epon 812/DDSA/NMA, as embedding plastic, XXXIX, 140
Epoxide
bond definition, LII, 228
in preparation of lipophilic Sephadex, XXXV, 380
Epoxide-glutathione *S*-transferase, genetic control, LII, 232
Epoxide hydrase, *see* Epoxide hydrolase
Epoxide hydratase, *see* Epoxide hydrolase
Epoxide hydrolase
activators and inhibitors, LII, 200
Ah locus, LII, 232
assay, LII, 193, 194, 416–418
benzo[*a*]pyrene metabolism, LII, 284
molecular weight, LII, 199
purification procedure, LII, 195–198
purity, LII, 199
stability, LII, 199, 417
substrate specificity, LII, 199, 200
3,4-Epoxybutene, structural formula, XLIV, 197
3′,4′-Epoxybutyl-β-D-glycoside, XLVI, 366
1,2-Epoxy-(β-glucopyranosyl)ethane, XLVI, 369
Epoxy glue, in anaerobic fittings, LIV, 116
2*R*-Epoxyhexitol-6-phosphate, synthesis, XLVI, 387
2*S*-Epoxyhexitol-6-phosphate, XLVI, 384–387
1,2-Epoxy-3-norepinephrine propane, XLVI, 592
16α,17α-Epoxyprogesterone, 16α-hydroxyprogesterone, XXXVI, 379
Epoxypropane, reaction with tributylphosphine, XLVII, 113

2′,3′-Epoxypropyl-2-acetamido-2-deoxy-β-D-glucopyranoside, XLVI, 366
synthesis, XLVI, 410
2′,3′-Epoxypropyl *O*-(2-acetamido-2-deoxy-β-D-glucopyranosyl)-(1→4)-2-acetamido-2-deoxy-β-D-glucopyranoside, XLVI, 407
2′,3′-Epoxypropyl-2-acetamido-3,4,6-tri-*O*-acetyl-2-deoxy-β-D-glucopyranoside, XLVI, 365
synthesis, XLVI, 409, 410
2′,3′-Epoxypropyl *O*-(2-acetamido-3,4,6-tri-*O*-acetyl-2-deoxy-β-D-glucopyranosyl)-(1→4)-2-acetamido-3,6-di-*O*-acetyl-2-deoxy-β-D-glucopyranoside, synthesis, XLVI, 407
(*R*,*S*)-2′,3′-Epoxypropyl β-D-glucopyranoside
affinity labeling, XLVII, 482
structure, XLVII, 482
2′,3′-Epoxypropyl β-D-glucopyranoside tetraacetate, synthesis, XLVII, 492
Epoxypropyl glycoside, XLVI, 365, 366
2′,3′-Epoxypropyl β-glycoside of *N*-acetyl-D-glucosamine, XLVI, 404–410
Eppenbach colloid mill, LIII, 233
Eppendorf centrifuge, modification, LVI, 223
Eppendorf Microcentrifuge Model 3200, XXXVI, 432, 433
in tRNA sequence analysis, LIX, 62
Eppendorf microliter pipette, XXXII, 113
Eppendorf photometer, LIV, 490, 491
EPR, *see* Electron paramagnetic resonance
EPRCAL/computer simulation program, XLIX, 422
Epsilon$_i$, evaluation, XLVIII, 110
Epstein-Barr virus, LVIII, 37, 42
effect on lymphocytoid cell lines, XXXII, 812, 814
Equilenin, biosynthesis, in perfused placenta, XXXIX, 251
Equilibrium
electrochemical, role of mediator-titrants, LIV, 396–398
in metabolic regulation, XXXVII, 282, 283
Equilibrium constant
adenylate kinase, LV, 232
of binding reaction, determination, LIV, 69
intrinsic, XLVIII, 112, 113
Equilibrium dialysis
measurement of antibiotic-ribosome binding, LIX, 862–866
of glutamate binding, LVI, 427
Equilibrium dialysis assay, XXXII, 422
Equilibrium dialysis cell, LIX, 863, 864, 866
Equilibrium model
adsorption with fixed binding constant, XXXIV, 141–147, *see also* Affinity method, theory
elution by change in K_1, XXXIV, 147–155, *see also* Affinity method, theory
by competitive inhibitors, XXXIV, 155–158
ER, *see* Estrogen receptor
Erabutoxin A, classification of tyrosyl residues, XLIX, 119
Erbium, magnetic circular dichroism, XLIX, 177
Ergosterol
biosynthesis of, aberration of in yeast, XXXII, 820
cytochromeless strains, LVI, 122
estimation, LVI, 563
heme-deficient mutants, LVI, 559, 560
liposome formation, XLIV, 227
plant membranes, XXXII, 543
in promitochondria, XXXI, 627, 628
Ergosterol 5,8-dioxygenase, LII, 19
Eriochrome blue, relaxation time, LVI, 307
Eriochrome blue SE
absorption spectra, LVI, 314, 315
magnesium ions, LVI, 313–317
relaxation time, LVI, 314
Erwinia aroideae
fungal penicillin acylase, XLIII, 722
penicillin acylase, XLIII, 718
Erythrina subrosa, lectin, XXXIV, 334
Erythrocruorin, LII, 16
prosthetic group, LII, 5

Erythrocyte, *see also* Red blood cell
adenylate cyclase, XXXVIII, 188
assay, XXXVIII, 170, 171
properties, XXXVIII, 173
purification, XXXVIII, 171, 172
bovine
hemolysate preparation, LIII, 388
superoxide dismutase, LIII, 383, 388, 389
contamination, LVIII, 492
countercurrent separation, XXXII, 637, 641, 643–646
2′,3′-cyclic nucleotide-3′-phosphohydrolase, XXXII, 127
density centrifugation, XXXI, 721
2,3-diphosphoglycerate mutase from human, assay, purification, and properties, XLII, 450–454
1,3-diphosphoglycerate phosphatase from human, preparation and purification, XLII, 419, 420
electrophoretic purification, XXXI, 760, 761
formalinization, L, 307
freezing, LVIII, 31
galactokinase, assay, purification, and properties, XLII, 47–53
ghost, XXXI, 75; XLVI, 75, 579
amidination, LVI, 623
ATPase, XXXII, 305, 306
cAMP derivatives, XXXVIII, 398
freeze fracturing, XXXII, 41
glycosphingolipids, XXXII, 345
IR spectroscopy, XXXII, 250, 253
isolation, XXXII, 350–352
negative staining, XXXII, 28
preparation, XXXI, 168–172; XXXII, 276; L, 204
impermeable and inside-out vesicles, XXXI, 172–180
sidedness assays, XXXI, 176–178
Raman spectroscopy, XXXII, 256, 257
sonicated, human, NADH-cytochrome b_5 reductase binding, LII, 108
assay, LII, 464, 465
properties, LII, 470, 471
purification procedure, LII, 467
spectrin isolation, XXXII, 275–277
glucose-6-phosphate dehydrogenase, from human, assay, purification, and properties, XLI, 208–214
glycolysis, XXXVII, 279
hemolysis, XXXII, 131, 132, 136
human
adenine phosphoribosyltransferase, LI, 568–574
adenosine deaminase, LI, 502–507
adenylate deaminase, LI, 497–502
adenylate kinase, LI, 466–473
enzyme isolation, general method, LI, 581–586
guanylate kinase, LI, 486–490
hemolysate preparation, LI, 146, 469, 498, 499, 503, 546, 570, 571, 582
from fresh blood, LI, 568, 569
hypoxanthine phosphoribosyltransferase from, LI, 543–549
lysis procedure, LI, 367, 583
micelle formation, XLVIII, 309
nucleoside diphosphokinase, LI, 376–386
nucleoside triphosphate pyrophosphohydrolase, LI, 282–284
preparation, XLII, 115
purine nucleoside phosphorylase, LI, 530–538
iodination, XXXII, 104, 105
IR spectroscopy, XXXII, 252
lipids
bilayer formation and properties, XXXII, 516, 518, 532–534, 537
extraction, XXXII, 132–135
in liver perfusion, XXXV, 598
media, XXXIX, 27, 36
long-term storage, XXXI, 3
lymphocyte separation, LVIII, 490, 491
lysis by ether, LII, 448
by hyptonic shock, LII, 456, 467, 488
by toluene, LII, 456
membranes, XLVI, 83, 87, 167, 344
chemical analyses, XXXII, 474
fluorescent probe studies, XXXII, 234

human
anion transport, XLIX, 14
labeling, with eosin derivatives, LIV, 50, 51
rotation of band, LIV, 3, 58, 59
thermal transition studies, XLIX, 4, 7, 13, 14
lipid, XXXII, 543
lipid bilayer, XXXII, 513
NMR spectra, XXXII, 206, 210
phospholipase effect, XXXII, 131–140
photosensitivity, LIV, 52
reconstitution, XXXII, 468–475
methemoglobin reduction system, LII, 463–473
microencapsulation, XLIV, 206, 210, 212, 214, 215
(Na^+ + K^+)-ATPase isolation, XXXII, 280
negative staining, XXXII, 31
oxidase, LII, 16
phosphofructokinase from human, assay, purification, and properties, XLII, 110–115
6-phosphogluconate dehydrogenase, from human, assay, purification, and properties, XLI, 220–226
phosphoglucose isomerase, from human, assay, isolation, and properties, XLI, 392–396, 399
3-phosphoglycerate kinase from human, assay, purification, and properties, XLII, 144–148
phosphoglycerate mutase from human, XLII, 438, 445
phospholipid exchange, XXXII, 140
protein coating, XXXII, 65, 66
pyruvate kinase from human, assay, purification, and properties, XLII, 182–186
rabbit, nucleoside triphosphate pyrophosphohydrolase, LI, 276–279
removal, LVII, 467, 470
scanning electron microscopy, XXXII, 52, 54
stroma
chromatography, XXXII, 471
isolation, L, 212
as matrix, XXXIV, 336
in transport studies, XXXIX, 505
triosephosphate isomerase, from human, assay, isolation and properties, XLI, 442–447
Erythrocyte cytochrome b_5 reductase, *see* NADH-dependent reductase
alternative names, LII, 464
Erythro-9-(2-hydroxy-3-nonyl)adenine, inhibitor of adenosine deaminase, LI, 507
Erythroleukemia cell, *see* Mouse, erythroleukemia cell
Erythromycin, LVI, 31, 33
binding to ribosomal particles, LIX, 865, 866
biosynthesis, XLIII, 487–498, *see also* *S*-Adenosylmethionine:erythromycin C *O*-methyltransferase
^{14}C-labeled, preparation, LIX, 864
effect on protein synthesis, LIX, 855, 860
gas-liquid chromatography, XLIII, 245–248
calculation, XLIII, 247
euteric coated tablet, XLIII, 246
high-pressure liquid chromatography, XLIII, 308, 309, 319
in media, LVIII, 112, 114
mitochondria, LVI, 32
in photoaffinity labeling studies, LIX, 804, 805
proton magnetic resonance spectroscopy, XLIII, 391
resistance, LVI, 140
gene mapping, LVI, 185
Streptomyces erythreus, media, XLIII, 20
thin-layer chromatography, XLIII, 198, 204
Erythromycin A, XLIII, 490
derivatization, XLIII, 494
partition chromatography, XLIII, 492, 493
structure, XLIII, 488
thin-layer chromatography, XLIII, 493, 494
Erythromycin B
paper chromatography, XLIII, 131
structure, XLIII, 488

Erythromycin C
partition chromatography, XLIII, 492, 493
spiroketal (6 → 9:12 → 9), XLIII, 490
structure, XLIII, 488
thin-layer chromatography, XLIII, 493, 494
Erythromycin methyltransferase, *see* *S*-Adenosylmethionine:erythromycin C *O*-methyltransferase
Erythropoietin
bioassay, XXXVII, 109, 110
mode of action, XXXVII, 115–121
plethoric mouse assay, XXXVII, 111, 112
starved rat assay, XXXVII, 110, 111
in vitro assay, XXXVII, 112–115
Erythrose-4-phosphate, substrate, transaldolase, XLIV, 495
Escherichia coli
N-acetylglucosamine-6-phosphate deacetylase, assay, purification, and properties, XLI, 497–502
adenine phosphoribosyltransferase, LI, 558–567
adenosine monophosphate nucleosidase, LI, 263, 267–271
adenylate cyclase, XXXVIII, 155
assay, XXXVIII, 156–158
properties, XXXVIII, 159, 160
purification, XXXVIII, 159
adenylosuccinate synthetase, LI, 212
aldolases, XLII, 258
alkaline phosphodiesterase, LI, 238
aminoacyl-tRNA synthetase, preparation, LIX, 314, 315
aminoglycosidic antibiotics, XLIII, 120
anaerobically grown
culture medium and growth conditions, LVI, 380
membrane vesicle preparation, LVI, 380, 381
L-arabinose isomerase, assay, purification, and properties, XLI, 454–458
aspartate transcarbamoylase, LI, 35–41
ATPase, LV, 298, 299, 341, 342
dissociation and reconstitution, LV, 800–810
inhibitor, LV, 402, 404, 405
subunits, LV, 795, 809, 810
ATP-P_1 exchange preparation
assay, LV, 358, 359
characteristics, LV, 360–363
preparation of membrane particles, LV, 359, 360
aurovertin-resistant mutants, LVI, 178, 179
selection procedure and properties, LVI, 179–182
as bacterial associate, XXXIX, 485
bacteriophage P1 kc, XLIII, 45–47
bacteriophage T2-induced deoxycytidylate deaminase, LI, 412–418
cAMP-receptor protein, XXXVIII, 367
assay, XXXVIII, 368
bacteria, XXXVIII, 368, 369
DNA preparation, XXXVIII, 372, 373
growth of bacteria, XXXVIII, 373, 374
ion-exchange resin preparation, XXXVIII, 371
mechanism of action, XXXVIII, 375, 376
properties, XXXVIII, 376
purification, XXXVIII, 369–371
transcription assay, XXXVIII, 372
carbamoyl-phosphate synthetase, LI, 21–29
carbodiimide-resistant ATPase mutants, LVI, 163, 164
biochemical analysis, LVI, 167–170
selection, LVI, 164–167
strain RF-7, LVI, 170–173
cell-envelope enzymes, XXXII, 91
cell extract preparation, LIII, 233, 234
cells, internal volume, LV, 551
chloramphenicol acetyltransferase, XLIII, 741, *see also* Chloramphenicol acetyltransferase
chloramphenicol hydrolase, XLIII, 735
chlorate resistant mutant, LIII, 355
CMP-dCMP kinase, LI, 332
coenzyme Q homologs, LIII, 593

colicin K, LV, 533
countercurrent distribution, XXXI, 768
crude extract preparation, LI, 215, 230, 265, 311, 414, 443, 477, 509, 510, 553, 562; LIX, 314
cultivation, LV, 165
culture, XLIV, 740
2′,3′-cyclic nucleotide-2′-phosphohydrolase in, XXXII, 127
cystine-binding protein, XXXII, 422–427
  physical properties, XXXII, 427
  purification, XXXII, 425–427
cytidine deaminase, LI, 401–405
cytidine triphosphate synthetase, LI, 79–83
cytochrome, LIII, 207, 208, 210, 211
cytochrome *b*-562, CD spectrum, LIV, 283
cytochrome b_1, LIII, 232–237
DCCD-binding protein
  preparation of labeled proteolipid, LV, 429–431
    of unlabeled proteolipid, LV, 431, 432
  purification of unmodified protein, LV, 432–434
differential scanning calorimetry, XXXII, 265
Dio-9, LV, 515
efrapeptin, LV, 495
elongation factor Tu, cGMP assay, XXXVIII, 85–90
enolase, assay, purification, and properties, XLII, 323–329
envelope, electrophoretic purification, XXXI, 759
everted membrane vesicles for measurement of calcium transport, LVI, 233–236
  assay of transport, LVI, 238–241
  preparation of vesicles, LVI, 236–238
formate dehydrogenase, LIII, 372
gentamicin adenylyltransferase, XLIII, 625
glucosamine-6-phosphate deaminase, assay, purification, and properties, XLI, 497–502
glucose-6-phosphate 1-epimerase, molecular weight, XLI, 493
glucosephosphate isomerase, anomerase activity, XLI, 57
glutamine-binding protein, XXXII, 422–427
  properties, XXXII, 425
  purification, XXXII, 424, 425
D-glycerate 3-kinase, assay, purification, and properties, XLII, 124–127
glycerol kinase, assay, purification, and properties, XLII, 148–156
L-glycerol-3-phosphate dehydrogenase, assay, purification, and properties, XLI, 249–254
glycosyl transferase complex reconstitution, XXXII, 449–459
GMP synthetase, LI, 213
growth, LI, 60, 214, 229, 230, 265, 311, 413, 414, 443, 509, 553, 561, 562; LIII, 233, 347, 520
  adenylate pool size, LV, 233
  for iron transport studies, LVI, 389
guanine phosphoribosyltransferase, LI, 549–558
guanylate kinase, LI, 473–482
hemin-permeable, LVI, 178
hydrogenase reaction, LIII, 287
hypoxanthine phosphoribosyltransferase, LI, 549–558
immobilization, XLIV, 740, 741
intact cells for P/O ratio determination
  assay, LV, 169–174
  equilibration of starved cells with $^{32}P_1$, LV, 169
  preparation of cells, LV, 169
isolation of enterochelin, LVI, 394, 395
  materials, LVI, 396–398
  principle, LVI, 395, 396
isomeric specificity of aminoacylation, LIX, 272–275
kanamycin acetyltransferase, XLIII, 623, 624
2-keto-4-hydroxyglutarate aldolase, assay, purification, and properties, XLII, 285–290

β-lactamase, XLIII, 672–677, *see also* β-Lactamase, IIIa
D-lactate dehydrogenase, LIII, 439, 440, 519–527
L-lactate dehydrogenase, LIII, 439
lipid mutants, XXXII, 819
measurement of ΔpH, LV, 563
membrane, optical probes, LV, 573
membrane isolation, XXXI, 643
membrane particle preparation, LI, 60, 61
membrane potential, LV, 452, 455
membrane vesicles
  anaerobic active transport, LVI, 385, 386
  transport using K^+ diffusion gradients, LV, 676–680
methylglyoxal synthetase, preparation, XLI, 505, 506
mutagenesis and primary screening, LVI, 107, 108
mutant
  affected in oxidative phosphorylation or quinone biosynthesis, LVI, 106–117
  choice of starting strain, LVI, 106, 107
  for membrane study, XXXII, 884, 890
mutarotase, substrate specificities, XLI, 487
neomycin phosphotransferase, XLIII, 626
nicotinamide nucleotide transhydrogenase, LV, 266, 270
  reconstitution, LV, 787–800
nitrate reductase, LIII, 643
number fluctuation studies, XLVIII, 451, 452
organotin, LV, 509
oxidase, LII, 15, 19
penicillin acylase, XLIII, 705–721; XLIV, 760, *see also* Penicillin acylase, from *Escherichia coli*
phage M_{13} coat protein, incorporation into liposomes, LV, 703
phosphocellobiase, XLII, 493
phosphodiesterase
  assay, XXXVIII, 249, 250
  culture, XXXVIII, 250, 251
  extract preparation, XXXVIII, 251
  properties, XXXVIII, 255, 256
  purification, XXXVIII, 251–255
phosphofructokinase, assay, purification, and properties, XLII, 84, 91–98
3-phosphoglycerate kinase, assay, purification, and properties, XLII, 139–144
phosphoglyceromutase, assay, purification, and properties, XLII, 139–144
  purification and properties, XLII, 438, 444, 446
phosphorylating vesicles, preparation, LV, 165, 166
pyruvate formate-lyase, assay, purification, and properties, XLI, 508–518
radiolabeled, in phagocytosis tests, XXXII, 652
rhamnulose-1-phosphate aldolase, XLII, 265
ribonucleoside diphosphate reductase, LI, 227–237, 246, 247
ribosomal subunits, LX, 333
  30 S, structure, XLVIII, 376
ribosomes, *see also* Ribosome, prokaryotic, 70 S
  cocentrifugation, LVI, 84
L-ribulose-5-phosphate 4-epimerase, assay, purification, and properties, XLI, 419–423
S30 extract, preparation, LIX, 297, 363, 445
S100 extract
  preparation, LIX, 357, 363, 818, 827, 828, 839
    RNA-free, LIX, 279
source of acetyl-CoA carboxylase, XXXV, 17–37
  of acyl carrier protein synthetase, XXXV, 95–101
  of elongation factors, LX, 596, 609
  of initiation factors, LX, 3, 4, 205, 206, 216, 225, 226, 231, 314, 344
  of long-chain fatty acyl thioesterases I and II, XXXV, 102–109
  of ribosomal protein S1, LX, 449, 456, 457
spin-labeled components, XXXII, 194

starved cell preparations, pertinent characteristics, LV, 174, 175
strain RF–7
properties
active transport, LVI, 171
genetic mapping, LVI, 171
site of inhibition by carbodiimide, LVI, 171
suppression of inhibition, LVI, 171, 172
streptomycin phosphotransferase, XLIII, 627
streptomycin-spectinomycin adenylyltransferase, XLIII, 625
substrate uptake, LV, 453
succino-AICAR synthetase, LI, 190
sulfite reductase, siroheme, LII, 437, 439–441, 443
thymidine kinase, LI, 354
for transduction, XLIII, 46
transformation of resistance plasmids, XLIII, 48
translation, initiation sites, LX, 416, 417
transport systems, XXXI, 702, 704, 707–709
ubi mutants
culture and maintenance, LIII, 600–605, 608
radiolabeling, LIII, 608, 609
ubiquinone biosynthesis intermediates, LIII, 600–609
UMP kinase, LI, 332
unsaturated fatty acid auxotroph, XXXII, 856–864
use in industrial production of L-aspartic acid, XLIV, 740–745
verification of carbodiimide-resistance, LVI, 166
vesicles
internal volume, LV, 551
membrane potential, LV, 555, 559
Escherichia coli A19, LIX, 444, 483
growth, LIX, 444
lysis, LIX, 483
Escherichia coli 0111a
electron microscopy, XXXI, 634
intracellular membrane isolation, XXXI, 630–642
large-scale method, XXXI, 639–641
Escherichia coli AB301/105λ⁻, growth, LIX, 830, 831
Escherichia coli C3
crude extract preparation, LIX, 827
ribonuclease III, LIX, 827–830
Escherichia coli D10, LIX, 817
Escherichia coli GI 238, ribosomes, LIX, 839
Escherichia coli K12, LIX, 297
Escherichia coli K12 S26, LIX, 363
Escherichia coli MRE600
crude extract preparation, LIX, 194
deuteration, LIX, 640–643
endogenous polysomes, purification, LIX, 353–356, 363–366
1F fraction, preparation, LIX, 363
growth conditions, LIX, 193, 354, 364, 365, 451, 552–554
lysis by alumina grinding, LIX, 451
by freeze-thaw-lysozyme, LIX, 365
by lysozyme treatment, LIX, 354, 355
in pressure cell, LIX, 194, 464
by solid CO_2 grinding, LIX, 365
protein labeling with [^{3}H]amino acids, LIX, 552
with $H_2[^{35}S]O_4$, LIX, 552, 553
ribosomal RNA labeling with [^{14}C]uracil, LIX, 553
with $H_3[^{32}P]O_4$, LIX, 553, 554
Escherichia intermedia, XLIV, 886
ESR, *see* Electron spin resonance
Ester, selective hydrolysis, XLIV, 835, 836
Esterase
activity
acrosin, XLV, 332
definition of, as opposed to lipase activity, XXXV, 182
staphylococcal protease, XLV, 472, 475
assay
fluorimetric, XLV, 18
Leucostoma peptidase A, XLV, 398
carbamylation, XXXV, 192, 202
cephalosporin, *see* Cephalosporin acetylesterase
gel electrophoresis, XXXII, 83, 88, 89
as marker enzymes, XXXI, 735

in plasma membrane, XXXI, 89
serine residues, reaction with diisopropylfluorophosphate, XLIV, 15
Esterolytic activity, assay, XLIV, 73
Estradiol, XXXVI, 126; XLVI, 75, 76
adsorption to various materials, XXXVI, 93, 94
assay, XXXIV, 677, 678
autoradiography, XXXVI, 139, 150, 151, 153
cytoplasmic receptors, XXXVI, 172, 210
in mammary tumor, XXXVI, 248–254
diazonium agarose linked, XXXIV, 105
effects on nuclear polymerase, XXXVI, 319–327
electron-capture detection, XXXVI, 64–67
logit-log assay, XXXVII, 9
mercurated, XXXVI, 422
metabolites, gas chromatographic and mass spectral properties, LII, 387
nuclear receptors, XXXVI, 269–275
assay, XXXVI, 283–286
protein complex, XXXVI, 275–283
PBP affinity, XXXVI, 125
phagocytosis, LVII, 493
radioimmunoassay, XXXVI, 16
receptor protein, XXXIV, 104
receptor-protein assay, XXXVI, 49–51
in vivo production and interconversion, XXXVI, 68
17α-Estradiol
antiestrogen, XL, 292
mercuric derivatives, antiestrogenic activity, XXXVI, 405–408
17β-Estradiol, XXXIV, 555, 675; XLVI, 57, 451, 453
antibodies to derivatives, XXXVI, 54
derivatives, XXXVI, 18, 19
effects on mercaptide formation, XXXVI, 389
on uterine protein synthesis, XXXVI, 445–455
medium containing, XXXIX, 300
mercuric derivatives, XXXVI, 376–378
estrogenic activity, XXXVI, 405–408
nuclear receptor binding, XXXVI, 319
protein-binding assay, XXXVI, 37
radioimmunoassay, XXXVI, 20, 21, 28, 29
testicular secretion, XXXIX, 272, 277
thin-layer chromatography, XXXIX, 262
Estradiol-agarose, XXXIV, 558, 671, 672
Δ^5-3-ketosteroid isomerase, XXXIV, 562
stability, XXXIV, 671, 672
Estradiol-(benzyl)cellulose, XXXIV, 670
17β-Estradiol-6-(*O*-carboxymethyl)oxime, XXXIV, 680
17β-Estradiol-6-(*O*-carboxymethyl)oxime-diaminodipropylaminoagarose, XXXIV, 673, 680
Estradiol dehydrogenase, XLVI, 56–58
transhydrogenase system, XLVI, 57
17β-Estradiol dehydrogenase, XXXIV, 3, 4, 555–557; XLVI, 145, 148, 150, 451, 456, 457
17β-Estradiol-3-*O*-hemisuccinate, XXXIV, 558, 682
17β-Estradiol-17-hemisuccinate, XXXIV, 675
17β-Estradiol-17-hemisuccinyl-albumin-agarose, XXXIV, 673, 681, 686
17β-Estradiol-17-hemisuccinyldiaminodipropylaminoagarose, XXXIV, 673, 679, 686
17β-Estradiol-17-hemisuccinyl-poly-L-(lysyl-DL-alanine)-agarose, XXXIV, 673, 681
Estradiol-polyvinyl(*N*-phenylenemaleimide), XXXIV, 671
Estradiol receptor, XXXIV, 670–688
adsorbents
adsorptive properties, XXXIV, 671
ineffective for purification, XXXIV, 686–688
selective, XXXIV, 679–681
washing, XXXIV, 683–684
assay, XXXIV, 677–679
elution, XXXIV, 684
leakage, XXXIV, 671, 672, 687, 688

macromolecular spacers, XXXIV, 676, 681
purification, XXXIV, 685–688
quantitation of ligand, XXXIV, 683
selective adsorbents, XXXIV, 679–681
stability, XXXIV, 671, 672
Estradiol-receptor complex, XXXIV, 678, 679
Estradiol-receptor protein, XXXIV, 104
17β-Estradiol-3-*O*-succinyldiaminodipropylaminoagarose, XXXIV, 558
Estriol
chromatographic separation, XXXVI, 487
effects on mercaptide formation, XXXVI, 389
formation in perfused placenta, XXXIX, 250, 251
Estriol 16-hemisuccinate, XXXIV, 4
Estriol 2α-hydroxylase, LII, 15
Estriol 3-methyl ether, XLVI, 449
Estrogen, XLVI, 580
cell lines dependent, XXXII, 559, 560
chromatographic separation, XXXVI, 488
column chromatography, XXXVI, 39
cytoplasmic and nuclear receptor proteins of, density gradient centrifugation, XXXVI, 166–176
extraction artifacts, XXXVI, 176–187
marker proteins, XXXVI, 170, 171
effects on uterine protein synthesis, XXXVI, 444–455
in monkey plasma, XXXIX, 197
nuclear receptors, XXXVI, 265–275
assay, XXXVI, 283–286
transformation, XXXVI, 271–274
photochemically active derivative, LVI, 660
receptor, XLVI, 75, 76, 84, 89, 90
Estrogen-binding protein, from cytosol, XXXVI, 336–349
Estrogen receptor, XXXIV, 670–688, *see also* Estradiol receptor
collection and preparation of tissue, XXXVI, 336
from cytosol, XXXVI, 336–349
different forms, XXXVI, 331–334
in mammary carcinoma, XXXVI, 248–254
molecular parameters, XXXVI, 365
purification, XXXVI, 331–365
basic guidelines, XXXVI, 334, 335
buffers, XXXVI, 352
with estradiol-containing gels, XXXIV, 96
receptor-protein assay, XXXVI, 49
sedimentation behavior, XXXVI, 363–365
sources, XXXVI, 335, 336
in vivo production and interconversion, XXXVI, 68
Estrogen receptor protein, nuclear
artifacts in extraction, XXXVI, 182–186
density gradient centrifugation, XXXVI, 166–176
preparation, XXXVI, 169, 170
Estrone, XLVI, 56, 75, 76, 455
electron-capture detection, XXXVI, 60, 64–67
formation, in perfused placenta, XXXIX, 250, 251
gravimetric assay, XXXVI, 460–467
receptor-protein assay, XXXVI, 50
in vivo production and interconversion, XXXVI, 68
Estrone-agarose, XXXIV, 555–557
concentration effect, XXXIV, 556
Estrone-aminocaproate-agarose, XXXIV, 555
Estrone-3-hemisuccinate-agarose, XXXIV, 564
Estrone-hemisuccinate-ethylenediamine-agarose, XXXIV, 555
Estrone sulfatase, multiple sulfatase deficiency, L, 474
Etalon, XLVIII, 455, 456
Etamycin
countercurrent distribution, XLIII, 336
peptide antibiotic lactonase 1ction, XLIII, 773, *see also* Peptide antibiotic lactonase
structure, XLIII, 768
Etching, in freeze fracturing, XXXII, 40, 41
ETF, *see* Electron-transferring flavoprotein

ETF dehydrogenase, iron-sulfur properties, EPR characteristics, LIV, 140, 141
ETF-ubiquinone oxidoreductase
  component stoichiometries, LIV, 145
  EPR signal properties, LIV, 143, 144
  EPR spectrum, LIV, 146, 147
Ethacrynate, as ATPase inhibitor, XXXII, 305, 306
Ethamoxytriphetol, antiestrogen, XL, 292
1,2-Ethanediol, XLIX, 243
  electron spin resonance spectra, XLIX, 383
Ethanediol ester, lipophilic Sephadex chromatography of, XXXV, 389
Ethanediol ether, lipophilic Sephadex chromatography, XXXV, 389
Ethane thiol, XLIV, 88; XLVII, 347
Ethanol, XLIV, 34, 44, 92–94, 102, 131, 204, 284, 285, 302, 385, 460, 530, 749, 756, 834, 862, 863, 865, 867, 868, 872
  in AICAR synthesis, LI, 189
  in 6-amino-1-hexanol phosphate preparation, LI, 253
  analysis of periodate-oxidated tRNA, LIX, 180
  assay
    of long-chain aldehydes, LVII, 190, 191
    of malate, LVII, 187
    of oxaloacetate, LVII, 187
    using immobilized enzyme rods, LVII, 211, 212, 214
    of quinones and hydroquinones, LIII, 586–589
    of translocation, LIX, 360
  benzoquinone activation, XLIV, 35
  chromatographic separation of nucleosides and nucleotides, LI, 300, 366
  concentration determination, with enzyme electrode, XLIV, 590, 591
  crystallization of transfer RNA, LIX, 7
  deoxynucleoside storage, LI, 338
  effect on EPR signals, LIV, 148, 149
  electron spin resonance spectra, XLIX, 383, 470
  in 7-ethoxycoumarin synthesis, LII, 374
  ferricyanide, LVI, 232, 233
  in 5-fluoro-2′-deoxyuridine 5′-(p-aminophenyl phosphate) synthesis, LI, 99, 100
  fractionation of chloroperoxidase, LII, 526
  hollow-fiber dialytic permeability, XLIV, 297, 302
  inhibitor, of epoxide hydrase, LII, 200
  inhibitor solutions, LVI, 32
  ionophores, LV, 448
  isolation of 2-octaprenyl-6-methoxyphenol, LIII, 606
  in media, LVIII, 70
  metabolism, pathways, in liver, LII, 355
  ninhydrin solution, LIX, 777
  nuclear magnetic resonance spectra, XLIX, 319
  oxaloacetate assay, LV, 207
  oxidation rate, XLIV, 316
  in oxygen consumption studies, LIV, 487
  Pasteur effect, LV, 294, 295
  polarography, LVI, 452, 467, 468, 478
  precipitation
    of orotidylate decarboxylase, LI, 75, 76
    purification, of adenine phosphoribosyltransferase, LI, 571
      of AIR synthetase, LI, 194
      orotate phosphoribosyltransferase, LI, 71
    of RNA, LIX, 120, 126–128, 132–134, 140, 141, 147, 149, 161, 170, 171, 180, 192, 210, 216, 220, 221, 230, 269, 270, 273, 299, 312, 343, 446, 452, 453, 555, 556, 832, 836
  preparation of chromophore-labeled sarcoplasmic reticulum vesicles, LIV, 51
    of dehydroluciferol, LVII, 25
    of eosin isothiocyanate, LIV, 49
    of protein-depleted 30 S subunits, LIX, 621
    of ribosomal RNA, LIX, 446

pressure-effect studies, LVII, 364
purification of 8-bromoadenosine 5′-monophosphate, LVII, 117
of complexes I–III and II–III, LIII, 57
of complex II, LIII, 23
of phenylalanine hydroxylase, LIII, 281
of superoxide dismutase, LIII, 388
radioassay of nucleotide incorporation, LIX, 182
recrystallization of ammonium sulfate, LIII, 56
of bile salts, LIII, 3, 74
of cholic acid, LIII, 56
of potassium cholate, LIII, 83
of *N,N,N′,N′*-tetramethyl-*p*-phenylenediamine, LIII, 470
release of ribosomal proteins, LIX, 818
solvent
of antimycin, LIII, 83
of 2,4-dinitro-5-fluorophenylazide, LIII, 173
of menadione, LIII, 20
of organic hydroperoxide, LII, 510
of phenylmethylsulfonylfluoride, LIX, 374, 484, 490
of 8-quinolinol, LIII, 637
of quinones, LIII, 581
of tetranitromethane, LIII, 172
of 2-thenoyltrifluoroacetone, LIII, 26
of ubiquinone-1, LIII, 13
of ubiquinone-2, LIII, 26, 39
reduced, LIII, 39
of water-insoluble antibiotics, LIX, 853
storage solution of [γ-^{32}P]ATP, LIX, 61
substrate
alcohol dehydrogenase, XLIV, 64
of reconstituted MEOS system, LII, 366, 367
synthesis of 2-carbethoxy-2-diazoacethydrazide, LIX, 811
of *S*-(3-carbethoxy-3-diazo)acetonyl-7-thiolincomycin, LIX, 814
of chemiluminescent aminophthalhydrazides, LVII, 430, 431, 433, 434, 435
of luciferyl sulfate analogs, LVII, 251
tissue extraction, LV, 202
as uncoupler solvent, LV, 464
uranyl acetate staining solution, LIX, 842, 843
yeast mitochondria, LV, 150, 151, 159
Ethanolamine, XLIV, 73, 266, 273, 375
for CO_2 collection, LI, 156
intermediate in phosphatidylethanolamine synthesis, XXXV, 461
isothiocyanato group blocking, XLVII, 322
Ethanolamine hydrochloride
blocking reagent, of activated support, XLIV, 493
coupling of cytochrome *c* to Sepharose 4B, LIII, 101
Ethanolamine oxidase, LII, 18; LVI, 474
Ethanolamine phosphate, XLVII, 487, 488
synthesis of adenosine derivatives, LVII, 119
Ethanol-ammonium acetate solvent system
chromatographic purification of adenosine derivatives, LVII, 117, 118
for luciferin, LVII, 24
Ethanol dehydrogenase, XXXI, 576
Ethanol-hexane, purification of *Diplocardia* luciferin, LVII, 377
Ethanol-triethylammonium bicarbonate
chromatographic separation of 8-bromoadenosine 5′-monophosphate, LVII, 117
purification of chemiluminescent aminophthalhydrazides, LVII, 434, 437
Ethanol-water
electron spin resonance spectra, XLIX, 383
synthesis of *N*-(3-carbethoxy-3-diazo)acetonylblasticidin S, LIX, 814
of *N*-(3-carbethoxy-3-diazo)acetonyl-gougerotin, LIX, 814

of *N*-(ethyl-2-diazomalonyl)blasticidin S, LIX, 814
1,*N*[6]-Ethenoadenosine 3′,5′-monophosphate
biological activities, XXXVIII, 428, 429
characterization, XXXVIII, 430
synthesis and purification, XXXVIII, 429, 430
1-*N*[6]-Ethenoadenosine triphosphate, interaction, with GMP synthetase, LI, 224
1-*N*[6]-Etheno-ATP, *see* 1-*N*[6]-Ethenoadenosine triphosphate
1,*N*[6]-Etheno-coenzyme A, fluorescent donor, XLVIII, 378
Ether, XLIV, 102, 130–133, 198, 204, 324, 690, 862, 863, 865; LII, 62
analysis of periodate-oxidized tRNA, LIX, 180
anhydrous
preparation of chicken liver acetone powder, LI, 198
purification of dihydroorotate dehydrogenase, LI, 65
for fatty acid extraction, LII, 320
preparation of hydrazine-substituted tRNA, LIX, 120
of labeled tRNA, LIX, 103
of MS2 RNA, LIX, 299
purification of *Diplocardia* luciferin, LVII, 377
radioassay of nucleotide incorporation, LIX, 182
as red blood cell hemolysing agent, LII, 448
removal of acetic anhydride, LIII, 458
siroheme solubility, LII, 437
synthesis of chemiluminescent aminophthalhydrazides, LVII, 435, 436
Ether deoxylysolecithin, LVI, 747
micellar properties, LVI, 745
structure, LVI, 735
Ether-ethanol, assay of cell-free protein synthesis, LIX, 449
Ether-hexane extraction, of TRF and LRF, XXXVII, 407
Ether-hexane-methanol, isolation of 3-octaprenyl-4-hydroxybenzoic acid, LIII, 605
Ether lecithin, intermediate in phosphatidylcholine synthesis, XXXV, 444–453
Ether-petroleum ether, purification of *N*-trifluoroacetylthyroxine, LVII, 432
Ether phosphatidylethanolamine, intermediate in phosphatidylethanolamine synthesis, XXXV, 458–460
Ethidium, as optical probe, LV, 573
Ethidium azide, XLVI, 647–649
Ethidium bromide, LV, 516; LVI, 31
binding to tRNA, principle, LIX, 111–114
spectral properties, LIX, 112
as curing agent, XLIII, 51
DAPI stained cells, LVI, 731
DNA extraction, LVIII, 408
as fluorescent probe, XXXII, 235, 238, 239
mutagenesis, LVI, 185
petite mutants, LVI, 157, 159
photochemically active derivative, LVI, 660
staining, LVI, 186, 188; LVIII, 412
in studies of protein-RNA interactions, LIX, 578
tritiated, synthesis, LVI, 662
in vivo administration, LVI, 37
Ethionine-agarose, XXXIV, 510
ETH 129 ionophore, LVI, 442
calcium complexes, LVI, 440, 441
ETH 149 ionophore, LVI, 442
selectivity, LVI, 441
ETH 157 ionophore, LVI, 442
selectivity, LVI, 441
ETH 227 ionophore, LVI, 442
selectivity, LVI, 441
ETH 1001 ionophore, LVI, 442
calcium complexes, LVI, 440, 441
ETH 1097 ionophore, LVI, 442
selectivity, LVI, 441
Ethoxazolamide, XLIX, 168
N-Ethoxycarbonyl-2-ethoxy-1,2-dihydroquinoline, XXXIV, 89; XLVII, 598
ATPase, LVI, 181
1-Ethoxycarbonyl-2-ethoxy-1,2-dihydroquinoline, LV, 515

7-Ethoxycoumarin
fluorescence spectra, LII, 375, 376
metabolism, LII, 66–69, 373
NMR spectral properties, LII, 374
preparation, LII, 373, 374
7-Ethoxycoumarin *O*-deethylase, *Ah*[b] allele, LII, 231
β-Ethoxyethyl methacrylate, *see* Nobecutan
Ethoxyformic anhydride
concentration determination, LVII, 175
modification of bacterial luciferase subunits, LVII, 175, 176
7-Ethoxyphenoxazone, *see* 7-Ethoxyresorufin
7-Ethoxyresorufin
fluorescence spectra, LII, 375, 376
microsomal metabolism, LII, 373
NMR spectral properties, LII, 374, 375
preparation, LII, 374, 375
Ethoxyresorufin *O*-deethylase, *Ah*[b] allele, LII, 231
7-Ethoxyumbelliferone, *see* 7-Ethoxycoumarin
DL-Ethyl-2-acetamido-3-oxo-4-octadecenoate intermediate in sphingomyelin synthesis, XXXV, 498
Ethyl acetate, XLIV, 132, 187, 838, 894; XLVII, 294, 305, 326
δ-aminolevulinic acid synthetase assay, LII, 354
determination of [^{3}H]puromycin specific activity, LIX, 360
for drug isolation, LII, 333, 334
effect on Tygon tubing, LVII, 460
electron spin resonance spectra, XLIX, 383
extraction of biphenyl metabolites, LII, 403, 405
of hemin, LII, 453
of styrene glycol, LII, 417
for glycol removal, LII, 193, 194
purification of 4-*N*-(3-chloro-2-hydroxypropyl)amino-*N*-methylphthalimide, LVII, 430
of *Diplocardia* luciferin, LVII, 377
of firefly luciferase, LVII, 9
siroheme solubility, LII, 437
solvent in peptide synthesis, XLVII, 558
synthesis of chemiluminescent aminophthalhydrazides, LVII, 432
of dithiodiglycolic dihydrazide, LIX, 385
Ethyl acetate-acetone
extraction of polycyclic aromatic hydrocarbons, LII, 284
salt accumulation, XLVII, 339
Ethyl acetate-cyclohexane, in steroid TLC, LII, 380
Ethyl acetate-ethyl alcohol-water, purification of *Cypridina* luciferin, LVII, 367, 368
Ethyl acetate-hexane, isolation of quinone intermediates, LIII, 606–608
Ethyl acetate-hexane-methanol, isolation of 3-octaprenyl-4-hydroxybenzoic acid, LIII, 605
Ethyl acetate-methanol-water, in peroxyoxalate chemiluminescence reactions, LVII, 458, 459, 460
Ethyl acetimidate, cross-linking, LVI, 623, 629
Ethyl-2-acetyl-3-oxo-4-octadecenoate, intermediate in sphingomyelin synthesis, XXXV, 494, 502
Ethyl alcohol, XLIX, 238
Ethylamine, measurement of ΔpH, LV, 562, 563
Ethyl-*erythro*-2-amino-3-hydroxy-4-octadecenoate intermediate in sphingomyelin synthesis, XXXV, 498, 499
4-*N*-Ethylamino-*N*-methylphthalimide
synthesis, LVII, 431
of 4-*N*-[(3-chloro-2-hydroxypropyl)-*N*-ethyl]amino-*N*-methylphthalimide, LVII, 431
Ethyl atropate, XLVI, 545
Ethyl 2-benzothiazolesulfonic acid, XLVI, 539
inhibitor, XLVI, 541
synthesis, XLVI, 540
N-(2-Ethylcellulose)-glycyl-D-arginine, XXXIV, 412
N-(2-Ethylcellulose)-glycyl-D-phenylalanine, XXXIV, 412, 418

Ethylcellulose N200, fibers, preparation, XLIV, 231
Ethyl-4-chloro-2-diazoacetoacetate
structural formula, LIX, 812
synthesis, LIX, 811
of ethyl-4-iodo-2-diazoacetoacetate, LIX, 811
Ethyl chloroformate
for protein insolubilization, XXXIV, 706
synthesis of 6-*N*-[(3-biotinylamido)-2-hydroxypropyl]amino-2,3-dihydrophthalazine-1,4-dione, LVII, 430
of *N*-trifluoroacetylthyroxinyl ethyl carbonic anhydride, LVII, 432
Ethylchloromalonylcymarin, LVI, 660
4′-(Ethylchloromalonyl)cymarin, XLVI, 525, 526
photoaffinity labeling, XLVI, 527–530
Ethyl citrate, transport, LVI, 248
Ethyl-2-diazoacetate
synthesis of ethyl-4-chloro-2-diazoacetoacetate, LIX, 811
of ethyl-2-diazomalonylchloride, LIX, 813
Ethyl diazomalonic acid, XLVI, 96
Ethyldiazomalonyladenosine 3′,5′-phosphate, LVI, 656
Ethyldiazomalonyladenosine 5′-phosphate, LVI, 656
N-(Ethyl-2-diazomalonyl)blasticidin S
structural formula, LIX, 812
synthesis, LIX, 814
Ethyl 2-diazomalonyl chloride, XLVI, 95
structural formula, LIX, 812
synthesis, XXXVI, 417; XLVI, 638, 639
of 2-carbethoxy-2-diazoacethydrazide, LIX, 811
of puromycin derivative, LIX, 809
[^{14}C]Ethyl-2-diazomalonylchloride
microscale synthesis, LIX, 811–813
synthesis of 2-[^{14}C]carbethoxy-2-diazoacethydrazide, LIX, 814
N^6-(Ethyl 2-diazomalonyl) cyclic adenosine monophosphate, XLVI, 579
preparation, XXXVIII, 391
labeled, XXXVIII, 392
$O^{2'}$-(Ethyl-2-diazomalonyl)-cyclic adenosine monophosphate
synthesis, XXXVIII, 389, 390
labeled, XXXVIII, 391, 392
Ethyldiazomalonylcymarin, LVI, 660
4′-(Ethyldiazomalonyl)cymarin, XLVI, 524–526
photoaffinity labeling, XLVI, 527–530
Ethyl diazomalonyl group, XLVI, 627
Ethyl-2-diazomalonyl-*N*-hydroxysuccinimide ester, synthesis, XLVI, 639
Ethyldiazomalonylphenylalanyl-tRNA, LVI, 656
N-(Ethyl-2-diazomalonyl)Phe-tRNA, XLVI, 638, 641–644
synthesis, XLVI, 638–640
N-Ethyl-2-diazomalonylpuromycin
inhibitor of peptidyltransferase activity, LIX, 809
structural formula, LIX, 812
N-(Ethyl-2-diazomalonyl)streptomycyl hydrazone
radiolabeled, microscale synthesis, LIX, 815
structural formula, LIX, 813
synthesis, LIX, 815
Ethyldimethylaminopropyl carbodiimide, XLVII, 297
dicyclohexylcarbodiimide-resistant mutants, LVI, 170
resistant mutants, selection, LVI, 167
N-Ethyl-*N*′-dimethylaminopropyl carbodiimide, XXXIV, 589
N-Ethyl-*N*′-(3-dimethylaminopropyl)carbodiimide, synthesis of affinity matrix, LII, 516, 517
1-Ethyl-3-(3-dimethylaminopropyl)carbodiimide, XXXIV, 89, 303, 355, 402, 515, 549, 553, 603, 612, 620, 679; XLIV, 61, 63, 77, 506, 507, 521, 524, 860, 863, 866, 867
in affinity gel preparation, LI, 32, 100, 367
1-Ethyl(3,3-dimethylaminopropyl)carbodiimide, XLVI, 222, 223
N-Ethyl(*N*′-dimethylaminopropyl)carbodiimide hydrochloride, XLVII, 282, 383
1-Ethyl-3(3-dimethylaminopropyl)carbodiimide-hydrogen chloride, as radioimmunoassay reagent, XXXVII, 25, 26

N-Ethyl-*N*′-(3-dimethylaminoprop-1-yl)-carbodiimide methiodide
proteolipid, LV, 429
Sepharose coupling to ADP, LV, 190
Ethyl-2,3-dioxo-4-octadecenoate-2-phenylhydrazone, intermediate in sphingomyelin synthesis, XXXV, 494, 498
Ethylene, XLIV, 85
1,1′-Ethylene 2,2′-bipyridylium dichloride, as mediator-titrant, LIV, 409, 410
Ethylene diacrylate, LIX, 833
as cross-linking agent, XLVI, 526, 527
Ethylenediamine, XXXIV, 37; XLIV, 86
aminolysis of chlorinated resin, XLVII, 268
derivative, XLIV, 284
succinylated, XXXIV, 404
synthesis of adenosine derivatives, LVII, 116, 117, 118
Ethylenediamine-cobalt(III) complex, XLVI, 315
Ethylenediaminetetraacetic acid, XXXI, 537; XLIV, 41, 42, 106, 111, 197, 313, 326, 337, 347, 494, 495, 508, 823, 865, 878, 881, 925; XLIX, 177
activator, of cytidine deaminase, LI, 411
adrenodoxin reductase purification, LII, 133, 134
aminoacylation of tRNA, LIX, 126, 127, 130, 221
δ-aminolevulinic acid synthetase assay, LII, 351
anaerobic TEA-maleate buffer, LIII, 366
assay of adenosine monophosphate nucleosidase, LI, 264
of adenylate kinase, LI, 468
of ADP, LVII, 61
of aminoacyl-tRNA synthetases, LIX, 231
of bacterial luciferase-coupled antibodies, LVII, 404
of catalase, LVII, 404
of complex I, LIII, 14
of complex II activities, LIII, 26, 27
of complex III, LIII, 91
of creatine kinase isoenzymes, LVII, 60, 61
of dCTP deaminase, LI, 418
of deoxythymidylate phosphohydrolase, LI, 286
of flavin compounds, LIII, 427
of hydrogen peroxide, LVII, 403
of *p*-hydroxybenzoate hydroxylase, LIII, 545
of hypoxanthine phosphoribosyltransferase, LI, 543
of luciferase, LVII, 30
of 5-methyluridine methyltransferase, LIX, 191
of NADH-cytochrome *c* reductase, LIII, 9, 10
of PAP, LVII, 253
of phosphoribosylglycinamide synthetase, LI, 180
of phosphoribosylpyrophosphate synthetase, LI, 3, 5
radiolabeled ribosomal particles, LIX, 779
of ribonucleoside diphosphate reductase, LI, 228
of ribosephosphate pyrophosphokinase, LI, 12, 13
of tRNA methyltransferase, LIX, 191
of salicylate hydroxylase, LIII, 530
of succinate dehydrogenase, LIII, 34
of superoxide dismutase, LIII, 383; LVII, 405
of thymidine phosphorylase, LI, 442
of uridine 5-oxyacetic acid methyl ester methyltransferase, LIX, 191
of *in vitro* energy transfer, LVII, 260
ATPase, LV, 201
in brush border isolation, XXXI, 124, 125
in calcium measurement, XXXIX, 515–520
crystallization of transfer RNA, LIX, 6
cyclic nucleotide immunoassay, XXXVIII, 105

cytochrome isolation, LII, 99, 110, 111, 118–121, 127, 128, 160
cytochrome *P*–450 assay, LII, 126, 138
determination of ferrocytochrome *c* oxidation, LIII, 45
in dissociation of carbamoyl-phosphate synthetase, LI, 25
E. coli cell lysis, LIX, 300, 354
effect on epoxide hydrase activity, LII, 200
electrophoresis buffer, LII, 325
electrophoretic analysis of ribosomal protein, LIX, 65, 506
of RNA fragments, LIX, 574
for endocrine tissue dissociation, XXXIX, 114, 115
in Enzyme Buffer, LI, 14
extraction of nucleic acids, LIX, 311, 831, 835
hemoglobin isolation, LII, 448
in HEPES buffers, XXXI, 10, 11, 14, 15
Hummel-Dreyer gel filtration, LIX, 326
in hydride transfer, stereospecificity experiments, LIV, 228
inactivation of nuclease P_1, LIX, 104
inactivator
of aminopeptidase, XLVII, 87, 88
of *Cypridina* luciferase, LVII, 371
inhibitor
of brain α-hydroxylase, LII, 317
of *Cypridina* luciferase, LVII, 353
of *Cypridina* luminescence, LVII, 349
of endogenous nucleotidases, LVII, 122
of formate dehydrogenase, LIII, 371
of lipid peroxidation, LII, 366
of nucleoside phosphotransferase, LI, 393
of orotate phosphoribosyltransferase, LI, 74
of protease contaminant, XLVII, 190
of rat liver acid nucleotidase, LI, 274
of streptomycin-6-phosphate phosphohydrolase, XLIII, 470
of uridine nucleosidase, LI, 294
iron chelator, LII, 305
isolation of mitochondria, LV, 4, 12, 13
ligand binding studies, LVII, 120
luciferase renaturation buffer, LVII, 178
in media, LVIII, 70
MgK_2 salt, purification of glycine reductase, LIII, 376, 377, 381
microsomal protein solubilization, LII, 93, 104, 195
microsome isolation, LII, 92, 98, 104, 190
mitochondrial incubation, LV, 7
mitochondrial protein synthesis, LVI, 21
in oxygen determination, LIV, 119
particles of mitochondria fluorescent probe studies, XXXII, 237
in PE buffer, LIII, 222
in PED buffer, LI, 137
peptide digestion, LIX, 301
periodate oxidation of tRNA, LIX, 269
phosphodiesterase, XXXVIII, 233, 243, 248, 256
as phospholipase inhibitor, XXXII, 138, 160, 161
photochemical reduction of flavin, LVII, 139
in plant-cell fractionation, XXXI, 505, 591, 592, 598, 603
potassium salt, reaction termination, LI, 550, 559
preparation of aequorin crude extract, LVII, 278, 280, 307
of apohydroxylase, LIII, 360
of beef liver mitochondria, LIII, 497
of denatured DNA, LIX, 840
of dialysis tubing, LVII, 10
of *E. coli* ribosomes, LIX, 647, 817
of luciferase subunits, LVII, 172, 173, 178
of mitochondrial membranes, LIII, 102, 103
of mitochondrial ribosomes, LIX, 422, 423

of protein from cross-linked ribosomal subunits, LIX, 541
of rat liver tRNA, LIX, 230
of ribosomal proteins, LIX, 484–486, 490, 492
of ribosomal RNA, LIX, 452, 556, 648
of ribosomal subunits, LIX, 411, 412, 416
of tRNA for NMR studies, LIX, 25
of synthetic nucleotide polymers, LIX, 826
of yeast submitochondrial particles, LIII, 113
as preservative, LVII, 150
purification of adenosine deaminase, LI, 509, 510, 512
of adenylate kinase, LI, 471
of aequorin, LVII, 281, 301, 307, 308, 309
of aminoacyl-tRNA synthetases, LIX, 258, 259
of AMP nucleosidase, LI, 265
of aspartate carbamyltransferase, LI, 53
of bacterial luciferase, LIII, 567; LVII, 146, 147, 148
of blue-fluorescent protein, LVII, 227
of carbamoyl-phosphate, synthetase, LI, 31–33
of complex III, LIII, 88
of CTP(ATP):tRNA nucleotidyltransferase, LIX, 123, 184
of cytochrome bc_1 complex, LIII, 115
of cytochrome oxidase, LIII, 62
of dCTP deaminase, LI, 420, 421
of deoxythymidylate phosphohydrolase, LI, 288
of dihydroorotate dehydrogenase, LI, 65, 66
of electron-transferring flavoprotein, LIII, 515, 516
of FGAM synthetase, LI, 197
of firefly luciferase, LVII, 7, 8, 9, 10, 31, 32, 35, 36, 62
of flavocytochrome b_2, LIII, 240
of GMP synthetase, LI, 215, 216, 221
of guanylate kinase, LI, 477, 478, 479, 490
of *p*-hydroxybenzoate hydroxylase, LIII, 545–547
of hypoxanthine phosphoribosyltransferase, LI, 545
of luciferin sulfokinase, LVII, 247, 249
of melilotate hydroxylase, LIII, 554
of mitochondrial tRNA, LIX, 204, 205
of nitrate reductase, LIII, 348, 642
of oxidoreductase, LVII, 203
of purine nucleoside phosphorylase, LI, 519, 521, 539
of pyrimidine nucleoside phosphorylase, LI, 434
of ribonucleoside diphosphate reductase, LI, 239
of ribonucleoside triphosphate reductase, LI, 252
of tRNA methyltransferase, LIX, 195, 197, 198, 200
of superoxide dismutase, LIII, 391
of thymidine kinase, LI, 356, 357, 368
of thymidine phosphorylase, LI, 439, 443, 444
of thymidylate synthetase, LI, 94, 95
of tryptophanyl-tRNA synthetase, LIX, 239, 241, 242
of uridine-cytidine kinase, LI, 311
of uridine phosphorylase, LI, 426, 427
in radioimmunoassay, XXXVII, 31
for reaction termination, LI, 569
reconstitution of ribosomal subunits, LIX, 447, 451, 654, 864
reductase purification, LII, 93, 94, 104
regeneration of aequorin, LVII, 289
removal, from purified aequorin, LVII, 311, 312
reverse salt gradient chromatography, LIX, 215
in reversible dissociation of flavoproteins, LIII, 431, 433, 434

ribosome dissociation, LVI, 84
ribosome removal from microsome subfractions, LII, 76
tRNA digestion mixture, LIX, 81, 82, 107, 833
separation of aminoacylated tRNA, LIX, 216
of bacterial ribosomal subunits, LIX, 398
in SKE buffer, LIII, 114
in sodium dihydrolipoate reagent, LI, 248
stabilizer of cytidine triphosphate synthetase, LI, 82
of phosphoribosylpyrophosphate synthetase, LI, 16
staining method, XL, 25
storage buffer of aequorin, LVII, 281
submitochondrial particles, LV, 106, 107
sucrose gradient buffers, LIX, 329
termination of complex III assay, LIII, 39
Titriplex, K^+, Mg^{2+} salt, LI, 550, 557, 559
Ethylene dichloride
determination of tyrosine, LIII, 279
magnetic circular dichroism, XLIX, 159
Ethylene dimethacrylate, XLIV, 67, 172–175
Ethylene glycol, XLIV, 91, 93, 320; XLVII, 89, *see also* Ethanediol
effect on EPR signals, LIV, 150
in enzyme storage, LIX, 244
inhibitor of deoxythymidine kinase, LI, 365
in low-temperature solvent mixture, LIV, 105, 107, 108
magnetic circular dichroism studies, XLIX, 159
in MCD studies, LIV, 293
phosphorescence studies, XLIX, 238
polymers, self-association, XLVIII, 110, 118
purification of luciferase oxygenated-flavin intermediate, LIII, 570
stabilizer of adenine deaminase, LI, 509, 510, 512
of carbamoyl-phosphate synthetase, LI, 118
of dCTP deaminase, LI, 420–422
storage of luciferase reaction intermediate, LVII, 197
for termination of periodate oxidation reaction, LIX, 269
Ethylene glycol bis(β-aminoethyl ether) *N,N'*-tetraacetic acid
assay of calcium with aequorin, LVII, 324
of modulator protein, LVII, 109, 110
brain adenylate cyclase, XXXVIII, 142
in brain homogenization, LII, 312
effect on brain α-hydroxylase, LII, 317
heart mitochondria, LV, 43, 46
in hepatocyte isolation buffer, LII, 62
inactivator, of *Cypridina* luciferase, LVII, 371
inhibitor, of *Cypridina* luminescence, LVII, 349
mitochondrial incubation, LV, 7
in opsonization studies, LVII, 487
phosphodiesterase activator, XXXVIII, 273
as phospholipase inhibitor, XXXII, 138
in plant cell fractionation, XXXI, 591, 592, 597, 598
preparation of mitochondria, LV, 667
in relaxing medium, LVII, 248
Ethyleneglycol dimethacrylate, XLIV, 172
Ethylene glycol-phosphate buffer, chromatographic separation of luciferase reaction intermediate, LVII, 196
Ethylene glycol-water, in ligand binding studies, LIV, 528
Ethyleneimine, XLIII, 36; XLIV, 196; XLVII, 115, 116, 126
alkylation of protein A, LIII, 379
reaction with tributylphosphine, XLVII, 113
toxicity, XLVII, 419
Ethylenemaleic anhydride, as carrier, XXXIV, 5
Ethylene oxide, XLIV, 70
Ethyl ester moiety, peptide synthesis, XLVII, 521

Ethyl ether, XLIX, 238
purification of *Cypridina* luciferin, LVII, 366, 367
Ethyl ether-hexane, purification of ubiquinone, LIII, 639
Ethyl ether/isopentane/ethyl alcohol, XLIX, 159
components, XLIX, 238, 383
electron spin resonance spectra, XLIX, 383
Ethyl formate, XLIV, 130
Ethyl[2-hydroxy-3-methoxycarbonyl-5-(4-methoxyphenyl)]valerate potassium enolate, LVII, 435
synthesis, LVII, 435
of 3,4-dihydronaphthalene-1,2-dicarboxylic acid anhydride, LVII, 435
4-(*N*-Ethyl-*N*-[2-hydroxy-3-(*N*-phthalimido)propyl]amino)-*N*-methylphthalimide
synthesis, LVII, 431
of 6-[*N*-(3-amino-2-hydroxypropyl)-*N*-ethyl]amino-2,3-dihydrophthalazine-1,4-dione, LVII, 431
6-(*N*-Ethyl-*N*-[2-hydroxy-3-(thyroxinylamido)propyl])amino-2,3-dihydrophthalazine-1,4-dione
chemiluminescence of ligand conjugates, LVII, 440
structural formula, LVII, 428
synthesis, LVII, 432
Ethyl iminobenzoate hydrochloride, intermediate in sphingomyelin synthesis, XXXV, 502
Ethyl iodide, in 7-ethoxycoumarin synthesis, LII, 374
Ethyl-4-iodo-2-diazoacetoacetate
structural formula, LIX, 812
synthesis, LIX, 811
Ethyl isocyanide, XLIV, 541–543
cytochrome *P*–450 difference spectra, LII, 273, 279
S-Ethylisothiourea, synthesis of *Cypridina* luciferin, LVII, 369
Ethyl-luciferase, XLVI, 538
[^{3}H]Ethylmaleimide, preparation of radiolabeled protein, LIX, 794, 795
N-Ethylmaleimide, XLIII, 502; XLVI, 62, 63, 67, 612; XLVII, 407; LV, 518
acetyl-CoA assay, LV, 210
carnitine-acylcarnitine translocase, LVI, 378
carrier sites, LVI, 251
cross-linked products, LVI, 636
deactivation of elongation factor eEF–2, LX, 254, 580, 665, 708
glutamate uptake, LVI, 247, 421, 427, 430
guanylate cyclase, XXXVIII, 199
inactivator
of L-3-hydroxyacyl-CoA dehydrogenase, XXXV, 128
of mitochondrial acetoacetyl-CoA thiolase, XXXV, 135
of tryptophanyl-tRNA synthetase, LIX, 250
inhibition of initiation, LX, 575–577
inhibitor, of amidophosphoribosyltransferase, LI, 178
of cytidine deaminase, LI, 411
of deoxythymidine kinase, LI, 365
of guanylate kinase, LI, 489
of thymidylate synthetase, LI, 96
of uridine phosphorylase, LI, 429
isolation of succinate dehydrogenase active site, LIII, 482, 483
measurement of proton translocation, LV, 638, 642, 646, 648
modification of bacterial luciferase subunits, LVII, 176, 177
phosphate transport, LVI, 342, 355
preservation of activity of heme-reversible translational inhibitor, LX, 485, 491–493
proline uptake, LV, 199
in protection method, LIX, 593
reaction with sulfhydryl groups, LX, 100, 424
transport, LVI, 248, 260, 292
Ethylmercurithiosalicylate, XLVII, 428
Ethyl methanesulfonic acid
as mutagen, XLIII, 666; LVIII, 312
mutagenesis, LVI, 98, 99
temperature-sensitive mutants, LVI, 133–135
Ethylmorphine
N-demethylation, determination, LII, 300–302
metabolism, LII, 66

as mixed-function oxidase substrate, XXXI, 233
substrate of cytochrome P–450LM forms, LII, 117
Ethylmorphine N-demethylase
Ah locus, LII, 232
inhibition by anti-NADPH-cytochrome c reductase, LII, 246, 247
in microsomal preparations, LII, 87
N-Ethylmorpholine, XLIV, 127; XLVII, 564
N-Ethylmorpholine acetate, XLVII, 42, 43
thermolysin hydrolysis, XLVII, 183–185
N-Ethylmorpholinium acetate, acetylation of flavin peptides, LIII, 459
Ethyl nitrosourea, as mutagen, LVIII, 312
N-Ethyl phenazonium ethosulfate, as redox mediator, LIV, 423, 429, 433
Ethyl α-phenylglycidate, XLVI, 545
N-Ethyl-5-phenylisoxazolium 3′-sulfonate, XLIV, 85
2-Ethyl-5-phenyloxazolium 3′-sulfonate, XLVII, 556, 557
4-[N-Ethyl-N-4-(N-phthalimido)butyl]amino-N-methylphthalimide
synthesis, LVII, 433
of 6-[N-aminobutyl-N-ethyl]amino-2,3-dihydrophthalazine-1,4-dione, LVII, 433
Ethyl thiotrifluoroacetate, modification of lysine, LIII, 142
6-[N-Ethyl-N-(4-thyroxinylamido)butyl]amino-2,3-dihydrophthalazine-1,4-dione
chemiluminescence of ligand conjugates, LVII, 440, 444
in heterogeneous competitive protein-binding assay, LVII, 442–444
structural formula, LVII, 429
synthesis, LVII, 433–434
7-[N-Ethyl-N-(4-thyroxinylamido)butyl]aminonaphthalene-1,2-dicarboxylic acid hydrazide
chemiluminescence of ligand conjugates, LVII, 440, 444
structural formula, LVII, 429
synthesis, LVII, 437
S-Ethyl trifluorothiol acetate, in N-trifluoroacetyl-6-amino-1-hexanol phosphate preparation, LI, 253
3-O-Ethyl-2,4,6-tri-O-methyl-D-galactose, L, 28
3-O-Ethyl-2,4,6-tri-O-methyl-D-glucose, L, 28
Ethylxanthate, interaction with cytochrome c oxidase, LIII, 195
5-Ethynouracil, XLVI, 164
5-Ethynyldeoxyuridine, substrate of thymidine kinase, LI, 370
17-Ethynyl-19-nortestosterone, molecular bioassay, XXXVI, 465
Etiolactone
isolation, XXXVI, 506, 507
synthesis, XXXVI, 508, 509
Etioluciferamine, synthesis of *Cypridina* luciferin, LVII, 369
Etioluciferin
Cypridina, structural formula, LVII, 346
methyl ether, in synthesis of luciferyl sulfate analogs, LVII, 251
in synthesis of *Cypridina* luciferin, LVII, 369
Etioplast
isolation, XXXI, 544–553
marker enzymes, XXXI, 735, 744, 745
in plant cells, XXXI, 493
function, XXXI, 495
Etioporphyrin manganese(III), axial mode assignments, LIV, 242
Etorphine, XXXIV, 622, 623; XLVI, 603
ETP, *see* Electron transport particle; Succinate dehydrogenase, electron-transferring flavoprotein
ETPH, *see* Electron transport particle, phosphorylating
Euactinomycete, aerobic, XLIII, 687, 688
Eubacteria
aminoglycoside-modifying mechanisms, XLIII, 611
antibiotic fermentations, XLIII, 16–18
media for maintaining, XLIII, 5, *see also* Media, for eubacteria
paper chromatography, XLIII, 135
Euchromatin, disperse, distribution, XL, 33
Euflavin, *in vivo* administration, LVI, 37, 38

Euglena, *see also* specific genera
  free cells, XXXII, 723
  glyoxylate cycle enzymes, XXXI, 566
  membrane mutants, XXXII, 865
  organelles, XXXI, 508, 513, 514
Euglena gracilis
  acyl carrier protein, XXXV, 110–114
  aldolase, XLII, 234
  disruption, LV, 138
Euglena gracilis var. *bacillaris*
  chloroplastic tRNA, LIX, 208–214
  culture conditions, LIX, 208
  plastid ribosomes, LIX, 434–437
Euonymus europaeus, lectin, XXXIV, 333
Euphausiid shrimp, photogenic organ, LVII, 331
Europium acetate, derivatization of transfer RNA, LIX, 14
Everett oxygenator, XXXIX, 278
Everninomicin, XLIII, 128, 129
EXAFS spectrum, *see* X-ray absorption fine structure spectrum, extended
Exchange
  electroneutral, LVI, 253
  measurement, LVI, 289, 290
  mitochondrial transport, LVI, 245, 246
Exchange reaction
  calculations, LV, 258–261
  corrections, LV, 259, 260
  uncouplers, LV, 463
Excimer emission, LVII, 502–504
Exciplex
  definition, LVII, 504
  emission, LVII, 504–508, 522
Excitation profile, definition, XLIX, 95
Excited state
  advantages, LVII, 494, 495
  electronically generated
    luminescence, LVII, 561–566
    perturbations, LVII, 566–568
  evolutionary significance, LVII, 561
Excited-state angular momentum of transition, XLIX, 171
Exciton splitting, LIV, 256, 263
Exclusion chromatography
  estrogen receptor assay, XXXVI, 348, 349
  of prolactin, XXXVII, 393
Exfoliatin, XLIII, 128
Exoaffinity labeling, XLVI, 12
  exo-alkylating agent, XLVI, 242
  reagents, XLVI, 115–130, *see also* specific substances
    active center, XLVI, 115
    properties, XLVI, 119–121
Exocrine tissue, XXXIX, 441–482
Exo-*cis*-3,6-endoxo-Δ^4-tetrahydrophthalic anhydride, *see* Phthalic anhydride
Exoglucosidase, in yeast cell wall hydrolysis, XXXI, 611
Exoglycosidase
  assay, L, 493, 494
  isozymes, L, 488
  from lysosomes, L, 488
  oligosaccharides, L, 269
  properties, L, 490, 491
λ-Exonuclease, XXXIV, 466
Exonuclease III, XXXIV, 469
Exopenicillinase, fluorine nuclear magnetic resonance, XLIX, 274
Explant
  plant cell, LVIII, 479
  skin, *see* Skin explant
Exponential curve, identification, LIV, 509
Exponential filtering
  for resolution enhancement, LIV, 161, 170
  for sensitivity enhancement, LIV, 160, 170
Extensin, potato lectin structure, L, 343
Extensor digitorum longus, of rat, XXXIX, 84
  in vitro studies, XXXIX, 91
Extensor digitorum soleus, of rat, XXXIX, 84
Extinction coefficient
  molar, apparent, definition, LIV, 312
  standard, for protein concentration measurements, XLIX, 28
Extraction, cold-solvent apparatus, LVII, 366
Extrusion, for plant-cell rupture, XXXI, 502
Eyring relation, LIV, 508, 524
Eyring theory, of absolute reaction rates, LVII, 363

F

F–420, in hydrogenase reactions, LIII, 287
FA, *see* Fluoranthene
$F(ab')_2$ antibody
hybridization, XXXII, 64
preparation, XXXII, 63, 64
Fab fragment, XXXIV, 711
and antigen localization, XLIV, 712
Fab piece, antibodies, L, 251, 253
Fabry's disease
enzymic diagnosis, L, 479–481
glycosphingolipids, L, 247
Fabry-Perot interferometer, XLVIII, 455
Facility, *see* Laboratory facility
Factor II, adsorption to barium citrate, XLV, 51
Factor V
bovine
assay, XLV, 107
one stage, XLV, 108
two stage, XLV, 110
calcium, interaction, XLV, 117
composition, XLV, 115
factor X_a, XLV, 116
phospholipids, XLV, 116
properties, XLV, 113
immunochemical, XLV, 119
physical, XLV, 114
proteolytic enzymes, XLV, 118
prothrombin, XLV, 116
purification, XLV, 111–113
purity, XLV, 114
thrombin, XLV, 117
human
anticoagulants, circulating, XLV, 122
assay, XLV, 119
catabolism, XLV, 122
half-life, XLV, 121
physical, XLV, 121
plasma levels, XLV, 121
platelet, XLV, 122
properties, XLV, 120
prothrombin assay, XLV, 125
purification, XLV, 120
purity, XLV, 121
stability, XLV, 120
synthesis, XLV, 121
Factor V–deficient plasma, XLV, 108
Factor VII, bovine
assay, XLV, 49, 50
composition, XLV, 53, 54
inhibitors, XLV, 55
properties, XLV, 53
proteolytic activators, XLV, 53
purification, XLV, 50, 51
role, XLV, 83
specificity, XLV, 55
stability, XLV, 53
Factor VIIa, composition, XLV, 54
Factor VIII, XXXIV, 337
bovine
assay, XLV, 85
properties, XLV, 89
purification, XLV, 86–88
Factor IX
barium citrate, adsorption, XLV, 51
bovine, assay, XLV, 76
conversion to factor IX_a, XLV, 81, 82
properties, XLV, 81–83
purification, XLV, 74, 77–81
role, XLV, 74
Factor IX_a, bovine
composition, XLV, 82
conversion from factor IX, XLV, 81, 82
hirudin, inactivated, XLV, 678
interaction with other factors, XLV, 83
Factor IX–deficient plasma, for assay, XLV, 76
Factor X, XXXIV, 592–594
bovine, *see also* Coagulant protein
activation, XLV, 95
assay, XLV, 100
barium citrate, adsorption, XLV, 51
cleavage sites, XLV, 99
pathways, XLV, 97
Russell viper venom, XLV, 191, 192
composition, XLV, 98
hydroxylamine cleavage, XLVII, 140, 142
interaction with other factors, XLV, 83
properties, XLV, 92

purification, XLV, 89–92, 101–103, 128
with prothrombin, XLV, 128, 130–132
role, XLV, 89
tissue factor assay, XLV, 45–47
Factor X_a, bovine
activation, cleavage sites, XLV, 106
assay, XLV, 101
proteolytic action, XLV, 105
prothrombin activation, XLV, 155
Factor X_1, bovine
composition, XLV, 94, 98
isolation, XLV, 89
Factor X_2, bovine, isolation, XLV, 89
Factor X_{14}, bovine
composition, XLV, 94, 98
preparation, XLV, 104
Factor $X_a\alpha$, bovine
composition, XLV, 94, 95, 98
preparation, XLV, 103
Factor $X_a\beta$, bovine
composition, XLV, 94, 98
preparation, XLV, 104
Factor X–agarose, XXXIV, 593
Factor X–deficient plasma, XLV, 90
Factor XI, bovine
assay, XLV, 67
composition, XLV, 73
properties, XLV, 73
purification, XLV, 66, 68
role, XLV, 65
Factor XI_a, role, XLV, 65
Factor XI–deficient plasma, XLV, 66
Factor XII
human
assay, XLV, 57–60
chromatography, XLV, 61
composition, XLV, 64
inhibitors, XLV, 65
properties, XLV, 62, 63
immunochemical, XLV, 65
purification, XLV, 60
purity, XLV, 63
stability, XLV, 62
zymogen form, XLV, 56
rabbit, amino acid composition, XLV, 64
Factor XIII
assay, XLV, 178–189
purification, XLV, 177
role, XLV, 177
zymogens
properties, XLV, 189
specific activity stain, XLV, 186
Factor XIII_a, *see* Fibrinoligase
FAD, *see* Flavin adenine dinucleotide
FAGLA, *see* Furacryloylglycyl-L-leucinamide
Falcon plastic tube, XXXII, 113
Falcon plastic vessel, XXXIX, 203
Falcon tissue culture dish, XXXII, 744
Falcon tissue culture plastic, XXXII, 609–611
Familial hypercholesterolemia, *see* Hypercholesterolemia, familial
Faraday balance, LIV, 385–387, 390, 391
Faraday conversion factor, LIV, 412
Faraday effect, XLIX, 149, 150
A term, LIV, 285, 286
B term, LIV, 286
C term, LIV, 286, 287
Faraday factor, membrane potential, LVI, 523
Farber's lipogranulomatosis, L, 461–465
ceramidase, L, 463–465
enzymic diagnosis, L, 463–465
Farghaley's minimal media
formulations, LVII, 154
preparation, LVII, 155
species supported, LVII, 165
Fast Analyzer, XXXI, 790–833
activity assays, using Fast Analyzer, XXXI, 807–815
analytical module, XXXI, 791–800
applications, XXXI, 805–833
colorimetric and spectrophotometric analysis, XXXI, 807
computer use, XXXI, 801–805
data system, XXXI, 800–805
enzyme activity assays, XXXI, 807–815
fluorescence assays, XXXI, 826–828
instrumentation, XXXI, 790–805
multiple-cuvette rotor, XXXI, 791–793
multiple parallel analysis, XXXI, 805–807

multiple wavelength scan of absorbance, XXXI, 824–826
optical systems, XXXI, 794–799
fluorometric, XXXI, 796–799
photometric, XXXI, 794–796
substrate analysis, XXXI, 815–824
synchronization, XXXI, 800
temperature control, XXXI, 799, 800
Fast Blue RR, as esterase stain, XXXII, 88
Fast Garnet GBC, as enzyme stain, XXXII, 89
Fast Red TR salt
as enzyme stain, XXXII, 90, 91
reagent, alkaline phosphatase staining, XLIV, 713
Fat cell, XXXIV, 6
brown, isolation, XXXV, 555–561
enzyme distribution, XXXI, 67, 68
hormonal regulation, XXXI, 68, 69
fractionation, XXXI, 61–71
biochemical characterization, XXXI, 64–67
packed cell volume, XXXI, 65
ghosts, preparation, XXXI, 109–111
suspending medium, XXXI, 110
glucose transport, hormone effects, XXXVII, 269–276
insulin binding, XXXVII, 193–198
proteolytic modification, XXXVII, 211–213
insulin response, XXXV, 560
isolation, XXXI, 103–114
microsomes, XXXI, 62, 63
perifused
lipolysis, XXXV, 607–612
albumin dependence, XXXV, 611, 612
epinephrine stimulation, XXXV, 609–612
preparation, XXXV, 609
preparation, XXXI, 60, 61
protein, XXXI, 64–66
white, isolation, XXXV, 555–561
Fatty acetyl-CoA dehydrogenase system, LIV, 143
Fatty acid, XXXIV, 6, *see also* specific type
analysis of positional distribution in glycerolipids, XXXV, 315–325
bioassay, LVII, 193
in biomembrane phospholipids, XXXII, 542, 544
biosynthesis, XXXI, 23; XLVI, 159
in yeast, XXXII, 820, 821
branching, determination of sites by mass spectrometry, XXXV, 344, 345
chromatography on lipophilic Sephadex, XXXV, 388, 392
on silicic acid-Florisil, XXXV, 411, 423
composition of yeast, LVI, 570
Tween-supplemented mutants, LVI, 572, 576
content of brown adipose tissue mitochondria, LV, 75
control studies on metabolism, XXXVII, 292, 295
desaturation, heme-deficient mutants, LVI, 560, 651
deuterated
gas chromatography, XXXV, 285–287, 340–342, 345
mass spectrometry, XXXV, 342–348
high resolution, XXXV, 347, 348
synthesis, XXXII, 201, 202; XXXV, 340
dichlorofluorescein, as indicator for thin-layer chromatography, XXXV, 319, 324, 325
effect on steroid binding, XXXVI, 95
estimation, LVI, 562, 563
free, inhibitor of dihydroorotate dehydrogenase, LI, 62
gas chromatography of methyl esters, XXXV, 282, 283, 285–287
in glycolipids, XXXII, 361, 362
in glyoxysome glyoxylate cycle, XXXI, 565, 566
hydroxylation, distribution in rodent tissues, LII, 319
α-hydroxylation, assays, LII, 311–315
ω-hydroxylation, LII, 318–324
ω-1-hydroxylation, LII, 318–324
incubation of mitochondria, LV, 8
O-isopropylidene derivatives of unsaturated fatty acids, XXXV, 347

lipolysis in fat pads, XXXV, 607
in perifused fat cells, XXXV, 607–712
long chain
α-hydroxylase, LII, 310–318
as uncouplers, LV, 472
mammary synthesis, assay, XXXIX, 453, 454
mass spectrometry of methyl esters, XXXV, 283–285, 340–348
methylation
for gas-liquid chromatography, XXXV, 76, 282, 319
procedure, LII, 321
methyl esters, separation, LII, 321–323
mitochondrial composition, function, LVI, 563
in NMR studies of lecithin, XXXII, 208
oxidation
by brown adipose tissue mitochondria, LV, 75, 76
in intestinal mucosa isolates, XXXII, 668, 669
Pasteur effect, LV, 296
β-oxidation enzymes, XXXV, 122–151
β-oxidation in microbodies, LII, 496, 502
in phosphoglycerides
analysis of positional distribution, XXXV, 317–319
pancreatic lipase hydrolysis, XXXV, 317
snake venom phospholipase hydrolysis, XXXV, 317, 323, 325
thin-layer chromatography to monitor deacylation, XXXV, 319, 325
of phospholipids of *E. coli* auxotrophs, XXXII, 861
photochemically active derivatives, LVI, 660
in plant cells, problems, XXXI, 521, 522, 524, 590–593
in plant extracts, XXXI, 526, 527
in plasma membranes, XXXI, 87
polyunsaturated, in media, LVIII, 54, 55
in promitochondria, XXXI, 627
separation, LII, 321
spin-labeled, coupling to atractyloside, LV, 527
spin-labeling studies, XXXII, 167, 177, 191, 193
synthesis
in biomembrane phospholipids, XXXII, 542, 544
in parenchyma cells, XXXII, 706
by perfused liver, XXXV, 597–607
rates, XXXV, 279–287, 597, 606, 607
relation to diet, XXXV, 598
to food intake, time course, XXXV, 597
substrate entry, XXXIX, 495
of thermophilic bacterium PS3, LV, 370, 371
thin-layer chromatography, XXXV, 397, 399, 400, 417, 423
in triglycerides
analysis of positional distribution, XXXV, 320–325
Grignard reagent hydrolysis, XXXV, 320, 322–324
pancreatic lipase hydrolysis, XXXV, 320–322, 324
thin-layer chromatography to monitor deacylation, XXXV, 320, 321, 324, 325
unsaturated
E. coli auxotroph, XXXII, 856–864
strains, XXXII, 859
detection by mass spectrometry, XXXV, 345
position assignment in glycerides, XXXV, 345, 346, 352
requirement, *cyd 1*, LVI, 121
stereochemistry determination, with *O*-isopropylidene derivatives, XXXV, 347
yeast requiring, XXXII, 820–823
Fatty acid-albumin complex, preparation, XXXIX, 44, 45
Fatty acid anhydride, intermediate in phosphatidic acid synthesis, XXXV, 440
Fatty acid hydroperoxide, formation, XXXI, 521

Fatty acid α-hydroxylase, LII, 20, 310–318
Fatty acid ω-hydroxylase, *see* Alkane 1-monooxygenase
Fatty acid (ω-1)-hydroxylase, LII, 14, 318–324
Fatty acid (ω-2)-hydroxylase, LII, 14
Fatty acid peroxide, in plant extracts, XXXI, 526, 527
Fatty acid synthase
  in assay of acetyl-CoA carboxylase, XXXV, 9
    of biotin carboxyl carrier protein, XXXV, 18, 19
    of long-chain fatty acyl-CoA, XXXV, 273, 274
  from chicken liver, XXXV, 59–65
    acetyl-CoA binding sites, XXXV, 65
    assay, XXXV, 60, 61
    dissociation and reassociation, XXXV, 63
    malonyl-CoA binding site, XXXV, 65
    products of reaction, XXXV, 63–65
    stability, XXXV, 63
    sulfhydryl content, XXXV, 65
  multienzyme complex, XXXV, 37, 45, 59, 84
  from *Mycobacterium phlei*, XXXV, 84–90
    activators, XXXV, 84, 85, 87–89
    assay for *de novo* synthesis, XXXV, 85, 86
    effect of long-chain acyl-CoA thioesterase on elongation activity, XXXV, 89, 90
      of primer length on elongation activity, XXXV, 89
    elongation of acyl-CoA derivatives, XXXV, 89, 90
    inhibitors, XXXV, 88, 89
    methylated polysaccharide activators, XXXV, 84–85, 87, 95
    phosphopantetheine content, XXXV, 87
    primer specificity, XXXV, 88
    products of *de novo* synthesis, XXXV, 84, 89
    Type I, XXXV, 84–90
    Type II, XXXV, 84, 90
  4′-phosphopantetheine content, XXXV, 42, 43, 58, 59, 87
  from pigeon liver, XXXV, 45–59
    acetyl-CoA binding sites, XXXV, 59
    assay of component reaction activities, XXXV, 49–52
      of covalent binding sites of acetyl and malonyl groups, XXXV, 51, 52
      of overall activity, XXXV, 46–48, 57
    effect of nutritional state, XXXV, 52, 54
    inactivation and dissociation, XXXV, 48, 55–57
    malonyl-CoA binding sites, XXXV, 59
    reaction mechanism, XXXV, 45, 46
    reactivation and reassociation, XXXV, 57
    stability, XXXV, 47, 48, 53–55, 59
    sulfhydryl content, XXXV, 58
  purification from *Euglena gracilis*, XXXV, 111
  from rabbit mammary gland, XXXV, 74–83
    acetylation with acetic anhydride, XXXV, 83
    assay, XXXV, 75, 76
    chain termination factor, XXXV, 83
    inhibitors, XXXV, 80
    products of reaction, XXXV, 81, 82
    stability, XXXV, 79
    substrate control of chain length, XXXV, 81, 82
    sulfhydryl content, XXXV, 81
  from rat liver, XXXV, 37–44
    assay, XXXV, 37, 38
    control of rate of synthesis, XXXV, 44
    immunochemical assay, XXXV, 43, 44
    stability, XXXV, 41, 42
    sulfhydryl content, XXXV, 42

from rat mammary gland, XXXV, 65–74
acetyl-CoA binding sites, XXXV, 73
assay, XXXV, 65–67
dissociation and reassociation, XXXV, 69, 70
effect of nutritional state, XXXV, 67
inhibitors, XXXV, 73
malonyl-CoA binding sites, XXXV, 73
products of the reaction, XXXV, 73
stability, XXXV, 70
substrate control of chain length, XXXV, 73
sulfhydryl content, XXXV, 73
Fatty acyl carnitine, tissue content, LV, 221
Fatty acyl-CoA dehydrogenase
brown adipose tissue mitochondria, LV, 76
in glyoxysomes, LII, 502
long-chain, LIII, 402
Fatty acyl-CoA desaturase, XXXV, 253; LII, 13
Fatty acyl-CoA synthetase, brown adipose tissue mitochondria, LV, 76
Fatty acyl coenzyme A
brown adipose tissue mitochondria, LV, 75
content
of rat heart, LV, 217
of rat liver, LV, 213
long-chain, assay, XXXV, 273–278
short-chain, separation from long-chain, XXXV, 275, 276
Fatty acyl thioesterase I, XXXV, 106, 107, *see also* Palmityl thioesterase II
Fatty acylthiokinase, in glyoxosomes, XXXI, 569
FCCP, *see* Carbonylcyanide *p*-trifluoromethoxyphenylhydrazone
70–F cell, plasma membrane isolation, XXXI, 161
F_c crystal, negative staining, XXXII, 29
Fd, *see* Ferredoxin
FDCD, *see* Circular dichroism, fluorescence-detected
F12/DME mixed medium, LVIII, 103, 104, 106
FDNB, *see* 1-Fluorodinitrobenzene
Feedback inhibition, in multistep enzyme systems, XLIV, 916–919
Feedback inhibitor, as ligand, XXXIV, 390
Feedback regulation
of pyrimidine nucleoside monophosphate kinase, LI, 331
of uridine-cytidine kinase, LI, 307, 314, 320
Feeder cell, Japanese quail, LVIII, 392
Feeder layer
epithelium, LVIII, 265
human fibroblast, LVIII, 358
mesenchymal cell, LVIII, 270
preparation, LVIII, 278
Fenwal bag, XXXIX, 363
Fermentation
ATP formation, LV, 293
in mutation program, XLIII, 25, 26
oxygen, LV, 290
thin-layer chromatography for study, XLIII, 201–206, *see also* Thin-layer chromatography, fermentation
Fermentor, mass culture, LVIII, 204, 205
Fermi resonance, XLIX, 118
Fermi shift, *see* Contact shift
Ferredoxin, *see also* Iron-sulfur protein, high-potential
amino acid composition, LIII, 627
assay, LIII, 617
from *Bacillus polymixa*, LIII, 628
from *Bacillus stearothermophilus*, LIII, 629
bacterial, EPR characteristics, LIII, 486
bovine adrenal cortex mitochondrial, *see* Adrenodoxin
from *Chromatium,* thermolysin hydrolysis products, XLVII, 176, 184
classification of iron proteins, LIII, 260
clostridial, magnetic circular dichroism, XLIX, 175, 176

from *Desulfovibrio desulfuricans*, LIII, 629
from *Desulfovibrio gigas*, LIII, 617, 623, 624, 627–629
electron paramagnetic resonance data, LIII, 630
electron paramagnetic resonance studies, XLIX, 514
4Fe-4S* type, Mössbauer studies, LIV, 370–372
(4Fe-4S) subgroup, *see also* Iron-sulfur protein, high potential
bacterial sources, LIII, 329, 330
optical properties, LIII, 332, 334
forms of spectral properties, LIII, 629
fructose 1,6-bisphosphatase system, activation, XLII, 397–405
in hydrogenase reactions, LIII, 287
iron-sulfur centers, characterization, LIV, 203
magnetic susceptibility studies, LIV, 395
molecular weight, LIII, 627, 629
partial specific volume, XLVIII, 29
physical properties, LIII, 262, 263
plant-type, Mössbauer spectroscopy, LIV, 369, 370
presence in *Desulfovibrio* strains, LIII, 632
prosthetic groups, LIII, 627
proton magnetic resonance studies, LIV, 204–206
purification, LIII, 623, 624
redox potentials, LIII, 630
reductive titration, LIV, 124
renal mitochondria, LV, 12
spinach, magnetic circular dichroism, XLIX, 175, 176
reduced, *g* value, XLIX, 523
from *Spirocheta aurantia*, LIII, 629
Ferredoxin-NADP reductase
assay of ω-hydroxylase, LIII, 356, 357
of rubredoxin, LIII, 341
Ferredoxin-NADP$^+$ reductase
in adrenal gland, LII, 124
in cytochrome *P*–450 activity assay, LII, 125, 126, 131, 139
kinetics, LII, 141
molecular weight, LII, 141
optical properties, LII, 141
prosthetic group, LII, 141
purification procedure, LII, 133, 134
adrenodoxin-substituted Sepharose column, LII, 134
solubilization, LII, 141
spectral properties, LII, 134
Ferric ammonium sulfate, XLIV, 290
Ferric chloride
in adrenodoxin chromophore reconstitution, LII, 137
determination of acid-labile sulfide, LIII, 276
test for hydrazones, XLVI, 211
Ferricenium cation, excited state, LVII, 99
Ferric-enterochelin esterase, function, LVI, 395
Ferric ion
buffers, LVI, 389
in dihydroorotate dehydrogenase, LI, 68
inhibitor
in hydrogen peroxide assay, LVII, 378, 380
of nucleoside phosphotransferase, LI, 393
of *Pholas* luciferin oxidation, LVII, 399
lipid peroxidation, LII, 305, 309
Ferric myoglobin, azide binding kinetics, LIV, 72
Ferric oxalate, as mediator-titrant, LIV, 408
Ferric thiocyanate complex, pressure-jump studies, XLVIII, 310
Ferricyanide, *see also* Potassium ferricyanide
assay of chemiluminescence, LVII, 427
of glucose, LVII, 452, 453
of succinate dehydrogenase, LIII, 466, 468–470
catalyst/cooxidant, in luminol chemiluminescence, LVII, 410, 448, 449, 450
complex I, LVI, 586, 587
for cytochrome oxidase oxidation, LIV, 108, 109
as electron acceptor, LVI, 386
for complex I, LIII, 13
for complex I–III, LIII, 8

for dihydroorotate dehydrogenase, LI, 59
extinction coefficient, LIII, 14, 21
in flavoprotein assays, LIII, 406
Keilin-Hartree preparation, LV, 126
in lactic acid determination, XLIV, 591, 592
millimolar extinction coefficient, LIII, 469
oxidant, of cytochrome c_1, LIII, 226
permeability, LVI, 231
in pipette calibration procedure, LIV, 128
polarography, LVI, 478
preparation of luciferin, LVII, 19
proline uptake by proteoliposomes, LV, 197
reduction
measurement, LV, 542
pathways, LVI, 231
spectrophotometric determination, LIII, 14, 21
spin probe measurements, LVI, 522, 523
stability, in solution, LVII, 453
substrate of complex II, LIII, 21
of NADH dehydrogenase, LIII, 18
Ferricyanide method, to determine sidedness of membrane-bound dehydrogenases, LVI, 229, 230
method, LVI, 231–233
principle, LVI, 230, 231
Ferricyanide reagent, XXXII, 787
Ferricytochrome *c*, *see* Cytochrome *c*; Cytochrome *c*, oxidized
Ferrihemoglobin fluoride, electronic absorption spectrum, XLIX, 135
Ferritin
antibody ratio in conjugates, LVI, 712, 713
conjugation to affinity antibody
assay of conjugates, LVI, 712–715
cytochrome oxidase antibody, LVI, 710, 711
general, LVI, 708, 709
α-globulin conjugation, XXXII, 61, 62
as immune reagent, XXXII, 61
as immunochemical marker, XXXVII, 133
immunoglobulin conjugation, LVI, 226, 227
electron microscopy, LVI, 228
as lectin label, XXXII, 616
in microsomal preparations, LII, 87, 88
molar extinction, LVI, 712, 713
monoamine oxidase antibody, LVI, 711, 712
preparation of ferritin, LVI, 709, 710
purification, XXXII, 61
tRNA mapping, LVI, 9
structure, LVI, 709
Ferritin tracer, in electron microscopy, XXXIX, 148, 149
Ferroammonium sulfate, in hydrogen peroxide assay, LII, 346, 347
Ferrocene, as mediator-titrant, LIV, 408
Ferrocene acetic acid, as mediator-titrant, LIV, 408
Ferrocene 1,1′-dicarboxylic acid, as mediator-titrant, LIV, 408
Ferrocenyl methyl trimethyl ammonium perchlorate, as mediator-titrant, LIV, 408
Ferrocyanide, determination by UV-spectrophotometry, XLI, 29
Ferrocytochrome *c*, *see also* Cytochrome *c*, reduced
absorption spectrum, XLIX, 129
complex IV, LVI, 586
electronic absorption spectrum, XLIX, 132
excitation profile, XLIX, 132–134
resonance Raman spectra, XLIX, 131
Ferrocytochrome c_2, LIV, 31
Ferrocytochrome *c*: oxygen oxidoreductase, *see* Cytochrome *c* oxidase
Ferroperoxidase, oxy form, spectrum, LII, 36
Ferrous acetate
in metallation of porphyrins, LII, 453
preparation, LII, 453
Ferrous ammonium sulfate
in anaerobic TEA-maleate buffer, LIII, 366
in assay of ribonucleoside diphosphate, LI, 238, 244

Ferrous ion
activator of deoxycytidine kinase, LI, 345
of thymidine kinase, LI, 370
of uridine-cytidine kinase, LI, 319
effect on ribonucleoside diphosphate reductase, LI, 244
inhibitor of deoxythymidine kinase, LI, 365
of nucleoside phosphotransferase, LI, 393
of phosphoribosylpyrophosphate synthetase, LI, 11
microsome aggregation, LII, 87
Pholas luciferin chemiluminescence, LVII, 399, 400
Ferrous myoglobin, O_2 binding kinetics, LIV, 72, 73
Ferrous sulfate
acrylamide polymerization, LIX, 68
assay of *Pholas* luciferin, LVII, 386
for porphyrin esterification, LII, 432, 452
in siroheme demetallation, LII, 443
Ferroxidase, LII, 10
oxy form, spectrum, LII, 36
prosthetic group, LII, 4
reaction mechanism, LII, 39
Fervenulin, XLIII, 336
Fetal calf serum, *see* Calf, serum, fetal
α-Fetoprotein, XXXIV, 6
Fetuin, XXXIV, 610, 611
desialylated, β-galactosidase, L, 482
gel electrophoresis, XXXII, 101
in media, LVIII, 88, 105
Fetuin-agarose, XXXIV, 611
Fetus, *see also* Embryo
blood circulation, XXXIX, 245
perfusion, XXXIX, 245–249
liver explant, XXXIX, 36–40
FGAM, *see* 2-Formamino-*N*-ribosylacetamidine 5′-phosphate
FGAM synthetase, *see* 2-Formamino-*N*-ribosylacetamidine 5′-phosphate synthetase
FGAR, *see* 2-Formamido-*N*-ribosylacetamide 5′-phosphate
FGAR amidotransferase, *see* 2-Formamino-*N*-ribosylacetamidine 5′-phosphate synthetase
Fiber, *see also* specific type
column packing, XLIV, 825
derivatization, XXXIV, 200–205
dry weight determination, XLIV, 232
fractionation, XXXIV, 198–255, 720
cell attachment, XXXIV, 208, 209
cell binding, XXXIV, 205–220
cell population studies, XXXIV, 220
cell separation, XXXIV, 205
observations, XXXIV, 199, 200
plucking, XXXIV, 203
rosette methods, XXXIV, 217–219
significance, XXXIV, 199
specificity, XXXIV, 199
gelatin coated, XXXIV, 208
laboratory reactor, XLIV, 233, 234
nitrogen content determination, XLIV, 232
physical properties, XLIV, 233–235
preparation, XLIV, 228–231
ultrastructure, XLIV, 235, 236
use, XLIV, 232, 233, 241–243
Fiberglass screening, XXXI, 116
Fiber-rosette method, XXXIV, 75, 217–219
Fibrinogen
batroxobin assay, XLV, 215
crotalase clotting assay, XLV, 229
electrophoretic light scattering properties, XLVIII, 435
factor V assay, XLV, 110
inert proteic matrix, XLIV, 268
thrombin assay, XLV, 157, 158
yeast electron microscopy, LVI, 720
Fibrinoligase
fibrinogen clotting, XLV, 35, 36
filter paper method for titration, XLV, 186
function, XLV, 177
substrates, synthetic, XLV, 179
Fibrin-stabilizing factor, *see* Factor XIII
Fibroblast
cell culture lines, XXXII, 584, 585
cell hybrids, XXXII, 578, 582
chicken embryo, LVIII, 55, 76, 84
freezing, LVIII, 392, 393
heart, media, LVIII, 90
media, LVIII, 59

microcarrier culture, LVIII, 188, 189
preparation, LVIII, 381–384
storage, LVIII, 392, 393
culture, L, 440–442
duck embryo, media, LVIII, 59
explant, LVIII, 446–448
fetal lung, LVIII, 50
freezing, LVIII, 448, 449
growth factor, LVIII, 78, 163
in media, LVIII, 99
human, LVIII, 55, 76
cells resembling, culture, XXXII, 799, 800, 806–811
culture, XXXII, 801–811
DNA content, LVIII, 148
embryo lung, microcarrier culture, LVIII, 190
as feeder cell, LVIII, 358
foreskin, LVIII, 51
microcarrier culture, LVIII, 190
interferon production, LVIII, 295
isolation, LVIII, 444–450
media, LVIII, 56, 58, 59
monolayer culture, LVIII, 132
preservation, LVIII, 31
propagation, LVIII, 444
human diploid
frozen storage, XXXII, 564, 565
monolayer culture, XXXII, 561–568
subcultivation, XXXII, 565
α-L-iduronidase activity, L, 443, 444
mouse
effect of hormone, LVIII, 101, 102
embryo, LVIII, 90
media for, LVIII, 58, 59
monolayer culture, LVIII, 132
Swiss 3T3, LVIII, 105, 106
mucopolysaccharide storage disorders, L, 439
plasma membranes, XXXI, 144
storage, LVIII, 448, 449
Fibronectin, LVIII, 267
Fibrosarcoma, LVIII, 376
Fibrous tissue, nuclei isolation, XXXI, 258
Ficaprenol, in phosphokinase assay, XXXII, 439, 440
Ficaprenol monophosphate, L, 405
Ficin, XXXIV, 532
azide binding, XLIV, 52
conjugate, assay, XLIV, 338, 339
in hormone receptor studies, XXXVII, 212
immobilization
on (carboxymethyl)cellulose, XLIV, 52, 338, 339
with collagen, XLIV, 244
property changes, XLIV, 53
kinetics, of substrate diffusion, XLIV, 437
Ficoll, XXXI, 733
gradient properties, XXXI, 508
isolation of chloroplasts, LIX, 208, 209, 434
of mitochondria, LV, 6
as plant-cell protectant, XXXI, 536, 549, 599
purification of monoamine oxidase, LIII, 498, 499
solution, preparation, XXXI, 399
Ficoll gradient, LVIII, 257, 258
Ficoll-Hypaque
leukocyte isolation, LVII, 470
solution, preparation, XXXII, 634, 635
Ficoll-Isopaque density gradients, XXXII, 760
FID, *see* Free-induction decay
Field-indicating absorption change, LIV, 61, 62
Field jump, LIV, 2
Fieser's solution, formulation, LIII, 299
Fig
enzyme studies, XXXI, 541
latex, peroxidase, LII, 514
organelles, XXXI, 533
Filipin, effect on liposomes, XXXII, 512
Film
autoradiography, LVIII, 287
for optical diffraction technique, XLIX, 45, 48
Filter, *see also* specific type
bacteriological, LVIII, 8
calcium transport assays, LVI, 239, 241
cell, LVIII, 514
high-efficiency particle air, LVIII, 10, 40

for HPLC solvents, LII, 282
low-frequency cutoff, XLVIII, 467
in nanosecond absorbance spectrometer, LIV, 37
optical, for photometer in bioluminescence assays, LVII, 260
phospholipid-impregnated
for measurement of penetrating ion concentration
apparatus, LV, 597, 598
applicability, LV, 600, 601
filters, LV, 599
incubation mixture, LV, 599, 600
penetrating ions, LV, 598, 599
principle, LV, 597
proteoliposomes associated, LV, 763, 766–768
photoaffinity labeling, LVI, 672
spatial, XLVIII, 187
sterilizing, LVIII, 75
for use with microliter volumes, LVII, 312, 313
Filter assay, for steroid-receptor binding, XXXVI, 234–239
Filter binding assay, XXXII, 423; LIX, 847–850
Filter cloth, *see also* Miracloth; Nitex
for chloroplast isolation, XXXI, 604, 605
Filtering apparatus, for bacteria, XXXII, 844
Filter membrane
lipid-impregnated, LV, 604, 605
assay of electrical potentials across, LV, 612
of photopotentials, LV, 608–611
electrical characteristics, LV, 607, 608
preparation, LV, 605–608
summary, LV, 613
Filter paper
bridge formation, XLIV, 166
for cytochrome *c* extraction, LIII, 129
disk, for radioassay, LIX, 216
Filter-Solve, LIX, 324
Filtration, *see also* specific type
of biological fluids, LVII, 72
centrifugal, LVI, 283–285
of environmental samples, for ATP determination, LVII, 80
isolation of mitochondria, LV, 5
rapid, for plant-cell extraction, XXXI, 541, 542
of subcellular particles, XXXII, 5–7
Filtration aid, *see* Celite 505; Hyflo Super-Cel
Filtration apparatus, iron transport studies, LVI, 391
construction and use, LVI, 392–394
Filtration system, supplier, LVIII, 16, 17
Fine-structure splitting, LIV, 349
Fingerprinting
alternative procedure, LIX, 66
of mononucleotides, LIX, 71–74
of oligonucleotides, LIX, 76–78, 93
standard procedure, LIX, 65, 66
Finnigan 1015-PDP 8/I gas chromatograph-mass spectrometer-computer system, LII, 333
Firefly
bioluminescent reaction, LIII, 559
luciferase, LIII, 560
Firefly lantern
drying procedures, LVII, 6
extract FLE–50, LVII, 61, 75, 87
Firefly luciferase, *see* Luciferase, firefly
Firefly luciferin, *see* Luciferin, firefly
Fischer's medium, LVIII, 56, 86
Fischer method, for amino acid esterification, XLVII, 523
Fish
blood collection, LVIII, 472
cell, LVIII, 467
corneal tissue culture, LVIII, 473
fin, culture, LVIII, 473
leukocyte, LVIII, 472
muscle, enolase, assay, purification, and properties, XLII, 329–334
Fisher Model 13-639-92 microcombination pH electrode, LIV, 404
FITC, *see* Fluorescein isothiocyanate
FITC-dextran, internal pH, LV, 566
Fitting
in anaerobic systems, LIV, 116
in freeze-quench techniques, LIV, 89
metal-to-metal compression-type, LIV, 116

Fixation
for DAPI staining, LVI, 730
electron microscopy, LVI, 719, 720
monolayer cultures, LVIII, 140
for scanning electron microscopy, XXXII, 55, 58
of subcellular particles, XXXII, 4, 5
for transmission electron microscopy, XXXIX, 133–138
Fixative
solution, metaphase preparation, LVIII, 171
for transmission electron microscopy, XXXIX, 136
F12K medium, LVIII, 59, 87, 155
Flame, chemical, as excitation source, LIV, 450–454, 464, 468
Flash activation, LIV, 43–45
stimulus generation, LIV, 44
Flash discharge tube, relative brightness, LIV, 95
Flash photolysis technique, LIV, 93–101
apparatus, LIV, 53, 54
applications, LIV, 47, 108
conversion of intensity changes to absorbance changes, LIV, 54, 55
data analysis, LIV, 54–57
data recording, LIV, 97–101, 108
experimental procedure, LIV, 106–108
integrating sphere, LIV, 98, 99
in ligand binding studies, LIV, 511, 512
light sources, LIV, 95, 96
limitations, LIV, 111
oxygen addition methods, LIV, 105, 106
principle, LIV, 94, 95
Flavacid, XLIII, 135
Flaveolin, XLIII, 336
Flavin
8α-aminoacyl, synthetic, properties, LIII, 465
in complex I, LIII, 12
in complex II, LIII, 24
in complex III, LIII, 89, 96, 97
in complex I–III, LIII, 7, 8
complex V, LVI, 580, 584
covalently bound, types, LIII, 449
in cytochrome bc_1 complex, LIII, 116
Keilin-Hartree preparation, LV, 123
M. phlei ETP, LV, 184, 185
in NADH dehydrogenase, LIII, 17, 18, 19
prosthetic group, in oxidases, LII, 4, 6
redox behavior, as mediator-titrants, LIV, 402
site of amino acid substitution, determination, LIII, 460
Flavin adenine dinucleotide, XLIV, 152, *see also* Adenosine 5′-monophosphate
complex II, LVI, 580
complex V, LV, 315
in cytochrome P–450_{cam} activity assay, LII, 156
in enzyme electrode storage buffer, XLIV, 590, 603
in ETF-ubiquinone oxidoreductase stoichiometry, LIV, 145
fluorescent acceptor, XLVIII, 378
inhibitor of brain α-hydroxylase, LII, 317
in NADPH-cytochrome P–450 reductase, LII, 90, 94
in oxidase determination, XLIV, 852
photochemically active derivative, LVI, 656
prosthetic group, XLIV, 389; LII, 4
as putidaredoxin reductase stabilizer, LII, 187
Flavin 8-azidoadenine dinucleotide, LVI, 656
Flavin compound, EPR characteristics, LIV, 134, 137, 138
Flavin enzyme, XLVI, 33, 160
Flavin mononucleotide, XLIV, 839, *see also* Riboflavin 5′-phosphate
activator of fatty acid synthase, XXXV, 84, 87
assay of luciferase, LVII, 137
of malate, LVII, 187
of oxaloacetate, LVII, 187
of oxidoreductases, LVII, 205
complex I, LVI, 580
dihydroorotate dehydrogenase, LI, 68
immobiliation, XLIV, 884, 885
inhibitor of brain α-hydroxylase, LII, 317
luciferase-mediated assays, LVII, 218, 221

as mediator-titrant, LIV, 409
in NADH-dehydrogenase, stoichiometry, LIV, 145
in NADPH-cytochrome P–450 reductase, LII, 90, 94
nonradiative energy transfer, LVII, 287, 288
phosphorescence quenching, XLIX, 247
prosthetic group, LII, 4
reduced
assay of bacterial luciferase-coupled antibodies, LVII, 404
of luciferase, LVII, 205
of protease, LVII, 200
autoxidation, LVII, 126, 133, 136
luciferase affinity assay, LVII, 170, 171
4α-peroxy adduct, luminescence reaction intermediate, LVII, 133–135, 152
Pholas luciferin chemiluminescence, LVII, 399
preparation, LVII, 138–140
by catalytic reduction, LVII, 138
by chemical reduction, LVII, 139
by enzymic reduction, LVII, 140
by photochemical reduction, LVII, 139
substrate, of bacterial luciferase, LVII, 126, 133, 151
solution, stability, LVII, 137
Flavin mononucleotide-agarose, XXXIV, 300, 302
Flavin mononucleotide-cellulose, XXXIV, 300–302
Flavin mononucleotide-cellulose phosphate, XXXIV, 300
Flavin peptide
criteria for purity, LIII, 456
degradation to aminoacyl flavins, LIII, 456, 457
8α-substituted, LIII, 449–465
naturally occurring, isolation, LIII, 452–457
synthesis, LIII, 457–460
from succinate dehydrogenase, LIII, 482
Flavin radical, neutral, interatomic distance calculations, XLIX, 331
Flavin reductase, in coupled assay, with luciferase, LVII, 136, 140
Flavipin, XLIII, 163
Flavobacterium, XLIII, 737
oxidase, LII, 20
Flavocytochrome b_2, LIII, 238–256
assay, LIII, 239, 240
cleaved crystalline form, LIII, 240–245
properties, LIII, 244
purification, LIII, 240–245
purity, LIII, 245
stability, LIII, 245
extinction coefficient, LIII, 240
from *Hansenula anomala*, LIII, 238, 249–256
intact noncrystalline form, LIII, 244–254
properties, LIII, 244
purification, LIII, 245–248, 249–254
purity, LIII, 248, 249, 254
stability, LIII, 248, 249, 254
kinetic properties, LIII, 244
molar activity, LIII, 244
molecular weight, LIII, 244
pH optimum, LIII, 244
photosensitivity, LIII, 254, 255
from *Saccharomyces cerevisiae*, LIII, 238, 240–249
spectral properties, LIII, 246
subunit structure, LIII, 244
Flavodoxin, LIII, 403
activity, LIII, 630
amino acid composition, LIII, 627
assay, LIII, 617
from *Desulfovibrio gigas*, LIII, 617, 624, 625, 630, 631
in *Desulfovibrio* strains, LIII, 632
dissociation kinetics, LIII, 430
holoenzyme reconstitution, LIII, 436
in hydrogenase reactions, LIII, 287
molecular weight, LIII, 627
from *Peptostreptococcus elsdenii*, redox potential, LIII, 400
prosthetic groups, LIII, 627, 630
purification, LIII, 622–625
redox potentials, LIII, 630

reversible dissociation, LIII, 433
spectral properties, LIII, 631
Flavofungin, XLIII, 136
Flavone derivative, enzyme induction, XXXIX, 39
Flavonoid, in plants, XXXI, 529, 530
Flavoprotein, LIII, 395–570, *see also* specific enzyme
assay methods, LIII, 403, 404, 421–429
biological types, LIII, 419
brown adipose tissue mitochondria, LV, 75
catalysis mechanisms, LIII, 401–403
chromatographic separation, LIII, 426, 427
classification, LIII, 397–399
complex, definition, LIII, 397
covalently bound, definition, LIII, 419, 449–451
definition, LIII, 397
determination, by enzymic analysis, LIII, 421–423
by fluorometric analysis, LIII, 423, 425–429
by spectrophotometric analysis, LIII, 421
fluorescence, LII, 57
hydroxylated, LIII, 419
inactivation by apoprotein modification, LIII, 441–443, 447, 448
by flavin modification, LIII, 438–441, 447
kinetic mechanism, LIII, 443–445
isoalloxazine ring, reactivity, LIII, 400
membrane-bound, LIII, 405–413
activity determination, LIII, 405–407
comparison to soluble forms, LIII, 407–413
extraction, LIII, 405
of microsomes, LII, 44, 45
with noncovalently bound flavin, LIII, 419–429
definition, LIII, 419
purification procedures, LIII, 420
quantitative determination, LIII, 421–429
oxygen transport, LII, 6, 8
reconstitution, LIII, 436, 437
redox potentials, range, LIII, 400
reversible dissociation, LIII, 429–437
by acid-ammonium sulfate treatment, LIII, 430–432
by calcium chloride treatment, LIII, 433–434
by dialysis against potassium bromide, LIII, 432, 433
by guanidine hydrochloride treatment, LIII, 433, 434
simple, definition, LIII, 397
suicide substrates, LIII, 437–448
versatility, factors, LIII, 399–401
Flavoprotein dehydrogenase, definition, LIII, 398
Flavoprotein-linked monooxygenase, LII, 15
Ah^b allele, LII, 231
assay, LII, 234–236, 372–377, 409, 410, 413–415
sensitivity, LII, 236
standard curves, LII, 238, 239
differences in basal and aromatic hydrocarbon-induced forms, LII, 235
distribution, LII, 202
for fatty acid hydroxylation, LII, 319–324
inhibitors, LII, 410
in liver microsomes, components, LII, 201
microassay, LII, 236–238
for steroid hydroxylation, LII, 377, 378
subcellular localization, LII, 238–240
Flavoprotein monooxygenase
definition, LIII, 398, 399
external, LIII, 399
internal, LIII, 399
reaction mechanism, LIII, 403
Flavoprotein oxidase, definition, LIII, 397, 398
Flavoprotein oxygenase, definition, LIII, 398, 399
Fletcher and Powell method, LVI, 268
Flexibility gradient, of hydrocarbons, spin-labeling studies, XXXII, 167, 168
Flight muscle, mitochondria, XXXI, 491

Flocculation, in brush-border preparation, XXXI, 132–134
Florisil
in enzyme analyses, XXXVII, 290
in glycolipid isolation, XXXII, 349
for steroid adsorption, XXXVI, 40, 47
Florisil chromatography
column preparation, LIII, 453
in isolation of flavin peptides, LIII, 453, 454
Flotation, of liver homogenate, XXXI, 80
Flotation equilibrium centrifugation, XXXI, 716
in plasma-membrane isolation, XXXI, 163
Flotation gradient centrifugation, preparation of mitochondria, LIX, 422, 423
Flotation method, density determinations, XLVIII, 23–29
Flow cytometer, LVIII, 235
suppliers, LVIII, 234
Flow cytometry, LVIII, 233–247
instruments, LVIII, 233, 234
principles, LVIII, 233–239
Flow dialysis
determination of ΔpH and active transport, LV, 680–682
calculation of ΔpH, LV, 686–688
method, LV, 682–686
measurement of ion distribution, LV, 556
of pH changes, LV, 564
transport studies, LVI, 387, 388
Flow experiment, immobilized mitochondria, LVI, 552–554
Flow microfluorometry, LVIII, 149, 252
Flow sorter
schematic, LVIII, 234
suppliers, LVIII, 234
Flow time, measurement, LVI, 494
Fluctuation scattering function, LIX, 677, 710
Fluoboric acid, purification of 2-amino-4-nitrophenol, LVI, 659
Fluoranthene
chemiluminescent reaction, LVII, 516, 517
structural formula, LVII, 498
Fluorene
internal standard, XLVII, 34
metabolite binding, reduced cytochromes, LII, 274, 275
Fluorescamine, XL, 115; XLVII, 8, 13
amino group determination, XLVII, 303
assay
procedure, LIX, 500, 501
proteolytic activity, XLV, 25
chain termination, XLVII, 601
chemical properties, XLVII, 236
measurement of amines, LVI, 624, 625
peptide visualization, XLVII, 201, 202, 236–243
preparation, XLVII, 17
protein amino group determination, L, 140
reagent, for tube analysis, XLVII, 237
salts, XLVII, 239
solid-phase peptide synthesis monitoring, XLVII, 599
spray reagent, for thin-layer plates, XLVII, 196, 201, 240
Fluorescein, XLVI, 563, 564, 567
conjugate, XL, 303
emission spectrum, XLVIII, 369
extinction coefficient
of free dye, LIX, 151
of tRNA-bound dye, LIX, 151
fluorescent quantum yield, LVII, 590
as immunochemical marker, XXXVII, 133
intensity readings, on digital, photon-counting fluorescence polarometer, XLVIII, 401
lectin labeling, XXXII, 616
as optical probe, LV, 573
preparation, XLVI, 565
product, in microfluorometry, XLIV, 374–376
Fluorescein isothiocyanate, LVIII, 176; LIX, 147, 157
determination of tRNA-ester modification, LIX, 161
incorporation into modified tRNA, LIX, 149
labeling, prothrombin, XLV, 134
modification of lysine, LIII, 141
of ribosomal proteins, LIX, 819

in preparation of eosin isothiocyanate, LIV, 49
Fluorescein mercuric acetate, XLVII, 408
for plasma membrane isolation, XXXI, 156–163
large scale, XXXI, 157, 158
small scale, XXXI, 158, 159
Fluorescence
amine uptake, LV, 564, 565
ATPase-bound aurovertin, LV, 301
aurovertins, LV, 485
cell, LVIII, 233, 234, 236
changes, nonspecific, LVI, 498
coefficient, definition, XLI, 55
conformation studies, XLIV, 362–366
definition, XLVIII, 348, 349
donor-acceptor pairs, XLVIII, 362, 363
donor quenching, XLVIII, 354, 355, 360, 364, 365
dye choice, XLVIII, 359
emission, polarization, XLVIII, 350–352
of energized vesicles, measurement, LV, 366
energy-linked processes, LVI, 496, 497
apparatus, LVI, 497, 498
measurements, LVI, 498–501
remarks, LVI, 498
1,N^6-ethenoadenosine 3′,5′-monophosphate, XXXVIII, 430
of ETP with ANS, LV, 183
high-pressure studies, XLIX, 16
indicator, calcium, LVI, 303
influence of local environment, XLVIII, 348–350
interference, in Raman spectroscopy, LIV, 248
for measurement of reaction equilibria and kinetics, XLVIII, 380–415
advantages and disadvantages, XLVIII, 381
molar
definition, XLVIII, 384
evaluation, XLVIII, 385
nonradiative energy transfer
efficiency, XLVIII, 354–356
measurement, XLVIII, 354–356
optimal conditions for monitoring, XLVIII, 360
random labeling case, data analysis, XLVIII, 373–375
rate, XLVIII, 353, 354
sample preparation, XLVIII, 363, 364
specific-site labeling case, data analysis, XLVIII, 368–373
theory, XLVIII, 352–355
nuclear, of mouse L cells, XL, 303
parameters, species concentration, XLVIII, 384–385
polarization, evaluation, XLVIII, 385
quantum yield, LVII, 569
absolute measurement, LVII, 589–595
relative measurement, LVII, 595–597
Raman scattering, XLIX, 83, 89, 94
reaction stoichiometry, XLVIII, 382, 383
sensitized emission, XLVIII, 355, 361
time range, LIV, 1
time scale, XLVIII, 348
types of labeling, XLVIII, 356–359, 364
Fluorescence detection
detector, XLVII, 4
o-phthalaldehyde, XLVII, 8
sensitivity, XLVII, 7, 13
Fluorescence polarization
applications, LIV, 47, 49
measurement
calculation of results, LX, 716–718
observations, LX, 715, 716
theoretical basis, LX, 713, 714
Fluorescence polarometer
analog, direct reading, XLVIII, 401–404, 406, 408
sensitivity, XLVIII, 402, 403
digital, photon-counting, XLVIII, 393–401
detector module components, XLVIII, 394, 396, 397
excitor module components, XLVIII, 397–402
operating procedure, XLVIII, 397–401
optical components, XLVIII, 394, 395

sensitivity, XLVIII, 400, 401
stopped-flow, XLVIII, 405, 407, 409–415
sensitivity, XLVIII, 415
Fluorescence quenching
data analysis, XLIX, 225–231, 236
effect of pH, XLIX, 234
of temperature, XLIX, 234
excitation wavelength, XLIX, 235
quencher selection, XLIX, 231–233
by solute, XLIX, 222–236
spectral shifts, XLIX, 231
Fluorescence spectroscopy, XLVIII, 348–356
for donor lifetime quenching technique, XLVIII, 367, 368
for donor quenching technique, XLVIII, 364, 365
ligand binding, XLVIII, 307
for sensitized emission technique, XLVIII, 365–367
single photon counting data, analysis, XLVIII, 367
Fluorescent probe, XXXII, 234–246, 506
ANS technique, XXXII, 234
apparatus, XXXII, 236, 237
attachment to tRNA, LIX, 146–156
examples, XXXII, 238, 239
fluorescence emission spectra, XXXII, 243–245
membrane preparations, XXXII, 237, 238
Fluorescent stain, XXXIV, 216
Fluorescent staining, *see* Staining, fluorescent
Fluoride
cAMP inhibitor, XXXVIII, 275
cyclase assay, XXXVIII, 119, 140, 145–147, 149, 160, 170, 173, 190, 191
as enzyme inhibitor, XXXVII, 294
inhibitor of chloroperoxidase, LII, 529
of salicylate hydroxylase, LIII, 541
interaction with cytochrome *c* oxidase, LIII, 194, 200, 201
phosphodiesterase assay, XXXVIII, 133
protein kinase assay, XXXVIII, 289
Fluoride ion, inhibitor of pyrophosphatases, LI, 172
Fluorimetric assay, XLIV, 528–538
Fluorimetry, in direct quantitative histone assay, XL, 114
Fluorine, in media, LVIII, 69
Fluorine-19
lock nuclei, XLIX, 348, 349; LIV, 189
nuclear magnetic resonance, XLIX, 270–295
resonance frequency, LIV, 189
Fluoroacetol phosphate, synthesis, XLVI, 141
2-Fluoroadenosine
inhibitor of adenosine deaminase, LI, 507
of GMP synthetase, LI, 224
Fluoroalcohol, as uncouplers, LV, 470
P_3-Fluoro-P^1-5′,8-azidoadenosine triphosphate, synthesis, LVI, 652
Fluorobenzamide, XLVI, 295
N^6-*o*-Fluorobenzoyladenosine 5′-phosphate, synthesis, XLVI, 296, 297
N^6-*o*-Fluorobenzoyladenosine 5′-triphosphate, XLVI, 295–299
synthesis, XLVI, 296, 298
N^6-*p*-Fluorobenzoyladenosine 5′-triphosphate, XLVI, 295–299
synthesis, XLVI, 296–298
Fluorocarbon, for liver perfusion media, XXXIX, 36
Fluorocarbon-43, hollow-fiber membranes, XLIV, 302, 303
Fluorocarbon FC–47, as hepatic perfusion, XXXIX, 36
2-Fluorocinnamic acid, pseudosubstrate of melilotate hydroxylase, LIII, 557
3-Fluorocinnamic acid, pseudosubstrate of melilotate hydroxylase, LIII, 557
4-Fluorocinnamic acid, pseudosubstrate of melilotate hydroxylase, LIII, 557
Fluorocitrate, as enzyme inhibitor, XXXVII, 294
9α-Fluorocortisol, effects on sodium transport, XXXVI, 444
5-Fluorocytidine, substrate of uridine-cytidine kinase, LI, 313
2-Fluorodeoxyadenosine, inhibitor of adenosine deaminase, LI, 507
5′-Fluoro-5′-deoxythymidine, substrate of nucleoside deoxyribosyltransferase, LI, 450

5-Fluorodeoxyuridine
as mutagen, LVIII, 316, 317
substrate of deoxythymidine kinase, LI, 364
of thymidine kinase, LI, 358
5-Fluoro-2′-deoxyuridine, DNA inhibitor, XL, 276
5-Fluoro-2′-deoxyuridine 5′-(*p*-aminophenyl phosphate)
coupling to agarose, LI, 100
preparation, LI, 99, 100
5-Fluoro-2′-deoxyuridine 5′-(*p*-nitrophenylphosphate), preparation, LI, 98, 99
5-Fluoro-2′-deoxyuridine 5′-phosphate, inhibitor of thymidylate synthetase, LI, 97, 98
5-Fluoro-2′-deoxyuridylate, XLVI, 307–312
radioactive, XLVI, 310–312
5-Fluoro-2′-deoxyuridylate 5,10-methylenetetrahydrofolate-thymidylate synthetase complex, XLVI, 309
synthesis, XLVI, 311, 312
Fluorodinitrobenzene, XLIV, 284; XLVI, 64, 65
1-Fluoro-2,4-dinitrobenzene, LV, 517
in histone assay, XL, 115
Fluorography
analysis of products of tRNA methyltransferase, LIX, 192
method, indirect, protein visualization, LVI, 71, 72
3-Fluoro-4-hydroxybenzoic acid, substrate of *p*-hydroxybenzoate hydroxylase, LIII, 549
Fluoroketone, XLVI, 131
Fluorometer
adaptations for solid surface method, XLIV, 621–624
adenylate cyclase assay, XXXVIII, 162, 163
cAMP assay, XXXVIII, 62
cGMP assay, XXXVIII, 82
cyanine dyes, LV, 690
enzymatic assays, LV, 200, 201
for fluorescent probe studies, XXXII, 236, 237
Fluorometric Fast Analyzer, principles and use, XXXI, 796–799
Fluorometry
L-asparaginase activity assay, XLIV, 690, 691
assay of hydroxylated biphenyls, LII, 399–407
calculations, LII, 402, 403
chymotrypsinogen tertiary structure, XLIV, 524–526
corticosterone assay, XXXIX, 345–347
determination, of dehydrogenase activity using resorufin, XLI, 53–56
of flavin site of substitution, LIII, 461, 462, 465
in D-galactose assay, XLI, 8, 10
of immobilized proteins, XLIV, 530, 531, 533, 534
melatonin determination, XXXIX, 382–384
for microanalysis, XLIV, 373–379
procedural advantages, XLIV, 618
pyridine nucleotide assay, LV, 267, 269
quantitative determination of flavins, LIII, 423–429
silicone rubber pad method, XLIV, 619–625
advantages, XLIV, 624, 625
preparation, XLIV, 619–624
procedure, XLIV, 624
solid surface methods, XLIV, 618–633
titration of active sites, XLIV, 391, 392
4-Fluoro-3-nitroaniline, XLIV, 284; XLVI, 264
Fluoro-2-nitro-4-azidobenzene
commercial source, XLIV, 283, 284
in photochemical immobilization, XLIV, 281, 282
preparation, XLIV, 284
4-Fluoro-3-nitrophenyl azide, XLVI, 264, 266, 267, 269; XLVII, 421
5-Fluoroorotate, inhibitor of orotate phosphoribosyltransferase, LI, 166
Fluoropa Premix, XLVII, 26, 27, 29, 31, *see also* Durrum resin
2-Fluorophenylalanine, substrate of phenylalanine hydroxylase, LIII, 284

3-Fluorophenylalanine, substrate of phenylalanine hydroxylase, LIII, 284
4-Fluorophenylalanine
  inhibitor of phenylalanine hydroxylase, LIII, 285
  substrate of phenylalanine hydroxylase, LIII, 284
Fluorostyrene, XLIV, 108, 112
3-Fluorostyrene, in copolymer, XLIV, 84
4-Fluorostyrene, in copolymer, XLIV, 84
Fluorosulfonyl benzamidine, XXXIV, 732
m-(2-[*o*-(*m*-Fluorosulfonylbenzamido)phenoxy]ethoxy)benzamidine, XLVI, 118
p-(3-[*p*-(*p*-Fluorosulfonylbenzamido)phenoxy]propoxy)benzamidine, XLVI, 128
3′-*p*-Fluorosulfonylbenzoyladenosine
  reaction with buffers, XLVI, 245
    with glutamate dehydrogenase, XLVI, 246
  synthesis, XLVI, 243, 244
5′-*p*-Fluorosulfonylbenzoyladenosine, XLVI, 241
  reaction with buffers, XLVI, 245
    with glutamate dehydrogenase, XLVI, 247, 248
    with pyruvate kinase, XLVI, 248
  synthesis, XLVI, 242, 243
m-Fluorosulfonylbenzoyl chloride, XLVI, 118
p-Fluorosulfonylbenzoyl chloride, XLVI, 243, 244
DL-5-Fluorotryptophan, substrate of tryptophanyl-tRNA synthetase, LIX, 249
DL-6-Fluorotryptophan, substrate of tryptophanyl-tRNA synthetase, LIX, 249
5-Fluorouracil
  inhibitor of orotate phosphoribosyltransferase, LI, 166
  substrate of orotate phosphoribosyltransferase, LI, 152
5-Fluorouridine, substrate of uridine-cytidine kinase, LI, 313, 319
5-Fluorouridine triphosphate, substrate of nucleoside diphosphokinase, LI, 385
Fluoxymesterone, antiglucorticoid, XL, 291, 292
Fluram, *see* Fluorescamine
Flux coefficient, in substrate entry studies, XXXIX, 497, 498
Flux equation, for asymptotic boundary profiles, XLVIII, 196–201
Flux measurement device, LIV, 388
Flux technique, for transport linked membrane potential
  preparation of cells, LV, 663, 664
  procedure, LV, 664–666
Fly, flight muscles, mitochondria, LV, 22–24
FMA, *see* Fluorescein mercuric acetate
F12 medium, *see* Ham's medium F 12
fMet-tRNA, *see* Formylmethionyl-transfer ribonucleic acid
FMIR, *see* Frustrated multiple internal reflectance
F12M medium, LVIII, 155
FNAB, *see* Fluoro-2-nitro-4-azidobenzene
Foam fractionation, of brain endothelial cells, XXXII, 717
FOCAL language, use with Fast Analyzer, XXXI, 804
Focus-forming assay, for transformed cells, XXXII, 591, 592
Förster critical distance, XLVIII, 354, 356
  range, calculation, XLVIII, 369–373
  values, for assembled systems, XLVIII, 378, 379
Förster mechanism, LIV, 25
Förster theory, of fluorescent chromophores, XXXII, 236
Folate deconjugase, in brush order, XXXI, 130
Folate reductase, fluorine nuclear magnetic resonance, XLIX, 274
Folch's partition method
  for glycolipids, XXXII, 349, 352
  for lipid, XXXV, 396, 397, 409, 421, 423, 549–551
Fold-over, NMR spectrum, LIV, 157, 160
  elimination, LIV, 159, 160, 168
Folic acid, XLIII, 520
  chemotactic potential, in amoebae, XXXIX, 490
  deficiency, LVIII, 53
  as eluent, XXXIV, 286

as ligand, XXXIV, 287
logit-log assay, XXXVII, 9
in media, LVIII, 54, 64
Folimycin, XLIII, 137
Folin–Ciocalteau reagent, XL, 111, 112
Folinic acid, in media, LVIII, 54, 64
Folin phenol reagent, measurement of protein, XL, 375
Folin reagent, XLIV, 110, 387
melanoma cell, LVIII, 568
Follicle
diameter, XXXIX, 186
granulosa cell isolation, XXXIX, 188–199
transformation to corpus luteum, XXXIX, 185
Follicle-stimulating hormone, XXXIV, 5, 338
autoradiography, XXXVI, 150; XXXVII, 165
desialylation, XXXVII, 321, 322
effect on interstitial cell function, XXXIX, 252, 253
human, purification, XXXVII, 382–386
logit-log assay, XXXVII, 9
LRF effects, XXXVII, 87
in luteinization, XXXIX, 184–185
in media, LVIII, 102
pineal effects, XXXIX, 381, 382
preparation, XXXVII, 365–370
ovine and bovine, XXXVII, 365–368
rat and rabbit, XXXVII, 369, 370
releasing factor, XXXIX, 382
secretion by cloned pituitary cells, XXXIX, 128–132
secretory granule use in study, XXXI, 418
sialic acid content, XXXVII, 323
tritium labeling, XXXVII, 321–326
unit of activity, XXXVII, 362
Follicle-stimulating hormone-releasing hormone, bioassay, XXXVII, 233–238
Follicular cell, rat, effect of hormone, LVIII, 100, 101
Fomecin A, XLIII, 163
Food additive, effect on liver microsomes of laboratory animals, LII, 272
Forceps, dissecting, for rat surgery, XXXIX, 171
Foreskin fibroblast, LVIII, 51
Forest tent caterpillar, media, LVIII, 457
Formaldehyde, XLIV, 248
alcohol oxidase, LVI, 483
^{14}C-labeled, in steroid receptor studies, XXXVI, 164
colorimetric assay, LII, 91, 96, 205, 298–302
in CsCl gradient analysis of ribosomal subunits, LIX, 416, 417
fixation, XL, 4
hollow-fiber membrane sterilization, XLIV, 308
hydrogen peroxide assay, LII, 349, 350
in (±)-L-methylenetetrahydrofolate preparation, LI, 91
microsomal reactions yielding, LII, 297, 298
protein labeling, XLVIII, 332
radioactive labeling, XLVII, 476
of initiation factors, LX, 34, 35, 55, 57, 68, 122, 208
in vitro, of ribosomal proteins, LIX, 586
reductive alkylation, XLVII, 469–472, 476
tissue fixation, XLIV, 712, 713
[^{14}C]Formaldehyde, preparation of radiolabeled ribosomal protein, LIX, 777, 783, 785, 786, 793
[^{3}H]Formaldehyde, preparation of radiolabeled ribosomal protein, LIX, 790
Formaldehyde-glutaraldehyde fixative, SEM, XXXII, 56, 58
Formalin, XLIV, 110
Formamide, XLIV, 127, 130
deionizing procedure, LIX, 65
denaturation, XL, 306
electron spin resonance spectra, XLIX, 383
in gel electrophoresis, LIX, 67
identification of iron-sulfur clusters, LIII, 269
in spreading solution, LIX, 842, 847
Formamidine, XLIII, 451, 461

Formamidine disulfide·2HCl, as inosamine-phosphate amidinotransferase inhibitor, XLIII, 458
2-Formamido-*N*-ribosylacetamide 5′-phosphate
substrate of FGAM synthetase, LI, 193
in succino-AICAR synthesis, LI, 189
2-Formamido-*N*-ribosylacetamide 5′-phosphate amidotransferase, reactive SH group, XXXVI, 413
2-Formamido-*N*-ribosylacetamide 5′-phosphate:L-glutamine amido-ligase, *see* 2-Formamino-*N*-ribosylacetamidine 5′-phosphate synthetase
2-Formamino-*N*-ribosylacetamidine 5′-phosphate, product of FGAM synthetase, LI, 193
2-Formamino-*N*-ribosylacetamidine 5′-phosphate synthetase, LI, 193–201
activity, LI, 193
alternate name, LI, 193
assay, LI, 193–197
by ADP determination, LI, 196, 197
colorimetric, LI, 194, 195
by glutamate determination, LI, 195, 196
from chicken liver, LI, 197–200
inhibitors, LI, 193, 201
kinetic properties, LI, 201
molecular activity, LI, 200, 201
molecular weight, LI, 200
properties, LI, 200, 201
purification, LI, 197–200
from *Salmonella typhimurium*, LI, 200
Formate acetyltransferase
in clostridiae, XLI, 518
from *Escherichia coli*
activator system, XLI, 508–518
assay, XLI, 510–513
chromatography, XLI, 515
molecular weight, XLI, 517
properties, XLI, 517
purification, XLI, 513–516
in *Streptococcus faecalis*, XLI, 518
Formate dehydrogenase, LIII, 360–372
activators, LIII, 371
activity, LIII, 360, 370
assay, LIII, 363–365
from *Clostridium acidurici*, LIII, 372
from *Clostridium formicoaceticum*, LIII, 372
from *Clostridium sticklandii*, LIII, 375, 376
from *Clostridium thermoaceticum*, LIII, 360–372
from *Escherichia coli*, LIII, 372
inhibitors, LIII, 371
kinetic properties, LIII, 370
metal content, LIII, 371, 372, 375, 376
molecular weight, LIII, 371
pH optimum, LIII, 370
purification, LIII, 365–369
radiolabeling, LIII, 369
reversible dissociation, LIII, 436
substrate specificity, LIII, 369
temperature effects, LIII, 367, 370, 371
turnover number, LIII, 369
Formate dehydrogenase:fumarate reductase, anaerobic electron transfer, LVI, 386
Formate dehydrogenase:nitrate reductase, anaerobic electron transfer, LVI, 386
Formazan granule, in interstitial cells, XXXIX, 258
Form factor, *see* Intramolecular scattering form factor
Formic acid
in acetate formation pathway, LIII, 361
activator of aspartate carbamyltransferase, LI, 49
of succinate dehydrogenase, LIII, 33
in fingerprinting procedure, LIX, 66
inhibitor of succinate dehydrogenase, LIII, 33
interaction with cytochrome *c* oxidase, LIII, 201
preparation of 8-bromoadenosine 5′-monophosphate, LVII, 117
of deuterated NAD^+, LIV, 227
protein cleavage, XLVII, 147, 340, 342

purification of adenosine derivative, LVII, 117
of flavin peptide, LIII, 453
for reaction termination, LI, 338, 346, 487
removal of *tert*-butyl-type protectors, XLVII, 562, 569
procedure, XLVII, 545, 546
synthesis of 8α-flavins, LIII, 460
thin-layer chromatography, XXXVIII, 29, 30
Formic acid buffer, protein elution, XLVIII, 343
Formic acid-hydrochloric acid, modification of tryptophan, LIII, 167
Formvar, for negative staining grids, XXXII, 24
Formycin, XLIII, 154, 156
Formycin A, substrate of adenosine deaminase, LI, 506, 507
Formycin B, inhibitor of purine nucleoside phosphorylase, LI, 537
N^{10}-Formylaminopterin, XXXIV, 48
Formylaminopterin acrylamide, XXXIV, 274
8-Formyl-2′,5′-anhydroriboflavin, breakdown product of 8α-aminoacylflavins, LIII, 465
Formylglycinamide ribonucleotide, *see* 2-Formamido-*N*-ribosylacetamide 5′-phosphate
Formylglycinamide ribonucleotide amidotransferase, XLVI, 36, 420, 424
5-Formyl-2′,3′-isopropylideneuridine, XLVI, 347
synthesis, XLVI, 348
Formylmethionine-transfer ribonucleic acid, XLVI, 182; LIX, 852, *see also* Methionyl-transfer ribonucleic acid
assay for initiation complexes, LX, 5, 120, 182, 183, 208–212, 226, 227
binding to 30 S ribosome subunit, LX, 209–211, 332–338, 340–343
to 70 S ribosomes, LX, 11, 15, 205, 431, 432
complex with initiation factors, LX, 14, 15, 120, 215, 217–224
from *Escherichia coli,* ester modification, LIX, 163, 164
from human placenta fingerprint, LIX, 76–78
mitochondrial ribosomes, LVI, 91
preparation, LIX, 841; LX, 113, 216, 316, 317, 325, 339
in studies of antibiotic action on initiation, LIX, 856–857
Formylmethionylpuromycin, mitochondrial ribosomes, LVI, 91
Formylmethionyl-transfer ribonucleic acid, *see* Formylmethionine-transfer ribonucleic acid
8-Formylriboflavin, breakdown product of 8α-aminoacylflavins, LIII, 464, 465
3-Formyl rifampin, XLIII, 305
Formyltetrahydrofolic acid, for preparing aminoacyl-transfer ribonucleic acids, LX, 317, 325, 598
10-Formyltetrahydrofolic acid, acetate formation pathway, LIII, 361
N-Formyl-L-tryptophan, inhibitor of tryptophanyl-tRNA synthetase, LIX, 252
α-5-Formyluridine 5′-monophosphate, synthesis, XLVI, 348, 349
β-5-Formyluridine 5′-monophosphate, synthesis, XLVI, 348, 349
α-5-Formyluridine 5′-triphosphate
binding site, XLVI, 351, 352
inhibition, XLVI, 350, 351
synthesis, XLVI, 347–350
β-5-Formyluridine 5′-triphosphate, synthesis, XLVI, 347–350
5-Formyluridine 5′-triphosphate, 346-352
Foromacidin, XLIII, 131
Forssman glycolipid
IR spectroscopy, XXXII, 366
isolation, XXXII, 355
in liposomes, XXXII, 503, 506, 510, 511
optical rotation, XXXII, 365
Forssman hapten, L, 247
Fortran
in computer analysis of radioligand assay data, XXXVII, 9
program, for morphometry, XXXII, 13, 16–20
Fosfomycin, XLVI, 543; XLVII, 409
Fourier transformation, of NMR data, LIV, 161, 162
Fourier transform envelope, of unit cell, XXXII, 218

Fourier-transform infrared
interferometry, LIV, 322, 323
Fourier transform nuclear magnetic
resonance spectroscopy, pulsed, LIX, 27
Fractionation, methods for green tissue, XXXI, 548–553
Fractionation experiment, XLVIII, 245
Fractionation medium, for digitonin method, LVI, 209
Fraction C, purification, LIII, 381
Fractophorator, LIII, 326
Fragment reaction, LIX, 819
as assay of reconstituted ribosomal particles, LIX, 779, 780
Framicetin, XLIII, 211
Franck-Condon factors, LVII, 495, 508, 510, 564, 566, 567
Franck-Condon overlap, LIV, 235
Free energy of activation, for electron-transfer chemiluminescent reactions, LVII, 509
Free-hand slice technique, of macroautoradiography, XXXVI, 138
Free-induction decay, LIX, 26
data, Fourier transformation, LIV, 161, 162
Free radical, XLIV, 37, 172, 179
sources, LIII, 387
Freeze-drying
advantages, LIX, 612
in autoradiography, XXXVI, 139, 143, 144, 153, 154
effects on plant cells, XXXI, 532
instrumentation, LIX, 613, 614
prior to scanning electron microscopy, XXXIX, 155
for scanning electron microscopy, XXXII, 56–58
Freeze-etch microscopy
of algae, XXXII, 875, 876
of membranes, XXXII, 54
Freeze-fracture, XXXII, 35–50
of electric tissue membranes, XXXII, 323
etching, XXXII, 40, 41
fracturing, XXXII, 39, 40
freezing, XXXII, 36–39
freezing solution, XXXII, 38
of lamellar phase lipid, XXXII, 47, 48
of lipid membranes, XXXII, 45–50
membrane studies, XXXII, 35–44
replica cleaning, XXXII, 43, 44
replica formation, XXXII, 42, 43
Freeze-quench technique, LIV, 85–93
apparatus, LIV, 86–91
associated problems, LIV, 91–93
data analysis, LIV, 91, 92
design criteria, LIV, 86
principle, LIV, 85
Freeze-stop, metabolite measurement, LVI, 202
Freeze-thaw cycle, for *E. coli* lysis, LIX, 365
Freeze-thawing, for cell lysis, LIII, 639
Freeze-thaw technique
dissociation of ATPase, LV, 806
reconstitution experiments, LV, 707
Freezing, *see also* specific cell type
adjuvants, LVIII, 30
cell damage, LVIII, 30
conditions, LVIII, 30
cooling rate, LVIII, 30, 31
method, LVIII, 33
of plant tissue, XXXI, 532
protocol, LVIII, 220, 221
recovery, LVIII, 448
storage, LVIII, 448
thawing, lipid peroxidation, LII, 305
Freezing mixture, LIII, 489
Freezing point, of biological material, XXXII, 36
Fremy's salt, *see* Peroxylamine disulfonate
French Press, XXXI, 636, 643
for plant-cell rupture, XXXI, 502
French pressure cell, XXXII, 826
for bacterial cell disruption, LI, 53, 60, 266
cell disruption, LV, 165, 360, 790
in cell extract preparation, LIII, 618, 619, 633, 636, 639, 643, 644
everted membrane preparation, LVI, 236
fractionation of mitochondria, LV, 90–93
membrane preparation, LVI, 167
Freon 22, in freeze-fracture method, XXXII, 38
Frequency-conformation correlation approach, principle, XLIX, 99

Frequency shift, in magnetic susceptibility determinations, LIV, 388, 389
Freund's adjuvant, XXXIV, 190, 724; LVIII, 175
  complete, LII, 242–244
    for stimulation of alveolar macrophage production, LVII, 471
  rabbit immunization, LVI, 224
Frictional coefficient, lack of knowledge, XLVIII, 4
Friend cell, *see* Mouse erythroleukemia cell
Friend leukemia virus, LI, 408
Fringe displacement
  absolute, XLVIII, 170
  numerical integration, XLVIII, 183, 184
Frog
  erythrocytes, adenylate cyclase, XXXVIII, 170–173
  haploid cells, isolation and culture, XXXII, 789–799
  oxyntic cell isolation, XXXII, 709
  skin
    in MSH assay, XXXVII, 126, 127
    preparation, XXXVII, 122
  ventricular muscle isolation, XXXIX, 60–63
β-Fructofuranosidase
  coimmobilization, XLIV, 472
  conjugate, kinetics, XLIV, 408
  efficiency, effect of substrate concentration, XLIV, 238
  fiber entrapment, XLIV, 230, 231
    activity, XLIV, 237
    industrial application, XLIV, 241
    stability, XLIV, 241
  immobilization
    by bead polymerization, XLIV, 194
    on collagen, XLIV, 255, 260
    by entrapment, XLIV, 176, 178
    in liposomes, XLIV, 224, 702
    with magnetic carrier, XLIV, 278, 279
    by microencapsulation, XLIV, 214
    photochemical, XLIV, 286, 287
  specific activity, XLIV, 255
β-D-Fructofuranosidase
  in intestinal mucosa, XXXII, 667
  from yeast
    external
      assay, XLII, 504, 505
      properties, XLII, 508
      purification, XLII, 506–508
    internal
      activators and inhibitors, XLII, 511
      assay, XLII, 509
      properties, XLII, 511
      purification, XLII, 509–511
β-D-Fructofuranoside fructohydrolase from yeast, *see* β-D-Fructofuranosidase
Fructokinase
  estimation, in crude tissue preparations, XLI, 61–63
  from *Leuconostoc mesenteroides*, XLII, 39–43
    assay, XLII, 39, 40
    chromatography, XLII, 42
    molecular weight, XLII, 43
    properties, XLII, 42, 43
    purification, XLII, 40–42
Fructose
  chromatographic constants, XLI, 17
  degradation on heating in boric acid/2,3-butanediol or in borax, XLI, 21
  from glucose, mechanism, XLIV, 810, 811
  production in reactor, XLIV, 768, 775, 776, 809–821
D-Fructose, D-fructose kinase in metabolism, in *Leuconostoc mesenteroides*, XLII, 39
Fructose-1,8-biphosphatase, *see* Fructose-bisphosphatase
Fructose 1,6-biphosphate, *see* Fructose 1,6-bisphosphate
Fructose-bisphosphatase, XXXIV, 164, 165
  from bovine liver, XLII, 363–368
    activators and inhibitors, XLII, 368
    assay, XLII, 363, 364
    molecular weight, XLII, 367
    properties, XLII, 367, 368
    purification, XLII, 364–367

from *Candida utilis*, XLII, 347–353
activators and inhibitors, XLII, 352
assay, XLII, 347
molecular weight, XLII, 352
properties, XLII, 351–353
purification, XLII, 348–350
from *Polysphondylium pallidum*, XLII, 360–363
assay, XLII, 360
chromatography, XLII, 362
properties, XLII, 362, 363
purification, XLII, 361, 362
from rabbit liver, XLII, 354–359, 369–374
activators and inhibitors, XLII, 358, 374
assay, XLII, 354, 355, 369, 370
chromatography, XLII, 356
distribution, XLII, 359
molecular weight, XLII, 357
properties, XLII, 357–359, 373, 374
purification, XLII, 355, 356, 370–373
site-specific reagent, XLVII, 481, 483
in spinach chloroplasts
activation and inhibition, XLII, 402
amino acids, XLII, 402
assay, XLII, 398–400
chromatography, XLII, 401
molecular weight, XLII, 402
properties, XLII, 402
purification, XLII, 400–402
from swine kidney, XLII, 375, 385–389
assay, XLII, 385, 386
inhibition, XLII, 389
molecular weight, XLII, 388
properties, XLII, 388, 389
purification, XLII, 386–388
Fructose-1,6-bisphosphatase, gluconeogenic catalytic activity, XXXVII, 281
Fructose-bisphosphatase system
ferredoxin-activated from spinach chloroplasts, XLII, 397–405
activators and inhibitors, XLII, 397–405
properties, XLII, 404, 405
fructose-1,6-diphosphatase component from spinach chloroplasts, XLII, 397–403
protein factor component from spinach chloroplasts
assay, XLII, 403
chromatography, XLII, 404
properties, XLII, 404
purification, XLII, 404
Fructose 1,6-bisphosphate, XLIV, 337
assay, LV, 221
enzymatic preparation of sedoheptulose 1,7-biphosphate, XLI, 77–79
inhibitor of adenylosuccinate synthetase, LI, 212
inorganic phosphate assay, LV, 212
Pasteur effect, LV, 291
substrate, aldolase, XLIV, 495
Fructose-bisphosphate aldolase
affinity labeling procedure, XLVII, 483, 484, 491–496
from blue-green algae, XLII, 228–234
assay, XLII, 228–230
chromatography, XLII, 232
molecular weight, XLII, 232
properties, XLII, 232–234
purification, XLII, 230–232
detection, by tetrazolium dye reduction, XLI, 66, 67
from *Drosophila melanogaster*, XLII, 223
from lobster muscle, XLII, 223–228
amino acids, XLII, 226, 227
assay, XLII, 223, 224
chromatography, XLII, 225
molecular weight, XLII, 226
properties, XLII, 226–228, 239
purification, XLII, 224–226
from mammalian tissues, XLII, 240–249
assay, XLII, 240–242
chromatography, XLII, 243, 244, 247
electrophoresis, XLII, 241, 242
purification, XLII, 242–249
of *Peptococcus aerogenes*, XLII, 249–258
amino acids, XLII, 256, 257

assay, XLII, 250, 251
chromatography, XLII, 253
properties, XLII, 255–257
purification, XLII, 252–255
from spinach, XLII, 234–239
assay, XLII, 235, 236
chromatography, XLII, 237
molecular weight, XLII, 238
properties, XLII, 238, 239
purification, XLII, 236–238
Fructose-1,6-bisphosphate aldolase, XLVI, 49–51
Fructose 1,6-diphosphatase, *see* Fructose-bisphosphatase
Fructose-1,6-diphosphatase phosphohydrolase, *see* Fructose-bisphosphatase
Fructose 1,6-diphosphate, *see also* Fructose-1,6-bisphosphate
Fructose-diphosphate aldolase, *see* Fructose-bisphosphate aldolase
D-Fructose kinase, *see* Fructokinase
Fructose 6-phosphate
in isolated renal tubule, XXXIX, 16
Pasteur effect, LV, 291
substrate, transaldolase, XLIV, 495
Fructose-6-phosphate amidotransferase, XLVI, 420
Fructose-6-phosphate kinase, phosphorylation of sedoheptulose 7-phosphate, XLI, 34–36
D-Fructose-1-phosphate kinase, *see* 1-Phosphofructokinase
Fruit fly, *see also* specific genera
media, LVIII, 464
suspension culture, LVIII, 454, 455
Fruit juice, debittering process, XLIV, 241
Frustrated multiple internal reflectance technique, LIV, 323
FSH, *see* Follicle stimulating hormone
FTC
fluorescent acceptor and donor, XLVIII, 362, 363
spectral properties, XLVIII, 362
structural formula, XLVIII, 362
FTIR, *see* Fourier-transform infrared interferometry
FTNMR, *see* Nuclear magnetic resonance, Fourier-transform
Fuchsin
basic, autoradiography prestain, LVIII, 286
for Schiff reagent, XXXII, 97
Fuchsin-sulfite solution, preparation, XXXII, 87, 88
Fuchs Rosenthal hemacytometer, XXXI, 581
Fucitol, paper electrophoresis, L, 51
Fuco-glycosphingolipid
blood group activity, L, 247
structures, L, 244–246
D-Fuconate dehydratase, from pseudomonad MSU-1, XLII, 305–308
assay, XLII, 305, 306
chromatography, XLII, 306
properties, XLII, 307, 308
purification, XLII, 306, 307
α-L-Fucopyranoside, XXXIV, 361, 363
Fucopyranoside-agarose, XXXIV, 370, 371, 372
O-α-L-Fucopyranosyl-polyacrylamide, XXXIV, 361
Fucose
chromatographic constants, XLI, 16, 17, 20
lectin purification, XXXIV, 336
in progesterone-binding protein, XXXVI, 124
β-L-Fucose, lectin purification, XXXIV, 347
D-Fucose
lectin, XXXIV, 333
metabolism in pseudomonad MSU-1, XLII, 269
D-fuconate dehydratase, XLII, 305
L-Fucose, L, 94
cultured cell growth, L, 190, 191
enzymic microassay, XLI, 5, 7–10
glycosyl linkage, L, 101
interconversion, in animal cells, L, 183
lectin, XXXIV, 332
purification from HeLa cells, L, 198–200
quantitation of cellular pools, L, 191–198
high voltage paper electrophoresis, L, 194, 195
paper chromatography, L, 194, 195

thin-layer chromatography, L, 196, 197
tritium labeling
of fibroblasts, L, 185
of HeLa cells, L, 186
L-Fucose-agarose derivative, XXXIV, 347
L-Fucose-binding protein, XXXIV, 328
L-Fucose dehydrogenase
in L-fucose microassay, XLI, 5
from sheep liver, XLI, 173–177
assay, XLI, 174
chromatography, XLI, 176
molecular weight, XLI, 176
properties, XLI, 176
purification, XLI, 174–176
Fucose glycolipid, of human adenocarcinoma, XXXII, 365
L-Fucose isomerase, *see* D-Arabinose isomerase
L-Fucose kinase
^{32}P-labeling of fucose, L, 191
quantitation of fucose, L, 200–204
L-Fucose-starch adsorbent, XXXIV, 329
α-L-Fucosidase
assay, L, 453
distribution, L, 453
fucosidosis, L, 453
isolation of 6′-galactosyllactose, L, 219
mucolipidoses II and III, L, 453
from rat liver lysosomes
assay, L, 506
properties, L, 508–510
purification, L, 506–508
Fucosidosis
α-L-fucosidase, L, 453
urinary oligosaccharides, L, 229–233
2-Fucosylgalactose, structure, L, 228
2-Fucosylglucose, human urine, L, 228
Fucosylinositol, human urine, L, 228
Fucosyl-*myo*-inositol, human urine, L, 227
Fucosyllacto-*N*-hexaose I, structure, L, 217
Fucosyllacto-*N*-neooctaose
isolation from human milk, L, 220
structure, L, 217
Fucosyllacto-*N*-octaose
isolation from human milk, L, 220
structure, L, 217
2′-Fucosyllactose
chromatography, L, 219, 220
structure, L, 163
3-Fucosyllactose
chromatography, L, 219, 220
structure, L, 163
O-α-L-Fucosyl polyacrylamide, XXXIV, 367
Ful-jak allihn condensor, XXXIX, 50
Fuller's earth, for steroid adsorption, XXXVI, 40
Fumarase, XXXIV, 410, 411; *see also* Fumarate hydratase
assay, XXXI, 742
in chromaffin granule, XXXI, 381
as mitochondria marker, XXXI, 406, 408, 735, 740, 742
Fumarate epoxidase, LII, 20
Fumarate hydratase, XLIV, 832
brain mitochondria, LV, 58
latency, intact mitochondria, LV, 143
toluene-treated mitochondria, LVI, 547
Fumaric acid
activator of adenylate kinase, LI, 465
of succinate dehydrogenase, LIII, 33
assay, LV, 207
content of brain, LV, 218
as electron acceptor, LVI, 386
inhibitor of succinate dehydrogenase, LIII, 33
product of adenylosuccinate AMP-lyase, LI, 202
of complex II, LIII, 21
of succinate dehydrogenase, LIII, 27
renal tubule permeability, XXXIX, 19
in succino-AICAR synthesis, LI, 189
Fume hood, LVIII, 5, 10, 40
Functional group density, XXXIV, 52
aliphatic amino group determination, XXXIV, 57
carboxyl group determination, XXXIV, 57
dry weight determination, XXXIV, 56
estimation, XXXIV, 56–58
hydrazide determination, XXXIV, 57
polyacrylamide, XXXIV, 33, 34, 36

Functional purification, XXXIV, 631, 632
Fungal contamination, *see* Contamination, fungal
Fungichromin, XLIII, 136
Fungizone, in media, LVIII, 523
Fungus, *see also* specific types
antibiotic fermentation by, XLIII, 14–16
double-stranded ribonucleic acid, LX, 549–554
filamentous, disruption, LV, 135, 139, 140
free cells, XXXII, 723
D(−)-lactate dehydrogenase assay, purification and properties, XLI, 293–298
media for maintaining, XLIII, 4, *see also* Media, for fungi
mycelium, homogenization with glass beads, LI, 65
oxidases, LII, 9–21
spheroplast and membrane vesicles, XXXI, 609–626
vacuole isolation, XXXI, 574
Furacryloylglycyl-L-leucinamide, substrate, of thermolysin, XLIV, 89, 95–97
Furacryloylglycyl-L-leucinamide hydrolase, activity, unit, definition, XLIV, 96
Furnace, *see* specific type
Furnace atomizer
disadvantages, LIV, 464
principles, LIV, 465, 466, 468, 470
Furoxan, XLVI, 100
Fusarium avenceum, penicillin acylase, XLIII, 723
Fusarium moniliforme, penicillin acylase, XLIII, 723
Fusarium semitectum, penicillin acylase, XLIII, 723–725, *see also Penicillium fusarium* acylase
Fusarum culmorum, vacuole isolation, XXXI, 574
Fuscin
glutamate transport, LVI, 421
transport, LVI, 248
Fusidic acid, XLIII, 159, 352; LVI, 31
effect on protein synthesis, LIX, 853, 854, 860, 862
inhibition of elongation, LX, 410, 414–416, 607–610, 710, 712, 722–724, 729–732, 736, 737
inhibitor of *in vitro* protein synthesis, LIX, 370
protein inhibitor, XL, 287, 288
sodium salt, partial specific volume, XLVIII, 21
Fusidine, XLIII, 204
Fusion, procedure, reconstitution studies, LV, 709
Fusion index, of cells, definition, XXXII, 582
Fuwa-Vallee long-path absorption tube, LIV, 462, 464

G

GABA aminotransferase, *see* γ-Aminobutyrate-α-ketoglutarate aminotransferase
Gabaculine, *see* 1,3-Cyclohexadienyl-5-aminocarboxylic acid
GABA transaminase, *see* γ-Aminobutyrate-α-ketoglutarate transaminase
Gabromycin, XLIII, 200
Gadolinium acetate, derivatization of transfer RNA, LIX, 14
D-Galactal, XLVI, 26
Galactitol
gas-liquid chromatography, of 2-deoxy-2-(*N*-methylacetamido) acetate derivatives, L, 6
paper electrophoresis of tritiated derivative, L, 51
Galactoglycerolipid, assay, XLI, 8
Galactokinase, XLIV, 698
from human erythrocytes, XLII, 47–53
assay, XLII, 48–51
chromatography, XLII, 51, 52
clinical significance, XLII, 47, 48
molecular weight, XLII, 53
properties, XLII, 52, 53
purification, XLII, 51, 52
from pig liver, XLII, 43–47
assay, XLII, 43, 44
chromatography, XLII, 45
properties, XLII, 45, 47
purification, XLII, 44–46

from yeast, antibody binding, L, 55
Galactolipase, enzymatic activity, XXXI, 522, 591
Galactolipid
in chloroplast, XXXII, 542
chromatography, on lipophilic Sephadex, XXXV, 388
enzymatic deacylation, XXXI, 522
Galactomannan, XXXIV, 337
Galactometasaccharinic acid lactone, preparation, XLI, 95, 96
L-Galactonolactone, oxidation, LVI, 474
α-D-Galactopyranoside, XXXIV, 363
α-D-Galactopyranosylamine, XXXIV, 347
β-D-Galactopyranosylamine, XXXIV, 323, 324
synthesis, XLVI, 400
1-*O*-β-D-Galactopyranosyl-D-erythritol, L, 25, 26
3-*O*-α-D-Galactopyranosyl-D-glucose, synthesis, L, 121
L-α-D-Galactopyranosyl polyacrylamide, XXXIV, 181
Galactosamine, in affinity gel preparation, LI, 367
D-Galactosamine, neoplastic cell growth, L, 190
D-Galactosamine hydrochloride, XLVII, 489
Galactose, XXXIV, 318
Golgi complex, transfer, XXXI, 23, 31
lectin purification, XXXIV, 336
in media, LVIII, 53, 66
polarography, LVI, 452, 465–467
product, of α-galactosidase, XLIV, 350
substrate, of β-galactose dehydrogenase, XLIV, 881
β-D-Galactose, XXXIV, 347
D-Galactose, XXXIV, 333, 334
2-amino-2-methoxyethylthioglycoside, liver membrane binding, L, 113
chromatographic constants, XLI, 16, 17, 20
^{14}C–labeling of BHK cells, L, 186
cultured cell growth, L, 190
enzymic microassay, XLI, 4, 7–10
fluorometric assay, XLI, 8, 10
gas-liquid chromatography, of 2,3,4,6-tetra-*O*-methyl derivative, L, 7, 9
interconversion, in animal cells, L, 182
L-Galactose, lectin, XXXIV, 332
D-Galactose-agarose derivative, XXXIV, 347
Galactose-binding lectin, *see* Electrolectin
Galactose binding protein
asialoglycoproteins, L, 287, 288
erythrocyte agglutination, L, 288
structure, L, 288
β-Galactose dehydrogenase, in enzyme reactor, XLIV, 880–882
D-Galactose dehydrogenase, in D-galactose microassay, XLI, 4
α-D-Galactose 1,6-diphosphate, preparation, XLI, 83
Galactose oxidase, LII, 10
assay, with *Diplocardia* bioluminescence, LVII, 381
immobilization, by entrapment, XLIV, 176
reaction catalyzed, LVI, 465
sources, LVI, 465
tritiation
of cell-surface glycoproteins, L, 204–206
of glycolipids, L, 204–206
uronic acid analysis, L, 31, 32
Galactosidase
assay, XXXI, 20
as diagnostic enzyme, XXXI, 19, 20, 31, 185, 186
α-Galactosidase, XXXIV, 347–358
conjugate, assay, XLIV, 350
purification, XXXIV, 349, 350
β-Galactosidase, XXXIV, 105, 110, 118, 119, 350–358, 731; XLVI, 26, 36, 156, 364, 365, 398–403
N-acetylglucosamine binding protein, L, 291
activation energy, XLIV, 796, 797
activity assay, XLIV, 793
agarose derivatives, XXXIV, 353–355
assay methods, XXXIV, 351, 352
from bovine testes
assay, L, 515
properties, L, 518–520
purification, L, 515–518
cAMP binding, XXXVIII, 367

conjugate
activity assay, XLIV, 269–271, 343, 352
kinetics, XLIV, 426
coupling efficiencies, XLIV, 795
cross-reacting protein, XXXIV, 351
from *E. coli,* antibody binding, L, 55, 56
entrapment in cellulose triacetate fibers, XLIV, 824
in enzyme thermistor, XLIV, 674, 675
in polyacrylamide gels, XLIV, 176–178
fiber entrapment, industrial application, XLIV, 241
fluorine nuclear magnetic resonance, XLIX, 273, 274
glutaraldehyde binding, XLIV, 52
for glycolipid structure studies, XXXII, 363
immobilization
on (aminoethyl)cellulose, XLIV, 52
by chemical aggregation, XLIV, 269, 270
on collagen, XLIV, 153, 255, 260
in hollow-fiber membrane device, XLIV, 313, 314
on inert protein, XLIV, 909
by microencapsulation, XLIV, 212, 214
in multistep enzyme system, XLIV, 456, 457
on porous ceramics, XLIV, 793
inactivation, XLVI, 401, 402
induction, DCCD, LV, 498
inhibition by metal ions, XLIV, 261
kinetics, in multistep enzyme system, XLIV, 465, 466
for lactose reduction of milk, XLIV, 822–830
in macrophage fractions, XXXI, 340
magnesium requirement, XLIV, 827
molecular weight, LIII, 350, 351, 353
mucolipidoses II and III, L, 455
from *Neurospora crassa*, XLII, 497–503
amino acid, XLII, 501, 502
assay, XLII, 498, 502
chromatography, XLII, 500, 501
molecular weight, XLII, 501, 503
properties, XLII, 501–503
purification, XLII, 498–501, 502
nicotinic acid receptors, XXXII, 320
pH activity profile, XLIV, 793, 795–798
polyacrylamide derivatives, XXXIV, 352, 353
properties, XXXIV, 358
purification, XXXIV, 355–358
rate equation, XLIV, 798, 799
soluble aggregate, activity assay, XLIV, 474
sources, XLIV, 793; XLVI, 401
spacers, XXXIV, 356, 357
specific activity, XLIV, 255
sphingolipidoses, L, 457
endo-β-Galactosidase, from *Diplococcus pneumoniae*
assay, L, 560, 561
properties, L, 565, 566
purification, L, 563–656
α-Galactosidase A
assay, L, 481
distribution, L, 480
Fabry's disease, L, 480
α-Galactosidase B, Fabry's disease, L, 481
β-Galactosidase chymotrypsin, XLVI, 155
β-Galactosidase inhibitor, XXXIV, 7
β-Galactoside
in brain fractions, XXXI, 471
transport, XLVI, 67
Galactoside-blockable lectin, *see* Electrolectin
β-D-Galactoside-specific hemagglutinin, *see* Electrolectin
β-D-Galactosylamine, synthesis, XLVI, 363
Galactosylceramidase
assay, L, 468–470
globoid cell leukodystrophy, L, 468
Galactosylceramide, L, 247
^{14}C–labeled, preparation, L, 469
globoid cell leukodystrophy, L, 468
thin-layer chromatography, XXXII, 356
Galactosylceramide sulfate, metachromatic leukodystrophy, L, 471
Galactosylceramide-I^3-sulfate, L, 247

Galactosylgalactosidase, XLVI, 27
Galactosylglyceride, thin layer chromatography, XXXV, 396, 402, 403
Galactosylhydroxylysine
human urine, L, 229
structure, L, 229
Galactosylhydroxylysine collagen, XL, 353
IV-α-Galactosyllactoneotetraglycosylceramide, L, 247
6′-Galactosyllactose
isolation from human milk, L, 219, 220
paper chromatography, L, 219, 220
structure, L, 217
Galactosyl pyrophosphate, as ligands, XXXIV, 482
β-D-Galactosyl-specific lectin, *see* Electrolectin
Galactosylsphingosine, globoid cell leukodystrophy, L, 470
Galactosylsphingosine sulfate, arylsulfatase A, L, 474
β-D-Galactosyl-β-thiogalactopyranoside, electrolectin inhibition, L, 292, 300
Galactosyltransferase, XXXIV, 4, 327, 328, 359, 360, 485, *see also* specific types
as diagnostic enzyme, XXXI, 25, 33, 36, 37
in plasma membrane, XXXI, 102, 180, 183, 184, 186–190
spin-label studies, XLIX, 458–463
D-Galacturonic acid, colorimetric assay, XLI, 30, 31
Galegine, *M. phlei* membrane systems, LV, 186
Gallbladder, cAMP, XXXVIII, 3
Gallium, atomic emission studies, LIV, 455
Gallium arsenide semiconductor injection laser, wavelength, XLIX, 85
Gal-repressor, XXXIV, 6
Gamma counter, XL, 294
Gamma irradiation, for cross-linking, XLIV, 37, 178
Gamma ray, radiation properties, XXXVII, 146, 147
Ganglion cell, LVIII, 581
Ganglioside, L, 247–250, *see also* Glycosphingolipid
antigenic determinants, L, 250
characterization, L, 236
column chromatography, XXXV, 417, 421
hormone receptors, L, 250
isolation, XXXII, 345, 346, 351, 352, 354, 355
partition, XXXII, 354, 355
structures, L, 248–250
SV40-transformed mouse cells, L, 247
Tay–Sachs disease, L, 247
thin-layer chromatography, XXXII, 356; XXXV, 397, 400, 406, 407, 421, 422
toxin receptors, L, 250
G_{M1}-Ganglioside, structure, L, 482
Ganglioside-agarose
adsorbents, XXXIV, 613–616
aldehyde derivatives, XXXIV, 612, 613
assay, XXXIV, 616
carboxyl groups, XXXIV, 611, 612
cholera toxin, XXXIV, 610–619
Ganglioside-albumin-agarose, XXXIV, 611
Ganglioside-diaminodipropylaminoagarose, XXXIV, 611
G_{M1}-Ganglioside β-galactosidase
assay, L, 482, 483
galactosylceramidase, L, 483
lactosylceramide, L, 471
Ganglioside methyl ester, acetylation, L, 139
Ganglioside-poly(L-lysine)-agarose, XXXIV, 611
Ganglioside-poly(L-lysyl-DL-alanine)-agarose, XXXIV, 611
G_{M1}-Gangliosidosis
clinical forms, L, 481, 482
enzymic diagnosis, L, 481–483
urinary oligosaccharides, L, 229–233
G_{M2}-Gangliosidosis, *see also* Juvenile G_{M2}-Gangliosidosis; Sandhoff's disease; Tay–Sachs disease
AB variant, L, 484
urinary oligosaccharides, L, 229–233
Gap junction, of plasma membranes, XXXI, 85
Gas, purification method, LIV, 404

Gas analysis, LIV, 119, 120
Gas bubble, formation, prevention, XLVII, 212
Gas chromatography
acetate determination, XXXV, 307–311
of alkanols, steric analysis as 2-D-phenylpropionate derivatives, XXXV, 329
in assay of fatty acid synthesis, XXXV, 76, 285–287
of long-chain fatty acids, XXXV, 73
of biphenyl metabolites, LII, 404, 405
chemical ionization mode, procedure, LII, 338–340
of deuterated fatty acids, XXXV, 285–287, 340–342, 345
deuterium incorporation into fatty acids, XXXV, 285–287
electron impact mode, procedure, LII, 335
fatty acid synthesis, rate measurement, XXXV, 282, 283, 285–287
ghost peaks in determination of acetate, XXXV, 307, 308
of hydroxy fatty acids, steric analysis as (−)-menthyloxycarbonyl and 2-D-phenylpropionate derivatives, XXXV, 328–331
instrument types, comparison, LII, 337
internal standards, LII, 340, 341
in nitrogen gas determination, LIII, 644
perchloric acid peaks, interference in acetate assay, XXXV, 309
precision and accuracy, LII, 341
of prostaglandins, XXXV, 363, 365–367, 377
for separation of fatty acid methyl esters, LII, 323, 324
steric analysis of hydroxy fatty acids and alkanols, XXXV, 328–331
trichloroacetic acid peaks, XXXV, 308
of triglycerides, XXXV, 331
of trimethylsilylated dipeptides, XLVII, 395–397
Gas chromatography–mass spectroscopy
of deuterated fatty acids, XXXV, 340–348
identification of steroids, LII, 381–387
noninterference, by lipophilic Sephadex, XXXV, 383–385
of prostaglandins, XXXV, 359–377
of triglycerides, XXXV, 358, 359
Gas-Chrom Q, XXXII, 360, 362
Gas contaminant, in bottled gas, LVIII, 202
Gas-liquid chromatography, XLIII, 213–256, *see also* Gas chromatography
of alditol acetate derivatives of neutral polysaccharides, L, 4
aminocyclitol antibiotics, XLIII, 217, 218
antimycin A, XLIII, 253
aromatic antibiotics, XLIII, 228–234
celestosaminide, XLIII, 253
chloramphenicol, XLIII, 228–232
clindamycin, XLIII, 239–242
clindamycin palmitate, XLIII, 243–245
cycloheximide, XLIII, 234–237
electron-capture techniques, XXXVI, 58–67
erythromycin, XLIII, 245–248
for ethanol metabolism studies, LII, 357
gentamicin, XLIII, 217, 218
glutarimide antibiotic, XLIII, 234–237
griseofulvin, XLIII, 232–234
internal standard, XLVII, 34, 40
kanamycin, XLIII, 218–220
lincomycin-clindamycin family, XLIII, 237–245
lividomycin, XLIII, 220
methylated polysaccharides, acetylated forms, L, 4–31
neomycin, XLIII, 220–228
paromomycin, XLIII, 228
penicillin, XLIII, 248–251
phosphonomycin, XLIII, 254
quantitation of cellular hexose pools, L, 193
spectinomycin, XLIII, 215–217
tetracycline, XLIII, 251–253
thiamphenicol, XLIII, 254
validamycin, XLIII, 254–265
Gas manifold, in low-temperature EPR spectroscopy, LIV, 118, 119

Gas regulator, leakage, LIV, 115, 116
Gastric mucosa, oxyntic cell isolation, XXXII, 707–717
Gastrin
derivatives, XLVI, 479
heterogeneity, XXXVII, 40, 41
photochemically active derivative, LVI, 660
precursor, XXXVII, 326
receptor, XLVI, 85, 89
Gastrointestinal tract, and immobilized enzymes, XLIV, 680, 681, 696
Gas vesicle
isolation from blue-green algae, XXXI, 678–686
storage, XXXI, 686
GAT_I, *see* Gentamicin acetyltransferase I
GAT_{II}, *see* Gentamicin acetyltransferase II
GAT_{III}, *see* Gentamicin acetyltransferase III
Gaucher's disease
clinical forms, L, 475
enzymic diagnosis, L, 475–479
glucocerebrosidase treatment, L, 532
glycosphingolipids, L, 247
Gaulin laboratory homogenizer, LIII, 650
Gaussian band, resolution, fourth derivative analysis, LVI, 505, 506, 510, 511
Gauss-Newton method, in curve fitting, XXXVII, 12
G–banding, *see* Giemsa banding
GDP, *see* Guanosine 5′-diphosphate
GDP-mannosyltransferase, in microsomes, XXXI, 201
Gel, *see also* specific type
analysis of mitochondrial products, LVI, 104, 105
drying, LVI, 65, 71, 606
enzyme electrodes, LVI, 486
enzyme immobilization, LVI, 460
fragmentation procedure, XLIV, 174
macroreticular, XLIV, 56, 57
pouring and electrophoresis, LVI, 62–64
scanning, LVI, 676
separating, composition, LVI, 70, 71
for separation of mtDNA fragments, LVI, 186
shelf solution, composition, LVI, 70, 71
stacking, composition, LVI, 70, 71
two-dimensional, LVI, 606
identification of spots, LVI, 611–613
isoelectric focusing, LVI, 608–610
SDS exponential gradient gel electrophoresis, LVI, 610, 611
solutions, LVI, 607, 608
Gel apparatus, analysis of labeled mitochondria, LVI, 61
Gelatin, XXXI, 537
assay of tryptophanyl-tRNA synthetase, LIX, 237, 238
dinitrophenylated, XXXIV, 203
in media, LVIII, 525, 526
nylon derivatives, XXXIV, 203
pretreatment, LIX, 236
surface treatment, LVIII, 577
Gelatinase, identification of luminous bacteria, LVII, 161, 162
Gel buffer
analysis of labeled mitochondria, LVI, 61, 62
for chromaffin granule isolation, XXXI, 383
Gel chromatography
column design, XLVII, 211
procedure of Hummel and Dreyer, XLVIII, 307
Geldanamycin
biosynthesis, XLIII, 409
paper chromatography, XLIII, 137
proton magnetic resonance spectroscopy, XLIII, 390, 391
structure, XLIII, 397, 398
Geldanamycin acetate, XLIII, 401, 403
Gel electrophoresis, *see also* Electrophoresis; specific type
in absence of detergents, XLVIII, 4
of ATPase, XXXII, 297
of camphor 5-monooxygenase system components, LII, 175
of chloroplast membrane polypeptides, XXXII, 878–880
of cross-linked ribosomal proteins, LIX, 538–550
diagonal polyacrylamide-sodium dodecyl sulfate
principles, LIX, 535–538

procedure, LIX, 538–540
of *E. coli* ribosomal proteins, LIX, 495–500
in estrogen receptor purification, XXXVI, 356, 357
of hemoproteins, LII, 324–331
running procedure, LII, 325–327
staining, LII, 327–338
high-pressure technique
apparatus, XLIX, 19, 20
sample loading, XLIX, 20, 21
of membrane proteins, XXXII, 70–81
molecular weight determinations, XLVIII, 3–10
of ($Na^+ + K^+$)-ATPase, XXXII, 288, 289
oxidizing solution, XXXII, 97
of photoaffinity-labeled ribosomes, LIX, 799–808
preparation of cylindrical gels, LII, 326
of slab gels, LII, 326
preparative
maintenance of constant conditions, LIX, 469
in separation of ribosomal fragments, LIX, 468, 469
in presence of denaturing detergents, XLVIII, 5–7
protein elution from diagonal gels, LIX, 544, 545
of protein-RNA interactions, LIX, 567
purification of *Cypridina* luciferase, LVII, 370, 371
of rat liver ribosomal proteins, LIX, 506–514, 520–525, 528–532
of ribosomal fragment components, LIX, 471, 472
of RNA fragments, LIX, 572–575
of 30 S pre-rRNA cleavage products, LIX, 833, 835
of 40 S and 60 S subunits, LIX, 506, 507
sample preparation, LII, 326, 327
of sarcoplasmic reticulum, XXXII, 292
in separation of ribosomal subunits, LIX, 397–402
sodium dodecyl sulfate-acrylamide, procedure, LIX, 507, 509
solutions, XLVIII, 8–10
of steroid cytoplasmic receptors, XXXVI, 222–228
styrofoam partition, XXXII, 78
of TeBG, XXXVI, 117, 118
three-dimensional, definition, LIX, 503
two-dimensional
gel preparation, LIX, 68, 69
polyacrylamide-urea, procedure, LIX, 506, 507, 544–546
Gel entrapment, *see* Entrapment
Gel filtration, XLVII, 97–107
aseptic conditions, XLVII, 106, 107
column assembly, XLVII, 101–103
cytochrome oxidase subunits, LVI, 606
eluent deaeration, XLVII, 104
of gonadal receptors, XXXVII, 184, 187, 188
hydroxamates, XLVII, 133
matrix
recovery, XLVII, 107
selection, XLVII, 98–100
swelling, XLVII, 100, 101
micelle size, LVI, 737
for molecular weight determinations, XLVIII, 6
in quantitative histone assay, XL, 120
in radioimmunoassay, XXXVII, 34
recycling, XLVII, 106
sample application, XLVII, 103, 104
separation monitoring, XLVII, 105, 106
in study of tRNA-protein interactions, LIX, 325–328, 565, 567
use of pumps, XLVII, 104
Gel filtration chromatography, *see also* BioGel A-1.5; Sephadex
of cytoplasmic steroid receptors, XXXVI, 213–222
DNA-steroid receptor binding assay, XXXVI, 311–313
in estrogen receptor purification, XXXVI, 355
Gel formation, with polyvalent ions and polyelectrolytes, XLIV, 179
Gelman Sepraphore III strip, LI, 37
Gel-permeation chromatography, computer simulation, XLVIII, 236

Gel tube, cleaning and coating, XXXII, 73
Gene
mitochondrial, mapping, LVI, 16
structural, identification, LVI, 196
for ubiquinone biosynthesis, map location, LVI, 110, 111
Gene conversion, mitochondrial, LVI, 152
Genetic complementation, *see* Complementation, genetic
Genetic disorder, *see* specific type
Genetic mapping
of carbodiimide resistance, LVI, 171
hybridization experiments, LVI, 191, 192
DNA-DNA plateau, LVI, 192
MA1 mutation, LVI, 181, 182
pitfalls and problems, LVI, 192–194
procedures
consolidation of map, LVI, 188–191
construction of overlap map of petite mtDNA's, LVI, 188
DNA isolation, LVI, 186
isolation of set of *rho* − petite mutants, LVI, 185, 186
restriction enzyme analysis, LVI, 186–188
Genetics, temperature-sensitive mutants, LVI, 138
Genetic system, nucleocytoplasmic mitochondrial components, LVI, 40
Genetic transmission, extrachromosomal, LVI, 155
Genome, mitochondrial, LVI, 3
elimination, LVI, 152
Gentamicin
biosynthesis, XLIII, 615–618, *see also* Aminoglycoside modifying enzymes
gas liquid chropatography, XLIII, 217–218
in media, LVIII, 112, 114, 116, 198
Micromonospora purpurea, media, XLIII, 21
relative mobilites, XLIII, 126
solvent system, XLIII, 119, 120
structure, XLIII, 613
thin-layer chromatography, XLIII, 200
Gentamicin acetyltransferase, XXXIV, 4
assay, XLIII, 615–618
from *Escherichia coli*, XLIII, 624
from *Providencia*, XLIII, 624
from *Pseudomonas aeruginosa*, XLIII, 624
Gentamicin acetyltransferase I, XLIII, 616
Gentamicin acetyltransferase II, XLIII, 616
Gentamicin acetyltransferase III, XLIII, 616, 625
Gentamicin adenylytransferase, XLIII, 616, 625
Gentamicin C1, XXXIV, 4
Gentamicin C complex, XLIII, 274
Gentamycin, *see* Gentamicin
Gentiobiose, metabolism in *Aerobacter aerogenes*
β-glucoside kinase, XLII, 3
phosphocellobiase, XLII, 494
Gentisate 1,2-dioxygenase, LII, 11
Gentisyl alcohol dehydrogenase, assay, XLIII, 544
Geodin, XLIII, 163
Geraniol hydroxylase, LII, 14
GERL theory, in electrophoresis, XXXI, 751
Germanium
atomic emission studies, LIV, 455
infrared spectral properties, LIV, 316
volatilization losses, in trace metal analysis, LIV, 481
Germanium chloride, phase shift studies, LIV, 344
Germinal cell, velocity sedimentation, XXXIX, 290, 291
Gettering furnace, for oxygen removal, LIV, 118
Gey's balanced salt solution, LVIII, 120, 121, 142
GFP, *see* Green-fluorescent protein
GHAT medium, LVIII, 352
GH cell, *see* Growth hormone cell
Ghost, *see* specific type
Giardia, rat infection, XXXII, 667, 668
Gibbs activation energy, LIV, 508, 510, 519–521, 530
Gibbs-Duhem equation, XLVIII, 85, 98, 124
Gibbs energy profile, LIV, 530

Gibbs reagent, LII, 403
Gibson mixer, efficiency, LIV, 89
Giemsa banding, LVIII, 324, 325, 334–336
 method, LVIII, 326
 protocol, LVIII, 326
Giemsa buffer, LVIII, 172, 173
Giemsa stain, LVIII, 172, 173, 285–287, 324, 333, 422
Gifford-Wood-Eppenbach colloid mill, LIX, 205
Gifford-Wood mill, XXXII, 826
Gilbert technique, XLVIII, 213, 223, 244
Gilford 2500 recording spectrophotometer, LII, 523
Gilmont Model S3200 ultraprecision micrometer syringe, LIV, 404
Gilson Model KM Oxygraph, LIII, 160, 530, 539
Gimmel factor, LVIII, 96, 101, 106
 preparation, LVIII, 104, 105
Glacial acetic acid, *see* Acetic acid, glacial
Gladiolic acid, XLIII, 163
 in modification of lysine, LIII, 140
Glancing angle diffraction, in membrane studies, XXXII, 219, 220
Glan-Taylor prism, LIV, 53
Glass, XXXIV, 699, 700, 717, 722, *see also* Aminopropyl glass; specific type
 activated, storage, XLVII, 322
 aldehyde carrier, XXXIV, 69
 amide linkage, XXXIV, 70, 71
 aminated, color, impurities, XLVII, 322
 aminoalkylation, XXXIV, 290
 antibody-producing cells, XXXIV, 722
 attachment, XLVII, 301
 azo linkage, XXXIV, 71, 72
 boron-free, calcium concentration, LVII, 282
 carbodiimides coupling, XXXIV, 21, 67
 carboxylate carriers, XXXIV, 70, 71
 as carrier, XXXIV, 4
 cleaning, XXXIV, 290
 constant volume, advantages, XLVII, 272
 controlled pore, XXXIV, 59–63, 288–294; XLIV, 149–169, *see also* Lipoamide glass
  bonding mechanisms, XLIV, 149–153
  comparative properties of types, XLIV, 158
  coupling through bifunctional bisdiazotized reagent, XLIV, 147
  cyanogen bromide coupling, XLIV, 146, 147
  immobilized enzyme reactors, LVI, 487, 488
  octadecylsilylated, binding mitochondria, LVI, 551, 552
  pH, operating range, XLIV, 163
  physical properties, XLIV, 135
  pore diameter, XLIV, 155, 156
  protein binding by adsorption, XLIV, 150–152
  relationship between pore size and surface area, XLIV, 136, 137
  relative cost, XLIV, 776, 777
  silanization, XLIV, 139, 140
  surface pH, XLIV, 155
  uricase immobilization, LVI, 491
  yeast cell breakage, LVI, 48, 61
 degradative efficiency, XLVII, 274, 275
 derivatization, XLVII, 266, 268
 diazotization, XXXIV, 71, 72
 DITC-activated, advantages, XLVII, 271
 durability, XXXIV, 63
 enzyme immobilization, XLVII, 41, 42
 glutaraldehyde derivative, XXXIV, 66
 hydrazide linkage, XXXIV, 69
 N-hydroxysuccinimide derivative, protein attachment, XLVII, 362
 infrared spectra, XXXIV, 61
 for large peptides, XLVII, 321
 lipoamide, XXXIV, 288–294, *see also* Lipoamide glass
 as matrix, XXXIV, 59–72
 matrix selection, XXXIV, 18
 nonspecific adsorption, XXXIV, 292, 293
 organic, supercooled, saturation-transfer studies, XLIX, 500, 501
 physical properties, XXXIV, 61

in picosecond continuum generation, LIV, 12
porous, XXXIV, 59–72
protein adsorption, LIX, 587
protein derivatives, XXXIV, 67–72, *see also* specific derivatives
reactive sites, XXXIV, 62
silane coupled, XXXIV, 63
silanization, XXXIV, 63–72; LII, 326
aldehyde derivative, XXXIV, 60
γ-aminopropyltriethoxysilane, XXXIV, 64
aqueous, XXXIV, 64
arylamine derivative, XXXIV, 65
carboxyl derivative, XXXIV, 66
organic, XXXIV, 64
triethoxysilane, XXXIV, 63
surface coating, XLVIII, 487
surface properties, XXXIV, 61–63
thiourea linkage, XXXIV, 68
use in spinning-cup sequenator, XLVII, 307–311, 317
Glass bead, XXXIV, 19, 59–72; LI, 8, 356, 402
activation, XLIV, 670
alkyl amine, XXXIV, 289
photochemical immobilization, XLIV, 284, 285
preparation, XLIV, 931, 932
aminoalkylation of, XXXIV, 290
arylamine, preparation, XLIV, 932
catecholamine immobilization, XXXVIII, 180, 181
alkylamine glass synthesis, XXXVIII, 181, 182
arylamine glass production, XXXVIII, 182, 183
catecholamine glass production, XXXVIII, 184
succinyl glass and long-chain alkylamine glass synthesis, XXXVIII, 183, 184
washing procedure, XXXVIII, 185, 186
for column sieving fractionation, XXXI, 551
commercial source, LVII, 214
for dehydrogenase purification, XXXII, 380, 384–389
density, XLIV, 721
estradiol adsorbents, XXXIV, 687, 688
with gel pellicle, XLIV, 179
homogenization, LV, 4
incorporation of thioester linkage, XLIV, 524
preparation, LIX, 205
sonication, XXXIV, 290
succinyl, preparation, XLIV, 506, 932
yeast cell breakage, LV, 152–155, 162, 163
Glass cuvette, *see also* Cuvette; Sample cell
for anaerobic assays, LIII, 364
Glass electrode
conditioning and storage, LVI, 361
construction, LVI, 361
hydrogen ion measurements, LVI, 310, 312
rapid proton-transfer reactions, LV, 615
Glass fiber, affinity chromatography, LVI, 425
Glass fiber filter, LVII, 80
assay of tRNA methyltransferases, LIX, 191
Glass joint, commercial source, LIV, 124, 126
Glass-Teflon homogenizer, XXXI, 181
Glass tubing
in anaerobic procedures, LIV, 116
in purge gas connections, LIV, 404
Glassware
cleaning, LV, 41
for determination of nucleoside phosphorylation, LI, 424
cleaning procedure, LVII, 425
in copper analysis, LIV, 444
with hydrochloric acid, 109n
in iron analysis, LIV, 438
for low-temperature EPR spectroscopy, LIV, 121, 122
for trace metal analysis, LIV, 477, 478
coating with nonwetting agent, LIX, 4
contamination, XLVII, 18
drying, LVIII, 6
for human cell culture, XXXII, 816–819
iron transport studies, LVI, 390

liver cells, LVIII, 541
sterilization, LVIII, 4–7, 10, 11
suppliers, LVIII, 16
for tissue culture, XXXIX, 110
Glebomycin
bioautography, XLIII, 42
paper chromatographic data, XLIII, 121
structure, XLIII, 429
Glial cell, isolation, XXXV, 561–579
Glioma, LVIII, 377
Glioma C_6 cell, rat, LVIII, 97
effect of hormone, LVIII, 103–105
media, LVIII, 57
Glioma cell line, in neurobiological studies, XXXII, 769, 770
Globin
human, fluorescent peptide map, XLVII, 30
millimolar extinction coefficient, LII, 449
preparation, LII, 448, 449, 488–490
recombination with modified hemes, LII, 454
Globoid cell leukodystrophy, enzymic diagnosis, L, 468–471
Globopentaglycosylceramide, L, 247
Globoside, L, 236
carbohydrate sequence, XXXII, 363
IR spectroscopy, XXXII, 366
isolation, XXXII, 345, 350, 355
optical rotation, XXXII, 365
Globotetraglycosylceramide, L, 247, *see also* Globoside
Globotriglycosylceramide, L, 247
Globulin, *see also* specific types
corticosteroid-binding, XXXIV, 6
synthesis, in hepatic explants, XXXIX, 39
γ-Globulin, *see also* Immunoglobulin G
cross-linking protein antigens, LVI, 225, 226
isolation, XXXII, 61
labeling with ^{125}I, LVI, 226
nonimmune, weak antigens, LVI, 224
as plant-cell protectant, XXXI, 536
preparation, LVI, 225
radiolabeled, in estrogen receptor studies, XXXVI, 170, 171, 174
γ-Globulin-Sepharose 4B, in ribonucleoside diphosphate reductase purification, LI, 231, 232
Glomerular basement membrane, *see* Basement membrane
Glomerular filtration, of perfused rat kidney, XXXIX, 7–9
Glomerulus, renal, isolation, XXXII, 653–658
Glo-Quartz heating rod, XXXII, 747
Glove bag, source, LIX, 811
Glucagon, XXXIV, 7, 106
adenylate cyclase, XXXVIII, 147–149, 173, 176, 178, 180
cAMP inhibitor, XXXVIII, 275
cyclic nucleotide levels, XXXVIII, 94
digestion by aspergillopeptidase C, XLV, 428, 429
by thermomycolin, XLV, 428, 429
effect on β-cell, XXXIX, 425
on fat-cell hormones, XXXI, 68
on hepatocytes, XXXV, 592, 593
on perfused liver, XXXIX, 34
on synthesis of fatty acid synthase, XXXV, 44
fat cell response, modification, XXXVII, 213
inactivation studies, XXXVII, 198–211
lipolysis stimulation in adipose tissue, XXXV, 181, 187
in perfused pancreas, XXXIX, 371
precursor, XXXVII, 326
protein kinase, XXXVIII, 364
radioiodination, XXXVII, 200, 201
structure-function studies, XLVII, 170
Glucagon-agarose, XXXIV, 654
Glucagon receptors, XXXIV, 654
Glucan, in cell wall, XXXI, 610
α-Glucan, support, for penicillin acylase, XLIV, 765
Glucan synthetase
chromatographic separation, XXXII, 400, 401, 405
as possible membrane marker, XXXII, 393, 403
D-Glucitol, XLVI, 383
chromatography, gas-liquid of 2-deoxy-2-(*N*-methylacetamido)acetate derivatives, L, 6

Glucoamylase
commercial source, XLIV, 777
conjugate, activity assay, XLIV, 778
fiber entrapment, industrial application, XLIV, 241
immobilization, XLIV, 42
by bridge formation, XLIV, 167
on cellulose, XLIV, 52, 167
by entrapment, XLIV, 178, 180
in liposomes, XLIV, 224
on porous ceramics, XLIV, 136, 137, 778
scale-up data, XLIV, 781
industrial production of glucose, XLIV, 776–792
kinetics, in enzyme system, XLIV, 433, 434
support for, choice, XLIV, 776, 777, 779–783
titanium chloride binding, XLIV, 52
Glucocerebrosidase, XLIV, 702
Gaucher's disease, L, 532
from human placenta
assay, L, 529, 530
properties, L, 532
purification, L, 530–532
Glucocorticoid
cell lines dependent, XXXII, 559
effect on glucose transport, XXXVI, 429
on pituitary hormones, XXXVII, 91
in steroid hormone assay, XXXVI, 36
Glucokinase, XXXIV, 328, 490
hormonal control, XXXVII, 293
as marker enzyme, XXXII, 733
plasma membrane, XXXI, 90
from rat liver, XLII, 31–39
assay, XLII, 31, 32
chromatography, XLII, 34
electrophoresis, XLII, 32, 35, 36
inhibition, XLII, 38
molecular weight, XLII, 38
properties, XLII, 38, 39
purification, XLII, 33–37
from yeast, XLII, 25–30
assay, XLII, 25, 26
chromatography, XLII, 27, 28
inhibitors, XLII, 30
molecular weight, XLII, 30
properties, XLII, 30
purification, XLII, 26–30
Gluconase, as cell wall hydrolyzing enzyme, XXXI, 610, 611
D-Gluconate dehydratase, from *Alcaligenes*, XLII, 301–304
assay, XLII, 301
distribution, XLII, 304
electrophoresis, XLII, 302, 303
inhibitors, XLII, 304
properties, XLII, 303, 304
purification, XLII, 302, 303
Gluconate-6-phosphate dehydrogenase, mammary uptake, XXXIX, 454
Gluconeogenesis
control studies, XXXVII, 292, 294, 295
epinephrine effects, XXXI, 70, 71
fructose-1,6-diphosphatase, XLII, 398
in isolated renal tubule, XXXIX, 13, 14
localization, LVI, 207
pathway, in hepatocyte cell lines, XXXII, 740
in perfused liver, XXXIX, 34
rates, XXXV, 597
Gluconic acid, product, of glucose oxidase, XLIV, 272
Gluconobacter, 5-keto-D-fructose produced, XLI, 84, 85
Gluconobacter albidus, 5-keto-D-fructose reductase, XLI, 128
Gluconobacter cerinus
aldohexose dehydrogenase, assay, purification, and properties, XLI, 142–147
5-keto-D-fructose reductase, assay, purification, and properties, XLI, 127–131
Gluconolactonase
actinomycin, XLIII, 763–767, *see also* Actinomycin lactonase
peptide antibiotic, XLIII, 767–773, *see also* Peptide antibiotic lactonase
Gluconolactone
inhibitor, of β-glucosidase, XLIV, 433, 918
in media, LVIII, 66
α-D-Glucopyranoside, XXXIV, 363
β-D-Glucopyranoside, XXXIV, 363

O-α-D-Glucopyranosyl-(1→4)-2-acetamido-2-deoxy-D-glucose, L, 104
α-D-Glucopyranosyl-L-altritol, *Klebsiella* capsular polysaccharide, L, 22
α-D-Glucopyranosyl-D-galactitol, *Klebsiella* type 37 capsular polysaccharide, L, 22
Glucosamine, XLVI, 75; XLVII, 89
internal standard, XLVII, 34, 40
in media, LVIII, 66
radioactive, microsome labeling, XXXI, 218
D-Glucosamine
animal cell growth and, L, 190
β-1,6-di-*N*-acetyl, synthesis, L, 112
enzymic microassay, XLI, 5–10
interconversion, in animal cells, L, 182
Glucosamine-6-phosphate deaminase
from *Escherichia coli*, XLI, 497–502
activators, XLI, 502
assay, XLI, 498
chromatography, XLI, 499
properties, XLI, 501, 502
purification, XLI, 498–500
in pig kidney, XLI, 497
in *Proteus vulgaris*, XLI, 497
Glucosaminephosphate isomerase, from house flies, XLI, 400–407
activators and inhibitors, XLI, 406
assay, XLI, 400–403
properties, XLI, 406
purification, XLI, 403–406
Glucose, XLVI, 167
active transport, XLIV, 921–924
2-amino-1-ethyl glycosides, XXXIV, 318
analogs, XLVI, 365
assay, LV, 221
of adenylate kinase, LI, 468
in continuous flow analyzer, XLIV, 638–641
of glucose oxidase, LVII, 403
using luminol chemiluminescence, LVII, 403, 446, 451–456
by microcalorimetry, XLIV, 673–675
of phenylalanine hydroxylase, LIII, 279
ATPase inhibitor, LV, 425
automated analysis, comparative discussion, XLIV, 636, 637
bacterial luminescence, LVII, 128, 164
catabolite repression, XXXVIII, 367
^{14}C-labeled, fatty acid synthesis, precursor, XXXV, 280
Crabtree effect, LV, 296, 297
disappearance, measurement, LV, 294, 295
fluorometric determination, XXXIX, 15
growth of *E. coli*, LV, 165
of fatty acid mutant, LVI, 571, 572
hem mutants, LVI, 175, 559
inorganic phosphate assay, LV, 212
isomerization, XLIV, 768–776, 809–821
α-lactalbumin conjugate, XXXIV, 360
mammary uptake, XXXIX, 454
mating efficiency, LVI, 141
in media, LVIII, 53, 66, 198
oxidation of TSH effects, XXXVII, 262–268
in oxygen scavenging system, LII, 223
periodate oxidation of tRNA, LIX, 186
as peroxide source, XXXII, 108
phosphorescence studies, XLIX, 238, 243
polarography, LVI, 451, 452, 461–465
preparation of labeled oligonucleotides, LIX, 75
product, in invertase assay, XLIV, 232
production, from corn syrup, XLIV, 776–792
purification of aminoacyl-tRNA synthetases, LIX, 258
in rat hepatocyte isolation, LII, 358
tRNA labeling procedure, LIX, 71
substrate, of glucose oxidase, XLIV, 198, 347, 476
suppression of carbodiimide-resistance by growth, LVI, 171, 172
transport, XLVI, 89, 167, 608
of plasma membrane, XXXI, 68, 69
trapped marker in lysosomes assay, XXXII, 501–513

utilization by *in vitro* placenta, XXXIX, 249

D-Glucose, XXXIV, 334
assay
of guanosine triphosphate, LVII, 89
using immobilized enzyme rods, LVII, 210, 211, 213
chromatographic constants, XLI, 16, 17, 20
^{13}C NMR spectroscopy, L, 40, 41
$\alpha(1\rightarrow6)$-disaccharide, *see* Isomaltose
enzymic microassay, XLI, 4, 7–10
gas-liquid chromatography of 2,3,4,6-tetra-*O*-methyl derivative, L, 7
in perfusion fluid, LII, 51
2,3,5,6-tetra-*O*-methyl derivative, L, 7
trisaccharide, *see* Panose

L-Glucose, cultured cell growth, L, 190

Glucose carrier, reconstitution, LV, 707

Glucose dehydrogenase
assay of phenylalanine dehydrogenase, LIII, 278, 279
in continuous flow analyzer, XLIV, 639, 646
immobilization, on nylon tube, XLIV, 640, 646

α-D-Glucose 1,6-diphosphate
characterization, XLI, 82, 83
preparation, XLI, 81, 82

α-D-[1-^{32}P]Glucose 1,6-diphosphate, preparation, XLI, 83

α-D-[6-^{32}P]Glucose 1,6-diphosphate, preparation, XLI, 83

Glucose isomerase
activity, XLIV, 161, 163, 164, 200, 237, 811–813
effect of metal ions, XLIV, 815, 816
efficiency
effect of pH, XLIV, 237, 238
of temperature, XLIV, 238, 239
immobilization
by adsorption, XLIV, 161
on collagen, XLIV, 770
on controlled-pore alumina, XLIV, 161–164
in cross-linked cell homogenates, XLIV, 812
on DEAE-cellulose, XLIV, 165
industrial use, XLIV, 47, 165, 241, 809–821
kinetics, XLIV, 769
microbial sources, XLIV, 770, 811
oxygen sensitivity, XLIV, 816
pH optimum, XLIV, 164
productivity, XLIV, 811
stability, XLIV, 811–813
use in reactor, XLIV, 769–776

Glucose monohydrate, substrate, of glucose oxidase, XLIV, 157

Glucose oxidase, XXXIV, 338; LII, 6, 18; LVI, 474
activity assay of immobilized enzyme, XLIV, 198–200, 347–349, 482, 483
of insoluble aggregate, XLIV, 268–272
by microcalorimetry, XLIV, 663, 664
of native protein, XLIV, 479–482
of soluble aggregate, XLIV, 474
assay
of glucose, LVII, 403, 452
with luminol, LVII, 403, 446
catalytic reaction, XLIV, 478
commercial source, LII, 222
conjugate
pH optimum shift, XLIV, 468, 469, 607
support surface area and activity, XLIV, 156
conjugation, with antibody, XLIV, 711
controlled-pore glass columns, LVI, 487, 488
in enzyme electrode, XLIV, 585, 597
holoenzyme reconstitution, LIII, 436
immobilization, LVI, 482, 483, 489
by carbohydrate residue activation, XLIV, 319, 320
in cellophane membrane, XLIV, 904, 905, 908; LVI, 487
by co-cross-linking, XLIV, 268–272, 908
on collagen, XLIV, 260
on controlled-pore ceramics, XLIV, 152, 153, 156–160
by copolymerization, XLIV, 198, 200

by entrapment, XLIV, 173–176, 908
on glass beads, XLIV, 347, 671
with magnetic carrier, XLIV, 278, 279
by microencapsulation, XLIV, 214
in multistep enzyme systems, XLIV, 181, 460, 461, 472–478, 917
on nylon tube, XLIV, 124; LVI, 490
immobilized, assay of glucose, LVII, 454
kinetics
on conjugate, XLIV, 201, 425, 426, 483–488
in multistep enzyme system, XLIV, 433, 434
loading capacity on collagen membrane, XLIV, 255
oxidation, periodate, XLIV, 318–320
in oxygen scavenging system, LII, 223
peroxidase inhibition, XLIV, 160
as peroxide source, XXXII, 108
photometry or fluorimetry, LVI, 461
polyacrylamide adduct, LVI, 486
properties, XLIV, 478, 479
quantitization, XLIV, 389
sources, LVI, 461
specific activity, XLIV, 255
specificity, XLIV, 583
spontaneous structuration studies, XLIV, 927, 928
stain for detection, XLIV, 713, 905
substrates, XLIV, 615
temperature-activity profile, XLIV, 259
triazole treatment, LVI, 490
unit cell dimensions, XLIV, 159
use in quantitative immunodiffusion, XLIV, 716, 717
Glucose-6-phosphatase
assay, XXXI, 20, 91, 93, 94; LV, 102
in brush borders, XXXI, 121
in chromaffin granule, XXXI, 381
as diagnostic enzyme, XXXI, 20, 31, 33, 35–38, 84, 262, 303, 327, 338, 361, 726, 729, 743
in endoplasmic reticulum, XXXI, 23, 24
gluconeogenic catalytic activity, XXXVII, 281
lipid peroxidation, LII, 304
in liver, XXXI, 97, 195
in microsomal fractions, LII, 83
in nuclear membrane, XXXI, 290–292
in plasma membranes, XXXI, 88, 91, 100, 185, 186
in rough microsomal subfractions, LII, 75, 76
site-specific reagent, XLVII, 483
Glucose 1-phosphate
AMP assay, LV, 209
inorganic phosphate assay, LV, 212
Glucose 6-phosphate
active transport, XLIV, 921–924
AMP assay, LV, 209
assay, XXXII, 397; LV, 221
cytochrome P–$450_{11\beta}$assay, LII, 126
dimethylaniline monooxygenase assay, LII, 143
enzyme electrode, XLIV, 879
formation, XXXVI, 429–431
inhibitor of uridine nucleosidase, LI, 294
inorganic phosphate assay, LV, 212
in isolated renal tubule, XXXIX, 16
as membrane marker, XXXII, 393
microsomal fraction, LII, 88
Pasteur effect, LV, 291
substrate, for glucose-6-phosphate dehydrogenase, XLIV, 340, 455
α-D-Glucose 1-phosphate, assay of guanosine triphosphate, LVII, 89
D-Glucose 6-phosphate, isomerization by glucosephosphate isomerase from baker's yeast, XLI, 57
Glucose-6-phosphate dehydrogenase, XXXIV, 164, 252, 478–479; XLIV, 342, 629; XLVI, 453
activity, LVII, 86
assay, LVIII, 167, 168
adenylate cyclase, XXXVIII, 162
of adenylate kinase, XLIV, 892; LI, 468
dimethylaniline monooxygenase, LII, 143
by Fast Analyzer, XXXI, 816
GTP, LV, 208
using immobilized enzyme rods, LVII, 210

inorganic phosphate, LV, 212
pyridine nucleotide, LV, 267
associated with cell fractions, XL, 86
from bovine mammary gland, XLI, 183–188
assay, XLI, 183, 184
chromatography, XLI, 186, 187
purification, XLI, 184–188
from *Candida utilis*, XLI, 205–206
activators and inhibitors, XLI, 208
assay, XLI, 205, 206
chromatography, XLI, 207
properties, XLI, 207
purification, XLI, 206–208
cell monitoring, LVIII, 167, 168
conjugate, activity assay, XLIV, 200, 271, 340, 341
in cortexolone formation, XXXVII, 310
from cow adrenal cortex, XLI, 188–196
assay, XLI, 188–190
chromatography, XLI, 191, 192
inhibitors, XLI, 195
properties, XLI, 195
purification, XLI, 190–194
in cytochrome *P*–450 assay, LII, 126, 139
deficiency, LVII, 489
determination of guanosine nucleotides, LVII, 86, 89, 93
in D-glucose microassay, XLI, 4
from human erythrocytes, XLI, 208–214
amino acids, XLI, 214
assay, XLI, 209
properties, XLI, 212, 213
purification, XLI, 210–212
immobilization in collodion membrane, XLIV, 906, 908
in multistep enzyme system, XLIV, 454–457
interference in malonyl-CoA assay, XXXV, 313
in intestinal mucosa, XXXII, 667
isoenzyme, LVIII, 167, 168
kinetics, in enzyme system, XLIV, 432, 433, 463–466
of phosphoribosylpyrophosphate synthetase, LI, 6
from *Leuconostoc mesenteroides*, XLI, 196–201
assay, XLI, 196, 197
chromatography, XLI, 198
inhibitors, XLI, 201
properties, XLI, 199–201
purification, XLI, 197–200
mammary uptake, XXXIX, 454
as marker enzyme, XXXI, 735
molecular weight, LVII, 210
in NADPH-generating system, XLIV, 850, 851, 865
from *Neurospora crassa*, XLI, 177–182
assay, XLI, 177–179
chromatography, XLI, 180
genetics, XLI, 182
properties, XLI, 181, 182
purification, XLI, 179–181
in parenchyma cells, XXXII, 706
from *Penicillium duponti*, XLI, 201–205
assay, XLI, 201, 202
chromatography, XLI, 203
properties, XLI, 204, 205
purification, XLI, 202–204
P/O ratio, LV, 226
reduction of $NADP^+$ analogs, XLIV, 874–876
stain, LVIII, 168, 169
stereospecificity determination, LIV, 229, 230
in studies of glucose-marked liposomes, XXXII, 512
substrate specificity, LVII, 93
Glucose-6-phosphate 1-epimerase
from baker's yeast, XLI, 488–493
assay, XLI, 488–491
chromatography, XLI, 492
inhibitors, XLI, 493
properties, XLI, 493
purification, XLI, 490–493
from *Escherichia coli*, molecular weight, XLI, 493
from potato tubers, molecular weight, XLI, 493
from *Rhodotorula gracilis,* molecular weight, XLI, 493

Glucosephosphate isomerase, XLIV, 811
affinity for calcium phosphate gel, LI, 584
affinity labeling procedure, XLVII, 493, 494
from *Bacillus stearothermophilus*, XLI, 383–387
assay, XLI, 383, 384
chromatography, XLI, 385
inhibitors, XLI, 387
properties, XLI, 387
purification, XLI, 384–386
conjugate
activity, XLIV, 271
kinetics, XLIV, 917
determination of anomerase activity, from baker's yeast, XLI, 57–61
from human erythrocytes and cardiac tissues, XLI, 392–400
assay, XLI, 392–394
isolation, XLI, 394–399
properties, XLI, 399
immobilization, on inert protein, XLIV, 909
from peas, XLI, 388–392
assay, XLI, 388, 389
chromatography, XLI, 391
properties, XLI, 391, 392
purification, XLI, 389–391
site-specific reagent, XLVII, 482, 483
Glucose-6-phosphate isomerase, XLVI, 23, *see also* Glucosephosphate isomerase
Glucose-1-phosphate uridylyltransferase
activity, LVII, 86
in determinations of guanosine nucleotides, LVII, 86, 89
inhibitor, LVII, 93
substrate specificity, LVII, 93
Glucose pulse, hormone effects, XXXVI, 429–432
Glucose tetrasaccharide, human urine, L, 228
Glucose transferase, in plasma membrane, XXXI, 88
Glucose transport system
in cell membrane, XXXVII, 213
control studies, XXXVII, 294
in fat cells, hormone effects, XXXVII, 269–276
Glucosidase, XLVI, 368–381
inhibition, XLVI, 369
with bromoconduritol, XLVI, 381
α-Glucosidase
immobilization, by microencapsulation, XLIV, 214
reactions, XLVI, 369, 370
β-Glucosidase
in amebic phagosomes, XXXI, 697
for glycolipid structure studies, XXXII, 363
reactions, XLVI, 369, 370, 377
β-D-Glucosidase
acyl azide coupling, XLIV, 103, 104
in amygdalin electrode, XLIV, 592, 604–607
assay, spectrophotometric, XLIV, 101
conjugate
activity, XLIV, 103, 104, 271, 272
stability, XLIV, 104
storage, XLIV, 103
immobilization in multistep enzyme system, XLIV, 473–475, 917
on polyacrylic type copolymers, XLIV, 103, 104
in porous matrix, XLIV, 270–272
kinetics, in enzyme system, XLIV, 433
pH optimum, XLIV, 607
source, XLIV, 101
β-Glucosidase A_3, XLVI, 379, 380
β-Glucoside, metabolism, phosphocellobiase, XLII, 494
D-Glucoside 3-dehydrogenase
from *Agrobacterium*, XLI, 153–158
assay, XLI, 153–155
chromatography, XLI, 155
electrophoresis, XLI, 156
properties, XLI, 156–158
purification, XLI, 155–157
in 3-ulose determination, XLI, 25, 26
β-Glucoside kinase
assay, XLII, 3, 4
chromatography, XLII, 5
molecular weight, XLII, 6
properties, XLII, 5, 6
purification, XLII, 4, 5

Glucosylceramidase
assay
with glucosylceramide, L, 476–478
with 4-methylumbelliferyl β-glucoside, L, 478, 479
Gaucher's disease, L, 476–479
Glucosylceramide
Gaucher's disease, L, 476
isolation, XXXII, 350
tritiation, L, 477, 478
Glucosylfucosylthreonine
human urine, L, 227, 229
structure, L, 229
D-Glucosyl isothiocynate, XLVI, 166
Glucosylsphingosine, glucosylceramidase, L, 479
Glucosyltransferase, XL, 353
Glucuronic acid, in media, LVIII, 66
D-Glucuronic acid, colorimetric assay, XLI, 30, 31
β-Glucuronidase, XXXIV, 4
assay, L, 451, 452
in brain homogenates, XXXI, 464, 466, 467, 469–471
distribution, L, 451
gel electrophoresis, XXXII, 83, 89
in lysosomes, XXXI, 330, 344
as marker enzyme, XXXI, 753
mucolipidosis II, L, 452, 455
mucolipidosis III, L, 452, 455
mucopolysaccharidosis VII, L, 452
preparation, L, 147
in snail gut juice, XXXI, 612
test for β-glucuronide, L, 147
β-D-Glucuronidase
from rat liver, L, 489
from rat preputial gland
assay, L, 510, 511
properties, L, 513, 514
purification, L, 511–513
β-D-Glucuronide, α-L-idosides, L, 147
Glucuronolactone, in media, LVIII, 66
Glucuronosyltransferase, as marker enzyme, XXXII, 733
Glucuronyltransferase, *see* Glucuronosyltransferase
Glusulase
for drug metabolite isolation, LII, 332, 334
yeast spheroplasts, LVI, 19, 20, 129
Glutamate, *see also* Glutamic acid
binding to mitochondrial fractions, LVI, 427, 428
carrier system, isolation from pig heart mitochondria, LVI, 419
basis of experimental approach, LVI, 420
choice and preparation of mitochondria, LVI, 420, 421
glutamate binding to fractions, LVI, 427, 428
main properties of carrier, LVI, 426, 427
procedure, LVI, 421–426
reconstitution of transport system, LVI, 428–430
concentration determination, with enzyme electrode, XLIV, 877–880
chemical properties, XLIV, 12–18
efflux, kinetic analysis, LVI, 268–271
exchange, LVI, 253, 254
loading for efflux kinetics, LVI, 258, 259
mitochondrial protein synthesis, LVI, 20, 22
product of amidophosphoribosyltransferase, LI, 171
of carbamoyl-phosphate synthetase, LI, 21, 29
of cytidine triphosphate synthetase, LI, 84
of FGAM synthetase, LI, 193
of GMP synthetase, LI, 213
tRNA synthetase, XXXIV, 170
transport, LVI, 248, 249, 256, 292
measurement, LVI, 421
uptake, inhibition, LVI, 247
Glutamate-aspartate, equilibrium, LVI, 255
Glutamate-aspartate transaminase, *see* Aspartate aminotransferase
Glutamate decarboxylase
assay, XXXII, 779–781
cultured cells, XXXII, 772, 773
labeling, XLVI, 139
in neurobiology, XXXII, 765, 768, 770
Glutamate dehydrogenase, XXXIV, 126; XLVI, 21, 83, 87, 90, 242, 245–248
allosteric effectors, XLIV, 512, 513

ammonia assay, LV, 209
assay of FGAM synthetetase, LI, 196
in nonaqueous tissue fractions, LVI, 204
from bovine liver, hydroxylamine cleavage, XLVII, 140, 142
chemical cleavage at aspartyl-prolyl bond, XLVII, 145
conjugate
assay, XLIV, 508
coenzyme binding, XLIV, 508–510
denaturation studies, XLIV, 510–512
elimination from mitochondria, LVI, 421
in enzyme electrode, XLIV, 877
enzymic oxidation of NADPH, XLIV, 865
histidine modification, XLVII, 433, 441
immobilization, LVI, 483, 484
by azide method, XLIV, 907
on collagen, XLIV, 907, 908
on porous glass beads, XLIV, 506
α-ketoglutarate assay, LV, 206
liposomes, LVI, 428, 429
in liver compartments, XXXVII, 288
localization, LVI, 205
as marker enzyme, LVI, 210–212, 217, 218, 222
mercuri-estradiol reaction, XXXVI, 389
NAD-specific, cyanylation, XLVII, 131
negative staining, XXXII, 29
from *Neurospora crassa*
arginine blocking, XLVII, 160, 161
staphylococcal protease hydrolysis, XLVII, 190, 191
pyridine nucleotide assay, LV, 267
reduction of $NADP^+$ analogs, XLIV, 874–876
self-assembly, XLVIII, 309
structure, XLIV, 507
Glutamate dehydrogenase (NAD), brain mitochondria, LV, 58
Glutamate dehydrogenase (NADP), brain mitochondria, LV, 58
Glutamate-glyoxylate aminotransferase, in microbodies, LII, 496
Glutamate γ-methyl ester, affinity chromatography, LVI, 425
Glutamate-oxalacetate aminotransferase, *see* Aspartate transaminase
Glutamate-oxaloacetate aminotransferase, in microbodies, LII, 496
Glutamate-oxaloacetate transaminase, *see* Aspartate aminotransferase
D-Glutamate oxidase, LII, 21; LVI, 474
L-Glutamate oxidase, LII, 19
Glutamate synthase, XLVI, 420
Glutamate synthetase, XXXIV, 135, 136
Glutamic acid, *see also* Glutamate
affinity labeling, XLVII, 482, 493, 494, 498
aminoacylation isomeric specificity, LIX, 274, 275, 279, 280
assay, LV, 221
brain mitochondria, LV, 57
γ-carboxyl group blocking, XLVII, 526, 528, 531, 532, 613, 614
cleavage, with staphylococcal protease, XLVII, 189–191
content
of brain, LV, 219
of hepatocytes, LV, 214
of kidney cortex, LV, 220
of rat heart, LV, 216
of rat liver, LV, 213
detection by fluorescence, XLVII, 242
in media, LVIII, 63
phenylthiohydantoin derivative, identification, XLVII, 348
reaction with diethylpyrocarbonate, XLVII, 440
renal gluconeogenesis, XXXIX, 15, 16
transport
colcin, LV, 536
by *M. phlei*, LV, 186–188, 193
ubiquinone redox state, LIII, 583
Glutamic acid dihydrazide, preparation, LIX, 179
Glutamic decarboxylase, XLVI, 445
Glutamic dehydrogenase
assay by Fast Analyzer, XXXI, 816
in lung lamellar bodies, XXXI, 423
as marker enzymes, XXXI, 760
in mitoplasts, XXXI, 323
Glutamic-oxaloacetic transaminase, *see* Aspartate aminotransferase

Glutaminase, XLVI, 35, 420
effect on immune response, XLIV, 705
phosphorylation potential determination, LV, 237
renal mitchondria, LV, 12
Glutaminase A, XLVI, 424
Glutaminase activity, of carbamoyl-phosphate synthetase subunits, LI, 26, 28, 29, 35
Glutamine
aminoacylation isomeric specificity, LIX, 274, 275, 279, 280
analogs, XLVI, 414–427
inhibition, XLVI, 420
properties, XLVI, 414–419
synthesis, XLVI, 414–419
use, XLVI, 419–423
assay, LV, 209
binding sites, XLVI, 414–427
carbodiimide activation, XLVII, 553, 554
carboxamide group blocking, XLVII, 614
content
of brain, LV, 219
of kidney cortex, LV, 220
dipeptides containing, identification, XLVII, 396
hydrogen exchange, XLIX, 25
identification, XLVII, 83, 84
isoxazolium salt activation, XLVII, 556
in media, LVIII, 53, 62, 89
problems in peptide synthesis, XLVII, 608
residue, effect on Raman spectra, XLIX, 116
Sepharose-bound, preparation, LI, 31, 32
substrate of amidophosphoribosyltransferase, LI, 171
of carbamoyl-phosphate synthetase, LI, 21, 29, 33, 106, 112, 114, 122
of cytidine triphosphate synthetase, LI, 80, 84, 85
of FGAM synthetase, LI, 193, 194, 196, 197
of GMP synthetase, LI, 213, 217, 219
transport by *M. phlei*, LV, 186–188, 193
L-[^{14}C]Glutamine, XLIII, 443
Glutamine-binding protein, from *E. coli*, isolation, XXXII, 422–427
L-Glutamine:keto-*scyllo*-inositol aminotransferase, XLIII, 439–443, 462
assay, XLIII, 440–443
L-[^{14}C]glutamine, XLIII, 443
myo-inositol 2-dehydrogenase, XLIII, 443
biological distribution, XLIII, 442
ion-exchange chromatography, XLIII, 441
properties, XLIII, 442, 443
reaction scheme, XLIII, 439
specificity, XLIII, 442, 443
stability, XLIII, 442
Glutamine-mercaptoethanol-Tris buffer, LI, 86
Glutamine phosphoribosyl pyrophosphate amidotransferase, hormone induction of biorhythms, XXXVI, 480
Glutamine synthetase, XLVI, 24, 25, 66
glutamine assay, LV, 209
phosphorylation potential determination, LV, 237
Glutamine synthetase adenylyltransferase, XXXIV, 135
γ-Glutaminyl-4-hydroxybenzene hydroxylate, LII, 20
Glutaminylpeptide γ-glutamyltransferase, XLVI, 151
fluorescent labeling, XLVIII, 358
Glutaminyl residue, fluorescent labeling, XLVIII, 358
Glutaminyl-tRNA synthetase, subcellular distribution, LIX, 233, 234
Glutamylalanyllysylchloromethane, XLVI, 206
Glutamyl γ-carboxylate, as active site, XLVI, 141
γ-Glutamylcysteine synthetase, XLVI, 420
γ-Glutamyl ester, substrates, of FGAM synthetase, LI, 201
γ-Glutamyl hydrazide, XLVI, 419

γ-Glutamyl hydrazine, substrate, of FGAM synthetase, LI, 201
γ-Glutamylhydroxamatase, activity, of carbamoyl-phosphate synthetase subunits, LI, 26, 28
γ-Glutamyl hydroxamate, XLVI, 419
 substrate, of FGAM synthetase, LI, 201
Glutamyl-tRNA synthetase, subcellular distribution, LIX, 233, 234
γ-Glutamylthioester, substrates, of FGAM synthetase, LI, 201
γ-Glutamyltransferase, in brush borders, XXXI, 122
Glutamyltranspeptidase
 assay by Fast Analyzer, XXXI, 816
 in microvillous membrane, XXXI, 130
Glutaraldehyde, XXXIV, 338
 antigen cross-linking to γ-globulin, LVI, 226
 in buffer, XXXI, 17
 coupling, density perturbation, XXXIV, 173
 for covalent binding of aminoethyl cellulose, XLIV, 48, 51
  after microencapsulation, XLIV, 212
  of cellophane and enzyme, XLIV, 905
  of ceramics, XLIV, 778, 851
  of enzyme aggregates, XLIV, 263, 473
  of inert protein and enzymes, XLIV, 21, 710–712, 903, 904, 912, 925
  of inorganic supports, XLIV, 140, 141
  of nylon, XLIV, 121–127
 enzyme/hair conjugation, XLIV, 561
 enzyme immobilization, LVI, 491
 enzyme polymerization onto electrode membrane, LVI, 486
 fixation, XXXI, 20, 21; XL, 4
  for electroscopy, XXXIX, 137
  of initiation complexes, LX, 333, 340–342
  of ribosome-DNA complexes, LIX, 841, 842
 in hybrid antibody assay, XXXII, 65
 in immune conjugate preparation, XXXII, 62, 63
 immunoglobulin conjugation to ferritin, LVI, 226, 227
 inhibition of initiation complex formation, LX, 263–265
 in Karnovsky fixative, XXXIX, 137
 lectin coupling, XXXIV, 341
 lectin purification, XXXIV, 336
 as negative stain interferent, XXXII, 27, 35
 polyacrylamide activation, XLIV, 60, 61, 86
 porous silica alumina pellet activation, XLIV, 480
 protein coupling, XXXIV, 56, 660
 protein crystal insolubilization, XLIV, 546, 548–555
  mechanism, XLIV, 551–555
 reagent, XXXII, 34
  for subcellular fraction studies, XXXII, 4, 5
 SDS polyacrylamide gel electrophoresis, L, 59
 Sepharose activation, XLIV, 458
 tanning of collagen membrane, XLIV, 248, 250, 254, 257
 tissue fixation, XLIV, 713
Glutaraldehyde-glass, XXXIV, 66
Glutaredoxin, assay, LI, 231
Glutaric anhydride, XXXIV, 38
 purification, XXXIV, 38
Glutarimide antibiotics, XLIII, 234–237
Glutaryl dichloride, agar gel cross-linker, XLIV, 22, 37
Glutarylhydrazide derivative, XXXIV, 38, 39
Glutathione, XXXIV, 536; XLVI, 135, 136, 141; XLVII, 489
 activator of cytidine deaminase, LI, 411
  of pyrimidine nucleoside monophosphate kinase, LI, 330
 aminoacylation of tRNA, LIX, 216
 coupling, XXXIV, 537
 in cytochrome P-450 isolation, LII, 362
 determination of aminoacylation isomeric specificity, LIX, 272
 dissociation of OPRTase-OMPdecase complex, LI, 141
 drug metabolism, LII, 69, 70

inhibitor of nucleoside phosphotransferase, LI, 393
in media, LVIII, 63
oxidized, XXXIV, 4
in plant-cell fractionation, XXXI, 505, 603
in preparation of deuterated NADPH, LIV, 232
of *E. coli* crude extract, LIX, 298
reduced, steroid reaction, XXXVI, 389, 392
reduction, XXXIV, 294
reverse salt gradient chromatography, LIX, 216
tRNA reconstitution, LIX, 140, 141
stabilizer of cytidine deaminase, LI, 398
of nucleoside triphosphate pyrophosphohydrolase, LI, 276–284
Glutathione-agarose, XXXIV, 536–538
Glutathione peroxidase, LII, 506–513
aggregates and charge forms, LII, 510–512
assay, LII, 506
from erythrocytes, LIII, 372
in hydroperoxide measurement, LII, 509, 510
isolation, LII, 506, 507
kinetic properties, LII, 513
molecular weight, LII, 512
specificity, LII, 513
subcellular distribution, LII, 513
Glutathione-2-pyridyl disulfide-agarose, XXXIV, 537, 538
capacity, XXXIV, 240, 241
Glutathione reductase, XXXIV, 4, 246
flavoprotein classification, LIII, 397
in preparation of deuterated NADPH, LIV, 232
pyridine nucleotide assay, LV, 264, 265, 267–269
transhydrogenase assay, LVI, 113
Glycera dibranchiata, hemoglobin, LIV, 275
Glyceraldehyde 3-phosphate, XLVI, 18
control studies on formation, XXXVII, 295
Pasteur effect, LV, 290, 291
phosphorylation potential, LV, 244, 245
D-Glyceraldehyde 3-phosphate, XLVI, 134, 143
Glyceraldehyde phosphate dehydrogenase, XXXIV, 34, 43, 123, 148, 165, 237, 246; XLVI, 18, 75, 87
affinity labeling, XLVII, 424, 425, 427
from *Bacillus stearothermophilus*, XLI, 268–273
amino acids, XLI, 272
assay, XLI, 268, 269
hydroxylamine cleavage, XLVII, 140
inhibition, XLI, 273
properties, XLI, 271–273
purification, XLI, 269–272
cGMP assay, XXXVIII, 106, 107
cyanylation, XLVII, 131
dissociation, XLVIII, 180
in erythrocyte ghost sidedness assay, XXXI, 176, 178, 179
from *Escherichia coli,* purification, XLII, 444
gluconeogenic catalytic activity, XXXVII, 281, 288
GTP assay, LV, 208
inorganic phosphate assay, LV, 212
ligand binding studies, XLVIII, 276–278
NAD binding, kinetics, LIV, 82
peptide separation, XLVII, 216, 218, 219
from rabbit muscle, XLI, 264–267
assay, XLI, 264, 265
properties, XLI, 267
purification, XLI, 265–267
in tRNA sequence analysis, LIX, 61
sedimentation equilibrium, XLVIII, 175
site-specific reagent, XLVII, 483
D-Glycerate dehydrogenase
from hog spinal cord, XLI, 289–293
assay, XLI, 289, 290
chromatography, XLI, 291, 292
properties, XLI, 292, 293
purification, XLI, 290–292
D-Glycerate 3-kinase
from *Escherichia coli*, XLII, 124–127
assay, XLII, 124
chromatography, XLII, 125

properties, XLII, 127
purification, XLII, 124–126
D-Glyceric acid, XLVI, 4
Glyceride, *see* Phosphoglyceride; Triglyceride
Glyceroacyltransferase, in *E. coli* auxotrophs, XXXII, 864
Glycerokinase, XXXIV, 246, 252, *see also* Glycerol kinase
Glycerol, XXXIV, 563; XLIV, 157; LVIII, 12
assay of carbamoyl-phosphate synthetase, LI, 112, 113, 122
in Buffer A, LI, 229
in cell freezing preservation, XXXIX, 121
control of microbial contamination, XLIV, 738
as cryoprotectant, XXXII, 36, 37
cytochrome b_5 assay, LII, 110
cytochrome *P*–450 purification, LII, 110, 140, 362, 363
detergent removal, LII, 113
effect on aspartate carbamyltransferase, LI, 48
on rat liver carbamoyl-phosphate synthetase, LI, 118–120
electron spin resonance studies, XLIX, 379
in electrophoresis buffer, LI, 342, 343
electrophoretic analysis of *E. coli* ribosomal proteins, LIX, 538
for enzyme storage, LI, 82, 303, 328, 344, 362, 363, 524; LIII, 500
enzyme storage solution, LIX, 61, 125, 126
in freezing, LVIII, 30, 31
gradient, ATPase purification, LV, 356
gradient properties, XXXI, 508
hem mutants, LVI, 175
high concentration polyacrylamide step gels, LVI, 603
petite mutants, LVI, 155
initiation factor buffer, LIX, 838
lipid extraction of microsome fraction, LII, 98
from lipolysis
in fat pads, XXXV, 607
in perifused fat cells, XXXV, 607–612
in liver preservation, XXXI, 3
lysis, of mycoplasmas, XXXII, 461
magnetic circular dichroism studies, XLIX, 159
in MCD studies, LIV, 293
microsomal disruption, LII, 360
periodate removal, LII, 92
phosphorescence studies, XLIX, 238
preparation of aminoacyl-tRNA synthetase, LIX, 220, 314, 315
of 3′-amino-3′-deoxy-ATP, LIX, 135
of CHO crude extract, LIX, 232
of *E. coli*
crude extract, LIX, 194
ribosomes, LIX, 818
of polysomes, LIX, 363, 368
of ribosomal subunits, LIX, 412
protein solubilization, LII, 93, 118–121, 195, 201, 202
purification of aminoacyl-tRNA synthetase, LIX, 259
of CPSase-ATCase-dihydroorotase complex, LI, 124
of CTP(ATP):tRNA nucleotidyltransferase, LIX, 185
of FGAM synthetase, LI, 197
of *Lactobacillus* deoxynucleoside kinases, LI, 348, 349, 351
of nitrogenase, LIII, 318
of purine nucleoside phosphorylase, LI, 522
of pyrimidine nucleoside phosphorylase, LI, 434
of tRNA methyltransferases, LIX, 194, 195, 197, 198, 200, 201
of RNase III, LIX, 826, 827
of thymidine kinase, LI, 367, 368
of uridine-cytidine kinase, LI, 303, 304
removal, by dialysis, LII, 197
renal gluconeogenesis, XXXIX, 14
restoration of growth on by a rho^0 tester strain, LVI, 102, 103
as stabilizer of cytochrome *P*–450, LII, 203
of reductase, LII, 93–95
of ribonucleoside diphosphate reductase, LI, 242
storage of mitochondria, LV, 31, 32

in storage reagent for enzyme rods, LVII, 204, 205, 207
for luminescent bacteria, LVII, 156
suppression of carbodiimide-resistance by growth, LVI, 171, 172
Glycerol-1′-alkenyl ether, intermediate in plasmalogen synthesis, XXXV, 477, 479–481
sn-Glycerol-3-benzyl ether, intermediate in plasmalogen synthesis, XXXV, 478, 479
Glycerol-1,2-cyclic carbonate, intermediate in plasmalogen synthesis, XXXV, 477
sn-Glycerol-2,3-cyclic carbonate, intermediate in plasmalogen synthesis
via *sn*-glycerol-3-benzyl ether, XXXV, 478, 479
via *sn*-glycerol-3-(2′,2′,2′-trichloroethylcarbonate), XXXV, 479
Glycerol dehydratase, from *Aerobacter aerogenes*, XLII, 315–323
activators and inhibitors, XLII, 321
assay, XLII, 316, 317
molecular weight, XLII, 320
properties, XLII, 320–323
purification, XLII, 317–320
Glycerol dehydrogenase, assay by Fast Analyzer, XXXI, 816
Glycerol density gradient, nuclear aldosterone complex studies, XXXVI, 289–291
Glycerol derivative, *see also* specific types
intermediate in synthesis of phosphatidylcholine analogues, XXXV, 514
partial specific volumes, XLVIII, 22
sn-Glycerol-1,2-diacylate, intermediate in synthesis of phosphatidylethanolamine analogues, XXXV, 510
Glycerol diether, intermediate in phosphatidylcholine synthesis, XXXV, 449–452, 454
DL-Glycerol-1,2-dihexadecyl ether-3-phosphorylcholine, intermediate in phosphatidylcholine synthesis, XXXV, 453
Glycerol dioleate, for lipid monolayer formation, XXXII, 552
Glycerol-ester hydrolase, enzymatic activity, XXXI, 522
Glycerol gradient centrifugation
isolation of DNA-ribosome complexes, LIX, 842, 843, 850
preparation of polysomes, LIX, 367
purification
elongation factors, LX, 675
eukaryotic elongation factor eEF–1, LX, 583
heme-reversible translational inhibitor, LX, 487, 488, 495
initiation factors, LX, 21, 23, 40, 42, 43, 46–48, 52, 55, 56, 106
of 5-methyluridine methyltransferase, LIX, 199
molecular weight determination, LX, 52, 492–494
Qβ ribonucleic acid replicase, LX, 635
ribosome dissociation factor, LX, 292–294
tRNA labeling procedure, LIX, 71, 103
RNase III digestion buffer, LIX, 837
storage solution for *E. coli* polysomes LIX, 355, 356
synthetase solvent, LIX, 288
tritium labeling studies, LIX, 342, 349
L-Glycerol-α-iodohydrin, intermediate in phosphatidic acid synthesis, XXXV, 435, 436
Glycerol-α-iodohydrin diester, intermediate in phosphatidylethanolamine synthesis, XXXV, 457–458
sn-Glycerol-1,2-isopropylidene ketal, *see also* Isopropylidene-D-glycerol
intermediate in phosphatidylcholine synthesis, XXXV, 454
Glycerol kinase
activators and inhibitors, XLII, 156
activity on ATP analogs, XLIV, 876
amino acids, XLII, 154
assay, XLII, 149–151
catalytic properties, XLII, 155
chromatography, XLII, 153
molecular weight, XLII, 154

in parenchyma cells, XXXII, 706
properties, XLII, 154–156
purification, XLII, 151–154
thermodynamics, XLII, 156
Glycerol methacrylate embedding plastic, XXXIX, 140–141
sn-Glycerol-1-monoalkyl-3-trityl diether, intermediate in phosphatidylcholine synthesis, XXXV, 447
Glycerol-α-monoether, intermediate in phosphatidylcholine synthesis, XXXV, 444–449
Glycerol monooleate, in lipid bilayers, XXXII, 516
Glycerol monopalmitate, in lipid bilayers, XXXII, 516
sn-Glycerol-1-*cis,cis*-(1′,9′-octadecadienyl)-ether, intermediate in plasmalogen synthesis, XXXV, 480, 481
sn-Glycerol-1-octadecyl ether, intermediate in phosphatidylcholine synthesis, XXXV, 446, 447
Glycerolphosphatase, in plasma membranes, XXXI, 88
Glycerol 3-phosphate, XXXIV, 4
cell fractionation, LVI, 210–212, 217, 218, 222
α-Glycerolphosphate, ferricyanide, LVI, 232, 233
Glycerol-3-phosphate acyltransferase, *see* Glyceroacyltransferase
Glycerol 3-phosphate dehydrogenase, XXXIV, 4; XLIV, 337, 495, 496
brown adipose tissue mitochondria, LV, 75–77
of chicken breast muscle, XLI, 245–249
assay, XLI, 245, 246
chromatography, XLI, 247
properties, XLI, 248, 249
purification, XLI, 246–248
in *E. coli* auxotrophs, XXXII, 864
flavin determination, LIII, 429
from honey bee, and other insects, XLI, 240–245
amino acids, XLI, 245
assay, XLI, 241
inhibition, XLI, 244
molecular weight, XLI, 244, 245
properties, XLI, 243–245
purification, XLI, 241–243
localization, LVI, 230
in microbodies, LII, 496, 501
in parenchyma cells, XXXII, 706
α-Glycerolphosphate dehydrogenase, associated with cell fractions, XL, 86
L-Glycerol-3-phosphate dehydrogenase
from *Escherichia coli*, XLI, 249–254
assay, XLI, 249, 250
chromatography, XLI, 251, 252
distribution, XLI, 254
inhibition, XLI, 254
properties, XLI, 253, 254
purification, XLI, 250–253
from pig brain mitochondria, XLI, 254–259
assay, XLI, 254, 255
inhibitors, XLI, 258
properties, XLI, 258, 259
purification, XLI, 255–258
α-Glycerolphosphate dehydrogenase:nitrate reductase, anaerobic electron transfer, LVI, 386
sn-Glycerol-3-phosphoric acid, intermediate in phosphatidic acid synthesis, XXXV, 439–441
sn-Glycerol-3-phosphorylcholine, intermediate in phosphatidylcholine synthesis, XXXV, 442
sn-Glycerol-3-(2′,2′,2′-trichloroethylcarbonate)
intermediate in plasmalogen synthesis, XXXV, 479
in synthesis of phosphatidylethanolamine analogues, XXXV, 509
sn-Glycerol-1-trityl ether, intermediate in phosphatidylcholine synthesis, XXXV, 451
Glycerol-water, solvent, in ligand binding studies, LIV, 528
α-Glycerophosphate
assay, LV, 221
measurement of H^+/site ratios, LV, 648
oxidation ubiquinone extraction, LIII, 578
phosphorylating vesicles, LV, 166
β-Glycerophosphate, substrate, of acid nucleotidase, LI, 274

α-Glycerophosphate dehydrogenase, LIII, 405, *see also* Glycerol-3-phosphate dehydrogenase
assay, XLI, 37
effect of extraction methods, LIII, 409
L-α-Glycerophosphate dehydrogenase
in bacterial membrane, XXXI, 649
from beef liver, XLI, 259–264
assay, XLI, 260
chromatography, XLI, 261, 262
properties, XLI, 262–264
purification, XLI, 260–263
α-Glycerophosphate oxidase body, LII, 501
L-α-Glycerophosphorylcholine, intermediate in phosphatidylcholine synthesis, XXXV, 443
Glycerophosphorylserine, intermediate in phosphatidylserine synthesis, XXXV, 463–465
Glyceryl dipalmitate, in lipid bilayers, XXXII, 516
Glyceryl distearate, in lipid bilayers, XXXII, 516
Glyceryl ether phospholipid
methanolysis, XXXV, 423
thin-layer chromatography, XXXV, 423
D-Glycidaldehyde, XLVII, 490
synthesis, XLVI, 387
L-Glycidaldehyde, synthesis, XLVI, 384–387
R,S-Glycidaldehyde, synthesis, XLVI, 383
Glycidol, XLVI, 382
Glycidol phosphate, XLVI, 144, 145, 381–387; XLVII, 425
affinity labeling, XLVII, 483, 497, 498
as inhibitor for triosephosphate isomerase from rabbit muscle, XLI, 453
preparation of, and complex formation with enolase, XLI, 121
structure, XLVII, 483
synthesis, XLVI, 382, 383
D-Glycidol phosphate, synthesis, XLVI, 387
L-Glycidol phosphate, synthesis, XLVI, 384–387
Glycidyl methacrylate, XLIV, 72
Glycinamide, LI, 101
Glycinamidocaproyl-hydrazide derivative, XXXIV, 41, 42
Glycine, XLVI, 94, 104
aggregation, XLIV, 272, 273
alanine carrier, LVI, 435
amidophosphoribosyltransferase assay and, LI, 172
aminoacylation isomeric specificity, LIX, 274, 275, 279, 280
δ-aminolevulinic acid synthetase assay, LII, 351
amounts bound to Spheron gels, XLIV, 73, 74
assay of adenosine monophosphate nucleosidase, LI, 264
of luciferase, LVII, 30
of phosphoribosylglycinamide synthetase, LI, 180
blocking agent, XLIV, 529, 905, 912, 920
cleavage rates, XLVII, 342–344
conformational state, XLIX, 103, 104
cyanogen bromide binding, XLIV, 62
derivative, XLIV, 284
in digestion of tRNA terminal mononucleotides, LIX, 140
effect on rat liver carbamoyl-phosphate synthetase, LI, 119
in electrophoresis buffer, LI, 357
in electrophoresis reagent, XXXII, 85
1-ethyl-3(3-dimethylaminopropyl) carbodiimide binding, XLIV, 62
glutaraldehyde binding, XLIV, 61
imido carbonate deactivation, XLIV, 30
in media, LVIII, 53, 63
peptide synthesis, XLVII, 507, 509
purification of deoxythymidylate phosphohydrolase, LI, 288
of luciferase, LVII, 31, 32
radiolabeled, in phosphoribosylglycinamide synthetase, LI, 181
reduction, energetics, LIII, 374
for removal of glutaraldehyde, LIX, 842
residue, cleavage site, LIX, 105
in tRNA reconstitution, LIX, 140, 141
tRNA synthetase, XXXIV, 170
solid-phase peptide synthesis, XLVII, 610

substituted-Sepharose preparation, LII, 133, 138
substrate of glycine reductase, LIII, 373
of phosphoribosylglycinamide synthetase, LI, 179
thermolysin hydrolysis, XLVII, 175
vinylketo deactivation, XLIV, 36
Glycine-arsenate buffer, assay of luciferase, LVII, 30
Glycine-bicarbonate buffer, for phosphodiesterase assay, XXXI, 93
Glycine buffer
in electrophoretic analysis of *E. coli* ribosomal proteins, LIX, 539, 545, 546
for enzyme assay, XXXII, 125
Glycine ethyl ester, in carboxyl group determination, XLIV, 384
Glycine-hydrochloride buffer, preparation of yeast submitochondrial particles, LIII, 114
Glycine max, lectin, XXXIV, 334
Glycine reductase
activity, LIII, 373, 374, 382
assay, LIII, 373–376
from *Clostridium sticklandii*, LIII, 372–382
purification, LIII, 376–382
selenium-dependent, LIII, 373–382
Glycine-sodium hydroxide buffer, XLVIII, 261, 262, 264
peroxidase-luminol coupled assays, LVII, 405
Glycinin
molecular weight, XLVIII, 96
osmometric studies, XLVIII, 94–96
Glycocalyx
description, XXXI, 123
staining, XXXIX, 143–145
Glycochenodeoxycholate, sodium salt, partial specific volume, XLVIII, 21
Glycocholate
micelles, ionic strength and pH effects, LVI, 747
sodium salt, partial specific volume, XLVIII, 21
Glycodeoxycholate
micelles, ionic strength and pH effects, LVI, 746
sodium salt, partial specific volume, XLVIII, 21
Glycoenzyme, immobilization, via carbohydrate residue activation, XLIV, 317–320
Glycogen, XXXIV, 338
activation with cyanogen bromide, XXXIV, 126
AMP assay, LV, 209
assay, LV, 221
cAMP inhibitor assay, XXXVIII, 279, 280
as electron microscopic tracer, XXXIX, 149
formation, Pasteur effect, LV, 295
gradient separation, XXXI, 735, 745, 746
isolation of mitochondria, LV, 37
in liver, XXXI, 16
phosphate assay, LV, 212
polarography, LVI, 477
storage, substrate entry, XXXIX, 495
synthesis, control studies, XXXVII, 292, 295
Glycogen-agarose, XXXIV, 126, 127
spacers, XXXIV, 126, 127
Glycogenolysis
control studies, XXXVII, 292, 294
in perfused liver, XXXIX, 33, 34
in situ perfusion method, XXXIX, 67–73
Glycogen phosphorylase, XXXIV, 131–134; XLIV, 501, 887
as ovarian viability criterion, XXXIX, 235, 236
Glycogen phosphorylase *b,* from rabbit muscle, reductive alkylation, XLVII, 474, 475
Glycogen staining
with thiosemicarbazide, XXXIX, 152
in transmission electron microscopy, XXXIX, 151, 152
Glycogen synthetase, XXXIV, 131, 132, 164, 338
control studies, XXXVII, 293
insulin effects, XXXIX, 99
protein kinase, XXXVIII, 328, 351
from swine kidney, XLII, 375–381
assay, XLII, 375–377
molecular weight, XLII, 380
properties, XLII, 380

purification, XLII, 377–381
testosterone activation, XXXIX, 98
Glycolaldehyde, acetals, L, 27
Glycolaldehyde phosphate, XLVI, 20
Glycolate oxidase, XXXIV, 302; LII, 18
in glyoxysomes, XXXI, 565, 569
as marker enzyme, XXXI, 735, 742
in peroxisomes, XXXI, 501
from pig liver, XLI, 337–343
assay, XLI, 338
chromatography, XLI, 339, 340
inhibitors and activators, XLI, 342
properties, XLI, 340–343
purification, XLI, 338–341
Glycol ether, XLIV, 117
Glycolic acid, XLVI, 4
inhibitor of succinate dehydrogenase, LIII, 33
Glycolic acid oxidase, *see* Glycolate oxidase
Glycolipid, XXXIV, 85, *see also* Glycosphingolipid
anomeric linkages, XXXII, 363, 364
carbohydrate components, XXXII, 358–361
carbohydrate sequence, XXXII, 363, 364
characterization, XXXII, 348
fatty acid determination, XXXII, 361, 362
glycosyl linkages, XXXII, 364
identification, XXXII, 358
isolation, XXXII, 346
neutral, resolution into components, XXXII, 355, 356
NMR spectroscopy, XXXII, 366, 367
physical characterization, XXXII, 365–367
sphingosine bases, XXXII, 361, 362
thin-layer chromatography, XXXII, 356–358
Glycolipid acid
complex with agarose, L, 140
with glass beads, L, 139, 140
with methylated bovine serum albumin, L, 140
preparation, L, 139
N-Glycolylneuraminic acid
N-acylneuraminate-9(7)-*O*-acetyltransferase, L, 381
crystallization, L, 70
determination, XXXII, 361
gas-liquid chromatography, L, 376
thin layer chromatography, L, 376
Glycolysis
in cell characterization, XXXI, 23
in cells, XXXII, 801
control studies, XXXVII, 292, 295
in intestinal mucosa isolates, XXXII, 668, 669
Pasteur effect, LV, 290, 291
pyruvate kinase, XLII, 182
rate, in perfused liver, XXXV, 606
Glycolytic enzyme, active site studies, XLVII, 479–498
Glycolytic intermediate, in isolated renal tubules, XXXIX, 16
Glyconolactone, XLVI, 21
Glycopeptide, XXXIV, 3, 332–341, *see also* specific glycopeptide
molecular weight determination, XLVII, 62, 65
Glycopeptide hormone, radioimmunoassay, XXXVII, 25
Glycophorin, *N*-acetylneuraminic acid, L, 53
Glycoprotein, XXXIV, 6, 610, 688–691; L, 273–279, *see also* Asialoglycoprotein receptor; specific types
acid hydrolysis, L, 52
analysis of monosaccharides, XLI, 19, 21
biosynthesis
2-deoxy-D-glucose, L, 188
fluoro-sugar analogs, L, 188
density gradient centrifugation, XXXVI, 165
desorption, XXXIV, 341
galactose oxidase tritiation, L, 204–206
gel electrophoresis of tritiated, L, 205, 206
group-specific separation, XXXIV, 331–341, *see also* specific glycoprotein
hydrazido-agarose derivatives, XXXIV, 85
identification in SDS-polyacrylamide gels specific lectin binding, L, 54–64

immobilization, XLIV, 320
insulin receptors, XXXIV, 668
lectin
covalent coupling, XXXIV, 337–339
polymerization, XXXIV, 335, 337, 338
precipitation, XXXIV, 335–337
luciferases, LIII, 562
in membranes
gel electrophoresis of, XXXII, 86–88
molecular weight determination by gel electrophoresis, XXXII, 92–102
molecular weight, by gel electrophoresis, XXXVI, 99
purification, XXXIV, 335–338
receptor, XXXIV, 221
sodium borotritide reduction, L, 205
steroid-binding, XXXVI, 91–104
isoelectric focusing, XXXVI, 98–102
sedimentation coefficients, XXXVI, 102–104
α_1-Glycoprotein, gel electrophoresis, XXXII, 101
β-Glycoprotein, XXXIV, 731
Glycoprotein hormone
desialylated, effect on activity, XXXVII, 321, 322
human, purification, XXXVII, 380–389
sialylated, tritium labeling, XXXVII, 321–326
Glycosaminoglycan, LVIII, 267, 268
differentiation, LVIII, 268, 269
preparation, LVIII, 274
types, L, 279, 280
Glycosidase, XXXIV, 4; XLVI, 22
carbohydrate structure, L, 94
isolation, L, 237
mammalian, L, 488–494
Glycoside, XXXIV, 4
as detoxification mechanism, XXXI, 529
oligosaccharide synthesis, L, 95
preparation from alcohols, L, 97
β-Glycoside, XXXIV, 334
β-L-Glycoside, XXXIV, 364, 365
D-Glycoside, XXXIV, 336
O-Glycoside, XXXIV, 318–323, *see also* specific glycoside
acetobromo derivatives, XXXIV, 321–323
Glycosphingolipid, XXXII, 345–367; L, 236–250, *see also* Carbohydrate, complex; Glycolipid
acetylation, XXXV, 419–421, 424; L, 209–211
analysis of tritiated, L, 206
borate complexes, XXXV, 424
characterization, L, 237
column chromatography, XXXV, 409–411, 417–421, 423
galactose oxidase tritiation, L, 204–206
general structure, L, 236
Iatrobead column chromatography, L, 241
indicators, on thin-layer chromatography, XXXV, 403–407, 412, 413, 415–417
infrared spectroscopy, XXXII, 366
ion-exchange chromatography, XXXV, 419
isolation, L, 237
flowsheet, XXXII, 346
quantitative, XXXII, 347, 349, 350
long-chain neutral glycolipid isolation, L, 241
neutral
acetylation, L, 139
fractionation, L, 208, 209
myelin formation, L, 247
structures, L, 242, 243
nomenclature, L, 236, 237
NMR spectroscopy, XXXII, 366, 367
optical rotatory studies, XXXII, 365
ozonolysis, L, 137
physical characterization, XXXII, 365–367
separation, L, 237
silicic acid column chromatography, L, 241
thin-layer chromatography, XXXV, 396–425
quantitative analysis, XXXV, 413, 416
types, L, 273

Glycosylceramide, *see also* Glycosphingolipid
column chromatography, XXXV, 416, 417, 421
thin-layer chromatography, XXXII, 356; XXXV, 396, 397, 400, 402, 403, 407, 416
Glycosyl halide
anomerization, L, 102, 103
inversion, L, 101, 102
Koenigs-Knorr reaction, L, 96–100
stabilization, L, 101
substituent effects, L, 103, 104
O-Glycosyl polyacrylamide
affinity chromatography, XXXIV, 366, 367
copolymers, XXXIV, 366
electrophoresis, XXXIV, 178–181
phytohemagglutinins, XXXIV, 361–367
preparation, XXXIV, 366
properties, XXXIV, 366
Glycosyltransferase
of bacterial cell envelopes, reconstitution, XXXII, 449–459
effect on cell glycosphingolipids, XXXII, 347
galactose binding protein, L, 288
in plasma membranes, XXXI, 102
Glycylamphenicol hydrochloride, XLVI, 704, 705
Glycylglycine, XLIII, 161
purification of firefly luciferase, LVII, 7, 31
Glycylglycine buffer
assay of ATP, LVII, 11
of cyclic nucleotide phosphodiesterases, LVII, 96, 97
of modulator protein, LVII, 109
of photophosphorylation, LVII, 55
Glycylglycyl-(*O*-benzyl)-L-tyrosylarginine, XXXIV, 5
Glycylglycyltyrosine derivative, XXXIV, 49
Glycyl-D-phenylalanine, XXXIV, 4
Glycyl-D-phenylalanine-agarose, XXXIV, 435
Glycylphenylalanyl-β-naphthylamide, substrate, of dipeptidyl aminopeptidase I, XLVII, 393
Glycyl-tRNA synthetase, subcellular distribution, LIX, 233, 234
Glycyl-L-tyrosine, XXXIV, 395
substrate, carboxypeptidase, Aα, XLIV, 557
Glyoxal, XLIV, 248; XLVI, 64
Glyoxalic acid, inhibitor of succinate dehydrogenase, LIII, 33
Glyoxylate cycle, LII, 501, 502
Glyoxylate oxidase, LII, 18, 19; LVI, 474
Glyoxylate reductase
in microbodies, LII, 496, 499
from *Pseudomonas*, XLI, 343–348
assay, XLI, 343
chromatography, XLI, 345
properties, XLI, 347, 348
purification, XLI, 344–347
Glyoxysome, LII, 494, 501, 502
enzymes, XXXI, 569, 743
in plant cells, XXXI, 496, 590
extraction, XXXI, 541, 552, 553, 565–571
function, XXXI, 565
large-scale isolation, XXXI, 570, 571
20-GM80 hollow-fiber device, XLIV, 293, 310
hydraulic permeability, XLIV, 298
GMP, *see* Guanosine 5′-monophosphate
GMP kinase, *see* Guanylate kinase
GMPPCP, XLVI, 660
GMPPNP, XLVI, 194, 660
GMP reductase, XLVI, 301
GMP synthetase, XLVI, 420; LI, 213–224
activity, LI, 213
alternate names, LI, 219
amino donor, LI, 213, 217
assay, LI, 213, 214
diffusion coefficient, LI, 218
from Ehrlich ascites cells, LI, 219–224
from *Escherichia coli*, LI, 213–218
extinction coefficient, LI, 217
inhibitors, LI, 217, 223, 224
kinetic properties, LI, 217, 223, 224
molecular weight, LI, 218, 223
partial specific volume, LI, 218
pH optimum, LI, 223
properties, LI, 216–218
purification, LI, 214–216

purity, LI, 216, 217
sedimentation coefficient, LI, 218
storage, LI, 223
substrate specificity, LI, 224
subunit structure, LI, 218
GMT buffer, formulation, LI, 86
Gnorimoschema operculella, media, LVIII, 462
Goat
erythrocyte, phosphoglycerate mutase, XLII, 438, 445
mammary gland, phosphofructokinase, purification, XLII, 110
Goblet cell, in small intestine, XXXII, 670
Gold
atomic emission studies, LIV, 455
in carbon coating, XXXIX, 155, 156
latensification, XL, 10
in negative staining studies, XXXII, 35
Goldman equation, diffusion potentials, LV, 559
Goldman-Hodgkin-Katz equation, XXXII, 536
membrane potential, LV, 692
Golgi apparatus
enzyme marker, XXXII, 393, 402, 403
α-mannosidase from membrane, XXXII, 83
membrane enzymes, XXXII, 86, 87
5′-nucleotidase, XXXII, 368
Golgi complex, XXXI, 102
characterization, XXXI, 20, 23–25, 31, 33, 183–184
diagnostic enzyme, XXXI, 20
differential separation, XXXI, 727, 730–732
electron microscopy, XXXI, 26, 30, 31
electrophoretic purification, XXXI, 754
lipids, XXXI, 29
from liver cells, XXXI, 180–191
of pancreas, XXXI, 41, 42
in plant cells, *see* Dictyosome
sensitivity, XXXI, 29, 31
Gonad, X-ray effects, XXXII, 215
Gonadal receptor
for LH and hCG, XXXVII, 167–193
assay of binding, XXXVII, 182–184
physical properties, XXXVII, 184, 185
Gonadotropin
antiprogestational assay, XXXVI, 458
binding studies, XXXIX, 256
chorionic, XXXIV, 6; LVIII, 184
cultured cells dependent, XXXII, 557, 558
effect on cellular respiration, XXXIX, 424
on granulosa cells, XXXIX, 423
on interstitial cell function *in vitro,* XXXIX, 252–271
enzyme regulation, XXXIX, 237
gonadal receptors, XXXVII, 167–193
human chorionic, XXXIV, 339
autoradiography, XXXVII, 165
binding to granulosa cells, XXXIX, 223, 224, 230
effect on estradiol hydroxylation, XXXIX, 251
on testicular metabolism and secretion, XXXIX, 252–271, 280, 281
gel electrophoresis, XXXII, 101
gonadal receptors, XXXVII, 167–193
binding studies, XXXVII, 174–181
extraction, XXXVII, 181–193
structural aspects, XXXVII, 178–181
iodination, XXXVII, 169, 170; XXXIX, 223, 224
labeling, XXXVII, 168–174
specific activity, XXXVII, 173, 174
logit-log assay, XXXVII, 9
purification, XXXVII, 170–173
radioligand assay, XXXVII, 176, 178
testis binding, XXXIX, 269–271
interstitial cell binding, XXXIX, 268–271
secretion by cell hybrids, XXXIX, 127
by cloned pituitary cells, XXXIX, 128–132

Gonadotropin releasing factor, solvent extraction, XXXVII, 404
Gonadotropin-releasing hormone, antisera, XXXVI, 17
Gonyaulax
  luciferase, LIII, 560
    structure, LVII, 258
  luciferin-binding protein, LVII, 258
Goodwin IPL–52 medium, LVIII, 460
  preparation, LVIII, 460, 461
Gordon arc, LIV, 456, 459
Gossypol, as extraction interferent, XXXI, 537, 540
G:O transaminase, *see* Aspartate aminotransferase
Gougerotin, synthesis of *N*-(3-carbethoxy-3-diazo)acetonylgougerotin, LIX, 814
Gout, treatment, XLIV, 704
Gouy-Chapman equation, ionic strength, LVI, 525
Gouy-Chapman potential, expression, LVI, 523, 524
Gouy force technique, LIV, 387, 388, 391
GpApUpU
  inhibition of phenylalanyl-transfer ribonucleic acid binding to 70 S ribosomes, LX, 727, 728
  synthesis, LX, 618
GPC Maltrin syrup, XLIV, 786, 787
G_1 phase
  cell preparation, XXXII, 592, 594, 595
  in measurement of termination point, XL, 50, 52
G_2 phase
  cell preparation, XXXII, 592, 595, 596
  in measurement of termination point, XL, 51, 54
GpUpUpU*, XLVI, 623
Grace medium, LVIII, 457–460
  preparation, LVIII, 457–460
Gracilis muscle, *in situ* perfusion method, XXXIX, 67–73
Gradient, *see also* specific type
  nonaqueous, composition, LVI, 203
  for plant-cell isolation, XXXI, 507–509
  properties, XXXI, 508
Gradient centrifugation, LVIII, 256–258
Gradient elution, XXXIV, 237–239
Gradient gel, preparation, LVI, 70, 71
Gradient mixer, XXXIX, 368
Grain size, effect on binding ability of support, XLIV, 108
Gramicidin
  amino acid transport, LVI, 406
  bacteriorhodopsin photovoltage, LV, 754, 755
  chromatophore planar membranes, LV, 756
  countercurrent distribution, XLIII, 324–327
  effects on membranes, XXXII, 882
  M. phlei ETP, LV, 186
  oxidative phosphorylation, LV, 451
  photovoltage, LV, 592, 612
  polymyxins compared, XLIII, 584
  solvent systems, XLIII, 336
Gramicidin A
  effect on bilayers, XXXII, 500, 552
  as ionophore, LV, 440–442
Gramicidin S
  biosynthesis, XLIII, 567, *see also* Gramicidin S synthetase
  molecular weight determination, XLIII, 328–330
  solvent system for countercurrent distribution, XLIII, 336
  spin-spin distance, XLIX, 424
  structure, XLIII, 567
  thin-layer chromatography, XLIII, 208, 209
Gramicidin S synthetase, XLIII, 567–579
  affinity chromatography, XLIII, 575
  amino acid-dependent ATP-[^{14}C]AMP exchange, XLIII, 569, 570
  amino acid-dependent ATP-$^{32}PP_i$ exchange, XLIII, 569, 576
  aminosulfate precipitation, XLIII, 573
  assay, XLIII, 568–571
  Bacillus brevis preparation, XLIII, 571, 572
  crude extract preparation, XLIII, 573
  DEAE Sephadex A–50 chromatography, XLIII, 573, 574
  heavy enzymes
    physical properties, XLIII, 577
    specificity, XLIII, 577
  incubation mixture, XLIII, 568, 569

inhibitors, XLIII, 578
kinetic constants, XLIII, 578
light enzyme, properties, XLIII, 577, 578
molecular weight, XLIII, 577
4′-phosphopantetheine, XLIII, 577
properties, XLIII, 576–579
purification, XLIII, 571–576
purity, XLIII, 576
racemization of phenylalanine, XLIII, 571
reaction scheme, XLIII, 568
separation of light and heavy enzymes, XLIII, 575
Sephadex G–200 chromatography, XLIII, 573, 574
specificity, XLIII, 577
stability, XLIII, 577
streptomycin sulfate precipitation, XLIII, 573
thio ester-bound amino acids, XLIII, 570
thiotemplate mechanism, XLIII, 568
Granulation
disadvantages, XLIV, 176, 191
procedure
by chopping, XLIV, 174
with syringe, XLIV, 174
using homogenizer, XLIV, 186
using 30-mesh nylon net, XLIV, 186
using metal sieve, XLIV, 198
Granule, hepatic, morphometry, XXXII, 8, 15
Granulocyte
countercurrent separation, XXXII, 637, 646
DNA content, LVIII, 148
growth factor, LVIII, 163
Granulocyte chemiluminescence assay
in studies of granulocyte phagocytic defects, LVII, 488, 489
of opsonization, LVII, 485–487
Granulomatous disease, chronic, phagocytic chemiluminescence, LVII, 465, 477, 481, 483, 488, 489
Granulosa cell
differentiation
hormonal stimulation, XXXIX, 227, 228
in vitro, XXXIX, 183–230
estrogen secretion, XXXIX, 219
human chorionic gonadotropin binding, XXXIX, 223, 224
isolation from ovarian follicle, XXXIX, 188–199
luteinization
hormonal control, XXXIX, 227, 228
inhibition, XXXIX, 228, 229
luteinized, morphology, XXXIX, 217
morphological assay, XXXIX, 211–216
succinoxidase activity, XXXIX, 423
tissue culture, XXXIX, 199–210
short-term, XXXIX, 210, 211
Graphite furnace atomization cell, LIV, 462, 464
Grass, tropical, pyruvate orthophosphate dikinase in leaves, XLII, 187, 199, 201
Grassfrog, *see Rana pipiens*
Grass stimulator apparatus, XXXIX, 62
Grating
holographic, LIV, 461
monochromator, XLIX, 88
Green-fluorescent protein
of *Aequorea,* isomeric forms, LVII, 287
assay of *in vitro* energy transfer, LVII, 260, 261
comparative structures, LVII, 287
copurification with aequorin, LVII, 308, 309
with luciferase, LVII, 265
effect on luciferase-luciferin reaction, LVII, 240, 241
fluorescence properties, LVII, 240, 260
interaction with luciferase, LVII, 263–267
after side chain modification, LVII, 264, 265
complex formation, LVII, 265–267
species specificity, LVII, 263, 264
isoelectric point, LVII, 308
molecular weight, LVII, 240, 260
nonradiative energy transfer, LVII, 287, 288
of *Renilla reniformis*, LVII, 240, 258, 260, 261, 287

Greening, of *Chlamydomonas reinhardi* cells, XXXII, 866
Green tissue, *see* Plant tissue
Grid, for negative staining, XXXII, 24
Grid staining, for transmission electron microscopy, XXXIX, 142, 143
Griffonia simplicifolia lectin, XXXIV, 333
Grignard reagent, deacylation of triglycerides, XXXV, 320, 322–324
Grinding
  for plant-cell rupture, XXXI, 502
  for subcellular fractionation, XXXI, 737
Grind mill, disruption of *N. crassa* cells, LV, 145, 146
Grisein, XLIII, 336
Griseofulvin
  chiroptical studies, XLIII, 352
  chromatographic date, XLIII, 163
  gas-liquid chromatography, XLIII, 232–234
  high-pressure liquid chromatography, XLIII, 307, 308, 319
  paper chromatography, XLIII, 161
  Penicillium media, XLIII, 15
  thin-layer chromatography, XLIII, 200
    in pharmacokinetic studies, XLIII, 212, 213
Griseolutein, XLIII, 336
Grisorixin, LV, 446
Ground-glass diffuser, LIV, 5
Groundnut chymotrypsin inhibitor, *see* Chymotrypsin inhibitor, groundnut
Groundnut trypsin inhibitor, *see* Trypsin inhibitor, groundnut
Ground-state depletion signal, LIV, 49
Group C anticarbohydrate antibody, XXXIV, 712
Group density, XXXIV, 52
Growth
  on glycerol, restoration by a rho^0 tester strain, LVI, 102–104
  rate and yield, of temperature-sensitive mutants, LVI, 137
  yield, *E. coli* mutants, LVI, 108, 109
Growth factor, for cell types, LVIII, 163
Growth hormone, *see also* specific types
  effects, XL, 320
    on levator ani muscle, XXXIX, 99
  heterogeneity, XXXVII, 40, 41
  immunological assay, XXXVII, 139
  in media, LVIII, 107
  from ovine pituitary, fragment, synthesis, XLVII, 589, 607
  pituitary hormone cell, LVIII, 527
  preparation, XXXI, 419; XXXVII, 376–378
  production, LVIII, 183
  radioimmunoassay, XXXVII, 34
  release, cAMP derivatives, XXXVIII, 401
  secretory granule use in study, XXXI, 410, 418
  unit of activity, XXXVII, 362
Growth hormone cell, LVIII, 528, 529, *see also* 96
  characteristics, LVIII, 529, 532–534
  culture, LVIII, 530
  growth rates, LVIII, 532
  hormone, effect, LVIII, 98–100
  hormone-producing, LVIII, 532–534
  karyotype, LVIII, 532
  media, LVIII, 530
  in somatic cell hybids, LVIII, 534
  stability, LVIII, 532
  subculture, LVIII, 532
Growth hormone release inhibiting factor, *see* Somatostatin
Growth kinetics, LVIII, 133, 214–217, 241, 245
Growth medium
  axenic, for amoeba, composition, XXXI, 687
  for human cell culture, XXXII, 802
Growth parameter, LVIII, 214–217
Growth phase, LVIII, 214
Growth promotion test, LVIII, 20, 21
Grundschlitten microtome apparatus, XXXVII, 150
GSH, *see* Glutathione
GTP, *see* Guanosine 5′-triphosphate
Guanidine
  activator of NADH dehydrogenase, LIII, 20
  effect of *Pholas* luciferin fluorescence, LVII, 392
  extraction of chick bone collagen, XL, 340

Guanidine hydrochloride, XLVII, 113, 137, 143, 147, 150, 322
aldolase dissociation, XLIV, 494
dissociation of F_1 ATPase subunits, LV, 354, 355
of TF_1, LV, 784
inhibitor of GMP synthetase, LI, 217
reconstitution studies with D-lactate dehydrogenase, LIII, 526
reversible dissociation of flavoproteins, LIII, 433, 434
separation of cytochromes, LIII, 85
of elongation factors EF-1β and EF-1γ, LX, 672, 673
Guanidine thiocyanic acid, purification of cytochrome oxidase subunits, LIII, 77, 78
Guanidinium chloride, XLIV, 370
dissociation, of lactate dehydrogenase, XLIV, 514
Guanidinium hydrochloride, XLVIII, 95
effect on osmometer membranes, XLVIII, 138, 139
on sedimentation equilibrium data, XLVIII, 175, 176
partial specific volumes, XLVIII, 42–44
protein volume changes, XLVIII, 38–42
solvent, in osmometric studies, XLVIII, 73, 75, 94
viscosity measurements, XLVIII, 53, 55
1,D-1-Guanidino-3-amino-1,3-dideoxy-*scyllo*-inositol, XLIII, 432
1,D-1-[^{14}C]Guanidino-3-amino-1,3-dideoxy-*scyllo*-inositol, XLIII, 463
1,L-1-[^{14}C]Guanidino-3-amino-1,3-dideoxy-*scyllo*-inositol, XLIII, 465
1,D-1-Guanidino-3-amino-1,3-dideoxy-*scyllo*-inositol 6-phosphate, XLIII, 432
L-[Guanidino-^{14}C]arginine, XLIII, 447
ATP:inosamine phosphotransferase assay, XLIII, 447
inosamine-phosphate pamidinotransferase assay, XLIII, 452–454
Guanidinodeoxy-*scyllo*-inositol, XLIII, 432
2-Guanidino-1-deoxy-*neo*-inositol, XLIII, 432
1-Guanidino-1-deoxy-*scyllo*-inositol 4-phosphate, XLIII, 432
1-[^{14}C]Guanidino-1-deoxy-*scyllo*-inositol 4-phosphate, XLIII, 460
2-Guanidino-2-deoxy-*neo*-inositol 5-phosphate, XLIII, 432
1-Guanidino-1-deoxy-*scyllo*-inositol 4-phosphate phosphohydrolase, XLIII, 459–461
assay, XLIII, 459–461
biological distribution, XLIII, 461
inhibitors, XLIII, 461
1,D-1-Guanidino-1-deoxy-3-keto-*scyllo*-inositol, XLIII, 432, 463
Guanidinoethylcellulose, XXXI, 538
Guanidino group, XXXIV, 103
Guanidium hydrochloride, preparation of radiolabeled ribosomal protein, LIX, 777, 783, 793
Guanine
absorption maximum, LI, 525
chromatographic separation, LI, 220, 551
electrophoretic mobility, LI, 551
hydrogen exchange, XLIX, 32
in media, LVIII, 67
molar extinction coefficient, LI, 552
product/substrate, of purine nucleoside phosphorylase, LI, 537
radiolabeled
determination, LI, 220
standardization, LI, 552
substrate of guanine deaminase, LI, 512, 516
of guanine phosphoribosyltransferase, LI, 549
of hypoxanthine phosphoribosyltransferase, LI, 543, 548
Guanine aminohydrolase, *see* Guanine deaminase
Guanine deaminase, XXXIV, 523–527; LI, 512–517
activity, LI, 512
assay, XXXIV, 526; LI, 512, 513
inhibitors, LI, 516
kinetic properties, LI, 515, 516
molecular weight, LI, 517
pH optimum, LI, 515, 516

purification, LI, 513–515
purity, LI, 515
from rabbit liver, LI, 512
reaction mechanism, LI, 517
stability, LI, 516, 517
Guanine nucleotide analog, XLVI, 289
Guanine phosphoribosyltransferase, LI, 549–558
assay, LI, 550–553
cation requirement, LI, 557
cellular localization, LI, 557
from enteric bacteria, LI, 549–558
inhibitors, LI, 557, 558
kinetic properties, LI, 557
physiological function, LI, 558
purification, LI, 553–555, 558
purity, LI, 555
reaction stoichiometry, LI, 555
regulation, LI, 550
stability, LI, 555
substrate specificity, LI, 549, 550, 557
Guanosine
absorption maximum, LI, 525
chromatographic mobility, LIX, 73
chromatographic separation, LIX, 188
inhibitor of adenosine deaminase, LI, 507
of cytosine deaminase, LI, 400
NMR spectrum, LIX, 35, 42
optical spectra, XLIX, 165
substrate of nucleoside phosphotransferase, LI, 391
of purine nucleoside phosphorylase, LI, 521, 529, 537
Guanosine-2′,3′-cyclic phosphate hydrolase, hydrolysis rate, XXXII, 128
Guanosine 5′-diphosphate
activator of thymidine kinase, LI, 358
assay for elongation factor EF-G, LX, 595–597, 607–610
binding
of elongation factor eEF–1, LX, 583, 584
of elongation factor EF–Tu, LX, 585–594, 636
initiation factor eIF–2, LX, 272, 273, 396, 583, 584
binding sites, XLVI, 649–656
brown adipose tissue mitochondria, LV, 16, 71–73
Co(III) complex, XLVI, 313
complex with elongation factor EF–G and 70 S ribosomes, LX, 729
with elongation factor EF–1α, LX, 663
determination, LVII, 85
effect on orotidylate decarboxylase, LI, 79
enzymic conversion to GTP, LVII, 91
inhibitor of adenylosuccinate synthetase, LI, 212
of biotin carboxylase, XXXV, 31
of cytosine deaminase, LI, 400
labeled, cGMP assay, XXXVIII, 106–112
as ligands, XXXIV, 482
photochemically active derivative, LVI, 656
preparation, XXXVIII, 86, 87
product of adenylosuccinate synthetase, LI, 207
of guanylate kinase, LI, 473, 483
ratio to guanosine triphosphate, LX, 578, 586, 589
regulation of initiation and elongation complex formation, LX, 589
removal from EF Tu, XXXVIII, 86
ribosome complex, XLVI, 633
stimulation, XXXVII, 248
substrate of nucleoside diphosphokinase, LI, 375
of pyruvate kinase, LVII, 82
of ribonucleoside diphosphate reductase, LI, 235, 236, 245
Guanosine diphosphate fucose
animal cell pools, L, 179
intracellular pool quantitation, L, 191–198
Guanosine diphosphate mannose
animal cell pools, L, 179
dolichol diphosphate-*N*-acetylglucosaminyl-*N*-acetylglucosaminyl-mannose, L, 422
dolichol intermediates, L, 418–426
Guanosine 3′,5′-monophosphate, *see* Cyclic guanosine monophosphate
Guanosine 5′-monophosphate
cyclic, LVI, 656

determination, LVII, 85–94
electrophoretic mobility, LIX, 64
enzymic conversion to GTP, LVII, 91
estimation, XXXVII, 248
M value range, LIX, 89
Guanosine 5′-monophosphate-agarose, XXXIV, 477
Guanosine monophosphate analog, XLVI, 299
Guanosine monophosphate kinase, cGMP assay, XXXVIII, 73, 78, 80, 85, 89, 106, 109–111
Guanosine nucleotide, *see also* specific types
assay for elongation factor EF–2, LX, 705, 706
binding to elongation factor EF–G, LX, 737–745
determinations
data analysis, LVII, 92
principles, LVII, 86
procedures, LVII, 88–92
sensitivity, LVII, 92
sources of error, LVII, 93, 94
extraction methods, LVII, 88
free vs. ribosome-bound EF–G, LX, 738–741
site in elongation factor EF–G, LX, 742–745
standard solutions, LVII, 89
Guanosine 2′(3′)-phosphate, electrophoretic separation, LIX, 177
Guanosine 5′-phosphate, *see* Guanosine 5′-monophosphate
Guanosinetriphosphatase, XLVI, 633, 659; LX, 117, 118, 705, 708, 710, 732, 740, 741, *see also* Guanosine triphosphate, hydrolysis
Guanosine 5′-triphosphate, XXXIV, 4; XLIII, 515; XLVI, 194, 658
activator of cytidine triphosphate synthetase, LI, 83, 84
of ribonucleoside diphosphate reductase, LI, 238, 245
of thymidine kinase, LI, 358
allosteric effector, of glutamate dehydrogenase, XLIV, 507, 512, 513
analog, XLVI, 633
γ-phosphate, XLVI, 658, 659
γ-phosphate-photoactivated reactivity with guanosinetriphosphatase, LX, 732
specificity of binding in ternary complex, LX, 731
structure, LX, 729
use to study binding to elongation factor EF-G, LX, 733–735, 738–741
photoactivated, XLVI, 656–660
ribose, XLVI, 656–660
ribose-photoactivated
specificity of binding in ternary complex, LX, 732
structure, LX, 729, 731
use to study binding to elongation factor EF-G, LX, 733, 734, 736–739, 742–745
assay, LV, 208, 221; LVII, 85–94
of cytidine triphosphate synthetase, LI, 80
data analysis, LVII, 91, 92
for elongation factor EF–G, LX, 595–597
for initiation factor
eIF–2, LX, 19, 32, 36, 37, 46, 58, 118, 119, 126, 144, 168, 169, 183, 184, 190, 203, 241, 251, 266, 267, 309, 381, 382, 468, 469
IF, LX, 5
for methionyl-transfer ribonucleic acid binding, LX, 256–265
of modulator protein, LVII, 109
of polypeptide chain elongation, LIX, 358
of polypeptide synthesis, LIX, 367, 459
procedure, LVII, 89–91
sources of error, LVII, 93, 94
for translocation, LIX, 359
ATP-P_1 exchange preparation, LV, 361
binding of elongation factor eEF-1 and eIF-2, LX, 583, 584
brown adipose tissue mitochondria, LV, 71
cell-free protein synthesis system, LIX, 300, 448, 459, 850

complex with elongation factor EF-2 and 70 S ribosomes, LX, 726–741
with elongation factors and aminoacyl-transfer ribonucleic acid, LX, 615, 616, 663, 698
complex V, LV, 314
F_1 ATPase, LV, 328
functional role, LX, 578, 726, 727, 761, 762
hydrolysis, LX, 61, 564, 574, 578, 579, 614, 616, 617, 684, 698, 708
EF-G plus ribosome-dependent, assay, LIX, 360–362, 460
ribosomal proteins, LIX, 823, 824
inhibitor of adenylate deaminase, LI, 496, 502
of aspartate carbamyltransferase, LI, 57
of cytidine deaminase, LI, 400
of cytosine deaminase, LI, 400
initiation factors, LX, 14, 15, 245, 246, 272, 273, 573
labeled
determination of specific activity, LX, 513–516
impurities, XXXVIII, 45, 46
preparation, LX, 497, 498, 512, 513
use in lysate phosphoprotein synthesis, LX, 511–516
phosphate donor, of thymidine kinase, LI, 358
for uridine-cytidine kinase, LI, 306, 308, 309, 313, 320
photoactivated analogs, *see* Guanosine 5′-triphosphate, analog
photoanalogs, XLVI, 659, 660
photochemically active derivative, LVI, 656
preparation of *E. coli* crude extract, LIX, 297
of ribosomal subunits, LIX, 404
of ribosome-DNA complexes, LIX, 841
in protein synthesis, XXXVII, 247, 248
ratio to guanosine diphosphate, LX, 578, 586, 589
regulation of initiation and elongation complex formation, LX, 589
regenerating systems, LX, 74, 582
requirement for binding of elongation factor EF–G to ribosomes, LX, 725–730
for polyphenylalanine synthesis, LX, 683, 684, 761–779
ribosome release, LIX, 862
role in cytidine triphosphate synthetase substrate binding, LI, 83, 90
separation, LV, 287, 288
site of binding on elongation factor EF–2, LX, 733–737, 739–745
substrate of adenylosuccinate synthetase, LI, 207
for biotin carboxylase, XXXV, 31
of luciferase, LVII, 82
of nucleoside triphosphate pyrophosphohydrolase, LI, 279
of phosphoribosylpyrophosphate synthetase, LI, 10
in succinyl-CoA generating system, LII, 352
ternary complex with initiation factor eIF-2 and methionyl-transfer ribonucleic acid, LX, 460, 461, 562–566, 570, 585, 586
utilization by casein kinases, LX, 503
in *in vitro* labeling of ribosomal proteins, LIX, 517, 527
[γ-^{32}P]Guanosine triphosphate, LIX, 819
assay of GTP hydrolysis, LIX, 361
Guanosine 5′-triphosphate-agarose, XXXIV, 477
Guanosine triphosphate-8-formylhydrolase, XLIII, 515–520
activators, XLIII, 517
ammonium sulfate fractionation, XLIII, 517
assay, XLIII, 516
cell-free extracts, XLIII, 516
DE–52 cellulose column chromatography, XLIII, 517
dissociation, XLIII, 519
inhibitors, XLIII, 517

Michaelis constant, XLIII, 519
molecular weight, XLIII, 519
purification, XLIII, 516–518
Sephadex G–200 column chromatography, XLIII, 517
specific activity, XLIII, 516
specificity, XLIII, 517
Streptomyces rimosus preparation, XLIII, 516
unit definition, XLIII, 516

Guanylate
chromatographic separation, LI, 550, 551
effect on orotidylate decarboxylase, LI, 79
electrophoretic mobility, LI, 551
extinction coefficient, LI, 220
inhibitor of cytosine deaminase, LI, 400
of orotate phosphoribosyltransferase, LI, 163
of orotidylate decarboxylase, LI, 163
product of GMP synthetase, LI, 213, 219
of hypoxanthine phosphoribosyltransferase, LI, 543
spectrophotometric determination, LI, 219, 220
substrate of guanylate kinase, LI, 473, 483, 488, 489

Guanylate cyclase
assay, general principles
determination of residual substrate, XXXVIII, 124, 125
divalent cations, XXXVIII, 120
enzyme concentration, XXXVIII, 117
labeled nucleoside triphosphates, XXXVIII, 118, 119
linearity of cyclic nucleotide production, XXXVIII, 122
nucleoside triphosphate concentration and volume, XXXVIII, 115–117
protection of cyclase activity, XXXVIII, 121, 122
purification, proof of identity and determination of product, XXXVIII, 123, 124
reduction of product degradation, XXXVIII, 120, 121
substrate degradation, XXXVIII, 119, 120
termination of cyclase reaction, XXXVIII, 122, 123
bovine lung
assay, XXXVIII, 192
properties, XXXVIII, 195
purification, XXXVIII, 192–195
human platelets
apparent activation with time, XXXVIII, 201, 202
assay, XXXVIII, 199, 200
distribution among blood cells, XXXVIII, 200
divalent cations, XXXVIII, 201
homogenate preparation, XXXVIII, 200
subcellular distribution, XXXVIII, 201
rod outer segments, XXXVIII, 154, 155
sea urchin sperm
assay, XXXVIII, 196–198
properties, XXXVIII, 198, 199

Guanylate kinase, LI, 473–490
activity, LVII, 86
affinity for calcium phosphate gel, LI, 584
assay, LI, 474–477, 483, 484
cation requirements, LI, 481, 489
contaminant, in GMP synthetase purification, LI, 222, 223
determination of cyclic GMP phosphodiesterase, LVII, 96–98
of guanosine nucleotides, LVII, 86, 91
from *E. coli*, LI, 473–482
from hog brain, LI, 485–490
from human erythrocytes, LI, 483
isoelectric variants, LI, 488
kinetic properties, LI, 488, 489
molecular weight, LI, 482, 488
pH optimum, LI, 482
purification, LI, 477–480, 485–487
purity, LI, 482, 488
from rat liver, LI, 485–490

reaction mechanism, LI, 490
stability, LI, 480, 481, 487, 488
substrate specificity, LI, 473, 474, 481, 488, 489; LVII, 93
Guanylate synthetase, *see* GMP-synthetase
Guanyl cyclase, *see* Guanylate cyclase
1-Guanyl-3,5-dimethylpyrazole nitrate, affinity labeling, XLVII, 485, 487
Guanylyl cyclase, *see* Guanylate cyclase
Guanyl-5′-yl imidodiphosphate
complex with elongation factor EF–Tu, LX, 617, 622, 624, 625
induction of translocation with elongation factor EF-G, LX, 761
interference with initiation processes, LX, 73, 74, 117, 123, 564
stabilization of preinitiation complex, LX, 62, 74, 75, 118, 123, 617
use for assay for elongation factors, LX, 116, 658, 662
Guanylyliminodiphosphate, preparation
principle, XXXVIII, 425, 426
procedure, XXXVIII, 426, 427
Guanyl-5′-yl methylenediphosphonate
competition with guanosine triphosphate, LX, 710, 711
complex with elongation factor EF–G and 70 S ribosomes, LX, 728, 729, 732, 736
induction of translation with elongation factor EF-G, LX, 761–764, 766–779
inhibition of protein synthesis, LX, 366, 369, 564
nonhydrolyzable analog of guanosine triphosphate, LX, 357, 366, 564, 574, 617, 726, 732, 761
Guillotine, for rats, XXXI, 8
Guinea pig
brown adipose tissue, LV, 66, 67
brush borders from intestines, XXXI, 134
liver, microsome isolation, XXXI, 199
oxyntic cell isolation, XXXII, 709
pancreas fractionation, XXXI, 43
seminal vesicle, inhibitor, XLV, 826–832
Guinier approximation, LIX, 685, 712
Guinier plot, LIX, 685
D-Gulono-γ-lactone oxidase, LII, 18
L-Gulono-γ-lactone oxidase, flavin linkage, LIII, 450
Gum arabic, in freeze-fracture method, XXXII, 39
Gummiferin, *see* Carboxyatractyloside
g value, *see also* Absorption anisotropy factor; Emission anisotropy factor
anisotropy, LIV, 385
electron paramagnetic spectra, XLIX, 526, 527
in metalloproteins, XLIX, 517
for α-chymotrypsin crystal, XLIX, 473
in crystal lattice, XLIX, 425
of magnetic field standards, XLIX, 523
measurement, XLIX, 373–375
for metalloprotein, XLIX, 514, 517
for nitroxide compound, XLIX, 468
Gypsy moth, media, LVIII, 457
Gyromagnetic ratio, XLIX, 282
of nucleus, LIV, 156, 188, 196

H

HABA, *see* 4-(Hydroxyazobenzyl)-2′-carboxylic acid
Haber-Weiss reaction, LVII, 463, 489
Hac, *see* 9-Hydrazinoacridine
Haemophilus influenzae
crude extract preparation, LI, 433
growth, LI, 433
pyrimidine nucleoside phosphorylase, LI, 432–437
type b polysaccharide, XXXIV, 715, 716
Haemosol, cleaning agent, LIV, 122
Hageman factor, XXXIV, 435, *see also* Factor XII
Hair, as elastic support, XLIV, 560, 561
binding procedure, XLIV, 561
Hair cell, viable, studies, XXXII, 724
Half cell, definition, LIV, 411
Halide, interference, in atomic spectroscopy, LIV, 466, 480, 481
Halistaura, bioluminescence, LVII, 275
Halle's medium SM–20, LVIII, 59, 90
Haloacetol phosphate, XLVI, 132, 139
affinity labeling, XLVII, 483, 497
assay, XLVI, 141

structure, XLVII, 483
3-Haloacetol phosphate, XLVI, 139–145
as inhibitors for triosephosphate isomerase from rabbit muscle, XLI, 453
Haloacetyl derivative, XLVI, 153–157, 627
labeling, XLVI, 554–561
Haloacid
interaction with organic sulfoxides, XLVII, 443, 444, 454
oxygen transfer, XLVII, 445, 457
Halobacterium
cell membrane isolation, XXXI, 667–678
purple membrane separation, XXXI, 671, 672
red membrane separation, XXXI, 672–678
Halobacterium halobium
bacteriorhodopsin, LIV, 58
cells, internal volume, LV, 551
Dio-9, LV, 515
envelope vesicles
assay of electrical potentials, LV, 612
external membrane potential probes, LV, 559
internal volume, LV, 551
preparation and properties, LVI, 401–404
growth, LVI, 399–401
membranes, optical probes, LV, 573
α-Halocarbonyl compound, fluorescent labeling, XLVIII, 359
Halogen, activation agent, XLIV, 36, 37
ω-Halogenalkylacetic acid, XLVI, 251
Halohydrin, XLIV, 32, 36
α-Haloketone, XLVI, 130–153
as reagents, disadvantages, XLVI, 131
Halomethyl ketone, XLVI, 197–208
as alkylating agents, XLVI, 197, 204
derivatives of amino acids and peptides, XLVI, 201
design, XLVI, 198
effect on disulfides, XLVI, 205
on enzymes, XLVI, 205
on histidine, XLVI, 205
of pH, XLVI, 204
on serine, XLVI, 205
heavy atom-containing, XLVI, 207
inhibition kinetics, XLVI, 203
for NMR, XLVI, 207
reaction site, XLVI, 204
reactivity series, XLVI, 199
as reporter groups, XLVI, 207
solubility, XLVI, 204
synthesis, XLVI, 199–203
Halomicin, XLIII, 128
Halophenol, as uncoupler, LV, 470
6-Halopurine riboside, XLVI, 328
Hamilton CR 700–20 automatic syringe, in aequorin assay apparatus, LVII, 294, 297
Hamilton microsyringe, LIX, 864
Hamilton syringe, XXXVI, 32
automatic dispenser, XXXII, 112
disadvantage, in chemiluminescent assays, LVII, 445
Ham's medium F10, XXXII, 741, 742, 744; LVIII, 56, 86, 155, 393, 522
composition, LVIII, 62–70
Ham's medium F12, LVIII, 53, 57, 58, 86, 89, 91, 95, 97, 99, 100, 106, 107, 130, 155, 524
composition, LVIII, 62–70
for endocrine cell culture, XXXIX, 113
modified, XXXII, 734
Hamster
brown adipose tissue, LV, 66, 67
brush border isolation from intestine, XXXI, 124, 125
cheek pouch, tissue removal, XXXIX, 116, 117
Hamster cell line
CPSase-ATCase-dihydroorotase complex, LI, 123–134
growth, LI, 124
Hamycin, XLIII, 136, 338
Hanks' balanced salt solution, XXXIX, 38; LVIII, 120, 121, 130, 142
composition, XXXI, 309
for haploid frog cell culture, XXXII, 798
for placental perfusion, XXXIX, 245
Hanks' solution, modified, XXXII, 734
Hanks' and Wallace balanced physiological solution, XXXII, 659

Hansch-type equation, uncouplers, LV, 466, 469
Hansen medium S–301, LVIII, 465, 466
Hansenula, cytochrome b_2, CD spectrum, LIV, 283
Hansenula anomala, culture procedure, LIII, 249, 250
Hansenula ciferri
  growth, XXXV, 245, 250, 251
  as source of acetyl CoA:amine acetyltransferase, XXXV, 247
    of acetyl CoA:long-chain base acetyltransferase, XXXV, 242
    of phytosphingosine and dihydrosphingosine, XXXV, 529
H–227 antibiotic, streptothricin-like, XLIII, 336
H–2 antigen
  cell surface distribution, XXXII, 66, 67
  cyanylation, XLVII, 132
Haploid frog cell
  androgenetic, isolation, XXXII, 790
  characteristics, XXXII, 797
  chromosome analysis, XXXII, 795–797
  dissociating solution, XXXII, 798
  gynogenetic, isolation, XXXII, 790
  isolation and culture, XXXII, 789–799
Haploid syndrome, XXXII, 791
Haplopappus, cell cultures, XXXII, 729
HAPTA technique, XL, 27
Hapten, *see also* specific types
  immobilizing, XXXIV, 710
Haptoglobin, XXXIV, 6
  human, reductive methylation, XLVII, 475
  interaction with hemoglobin, XLVII, 188
Harding-Passey cell line, LVIII, 564
Harmine, as optical probe, LV, 573
Harmonic force constant, LIV, 313
Hartree modification, of Lowry measurement of protein, XL, 376
Hartridge-Roughton flash photolysis experiment, LIV, 94
Harvard withdrawal-infusion pump, XXXIX, 275
Harvesting, *see* specific type
Hastings-Weber standard source, LVII, 386
Hatching enzyme, XLV, 371–373, *see also* Sea urchin, hatching enzyme
HAT medium, LVIII, 351–353
HAT selective medium
  for cell hybridization, XXXII, 575–579
  for somatic cell hybridization, XXXIX, 122–124
    half-selection procedure, XXXIX, 124–126
HBA, *see* 2-Hydroxy-3-butynoate
hCG, *see* Gonadotropin, human chorionic
HDZ, *see* Hydroxylalkyl methacrylate gel, hydrazide derivative
Head holder, for rat surgery, XXXIX, 167, 168
Heart, XXXIX, 41–63, *see also* specific types
  beef, isolation of heme A, LII, 423–427
  biopsy, XXXVIII, 4, 5
  cAMP-receptor, XXXVIII, 380
  cannulation, XXXIX, 45, 46
  carnitine, LV, 221
  cell
    beating, isolation, XXXII, 632
    isolation and cultivation, XXXII, 740–745
    media, LVIII, 59, 90
  dispersion, LVIII, 123
  L-3-hydroxyacyl-CoA dehydrogenase, XXXV, 122–128
  isolated, perfusion, XXXIX, 43–60
  Keilin-Hartree preparation, LV, 118, 119, 124–127
    preparation, LV, 119–121
    properties, LV, 121–124
    yield, LV, 121
  lipofuscin isolation from cells, XXXI, 427–430
  mechanical performance, XXXIX, 56–59
    ischemic hearts, XXXIX, 52–56
    working hearts, XXXIX, 49–52
  metabolic contents, LV, 216
  mince, preparation of electron-transport complexes, LIII, 58, 61
  mitochondria, LV, 13, 14
    ATPase, LV, 298, 299, 300

human, characteristics, LV, 45
isolation from small amounts of tissue, LV, 39–46
oligomycin-sensitive ATPase, LV, 303
properties, LV, 14
storage, LV, 31
muscle
cloned cells, XXXII, 770
1,3-diphosphoglycerate, preparation and purification, XLII, 418, 419
phosphoglycerate mutase from human, XLII, 438, 445
triosephosphate isomerase, XLI, 446
nuclei isolation from cells, XXXI, 258–260
optical probes, LV, 573, 580
phosphodiesterase
assay, XXXVIII, 218, 219
properties, XXXVIII, 222, 223
purification, XXXVIII, 219–221
phospholipid exchange enzyme, XXXV, 262
preparation for perfusion, XXXIX, 45
protein kinase, XXXVIII, 298, 365, 366
assay, XXXVIII, 308–310
properties, XXXVIII, 313–315
purification, XXXVIII, 310–313
submitochondrial vesicles, XXXI, 292–299
succinate dehydrogenase isolation, XXXI, 777
tissue
EPR spectrum, LIV, 146, 147
phosphoglucose isomerase, assay isolation, and properties, XLI, 392–394, 397–400
ventricular muscle isolation, XXXIX, 60–63
Heat
for dissociation of ribosomal cleavage products, LIX, 473, 474
precipitation, of liver proteins, LII, 507
Heat change, measurement, LVII, 541–549
Heat Systems–Ultrasonics Sonifier Cell Disruptor Model W185, LI, 92; LII, 208; LIII, 222, 533
Heat treatment
ATPase inhibitor, LV, 410, 411
purification of adenosine monophosphate nucleosidase, LI, 266, 268
of adenylosuccinate synthetase, LI, 209
of amidophosphoribosyltransferase, LI, 173, 175
of cytidine deaminase, LI, 406
of dCTP deaminase, LI, 421
of FGAM synthetase, LI, 199
of flavins, LIII, 420
of formate dehydrogenase, LIII, 366, 367
of hydrogenase, LIII, 288, 311
of *p*-hydroxybenzoate hydroxylase, LIII, 547
of hypoxanthine phosphoribosyltransferase, LI, 545, 546, 547, 548
of nitrate reductase, LIII, 349
of nucleoside deoxyribosyltransferase, LI, 449
of nucleoside triphosphate pyrophosphohydrolase, LI, 280
of OPRTase-OMPdecase complex, LI, 146
of orotate phosphoribosyltransferase, LI, 71
of phosphoribosylpyrophosphate synthetase, LI, 15
of purine nucleoside phosphorylase, LI, 522, 528
of superoxide dismutase, LIII, 391
of uridine phosphorylase, LI, 427
Heavy atom labeling, in small-angle X-ray studies, XXXII, 218
Heavy metal
effect on luciferin light emission, LVII, 3
interference in *Rhizobium P*–450 purification, LII, 159
Heavy-metal ion, inhibitor of hydrogenase, LIII, 314

Heidenhain hematoxylin, for SEM fixation, XXXII, 55
HeLa cell, LVIII, 77, 84, 96
chromosome, LVIII, 174
contamination, LVIII, 439
DNA replication, nuclear system, XL, 41
effect of hormone, LVIII, 100
fucose production, L, 184, 185
growth and labeling, LVI, 68
media, LVIII, 56, 84, 85
microcarrier culture, LVIII, 189–191
mitochondrial protein biogenesis, LVI, 66, 67
method of procedure, LVI, 68–73
results, LVI, 73–79
in vitro, LVI, 73, 79
monitoring, LVIII, 166
monolayer culture, LVIII, 132, 133
plasma membranes, XXXI, 144
protein content, LVIII, 145
protein kinase, XXXVIII, 365, 366
serum-free, LVIII, 99
subcellular fractionation, LVI, 68, 69
HeLa culture × tobacco mesophyll, LVIII, 365
HeLa–S cell, effect of hormone, LVIII, 107–109
Helicase
in spheroplast isolation, XXXI, 572
yeast spheroplasts, LVI, 19
Heliothis zea, media, LVIII, 456, 457, 460
Helium, in anaerobic procedures, LIV, 114, 115
α-Helix
topographical structure
definition, XLIX, 104, 105
dihedral angles, XLIX, 104
hydrogen-bonding patterns, XLIX, 106–109
left-handed, XLIX, 113
occurrence, XLIX, 108
vibrational frequencies, XLIX, 112–114
vacuum ultraviolet circular dichroism, XLIX, 218
Helix-coil exchange rate, determination, LIX, 50–52
Helix pomatia, XLVIII, 245, 246, 248, 265, 266
trypsin-kallikrein inhibitors, *see* Trypsin-kallikrein inhibitor, snail
Hellebrigenin, XLVI, 523
Hellium-neon gas laser, wavelength, XLIX, 85
Helminthosporium No. 215, 3′-amino-3′-deoxyadenosine, LIX, 135
Helper virus
mouse sarcoma virus, LVIII, 413
Rous sarcoma virus, LVIII, 380
transformation, LVIII, 369
Helvolic acid, XLIII, 159
HEMA, *see* 2-Hydroxyethyl methacrylate
Hemagglutination, in assay of hybrid antibodies, XXXII, 65
Hemagglutinin, XXXIV, 328–331
Hematin
chemiluminescence assays, LVII, 426
stock solution, preparation, LVII, 425
Hematocrit, LVIII, 149
Hematopoietic precursor cell, growth factor, LVIII, 163
Hematoporphyrin, heme-deficient mutants, LVI, 559, 560
Hematoporphyrin-imidazole, light-driven electron transfer, LV, 537
phosphorylation, LV, 540
Hematoporphyrin IX, in active site, prosthetic group, LII, 4
Hematoside, *see also* Glycosphingolipid
glass bead complex, L, 137, 138
isolation, XXXII, 345, 350, 354, 355
thin-layer chromatography, XXXII, 356
thin-layer chromatography, XXXV, 397, 407, 410, 424
Hematoxylin
autoradiography, poststain, LVIII, 287
for SEM fixation, XXXII, 55
Heme, *see also* Dipyridine heme; specific types
biosynthetic pathway, LVI, 173
cleavage, XLVII, 466
ethyl acetate removal, XLVII, 465
infrared spectral properties, LII, 430
Keilin-Hartree preparation, LV, 123
mode assignments, LIV, 239–242

oxidative degradation, LII, 45
phosphorescence quenching, XLIX, 247
prosthetic group, in oxidases, LII, 5, 6
resonance Raman bands, LIV, 239
structure, XLIX, 128
transition, characteristic spectral regions, LIV, 254

Heme A
dipyridine derivatives, spectral properties, LII, 427, 428
isolation, LII, 423–427
structure, LII, 422

Heme *a*
in cytochrome *a,* definition, LIII, 203
cytochrome oxidase, LIII, 64, 76; LVI, 693
hemoprotein subunit, LIII, 78, 79
ligand binding, LIV, 108–110
prosthetic group, LII, 4
reduced, extinction coefficient, LIII, 43
Soret CD spectrum, LIV, 264

Heme a_2, prosthetic group, LII, 4

Heme a_3, ligand binding, LIV, 108–110

Heme B
concentration determination, LII, 159
separation on Celite column, LII, 425
structure, LII, 421
thin-layer chromatographic characteristics, LII, 442

Heme *b,* as measure of purification of complex III, LIII, 95

Heme-binding peptide, of complex III, LIII, 82–84

Heme C, structure, LII, 422

Heme *c*
midpoint potential, LIII, 656
in *Pseudomonas* cytochrome oxidase, LIII, 647, 655, 656
spectral properties, LIII, 126

Heme *c* peptide, CD spectra, LIV, 262, 263

Heme d_1
midpoint potential, LIII, 656
in *Pseudomonas* cytochrome oxidase, LIII, 647, 655, 656

Heme *h*, prosthetic group, LII, 4

Heme-α-methenyl oxygenase, LII, 15

Heme oxygenase
assay, LII, 368–370
distribution, LII, 367, 372
homogeneity, LII, 370
inhibitors, LII, 371
kinetics, LII, 371
molecular weight, LII, 370
reactivity, LII, 367, 368
specificity, LII, 371
stability, LII, 370, 371

Heme peptide, XXXIV, 6
of cytochrome *c*, XXXIV, 194

Heme-polylysine complex, CD spectrum, LIV, 260

Heme precursor, single administration, in measurement of protein degradation, XL, 257

Heme protein, *see also* Hemoprotein
in microsomes, LII, 45–47

Heme-regulated translational inhibitor
irreversible
assay, LX, 287–290, 470
functional role, LX, 459–461
inactivation of initiation factor eIF-2, LX, 481–484
inhibition of methionyl-transfer ribonucleic acid binding, LX, 287–290
of protein synthesis, LX, 287, 470, 475, 476, 478, 480–484
molecular weight, LX, 477, 484
phosphorylation of initiation factor eIF-2 subunit, LX, 287–290, 467, 468, 470, 476, 477, 496, 532
properties, LX, 460
purification, LX, 467, 471–475, 478
relation to reversible form, LX, 494, 495
specificity, LX, 479
reversible
assay of purity, LX, 488, 489
inhibition of protein synthesis, LX, 490
molecular weight, LX, 492
phosphorylation of initiation factor eIF-2 subunit, LX, 490, 492, 493, 496
properties, LX, 489, 490

purification, LX, 485–488, 493–495
relation to irreversible form, LX, 494, 495
self-phosphorylation, LX, 490–494
specificity, LX, 491
ternary complex dissociation, LX, 49, 50
Hemerythrin, LII, 13
dissociation studies, XLVIII, 321
magnetic circular dichroism, XLIX, 175, 176
oxy form
resonance frequencies, LIV, 241
spectrum, LII, 36
structure, LII, 4, 5
Heme S, LII, 422
Heme species
high-spin, EPR characteristics, LIV, 134
low-spin, EPR characteristics, LIV, 134
Hemiacetal, XLVI, 221
Hemicorpus, perfusion method, XXXIX, 73–82
Hemidiaphragm, of rats and mice, XXXIX, 84, 85
Hemiketal, XLVI, 28
Hemin
catalase, as source, LVII, 445
cytochromeless mutants, LVI, 122
electron transport reconstruction, LVI, 176, 177
infrared spectral properties, LII, 430
in luminol chemiluminescence reactions, LVII, 448
preparation of metal-free porphyrin ester, LII, 430–432
protein synthesis, LX, 66, 352, 380, 390–392, 513, 516, *see also* Heme-regulated translational inhibitor
requirement, LVI, 172–175
solutions, LVI, 174
Hemin B
conversion to μ-oxobishemin B dimethyl ester, LII, 429, 430
to protoporphyrin IX dimethyl ester, LII, 432
paramagnetic resonance spectrum, LII, 435
Hemin chloride, in porphyrin ester synthesis, LII, 432
Hemin-controlled repressor, *see* Heme-regulated translational inhibitor
Hemisuccinate, of steroids, XXXVI, 18, 19, 22
Hemithioacetal, XLVI, 221
stability, XLVI, 221
Hemocyanin, XLVI, 491; LII, 10
boundary patterns, XLVIII, 212
coupling to succinylated cyclic nucleotides, XXXVIII, 98
electronic absorption spectra, XLIX, 139
excitation profiles, XLIX, 139
fractionation results, XLVIII, 269
ligand-mediated dimerization, XLVIII, 243
magnetic circular dichroism, XLIX, 175
from *M. trunculus,* translational diffusion coefficient, XLVIII, 420
oxy form
Raman spectra, XLIX, 137–139
spectrum, LII, 36
phosphorescence studies, XLIX, 239, 241, 245, 246, 247
properties, XLIX, 137
prosthetic group, LII, 4
reciprocal relaxation times, XLVIII, 264
resonance Raman spectra, XLIX, 140
reversible dimerization and, XLVIII, 260–270
as visual marker, XXXII, 63
α-Hemocyanin
of *Helix pomatia*, XLVIII, 245, 246, 248, 265, 266
pressure-jump studies, XLVIII, 310, 314
Hemocytometer, counting, LVIII, 188
protocol, LVIII, 143
Hemodialysis, microcapsules, XLIV, 694–696
Hemoglobin, XXXIV, 5, 6, 706; XLIV, 684; LII, 16, *see also* Oxyhemoglobin
β chain, carbon monoxide binding, data evaluation, LIV, 515, 523
absorption spectra, LIV, 274

Aeromonas neutral protease, XLV, 404, 406
by chloroform-methanol denaturation, LI, 469, 470
concentration determination, XLVIII, 332, 333, 337–338
  by pyridine hemochromagen method, LII, 457
conformational composition, LIV, 268, 270, 271
conjugate, ligand equilibria, XLIV, 540–543
deoxy form
  hyperfine shifted proton, resonances, LIV, 202
  Raman frequencies and intensities, XLIX, 136
  temperature dependence of line width, LIV, 203
derivatives
  preparation, LII, 457–459
  spectral properties, LII, 459–463
determination, using Austin and Drabkin method, LI, 276
dissociation by adsorption, XLIV, 153
electron paramagnetic resonance studies, XLIX, 514
elimination from microsomal preparations, LII, 201, 202
gel electrophoresis, XXXII, 101
gene expressions, LVIII, 507
human
  α-chain, cyanylation and, XLVII, 131
  preferred oxygen donor, in oxygen consumption studies, LIV, 487
  reaction with oxygen, kinetics, LIV, 73, 486
hydrogen exchange, XLIX, 37, 38
immobilization
  electrostatic, XLIV, 542, 543
  on Sephadex, XLIV, 542
inducibility, LVIII, 509, 510
inert proteic matrix, XLIV, 268, 271
interference in difference spectrum measurements, LII, 213–215
  in spectral analysis of microsomal hemoproteins, LII, 85
IR difference spectra, LIV, 306, 314
IR spectra, LIV, 305, 306, 309
isolated chains
  ligand binding studies, LIV, 530, 531
  reaction with oxygen, kinetics, LIV, 486
isolation, LII, 448, 449, 456, 457, 487, 488
Leucostoma peptidase A, XLV, 397
ligand binding, LIV, 73, 74, 82–84, 94, 95, 99, 100, 530, 531
magnetic circular dichroism, XLIX, 170
magnetic susceptibility studies, LIV, 393, 394
maleimide-labeled
  saturation-recovery time, XLIX, 502
  saturation-transfer spectra, XLIX, 507
MCD studies, LIV, 295
membrane osmometry, XLVIII, 74
metal-substituted, reconstitution procedure, LII, 490–492
microencapsulation, XLIV, 204–208
microspectrophotometry, XXXIX, 414, 415, 419
Mössbauer spectrum, LIV, 355
mouse, phthalylation, XLVII, 155
mutant, NMR studies, LIV, 217–222
opossum, carbonyl form, IR difference spectrum, LIV, 321
oxy form
  electronic absorption spectrum, XLIX, 133
  excitation profile, XLIX, 133, 134
  Raman frequencies and intensities, XLIX, 136
oxygen association kinetics, XLVIII, 274–276
perfusion fluid, LII, 49
as plasma membrane contaminant, XXXI, 85
preparation, LIV, 488
properties, LIV, 214
prosthetic group, LII, 5
proton magnetic resonance studies, LIV, 194, 195, 214–222
rabbit, carbon monoxide binding, IR studies, LIV, 310
reaction mechanism, LII, 38
reconstituted, purification, LII, 455

reconstitution procedure, LII, 454
removal, LI, 581–585
by DEAE-cellulose chromatography, LII, 360
resonance Raman spectra, LIV, 237, 241, 243
saturation-transfer studies, XLIX, 489–492
SDS complex, properties, XLVIII, 6
sickle-cell, saturation-transfer studies, XLIX, 495, 496
Soret CD spectrum, LIV, 260, 261, 273–276
spectrophotometry, LVIII, 511
spin-labeling studies, XXXII, 195
staining, LVIII, 510
structure, XLIX, 128
substrate
of chymotrypsin, XLIV, 73–75
of heme oxygenase, LII, 371
of papain, XLIV, 82
of pepsin, XLIV, 77–79
synthesis, measurement methods, XXXVII, 118–121
X-ray studies, XXXII, 218; LIV, 343
yeast, reaction with oxygen, kinetics, LIV, 486, 504

Hemoglobin A, carbonyl form
for cell path length determination, LIV, 318
IR studies, LIV, 310, 319

Hemoglobin-bound ligand
integrated intensities, LIV, 311
isotope shifts, LIV, 315
stretching frequencies, LIV, 311

Hemoglobin Chesapeake, LIV, 217, 219

Hemoglobin-CPA
in oxygen consumption assays, LIV, 492
reaction with oxygen, kinetics, LIV, 486

Hemoglobin-CPB, reaction with oxygen, kinetics, LIV, 486

Hemoglobin Iwate, LIV, 219, 220

Hemoglobin Kempsey, LIV, 219

Hemoglobin M_{Emory}, carbon monoxide binding, IR studies, LIV, 310

Hemoglobin Malmö, LIV, 219

Hemoglobin M Milwaukee, LIV, 221, 222

Hemoglobin Rainier, CD spectrum, LIV, 275

Hemoglobin Yakima, LIV, 219

Hemoglobin Zurich, carbonyl form, IR difference spectra, LIV, 310, 320, 321

Hemopexin heme carbonyl, carbon monoxide binding, IR studies, LIV, 310

Hemoprotein, *see also* specific type
band
α-band, LIV, 236, 254
β-band, LIV, 236, 254
γ-band, LIV, 236
Δ-band region, LIV, 254
B–band region, LIV, 236
D–band region, LIV, 254
N–band region, LIV, 254
near-IR region, LIV, 254
Q_0 band, LIV, 236
Q_v band, LIV, 236
random coil form, LIV, 268, 269
Soret band, LIV, 236, 254
biosynthesis, in liver, LII, 350
CD spectrum
heme optical activity, LIV, 259–265, 273
in intrinsic absorption region, LIV, 266–272
classification, LIII, 260
detection in SDS-polyacrylamide gels, LII, 324–331
distance measurements, LIV, 188, 190
electron paramagnetic studies, XLIX, 514
in ferric state, MCD spectra, LIV, 296–299
in ferrous state, MCD spectra, LIV, 296, 297
interference, in cytochrome difference spectrum measurements, LII, 125, 213
labeling, LVI, 45
ligand binding, principles, LIV, 506–532
MCD spectra, general characteristics, LIV, 287, 288
Mössbauer spectroscopy, LIV, 358
nomenclature for band regions, LIV, 254

oxidation state determination, from Raman bands, LIV, 243
oxygenated forms, spectra, LII, 36
proton magnetic resonance studies, LIV, 209–222
relaxation kinetics, LIV, 64–84
resonance enhancement, LIV, 236, 237
resonance Raman spectra, LIV, 233–249
spin coupling, MCD studies, LIV, 300–302
spin state determination, from Raman spectra, LIV, 243–245
spin-state equilibria, LIV, 299, 300
structural interpretations, from Raman bands, LIV, 242–245
Hemoprotein conjugate, properties, XLIV, 538, 539
Hemostasis, platelet function, XXXI, 149
Hemp, aleurone grain, XXXI, 576
Henry equation, XLVIII, 434
HEPA filter, *see* Filter, high-efficiency particle air
Heparan *N*-sulfatase, *see* Heparin sulfamidase
Heparan sulfate
multiple sulfatase deficiency, L, 474
repeating disaccharide structure, L, 280
Heparin, XXXIV, 4, 6; XLIV, 680; LII, 54, 62, 66; LVIII, 268
bound to Sepharose 4B, purification of elongation factor EF-1, LX, 692, 694
to Sepharose 6B, purification of initiation factors, LX, 124, 125, 127, 135, 167, 170–180, 196
butyl nitrite degradation, L, 150–152
cofactor assay, XLV, 657
effect on estrogen receptor studies, XXXVI, 173, 181
isolation of mitochondria, LV, 4, 13
mitochondrial ribosome dissociation, LVI, 83, 85
in preparation of hemoglobin, LIV, 488
in radioimmunoassay, XXXVII, 31, 32
Heparin sulfamidase
assay, L, 447, 448
distribution, L, 447
multiple sulfatase deficiency, L, 453, 454
Sanfilippo A syndrome and, L, 448, 449
specificity, L, 448
Heparin sulfate, LVIII, 268
Hepatitis
β-antigen, XXXIV, 6, 338
serum asialoglycoprotein levels, L, 288
Hepatocyte
culture, XXXII, 733–740
distribution of pyridine nucleotide, LV, 215
of fetal liver, XXXIX, 36
incubation procedures, LII, 66
isolation procedures, LII, 61–64, 358
media, LVIII, 53
metabolic distribution, LV, 214
from rat liver, XXXV, 579–594
effect of Ca^{2+}, XXXV, 585
of collagenase, XXXV, 585
gluconeogenesis, XXXV, 587–590
hormonal effects, XXXV, 592–594
incorporation of amino acids into protein, XXXV, 590–592
isolation and purification, XXXV, 580–584
microscopic examination, XXXV, 586
viability tests, LII, 64, 65
Hepatoma
large-scale growth, LVIII, 211
plasma membrane isolation, XXXI, 75–90
rat, LVIII, 544
tissue culture cell, LVIII, 544–552
cloning, LVIII, 547–549
cryopreservation, LVIII, 551, 552
effect of steroids, LVIII, 544, 545
media, LVIII, 545, 546
composition, LVIII, 546
monolayer culture, LVIII, 545
propagation, LVIII, 545
routine culture, LVIII, 546, 547
serum-free growth, LVIII, 549, 550

for steroid-receptor binding studies, XXXVI, 235–248
storage, LVIII, 551, 552
suspension culture, LVIII, 550, 551
Hepatoma cell
cell hybrids, XXXII, 582
lectin binding, XXXII, 624
HEPES, *see* *N*-2-Hydroxyethylpiperazine-*N*′-2-ethanesulfonic acid
HEPES buffer, *see* *N*-2-Hydroxyethylpiperazine-*N*′-2-ethanesulfonic acid
L-Heptadecylamine, XLIV, 151, 227
Heptaene, XLIII, 136
Heptafluorobutyric acid, testosterone derivative, EC detection, XXXVI, 59, 61
n-Heptafluorobutyric acid, XLVII, 307, 309, 327, 336, 340
removal, XLVII, 376
n-Heptane-ethanol, solvent, of quinones, LIII, 581
Heptane-ethyl acetate, XLVII, 346
advantage, XLVII, 336
Heptane 2-hydroxylase, LII, 20
n-Heptane-isoamyl alcohol, for extraction of hydroxylated biphenyls, LII, 400, 401
Heptane-pentane, electron spin resonance spectra, XLIX, 383
N-Heptylfluorobutyrylimidazole, in melatonin assay, XXXIX, 386
2-Heptyl-4-hydroxyquinoline *N*-oxide, inhibitor of complex III, LIII, 38
n-Heptylquinoline *N*-oxide, cytochrome b-c_1 region, LV, 461, 462
Herbicide, *see* specific substance
Heroin, logit-log assay, XXXVII, 9
Herpes simplex virus, thymidine kinase, LI, 369
Herpesvirus, bovine, LVIII, 28
Herpesvirus simiae, LVIII, 37, 43
Hetacillin, XLIII, 362, 363
Heteroadduct, XLVI, 174
Heterocarpus, coelenteramide, LVII, 344
Heterodyne detection, XLVIII, 423, 424
Heterodyne spectroscopy, XLVIII, 482–485
Heterokaryon, LVIII, 345
Heteroploid cell line, *see* Cell line, heteroploid
Heteropolypeptide, random-coil, hydrogen-exchange kinetics, XLVIII, 323
Hewlett Packard HP9825A desk-top computer, LIV, 54
Hewlett Packard HP9862A plotter, LIV, 54
Hexachloroplatinic acid, XLIV, 654
Hexacyanoferrate(III), XLVI, 49–53
Hexadecafluoronanoic acid, testosterone derivative, EC detection, XXXVI, 59–61
Hexadecanal, quantum yield, in bioassay, LVII, 192
[^{14}C]Hexadecane, photometer calibration and, LVII, 300
Hexadecanoate, *see* Hexadecanoic acid; Palmitic acid
Hexadecanoic acid, quantum yield, in bioassay, LVII, 192
trans-2-Hexadecenoic acid, intermediate in sphingomyelin synthesis, XXXV, 493, 494
trans-2-Hexadecenoyl chloride, intermediate in sphingomyelin synthesis, XXXV, 494
2-Hexadecyloxy-3-octadecyloxy-1-propanol, intermediate in phosphatidylcholine synthesis, XXXV, 449, 450
rac-2-Hexadecyloxy-3-octadecyloxypropylphosphonic acid, intermediate in synthesis of phosphatidic acid analogues, XXXV, 506 507
DL-2-Hexadecyloxy-3-octadecyloxypropyl-[2-(trimethylammonium)ethyl] phosphinate, intermediate in synthesis of phosphatidylcholine analogues, XXXV, 522–525
2-Hexadecyloxy-3-octadecyloxypropylphosphonic acid, intermediate in synthesis of phosphatidylethanolamine analogues, XXXV, 513
Hexadecyltrimethylammonium bromide, from microsome isolation, XXXI, 216, 224
1,5-Hexadiene, electron spin resonance spectra, XLIX, 383

Hexafluoroacetone trihydrate
coupling agent, XLVII, 363
distillation, XLVII, 359
Hexafluorobenzene, as lock nuclei source, LIV, 189
Hexafluoroisopropanol, XXXIV, 656; XLIX, 218
advantage, XLVII, 337
hydrogen peroxide stabilization, LVI, 459
Hexahelicine, LIV, 253, 254
Hexamethonium, cyclic nucleotide levels, XXXVIII, 94
Hexamethylene bisiodoacetamide, XLIV, 263
Hexamethylenediamine, XXXIV, 4; XLIV, 70, 77, 207; XLVII, 42
as spacer, XXXIV, 80
Hexamethylenediamine-agarose, XXXIV, 510
as spacer, XXXIV, 72
Hexamethylene diisocyanate, XLIV, 263
Hexamethylphosphoramide, XLVII, 570
electron-transfer chemiluminescence, LVII, 499
identification of iron-sulfur centers, LIII, 269, 272
solvent, in chemiluminescent reaction, LVII, 524
for highly apolar substances, XLVII, 563
thioacetylation, XLVII, 292
Hexamethyl phosphoric triamide, XLVI, 242, 243
guanylyliminodiphosphate preparation, XXXVIII, 426, 427
tRNA base replacement reactions, LIX, 114, 120
solvent, for luminol chemiluminescence reactions, LVII, 410
Hexamethylsilazane, for steroid silylation, LII, 380
Hexane, XLIV, 894
electron spin resonance spectra, XLIX, 383
purification of ethyl-4-iodo-2-diazoacetoacetate, LIX, 811
of ubiquinone, LIII, 639
as solvent for benzo[*a*]pyrene, LII, 414
for silyl ethers of steroids, LII, 381
for styrene oxide, LII, 416
n-Hexane
IR bandwidth, LIV, 311
in picosecond continuum generation, LIV, 12
Hexane-acetone, for aromatic hydrocarbon extraction, LII, 236–238
Hexane-ether, isolation of quinone intermediates, LIII, 607
Hexane-isopropanol, for HPLC separation of vitamin D metabolites, LII, 392, 395
Hexanoic acid hydrazide, XLIV, 126
1-Hexanol
in copper analysis, LIV, 444, 445
electron spin resonance spectra, XLIX, 383
2-Hexene, electron spin resonance spectra, XLIX, 383
Hexobarbital
commercial source, LII, 222
as endogenous substrate, LII, 271
interference, in hydrogen peroxide assay, LII, 349
metabolism, LII, 65, 66
as mixed-function oxidase substrate, XXXI, 233
optical-difference spectrum produced, LII, 259, 266
oxidation, difference spectrum measurements, LII, 216–218
stabilizer, of cytochrome *P*–450, LII, 215
type of binding reaction with cytochrome *P*–450, LII, 264
Hexobarbital monooxygenase, *Ah* locus, LII, 232
Hexokinase, XXXIV, 246, 252, 328; XLVI, 287
active transport studies, XLIV, 921–924
activity, LVII, 86
on ATP analogs, XLIV, 876
assay, LVII, 210
of adenylate kinase, XXXVIII, 162; XLIV, 892; LI, 468
AMP, LV, 209

by Fast Analyzer, XXXI, 816
GTP LV, 208
inorganic phosphate, LV, 212
oxidative phosphorylation, LV, 723, 724, 747
succinyl-CoA, LV, 210
associated with cell fractions, XL, 86
in ATP regeneration, XLIV, 698
coentrapment, XLIV, 475–478
conjugate, activity, XLIV, 200, 271, 342
determinations of guanosine nucleotides, LVII, 86, 89, 93
in D-glucosamine microassay, XLI, 5
hormonal control, XXXVII, 293
immobilization
on inert protein, XLIV, 908
by microencapsulation, XLIV, 214
in multistep enzyme system, XLIV, 454–457
in intestinal cells, XXXII, 673
kinetics, in multistep enzyme system, XLIV, 432–435, 463–466
in D-mannose microassay, XLI, 3
in parenchyma cells, XXXII, 706
Pasteur effect, LV, 291
in phosphoribosylpyrophosphate synthetase kinetics studies, LI, 6
plasma membrane, XXXI, 90
P/O ratio, LV, 226
of rat brain, XLII, 20–25
amino acids, XLII, 24
assay, XLII, 20, 21
chromatography, XLII, 22
molecular weight, XLII, 23
properties, XLII, 23–25
purification, XLII, 21–23
site-specific reagent, XLVII, 482
in studies of glucose-marked liposomes, XXXII, 512
substrate specificity, LVII, 93
from yeast, XLII, 6–20
assay, XLII, 8
chromatography, XLII, 11–14
isoenzymes A, B, and C, XLII, 7, 16–20
molecular weight, XLII, 18
preparation of labeled oligonucleotides, LIX, 75
properties, XLII, 16–20
purification, XLII, 9–16
in tRNA labeling procedure, LIX, 71
in tRNA sequence analysis, LIX, 61
solution, stability, LIX, 61
Hexokinase B, yeast, self-association, XLVIII, 117
Hexokinase C, yeast, self-association, XLVIII, 117
L-Hexonate dehydrogenase
in mammalian tissues, XLI, 159–162
chromatography, XLI, 162
Hexopyranoside:cytochrome *c* oxidoreductase, *see* D-Glucoside 3-dehydrogenase
Hexosamine
analysis, XLI, 3–10
in glycolipids, XXXII, 358–360, 365
β-Hexosaminidase, mucolipidoses, L, 455
Hexosaminidase A, from human placenta
assay, L, 548
immunochemical determination, L, 554, 555
properties, L, 551–554
purification, L, 549–551
β-Hexosaminidase A
assay, L, 485, 486
lysosomes, L, 490
Sandhoff's disease, L, 484
Tay–Sachs disease, L, 484
Hexosaminidase B, from human placenta
assay, L, 548
immunochemical determination, L, 554, 555
properties, L, 551–554
purification, L, 549–551
β-Hexosaminidase B
assay, L, 485, 486
lysosomes, L, 490
Sandhoff's disease, L, 484
Tay–Sachs disease, L, 484
Hexosaminidase C, L, 490
Hexose
cellular, methodology for quantitation, L, 192, 193
in glycolipids, XXXII, 358–360
in plasma membranes, XXXI, 87
Hexose isomerase, in plants, XXXI, 740

Hexose monophosphate shunt, control studies, XXXVII, 292, 295
Hexose oxidase, LII, 10
Hexuronic acid, colorimetric assays, XLI, 29–31
Hexylamine, XLIV, 126
measurement of ΔpH, LV, 562
n-Hexylbromide, XLIV, 44
HFB, *see* Heptafluorobutyric acid
HFN, *see* Hexadecafluoronanoic acid
Hfs, *see* Hyperfine shifted resonance
hGH, *see* Human growth hormone
H–GPRT, *see* Hypoxanthine-guanine phosphoribosyltransferase
High energy compound, in protein turnover, XXXVII, 247, 248
High-performance liquid chromatography, *see* High-pressure liquid chromatography
High-pressure liquid chromatography, XLIII, 278, 300–320; LII, 279–296; *see also* Reversed phase chromatography
actinomycin, XLIII, 313, 319
ampicillin, XLIII, 309–311, 319
aromatic antibiotics with nitrogen, XLIII, 301–305
aureovertins, LV, 480
bacitracin, XLIII, 314, 319
of biphenyl metabolites, LII, 405–407
cephalosporin C, XLIII, 311, 312, 319
clindamycin phosphate, XLIII, 306, 307, 319
daunomycin, XLIII, 318, 319
erythromycin, XLIII, 308, 309, 319
experimental considerations, LII, 281, 282
griseofulvin, XLIII, 307, 308, 319
kanamycin, XLIII, 307, 308, 319
leucomycin, XLIII, 309, 319
novobiocin, XLIII, 301–304, 319
oxytetracycline, XLIII, 318, 319
penicillin G, XLIII, 312, 313, 319
penicillin V, XLIII, 313, 319
principles, LII, 280–283
PTMA-amino acid indentification, XLVII, 381, 382
rifampin, XLIII, 304, 305, 319
for separation of fatty acid methyl esters, LII, 322, 323
tetracycline, XLIII, 315–319
virginiamycin, XLIII, 305, 319
of vitamin D metabolites, LII, 392
High-pressure techniques, XLIX, 14–24
gel electrophoresis, XLIX, 19–21
for hydrogen out-exchange, XLIX, 33
optical measurements, XLIX, 21–24
volume measurements, XLIX, 16–19
High-voltage electron microscopy, XXXIX, 156, 157
High-voltage electrophoresis
purification of 8α-flavins, LIII, 458–460
in tRNA sequence analysis, LIX, 65
for separation of adenine and adenylate, LI, 568–570
of adenosine phosphates, LI, 460
High-voltage shutdown circuit, LVII, 540
diagram, LVII, 537
High-voltage supply, in flash photolysis apparatus, LIV, 54
Higuchi's medium, LVIII, 57
composition, LVIII, 62–70
Hikizimycin, XLIII, 153
Hill coefficient, XLVIII, 274, 276–278
cooperativity, XLVIII, 292, 293
evaluation, XLVIII, 278, 279, 287–289, 297–299
second moment of bound-state distribution, XLVIII, 289
Hill plot, XXXVI, 10–12, 14, 15
binding site heterogeneity, XLVIII, 287
cooperativity, XLVIII, 287–293
of hemoglobin-oxygen binding, XLVIII, 274–276
meaning, XLVIII, 272, 273
qualitative relationship to Scatchard plot, XLVIII, 278, 279
Hindrance, steric, XXXIV, 118
Hink medium, LVIII, 461
HIOMST, *see* Hydroxyindole-*O*-methyltransferase
HIPA solution, *see* 2-Hydrazino-3-(4-imidazolyl)propionic acid solution
HiPIP, *see* Iron-sulfur protein, high-potential
Hippuryl-L-phenylalanine, potato carboxypeptidase inhibitor substrate, XLV, 736
Hirt procedure, extraction, LVIII, 406–409

Hirudin, XXXIV, 445
alkylation, XLV, 676
amino acid sequence, XLV, 675
assay, XLV, 677
bdellins, separation, XLV, 799
characteristics, XLV, 669
inhibitor, of thrombin, XLV, 174
preparation, XLV, 670
properties, XLV, 674
physical, XLV, 677
protease inhibitor, XLV, 669
purity, XLV, 674
reduction, XLV, 676
structure-activity relationships, XLV, 676
Hirudo medicinalis, preparation of hirudin, XLV, 670
Histamine
adenylate cyclases, XXXVIII, 147, 148, 176, 178, 180
cyclic nucleotide levels, XXXVIII, 94
Histamine polyacrylamide, XXXIV, 47
Histidine
acylation procedure, XLVII, 516
affinity labeling, XLVII, 483, 494, 495
aminoacylation isomeric specificity, LIX, 274, 275, 279, 280
assay of phosphoribosylglycinamide synthetase, LI, 181
in buffer for isolation of electron-transport carriers, LIII, 57
chemical properties, XLIV, 12–18
in chymotrypsin active site, XXXVI, 390
cleavage rates and, XLVII, 342–345
detection by radiolabeling, XLVII, 259
dipeptides containing, identification, XLVII, 396
EDTA-Tris buffer, preparation of cytochrome *c* reductase complexes, LIII, 57
enzyme aggregation, XLIV, 263
imidazole chain, hydrogen exchange, XLIX, 25
imidazole group blocking, XLVII, 527, 539–541, 611
in media, LVIII, 53, 62
modification by diethylpyrocarbonate, XLVII, 431–442
N-o-nitrophenylsulfenylation, XLVII, 518, 519
peptide electropheretic mobility, XLVII, 58, 59
peptide synthesis, XLVII, 509, 527
in plasmin, XLV, 270
inhibition and radiolabeling, XLV, 263
in proteins, iodination, XXXII, 103
purification of complex III peptides, LIII, 84
of ribonucleoside diphosphate reductase, LI, 241
radiolabeled, in protein turnover studies, XXXVII, 242
residue, modification, LIII, 137, 149
tRNA synthetase, XXXIV, 170
thermolysin hydrolysis, XLVII, 175
in thrombin, XLV, 173
transport of, in bacteria, XXXII, 843–849
L-Histidine
in affinity labeling of steroids, XXXVI, 386, 387
enzyme active site studies, XXXVI, 390
in protein-tRNA interaction studies, LIX, 323
L-Histidine ammonia-lyase, urocanic acid production, XLIV, 745
Histidine-binding protein, XXXIV, 133
Histidine binding protein J
deficiency in bacteria, XXXII, 849
fluorine nuclear magnetic resonance, XLIX, 274
Histidine decarboxylase, in neurobiological studies, XXXII, 769
Histidine dehydrogenase, from yeast, antibody binding, L, 55, 56
Histidine permease, deficiency in bacteria, XXXII, 852–854
Histidinol, resistant strains, LVIII, 314
8α-[N-(1)-Histidyl]-2′,5′-anhydroriboflavin, properties, LIII, 465
8α-[N-(3)-Histidyl]-2′,5′-anhydroriboflavin, properties, LIII, 465
Histidyl residue, modification with ethoxyformic anhydride, LVII, 176
8α-N(1)-Histidylriboflavin
properties, LIII, 465
purification, LIII, 459

8α-N(3)-Histidylriboflavin
properties, LIII, 465
purification, LIII, 459
8α-N-Histidylriboflavin isomer
fluorescence quenching values, LIII, 461
methylation of imidazole nucleus, LIII, 459, 460
separation, LIII, 458, 459
synthesis, LIII, 458
L-[^{3}H]Histidyl-tRNA^{His}, in protein-tRNA interaction studies, LIX, 322, 323, 328, 329
Histidyl-tRNA synthetase, subcellular distribution, LIX, 233, 234
Histocompatibility antigen, sequence analysis, XLVII, 331
Histone, XXXIV, 549–552; LVIII, 49, *see also* Cysteine-containing protein
AKP, XL, 104
analysis
cytochemical measurements, XL, 136
methods, XL, 102
arginine and glycine-rich, structure-function relationships, XLVII, 187
calf thymus fractions, amino acid composition, XL, 105
in chromatin characterization, XXXI, 278, 279
in chromatin fractions, XL, 101
composition, XL, 102
coupling with human serum albumin, XL, 191
with nucleic acids, XL, 192
with phosphorylated bovine serum albumin, XL, 192
dialysis, XL, 108
electrophoresis, Bonner method, XL, 127
fractionation
chemical, XL, 116
method, XL, 131
quantitative, XL, 109, 116
heterogeneity, XL, 102
immunochemical characteristics, XL, 191
immunochemical reactions, XL, 137
isolation, XL, 105
KSA, XL, 104
labeled
determination of specific activity, LX, 524
preparation by phosphorylation, LX, 524
substrate for acetylation, LX, 535, 537–539
mammary synthesis and acetylation, XXXIX, 454
nomenclature, XL, 103, 104
phosphorylation
assay for given histone fraction, XL, 139
cGMP assay, XXXVIII, 90, 92, 93
detailed analysis, within given histone fraction, XL, 142
protein kinase assay, XXXVIII, 287, 288, 313, 316, 317, 322, 324, 327, 328, 340–342, 357, 359
cGMP-dependent, XXXVIII, 330, 340–342, 348
proteolytic degradation, XL, 109
quantitative determination, XL, 110
chromatography, XL, 117
radioactivity, determination, XL, 125, 129
from rat liver, analysis, LIX, 520–522
selective dissociation from chromatin in circular dichroism, XL, 228
selective phosphorylation, assessment methods, XL, 138
separation, method, XL, 133
staining, XL, 19
Histone–dye interaction, in direct quantitative determination, XL, 115
Histone F3, XXXIV, 551
Histone F2al, XXXIV, 551
Histone F2b, osmometric studies, XLVIII, 75
HL–A antigen
cell surface studies, XXXII, 67
cyanylation, XLVII, 132
H medium, for granulosa cell culture, XXXIX, 205
HMG-CoA lyase, *see* 3-Hydroxy-3-methylglutaryl-CoA lyase
HMG-CoA synthase, *see* 3-Hydroxy-3-methylglutaryl-CoA synthase
HMP, *see* Keilin-Hartree heart muscle preparation
HMPA, *see* Hexamethylphosphoramide

HMPT, *see* Hexamethylphosphoric triamide
HN2, *see* Methyl bis(3-chloroethyl)amine
^{1}H NMR technique, *see* Nuclear magnetic resonance spectroscopy, hydrogen-1
33258 Höechst dye, as sensitizer, LVIII, 317
Hofstee plot, in enzymology, XXXVI, 4
Hog, *see also* Pig; Swine
 brain, guanylate kinase, LI, 485–490
 kidney, *N*-acetylglucosamine 2-epimerase, assay, purification, and properties, XLI, 407–411
 spinal cord
  D-glycerate dehydrogenase, assay, purification, and properties, XLI, 289–293
  D-3-phosphoglycerate dehydrogenase, assay, purification, and properties, XLI, 282–285
 spleen, *N*-acetylglucosamine kinase, assay, purification, and properties, XLII, 58–62
Hogben index, in melatonin bioassay, XXXIX, 384, 385
HOL–1, high-potential iron-sulfur protein, LIII, 330
Holey-carbon film, for negative staining grids, XXXII, 24, 26
Hollow-fiber device, XLIV, 292–295
Hollow-fiber membrane, XLIV, 291–317
 cleaning procedure, XLIV, 308–310
 commercial sources, XLIV, 292–294
 immobilization, XLIV, 303–317
  advantages, XLIV, 291
  enzyme slurry, XLIV, 310–312
  enzyme system, XLIV, 314–317
  modified soluble enzyme, XLIV, 310
  soluble enzyme, XLIV, 306–310
 leak check, XLIV, 308
 liquid membrane impregnation, XLIV, 303
 permeability changes, XLIV, 307
 permeability measurement, XLIV, 301, 302
 permeation selectivity, XLIV, 302, 303
 preparation, XLIV, 302, 303
 selection factors, XLIV, 295–300
 sterilization, XLIV, 307, 308
Hollyhock, crown gall studies, XXXI, 557
Holmium, magnetic circular dichroism, XLIX, 177
Holter medium, XXXIX, 404
Homarus americanus
 hemocyanin studies, XLVIII, 260–270
 nerve fiber, spin-label studies, XXXII, 177
Homoarginine, XLVI, 552, 553
 inhibitor, of clostripain, XLVII, 167
 internal standard, XLVII, 34, 38
Homobiotinyl *p*-nitrophenyl ester, XLVI, 615
Homochromatography
 in tRNA sequence analysis, LIX, 63, 64
 two-dimensional, mobility shift analysis, LIX, 82–88, 92, 94, 95, 104–106
Homocitrulline, internal standard, XLVII, 34, 39
Homodyne detection, XLVIII, 423, 457
Homogeneity spoil pulse, XLIX, 258, 268, 289, 337
Homogenization
 of adrenal glands, LII, 135
 of adrenocortical mitochondria, LII, 139
 of arteries and veins, LV, 61
 of brain tissue, LII, 312
 continuous-flow method, LII, 145
 of heart tissue, LV, 42
 of hepatoma tissue, LV, 86
 isolation of mitochondria, LV, 4, 5
 lipid peroxidation, LII, 305
 of liver, XXXI, 4, 5, 7, 8, 95; LII, 71, 72, 92, 145, 195, 201
 with Polytron tissue homogenizer, LII, 235
 with Potter-Elvehjem Teflon-glass homogenizer, LII, 139, 201, 235
 preparation of mitochondria, LV, 116, 117
 rapid, in plant-cell extraction, XXXI, 540, 541
 of rat liver, LV, 34, 35
  mitochondria, LVI, 18
 technique, LIX, 413

of tissue from freeze-stop, LVI, 202, 203
in Waring Blendor, LII, 92, 135, 195, 201, 507
Homogenizer, *see also* Sonifer
Braun, LV, 152, 154, 162, 416
Dounce, LV, 4, 52, 54
Dyno Mill KDL, LV, 155, 156
freeze-stop tissue, LVI, 203
Gaulin-Mantin, LV, 416
HeLa cell disruption, LVI, 68
Kontes Glass Co., LV, 47
no-clearance type, XXXI, 153
Polytron, LV, 4, 61, 202
Potter-Elvehjem, LV, 4, 34, 54, 56, 61, 62, 99, 155
Tekmar, LV, 385
Waring Blendor, LV, 158, 161
yeast cell breakage, LVI, 46
Homogentisate 1,2-dioxygenase, LII, 11
Homomycin, XLIII, 338
Homopeptide, XLIII, 145, 146
Homoserine, XLVI, 156, 414
attachment, XLVII, 283, 306, 309
internal standard, XLVII, 34
Homovanillic acid, XLIV, 632
Honda medium, composition, XXXI, 546
Hondamycin, XLIII, 137
Honey bee
glycerol-3-phosphate dehydrogenase, assay, purification, and properties, XLI, 240–245
triosephosphate dehydrogenase, assay, purification, and properties, XLI, 273–278
Hood, for tissue culture, XXXIX, 110
Hooker-Forbes progestin assay, XXXVI, 457
Hoplophorus, *see Oplophorus*
Hormone, *see also* Antihormone, specific type
adenylate cyclase stimulation, XXXI, 111
antagonist, from adipocytes, XXXVII, 431–438
bioassay, XXXVII, 433–435
characterization, XXXVII, 435–438
assay, XXXVII, 1–130
in perfused liver, XXXIX, 34, 35
biological effects, XXXVII, 221–310
bound, activity studies, XXXVII, 208
cell cultures dependent, XXXII, 557–561
dialysis, assays based, XXXVII, 202, 203
effect on calcium fluxes, *in vivo*, XXXIX, 513–573
on central nervous system, XXXIX, 429–440
on collagens, XL, 307
on embryonal carcinoma cells, LVIII, 105
experimental procedures, XL, 319
on fetal liver explants, XXXIX, 36–40
on GH_3 cells, LVIII, 98–100
on growth, LVIII, 94–109
on HeLa cells, LVIII, 100
on HeLa–S cells, LVIII, 107–109
on melanoma cells, LVIII, 102
metabolic studies, XL, 324
on mouse fibroblasts, LVIII, 101, 102
Swiss 3T3, LVIII, 105, 106
on neuroblastoma, LVIII, 97, 98
on perfused livers, XXXIX, 34
preparation, dosage, routes of administration, XL, 319
on protein turnover in perfused organs, XXXVII, 238–250
on rabbit intimal cells, LVIII, 106, 107
on rat follicular cells, LVIII, 100, 101
on rat glial C_6 cells, LVIII, 103–105
on testicular cells, LVIII, 103
in fat-cell enzyme regulation, XXXI, 68, 69
immunoreactivity, assays, XXXVII, 204, 205
inactivation at receptor sites, XXXVII, 198–211
assays, XXXVII, 199–208
specificity, XXXVII, 209, 210
structural studies, XXXVII, 210
levator ani test for action, XXXIX, 94–101
macroautoradiography, XXXVII, 148–155

microautoradiography, XXXVII, 155–167
microencapsulation, XLIV, 215
precipitation and adsorption, assays, XXXVII, 201, 202
purification and synthesis, XXXVII, 311–438
radioimmunoassay, XXXVII, 22–38
radioiodination, XXXVII, 200, 201
radioligand assay, XXXVII, 3–22
in serum, LVIII, 78, 83
studies using embryos, XXXIX, 297–302
tissue incubation, XXXVIII, 362
ultrastructural localization, XXXIX, 153, 154
Hormone-binding protein, XXXVI, 1–88
Hormone growth factor, LVIII, 78
Hormone receptor, XXXVII, 131–219
autoradiographic methods, XXXVII, 145–167
peroxidase-labeled antibody method, XXXVII, 133–144
Horse
liver
1,3-diphosphoglycerate phosphatase, preparation and purification, XLII, 413, 415, 416
triosephosphate isomerase, assay, purification, and preparation, XLI, 430–434
muscle
1,3-diphosphoglycerate phosphatase, purification, XLII, 412–414
3-phosphoglycerate kinase, purification, XLII, 129, 131, 133
serum, LVIII, 95, 213
in media, LVIII, 523, 530
Horseradish peroxidase, *see also* Peroxidase
assay of superoxide dismutase, LIII, 387
binding properties, LII, 516
carbon monoxide binding, IR studies, LIV, 310
CD studies, LIV, 269, 271, 272, 283, 284
Compound II, oxidation state determination, LIV, 243
conformational composition, LIV, 269, 271, 272
coupled assays with luminol, LVII, 403–405
crude, chromatography, LII, 518
in hydrogen peroxide assay, LII, 347, 348
inhibitor, LII, 349
magnetic circular dichroism, XLIX, 170
MCD studies, LIV, 296
Mössbauer spectra, LIV, 355
reaction with *Pholas* luciferin, LVII, 399, 402, 403
subcellular localization, LII, 251
Horseshoe crab, D-lactate dehydrogenase, assay, purification, and properties, XLI, 313–318
House fly, glucosamine-phosphate isomerase, assay, purification, and properties, XLI, 400–407
Howland current pump, XXXIX, 430
HPLC, *see* High-pressure liquid chromatography
HQS, *see* Lithium 8-hydroxy-5-quinolinesulfonate
HRP, *see* Horseradish peroxidase
HSA, *see* Human serum albumin
HTA Br, *see* Hexadecyltrimethylammonium bromide
HTC cell, *see* Hepatoma, tissue culture cell
Hughes press, XXXI, 656
description, XXXI, 660
properties, XXXI, 663
Human chorionic gonadotropin, *see* Gonadotropin, human chorionic
Human erythrocyte, *see* Erythrocyte
Human Genetic Mutant Cell Repository, LVIII, 441, 442
Human growth hormone
desialylation, XXXVII, 321, 322
iodination, XXXVII, 226–228, 325
logit-log assay, XXXVII, 9
in prolactin purification, XXXVII, 398
radioreceptor assay, XXXVII, 70–77
sialic acid content, XXXVII, 323

tritium labeling, XXXVII, 321–326
Human heart, *see* Heart muscle
Human liver, *see* Liver
Human mutant cell, isolation and culture, XXXII, 799–819
Human seminal plasma inhibitor, *see* Seminal plasma inhibitor, human
Human serum albumin
coupling of histones, XL, 191
to succinylated cyclic nucleotides, XXXVIII, 98
steroid bonding, XXXVI, 93–96
Humidity, effect on culture, LVIII, 73, 74, 136
Hummel-Dreyer gel filtration, LIX, 325–328, 331
Humphrey Electronics step scanner, XXXII, 213
Hunter syndrome, iduronate sulfatase, L, 446, 447
Hurler/Scheie syndrome, α-L-iduronidase activity, L, 443
Hurler syndrome, α-L-iduronidase activity, L, 141, 443
Hutner's Metals 44, modified in *Pseudomonas* growth medium, LIII, 529
HX, *see* Hydrogen exchange
Hyaluronic acid, LVIII, 268
aggregation, LVIII, 268
bovine, osmotic pressure studies, XLVIII, 74
repeating disaccharide structure, L, 279
Hyaluronidase
in mucosal cell isolation, XXXII, 665
in oxyntic cell isolation, XXXII, 712
in parenchymal cell isolation, XXXII, 625–627, 632
in renal tubule isolation, XXXIX, 12
tissue culture, LVIII, 125
Hyamine hydroxide, in α-hydroxylase assay, LII, 312
Hyamine hydroxide reagent, XXXVII, 263
Hyamine–10X hydroxide reagent, XXXII, 655
Hybrid
mouse × human, LVIII, 347, 357
somatic, LVIII, 359, 360
Hybrid cell, LVIII, 324, 345
Hybrid colony frequency, of cells, definition, XXXII, 582
Hybrid fusion, plants, LVIII, 364–366
Hybridimycin, XLIII, 119, 120
Hybridization
cytological, LVIII, 326
in genetic studies, XXXIX, 127
identification of petite mtDNA fragments, LVI, 191, 192
mitochondrial RNA, LVI, 162
Hybrid plant, cell fusion, LVIII, 364–366
Hybrid selection, procedures, LVIII, 351–353
Hydraulic permeability, definition, XLIV, 298
Hydraulic press, cell lysis, LVI, 237
Hydrazide, XLVI, 210
chemical structures, LIX, 173
chemiluminescence, mechanism, LVII, 416–422
derivatives, XXXIV, 36, 37, 50, 51
gel, dinitrophenylation and trinitrophenylation, XXXIV, 48, 49
group determination, XXXIV, 57
as inhibitors, XLVI, 38
linkage, glass, XXXIV, 69
in modification of tRNAs, LIX, 175
synthesis, XLVI, 210; LIX, 174, 179
test, XXXIV, 33; XLVI, 211
Hydrazide-agarose, XXXIV, 475, 476, 490, 491
Hydrazide Bio-Gel P, XLIV, 56
Hydrazide group, titration, XLIV, 446
Hydrazido-agarose, XXXIV, 84, 90–93
preparation of derivatives, XXXIV, 91, 92
Hydrazido-albumin, XXXIV, 92
(Hydrazido)cellulose, XXXIV, 84, 85
Hydrazidosuccinylagarose, XXXIV, 83, 84
Hydrazine, XLIV, 120, *see also* Hydrazide
for hemin reduction, LII, 433
interaction with cytochrome *c* oxidase, LIII, 195
peptide cleavage from support, XLVII, 604
removal, XLVII, 284

synthesis of 6-*N*-(4-aminonaphthalene)-1,2-dicarboxylic acid hydrazide, LVII, 437
of 6-[*N*-(4-aminobutyl)-*N*-ethyl]amino-2,3-dihydrophthalazine-1,4-dione, LVII, 433
of 7-[*N*-(4-aminobutyl)-*N*-ethyl]aminonaphthalene-1,2-dicarboxylic acid hydrazide, LVII, 437
of 6-*N*-(3-amino-2-hydroxypropyl)amino-2,3-dihydrophthalazine-1,4-dione, LVII, 430
of 6-[*N*-(3-amino-2-hydroxypropyl)-*N*-ethyl]amino-2,3-dihydrophthalazine, LVII, 431
of 2-carbethoxy-2-diazoacethydrazide, LIX, 811
of hydrazides, LIX, 174
transhydrogenase, LV, 268, 270
Hydrazine derivative, LIX, 117
incorporation
in tRNA, LIX, 114–121
into wybutine position, LIX, 120
Hydrazine hydrate, XLIV, 446; XLVII, 564, 565, 566
synthesis of 2-[^{14}C]carbethoxy-2-diazoacethydrazide, LIX, 814
of dithiodiglycolic dihydrazide, LIX, 385
9-Hydrazinoacridine, LIX, 110
incorporation in tRNA, LIX, 114, 115, 118
2-Hydrazino-3-(4-imidazolyl)propionic acid solution, for bacteria growth, XXXII, 850
Hydrazinolysis, of arginine peptide, XLVII, 284
3-Hydrazinoquinoline, incorporation in tRNA, LIX, 118
Hydrazoic acid, absorption maximum, LVI, 664
Hydrazone, XXXIV, 50, 51
test, XLVI, 211
Hydride ion, inactivator of flavoprotein enzyme, LIII, 440
Hydride transfer stereospecificity, *see* Stereospecificity, of hydride transfer
Hydrindantin, XLVII, 17, 223
Hydrobromic acid, preparation of dehydroluciferin, LVII, 25
Hydrocarbon
aromatic, monooxygenase induction, LII, 231, 232
chain, flexibility studies by spin-labeling, XXXII, 166–172
polarography, LVI, 477
polycyclic aromatic
as carcinogens, LII, 279
inducer, of cytochrome *P*–450, LII, 270, 271
metabolism, in rat liver, LII, 412, 413
nucleoside adducts, chromatographic separation, LII, 291, 292
phenols and dihydrodiols, chromatographic properties, LII, 293–296
Hydrocarbon-coated agarose, *see* Agarose, hydrocarbon coated
Hydrochloric acid
anilinothiazolinone hydrolysis, XLVII, 370, 371
in ashing procedure, LIV, 481
assay of GMP synthetase, LI, 220
for cleaning capillary tubes, LIX, 8
cleavage, XLVII, 340–343
dissociation of flavoproteins, LIII, 430
drying, XLVII, 341, 354
identification of substituted flavins, LIII, 463, 464
interaction with organic sulfoxides, XLVII, 443, 444, 454
metal-free, preparation, LIV, 476
methylthiazolinone hydrolysis, XLVII, 294
peptidyl thiohydantoin hydrolysis, XLVII, 363
preparation of aniline hydrochloride, LIX, 62
of (carboxymethyl)cellulose azide, LIX, 385
of Chelex 100, LVII, 302
of matrix-bound RNase, LIX, 463
of immobilized enzyme rods, LVII, 204
of matrix-bound RNase, LIX, 463
of zinc amalgam, LVII, 434
in prewashing of ion-exchange resins, LIX, 558

purification of *Cypridina* luciferin, LVII, 367
quality check, XLVII, 14, 15, 18
removal, LIX, 313
separation of protein from polyacrylamide gels, LIX, 313
for stopping biphenyl hydroxylation, LII, 401
synthesis of dithiodiglycolic dihydrazide, LIX, 384
Hydrochloric acid/dioxane, cleavage, XLVII, 342
Hydrocinnamaldehyde, XLVI, 125
Hydrocinnamic acid, XLVI, 123
Hydrocorticosterone, effects, $(Na^+ + K^+)$-ATPase, XXXVI, 435
Hydrocortisone
chromatographic separation, XXXVI, 487
effect on mammary epithelial cells, XXXIX, 445, 446
on melanin, XXXVII, 129
enzyme induction, in liver explants, XXXIX, 39
in media, LVIII, 100, 101, 109
radioimmunoassay, XXXVI, 21
Hydrofluoric acid, XLVI, 201
solvent, of silica gel, LII, 404
Hydrogen
assay of hydrogenase, LIII, 290, 292
exchange in 8-position
mechanism, LIX, 334, 335
microenvironment, LIX, 337–339
rate constants, LIX, 335–337
temperature dependence, LIX, 337
exchange reaction, apparatus, LIII, 300–303
gas
in anaerobic procedures, LIV, 114, 115
in reduction of FMN, LVII, 138
para, *see* Para hydrogen
product of hydrogenase, LIII, 297
proton
lock nuclei, LIV, 189
resonance frequency, LIV, 189
substrate of hydrogenase, LIII, 287, 297
Hydrogen-1, lock nuclei, XLIX, 348
Hydrogen-2, lock nuclei, XLIX, 348
Hydrogen-3
lock nuclei, XLIX, 348
in steroid autoradiography, XXXVI, 153
Hydrogenase, LIII, 286–314
activity, LIII, 287, 296
amino acid composition, LIII, 294
assay, LIII, 290–292
bacterial sources, LIII, 287, 297
from *Clostridium pasteurianum*, LIII, 286–296
composition, LIII, 292, 314
from *Desulfovibrio gigas*, LIII, 627
effect of pH, LIII, 296
inhibitors, LIII, 296, 314
iron content, LIII, 290
iron-sulfur centers
characterization, LIII, 292–294
EPR characteristics, LIII, 265, 266, 271, 292, 295
kinetic properties, LIII, 294, 295
molecular weight, LIII, 294, 313
multiple isoelectric forms, LIII, 313
oxygen-stable, LIII, 296–314
assay, LIII, 297–309
properties, LIII, 313, 314
purification, LIII, 310–313
pH optimum, LIII, 314
purification, LIII, 288–290
sedimentation coefficient, LIII, 313
spectral properties, LIII, 292–294, 313, 314
substrate specificity, LIII, 314
Hydrogenase/thiosulfate-sulfite reductase
assay of cytochrome c_3, LIII, 616
preparation of extract containing, LIII, 618, 619
Hydrogenation, XLVI, 200
catalytic, XLVII, 526, 530, 532, 534
procedure, XLVII, 547, 548
of prostaglandins, XXXV, 372
Hydrogen bacteria, ribulose 1,5-diphosphate carboxylase, assay, purification, and properties, XLII, 461–468
Hydrogen bond, XLIV, 149, 152
ionophores, LVI, 439

Hydrogen bonding
hydrogen exchange studies, XLIX, 26–29, 34, 35
in lipid monolayers, XXXII, 544
NMR studies, LIX, 22
pattern, in extended topographical structures, XLIX, 109, 110
in α-helical topographical structures, XLIX, 106–109
in plant tissue, XXXI, 529
states, XLIX, 107, 108
in transfer RNA, LIX, 17, 20
Hydrogen bromide
peptide cleavage, XLVII, 459–469, 546, 547, 566, 567
peptide deprotection, XLVII, 603
storage, XLVII, 463, 464
Hydrogen chloride, preparation of pyridine hydrochloride, LVII, 21
Hydrogen chloride-acetone method, XLVII, 549
Hydrogen cyanide, XLVII, 517
Hydrogen exchange, XLIX, 24–39, *see also* Tritium exchange
breathing model, XLIX, 35
conformation studies, XLIV, 369
data requirements, XLVIII, 330, 331
difference method, XLIX, 37–39
evaluation, XLVIII, 330
exchange rate constant determination, XLIX, 27
for free peptide, measurement, XLIX, 29, 30
H bond studies, XLIX, 26–30
kinetics, XLVIII, 322–325
methods, XLVIII, 325–327
of peptide group protons, XLIX, 26–30
in polynucleotides, XLIX, 30–33
technique reviews, XLIX, 24, 25
Hydrogen fluoride
catecholamine glass beads, XXXVIII, 186
deprotection, XLVII, 575, 576, 602, 603
reaction vessels, XLVII, 605
removal of iron from cytochrome *c*, LIII, 176
Hydrogen half cell, LIV, 412
redox potentials, at various temperatures, LIV, 413
Hydrogen iodide
anilinothiazolinone hydrolysis, XLVII, 371, 372
solid-phase hydrolysis, XLVII, 297
Hydrogen ion, *see also* Proton
bacteriorhodopsin, LV, 592, 594
calcium-selective electrodes, LVI, 348
concentration, effect on self-association, XLVIII, 105
electrochemical potential outside membrane, LV, 227
gradient, evaluation, LV, 227–229
indicators, LVI, 310–313
membrane charge density, LVI, 525
oxidative phosphorylation, LV, 226, 227
Pasteur effect, LV, 295
release, membrane orientation, LVI, 403
stereospecific transfer between pyridine nucleotides, LV, 274, 275
transport, LVI, 250, 336–338
Hydrogen nucleus
coherent scattering length, LIX, 630, 672
scattering cross section, LIX, 672
Hydrogenomonas eutropha, ribulose-1,5-diphosphate carboxylase
chromatography, XLII, 466
purification and properties, XLII, 465–468
Hydrogenomonas facilis, ribulose-1,5-diphosphate carboxylase
chromatography, XLII, 465
purification and properties, XLII, 464–468
Hydrogen peroxide, XLVI, 49–51, 315; LII, 3, 6, 342–350
acatalasemia, XLIV, 684–688
amperometric determination, XLIV, 586, 590, 591, 613, 614
in ashing procedure, LIV, 481
assay
chemical, LII, 346, 347
enzymic, LII, 347–349
interference, LII, 349, 350
with luminol, LVII, 403, 445–462
of *Pholas* luciferase, LVII, 386
using *Diplocardia* bioluminescent system, LVII, 377–381

in chloroperoxidase assay, LII, 523
in copper analysis, LIV, 444
decomposition and stabilization, LVI, 459
determination, LII, 345
dissolution of SDS gels, LVII, 176
elimination of chemiluminescence, LI, 122
formation, by microsomes, LII, 344, 345
heme-deficient mutants, LVI, 559
hydrogen donor, for cytochrome P–450 peroxidase activity, LII, 411
indole derivative chemiluminescence, LVII, 348
inhibitor, of brain α-hydroxylase, LII, 317
of thymidylate synthetase, LI, 96
interference, in oxygen consumption studies, LIV, 486, 487
in iron analysis, LIV, 441
means of formation, LVI, 461
measurement, LVI, 448, 449
methionine oxidation, XLVII, 456
microcalorimetric determination, XLIV, 674, 675
modification of methionine, LIII, 139
of tyrosine, LIII, 148, 149
oxidant, of cytochrome c_1, LIII, 226
oxygen donor, to cytochrome P–450, LII, 407
phagocyte chemiluminescence, LVII, 462–466, 488, 489
polarographic monitoring, XLIV, 617
polarography
general rationale, LVI, 448–450
medical and other applications, LVI, 450–452
preparation of cross-linked ribosomal subunits, LIX, 537
product of flavoprotein oxidases, LIII, 398
of melilotate hydroxylase, LIII, 557
of salicylate hydroxylase, LIII, 528, 537, 538
reagent, peroxidase stain, XLIV, 713
stock solutions, preparation and titration, LVII, 378
substrate
in luminol chemiluminescence reaction, LVII, 447, 448
of *Pholas* luciferase, LVII, 398
peroxidase, XLIV, 716
in uric acid assay, XLIV, 631, 632, 674
Hydrogen peroxide electrode, measurement of hydrogen peroxide, LVI, 456–459
Hydrogen peroxide-hematin system
chemiluminescence assay
disadvantages, LVII, 444, 445
procedure, LVII, 426
sensitivity, LVII, 438, 439, 444
competitive protein-binding assay, LVII, 443
Hydrogen peroxide-lactoperoxidase system
chemiluminescence assay
procedure, LVII, 427
sensitivity, LVII, 438
competitive protein-binding assay, LVII, 441, 442
Hydrogen peroxide-microperoxidase system
chemiluminescence assay
data reproducibility, LVII, 441
procedure, LVII, 426
sensitivity, LVII, 438, 439, 444
Hydrogen peroxide-potassium ferricyanide system
chemiluminescence assay
procedure, LVII, 427
sensitivity, LVII, 438
Hydrogen sulfide, XLIV, 231
in adrenodoxin chromophore reconstitution, LII, 137
interaction with cytochrome c oxidase, LIII, 201
preparation of dehydroluciferin, LVII, 25
of luciferin, LVII, 17
quantitative determination, LIII, 277
Hydrolase
assay by Fast Analyzer, XXXI, 816
in pancreas, XXXI, 47
in plant cells, XXXI, 501
use for plant-cell rupture, XXXI, 503

Hydrolysis
in combination with reduction-sulfopropylation, XLVII, 121, 122
fragment, XLIII, 260–263
by immobilized enzymes, XLVII, 40–45
of phenylthiohydantoins, XLVII, 351, 371
Hydroperoxide, quantitative determination, LII, 509, 510
Hydrophilicity, XLIV, 19, 20, 84
Hydrophobic chromatography, XXXIV, 114, 126–140
hydrocarbon-coated agarose, XXXIV, 130, 131, 137–139
hydrophobic interactions, XXXIV, 134
resolution test kit, XXXIV, 134–136
Hydrophobicity, XLIV, 84
of amino acid residues, XLIV, 13, 14
on glass surface, procedure, XLIV, 194
of mitochondrial proteins, LVI, 76
Hydrophobic probe, photoaffinity labeling, XLVI, 191
Hydroquinone, cofactor for dihydroorotate dehydrogenase, LI, 63
Hydroxamate
assay
carnitine palmitoyltransferase, LVI, 372
of long-chain fatty acyl-CoA synthase, XXXV, 117
determination, XLVII, 138
test, LII, 517
Hydroxamate method, LIX, 238
Hydroxamic acid, substrate of succino-AICAR synthetase, LI, 193
Hydroxamic acid-affinity matrix, synthesis, LII, 515, 516
Hydroxide ion, cyanogen bromide activation, XLIV, 28
Hydrox Purifier, Model 8301, LIV, 118
D-2-Hydroxy acid, assay with 2-hydroxy acid dehydrogenase, XLI, 328
D-α-Hydroxyacid dehydrogenase, flavoprotein classification, LIII, 397
D-2-Hydroxyacid dehydrogenase
from animal tissue, assay, XLI, 323–329
from rabbit kidney, XLI, 325–329
inhibitors, XLI, 328
preparation and purification, XLI, 325–327
properties, XLI, 327–329
α-Hydroxyacid oxidase, in microbodies, LII, 496–498, 503
D-α-Hydroxyacid oxidase, inactivation, LIII, 439
L-α-Hydroxyacid oxidase
inactivation, LIII, 439, 443, 446, 447
as marker enzyme, XXXI, 742
in peroxisomes, XXXI, 364, 367
β-Hydroxyacyl-CoA dehydrogenase
in assay of crotonase, XXXV, 137
of α,β-unsaturated acyl-CoA derivatives, XXXV, 140
in glyoxysomes, XXXI, 569
L-3-Hydroxyacyl-CoA dehydrogenase, XXXV, 122–128
assay, XXXV, 123
inhibitors, XXXV, 127, 128
stability, XXXV, 127
Hydroxyalkyl methacrylate gel, XLIV, 66–83
activated
preparation steps, XLIV, 70
storage, XLIV, 70
amino derivatives, preparation, XLIV, 76, 77
anhydride derivative, XLIV, 83
carboxyl derivatives, preparation, XLIV, 76, 77
chemical modifications, methods, XLIV, 69, 70
chemical properties, XLIV, 69
copolymerization with monomers containing functional groups, XLIV, 71
containing functional group precursors, XLIV, 71, 72
hydrazide derivative, XLIV, 82, 83
mechanical stability, XLIV, 66
physical properties, XLIV, 68, 69
polymer analogous transformation, XLIV, 70
preparation, XLIV, 66–68
reaction mechanism, XLIV, 66, 67
surface modification, XLIV, 70
swelling capacity, XLIV, 66, 69

3β-Hydroxy-5α-androstane 17α-acetate, in blood, XXXVI, 73
3α-Hydroxy-5α-androstan-17-one, chromatographic separation, LII, 379
17β-Hydroxyandrostan-3-one, PBP affinity, XXXVI, 125
17β-Hydroxy-5α-androstan-3-one, *see also* 5α-Dihydrotestosterone
  chromatographic separation, LII, 379
17β-Hydroxyandrostan-3-yl succinamide-*p*-aminobenzene, in affinity chromatography, XXXVI, 114, 115
3β-Hydroxy-5-androstene-17β-carboxylic acid, XLVI, 477
17β-Hydroxyandrost-4-en-3-one, *see* Testosterone
3-Hydroxyanthranilate 3,4-dioxygenase, LII, 11
3-Hydroxyanthranilate oxidase, LVI, 475
Hydroxyapatite *see also* Hydroxylapatite
  adenylate cyclase, XXXVIII, 165
  in chromatin fraction, XL, 160
  in chromatin separation, XL, 162
  chromatography
    purification of complex III, LIII, 88, 94
      of CTP(ATP):tRNA nucleotidyltransferase, LIX, 125
      of cytochrome oxidase subunits, LIII, 77
      of *Diplocardia* luciferase, LVII, 376
      of flavocytochrome b_2, LIII, 247, 248
      of hydrogenase, LIII, 289
      of *p*-hydroxybenzoate hydroxylase, LIII, 547
      of nitrate reductase, LIII, 643
      of peptide constituents of cytochrome bc_1 complex, LIII, 107–109
      of tRNA methyltransferase, LIX, 197, 198
  cleanup for use, LX, 103
  protein kinase, XXXVIII, 312, 318–321, 326
    regulatory subunit, XXXVIII, 327
  purification
    casein kinases, LX, 501, 502
    elongation factors, LX, 142, 582, 583, 639, 641–645, 654–656, 675, 691, 693, 707, 709
    of F_1 ATPase subunit, 1, LV, 355
      of subunit, 2, LV, 355
    of heme-regulated translational inhibitor, LX, 471, 472, 486, 487, 495
    of initiation factors
      Co-EIF-1, LX, 40, 54, 56
      eIF, various, LX, 106, 187, 189, 242, 243
      EIF-1, LX, 42, 44
      IF_{EMC}, LX, 96
      $IF–M_{\alpha}$, LX, 99
    of long-chain fatty acyl-CoA synthase, XXXV, 120
    of messenger ribonucleic acid-binding protein, LX, 396, 397
    of palmityl thioesterase I, XXXV, 104
    of protein kinases, LX, 509
  thermal elution, XL, 94
  transhydrogenase, LV, 278, 279
Hydroxyapatite-Sephadex G–25 column
  preparation, LIII, 316
  purification of nitrogenase, LIII, 316
2-L-Hydroxyarachidic acid methyl ester, (−)-menthyloxycarbonyl derivative, gas chromatography, XXXV, 328
4-(Hydroxyazobenzyl)-2′-carboxylic acid, as matrix, XXXIV, 265, 266
2-Hydroxybehenic acid
  in α-hydroxylase assay, LII, 311
  in saponification mixture, LII, 312
  synthesis, LII, 311
m-[^{14}C]Hydroxybenzaldehyde, XLIII, 544
p-Hydroxybenzoate hydroxylase, *see also* 4-Hydroxybenzoate-3-monooxygenase
  oxy form, spectrum, LII, 36
  reaction mechanism, LII, 39
3-Hydroxybenzoate 4-hydroxylase, *see* 3-Hydroxybenzoate 4-monooxygenase
3-Hydroxybenzoate 6-hydroxylase, *see* 3-Hydroxybenzoate 6-monooxygenase
4-Hydroxybenzoate 3-hydroxylase, *see* 4-Hydroxybenzoate 3-monooxygenase

3-Hydroxybenzoate 4-monooxygenase, LII, 16
3-Hydroxybenzoate 6-monooxygenase, LII, 17
4-Hydroxybenzoate 3-monooxygenase, LII, 17; LIII, 543–552
  activity, LIII, 544
  assay, LIII, 545
  crystallization, LIII, 547
  flavin content, LIII, 548
  inhibitors, LIII, 551
  molecular weight, LIII, 548
  oxygenated-flavin intermediates, LIII, 550, 551
  from pseudomonads, LIII, 544
  from *Pseudomonas fluorescens*, LIII, 544–552
  purification, LIII, 545–547
  reaction mechanism, LIII, 551, 552
  reconstitution, LIII, 549
  reduction by NADPH, LIII, 549, 550
  reversible dissociation, LIII, 433
  stability, LIII, 547, 548
  substrate specificity, LIII, 548, 549
3-Hydroxybenzoic acid
  pseudosubstrate of melilotate hydroxylase, LIII, 557
4-Hydroxybenzoic acid
  accumulation by mutants, LVI, 110
  intermediate in ubiquinone biosynthesis, LIII, 600, 601
  purification of *p*-hydroxybenzoate hydroxylase, LIII, 546
  structural formula, LIII, 601
  substrate
    of *p*-hydroxybenzoate hydroxylase, LIII, 55
    of salicylate hydroxylase, LIII, 538, 539
4-Hydroxy[U-^{14}C]benzoic acid, for radiolabeling of quinone intermediates, LIII, 608
3-Hydroxybenzo[*a*]pyrene
  chromatographic separation, LII, 285–287
  fluorometric assay, LII, 234–236
  spectrofluorometric determination, LII, 410
  for standard curve preparation, LII, 238, 239
9-Hydroxybenzo[*a*]pyrene, chromatographic separation, LII, 285–287
D-(−)-2-(6′-Hydroxy-2′-benzothiazolyl)-Δ^{2}-thiazoline-4-carboxylic acid, *see* Luciferin
1-Hydroxybenzotriazole, XLVII, 540, 568, 570, 597
  carbodiimide activation, XLVII, 282
1-Hydroxybenzotriazole monohydrate, XLVII, 383
  recrystallization, XLVII, 382
m-Hydroxybenzyl alcohol dehydrogenase, XLIII, 540–548
  assay, XLIII, 544
  calculation, XLIII, 544
  cell-free extract, XLIII, 543
  gentisyl alcohol dehydrogenase assay, XLIII, 544
  m-[1-^{14}C]hydroxybenzaldehyde assay, XLIII, 544
  inhibitors, XLIII, 546
  kinetic properties, XLIII, 546
  Michaelis-Menten plots, XLIII, 546
  molecular weight, XLIII, 546
  Penicillium urticae preparation, XLIII, 542
  pH optimum, XLIII, 546
  Polyclar AT treatment, XLIII, 543, 545
  polyketides, XLIII, 541
  properties, XLIII, 545–548
  purification, XLIII, 544, 545
  reaction scheme, XLIII, 541
  specificity, XLIII, 547
  unit definition, XLIII, 544
3-Hydroxybenzyl alcohol:NADP oxidoreductase, *see m*-Hydroxybenzyl alcohol dehydrogenase
Hydroxybenzylpindolol, XLVI, 591
2-Hydroxybiphenyl
  chromatographic properties, LII, 405
  maximum fluorescence wavelength, LII, 400
  metabolite, of biphenyl, LII, 399
3-Hydroxybiphenyl, chromatographic properties, LII, 405
4-Hydroxybiphenyl
  chromatographic properties, LII, 405

maximum fluorescence wavelength, LII, 400
metabolite, of biphenyl, LII, 399
5-*exo*-Hydroxy-2-bornanone, *see* 5-*exo*-Hydroxycamphor
2-Hydroxy-3-butenoic acid, XLVI, 39, *see also* Vinyl glycolate
9-(4′-Hydroxybutyl)adenine, inhibitor of tryptophanyl-tRNA synthetase, LIX, 252
9-(4′-Hydroxybutyl)adenine triphosphate, inhibitor of tryptophanyl-tRNA synthetase, LIX, 252
2-Hydroxy-3-butynoic acid, XLVI, 162
inactivator of flavoprotein enzymes, LIII, 439–441
suicide substrate of D-lactate dehydrogenase, LIII, 525
synthesis, scheme, LIII, 438
3-Hydroxybutyrate, *see* 3-Hydroxybutyric acid
D-β-Hydroxybutyrate-acetoacetate ratio, in perfusion fluid, LII, 51
D-β-Hydroxybutyrate apodehydrogenase
adsorption chromatography, XXXII, 389, 390
assay, XXXII, 378, 379
glass-bead chromatography, XXXII, 384–389
lecithin requirement, XXXII, 374–391
precipitation, XXXII, 384
properties, XXXII, 391
protein assays, XXXII, 378
release from mitochondria, XXXII, 381–383
specific activity, XXXII, 383
3-Hydroxybutyrate dehydrogenase, XXXIV, 238
hepatoma mitochondria, LV, 82
in liver compartments, XXXVII, 288
stereospecificity determination, LIV, 229
3-Hydroxybutyric acid, XXXI, 39
brain mitochondria, LV, 57
hepatoma mitochondria, LV, 81–83
oxidation, LVI, 285
characteristic → H^+/O values, LV, 638
β-Hydroxybutyryl-*N*-acetylcysteamine
in assay of dehydrase activity of fatty acid synthase complex, XXXV, 50, 51
synthesis, XXXV, 48
5-*exo*-Hydroxycamphor
product of camphor 5-monooxygenase system, LII, 166, 168
of cytochrome P–450_{cam}, LII, 156
3-D-Hydroxycapric acid methyl ester, (−)-menthyloxycarbonyl derivative, gas chromatography, XXXV, 328
10-Hydroxycaproic acid, synthesis, LII, 320
5-Hydroxy-7-chlorotetracycline, XLIII, 342
25-Hydroxycholecalciferol, kidney mitochondria, LV, 11, 12
20α-Hydroxycholesterol
effect on cytochrome P–450_{scc}, LII, 131
substrate of cytochrome P–450_{scc}, LII, 132
22*R*-Hydroxycholesterol
effect on cytochrome P–450_{scc}, LII, 131
substrate of cytochrome P–450_{scc}, LII, 132
2-Hydroxycinnamic acid, substrate of melilotate hydroxylase, LIII, 557, 558
3-Hydroxycinnamic acid, pseudosubstrate of melilotate hydroxylase, LIII, 557
17-Hydroxycorticosteroid, effect on collagens, preparation, dosage, route of administration, XL, 319
Hydroxycorticosterone, in superfused adrenal tissue, XXXIX, 307–314
18-Hydroxycorticosterone, metabolites, synthesis and isolation, XXXVI, 503–512
6β-Hydroxycortisol, separation, XXXVI, 501–503
7-Hydroxycoumarin
in 7-ethoxycoumarin synthesis, LII, 373
fluorescence spectra, LII, 375, 376
production, LII, 67
β-Hydroxydecanoyl thioester dehydrase, XLVI, 158
deficiency of, in *E. coli* mutants, XXXII, 859

Hydroxydecanoyl thiolester dehydrase, inactivator, LIII, 438
18-Hydroxy-11-dehydrocorticosterone
  biosynthesis, XXXVI, 509, 510
  oxidation, XXXVI, 510, 511
18-Hydroxydeoxycorticosterone
  metabolism, in adrenal gland, LII, 124
  R_f value, LII, 126
  in superfused adrenal tissue, XXXIX, 307–314
m-Hydroxydiphenyl, in colorimetric assay of hexuronic acids, XLI, 31
Hydroxydodecanoic acid, in mass spectrometry of triglycerides, XXXV, 355
6-Hydroxydopamine
  autoxidation, as source of free radicals, LIII, 387
  destruction of adrenergic terminals, XXXIX, 435
9-(2′-Hydroxyethyl)adenine, inhibitor of tryptophanyl-tRNA synthetase, LIX, 252
9-(2′-Hydroxyethyl)adenine triphosphate, inhibitor of tryptophanyl-tRNA synthetase, LIX, 252
Hydroxyethylcellulose, osmometric studies, XLVIII, 73, 75
2-Hydroxyethyl disulfide, inhibitor of thymidylate synthetase, LI, 96
N-(2-Hydroxyethylenediamine)-*N*,*N*′,*N*′-triacetic acid, as calcium buffer, LVI, 352
Hydroxyethylferrocene, as mediator-titrant, LIV, 408
2-Hydroxyethyl methacrylate, XLIV, 57, 67, 172–175
2-Hydroxyethyl methacrylate gel
  hydrolytic stability, XLIV, 69
  specific surface area, XLIV, 68
  structural formula, XLIV, 67
N-2-Hydroxyethylpiperazine-*N*′-2-ethanesulfonic acid, XXXII, 375, 376; XXXVII, 84; LII, 66, 205, 222
  in activation of succinate dehydrogenase, LIII, 474, 475
  aminoacylation of tRNA, LIX, 216
  in antibody studies with nitrate reductase, LIII, 352, 353
  assay of adenylosuccinate synthetase, LI, 207
    of carbamoyl-phosphate synthetase, LI, 112, 113, 122
    of 5-methyluridine methyltransferase, LIX, 191
    of ribonucleoside diphosphate reductase, LI, 228, 238
    of PAP, LVII, 253
    of uridine 5-oxyacetic acid methyl ester methyltransferase, LIX, 191
  cell-free protein synthesis system, LIX, 300
  determination of aminoacylation isomeric specificity, LIX, 272
  in granulosa cell culture media, XXXIX, 205, 208, 209
  for HGH assay, XXXVII, 70–77
  inhibitor in hydrogen peroxide assay, LVII, 380
  isolation of chloroplastic tRNA, LIX, 208, 209
    of chloroplasts, LIX, 434
  Krebs Ringer, XXXVII, 193
  for liver fractionation, XXXI, 8, 10, 11, 14
  for liver preservation, XXXI, 4, 5
  in media, LVIII, 70, 72, 73
  for plasma membrane isolation, XXXII, 119
  purification of luciferin sulfokinase, LVII, 249
    of 50 S ribosomal proteins, LIX, 490, 502
  reductive methylation reaction, LIX, 784
  reverse salt gradient chromatography, LIX, 216
  for sarcoplasmic reticulum isolation, XXXI, 244, 245
  in seawater complete medium, LVII, 155
  for tissue culture, XXXIX, 111
Hydroxy fatty acid, thin-layer chromatography, XXXV, 403
α-Hydroxy fatty acid
  copper chelate, preparation, LII, 314
  standards, preparation, LII, 320

2-Hydroxy fatty acid
(–)-menthyloxycarbonyl derivatives
gas–liquid chromatography, for resolution of stereoisomers, XXXV, 328–330
mass spectra, XXXV, 331
preparation, XXXV, 326, 327
thin-layer chromatography, XXXV, 327
stereoisomers, resolution as (–)-menthyloxycarbonyl derivatives, XXXV, 326–331
3-Hydroxy fatty acid
2-D-phenylpropionate derivatives
gas liquid chromatography, for resolution of stereoiosmers, XXXV, 328–330
mass spectra, XXXV, 331
preparation, XXXV, 328
thin-layer chromatography, XXXV, 328
stereoisomers, resolution as 2-D-phenylpropionate derivatives, XXXV, 326–331
C(4a)-Hydroxyflavin, reaction intermediate, LIII, 551
Hydroxyindole-*O*-methyltransferase
assay, XXXIX, 391, 392
in melatonin biosynthesis, XXXIX, 379, 380
2,2-Hydroxy-3-isopropylaminopropoxyiodobenzene, XLVI, 579
Hydroxylamine
arginine blockage, XLVII, 159
assay, LIII, 637
of carbamoyl-phosphate synthetase, LI, 30
of superoxide dismutase, LIII, 387
of tryptophanyl-tRNA synthetase, LIX, 238
for α,β-unsaturated acyl-CoA derivatives, XXXV, 140
decarbethoxylation, XLVII, 431, 435, 438, 439
determination, XLVII, 138
formation in spin-labeling studies, XXXII, 185
inactivator of protein B, LIII, 381
inhibitor, of catalase, LII, 412
of complex IV, LIII, 45
of FGAM synthetase, LI, 201
of GMP synthetase, LI, 217
interaction with cytochrome *c* oxidase, LIII, 194, 199
M. phlei membrane systems, LV, 186
as mutagen, XLIII, 35
oxidation by *Nitrosomonas europaea*, LIII, 637
penicillin analysis, XLIV, 760
reactivation of tryptophanyl-tRNA synthetase, LIX, 251
salt-free, preparation, LIX, 236
specific protein cleavage, XLVII, 132–145
thioester bond cleavage, XLIV, 524
Hydroxylamine-cytochrome *c* reductase, from *Nitrosomonas europaea*, LIII, 637
Hydroxylamine hydrochloride, XLIV, 105, 385, 532
in copper analysis, LIV, 444
synthesis of affinity matrix, LII, 517
Hydroxylamine oxidase, LII, 21; LVI, 475
Hydroxylapatite, *see also* Hydroxyapatite
BKA-protein complex, LVI, 414
CAT-protein complex, LVI, 409, 412
column preparation, LI, 114, 115
purification of adenosine deaminase, LI, 511
of adenosine monophosphate nucleosidase, LI, 266, 267
of adenylosuccinate AMP-lyase, LI, 204
of adenylosuccinate synthetase, LI, 209, 210
of aspartate carbamyltransferase, LI, 44, 53
of CPSase-ATCase complex, LI, 115, 116
of cytidine deaminase, LI, 403
of deoxythymidylate phosphohydrolase, LI, 287
of guanine deaminase, LI, 514, 515
of orotate phosphoribosyltransferase, LI, 72
of orotidylate decarboxylase, LI, 76, 78
of purine nucleoside phosphorylase, LI, 527, 528

of pyrimidine nucleoside phosphorylase, LI, 434
of succino-AICAR synthetase, LI, 192
of thymidine phosphorylase, LI, 440, 441, 444
of thymidylate synthetase, LI, 94, 95
of uridine-cytidine kinase, LI, 318
of uridine nucleosidase, LI, 293
of uridine phosphorylase, LI, 428
Hydroxylapatite-cellulose column, preparation, LII, 113
Hydroxylapatite chromatography
of CBG, XXXVI, 108, 109
of cytoplasmic progesterone receptors, XXXVI, 204
in epoxide hydrase purification, LII, 197, 198
in NADH-cytochrome b_5 reductase purification, LII, 105
in *Rhizobium* *P*–450 purification, LII, 163, 164
Hydroxylapatite-silica gel column, preparation, LII, 112
Hydroxylase, for steroid hydroxylation, LII, 6, 383, 384
α-Hydroxylase
activators and inhibitors, LII, 316, 317
in brain, assay, LII, 310–316
effects of aging, LII, 318
kinetic properties, LII, 317
in liver, LII, 317
reaction mechanism, LII, 317
specificity, LII, 317, 318
stability, LII, 316
subcellular localization, LII, 316
ω-Hydroxylase, *see* Alkane 1-monooxygenase
11β-Hydroxylase, XLIV, 184, 471; LII, 6
activity assay, XLIV, 187
cytochrome P_{450}, XXXVII, 309
effect of adrenodoxin antibody, LII, 244
17α-Hydroxylase, lack, in adrenals, XXXVII, 299
18-Hydroxylase, cytochrome P_{450}, XXXVII, 309
11β-Hydroxylation, in mitochondria, XXXIX, 163
ω-Hydroxylation reaction, schematic, LIII, 346
11-Hydroxylauric acid
methyl esters, chromatographic separation, LII, 321–323
synthesis, LII, 320
12-Hydroxylauric acid
methyl esters, chromatographic separation, LII, 321–323
synthesis, LII, 320
Hydroxyl group
acetylation, XLVII, 546
activation, XLIV, 542
blocking, XLVII, 527, 528, 543, 544
2-L-Hydroxylignoceric acid methyl ester, (−)-menthyloxycarbonyl derivative, gas chromatography, XXXV, 328
Hydroxyl ion
exchange, LVI, 253
relative competitiveness in hydrogen exchange experiments, XLIX, 29
3-Hydroxyl-2-naphthoic acid, chemiluminescence, LVII, 423
Hydroxyl radical, role in phagocyte chemiluminescence, LVII, 463, 464
Hydroxylysine
^{14}C-labeled, measurement, XL, 353
total aglycosylated plus nonglycosylated, XL, 357
detection, XLVII, 17
glycosylated, separation, XL, 358
internal standard, XLVII, 34, 39
Hydroxylysine-linked carbohydrate, XL, 353
2-Hydroxy-4-maleimidobenzoylazide, XLVII, 421–422
6-Hydroxymelatonin, as melatonin metabolite, XXXIX, 381
p-Hydroxymercuribenzoate, XXXIV, 553; XLVII, 428
effect on epoxide hydrase, LII, 200
guanylate cyclase, XXXVIII, 199
inhibitor of adenine phosphoribosyltransferase, LI, 574
of amidophosphoribosyltransferase, LI, 178
of carboxypeptidase Y, XLVII, 89
of CPSase-ATCase complex, LI, 118
of cytidine deaminase, LI, 411

of deoxythymidine kinase, LI, 365
of guanylate kinase, LI, 489
of rat liver acid nucleotidase, LI, 274
phosphodiesterase, XXXVIII, 235
as spacer, XXXIV, 72
transport, LVI, 292
Hydroxymercurihydroquinone *O,O*-diacetate, in derivatization of transfer RNA, LIX, 14
p-Hydroxymercuriphenyl sulfonate
inhibitor of adenylate kinase, LI, 465
of rat liver acid nucleotidase, LI, 274
β-Hydroxymethethyl NAD^+ analog, XLVI, 74
7-Hydroxy-6-methoxycoumarin, *see* Scopoletin
4-Hydroxy-3-(3-methyl-2-butenyl)benzoic acid, XLIII, 506
(Hydroxymethyl)cellulose, LVIII, 580
Hydroxymethyldeoxycytidine triphosphate
activator of thymidine kinase, LI, 358
substrate of deoxycytidylate deaminase, LI, 417
3-Hydroxy-3-methylglutaryl-CoA lyase, interference in assay of HMG-CoA synthase, XXXV, 157, 162
3-Hydroxy-3-methylglutaryl-CoA synthase
assays, XXXV, 155–157, 161, 162, 174
cytosolic, from chicken liver, XXXV, 160–167
mitochondrial, from chicken liver, XXXV, 155–160
from yeast, XXXV, 173–177
5-Hydroxymethyl-2′,3′-*O*-isopropylidene-uridine, synthesis, XLVI, 347, 348
3*S*-5-Hydroxy-3-methylpentanoic acid lactone, XLIV, 839, 840
6-Hydroxymethyltetrahydropterin, substrate of phenylalanine hydroxylase, LIII, 285
α-Hydroxymyristic acid, quantum yield, in bioassay, LVII, 192
13-Hydroxymyristic acid, synthesis, LII, 320
14-Hydroxymyristic acid, synthesis, LII, 320
β-Hydroxymyristyl coenzyme A, in assay for palmityl thioesterase II, XXXV, 107
5-Hydroxynaphthalene-1,2-dicarboxylic anhydride, LV, 515
7-Hydroxynaphthalene-1,2-dicarboxylic anhydride
synthesis, LVII, 436
of dimethyl 7-aminonaphthalene-1,2-dicarboxylate, LVII, 436
1-Hydroxy-2-naphthoic acid, substrate of salicylate hydroxylase, LIII, 538
3-Hydroxy-2-naphthoic acid, substrate of salicylate hydroxylase, LIII, 538
2-Hydroxy-1,4-naphthoquinone
as mediator-titrant, LIV, 408, 423, 429, 433
in redox titration, LIII, 491
N-(2-Hydroxy-3-naphthyloxypropyl)-*N*′-bromoacetylethylenediamine, XLVI, 578, 597
N-(2-Hydroxy-3-naphthyloxypropyl)ethyl-enediamine, XLVI, 596, 597
D-6-Hydroxynicotine oxidase, LII, 18; LVI, 475
flavin linkage, LIII, 450
flavin peptide, amino acid analysis, LIII, 457
L-6-Hydroxynicotine oxidase, LII, 18; LVI, 475
6-Hydroxynicotinic acid, effector of *p*-hydroxybenzoate hydroxylase, LIII, 550
4-Hydroxy-3-nitroacetophenone, preparation, XLV, 8
2-Hydroxy-5-nitrobenzylbromide, in modification of tryptophan, LIII, 170
(2-Hydroxy-*S*-nitrobenzyl)sulfonium bromide, modification of tryptophans in enzyme, XXXV, 128
8-Hydroxyoctanoic acid, synthesis, LII, 320
2-Hydroxypalmitic acid, in saponification mixture, LII, 312
16-Hydroxypalmitic acid, synthesis, LII, 320
2-L-Hydroxypalmitic acid methyl ester, (−)-menthyloxycarbonyl derivative, gas chromatography, XXXV, 328
p-Hydroxyphenethyl derivatives, XXXIV, 49
4-Hydroxyphenylacetate oxidase, LII, 17

3-(2-Hydroxyphenyl)propionic acid, *see* Melilotate
p-Hydroxyphenylpyruvate hydroxylase, LII, 38
p-Hydroxyphenylpyruvate oxidase, LII, 12
Hydroxypolyethoxydodecane, cholesterol emulsions, LVI, 472
17α-Hydroxypregn-4-ene-3,20-dione, *see* 17-Hydroxyprogesterone
21-Hydroxypregn-4-ene-3,20-dione, *see* Deoxycorticosterone
3β-Hydroxypregn-5-en-20-one, *see* Pregnenolone
20α-Hydroxypregn-4-en-3-one
  chromatographic separation, XXXVI, 487
  PBP affinity, XXXVI, 125
20β-Hydroxypregn-4-en-3-one
  chromatographic separation, XXXVI, 487
  in corpus luteum, XXXIX, 240, 242, 243
  as 5α-oxidoreductase substrate, XXXVI, 473
  oxime derivative, XXXVI, 18
  PBP affinity, XXXVI, 125
11α-Hydroxyprogesterone, radioimmunoassay, XXXVI, 20, 21
11β-Hydroxyprogesterone, radioimmunoassay, XXXVI, 20, 21
16α-Hydroxyprogesterone, preparation, XXXVI, 379
17α-Hydroxyprogesterone
  antiglucocorticoid, XL, 291, 292
  chromatographic separation, XXXVI, 487
  cytochrome P_{450} effects, XXXVII, 310
  derivatives, XXXVI, 20
  logit-log assay, XXXVII, 9
  molecular bioassay, XXXVI, 465
  as 5α-oxidoreductase substrate, XXXVI, 473
  protein-binding assay, XXXVI, 34, 36, 46–48
  radioimmunoassay, XXXVI, 20, 21
  as testosterone precursor, XXXIX, 256
  thin-layer chromatography, XXXIX, 262
20α-Hydroxyprogesterone
  molecular bioassay, XXXVI, 465
  synthesis, LVIII, 571
6β-Hydroxyprogesterone ethyl diazomalonate, synthesis, XXXVI, 418
11α-Hydroxyprogesterone ethyl diazomalonate, synthesis, XXXVI, 417
Hydroxyproline
  ^{14}C-labeled, Juva and Prockop procedure, XL, 371
  detection system, XLVII, 8
  internal standard, XLVII, 34
  measurement
    paper chromatography, XL, 374
    procedures, XL, 367, 369, 371
    in urine samples, XL, 370
  in media, LVIII, 53, 63
9-(3′-Hydroxyprolyl)adenine, inhibitor of tryptophanyl-tRNA synthetase, LIX, 252
9-(3′-Hydroxyprolyl)adenine triphosphate, inhibitor of tryptophanyl-tRNA synthetase, LIX, 252
3-Hydroxypropanesulfonic acid γ-sultone, *see* 1,3-Propane sultone
6-Hydroxypurine, phosphodiesterase, XXXVIII, 244
Hydroxypyridine hydroxylase, LII, 20
4-Hydroxypyridine 3-hydroxylase, LII, 17
Hydroxypyruvaldehyde phosphate, XLVI, 49
Hydroxypyruvate reductase
  in glyoxysomes, XXXI, 569
  in microbodies, LII, 496, 499
8-Hydroxyquinoline
  inhibitor of *Cypridina* luciferase, LVII, 352
    of NADPH-cytochrome *P*–450 reductase, LII, 96
    of nucleoside phosphotransferase, LI, 393
    of phenylalanine hydroxylase, LIII, 285
    of *Pholas* luciferase, LVII, 398
  removal of iron, LVI, 390
Hydroxyquinoline *N*-oxide, interactions with bacterial cytochromes, LIII, 206

15-Hydroxystearic acid, steric analysis of 2-D-phenylpropionate derivatives by gas-liquid chromatography, XXXV, 329
16-Hydroxystearic acid, steric analysis, XXXV, 329
17-Hydroxystearic acid
steric analysis, XXXV, 329
synthesis, LII, 320
18-Hydroxystearic acid, synthesis, LII, 320
Hydroxysteroid dehydrogenase, XXXIV, 564–566
3α-Hydroxysteroid dehydrogenase, XXXIV, 558, 564
3β-Hydroxysteroid dehydrogenase, XXXIV, 558, 564
in luteinization, XXXIX, 215
17β-Hydroxysteroid dehydrogenase, XXXIV, 558, 564
20β-Hydroxysteroid dehydrogenase
active site studies, XXXVI, 390–405
modified amino acid, XXXVI, 400–405
affinity labeling, XXXVI, 387
3β-Hydroxysteroid oxidase, LII, 18
Hydroxystreptomycin, XLIII, 121
N-Hydroxysuccinimide, XLIV, 146, 894; XLVI, 157, 227, 437, 438, 568; XLVII, 488, 560–562, 568, 570; LIX, 157
preparation of bromoacetamidoethyl-Sepharose, LI, 310
N-Hydroxysuccinimide [^{14}C]bromoacetate, XLVII, 488
Hydroxysuccinimide ester, fluorescent labeling, XLVIII, 359
N-Hydroxysuccinimide ester, XXXIV, 85–90; XLVI, 96, 104, 491
carbodiimides, XXXIV, 87
1-cyclohexyl-3-(2-morpholinethyl) carbodiimide metho-*p*-toluenesulfonate, XXXIV, 89
N-ethoxycarbonyl-2-ethoxyl-1,2-dihydroquinoline, XXXIV, 89
1-ethyl-3-(3-dimethylaminopropyl) carbodiimide, XXXIV, 89
in modification of tRNA, LIX, 156
molecular sieves, XXXIV, 88
preparations, LIX, 158–160
reactions with modified tRNA, LIX, 160–165
spacer, XXXIV, 86
5-Hydroxytetracycline, in media, LVIII, 112, 114
Hydroxytoluene, butylated
as antioxidant, XXXII, 49
lipid peroxidation, LII, 304
in liposome preparation, LII, 189, 208
in microsomal protein solubilization, LII, 111
polyoxyethylene autoxidation, LVI, 742
prevention of cytochrome *P*–450 inactivation, LII, 306, 309
5-Hydroxytryptamine, stimulation of salivary glands, XXXIX, 467, 468
2-Hydroxytryptophan, *see* Oxindolylalanine
5-Hydroxytryptophan
inhibitor of tryptophanyl-tRNA synthetase, LIX, 251
as melatonin precursor, XXXIX, 379
5-Hydroxytryptophan 2,3-dioxygenase, LII, 16
5-Hydroxytryptophan pyrrolase, LII, 20
Hydroxyurea
DNA inhibitor, XL, 276
resistant strains, LVIII, 314
synchrony, LVIII, 259–261
5-Hydroxyuridine, substrate of uridine-cytidine kinase, LI, 313
25-Hydroxyvitamin D_2, assay, using HPLC, LII, 394–397
25-Hydroxyvitamin D_3
assay, using HPLC, LII, 394–397
normal serum level, LII, 398
vitamin D metabolism, LII, 388–391
25-Hydroxyvitamin D_3 1α-hydroxylase, LII, 15
Hyflo–Super–Cel, XLIV, 762, 765
purification of cytochrome *c*, LIII, 129
Hygienic practice, LVIII, 39
Hygromycin, XLIII, 338
Hypatite C, *see* Hydroxyapatite
Hypercholesterolemia, familial, LVIII, 444
Hyperfine coupling constant, LIV, 196, 204
Hyperfine interaction, principle, LIV, 193

Hyperfine shift, pseudocontact, LIV, 167, 175, 194, 196, 197
Hyperfine shifted resonance, *see also* Contact shift; Pseudocontact shift
definition, LIV, 194
differentiation from diamagnetic shifts, LIV, 197
observation, LIV, 166–171, 175, 176
by long-pulse method, LIV, 167–169
by rapid-scan correlation method, LIV, 169–171
by saturation transfer, LIV, 175, 176
Hyperfine splitting
for immobilized nitroxide molecules, XLIX, 429, 475
isotropic, XLIX, 422, 423
substrate orientation, XLIX, 476–478
Hyperfine tensor, measurement, XLIX, 373–375
Hyperparathyroidism, kidney, XXXVII, 42
Hyperuricemia, XLIV, 695
Hypervariable region, photoaffinity labeling, XLVI, 86
Hypochlorate-benzidine reagent, as ceramide stain, XXXII, 357
Hypochlorite, XLVII, 8, 14
oxidizing agent, in luminol chemiluminescence reactions, LVII, 410
Hypochlorite-cobalt chloride chemiluminescence assay
procedure, LVII, 427
sensitivity, LVII, 438
Hypochlorous acid, phagocyte chemiluminescence, LVII, 462, 464
Hypophosphite, inhibitor of formate dehydrogenase, LIII, 371
Hypophyseal portal vessel, cannulation, operating board, XXXIX, 169
Hypophysectomy
transaural, XXXIX, 339–340
transsphenoidal, XXXIX, 337–339
Hypophysiotropic peptide
from hypothalamus, extraction, XXXVII, 402–407
pituitary hormone secretion assay, XXXVII, 82–93
Hypophysis, oxidase, LII, 13
Hypothalamic peptide, regulatory effects, LVIII, 527
Hypothalamic releasing factor
pituitary receptor binding assay, XXXVII, 213–219
secretory studies, XXXIX, 179
Hypothalamus
hypophysiotropic peptides, extraction, XXXVII, 402–407
methods for study, XXXIX, 166–182
nuclear estrogen receptors, XXXVI, 286
Hypoxanthine, XXXIV, 525
chromatographic separation, LI, 551
electrophoretic mobility, LI, 551
in media, LVIII, 54, 67
molar extinction coefficient, LI, 552
product of nucleoside deoxyribosyltransferase, LI, 447
product/substrate of purine nucleoside phosphorylase, LI, 517, 537
radiolabeled standardization, LI, 552
substrate of hypoxanthine-guanine phosphoribosyltransferase, LI, 13
of hypoxanthine phosphoribosyltransferase, LI, 543, 548, 549
Hypoxanthine guanine phosphoribosyltransferase, XXXIV, 731
deficiency in cell hybridization, XXXII, 576, 577; XXXIX, 123, 124
hybrid selection, LVIII, 352
in phosphoribosylpyrophosphate synthetase assay, LI, 5, 13
Hypoxanthine oxidase, *see* Xanthine oxidase
Hypoxanthine phosphoribosyltransferase, LI, 543–558
activity, LI, 543, 549
amino acid composition, LI, 549
assay, LI, 543, 544, 550–553
cation requirement, LI, 557
cellular localization, LI, 557
from Chinese hamster brain, LI, 543–549
from enteric bacteria, LI, 549–558
from human erythrocytes, LI, 543–549
inhibitors, LI, 557, 558

isozymes, LI, 548
kinetic properties, LI, 548, 557
molecular weight, LI, 548
pH optimum, LI, 548, 557
physiological function, LI, 558
purification, LI, 545–548, 553–555, 558
purity, LI, 549, 555
reaction stoichiometry, LI, 555
regulation, LI, 550
sedimentation coefficient, LI, 548
stability, LI, 548, 555
substrate specificity, LI, 548–550, 557
subunit structure, LI, 548

I

IA-eosin, *see* Iodoacetamidoeosin
I–cell disease, *see* Mucolipidosis II
ICSH, *see* Interstitial cell-stimulating hormone
L-Iditol-6-phosphate, XLVI, 383
O-(α-L-Idopyranosyluronic acid)-(1→4)-2,5-anhydro-D-[^{3}H]mannitol, preparation, L, 154
O-(α-L-Idopyranosyluronic acid-2-sulfate)-(1→4)-2,5-anhydromannose-6-sulfate, isolation from heparin, L, 150
IDPase, *see* Inosinediphosphatase
Iduronate sulfatase
assay, L, 444, 445
distribution, L, 444
Hunter syndrome, L, 446, 447
α-L-iduronidase, L, 444, 447
multiple sulfatase deficiency, L, 453, 454
radioactive substrates, L, 150–154
specific activity, L, 445
L-Iduronic acid, colorimetric assay, XLI, 31
α-L-Iduronidase
assays, L, 148–150
colorimetric assay, L, 442, 443
mucopolysaccharidosis I, L, 443
radioactive assay, L, 443, 444
radioactive substrates, L, 150–154
synthetic substrates, L, 141–150
i factor, LX, 337, 418
IgE, *see* Immunglobulin E
Igepal CO–630 detergent, LVI, 741
Igepal CO–712 detergent, LVI, 741
IgG, *see* Immunoglobulin G
IgM, *see* Immunoglobulin M
Ikutamycin, XLIII, 137
Ilford emulsion, properties, XXXVII, 160
Ilford G5-plate, XXXVII, 154
Illumination, sources, LVI, 404
Image intensification system, LVII, 320, 321
Image processing, XLIX, 40–42, 59
Image reconstruction, XLIX, 39–63
Imaging, LIX, 10, 11
algorithm, LIX, 11
IMEM–ZO medium, LVIII, 58, 87
composition, LVIII, 62–79
Imidate, reactivity, LVI, 614
Imidate salt, of nylon, XLIV, 122–127
Imidazole, XXXIV, 102, 103; XLVII, 434
as ligands, XXXIV, 107, 108
NH proton, NMR spectroscopy, LIV, 186
nucleus, methylation, LIII, 459
phosphodiesterase, XXXVIII, 223, 244
uptake, pH change, LV, 564
Imidazole acetate
assay of creatine kinase isoenzymes, LVII, 60, 61
buffer, preparation, LI, 283
Imidazole buffer, hydrolysis of tRNA adducts, LIX, 164, 165
Imidazoleglycylglycine, in spectrin purification, XXXII, 277
Imidazole-HCl buffer, *see* Imidazole-hydrochloric acid buffer
Imidazole-hydrochloric acid buffer
assay of adenylate deaminase, LI, 491, 498
of uridine-cytidine kinase, LI, 316
interference by, in OPRTase-OMPdecase complex assay, LI, 167
purification of adenylate deaminase, LI, 499, 500
Imidazole-sulfate solution
preparation, LIII, 83
purification of complex III peptides, LIII, 83
Imidazolium picrate method, XLVII, 591, 599, 600

Imidazolylacetate hydroxylase, LII, 17
Imido carbonate, XXXIV, 723
formation in support, XLIV, 27
hydrolysis, XLIV, 28
reactive, on Sepharose, XLIV, 17
Imidoester compound
reactivity, XLIV, 321
stability, XLIV, 321, 322
synthesis, XLIV, 322
Imidoester-containing polymer, XLIV, 118–134, 320–324
Imine *N*-hydroxylase, LII, 15
2-Imino-3-butenoic acid, LIII, 443
2-Imino-4-pentynoic acid, LIII, 442
Iminothazolidine carboxylate group, removal, XLVII, 132
2-Iminothiolane
activity, LIX, 534–536
in modification of 50 S ribosomal subunits, LIX, 535–537, 550
Imipramine, in studies of phagocytosis, LVII, 493
Immobilization, *see also* specific techniques
allosteric regulation studies, XLIV, 504–506
anhydrous, XLIV, 288–290
choice of technique, for industrial applications, XLIV, 725–729
decrease in activity, XLIV, 24
functional groups involved, comparative data, XLIV, 929, 930
interpenetrating network, polyacrylamide, XLIV, 180
methods, classification, XLIV, 10, 11, 330–332
photochemical, XLIV, 280–288
materials, XLIV, 283, 284
procedure, XLIV, 284–287
theoretical considerations, XLIV, 280–283, 287, 288
Immune complex
aggregates, XLVI, 505–508
protein A, LVI, 50
Immune conjugate, assay, XXXII, 64–66
Immune serum, preparation
enzyme polymerization, LVI, 696, 697
general, LVI, 696
immunization and collection of sera, LVI, 697
Immunization, XXXIV, 723–725
bleeding schedule, XXXIV, 724
booster, XXXIV, 724
procedure for proteins, XL, 242
chromosomal, XL, 193
of rabbits, LVI, 224, 225
schedule, cyclic nucleotide assay, XXXVIII, 98
in steroid radioimmunoassay, XXXVI, 22, 23
Immunoabsorbent chromatography
agarose based, L, 57, 58
Sepharose based, L, 57
Immunoadsorbent, XXXIV, 703–722, *see also* specific immunoadsorbents
active-site peptides, XXXIV, 711, 712
antibody measurements, XXXIV, 709–711
for cells, XXXIV, 716–722
cellular, XXXIV, 720
enzymatically degradable, XXXIV, 719
leakage, XXXIV, 703
polysaccharide, XXXIV, 712–716
preparation, XXXIV, 725, 726
protein, XXXIV, 703–712
measurement, XXXIV, 708
spacers, XXXIV, 704
Immunoadsorbent column, LI, 229
Immunoaffinity chromatography, XXXIV, 723–731
antibody preparation, XXXIV, 723–728
column preparation, XXXIV, 726–728
immunization, XXXIV, 723–725
immunoadsorbent preparation, XXXIV, 725
immunoglobulin purification, XXXIV, 725
specific systems, XXXIV, 728–731
Immunochemical method, mitochondrial membrane proteins, LVI, 45, 49, 50
Immunocompetent cell, XXXIV, 716
Immunodiffusion, XL, 246
quantitative, and enzyme-labeled antibodies, XLIV, 716, 717
Immunoelectrophoresis, XL, 246
affinity antibody purity, LVI, 703

in assay of hybrid antibodies, XXXII, 64–66
ferritin-affinity antibody-conjugates, LVI, 712
Immunofluorescence, indirect, XL, 307
Immunogen, chromosomal protein, XL, 191
testing of nuclear origin, XL, 196
Immunoglobin, XLVI, 479, 480, 482, 512, 513, 515, 531
cross linking, XLVI, 501–504
Immunoglobulin, XXXIV, 725, *see also* specific type
antipolysaccharide specific, L, 316–323
conjugation to ferritin, LVI, 226, 227
human cells synthesizing, XXXII, 799, 800
human electrophoretic mobilities of peptides, XLVII, 57
light chain, rabbit, chemical cleavage, XLVII, 147, 148
molecular weight, by gel electrophoresis, XXXVI, 99
opsonin activity, LVII, 482, 485, 486
peroxidase conjugation, XXXVII, 134–136
radioactively labeled, binding to membranes, LVI, 228
receptor, XXXIV, 219
regional unfolding, XLIX, 4
in steroid hormone assay, XXXVI, 34
Immunoglobulin A
affinity chromatography, L, 319–322
chromatographic purification, L, 319
Immunoglobulin E
antiallergen, XXXIV, 709
measurement, XXXIV, 708–711
Immunoglobulin G, XXXIV, 337
as carrier, XXXIV, 7
$F(ab')_2$ isolation, XXXII, 63
heavy chain, SDS complex, properties, XLVIII, 6
hollow-fiber retention data, XLIV, 296
hybrid antibodies, XXXII, 63, 70
preparation from whole immune serum, LVI, 697, 698
rabbit, goat antibody, LVI, 700, 701
testosterone-antiserum preparation, XXXIX, 262, 263
use in antibody quantitation, XLIV, 714–717
Immunoglobulin G–agarose, XXXIV, 725, 726, 733, 742, 745, 747
Immunoglobulin G–*p*-azobenzamidoethyl-agarose, XXXIV, 741, 742
Immunoglobulin M, removal, XXXIV, 746
Immunological technique, β-lactamase, XLIII, 86–100
antisera
inoculation program with crude enzyme, XLIII, 89, 90
preparation, XLIII, 86–90
antistaphylococcal penicillinase, XLIII, 87
crude enzyme preparation, XLIII, 87–90
isolation of β-lactamase-less mutants, XLIII, 89
N-methyl-*N*-nitro-*N*-nitrosoguanidine, XLIII, 89
neutralization analysis, XLIII, 90–97
mixtures of β-lactamases, analysis of composition, XLIII, 96
molecular variants, XLIII, 94, 95
mutations, laboratory-induced, XLIII, 96
natural variants, detection, XLIII, 94, 95
neutralization curve, XLIII, 92
time course of reaction, XLIII, 92
precipitation analysis, XLIII, 98–100
Bacillus cereus, XLIII, 98
gel precipitation, XLIII, 99
immune electrophoresis, XLIII, 99
purified enzyme preparation, XLIII, 86, 87
staphylococcal penicillinase, XLIII, 87
Immunopoietic cell, XXXIV, 718
Immunoprecipitate
protein spot identification, LVI, 611–613
specificity, XL, 249
Immunoprecipitation, XLVI, 632
of ATPase complex, LV, 346, 347
for isolation of labeled protein, XL, 246
controls, XL, 250
quantitative, XL, 245

isolation of mitochondrial components, LV, 148
photoaffinity labeling studies, LIX, 803, 807
Immunotitration, of enzyme activity, XL, 243
IMP, *see* Inosine 5′-monophosphate
Impregnation, definition, XLIV, 149, 150
IMR–91 cell, protein content, LVIII, 145
Inactivation
kinetics, XLVI, 256
protein, XLVI, 325, 326
syncatalytic, *see* Syncatalytic inactivation
without affinity labeling, XLVI, 11
Incorporation procedure, reconstitution studies
with detergents, LV, 707, 709
without detergents, LV, 708, 709
Incubation
for aspartate efflux kinetics, LVI, 260, 261
in bone collagen metabolic studies, XL, 326
for cell steroid dynamic studies, XXXVI, 75–88
condition
mitochondrial protein synthesis, LVI, 22, 23
mammalian mitochondria, LVI, 20–22
yeast mitochondria, LVI, 22
for transport studies, LVI, 281, 282
of endocrine cells, XXXIX, 165
enzymatic, XL, 37
for glutamate efflux kinetics, LVI, 259, 260
mutant isolation, LVIII, 315, 316
small-scale, for glucose pulse studies, XXXVI, 432, 433
in soft tissue collagen metabolism studies, XL, 324
temperature
invertebrate cell line, LVIII, 450, 451
poikilotherm vertebrate cells, LVIII, 470–472
of tissue, for transport studies, XXXIX, 511, 512
vessel, large-scale, LVIII, 213, 214
Incubation medium, in cell-adhesion studies, XXXII, 601
Incubation mixture, *in vitro* HeLa cell mitochondrial protein synthesis, LVI, 73
Incubator, LVIII, 11
supplier, LVIII, 16
for tissue culture, XXXIX, 110
Indicator, in media, LVIII, 70
Indigo disulfonic acid, as mediator-titrant, LIV, 408
Indigo tetrasulfonic acid, as mediator-titrant, LIV, 408
Indigo White, in uric acid fluorometric assay, XLIV, 632
Indium, atomic emission studies, LIV, 455
Indium oxide, in semiconductor electrodes, LIV, 405
Indole
chromophore, circular dichroism, XLIX, 162, 164
inhibitor, of α-chymotrypsin, XLIX, 447–449
metabolism, in organ culture, XXXIX, 392, 394
nucleus, extinction coefficient, XLVII, 445
synthesis of *Cypridina* luciferin, LVII, 369
Indoleacetic acid, as plant growth regulator, XXXII, 731
3-Indoleacetic acid, as auxin, LVIII, 478
Indoleamine 2,3-dioxygenase, LII, 13
Indole derivative, chemiluminescence, LVII, 348
Indole dioxygenase, LII, 9
Indole ethanol oxidase, LII, 21
Indolepyruvate methyltransferase
ammonium sulfate treatment, XLIII, 500, 501
assay, XLIII, 499
Bio-Gel A–5m chromatography, XLIII, 501
crude extract, XLIII, 500
DEAE-Sephadex chromatography, XLIII, 501
inhibitors, XLIII, 512
kinetic properties, XLIII, 502
molecular weight, XLIII, 502
pH effect, XLIII, 501, 502

purification, XLIII, 500, 501
reaction scheme, XLIII, 498
Sephadex chromatography, XLIII, 501
specificity, XLIII, 502
Streptomyces griseus preparation, XLIII, 500
unit definition, XLIII, 500
Indolmycin, biosynthesis, XLIII, 498–502, *see also* Indolepyruvate methyltransferase
β-Indolylacetic acid, inhibitor of tryptophanyl-tRNA synthetase, LIX, 252
2-Indolylglyoxal, synthesis of *Cypridina* luciferin, LVII, 369
3-Indolylpropionic acid, XLVII, 449, 452, 455, 458
inhibitor of tryptophanyl-tRNA synthetase, LIX, 252
β-Indolylpyruvic acid, inhibitor of tryptophanyl-tRNA synthetase, LIX, 252
Induction, LVIII, 292–296
mutant, *see* Mutant
transformant, LVIII, 301
Infection
abortive, LVIII, 404
human, LVIII, 37, 38
laboratory-acquired, LVIII, 38
Inferon, as electron microscopic tracer, XXXIX, 148
Influenza virus, XXXIV, 7, 106
Information retrieval, fourth derivative analysis of spectra, LVI, 508, 509
Infrared absorption method, hydrogen-exchange, XLVIII, 325
Infrared energy, percent reflected from cell surface, computation, LIV, 314
Infrared spectroscopy, XXXII, 247–257; LIV, 302–323
apparent molar extinction coefficient, definition, LIV, 312
in aqueous media, LIV, 302–304
band parameters, LIV, 307–314
biomembrane absorption bands, XXXII, 249
conformation studies, XLIV, 368, 369
data analysis
computer techniques, LIV, 319–321
by triangulation, LIV, 313
experimental parameters, LIV, 303, 304, 321, 322
film preparation, XXXII, 250
of glycosphingolipids, XXXII, 366
instrumentation, LIV, 321, 322
integrated intensity, LIV, 312, 313
of heme protein carbonyls, LIV, 310
of hemoglobin-bound ligands, LIV, 311
results on various substances, XXXII, 251, 252
sample cell
absorption, LIV, 303
pathlength determination, LIV, 317, 318
window materials, LIV, 314–317
sample preparation, LIV, 318
spectral calibration, LIV, 308
spectral regions, characterization, LIV, 304–307
of steroid hormone receptors, XXXVI, 412
Inheritance, uniparental, LVI, 151, 152
Inhibition
competitive, XLVI, 4
density-dependent, LVIII, 47, 48
in diffusionally constrained systems, XLIV, 427–431
effect of immobilization, XLIV, 568–570
kinetics, XLVI, 203, 465, 466
Inhibitor, *see also* specific substance; Uncoupler
administration
drinking water, LVI, 35
gastric intubation, LVI, 35
ATPase assays, LV, 329
E. coli ATPase, LV, 796
kinetics, XLVI, 160
k_{cat}, XLVI, 28–41, 369
chloro and azido derivatives as, XLVI, 327
mechanism-based, XLVI, 7, 28–41
metabolic, *in vivo* cell cycle analysis, XL, 61
miscellaneous, LV, 515–518
of mitochondrial protein synthesis, LVI, 29–32
for selection of mutants, LVI, 33

use
in vitro, LVI, 32, 33
in vivo, LVI, 33–39
of mitochondrial transport, LVI, 248, 282
of *M. phlei* membrane systems, LV, 185, 186
in study of hormone mechanisms in cell culture, XL, 273
transition-state analogs for enolase, XLI, 120–124
Inhibitor-stop technique
for calcium influx, LVI, 346
general considerations, LVI, 343–345
method, LVI, 345, 346
transport studies, LVI, 290, 291
back exchange, LVI, 294
calculation of rate, LVI, 295, 296
procedure for uptake, LVI, 291–294
Initiation, action of antibiotics, LIX, 852–860
Initiation cofactor Co–EIF–1
assay, LX, 53, 58
complexing with EIF-1, LX, 57–61
electrophoretic purity, LX, 56, 57
functional role, LX, 36, 44, 45, 53, 57–60, 245
labeling by reductive alkylation, LX, 55, 57
molecular weight, LX, 56, 57
purification, LX, 53–56
Initiation cofactor Co–eIF–2
assay, LX, 184
preparation, LX, 191–193
wheat germ supernatant, LX, 182
Initiation complex, *see also* Ribosome, subunit
antibiotic action and formation, LIX, 856, 857
translation, LIX, 857–859
isolation, LIX, 857, 858
Initiation factor, XLVI, 182, 713–715
activity in polysomes, LIX, 368–370
assay, LIX, 367, 368
crude fraction, preparation, LIX, 363, 839, 840
eukaryotic, *see also* specific initiation factors
Krebs ascites cells, LX, 87–101
assay, LX, 93, 94
isolation, LX, 89–93
purification, LX, 94–100
pig liver, LX, 240–246
rabbit reticulocyte
assay
amino acid incorporation, LX, 126
globin synthesis, LX, 126, 127, 170, 202, 203
methionylpuromycin synthesis, LX, 406, 407
polyphenylalanine synthesis, LX, 115
complexes, LX, 32, 33, 71, 72, 75–78, 82–84, 122, 123
distribution in lysate, LX, 80–82
electrophoretic purity, LX, 28–30, 159, 204
functional role, LX, 32, 33, 63, 72, 77, 108–110, 136–138, 147, 161–163, 166, 334, 335
identification by gel electrophoresis, LX, 175–180, 521
molecular weight, LX, 28, 29, 63, 159–161, 166
nomenclature, LX, 63, 88, 108, 109, 137, 138
phosphorylation, LX, 505–511
pool size, LX, 78–80
protein content, LX, 31
purification, LX, 19–29, 67, 101–108, 124, 125, 128, 129, 137–143, 146–159, 167–181, 257, 258, 283, 599
radiolabeled, LX, 34, 35, 68, 69, 75–80, 122, 123, 208
wheat germ, LX, 181–204
Initiation factor EIF-1, *see* Initiation factor eIF-2

Initiation factor eIF-1, *see also* Initiation factor, eukaryotic
assay
amino acid incorporation, LX, 116–118
globin synthesis, LX, 17, 18, 31–33, 94, 101
functional role, LX, 565
molecular weight, LX, 100, 565
purification, LX, 20, 21, 28, 29, 91, 92, 95, 100, 132–135, 156–159, 562, 563
Initiation factor EIF-2
assay, LX, 37, 46–48
functional role, LX, 36, 37, 49–52, 60
molecular weight, LX, 48
separation and purification, LX, 39, 40, 42, 43
Initiation factor eIF-2, *see also* Initiation factor, eukaryotic
affinity for guanosine nucleotides, LX, 583, 584, 589
binding to messenger ribonucleic acid, LX, 381, 395, 396, 398–401
complex with cofactor Co-EIF-1, LX, 57–61
dephosphorylation, LX, 527
electrophoretic purity, LX, 46, 189, 267
functional role, LX, 46–52, 58, 59, 74, 181, 182, 265, 266, 380, 381
inactivation by heme-regulated translational inhibitor, LX, 481–484
molecular weight, LX, 45, 245, 395, 396
phosphorylation, LX, 267, 274, 275, 468, 470, 476–479, 490, 492, 493, 525, 532
pool size, LX, 78–80
preparation
from *Artemia salina* cysts, LX, 310, 311
by gel electrophoresis, LX, 267–275, 399, 400, 470
including heparin-Sepharose chromatography, LX, 131, 132
including ribonucleic acid-cellulose chromatography, LX, 247–255
from pig liver, LX, 240–244
from rabbit reticulocyte lysate, LX, 20, 21, 24, 25, 39–42, 47, 48, 107, 154–156, 406, 468–470, 581
from wheat germ, LX, 187–191, 199–201
subunits, dephosphorylation, LX, 532, 533
ternary complex
binding to 40 A ribosomal subunit, LX, 275–280, 461, 563, 586–588
dissociation by magnesium ion, LX, 562–564, 570
formation, LX, 460, 461, 565, 585, 586
Initiation factor EIF-3, *see* Initiation factor eIF-5
Initiation factor eIF-3, *see also* Initiation factor, eukaryotic
assay by globin synthesis, LX, 17, 18, 31–33, 93, 94, 121, 122
by methionylpuromycin synthesis, LX, 68
functional role, LX, 408, 409, 533
nomenclature, LX, 88
preparation from Krebs ascites cells, LX, 91, 92, 95, 97, 98
from rabbit reticulocytes, LX, 20–23, 107, 127–131, 149, 152–154
from wheat germ, LX, 198, 199
properties, LX, 88
subunits
acetylation, LX, 539–541
dephosphorylation, LX, 533, 534
phosphorylation, LX, 526, 527
Initiation factor eIF-5, *see also* Initiation factor, eukaryotic
assay
catalysis of complex formation, LX, 38
of guanosine triphosphate hydrolysis, LX, 108, 110, 117, 118
methionylpuromycin synthesis, LX, 18, 19, 31–33, 38, 68, 170
functional role, LX, 36, 38, 50, 51, 60, 74, 574
molecular weight, LX, 52

purification, LX, 20, 21, 24–26, 39–43, 131, 132, 156–159
Initiation factor eIF-2A
assay
aminoacyl-transfer ribonucleic acid binding, LX, 116, 182, 183
methionylpuromycin synthesis, LX, 119, 120
phenylalanylpuromycin synthesis, LX, 121
functional role, LX, 181, 182
molecular weight, LX, 189, 191
separation and purification, LX, 107, 164, 165, 187–191
Initiation factor eIF-4A, *see also* Initiation factor, eukaryotic
assay by globin synthesis, LX, 17, 18, 31–33, 93, 121, 122, 170
purification, LX, 20, 21, 26, 27, 94–97, 107, 132–135, 156–159
Initiation factor eIF-4B, *see also* Initiation factor, eukaryotic
assay by globin synthesis, LX, 17, 18, 31–33, 121, 122
preparation, LX, 20–22, 127–131, 149, 152–154
Initiation factor eIF-4C, *see also* Initiation factor, eukaryotic
assay, LX, 18, 19, 31–33, 68, 94, 101, 170
molecular weight, LX, 100
preparation, LX, 20, 21, 27, 91, 92, 95, 98–100, 132–135, 156–159, 164
Initiation factor eIF-4D, *see also* Initiation factor, eukaryotic
assay by methionylpuromycin synthesis, LX, 18, 19, 31–33, 68, 69
functional role, LX, 164–165
preparation, LX, 20, 21, 27, 28, 132–135
Initiation factor eIF-MI, *see* Initiation factor eIF-2A
Initiation factor IF-1
assay, LX, 5, 6, 209–215
catalysis of exchange of ribosomal subunits, LX, 212–215
functional role, LX, 20, 204, 205
interaction with ribosomal particles, LX, 13, 15
isotopically labeled, LX, 13
radiolabeled
preparation, LIX, 790, 791
ribosomal binding activity, LIX, 792
separation and purification, LX, 6–8, 206, 207
stimulation of formylmethionyl-transfer ribonucleic acid binding to 30S ribosomal subunits, LX, 209–211
of recycling of intiation factor IF-2, LX, 211, 212
yield and purity, LX, 8–11
Initiation factor IF–2
assay, LX, 5–6, 226, 227
binding of formylmethionyl-transfer ribonucleic acid, LX, 14, 15, 215, 217–222
molecular weight, LX, 229, 245
radiolabeled, LX, 208
preparation, LIX, 790, 791
recycling, LX, 211, 212
separation and purification, LX, 6, 7, 206, 207, 225–230, 339
yield and purity, LX, 8–11, 206, 229
Initiation factor IF–3
assay, LX, 5, 6, 230–239
binding sites in MS-2 ribonucleic acid, LX, 343–350
content of *E. coli* cells, LX, 237, 238
from *Escherichia coli,* reductive alkylation, XLVII, 475
functions, LX, 13–15, 230, 437
for photoaffinity labeling, LIX, 798
radiolabeling, LIX, 784–787, 790, 791, 794, 795; LX, 208
recycling, LX, 221–224
separation and purification, LX, 6–8, 206, 207, 234, 338, 339, 344, 345
ternary complex with 30 S ribosomal subunits and MS-2 ribonucleic acid, LX, 443–446
yield and purity, LX, 8–11, 206
Initiation factor IF_{EMC}, *see* Initiation factor eIF-4A
Initiation factor IF–M_3, *see* Initiation factor eIF-3

Initiation factor IF–M1, *see* Initiation factor eIF-2A
Initiation factor IF–Mβ, *see* Initiation factor eIf-1
Initiation factor TDF, *see* Initiation factor EIF-2
Initiator tRNA, LIX, 228
from *Neurospora crassa,* digest, LIX, 108
preparation, LIX, 207
Ink
^{14}C-labeled, LIX, 192
^{35}S-labeled, preparation, LIX, 70
Inoculation
intracerebral, Chikugunya virus, LVIII, 293
mutant isolation, LVIII, 315, 316
Inorganic salt, coprecipitation
cAMP, XXXVIII, 38, 39
reagents and procedure, XXXVIII, 39
results and discussion, XXXVIII, 39–41
Inosamine derivative, chemically phosphorylated, XLIII, 453, 454
Inosamine kinase, XLIII, 444–451
aminodeoxy-*scyllo*-inositol, XLIII, 448
L-arginine:inosamine-P amidinotransferase assay, XLIII, 447
assay
method I, XLIII, 445–447
extract preparation, XLIII, 446
mycelia growth, XLIII, 446
L-ornithine in incubation mixture, XLIII, 445
paper chromatography, XLIII, 446, 447
method II, XLIII, 447–450
L-arginine:inosamine-P amidinotransferase, XLIII, 447
radiochemical method, XLIII, 451
[γ-^{32}P]ATP assay, XLIII, 451
biological distribution, XLIII, 450
2-deoxystreptamine from kanamycin, XLIII, 449, 450
equations, XLIII, 444
inhibitors, XLIII, 451
monoamidinated streptamines, XLIII, 449
paper chromatography, XLIII, 446, 447
properties, XLIII, 450, 451
purification, XLIII, 450
specificity, XLIII, 450
streptamine and streptidine from dihydrostreptomycin, XLIII, 448, 449
Inosinate
activator of carbamoyl-phosphate synthetase, LI, 22, 27, 33
for AICAR synthesis, LI, 189
assay of phosphoribosylglycinamide synthetase, LI, 180
chromatographic separation, LI, 550, 551
electrophoretic mobility, LI, 551
inhibitor of orotate phosphoribosyltransferase, LI, 163
of orotidylate decarboxylase, LI, 163
product of adenylate deaminase, LI, 490, 497
of hypoxanthine-guanine phosphoribosyltransferase, LI, 13
of hypoxanthine phosphoribosyltransferase, LI, 543
substrate of adenylosuccinate synthetase, LI, 207
Inosine
absorption maximum, LI, 525
chromatographic separation, LI, 508
inhibitor of adenosine deaminase, LI, 507
product of adenosine deaminase, LI, 502
substrate of purine nucleoside phosphorylase, LI, 517, 518, 521, 529, 530, 537
Inosine 3′,5′-cyclic monophosphate, *see* Cyclic inosine monophosphate
Inosinediphosphatase
chromatographic separation, XXXII, 401, 402, 404
as membrane marker, XXXII, 393, 403
in microsomal fractions, LII, 83, 88
in plasma membranes, XXXI, 88

Inosine 5′-diphosphate
activator of succinate dehydrogenase, LIII, 33
Co(III) complex, XLVI, 313
substrate of nucleoside diphosphokinase, LI, 375
of pyruvate kinase, LVII, 82
Inosine 5′-monophosphate, estimation, XXXVII, 248
Inosine 5′-monophosphate analog, XLVI, 299
Inosine-5′-phosphate dehydrogenase, XLVI, 301, 302
Inosine 5′-triphosphate
activator of succinate dehydrogenase, LIII, 33
ATP-P_1 exchange preparation, LV, 361, 362
Co(III) complex, XLVI, 313
complex V, LV, 314
F_1 ATPase, LV, 328
phosphate donor, of pyrimidine nucleoside monophosphate kinase, LI, 329
of thymidine kinase, LI, 358
of uridine-cytidine kinase, LI, 306
photochemically active derivative, LVI, 656
in pyrophosphate determination, LI, 275
separation, LV, 287
substrate of nucleoside triphosphate pyrophosphohydrolase, LI, 279
of phosphoribosylpyrophosphate synthetase, LI, 10
Inosinicase, absence, in *E. coli* mutant, LI, 214
Inositol, L, 413, *see also* Phosphatidylinositol
chromatography, XLIII, 432
in media, LVIII, 54, 68
(−)-*chiro*-Inositol, XLVI, 378
(+)-*chiro*-Inositol, XLVI, 370, 378
myo-Inositol, XLVI, 371, 372, 374
bromination, XLVI, 375, 376
intermediate in phosphatidylinositol synthesis, XXXV, 472
in media, LVIII, 478
scyllo-Inositol, XLVI, 379
myo-Inositol 2-dehydrogenase, XLIII, 433–439, 443
myo-Inositol hexaacetate, chromium trioxide oxidation, L, 23
Inositol hexaphosphate, hemoglobin ligand binding, LIV, 216–220
Inositol-less death, enrichment for *ts* mutations, LVI, 135, 136
myo-Inositol:NAD^+ 2-oxidoreductase, *see* *myo*-Inositol 2-dehydrogenase
assay, XLIII, 434–438
aminodeoxy-*scyllo*-inositol, XLIII, 434
extract preparations, XLIII, 436
high-voltage paper electrophoresis, XLIII, 436, 437
myo-inositol
labeled, XLIII, 434
nonlabeled, XLIII, 437, 438
mycelia growth, XLIII, 435
reagents, XLIII, 434, 438
biological distribution, XLIII, 438
properties, XLIII, 438
purification, XLIII, 439
reaction scheme, XLIII, 433
specificity, XLIII, 439
Inositol oxidase, LII, 12
scyllo-Inosose, XLVI, 379
Insect, *see also* specific genera
flight, metabolism, glycerol-3-phosphate dehydrogenase, XLI, 240, 245
flight muscles, mitochondria, LV, 22–24
oxidases, LII, 13, 17, 21
salivary glands, in hormone studies, XXXIX, 466–476
Insect cell line, media, LVIII, 455–466
Insecticide, *see also* specific substance
automated analysis, XLIV, 637, 647–658
Insta-Gel, LI, 229; LIX, 449
Instrumentation, absorbance indicators, LVI, 309
Insulin, XXXIV, 5, 6, 730, 746–750
A chain, reduction of *S*-sulfo derivative, XLVII, 125, 126
adenylate cyclase, XXXVIII, 173
alkaline proteases, hydrolysis, XLV, 432
aminoethylation, XLVII, 115, 116
aspergillopeptidase C, reaction, XLV, 428–430

batch method, XXXIV, 749
B chain
sequence analysis, XLVII, 388, 389
synthetic, XLVII, 573, 575, 576
binding studies on fat-cell fractions in study, XXXI, 69–71
biosynthesis, peptide precursors, XXXVII, 326–345
circulating hormone precursors, XXXVII, 340–344
classification of tyrosyl residues, XLIX, 119
coupling, XXXIV, 658–660
disulfide cleavage, XLVII, 112
double antibody radioimmunoassay, XXXIX, 376
effect on collagens, preparation, dosage, route of administration, XL, 320
on fat cells, XXXV, 560
on glucose transport, XXXI, 69
in fat cells, XXXVII, 269–276
on hepatocytes, XXXV, 592, 593
on levator ani muscle, XXXIX, 99
on mammary epithelial cells, XXXIX, 445, 446, 449
on melanoma cells, LVIII, 567
on perfused liver, XXXIX, 34
on protein synthesis in perfused muscle, XXXIX, 81, 82
enzyme hydrolysis, XLVII, 41
fat-cell binding, XXXVII, 193–198
maximal binding capacity, XXXVII, 196, 197
proteolytic modification, XXXVII, 211–213
fragment synthesis, XLVII, 589
heterogeneity, XXXVII, 40, 41
induction of rat liver fatty acid synthase, XXXV, 44
iodinated, *see* Iodoinsulin
logit-log assay, XXXVII, 9
in media, LVIII, 70, 78, 99–107, 109
osmotic pressure studies, XLVIII, 73
in perfused pancreas, XXXIX, 371, 372
proinsulin from islet fractionation, XXXI, 374–376
protein-binding assay, XXXVI, 35
protein kinase, XXXVIII, 364, 365
radioimmunoassay, XXXVII, 22, 34
radioreceptor assay, XXXVII, 78–81
Raman spectra, XLIX, 121–127
reduction, XLVII, 115, 116
release, superfusion studies, XXXIX, 373–376
sequence analysis, XLVII, 318
standard, XXXIX, 210
structure capable of binding, XXXIV, 653
thermolysin cleavage sites, XLVII, 176, 177, 188
thermomycolin, reaction, XLV, 428–430
Insulin-agarose, XXXIV, 747
derivatives, XXXIV, 653, 654, 657–660
Insulin-antibody-agarose, XXXIV, 747
Insulinase, inactivation, XXXVII, 213
Insulin-free plasma, XXXIV, 747–750
Insulin-protamine reaction, mixed association, XLVIII, 122
Insulin receptor, XXXIV, 653–670; XLVI, 85
adsorbents, XXXIV, 668–670
affinity adsorbents preparation, XXXIV, 657–660
assay, XXXIV, 654, 655
detergents for solubilization, XXXIV, 655, 656
fat cells, XXXIV, 655
liver, XXXIV, 654
macromolecular agarose derivatives, XXXIV, 658–660
membranes preparation, XXXIV, 654, 655
placenta, XXXIV, 655
purification, XXXIV, 661–668
spacers, XXXIV, 663
Interaction, *see* specific type
Interaction energy, XLVIII, 275
Interaction kinetics, macromolecular, pressure-jump light-scattering observation, XLVIII, 308
Interference, absorption indicators, LVI, 307
Interference cross term, LIX, 631–633, 662, 668, 669
Interference factor i*a*, *see* Protein, ribosomal, S1
Interference filter, XLIX, 87

Interference fringe pattern, in sample cell path length determination, LIV, 317
Interference fringe shift, XLVIII, 170
Interference optics, XLVIII, 185
Interference ripple, LIX, 631
  computation, LIX, 703, 704
  function of shape and subunit orientation, LIX, 680, 681, 704, 705
Interferogram, LIV, 322
Interferon, XXXIV, 7, 338, 730
  assay, LVIII, 296
  formation in microcarrier culture, LVIII, 192
  inducers, LVIII, 293, 294, *see also* specific types
  induction, LVIII, 292–296
  production, LVIII, 292–296
    priming, LVIII, 296
  from RSTC-2 cells, LVIII, 293
  superinduction, LVIII, 295
Internal conversion electron, radiation properties, XXXVII, 146
Internal magnetic field, at nucleus, LIV, 350
Internal standard, for amino acid analysis, XLVII, 31–40
Interstitial cell
  enzymatic dispersion, XXXIX, 257–260
  hormone effects, XXXIX, 424
  isolation and culture, XXXIX, 295, 296
  of testis, *in vitro* function, XXXIX, 252–271
Interstitial cell-stimulating hormone, stress effects, XXXIX, 279
Inter-α-trypsin inhibitor
  amino acid composition, XLV, 772
  assay, XLV, 762
    in whole serum, XLV, 760
  distribution, XLV, 769
  kinetics, XLV, 769
  other inhibitors, XLV, 770
  properties, XLV, 766
    chemical, XLV, 768
    physical, XLV, 768
  specificity, XLV, 769
  stability, XLV, 766
Intestinal mucosal cell
  fractionation studies, XXXII, 670–673
  isolation, XXXII, 665–673
Intestine
  brush-border isolation, XXXI, 123–134
  epithelial cell, protein content, LVIII, 145
  isolated, perfusion, LII, 59
  mammalian, oxidase, LII, 11, 13, 20, 21
Intimal cell
  rabbit
    aortic, LVIII, 96, 97
    effect of hormone, LVIII, 106, 107
Intramer, in tryptophanyl-tRNA synthetase, LIX, 245
Intrinsic factor, XXXIV, 305, 312
Inulin
  autoradiography, XXXVI, 150
  hollow-fiber dialytic permeability, XLIV, 297
  hollow-fiber retention data, XLIV, 296
  in kidney function studies, XXXIX, 7–9
    as extracellular water marker, XXXIX, 19, 20
  radiolabeled, in protein turnover studies, XXXVII, 246
Inversion recovery method
  advantages, XLIX, 339
  principle, XLIX, 333
  procedure, XLIX, 336, 337
  spectrometer requirements, XLIX, 333–336
Invertase, *see* β-D-Fructofuranosidase
Invertebrate, oxidases, LII, 13, 16
Invertebrate cell, LVIII, 450–466
  culture methods, LVIII, 451, 452
  effect of atmosphere, LVIII, 450, 451
    of temperature, LVIII, 450, 451
  media, LVIII, 454–466
  storage, LVIII, 453
  subculture, LVIII, 451, 452
  suspension culture, LVIII, 454, 455
Iodide
  activator of succinate dehydrogenase, LIII, 33
  amino acid oxidase, LVI, 485

calcium-sensitive electrodes, LVI, 347
cholesterol oxidase, LVI, 485
fluorescence quenching, XLIX, 222, 225, 231, 235
glucose electrode, LVI, 483
inhibitor of firefly luciferase, LVII, 15
of *p*-hydroxybenzoate hydroxylase, LIII, 551
of salicylate hydroxylase, LIII, 541
thyroidal, TSH effects on transport, XXXVII, 256–262
in vitro, XXXVII, 258–261
in vivo, XXXVII, 257, 258
transport of, in thyroid cells, XXXII, 756–758
Iodination
with chloramine-T, XXXVII, 225, 226
of glycoproteins, XXXVII, 325
of human chorionic gonadotropin, XXXIX, 223, 224
with lactoperoxidase, XXXVII, 229, 232, 233
of peptide hormones, XXXVII, 224–233
of proteins, lactoperoxidase catalysis, XXXII, 103–109
protein spot identification, LVI, 612, 613
site of action, XLVII, 66
Iodine, XLVI, 64; XLVII, 138, 538
γ-globulin labeling, LVI, 226
as inactivator of β-lactamase, XLIII, 662
inhibitor of nucleoside phosphotransferase, LI, 393
in media, LVIII, 69
vapor
in glycolipid identification, XXXII, 358
indicator for lipids, XXXV, 321, 405
Iodine-125
in chemical protection method, LIX, 593–595
as hormone label, XXXVII, 29, 32, 49, 50, 145–147
as insulin label, in binding studies, XXXI, 69, 70
labeling
antitoxin to *Pseudomonas* toxin A, LX, 790, 791
messenger ribonucleic acid, LX, 384, 385, 404, 405
in liquid scintillation spectrometry, XXXVI, 164
properties, XXXVII, 146
in protein labeling, XXXII, 105
for steroid labeling, XXXVI, 24
in studies of perfused thyroid gland, XXXIX, 364
Iodine-127, lock nuclei, XLIX, 349; LIV, 189
Iodine-131
contaminants, XXXII, 107
as hormone label, XXXVII, 29, 32, 145–147
in liquid scintillation spectrometry, XXXVI, 164
in pineal-thyroid studies, XXXIX, 382
properties, XXXVII, 146
for steroid labeling, XXXVI, 24
Iodoacetamide, XLVI, 66, 391, 612, 667; XLVII, 407
alkylation of protein A, LIII, 379
effect on epoxide hydrase, LII, 200
on orotidylate decarboxylase, LI, 79
electrophoretic analysis of *E. coli* ribosomal proteins, LIX, 538, 539, 542, 544
inactivator of acetoacetyl-CoA thiolases, XXXV, 134, 135
of complex III, LIII, 82
of fatty acid synthase, XXXV, 81
of β-hydroxyacyl-CoA dehydrogenase, XXXV, 128
of palmityl thioesterase II, XXXV, 109
inhibitor of deoxythymidine kinase, LI, 365
of FGAM synthetase, LI, 201
of guanine deaminase, LI, 516
of 6-methylsalicylic acid synthetase, XLIII, 528
of thymidylate synthetase, LI, 96
of uridine phosphorylase, LI, 429
labeling of elongation factor EF-G, LX, 743, 744

preparation of cross-linked ribosomal subunits, LIX, 537
of protein from cross-linked ribosomal subunits, LIX, 537, 541
[1-^{14}C]Iodoacetamide, in fibrinoligase titration, XLV, 188
Iodoacetamide spin label
structure, XLIX, 492
tropomyosin denaturation, XLIX, 11
5-Iodoacetamidoeosin, LIV, 49, 50
in erythrocyte membrane studies, LIV, 59
extinction coefficient, LIV, 52
structural formula, LIV, 50
5-Iodoacetamidosalicylic acid, properties, XLVII, 415
Iodoacetic acid, XLIV, 392; XLVI, 68; XLVII, 407; LV, 517
alkylation, of coenzymes, XLIV, 859–867
of protein A, LIII, 379
in carboxymethylation of sulfhydryls, XXXII, 72
effect on orotidylate decarboxylase, LI, 79
as enzyme inhibitor, XXXVII, 294
inactivator of β-hydroxyacyl-CoA dehydrogenase, XXXV, 128
inhibitor of adenine phosphoribosyltransferase, LI, 574
of amidophosphoribosyltransferase, LI, 178
of brain α-hydroxylase, LII, 317
of FGAM synthetase, LI, 201
of guanine deaminase, LI, 516
of uridine phosphorylase, LI, 429
for methionine modification, LIII, 138
for removal of dithiothreitol, LI, 160
Iodoacetol phosphate, XLVI, 140
Iodoacetoxy-3-estrone, XLVI, 457
N-(Iodoacetylaminoethyl)-5-naphthylamine-1-sulfonic acid, XLVII, 417, 418
(Iodoacetyl)cellulose, preparation, XLIV, 749
N-Iodoacetyl-3,4-dimethoxy-5-hydroxyphenylethylamine, XLVI, 558–561
N-Iodoacetyl-3,5-dimethoxy-4-hydroxyphenylethylamine, XLVI, 558–561
Iodoacetyl-*N*-hydroxysuccinimide ester, synthesis, XLVI, 157
N-Iodoacetylpuromycin, XLVI, 668, 669
α-Iodo-β,γ-acyloxypropane, intermediate in phosphatidic acid synthesis, XXXV, 430
5-Iodo-5′-amino-2′,5′-dideoxyuridine, inhibitor of thymidine kinase, LI, 359, 360
Iodobenzene, photochemically active derivative, LVI, 660
Iodobenzene dichloride, XLVII, 457
p-Iodobenzoate, binding studies, XLVIII, 280–282
Iodobenzyl malonate, transport, LVI, 248
4-Iodobutyl methacrylate, XLIV, 84
5-Iodocytidine, substrate of cytidine deaminase, LI, 411
5′-Iodocytidine triphosphate, allosteric inhibitor, aspartate transcarbamylase, XLIV, 557
5-Iododeoxycytidine
substrate of cytidine deaminase, LI, 411
of thymidine kinase, LI, 370
5-Iododeoxycytosine triphosphate, activator of thymidine kinase, LI, 358
5-Iododeoxyuridine
substrate of deoxythymidine kinase, LI, 364
of thymidine kinase, LI, 358, 370
of thymidine phosphorylase, LI, 441
5-Iodo-2′-deoxyuridine, XLVI, 89
5′-Iodo-5′-deoxyuridine, inhibitor of uridine-cytidine kinase, LI, 313
5-Iododeoxyuridine triphosphate, activator of thymidine kinase, LI, 359
2-Iodoethyl methacrylate, XLIV, 84
α-Iodoglycerol, intermediate in phosphatidic acid synthesis, XXXV, 431–441
Iodohexestrol, XLVI, 89
Iodoinsulin
fat cell binding, XXXVII, 193–198
preparation, XXXVII, 194
properties, XXXVII, 230
Iodoketone, XLVI, 131

Iodolactic acid phosphate, preparation and reaction with enolase, XLI, 121
Iodomethyl ketone, XLVI, 199
Iodometric assay, of lipid hydroperoxides, LII, 306, 307, 309, 310
4-Iodophenyl azide, XLVI, 86
α-Iodo-L-propylene glycol, *see* L-Glycerol-α-iodohydrin
Iodosobenzene, oxygen donor, to cytochrome *P*–450, LII, 407
Iodosobenzoic acid, XLVII, 409
 inhibitor of uridine phosphorylase, LI, 429
3-Iodotyrosine, XXXIV, 6
 in thyroid cells, XXXII, 755, 756
5-Iodouracil, substrate of orotate phosphoribosyltransferase, LI, 152
Ion
 distribution
  determination of ΔpH across membranes, LV, 561–563
  without separation, LV, 556
 equilibrium distribution, determination by separation techniques, LV, 552–556
 permeable, electrochemical equilibrium distribution, LV, 548, 549
Ion exchange, property, XXXIV, 74
Ion-exchange cartridge, for water purification, LIV, 473
Ion exchange chromatography, XXXVI, 211, 212; XLIII, 256–278, *see also* specific substance
 Amberlite CG–50 resin, XLIII, 268, 270, 272, 274
 Amberlite IRC–50 resin, XLIII, 266, 269
 Amberlite IR–120 resin, XLIII, 266, 272, 273, 277
 Aminex A–27, A–28, XLIII, 266, 275
 4-amino-4-deoxy-α,α-trehalose, XLIII, 269, 270
 aminoglycoside antibiotics, XLIII, 263–278, *see also* Aminoglycoside antibiotics; specific antibiotic
 ammonia gradient, XLIII, 270
 bromelain, XLV, 744
 (carboxymethyl)cellulose, XLIII, 273, 274
 cation interference, XLIII, 266
 cellulose exchanger, XLIII, 273–275
 CM–Sephadex C–25, XLIII, 266, 273–275
 column design, XLVII, 211, 225–227
 cyclic nucleotides
  Dowex 50, XXXVIII, 10–13
  (polyethyleneimine)cellulose, XXXVIII, 16–20
  quaternary (aminoethyl)cellulose, XXXVIII, 13–16
 developers, XLVII, 214
 3′,4′-dideoxykanamycin B, XLIII, 271, 272
 Dowex 50, XLIII, 272, 273
 Dowex 1–X2, XLIII, 267, 275–277
 Duolite c–62, XLIII, 272
 elution rates, XLVII, 214, 215
 gentamicin C complex, XLIII, 274
 high pressure
  operation of equipment, XXXVIII, 20–22
  preliminary prepurification, XXXVIII, 24–26
  procedures, XXXVIII, 22–24
 high-pressure liquid chromatography, XLIII, 278
 history, XLVII, 3
 instrumentation, XLVII, 4–7
 internal standards, XLVII, 32–34
 kanamycin, XLIII, 268, 269
 kasugamycin, XLIII, 273
 Lewatit SP–120, XLIII, 272
 lividomycin, XLIII, 275
 nebramycin, XLIII, 271
 nonionic adsorption chromatography, XLIII, 275–280, *see also* Adsorption chromatography, nonionic
 in peptide purification, XXXVII, 316, 317
 phosphonic acid resin, XLIII, 272
 SE–Sephadex C–25, XLIII, 266, 273, 274
 streptothricin-like antibiotics, XLIII, 256–263, *see also* Streptothricin-like antibiotics
  hydrolysis fragments, XLIII, 260–262
  intact antibiotics, XLIII, 258–260
  sulfonic acid resins, XLIII, 272
 validamycins, XLIII, 273

Ion-exchange thin-layer chromatography, XLIII, 180, 181
Ion flux
in perfused liver, XXXIX, 34
substrates, XXXIX, 493–573
Ionic bond, in adsorption, XLIV, 149–153
Ionic strength
activity coefficients, LVI, 360
detergent micelles, LVI, 738, 740, 746
luminescence, LVI, 543
spin probes, LVI, 525
Ionic strength effect, XXXIV, 138, 139
Ionization chamber, LIV, 330
Ionizing radiation, mutation induction, XLIII, 31, 32
Ionophore, *see also* specific type
activity, assessment, LV, 438–440
bioenergetic systems, LV, 450–454
carboxylic, cation fluxes, LV, 674, 675
classification and cation selectivity, LV, 440–448
flexibility, LVI, 439
general use, LV, 448–450
model transmembrane redox reactions, LV, 542–544
nature, LV, 435–438
neutral
applications, LVI, 441–448
design features, LVI, 439–441
to measure transmembrane electrical potentials, LVI, 446–448
nonpolar groups, LVI, 439
pet9 mitochondria, LVI, 127
polar groups, LVI, 439
selectivity, LVI, 439, 440
synthetic, LV, 443
as uncouplers, LV, 471, 472
Ionophore A23187, stimulant, of neutrophile chemiluminescence, LVII, 466, 492
Ion permeability, brown adipose tissue mitochondria, LV, 72
Iontophoresis, problems in use, XXXIX, 438–440
Ion transport, by smooth muscle mitochondria, LV, 65
Ipegal CO–710, partial specific volume, XLVIII, 19, 20
Iproniazid, effect on melanin, XXXVII, 129
Iron, *see also* Ferric ion; Ferrous ion
analysis, LIV, 436–443
acid-extraction method, LIV, 442, 443
glassware, LIV, 437–439
procedure, LIV, 439–443
sources of error, LIV, 442
standard curve, LIV, 441, 442
assay of transport into bacteria
construction and use of filtration apparatus, LVI, 392–394
materials and methods, LVI, 389–391
principle, LVI, 388, 389
procedure, LVI, 391, 392
atom, possible electronic configurations, LIV, 198
atomic emission studies, LIV, 455
ATPase, LV, 301
colloidal, XL, 21
as stain, XXXIX, 152
complex V, LV, 315; LVI, 580, 584
in cytochrome oxidase, LIII, 64
inhibitor, of brain α-hydroxylase, LII, 317
in media, LVIII, 52, 69
in prosthetic groups, LII, 4
radioactive, choice of nuclide, LVI, 390
solutions, LVI, 391
volatilization losses, in trace metal analysis, LIV, 480
Iron(II)
correlation time evaluation, XLIX, 346
electron paramagnetic resonance studies, XLIX, 514
interatomic distance calculations, XLIX, 331
Iron(III)
correlation time evaluation, XLIX, 346
electron paramagnetic resonance studies, XLIX, 514
Iron-55
in membrane studies, XXXII, 884–893
properties, XXXII, 884

Iron-57, for isotopic enrichment in Mössbauer studies, LIV, 374
Iron-59, in blood-cell separation studies, XXXII, 644
Iron(II)-bathophenanthroline chelate, LV, 517
Iron-deuteroheme IX, substrate of heme oxygenase, LII, 371
Iron dextran-complex tracer, for electron microscopy, XXXIX, 148
Iron-dihydroporphyrin, *see* Chlorin
Iron-ethylenediaminetetraacetic acid, as mediator-titrant, LIV, 408
Iron(III)-ethylenediaminetetraacetic acid, *g* value, XLIX, 523
Iron(II)-ethylenediammonium sulfate, as iron standard, LIV, 436
Iron-manganese oxide particles, magnetic, commercial source, XLIV, 278
Iron mesoporphyrin IX, substrate of heme oxygenase, LII, 371
Iron-mesoporphyrin IX analog, structure-sensitive Raman bands, frequency pattern, LIV, 242
Iron(III)-octaethylporphyrin, resonance Raman scattering studies, LIV, 242
Iron porphyrin, esterification, LII, 429
Iron-porphyrin dimer, magnetic susceptibility studies, LIV, 395
Iron protein
  classification, LIII, 260
  of nitrogenase, *see* Azoferredoxin
Iron-protoporphyrin IX, LII, 5
  CD spectrum, LIV, 259, 260
  function, LII, 5
  substrate of heme oxygenase, LII, 371
Iron standard, preparation, LIV, 436
Iron-sulfur center
  displacement, LIII, 270–273
  EPR characteristics, LIV, 134, 138–143
  EPR studies, LIII, 486–490
  4Fe-4S cluster, oxidation states, LIII, 330
  in ferredoxin, 335-340, LIII, 627
  identification, LIII, 268–274
    by absorbancy ratio method, LIII, 269
    by EPR spectroscopy, LIII, 273, 274
  oxidation states, LIII, 264
  of rubredoxin, LIII, 344, 346
  of succinate dehydrogenase, LIII, 486–495
    EPR characteristics, LIII, 494, 495
    midpoint potentials, LIII, 490, 491
    thermodynamic parameters, LIII, 490–493
  transfer among proteins, LIII, 273
  types, LIII, 259, 261, 268
Iron-sulfur flavoprotein, fatty acid oxidation, LVI, 584
Iron-sulfur protein, LIII, 259–274, *see also* Iron-sulfur centers
  acid-labile sulfide, LIII, 275–277
  classification, LIII, 259, 260
  complex classes, LIV, 203
  of complex I, LVI, 588
  of complex III
    characterization, LIII, 82, 83
    isolation, LIII, 84
    molecular weight, LIII, 36
  in cytochrome bc_1 complex, LIII, 116
  definition, LIII, 259
  as electron carriers, assays, LIII, 264
  electron paramagnetic resonance characteristics, LIII, 265, 266, 268, 269, 273, 274
  high potential, LIII, 329–340
    assay, LIII, 331, 334
    bacterial sources, LIII, 329, 330
    chemical properties, LIII, 332, 333
    crystallization, LIII, 339
    detection methods, LIII, 331
    EPR characteristics, LIII, 486
    purification, LIII, 335–339
    spectral properties, LIII, 332, 334
    storage, LIII, 339, 340
  high potential type
    EPR characteristics, soluble center, LIV, 140, 141
    Mössbauer studies, LIV, 371
  Mössbauer studies, LIV, 369–372
  NMR studies of, LIV, 202–209
  nomenclature, LIII, 261, 265
  physical properties, LIII, 262, 263
  purification, LIII, 266, 267
  recognition, LIII, 261
  stability, LIII, 266, 267

two-iron compounds, structural model, LIV, 207
Iron(III)-tetraphenylporphyrin, resonance Raman scattering studies, LIV, 242
Iron-tris dithiocarbamate, phase shift studies, LIV, 343
Irradiance, relationship to photon irradiance, LVII, 572
Irradiation
cAMP diazomalonyl derivatives, XXXVIII, 393, 394
cross linking, XLVI, 170–172
damage, XLVI, 173
of *M. phlei* ETP, LV, 185
IRTRAN window, for IR spectroscopy, XXXII, 251
Ischemia, rat heart metabolites, LV, 217
Ischemic heart, perfusion, XXXIX, 52–56
ISCO Density Gradient Former, XXXI, 136
ISCO mixing device, XXXVI, 160
ISCO Model 380 Dialagrad programmed gradient pump, LIX, 373
ISCO Model UA–5 absorbance monitor, LII, 174, 328
3-Isoadenosylcobalamin, effector of ribonucleoside triphosphate reductase, LI, 258
Isoamyl alcohol
determination of *in vivo* levels of aminoacylated tRNA, LIX, 271
as emulsification retardant, LII, 400
in iron analysis, LIV, 437
Isobestic point, in difference spectra, LII, 219, 220, 266
Isobutyl chloroformate, XXXIV, 612; XLVI, 202; XLVII, 556
4-Isobutyl-2,6-dinitrophenol, LV, 470
Isobutylene, XLVII, 525, 526, 530
liquid, preparation, XLVII, 526
Isobutyral, XLIV, 128, 129, 131, 132
Isobutyramide, conformation studies, XLIX, 356
Isobutyrate, activator of aspartate carbamyltransferase, LI, 49
Isobutyric acid-ammonia, separation of RNase T2 digestion products of tRNA, LIX, 169, 170
Isobutyric acid-water-ammonium hydroxide
chromatographic separation of oligonucleotides, LIX, 343
in tRNA sequence analysis, LIX, 62
N-Isobutyrylphenylalanine, XLVI, 217
Isocaproic aldehyde, in adrenal gland, LII, 124
Isochlorotetracycline, XLIII, 199
Isocitrate, *see also* Isocitric acid
activator of adenylate kinase, LI, 465
assay, improvements, LV, 206
content
of brain, LV, 218
of kidney cortex, LV, 220
of rat heart, LV, 216
ferricyanide, LVI, 232
intact mitochondria, LV, 143
measurement of proton translocation, LV, 638
Isocitrate dehydrogenase, XXXIV, 164, 246
assay by Fast Analyzer, XXXI, 816
associated with cell fractions, XL, 86
from beef heart, affinity labeling, XLVII, 428
citrate assay, LV, 211
cyanylation, XLVII, 131
in heme oxygenase activity, LII, 368
inhibition, LV, 206
in liver compartments, XXXVII, 288
in NADPH-generating system, LII, 274, 359, 362, 378, 380, 401, 414
pyridine nucleotide assay, LV, 264, 265, 267–269
in sodium laurate hydroxylation, LII, 320
threo-D_s-Isocitrate dehydrogenase, reduction of $NADP^+$ analogs, XLIV, 874–876
Isocitrate dehydrogenase (NAD)
brain mitochondria, LV, 58
toluene-treated mitochondria, LVI, 547
Isocitrate dehydrogenase (NADP)
brain mitochondria, LV, 58
toluene-treated mitochondria, LVI, 547
Isocitrate lyase
in glyoxysomes, XXXI, 565, 568–570, 743

labeling, XLVI, 139
as marker enzyme, XXXI, 735, 743
in microbodies, LII, 496
Isocitric acid, LII, 401, 414, *see also* Isocitrate
in heme oxygenase activity, LII, 368
in isolated renal tubule, XXXIX, 16, 18
DL-Isocitric acid, in sodium laurate hydroxylation, LII, 320
Isocitric dehydrogenase, *see* Isocitrate dehydrogenase
1-Isocyanato-4-isothiocyanatobenzene, XLIV, 116
Isocyanide, cytochrome *P*–450 difference spectra, LII, 273
Isocyanide group
assay, titrimetric, XLIV, 131, 132
indirect coupling, XLIV, 38–40
Isodeacetylcolchicine, XLVI, 568
synthesis, XLVI, 569
Isoelectric focusing
of ATPase inhibitor, LV, 424, 425
of progesterone receptors, XXXVI, 208
proline carrier protein, LV, 196
protein kinases, XXXVIII, 325
elongation factors, LX, 142, 714, 715
heme-regulated translational inhibitor, LX, 471, 473, 474
purification of adenine phosphoribosyltransferase, LI, 576, 577
of adenylate kinase, LI, 461
of bacterial cytochrome *c*, LIII, 229–231
of deoxycytidylate deaminase, LI, 417
of deoxythymidylate phosphohydrolase, LI, 288
of guanylate kinase, LI, 487
of hydrogenase, LIII, 313
of luciferase, LVII, 62
of 5-methyluridine methyltransferase, LIX, 199, 200
of purine nucleoside phosphorylase, LI, 542
separation
of cytochrome *c* derivatives, LIII, 149
initiation factor eIF-2 subunits, LX, 265–270, 272–275, 396
messenger ribonucleic acid-binding proteins, LX, 397–399
for soluble enzyme aggregate characterization, XLIV, 473
of steroid-binding glycoproteins, XXXVI, 98–102
of steroid cytoplasmic receptors, XXXVI, 228–234
of TeBG, XXXVI, 115–117
two-dimensional gels, LVI, 608, 609
determination of pH gradient, LVI, 609, 610
preparation of gel for SDS gel electrophoresis, LVI, 609
Isoelectric focusing gel, XL, 157
Isoelectric focusing-gel electrophoresis, two-dimensional, mitochondrial proteins, LVI, 72, 73
Isoelectric point
adsorption, XLIV, 151, 152, 154
of ATPase inhibitors, LV, 402, 413, 414, 426
Isoenzyme, *see also* Isozyme
determination of levels, in lactate dehydrogenase, XLI, 47–52
Isoenzyme X, XXXIV, 605
Isoflavone, in plant-cell extraction, XXXI, 531
Isogriseofulvin, XLIII, 234
Isoinhibitor
cuttlefish, *see* Cuttlefish, isoinhibitors
garden bean trypsin, XLV, 710, 714, 715
Isolation buffer, for plant-nuclei extraction, XXXI, 559
Isolectin, from *Bandeiraea simplicifolia*
assay, L, 346
properties, L, 349
purification, L, 346–349
specificity, L, 345
Isoleucine, XXXIV, 5
aminoacylation isomeric specificity and, LIX, 274, 279, 280
deprivation, synchrony, LVIII, 258, 259
esterification, XLVII, 522

in media, LVIII, 52, 53
misactivation, LIX, 289
peptide synthesis, XLVII, 509
N^{α}-phenylthiocarbamyl methylamide derivative, identification, XLVII, 377
phenylthiohydantoin derivative, identification, XLVII, 334
solid-phase peptide synthesis, XLVII, 610
thermolysin hydrolysis, XLVII, 175–179
L-Isoleucine bromomethyl ketone, XLVII, 66
L-Isoleucine–transfer-RNA ligase, *see* Isoleucyl-tRNA synthetase
Isoleucyl-tRNA synthetase, XXXIV, 4, 170; XLVI, 90, 152
active site, XLVII, 66
in protein-tRNA interaction studies, LIX, 342
purification, LIX, 262, 266
subcellular distribution, LIX, 233, 234
in tritium labeling experiments, LIX, 342
Isoluminol
labeling compound, advantages, LVII, 424
stability, LVII, 425
storage, LVII, 425
structural formula, LVII, 428
Isomaltase, in brush borders, XXXI, 129, 130
Isomaltose
chromatographic constants, XLI, 16
excretion after injury, L, 233
polarography, LVI, 477
synthesis, L, 101, 102
Isomerase, XLVI, 132, 158
calculation of activity, XLI, 64
Isomeric shift, *see* Chemical shift
Isomerization, reaction kinetics, light scattering studies, XLVIII, 446–449
Isonicotinic acid hydrazide, incorporation in tRNA, LIX, 118
[^{14}C]Isonicotinic acid hydrazide, LIX, 178
Isonitrile, interaction with cytochrome *c* oxidase, LIII, 201
Isonovobiocin, XLIII, 304
Isooctane
in heme isolation, LII, 425
in hemin derivative crystallization, LII, 429
Isooctane-acetonitrile-isoamyl alcohol, elution of biphenyl metabolites, LII, 406
Isooctane-isoamyl alcohol, for extraction of biphenyl metabolites, LII, 406
4-Isooctyl-2,6-dinitrophenol, LV, 470
Isopentane, XLIX, 238
in freeze-quench technique, LIV, 85, 88
in freezing mixture, LIII, 489
in low-temperature EPR spectroscopy, LIV, 120
N^6-Isopentenyladenine, as cytokinin, LVIII, 478
6-Isopentyladenosine
identification, LIX, 92
NMR studies, LIX, 57
1-Isoprenaline, XLVI, 579
Isoprenoid-alcohol kinase
amino acid composition, XXXII, 446
assay, XXXII, 439
properties, XXXII, 444–446
purification, XXXII, 440–444
Isopropanol, *see* 2-Propanol
Isopropanol-acetic acid, in protein staining procedure, LVII, 176
Isopropanol-ammonium hydroxide-water, separation of 3′-amino-3′-deoxyadenosine, LIX, 135
Isopropylamine, electron spin resonance spectra, XLIX, 383
Isopropyl 2-benzothiazolesulfonate, XLVI, 539
2′,3′-*O*-Isopropylidene adenosine 5′-phosphate, in 5′-nucleotidase assay, XXXII, 368, 369
Isopropylideneadenosylcobalamin, effector of ribonucleoside triphosphate reductase, LI, 258
2′,3′-*O*-Isopropylidene-6-chloropurine ribonucleoside, XLVI, 299
O-Isopropylidene derivative of unsaturated fatty acids, XXXV, 347
2′,3′-*O*-Isopropylidene-*S*-(dinitrophenyl)-6-mercaptopurine riboside, synthesis, XLVI, 289, 290

Isopropylidene-D-glycerol, intermediate in phosphatidic acid synthesis, XXXV, 431–435, 444
2,3-Isopropylidene-*sn*-glycerol-(benzyl) phosphate
barium salt, XXXV, 466
silver salt, XXXV, 466
2,3-Isopropylidene-*sn*-glycerol-1-dibenzyl phosphate, intermediate in phosphatidylglycerol synthesis, XXXV, 466
1,2-Isopropylidene-*sn*-glycerol-3-(β,β,β-trichloroethylcarbonate), intermediate in synthesis of phosphatidylethanolamine analogues, XXXV, 509
2′,3′-Isopropylideneuridine, XLVI, 347
N^{ε}-Isopropyllysine, identification, XLVII, 477
α-Isopropylmalate synthetase, XXXIV, 4
Isopropylnorepinephrine, adenylate cyclases, XXXVIII, 147–149
1-(2-Isopropylphenyl)imidazole, activator of epoxide hydrase, LII, 200
Isopropyl-β-thiogalactoside, XXXIV, 351
Isoproterenol
adenyl cyclase, XXXVIII, 173
cyclic nucleotide levels, XXXVIII, 94
derivativization to glass beads, XXXVIII, 189
Isoproterenol-glass, XXXIV, 699, 700
Isopycnic centrifugation
of bacterial membranes, XXXI, 647–653
definition, XXXI, 715
of plant cells, XXXI, 515–518, 545, 549
of subcellular organelles, XXXI, 734–746
Isopycnic temperature method, for density determinations, XLVIII, 23–29
Isorhodopsin, formation, XXXII, 309
Isothiocyanate, XLIV, 38, 132, 141
from aryl amines, XXXIV, 53, 54
derivatives, XXXIV, 55
fluorescent labeling, XLVIII, 358
organic, XLVI, 164–168
reactions, XLVI, 167, 168
synthesis, XLVI, 165–167
Isothiocyanate-glass, XXXIV, 68
p-Isothiocyanatobenzoyl-DL-homoserine, lactone coupling procedure, XLVII, 286
Isothiocyanato glass
capacity, XLVII, 269
synthesis, XLVII, 268
Isothiocyanato polydimethylacrylamide, synthesis, XLVII, 268
3-Isothiocyanatostyrene, XLIV, 108, 113
in copolymer, XLIV, 84
4-Isothiocyanatostyrene, XLIV, 108
3-Isothiocyano-1,5-naphthalene disulfonate, XLVII, 261
Isotope
calcium distribution, LVI, 302
continuous, method of protein degradation determination, XL, 257
double, determination of heterogeneity of protein degradation, XL, 258
dual, in differential labeling, XLVI, 66, 67
effect, XLVI, 30
acetylenic inhibitors, XLVI, 159
exchange, carnitine, palmitoyltransferase assay, LVI, 370, 371
labeling, *in vivo* cell cycle analysis, XL, 59
reutilization, in measurement of protein degradation, XL, 254
short-term incorporation into enzyme, XL, 251
single, method of protein degradation determination, XL, 253
Isotope dilution
in progesterone-metabolites identification, XXXVI, 496
quantitation of cellular hexose pools, L, 192
Isotope dilution technique, assay of L-amino acids, LIX, 314–316
Isotope effect, LIV, 527
Isotope shift
in IR spectroscopy, LIV, 313, 314
in spectra of hemoglobin-bound ligands, LIV, 315
N-Isovaleryl-3-aminopropanal, *see* Luciferin, *Diplocardia*

Isovaleryl chloride, synthesis of K Diplocardia luciferin, LVII, 377 Kabikinase, streptokinase, purification, Isovaline, internal standard, XLVII, 34, XLV, 248 Kallikrein, XXXIV, 5, 432; XLVI, 206 Isoxazole, XLVI, 415, 418 human Isoxazolium salt, XLIV, 48 assay, XLV, 304-306 Isoxazolium salts method, carboxyl group inhibitors, XLV, 314 activation, XLVII, 556, 557 molecular weight, XLV, 311 Isozyme, XXXIV, 595-597 pH, optimum, XLV, 311 carbonic anhydrase, XXXIV, 595 from prekallikrein, activation, cell monitoring, LVIII, 166-170 XLV, 315 extract preparation, LVIII, 167 properties, XLV, 309 staining, LVIII, 168–170 physical, XLV, 311 Itran-2, infrared spectral properties, purification, XLV, 306-309 LIV, 316 alcohol fractionation, XLV, 307 chromatography, XLV, 308 J role, XLV, 303 stability, XLV, 309 Jakoby procedure, LI, 87 substrates Japanese quail natural, XLV, 312 feeder cell, LVIII, 392 synthetic, XLV, 313 Rous sarcoma virus, LVIII, 395 Kallikrein A, pig pancreatic Japanese radish, peroxidase, LII, 515 assay, XLV, 300-302 CD studies, LIV, 271, 272, 283, 284 spectrophotometric, XLV, 303 JASCO-J-40 CD spectrophotometer, with Technician autoanalyser, LIV, 251 XLV, 300 JEG-3 trophoblast line, in cell hybrid, composition, XLV, 294 XXXIX, 127 amino acid, XLV, 295, 296, 299 Jena UG-11 filter, XLIX, 192 carbohydrate, XLV, 295 Jena UG-12 filter, XLIX, 192 kinetics, XLV, 298, 299 Jensen sarcoma cell molecular weight, XLV, 297 multilayered culture, XXXII, 571-574 properties, XLV, 293, 294, 299 protein content, LVIII, 145 purification, XLV, 289-293 JEOLPFT-100 spectrometer, LIV, 190 purity, XLV, 294 Johnson Foundation dual-beam role, XLV, 289 spectrophotometer, LVII, 55 specific activity, XLV, 293 Johnston-Ogston effect, correction, stability, XLV, 293 XLVIII, 268 structure, XLV, 294 Joule heating, electrophoretic light Kallikrein B, pig pancreatic scattering, XLVIII, 487–489 assay, XLV, 300-302 J protein, deficiency in bacteria, XXXII, spectrophotometric, XLV, 303 849, 852, 854–856 JRB Model 3000 ATP photometer, see with Technician autoanalyser, ATP photometer XLV, 300 Juice extractor, for plant-cell composition, XLV, 294 maceration, XXXI, 541 amino acid, XLV, 295, 296, 299 Juvenile G_{M2} -gangliosidosis carbohydrate, XLV, 295 characteristics, L, 484 kinetics, XLV, 298, 299

molecular weight, XLV, 297

properties, XLV, 293, 294, 299

enzymic determination, L, 485–487

Tay-Sachs disease, L, 484

purification, XLV, 289–293
purity, XLV, 294
role, XLV, 289
specific activity, XLV, 293
stability, XLV, 293
structure, XLV, 294
Kanamycin
biosynthesis, XLIII, 615–618, *see also* Aminoglycoside modifying enzyme
biotransformation, XLIII, 208
carrier-free continuous electrophoresis, XLIII, 289, 290
2-deoxystreptamine, XLIII, 449, 450
for frog haploid cell culture, XXXII, 798, 799
gas liquid chromatography, XLIII, 218–220
hem mutants, LVI, 172–174
high-pressure liquid chromatography, XLIII, 278, 307, 308, 319
ion-exchange chromatography, XLIII, 268, 269, 272
Amberlite–IR–120 resin, XLIII, 272
Duolite C–62, XLIII, 272, 273
in media, LVIII, 112, 114, 116
nonionic adsorption chromatography, XLIII, 276, 277
solvent system, XLIII, 119, 120
stock solutions, LVI, 174
thin-layer chromatography, XLIII, 211
Kanamycin A
kinetic data, XLIII, 623
structure, XLIII, 612
Kanamycin acetyltransferase, XLIII, 616
assay, XLIII, 615–618
from *Escherichia coli*
activity, XLIII, 623
buffer, XLIII, 623
kinetic data, XLIII, 623
pH optimum, XLIII, 623
stability, XLIII, 623
from *Pseudomonas aeruginosa*, XLIII, 624
Kaolin suspension, reagent
factor VIII, XLV, 85
factor IX, bovine, XLV, 76
factor XI, XLV, 67
Kapton, in X-ray absorption spectroscopy, LIV, 332
Karnovsky high osmolarity fixative, XXXIX, 137
Karnovsky osmium tetroxide-ferrocyanide stain, XXXIX, 143, 144
Karnovsky paraformaldehyde-glutaraldehyde fixative, XXXIX, 134, 137, 200
diluted, XXXIX, 137
Karnovsky procedure, for SEM fixation, XXXII, 56
Karyotyping, LVIII, 170–174, 220, 322–344
characteristics, LVIII, 342
CHO cells, LVIII, 340–342
Drosophila melanogaster, LVIII, 342
GH cells, LVIII, 532
of haploid frog cells, XXXII, 796, 797
human, LVIII, 340–342
modifications, LVIII, 333, 334
mouse, LVIII, 341, 342
mutants, LVIII, 309
problems, LVIII, 330
rat, LVIII, 342
standardization, LVIII, 339
Y1 adrenal cell line, LVIII, 571
Kasugamycin, XLIII, 272, 273
effect on protein synthesis, LIX, 855, 858
KAT, *see* Kanamycin acetyltransferase
μ-kat, *see* Microkatal
Katsuta medium, DM, LVIII, 57, 62–70, 88
Kayser unit, LVII, 572
KB cell, plasma membrane isolation from, XXXI, 161
KB cell line
infection, LVIII, 425, 431
monolayer culture, LVIII, 430
properties and uses, XXXII, 585
suspension culture, LVIII, 430
K12 cell, synchronization, LVIII, 252–256
KDO, *see* 2-Keto-3-deoxyoctonate
KDP crystal, LIV, 7, 8, 21, 23
KED buffer, XXXVI, 469, 470
Keilin-Hartee particle, LIII, 155, 410
assay of cytochrome *c*, LIII, 160, 161

preparation of cytochrome c_1, LIII, 183
of succinate dehydrogenase, LIII, 476, 477
Keilin-Hartree preparation, heart muscle
comments, LV, 124–127
preparation, LV, 119–121
properties, LV, 121–124
stability, LV, 121
yield, LV, 121
Kel-F, as sample cell material, LIV, 403
Kel–F glassware, LIII, 176
Kel–F Kontes pestle, XXXI, 62
Kerasine, optical rotation, XXXII, 365
Keratinocyte, human epidermal, media, LVIII, 90
Kerry Vibrason System, XXXII, 207
Ketene, XLVI, 72
reactivity, LVI, 657
Kethoxal, in studies of protein-RNA interactions, LIX, 349, 577
Ketimine, XLVI, 44, 137
α-Ketoacid
decarboxylation, LII, 38
β,γ-unsaturated, XLVI, 32
3-Ketoacyl-CoA thiolase, *see* Acetyl-CoA acyltransferase
3-Ketoacyl coenzyme A
absorption at 303 nm (of Mg^{2+} complex), XXXV, 129
synthesis, XXXV, 129
3-Keto-*p*-arbutine, quantitative determination, XLI, 25
3-Ketobionic acid, quantitative determination, XLI, 25
2-Ketobutenoic acid, LIII, 445
α-Ketobutyrate, adenylate cyclase, XXXVIII, 169
Ketocarbene, XLVI, 485
Δ^4-3-Keto chromophore, XLVI, 477
3-Ketodecanoyl coenzyme A, in assay for 3-ketoacyl-CoA thiolase, XXXV, 130
2-Keto-3-deoxy-L-arabonate, equilibrium constant, XLII, 272
2-Keto-3-deoxy-L-arabonate aldolase, *see* 2-Keto-3-deoxy-L-pentonate aldolase
L-2-Keto-3-deoxyarabonate dehydratase
assay, XLII, 308, 309
chromatography, XLII, 311
electrophoresis, XLII, 313
inhibitors, XLII, 313
molecular weight, XLII, 313
properties, XLII, 312, 313
from *Pseudomonas saccharophila*, purification, XLII, 309
purification, XLII, 309–312
DL-2-Keto-3-deoxyarabonic acid, potassium salt, preparation and assay, XLI, 101, 103
L-2-Keto-3-deoxyarabonic acid, potassium salt, preparation and assay, XLI, 102, 103
2-Keto-3-deoxy-D-fuconate, equilibrium constant, XLII, 272
2-Keto-3-deoxygluconate-6-phosphate aldolase, XLVI, 132, *see also* Phospho-2-keto-3-deoxygluconate aldolase
2-Keto-3-deoxy-D-gluconic acid, preparation, calcium and potassium salts, XLI, 99–101
2-Keto-3-deoxyoctonate, in *E. coli* cell walls, XXXI, 642
2-Keto-3-deoxy-L-pentonate aldolase, from pseudomonad MSU-1, XLII, 269–272
assay, XLII, 269, 270
chromatography, XLII, 271
properties, XLII, 271, 272
purification, XLII, 270, 271
2-Keto-3-deoxy-6-phosphogluconic aldolase, *see* Phospho-2-keto-3-deoxygluconate aldolase
Ketoester, as fatty acid precursors, synthesis, XXXII, 191, 192
6-Ketoestradiol, XXXVI, 21
5-Keto-D-fructose
bisphenylhydrazone and bis-*p*-nitrophenylhydrazone, XLI, 86
chromatography, XLI, 86
colorimetric analysis, XLI, 87
enzymic assay, XLI, 87
isolation and purification, XLI, 85
production, and phosphates, XLI, 84–91
5-Keto-D-fructose 1,6-bisphosphate, preparation and properties, XLI, 89, 90
5-Keto-D-fructose 1-phosphate, preparation and properties, XLI, 88, 89

5-Keto-D-fructose reductase
from *Acetobacter melanogenum*, XLI, 128, 137
from *Gluconobacter albidus*, XLI, 128
from *Gluconobacter cerinus*, XLI, 127–131
assay, XLI, 127
properties, XLI, 131
purification, XLI, 128–130
from yeast, XLI, 132–138
assay, XLI, 132
chromatography, XLI, 134
inhibitors, XLI, 137
properties, XLI, 136–138
purification, XLI, 133–136
Ketogenesis
control studies, XXXVII, 292
rate, by perfused liver, XXXV, 597, 606
α-Ketoglutarate dehydrogenase, *see also* 2-Oxoglutarate dehydrogenase
brain mitochondria, LV, 58
coenzyme A assay, LV, 209
α-Ketoglutarate tyrosine aminotransferase, induction by hydrocortisone, XXXIX, 39
α-Ketoglutarate tyrosine transaminase, *see* α-Ketoglutarate tyrosine aminotransferase
α-Ketoglutaric acid
activator of adenylate kinase, LI, 465
assay, LV, 221
improvement, LV, 206
content
of brain, LV, 218
of hepatocytes, LV, 214
of kidney cortex, LV, 220
of rat heart, LV, 216, 217
of rat liver, LV, 213
exchange, LVI, 253
in isolated renal tubules, XXXIX, 18
rat brain mitochondria, LV, 57
renal gluconeogenesis, XXXIX, 13, 14
substrate, glutamate dehydrogenase, XLIV, 508, 690
transport, LVI, 248, 249, 292
yeast mitochondria, LV, 150
Ketoheptose, qualitative analysis, XLI, 18
3-Ketohexanoyl coenzyme A, in assay for 3-ketoacyl-CoA thiolase, XXXV, 129, 130
3-Ketohexosamine, quantitative determination, XLI, 25
Ketohexose, qualitative analysis, XLI, 18
DL-2-Keto-4-hydroxyglutarate
assay, XLI, 118
preparation, XLI, 115–118
2-Keto-4-hydroxyglutarate aldolase
from bovine liver, XLII, 280–285
assay, XLII, 280, 281
chromatography, XLII, 282–284
equilibrium constant, XLII, 285
inhibition, XLII, 284
molecular weight, XLII, 284
properties, XLII, 284, 285
purification, XLII, 281–284
from *Escherichia coli*, XLII, 285–290
assay, XLII, 286
chromatography, XLII, 288
molecular weight, XLII, 289
properties, XLII, 289, 290
purification, XLII, 286–289
Keto-*scyllo*-inositol, XLIII, 432, 440, 441, *see also* L-Glutamine:keto-*scyllo*-inositol aminotransferase
L-guanidino-3-keto-*scyllo*-inositol aminotransferase assay, XLIII, 465
myo-inositol:NAD$^+$ 2-oxidoreductase assay, XLIII, 434, 437, 438
1,D-4-Keto-*myo*-inositol, XLIII, 442, 443
3-Ketolactose, polarographic identification, XLI, 23–25
3-Ketomethylthiohexoside, quantitative determination, XLI, 25
Ketone
aromatic, derivatives of aminoacyl-tRNA, XLVI, 678–683
as inhibitors, XLVI, 35, 36
photochemical reactions, XLVI, 470–479
Ketone body
brown adipose tissue mitochondria, LV, 76
liver mitochondria, LV, 21
α-Keto nitrene, XLVI, 78
2-Keto-3,4-pentadienoic acid, LIII, 442
3-Ketoreductase, in mouse kidney, XXXIX, 458

3-Ketosalicine, quantitative determination, XLI, 25
Δ^4-3-Ketosteroid, XLVI, 461, 469
Δ^4-3-Keto-C_{21}-steroid, synthesis, LVIII, 571
Δ^5-Ketosteroid, XLVI, 335, 469
Δ^{10}-3-Ketosteroid, XLVI, 469
3-Ketosteroid-Δ^1-dehydrogenase, XLIV, 184–190
  activity assay, XLIV, 186
  cofactor system, XLIV, 188
$\Delta^{3,5}$-Ketosteroid isomerase, XLVI, 159
Δ^5-3-Ketosteroid isomerase, XXXIV, 558; XLVI, 89, 461–468
  affinity labeling, XLVI, 476–478
  labeling by photoexcited steroid ketones, XLVI, 469–479
  purification, XXXIV, 560–564
3-Ketosucrose, polarographic identification, XLI, 24
Keto-sugar, *see also* specific type
  colorimetric assay, XLI, 29–31
3-Keto sugar, polarographic determination, XLI, 22–27
3-Ketotrihexose, quantitative determination, XLI, 25
α-Ketovalerate, adenylate cyclase, XXXVIII, 169
Key intermediary protein estradiol receptors, XXXVI, 320
Kidney, XXXIX, 1–22, *see also* specific type
  adenocarcinoma, LVIII, 376
  adenylate cyclase, XXXVIII, 150–152
  androgen metabolism and receptor activity, XXXIX, 454–460
  artificial, XLIV, 211
  brush-border isolation, XXXI, 115–123
  cAMP-receptor, XXXVIII, 380
  cell, LVIII, 377
    calcium efflux studies, XXXIX, 548, 549, 552
    calcium influx studies, XXXIX, 536, 537
    calcium measurements, XXXIX, 515–520
    cat, media, LVIII, 58
    monkey
      dissociation, LVIII, 127
      microcarrier culture, LVIII, 190
      suspension culture, LVIII, 203
      SV40, LVIII, 404
    preparation, XXXIX, 20–23
  cortex
    metabolite content, LV, 220
    mitochondria, LV, 11
      properties, LV, 11, 12
    mutarotase, assay, purification, and properties, XLI, 471–484
  dispersion, LVIII, 123
  epithelial cell, LVIII, 552–560
  glomeruli isolation, XXXII, 653–658
  isolated, perfusion, LII, 59
  3-ketoreductase activity, XXXIX, 458, 459
  lysosome isolation from cells, XXXI, 330–339
  mammalian, oxidase, LII, 10–12, 18
  microsomes, LII, 89
  $(Na^+ + K^+)$-ATPase isolation, XXXII, 281–285
  nuclear steroid receptors in, extraction, XXXVI, 287–289
  perfusion, XXXIX, 3–11
    biochemical parameters, XXXIX, 9, 10
    functional behavior, XXXIX, 7–9
  protein kinases, XXXVIII, 298
  removal procedure, XXXIX, 6
  tubule isolation, XXXII, 304, 305, 658–664; XXXIX, 11–20
    incubation procedure, XXXII, 661, 662
    mammalian, XXXIX, 11–20
    sampling procedure, XXXII, 663, 664
Kikumycin A, XLIII, 145, 146
Kimax tube, calcium concentration, LVII, 282
Kinase, XXXIV, 229, 246; XLII, 3–219; XLVI, 240–249, *see also* specific types
Kinetic analysis
  of calcium efflux, XXXIX, 547–555
  of calcium influx, XXXIX, 528–546
Kinetic isotope effect, LIV, 182
Kinetics
  of artificial enzyme membranes, XLIV, 913–919
  experimental data for parameter determination, XLIV, 438–443

of growth, LVIII, 133, 214–217, 241, 245
of irreversible monoenzyme system, XLIV, 913–916
measurement
general principles, LIV, 1–3
for immobilized enzymes, XLIV, 3
of proteoliposome incorporation into planar membranes, LV, 768
steady state, of enzyme level, XL, 252
theoretical, XLIV, 397–443
transition state, XLVI, 15–17
Kinetin, as cytokinin, LVIII, 478
Kininogen, substrate
of kallikrein, XLV, 312
of prekallikrein, XLV, 316
KIT–1 quartz oscillator, LIX, 615
Klebsiella aerogenes
β-lactamase in, XLIII, 673, 678
in D-ribulose preparation from ribitol, XLI, 104–106
Klebsiella pneumoniae
crude extract preparation, LIII, 323
growth, LI, 182; LIII, 323
nitrogenase, LIII, 319, 323, 324
Mössbauer studies, LIV, 371
phosphoribosylglycinamide synthetase, LI, 182
Kluyveromyces lactis, mitochondrial mutants, LVI, 120, 121
Knife, for freeze-fracturing, XXXII, 39, 40
Knob, globular, in plasma membranes, XXXI, 85, 86, 90
Kodak AR–10 Stripping film plate, XXXVII, 153
Kodak No-Screen Medical X-ray film, LIX, 70
Kodak Photoflo, LIX, 843
Kodak Royal X–O–Mat Film, LIX, 70, 301
Kodak Written filter, LIV, 500
Kodak XR–5 film 192, LIX, 834
Koenigs–Knorr reaction
disaccharide synthesis, L, 96–100
kinetics, L, 98
mechanism, L, 97, 98
peracylglucopyranosyl bromides, L, 105
stereochemistry, L, 99, 100
Kojate oxidase, LII, 13
Kojic acid, XLIII, 154
Koshland, Nemethy, Filmer model of cooperativity, XLVIII, 289, 290, 294
Koshland reagent, XLVII, 453
Krabbe's disease, *see* Globoid cell leukodystrophy
Krait, α-bungarotoxin, XXXII, 311–313
Kramers-Heisenberg-Dirac dispersion equation, LIV, 235
Kramers system, LIV, 346, 350, 355–365
Kratky camera, XXXII, 212
Kratky X-ray tube, XXXII, 212
Krebs cycle
control studies, XXXVII, 292, 295
enzymes, toluene-treated mitochondria, LVI, 546, 547
intermediates, renal tubule permeability, XXXIX, 11, 15–20
substrates, transport, LVI, 248, 249
Krebs-Henseleit bicarbonate buffer
as heart perfusion medium, XXXIX, 44
for hemicorpus perfusion, XXXIX, 74, 75
as kidney perfusate, XXXIX, 6
as liver perfusate, XXXIX, 27
Krebs phosphosaline buffer, XXXVII, 97
Krebs-Ringer bicarbonate buffer, XXXII, 688, 689, 696; XXXVII, 212; LII, 356
with albumin, XXXVII, 269, 270
in interstitial cell isolation, XXXIX, 257, 260
modified, for embryo culture, XXXIX, 298, 299
for muscle studies, XXXIX, 92
for pancreas superfusion, XXXIX, 374
for placental perfusion, XXXIX, 245
for renal tubule incubation, XXXIX, 14
Krebs-Ringer-Henseleit glucose buffer, composition, XXXVI, 269
Krebs-Ringer HEPES buffer, XXXVII, 193
Krönig fine structure, LIV, 340
Krogh's diffusion constant, XXXIX, 83
Krogh model, oxygen gradient studies, LVII, 226
KRS–5 cell, LIV, 303, 314
infrared spectral properties, LIV, 316

refractive index, LIV, 314, 316
Krypton gas laser, available wavelength, XLIX, 85, 86
Krypton ion laser, use in resonance Raman spectroscopy, LIV, 246
Kuhn dissymmetry factor, XLIX, 202
Kuhn-Kirkwood mechanism, LIV, 254
Kunitz soybean trypsin inhibitor, XXXIV, 5
active fragment, reconstitution, XLV, 702
amino acid sequence, XLV, 701
reactive site, alterations, XLV, 700, 703
kinetics, XLV, 700
properties, XLV, 700
Kupffer cell, *see also* Macrophage
of liver, XXXIX, 36
Kwik-fil glass, LVII, 313
Kynurenate 7,8-dihydroxylase, LII, 11
Kynurenine 3-hydroxylase, *see* Kynurenine 3-monooxygenase
Kynurenine 3-monooxygenase, LII, 17
assay, LV, 102
in outer mitochondrial membranes, XXXI, 319
KYST multidimensional scaling program, LIX, 609, 610
K-II zonal rotor, LIX, 206, 209

L

Labar technique, XLVIII, 171
Labeling, *see also* specific types
conditions for DNA localization, XL, 12
for protein localization, XL, 15
for RNA localization, XL, 13
electrophilic, XLVI, 70
iterative
column photolysis, XXXVIII, 395
dialysis photolysis, XXXVIII, 396, 397
principle, XXXVIII, 394, 395
nonspecific, XLVI, 111, 112
radioactive, XLVI, 110
scavengers, XLVI, 113
Labeling index analysis, in exponential cell cultures, XL, 48
Label triangulation, LIX, 676, 678–683
Labile sulfide
complex I, LVI, 580, 581
complex II, LVI, 580, 581
complex III, LVI, 580, 582, 583
complex V, LVI, 580, 584
Laboratory equipment, suppliers, LVIII, 15–17, 196
Laboratory facility
cleaning, LVIII, 3–7
design, LVIII, 4–7
sterilization, LVIII, 3–7
storage, LVIII, 3, 7–9
Laboratory operation, LVIII, 3
Laboratory safety, LVIII, 36–43
Lab-Tek Culture Chamber, XXXIX, 202
Lab-Tek tissue culture chamber slide, XXXIX, 300
Laburnum alpinum, lectin, XXXIV, 334
Laccase, LII, 10
immobilization, by microencapsulation, XLIV, 214
oxy form, spectrum, LII, 36
prosthetic group, LII, 4
Raman frequencies and assignments, XLIX, 144
reaction mechanism, LII, 39
resonance Raman spectrum, XLIX, 142
L251A cell line, LVIII, 219
Lac operator, XLVI, 172, 175
5-bromo-2′-deoxyuridine-substituted, XLVI, 89
Lac operon polysome, XXXIV, 7
Lac repressor, XXXIV, 368; XLVI, 89, 172, 175
lac repressor protein, fluorine nuclear magnetic resonance, XLIX, 273, 274
α-Lactalbumin, XXXIV, 4, 359; XLVI, 366
binding site, XLVI, 75
mammary epithelial cell synthesis, XXXIX, 450
assay, XXXIX, 453
α-Lactalbumin-agarose, XXXIV, 359, 360
β-Lactam antibiotic, *see also* specific antibiotic
chiroptical studies, XLIII, 352
CMR spectroscopy in biosynthesis of, XLIII, 410–425

DD-carboxypeptidase-transpeptidase, interaction, XLV, 624, 625
inhibition, XLV, 635
paper chromatography of, XLIII, 110–119
screening, XLIII, 52
titration, XLV, 627
β-Lactamase, XLVI, 531–537, *see also* Penicillinase
from Actinomycetes, XLIII, 687–698, *see also* β-Lactamase, from *Streptomyces*
affinity labeling, XLVI, 533–536
antisera preparation, XLIII, 86–90
crude enzyme preparation, XLIII, 87, 88
inoculation program for crude enzyme, XLIII, 89, 90
isolation of β-lactamase-less mutants, XLIII, 89
purified enzyme preparation, XLIII, 86, 87
antisera specific, XLIII, 90–100
assay, XLVI, 533
acidimetric methods, XLIII, 77–80, 83, 84
alkimetric methods, XLIII, 77–80
biological, XLIII, 81, 82
of DD-carboxypeptidase-transpeptidase, XLV, 613
hydroxylamine, XLIII, 81, 85
indicator method, XLIII, 78, 79
iodometric method, XLIII, 74–77, 83, 84
macroiodometric determination, XLIII, 74–76
macroiodometric method of Perret, XLIII, 83
manometric measurement of CO_2, XLIII, 79, 80
microbiological, XLIII, 81–83
microiodometric determination, XLIII, 76, 77
pH stat titration method, XLIII, 79
from *Bacillus cereus*, XLIII, 640–652, *see also* β-Lactamase I; β-Lactamase II
concentration in culture, XLIII, 643, 644
culture preparation, XLIII, 642, 643
β-lactamase II isolated from β-lactamase I, XLIII, 640, 641, 644
purification, XLIII, 641–644, 646
specific activity, XLIII, 642
substrate profile of extracellular, XLIII, 647
from *Bacillus licheniformis*, XLIII, 653–664
affinity chromatography, XLIII, 654
amino acid sequence, XLIII, 661
Bacillus licheniformis preparation, XLIII, 654
cell-bound, purification, XLIII, 655–658
chemical properties, XLIII, 660
culture preparation, XLIII, 654, 655
extracellular, purification of, XLIII, 655
heterogeneity, XLIII, 661
hydrolysis of penicillin and cephalosporin derivatives, XLIII, 659
immunology, XLIII, 662
iodine reaction, XLIII, 662
kinetic properties, XLIII, 658, 659
Michaelis constants, XLIII, 659
modification, XLIII, 662–664
molecular weight, XLIII, 662
physical properties, XLIII, 662
physiological efficiency, XLIII, 660
plasma membrane bound, purification, XLIII, 663
purification, XLIII, 653–658
secretion, XLIII, 663
specific activity, XLIII, 659
stability, XLIII, 658
tetranitromethane modification reaction, XLIII, 662
trypsin release, XLIII, 663
effect of pH, XLVI, 534, 535
from *Enterobacter*, XLIII, 678–687
antiserum, XLIII, 684
assay, XLIII, 679, 680
cross-reaction, XLIII, 684
dialysis and concentration, XLIII, 681

Enterobacter cloacae, preparation, XLIII, 680
homogeneity, XLIII, 685
hydrolysis of cephaloridine, XLIII, 687
Ia type, XLIII, 673
induction, XLIII, 684
inhibition, XLIII, 686, 687
isoelectric focusing patterns, XLIII, 685
kinetic properties, XLIII, 686
molecular weight, XLIII, 685
polyacrylamide disc electrophoresis, XLIII, 683
properties, XLIII, 683–687
purification, XLIII, 680–683
QAE–Sephadex chromatography, XLIII, 681, 682
Sephadex G–50 chromatography, XLIII, 681
specific activity, XLIII, 679
substrate specificity, XLIII, 686
temperature effect, XLIII, 686
ultrasonic disruption in purification, XLIII, 681
unit definition, XLIII, 679
from *Escherichia coli*, XLIII, 672–677, *see also* β-Lactamase, type IIIa
excretion by *Streptomyces* strain R39, XLV, 614
by *Streptomyces* strain R61, XLV, 615
fluorescence for testing antibiotics resistance to, XLIII, 208
immunological techniques, XLIII, 86–100, *see also* Immunological techniques
kinetics, XLVI, 535
molecular variants, detection of, XLIII, 94–95
neutralization analysis, XLIII, 90–97, *see also* Immunological technique
P99, XLIII, 678
in penicillin acylase preparations, XLIII, 700–701
precipitation analysis, XLIII, 98–100
reaction scheme, XLIII, 70
R_{TEM}–mediated, XLIII, 678
from *Staphylococcus aureus*, XLIII, 664–672
activity, XLIII, 670, 671
6-aminopenicillanic acid induction, XLIII, 666
cell-bound, XLIII, 670
cloxacillin induction, XLIII, 666
constitutive mutants, XLIII, 666, 667
culture preparation, XLIII, 665, 666
ethylmethanesulfonate as mutagen, XLIII, 666
N-methyl-*N*-nitro-*N*-nitrosoguanidine as mutagen, XLIII, 666
molecular properties, XLIII, 671, 672
natural variants from studies one, XLIII, 94, 95
purification, XLIII, 667–670
specificity, XLIII, 671
spot tests, XLIII, 670
stability, XLIII, 670
variants, XLIII, 664, 665
from *Streptomyces albus*, XLIII, 687–698
assay, XLIII, 689
concentration of enzyme preparations, XLIII, 693
crude extract preparation, XLIII, 692
culture media, XLIII, 688
hydrolysis of β-lactam antibiotics, XLIII, 689, 690, 697
inhibitors, XLIII, 697
iodine, sensitivity, XLIII, 696
isolation from strain Albus G, XLIII, 695, 696
from strain R 39, XLIII, 693, 695
kinetic properties, XLIII, 689, 690
K_m and V_{max} values, XLIII, 689, 690
metal ion requirements, XLIII, 696
pH optimum, XLIII, 696
phosphate buffer, XLIII, 693
physical properties, XLIII, 696, 697
polyacrylamide gel electrophoresis, XLIII, 693

properties, XLIII, 696, 698
purification, XLIII, 692–696
SH group reagents, sensitivity to, XLIII, 696
specific activity, XLIII, 693
specificity, XLIII, 697
stability, XLIII, 696
unit definition, XLIII, 689
synthesis, XLVI, 533
type Ia, XLIII, 673
type IIIa, XLIII, 672–677
centrifugation steps, XLIII, 675
DEAE-cellulose column chromatography, XLIII, 676
EDTA in sucrose purification, XLIII, 673
Escherichia coli K12, growth medium, XLIII, 673
G–75 Sephadex column, XLIII, 676
location in cells, XLIII, 672, 673
molecular weight, XLIII, 676, 677
properties, XLIII, 676, 677
Pseudomonas aeruginosa, growth medium, XLIII, 674
purification, XLIII, 673–676
R factor-mediated, XLIII, 673
specificity, XLIII, 676, 677
stability, XLIII, 676
ultrasonic treatment, XLIII, 675
β-Lactamase I
from *Bacillus cereus*, XLIII, 678
amino acid composition of, XLIII, 648
chemical properties, XLIII, 647, 648
conformational changes, XLIII, 649, 650
kinetic properties, XLIII, 647
modification reaction, XLIII, 649
molecular weight, XLIII, 649
properties, XLIII, 644–650
physical, XLIII, 644–650
purification, XLIII, 641–644, 646
stability, XLIII, 644, 647
β-Lactamase II
from *Bacillus cereus*
amino acid composition of, XLIII, 651
chemical properties, XLIII, 650–652
inhibitors, XLIII, 652
kinetic properties, XLIII, 650
modification, XLIII, 652
molecular weight, XLIII, 652
physical properties, XLIII, 652
purification, XLIII, 642–644, 646
stability, XLIII, 650
substrate profile of extracellular, XLIII, 651
thiol group modification of, XLIII, 652
β-Lactamase-less mutant, XLIII, 89
Lactase, *see also* β-D-Galactosidase
in microvillous membrane, XXXI, 130
L-Lactate, chemiluminescent assay of lactate dehydrogenase, LVII, 461
D-Lactate-cytochrome *c* reductase, flavoprotein classification, LIII, 397
L-Lactate cytochrome *c* reductase, *see* Flavocytochrome b_2
Lactate dehydrogenase, XXXIV, 165, 237–239, 246, 249, 252, 491, 598–605; XLVI, 20, 21, 145, 162, *see also* Ligand affinity chromatography
activity assay, XLIV, 271, 458–460, 664
affinity for calcium phosphate gel, LI, 584
affinity labeling, XLVII, 424
assay, LVIII, 169, 170
of adenylosuccinate synthetase, LI, 213
AMP, LV, 208
ATPase, LV, 313, 320, 329
inhibitor, LV, 408
associated with cell fractions, XL, 86
for biotin carboxylase, XXXV, 26, 28
bovine, indefinite, self-association, XLVIII, 112
brain mitochondria, LV, 58, 59
cell monitoring, LVIII, 168–170
cGMP, XXXVIII, 74, 78, 81, 82
of FGAM synthetase, LI, 197
fluorometric, XLIV, 627, 628
of GMP synthetase, LI, 219
of guanylate kinase, LI, 475, 483
using immobilized enzyme rods, LVII, 210, 211

of nucleoside diphosphokinase, LI, 377
using peroxyoxalate chemiluminescence, LVII, 455, 460, 461
pyridine nucleotide, LV, 264, 265, 267–269, 273, 276
of pyrimidine nucleoside monophosphate kinase, LI, 323
of uridine-cytidine kinase, LI, 300, 309, 316
conjugate
concentration determination, XLIV, 392, 393
ligand binding, XLIV, 514, 515
oligomer dissociation, XLIV, 514
physicochemical properties, comparative, XLIV, 934–936
determination of isoenzyme levels, XLI, 47–52
as diagnostic enzyme, XXXI, 24
in enzyme electrode, XLIV, 591, 877
gluconeogenic catalytic activity, XXXVII, 281, 288
histidine modification, XLVII, 433, 436, 437, 441
immobilization
by adsorption, effects on activity, XLIV, 44
on cellulose sheets, XLIV, 906, 908
by diazotization procedure, XLIV, 933
on (diethylaminoethyl)cellulose, XLIV, 52
on glass beads, XLIV, 932–934
by glutaraldehyde cross-linking, XLIV, 933, 934
on inert protein, XLIV, 903, 908
in multistep enzyme system, XLIV, 181, 457–460, 471, 472
on nylon tube, XLIV, 120, 641
on Sepharose 4B, XLIV, 930, 931
using soluble carbodiimide, XLIV, 932, 933
in intestinal mucosa, XXXII, 671
isoenzyme, XXXIV, 599, 605; LVIII, 169, 170
kinetics
of ihibition, XLIV, 428
in multistep enzyme system, XLIV, 467, 468
in lactic acid assay, XLI, 42–44
in D-mannose microassay, XLI, 4
as marker enzyme, LVI, 210, 211, 217, 218, 222
membrane osmometry, XLVIII, 75
molecular weight, LVII, 210
polarography, LVI, 478
reduction of NAD^+-analogs, XLIV, 873, 874
rhein, LV, 459
SDS complex, properties, XLVIII, 6
sialic acid determination, L, 75, 76
sialic acid *O*-lactyl group determination, L, 78
stain, LVIII, 169
staphylococcal membranes, LVI, 176
s-triazine binding, XLIV, 52
viability test, LII, 64
Lactate dehydrogenase-H_4, XXXIV, 600
Lactate dehydrogenase-M_4, XXXIV, 600
D-Lactate dehydrogenase, XXXI, 742; LIII, 519–527
activity, LIII, 519
assay, LIII, 519, 520
by Fast Analyzer, XXXI, 816
in bacterial membrane, XXXI, 649
biological role, LIII, 519
as cytosol marker, XXXI, 406, 408, 410, 740
from *Escherichia coli*, LIII, 519–527
flavin content, LIII, 524
from horseshoe crab, XLI, 313–318
assay, XLI, 313
chromatography, XLI, 314–316
inhibitors, XLI, 316
molecular weight and subunit structure, XLI, 316
properties, XLI, 316, 318
purification, XLI, 313–317
immunochemical studies, LIII, 526, 527
inactivation, LIII, 439, 440
inhibitors, LIII, 525
kinetic properties, LIII, 524, 525
molecular weight, LIII, 524
from *Peptostreptococcus elsdenii*, XLI, 309–312
activators and cofactors, XLI, 312

assay, XLI, 309, 310
chromatography, XLI, 311
properties, XLI, 312
purification, XLI, 310–312
pH optimum, LIII, 524
purification, LIII, 520–524
purity, LIII, 524
reconstitution studies, LIII, 525, 526
spectral properties, LIII, 524
substrate specificity, LIII, 524, 525
D(−)-Lactate dehydrogenase
from *Butyribacterium rettgeri*, XLI, 299–303
activators and inhibitors, XLI, 303
assay, XLI, 299
properties, XLI, 302, 303
purification, XLI, 300–303
from fungi, XLI, 293–298
assay, XLI, 294, 295
chromatography, XLI, 296
inhibitors and regulation, XLI, 298
molecular weight, XLI, 298
properties, XLI, 297, 298
purification, XLI, 295–297
L-Lactate dehydrogenase, XXXIV, 239, 595, 599
from *Bacillus subtilis*, XLI, 304–309
assay, XLI, 304
chromatography, XLI, 306
properties, XLI, 307–309
purification, XLI, 305–307
inactivation, LIII, 439
isozyme, XXXIV, 595
yeast, phosphorescence studies, XLIX, 247
L(+)-Lactate dehydrogenase, from animals, plants, and microorganisms, XLI, 293
Lactate dehydrogenase–X
from spermatozoa, chromatography, XLI, 322
purification, XLI, 322, 323
from testes of mouse, rat, and bull, XLI, 318–322
assay, XLI, 318, 319
inhibitors, XLI, 321
properties, XLI, 320, 321
purification, XLI, 319–321
L-Lactate monooxygenase, inactivation, protection, LIII, 447
Lactate oxidase, apoenzyme, assay of FMN, LIII, 422
L-Lactate oxidase, inactivation, LIII, 439
Lactate oxygenase, from *Mycobacterium phlei*, XLI, 329–333
assay, XLI, 329, 330
molecular weight, XLI, 332
preparation, XLI, 330
properties, XLI, 332, 333
purification, XLI, 330–332
L-Lactate/pyruvate ratio, in perfusion fluid, LII, 50
Lactating tissue, mitochondria isolation, XXXI, 306, 307
Lactation, mice, mammary gland parenchymal cells, XXXII, 693–706
Lactic acid
alcohol oxidase, LVI, 467
assay, LV, 221, 294
automated analysis, XLIV, 636
content of kidney cortex, LV, 220
as electron donor, LVI, 386
enzymic determination, XLI, 41–44
ferricyanide, LVI, 232, 233
flow dialysis experiments, LV, 683
oxidation, LVI, 474
oxidative decarboxylase, LII, 16
Pasteur effect, LV, 290, 294, 295
phosphorylating vesicles, LV, 166
polarography, LVI, 451, 452, 478
renal gluconeogenesis, XXXIX, 13–16
use by perfused liver, XXXIX, 34
D-Lactic acid, XLVI, 86
inhibitor of flavocytochrome b_2, LIII, 240
substrate of D-lactate dehydrogenase, LIII, 519
DL-Lactic acid
assay of flavocytochrome b_2, LIII, 240
isolation, XLVII, 491
purification of flavocytochrome b_2, LIII, 240, 246, 247
L-Lactic acid, XLIV, 246, 249
assay of flavocytochrome b_2, LIII, 239
enzyme electrode, XLIV, 585, 591, 592
substrate, lactate dehydrogenase, XLIV, 881

Lactic dehydrogenase, *see* Lactate dehydrogenase
Lactobacillus acidophilus
cell extract preparation, LI, 347
deoxynucleoside kinases from, LI, 346–354
growth, LI, 347
Lactobacillus brevis, D-xylose isomerase, assay, purification, and properties, XLI, 166, 471
Lactobacillus casei var. *rhamnosus*
growth conditions, LI, 92
thymidylate synthetase, LI, 90–97
Lactobacillus delbrueckii, nucleoside deoxyribosyltransferase, LI, 452
Lactobacillus gayonii, L-arabinose isomerase, assay, purification, and properties, XLI, 458–461
Lactobacillus helveticus
crude extract preparation, LI, 448
gramicidin S, XLIII, 577
growth, LI, 448
nucleoside deoxyribosyltransferase, LI, 446–455
Lactobacillus leichmannii
growth, LI, 250, 251
ribonucleoside diphosphate reductase, LI, 227, 237, 242
ribonucleoside triphosphate reductase, LI, 246–259
Lactobacillus plantarum, L-ribulose-5-phosphate 4-epimerase, properties, XLI, 423
Lacto-*N*-difucohexaose I, structure, L, 163
Lacto-*N*-difucohexaose I antibody, purification, L, 175
Lactodifucotetraose
α-fucosidases, L, 219
isolation of 6′-galactosyllactose, L, 219
structure, L, 227
Lactoferrin, in PMN granules, XXXI, 345
Lacto-*N*-fucopentaose I, structure, L, 163
Lacto-*N*-fucopentaose II, structure, L, 163
Lacto-*N*-fucopentaose III, structure, L, 163
Lacto-*N*-fucopentaose V
isolation from human milk, L, 220
structure, L, 217
Lactogen, placental, effect on mammary epithelial cells, XXXIX, 445, 446
β-Lactoglobulin
computer-simulated sedimentation, XLVIII, 236
crystal immobilization, XLIV, 548
membrane osmometry, XLVIII, 75
SDS complex, properties, XLVIII, 6
solute diffusion studies, XLIV, 557, 558
γ-Lactoglobulin, XLIV, 552, 554
β-Lactoglobulin A, self-association, XLVIII, 110
β-Lactoglobulin C, self-association, XLVIII, 110
Lactonase, *see* Gluconolactonase
γ-Lactone, XLIII, 72
Lactone antibiotic, macrocyclic, XLIII, 129–137, *see also* specific antibiotic
Lacto-*N*-neotetraose, structure, L, 163
Lacto-*N*-neotrifucoheptaose II
methylation analysis, L, 10, 11, 13
partial acid hydrolysis, L, 13
structure, L, 10, 11
Lactoperoxidase
antibody iodination, L, 58
chemiluminescence assay, LVII, 427
in determination of membrane proteins, XXXII, 103–109
hormone iodination, XXXVII, 229, 232, 233, 325
in hormone labeling, XXXVII, 147
as lectin label, XXXII, 616
for tyrosine modification, LIII, 137, 148, 149
Lactose, XXXIV, 121, 334; XLVI, 86
antibody, XLVI, 516–523
chromatographic constants, XLI, 17
excretion after head injury, L, 233, 234
fluorescence enhancement of electrolectin, L, 298
hydrolysis, kinetics, XLIV, 798, 799
intolerance, XLIV, 822
mammary synthesis, assay, XXXIX, 453
measurement, LVI, 467, 477
pallidin inhibition, L, 315

substrate, of β-galactosidase, XLIV, 314, 457
synthesis, Koenigs–Knorr reaction, L, 108, 109
transport, XLVI, 86, 608
in whey, commercial use, XLIV, 792–809
Lactose permease, fluorine nuclear magnetic resonance, XLIX, 274
Lactose synthetase, XXXIV, 4
β-Lactoside, XXXIV, 363
β-Lactoside hapten, XXXIV, 7
Lactosylceramide
β-galactosidase, L, 482
globoid cell leukodystrophy, L, 470, 471
isolation, XXXII, 350
optical rotation, XXXII, 365
Lactosylceramide sulfate
arylsulfatase A, L, 473
metachromatic leukodystrophy, L, 472
Lactosyl ceramidosis, L, 494
Lactosylphenethylamine, chromatography, L, 173, 174
O-β-Lactosyl polyacrylamide, XXXIV, 366, 367
Lacto-*N*-tetraose, structure, L, 163
LAHPO, *see* Linoleic acid hydroperoxide
Laidlomycin, LV, 446
Lamb serum, for granulosa cell culture, XXXIX, 208
Lamellar body
enzymatic activity, XXXI, 423
from lung, isolation, XXXI, 419–425
phospholipids, XXXI, 424
Lamellar phase lipid freeze-fracture, *see* Freeze-fracture
Lamelopodia, transformed cells, LVIII, 368
Laminar flow hood, XXXIX, 37; LIV, 479
Lamprey, hemoglobin, CD spectrum, LIV, 275
Langendorff heart perfusion method, XXXIX, 43, 46–52
apparatus, XXXIX, 47, 48
for kidney perfusion, XXXIX, 3
Langerhans islet
fractionation, XXXI, 377, 378
incubation, XXXVII, 331, 332
isolation, XXXVII, 330, 331
from pancreas, XXXI, 374–376
Langmuir-Adam surface balance, XXXII, 540, 541
Langmuir adsorption isotherm, XXXIV, 145
Lanosterol, heme-deficient mutants, LVI, 558–560
Lanthanide, XXXIV, 592
Lanthanide ion
calcium-binding proteins, XLV, 198
magnetic circular dichroism, XLIX, 177
Lanthanum
atomic emission studies, LIV, 455
calcium efflux, LVI, 350
calcium uptake, LVI, 339, 342
as electron microscopic tracer, XXXIX, 148
Lanthanum ion, bacteriorhodopsin photovoltage, LV, 754, 755, 768
Lanthanum stain, XXXIX, 144
Lapachol, *M. phlei* membrane systems, LV, 186
Large external transformation substance, *see* Fibronectin
Large-scale growth, *see* Culture, large-scale
Lasalocid
code number, LV, 445
homologs, LV, 446, 447
as ionophore, LV, 443, 446
solutions, LV, 448
structure, LV, 437
Laser, *see also* specific types
amplifier rod, LIV, 5, 7, 23
available wavelengths, XLIX, 85, 86
continuous wave
commercial availability, 36n
stabilization, LIV, 35
eye protection, XLIX, 86
oscillator rod, LIV, 5
power, measurement, XLIX, 94
Q-switching, LIV, 44
Raman scattering, XLIX, 84–86
for resonance Raman spectroscopy, LIV, 246, 247
Laser light scattering, in studies of protein-RNA interactions, LIX, 579

Laser light source, XLVIII, 453–456
argon, mode pattern, XLVIII, 456
coherence of light, XLVIII, 188, 191
electric field, XLVIII, 425
focusing optics, XLVIII, 456–458
helium-neon
advantages over mercury arc, XLVIII, 189
installation, XLVIII, 185, 186
interference optics, XLVIII, 185–191
modulation circuit, XLVIII, 188
wavelength of incident light *in vacuo*, XLVIII, 417, 419
optical path length determination, XLVIII, 484
wavelengths and powers, XLVIII, 455
Laser microspectrophotometer, XLIV, 370
Laser Raman spectroscopy, XXXII, 247–257
Laser spectrophotometer, double-beam, XLIV, 912, 919
Latex agglutination, XXXIV, 734, 735
Latex bead, use in amoebic membrane isolation, XXXI, 686–698
Latex particle, use in phagolysosome isolation, XXXI, 342, 343
Lathosterol oxidase, LVI, 474
Lathyrogenic compound, XL, 321
Lathyrus sativus L., lectin, XXXIV, 334
LATS, *see* Long-acting thyroid stimulator
Lauric acid
deuterated, mass spectrum of, XXXV, 342
hydroxylation by microsomal fractions, characteristics, LII, 323
methyl esters, chromatographic separation, LII, 321–323
radiolabeled, in plasma membrane, XXXII, 202
Lauric acid hydroxylase, activity in hamster microsomal fractions, LII, 319
1-(1′-Lauroyl-2′-oleyl-*sn*-glycerol-3′-phosphoryl)-*rac*-myoinositol, intermediate in phosphatidylinositol synthesis, XXXV, 475, 476
Lauryl aldehyde, assay of bacterial luciferase-coupled antibodies, LVII, 404
Lauryl coenzyme A, substrate for palmityl thioesterase I, XXXV, 106
Lauryldimethylamine oxide, bacteriochlorophyll proteoliposomes, LV, 760
as ionophore, LV, 443, 448
Lauryl sulfate, micelle formation, XLVIII, 309
Lauths violet, as mediator-titrant, LIV, 408
Lavage procedure, LVII, 472
Layer filtration, centrifugal, transport studies, LVI, 285, 286
L cell
large-scale growth, LVIII, 211
mouse, LVIII, 77, 84
cloning, LVIII, 163
interferon production, LVIII, 295
medium, LVIII, 84, 85
plasma membrane isolation, XXXI, 156–162
L-S cell, protein content, LVIII, 145
LCL, *see* Lymphocytoid cell line
Le^2 active glycolipid, sequential degradation, XXXII, 363
Lead
atomic emission studies, LIV, 455
effect on luciferin light emission, LVII, 3
urine analysis, XLIX, 175
volatilization losses, in trace metal analysis, LIV, 480
Lead(II), inhibitor of nucleoside phosphotransferase, LI, 393
Lead-207, lock nuclei, XLIX, 349; LIV, 189
Lead acetate, XLIV, 654
hydrogen sulfide trap, LVII, 17
Lead citrate, as stain, XXXI, 23; XXXIX, 142, 143
Lead citrate reagent, for algae membrane sectioning, XXXII, 872, 873
Lead nitrate, phosphodiesterase assay, XXXVIII, 260, 261
Lead salt, in ribonucleoprotein staining, XL, 24

Leaf
inoculation, of crown gall, XXXI, 555
mesophyll cell, LVIII, 360
pyruvate P_1 dikinase, assay, purification, and properties, XLII, 212–219
ribulose-1,5-diphosphate carboxylase, extraction, purification, and assay, XLII, 481–484
Leafhopper, media, LVIII, 462
Leakage, XXXIV, 58, 73, 93–102
comparison of ligands, XXXIV, 663
cyanobromide, XXXIV, 93–102
estradiol receptors, XXXIV, 671, 672, 675, 687, 688
immunoadsorbents, XXXIV, 703
nylon derivatives, XXXIV, 204
Lecithin, XLIV, 222, 226, 227, *see also* Egg lecithin; Phosphatidylcholine
adenylate cyclase, XXXVIII, 177, 180
analogue, XXXV, 514–525, *see also* Phosphatidylcholine analogue
bacteriorhodopsin sheets, LV, 601
cholesterol interactions, XXXII, 208, 209
derivatives, phase transition studies, XXXII, 172–174
egg, saturation-transfer studies, XLIX, 497, 498
electron spin resonance spectra, XLIX, 384, 399–401, 407
freeze-fracture, XXXII, 47–49
D-β-hydroxybutyrate apodehydrogenase requirement, XXXII, 374–391
IR spectroscopy, XXXII, 253
in lipid bilayers, XXXII, 515, 516, 530
for lipid monolayer formation, XXXII, 552
in liposomes, XXXII, 503, 505
in media, LVIII, 68, 500
NMR studies, XXXII, 207, 208, 211
as phospholipase C substrate, XXXII, 159, 160
preparation from lysolecithin, XXXI, 48
spin-labeling studies, XXXII, 167–172
synthetic, monolayers, XXXII, 543, 544
unsaturated, transhydrogenase, LV, 814, 815
Lecithin:cholesterol acyltransferase, XXXIV, 747
Lecithin-cholesterol mixture
bilayer formation, XXXII, 490–492
freeze-fracture, XXXII, 46, 48
Lecithin-cholesterol vesicle, spin-labeled, anesthesia, XXXII, 177, 179
Lectin, XXXIV, 7, 331–341, *see also* Agglutinin; specific types
adsorbents, XXXIV, 668–670
agglutination assay using, XXXII, 615–621
from *Arachis hypogaea*
applications, L, 367
assay, L, 361–363
properties, L, 364–367
purification, L, 363, 364
from *Bandeiraea simplicifolia*, L, 345–354, *see also* Isolectin, from *Bandeiraea simplicifolia*
carbohydrate–protein conjugates, L, 161
from *Bauhinia purpurea*
assay, L, 368
carbohydrate inhibition, L, 371
properties, L, 370–372
purification, L, 368, 369
specificity, L, 367, 368
binding-site measurement, XXXII, 616, 617
carbohydrate-protein conjugates, L, 161
from castor bean, L, 330–335, *see also* Ricin; *Ricinus*, agglutinin
immunization, L, 334
properties, L, 334, 335
toxicity, L, 330
cell-binding assay, XXXII, 621–625
cell transformation, LVIII, 368
from *Dictyostelium discoideum*
N-acetyl-D-galactosamine inhibition, L, 311
affinity chromatography, L, 309, 310
biological significance, L, 311, 312
hemagglutination assay, L, 307, 308
direct polymerization, XXXIV, 335, 337, 338

divalent cations, L, 60, 63, 64
glycoprotein binding, L, 54–64
O-glycosyl polyacrylamide derivatives, XXXIV, 362
iodination, L, 57, 58
from *Lens culinaris,* iodination, L, 57
from *Lotus tetragonolobus,* iodination, L, 57–58
from potato, *see also Solanum tuberosum*
assay, L, 340
hemagglutinating activity, L, 340, 344
properties, L, 342–344
purification, L, 340–342, 344
from red kidney bean, *see* Red kidney bean agglutinin
resistance, LVIII, 309, 314
from *Ricinus communis*, *see also* Ricin; *Ricinus,* agglutinin
carbohydrate-protein conjugates, L, 161
iodination, L, 57, 58
from *Ulex europaeus,* affinity chromatography, L, 58
from *Vicia faba*
affinity chromatography, L, 338
assay, L, 336
carbohydrate inhibition, L, 336
hemagglutinating activity, L, 335
properties, L, 338, 339
purification, L, 336–338
from wheat germ, *see* Wheat germ agglutinin
Lectin II
from *Bandeiraea simplicifolia*
assay, L, 350, 351
properties, L, 353, 354
purification, L, 352–353
specificity, L, 350
Lectin-nylon, XXXIV, 211
Leech, *see also Hirude medicinalis*
trypsin-plasmin inhibitors, bdellins, XLV, 797, *see also* Bdellin
Leghemoglobin, LII, 16
carbon monoxide binding, IR studies, LIV, 310
prosthetic group, LII, 5
Legume
oxidase, LII, 16
spherosome isolation, XXXI, 578
Legume seed, proteinase inhibitors, XLV, 697
Leibovitz's medium L–15, XXXII, 742; LVIII, 56, 87, 91
composition, LVIII, 62–70, 463
preparation, LVIII, 463, 464
Leighton tube, for granulosa cell culture, XXXIX, 200, 201, 206, 209
Leioptilus, coelenteramide, LVII, 344
Leitz in-line mirror monochromator, LIV, 53
Leitz Wetzlar micromanipulator, XXXIX, 182, 183
Lemacidin B_1, XLIII, 123
Lenoremycin, LV, 446
code number, LV, 445
Lens, isolation of cells, XL, 314
Lens culinaris, lectin, XXXIV, 334
Lens paper, siliconized, preparation, XXXIX, 446
Lentil, organelles, XXXI, 491
Lenz law, magnetic susceptibility determinations, LIV, 384
Lesch-Nyhan syndrome
cells from, in cell fusion studies, XXXII, 577
fibroblast, LVIII, 444
Lesion, pleiotropic, explanation of in heme mutants, LVI, 560
Lettuce
chloroplast
isolation, XXXI, 602
oxidase, LII, 20
Leucine, XXXIV, 4
aminoacylation isomeric specificity, LIX, 274, 275, 279, 280
codons, in MS2 proteins, LIX, 302–305, 309
free radical sites, XXXVII, 318
in media, LVIII, 53, 62
misactivation, LIX, 289, 291
mitochondrial protein synthesis, LVI, 21
phenylthiohydantoin derivative, identification, XLVII, 334
radioactive, microsome labeling, XXXI, 217–219
radiolabeled
cytochrome oxidase purification, LIII, 68

for labeling of cytochrome *b* polypeptides, LIII, 108
in pulse labeling of parathyroid tissue, XXXVII, 350
release, assay, XLVII, 86
solid-phase peptide synthesis, XLVII, 610
thermolysin hydrolysis, XLVII, 175–179
[^{14}C]Leucine, for *in vitro* labeling of ribosomal protein, LIX, 527
[^{3}H]Leucine
for *in vitro* labeling of ribosomal protein, LIX, 527
for *in vivo* labeling of ribosomal protein, LIX, 525
L-[^{3}H]Leucine, in aminoacylation of tRNA, LIX, 216
L-Leucine-agarose, XXXIV, 440
Leucine aminopeptidase, XLVI, 28, 431, *see also* Aminopeptidase, cytosol
crystalline, *see* Aminopeptidase, cytosol
L-Leucine-*N*-carboxyanhydride, XXXIV, 338
coupling, lectin purification, XXXIV, 336
Leucine chloromethyl ketone, XLVI, 609, 611, 612
Leucine-glyoxylate aminotransferase, in microbodies, LII, 496
Leucine-*p*-nitroanilide, assay, of aminopeptidase, XLVII, 77
Leucine-tRNA synthetase, XXXIV, 170
Leucinostatin
availability and preparation, LV, 513
biological properties, LV, 513, 514
physical properties, LV, 477
structure, LV, 512, 513
toxicology, LV, 513
uncoupling, LV, 506
Leucomycin, XLIII, 134, 309, 319
Leuconostoc mesenteroides
in L-aspartic acid assay, XLIV, 739
D-fructose kinase and D-mannose kinase, assay, purification, and properties, XLII, 39–43
glucose-6-phosphate dehydrogenase, XLIV, 851, 852; LIV, 229
assay, purification, and properties, XLI, 196–201
D-mannitol dehydrogenase, assay, purification, properties, and sources, XLI, 138–142
D-xylulose-5-phosphate phosphoketolase, purification, XLI, 413, 414
Leucoplast, in plant cells, types, XXXI, 495
Leucostoma, peptidase A
assay, XLV, 397, 398
arylamidase, XLV, 398
esterase, XLV, 398
by hemoglobin, XLV, 397
characteristics, XLV, 397
metal content, XLV, 403
pH optimum, XLV, 403
properties, XLV, 401
physical, XLV, 403
purification, XLV, 399–401
purity, XLV, 401
specificity, XLV, 403
stability, XLV, 403
Leucylnaphthyl amidase
in brush borders, XXXI, 125
in microvillous membrane, XXXI, 130
L-Leucyl-β-naphthylamidase, gel electrophoresis, XXXII, 83, 89
Leucyl β-naphthylamide, XXXIV, 449
L-Leucyl-β-naphthylamide
Aeromonas, aminopeptidase, substrate, XLV, 542
Leucostoma, peptidase A, substrate, XLV, 398
L-Leucyl-*p*-nitroanilide, substrate
Aeromonas aminopeptidase, XLV, 542
Aeromonas neutral protease, XLV, 405
thermophilic aminopeptidase I assay, XLV, 523
[^{3}H]Leucyl-tRNA^{Leu}, in study of protein-tRNA interactions, LIX, 326, 330
Leucyl-tRNA^{Leu} isoaccepting species
deacylation rates, LIX, 294
leucine transfer among, LIX, 294, 295
Leucyl-tRNA synthetase
purification, LIX, 262
subcellular distribution, LIX, 233, 234

Leukemia
acute myelocytic, blast cell
crude extract preparation, LI, 367
thymidine kinase, LI, 367
antigen, murine thymus, LVIII, 219
cell
granulocytic, DNA content, LVIII, 148
human, protein content, LVIII, 145
media, LVIII, 56
humans, cell lines, XXXII, 812
risk from X-rays, XXXII, 215
virus, *see* Murine leukemia virus
Leukocyte
cloning, LVIII, 155
human, LVIII, 163
countercurrent separation, XXXII, 637, 641, 643–645
isolation, L, 441; LVIII, 487–489
microencapsulation, XLIV, 215
mucopolysaccharide storage diseases, L, 440
mucopolysaccharidosis I, L, 443
peripheral, isolation, LVII, 467–469
poikilotherm, LVIII, 472
polymorphonuclear
assay of oxygen consumption, LIV, 493, 494, 498
granule isolation, XXXI, 345–353
Leukodystrophy, *see also* specific type
of brain, lipopigments, XXXI, 483
Leukopheresis, LVIII, 493, 494
Leupeptin
activity, XLV, 682
assay, by fibrinogen, XLV, 680
cathepsin B, XLV, 679
kinetics, XLV, 605
papain, XLV, 679
plasmin, XLV, 679
properties, XLV, 682
purification, XLV, 681
in *Streptomyces*, XLV, 679
trypsin, XLV, 679
Levator ani
in bioassay, XXXIX, 100
denervation, XXXIX, 99
as endocrine test organ, XXXIX, 94–101
incubation, XXXIX, 97
tension and contraction, effects, XXXIX, 100
Levomycin, XLIII, 338
Levorin, XLIII, 136, 338
Levorphanol, XLVI, 85
Lewis lung carcinoma, LVIII, 376
Leydig cell
2′,3′-cyclic nucleotide-3′-phosphohydrolase, XXXII, 128
in interstitial cell cultures, XXXIX, 295
luteinizing hormone effects, XXXIX, 252–256
succinoxidase activity, XXXIX, 424
LH, *see* Luteinizing hormone
D-LH_2, *see* Luciferin
Licheniformin, XLIII, 128, 548, *see also* Bacitracin synthetase
Lidocaine
cytochrome binding, LII, 276, 277
metabolism, LII, 65
type of binding reaction with cytochrome *P*–450, LII, 264
Ligand, XXXIV, 163, 229–253, *see also* specific ligands
active-site, XXXIV, 116
aldehyde, XXXIV, 69
alkylation by bromoacetyl polyacrylamide, XXXIV, 49, 50
amino group, XXXIV, 46–48
attachment to inorganic carriers, XXXIV, 58–72, *see also* specific carrier
attachment of specific, XXXIV, 45–51
axial, in heme proteins, XLIX, 172, 173
binding to hemoglobin conjugate, XLIV, 540–543
binding sites, XLVI, 188, 189
carboxyl group, XXXIV, 48
chromatography, *see* Ligand affinity chromatography, general, specific
concentration, XXXIV, 248
enzyme interaction, quantitation, XXXIV, 165
α-galactosidase inhibition, XXXIV, 350
group-specific, XXXIV, 250
hydrolysis, XXXIV, 58, 73
leakage, XXXIV, 96–102
control, XXXIV, 101, 102

rate, XXXIV, 99–102
macromolecular, XLVI, 183
polydentate, stability, LVI, 440
promoter, XXXIV, 599
protective, in differential labeling, XLVI, 63
proteins, XXXIV, 51–56, *see also* Protein, as ligands
quantitation of bound, XXXIV, 349
saturation, percent, XXXIV, 146
selection, XXXIV, 246
spacer attachment, XXXIV, 30–58
stability
cyanobromide coupled, XXXIV, 99–101
estrogen receptors, XXXIV, 96
leakage, XXXIV, 96–102
periodate oxidation method of coupling, XXXIV, 102
Ligand affinity chromatography
general, XXXIV, 598
specific, XXXIV, 598–605
chromatographic procedures, XXXIV, 604, 605
gel
preparation, XXXIV, 602, 603
stability, XXXIV, 602, 603
for lactate dehydrogenase, XXXIV, 598–605
Ligand-ATP conjugate
binding studies, LVII, 113–122
structure, LVII, 114, 115
Ligand binding, *see also* Association reaction; Binding; Cooperativity; Dissociation; Self-association
data evaluation
assuming conformational relaxation, LIV, 524, 525
assuming tunneling, LIV, 526–529
considering multiple energy barriers, LIV, 512–518
for nonexponential kinetics, LIV, 518–524
using multiple excitation, LIV, 525, 526
to discrete binding sites, XLVIII, 270–295
dissociation measurements, XLVIII, 321
experimental approach, using flash photolysis, LIV, 511, 512
experimental temperature range, LIV, 506
fluorometry, XLVIII, 307
to heme proteins, principles, LIV, 506–532
heterogeneous site, XLVIII, 302
interaction, XLVIII, 242–248
to linear lattice of binding sites, XLVIII, 295–299
measurement, preferred systems, XLVIII, 307
mediated reactions, *see* Dimerization
nonexponential, LIV, 518–524
photodissociation, LIV, 510–512
power law, LIV, 509, 514, 518, 523, 524
Ligand complex, boundary patterns, XLVIII, 212
apparatus, XLVIII, 308
autocorrelator operation, XLVIII, 469–473
collecting optics, XLVIII, 460–462
data analysis, XLVIII, 473–479
depolarized
for flexible chains, XLVIII, 443, 444
for rodlike molecules, XLVIII, 440–442
from dilute solution, XLVIII, 425–427
focusing optics, XLVIII, 456–458
measurements, XLVIII, 308–320
multicomponent biopolymer system, XLVIII, 430
optical apparatus, XLVIII, 452–462
photomultiplier, XLVIII, 462–466
reference beam light source, XLVIII, 483–485
sample cell, XLVIII, 458–460
signal amplification
analog mode, XLVIII, 466, 467
photon-counting mode, XLVIII, 467–469
preamplifier, XLVIII, 467
spectrum analysis, XLVIII, 479–482
heterodyne spectroscopy, XLVIII, 482–485
signal-to-noise ratio, XLVIII, 482
temperature control, XLVIII, 459
two component biopolymer system, XLVIII, 428, 429, 435
Ligand-matrix stability, XXXIV, 671

Ligand saturation, degree, measurement, LIV, 216, 217
Light
activation, XLVI, 478
circular dichroism analysis
circularly polarized, XL, 212
plane polarized, XL, 212
effect on growth, LVIII, 136
on pineal activity, XXXIX, 380
incident, frequency, polarizability, XLIX, 81
phosphodiesterase, XXXVIII, 154, 155
rod outer segment isolation, XXXVIII, 153, 154
Light detection, *see also* Photometer
calibration techniques, LVII, 300, 386, 543–545, 560–600
circuitry for calorimeter modification, LVII, 543
instrumentation, *see* Liquid scintillation counter; Photometer
with photomultiplier tube, advantages, LVII, 530
Light intensity, units, LVII, 571–576
Light microscopic autoradiography, of steroids, XXXVI, 138–148
Light microscopy, *see* Microscopy, light
Light-organ symbiosis, LVII, 129, 153
Light petroleum, in extraction of quinone intermediates, LIII, 605
Light quanta, equivalents, LVI, 672
Light scattering, LVIII, 238, *see also* Electrophoretic light scattering
Crabtree effect, LV, 297
instrumentation, LIX, 755, 756
intensity
for depolarized scattering from anisotropic molecules, XLVIII, 441
for dilute solution, XLVIII, 426
photocurrent, XLVIII, 422
for two-component biopolymer solution, XLVIII, 428
parameter in circular dichroism, XL, 230
theory, LIX, 756–759
transport studies, LVI, 255, 256
Light source, *see also* Laser light source
for flash photolysis, LIV, 95–97
for interference optics, XLVIII, 185–191
in nanosecond absorbance spectroscopy, LIV, 33–36
Light standard, *see also* specific standard lamps
radioactive, in evaluation of photometers, LVII, 300, 556–558
Lignoceric acid, radiolabeled
in α-hydroxylase assay, LII, 311
synthesis, LII, 315
Lignoceronitrile, lignoceric acid synthesis, LII, 311
Ligustrum, free cells, XXXII, 724
Lima bean inhibitor, XXXIV, 4
Lima bean trypsin inhibitor, *see also* Trypsin inhibitor, lima bean
in adrenal cortex cell studies, XXXII, 689, 690
in interstitial cell isolation, XXXIX, 295
Limulin
hemagglutination assay, L, 303
from *Limulus polyphemus*, L, 302–305
properties, L, 305
purification, L, 303–305
Limulus polyphemus
agglutinin, *see* Limulin
D-lactate dehydrogenase, assay, purification, and properties, XLI, 313–318
lectin, XXXIV, 335
Lincomycin, LVI, 31
in media, LVIII, 112, 114
in photoaffinity labeling studies, LIX, 804, 805
as protein synthesis inhibitor, XXXII, 869
Lincomycin group
bioautography, XLIII, 128
clindamycin, XLIII, 756
gas liquid chromatography, XLIII, 237–239
paper chromatography, XLIII, 128
proton magnetic resonance spectroscopy, XLIII, 390
solvent systems for countercurrent distribution, XLIII, 338
thin-layer chromatography in pharmacokinetic studies, XLIII, 211

Lincomycin tetrakistrimethylsilyl derivative, XLIII, 239
Lincosaminide, XLIII, 755, *see also* Clindamycin phosphotransferase
Lindahl equation, for cell radius, XXXIX, 325, 326
Line broadening
causes, LIV, 201
principle, LIV, 193, 194, 197–199
Lineweaver-Burk equation, XXXVI, 11, 14
Lineweaver and Burk graphical method, LII, 264; LVI, 265, 266
Line width, LIV, 197–199
temperature dependence, LIV, 201, 203
Linewidth method, for T_2 evaluation, XLIX, 344
Linoleic acid
in biomembrane phospholipids, XXXII, 542
effect on plant cells, XXXI, 520, 521
inhibitor, of myristic acid bioassay, LVII, 192
in linoleic acid hydroperoxide synthesis, LII, 412
in media, LVIII, 54, 68, 90
quantum yield, in bioassay, LVII, 192
Linoleic acid hydroperoxide
hydrogen donor, for cytochrome *P*-450 peroxidase activity, LII, 411
preparation, LII, 412
Linolenic acid
in biomembrane phospholipids, XXXII, 542
in mass spectrometry of triglycerides, XXXV, 354
Lipase, *see also* specific type
bioassay, LVII, 193, 194
enzymatic activity, XXXI, 521, 522
identification of luminous bacteria, LVII, 161, 162
immobilization
by microencapsulation, XLIV, 214
on polyacrylamide, XLIV, 447
inhibition, XXXI, 524
lipoprotein, XXXV, 181
microsome damage, XXXI, 195
mitochondria, XXXI, 591, 593
modification, effect on activity, XLIV, 447–450
monoglyceride, XXXV, 181
in NMR studies of membranes, XXXII, 211
pancreatic, XXXI, 47, 48; XXXV, 181–189, 273, 317, 320–322, 324
assay, XXXV, 182–185
of long-chain fatty acyl-CoA, XXXV, 273
in plasma membranes, XXXI, 88
triglyceride, hormone sensitive, XXXV, 181–189
activation by cyclic AMP-dependent protein kinase, XXXV, 187–189
assay, XXXV, 182–185
inhibitors, XXXV, 189
phosphorylation, XXXV, 188, 189
purification, XXXV, 185–187
species differences, XXXV, 189
Lipid, *see also* specific types
acetone extraction, LII, 98, 104
acetylation, XXXV, 419–421, 424
analysis of positional distribution of fatty acids in glycerolipids, XXXV, 315–325
bilayer
charge on surfaces, XLIV, 220
planar membranes, XXXII, 513–539
electrical studies, XXXII, 527–530
lipid constituents, XXXII, 516, 517
properties, XXXII, 531–539
solvents, XXXII, 518
thickness, XXXII, 530, 531
black film, XXXII, 546
bilayers, XXXII, 553, 554
column chromatography, XXXV, 409–411, 417–421, 423
complex I, LVI, 580
complex II, LVI, 580
complex III, LVI, 580, 582
complex IV, LIII, 44; LVI, 580
complex V, LVI, 580, 584
differential scanning calorimetry, XXXII, 268
effect on steroid binding, XXXVI, 95

extraction
of microsomes, XXXV, 118, 119
from plants, XXXI, 526
Folch partition, XXXV, 396, 397, 409, 421, 423, 549–551
indicators, on thin-layer chromatography, XXXV, 403–408, 412, 413, 415–417
ion-exchange chromatography of, XXXV, 419
IR spectroscopy, XXXII, 252, 253
Keilin-Hartree preparation, LV, 123
lipophilic Sephadex chromatography, XXXV, 386, 388–391
in liver fractions, XXXI, 29, 35
measurement of, in algae, XXXII, 866
in media, LVIII, 68
membranes
artificial, freeze-fracturing, XXXII, 45–50
enzymatic breakdown of, XXXI, 520
spin-label studies, XXXII, 174
X–ray studies, XXXII, 219
metabolism, in perfused liver, XXXIX, 34
monolayers
biomembranes, XXXII, 545–554
electrical properties, XXXII, 551–553
of ox brain, in bilayer formation, XXXII, 491
partial specific volumes, XLVIII, 22
peroxidation, LII, 45
assays, LII, 306
inhibition, LII, 303, 304
in microsomes, LII, 302–310
piericidin A binding, LV, 457
in plasma membranes, XXXI, 87
precipitation, LII, 424
Raman spectroscopy, XXXII, 257
removal by cheesecloth filtration, LII, 98, 111
in dihydroorotate dehydrogenase purification, LI, 65
by filtration, LIII, 115
rotenone binding, LV, 455
spin labels, XXXII, 186–191
synthesis, in endoplasmic reticulum, XXXI, 23
thin-layer chromatography, XXXV, 396–425
total, in microsomal fractions, LII, 88
Lipid mutant, of yeast, isolation and culture, XXXII, 819–843
Lipid peroxidase, in microsomal fractions, LII, 88
Lipid vesicle, thermal transitions, XLIX, 11, 12
Lipmann-Tuttle test, LII, 517
Lipoamidase, XXXIV, 292
Lipoamide, in preparation of deuterated NADH, LIV, 226
Lipoamide dehydrogenase, XXXIV, 288–294, *see also* Dihydrolipoamide reductase; Lipoamide glass
Lipoamide glass, XXXIV, 288–294
alkyl amine glass beads, XXXIV, 289
aminoalkylation, XXXIV, 290
cleaning of beads, XXXIV, 290
derivatization, XXXIV, 289, 290
lipoamide dehydrogenase purification, XXXIV, 292
nonspecific adsorption, XXXIV, 292, 293
preparation procedure, XXXIV, 291–294
as reducing agent, XXXIV, 293, 294
reduction, XXXIV, 294
sonication, XXXIV, 290
Lipofuchsin, see Lipofuscin
Lipofuscin
isolation from brain, XXXI, 425–432, 478–485
in neurons from adult human brains, XXXV, 575
Lipofusion pigment, reaction with malondialdehyde, LII, 308
Lipogenesis, *see also* Fatty acid, synthesis
glycolysis, LV, 295
Lipoic acid, XXXIV, 4
α-Lipoic acid, in media, LVIII, 54, 64
d,l-α-Lipoic acid, activator of pyrimidine nucleoside monophosphate kinase, LI, 330
Lipolysis, XXXI, 103
in fat pads, XXXV, 607
in perifused fat cells, XXXV, 607–612
albumin dependence, XXXV, 611, 612

epinephrine stimulation, XXXV, 609–612
Lipophilic Sephadex, *see* Sephadex, lipophilic
Lipopigment, of brain, isolation, XXXI, 478–485
Lipopolysaccharide, XXXV, 91
Klebsiella O-group 9, partial acid hydrolysis of methylated, L, 12
as mitogen, LVIII, 351
Salmonella typhimurium 395 MS
acid hydrolysate methylation, L, 12
partial acid hydrolysis, L, 12
Shigella flexneri variant Y, L, 18
Lipopolysaccharide-phospholipid-transferase enzyme complex, reconstitution, XXXII, 449–459
Lipoprotein, XXXIV, 746, *see also* specific types
activator requirement by lipoprotein lipase, XXXV, 181
cardiovascular disease, LVI, 468, 470
flotation separation, XXXI, 716
fluorescence quenching, XLIX, 222
inhibitor of phospholipase D, XXXV, 231
negative staining, XXXII, 29
phospholipid exchange, XXXII, 140
plasma, properties, LVI, 469
synthesis, in perfused liver, XXXIX, 35
Lipoprotein lipase, XXXIV, 4
muscle mitochondria, LV, 13
Liposome, XLVI, 86; LVIII, 54
Arthus reaction, XLIV, 709
assay of enzyme activity, XLIV, 225, 226
assay reagent
complete, XXXII, 507–512
incomplete, XXXII, 507–512
autoxidation, LII, 309
carrier protein, LVI, 252
clearance, XLIV, 699, 700, 705–707
composition, XXXII, 503
definition, XLIV, 69
effect of surface charge, XLIV, 688, 689, 703, 706, 708, 709
enzyme entrapment, XLIV, 218–227
external membrane potential probes, LV, 559
freeze-fracture, XXXII, 48, 49
immunological adjuvant, XLIV, 227, 700, 705
ionophores, LV, 439
lipid sources, XLIV, 708
lipid structure, XXXII, 542
in model system for transmembrane redox reactions, LV, 541, 542
P basic protein interaction, XXXII, 345
phosphatidylcholine, for stabilizing NADH-cytochrome b_5 reductase, LII, 107
phospholipid exchange, XXXII, 146
physical characteristics, XLIV, 218–220
preparation, XXXII, 142, 501–513; LII, 208; LVI, 428
proteolipid, preparation, LV, 418
reconstitution of glutamate transport system, LVI, 428–430
reconstitution studies, LV, 707, 708
reductase binding, LII, 108
sterilization, XLIV, 707
substrate for phospholipid exchange enzyme, XXXV, 264
therapeutic application, XLIV, 688, 689, 698–709
toxicity, XLIV, 708
trapped glucose marker in, assay, XXXII, 501–513
ultrastructure, XLIV, 219
Liposome-entrapped enzyme, assay, XLIV, 225, 226
Liposome-protein complex, properties, LII, 210, 211
Lipoteichoic acid carrier, L, 388
isolation from *Staphlococcus aureus* H, L, 391, 392
radioactive labeling, L, 391, 392
β-Lipotropin, ovine, fragment, synthesis, XLVII, 616
Lipoxygenase, XXXIV, 4; LII, 11
enzymatic activity, XXXI, 522, 523
inhibition, XXXI, 524
in plants, overcoming problems, XXXI, 520–528
Lipoyl chloride, XXXIV, 291
Lipoyl dehydrogenase, *see* Dihydrolipoamide reductase
Liquemin, *see* Heparin

Liquid chromatography, high-performance
of amino acid derivatives, XLVII, 3–18, 45–51
instrumentation, XLVII, 46
Liquid culture, XLIII, 11–14, *see also* Antibiotic, production in liquid culture
Liquid-diffusion method, LIX, 5
Liquid-liquid chromatography, for steroid hormone separation, XXXVI, 485–489
Liquid membrane electrode, construction, LVI, 362
Liquid nitrogen
in freezing, LVIII, 30
storage, LVIII, 33
Liquid nitrogen refrigerator, for liver preservation, XXXI, 4
Liquid scintillation counter
assay of bioluminescence, LVII, 216, 217, 219
of malate dehydrogenase, LVII, 182
measurement of phagocyte chemiluminescence, LVII, 474–476
Liquid scintillation counting, XL, 295
Liquid-surfactant membrane, definition, XLIV, 328
Liquifluor scintillation fluid, LI, 569
Lithium, atomic emission studies, LIV, 455
Lithium-7
lock nucleus, XLIX, 348, 349; LIV, 189
resonance frequency, LIV, 189
Lithium acetate, XLIX, 459
Lithium acetate buffer, XLVII, 16
Lithium acetylphosphate, phosphoribosylglycinamide synthetase assay, LI, 180
Lithium aluminum hydride, preparation of dehydroluciferol, LVII, 25
Lithium borate, chromatographic separation of nucleosides, LI, 509
Lithium bromide, cyclase preparation, XXXVIII, 145
Lithium carbamyl phosphate, assay of aspartate carbamyltransferase, LI, 51
Lithium chloride, XLIV, 550, 863
bentonite-charcoal treatment, LIX, 482
in chromatographic separation of modified nucleosides, LIX, 181
of nucleotides, LI, 158, 315, 333, 355
of protein from cross-linked ribosomal subunits, LIX, 537, 540–542
cyclic nucleotide separation, XXXVIII, 30, 31, 35, 36
as lock nuclei source, LIV, 189
preparation of 3′-amino-3′-deoxy-ATP, LIX, 136
of protein from cross-linked ribosomal subunits, LIX, 537, 540–542
of protein-depleted 30 S subunits, LIX, 621, 622
of ribosomal proteins, LIX, 440, 470, 484–488, 490–494, 648
of ribosomal RNA, LIX, 556, 557, 625, 648
of 16 S RNA, LIX, 648
SDS method, LIX, 556, 557
urea method, LIX, 440, 470, 625, 648
Lithium chloride solution, electron spin resonance spectra, XLIX, 383
Lithium hydroxide, XLIV, 861–867; XLVII, 137, 143
in 6-amino-1-hexanol phosphate preparation, LI, 253
in deacylation, LIX, 270
in hydride transfer stereospecificity experiments, LIV, 227
Lithium 8-hydroxy-5-quinolinesulfonate, effect on ribonucleoside diphosphate reductase activity, LI, 244
Lithium ion
activator of guanylate kinase, LI, 489, 501
of ribonucleoside triphosphate reductase, LI, 258
amino acid transport, LV, 187
Lithium lactate, substrate, lactate dehydrogenase, XLIV, 628
Lithium D-lactate, in assay of D-lactate dehydrogenase, LIII, 519
Lithium salicylate, activator, peroxyoxalate chemiluminescence reaction, LVII, 457

Litmocidin, XLIII, 168
Littoral cell, *see* Macrophage
Liver, XXXIX, 23–40, *see also* Hepatocyte; specific type
acetoacetyl-CoA thiolase, XXXV, 167–173
acetyl coenzyme A carboxylase, XXXV, 3
cAMP-receptor, XXXVIII, 380
carboxylesterase, XXXV, 190, 208
carnitine, LV, 221
cell
bilirubin conjugation, LVIII, 184
buffer solutions, LVIII, 540
cholestorol, XXXII, 543
cloning, LVIII, 163
counting, LVIII, 543, 544
culture, XXXII, 733–740
differentiation, LVIII, 541
digestion buffer, LVIII, 542, 543
establishment of cell lines, LVIII, 536–544
isolation, LVIII, 543
media, LVIII, 59, 537, 542, 543
microencapsulation, XLIV, 215
plating, LVIII, 538, 539, 543
rat
adult, LVIII, 539
fetal, LVIII, 538
media, LVIII, 90
cell-fraction analysis, XXXI, 19–23, 209
characterization, XXXI, 23–39
by differential centrifugation, XXXI, 504, 724, 725
crotonase, XXXV, 136–151
cytoplasmic phosvitin kinase, XXXVIII, 328, 329
cytosol protein kinase
assay, XXXVIII, 323, 324
mechanism of cAMP action, XXXVIII, 323
properties, XXXVIII, 327, 328
purification, XXXVIII, 324–327
desaturation system, XXXV, 253
dispersion, LVIII, 123
enzymes, intracellular, XXXI, 97
fatty acid synthase, XXXV, 37, 45, 59
fetal explant, in hormone studies, XXXIX, 36–40
from fetal mouse, in erythropoietin studies, XXXVII, 117, *see also* specific types
Golgi apparatus isolation, XXXI, 180–191
homogenate, fractionation, after storage, XXXI, 39
homogenization, XXXI, 95
human
alcohol dehydrogenase, assay, purification, and properties, XLI, 369–374
crude extract preparation, LI, 406
cytidine deaminase, LI, 405–407
superoxide dismutase, LIII, 389–393
triosephosphate isomerase, assay, purification, and properties, XLI, 430–434
3-hydroxy-3-methylglutaryl-CoA synthase, XXXV, 155, 160
lipofuscin isolation from cells, XXXI, 430–432
liposome uptake, XLIV, 699, 701, 706
long-chain fatty acyl-CoA synthase, XXXV, 117
long-term preservation, XXXI, 3–6
lysosome isolation, XXXI, 323–329
metabolic contents, LV, 213
microsome isolation, XXXI, 195–199
mitochondria
ATPase, LV, 299, 300
properties, LV, 20, 21
storage, LV, 31
mouse, homogenization, LII, 235
($Na^+ + K^+$)-ATPase isolation, XXXII, 280
nuclei, isolation, XXXI, 253–262
5′-nucleotidase isolation, XXXII, 368–374
oxidases, LII, 10–21
parenchymal cell isolation, XXXII, 625–632
parenchyma-like cell, plasma-membrane isolation, XXXI, 163
perfused
apparatus, XXXV, 598–604
fatty acid synthesis rate, XXXV, 597, 598, 604, 606, 607
albumin requirement, XXXV, 598

gluconeogenesis rate, XXXV, 597
glycolysis rate, XXXV, 606
ketogenesis rate, XXXV, 597, 606
technique, XXXV, 604–606
peroxisome isolation, XXXI, 356–368
phosphatidylcholine exchange protein, XXXII, 140–146
phosphodiesterase, of multiple forms
alternate preparations, XXXVIII, 258, 259
assay, XXXVIII, 257
chromatography, XXXVIII, 258
properties, XXXVIII, 259
tissue extract preparation, XXXVIII, 257
phosphofructokinase from human, purification, XLII, 107
phospholipid exchange enzyme, XXXV, 262, 269
pig, preparation of microsomes, LII, 143–145
prostaglandin binding, XXXII, 109, 110
protein kinases, XXXVIII, 298, 365, 366
rat
homogenization, LII, 92, 98, 104, 353
microsome isolation, LII, 85, 86
perfusion, XXXIX, 25–36; LII, 48–58; LVIII, 128, 129
media, XXXIX, 27
preparation, LII, 54, 55
for hepatocyte isolation, LII, 62, 63
for whole organ perfusion, LII, 54, 55
subcellular fractionation, XXXI, 6–41
flow diagram, XXXI, 10
sample storage, XXXI, 18, 19
subcellular particles
morphometry, XXXII, 8, 9
negative staining, XXXII, 31
triglyceride, analysis by gas chromatography–mass spectrometry, XXXV, 359
Liver aminopeptidase, *see* Aminopeptidase, human liver
Liver membrane, XXXIV, 7
insulin receptors, XXXIV, 664
in receptor purification, XXXIV, 654
Lividomycin
gas-liquid chromatography, XLIII, 220
ion-exchange chromatography
Amberlite CG–50 column, XLIII, 269
CM–Sephadex, XLIII, 275
solvent system, XLIII, 119, 120
structure, XLIII, 614
Lividomycin A, XLIII, 278
Lividomycin B, XLIII, 269
Lividomycin phosphotransferase, XLIII, 627
LKB Ampholine column, LIX, 199
LKB batch isothermal calorimeter, modified for light detection, LVII, 548
LKB 8100–10 column, LVII, 62
LKB 9000-PDP/12 gas chromatograph-mass spectrometer-computer system, LII, 333
LKB Ultrorac 7000 fraction collector, LIX, 373
LKB Ultrotome, XXXI, 23
LL–AB664 antibiotic, XLIII, 258
LL–AC541 antibiotic, XLIII, 258–260
Lloyd's reagent, for steroid adsorption, XXXVI, 40
LM medium, formulation, LVII, 154, 155
Lobster
cGMP-dependent protein kinase, XXXVIII, 330–335
muscle, fructose-diphosphate aldolase, assay, purification, and properties, XLII, 223–228
nerve, sample holder, XLIX, 392
Lock nucleus, table, XLIX, 348, 349, 351; LIV, 189
Locust, mitochondria, LV, 24
Locusta migratoria
cytochrome *b*, LIII, 213, 221
cytochrome oxidase, LIII, 66
Logit-log method
in curve fitting, XXXVII, 6, 7
graph paper, XXXVII, 8
Log-lin plot, in ligand binding data evaluation, LIV, 509, 518
Log-log plot, in ligand binding data evaluation, LIV, 509, 518
L-OH, *see* Dehydroluciferol
Loligo vulgaris, *see* Cuttlefish

Lombricine kinase, XXXIV, 546
Long-chain acyl-CoA thioesterase, effect on elongation activity of fatty acid synthase, XXXV, 89, 90
Longitudinal relaxation rate, hemoprotein distance measurements, LIV, 188, 190
Long-pulse method, LIV, 167–169, 184
 for elimination of water signal, LIV, 184–187
Lonomycin, LV, 445, 446
Lorentzian band, resolution, fourth derivative analysis, LVI, 505, 506, 510, 511
Lorentzian term, XLVIII, 419, 424, 428, 432
Lotus tetragonolobus, lectin, XXXIV, 332
Low-phosphate medium, formulation, LIX, 831
Lowry protein assay, LVIII, 144
 for cell cultures, XXXII, 568
 interference, LVIII, 146
 original, of measurement of protein with Folin phenol reagent, XI, 375
 for plant protein determination, XXXI, 542–544
 of quantitative determination of proteins, XL, 110
Low-temperature spectroscopy, *see* Electron paramagnetic resonance spectroscopy, low-temperature
L. pomonella, media, LVIII, 457
LRF, *see* Luteinizing hormone releasing factor
LSC, *see* Liquid scintillation counter
LS cell, LVIII, 89
LSD, as optical probe, LV, 573
LS-tetrasaccharide a, structure, L, 225
LS-tetrasaccharide b, structure, L, 225
LS-tetrasaccharide c, structure, L, 225
LTH, *see* Luteotropic hormone
Lubrol–PX surfactant, XXXIV, 656
 adenylate cyclases, XXXVIII, 137–140, 149, 174, 175
 partial specific volume, XLVIII, 19, 20
Lubrol–WX surfactant, XXXI, 314; LIV, 493
 ATPase isolation, LV, 331
 cyclase, XXXVIII, 140
 partial specific volume, XLVIII, 19, 20
 purification of NADH dehydrogenase, LIII, 418
Lucibacterium, *see Beneckea*
Luciferase, XLVI, 151, 287, 537–541; LII, 17, 20; LIII, 403
 activity, LIII, 558
 age, LVI, 537
 assay, LIII, 564–566
 AMP, LV, 209
  ATP, LV, 539
  of FMN, LIII, 422
  mediated by, *see* Luciferase-mediated assay
 bacterial
  absorbance coefficient, LVII, 172
  active site, LVII, 179–181
  activity units, LVII, 150
  anion stabilizers, LVII, 199
  assay
   aldehyde concentration, LVII, 138, 139
   coupled, LVII, 140
   of malate, LVII, 187
   of oxaloacetate, LVII, 187
   photometer, LVII, 137
   principles, LVII, 135, 136
   using immobilized enzyme rods, LVII, 205
  chemical modification, LVII, 132, 133
  classification, LVII, 136
  effect of pH, LVII, 220
  extinction coefficient, LVII, 149
  glycoprotein, LVII, 132, 149
  hybrid molecules, preparation, LVII, 178, 179
  immobilization, LVII, 204, 205
  immobilized, assay, LVII, 205
  inhibitors, LVII, 151, 152
  kinetics, for identification of luminous species, LVII, 161–163
  ligand binding studies, LVII, 132, 133
  molecular weight, LVII, 149
  photoexcitable, LVII, 140, 141
  protease sensitivity, LVII, 198

purification, LVII, 143–149
from *Beneckea harveyi*, LVII, 143–147
from *Photobacterium fischeri*, LVII, 147, 148
from *Photobacterium phosphoreum*, LVII, 149
quantum yield, LVII, 151
reaction, LIII, 562, 563
reaction intermediate, LVII, 127, 133–135, 152, 195–197
absorption spectra, LVII, 197
assay, LVII, 196
fluorescence spectra, LVII, 197
isolation, LVII, 195, 196
quantum yield, LVII, 197
stability, LVII, 196, 197
reaction sequence, LVII, 126, 133–135
reconstitution, LVII, 174
specific activities, LVII, 150, 172
stability, LVII, 149, 150
structural alterations, mutation, LVII, 168–171
structure, LVII, 258
substrate specificity, LVII, 151
subunit
absorbance coefficients, LVII, 173
chemical modification, LVII, 174–181
hydrolysis by proteases, LVII, 201
molecular weights, LVII, 171
preparation, LVII, 172, 173
properties, LVII, 173, 174
role in activity, LVII, 132, 179
structures, LVII, 132, 149, 171
α-subunit, LVII, 132, 133
β-subunit, function, LVII, 132
thermal stability, tests, LVII, 168, 169
turnover rate, LVII, 126, 127, 132, 136, 195, 223
from *Beneckea harveyi*, LIII, 566–570
bioluminescence quantum yields, LIII, 570
comparative molecular weights, LIII, 560, 567
Cypridina
assay, LVII, 371, 372
cation requirement, LVII, 371
content, in living *Cypridina*, LVII, 337
diffusion constant, LVII, 352, 353
isoelectric point, LVII, 352
kinetics, LVII, 353, 355–358
molecular weight, LVII, 350, 371
pH optimum, LVII, 352, 353
purification, LVII, 350–352, 366, 367, 369–371
salt concentration, LVII, 350, 351
sedimentation constant, LVII, 352, 353
Diplocardia
assay of hydrogen peroxide, LVII, 378
copper stimulation, LVII, 380, 381
purification, LVII, 376, 377
quantum yield, LVII, 376
firefly
acetone powder, preparation, LVII, 6, 7
activity, LVII, 3, 37, 58, 74, 86, 108, 125, 126
assay, LVII, 29, 30, 76, 77
adenylate charge, LVII, 73–85
of ATP, LVII, 36–50
of ATPase, LVII, 50–56
of bacterial numbers, LVII, 65–72
of biomass, LVII, 73–85
of creatine kinase isoenzymes, LVII, 56–65
of cyclic nucleotide phosphodiesterases, LVII, 94–106
of guanosine nucleotides, LVII, 85–94
of photophosphorylation, LVII, 50–56
commercial sources, LVII, 37
continuous measurement of adenosine triphosphatase, LVI, 530, 531
materials and instrumentation, LVI, 531–533
precautions, LVI, 541–544
qualitative aspects, LVI, 533
quantitation, LVI, 534–537

rapid kinetics, LVI, 540, 541
rate of change of ATP concentration, LVI, 537–540
crude extract preparation, LVII, 6–8
crystalline
preparation, LVII, 8–11
stability, LVII, 11
effect of arsenate, LVII, 13, 14
of salt concentration, LVII, 14, 15
enzyme classification, LVII, 126
extinction coefficient, LVII, 11
inhibitors, LVII, 14, 70
light reaction, kinetics, LVII, 11–13
molecular weight, LVII, 11
pH optimum, LVII, 15
purification, LVII, 30–36, 51, 61, 62, 82
reaction catalyzed, LVI, 531
reaction kinetics, LVII, 45, 49
theoretical treatment, LVII, 39–41
stability, LVII, 6, 7, 13
structure, LVII, 258
in studies, of ATP synthesis in mitochondria, LVII, 39–50
of protein-ligand binding, LVII, 113–122
of rapid kinetics of ATP synthesis, LVII, 45–50
substrate specificity, LVII, 14, 50, 82
flavin specificity, LIII, 569
generic definition, LIII, 558
as glycoproteins, LIII, 562
kinetic properties, LIII, 570
light measurement, LIII, 564
membranes, LIII, 562
oxy form, spectrum, LII, 36
oxygenated-flavin intermediate, LIII, 570
Pholas
amino acid analysis, LVII, 388, 390
assay, LVII, 386
complex formation with luciferin, LVII, 393–397
copper content, LVII, 388
electrochemical oxidation, LVII, 401
kinetics, LVII, 394–397
molecular weight, LVII, 388, 391
optimum ionic strength, LVII, 393
as peroxidase, LVII, 397, 398
pH optimum, LVII, 391, 393
physical properties, LVII, 388, 391
purification, LVII, 387, 388
stability, LVII, 391
sugar content, LVII, 388, 390
photoexcitability, LIII, 569
purification, LIII, 566–569
reconstitution, LIII, 567, 569
Renilla
activity, LVII, 246
alcyonarian coelenterate sources, LVII, 248
copurification with blue-fluorescent protein, LVII, 227, 229, 230
with luciferin sulfokinase, LVII, 238, 249
enzyme classification, LVII, 239
glycoprotein, LVII, 238
hydrophobicity value, LVII, 238
interaction with GFP
complex formation, LVII, 265–267
species specificity, LVII, 263, 264
molecular weight, LVII, 238, 260
properties, LVII, 238
purification, LVII, 249, 250
turnover number, LVII, 238
specificity, XLIV, 583
subunits, LIII, 567
Luciferase-luciferin
choice of concentrations, LVI, 541, 543
sources, LVI, 531
Luciferase-luciferin enzyme preparations
commercial, sensitivity comparison, LVII, 554, 555, 559
endogenous light, decay pattern, LVII, 552
Luciferase-luciferin sulfokinase, specific activity, LVII, 250
Luciferase-mediated assay
of adenylate charge, LVII, 73–85

of antibody binding sites, LVII, 402, 403
of ATP, LVII, 36–50
of ATPase, LVII, 50–56
of biomass, LVII, 73–85
of creatine kinase isoenzymes, LVII, 56–65
of cyclic nucleotide phosphodiesterases, LVII, 94–106
of FMN, LVII, 215–222
guanosine nucleotides, LVII, 85–94
of hydrogen peroxide, LVII, 378
integrated light-flux method, LVII, 77, 90
of ligand binding, LVII, 113–122
of malate, LVII, 187
of malate dehydrogenase, LVII, 181–188
of modulator protein, LVII, 107–112
of myristic acid, LVII, 189–194
of NADH, LVII, 215–222
of NADPH, LVII, 215–222
of oxaloacetate, LVII, 187
peak height method, LVII, 77, 89, 90
of photophosphorylation, LVII, 50–56
of protease, LVII, 200

Luciferase-oxyluciferin complex, of *Cypridina,* light-emission, LVII, 360, 361

Luciferase reagent
preparation, LVII, 55, 76, 88
stability, LVII, 218

Luciferin
bacterial
photagogikon derivative, emission wavelength, LVII, 341
structural formula, LVII, 341
Cypridina
assay, LVII, 371, 372
biological distribution, LVII, 340, 342–344
content, in living *Cypridina*, LVII, 337
extinction coefficients, LVII, 368
inhibition, by cyanide, LVII, 339
light emission by complex, LVII, 360, 361
molecular weight, LVII, 340, 352, 368
oxidation, LVII, 344–350
photagogikon derivative
emission wavelength, LVII, 341
structural formula, LVII, 341
purification methods, LVII, 338, 339, 366–368
quantum yield, LVII, 358–360
spectral properties, LVII, 368
structural formula, LVII, 341
synthesis, LVII, 369
Diplocardia
assay of hydrogen peroxide, LVII, 378
photagogikon derivative, emission wavelength, LVII, 341
purification, LVII, 376
structural formula, LVII, 341
synthesis, LVII, 377
ethylation, XLVI, 537–541
firefly
anaerobic solution preparation, LVII, 109, 110
analogs, LVII, 25–28
structure, LVII, 27
synthesis, LVII, 25–28
assay of firefly luciferase, LVII, 11, 12, 30
of photophosphorylation, LVII, 55
chromatographic properties, LVII, 26
concentration determination, LVII, 37
criteria for purity, LVII, 24, 25
electrophoretic mobility, LVII, 24
light emission, pH, LVII, 3
molar extinction coefficient, LVII, 110
oxidation mechanism, LVII, 347, 348
photagogikon derivative
emission wavelength, LVII, 341
structural formula, LVII, 341
purification on Sephadex gels, LVII, 28, 31, 35, 36
spectral properties, LVII, 22–24, 26
stability, LVII, 22
structural formula, LVII, 4, 16, 17, 18, 27, 341

synthesis
pathways, LVII, 17
principles, LVII, 15, 16
procedure, LVII, 16–22
fish, structural formula, LVII, 341
generic definition, LIII, 559
Latia
photagogikon derivative, emission wavelength, LVII, 341
structural formula, LVII, 341
Pholas
amino acid analysis, LVII, 390
assay, LVII, 385, 386
chemiluminescent oxidation, LVII, 399–401
molecular weight, LVII, 391
in peroxidase-antibody binding studies, LVII, 406
peroxide intermediate, effect of ionic strength, LVII, 400
physical properties, LVII, 391
prosthetic group, LVII, 392, 393
purification, LVII, 388, 390
quantum yield, LVII, 391
reaction with horseradish peroxidase, LVII, 399
spectral properties, LVII, 392, 393
sugar content, LVII, 390
Renilla
kinetics, LVII, 238, 239
structural formula, LVII, 246
substrate, of luciferase, LVII, 238
Luciferin alcohol, synthesis, LVII, 377
Luciferin-binding protein, calcium-triggered, LVII, 242–244, 258
Luciferin-luciferase, cAMP assay
enzyme decontamination, XXXVIII, 64, 65
instrumentation, XXXVIII, 62
procedure, XXXVIII, 63, 64
reagents, XXXVIII, 63
Luciferin-luciferase complex, of *Pholas dactylus,* emission mechanism, LVII, 393–398
Luciferin sulfokinase
activity, LVII, 237, 246
alcyonarian coelenterate sources, LVII, 248
copurification with luciferase, LVII, 238, 249
purification, LVII, 249, 250
stability, LVII, 247
Luciferyl adenylate
chromatographic properties, LVII, 26
spectral properties, LVII, 26
stability, LVII, 28
structural formula, LVII, 27
synthesis, LVII, 25, 27, 28
Luciferyl sulfate
analogs, LVII, 250
assay of PAP, LVII, 253, 254
of PAPS, LVII, 245
inhibitor of luciferase, LVII, 254
structural formula, LVII, 238, 246
substrate of luciferin sulfokinase, LVII, 237, 246
Lucigenine, chemiluminescence
detection system, LVII, 445
mechanism, LVII, 563
Lucite plastic, for neutron beam attenuation, LIX, 667
Lucite sample cell, disadvantages, LIV, 376, 403
Lucite spacer, in NMR tube, for elimination of light scattering, LIV, 217
Ludox
colloidal silica, XLVIII, 394
gradient properties, XXXI, 508
Luft's ruthenium red stain, XXXIX, 144
Lumase kit, LVII, 219
Lumase-P kit, LVII, 219
Lumbricus terrestri, hemoglobin, CD spectrum, LIV, 275
Lumichrome, LIII, 420
Lumiflavin, LIII, 420
quantitative determination, LIII, 424, 425
Lumiflavin 3-acetate, in oxygen determination, LIV, 119
Luminescence, *see also* specific types
bacterial
reaction sequence, LIV, 499
use, for oxygen determination, LIV, 499–505
chemically initiated electron exchange, LVII, 521, 524, 565
Luminescence biometer, LVII, 75

Luminol
amplifier, of phagocyte chemiluminescence, LVII, 476–479
assay of catalase, LVII, 404
of glucose, LVII, 403, 452–456
glucose oxidase, LVII, 403, 404
hydrogen peroxide, LVII, 403, 441, 445–462
superoxide dismutase, LVII, 404, 405
systems, sensitivity, LVII, 402, 403
calibration of light-detection system, LVII, 544, 545
chemiluminescence, LVII, 409–423
catalyst/cooxidants, LVII, 447–450
disadvantages, LVII, 461
effect of amino residue alkylation, LVII, 424
of luminol concentration, LVII, 451
of pH, LVII, 449–451
emission spectrum, LVII, 413
hydrogen peroxide detection limit, LVII, 454
inhibitors, LVII, 401
light emitter, LVII, 412–417
persulfate/hydrogen peroxide-driven, LVII, 419
photocalorimetry experiments, LVII, 547, 548
reaction conditions, LVII, 410, 411
reaction mechanism, LVII, 417–422, 474, 563
concentration, effect on phagocyte chemiluminescence, LVII, 477, 478
in dimethyl sulfoxide
molar extinction coefficient, LVII, 600
quantum yield, LVII, 589, 600
fluorescence emission spectra, LVII, 413, 415
quantum yield, LVII, 412
solubility, LVII, 476, 477
stability, in solution, LVII, 453
standard, for chemiluminescent quantum yields, LVII, 228, 250, 253, 600
structural analogs, chemiluminescence, LVII, 411, 412
structural formula, LVII, 409, 447
water-soluble salt, preparation, LVII, 477
Lumisome, LVII, 243, 244
Lung, *see also* specific type
carcinoma, LVIII, 376
cell
fibroblast
fetal, LVIII, 50
human embryo, microcarrier culture, LVIII, 190
human embryo, protein content, LVIII, 145
mink, LVIII, 414
dispersion, LVIII, 123
guanylate cyclase
assay, XXXVIII, 192
properties, XXXVIII, 195
purification, XXXVIII, 192–195
isolated, perfusion, LII, 59
lamellar body isolation, XXXI, 419–425
macrophages, isolation, XXXII, 762
mammalian, oxidase, LII, 13
microsomes, LII, 89
mitochondria, LV, 19, 20
properties, LV, 20
tissue, scanning electron microscopy, XXXII, 56
Lupanine hydroxylase, LII, 21
Lupersol 225, XLIV, 175
Luteal cell, ultrastructure, XXXIX, 185
Luteinization
of granulosa cells, XXXIX, 184, 185
assay, XXXIX, 211–227
induction, summary, XXXIX, 229, 230
Luteinizing hormone, XXXIV, 7, 338
autoradiography, XXXVI, 150
cation-exchange chromatography, XXXVII, 371–373
in corpus luteum, biogenesis, XXXIX, 238–244
cultured cells dependent, XXXII, 558
desialylation, XXXVII, 321, 322
effect on granulosa cells, XXXIX, 423
on Leydig cell, XXXIX, 252–256

on testicular metabolism, XXXIX, 252–271
gonadal receptors, XXXVII, 167–193
extraction, XXXVII, 181–193
structural aspects, XXXVII, 178–181
granulosa cell luteinization, XXXIX, 184, 186
high purification, XXXVII, 374, 375
human, purification, XXXVII, 382, 384–388
iodination, XXXVII, 325
logit-log assay, XXXVII, 9
LRF effects, XXXVII, 87, 89
in media, LVIII, 106
molecular weight, by electrophoresis, XXXVI, 99
ovine
radioactive labeling procedure, XLVII, 476
reductive methylation, XLVII, 475
pineal effects, XXXIX, 381, 382
preparation, XXXVII, 370–375
ovine and bovine, XXXVII, 370–375
rat and rabbit, XXXVII, 370
radioligand assay, XXXVII, 176–178
releasing factor, XXXIX, 382
secretion by cloned pituitary cells, XXXIX, 128–132
sialic acid content, XXXVII, 323
tritium labeling, XXXVII, 321–326
TSH activity separation, XXXVII, 373, 374
unit of activity, XXXVII, 362
Luteinizing hormone releasing factor, LVIII, 102
amino acid analysis, XXXVII, 422
biological activity
comparison of various samples, XXXVII, 423
estimation, XXXVII, 233–238
commercial source, XXXVII, 415
effect on luteinizing hormone, XXXVII, 87, 90, 91
inhibitors, synthesis, XLVII, 587
iodinated, properties, XXXVII, 230
pituitary receptor binding assay, XXXVII, 214–219
radiolabeled, XXXVII, 214
solid-phase synthesis, XXXVII, 416–424
solvent extraction, XXXVII, 402–407
synthesis, XLVII, 588
Luteomycin, XLIII, 338
Luteotropic hormone, radioimmunoassay, XXXIX, 196, 197
Lutetium acetate, derivatization of transfer RNA, LIX, 14
Lutetium chloride, derivatization of transfer RNA, LIX, 14
Lutidine, XLVII, 488
2,6-Lutidine-water-ammonia, chromatographic identification of siroheme, LII, 442
Lyase, assay by Fast Analyzer, XXXI, 816
Lymantria dispar, media, LVIII, 457
Lymph node, lymphocyte preparation, LVIII, 124
Lymphoblast
human, slide preparation, LVIII, 328–330
media, LVIII, 89
Lymphoblastoid cell, LVIII, 376
media, LVIII, 56
murine, LVIII, 218
EL4, LVIII, 223
Lymphocyte, XXXIV, 716, 750–755, *see also* specific types
activated, culture conditions, LVIII, 492
activation, LVIII, 487
AHH assay, LII, 236, 237
antigen, XXXII, 68, 69
screening, XXXIV, 223, 224
antihapten, XXXIV, 718
antihapten specific, XXXIV, 7
bone marrow-derived, *see* B cell
cloning, LVIII, 155
contaminants, LVIII, 492, 493
countercurrent separation, XXXII, 637, 645
differential binding, XXXIV, 215
fractionation apparatus, XXXIV, 210
for hHG assay, XXXVII, 70
homing, L, 94
human, LVIII, 486–494
cells resembling, culture, XXXII, 799, 800, 811

culture conditions, LVIII, 491, 492
DNA content, LVIII, 148
immunofluorescence, XXXIV, 215
irradiated, XXXIV, 219
isolation, XXXI, 509
large-scale growth, LVIII, 211
large scale isolation, XXXIV, 210
membrane glycoprotein, XXXIV, 338
optical probes, LV, 573
from peripheral blood, purification, XXXII, 633–636
platelet removal, LVIII, 490
preparation, LVIII, 487–493
from leukocytes, LVIII, 489, 490
preservation, LVIII, 31
quantitation, LVIII, 491
separation, L, 441
2-amino-2-methoxyethylthioglycosides, L, 113
cell wall peptidoglycan, L, 401
from erythrocytes, LVIII, 490, 491
suspension, XXXIV, 174; LVIII, 124
thymus-derived, *see* T cell
transformed, XXXIV, 212
viability, XXXIV, 209
yield, from blood, LVIII, 491

Lymphocytoid cell lines
culture, XXXII, 811–816
initiation, XXXII, 812–814

Lymphoid cell, XXXIV, 205
antigen-binding cells, XXXIV, 197
H-2 antigen distribution, XXXII, 66, 67
biohazard, LVIII, 37, 42
transformed, XXXIV, 207

Lymphoid tissue, lysosome isolation, XXXI, 353–356

Lymphoma, LVIII, 376
cell, plasma membrane isolation, XXXI, 161
mouse S49, LVIII, 238
cell cycle phase, LVIII, 242
DNA content, LVIII, 244

Lymphosarcoma, mouse MB (T-86157) suspension culture, LVIII, 203

LYO, *see* Lyophilization medium

Lyophilization
of activated agarose gels, XLIV, 33
effect on immobilized-enzyme activity, XLIV, 76
metabolite measurement, LVI, 202
preservation of cultures, XLIII, 3
protein denaturation, LIX, 499
tissue extraction, LV, 202

Lyophilization medium, formulation, LVII, 154

Lysergic acid diethylamine, salivary secretion effects, XXXIX, 475

Lysine
aminoacylation, isomeric specificity, LIX, 274, 279, 280
ε-amino group, XLIV, 12
coupling with activated Sepharose, XLIV, 17
ε-amino group blocking, XLVII, 526, 534–537, 610
assay, with enzyme electrode, XLIV, 277, 278
^{14}C-labeled, separation from ^{14}C-labeled hydroxylysine, XL, 357
cleavage, in presence of *S*-2-aminoethylcysteine, XLVII, 172, 173
derivation, with sulfonated phenylisothiocyanate, XLVII, 216, 262
detection, XLVII, 17
determination of 3′-terminal modification, LIX, 181
enzyme aggregation, XLIV, 263–273
esterification, XLVII, 522, 523
fluorescent labeling, XLVIII, 358
isoelectric focusing of cytochrome *c* derivatives, LIII, 149
in media, LVIII, 53, 62
mobility reference, XLVII, 57
modification, XLVII, 483, 485, 493–496
peptide attachment, XLVII, 283, 309
peptide synthesis, XLVII, 510
percent in protein, XLIV, 13
quinone binding, XXXI, 537
reaction with anhydrides, XLIV, 18
with diazonium compounds, XLIV, 18
with diethylpyrocarbonate, XLVII, 440
with glutaraldehyde, XLIV, 552, 713, 904
with imidoester compounds, XLIV, 321

reductively methylated derivatives, identification, XLVII, 476–478
residue
cross-linking, LVI, 635
enzyme immobilization, XLIV, 930
modification, LIII, 139–143, 151–153
reversible blockage by phthalic anhydride, XLVII, 149–155
side chain reactivity, XLIV, 14, 15
thermolysin hydrolysis, XLVII, 175
L-β-Lysine, XLIII, 123
L-Lysine-agarose, XXXIV, 425–427, 712, 713
Lysine buffer, in deacylation, LIX, 270
Lysine chloromethyl ketone, XLVI, 609, 611, 612
Lysine decarboxylase
conjugate, activity, XLIV, 271
in lysine assay, XLIV, 277, 278
L-Lysine hydrochloride, tRNA hydrolysis, LIX, 186
Lysine oxidative decarboxylase, LII, 16
Lysine oxygenative decarboxylase, oxy form, spectrum, LII, 36
Lysine-polyacrylamide, XXXIV, 427
affinity adsorbent for isolation of plasminogen, XXXIV, 46
Lysine-tRNA synthetase, XXXIV, 170
Lysine-Sepharose, preparation of ribosomal RNA, LIX, 557
Lysine-Sepharose affinity chromatography, factor XII, XLV, 61
Lysine–tryptophan ratio, determination, in histone analysis, XL, 137
Lysing medium, for fat-cell ghost preparation, XXXI, 110
Lysocellin, as ionophore, LV, 446
Lysoglycerophospholipid, *see also* specific compounds
method of synthesis, XXXV, 484–493
Lysolecithin
activator of phenylalanine hydroxylase, LIII, 284
adenylate cyclase, XXXVIII, 177
cell fusion, XXXIX, 123, 125; LVIII, 345
freeze-fracture, XXXII, 47, 48
intermediate in lysoglycerophospholipid synthesis, XXXV, 493
in phosphatidylcholine synthesis, XXXV, 443
reconstitution studies, LV, 707, 708
structure, LVI, 735
Lysophosphatidic acid phosphatase, in bacterial membrane, XXXI, 649
Lysophosphatidylcholine, transhydrogenase assay, LV, 276
Lysophosphatidylethanolamine, formation in phosphatidylethanolamine synthesis, XXXV, 460
Lysophospholipid, uses and properties, LVI, 745–747
Lysosomal enzyme, *see* Enzyme, lysosomal
Lysosome, XXXI, 494, 521
autofluorescence and fluorochroming, XXXI, 471–475
from brain cells
isolation, XXXI, 457–477
ultrastructure, XXXI, 475–477
characterization, XXXI, 23, 24, 31, 35, 39, 326–329
by enzyme assay, XXXI, 409
diagnostic enzymes, XXXI, 20, 187, 326, 327, 330, 735, 743, 744
differential sedimentation, XXXI, 727, 732
electron microscopy, XXXI, 26, 34
electrophoretic purification, XXXI, 753, 754
enzymes, bovine serum albumin, LV, 116
internal volume, LV, 551
isolation from kidney, XXXI, 330–339
from liver, XXXI, 10, 323–329
from lymphoid tissue, XXXI, 353–356
from macrophages, XXXI, 339–345
from rat liver, L, 491–493
liposome, interaction, XLIV, 701, 702
market enzyme, LV, 101
mitochondria, LV, 117
morphometry, XXXII, 15, 17, 18
purification, XXXI, 40, 41, 717
of enzymes, L, 491–493
purity criteria, XXXI, 326–329

rat liver, acid nucleotidase, LI, 271
Triton uptake, XXXI, 40
vacuole similarity, XXXI, 575
Lysostaphin, XLIII, 621, 752
spheroplast preparation, LVI, 175
Lysozome transfer, LVIII, 359
Lysozyme, XXXIV, 4, 5, 639–645, 705; XLVI, 403–414
acylation techniques, XLVII, 149
adenylate cyclase extraction, XXXVIII, 163, 164
affinity labeling, XLVI, 410–413
in algal lysis, XXXI, 682
assay by Fast Analyzer, XXXI, 816
in bacterial cell disruption, LI, 287
for bacterial lysis, LII, 153
binding site, XLVI, 75, 411, 412
in cell-wall degradation, XXXI, 609
from *Chaloropsis,* sequence analysis, XLVII, 169, 170
circular dichroic spectrum, XLIX, 161
classification of tyrosyl residues, XLIX, 118, 119
conjugate
activity assay, XLIV, 290
activity correlations, XLIV, 444–450
crystal
immobilization, XLIV, 548
X-ray analysis, XLIV, 555
diazotization, XLIV, 445, 446
from egg white, interaction with chitotriose, XLVIII, 272, 273
electron-nuclear distance, XLIX, 423
Escherichia coli, XLIII, 621
lysis, LIX, 300, 354, 365
fluorescence quenching, XLIX, 222
glutaraldehyde insolubilization, XLIV, 553
from hen egg white
activity, modification, XLVII, 160
modification with DMSO/HCl, XLVII, 448–451
reductive alkylation, XLVII, 474, 475
solid-phase sequencing, XLVII, 269
immobilization
on collagen, XLIV, 153, 255, 260
by microencapsulation, XLIV, 214
on polyacrylamide beads, XLIV, 444, 445
on polystyrene, XLIV, 288–290
infrared spectroscopy, XXXII, 252
light scattering properties, XLVIII, 442
from mare milk, chemical cleavage, XLVII, 147
membrane preparation, LV, 782
methionine residues, XXXIV, 192
modification, effect on activity, XLIV, 448–450
in mucosal cell isolation, XXXII, 665
native, XXXIV, 642, 643
nitrotyrosine peptides, XXXIV, 189
physical properties, XLVI, 404
in PMN granules, XXXI, 345
protoplast ghost preparation, LV, 181
purification by chitin, XXXIV, 164
of glycine reductase, LIII, 376
of nitrate reductase, LIII, 348
of nitrogenase, LIII, 316, 318
reductive denaturation studies, XLIV, 517
SDS complex, properties, XLVIII, 6
self association, XLVIII, 110
sequence analysis, XLVII, 252–255
high-sensitivity, XLVII, 331
soluble, activity assay, XLIV, 446
specific activity, XLIV, 255
spheroplast preparation, LVI, 175, 380, 382, 383
standard, in fluorescamine assay, LIX, 501
substrate specificity, XLVI, 404
synthetic, XXXIV, 644, 645
translational diffusion coefficient, XLVIII, 420
tryptophan content, XLIX, 161, 162
Lysozyme-cellulose, XXXIV, 705
Lysylalanyllysylchloromethyl ketone, XLVI, 206
Lysylchloromethyl ketone, XLVI, 201
L-Lysyl-D-glutamyl-D-alanine, XLIII, 697
Lysyl hydroxylase, LII, 12
Lysyl oxidase, XL, 321
activity of enzymes, XL, 347
Lysyl protocollagen oxidase, LII, 10
Lysyl residue, modification, with succinic anhydride, LVII, 177

Lysyl-tRNA synthetase, subcellular distribution, LIX, 233, 234
Lytechinus pictus, hatching enzymes, XLV, 372
Lyxose, chromatographic constants, XLI, 16, 17

M

MAB 87/3 medium, LVIII, 57, 88
  composition, LVIII, 62–70
Macaloid
  nucleic acid extraction, LIX, 311
  treatment of gelatin, LIX, 236
MA-134 cell line, properties and uses, XXXII, 585
Macerozyme, in protoplast isolation, XXXI, 579, 580, 583
Machery-Nagel Polygram plate, XXXVI, 467
Macroautoradiography, *see also* specific techniques
  of steroid hormones, XXXVI, 137, 138
Macrocyclic antibiotic, *see* Antibiotic, macrocyclic
α_2-Macroglobulin
  antibody binding, L, 55
  human
    amino acid composition, XLV, 651
    assay, by N^α-benzoyl-DL-arginine-p-nitroanilide, XLV, 641
    carbohydrate composition, XLV, 652
    characteristics, XLV, 639
    enzyme binding properties, XLV, 649
    inhibitor of kallikrein, XLV, 314
    properties, XLV, 641
      physical, XLV, 646
    purification, XLV, 641, 642
    purity, XLV, 646
    serum concentration, XLV, 639
    specificity, XLV, 647
    stability, XLV, 644
    subunit structure, XLV, 647
Macroion
  definition, XLVIII, 3
  electrophoretic mobility, XLVIII, 4
  molecular dimensions, XLVIII, 3
Macrolide, LVI, 31
  impermeability, LVI, 32
Macrolide antibiotic, *see also* specific antibiotic
  aglycosidic nonpolyene, XLIII, 136, 137
  chiroptical studies, XLIII, 352
  gas-liquid chromatography, XLIII, 245–248
  high-pressure liquid chromatography, XLIII, 308, 309, 319
  thin-layer chromatography, XLIII, 193, 198
Macromolecule
  agarose derivative, XXXIV, 658–660
  spacer, XXXIV, 676, 681
  synthetic, XXXIV, 631–649, *see also* specific types
Macrophage, LVIII, 492, 494–506
  activation, LVIII, 494
  adherence, LVIII, 497, 498
  alveolar, LVIII, 496, 497
    isolation, LVII, 471, 472
    phagolysosome isolation, XXXI, 343, 344
  centrifugation, LVIII, 497
  chemiluminescence, LVII, 465
  continuous cell lines, LVIII, 503, 504
  culture, XXXII, 762–765; LVIII, 506
    protocol, LVIII, 504–506
  detachment, LVIII, 501
  effect of serum, LVIII, 501
  in exudates, LVIII, 502
  fixing, LVIII, 502
  formation, XXXII, 758, 759
  functional activities, XXXII, 652
  growth factor, LVIII, 163
  histochemistry, LVIII, 498
  identification, LVIII, 498
  isolation from liver, XXXII, 647–663
  light and electron microscopy, XXXII, 650–652
  long-term culture, LVIII, 502
  lysis, LVIII, 502, 503
  lysosome isolation, XXXI, 339–345
  marker, LVIII, 499
  media, LVIII, 499–502
  morphology, LVIII, 498, 502

peritoneal, XXXII, 760–762; LVIII, 495, 496
isolation, LVII, 472, 473
phagocytic activity, XXXII, 649, 650
phagocytic uptake, LVIII, 498, 499
properties, XXXII, 765
purification, LVIII, 497, 498
separation, LVIII, 497, 498
sources, LVIII, 495–497
staining, LVIII, 52
viability, LVIII, 501, 502
Macroradioautography, of protein hormones, XXXVII, 148–155
Macroreticular resin chromatography, XLIII, 296–299
adsorption, XLIII, 298
cephalosporin C isolation from fermentation broth, XLIII, 299
clean-up of new resin, XLIII, 298
desorption of compounds, XLIII, 298
properties of resin, XLIII, 297
regeneration of resin, XLIII, 298
rehydration of resin, XLIII, 298
XAD–2, XLIII, 297
Macrotetralide, as ionophores, LV, 442, 444
Macula adhaerens, of plasma membranes, XXXI, 85
Maden Darby canine kidney cell line, LVIII, 552–560
blister formation, LVIII, 553
growth, LVIII, 554
kidney function studies, LVIII, 553
maintenance, LVIII, 554
media, LVIII, 554
as model of distal tubule, LVIII, 559, 560
response to antidiuretic hormones, LVIII, 559
transport studies, LVIII, 553, 554
Magnamycin, XLIII, 131, 338
Magnesium
in ACTH mitochondria, XXXVII, 303, 304
activation of acetyl-CoA carboxylase, XXXV, 3, 9
deficiency, XL, 321
isotopes, LVI, 302
in media, LVIII, 68
in mitochondrial matrix, LVI, 254
in ribosome studies, XXXVII, 249, 250
transport, parathyroid hormone effects, XXXIX, 20
Magnesium-28
in membrane studies, XXXII, 884–893
properties, XXXII, 884
Magnesium acetate, XLIV, 162
aminoacylation of tRNA, LIX, 216
assay of adenylosuccinate synthetase, LI, 207
of creatine kinase isoenzymes, LVII, 60
of GTP hydrolysis, LIX, 361
of modulator protein, LVII, 109
of photophosphorylation, LVII, 55
of polypeptide chain elongation, LIX, 357
of polypeptide synthesis, LIX, 367
of radiolabeled ribosomal particles, LIX, 779
for translocation, LIX, 359
cell-free protein synthesis system, LIX, 300, 448, 459, 850
crystallization of transfer RNA, LIX, 6
determination of aminoacylation, isomeric specificity, LIX, 272
digestion of tRNA terminal mononucleotides, LIX, 140
electrophoretic analysis of protein-RNA complexes, LIX, 567
extraction of nucleic acids, LIX, 311
of ribosomal proteins, LIX, 431
filter binding assay, LIX, 848
initiation factor buffer, LIX, 838
isolation of chloroplasts, LIX, 210, 434
of chloroplastic ribosomes, LIX, 434, 435
of mitochondria, LIX, 206
preparation of aminoacylated tRNA isoacceptors, LIX, 298
of chloroplastic tRNA, LIX, 210, 211
of *E. coli* crude extract, LIX, 297, 354
of *E. coli* ribosomes, LIX, 445, 451, 647, 817, 818, 838
of mitochondrial tRNA, LIX, 206

of polysomes, LIX, 363
of protein-depleted 30 S subunits, LIX, 621
of radiolabeled ribosomal protein, LIX, 783
of ribosomal fragments, LIX, 472
of ribosomal proteins, LIX, 446, 447, 453
of ribosomal RNA, LIX, 557
of ribosomal subunits, LIX, 411, 413, 415, 416
of ribosome-DNA complexes, LIX, 841, 842
of samples for neutron scattering, LIX, 655
of S100 extract, LIX, 357
of 30 S subunits, LIX, 618
purification of chloroplastic phenylalanine tRNA, LIX, 213
of cytidine deaminase, LI, 402
of *E. coli* endogenous polysomes, LIX, 353, 354, 355
reconstitution of ribosomal subunits, LIX, 447, 448, 451, 654, 864
in relaxing medium, LVII, 248
reverse salt gradient chromatography, LIX, 216
tRNA reconstitution, LIX, 140, 141
studies of protein-RNA interactions, LIX, 586
Magnesium acetate-ethylenediaminetetraacetic acid, isolation of ribosomes, LIX, 373
Magnesium-adenosine triphosphate ion, allosteric activator, of adenosine monophosphate nucleosidase, LI, 263, 270, 271
Magnesium atomic emission studies, LIV, 455
Magnesium chloride, XLIV, 313, 340, 342, 454, 456, 476, 549, 653, 793, 865, 890, 892
activator of nucleoside triphosphate pyrophosphohydrolase, LI, 279
adenylate kinase, LV, 285
aminoacylation of tRNA, LIX, 216, 221, 818
assay of adenine phosphoribosyltransferase, LI, 559, 569, 574
of adenosine monophosphate nucleosidase, LI, 264
of adenylate kinase, LI, 460, 468
of amidophosphoribosyltransferase, LI, 172
of aminoacyl-tRNA synthetases, LIX, 231
of carbamoyl-phosphate synthetase, LI, 30, 106, 112, 122
of cyclic nucleotide phosphodiesterases, LVII, 96, 97
of CTP(ATP):tRNA nucleotidyltransferase, LIX, 123
of cytidine triphosphate synthetase, LI, 80, 85
of dCTP deaminase, LI, 418
of deoxycytidylate deaminase, LI, 413
of deoxythymidine kinase, LI, 361
of deoxythymidylate phosphohydrolase, LI, 285
of FGAM synthetase, LI, 195, 196, 197
of GMP synthetase, LI, 213
of guanosine triphosphate, LVII, 89, 91
of guanylate kinase, LI, 475, 476
of hypoxanthine phosphoribosyltransferase, LI, 543, 550
of nucleoside diphosphokinase, LI, 377
of OPRTase-OMPdecase complex, LI, 135, 156
of orotate phosphoribosyltransferase, LI, 70, 144
of phosphoribosylglycinamide synthetase, LI, 180
of phosphoribosylpyrophosphate synthetase, LI, 3, 4
of purine nucleoside phosphorylase, LI, 539
of pyrimidine nucleoside monophosphate kinase, LI, 322, 323
of ribonucleoside diphosphate reductase, LI, 228, 238
of ribosephosphate pyrophosphokinase, LI, 12
of RNase III, LIX, 825

of succino-AICAR synthetase, LI, 187
of thymidine kinase, LI, 355, 365
of tryptophanyl-tRNA synthetase, LIX, 237, 238
of uridine-cytidine kinase, LI, 300, 308, 309, 315, 316
bentonite-charcoal treatment, LIX, 482
brain homogenization, LII, 312
carbohydrazide modification of tRNA, LIX, 147
cell-free translation system, LIX, 387
in chromatin fractionation, XL, 98
in coupling buffer, XLVII, 44
crystallization of transfer RNA, LIX, 6, 8
cytochrome *P*–450LM assay, LII, 205
determination of deuterium content of RNA, LIX, 647
in electrophoresis buffer, LI, 357
enhancer of aromatic hydrocarbon solubility, LII, 236
in Enzyme Buffer, LI, 14
enzymic conversion of AMP and ADP to ATP, LVII, 78
extraction of radiolabeled RNA, LIX, 832, 834, 836
fluorescein labeling of tRNA, LIX, 149
heme oxygenase assay, LII, 368
in HEPES buffers, XXXI, 10, 11, 14
hydrolysis of tRNA adducts, LIX, 164, 165
incorporation of terminal nucleotide analogs, LIX, 188
microsome peroxidase incubation, LII, 344
mitochondria, LV, 7
in mitochondrial incubation medium, LVII, 38
NADPH-generating system, LII, 414
periodate oxidation of tRNA, LIX, 269
preparation of aminoacylated tRNA, LIX, 126, 130, 133, 134
of aminoacyl tRNA synthetase, LIX, 314, 315
of 3′-amino-3′-deoxy-ATP, LIX, 135
of *Bacillus* crude extract, LIX, 438
of *Bacillus* ribosomes, LIX, 438
of CHO crude extract, LIX, 219, 232
of cross-linked ribosomal subunits, LIX, 535
of *E. coli* crude extract, LIX, 194
of *E. coli* ribosomal subunits, LIX, 444, 554, 752
of *E. coli* ribosomes, LIX, 444, 483, 554, 752
of labeled oligonucleotides, LIX, 75
of mitochondrial ribosomes, LIX, 425, 426
of modified tRNA, LIX, 170, 171, 179, 180
of postmitochondrial supernatant, LIX, 515
of protein from cross-linked ribosomal subunits, LIX, 542
of protein-depleted 30 S subunits, LIX, 621
of radiolabeled ribosomal protein, LIX, 777
of rat liver crude extract, LIX, 403
of rat liver ribosomes, LIX, 504
of rat liver tRNA, LIX, 230
of ribosomal proteins, LIX, 484, 490, 518, 837
of ribosomal RNA, LIX, 556, 557, 648
of ribosomal subunits, LIX, 403, 404, 440, 483, 505
of RNase III, LIX, 826
of *T. vulgare* ribosomes, LIX, 752
protein binding buffer, LIX, 322, 328
purification of adenine phosphoribosyltransferase, LI, 570, 571, 572
of AMP nucleosidase, LI, 268
of chloroplastic phenylalanine tRNA, LIX, 213
of CPSase-ATCase-dihydroorotase complex, LI, 124
of CTP(ATP):tRNA nucleotidyltransferase, LIX, 123
of deoxycytidylate deaminase, LI, 414, 415, 416
of flavin peptides, LIII, 455

of hypoxanthine phosphoribosyltransferase, LI, 545
of nitrate reductase, LIII, 348, 642
of nitrogenase, LIII, 316, 318, 322–324, 327
of nucleoside triphosphate pyrophosphohydrolase, LI, 276, 277, 278
of ribonucleoside diphosphate reductase, LI, 239
of tRNA methyltransferases, LIX, 194, 197
of shortened tRNAs, LIX, 186
of thymidylate synthetase, LI, 95
in pyrophosphate determination, LI, 275
reconstitution of ribosomal subunits, LIX, 441, 652, 837
of tRNA, LIX, 127, 128, 134
reverse salt gradient chromatography, LIX, 215
tRNA end-group labeling, LIX, 103
RNase III digestion buffer, LIX, 833, 834
separation of aminoacylated tRNA, LIX, 216
of bacterial ribosomal subunits, LIX, 398, 399
of mitochondrial tRNAs, LIX, 207
of tRNA isoacceptors, LIX, 221
sodium laurate hydroxylation, LII, 320
studies of protein-RNA interactions, LIX, 562, 563
succinyl-CoA generating system, LII, 352
transfer RNA NMR spectra, LIX, 25
tritium labeling studies, LIX, 342
Magnesium hydrate, binding, to transfer RNA, LIX, 17, 18
Magnesium ion
A23187, LV, 449
activator of adenine phosphoribosyltransferase, LI, 566, 570, 573
of brain α-hydroxylase, LII, 317
of CPSase-ATCase-dihydroorotase complex, LI, 118
of cytidine triphosphate synthetase, LI, 79
of dCTP deaminase, LI, 422
of deoxycytidine kinase, LI, 338, 345
of deoxycytidylate deaminase, LI, 417
of deoxythymidine kinase, LI, 360, 361
of deoxythymidylate phosphohydrolase, LI, 290
of guanine phosphoribosyltransferase, LI, 557
of guanylate kinase, LI, 481, 489
of hypoxanthine phosphoribosyltransferase, LI, 543, 557
of *Lactobacillus* deoxynucleoside kinases, LI, 346
of nucleoside diphosphokinase, LI, 371, 372, 375, 385
of nucleoside phosphotransferase, LI, 393
of orotate phosphoribosyltransferase, LI, 74, 140, 152
of phosphoribosylglycinamide synthetase, LI, 185
of phosphoribosylpyrophosphate synthetase, LI, 11
of pyrimidine nucleoside monophosphate kinase, LI, 321, 329
of ribonucleoside diphosphate reductase, LI, 227
of succino-AICAR synthetase, LI, 193
of thymidine kinase, LI, 358, 370
of UMP-CMP kinase, LI, 331, 332
of uridine-cytidine kinase, LI, 299, 305, 308, 314, 319
adenylate kinase, LV, 232
antipyrylazo III, LVI, 326, 328
arsenazo III, LVI, 319, 320, 329
ATPases, LV, 301
in bacterial cell wall isolation, XXXI, 646, 647
calcium influx measurement, LVI, 344–346
calcium-selective electrodes, LVI, 348
in chloroplast isolation, XXXI, 603

dissociation of ternary initiation complex, LX, 562–564, 570
effect
aminoacyl-transfer ribonucleic acid binding by 70 S ribosomes, LX, 625, 626
on calcium-aequorin assay, LVII, 324, 326
on liver fractionation, XXXI, 8, 9
polyuridylate binding by 30 S ribosomal subunits, LX, 433, 436
preribosomal complex formation, LX, 217, 219, 239
on transfer RNA, LIX, 53, 54
in vitro chain elongation, LIX, 370
elution of translating ribosomes from matrix-bound polyuridylate, LX, 759, 760
energy charge, LV, 233, 234
glucose isomerase activation, XLIV, 815
indicators, LVI, 313–317
inhibitor of *Cypridinium* luminescence, LVII, 349, 350
of immobilized lactase, XLIV, 261
initiation of protein synthesis, LX, 12, 13, 39, 41, 47, 48, 184, 185
microsome aggregation, LII, 87
mitochondrial protein synthesis, LVI, 21
mitochondrial ribosome preparation, LIX, 423
mitochondrial ribosomes, LVI, 86
murexide, LVI, 329
in opsonization studies, LVII, 487
Pasteur effect, LV, 291
phosphorylation potential, LV, 235, 236, 238, 244
in plant-cell fractionation, XXXI, 504
preparation of ribosomal fragment proteins, LIX, 470
in reconstituted hemoglobin, LII, 491
reconstitution studies, LV, 708
required by β-galactosidase, XLIV, 827
ribosomal protein synthesis, LVI, 27
ribosomal RNA digestion, LIX, 571
ribosomal RNA storage, LIX, 586
ribosome dissociation, LIX, 379, 380, 381
ribosome isolation, LVI, 25, 26
role in mitochondria, LV, 453
in sarcoplasmic reticulum, XXXI, 241
smooth muscle mitochondria, LV, 62, 64, 65
stabilization of ternary initiation complex, LX, 741
in *in vitro* RNA synthesis, LIX, 847
Magnesium oxide, pore properties, XLIV, 137
Magnesium phosphoribosylpyrophosphate solution, standardization, LI, 552, 560
Magnesium sulfate, XLIV, 132, 163, 327, 823, 824–836, 839, 894
aminoacylation of tRNA, LIX, 288
assay of AMP formation during aminoacylation, LIX, 287
of firefly luciferase, LVII, 11, 30
of GMP synthetase, LI, 219
of nucleotide incorporation into tRNA, LIX, 182, 183
of ATP, LVII, 11, 12, 98
in luciferase reagent, LVII, 70, 76, 88
misactivation studies, LIX, 286
of phosphoribosylpyrophosphate synthetase, LI, 13
purification of aminoacyl-tRNA synthetases, LIX, 258
of luciferase, LVII, 31, 32, 33, 36
of phosphoribosylglycinamide synthetase, LI, 184
Magnet, superconducting, for magnetic circular dichroism, XLIX, 153
Magnetic balancing, XLVIII, 29–68
density-volume changes, XLVIII, 30–50
osmotic pressure, XLVIII, 61–68
sensitivity, XLVIII, 31, 32
viscosity, XLVIII, 50–61
Magnetic circular dichroism, XLIII, 347; XLIX, 149–179, *see also* Spectropolarimetry
applications, XLIX, 159–179
A term, XLIX, 150, 151
B term, XLIX, 151, 152
calibration, XLIX, 156–159
C term, XLIX, 151, 152, 158
data analysis, XLIX, 154, 155

experimental considerations, XLIX, 153–159
low-temperature studies, XLIX, 158, 159
sample cell, XLIX, 157, 158
theoretical considerations, XLIX, 150–153
units, XLIX, 155, 156
Magnetic circular dichroism spectroscopy, LIV, 284–302
calibration, LIV, 290
data analysis, LIV, 290–292
experimental parameters, LIV, 291
with heme proteins, LIV, 295–302
instrumentation, LIV, 288, 289, 294
principles, LIV, 284–288
sample handling, LIV, 291, 293–295
sign convention, LIV, 289
solvent systems, LIV, 293
units of measurement, LIV, 289, 290
Magnetic ellipticity, tryptophan content, XLIX, 162
Magnetic field
calibration, standards, XLIX, 523
measurement, XLIX, 521, 522
Magnetic field effect on electrochemiluminescence, LVII, 518–520
Magnetic field modulation, XLIX, 484
Magnetic field region, LIV, 133
Magnetic field strength
dimensions, LIV, 289
measurement, LIV, 290
Magnetic float, applications, XLVIII, 18–21
Magnetic support, XLIV, 324–326
iron filings, XLIV, 216
Magnetic susceptibility, LIV, 85, 346, 379–396
definition, LIV, 380
experimental error, sources, LIV, 389–393
experimental methods, LIV, 386–393
Faraday force technique, LIV, 386, 387
force techniques, LIV, 386–388
Gouy technique, LIV, 387, 388
measurement, physical basis, LIV, 380–385
nuclear magnetic resonance techniques, LIV, 388
sample positioning, LIV, 390
spatial anisotropy, LIV, 385
temperature, LIV, 386, 389, 390, 394
Magnetite, XLIV, 325, 326
Magnetization
computation, LIV, 381, 386
definition, LIV, 180
Maize
leaf, *see* Leaf
organelles, XXXI, 509
vacuole isolation, XXXI, 574, 575
Maize × oat root, LVIII, 365
Malacosoma distria, media, LVIII, 457
Malate buffer, for desorption membrane-bound ATP, LVII, 70, 71, 72
Malate dehydrogenase, XXXI, 642, 768; XXXIV, 246, 252; XLVI, 21, 66
activity, LVII, 181
activity assay, XLIV, 458–460, 474
assay, LVII, 181, 210
of acetate, XXXV, 299, 302
L-asparaginase, XLIV, 690
of carboxyltransferase, XXXV, 32–34
citrate, LV, 211
coenzyme A, LV, 210
α-ketoglutarate, LV, 206
of long-chain fatty acyl-CoA, XXXV, 273
of malate, LVII, 187
of oxaloacetate, LV, 207; LVII, 187
in enzyme electrode, XLIV, 595
in glyoxysomes, XXXI, 568, 569
in hormone receptor studies, XXXVII, 212
immobilization
in multistep enzyme system, XLIV, 181, 182, 457–460, 472–475
on nylon tube, XLIV, 120
kinetics, in multistep enzyme system, XLIV, 434, 466–468
matrix, LV, 101, 104
assay, LV, 102
mitochondrial subfractions, LV, 94, 99
in mitoplasts, XXXI, 321, 322

reduction of NAD^+ analogs, XLIV, 873, 874
rhein, LV, 459
toluene-treated mitochondria, LVI, 547
Malate dehydrogenase (NAD), brain mitochondria, LV, 58
Malate oxidase, LII, 21; LVI, 474
Malate synthase, in microbodies, LII, 496
Malate synthetase, *see also* Malate synthase
in glyoxysomes, XXXI, 565, 569, 743
as marker enzyme, XXXI, 735, 743
Malate-vitamin K reductase, *M. phlei* ETP, LV, 184, 185, 186
Malayan pit viper venom, *see* Ancrod
Malbranchea pulchella var. *sulfurea*
growth characteristics, XLV, 416, 417
thermomycolin, XLV, 415, *see also* Thermomycolin
4-(*N*-Maleamido)benzyldimethylamine, XLVI, 583
Maleamidobenzyltrimethylammonium iodide, XLVI, 585
Maleate buffer, XXXIX, 136
Maleic acid
inhibitor of succinate dehydrogenase, LIII, 33
irreversible inactivation and dissociation of fatty acid synthase, XXXV, 55, 56
in TEA buffer, LIII, 364
Maleic acid anhydride, XLIV, 85, 108, 114, 179
structural formula, XLIV, 197
Maleic acid azide, structural formula, XLIV, 197
Maleic acid buffer
interference by, in OPRTase-OMPdecase complex assay, LI, 167
for lysozyme immobilization, XLIV, 289, 290
Maleic anhydride, acylation, XLVII, 149
Maleic anhydride/ethylene copolymer, XLIV, 85
Maleic anhydride/vinyl methyl ether copolymer, XLIV, 179
Maleimide, XLVI, 578
Maleimide nitroxide spin label, structure, XLIX, 489
4-(*N*-Maleimido)benzyldimethylamine, XLVI, 584
4-(*N*-Maleimido)benzyltrimethylammonium iodide, XLVI, 578, 582
hydrolysis, XLVI, 585, 586
kinetics, XLVI, 586, 587
synthesis, XLVI, 583–585
Malic acid
activator of adenylate kinase, LI, 465
assay, LV, 206, 207, 221; LVII, 181–188
cell fractionation by digitonin method, LVI, 213
coenzyme A assay, LV, 210
content
of brain, LV, 218
of hepatocytes, LV, 214
of kidney cortex, LV, 220
of rat heart, LV, 216, 217
of rat liver, LV, 213
as electron donor, LVI, 386
exchange, LVI, 253
inhibitor of rat liver acid nucleotidase, LI, 274
intact mitochondria, LV, 143
in isolated renal tubules, XXXIX, 16, 18
in liver compartments, XXXVII, 287
of malate dehydrogenase, LVII, 187
mitochondrial incubation, LV, 8
rat brain mitochondria, LV, 57
renal gluconeogenesis, XXXIX, 13, 14
succinate dehydrogenase, LVI, 586
transport, LVI, 248–250, 292
ubiquinone redox state, LIII, 583
yeast mitochondria, LV, 150
D-Malic acid, XLVI, 4
as enzyme inhibitor, XXXVII, 294
Malic acid medium, growth of *Chromatium vinosum*, LIII, 310
Malic enzyme
labeling, XLVI, 139
in parenchyma cells, XXXII, 706
steroid hydroxylation, LV, 10
Malignancy, cell fusion, LVIII, 348
Malondialdehyde
determination of lipid peroxidation, LII, 306, 308, 309

in vivo metabolism, LII, 308
Malonic acid
activation of succinate dehydrogenase, LIII, 33, 472–474
complex II, LVI, 586
as enzyme inhibitor, XXXVII, 294
inhibitor of succinate dehydrogenase, LIII, 33
submitochondrial particles, LV, 107
transport, LVI, 249, 292
ubiquinone redox state, LIII, 583
Malonic acid dihydrazide, preparation, LIX, 179
Malononitrile, purification, LV, 470
Malonyl-CoA decarboxylase, XXXV, 75, 81
interference with assay for fatty acid synthase, XXXV, 75
Malonyl coenzyme A
assay, with fatty acid synthase, XXXV, 312–315
carbon dioxide exchange reaction, XXXV, 49, 96
inhibitor of acetyl-CoA carboxylase, XXXV, 3, 11
6-methylsalicylic acid biosynthesis, XLIII, 521
patulin biosynthesis, XLIII, 521, 540
purification, XXXV, 74
stability, XXXV, 75
synthesis, XXXV, 48, 66, 74
tissue content, XXXV, 315
use in assay of acyl carrier protein synthase, XXXV, 96
of biotin carboxyl carrier protein, XXXV, 18, 19
of carboxyl transferase, XXXV, 32–34
covalent binding sites of malonyl groups of fatty acid synthase, XXXV, 51, 52
of fatty acid synthase, XXXV, 37, 60, 66, 75, 85
of malonyl-CoA pantetheinetransacylase, XXXV, 49
Malonylpantetheine
in assay for acyl carrier protein synthetase, XXXV, 96
carbon dioxide exchange reaction, XXXV, 110
synthesis, XXXV, 96
Maltase, in brush borders, XXXI, 122, 129, 130
Maltooligosaccharide, preparation, L, 412
Maltose
chromatographic constants, XLI, 17
degradation on heating in boric acid/2,3-butanediol or in borax, XLI, 21
paper chromatography, L, 417
polarography, LVI, 477
β-Maltose octacetate, XLIV, 460
Maltotriose, chromatographic constants, XLI, 17
Mammalian tissue, *see also* specific types
aldose reductase, assay, purification and properties, XLI, 159–165
fructose-diphosphate aldolases, assay, purification and properties, XLII, 240–249
mitochondria, isolation, LVI, 18, 19
monoamine oxidase, LIII, 450
mutarotase, XLI, 472
phosphofructokinase, assay, purification, and properties, XLII, 99–110
D-3-phosphoglycerate dehydrogenase, XLI, 284
phosphoglycerate mutase, properties, XLII, 447
Mammary gland
bovine, glucose-6-phosphate dehydrogenase, assay and purification, XLI, 183–188
mitochondria, LV, 19
mitochondria isolation, XXXI, 305–310
from tumor tissue, XXXI, 307, 308
organ culture, XXXIX, 443–449
parenchyma cell isolation, XXXII, 693–706
source of acetyl coenzyme A carboxylase, XXXV, 11–17
of fatty acid synthase, XXXV, 65–83
tumor cells of, hormone dependence, XXXII, 558, 559
Mammary tumor, estrogen receptor, XXXVI, 177, 178, 248–254
Mandelate racemase, XLVI, 541–548
effect of metal ion, XLVI, 546, 547

residue identification, XLVI, 547, 548
stoichiometry of binding, XLVI, 545, 546
Manduca sexta, media, LVIII, 457
Manganese
atomic emission studies, LIV, 455
as inhibitor, LVIII, 99
in media, LVIII, 52, 69
in superoxide dismutase, LIII, 393
Manganese(II)
correlation time evaluation, XLIX, 346
electron paramagnetic resonance studies, XLIX, 514
interatomic distance calculations, XLIX, 331
Manganese(III), electron paramagnetic resonance studies, XLIX, 514
Manganese-54
in membrane studies, XXXII, 884–893
properties, XXXII, 884
Manganese-55, lock nuclei, XLIX, 348; LIV, 189
Manganese chloride
in benzo[*a*]pyrene monooxygenase assay, LII, 414
mutagenesis, LVI, 98
in NADPH-generating system, LII, 378, 414
purification of hydrogenase, LIII, 311
requirement for dephosphorylation, LX, 529–531
Manganese ion
antipyrylazo III, LVI, 326, 328
binding to ATPase, LV, 301
eriochrome blue, LVI, 315
inhibitor of *Cypridina* luminescence, LVII, 349, 350
oxygen metabolism, LII, 8
phosphodiesterase assay, XXXVIII, 216, 217
in reconstituted hemoglobin, LII, 490, 491
stimulator of *Pholas* luciferase oxidation, LVII, 399
Manganese-substituted cytochrome *c*
preparation, LIII, 177
properties, LIII, 178–181
Manganin wire, in temperature-sensing devices, LIV, 377
Manganous chloride
assay of amidophosphoribosyltransferase, LI, 172
crystallization of transfer RNA, LIX, 8
purification of adenylosuccinate AMP-lyase, LI, 204
Manganous ion
activator of adenine phosphoribosyltransferase, LI, 566, 573
of cytidine deaminase, LI, 411
of dCTP deaminase, LI, 422
of deoxycytidine kinase, LI, 345
of deoxycytidylate deaminase, LI, 417
of guanine phosphoribosyltransferase, LI, 557
of guanylate kinase, LI, 481, 489
of hypoxanthine phosphoribosyltransferase, LI, 557
of nucleoside diphosphokinase, LI, 385
of phosphoribosylpyrophosphate synthetase, LI, 11, 16
of pyrimidine nucleoside monophosphate kinase, LI, 329
of succino-AICAR synthetase, LI, 193
of thymidine kinase, LI, 358, 370
of uridine-cytidine kinase, LI, 319
inhibitor of deoxythymidine kinase, LI, 365
of nucleoside phosphotransferase, LI, 393
of uridine nucleosidase, LI, 294
Mannan, XXXIV, 338
antigenic determinant, L, 166
in cell wall, XXXI, 610, 611
Mannitol
autoradiography, XXXVI, 150
in fungal mitochondrial membrane preparation, LI, 65
isolation of mitochondria, LV, 3
in oxygen determination with luminescent bacteria, LIV, 501
paper electrophoresis, L, 51

in plant-cell fractionation, XXXI, 596, 597
as protoplast stabilizer, LVIII, 360
reagent, for liver preservation, XXXI, 4
slat substitute, for luminous bacteria, LVII, 223
D-Mannitol, XLVI, 384–386
L-Mannitol, intermediate in phosphatidylglycerol synthesis, XXXV, 467–469
D-Mannitol dehydrogenase, from *Leuconostoc mesenteroides*, XLI, 138–142
assay, XLI, 138
chromatography, XLI, 140
molecular weight, XLI, 141
properties, XLI, 141, 142
purificaton, XLI, 139–141
sources, XLI, 139
Mannitol procedure, XXXV, 431, 433, 434
Mannoheptulose, chromatographic constants, XLI, 16
Mannokinase
from *Leuconostoc mesenteroides*, XLII, 39–43
assay, XLII, 39, 40
chromatography, XLII, 42
molecular weight, XLII, 43
properties, XLII, 42, 43
purification, XLII, 40–42
α-D-Mannopyranoside, XXXIV, 363
Mannosamine
interconversion, in animal cells, L, 183
neoplastic cell growth, L, 190
Mannose
chromatographic constants, XLI, 16, 17, 20
degradation on heating in boric acid/2,3-butanediol or in borax, XLI, 21
enzymic microassay, XLI, 3, 6, 8–10
lectin purification, XXXIV, 336
D-Mannose, XXXIV, 334
chromatography, gas-liquid of 2,3,4,6-tetra-*O*-methyl derivative, L, 4, 7
cultured cell growth, L, 190
glycosyl linkage, L, 101
interconversion, in animal cells, L, 182
D-mannose kinase in metabolism, in *Leuconostoc mesenteroides*, XLII, 39
L-Mannose, intermediate in phosphatidylglycerol synthesis, XXXV, 467, 468
α-D-Mannose 1,6-diphosphate, preparation, XLI, 83, 84
D-Mannose kinase, *see* Mannokinase
α-D-Mannosidase
activity, L, 452
assay, L, 452
distribution, L, 452
gel electrophoresis, XXXII, 83, 90, 91
mannosidosis, L, 452–453
mucolipidoses II and III, L, 455
from rat liver, L, 489
golgi
assay, L, 501
properties, L, 504, 505
purification, L, 501–504
lysosomes
assay, L, 495, 496
properties, L, 498–500
purification, L, 496–499
Mannosidosis
α-mannosidase, L, 452–453
urinary oligosaccharides, L, 229–233
Mannosidostreptomycin
biosynthesis, XLIII, 637–640, *see also* Mannosidostreptomycin hydrolase
paper chromatographic data, XLIII, 121
structure, XLIII, 614
Mannosidostreptomycin hydrolase, XLIII, 637–640
assay, XLIII, 638, 639
inhibitors, XLIII, 639
location, XLIII, 639
metal requirement, XLIII, 639
pH optimum, XLIII, 639
reaction scheme, XLIII, 637
specificity, XLIII, 639
stimulators, XLIII, 639, 640
substrate source, XLIII, 639
unit definition, XLIII, 639
Mannosylglucosaminide, XLIII, 272

[^{14}C]Mannosyl-oligosaccharide, protein attachment, L, 424–426
O-α-D-Mannosyl polyacrylamide, XXXIV, 181, 367
D-Mannuronic acid, colorimetric assay, XLI, 30, 31
Manometer, 10-mm, LIII, 304
Manton-Gaulin homogenizer
  lysis of yeast, LIII, 114
  preparation of bacterial cell extract, LI, 6, 32, 137, 414; LIII, 546
  yeast cell breakage, LVI, 46
MAO, *see* Monoamine oxidase
MAP 954/1 medium, LVIII, 57, 89
Marburg virus, LVIII, 37
Mariotte bottle, LIII, 555
Marker enzyme, *see also* specific types
  for liver cells, XXXI, 23–26, 31–39
  for particles separated on sucrose gradients, XXXI, 735, 739, 740
  in subcellular particle studies, XXXII, 15
Markham technique, for image reconstruction, XLIX, 40
Maroteaux–Lamy syndrome, arylsulfatase B activity, L, 451
Marquardt-Levenberg iteration, in curve-fitting, XXXVII, 12
Marquardt-Levenberg method, XXXVI, 14
Marrow seed, enzyme studies, XXXI, 521
Mask
  aperature shape, XLIX, 49, 52
  aperature size, XLIX, 52–55
  design criteria, XLIX, 45
  preparation, XLIX, 46–48
  principle, XLIX, 41
  rectangular, XLIX, 52
Mass culture, *see* Cell culture, large-scale
Mass spectroscopy
  deuterated fatty acids, XXXV, 342–348
    branching, determination of sites, XXXV, 344, 345
    double bonds, detection and position assignment, XXXV, 345, 346
    high resolution, XXXV, 347, 348
  deuterium incorporation into fatty acids, XXXV, 283–285, 342–348
  dipeptide identification, XLVII, 397–399
  fatty acid synthesis rate measurement, XXXV, 283–285
  2-hydroxy fatty acids, (−)-menthyloxycarbonyl derivatives, XXXV, 331
  3-hydroxy fatty acids, 2-D-phenylpropionate derivatives, XXXV, 331
  identification of coenzyme Q analogs, LIII, 595, 596
  methylated polysaccharides, acetylated forms, L, 4–31
  prostaglandins, XXXV, 360–364, 367–377
  triglycerides, XXXV, 348–359
    analysis of mixtures, XXXV, 355–359
    applications, XXXV, 358
    double bond position determination, XXXV, 352
    molecular weight distribution determination, XXXV, 357, 358
    order of fatty acids, XXXV, 353
Mass spectroscopy-gas chromatography, for identification of steroids, LII, 381–387
Mast cell
  isolation, XXXI, 397–399
  secretory granule isolation, XXXI, 395–402
  tumor lines, in neurobiological study, XXXII, 769
Masterflex rotary pump, XXXIX, 47, 52
Mastocytoma, large-scale growth, LVIII, 211
Mathematics of affinity procedure, XXXIV, 140–162, *see also* Affinity method, theory
Mating, yeast, mitochondrial markers, LVI, 141, 142
Matrix, *see also* Support
  membrane, marker enzyme, LV, 101
Matrix selection, XXXIV, 18, 19
Matrix space, calculation, LVI, 261, 262
MB 752/1 medium, LVIII, 57, 88, 89
  composition, LVIII, 62–70

McCoy's medium 5A, LVIII, 56, 86, 91, 208, 423
composition, LVIII, 62–70
for endocrine cell culture, XXXIX, 113
suspension culture, LVIII, 208
MCD, *see* Magnetic circular dichroism
MCDB medium, for specific species, LVIII, 87
MCDB 104 medium, LVIII, 59, 87, 91, 155
MCDB 105 medium, LVIII, 59, 76, 87
composition, LVIII, 62–70
MCDB 202 medium, LVIII, 59, 76, 87, 91
composition, LVIII, 62–70
MCDB 301 medium, LVIII, 58, 89
composition, LVIII, 62–70
MCDB 401 medium, LVIII, 59, 87, 90, 91
composition, LVIII, 62–70
MCDB 411 medium, LVIII, 57, 89
composition, LVIII, 62–70
MCDB 501 medium, LVIII, 59, 87
composition, LVIII, 62–70
McElroy's medium, formulation, LIV, 500; LVII, 224
McGhee and vonHippel model of protein-nucleic acid interaction, XLVIII, 295–299
McGinty progestin assay, XXXVI, 457
McIlvaine buffer, XLIV, 114
MCIM cell, plasma membrane isolation, XXXI, 161
McNeil 4308, as ionophore, LV, 443, 448
MCR, *see* Metabolic clearance rate
MDCK cell, *see* Maden Darby canine kidney cell line
MD 705/1 medium, LVIII, 57, 88
M–D–SCAL multidimensional scaling program, LIX, 609, 610
Mean residue molecular weight, LIV, 252
Mean square angular displacement, XLIX, 482, 483
Meat, microbiological test, resazurin and resorufin, XLI, 56
Mechanical oscillator densimeter, XLVIII, 15, 18–22
Mechanochemistry, XLIV, 558–576
enzyme activity, XLIV, 558–571
protein-protein interaction, XLIV, 571–576
Mediator-titrant, LIV, 406–410, *see also* Redox mediator
redox potentials, table, LIV, 408, 409, 423
reduction, LIV, 407
Medium, XXXIX, 199, 445; LVIII, 44–93, *see also* specific type
A2 + APG, LVIII, 57
for actinomycetes, XLIII, 5–7
N-Z amine-starch-glucose medium, XLIII, 7
Bennett's agar, XLIII, 6
oatmeat agar, XLIII, 6
tomato paste oatmeal agar, XLIII, 6
trypticase-yeast extract agar, XLIII, 6
yeast extract agar, XLIII, 5
aeration effect, XLIII, 7
alpha-MEM, LVIII, 56
AMBD 647/3, LVIII, 59
5A (McCoy), *see* McCoy's medium 5A
basal (Eagle), *see* Eagle's basal medium
Birch and Pirt's, LVIII, 58
for centrifugation, XXXI, 719, 720
chemically defined, LVIII, 82
choice, LVIII, 81, 90–93
CMRL 1415, LVIII, 58, 62–70, 87, 90, 91
CMRL 1969, LVIII, 59, 62–70, 87, 91
CMRL 1415–ATM, LVIII, 58, 87
CMRL 1066 (Parker), LVIII, 57, 62–70, 88, 155
commercial, LVIII, 84, 196, 197
human adenovirus, LVIII, 427
composition
amino acid derivatives, LVIII, 63
buffers, LVIII, 70
carbohydrates, LVIII, 66
chelating agents, LVIII, 70
detergents, LVIII, 70
essential amino acids, LVIII, 62
indicators, LVIII, 70
inorganic ions, bulk, LVIII, 68
lipids, LVIII, 68
nonessential amino acids, LVIII, 63

nucleic acid derivatives, LVIII, 66, 67
polymers, LVIII, 70
proteins, LVIII, 70
solvent, LVIII, 70
trace elements, LVIII, 69
vitamins, LVIII, 64, 65
conditioned, LVIII, 47, 153, 524, 581
preparation, LVIII, 161–164
costs, LVIII, 204–206
7C's, LVIII, 58, 89
defined, LVIII, 88–90
development, LVIII, 91–93
DM, LVIII, 59
DM 120, LVIII, 57, 88
DM 145, LVIII, 57, 62–70, 88
DM 160, LVIII, 57, 88
Dulbecco's, *see* Dulbecco's modified Eagle's medium
EM–1, LVIII, 105
for eubacteria, XLIII, 5
nutrient agar, XLIII, 5
tryptone glucose yeast extract agar, XLIII, 5
F10(Ham), *see* Ham's medium F10
F12(Ham), *see* Ham's medium F12
F12 + hormones, LVIII, 57
Fischer's, *see* Fischer's medium
F12K, LVIII, 59, 87, 91, 155
F12M, LVIII, 155
for fungi, XLIII, 4
Czapek's solution agar, XLIII, 4
malt extract agar, XLIII, 5
potato dextrose agar, XLIII, 4
GHAT, LVIII, 352
Goodwin IPL–52, *see* Goodwin IPL–52 medium
Grace, *see* Grace medium
growth of *C. arbuscula*, LV, 479
of *E. coli*, LV, 790, 802
of *fep*$^-$ mutant, LVI, 396, 397
of *M. phlei*, LV, 178, 179
of *S. mobaraensis*, LV, 457, 458
of *Staphylococcus aureus*, LVI, 173, 174
half-selective hybrid, LVIII, 352
Hansen S–301, *see* Hansen medium S–301
HAT, LVIII, 351–353
Higuchi's, *see* Higuchi's medium
Hink, *see* Hink medium
history, LVIII, 84, 85
hormone-supplemented, LVIII, 94–109
IMEM–ZO, LVIII, 58, 62–70, 87
inositol-free, LVI, 136
insect cell, LVIII, 455–466
in interstitial cell isolation, XXXIX, 257
isolation
of aurovertin-resistant mutants, LVI, 179, 180
manipulation of membrane lipid composition, LVI, 570, 571
of temperature-sensitive mutants, LVI, 132
of mitochondria
possible additions, LV, 4
principle osmotic support, LV, 3, 4
light effect, XLIII, 7, 8
liquid culture, XLIII, 11–14
fermentation, XLIII, 12, 13
inoculum, XLIII, 11
listing, LVIII, 56–59
liver cells, LVIII, 537
L–15 (Leibovitz), LVIII, 56, 62–70, 87, 91, 463
MAB 87/3 (Gorham and Waymouth), LVIII, 57, 62–70, 88
macrophage, LVIII, 499–502
maintenance, LVIII, 87
MAP 954/1 [Waymouth (Donta)], LVIII, 57, 89
MB 752/1 (Waymouth), LVIII, 57, 62–70, 88, 89, 90
McCoy's 5A, *see* McCoy's medium 5A
MCDB 104, LVIII, 59, 87, 91, 155
MCDB 105, LVIII, 59, 62–70, 76, 87
MCDB 202, LVIII, 59, 62–70, 76, 87, 91
MCDB 301, LVIII, 58, 62–70, 89
MCDB 401, LVIII, 59, 62–70, 87, 90, 91
MCDB 411, LVIII, 57, 62–70, 89
MCDB 501, LVIII, 59, 62–70, 87
MD 705/1 (Kitos), LVIII, 57, 88
Medium 199, LVIII, 56, 62–70, 87, 90, 91, 129, 155

for microorganismal mitochondria, LV, 142
minimum essential (Eagle), *see* Eagle's minimum essential medium
Mitsuhashi and Maramorosch, *see* Mitsuhashi and Maramorosch medium
mixture, LVIII, 97, 99, 102–104, 106
four factors, LVIII, 98
Mohberg and Johnson's, LVIII, 57
MPNL65/C, LVIII, 59
N16, LVIII, 56
Nagle and Brown's, LVIII, 58, 59
for *N. crassa* culture
for flux and electrical studies, LV, 658
for H^+-flux studies, LV, 658
minimal, LV, 657
starvation, LV, 657, 658
NCTC–109, LVIII, 57
NCTC 135 (Evans), LVIII, 57, 62–70, 88
Parsa, LVIII, 90
pH, LVIII, 199, 200
effect, XLIII, 7
preparation, LVIII, 7–9, 94–109, 198, 199
products, LVIII, 225–228
quality control, LVIII, 453, 454
RPMI 1640, LVIII, 56, 62–70, 89, 91, 155, 213
S77, LVIII, 546
Schneider, *see* Schneider medium
selection of carbodiimide-resistant mutants, LVI, 164
selection and isolation of heme-deficient mutants, LVI, 559, 560
selective hybrid, LVIII, 351, 352
serumless, LVIII, 57, 62–70, 94–109
serum supplements, LVIII, 122
Sinclair's, LVIII, 58
SM–201, LVIII, 59
SM–20 (Halle), LVIII, 59, 90
sterility testing, LVIII, 21
storage, LVIII, 7–9
supplements, LVIII, 122
suppliers, LVIII, 15, 16
temperature effect, XLIII, 7
testing, LVIII, 91
toxicity, LVIII, 117
toxicity testing, LVIII, 91, 453, 454
V–605, LVIII, 56
virus, LVIII, 225–228
White's, LVIII, 56
Williams E, LVIII, 59, 90
Williams G, LVIII, 59, 62–70, 90
WO_5, LVIII, 57
Wollenberger, LVIII, 90
Yamane's, *see* Yamane's medium
yeast growth, LVI, 43, 59
yeast mutant selection, LVI, 97, 98
Yunker *et al.*, *see* Yunker, Vaughn and Cory medium
Medium A, for lymphocytoid cell line culture, XXXII, 813
Medium A2 + APG, LVIII, 57
Medium alpha-MEM, LVIII, 56
Medium AMBD 647/3, LVIII, 59
Medium 5A (McCoy), *see* McCoy's medium 5A
Medium 7C's, LVIII, 58, 89
Medium CMRL 1066, LVIII, 57, 88, 155
composition, LVIII, 62–70
Medium CMRL 1415, LVIII, 58, 87, 90, 91
composition, LVIII, 62–70
Medium CMRL 1969, LVIII, 59, 87, 91
composition, LVIII, 62–70
Medium CMRL 1415-ATM, LVIII, 58, 87
Medium DM, LVIII, 59
Medium DM 120, LVIII, 57, 88
Medium DM 145, LVIII, 57, 88
composition, LVIII, 62–70
Medium DM 160, LVIII, 57, 88
Medium E, for bacteria, XXXII, 844
Medium EM-1, LVIII, 105
Medium E (Williams), LVIII, 59, 90
Medium F10 (Ham), LVIII, 56, 86, 155
Medium F12 (Ham), LVIII, 53, 57, 58, 86, 89, 91, 155
composition, LVIII, 62–70
Medium F12K, LVIII, 59, 87, 91, 155
Medium F12M, LVIII, 155
Medium GHAT, LVIII, 352
Medium G (Williams), LVIII, 59, 90
composition, LVIII, 62–70
Medium HAT, LVIII, 351–353
Medium IMEM–ZO, LVIII, 58, 87
composition, LVIII, 62–70

Medium IPL-52, LVIII, 460
Medium L–15 (Leibovitz), LVIII, 56, 91
  composition, LVIII, 62–70, 463
Medium MAB 87/3, LVIII, 57, 88
  composition, LVIII, 62–70
Medium MAP 954/1, LVIII, 57, 89
Medium MB 752/1, LVIII, 57, 88, 89, 90
  composition, LVIII, 62–70
Medium MCDB, for specific species, LVIII, 87
Medium MCDB 104, LVIII, 59, 87, 91, 155
Medium MCDB 105, LVIII, 59, 76, 87
  composition, LVIII, 62–70
Medium MCDB 202, LVIII, 59, 76, 91
  composition, LVIII, 62–70
Medium MCDB 301, LVIII, 58, 59
  composition, LVIII, 62–70
Medium MCDB 401, LVIII, 59, 87, 90, 91
  composition, LVIII, 62–70
Medium MCDB 411, LVIII, 57, 89
  composition, LVIII, 62–70
Medium MCDB 501, LVIII, 59, 87
  composition, LVIII, 62–70
Medium MD 705/1, LVIII, 57, 88
Medium 199 (Morgan), LVIII, 56, 84, 87, 90, 91, 129, 155
  composition, LVIII, 62–70
Medium MPNL 65/C, LVIII, 59
Medium N16, LVIII, 56
Medium NCTC 109, LVIII, 57, 89
Medium NCTC 135, LVIII, 57, 88
  composition, LVIII, 62–70
Medium RPMI 1640, LVIII, 56, 89, 91, 155, 213
  composition, LVIII, 62–70
Medium S77, LVIII, 546
Medium SM–201, LVIII, 59
Medium SM–20 (Halle), LVIII, 59
Medium V–605, LVIII, 56
Medium WO_5, LVIII, 56
Medusa aequorea, *see Aequorea aequorea*
Meganyctiphanes, photoprotein system, LVII, 355
Meicelase, in protoplast isolation, XXXI, 580, 581
MEK factor, *see* Molday, Englander and Kallen factor
MEK method, *see* Methyl ethyl ketone method
Melanin
  autotoxicity, LVIII, 567
  dispersion, cAMP derivatives, XXXVIII, 401
  in melanocytes, LVIII, 564, 565
  in pigmentary hormone assay, XXXVII, 121–130
  precursors, LVIII, 567
Melanin granule
  electron microscopy, XXXI, 393
  isolation, XXXI, 389–394
Melanocyte, LVIII, 564–570
Melanocyte-stimulating hormone
  bioassay, XXXVII, 121–130
  cAMP levels, LVIII, 567
  in melatonin bioassay, XXXIX, 384
  pineal effects, XXXIX, 382
  radioimmunoassay, XXXVII, 27, 32
α-Melanocyte-stimulating hormone
  bioassay, XXXVII, 129
  radioimmunoassay, XXXVII, 25, 28
  synthetic, use in hormone assay, XXXVII, 125, 126
β-Melanocyte-stimulating hormone
  bioassay, XXXVII, 129
  radioimmunoassay, XXXVII, 25, 34
  synthesis, XLVII, 584
Melanoma, LVIII, 376, 564–570
  amelanotic, LVIII, 566
  cell, 2′,3′-cyclic nucleotide-3′-phosphohydrolase, XXXII, 128
  cell line PG19, mouse, tumorigenicity testing, LVIII, 374
  cell lines, in neurobiological studies, XXXII, 769
  cloning, LVIII, 566
  effect of hormone, LVIII, 102, 103
  Folin reagent, LVIII, 568
  maintenance, XXXI, 390, 391
  media, LVIII, 565
  melanin formation, LVIII, 565
  mutants, LVIII, 565, 566
  pigment formation, LVIII, 565
Melanotropin, *see* Melanocyte-stimulating hormone
Melatonin
  bioassay, XXXIX, 384, 385
  biological effects, XXXIX, 377–382

biosynthesis and metabolism, XXXIX, 377–382
determination, XXXIX, 382–387
effect on melanin, XXXVII, 129
fluorometric determination, XXXIX, 382–384
gas chromatography-mass spectrometry, XXXIX, 385–387
in pineal, amounts, XXXIX, 380, 381
as pineal hormone, XXXIX, 376
radiolabeled, XXXIX, 380, 381
synthesis and metabolism, XXXIX, 395
synthesis, XXXIX, 376
in urine, isolation, XXXIX, 396, 397
Melibionate–albumin conjugate, preparation, L, 162
Melibiose, XXXIV, 333
chromatographic constants, XLI, 16
polarography, LVI, 477
Melilotate hydroxylase, *see* Melilotate 3-monooxygenase
Melilotate 3-monooxygenase, LII, 16
activity, LIII, 552, 556
from *Arthrobacter*, LIII, 552
assay, LIII, 552, 553
effectors, LIII, 557
flavin content, LIII, 556
kinetic properties, LIII, 552
optical properties, LIII, 556
oxy form, spectrum, LII, 36
oxygenated-flavin intermediates, LIII, 557, 558
from *Pseudomonas*, LIII, 552–558
purification, LIII, 553–556
purity, LIII, 556
reconstitution, LIII, 555
reduction, LIII, 557
stereospecificity determination, LIV, 229
substrate specificity, LIII, 556, 557
subunits, LIII, 556
uncoupling, LIII, 557
Melilotic acid
pseudosubstrate of melilotate hydroxylase, LIII, 557
substrate of melilotate hydroxylase, LIII, 552, 557
synthesis, LIII, 553
Melilotic lactone, *see* Dihydrocoumarin
Melilotus alba, oxidase, LII, 20
Melinacidin, XLIII, 144
Melizitose, chromatographic constants, XLI, 17
MEM, *see* Eagle's minimum essential medium
Membrane, XXXIV, 171–177, *see also* Biomembrane; specific type
affinity labeling, problems, LVI, 654, 655
alteration
in *E. coli* auxotroph, XXXII, 856–864
in yeast, XXXII, 838–843
antibodies in studies, XXXII, 60–70
autoxidation, proposed mechanism, LII, 305
bilayer, ionophores, LV, 439
biological, ionophores, LV, 439, 440
biophysical techniques in study, XXXII, 161–271
brain, XXXIV, 622, 623
carbodiimide-resistant mutants, preparation, LVI, 167, 168
cation and antibiotic effects, XXXII, 881–893
chaotropic ion destabilization, XXXI, 770–790
characterization, XXXII, 1–272
of *Chlamydomonas reinhardi,* modulation of, XXXII, 865–871
collagen, preparation, XLIV, 246, 247
component isolation, XXXII, 272–445
animals, XXXII, 275–391
plants, XXXII, 391–422
unicellular organisms, XXXII, 422–445
components other than ATPase, inhibitors interacting, LV, 495–515
detergents, LVI, 734, 736, 737, 748, 749
differential scanning calorimetry, XXXII, 262–272
effect
on electrode current, LVI, 458
of mechanical deformation, XLIV, 576

enzyme
artificial, XLIV, 901–929
experimental methods using, XLIV, 907–913
kinetics, XLIV, 913–919
oscillatory behavior, XLIV, 926, 927
preparation, XLIV, 902–907
gel electrophoresis, XXXII, 82–85
enzyme markers, XXXII, 393
epinephrine-sensitive, preparation, XXXVIII, 188, 189
exposed proteins, lactoperoxidase in determination, XXXII, 103–109
extraction of TF_1, LV, 782
fat cells, XXXIV, 655
fluorescent probe studies, XXXII, 234–246
fractionation, XXXIV, 177
fragment, of *M. phlei*, LV, 195, 196
freeze-fracture studies, XXXII, 35–44
glycoproteins, molecular weight, XXXII, 92–102
glycosphingolipids, XXXII, 345–367
hybrid, LVI, 234
impregnation method, XLIV, 246, 247
IR spectroscopy, XXXII, 252
lipid composition
manipulation in batch cultures, LVI, 570–573
in glucose-limited chemostat cultures, LVI, 573–577
lipid peroxidation, LII, 304–306
liver, XXXIV, 654, 689
marker enzymes, XXXII, 124–131
in microsomes, XXXI, 25
mitochondria, XXXIV, 566, 571
separation, LV, 88–98
model, XXXII, 483–554
using insolubilized protein crystals, XLIV, 558
for microencapsulation, XLIV, 203
system, electron spin resonance spectra, XLIX, 399–401, 405, 406
morphometry, XXXII, 1–20
mosaic, origin, LVI, 234
negative staining studies, XXXII, 20–35
nerve, XXXIV, 571
NMR studies, XXXII, 198–211
optical rotation data, XXXII, 220–233
optimal ratio to probes, LV, 580
for osmometer
commercial source, XLVIII, 137
pretreatment procedures, XLVIII, 137–139
particle, ATPase-depleted, preparation, LV, 796, 797
penicillin-binding proteins, XXXIV, 402
phase transitions, electron spin resonance studies, XLIX, 384, 385
phospholipids, XXXII, 131–271
placenta, XXXIV, 655
planar
chromatophore photosynthetic redox chain, LV, 756
cytochrome oxidase containing, LV, 755, 756
mixture of phospholipid, protein and decane, LV, 751, 752
proteoliposomes, LV, 756, 757
planar black, association of bacteriorhodopsin sheets, LV, 772
planar phospholipid, proteoliposomes associated, LV, 763, 766–768
polarographic determination of phase changes, XXXII, 258–262
preparation
fluorescence studies, LVI, 497, 498
respiratory deficient, LVI, 175, 176
unc mutants, LVI, 112
prostaglandin E receptor, XXXII, 109–123
purple, synthesis, LVI, 399, 400
radioiodination, XXXIV, 175
reconstitution, XXXII, 447–481
release of ATPase, LV, 803
residue, solubilization, LVI, 423–425
saturation-transfer studies, XLIX, 493, 494, 497–500
scanning electron microscopy, XXXII, 50–60
separation
by disc gel electrophoresis, XXXII, 105, 106

from endoplasmic reticulum, XXXIV, 175
sidedness of components, antibodies, LVI, 223–228
small-angle X-ray diffraction, XXXII, 211–220
sodium channels, XLVI, 288
spin-label measurements, XXXII, 161–198; XLIX, 422, 423, 427
subcellular, calcium in isolation, LII, 86, 87
thermal transition studies, XLIX, 12–14
thickness, equation, LV, 608
thyroid plasma, XXXIV, 694
topology, XXXIV, 177
use of mtDNA mapping in research, LVI, 194–197
Membrane electrode, main functions, LVI, 454, 455
Membrane factor
covalent binding to collagen membrane and reassociation with F_1, LV, 744–746
preparation, LV, 742–744
Membrane filtration
apparatus, LIX, 584
studies of protein-RNA interactions, LIX, 322–324, 331, 568, 569, 577, 578, 583–591
Membrane-ligand complex, XXXIV, 172
Membrane particle preparation, LI, 60, 61
Membrane potential, XLIV, 924, 925
brown adipose tissue mitochondria, LV, 72
definition, LV, 547
determination, LV, 556, 557
extrinsic probes, LV, 558, 559
intrinsic probes, LV, 557, 558
generation, LV, 548
principle, LV, 678
procedure, LV, 679
reagents, LV, 678
light-dependent, assay, LV, 779, 780
light-induced changes, LVI, 406, 407
maintenance, LVI, 254
measurement, LV, 451, 452
with amphiphilic spin labels, LVI, 515–517
experimental, LVI, 516–518
principle, LVI, 517
sample calculation, LVI, 526
neutral ionophores, LVI, 446–448
optical probes as monitors, LV, 569, 570
calibration, LV, 581–585
conclusions, LV, 585, 586
probe selection, LV, 570–574
probe sources, LV, 574
technique and instrumentation, LV, 574–581
protein-produced
apparatus, LV, 588, 589
applicability to natural membrane vesicles, LV, 595, 596
to reconstituted systems, LV, 595
electrolyte solutions, LV, 589
estimation of electromotive force and equivalent electric scheme, LV, 592–595
filters, LV, 589
phospholipid treatment of filters, LV, 589
principle, LV, 588
protein incorporation, LV, 589–591
rate of potential generation, LV, 592–595
stability of system, LV, 591, 592
reconstitution, LV, 586–603
synthetic penetrating ions as probes in proteoliposomes, LV, 763
transport-linked, LV, 656, 657
culture media, LV, 657, 658
electrophysiological techniques, LV, 658–662
flux techniques, LV, 663–666
typical values, LV, 555
use of cyanine dyes for determination, LV, 689, 690
calibration, LV, 691, 692
comments, LV, 694
instruments, LV, 690
optimization of procedure, LV, 690, 691
other classes of dyes, LV, 694, 695
precautions, LV, 692–694
reagents, LV, 690
reversibility, LV, 691
studies with single cells, LV, 695

Membrane process, reconstitution procedures, LV, 699–701
cholate-dialysis, LV, 701–703
detergent-dilution, LV, 703, 704
fusion, LV, 709
incorporation, LV, 707–709
regeneration systems, LV, 710, 711
sonication, LV, 704–707
Membrane protein, XXXII, 70–131
integral, orientation, LIV, 56
intrinsic
M. phlei, LV, 188
purification, LV, 195
molecular weight, XXXII, 92–102
outer, iron-sulfur center
EPR characteristics, LIV, 140, 141, 144, 145
EPR spectrum, LIV, 148–150
stoichiometry, LIV, 145
problems in study, XXXII, 407–415
purification, XXXII, 406–422
reconstitution of electrogenic function
association of bacteriorhodopsin sheets with planar black membranes, LV, 772
planar membranes made of mixture of phospholipid, protein, and decane, LV, 751–772
proteins situated on hydrocarbon water interface, LV, 773–776
Membrane receptor, XXXIV, 171–177
Membrane resistance, measurement, LV, 661
Membrane transport, XLVI, 167
in perfused liver, XXXIX, 34
Membrane vesicle, XXXIV, 173
bacterial, binding of D-lactate dehydrogenase, LIII, 525, 526
capable of alanine transport, reconstitution, LVI, 433, 434
colicins, LV, 536
E. coli, preparation, LV, 677
Halobacterium
preparation, LVI, 401, 402
properties and handling, LVI, 402–404
potassium-loaded
principle, LV, 677, 678
procedure, LV, 678
reagents, LV, 678
preparation, LV, 359, 360
for measurement of calcium transport, LVI, 233–241
reacted with ferritin-conjugated immunoglobulin, electron microscopy, LVI, 228
Membrane viscosity, protein concentration, LIV, 61
MEM Eagle, Spinner modified medium, XXXIX, 21
Memory, in artificial enzyme membranes, XLIV, 925, 926
Menadione, XLIV, 188; XLVI, 76, *see also* Vitamin K
electron acceptor, for dihydroorotate dehydrogenase, LI, 64
photochemically active derivative, LVI, 656
substrate, of NADH dehydrogenase, LIII, 18, 20
Menaquinone
accumulation, LVI, 110
biosynthesis, deficient mutants, LVI, 107, 108
cofactor for dihydroorotate dehydrogenase, LI, 63
difference extinction coefficient, LIII, 587, 588
extinction coefficient, LIII, 587
M. phlei ETP, LV, 184, 185
protoplast ghosts, LV, 182
redox reactions, determination in extracts, LIII, 585–589
in membranes, LIII, 589–591
thin-layer chromatography, LVI, 109
Meningococcal group B polysaccharide, XXXIV, 716
Meningococcal polysaccharide, *see* Polysaccharide, meningococcal
(−)-Menthyl chloroformate
2-hydroxy fatty acids, menthyloxycarbonyl derivatives, XXXV, 326–331
preparation, XXXV, 326
MEOS, *see* Microsome, ethanol oxidizing system
MERCAP, *see* 2-Mercaptobenzothiazole
Mercaptan, XLVI, 317

Mercaptoacetic acid, XLVII, 458, *see also* Thioglycolic acid
Mercapto-agarose gel
preparation, XLIV, 41, 42
storage, XLIV, 41
Mercaptoalbumin, from bovine plasma, protein interaction kinetics, XLVIII, 308
p-Mercaptobenzoic acid, substrate of *p*-hydroxybenzoate hydroxylase, LIII, 549, 551
Mercaptobenzothiazole, as antioxidant, XXXI, 538
2-Mercaptobenzothiazole, XLVI, 540
in plant-cell fractionation, XXXI, 592, 596
plant mitochondria, LV, 24
4-Mercaptobutyrimidate, *see* 2-Iminothiolane
8-Mercapto-cyclic adenosine monophosphate, activity, XXXVIII, 401
5-Mercaptodeoxyuridine, substrate of thymidine kinase, LI, 358
Mercaptoethanesulfonic acid, XLVII, 452, 456
Mercaptoethanol, XLVI, 402, 451; LIX, 235
aminoacylation of tRNA, LIX, 216, 298
analysis of ribosomal proteins, LIX, 507, 509
assay of malate, LVII, 187
of oxaloacetate, LVII, 187
cell-free protein synthesis system, LIX, 300, 448, 459
disadvantage, purification of tryptophanyl-tRNA synthetase, LIX, 242
electrophoretic separation of ribosomal proteins, LIX, 431
elution of protein from gels, LIX, 544, 545
in gel electrophoresis, XXXII, 79, 97
inhibitor of formate dehydrogenase, LIII, 371
of *Pholas* luciferin oxidation, LVII, 399, 400
initiation factor buffer, LIX, 838
isolation of chloroplastic ribosomes, LIX, 434, 435, 437
of chloroplasts, LIX, 434
ribosomes, LIX, 373
modification of cytochrome *c*, LIII, 175
periodate oxidation of tRNA, LIX, 269
phosphodiesterase, XXXVIII, 234, 235
in plant-cell fractionation, XXXI, 505, 603, 612
preparation of aminoacyl tRNA synthetase, LIX, 314, 315
of *Bacillus* ribosomes, LIX, 438
of chloroplastic tRNA, LIX, 210
of cross-linked ribosomal subunits, LIX, 535
of dialysis tubing, LVII, 10
of *E. coli* crude extract, LIX, 297
of *E. coli* ribosomal subunits, LIX, 444, 554
of *E. coli* ribosomes, LIX, 444, 445, 451, 554, 647, 752, 817, 838
of labeled oligonucleotides, LIX, 75
of mitochondrial tRNA, LIX, 206
of postmitochondrial supernatant, LIX, 515
of radiolabeled ribosomal protein, LIX, 777, 783
of ribosomal protein, LIX, 453, 471, 484, 490, 557, 559, 649, 837
of ribosomal subunits, LIX, 403, 404, 440, 505
of RNase III, LIX, 826, 827
of S100 extract, LIX, 357
of 16 S RNA, LIX, 648
purification of blue-fluorescent protein, LVII, 227, 230
of chloroplastic phenylalanine tRNA, LIX, 213
of CTP(ATP):tRNA nucleotidyltransferase, LIX, 123, 124, 125
of cytochrome c_1, LIII, 183, 186, 223
of cytochrome oxidase subunits, LIII, 79

of *E. coli* endogenous polysomes, LIX, 354
of glycine reductase, LIII, 379
of high-potential iron-sulfur protein, LIII, 337
of hydrogenase, LIII, 311, 313
of ω-hydroxylase, LIII, 358
of D-lactate dehydrogenase, LIII, 520, 522
of luciferin sulfokinase, LVII, 249
of nitric oxide reductase, LIII, 645
of peptide constituents of cytochrome bc_1 complex, LIII, 106, 107
reactivation of bacterial luciferase, LVII, 176, 180
reconstitution of ribosomal subunits, LIX, 441, 447, 451, 652, 654, 837, 864
reduction of cross-linked ribosomal proteins, LIX, 539
regeneration of aequorin, LVII, 289
removal, LIX, 387
of protein from polyacrylamide gels, LIX, 313
reverse salt gradient chromatography, LIX, 215, 216
separation of aminoacylated tRNA, LIX, 216
stabilizer, of luciferin sulfokinase, LVII, 247
studies of protein-RNA interactions, LIX, 562, 563
2-Mercaptoethanol, XXXIV, 547, 554; XLIV, 81, 457, 494, 845; XLVII, 17, 115, 458; XLVIII, 53, 55, 75
activator of orotidylate decarboxylase, LI, 74, 75
of pyrimidine nucleoside monophosphate kinase, LI, 330, 331
assay of dCTP deaminase, LI, 418, 419
of cytidine triphosphate synthetase, LI, 85
of deoxycytidylate deaminase, LI, 413
of nucleoside diphosphokinase, LI, 372
of orotate phosphoribosyltransferase, LI, 70
of uridine phosphorylase, LI, 425
in buffers, for camphor 5-monooxygenase system isolation, LII, 172, 183
dissociation of aspartate transcarbamoylase, LI, 36
disulfide bond reduction, XLVII, 198, 199
electrophoresis buffer, LI, 342, 357
Enzyme Buffer, LI, 14
for enzyme storage, LI, 82, 362
inhibitor of GMP synthetase, LI, 223
interference, in hemoprotein staining, LII, 330
in (±)-L-methylenetetrahydrofolate preparation, LI, 91
protein determination, LI, 339
purification, XLVII, 393
of adenine phosphoribosyltransferase, LI, 575, 576
of adenylate deaminase, LI, 492, 495, 499, 500
of adenylate kinase, LI, 461
of adenylosuccinate AMP-lyase, LI, 205
of amidophosphoribosyltransferase, LI, 174, 175, 176
of aspartate carbamyltransferase, LI, 45, 53
of cytidine triphosphate synthetase, LI, 81, 86
of dCTP deaminase, LI, 420, 421, 423
of deoxycytidine kinase, LI, 339, 340, 341
of deoxycytidylate deaminase, LI, 414–417
of deoxythymidine kinase, LI, 362, 363
of deoxythymidylate phosphohydrolase, LI, 287, 288
of erythrocyte enzymes, LI, 584
of guanine deaminase, LI, 513, 514, 515, 517
of phosphoribosyltransferases, LI, 555
of purine nucleoside phosphorylase, LI, 521, 523, 524, 527, 528, 530

of pyrimidine nucleoside monophosphate kinase, LI, 324, 325, 326, 328
of ribonucleoside triphosphate reductase, LI, 251, 252
of thymidine kinase, LI, 357, 358
of thymidine phosphorylase, LI, 440, 441, 442
of thymidylate synthetase, LI, 92, 94, 95, 101
of uridine phosphorylase, LI, 427, 428, 429
as putidaredoxin stabilizer, LII, 172, 183, 184, 187
reductive denaturing agent, XLIV, 519, 522
removal, LI, 326, 328
substrate, of dimethylaniline monooxygenase, LII, 151
2-Mercaptoethylamine, in hybrid antibody preparation, XXXII, 64
2-Mercaptoimidazole, inhibitor of horseradish peroxidase, LVII, 404
3-Mercaptopropioimidate, XLIV, 42
β-Mercaptopropionic acid, purification of purine nucleoside phosphorylase, LI, 523, 524
γ-Mercaptopropyltriethoxysilane, for support of immobilized enzymes, LVI, 490, 491
γ-Mercaptopropyltrimethoxysilane, XLIV, 146
6-Mercaptopurine
analogs of nucleotides, XLVI, 289
substrate of adenine phosphoribosyltransferase, LI, 572
of hypoxanthine phosphoribosyltransferase, LI, 557
6-Mercaptopurine ribonucleoside, XLVI, 300
6-Mercaptopurine riboside 5′-monophosphate, XXXIV, 242
2-Mercapto-1-(β-4-pyridethyl)benzimidazole, RNA inhibitor, XL, 282
2-Mercaptosuccinic acid, in cytochrome *m* crystallization, LII, 183
Merck DC Fertigplatten Kieselgel F 254, LIV, 50
Merck gel 2000, XLVII, 83
Mercurial, LIII, 20
complex I or V, LVI, 586
inhibitors of complex I, LIII, 13
of complex I–III, LIII, 8
of NADH dehydrogenase, LIII, 20
transport, LVI, 248
Mercurial-agarose, XXXIV, 244, 546, *see also* Organomercurial-agarose
p-Mercuribenzoate, XLVI, 58
in hemoglobin-ligand binding studies, LIV, 531
inhibitor of aspartate carbamyltransferase, LI, 35, 40, 48
of cytidine deaminase, LI, 401
of cytosine deaminase, LI, 399
of uridine phosphorylase, LI, 429
phosphate transport, LVI, 353–355, 357
Mercuribenzoate-agarose column, purification of glycine reductase, LIII, 379
Mercuric chloride, XLIV, 845
for dithiol removal, LI, 250
in Parducz fixative, XXXII, 55
preparation of zinc amalgam, LVII, 434
S-Mercuric *N*-dansylcysteine, XLVII, 419, 420
Mercuric derivative steroid, XXXVI, 422–426
Mercuric ion, XLVII, 408
inhibitor of adenine phosphoribosyltransferase, LI, 573
of aspartate carbamyltransferase, LI, 48
of cytidine deaminase, LI, 411
of deoxythymidine kinase, LI, 365
of deoxythymidylate phosphohydrolase, LI, 289
of orotidylate decarboxylase, LI, 79
of rat liver acid nucleotidase, LI, 274
4-Mercuriestradiol-agarose, XXXIV, 687
Mercuriestradiol derivative, XXXIV, 682
estrogenic activity, XXXVI, 405–408
synthesis, XXXVI, 376–378
uses, XXXVI, 388–390, 393

Mercurinitrophenol, fluorescent acceptor, XLVIII, 378
p-Mercuriphenyl sulfonate
inhibitor of uridine phosphorylase, LI, 429
storage, LVI, 615
Mercurous ion, microsome aggregation, LII, 87
Mercury
as anode, LVI, 453, 454
effect on luciferin light emission, LVII, 3
as immunochemical marker, XXXVII, 133
volatilization losses, in trace metal analysis, LIV, 466, 480
Mercury-199, lock nuclei, XLIX, 349; LIV, 189
Mercury arc, LIV, 34
interference optics, XLVIII, 189, 190
Mercury lamp, LIII, 590
photoaffinity labeling and, LVI, 664, 665
Meristem, vacuoles, isolation, XXXI, 574, 575
Merocyanine, membrane potentials, LV, 559, 571, 573
Merocyanine-540, membrane potential, LV, 694
Meromyosin, XXXIV, 479, 490; XLIV, 882; XLVI, 275, 276, 318–321
Merrifield postulate, LVII, 518
Merrifield resin, *see* Polystyrene, chloromethylated
Merrifield solid phase method of hormone synthesis, XXXVII, 416, 417
Mersalyl, *see also* Sodium mersalyl
carnitine-acylcarnitine translocase, LVI, 374–376
cell fractionation, LVI, 217
inhibitor of aspartate carbamyltransferase, LI, 48
of CPSase-ATCase complex, LI, 118
of NADH dehydrogenase, LIII, 20
NADH dehydrogenase spectral properties, LIII, 19
transport, LVI, 292
Merthiolate, preservative, XLIV, 716
MES buffer, *see* 2-(*N*-Morpholino)ethanesulfonic acid
Mescaline, salivary secretion effects, XXXIX, 475
Mesenchymal cell, feeder layer, LVIII, 270
Mesitylenecarboxylic acid, inhibition studies of tryptophanyl-tRNA synthetase, LIX, 255
Mes-Na^+ buffer, *see* 2-(*N*-Morpholino)ethanesulfonate-sodium hydroxide buffer
Me_2SO, *see* Dimethyl sulfoxide
Mesophyll cell
leaf, LVIII, 360
pea, LVIII, 361
of plants, XXXI, 493, 499
Messenger ribonucleic acid, XXXIV, 299
analogs, XLVI, 621–626
binding site, XLVI, 621–626
eukaryotic
binding to initiation factor eIF-2, LX, 387–389, 395, 396, 399–401
to protein, LX, 395
completeness, LX, 398
specificity, LX, 398, 399
to ribosomes, LX, 351–360, 408–410
assay, LX, 355–360, 412–414
number of sites, LX, 410–417
to 40 S ribosomal subunits, LX, 408–410
bonded to cellulose, purification of initiation factor eIF-2, LX, 247–255
capped messages, LX, 248, 361, 363, 364, 372–375, 381, 382, 390
distribution in reticulocyte lysate, LX, 85–87
globin
assay of initiation factors, LX, 17, 18, 32, 33, 121, 122
isolation, LX, 383, 384, 402
purification, LX, 69, 78, 125, 402, 403
translation, stimulation by polyamines, LX, 555

histone, binding to ribosomes, LX, 353, 357, 358
hybridization with translational control ribonucleic acid, LX, 548
Mengo virus, LX, 382, 385, 389, 390
message specificity
analysis, LX, 375–380
effect of nonspecific initiation factor on concurrent translation, LX, 375–378
initiation vs. elongation, mathematical treatment, LX, 376–378
use of elongation inhibitors, LX, 379
ovalbumin, purification, LX, 403
polyadenylate-rich, preparation, LX, 394, 395
preparation of labeled
iodine-125, LX, 383–387, 404, 405
phosphorus-32, LX, 362–364
tritium, LX, 353–355
protected fragments as binding sites, LX, 360, 361
confirmation of activity, LX, 373–375
fractionation, LX, 369
preparation, LX, 364–369, 372
sequence analysis, LX, 369–373
vesicular stomatitis virus, LX, 353–355, 358–360
fiberglass filters, XXXIV, 470
globin, XXXIV, 472
isolation with poly(U)-agarose, XXXIV, 496–499
mitochondrial, LVI, 9, 10
oligothymidylic acid-cellulose, XXXIV, 472
polyacrylamide, XXXIV, 471
polynucleotide-cellulose, XXXIV, 472
polysomal, XXXIV, 496
prokaryotic
aminoacylation, LX, 62, 622, 623
binding to 30 S ribosome subunit, LX, 14, 15, 218–221
fractionation, LX, 324
70 S initiation complex from, LX, 313, 317, 325–328
MS2
assay for ribosome protein S1, LX, 453, 454
control of formylmethionyl-transfer ribonucleic acid binding, LX, 431
preparation of ^{32}P-labeled, LX, 315, 316, 322–324, 345
protected fragments as binding sites, LX, 311–313, 318–321, 343, 344
characterization, LX, 347–350
fractionation, LX, 328–331
preparation, LX, 345–347
sequence analysis, LX, 320, 321, 329
sites binding initiation factor IF-3, LX, 343–350
ternary complex with initiation factor IF-3 and 30 S ribosomal subunits, LX, 443–446
in protein turnover, XXXVII, 248
R17
effect on initiation complex formation, LX, 333–338, 340–343
initiation factor assay, LX, 9, 11
preparation, LX, 339
stimulation of binding, XLVI, 630
in studies of codon-anticodon recognition, LIX, 292–309
Messenger RNA, *see* Messenger ribonucleic acid
Mesyl chloride, *see* Methanesulfonyl chloride
Metabisulfite, as antioxidant, XXXI, 538, 592
Metabolic clearance rate, of blood steroids, XXXVI, 69–71
Metabolic cooperation, *see* Contact feeding
Metabolic Shaker, XXXII, 734
Metabolism
insect flight, glycerol-3-phosphate dehydrogenase role, XLI, 240, 245
regulation, application of digitonin method and cavitation procedure, LVI, 220, 221

Metabolite
assay in nonaqueous tissue fractions, LVI, 204, 205
calculation of content of subcellular fractions, LVI, 205, 206
averaging of results, LVI, 206
intramitochondrial, labeling, LVI, 281
uptake, measurement, LVI, 288, 289
Metabolite indicator method, LVI, 214
cellular compartments, LVI, 207
Metachromatic leukodystrophy
arylsulfatase A, L, 471–474
enzymic diagnosis, L, 471–474
glycosphingolipids, L, 247
lactosylceramide sulfate, L, 472
multiple sulfatase deficiency, L, 454
Metal, heavy, *see* Heavy metal
Metal ion, *see also* specific metal
ATPases, LV, 301
indicators, criteria for selection
affinity, LVI, 304, 305
binding and penetration, LVI, 305, 306
interferences, LVI, 307
response time, LVI, 307
side effects, LVI, 306
specificity and selectivity, LVI, 306, 307
inhibitor
of carboxypeptidase Y, XLVII, 89
of collagen-bound enzyme, XLIV, 261
site, XLVI, 548
support pretreatment, XLIV, 155
Metal-linked enzyme, XLIV, 166–169
Metallochromic indicator, calcium, LVI, 303, 304, 445, 446
Metallocytochrome *c*, *see also* specific type
preparation, LIII, 177, 178
properties, LIII, 178–181
Metalloendopeptidase, XXXIV, 435–440, *see also* Thermolysin
neutral, XXXIV, 435–440
Metalloenzyme, magnetic circular dichroism, XLIX, 166–179
Metalloporphyrin
preparation, LII, 490
resonance Raman spectroscopy, LIV, 238
Metalloprotein, *see also* specific protein
nonheme, Raman spectra, XLIX, 137–143
Metaphase cell
evaluation, LVIII, 172–174
preparation, LVIII, 170
staining, LVIII, 172–174
Methacrylanilide, component of copolymers, XLIV, 71
Methacrylic acid
in copolymer, XLIV, 57, 84, 107, 108, 112, 113
in tissue embedding, XXXVII, 139, 144
Methacrylic acid *m*-fluoranilide, in copolymer, XLIV, 84, 107, 108, 112
Methacrylic acid/3-fluorostyrene copolymer
nitration, XLIV, 113
papain immobilization, XLIV, 113
polymerization, XLIV, 113
Methacrylic acid/3-fluorostyrene/divinyl benzene copolymer, XLIV, 84
Methacrylic acid/4-fluorostyrene/divinyl benzene copolymer, XLIV, 84
Methacrylic acid/*m*-isothiocyanatostyrene/divinyl benzene copolymer, XLIV, 84
Methacrylic acid/methacrylic acid 3-fluoroanilide copolymer
nitration, XLIV, 112
papain immobilized, XLIV, 112
polymerization, XLIV, 112
Methacrylic acid/methacrylic acid *m*-fluoroanilide/divinyl benzene copolymer, XLIV, 84
Methacycline hydrochloride, XLIII, 385
Methane, as contaminant, LVIII, 202
Methane oxidase, LII, 20
Methanesulfonic acid, XLVII, 452, 455
for gel solubilization, LII, 517
intermediate in phosphatidylcholine synthesis, XXXV, 444, 450
peptide hydrolysis, XLVII, 616
Methanesulfonyl chloride, affinity labeling, XLVII, 428
Methanobacterium, hydrogenase reaction, LIII, 287
Methanococcus vannielii, formate dehydrogenase, LIII, 372

Methanol, XLIV, 58, 112, 113, 116, 121, 122, 127, 130, 133, 146, 175, 186, 231, 324, 385, 386, 446, 474, 716, 834; XLVII, 89
in affinity gel preparation, LI, 366
alcohol oxidase, LVI, 467, 483
benzo[*a*]pyrene recrystallization, LII, 237
for chromatographic separation of nucleosides and nucleotides, LI, 308, 372, 373, 419, 509
crystallization of quinone intermediates, LIII, 605–607
cytochrome *P*-450 assay, LII, 125, 126
effect on EPR signals, LIV, 150
on Tygon tubing, LVII, 460
electron spin resonance spectra, XLIX, 383, 401, 402
elution of polycyclic aromatic hydrocarbons, LII, 283–285
7-ethoxycoumarin crystallization, LII, 374
extraction nucleic acids, LIX, 311
extraction of NAD, LIII, 582
in 5-fluoro-2′-deoxyuridine 5′-(*p*-aminophenyl phosphate) synthesis, LI, 99
formaldehyde binding, LII, 344
gas chromatography, LII, 332
gel staining solution, LIX, 540, 560
ionophores, LV, 448
in low-temperature solvent mixture, LIV, 105, 108
in oxygen consumption studies, LIV, 487, 492, 493
phosphorescence studies, XLIX, 238
phosphorylating vesicles, LV, 166
porphyrin ester synthesis, LII, 432
postcoupling wash, XLVII, 326
preparation of carbamoylthiocarbonylthioacetic acid, LVII, 17
of ribosomal RNA, LIX, 446
proteolipid extraction, LV, 416
purification of *Cypridina* luciferin, LVII, 367, 368
in scintillation cocktail, LI, 4
solvent of *p*-anisidine, LVII, 19
of *p*-azidophenacyl bromide, LIII, 175
of 2-carbamoyl-6-methoxybenzothiazole, LVII, 19
2-cyano-6-hydroxybenzothiazole, LVII, 21
of duroquinone, LIII, 90
luciferin, LVII, 110
of organic hydroperoxide, LII, 510
of phenylmethylsulfonyl fluoride, LIII, 520
trichloroacetamide, LVII, 17
of ubiquinone isoprenologs, LIII, 14
in solvent system of peroxyoxalate chemiluminescence reactions, LVII, 458, 460
for storage of cellogel strips, LIX, 65
support swelling, XLVII, 272, 298
synthesis of chemiluminescent aminophthalhydrazides, LVII, 431, 435, 436, 437
of dithiodiglycolic dihydrazide, LIX, 383, 384
thioacetylation, XLVII, 292, 298
wash solution, for gel electrophoresis, XXXII, 96
as water space marker, LV, 555
Methanol-acetic acid
for destaining, LIX, 509
in staining solution, LIX, 560
Methanol-benzene, in carboxyl group titration, XLIV, 384
Methanol-chloroform
isolation of radiolabeled quinone intermediate, LIII, 609
purification of *Diplocardia* luciferin, LVII, 377
Methanol/ethyl acetate, solvent system, for luciferin, LVII, 24
Methanol-hydrochloride, synthesis of luciferyl sulfate analogs, LVII, 251
Methanol-light petroleum, extraction of quinones, LIII, 575, 580, 581
Methanol oxidase, in microbodies, LII, 496
Methanol-phosphate buffer, isolation of benzyl luciferyl disulfate, LVII, 252
Methanol-sulfuric acid, in sirohydrochlorin esterification, LII, 444, 445

Methanol-water, for extraction of linoleic acid hydroperoxide, LII, 412
Methemalbumin, substrate of heme oxygenase, LII, 368, 369, 371
Methemoglobin
membrane osmometry, XLVIII, 75, 76
preparation, LII, 459
reduction system, in erythrocytes, LII, 463
spin state determination, from resonance Raman bands, LIV, 243, 244
substrate of heme oxygenase, LII, 371
X–ray absorption spectroscopy, LIV, 332
Methemoglobin reductase system, LIV, 199
5,10-Methenyltetrahydrofolic acid, acetate formation pathway, LIII, 361
Methicillin, XLIII, 243; XLVI, 535, 536
Methimazole, substrate of dimethylaniline monooxygenase, LII, 142, 143
Methionine, XLIV, 12–15
aminoacylation, isomeric specificity, LIX, 274, 275, 278–280
carbonium ion scavenger, XLVII, 604
cytochromeless strains, LVI, 122
heme-deficient mutants, LVI, 558–560
hydrogen bromide cleavage, XLVII, 547
hydrophobicity, XLIV, 14
in media, LVIII, 53, 62
misactivation, LIX, 289, 291
modification, XLVII, 445, 448, 453–459, 614, 615
peptides, XXXIV, 184
percent in protein, XLIV, 13
polarography, LVI, 477
radioactive aureovertin, LV, 480, 481
radiolabeled
binding by reticulocyte lysate components, LX, 75–80, 86
monitoring of protein synthesis, LX, 391
transfer ribonucleic acid charging, LX, 4, 5, 17, 92, 93, 112, 140, 216, 240, 258, 276, 277, 282, 316, 339, 561, 562, 567, 568, 579, 580
recovery, XLVII, 371
residue, modification, LIII, 138, 139, 149–151, 171
side chain reactivity, XLIV, 14
solid-phase peptide synthesis, XLVII, 614, 615
thermolysin hydrolysis, XLVII, 175
[^{3}H]Methionine, for *in vivo* labeling of ribosomal proteins, LIX, 517
L-Methionine, XLIII, 487
in affinity labeling of steroids, XXXVI, 386, 387
enzyme active site studies, XXXVI, 390
isolation, XLIV, 757
[^{35}S]Methionine, for *in vitro* labeling of ribosomal proteins, LIX, 517
Methionine carboxymethyl sulfonium salt, XLVI, 156
Methionine-nitrophenylsulfonyl derivative, XXXIV, 510
Methionine peptide, XXXIV, 192
Methionine tRNA synthetase, sequence analysis, XLVII, 331
Methionine sulfone, XLVII, 455
Methionine sulfoxide
determination, XLVII, 454–456
reduction, XLVII, 458, 459
tryptophan modification, XLVII, 445, 452
Methionine sulfoximine phosphate, XLVI, 26
Methionylhexamethylenediamine-agarose, XXXIV, 506
Methionylpuromycin
assay of initiation factors, LX, 18, 19, 38, 41, 119, 120, 169, 170, 208, 571
formation, LX, 575, 576
Methionyl-tRNA deacylase, LX, 261, 262
Methionyl-tRNA synthetase, XXXIV, 506–513
purification procedures, XXXIV, 511, 512
spacer, XXXIV, 508
subcellular distribution, LIX, 233, 234
Methionyl-tRNA transferase, atomic emission studies, LIV, 459

Methionyl-transfer ribonucleic acid
  assay
    binding to 40 S and 80 S ribosomes, LX, 256, 258–265, 562, 563, 569–571, 574
    initiation factors eIF, LX, 18, 19, 33, 37–39, 93, 94
  complex, 48 S, with initiation factors in presence of edeine, LX, 79, 82, 84, 86, 87
  methionylpuromycin assay for initiation factor eIF-2, LX, 119, 120, 169, 170
  preparation, LX, 69, 112, 140, 216, 240, 258, 276, 277, 282, 561, 567, 568, 579, 580
  reticulocyte lysate, LX, 75–79
  ternary complex
    with adenyluridylyl guanylate and 40 S ribosome subunits, LX, 37, 41, 49, 60, 61
      assay for EIF-2, LX, 37, 38
        for eIF-2, LX, 184, 185, 190
        for eIF-5, LX, 38, 50–52
        for eIF-2A, LX, 183, 190
    with guanosine triphosphate and initiation factor eIF-2, LX, 32, 33, 36, 41, 58, 59, 245–247, 273, 399–401, 562–565, 570
      adenylate energy charge requirement, LX, 585–586
      binding to ribosomes, LX, 285, 286, 289, 290, 573
        to 40 S ribosome subunit, LX, 275–280, 283–285, 309, 310, 586, 587
      stability, LX, 74–76
      use in assay for initiation cofactor Co-EIF-1, LX, 53, 184
        for initiation factor EIF-2, LX, 37, 43, 46–48
        for initiation factor eIF-2, LX, 36, 37, 39, 118, 119, 144, 168, 169, 183, 184, 190, 249, 251, 309, 381, 382, 468, 469
Methocel, *see* (Hydroxymethyl)cellulose
Methotrexate, resistance, LVIII, 309, 314
Methotrexate-agarose, XXXIV, 272–281
Methotrexate-AH-agarose, XXXIV, 276
5-Methoxy-*N*-acetyltryptamine, *see* Melatonin
p-Methoxybenzenethiol, identification of iron-sulfur center, LIII, 273
4-Methoxybenzoate *O*-methyl hydroxylase, LII, 12
p-Methoxybenzyl ester moiety, peptide synthesis, XLVII, 522
2-Methoxyethylmercaptan, XLVII, 393
Methoxyhydroxybenzoic acid, in enzyme assay, XXXII, 778
Methoxyindole, melatonin as prototype, XXXIX, 377
5-Methoxyindole-2-carboxylic acid, as enzyme inhibitor, XXXVII, 294
6′-Methoxyluciferin, structural formula, LVII, 27
2,3-Methoxy-5-methyl-6-pentyl benzohydroquinone, substrate of cytochrome bc_1 complex, LIII, 117
7-Methoxynaphthalene-1,2-dicarboxylic anhydride
  synthesis, LVII, 435, 436
    of 7-hydroxynaphthalene-1,2-dicarboxylic anhydride, LVII, 436
4-[4-Methoxyphenyl]butyric acid
  synthesis, LVII, 435
    of methyl 4-(4-methoxyphenyl)butyrate, LVII, 435
4-(4-Methoxyphenyl)-4-oxobutyric acid
  synthesis, LVII, 434
    of 4-[4-methoxyphenyl]butyric acid, LVII, 434
8-Methoxypsoralen, XLIII, 36
4-Methoxythiooxanilamide
  chromatographic properties, LVII, 26
  preparation of luciferin, LVII, 16
  spectral properties, LVII, 26
  structural formula, LVII, 18
(−)-α-Methoxy-α-trifluoromethylphenylacetyl chloride, in BP metabolite optical purity studies, LII, 290, 291
Methoxytryptamine
  fluorometric assay, XXXIX, 384
  gas chromatography-mass spectrometry, XXXIX, 386, 387
5-Methoxytryptophol, pineal synthesis, XXXIX, 376

Methsuximide, detection in biological samples, LII, 336
N-Methylacetamide, Raman bands for amide groups, XLIX, 100
2-(N-Methylacetamido)-2-deoxy-1,3,4,5,6-penta-O-methyl-D-glucitol, L, 21
Methyl acetate, permeability, XLIV, 302
Methylacetimidate, amidination, LVI, 629
N-Methylacridine, acceptor, in chemiluminescent reaction, LVII, 563
N-Methylacridinium-agarose, XXXIV, 573
Methyladenosine, chromatographic mobility, LIX, 73
N^6-Methyladenosine, inhibitor of adenosine deaminase, LI, 507
N'-Methyladenosine, inhibitor of adenosine deaminase, LI, 507
1-Methyladenosine
  chromatographic mobility, LIX, 73
  conversion to 6-methyladenosine, LIX, 96–98
  electrophoretic mobility, LIX, 91
  identification, LIX, 87, 92, 94, 95, 98
2′-O-Methyladenosine, inhibitor of tryptophanyl-tRNA synthetase, LIX, 252
3′-O-Methyladenosine, inhibitor of tryptophanyl-tRNA synthetase, LIX, 251
6-Methyladenosine
  chromatographic mobility, LIX, 73
  electrophoretic mobility, LIX, 91
  identification, LIX, 98
  NMR studies, LIX, 57
1-Methyladenosine 5′-monophosphate, thin-layer chromatographic mobility, LIX, 73
3′-O-Methyladenosine triphosphate, inhibitor of tryptophanyl-tRNA synthetase, LIX, 251
Methylamine
  alkylamine conversion, XLVII, 375–377, 384, 385
  measurement of ΔpH, LV, 562, 563
  preparation of ribosomal protein, LIX, 557, 649
  source, XLVII, 382
  synthesis of N-methyl-4-nitrophthalimide, LVII, 427
Methylamine oxidase, LII, 21
N-Methylamino-acid oxidase, LII, 18; LVI, 475
p-Methylaminobenzoate, XXXIV, 428
1-Methyl-9-[N^{β}-(ε-aminocaproyl)-β-aminopropylamino]acridinium bromide, XXXIV, 574
5-Methylaminomethyl-2-thiouracil, occurrence, LIX, 223, 224
6-Methylaminopurine ribonucleoside, substrate of adenosine deaminase, LI, 511
17α-Methylandrosta-1,4,6-trien-17β-ol-3-one, synthesis, XXXVI, 424, 425
17α-Methylandrost-5-ene-3β,17β-diol, cytochrome P–450 difference spectra, LII, 268
9-Methylanthracene, chemiluminescence reaction, LVII, 507
N-Methylanthranilic acid, modification of tRNA, LIX, 156
Methylation, L, 3–33
  of ceramide-1-phosphoryl-N,N-dimethylethanolamine, XXXV, 540
  of phosphatidyl-N,N-dimethylcholine, XXXV, 536
  reductive
    of initiation factor IF-3, LIX, 784–787, 790, 791
    labeling initiation factors, LX, 13, 34, 35, 68, 122
    optimum experimental parameters, LIX, 784–789
    of ribosome, protein, LIX, 777–779, 783–787, 790, 791
N-Methyl-N-(2-p-azidophenylethyl)norlevorphanol, XLVI, 604, 606
N-Methyl benzothiazolone hydrazone hydrochloride, XL, 362
L(+ or −)-2-α-Methylbenzylamine, XLIX, 437
Methylbenzyl group, sulfhydryl group blocking, XLVII, 539, 614
N-Methyl-N-benzylpropynylamine, *see* Pargyline
Methyl bis(3-chloroethyl)amine, XLIII, 33, 34
Methyl bromopyruvate, preparation of dehydroluciferin, LVII, 25

2-Methyl-2-butanol, electron spin resonance spectra, XLIX, 383
3-Methyl-1-butanol, *see* Isoamyl alcohol
Methyl butyrimidate, amidination, LVI, 629
O-Methylcaprolactim, XLIV, 126
3-Methylcarboxin, kinetic studies of succinate dehydrogenase, LIII, 411
4-Methylcatechol, in polyphenol oxidase assay, XXXI, 534
Methyl Cellosolve
  phthalic anhydride solvent, XLIV, 632
  as uncoupler solvent, LV, 464
Methylcellulose, in media, LVIII, 70
3-Methylcholanthrene
  activities induced, LII, 65, 67, 68, 73, 203, 328, 329, 373, 377
  enzyme induction, XXXIX, 39
  injections, for liver enzyme induction, LII, 118
  metabolism, LII, 65, 67
3-Methylcholanthrene 11,12-oxide, substrate of epoxide hydrase, LII, 199, 200
Methylcyclohexane
  in freezing mixture, LIII, 489
  substrate of ω-hydroxylase, LIII, 360
4-*N*-Methylcytidine, substrate of cytidine deaminase, LI, 404
Methylcytosine
  chromatographic mobility, LIX, 73
  electrophoretic mobility, LIX, 91
3-Methylcytosine, identification, LIX, 87
5-Methylcytosine
  chromatographic mobility, LIX, 73
  electrophoretic mobility, LIX, 91
  in media, LVIII, 67
  NMR studies, LIX, 57
  substrate of cytosine deaminase, LI, 395, 399
5-Methyldeoxycytidine
  in media, LVIII, 67
  substrate of cytidine deaminase, LI, 411
5-Methyldeoxycytidylate, substrate of deoxycytidylate deaminase, LI, 417
Methyl (±)-1,2-dihydronaphtho[2,1-*b*]furan-2-carboxylate, XLIV, 834
Methyl 4-dimethylamino-3-nitro(α-benzamido)cinnamate
  electronic absorption spectrum, XLIX, 146
  resonance Raman spectrum, XLIX, 147
  structure, XLIX, 145
  substrate, of papain, XLIX, 143
Methyldithioacetate
  reactivity toward amino groups, XLVII, 291, 292
  sequence analysis, XLVII, 290–293
  synthesis, XLVII, 290, 291
Methyl-4,4′-dithiobisbutyrimidate, XLIV, 556
Methyl donor, in 5-methyluridine synthesis, LIX, 202
β,γ-Methyleneadenosine triphosphate, adenylate cyclase, XXXVIII, 169
N,N′-Methylene bisacrylamide, XLIV, 56, 57, 86, 91, 171–176, 185, 194–196, 198, 273, 455, 458, 476, 572, 600, 662, 741, 750
  in acrylamide solution, XXXII, 74
  concentration effects, XLIV, 174, 175
  electrophoretic analysis of protein-RNA complexes, LIX, 567
    of RNA fragments, LIX, 65–69, 574
  electrophoretic separation of *E. coli* ribosomal proteins, LIX, 539, 544
  for gel electrophoresis, XXXII, 79, 94, 95
  in gel preparation, LI, 342, 343; LII, 325
  SDS-acrylamide gel, LIX, 509
Methylene blue, XLIV, 132; XLVI, 563; LIV, 4
  assay of hydrogenase, LIII, 292
  chemiluminescent assay of lactate dehydrogenase, LVII, 460, 461
  electron acceptor, of complex I–III, LIII, 8
  inhibitor of palmityl thioesterase II, XXXV, 109
  as mediator-titrant, LIV, 407, 408
  method, for determination of acid-labile sulfide, LIII, 276, 277
  M. phlei membrane systems, LV, 186
  photooxidation of purine nucleoside phosphorylase and, LI, 530

polarography, LVI, 478
reduction by complex I, LIII, 13
for staining RNA, LIX, 458, 557, 567
staining for viable cells, LVI, 135
substrate of hydrogenase, LIII, 314
of NADH dehydrogenase, LIII, 18
Methylene chloride, *see* Dichloromethane
Methylene monooxygenase system, LII, 166–168
2-Methyleneoxide-(1,3,5/4,6)-pentahydroxycyclohexane, XLVI, 369
Methylene succinate, inhibitor of succinate dehydrogenase, LIII, 33
(±)-L-Methylenetetrahydrofolate
preparation, LI, 91
in thymidylate synthetase assay, LI, 91, 101
5,10-Methylenetetrahydrofolic acid, XLVI, 307
in acetate formation pathway, LIII, 361
Methyl ester moiety, peptide synthesis, XLVII, 521
Methyl ethyl ketone method
apomyoglobin preparation, LII, 478
globin preparation, LII, 448, 449
Methyl ethyl sulfide, cation scavenger in peptide synthesis, XLVII, 547, 615
N-Methylformamide, electron spin resonance spectra, XLIX, 383
3-*O*-Methylglucosamine
affinity chromatography of *Vicia faba* lectin, L, 336, 337
preparation, L, 337
Methylglucose lipopolysaccharide, *see* Mycobacterial polysaccharide
O-Methyl glucose pulse, hormone effects, XXXVI, 431
α-Methylglucoside phosphotransferase, in bacterial membrane, XXXI, 649
Methylglyoxal synthetase
assay, XLI, 502, 503
distribution, XLI, 508
from *Escherichia coli*, XLI, 504–508
preparation, XLI, 504
properties, XLI, 506, 508
purification, XLI, 507
inhibition, XLI, 508
molecular weights, XLI, 508
from *Pseudomonas saccharophila*, XLI, 503–505
preparation, XLI, 503, 504
purification, XLI, 505
Methyl Green
mobility reference, XLVII, 57
as plant nuclei stain, XXXI, 560
Methylguanidine sulfate, inhibitor of deoxythymidine kinase, LI, 365
Methylguanosine
chromatographic mobility, LIX, 73
electrophoretic mobility, LIX, 91
1-Methylguanosine
chromatographic mobility, LIX, 73
electrophoretic mobility, LIX, 91
2-Methylguanosine
chromatographic mobility, LIX, 73
electrophoretic mobility, LIX, 91
NMR studies, LIX, 57
7-Methylguanosine
chromatographic mobility, LIX, 73, 169
conversion to 5-(*N*-methyl)formamido-6-ribosylaminocytosine, LIX, 99
electrophoretic mobility, LIX, 91
hydrogen bond, LIX, 39–42
identification, LIX, 87, 92, 94, 95
NMR spectrum, LIX, 42
replacement by ethidium bromide, LIX, 113
site of chemical cleavage of tRNA, LIX, 100–102
4-*O*-Methyl-D-*arabino*-heptulosonic acid, estimation with thiobarbituric acid, XLI, 33
N-Methylhydrazine demethylase, LII, 15
N-Methyl-1-hydroxyphenazonium methosulfate, as redox mediator, LIV, 423
N-Methyl-4-*N*-[2-hydroxy-3-(*N*-phthalimido)propyl]aminophthalimide
synthesis, LVII, 430
of 6-*N*-(3-amino-2-hydroxypropyl)amino-2,3-dihydrophthalazine-1,4-dione, LVII, 430
Methylhydroxypyridine-carboxylate dioxygenase, LII, 16
flavoprotein classification, LIII, 399

2-Methylimidazole, in Raman studies of hemoproteins, LIV, 244
N-Methylimidazole-2-sulfinic acid, product of dimethylaniline monooxygenase, LII, 143
3-Methylindole, *see* Skatole
3-Methylindole pyrrolooxygenase, LII, 11
3-Methylindolepyruvate, XLIII, 500
N-Methylindoxyl acetate, substrate, cholinesterase, XLIV, 626
Methyl iodide, XLVII, 291
  protein labeling, XLVIII, 332
  synthesis of 8α-flavins, LIII, 459
Methyl isobutyl ketone, XLIV, 767
1-Methyl-3-isobutylxanthine
  cyclase assays, XXXVIII, 72, 121, 154
  in interstitial cell metabolism studies, XXXIX, 259
  ovarian follicle isolation, XXXIX, 194–198
  as phosphodiesterase inhibitor, XXXIX, 253, 256
  in vivo hormonal status, evaluation, XXXIX, 195–198
Methyl isothiocyanate, XLVII, 350
  excess amino group blocking, XLVII, 286, 287
  polystyrene derivatization, XLVII, 361
O-Methylisourea, modification of lysine, LIII, 141
Methyl β-3-ketoglucoside, quantitative determination, XLI, 25
Methyl ketone, XLVI, 199
α-Methyl ketone, XLVI, 135
cis-5-Methylluciferin, structural formula, LVII, 27
trans-5-Methylluciferin, structural formula, LVII, 27
N^{ϵ}-Methyllysine, identification, XLVII, 477
N-Methyllysine oxidase, LII, 21
N^6-Methyl-L-lysine oxidase, LVI, 475
Methylmalonyl-CoA carboxytransferase, XLVI, 21
  conformational study, XLIX, 354, 356
Methyl α-D-mannopyranoside, XXXIV, 669, 670
Methylmannose polysaccharide, *see* Polysaccharide, mycobacterial
Methyl α-mannoside, XXXIV, 207
Methyl α-D-mannoside, XXXIV, 334
Methyl 4-mercaptobutyrimidate hydrochloride, modification of cytochrome *c*, LIII, 175
N-Methyl-2-mercaptoimidazole, *see* Methimazole
1-Methyl-2-mercaptoimidazole, in thyroid cell studies, XXXII, 756
Methyl 3-mercaptopropionimidate, LIX, 550
Methyl methacrylate, XLIV, 84
Methyl methacrylate/4-iodobutyl methacrylate copolymer, XLIV, 84
Methyl methacrylate/2-iodoethyl methacrylate copolymer, XLIV, 84
Methyl methanesulfonate, as mutagen, LVIII, 312
Methyl methanethiolsulfonate
  active group titration, XLVII, 427
  reaction conditions, XLVII, 426
Methyl 4-(4-methoxyphenyl)butyrate
  synthesis, LVII, 435
    of ethyl[2-hydroxy-3-methoxycarbonyl-5-(4-methoxyphenyl)]valerate potassium enolate, LVII, 435
3-*O*-Methyl-2-(*N*-methylacetamido)-2-deoxy-D-glucose, L, 33
N-Methylmorpholine, XLVI, 202, 203, 224; XLVII, 279, 284, 322, 564; XLIX, 461
  hydrazine removal, XLVII, 284
  methylthiazolinone analysis, XLVII, 294
N-Methylmorpholine-HCl, XLVI, 311
Methyl myristate ester, quantum yield, in bioassay, LVII, 192
2-Methylnaphthoquinone, *see* Menadione
2-Methyl-1,4-naphthoquinone, XLIV, 188
1-Methyl-2,4-naphthoquinone-3-thioglycolyldiazoketone, LVI, 656
N^1-Methylnicotinamide, oxidation, LVI, 474
Methyl 4-nitrobenzenesulfonate, XLVI, 537, 539
N-Methyl-*N*′-nitro-*N*-nitrosoguanidine
  carbodiimide-resistant mutants, LVI, 165
  as mutagen, XLIII, 34, 35, 89, 666; LVII, 167; LVIII, 312
  mutagenesis, LVI, 107–108

N-Methyl-4-nitrophthalimide
synthesis, LVII, 427
of 4-amino-*N*-methylphthalimide, LVII, 427
N-Methyl-*N*-nitroso-*p*-toluenesulfonamide, in ethereal diazomethane synthesis, LII, 444
17α-Methylnortestosterone, antihormone, XL, 291, 292
Methylococcus, oxidase, LII, 20
Methyl orange, titration of TEAB, LIX, 61
17α-Methyl-2-oximino-5α-androstan-17β-ol-3-one, synthesis, XXXVI, 416
5-*N*-Methyl 1-oxyphenazine, as mediator-titrant, LIV, 408, 423
2-Methylpentane-2,4-diol
crystallization of transfer RNA, LIX, 7
preparation of crystals, LIX, 5
3-Methylpentane-1,5-diol, XLIV, 839
Methylpentose, in glycolipids, XXXII, 358–360
N-Methylphenazine, ferricyanide, LV, 543, 544
N-Methyl phenazonium methosulfate, as redox mediator, LIV, 423, 429, 431–433
10-Methylphenothiazine
chemiluminescent reaction, LVII, 516, 517
structural formula, LVII, 498
3-Methyl-7-(2-phenoxyacetamido)-3-cephem, XLIII, 411, 412
N-Methyl-L-phenylalanine, XLVI, 227
p-Methylphenylalanine
internal standard, XLVII, 34
substrate of phenylalanine hydroxylase, LIII, 284
Methyl phosphonodichloridate, XXXIV, 587
N-Methyl-4-*N*-[4-(*N*-phthalimido)butyl]aminophthalimide
synthesis, LVII, 433
of 6-*N*-(4-aminobutyl)amino-2,3-dihydrophthalazine-1,4-dione, LVII, 433
of 6-[*N*-(4-aminobutyl)-*N*-ethyl]amino-2,3-dihydrophthalazine-1,4-dione, LVII, 433
N-Methyl propionamide, electron spin resonance spectra, XLIX, 383
Methyl propionic acid, electron spin resonance spectra, XLIX, 383
6-Methylpterin, substrate of phenylalanine hydroxylase, LIII, 284
7-Methylpterin, substrate of phenylalanine hydroxylase, LIII, 284
4-Methylpyrazole, inhibitor of alcohol dehydrogenase, LII, 356
N-Methylpyridinium-agarose, XXXIV, 573
Methyl salicylic acid, substrate of salicylate hydroxylase, LIII, 538, 539
6-Methylsalicylic acid
biosynthetic, XLIII, 533–535
chemical synthesis, XLIII, 532, 533
^{14}C–labeled, XLIII, 531, 533, 535
decarboxylase activity, XLIII, 530
patulin biosynthesis, XLIII, 520, 521, 540
preparation, XLIII, 535
reaction sequence, XLIII, 520, 521
6-Methylsalicylic acid decarboxylase, XLIII, 530–540
ammonium sulfate fractionation, XLIII, 536, 537
assay, XLIII, 531, 532
DEAE–Sephadex A–50 column chromatography, XLIII, 437
fluorescence assay, XLIII, 531
homogeneity, XLIII, 538
hydroxyapatite column chromatography, XLIII, 437, 438
inhibitors, XLIII, 540
isotope effects, XLIII, 539
kinetic properties, XLIII, 539
6-methylsalicylic acid
biosynthesis
isolation, XLIII, 533
preparation, XLIII, 535
chemical synthesis, XLIII, 532, 533
molecular weight, XLIII, 539
Penicillium patulum preparation, XLIII, 534–536
properties, XLIII, 538–540
purification, XLIII, 536–538

radioactive assay, XLIII, 531, 532
Sephadex G–100 gel filtration, XLIII, 437
specificity, XLIII, 539
stability, XLIII, 538
ultracentrifugation, XLIII, 536
unit definition, XLIII, 532

6-Methylsalicylic acid synthetase, XLIII, 520–530
ammonium sulfate fractionation, XLIII, 525
assay, XLIII, 521–523
cell breakage, XLIII, 524
DEAE-cellulose chromatography, XLIII, 526, 527
fluorometric assay, XLIII, 521–523
homogeneity, XLIII, 528
hydroxypatite chromatography, XLIII, 525, 526
inhibitors, XLIII, 528
kinetic properties, XLIII, 528
molecular weight, XLIII, 528
Penicillium patulum, preparation, XLIII, 523
polyethylene glycol 1500 fractionation, XLIII, 524, 525
polyethylene glycol 6000 precipitation, XLIII, 524
properties, XLIII, 528–530
purification, XLIII, 524–527
reaction scheme, XLIII, 529, 530
Sepharose 6B, gel filtration, XLIII, 526
shake culture, XLIII, 524, 534
specificity, XLIII, 529
stability, XLIII, 528
unit definition, XLIII, 523

6-Methylselenopurine ribonucleoside, substrate of adenosine deaminase, LI, 507

L-Methyl succinate, LVI, 586

17α-Methyltestosterone, antiglucocorticoid, XL, 291, 292

5-Methyltetrahydrofolic acid, in acetate formation pathway, LIII, 361

2-Methyltetrahydrofuran, solvent, in radical ion chemiluminescence reactions, LVII, 506

Methylthiazolinone
O-acetylation, XLVII, 294, 295
ammonolysis, XLVII, 295, 296

17β-Methylthioandrost-4-en-3-one, synthesis, XXXVI, 422

Methylthiocarbamoyl γ-amino-*n*-butyric acid polymer
peptide attachment, XLVII, 362
preparation, XLVII, 360, 361
swelling ratios, XLVII, 368

Methyl thiocyanate, XLVII, 455

Methylthioethylidene morpholidium iodide, synthesis, XLVII, 291

1-Methyl-6-thiohypoxanthine, XLIX, 165

9-Methyl-6-thiohypoxanthine, XLIX, 165

6-Methylthioinosine, inhibitor of adenosine deaminase, LI, 507

Methylthioinosinedicarboxyaldehyde, XLVI, 353, 355, 356
synthesis, XLVI, 353–355

2-Methylthio-N^6-(2-isopentenyl)adenosine
chromatographic behavior, LIX, 169
electrophoretic mobility, LIX, 91
identification, LIX, 92

2-Methylthio-N^6-(2-isopentenyl)adenosine 5′-phosphate, identification, LIX, 89

Methyltransferase, *see* specific types

3-*O*-Methyl-4-*O*-trideuteriomethyl-2-(*N*-methylacetamido)-2-deoxy-D-glucose, L, 32

Methyl tri-*n*-octyl ammonium salt, XXXIV, 484

β-Methyl-DL-tryptophan, inhibitor of tryptophanyl-tRNA, LIX, 251

5-Methyltryptophan, inhibitor of tryptophanyl-tRNA synthetase, LIX, 251

6-Methyltryptophan, inhibitor of tryptophanyl-tRNA synthetase, LIX, 251

Methylumbelliferol ester, of *N*-benzyloxycarbonyl-L-lysine, substrate for trypsinlike enzymes, XLV, 18

4-Methylumbelliferone, XLIV, 391
fluorescence, XLV, 15

Methylumbelliferyl *p*-guanidinobenzoate, titration, of trypsinlike enzymes, XLV, 13, 14

4-Methylumbelliferyl α-L-iduronide, L, 141
α-L-iduronidase assay, L, 149, 150

synthesis, L, 146, 147
4-Methylumbelliferyl *p*-(*N,N,N*-trimethylammonium)cinnamate, XLIV, 391
N-Methylurea, substrate, urease, XLIV, 595
Methyluridine, chromatographic mobility, LIX, 73
5-Methyluridine, LIX, 193
chromatographic behavior, LIX, 73, 169
electrophoretic mobility, LIX, 91
NMR studies, LIX, 57
substrate of pyrimidine nucleoside phosphorylase, LI, 436
of uridine-cytidine kinase, LI, 313
of uridine nucleosidase, LI, 294, 295
5-Methyluridine methyltransferase
assay, LIX, 191–193
inhibition, LIX, 203
Methyl viologen
assay of formate dehydrogenase, LIII, 363, 364
of hydrogenase, LIII, 290, 291, 294, 295
of nitrate reductase, LIII, 348, 350
extinction coefficient, LIII, 363
as mediator-titrant, LIV, 409, 410, 423
in redox titration, LIII, 491
substrate, of hydrogenase, LIII, 314
Methylxanthine
effect on melanin, XXXVII, 129
phosphodiesterase, XXXVIII, 223, 244
Methymycin, XLIII, 131
Metmyoglobin
circular dichroism, XXXII, 221
quenching-time determination, LIV, 92
Metrizamide, gradient contrifugation, LII, 77–79
Met-$tRNA_f^{met}$, XLVI, 627–629, 632
Metrohn recording titrator, iodination, LIII, 149
Mets-Bogorad system, modified, for separation of ribosomal proteins, LIX, 429–433
Metyrapone
activator of epoxide hydrase, LII, 200
cytochrome binding, LII, 273, 275, 276
effect on cytochrome P–$450_{11\beta}$, LII, 131, 132
inhibitor of cytochrome P–450-linked monooxygenase activities, LII, 231
Meyerhof quotient, Pasteur effect, LV, 292, 293
Mg^{2+}-ATP, *see* Adenosine 5′-triphosphate, Mg^{2+}-activated
Mg^{2+}-ATPase, *see* Adenosinetriphosphatase, Mg^{2+}-activated
MgK_2-EDTA, *see* Ethylenediaminetetraacetic acid, MgK_2 salt
Micelle, critical concentration and size, LVI, 736, 737
Michael acceptor, XLVI, 31, 158
Michaelis buffer, XXXI, 21, 22
factor X, bovine, XLV, 90
Michaelis constant
determination by isomerase assay, XLI, 66
luminescent determination of ATP, LVI, 535
spectral dissociation, LII, 261, 262
Michaelis-Menten complex, XLVI, 6, 55
Michaelis-Menten equation, XXXI, 819, 821
rate of change of ATP concentration, LVI, 537, 538, 540
translocator data, LVI, 265
Michler hydrol, *see* 4,4′-Bisdimethylaminodiphenylcarbinol
Mickle Shaker apparatus, XXXI, 104, 653, 656, 665
properties, XXXI, 662
Microassay, *see* specific type
Microautoradiography
of protein hormones, XXXVII, 155–167
surface binding type, XXXVII, 164–167
Microbiological contamination, resazurin as indicator, XLI, 56
Microbody, *see also* specific types
alternative names, LII, 495
catalase localization, XXXI, 495, 496, 566, 735

electrophoretic purification, XXXI, 754
marker enzymes, XXXI, 735, 742, 743, 746
of plant cells, XXXI, 495, 496, 499
isolation, XXXI, 544, 549, 552, 553
Microcalorimetry, XLIV, 659–667
glucose oxidase assay, XLIV, 663, 664
instrumentation, XLIV, 661
theory, XLIV, 659–661
trypsin assay, XLIV, 665
Microcannula, for hypophyseal portal vessel, XXXIX, 180
Microcapsule, *see also* Microencapsulation
biodegradable, XLIV, 698
surface area, XLIV, 691
Microcarrier culture, LVIII, 184–194
cell counting, LVIII, 188
cell yield, LVIII, 210
culture initiation, LVIII, 187
growth cycle, LVIII, 187–190
harvesting, LVIII, 190, 191
interferon formation, LVIII, 192
large-scale, LVIII, 193
preparation, LVIII, 185–187
primary cells, LVIII, 207, 208
seeding, LVIII, 187, 188
source, LVIII, 185–187
synthesis, LVIII, 185–187
toxicity, LVIII, 184, 185
Micrococcin, XLIII, 338
effect on protein synthesis, LIX, 853, 854, 862
Micrococcus
cell-wall isolation, XXXI, 659
hydrogenase reaction, LIII, 287
oxidase, LII, 21
Micrococcus aerogenes, rubredoxin, LIII, 344
Micrococcus denitrificans, *see Paracoccus denitrificans*
Micrococcus flavus, bacitracin synthetase, XLIII, 551
Micrococcus lysodeikticus
ATPase, LV, 341
b-type cytochrome, LIII, 208
cell-wall isolation, XXXI, 655, 658, 665
IR spectroscopy, XXXII, 253
lysis, XLIV, 445, 446
membranes, differential scanning, calorimetry, XXXII, 265
Micrococcus pyrogenes, oxidase, LII, 15
Micrococcus pyrogenes var. *albus,* cytochrome *o*, LIII, 207
Micrococcus roseus, penicillin acylase, XLIII, 719
Micrococcus rubens, oxidase, LII, 18
Microcolumn, design, LVII, 281
Microculture, neurons, LVIII, 582, 583
Microcystis, lysis, XXXI, 681
Microdiver, for microgasometric measurements, XXXIX, 404–413
Microelectrode
electrophysiological techniques, LV, 660, 661
insertion into hyphae, LV, 661, 662
ion selective, LVI, 362, 363
Microencapsulation, XLIV, 201–218
advantages, XLIV, 218
in cellulose nitrate, XLIV, 203–207
in cellulose polymer material, XLIV, 210
chemical method, XLIV, 207–209
in fibers, XLIV, 211, 227–242
with heparin-complexed membrane, XLIV, 212
with lipid membranes, XLIV, 212
with lipid-protein membrane, XLIV, 212
in liposomes, XLIV, 212, 218–227, 698–709
in liquid membrane emulsions, XLIV, 211
magnetic preparations, XLIV, 216
membrane surface charge, XLIV, 212
in microspheres, XLIV, 201–218
of multistep enzyme systems, XLIV, 213–217
in nylon, XLIV, 207–209
by drop technique, XLIV, 207, 208
by emulsification technique, XLIV, 207–209
physical method, XLIV, 203–207
with protein membranes, XLIV, 212
in red blood cells, XLIV, 212
in silicone rubber, XLIV, 210
theory, XLIV, 202, 203, 210

therapeutic application, XLIV, 676–709
Microenvironment, kinetic effects, XLIV, 399, 400
Microferm fermenter, XXXI, 628, 629, 668
β_2-Microglobulin, sequence analysis, XLVII, 331
Microhemadsorption, in lectin-binding studies, XXXII, 616, 617
Microinjection, instrumentation, LVII, 312–318
Microiontophoretic circuitry, design, XXXIX, 429, 430
Microkatal, definition, XLIV, 3
Micromanipulator
for crystal mounting, LIX, 9
electrophysiological techniques, LV, 660
for pituitary stalk, XXXIX, 179
Micromedics automatic pipetting apparatus, XXXVI, 32
Micro–Mill MV–6–3, yeast cell breakage, LVI, 47
Micromonospora
N–Z amine-starch-glucose medium, XLIII, 7
streptzyme, XXXI, 613
Micromonospora purpurea
aminoglycoside antibiotics, XLIII, 265
gentamicin, media, XLIII, 21
Microorganism, *see also* specific type
care of enzyme-activated electrodes, LVI, 472, 473
contamination of glucose reactor, XLIV, 784, 785
of lactose reactor, XLIV, 804–826
immobilization, in polyacrylamide, XLIV, 183–190
isolation and characteriazation of mitochondria
controlled breakage of organisms, LV, 135–142
criteria of integrity, LV, 143, 144
preparation of subcellular fractions, LV, 142, 143
Microperoxidase
chemiluminescence assays, LVII, 426
stock solution, preparation, LVII, 425
Micropipette
beveling, LVII, 314
holder design, LVII, 315, 316
preparation, LVII, 313, 314
Micropipette puller, XXXIX, 180
Microsac, in eel tissue, XXXII, 317
Microscope, LVIII, 11, 12
for microspectrophotometer, XXXIX, 418
suppliers, LVIII, 17
Microscope lens, light recording apparatus, LVII, 319
Microscopy
chromosome study, LVIII, 336–339
light, and localization of cellular constituents, XLIV, 712, 713
monolayer cultures, LVIII, 139, 140
Microsome, XXXI, 521
aggregation
with guanidine hydrochloride, LII, 146
testing, LII, 79
from beef liver, LII, 98
cAMP receptor, XXXVIII, 380, 381
components, physical properties, XXXI, 227
definition, XXXI, 16
differential centrifugation, XXXI, 727, 728
differential scanning calorimetry, XXXII, 265
disruption, by continuous sonication, LII, 80–82
drug metabolism, lipid peroxidation, LII, 304
electron microscopy, XXXI, 29, 32
electron-transport reactions, LII, 43–71
schematic, LII, 44
electrophoretic purification, XXXI, 754, 755
enzymes, XXXII, 91
immunological studies, LII, 247–249
marker, XXXII, 671
ethanol oxidizing system
assay, in isolated hepatocytes, LII, 357, 358
in liver microsomes, LII, 358–362
in liver slices, LII, 356, 357
in reconstituted system, LII, 362–367

isolation from liver microsomes, LII, 359–362
reconstituted, activity properties, LII, 366
substrates, LII, 356
from fat cells, XXXI, 62, 63, 67
flavoproteins, LII, 44, 45
fraction I, isolation procedure, LII, 72
fraction II, isolation procedure, LII, 72, 73
fraction III, isolation procedure, LII, 72
fraction IV, isolation procedure, LII, 72
heme protein, LII, 45–47
from hen liver, XXXV, 256, 257
from house fly abdomen, LII, 89
isolation, XXXVI, 491; LV, 101
by calcium sedimentation, LII, 85, 86, 98
by centrifugation, LII, 104, 118
by density gradient techniques, LII, 71–83
homogenization procedure, LII, 71, 72
using EDTA and PMSF, LII, 93
lipids, XXXI, 35
from liver, preservation and purification, XXXI, 6, 15–18
marker enzymes, XXXI, 735, 743; LV, 101
membrane-bound components, separation, XXXI, 219–225
aggregation and pelleting effects, XXXI, 235, 236
membranes, thermal transition studies, XLIX, 13
millimolar extinction coefficient, LII, 369
morphometry, XXXII, 9, 15
from mouse liver, preparation, LII, 235
from nonhepatic tissue of mouse, preparation, LII, 235
5′-nucleotidase, XXXII, 368
optical-difference spectroscopy, LII, 263–266
oxidases, LII, 11–15
from pancreas
chemistry, XXXI, 46, 47
electron microscopy, XXXI, 56, 58
isolation, XXXI, 43–46, 50, 51, 199, 200
smooth and rough subfractions, XXXI, 45
phospholipid exchange, XXXII, 140
from pig liver, preparation, LII, 143–145
in plants, isolation, XXXI, 584, 585
preparation, XXXII, 141, 142; LVIII, 224
protein solubilization, *see* Protein, solubilization
from rabbit liver
NADH-cytochrome b_5 reductase binding, LII, 108
preparation, LII, 111, 201
from rat liver, XXXV, 118
multiple forms of cytochrome *P*-450, LII, 328, 329
preparation, LII, 118, 190, 194, 195, 359, 370; LIX, 403
relative enzyme activities in calcium-sedimented preparations, LII, 87, 88
ribosomal and membranous component isolation, XXXI, 201–215
electron microscopy, XXXI, 237
ribosome removal by EDTA treatment, LII, 76
rough, subfractionation, LII, 74–78
rough and smooth
isolation, XXXI, 191–201, 215
on three-layered discontinuous sucrose gradient, LII, 72, 73
sedimentation, XXXI, 716
smooth, subfractionation, LII, 78, 79
from southern armyworm midgut, LII, 89
spin-labeled phosphatidic acid, XXXII, 193
stability, XXXI, 194, 195
storage 93
surface charge, XXXI, 193, 194
vesicle-contents release, XXXI, 215–225
Microspectrophotometry
calcium transport, LVI, 323–326

of endocrine cells, XXXIX, 413–422
semimicromodification, XXXIX, 420–422
ultramicromodification, XXXIX, 422
Microtest Tissue Culture Plate, XXXIX, 131; LVIII, 156
Microtomy, in freeze fracture, XXXII, 39
Microtubule, XLVI, 567
disruptive drugs, XL, 322
membrane, energy transfer studies, XLVIII, 378
of plant cells, XXXI, 493
function, XXXI, 494
protein, XXXIV, 623
Microtubule wall, *see* Tubulin
Microvillus
in brush borders, XXXI, 115, 123, 124
membrane, properties, XXXI, 129–132
scanning electron microscopy, XXXII, 52, 53
transformed cells, LVIII, 368
Microwave cavity, XLIX, 391
Microwave frequency, measurement, XLIX, 521
Microwave generator, for tritium labeling, XXXVII, 314
Microwave plasma source, LIV, 457–459
Microwave power saturation, XLIX, 522
Microwave radiation, rapid tissue fixation, XXXVIII, 5, 6
Microzone Electrophoresis Cell, LI, 37
MIF, *see* Migration inhibition factor
Migration inhibition factor, A1 protein activity, XXXII, 341
Milk
assay for bacteria in, resazurin and resorufin, XLI, 56
continuous sterilization, XLIV, 824
lactose reduction, XLIV, 822–830
oxidase, LII, 17
ribonuclease, XXXII, 109
Miller liver perfusion system, XXXII, 626
Millipore assay, of nectin, XXXII, 434–436
Millipore cell, XXXII, 3, 6
Millipore filter
lipid-impregnated
assay of ionophore activity, LV, 419, 420
preparation, LV, 605–608
presoak, LIX, 848
reconstitution and membrane potential, LV, 589
in ribosome binding assay, LIX, 847–849
in ribosome release assay, LIX, 861
separation of cells from medium, LV, 204
of organelles, LV, 556
Millipore Immersible Molecular Separator, LIII, 652, 653
Millipore Milli-Q system, LII, 282; LIV, 474
Millonig phosphate buffer, XXXI, 402
Milton Roy Model 396 minipump, LIX, 221
Mineral, yeast growth media, LVI, 59
Mineralight UVSL-25 long-wavelength lamp, LII, 443
Mineralocorticoid, binding proteins, XXXVI, 286
Minibeaker, XLIV, 293
Minicon concentrator, LII, 156
Mini-Escargo Fractionator, XXXIX, 368
Minigrid, as optically transparent electrodes, LIV, 405, 406
Minimal Medium M9, composition, XXXI, 643, 644
Minimal salts medium 56, culture of *E. coli* ubiquinone mutants, LIII, 602
Mini-mill, XXXII, 826
spheroplast rupture, LVI, 129
Minimum essential medium, *see* Eagle's minimum essential medium
Minimum Essential Medium, Earle's salt, XXXVII, 350
Miniplant, XLIV, 293
MINISSA 1 multidimensional scaling program, LIX, 609
Minitube, XLIV, 293, 307, 308
Mink, lung cell, LVIII, 414
Miracloth
filtration cloth, XXXI, 584, 589, 593, 604, 605, 737
for lipid removal, LIII, 115

in washing of DEAE-cellulose, LVII, 145, 146
Mirror
100% dielectric, LIV, 10
in optical diffractometer, advantages and disadvantages, XLIX, 43, 44
100% reflective, LIV, 9
Mirror image subtraction, LIV, 319, 320
Mistral centrifuge, XXXII, 376
Mithramycin, XLIII, 127
DNA determination, LVIII, 147
Mitochondria, XXXI, 521; XLVI, 84, 553, 554
activation of succinate dehydrogenase, LIII, 475
of adipose tissue
brown, LV, 16, 17
white, LV, 15, 16
of adrenals, preparation, XXXVII, 297, 298
adrenocortical, XXXIX, 162–164; LV, 9, 10
cytochrome isolation, LII, 124–132, 139, 140
oxidase, LII, 14
properties, LV, 10, 11
rupture by sonication, LII, 134
adrenocorticotropin effects, XXXIX, 162, 163
alteration of, in petite-negative yeasts, XXXII, 838–843
amidination, LVI, 624
analysis of preparations, LVI, 624, 625
osmotic sensitivity, LVI, 625, 626
polyacrylamide gel electrophoresis, LVI, 626–629
analysis of products by slab gel electrophoresis, LVI, 104, 105
antimycin loading, LIII, 93
ATPase, *see also* Adenosinetriphosphatase
assay, LVI, 101
ATPase inhibitors, LV, 408–414, 421–426
applications, LV, 407
properties, LV, 399–407
ATP-dependent functions, resolution and reconstitution, LV, 736–741
ATP synthesis, potassium ion gradients, LV, 666–675
aureovertin inhibitory effects
energy-conserving systems, LV, 482, 483
soluble ATPase, LV, 481, 482
2-azido-4-nitrophenol binding, LVI, 666–668
beef heart, XXXI, 292; XXXV, 264
ATPase inhibitor, LV, 402, 404, 405
aureovertin, LV, 482
DCCD-binding protein, LV, 428, 429
improved method for ATPase isolation, LV, 317–319
isolation, LV, 385
preparation of ATPase, LV, 304–308
by chloroform method, LV, 338–342
in high yield, LV, 46–50
of rutamycin-sensitive ATPase, LV, 315–317
purification of transhydrogenase, LV, 276–283
reconstitution of ATP synthetase, LV, 711–715
of pyridine nucleotide transhydrogenase, LV, 811–816
brain
nonsynaptic, LV, 17, 18
synaptosomal, LV, 18
of brown adipose tissue, LV, 65, 66
appearance, LV, 67, 68
ATP synthesis, LV, 77, 78
BSA, LV, 70
carnitine and ATP effects, LV, 74
coupling, LV, 70
fatty acid content, LV, 75
fatty acid oxidation, LV, 75, 76
glycerol 3-phosphate oxidation, LV, 76, 77
isolation procedure, LV, 68, 69
location of tissue, LV, 66, 67
other substrates, LV, 77
pattern of substrate oxidation, LV, 75
pH effects, LV, 70
phospholipid content, LV, 78
purine nucleotides, LV, 70–74
requirement for KCl, LV, 69, 70

source of tissue, LV, 66
calcium depletion, LVI, 351
calcium transport, measurement, LVI, 320, 321, 327, 329
calcium uptake, XXXIX, 520
calculation of → H^+/O values and essential controls, LV, 635–637
characterization, XXXI, 23, 24, 33, 36, 38
by enzyme assay, XXXI, 409
cholesterol, XXXI, 29
complex III, XLVI, 84
computer methodology for reconstruction
alignment, LVI, 721–723
allocation of profile label, LVI, 723
data entry, LVI, 721
three-dimensional reconstructions, LVI, 723
computer system for reconstruction
computer, LVI, 727
coordinate digitizer and stylus, LVI, 723, 726
data retrieval and display, LVI, 727, 728
data storage, LVI, 727
condensed, with broken outer membranes, LVI, 692
countercurrent distribution, XXXI, 761
Crabtree effect, LV, 297
cristae, negative staining, XXXII, 31, 33
cross-linking agents, LVI, 622–629
cytochrome *c*-depleted, LIII, 155
of cytochromeless mutants, LVI, 125
cytochrome *P*–450 isolation, LII, 139, 140
defective membranes, biochemical characterization, LVI, 138
deoxyribonucleic acid
cloning, LVI, 5
replication, LVI, 6
size, LVI, 4
transcripts, LVI, 7
diagnostic enzymes, XXXI, 20, 165, 185, 186, 735, 742
differential scanning calorimetry, XXXII, 265
differential sedimentation, XXXI, 727, 742
Dio-9, LV, 514, 515
DNA, LV, 9
electron microscopy, XXXI, 26, 27
electrophoretic purification, XXXI, 752, 753
membranes, XXXI, 757–759
enzyme
in membranes, XXXII, 91
purity, LVI, 692–696
toluene treatment, LVI, 546–548
enzyme marker, XXXII, 393
EPR spectra, LIV, 146, 147
estimation of proton-motive force, LV, 567–569
extraction of lipid, LVI, 428
from fat cells, XXXI, 63, 64, 67
in flight muscle, XXXI, 491
fluorescence changes of ANS, LVI, 498, 499
fragments, LVI, 500, 501
intact mitochondria, LVI, 499, 500
fluorescent probe studies, XXXII, 237, 240–243
fraction, preparation, LVII, 38
freeze-etching, XXXII, 54
function, effect of changed sterol composition, LVI, 563–566
gel electrophoresis, XXXII, 79
general properties
criteria of integrity, LV, 8, 9
incubation medium, LV, 7, 8
substrate permeases, LV, 9
genes, mapping, LVI, 16
genome, LVI, 3
heavy, beef-heart, submitochondrial particles, LIII, 574
heavy and light fractions, XXXII, 377
hepatic, NADH-cytochrome b_5 reductase binding, LII, 108
high-yield and large scale preparation
isolation from nuclear fraction, LV, 35–37
preparation of rat liver homogenate, LV, 33–35
purification from postnuclear supernatant, LV, 37, 38
storage, LV, 38, 39
yield and characterization, LV, 39
human heart, characteristics, LV, 45

D-β-hydroxybutyrate apodehydrogenase, XXXII, 374–391
immobilized, functional properties
binding to octadecylsilylated beads, LVI, 551, 552
calibration of electrode, LVI, 554, 555
flow experiments, LVI, 552–554
preparation of mitochondria, LVI, 551
of solid support, LVI, 550, 551
representative results, LVI, 556, 557
incubation medium, LVII, 38
inhibitory effects of citreoviridin, LV, 487, 488
inner membrane-matrix fraction, preparation, LVI, 687–689
inner membranes
amidination, LVI, 624
distribution of label, LVI, 616, 617
effect of amidination on enzymes, LVI, 629
of labeling on enzymes, LVI, 619
isolation, XXXI, 310–323
labeling, with DABS or PMPS, LVI, 616
by ferritin-antibody, LVI, 715, 716
permeability, LVI, 245
polypeptides, LVI, 12, 13, 40
preparation, LVI, 358, 615
SDS-polyacrylamide gel electrophoresis of labeled membranes, LVI, 617–619
surface labeling, LVI, 613–615
of insect flight muscle, LV, 22–24
internal volume, LV, 551
measurement, LVI, 287, 288
inverted inner membrane vesicles
phosphate transport, LVI, 358, 359
preparation, LVI, 689
ionophores, LV, 440, 447, 449–453
isolation, LVIII, 222
from brain, materials and methods, LV, 51, 52
characterization from microorganisms
controlled breakage of organisms, LV, 135–142
criteria of integrity, LV, 143, 144
preparation of subcellular fractions, LV, 142, 143
from fatty acid depleted cells, LVI, 577
of glutamate carrier system, LVI, 419–430
postnuclear supernatant, LV, 37, 38
relevant properties, K^+ gradients, LV, 667, 668
from small amounts of heart tissue, LV, 39, 40
characterization, LV, 45, 46
cleaning of glassware, LV, 41
equipment, LV, 40
glassware and tools, LV, 40, 41
handling of tissue, LV, 41, 42
homogenization, LV, 42
isolation medium, LV, 41
separation and washing, LV, 42–45
stock solutions, LV, 41
kidney, oxidase, LII, 15
of kidney cortex, LV, 11
properties, LV, 11, 12
kinetics of glutamate and aspartate efflux
aspartate loading, LVI, 260
calculation of dilution factors, sucrose and matrix spaces, LVI, 261
example of specimen calculation, LVI, 261, 262
glutamate loading, LVI, 258, 259
incubation conditions for aspartate efflux, LVI, 260, 261
for glutamate efflux, LVI, 259, 260
labeled
analysis on exponential gradient gel slabs, LVI, 61–66
small scale isolation, LVI, 60, 61
leucinostatin, LV, 513
lipids, XXXI, 35

of liver, LV, 20, 21
isolation, XXXI, 10–15
rapid technique, XXXI, 299–305
preservation and purification of, XXXI, 5, 6, 9, 10
loading
with carnitine, LVI, 375, 376
with metabolites, LVI, 280, 281
long-term storage
assay and characterization, LV, 30, 31
comments, LV, 31, 32
principle, LV, 28, 29
procedure, LV, 29, 30
of lung, LV, 19, 20
magnesium efflux, LVI, 315, 316
of mammary gland, XXXI, 305–310; LV, 19
manipulation of unsaturated fatty acid composition, LVI, 571–577
marker segregation in yeast
estimating parameters for random segregation model, LVI, 150, 151
gene conversion, LVI, 152, 153
mating, LVI, 141, 142
mutants, LVI, 140, 141
pedigree studies, LVI, 147–150
random diploid analysis, LVI, 142–144
segregation analysis, LVI, 144–146
segregation problem, LVI, 146, 147
segregation rates, LVI, 153, 154
uniparental inheritance, LVI, 151, 152
zygote clone analysis, LVI, 144
measurement of Ca^{++}/site ratio, LV, 649, 650
isotopic Ca^{++}-jump procedure, LV, 650–652
simple steady-state rate method, LV, 652–655
simultaneous determination of Ca^{++}/site, H^{+}/site and H^{+}/solidus Ca^{++} ratios, LV, 655, 656
of ΔpH, LV, 563
membrane, XXXI, 24; XLVI, 86, *see also* Submitochondrial particle
ATPase, XXXIV, 566
cholesterol, LVI, 208
cytochrome *c*-depleted, preparation, LIII, 102
definition, 68n
electron microscopy, XXXII, 20–35
genetic modification, LVI, 117–131
IR spectroscopy, XXIII, 250, 252
NMR studies, XXXII, 210
optical probes, LV, 573
polarographic determination of phase changes, XXXII, 258–262
surface potential change with ATP, LVI, 526
thermal transition studies, XLIX, 13
membrane assembly, lipid depletion, LVI, 565, 566
membrane enzymes
cross-reactivity, LVI, 705–707
inhibition by affinity antibody, LVI, 707, 708, 713
labeling with ferritin-antibody probes and preparation for electron microscopy, LVI, 715–717
membrane potentials, LV, 555
extrinsic probes, LV, 559, 560
membrane proteins, biogenesis in *Neurospora*, LVI, 51–58
messenger RNA, LVI, 9, 10
metabolite measurement, methods, LVI, 202–206
metabolite transport systems, LVI, 245
general methodology, LVI, 245–247
isolation of carrier proteins, LVI, 251, 252
molecular approach to carriers, LVI, 250, 251
regulation, LVI, 250
survey, LVI, 247–249
Morris hepatoma, isolation and properties, LV, 79–88
mutations affecting, LVI, 13–16
negative staining, XXXII, 28, 31, 33
of neoplastic tissues, LV, 22

of *Neurospora crassa*, LV, 25, 26
ATPase complex, LV, 344–351
isolation, LIX, 205, 206, 422, 423
preparation, LV, 144–148
nicotinamide nucleotide transhydrogenase, LV, 263
nonaqueous fractionation, LVI, 205
nonsynaptic, preparation, LV, 52, 53
5′-nucleotidase, XXXII, 368
organotins, LV, 508, 509
ossamycin, LV, 512
outer membrane
isolation, XXXI, 310–323
labeling with ferritin-antibody, LVI, 716, 717
monoamine oxidase, LVI, 684
preparation of vesicles, LVI, 689, 691, 692
oxidative phosphorylation, LIII, 3, 4, 6
of pancreas, XXXI, 42
permeability with toluene
applications, LVI, 549, 550
reagents and solutions, LVI, 544–549
peroxide generation, XXXII, 108
pet9 mutants
preparation, LVI, 128–130
properties, LVI, 127, 128
phosphate content, LVI, 354
phosphate transport
analytical methods, LVI, 354
assay:net uptake, LVI, 354, 355
assay:$^{32}P_i$ exchange, LVI, 355–357
comparison with other methods, LVI, 357, 358
isolation of mitochondria, LVI, 354
phospholipid exchange, XXXII, 140, 146
phosphorylation potential measurement, LV, 237–239
pig brain, L-glycerol-3-phosphate dehydrogenase assay, purification, and properties, XLI, 254–259
pig heart, choice and preparation, LVI, 420, 421
plant, XXXI, 570; LV, 24, 25
isolation, XXXI, 501, 505, 506, 508–510, 516, 517, 520, 524, 533, 534, 541, 542, 544, 545, 549, 550, 552, 553, 589–600
phenol effects, XXXI, 530, 531
protectants, XXXI, 536–539
poky mutants, LVI, 11, 12
polarity, locus *omega*, LVI, 143
preparation
from beef liver, LIII, 498
general, LVI, 685, 687
measurement of proton translocation, LV, 633, 634
from *N. crassa*
cultivation of cells, LV, 144, 145
disruption of cell walls, LV, 145–147
isolation, LV, 147
properties, LV, 147, 148
from pig liver, LIII, 509, 512
of outer membrane from pig heart, LV, 98, 99
controls of purity, LV, 101–104
principle, LV, 99
procedure, LV, 99–101
from yeast, LV, 149–151, 160, 161
enzymatic method, LV, 157–159
large scale method, LV, 155–157, 161, 162
small scale methods, LV, 152–155, 162, 163
preparation techniques
differential centrifugation, LV, 5, 6
filtration, LV, 5
gradient centrifugations, LV, 6
homogenization, LV, 4, 5
isolation medium, LV, 3, 4
physiological state of organism, LV, 3
storage, LV, 6, 7
preservation, XXXI, 3
protein, resolution, LVI, 602, 606–613
protein synthesis, LVI, 17, 18
assay *in vivo*, LVI, 99–101
incubation conditions, LVI, 20–23

inhibitors, LVI, 29–32
selection of mutants, LVI, 33
use *in vitro*, LVI, 32, 33
in vivo, LVI, 33–39
preparation of mitochondria, LVI, 18–20
of protein for radioactivity determination, LVI, 23
proton/site ratio of electron transport, LV, 641, 642, 645, 646
calculations, LV, 645
modified oxygen pulse procedure, LV, 642, 643
procedure, LV, 643–645
with Ca^{++}, LV, 648, 649
using K^{+}, plus valinomycin, LV, 647, 648
reagents and test medium, LV, 643
steady-state method, LV, 646, 647
purification, XXXI, 40, 41
quinone redox reactions, LIII, 589–591
rapid separation from medium, LV, 228, 229
rat brain
preparation of two populations of synaptic and one population of nonsynaptic, LV, 53–56
properties of four populations of synaptic and nonsynaptic, LV, 56–60
rat heart, isolation, LVI, 546
rat liver
ATPase inhibitor, LV, 402, 405, 408–414
aureovertin, LV, 482
electrogenicity of calcium translocation, LV, 639, 640
internal volume, LV, 551
isolation, LVI, 545, 546
of oligomycin-sensitive ATPase, LV, 328–333
membrane potential, LV, 555
purification of F_1 ATPase, LV, 320–328, 333–337
reconstitution of ATP-dependent functions, LV, 736–741
red-green separation, LIII, 5, 6, 42
respiratory carriers, measurements of proton translocation, LV, 623–627
respiratory enzymes, assay, LVI, 101, 102
respiratory measurements, XXXI, 599, 600
ribosomal proteins, analysis, LVI, 91
ribosomal RNA, LVI, 7, 8
deletion mapping, LVI, 162
ribosomes, LVI, 10, 11, *see also* Ribosome, mitochondrial
analysis of function, LVI, 91
components of RNA, LVI, 90, 91
electron microscopy, LVI, 86–88
immunological studies, LVI, 88–90
isolation, LVI, 80–84
proteins, LVI, 91
sucrose density gradient centrifugation, LVI, 84–86
sedimentation behavior, XXXI, 716, 717
segregating units, numbers, LVI, 151
separation, LVI, 130
into acid, LV, 204, 205
into alkali, LV, 205
of intermembrane components, LV, 91–93
of membranes
analyses, LV, 90, 91
general principles, LV, 90
reagents and solutions, LV, 90
of outer membrane components, LV, 91
smooth muscle, LV, 14, 15
content and yield, LV, 62
isolation, LV, 60, 61
properties, LV, 62–65
spectra, fourth derivative analysis, LVI, 501–503
of spermatozoa, LV, 21, 22
of spleen, LV, 21
stabilization of function, LV, 115, 116
criteria of purity, LV, 117, 118
experimental techniques, LV, 116, 117
storage, LV, 6, 7, 31
substrate for phospholipid exchange enzyme, XXXV, 264
systems studied, LVI, 3, 4

testicular, LV, 11
of thymus gland, LV, 19
of thyroid gland, LV, 18
transfer RNA, LVI, 8, 9
translation products, LVI, 11, 12
in vivo labeling, LVI, 59, 60
ultrafiltration, XXXII, 5
ultrastructure of toluene-treated, LVI, 548, 549
uncoupler binding site, photoaffinity labeling, LVI, 673–683
volume, XXXII, 15, 17, 18
volume changes, recording, LV, 668
wash, LIII, 5, 42
yeast
ATPase isolation by chloroform method, LV, 341, 342
enzymatic preparation, LV, 157–159
isolation, XXXII, 825–829; LV, 26, 27
large scale preparation, LV, 161, 162
mechanical preparation, LV, 151–157
promitochondria, LV, 28
properties, LV, 27, 28
small scale preparation, LV, 162, 163
Mitochondrial paste, preparation, LIII, 61, 62, 113, 114
Mitogen
dosage, LVIII, 492
effect on lymphocyte, LVIII, 486
from *Phytolacca americana*
biological activities, L, 359–361
characterization, L, 357–359
lymphocyte specificity, L, 354
preparation, L, 355, 356
Streptococcus pneumoniae, L, 359
Mitomycin, as curing agent, XLIII, 52
Mitomycin C
colicin K preparation, LV, 534
DNA, RNA inhibitor, XL, 281
Mitomycin C group, XLIII, 152, 153
Mitoplast
electron microscopy, XXXI, 320
isolation, XXXI, 317, 318
physically prepared characteristics, LV, 98
properties, XXXI, 319–321
rat
heart, structure, LV, 97
liver, structure, LV, 96
separation, LV, 91
structure, LV, 95
Mitosis
hormone induction of biorhythms, XXXVI, 480
in measurement of termination point, XL, 49, 52
peak activity, LVIII, 254–256
Mitotic apparatus, XL, 83
isolation, XL, 84
problems, XL, 88
stability, XL, 87
parallel isolation procedures with metaphase chromosomes and nuclei, XL, 75
Mitotic cell
increase, LVIII, 255
preparation, XXXII, 59, 594
selection, LVIII, 192, 193
Mitotic cell accumulation function, rate of increase, XL, 56
Mitotic detachment, synchronization, LVIII, 252–256
Mitotic index, LVIII, 233, 327
Mitotic index analysis, in exponential cell cultures, XL, 47
Mitotic spindle inhibitor, LVIII, 327
Mitsuhashi and Maramorosch medium, LVIII, 454, 462
preparation, LVIII, 462
MIX, *see* 1-Methyl-3-isobutylxanthine
Mixed anhydride analog, XLVI, 302–307
Mixed anhydride method, XLVI, 200, 210
carboxyl group activation, XLVII, 554
Mixed culture, *see* Cell culture, mixed
Mixed-disulfide matrix, XXXIV, 533–536
Mixed-function amine oxidase, *see* Amine oxidase, mixed function
Mixed function oxidase
bacterial luciferase, LIII, 562
microsomal, drug substrate studies, XXXI, 233, 236
Mixed-function oxidase system, *see* Flavoprotein-linked monooxygenase
Mixed media, *see* Media, mixture

Mixer, four-jet tangential, LIV, 89
Mixer 66, for plant-cell rupture, XXXI, 503
Mixing, volume change, XLVIII, 33–38
Mixing-bath loading, *see* Shaking-bath loading adsorption method
Mixing chamber, continuous-flow pH measurement, LV, 621
M9 mineral salts medium, LIX, 640
MMTS, *see* Methyl methanethiolsulfonate
Mnemiopsis, luciferase, LIII, 560
MNNG, *see* *N*-Methyl-*N*-nitro-*N*′-nitrosoguanidine
MN Polygram Cel 300 apparatus, XXXII, 780
M12NS, *see* 2,2-Dimethyl-5-hexylmethylundecanoate-*N*-oxyloxazolidine
Mobility shift analysis, LIX, 80–88
MODELAIDE computer program, XXXVI, 10
Model enzyme system, XLIV, 63, 64, 558
Mode-locking dye cell, LIV, 5
Mode-locking medium, LIV, 7
Modification buffer, for thin-layer peptide mapping, XLVII, 196
Modulation, selective, of polarized light, XLIX, 188
Modulation spectroscopy, XLIX, 484, 485
Modulator protein, calcium-dependent
  activity, LVII, 108
  microassay, LVII, 107–112
Mössbauer spectroscopy, LIV, 346–379
  combination with low-temperature EPR spectroscopy, LIV, 132, 363–365
  of complex systems, LIV, 365–369
  effects of freezing, LIV, 377, 378
  isotopic enrichment, LIV, 374, 375
  sample cells, LIV, 375, 376
  short-lived species and, LIV, 379
  temperature and, LIV, 373, 376, 377
  theoretical, background, LIV, 347–351
Mohberg and Johnson's medium, LVIII, 57
Moiré pattern, in negative staining, XXXII, 34
Molar ellipticity
  computation, LIV, 252
  equivalents, LIV, 290
Molday, Englander, and Kallen factor, XLIX, 27, 28
Moldicin B, XLIII, 135
Molecular dipstick, XLIX, 464
Molecular motion, *see* Chemical bond vibrations; Molecular rotations
Molecular polarizability derivative, LIV, 235
Molecular rotation, time range, LIV, 2
Molecular sieve, XXXIV, 522
  catalyst, for oxygen removal, LIV, 117
Molecular Sieve 13X, LIII, 644
Molecular transport, of reversibly reacting systems, XLVIII, 195–212
Molecular weight
  average, apparent weight, XLVIII, 74
  computation
    linearization technique, XLVIII, 179–184
    from sedimentation analysis data, XLVIII, 179–184
    two-species plot, XLVIII, 176, 177
  dalton preference, XXXII, 35
  determination
    electrophoretic mobilities, XLVII, 61, 62
    elution from sorbents, LX, 189, 397, 477, 529, 702, 708
    by gel electrophoresis, XLVIII, 3–10
    magnetic suspension, XLVIII, 29–69
    osmotic pressure, XLVIII, 82–87
    by sedimentation equilibrium, XLVIII, 163
    gradient centrifugation, LX, 52, 56, 294, 492–494
    by neutron scattering, LIX, 689, 690, 712
    polyacrylamide gel electrophoresis, *see* Polyacrylamide gel electrophoresis
  electrophoretic determination, XXXII, 77, 78
  elongation factors, *see* specific elongation factors
  function of osmotic pressure, XLVIII, 82–87

initiation factors, LX, 28, 29, 63, 159–161, *see also* specific initiation factors
of membrane protein and glycoprotein, by gel electrophoresis, XXXII, 92–102
calculations, XXXII, 100, 101
number average, XLVIII, 124
concentration dependence, for self-associating species, XLVIII, 104
definition, XLVIII, 98
evaluation, XLVIII, 107, 108
primary charge effect, XLVIII, 89
relation to radius of gyration, LIX, 720, 721
weight average, definition, XLVIII, 108
Z average, XLVIII, 179
Mollusk, oxidase, LII, 10
Moloney murine leukemia virus, microcarrier culture, LVIII, 192
Molybdate, in negative stains, XXXII, 22, 24, 29, 30
Molybdenum
atomic emission studies, LIV, 455
formate dehydrogenase, LIII, 371
interference, in iron analysis, LIV, 440
in media, LVIII, 52, 69
in nitrate reductase, LIII, 352
role in biological oxidations, LIII, 401
Molybdenum(III), electron paramagnetic resonance studies, XLIX, 514
Molybdenum(IV), electron paramagnetic resonance studies, XLIX, 514
Molybdenum(V), electron paramagnetic resonance studies, XLIX, 514
Molybdenum(VI), electron paramagnetic resonance studies, XLIX, 514
Molybdenum cyanide ion, as mediator-titrant, LIV, 408
Molybdenum species, EPR characteristics, LIV, 134
Molybdoferredoxin, *see also* Nitrogenase
crystallization, LIII, 325
purification, LIII, 317, 320, 322–325, 327
specific activity, LIII, 319
Momordia charantia, lectin, XXXIV, 334
Monactin
as ionophore, LV, 442
structure, LV, 436
transport, LV, 680
Monamycine, XLIII, 338
Monazomycin, XLIII, 338
as ionophore, LV, 441, 442
Monensin
code number, LV, 445
as ionophore, LV, 442, 446
solutions, LV, 448
Monitoring
antigen-antibody reaction, LVIII, 174–178
with antiserum, LVIII, 176–178
cell characteristics, LVIII, 164–178
by chromosomal examination, LVIII, 170–174
by isozymes, LVIII, 166–170
Monitoring film badge, for X-ray workers, XXXII, 215
Monkey kidney cell
dissociation, LVIII, 127
suspension culture, LVIII, 203
Monkey serum, for granulosa cell culture, XXXIX, 207
Mono-*N*-acetylneomycin C, XLIII, 227
Monoacylhydrazide, chemiluminescence, LVII, 422
Monoamine oxidase, XLVI, 38, *see also* Amine oxidase, flavin-containing
activity assay, LIV, 492, 497
assay, XXXIX, 388–389
in melatonin biosynthesis, XXXIX, 379
in mitochondrial membranes, XXXII, 91
in outer mitochondrial membranes, XXXI, 319
in pineal, XXXIX, 396
plasma membrane, XXXI, 90
Monoaminopeptidase, activity assay, XLVII, 393
Monoazidoethidium bromide, XLVI, 84
Mono(*p*-azobenzenearsonic acid)-*N*-*t*-butyloxycarbonyl-L-tyrosine, XLVI, 494, 495
Mono(*p*-azobenzenearsonic acid)-L-tyrosine, XLVI, 494

N^6-Monobutyryl-cyclic adenosine monophosphate, cAMP inhibitor, XXXVIII, 275, 282, 283
2′-*O*-Monobutyryl-cyclic inosine monophosphate, preparation, XXXVIII, 408
2′-*O*-Monobutyryl-cyclic isoadenosine monophosphate, preparation, XXXVIII, 409
2′-*O*-Monobutyryl-cyclic uridine monophosphate, preparation, XXXVIII, 407
Monochloroacetic acid, in preparation of luciferin, LVII, 19
Monochlorodimedone, preparation, LII, 523
Monochlorosuccinic acid, XXXIV, 406
Monochloro-*s*-triazinyl derivative, XLIV, 50
Monochromator, XLIX, 87, 88
  in flash photolysis apparatus, LIV, 53
  for microspectrophotometer, XXXIX, 417
  in nanosecond absorbance spectroscopy, LIV, 36, 37
  for X-ray spectroscopy, LIV, 328, 329
    calibration, LIV, 329
Monocyte, LVIII, 492, 494, 495
  chemiluminescence, LVII, 465
  formation, XXXII, 758
  isolation, XXXII, 759–762; LVII, 467, 470
Monod, Wyman and Changeux model of cooperativity, XLVIII, 289, 290, 294
Monofluorosuccinate, LVI, 586
Monogalactosyldiglyceride, globoid cell leukodystrophy, L, 470
Monoglyceride, *see also* Triglyceride
  thin layer chromatography, XXXV, 400
Monolayer, use and interaction, XXXII, 539–544
Monolayer culture
  HTC cells, LVIII, 545
  human adenovirus, LVIII, 426
  media, LVIII, 88, 89
  microscopic evaluation, LVIII, 139
  morphology, LVIII, 139, 140
  poikilotherm vertebrate cells, LVIII, 474–476
  properties, LVIII, 133–135
  protocol, LVIII, 138, 139
  substrate for attachment, LVIII, 137, 138
  techniques, LVIII, 132–140
Monomethoxyltropylium ion
  identification by mass spectroscopy, LIII, 596
  structural formula, LIII, 596
(Monomethoxytrityl)cellulose, XXXIV, 646
ε-Monomethyllysine, chromatographic separation, LIX, 788
Mononucleotide, electrophoretic mobilities, LIX, 73
Monooxygenase, *see also* specific type
  activities associated with *Ah* locus, LII, 231, 232
  definition, LII, 6
  external
    definition, LIII, 399
    luciferase, LIII, 562
    salicylate hydroxylase, LIII, 527, 543, 544
  internal
    definition, LIII, 399
    luciferases, LIII, 558
  system, schematic, LII, 227
Monophenol monooxygenase, XLIV, 886
  melanocytes, LVIII, 564
Monosaccharide, XXXIV, 317–328
  cross-linked, XXXIV, 329–331
  cultured cell growth, L, 189–191
  interconversion, in animal cells, L, 182, 183
Monosaccharide-agarose, XXXIV, 317–328
  adsorbents, XXXIV, 327
Monosialoganglioside, XXXIV, 611
  isolation, XXXII, 354, 355
Monosialyllacto-*N*-hexaose, structure, L, 224
Monosialyllacto-*N*-neohexaose, structure, L, 225
Monosialylmonofucosyllacto-*N*-neooctaose, structure, L, 224
Monosialylmonofucosyllacto-*N*-octaose, structure, L, 224
Monostable photometer device, LVII, 534, 535
  in autoranging instrument amplifier, LVII, 538

in digital integrator, LVII, 539
Monovalent anion
for displacement of oxaloacetate, LIII, 472
inhibitor of *p*-hydroxybenzoate hydroxylase, LIII, 551
of salicylate hydroxylase, LIII, 541
Monoxime reagent, in dihydroorotase, LI, 123
Montal-Mueller apparatus, XXXII, 526, 527
Monte Carlo method, in error estimation, XXXVI, 14, 15
Moore's medium RPMI 1640, LVIII, 56, 62–70, 89, 91
MOPC, *see* Myeloma protein, dinitrophenol-binding
MOPC 315, mouse tumor, XLVI, 503
MOPC 460, mouse tumor, XLVI, 83
MOPS buffer, *see* 3-(*N*-Morpholino)propanesulfonic acid
Moraxella, oxidase, LII, 11
Mordant, for Parducz fixation, XXXII, 55
Morgan-Elson reaction
in glycolipid characterization, XXXII, 348, 364, 365
modified, XLVII, 490
Morphine, XXXIV, 619–623
antibodies, XXXIV, 621, 622
antisera, XXXVI, 17
logit-log assay, XXXVII, 9
photochemically active derivatives, LVI, 660
Morphine-agarose, XXXIV, 619–623
antibodies, XXXIV, 621–622
Morphine-glass, XXXIV, 619
Morphine receptor, XXXIV, 622
Morphogenesis, in amoebae, effects, XXXIX, 491, 492
Morphol dioxygenase, LII, 11
Morpholine, uptake, pH change, LV, 564
4-Morpholine *N,N*′-dicyclohexylcarboxamidine, XLVI, 252
Morpholine-hydrochloride buffer, preparation of tRNA-C-C-dA, LIX, 128, 134
2-(*N*-Morpholino)ethanesulfonate-sodium hydroxide buffer, preparation of tRNA-C-C-dA, LIX, 128, 134
2-(*N*-Morpholino)ethanesulfonic acid, XXXII, 718; XLVII, 186
activation of succinate dehydrogenase, LIII, 474
in deoxythymidylate phosphohydrolase assay, LI, 285
determination of ferrocytochrome *c* oxidation, LIII, 45
purification of nitrogenase, LIII, 322
3-(2-Morpholinoethyl)carbodiimide, XLIV, 144
3-(*N*-Morpholino)propanesulfonic acid
assay of ATP, LVII, 98
in bacterial growth medium, LVII, 190
interference by, in OPRTase-OMPdecase complex assay, LI, 167
luciferase-mediated binding assays, LVII, 114
preparation of ribosomal subunits, LIX, 412, 419
purification of complex III, LIII, 34, 93
Morphometry
biochemical data correlation, XXXII, 15
FORTRAN program, XXXII, 13, 16–20
of subcellular fractions, XXXII, 3–20
Morris hepatoma, LVIII, 544
isolation of mitochondria
using homogenization
method, LV, 86, 87
principle, LV, 84–86
properties, LV, 87, 88
using proteolysis
assay procedures, LV, 81
enzymatic properties, LV, 82–84
principle, LV, 79, 80
procedure, LV, 80, 81
Mortimore liver perfusion system, XXXII, 626
Moscona's saline, LVIII, 138, 139
Mosquito cell
media, LVIII, 457, 462
suspension culture, LVIII, 454
Moth, *see also* specific type
cell line, LVIII, 450
media, LVIII, 456, 462

suspension culture, LVIII, 455
Motility, bacterial, quasi-elastic light scattering, XLVIII, 421
Motor, continuous-flow pH measurement, LV, 619
Mouse
acatalasemic, XLIV, 684–689
Ah locus, LII, 228, 229
bone and skin, procollagenase, XL, 352
brain, metabolite content, LV, 218, 219
breeding, LII, 230, 231
efficiency, LII, 230
cell
nontransformed, media, LVIII, 56
tumorigenic, LVIII, 376
embryo
carcinoma, LVIII, 97
cooling rate, LVIII, 31
fibroblast
media, LVIII, 58, 59, 90
monolayer culture, LVIII, 132
media, LVIII, 87
preparation, XXXIX, 297–302
erythroleukemia cell, LVIII, 346, 506–511
cloning, LVIII, 509
growth conditions, LVIII, 507–509
hemoglobin induction, LVIII, 509, 510
media, LVIII, 508
serum
requirements, LVIII, 508
testing, LVIII, 508, 509
fibroblast
effect of hormone, LVIII, 101, 102
Swiss 3T3, effect of hormone, LVIII, 105, 106
implanted lymphosarcoma, XLIV, 689–694, 703
inbred strains, availability, LII, 230
karyotype, LVIII, 341, 342
L cell
cloning, LVIII, 163
interferon production, LVIII, 295
microscopy, XL, 304
nuclear fluorescence, XL, 303
leukemia, treatment, XLIV, 703
liver
microsome isolation, XXXI, 199
plasma membrane isolation, XXXI, 75–90
lymphoblastoid cell, LVIII, 218
EL4, LVIII, 223
lymphocyte cell, in enzyme binding assays, LVII, 403, 406
lymphoma cell, S49
cell cycle phase, LVIII, 242
DNA content, LVIII, 244
fluorescence spectrum, LVIII, 238
lymphosarcoma MB(T-86157), suspension culture, LVIII, 203
marrow cell, growth factor, LVIII, 163
mast cell tumor, uridine-cytidine kinase from, LI, 314–321
melanoma cell line PG19, tumorigenicity testing, LVIII, 374
neuroblastoma cell
C1300, LVIII, 89
media, LVIII, 57
nude, *see* Nude mouse
pancreatic cell, media, LVIII, 59
pituitary cell, characteristics, LVIII, 531
plasmacytoma, cell fusion, LVIII, 349
spleen, leukemic
crude extract preparation, LI, 409
cytidine deaminase from, LI, 408–412
testes, lactate dehydrogenase–X, assay, purification, and properties, XLI, 318–322
testicular cell TM4, LVIII, 96, 97
thymusless, LVIII, 370
thymus leukemia antigen, LVIII, 219
virus, *see* specific type
Mouse × human hybrid, LVIII, 347, 357
10-MP, *see* 10-Methylphenothiazine
MPD, *see* 2-Methylpentane-2,4-diol
MPNL 65/C medium, LVIII, 59
MPO, *see* Myeloperoxidase
mRNA, *see* Messenger ribonucleic acid
MRW, *see* Mean residue molecular weight
6-MSA, *see* 6-Methylsalicylic acid
MSE ultrasonic disintegrator, LIII, 358

MSH, *see* Melanocyte-stimulating hormone
ms^2i^6A, *see* 2-Methylthio-N^6-(2-isopentenyl)adenosine
MT80 cell, plasma membrane isolation, XXXI, 161
MTPA-C1, *see* (−)-α-Methoxy-α-trifluoromethylphenylacetyl chloride
MTT-tetrazolium salt, *see* 3(4,5-Dimethylthiazolyl-2)-2,5-diphenyltetrazolium bromide
Mucolipidosis II
  acid phosphatase, L, 454, 455
  arylsulfatase A, L, 455
  assay, L, 455
  α-L-fucosidase, L, 453
  β-galactosidase, L, 455
  β-glucuronidase, L, 452, 455
  β-hexosaminidase, L, 455
  iduronate sulfatase, L, 446, 455
  α-L-iduronidase activity, L, 444, 454
  α-mannosidase, L, 455
  multiple sulfatase deficiency, L, 454
Mucolipidosis III
  acid phosphatase, L, 454, 455
  arylsulfatase A, L, 455
  assay, L, 455
  α-L-fucosidase, L, 453
  β-galactosidase, L, 455
  β-glucuronidase, L, 452, 455
  β-hexosaminidase, L, 455
  iduronate sulfatase, L, 446, 455
  α-L-iduronidase activity, L, 444, 454
  α-mannosidase, L, 455
  multiple sulfatase deficiency, L, 454
Mucopeptide, bacterial polysaccharides, L, 253
Mucopolysaccharide, LVIII, 267, 268
  effect on estrogen receptor studies, XXXVI, 173
  elimination, from nitrogenase preparation, LIII, 316, 317
  human cells synthesizing, XXXII, 799
  storage disease, LVIII, 444
    enzymic diagnosis, L, 439–456
Mucopolysaccharidosis I, *see also* Hurler/Scheie syndrome; Hurler syndrome; Scheie syndrome
  α-L-iduronidase activity, L, 443
Mucopolysaccharidosis VII, β-glucuronidase, L, 452
Mucus, *see* specific type
Mueller-Rudin brush technique for bilayer formation, XXXII, 524
Multicell culture, LVIII, 264
Multidimensional scaling
  computer programs, LIX, 609–611
  of immunoelectron microscopic data, LIX, 611
  ribosomal models, LIX, 602–611
Multiple-cuvette rotor, for Fast Analyzer, XXXI, 791–793
Multiple isomorphous replacement method, LIX, 3, 11–13
Multiple sclerosis, A1 protein, XXXII, 324
Multiple sulfatase deficiency
  arylsulfatase A, L, 453
  arylsulfatase B, L, 453
  arylsulfatase C, L, 454
  assay, L, 454
  enzymic diagnosis, L, 474, 475
  heparin sulfamidase, L, 453
  iduronate sulfatase, L, 453
  metachromatic leukodystrophy, L, 471
  steroid sulfatases, L, 454
Multiplication stimulating activity, LVIII, 79, 96, 107, 162, 163
Multistep enzyme system
  application, XLIV, 468–473
  assay, XLIV, 455, 456, 462
  asymmetric for active site distribution, XLIV, 921–924
  bound protein determination, XLIV, 462, 463
  cofactor regeneration, XLIV, 698
  coimmobilization, XLIV, 454–478
  definition, XLIV, 453
  enzyme activity ratio, XLIV, 461, 462
  immobilization
    in hollow fiber membranes, XLIV, 312–314
    on inert protein, XLIV, 917–919
    technique, choice, XLIV, 461
  kinetics, XLIV, 432–435, 463–468, 917–919
  perturbed pH optima, XLIV, 468, 469
  physical characterization, XLIV, 370

polarimetric assay, XLIV, 350
soluble aggregates, XLIV, 473–475
support, choice, XLIV, 461
Mung bean
organelles, XXXI, 533
polyribosome isolation, XXXI, 589
Muramic acid, XXXIV, 333
Muramidase, *see also* Lysozyme
lysozyme, use in algal lysis, XXXI, 682
Murashige and Skoog inorganic salt, LVIII, 480
Murexide
calcium, LVI, 305, 323, 331, 332
comparison to other calcium indicators, LVI, 327, 329–332
properties, LVI, 330
Murine cl 1 cell, microcarrier culture, LVIII, 190
Murine leukemia virus, LVIII, 225, 226
assay, LVIII, 412–424
enzymic, LVIII, 416–421
immunological, LVIII, 416–421
virological, LVIII, 421–424
cells for propagation, LVIII, 414
host range, LVIII, 413
purification, LVIII, 412–424
S^+L^- test, LVIII, 423, 424
source, LVIII, 414
XC test, LVIII, 421–423
Murine lymphoma virus, transformation, LVIII, 369
Murine sarcoma virus, LVIII, 413
cells transformed, plasma membrane isolation, XXXI, 163
transformation, LVIII, 369
Murine type C virus, LVIII, 412
Muscle, *see also* specific types
adenylate cyclase
assay, XXXVIII, 143, 144
preparation, XXXVIII, 144–146
properties, XXXVIII, 146–149
solubilization, XXXVIII, 149
cAMP-receptor, XXXVIII, 380
cell dissociation, LVIII, 513–516
extract, immobilization, by microencapsulation, XLIV, 214
human, enolase, assay, purification, and properties, XLII, 335–338
$(Na^+ + K^+)$-ATPase isolation, XXXII, 280
oxidase, LII, 16, 18
perfusion methods, XXXIX, 65–106
phosphodiesterase
assay, XXXVIII, 240, 241
properties, XXXVIII, 243, 244
purification, XXXVIII, 241–243
phosphofructokinase, cAMP derivatives, XXXVIII, 396–398
protein kinases, XXXVIII, 298–301, 366
catalytic subunits, XXXVIII, 306–308
holoenzyme, XXXVIII, 301–306
inhibitor, XXXVIII, 353–355
smooth
mitochondria, LV, 14, 15, 60–65
perfusion methods, XXXIX, 65–106
striated
clones, XXXII, 770
mitochondria, LV, 12, 13
Mustard-green leaf, peroxidase in, binding properties, LII, 516, 521
Mutagen, *see also* specific substance
care in handling, LVIII, 312, 313
toxicity, LVIII, 312, 313
Mutagenesis, XLVI, 644
of *E. coli*, LVI, 107, 108
effect of medium, LVIII, 310, 311
of serum, LVIII, 310
of luminous bacteria, LVII, 167
petite mutants, LVI, 159
possible mechanism, LII, 228
procedures, LVI, 98, 99
sensitizer, LVIII, 317
temperature-sensitive mutants, LVI, 133–135
Mutagenic technique, XLIII, 29–36, *see also* Mutation; Mutagen-treated population
Mutagen-treated population, selection, XLIII, 36–41
auxotrophs, XLIII, 38
morphological differences, XLIII, 37
mutants
producing one antibiotic entity, XLIII, 39, 40

resistant to toxic analogs of precursors, XLIII, 39
overlay techniques, XLIII, 37, 38
prototrophs, XLIII, 38, 39
random isolation, XLIII, 37
Mutant
affected in oxidative phosphorylation
characterization, LVI, 112–114
genetic classification, LVI, 114, 115
amelanotic, isolation, LVIII, 565
antibiotic resistance, LVI, 14
ATP-P_i exchange, LV, 362, 363
aurovertin-resistant, LVI, 178, 179
selection procedures and properties, LVI, 179–182
auxotroph, LVIII, 316
carbodiimide-resistant
principle, LVI, 164
properties, LVI, 169, 170
reproducibility of plating procedure, LVI, 166
screening of resistant candidates, LVI, 165, 166
selection of EDC-resistant mutants, LVI, 167
selection procedure, LVI, 164, 165
verification of DCCD-resistant growth, LVI, 166
carrying *rec* A allele, preparation, LVI, 116
cell, human culture, XXXII, 799–819
cell cycle action, LVIII, 316
classification
meiotic segregation, LVI, 103
mitotic segregation, LVI, 103
as nuclear or mitochondrial, LVI, 102, 103
cloning, LVIII, 321
colony isolation, LVIII, 321
conditional, bioluminescence studies, LVII, 131
conditionally lethal, LVIII, 316–320
cyc 4, deficiency, LVI, 559
cyd 1, deficiency, LVI, 558, 559
cytoplasmic petite, gene mapping, LVI, 184, 185
dark, LVII, 130, 131
drug-resistance, LVIII, 313–316
fep$^-$, use, LVI, 395, 396
formation, LVIII, 308
frequencies, LVIII, 309
heme-deficient
estimation of fatty acids and sterols, LVI, 562, 563
explanation of pleiotropic lesions, LVI, 560
isolation and characterization, LVI, 174, 559, 560
manipulation of composition of cells, LVI, 560–562
nature and properties, LVI, 558, 559
induction, LVIII, 312, 313
isolation, LVIII, 308–322
effect of ploidy, LVIII, 308
incubation, LVIII, 315, 316
inoculation, LVIII, 315, 316
method, LVIII, 311
protocol, LVIII, 311
strategy, LVIII, 311
timing, LVIII, 311
karyotyping, LVIII, 309
melanoma cell, LVIII, 566
mit$^-$, LVI, 14–16
further characterization, LVI, 106
nature, LVI, 95, 96
mitochondrial, LVI, 140, 141
of mitochondrial membranes, selecting and maintaining, LVI, 119–121
multistep selections, LVIII, 316
mutagen use, LVIII, 312
need for isogenic controls, LVI, 119
new *unc* alleles, complementation test, LVI, 116, 117
ole 3
characterization, LVI, 559
manipulation of cell composition, LVI, 560–562
petite
conservation of purified strains, LVI, 161
cytoplasmic, LVI, 155, 156
detection of mitochondrial markers, LVI, 159, 160
nature, LVI, 95, 96, 155, 156
ole mutants, LVI, 576, 577
purification by subcloning, LVI, 160, 161

selection of other gene regions, LVI, 161, 162
types, LVI, 193
phenotype, LVIII, 313
presumptive oxidative phosphorylation deficient
mapping by transduction, LVI, 112
preparation of membranes, LVI, 112
presumptive ubiquinone-deficient, extraction of quinones and polyisoprenoid intermediates, LVI, 109, 111, 112
protein synthesis, LVIII, 317–319
protein synthesis negative, resolution into *syn*⁻ and *rho*⁻, LVI, 103, 104
requiring δ-aminolevulinate
growth media, LVI, 123, 124
nutrition, LVI, 122
strain variation in cytochrome spectra, LVI, 122, 123
synergistic and temperature effects, LVI, 121, 122
types, LVI, 121
rho⁻
growth at nonpermissive temperature, LVI, 137
nature, LVI, 95, 96
rho⁻ petite, isolation, LVI, 185, 186
screening, LVIII, 319, 320
temperature-sensitive, LVIII, 319
selection, inhibitors, LVI, 33
strains, source, LVI, 117
syn⁻, nature, LVI, 95, 96
temperature sensitive
enrichment for by inositol-less death, LVI, 135, 136
further screening, LVI, 137, 138
general strategy, LVI, 132
isolation, LVI, 132, 133
mutagenesis, LVI, 133–135
revertants, LVI, 136, 137
temperature-sensitive transformation, LVIII, 370
transformation-defective, of RSV, LVIII, 390
unc alleles, method for incorporation into episome, LVI, 115, 116
yeast
selection, LVI, 96, 97
stability and storage, LVI, 105, 106
Mutarotase, *see* Aldose 1-epimerase
Mutation, XLIII, 24–41, *see also* Mutagen-treated population
alkylating agents, XLIII, 32–35
assay, XLIII, 26, 27
auxotrophy, LVIII, 309, 316–320
chemical, LVIII, 312, 313
conditional, temperature-sensitive, LVIII, 309
enzymes, LVIII, 309
fermentation, XLIII, 25, 26
induced by combined action of mutagens, XLIII, 36
induction techniques, XLIII, 27–36
ionizing radiation, XLIII, 31, 32
isogenic control, need, LVI, 119
lectin resistance, LVIII, 309, 314
N-methyl-*N*′-nitro-*N*-nitrosoguanidine, XLIII, 34–35
mitochondria, LVI, 13–16
nitrogen mustard, XLIII, 33, 34
nuclear, mitochondria, LVI, 13
recessive, LVIII, 309
in CHO cells, LVIII, 309
recessive nuclear, maintenance, LVI, 120
rho⁻ state, LVI, 120
secondary, circumvention, LVI, 120, 121
selection of colonies, XLIII, 36–41
suspension preparation, XLIII, 26
types affecting mitochondrial membranes, LVI, 118, 119
ultraviolet light irradiation, XLIII, 29–31
Mycaminose, XLIII, 134
L-Mycarose, XLIII, 487
Mycarosyl erythronolide B, XLIII, 495
Mycelium, fungal, LI, 65
Mycetin, XLIII, 145
Mycobacillin, XLIII, 338
Mycobacterium
cell-wall isolation, XXXI, 655, 659
oxidase, LII, 18
Mycobacterium phlei
cultivation, XXXV, 91, 92

cytochrome a_3, LIII, 207, 208
fatty acid synthase, XXXV, 84–90
lactate oxygenase, assay, purification, and properties, XLI, 329–333
membrane isolation, XXXI, 701
transport systems, XXXI, 704
membrane structure, LV, 175–178
assay, for coupling factor activity, LV, 190, 191
for latent ATPase activity, LV, 191, 192
of transport with intact cells, LV, 179, 180
amino acid transport in proteoliposomes, LV, 197–200
comparison of active transport by ghosts and ETP, LV, 186, 187
growth media, LV, 179–189
inhibitors, LV, 185, 186
membrane fragments, LV, 195, 196
preparation and properties of depleted ETP, LV, 187, 188
of detergent-treated ETP, LV, 192, 193
of ETP, LV, 182–185
of protoplast ghosts, LV, 180–182
purification of BCF_1 by affinity chromatography, LV, 189, 190
of coupling factor-latent ATPase, LV, 189
of intrinsic membrane protein, LV, 195
of proline carrier protein, LV, 196
reconstruction of proline transport, LV, 197
solubilization and purification of DCCD-sensitive latent ATPase, LV, 193–195
summary
of reconstitution of functions, LV, 199
of types, LV, 198
methylated polysaccharide activators of fatty acid synthase, XXXV, 84, 85, 90–95
oxidase, LII, 16

Mycobacterium smegmatis, L-lactate oxidase, LIII, 439
Mycophenolic acid, XLIII, 154
Mycoplasma, LVIII, 21
autoradiography, LVIII, 27
biochemical tests, LVIII, 25
broth, LVIII, 22
colony, LVIII, 24
cultivation, XXXII, 459, 460
DNA stain, LVIII, 25
immunofluorescence, LVIII, 24, 25
infection, testing, LVIII, 375
isolation from medium, LVIII, 227
large-scale production, LVIII, 227
lipids of, X-ray studies, XXXII, 217
membranes
reconstitution, XXXII, 459–468
studies, XXXII, 864
microscopy, LVIII, 25
pseudo-colony, LVIII, 24
ribosomal RNA, LVIII, 27
scanning electron microscopy, LVIII, 28
in serum, LVIII, 23
testing, LVIII, 22–27
uracil uptake, LVIII, 26, 27
uridine phosphorylase, assay, LVIII, 25, 26
Mycoplasma hyorhinis, immunofluorescence, LVIII, 24, 25
Mycoplasma laidlawii, membranes, thermal transition studies, XLIX, 13
Mycoplasmal contamination, *see* Contamination, mycoplasmal
Mycoplasma pneumoniae, membrane reconstitution, XXXII, 464
Mycostatin, in media, LVIII, 198
Myelin
A1 protein isolation, XXXII, 330–335, 341
basic proteins, XXXII, 323–341, 344
2,3-cyclic nucleotide-3′-phosphohydrolase, XXXII, 127
electric-tissue postsynaptic membranes, XXXII, 317, 318
globules resembling *M. pneumoniae* glycolipids, XXXII, 464
IR spectroscopy, XXXII, 253

isolation from nerve tissue, XXXI, 435–444
lipid composition, XXXI, 444
marker enzyme, XXXII, 128
membrane, X-ray studies, XXXII, 216, 218, 219
occurrence and function, XXXI, 435, 436
from peripheral nerves, basic proteins, XXXII, 341–345
preparation of brain mitochondria, LV, 55
purity criteria, XXXI, 443, 444
Myeloma, LVIII, 376
cell, plasma membranes, XXXI, 144
hybrid, LVIII, 349
IgG, XXXIV, 337
protein, XXXIV, 106, 185, 703
bacterial polysaccharides, L, 252
dinitrophenol-binding, XXXIV, 183
Myeloperoxidase
assay, XXXI, 347, 348
phagocytic cell chemiluminescence, LVII, 462–465, 488, 489
in PMN granules, XXXI, 346, 352
Myeloperoxidase-hydrogen peroxide-halide reaction, LVII, 462, 463, 465
Mylar capacitor, LVII, 539
Myoblast, skeletal, LVIII, 377
cell course, LVIII, 512
culture, LVIII, 511–527
differential release, LVIII, 517
differentiation from fibroblasts, LVIII, 517
establishment of culture, LVIII, 512
inoculum size, LVIII, 518–520
media, LVIII, 520–524
composition, LVIII, 518–520
conditioned, LVIII, 524, 525
secondary suspension, LVIII, 516–518
synchrony, LVIII, 511, 518–520
Myoblast fusion, electrolectin, L, 300
Myocardial cell, calcium efflux studies, XXXIX, 549
Myocardial infarction
acute, effect on creatine kinase B activity, LVII, 59, 63
sialic acid-rich oligosaccharide excretion, L, 233
Myocardium, *see also* Heart
adenylate cyclase, solubilization and role of phospholipids, XXXVIII, 174–180
Myofibril, ATPase inhibitor, LV, 402
Myoglobin, LII, 16, 473–486
absorption spectrum, LIV, 287
acylation techniques, XLVII, 149
from California gray whale, staphylococcal protease hydrolysis, XLVII, 190
carbon monoxide binding
data evaluation, LIV, 514, 515, 523, 529–531
IR studies, LIV, 310
conformational composition, LIV, 268, 271, 272
conjugate, conformational transitions, XLIV, 539, 540
as electron microscopic tracer, XXXIX, 147, 148
as electrophoretic marker, XXXII, 80
electrophoretic properties, LII, 478, 479
free peptide hydrogen exchange, XLIX, 30, 34, 35
histidine titration, LII, 481, 482
hollow-fiber retention data, XLIV, 296
immobilization, on Sephadex, XLIV, 540
intramolecular diffusion studies, LII, 484–486
intrinsic CD spectrum, LIV, 270, 272
IR studies, LIV, 305, 309, 310
isolation, LII, 474, 475
Keilin-Hartree preparation, LV, 119, 120
magnetic circular dichroism, XLIX, 170, 172; LIV, 287, 288, 295, 297, 298
magnetic susceptibility studies, LIV, 393
metal substitutions, LII, 492, 493
oxygen-binding equilibrium, LII, 483, 484
as oxygen donor, in oxygen consumption studies, LIV, 488
phthalylation, XLVII, 155

primary structure, LII, 479, 480
prosthetic group, LII, 5
purification, LII, 475, 476
reaction with oxygen, kinetics, LIV, 486
Soret CD spectrum, LIV, 260, 261, 272, 273
spectral properties, LII, 480, 481
from sperm whale
analysis, XLVII, 47, 48
solid-phase sequencing, XLVII, 269
spin state determination, from resonance Raman bands, LIV, 243, 244
stability, LII, 482, 483
unfolding, XLIX, 4
Myoinositol, *see* *myo*-Inositol
Myokinase, XXXIV, 246, *see also* Adenylate kinase
capacity, XXXIV, 249
preparation of 3′-amino-3′-deoxy-ATP, LIX, 135
Myosin, XXXIV, 490; XLIV, 882; XLVI, 318–321
antibody binding, L, 55
cGMP assay, XXXVIII, 106, 109
cyanylation, XLVII, 131, 132
in heart cells, XXXII, 741, 744, 745
heart mitochondria, LV, 47
heavy chain, SDS complex, properties, XLVIII, 6
rabbit muscle, viscosity, XLVIII, 54
removal from sarcoplasmic reticulum, XXXII, 292
saturation-transfer studies, XLIX, 492–495
Myosin adenosinetriphosphatase, XLVI, 86, 260, 287, 318
reaction with arylazido-β-alanine ATP, XLVI, 273–276
Myosin ATPase, *see* Myosin adenosinetriphosphatase
Myotonic dystrophy membrane, saturation-transfer studies, XLIX, 499, 500
Myristic acid, assay, LVII, 189–194
Myristoleic acid, quantum yield, in bioassay, LVII, 192
Myristyl alcohol, quantum yield, in bioassay, LVII, 192
Myristyl coenzyme A, substrate for palmityl thioesterase I, XXXV, 106
Myrothecium, oxidase, LII, 11
Myxomycete, free cells, XXXII, 723

N

N16, medium, LVIII, 56
4N27, *see* 2,2-Dimethyl-5,5-ditridecane-*N*-oxyloxazolidine
6N11, *see* 2,2-Dimethyl-5,5-dipentyl-*N*-oxyloxazolidine
Nactrin, homologs, as ionophores, LV, 444
NAD, *see* Nicotinamide adenine dinucleotide
NAD^+, *see* Nicotinamide adenine dinucleotide
NAD-agarose, XXXIV, 229, 232, 233, 302, 477
elution, XXXIV, 236
NAD^+-N^6-[*N*-(6-aminohexyl) acetamide], XXXIV, 229
NAD^+-N^6-[*N*-(6-aminohexyl) acetamide]-agarose, XXXIV, 233–235
NADH, *see* Nicotinamide adenine dinucleotide, reduced
NADH-CoQ reductase, XLVI, 284, 285; LIII, 407
assay and inhibitors, LVI, 585, 586
components, LVI, 580, 581
NADH-cytochrome b_5 reductase
absorption spectrum, LII, 107
activity, LII, 107
antibody against, purification, LII, 241, 242
assay, LII, 103
contaminant, of cytochrome *P*–450LM forms, LII, 117
in cytochrome b_5 assay, LII, 97
dispersion state, LII, 106
distribution, LII, 102
effect of extraction methods on properties, LIII, 408, 409
of protease treatment, LII, 108
flavin content, LII, 107
inhibition, by phosphate anions, LII, 244
membrane binding, LII, 108
in microsomal fractions, LII, 88
molecular weight, LII, 106

purification, LII, 104, 105
requirement for regeneration of stearyl-CoA desaturase activity, XXXV, 258, 259
stability, LII, 107
storage, LII, 105

NADH:cytochrome *c* oxidoreductase, *see also* NADH-cytochrome *c* reductase
in erythrocyte ghost sidedness assay, XXXI, 176–178, 179

NADH-cytochrome *c* reductase
assay, XXXI, 20; XXXII, 397, 398, 400; LII, 244; LIII, 9, 10, 156; LVI, 102
chromatographic separation, XXXII, 401–404
as diagnostic enzyme, XXXI, 20, 24, 33, 36, 37, 148
effect of DABS labeling, LVI, 621
in endoplasmic reticulum, XXXI, 165, 185–187
inhibitors, LIII, 8, 9
in intestinal mucosa, XXXII, 667, 671
as marker enzyme, XXXI, 262, 746
as membrane marker, XXXII, 393, 402
in microbodies, LII, 496
in microsomal fractions, LII, 88
in microsomes, XXXI, 65, 67, 225
as mixed-function oxidase substrate, XXXI, 233
in nuclear membrane, XXXI, 290–292
properties, LIII, 7–9
purification, LIII, 5–7
of complex III, LIII, 86
rotenone-insensitive, outer membrane, LV, 101–104
rotenone-sensitive, inner membrane, LV, 101–104
submitochondrial particles, LV, 109
toluene-treated mitochondria, LVI, 547

NADH-cytochrome *P*-450 reductase, assay, LII, 244

NADH dehydrogenase, LIII, 58, 397, 405, 406, 408, 413–418, *see also* Erythrocyte cytochrome b_5 reductase
activators, LIII, 20
activity, LIII, 17, 18; LVI, 587
assay, LIII, 20, 21, 406
chaotropic ion solubilization, XXXI, 778–782
component stoichiometries, LIV, 145
composition, LIII, 17
effect of DABS or PMPS labeling, LVI, 621
energy conservation site, LIII, 1, 413
EPR signal characteristics, LIV, 138, 140, 141, 143, 144
EPR spectrum, effect of alcohols, LIV, 150
high molecular weight form, LIII, 414–417
inhibitors, LIII, 20
iron-sulfur center, EPR spectra, LIV, 146, 147
Keilin-Hartree preparation, LV, 126, 127
low molecular weight form, LIII, 413, 415–418
molecular weight, LIII, 18, 19
pH effects, LIII, 20
polypeptides, LVI, 588, 589
purification, LIII, 15–17
purity, LIII, 18
reductant, of azurin, LIII, 661
of cytochrome *c*-551, LIII, 661
reversible dissociation, LIII, 431, 433, 436
rotenone binding, LV, 455
rotenone sensitivity, LIII, 413
soluble, properties, LIII, 413
spectral properties, LIII, 19
stability, LIII, 17
substrate specificity, LIII, 17

NADH-dependent ferricyanide reductase, in column eluate, LII, 93

NADH-dependent reductase, *see* Erythrocyte cytochrome b_5 reductase

NADH-ferricyanide reductase
assay, LII, 103, 244; LIII, 13, 14
stability, LII, 107

NADH:FMN oxidoreductase
assay, LVII, 205
immobilization, LVII, 204, 205
purification, LVII, 203

NADH:hydroxypyruvate oxidase, as marker enzyme, XXXI, 742

NADH-hydroxypyruvate reductase, as marker enzyme, XXXI, 735

NADH-linked electron transport enzyme enrichment, LII, 82, 83
NADH-menadione reductase, halophile membranes, LVI, 403
NADH-methemoglobin reductase, *see* Erythrocyte cytochrome b_5 reductase
NADH oxidase
in bacterial membrane, XXXI, 649
in *E. coli* membrane, XXXI, 642
effects of DABS or PMPS labeling, LVI, 621
inhibition, LV, 266
Keilin-Hartree preparation, LV, 121, 124, 125
in microsomes, XXXI, 65
submitochondrial particles, LV, 109–112
system, reconstitution, LIII, 49
ubiquinone extraction, LIII, 573, 575–578
vesicles, LV, 168
NADH-rubredoxin reductase, from *Pseudomonas oleovorans*, LIII, 356
NADH-tetrazolium oxidoreductase, in *E. coli*, XXXII, 91
NADH:ubiquinone oxidoreductase, *see* Complex I
NADH-ubiquinone reductase
assay, LIII, 13, 14
EPR spectrum, LIV, 146, 148
NAD-linked enzyme, *see* Enzyme, NAD-linked
NAD-malate dehydrogenase, in microbodies, LII, 496, 499
NAD(P):menadione oxidoreductase, *Ah* locus, LII, 232
NAD nucleotidase, in plasma membrane, XXXI, 89
NADP, *see* Nicotinamide adenine dinucleotide phosphate
$NADP^+$, *see* Nicotinamide adenine dinucleotide phosphate
NADP-agarose, XXXIV, 123, 477
NADPH, *see* Nicotinamide adenine dinucleotide phosphate, reduced
NADPH-adrenoxin reductase, adrenal cortex mitochondria, LV, 10
NADPH-cytochrome b_5 reductase, lipid peroxidation, LII, 305
NADPH-cytochrome *c* reductase, XXXIV, 302; LII, 45
Ah locus, LII, 232
antibody against, purification, LII, 241, 242
assay, XXXI, 743; XXXII, 397, 398, 400; LII, 243, 362
detergent-solubilized, properties, LII, 241, 245, 246
immunological screening of microsomal and mitochondrial extracts, LII, 248
inhibition, by anti-NADPH-cytochrome *c* reductase, LII, 246, 247
in isolated MEOS fraction, LII, 363
lipid peroxidation, LII, 303
in liver, XXXI, 97
as marker enzyme, XXXI, 743
as membrane marker, XXXII, 393
in microsomal fractions, LII, 88, 89
in nuclear membrane, XXXI, 290–292
partial purification, LII, 364
in plasma membrane, XXXI, 89, 91, 100, 225
protease-solubilized, properties, LII, 241, 245, 246
removal, from microsomal protein, LII, 362
in rough microsomal subfractions, LII, 75, 76
NADPH-cytochrome *P*–450 reductase, *see also* NADPH dehydrogenase
effect of extraction methods on properties, LIII, 408, 409
flavin determination, LIII, 429
as mixed-function oxidase substrate, XXXI, 233
NADPH-cytochrome reductase, LIII, 404–406
microsomes, LV, 101, 103
assay, LV, 102
outer membranes, LV, 103, 104
NADPH dehydrogenase, LII, 19, 45, 46
activity, LII, 95
Ah locus, LII, 232
assay, LII, 90–92, 243
by dual-wavelength stopped-flow spectroscopy, LII, 221–226
biphenyl hydroxylation, LII, 399

chromatographic purification, LII, 93, 94
contaminant, of cytochrome *P*–450LM forms, LII, 117
in cytochrome *P*–450 assay, LII, 110, 205
degree of purity, LII, 96
detergent-solubilized, properties, LII, 241, 245, 246
dissociation kinetics, LIII, 430
flavin content, LII, 90, 94, 95
holoenzyme reconstitution, LIII, 436
inhibitors, LII, 96
in microsomal fractions, LII, 88, 201
molecular weight, LII, 90, 95
optimal purification method for active preparation, LII, 203, 204
purification, LII, 89, 90, 92–94
requirement by mitochondrial P_{450}, XXXVII, 304
reversible dissociation, LIII, 429, 433
stability, LII, 95

NADPH-dichlorophenolindophenol reductase, *see* Dihydrolipoamide reductase

NADPH:ferricytochrome oxidoreductase, *see* NADPH-cytochrome *P*–450 reductase

NADPH:FMN oxidoreductase
assay, LVII, 205
immobilization, LVII, 204, 205
purification, LVII, 203

NADPH-generating system, LII, 274, 359, 378, 414

NADPH glyoxylate reductase
assay, XXXI, 745
as marker enzyme, XXXI, 744, 745

NADPH:Δ^4-3-ketosteroid 5α-oxidoreductase, XXXVI, 466–474
assay, XXXVI, 466–468
preparation, XXXVI, 468–470
properties, XXXVI, 470–474

NADPH-linked electron transport enzymes, enrichment, LII, 81, 83

NADPH-neotetrazolium oxidoreductase, in microsomes, XXXII, 91

NADPH oxidase, LII, 17
in microsomal fractions, LII, 89
plasma membrane and, XXXI, 90
role in phagocyte respiratory burst, LVII, 464, 488

NADP-isocitrate dehydrogenase, in microbodies, LII, 496, 499

NADP-protection of SH protein, XXXIV, 552–554

NADP:tetracycline 5a(11a)-dehydrogenase, XLIII, 606, 607

NAD pyrophosphorylase, in nuclei, XXXI, 253, 259, 261, 262

NAD^+-pyruvate adduct, XXXIV, 239

NAD:rubredoxin oxidoreductase, *see* Rubredoxin:NAD oxidoreductase

NAD synthetase, XLVI, 420

NAD(P) transhydrogenase
reconstitution
from beef heart mitochondria
assay, LV, 811
comments, LV, 815, 816
enzyme preparation, LV, 811
procedure, LV, 812–815
from *E. coli*, LV, 787, 788
enzyme assays, LV, 788–790
preparation of ATPase, LV, 790–796
of stripped membrane particles, LV, 796, 797
procedure for reconstitution, LV, 797–800

Naficillin, XLIV, 615

Nafoxidine
antiestrogen, XL, 292
in estrogen receptor studies, XXXVI, 270

NAG, *see* *N*-Acetylglucosamine

Nagarse
in brain lipopigment isolation, XXXI, 484
mitochondria, LV, 8, 13–15, 41, 42, 61, 80
preparation of mitochondria, LVI, 375
rat heart mitochondria, LVI, 546

Nagle and Brown's medium, LVIII, 58, 59

Naja naja
neurotoxin, XXXIV, 6
phospholipase A_2, XXXII, 137; LI, 63, 66
venom, assay of succinate dehydrogenase, LIII, 468

Naja naja siamensis, neurotoxin, XXXII, 313, 314

(Na$^+$ + K$^+$)-ATPase, *see* Adenosinetriphosphatase, Na$^+$,K$^+$-activated
Naldixic acid
  plasmid transfer, LVI, 115, 116
  resistance, as genetic marker, XLIII, 44
Naloxone, XXXIV, 622; XLVI, 603, 604
Naltrexone, XLVI, 603
NANA, *see* *N*-Acetylneuraminic acid
Nanosecond absorbance spectroscopy, LIV, 2, 32–46
  detector system, LIV, 39–43
  light source, LIV, 33–36
  monochromator, LIV, 36, 37
  optics, LIV, 37
  sample cells, LIV, 37–39
  sample density, LIV, 37–39, 45, 46
  stimulus, LIV, 43–45
Naphthacene, activator, oxidation potential, LVII, 522
Naphthacyl-α-chymotrypsin, XLVI, 89
Naphthalene
  photochemically active derivative, LVI, 660
  in scintillation cocktail, LII, 392
Naphthaleneacetic acid, as plant growth regulator, XXXII, 731, 732
1-Naphthaleneacetic acid, as auxin, LVIII, 478
Naphthalene-dioxane, in scintillation cocktail, LI, 52
Naphthalenedisulfonic acid
  extraction of nucleic acids, LIX, 311
  storage, LIX, 311
Naphthalene epoxidase, LII, 15
Naphthalene monooxygenase, Ah^b allele, LII, 231
Naphthalene 1,2-oxide, substrate of epoxide hydrase, LII, 199, 200
Naphthalenesulfonic acid
  as optical probes, LV, 573
  purification, LV, 574
2,3-Naphthalic hydrazide
  chemiluminescence, LVII, 420
  reaction intermediate in luminol chemiluminescence, LVII, 421, 423
α-Naphthoflavone
  inhibitor of cytochrome *P*-450-linked drug metabolism, LII, 67, 68
    of monooxygenases, LII, 231
β-Naphthoflavone, as monooxygenase inducer, LII, 233
2-Naphthohydroxamic acid-peroxidase complex, apparent dissociation constants, LII, 516
α-Naphthol, indicator for glycolipid, XXXV, 407
β-Naphthol, color reaction with diazo groups, XLIV, 83
Naphthol-AS-BI-glucuronic acid, as enzyme substrate, XXXII, 89, 90
Naphthol AS–BI phosphate, substrate, alkaline phosphatase, XLIV, 626, 627
Naphthol AS–Mx phosphoric acid disodium salt, reagent, alkaline phosphatase stain, XLIV, 713
Naphthol blue-black, protein stain, LI, 141
2-Naphthol-6,8-disulfonic acid, in plant-cell fractionation, XXXI, 504
1-Naphthol 2-sulfonate indophenol, as mediator-titrant, LIV, 408
Naphthoquinone, as mediator-titrant, disadvantages, LIV, 407
1,4-Naphthoquinone, chromatophore planar membranes, LV, 756
Naphthoquinone-β-sulfonate, for oxygen removal, LIV, 117
1,4-Naphthoquinone 2-sulfonate, reduced, proton translocation, LV, 751
Naphthoxyacetic acid, for modification of tRNA, LIX, 156
α-Naphthylamine, XLVII, 138
β-Naphthylamine, fluorometric detection, XLVII, 393
1-Naphthylamine-7-sulfonic acid, *see* Cleves acid
(Naphthylcarbamoyl)cellulose, XXXIV, 646–649
β-Naphthylcarbinol oxidase, LVI, 474
β-Naphthylcarbonyl oxidase, LVI, 474
α-Naphthyl ester, as esterase substrates, XXXII, 88
N-(1-Naphthyl)ethylenediamine, measurement of pH changes, LV, 564, 565

N-1-(Naphthyl)ethylenediamine dihydrochloride
assay of FGAM synthetase, LI, 195
of nitrate reductase, LIII, 348, 350
of phosphoribosylglycinamide synthetase, LI, 180
nitrite determination, LIII, 638
Naphthylhydrazide, stability, LVII, 425
1-Naphthyl isocyanate, XXXIV, 647
α-Naphthyl phosphate, substrate of acid nucleotidase, LI, 274
Narasin
code number, LV, 445
as ionophore, LV, 443
Narbomycin, XLIII, 198
Narcine, nicotinic receptors, XXXII, 315–318, 322, 323
Naringin, XLIV, 241
Nash reaction, principle, LII, 298
Nash reagent, LII, 299
in hydrogen peroxide determination, LII, 345
Nasopharynx, carcinoma, LVIII, 376
National Institute on Aging Cell Repository, LVIII, 442
National Science Foundation Cell Culture Centers, LVIII, 443
NBD chloride, *see* 7-Chloro-4-nitrobenzo-2-oxa-1,3-diazole
NBD derivative
fluorescent donor, XLVIII, 362, 363
spectral properties, XLVIII, 362
structural formula, XLVIII, 362
Nbf-Cl, *see* 4-Chloro-7-nitrobenzofurazan
NBS, *see* *N*-Bromosuccinimide
NBT stain, *see* Nitrotetrazolium Blue stain
NCCD, *see* Carbodiimide
NCS solubilizer, LIX, 521
determination of radioactivity in slab gels, LIX, 521
NCTC 109 medium, XXXIX, 445; LVIII, 57, 89
NCTC 135 medium, LVIII, 57, 88
composition, XXXIX, 400; LVIII, 670
NDP kinase, *see* Nucleoside diphosphate kinase
NDSA, *see* Naphthalenedisulfonic acid
Neamine, XLIII, 126, 338
Nebramycin, XLIII, 271
Nebularylcobalamin, effector of ribonucleoside triphosphate reductase, LI, 258
Nectin
chromatography, XXXII, 436–438
Millipore assay, XXXII, 434–436
pelleting assay, XXXII, 434, 435
properties, XXXII, 438, 439; LV, 397
from *S. faecalis* membrane, XXXII, 428–439
Nectria, fungal penicillin acylase from, XLIII, 721
Necturus, gastric mucosa isolation, XXXII, 707–712
Negative stain, for transmission electron microscopy, XXXIX, 143
NE 901 glass, XXXIX, 531, 532
Neisseria meningitidis
2-acetamido-2-deoxymannopyranose 1-phosphate homopolysaccharide of serogroup A, L, 48
polysaccharide antigens, L, 89
carbon-13 nuclear magnetic resonance spectra, L, 41–46
Neisser's Stain A, in cell count method, XXXII, 677
Nelder-Mead method, XXXVI, 14
Nembutal, *see* Pentobarbital
Neodymium, magnetic circular dichroism, XLIX, 177
Neodymium glass laser, LIV, 7, 23, 96
energy gap, LIV, 10
output, LIV, 45
relative brightness, LIV, 95
Neohydrin, purification, LI, 41
Neomethymycin, XLIII, 131
Neomycin
N-acetylneomycins, relative mobilities, XLIII, 126
biosynthesis, XLIII, 618, *see also* Aminoglycoside modifying enzymes
^{13}C nuclear magnetic resonance studies, XLIII, 409
gas-liquid chromatography, XLIII, 220–228
apparatus, XLIII, 222
bulk powder, XLIII, 224
calculation, XLIII, 226
column preparation, XLIII, 222, 223

freeze-dry procedure, XLIII, 225
petrolatum-based ointments, XLIII, 224
problems, XLIII, 228, 229
ion-exchange chromatography, XLIII, 269
in media, LVIII, 112, 114, 116
nonionic adsorption chromatography, XLIII, 277
paper chromatography, XLIII, 120
proton magnetic resonance spectroscopy, XLIII, 391
solvent systems, XLIII, 119, 120, 338
stimulator of *in vitro* polypeptide synthesis, LIX, 850, 851
Streptomyces fraciae, media, XLIII, 22
structure, XLIII, 614
thin-layer chromatography in pharmacokinetic studies, XLIII, 211

Neomycin phosphotransferase
assay, XLIII, 616, 618
from *Escherichia coli*, XLIII, 626
from *Pseudomonas aeruginosa*, XLIII, 626
from *Staphylococcus aureus*, XLIII, 627

Neomycin phosphotransferase I, XLIII, 616

Neomycin phosphotransferase II, XLIII, 616

Neomycin sulfate, XLIII, 387

Neon gas laser, wavelength, XLIX, 85, 86

Neopterin, XLIII, 520

Nephron, intact, isolation, XXXII, 632

Nernst curve, XLVIII, 270, *see also* Hill plot
pH dependence, LIV, 415, 416
for redox couples, LIV, 414–416
slope, value, LVI, 361, 362
succinate dehydrogenase iron-sulfur centers, LIII, 491

Nernst equation, LIV, 413
data
acquisition, LIV, 398, 399
interpretation, LIV, 401, 402
method of mixtures, LIV, 398, 399
diffusion potentials, LV, 558, 559
optical probes, LV, 583

Nerve
acetylcholine in impulse, XXXII, 309
fluorescent probe studies, XXXII, 240
growth factor, LVIII, 580
in media, LVIII, 102
thermolysin hydrolysis products, XLVII, 176
spin-label studies, XXXII, 171, 177

Nervous system, peripheral, myelin isolation, XXXI, 442, 443

NESA semiconductor, LIV, 405

NESATRON semiconductor, LIV, 405

Netropsin, XLIII, 145, 146

Neumann and Tytell medium, XXXIX, 205, 209; LVIII, 57
composition, LVIII, 62–70

Neural tissue, XXXIX, 427–440
myelin isolation, XXXI, 435–444
subcellular fractions derived, XXXI, 433–485

Neuraminic acid, derivatives, L, 64–66

Neuraminic acid β-methylglycoside
crystallization, L, 70
preparation, L, 68
purification, L, 69

Neuraminidase, XXXIV, 5, 337, 689
N-acetylglucosamine binding protein, L, 291
N-acylneuraminate-9(7)-*O*-acetyltransferase assay, L, 383
azo linkage, XXXIV, 106
glycoprotein tritiation with galactose oxidase, L, 204–206
I–cell disease, L, 229–231
immobilization, in liposomes, XLIV, 223
interference in purification of TBG, XXXIV, 389
in kallikrein purification, XLV, 291
peanut agglutinin, L, 361, 362
in phospholipase studies, XXXII, 137
potato lectin hemagglutination activity, L, 344
from rat liver, L, 490
sphingolipidoses, L, 457
in testis hormone receptor studies, XXXVII, 186, 187

Neurinoma cell line, in neurobiological studies, XXXII, 769

Neurite
- enzyme studies, XXXII, 768
- separation from soma, LVIII, 30307

Neurobiology, cultured cell systems, XXXII, 765–788

Neuroblastoma cell
- C1300, LVIII, 585
  - media, LVIII, 89
  - neurobiological enzyme studies, XXXII, 766–774
- 2′,3′-cyclic nucleotide-3′-phosphohydrolase, XXXII, 128
- growth, LVIII, 97, 98
- media, LVIII, 57, 89
- rat, LVIII, 96

Neuron, *see also* specific types
- action potential, LVIII, 307
- adrenergic, *see* Adrenergic neuron
- chemical composition, XXXV, 578
- co-culture with nonneuronal cells, LVIII, 581
- electrical stimulation, LVIII, 307
- fixation for electron microscopy, XXXV, 577
- isolation, XXXV, 561–579
- morphology, XXXV, 576, 577
- plating, LVIII, 306
- sympathetic, *see* Sympathetic neuron
- transmitter function, LVIII, 307

Neuronal cell
- cloning, LVIII, 588, 589
- establishment of cell line, LVIII, 587, 588
- isolation, LVIII, 585
- large-scale growth, LVIII, 590
- media, LVIII, 586, 587
- from mouse hypothalamus, LVIII, 585
- from neoplastic tissue, LVIII, 587, 588
- primary culture, LVIII, 585
- recovery, LVIII, 589
- from rodent neoplasm, LVIII, 584–590
- routine passage, LVIII, 590
- selection, LVIII, 588, 589
- serum, LVIII, 587
- storage, LVIII, 589
- transformation, LVIII, 585
- tumor induction, LVIII, 586

Neuronal ceroid lipofuscinosis, ceroid isolation, XXXI, 483

Neuronal culture, LVIII, 574

Neuronal membrane, 2′,3′-cyclic nucleotide-3′-phosphohydrolase, XXXII, 127

Neurophysin, XXXIV, 7
- electrophoresis, XXXI, 409, 410

Neurosecretory granule
- bioassay, XXXI, 409
- isolation from posterior pituitary, XXXI, 403–410

Neurospora, XLVI, 612
- calibration of optical probe response, LV, 581, 582
- choline requirers, XXXII, 822, 823
- fatty acid mutants, XXXII, 821
- glyoxylate cycle enzymes, XXXI, 566
- growth and isolation of mitochondria, LVI, 80
- mitochondria, LVI, 3
  - size of DNA, LVI, 5
- oxidase, LII, 12
- spin labeling, XXXII, 830

Neurospora crassa
- biogenesis of mitochondrial membrane proteins, LVI, 50
  - pulse-labeling of proteins with leucine, LVI, 51–54
  - *in vivo* incorporation of leucine after inhibition of mitochondrial or cytoplasmic protein synthesis, LVI, 54–58
- carbamyl-phosphate synthase (glutamine)-aspartate carbamyltransferase complex, LI, 105–111
- coenzyme Q homolog, LIII, 593
- complex III, LIII, 98–112
- culture conditions, LIX, 204
- cytochrome *b*, LIII, 212–221
- cytochrome oxidase, LIII, 66–73
- dihydroorotate dehydrogenase, LI, 63–69
- disruption, LV, 139, 145–147
- β-galactosidases, assay, purification, and properties, XLII, 494–500
- glucose-6-phosphate dehydrogenase, assay, purification, and properties, XLI, 177–182
- growth conditions, LI, 107

growth method, XXXI, 616–620
mitochondria, LV, 25, 26
cultivation of cells, LV, 144, 145
disruption of cell walls, LV, 145–147
isolation, LV, 147
properties, LV, 147, 148
mitochondrial ATPase complex
immunoprecipitation, LV, 346, 347
preparation of antiserum, LV, 345, 346
principle, LV, 344
properties, LV, 347–351
purification, LV, 344, 345
mitochondrial initiator tRNA sequence analysis, LIX, 107–109
mitochondrial ribosomes, isolation, analysis and use, LVI, 79–92
mitochondrial tRNA, LIX, 203–207
mycelial growth conditions, LI, 64, 65
nitrate reductase, siroheme in, LII, 440, 443
6-phosphogluconate dehydrogenase, assay, purification, and properties, XLI, 227–231
$tRNA^{Phe}$, sequence analysis, LIX, 100
spheroplast isolation, XXXI, 609–626
spin-labeled fatty acids, XXXII, 194
succino-AICAR synthetase, LI, 188, 190
transport linked to membrane potential, LV, 656, 657
culture media, LV, 657, 658
electrophysiological techniques, LV, 658–662
flux techniques, LV, 663–666
vacuole isolation, XXXI, 574
Neurotensin, bovine, XLVII, 67
α-Neurotoxin, XXXII, 309–323; XLVI, 587, *see also* specific type
Neutralization, analysis, β-lactamase, XLIII, 90–97, *see also* Immunological technique, β-lactamase
Neutral red, XLIV, 132, 431
as internal pH indicator, LV, 565, 566
as mediator-titrant, LIV, 409
Neutrapen, XLIII, 642
Neutron, wavevector, LIX, 671
Neutron scattering, LIX, 629–775, *see also* Protein pair scattering
data interpretation, LIX, 636–638, 684–706, 711–717
by ensemble of atoms, LIX, 673, 674
experimental feasibility, LIX, 633–638
experimental techniques, LIX, 675, 676
forward scatter, LIX, 685–690
instrumentation, LIX, 683, 684, 754, 755
isomorphous replacement technique, LIX, 675, 676
by macromolecules in solution, LIX, 674–676
model, macromolecules, LIX, 714–714
molecular weight determination, LIX, 689, 690
preparation of materials, LIX, 639–655
protein triangulation, LIX, 702–706
ribosome structure, LIX, 629–638
shape analysis, LIX, 700–702, 714–717, 756–759
theory, LIX, 671–673, 708–711
other scattering techniques, LIX, 756–759
Neutron scattering length
of biological nuclei, LIX, 630, 672
definition, LIX, 630
Neutron scattering profile, definition, LIX, 631
Neville-Eyring theory of enzyme conformation, *Cypridina* luciferase, LVII, 351, 352
Newcastle disease virus, interferon producing, LVIII, 294
Newton-Raphson method, XXXVI, 9, 14
in curve-fitting, XXXVII, 12
NGNA, *see* *N*-Glycolylneuraminic acid
Niacinamide, in media, LVIII, 54, 56
Nichol-Winzor method, for analysis of mixed associations, XLVIII, 125
Nickel
atomic emission studies, LIV, 455
in media, LVIII, 69
Nickel(II)
correlation time evaluation, XLIX, 346

interatomic distance calculations, XLIX, 331

Nickel-63

in membrane studies, XXXII, 884–893

properties, XXXII, 884

Nickel amine reaction, nuclear magnetic resonance, XLIX, 296, 312, 320, 321

Nickel(II)-bathophenanthroline chelate, LV, 517

Nickel ion, *see also* Raney nickel

activator of adenine phosphoribosyltransferase, LI, 573

of deoxycytidine kinase, LI, 345

of deoxythymidylate phosphohydrolase, LI, 289

of pyrimidine nucleoside monophosphate kinase, LI, 329

of uridine-cytodine kinase, LI, 319

identification of modified arginyl residues, XLVII, 160

inhibitor, of deoxythymidine kinase, LI, 365

of phosphoribosylpyrophosphate synthetase, LI, 11

of uridine nucleosidase, LI, 294

Pholas luciferin chemiluminescence, LVII, 400

Nickel-substituted cytochrome *c*

preparation, LIII, 177, 178

properties, LIII, 178–181

Nickel sulfate, optical filter, XLIX, 192

Nickel-tetraphenylporphyrin, X-ray absorption spectroscopy, LIV, 332

Nicolsky-Eisenman equation, ion selective electrodes, LVI, 441–443

Nicotiana, cell cultures, XXXII, 729

Nicotiana glauca × *N. langsdorffi* mesophyll, LVIII, 365

Nicotiana tabacum L., callus culture, LVIII, 479

Nicotinamide

in cytochrome *P*–450 isolation, LII, 362

in media, LVIII, 64

Nicotinamide adenine dinucleotide, XXXIV, 2, 4, 243–253, 486; XLVI, 240, *see also* Pyridine nucleotide

3-acetylpyridine analogue, in FGAM synthetase assay, LI, 196

alkylation, XLIV, 862, 863

amino acid transport, LV, 187

ammonia assay, LV, 209

AMP assay, LV, 209

analogs, XXXIV, 125, 126; XLVI, 241, 249, 250; XLVII, 423, 424

characterization, XLIV, 868, 869

coenzyme activity, XLIV, 873–876

conformation studies, XLIX, 355, 356, 357

haloketone, XLVI, 145–153

immobilization, XLIV, 869–873, 882–884

synthesis, XLIV, 862–866

in ANS fluorescence measurements, XXXII, 237

assay, LV, 221

of malate, LVII, 187

of malate dehydrogenase, LVII, 181

azo-linked, XXXIV, 105, 122–126

alcohol dehydrogenase, XXXIV, 122

dehydrogenases, XXXIV, 122–126, *see also* specific dehydrogenases

binding, to immobilized lactate dehydrogenase, XLIV, 514, 515

binding studies, XLVIII, 276

brown adipose tissue mitochondria, LV, 69

chemiluminescent assay of lactate dehydrogenase, LVII, 460, 461

chromatography, XLVI, 281

citrate assay, LV, 211

coenzyme A assay, LV, 210

content

of brain, LV, 219

of rat heart, LV, 217

of rat liver, LV, 213

dehydrogenase coupling to platinum electrode, LVI, 477, 478

derivative, chromatography, XXXIV, 246

deuterated, preparation, LIV, 226, 227
as electron donor, LVI, 386
expensive cofactor, XLIV, 837
for horse liver alcohol dehydrogenase, XLIV, 837–840
extraction, LIII, 582
fiber entrapment of coenzyme, XLIV, 243
fluorescence changes, LVI, 501
as general ligand, XXXIV, 229–253, *see also* specific ligands
hollow-fiber dialytic permeability, XLIV, 297, 302
immobilization
on Dextran, XLIV, 872
in enzyme system, XLIV, 216, 243, 315, 316
immobilized, XXXIV, 243–253
ligand selection, XXXIV, 246, 247
properties, XXXIV, 244, 245
liposomes, reduction, LVI, 429
mammalian mitochondria, LV, 8
in media, LVIII, 64
M. phlei
ETP, LV, 184, 185
membrane fragments, LV, 196
oxaloacetate assay, LV, 207
oxidation
characteristic → H^+/O values, LV, 638
inhibitors, LV, 455–460
oxidized, in assay for elongation factor EF-2, LX, 706, 710
phosphorylating vesicles, LV, 167
phosphorylation potential, LV, 240
photochemically active derivative, LVI, 656
product of NADH:cytochrome *c* oxidoreductase, LIII, 5
of NADH:ubiquinone oxidoreductase, LIII, 11
pseudobase formation, XLIV, 842
reduced, XLVI, 20, 57, 66; XLVII, 441
activation of succinate dehydrogenase, LIII, 472
activator of brain α-hydroxylase, LII, 317
in AHH assay, LII, 236
analogs, XLVI, 241, 249, 250
synthesis, XLVI, 253, 254
assay, LVII, 182, 446
of adenylosuccinate synthetase, LI, 213
applications, LVII, 206–214
with bacterial luciferase, LIII, 564
coupled, LVII, 140
of complex I, LIII, 14
cytochrome b_5, LII, 97, 207
cytochrome b_5 reductase, LII, 207
cytochrome P–450_{cam} activity, LII, 156
of FGAM synthetase, LI, 197
of GMP synthetase, LI, 219
of guanylate kinase, LI, 474, 475, 484
of malate dehydrogenase, LVII, 188
of melilotate hydroxylase, LIII, 553
of NADH-cytochrome *c* reductase, LIII, 9, 10
of nitric oxide reductase, LIII, 645
of nucleoside diphosphokinase, LI, 376, 377
of oxaloacetate, LVII, 187
procedure, LVII, 205, 206
for *Pseudomonas* toxin A, LX, 788, 789
of pyrimidine nycleoside monophosphate kinase, LI, 323
of salicylate hydroxylase, LIII, 530
in stearyl-CoA desaturase activity, LII, 189
of uridine-cytidine kinase, LI, 300, 309, 316
of uridylate-cytidylate kinase, LI, 333, 334
binding, to immobilized glutamate dehydrogenase, XLIV, 508–510
in camphor 5-monooxygenase system, LII, 168
commercial source, LII, 244
deuterated, preparation, LIV, 226

as electron donor, XXXI, 701
of erythrocyte cytochrome b_5 reductase, LII, 465
for stearyl-CoA desaturation, XXXV, 253, 260, 261
extinction coefficient, LIII, 10, 14, 21
hydride transfer, XLVI, 20
lipid extraction of microsome fraction, LII, 98, 104
lipid peroxidation, LII, 305
oxidation
determination, LI, 300, 309, 316, 323, 333, 334, 376, 377
by 2,6-dichloroindophenol, LIII, 20, 21
enzyme, XLIV, 863
by ferricyanide, LIII, 21
by menadione, LIII, 20, 21
spectrophotometric determination, LIII, 10, 14, 20, 21
by ubiquinone-1, LIII, 20, 21
oxidation rate
for hepatocyte viability test, LII, 64
in NADH-cytochrome b_5 reductase assay, LII, 103
as redox titrant, LIV, 409
solution, stability, LVII, 218
stereospecifically deuterated, NMR spectra, LIV, 229, 230
substrate for fatty acid synthase, XXXV, 84, 87
of melilotate hydroxylase, LIII, 552
of NADH:cytochrome *c* oxidoreductase, LIII, 5
of NADH:ubiquinone oxidoreductase, LIII, 11
of salicylate hydroxylase, LIII, 528
reduction
ATPase inhibitor, LV, 407, 412
coupling factor B, LV, 384
ribosyl-linked, XXXIV, 120, 121
substrate for L-3-hydroxyacyl-CoA dehydrogenase, XXXV, 122

Nicotinamide adenine dinucleotide dehydrogenase, *see* NADH dehydrogenase
Nicotinamide adenine dinucleotide phosphate, XLVI, 240
adenylate cyclase assay, XXXVIII, 162
assay, LV, 221
of AMP, LV, 209
of citrate, LV, 211
of cytochrome *P*–450, LII, 126, 139
of dimethylaniline monooxygenase, LII, 143
of formate dehydrogenase, LIII, 364
of glutamine, LV, 209
of GTP, LV, 208
of inorganic phosphate, LV, 212, 215
of phenylalanine hydroxylase, LIII, 278, 279
content of rat liver, LV, 213
effect on adrenal steroidogenesis, XXXII, 691
enzyme reduction, XLIV, 865
extinction coefficient, LIII, 363
immobilization, on Dextran, XLIV, 872, 873
in media, LVIII, 64
NADH dehydrogenase, LVI, 586, 587
NADPH-generating system, LII, 359, 378, 380, 414
pyridine 4 proton, resonance frequency, LIV, 232
in reactivation of dissociated pigeon liver fatty acid synthase, XXXV, 57
reduced, XXXIV, 273; XLVI, 57
activator of brain α-hydroxylase, LII, 317
in AHH assay, LII, 236
applications, LVII, 206–214
assay of *p*-hydroxybenzoate hydroxylase, LIII, 545
of ω-hydroxylase, LIII, 356
binding, to immobilized glutamate dehydrogenase, XLIV, 508–510
commercial source, LII, 222, 244

in corpus luteum, XXXIX, 244
effect, XXXIX, 241
in cytochrome P–450 assay, LII, 125, 131, 205
for depletion of endogenous substrates, LII, 271
effect on cytochrome b_5 concentration measurement, LII, 215, 216
electron donor for stearyl-CoA desaturation, XXXV, 260, 261
extinction coefficient, LII, 91
fluorescence, LII, 57
in fluorometric analysis of mixed-function oxidase activity, LII, 375–377
generation
buffer, LII, 216, 217
drug effects, LII, 69
in α-hydroxlyase assay, LII, 311
lipid peroxidation, LII, 305
in microsomal peroxide production system, LII, 344
oxidation
determination, LI, 227, 228, 249, 250; LIII, 357
enzymic, XLIV, 865
procedure, LII, 92
in reductase assay, LII, 91
procedure, LVII, 205, 206
pyridine protons, resonance frequencies, LIV, 232
in reactivation of dissociated pigeon liver fatty acid synthase, XXXV, 57
reductant, for pig liver oxidase, XLIV, 850–852
stereospecifically deuterated, preparation, LIV, 232
substrate, of complex I–III, LIII, 8
for fatty acid synthase, XXXV, 45, 87, 312
of *p*-hydroxybenzoate hydroxylase, LIII, 544, 548
of NADH dehydrogenase, LIII, 18
of NADH:ubiquinone oxidoreductase, LIII, 11
reduction
assay in mutants, LVI, 113, 114
determination, LI, 468; LIII, 363, 364
spin labeled, XLVI, 285, 286
substrate, of formate dehydrogenase, LIII, 360, 369
Nicotinamide 8-azidoadenine dinucleotide, LVI, 656
Nicotinamide [5-(bromoacetyl)-4-methylimidazol-1-yl]dinucleotide, XLVI, 147, 150
Nicotinamide-(*S*-methylmercury-thioinosine) dinucleotide, reaction conditions, XLVII, 424
Nicotinamide mononucleotide, chromatography, XLVI, 281
Nicotinamide nucleotide
intact mitochondria, LV, 143
N. crassa mitochondria, LV, 147
Nicotinamide nucleotide transhydrogenase, *see also* Transhydrogenase
assay, LV, 276
principle, LV, 261–263
procedure, LV, 263–275
energy-linked
assay, LV, 726–728
properties, LV, 728, 729
reconstitution, LV, 716
enzyme preparations, LV, 263
localization, LV, 261
purification
procedure, LV, 277–282
properties, LV, 282, 283
reagents, LV, 276
Nicotine 2-hydroxylase, LII, 17
Nicotinic acetylcholine receptor, XXXII, 309–323
Nicotinic acid, in media, LVIII, 64
Nicotinic receptor
assay, XXXII, 319
purification and properties, XXXII, 319–323
solubilization and properties, XXXII, 318, 319
Niddamycin, XLIII, 131, 495
Niemann–Pick disease
enzymic diagnosis, L, 465–468
sphingomyelinase deficiency, L, 465
Nigericin
anion transport, LV, 453

cation fluxes, LV, 674, 675
chemiosmotic hypothesis, LV, 451
code number, LV, 445
complex V, LV, 315
flow dialysis experiments, LV, 683
as ionophore, LV, 442, 446
measurement of proton translocation, LV, 637
model redox reactions, LV, 543
M. phlei ETP, LV, 186
phosphate efflux, LVI, 354
photophosphorylation, LV, 452
regenerating systems, LV, 710
solutions, LV, 448
structure, LV, 437
NIH–LH, in media, LVIII, 101, 102, 104
NIH–LH-B9, in media, LVIII, 106, 107
Nikon 6-C, XLIX, 47
NIL2 cell line, properties and uses, XXXII, 584, 590
Ninhydrin, XLIII, 117
assay of carboxypeptidase C, XLVII, 77
of carboxypeptidase Y, XLVII, 86
colorimetric assay
estimation of amino acid components, XLIV, 94
of L-methionine, XLIV, 747
detection, XLVII, 6–9
error sources, XLVII, 12
of lysine derivatives, XLVII, 477
reagent dilution, XLVII, 8
sensitivity, XLVII, 13
reagent preparation, XLVII, 16, 223
solid-phase peptide synthesis monitoring, XLVII, 599
thermolysin hydrolysis products, XLVII, 185
for visualization of ribosomal proteins, LIX, 778
Ninhydrin-cadmium reagent, peptide staining properties, XLVII, 60
Niobium, atomic emission studies, LIV, 455
Niobium-93, lock nuclei, XLIX, 349; LIV, 189
NIPEX nylon mesh, LII, 525
Nisin
fragments, XLIII, 338
solvent systems, XLIII, 338
structure, XLIII, 330
Nitella, free cells, XXXII, 723
Nitex bolting cloth, XXXII, 120, 771
Nitex filter cloth, XXXI, 605
p-Nitraniline
diazotized, ubiquinone intermediates, LVI, 109, 111
identification of quinone intermediates, LIII, 603, 605, 606
Nitrate, *see also* Nitric acid
activation of succinate dehydrogenase, LIII, 33, 472, 474, 475
calcium-sensitive electrodes, LVI, 347
as electron acceptor, LVI, 386
fluorescence quenching agent, XLIX, 94
inhibitor, of firefly luciferase, LVII, 15
in media, LVIII, 68
oxidation, by *Nitrobacter*, LIII, 634, 640, 641
Nitrate reductase
antibody against
preparation, LIII, 352
subunit studies, LIII, 352–355
assay, LIII, 349, 350
from *Escherichia coli*, LIII, 347–355
immobilization, by microencapsulation, XLIV, 214
isoelectric point, LIII, 643
metal content, LIII, 352, 355
from *Micrococcus halodenitrificans*, LIII, 643
molecular weight, LIII, 350, 351, 643
from *Paracoccus denitrificans*, LIII, 642
pH optimum, LIII, 643
from *Pseudomonas perfectomarinus*, LIII, 642
purification, LIII, 348, 349, 355, 642, 643
radiolabeling, LIII, 354
respiratory-deficient cells, LVI, 175, 176
sedimentation coefficient, LIII, 351
subunit composition, LIII, 351, 352, 355
synthesis, LIII, 355

Nitration
with HNO_3/H_2SO_1, XLIV, 112
site of action, XLVII, 66, 67
Nitrene, XLVI, 10, 340, 580
in photoaffinity labeling, XLVI, 78, 645
reactivity, LVI, 614, 658
8-Nitrene-cAMP, *see* 8-Nitrene cyclic adenosine monophosphate
8-Nitrene cyclic adenosine monophosphate, XLVI, 340
Nitric acid, XLIV, 112; *see also* Nitrate
in determination of tyrosine, LIII, 279
extraction of adenine nucleotides, LVII, 81
of bacterial ATP, LVII, 68, 70, 71, 72
glass bead activation, XLIV, 931
for glassware cleaning, LVII, 425
in hydride transfer stereospecificity, experiments, LIV, 227
in iron analysis, LIV, 439
metal cleaning, XLIV, 654
metal-free, preparation, LIV, 476, 477
polystyrene nitration, XLVII, 265
in wet digestion procedure, LIV, 481
Nitric oxide
determination, LIII, 643, 644
inhibitor, of *Pseudomonas* cytochrome oxidase, LIII, 657
stretching frequency, LIV, 311
Nitric-oxide reductase
assay, LIII, 645
from *Pseudomonas perfectomarinus*, LIII, 644, 645
purification, LIII, 644, 645
Nitrification, principle, LIII, 634, 635
Nitrile, XL, 321
aromatic, XLVI, 117
Nitrilotriacetic acid
as calcium buffer, LVI, 352
for inactivation of phosphatase, LIX, 62, 74, 75, 100, 102
iron solutions, LVI, 391
2,2′,2′′-Nitrilotriethanol hydrochloride, *see* Triethanolamine
Nitrite
assay, LIII, 638
oxidation, assay, LIII, 640, 641
reaction with *Pseudomonas* cytochrome oxidase, LIII, 656
Nitrite oxidase, in *Nitrobacter*, LIII, 641
Nitrite reductase, *see also Pseudomonas*, cytochrome oxidase
assay, LIII, 643, 649
inhibitor, LIII, 646
from *Paracoccus denitrificans*, LIII, 643
purification, LIII, 643
4-Nitroacrylanilide, XLIV, 98
Nitroaliphatic compound, oxidation, LVI, 475
Nitroalkane
anion, inactivator of flavoprotein enzymes, LIII, 440, 441
metabolite binding to reduced cytochrome, LII, 274
N-Nitroamide, XLVI, 216
p-Nitroaniline
as mixed-function oxidase subtrate, XXXI, 233
release, measurement, XLVII, 77, 78
p-Nitroanisole, substrate of cytochrome *P*–450LM forms, LII, 117
p-Nitroanisole *O*-demethylase
Ah^b allele, LII, 231
in microsomal fractions, LII, 89
N^ω-Nitroarginine, stability, XLVII, 532, 533
N^G-Nitroarginine, XLVI, 231
Nitroaryl azide, XLVI, 80
2-Nitro-4-azidobenzoylhydrazide, XLVI, 656, 658, 660, 661
2-Nitro-4-azidobenzoyl hydrazone, of GTP, XLVI, 656–658
2-Nitro-4-azidocarbonylcyanide phenylhydrazone, LVI, 660
N-[N^α-(2-Nitro-4-azidophenyl), N^ε-dansyllysyl]-*p*-aminophenyl-β-lactoside, XLVI, 520
2-Nitro-4-azidophenyl-β-D-glycoside, XLVI, 366
2-Nitro-4-azidophenyl group, in photochemical immobilization, XLIV, 281–288
Nitrobacter
culture, LIII, 640
nitrification, LIII, 634, 635
m-Nitrobenzamidine hydrochloride, XXXIV, 441

p-Nitrobenzamidoagarose, XXXIV, 658
p-Nitrobenzamidoethylagarose, XXXIV, 625
6-*p*-Nitrobenzamidohexan-1-ol, XXXIV, 521
m-(3-[*m*-(*p*-Nitrobenzamido)phenoxy]propoxy)benzamidine, XLVI, 127
Nitrobenzene, XLVII, 360
  fluorescence quenching agent, XLIX, 94
  synthesis of 4-(4-methoxyphenyl)-4-oxobutyric acid, LVII, 434
m-Nitrobenzenediazonium fluoroborate, XXXIV, 188
p-Nitrobenzenediazonium fluoroborate, reaction conditions, XLVII, 413
N-(*o*-Nitrobenzenesulfenyl)ethanolamine, intermediate in phosphatidylethanolamine synthesis, XXXV, 461
7-Nitrobenzofurazan, derivatives, properties, LV, 489
o-Nitrobenzoic acid, substrate of salicylate hydroxylase, LIII, 538
p-Nitrobenzoyl azide, XXXIV, 658, 742
p-Nitrobenzoyl chloride, XXXIV, 65; XLIV, 86, 142
  arylamine glass production, XXXVIII, 183
  preparation of arylamine glass, XLIV, 932
9-(*p*-Nitrobenzyl)adenine, synthesis, XLVI, 329–332
p-Nitrobenzyl alcohol, XLVII, 529
p-Nitrobenzyl bromide, XLVII, 524
p-Nitrobenzyl ester moiety, peptide synthesis, XLVII, 522
4-Nitrobenzyl-4-guanidinobenzoate, inhibitor of urokinase, XLV, 242
α-(*p*-Nitrobenzyloxycarbonyl)arginine, XLVI, 230
N^{α}-*p*-Nitrobenzyloxycarbonyl-L-arginine chloromethyl ketone, XLVI, 229
  synthesis, XLVI, 230, 231
α-(*p*-Nitrobenzyloxycarbonyl)arginyl chloride, XLVI, 231
Nitro blue tetrazolium
  assay of D-lactate dehydrogenase, LIII, 520
    of superoxide dismutase, LIII, 387
  isoenzyme staining, LVIII, 168, 169
2-Nitro-4-carboxyphenyl-*N,N*-diphenyl carbamate, inhibitor of chymotrypsin, LIX, 598
S-(2-Nitro-4-carboxyphenyl)-6-mercaptopurine riboside, synthesis, XLVI, 291, 292
S-(Nitro-4-carboxyphenyl)-6-mercaptopurine riboside 5′-monophosphate, synthesis, XLVI, 291–293
Nitrocellulose, fibers of, preparation, XLIV, 230
Nitrocellulose-cellulose platelet, in liquid scintillation counting techniques, LIX, 62
Nitrocellulose disk, LIX, 254
Nitrocellulose membrane, LIX, 236, *see also* Membrane filtration
  prewetting, LIX, 584
p-Nitro-α-dimethylaminotoluene, XLVI, 583
4-Nitroestrone, XLVI, 455
4-Nitroestrone methyl ether, XLVI, 455, 456
Nitroethane, inactivator of flavoprotein enzymes, LIII, 440
Nitrogen
  agarose derivatization, XLIV, 39
  analysis, for bound protein determination, XLIV, 387, 388
  assay, in fibers, XLIV, 232
  deoxygenation, LIII, 364; LVI, 384
  determination, LIII, 643, 644
  fixation, assay of nitrogenase, LIII, 329
  gaseous, XLIV, 50, 58, 92, 102, 110, 111, 117, 174, 175, 177, 185, 198, 230, 346, 392, 455, 476, 482, 524, 600, 662, 862, 912
    in anaerobic procedures, LIV, 114, 115
    for deoxygenation of oxyhemoglobin, LIV, 489, 491–493
    level of oxygen contamination, LIV, 501
    as purge gas, purification, LIV, 404
  as inert atmosphere, XXXI, 540
  liquid
    in low-temperature EPR spectroscopy, LIV, 120

polyethylene bottles for sample storage, XXXII, 121
for storage of ribosomes, LIX, 441
tissue freezing, XXXVIII, 4
preparation of luciferin, LVII, 17, 19
prepurified, for yeast growth, XXXI, 629
product
of luminol oxidation, LVII, 410, 412
of nitrous oxide reductase, LIII, 646
in photochemical immobilization, XLIV, 282, 285, 287
Nitrogen-14, nucleus
coherent scattering length, LIX, 672
scattering cross section, LIX, 672
Nitrogen-15, lock nuclei, XLIX, 348
Nitrogenase
from *Anabaena cylindrica*, LIII, 328
assay, LIII, 329
from *Azotobacter chroococcum*, LIII, 319, 326–328
from *Azotobacter vinelandii*, LIII, 319, 324–326
from *Bacillus polymyxa*, LIII, 315–319
from *Clostridium pasteurianum*, LIII, 319, 321, 322
component 1, LIII, 314
component 2, LIII, 314
electron paramagnetic resonance studies, XLIX, 514
EPR characteristics, LIII, 265, 272
freezing effects, XXXI, 532
from *Klebsiella pneumoniae*, LIII, 319, 323, 324
low-temperature EPR studies, LIV, 123, 131
Mo-Fe protein, *see* Molybdoferredoxin
Mössbauer spectroscopy, LIV, 364, 365, 371
protein components, LIII, 314
specific activities, LIII, 319
X-ray absorption spectroscopy, LIV, 332, 338
from *Rhodospirillum rubrum*, LIII, 318–321
from soybean, LIII, 328
from *Spirillum lipoferum*, LIII, 328
Nitrogen flowing gas laser, wavelength, XLIX, 85, 86
Nitrogen gas laser, use in resonance Raman spectroscopy, LIV, 246
Nitrogen mustard, XLIII, 33, 34, *see also* specific substance
Nitrogenous base
cytochrome *P*–450 difference spectra, LII, 273
metabolism, LII, 69
synthesis of affinity matrix, LII, 516, 517
Nitro-γ-globulin, XXXIV, 189
2-Nitro-5-mercaptobenzoic acid, in affinity steroid labeling, XXXVI, 391
Nitromethane, XLVII, 340
anion, inactivator of flavoprotein enzymes, LIII, 440
coupling yields, XLVII, 270
solvent in carboxyl group activation, XLVII, 556
Nitrophenol
solubility, LV, 463
as uncouplers, LV, 470
p-Nitrophenol, XLVI, 208, 614; XLIX, 437
concentration determination, by spectroscopy, XLIV, 101
determination, LI, 272
ester derivatives of hydroxyalkyl methacrylate gels, XLIV, 80–82
capacity, XLIV, 80, 81
p-Nitrophenolic ester, fluorescent labeling, XLVIII, 359
p-Nitrophenyl-*O*-(2-acetamido-2-deoxy-β-D-glucopyranosyl)-(1→3)-*O*-(2-acetamido-2-deoxy-β-D-glucopyranosyl)-(1→6)-2-acetamido-2-deoxy-β-D-glucopyranoside, preparation, L, 118, 119
p-Nitrophenyl acetate, XLIV, 574
hydrolysis, rapid mixing device, LV, 246
in photoaffinity labeling studies, LIX, 803
substrate, of acetylesterase, XLVII, 78
4-Nitrophenyladenosine 5′-phosphate, XLVI, 670–672, 673
4-Nitrophenylalanine peptide, XLVI, 90
3-Nitro-5-phenylbenzofuroxan, XLVI, 80

p-Nitrophenyl-N^{α}-benzoylcarbonyl-L-lysinate
bromelain A titration, XLV, 9
protease inhibitors, pineapple stem, XLV, 740
p-Nitrophenyl-*N*-benzyloxycarbonyl glycinate, assay, of papain, XLV, 6
4-[4-(4-Nitrophenyl)butanamido]phenyl-β-D-fucopyranoside, XXXIV, 370
4-[4-(4-Nitrophenyl)butanamido]phenyl-tri-*O*-acetyl-β-D-fucopyranoside, XXXIV, 369
p-Nitrophenylcarbamyl group, XLVI, 627
p-Nitrophenyl chloroacetate, XLIV, 263
p-Nitrophenyl chloroformate, XLVI, 40, 118, 209, 210; XLVII, 536
Nitrophenyl derivative, reduction, XLVI, 673
p-Nitrophenyl diazoacetate, XLVI, 36
synthesis, XLVI, 40
p-Nitrophenyl diazonium fluoroborate, XLVI, 578
p-Nitrophenyldiethyl phosphate, for active site titration of carboxylesterase, XXXV, 192, 202
Nitrophenyldimethylcarbamate
active site titration of carboxylesterase, XXXV, 191, 192
acyl-enzyme formation, XXXV, 200–202
Nitrophenyl ester, XLVI, 94, 96
p-Nitrophenyl ester, XXXIV, 258, 259; XLVI, 209
O-Nitrophenyl-β-D-galactopyranoside, substrate, of β-galactosidase, XLIV, 269, 270, 313, 823
o-Nitrophenyl-β-D-galactopyranoside 6-phosphate, preparation and enzymic assay, XLI, 119, 120
p-Nitrophenyl β-D-glucopyranoside, substrate, of β-D-glucosidase, XLIV, 101, 474
p-Nitrophenyl *p*′-guanidinobenzoate, XLIV, 391
in acrosin assay, XLV, 334
titration, of trypsinlike enzymes, XLV, 13
p-Nitrophenyl-α-D-mannopyranoside, XLIII, 638
p-Nitrophenyl methyl phosphorochloridate, XXXIV, 587
O-(4-Nitrophenyl)-*O*′-phenylthiophosphate, XXXIV, 606, 607
Nitrophenylphosphatase, in plasma membranes, XXXI, 88, 148
p-Nitrophenyl phosphate
in acid nucleotidase assay, LI, 272
phosphate donor of nucleoside phosphotransferase, LI, 391, 392
substrate of acid nucleotidase, LI, 274
of alkaline phophatase, XLIV, 308, 309
5′-(4-Nitrophenylphospho)adenylyl(3′,5′)-uridine, XLVI, 673
5′-(4-Nitrophenylphosphoryl)uridine 2′(3′)-phosphate, XXXIV, 492, 493
5′-(4-Nitrophenylphospho)uridylyl 5′-phosphate, XLVI, 671
p-Nitrophenylserinol, XLIII, 735, 737
p-Nitrophenylsulfate
assay of PAP, LVII, 254
of PAPS, LVII, 245
O-Nitrophenylsulfenyl, XXXIV, 508, 513
Nitrophenylsulfenyl amino acid, XXXIV, 508, 513
2-Nitrophenylsulfenyl chloride, XLVII, 453, 466, 539
N-*o*-Nitrophenylsulfenyl group
removal, XLVII, 517, 538, 539, 548, 549
stability, XLVII, 517
2-(2-Nitrophenylsulfenyl)-3-methyl-3-bromoindolenine, XLVII, 443, 456, 460, 468
stability, XLVII, 468
p-Nitrophenylthiol diazoacetate, XLVI, 40
O-(4-Nitrophenyl)thiophosphoryl chloride, XXXIV, 607
p-Nitrophenylthymidine 5′-phosphate, as enzyme substrate, XXXII, 90
4-Nitrophenyl-tri-*O*-acetyl-β-D-fucopyranoside, XXXIV, 360
p-Nitrophenyl trimethylacetate, acylating agent, XLIV, 562, 567, 572
m-(3-[*m*-(*p*-Nitrophenylureido)phenoxy]-propoxy)benzamidine, XLVI, 127
4-Nitrophthalic acid, synthesis of 4-nitrophthalic anhydride, LVII, 427

4-Nitrophthalic anhydride, synthesis of *N*-methyl-4-nitrophthalimide, LVII, 427
Nitroprusside reaction, XXXIV, 268
N-Nitroso-*N*-benzyl amide, of *N*'-isobutyrylphenylalanine, XLVI, 217
Nitrosococcus, nitrification, LIII, 634, 635
Nitrosococcus oceanus
  ammonia oxidase, LIII, 636
  culture, LIII, 636
Nitroso compound, cytochrome *P*–450 difference spectra and, LII, 273, 275
Nitrosocystis, *see Nitrosococcus*
N-Nitroso-1,4-dihydro-6,7-dimethoxy-3(2*H*)-isoquinolone, XLVI, 218, 219
Nitrosoguanidine
  aurovertin-resistant mutants and, LVI, 180
  mutagenesis, LVI, 98, 99
N-Nitrosolactam, as inhibitors, XLVI, 39
Nitrosomonas
  nitrification, LIII, 634, 635
  oxidase, LII, 21
Nitrosomonas europaea
  ammonia oxidase, LIII, 637
  culture, LIII, 636
1-Nitroso-2-naphthol, determination of tyrosine, LIII, 279
N-Nitrosovalerolactam, XLVI, 218
Nitrosyl ion
  inhibitor of cytochrome *c* oxidase, LIII, 200
  interaction with cytochrome *c* oxidase, LIII, 194, 199, 200
  preparation, LIII, 200
Nitrotetrazolium Blue stain, in oxyntic cell studies, XXXII, 712
2-Nitro-5-thiocyanatobenzoic acid, inhibitor of NADPH-cytochrome *P*-450 reductase, LII, 96
2-Nitro-5-thiocyanobenzoate, XLVII, 411
  cysteine cyanylation, XLVII, 129–132
3-Nitrotyrosine
  internal standard, XLVII, 34, 36, 37
  purification procedure, XLVII, 37
Nitrotyrosine-albumin, XXXIV, 188
Nitrotyrosine peptide, XXXIV, 188, 189
  antibodies and antigens, XXXIV, 188, 189
  from lysozyme, XXXIV, 189
Nitrotyrosyl lysozyme, peptides, XXXIV, 184
Nitrotyrosyl peptide, XXXIV, 184
Nitrous acid, XLIV, 88, 99, 125
  as mutagen, XLIII, 35
Nitrous oxide
  as contaminant, LVIII, 202
  determination, LIII, 643, 644
  in IR spectroscopy, LIV, 304, 309, 311
Nitrous oxide/acetylene flame, LIV, 451, 454, 455, 462, 464
Nitroxide, in spin-label studies, XXXII, 162
Nitroxide compound, *see also* specific compound
  commercial availability, XLIX, 434, 435
  correlation time, rotational calculation, XLIX, 467–470
  electron spin resonance spectra, XLIX, 468, 475
  polarity, XLIX, 422, 423
  principal *A* and *g* values, XLIX, 375, 376
  properties, XLIX, 419, 433, 434
  stability, XLIX, 434
  structure, XLIX, 375, 376, 434, 441, 467, 489, 492, 493, 500, 501
  symmetry and orientation, XLIX, 425–427
  synthesis, XLIX, 370, 371, 434, 435
Nitroxide free radical
  electron spin resonance, XLIX, 372–389
  interatomic distance calculations, XLIX, 331, 350, 351
N16 medium, composition, XXXIX, 399
NMR, *see* Nuclear magnetic resonance
Nobecutan, XXXIX, 420
Noble agar, LVIII, 23
Nocardia, penicillinacylase, XLIII, 720
Nocardia erythropolis, oxidase, LII, 18
Nocardia globerula, oxidase, LII, 17
Nocardia restrictus, LII, 11
Noise, *see also* Shot noise
  in derivation of optimum sample density, LIV, 46
  fourth derivative spectral analysis, LVI, 511, 512
    digital filtering, LVI, 512
    random noise, LVI, 513–515

using different Δλ intervals, LVI, 512, 513
granular, XLIX, 39, 40, 53
phototube, dark current, XLVIII, 464, 465
relation to light intensity, LIV, 34
shot, fluorescence-detected circular dichroism, XLIX, 211
source, in electron microscopy, XLIX, 39
Nojirimycin, XLIII, 272
Nomarski imaging, cell-shape studies, XXXII, 53, 54
Nonactin, effect on bilayers, XXXII, 500, 552
Nonadecanoic acid, deuterated, mass spectrum, XXXV, 342
Nonanal, quantum yield, in bioassay, LVII, 192
Nonane-4,6-dione, LV, 517
Nonheme iron
in complex I, LIII, 12; LVI, 580, 581
in complex II, LIII, 24; LVI, 580, 581
in complex III, LIII, 89, 96, 97; LVI, 580, 582, 583
in complex I–III, LIII, 8
in cytochrome bc_1 complex, LIII, 116
in hydrogenase, LIII, 292, 314
in ω-hydroxylase, LIII, 360
Keilin-Hartree preparation, LV, 123
M. phlei ETP, LV, 185
in NADH dehydrogenase, LIII, 17
in nitrate reductase, LIII, 352
in succinate dehydrogenase, LIII, 30, 476, 484; LV, 700
Nonheme iron prosthetic group, in oxidases, LII, 4, 5
Nonheme iron protein, oxygenated form, spectra, LII, 36
Nonhistone chromosomal protein, XL, 144
analysis, XL, 165
chemical techniques, XL, 158, 161
disc gel electrophoresis, XL, 154
fractionation
ion exchange chromatography, XL, 160
QAE-Sephadex, XL, 164
immunochemical characteristics, XL, 192
isolation and characterization, XL, 160
general strategies, XL, 145
phosphorylated, column chromatography, XL, 200
phosphorylation *in vitro*, XL, 164
SDS isolation, XL, 165
SE preparation, XL, 150
controls and reproducibility, XL, 153
Nonidet P–40 detergent, XXXVII, 119; LVI, 741
lysis of Chinese hamster ovary cells, LIX, 220, 232
of mitochondria, LIX, 425
preparation of ribosomal subunits, LIX, 412
Nonidet P–42 detergent, activator of dihydroorotate dehydrogenase, LI, 62
Non-Kramers system, LIV, 346, 350–355
Nonneuronal cell
co-culture with neurons, LVIII, 581
removal, LVIII, 580, 581
Nonpolyene macrolide
aglycosidic, XLIII, 136, 137
glycosidic, XLIII, 129–134
2-Nonyl-4-hydroxyquinoline *N*-oxide, inhibitor of complex III, LIII, 38
2-*n*-Nonylhydroxyquinoline *N*-oxide, *M. phlei* membrane systems, LV, 185
Nopalcal 6–L detergent, LVI, 741
Noradrenochrome, XLIV, 256
19-Nor-4-androstene-3,17-dione, XLVI, 457–459
Norbiotinylglycyl-*p*-nitrophenyl ester, XLVI, 615
Norbiotinyl *p*-nitrophenyl ester, XLVI, 615
Norelevorphanol, synthesis, XLVI, 605
Norepinephrine, XXXIV, 105; XLIV, 256, 257; XLVI, 336, 591, 592
adenylate cyclases, XXXVIII, 147–149, 173, 179, 180
in chromaffin granule, XXXI, 381
effect on Purkinje cell, XXXIX, 435
on renin production, XXXIX, 22
inhibitor of phenylalanine hydroxylase, LIII, 285
in pineal, XXXIX, 395–396
release, XXXIX, 376

Norepinephrine-agarose, XXXIV, 697
Norepinephrine-glass, XXXIV, 699
Norit
assay of guanylate kinase, LI, 476, 477
of phosphoribosylpyrophosphate synthetase, LI, 4
of thymidylate synthetase, LI, 91
washing procedure, LI, 12, 13
Norit A, LIX, 235
assay of tryptophanyl-tRNA synthetase, LIX, 238
recrystallization of bile sales, LIII, 4, 56, 74
for steroid adsorption, XXXVI, 40
Norleucine, internal standard, XLVII, 32, 34, 36, 81
19-Norprogesterone, molecular bioassay, XXXVI, 465
19-Norsteroid, bioassay, XXXVI, 458
19-Nortestosterone, XLVI, 476
19-Nortestosterone acetate, XLVI, 89, 476–478
Nortestosterone-agarose, XXXIV, 558
Δ^5-3-ketosteroid isomerase, XXXIV, 560–564
19-Nortestosterone 17-hemisuccinate, XXXIV, 675
19-Nortestosterone 17-*O*-hemisuccinate, XXXIV, 558, 559
19-Nortestosterone 17-*O*-succinate, XXXIV, 680
19-Nortestosterone-17-*O*-succinyldiaminodipropylaminoagarose, XXXIV, 558, 673, 680
Norvaline, internal standard, XLVII, 34, 39
Noryl cap, on centrifuge tubes, XXXI, 18
Nossal cell disintegrator, XXXI, 658
properties, XXXI, 662
Nostoc, lysis, XXXI, 681
Nostoc muscorum, aldolase, XLII, 233
Novenamine, XLIII, 304
Novikoff ascites cell
adenylosuccinate synthetase, LI, 212
collection, LI, 301
culture procedure, LI, 326
homogenization, LI, 301
pyrimidine nucleoside monophosphate kinase, LI, 321–331
ribonucleoside diphosphate reductase, LI, 227, 245
uridine-cytidine kinase, LI, 299
Novobiocic acid synthetase, XLIII, 502–508
3-amino-4,7-dihydroxy-8-methylcoumarin, XLIII, 506
assay, XLIII, 503–505
paper chromatographic, XLIII, 504
thin-layer chromatographic, XLIII, 504
cell-free extract, XLIII, 507
hydrolysis of novobiocin, XLIII, 505
4-hydroxy-3-(3-methyl-2-butenyl)benzoic acid, XLIII, 506
pH optimum, XLIII, 508
specificity, XLIII, 507
stability, XLIII, 508
Streptomyces niveus, XLIII, 502, 507
unit definition, XLIII, 504
Novobiocin
biosynthesis, XLIII, 502–508, *see also* Novobiocic acid synthetase
high-pressure liquid chromatography, XLIII, 301–304
paper chromatography, XLIII, 160, 161, 163
structure, XLIII, 502, 503
NPAC gel, *see* *p*-Nitrophenol, ester derivative of hydroxyalkyl methacrylate gel
NPGB, *see* *p*-Nitrophenyl *p*′-guanidinobenzoate
NR, *see* Nitrate reductase
4NS, *see* 2,2-Dimethyl-5-tetradecane-5-propionic acid-*N*-oxyloxazolidine
NSF, *see* National Science Fountation
NTA, *see* Nitrilotriacetic acid
NTB emulsion, properties, XXXVII, 160
NTCB, *see* 2-Nitro-5-thiocyanobenzoate
NTPH, *see* Nucleoside triphosphate pyrophosphohydrolase
Nuclear-cytoplasmic interaction, XLVI, 644
Nuclear emulsion, *see* Emulsion, nuclear
Nuclear envelope
chemical composition, XXXI, 288–290
enzymes, XXXI, 290–292
isolation and properties, XXXI, 279–292
morphology, XXXI, 286–288

Nuclear fraction, isolation of mitochondria
procedure, LV, 35–37
reagents, LV, 35
Nuclear Hamiltonian, LIV, 347–349
Nuclear magnetic moment, LIV, 350
Nuclear magnetic resonance, XLIX, 253–369; LIV, 151–223
carbon
chemical shift range, XLIX, 271, 334
sample size, XLIX, 271, 336
carbon–13
^{13}C-enriched precursor, XLIII, 407, 411, 417
cephalosporin C biosynthesis, XLIII, 408–409, 411–417, 418–425
chemical shifts
in cephalosporin models, XLIII, 415
in spectra, XLIII, 415–417
experimental conditions, XLIII, 406–408
fermentation, XLIII, 411, 414
Fourier transform method, L, 39
high field spectrometer, XLIII, 405
instrumentation, XLIII, 405, 405, L, 48, 49
isolation, XLIII, 411, 414
labeling conditions, XLIII, 407–409, 411, 414
β-lactam antibiotics biosynthesis, XLIII, 410–425
model compounds synthesis, XLIII, 411
monosaccharide signal assignment, L, 40, 41
natural abundance CMR spectrum, XLIII, 407
nuclear Overhauser effect, XLIII, 407
Overhauser enhancements, L, 50
penicillin V biosynthesis, XLIII, 414, 418–425
pH dependence of resonance in cephalosporin C, XLIII, 415, 416
polysaccharide conformation, L, 46–48
polysaccharide signal assignment, L, 41–45
polysaccharide spectra solutions, L, 49, 50
polysaccharide structure determination, L, 39–50
procedures, L, 48–50
polysaccharide substituent effects, L, 45, 46
preliminary ^{14}C experiment, XLIII, 407
proton noise decoupling, XLIII, 406
pulsed Fourier transform, XLIII, 406
satellite method, XLIII, 406
spectrum
assignments of chemical shifts, XLIII, 415, 416
integration, L, 50
recording, XLIII, 414, 415
characterization of iron-sulfur proteins, LIII, 274
chemical shift reference compounds, XLIX, 256
chemical shift units, XLIX, 256
computer amplitude correction, XLIX, 265
computer phase correction, XLIX, 264, 265, 267
conformation studies, XLIV, 366
continuous-wave method, XXXII, 199, 200–203; XLIX, 323, 343, 344; LIV, 151, 152, 169–171, 187
correlation, XLIX, 259–264
crystal filtering, XLIX, 262
data analysis, LIV, 160–162
decoupling
time-shared, XLIX, 268
underwater, XLIX, 262, 268
determination of flavin site of substitution, LIII, 462, 463
digital timing system, XLIX, 267
effective field formalism, XLIX, 264
flow cell designs, XLIX, 304–314
fluorine, XLIX, 270–295
advantages, XLIX, 271, 272
chemical shift, XLIX, 272, 288, 289, 292–294
computer delay, XLIX, 284, 286
computer word length, XLIX, 284

data scaling, XLIX, 284
digitizer resolution, XLIX, 284, 285
field/frequency lock, XLIX, 288
filtering, XLIX, 283, 284
Fourier-transform time-averaging, XLIX, 275
labeling method, XLIX, 273, 275
linewidths, calculated, XLIX, 277, 278, 294, 295
magnetic field strength, XLIX, 277–279
nuclear Overhauser enhancement, XLIX, 294, 295
principles, XLIX, 270–275
progressive saturation experiment, XLIX, 289
proton noise decoupling, XLIX, 286, 287
receiver gating, XLIX, 284, 286
relaxation time measurements, XLIX, 289, 290, 294, 295
sample size, XLIX, 279, 280
spectrometer, choice, XLIX, 277–290
spectrum interpretation, XLIX, 290–295
two-parameter nonlinear least squares program, XLIX, 290, 291
Fourier-transform, XLIX, 254–258, 323, 332, 333
data parameters, XLIX, 280–283
difference spectroscopy technique, XLIX, 302, 303
free induction decay acquisition time, XLIX, 280
principles, XLIX, 275
pulse width, XLIX, 281–283
sweep width, XLIX, 280, 281
tipping angle, XLIX, 282
frequency-scan, XLIX, 258, 259
gating, XLIX, 267
geometry of enzyme-bound substrates, XLIX, 322–359
of glycosphingolipids, XXXII, 366–367
high-resolution type, XXXII, 205–211
hydrogen-1, XLIII, 338–404; XLIV, 366, 367; LIV, 223–232
acetone-d_6, XLIII, 394
acetonitrile-d_2, XLIII, 393
acetylation, XLIII, 403
anisotropic solvent, XLIII, 393, 401
applicability, XLIII, 390–392
area measurement, XLIII, 394
aromatic coupling, XLIII, 400
carbamation, XLIII, 403
chemical shift measurement, XLIII, 396
computer programs for molecular geometry, XLIII, 402
coupling constants, XLIII, 398–400
data analysis, LIV, 229–231
deuteriochloroform as solvent, XLIII, 393
deuterium oxide, XLIII, 393
dimethyl-d_7 formamide, XLIII, 393
dimethyl-d_6 sulfoxide, XLIII, 393
geminal coupling, XLIII, 398
hydrogen types, XLIII, 397, 398
INDOR sweep, XLIII, 396
instrumentation, XLIII, 392
long-range couplings, XLIII, 400
methanol-d_4, XLIII, 394
molecular structure of antibiotics, XLIII, 388–404
nuclear Overhauser effect, XLIII, 402
paramagnetic organometallic complex as shift reagent, XLIII, 401, 402
polysaccharide structure determination, L, 39
resonance standard, LIV, 200
sample
changing, XLIII, 403, 404
requirements, XLIII, 392–394
sample preparation, LIV, 228
shift interpretation, XLIII, 397, 398
sodium 4,4-dimethyl-4-silapentane-1-sulfonate, XLIII, 393
sodium 3-trimethylsilyltetradeuteriopropionate, XLIII, 393
solvent, XLIII, 393
spectral analysis, XLIII, 395, 396
changing, XLIII, 400–402
deuterium, XLIII, 400–410

interpretation, XLIII, 394–396
spectrometer specifications, LIV, 228
structure proposal, XLIII, 396–400
trifluoroacetic acid, XLIII, 394
vacinal coupling, XLIII, 399
van der Waals effect, XLIII, 400
vinyl coupling, XLIII, 399, 400
identification of coenzyme Q analogs, LIII, 596
of iron-sulfur centers, LIII, 269
line-broadening measurements, LIV, 178–180
longitudinal relaxation rate, theoretical considerations, XLIX, 324
long-pulse, XLIX, 264–269
magnetic susceptibility determinations, LIV, 388, 392, 393
of membrane systems, XXXII, 198–211
observation frequency, choice, LIV, 201
^{18}O measurements, LV, 252
Overhauser enrichment, XXXII, 200
paramagnetic contamination, elimination, XLIX, 336
phosphorus, sample, XLIX, 336
probe, flow system, XLIX, 314, 315
protein concentration, XLIX, 271
proton
in aqueous solutions, XLIX, 253–270
data acquisition, LIX, 25
for determination of transfer RNA structure, LIX, 21–57
elimination of H_2O peak, XLIX, 253–255
instrumentation, LIX, 23, 24
pH measurements, XLIX, 255, 256
protein molecular weight, XLIX, 271
pulsed methods, XLIX, 323
sample preparation, LIX, 24, 25
sample size, XLIX, 271
saturation, LIX, 26
solvent saturation methods, XLIX, 256–258, 342
spectral width, XLIX, 271
use of deuterated solvent, XLIX, 255, 256
pulsed, XXXII, 199, 203–205, 208–209
pulse Fourier-transform method, LIV, 151–192
advantages, LIV, 152, 153
in aqueous solutions, LIV, 182–187
data acquisition, LIV, 157–160
experimental conditions, LIV, 155–166
kinetic isotope effects, LIV, 182
long-pulse methods, LIV, 167–169, 184–187
principle, LIV, 151, 155
pulse parameters, selection, LIV, 156, 157, 163
of redox proteins, LIV, 166–190
ring current shifts, LIV, 167
sample parameters, LIV, 154, 155, 157
saturation transfer experiments, LIV, 171–178
spectrometer design, LIV, 153, 154
steady-state techniques, LIV, 163–166
symbol glossary, LIV, 190–192
quadrature detection, XLIX, 262, 265, 267, 280
rapid-scan correlation method, *see* Nuclear magnetic resonance, continuous-wave method
relative signal intensities, XLIX, 271
relaxation rates, theoretical paramagnetic effects, XLIX, 326–332
sample preparation, LIV, 199, 200
sample tubes, XXXII, 200
sensitivity, XLIX, 253, 322
shift reagents, XLIX, 269
signal averaging, XXXII, 199, 200
single sideband crystal filtering, XLIX, 280
spectra
glutaraldehyde, XLIV, 552, 553
phase display convention, XLIX, 373
of protein, effects of paramagnetic center, LIV, 193

spectrometer
lock channel, XLIX, 335, 347
locking frequency, source, XLIX, 347
proton decoupling channel, XLIX, 335
for stopped-flow method, XLIX, 316–318
for T_1 measurement, XLIX, 333–336, 340
of spin labels, XXXII, 193
of steroid hormone receptors, XXXVI, 412
stopped-flow, XLIX, 295–321
continuous-wave, XLIX, 296–299
flow-caused transient phenomena, XLIX, 318–321
Fourier-transform, XLIX, 299–303
sensitivity, XLIX, 301
pulse length, XLIX, 318
syringe blocks, XLIX, 315, 316
techniques, LIV, 200
tube, light scattering, elimination, LIV, 216, 217
units, LIV, 200
variable-frequency observations, LIV, 187, 188
wide-line, XXXII, 199, 200–205
wiggle signal, XLIX, 259, 260
Nuclear magnetic resonance spectroscopy, *see* Nuclear magnetic resonance
Nuclear membrane
AHH activity, LII, 239
in liver preparations, XXXI, 7
Nuclear Overhauser effect, XLIII, 402, 407
Nuclear polymerase I, estradiol effects, XXXVI, 319
Nuclear 5 S receptor, purification, XXXVI, 357–362
Nuclear steroid hormone receptor, XXXVI, 265–327
for androgens, XXXVI, 372, 373
assay of complex, XXXVI, 283–286
differential extraction, XXXVI, 286–292
estrogen interaction with target tissue, XXXVI, 267–275
interactions with nuclear constituents, XXXVI, 292–313
protein complex, XXXVI, 275–283
5 S receptor, purification, XXXVI, 357–362
Nuclear transfer technique, XXXII, 789
Nuclease, XXXIV, 492–496; XLVI, 175, 358–362, *see also* specific types
affinity labeled peptides, XXXIV, 187
circularly polarized luminescence, XLIX, 186
digestion of proteins, XL, 35
inhibition, by diethylpyrocarbonate, XLVII, 441
plant, removal, XXXI, 579
staphylococcal, XXXIV, 495, 496
circular dichroic spectrum, XLIX, 161
thermolysin hydrolysis, XLVII, 188
Nuclease P1
in oligonucleotide digestion, LIX, 79
from *Penicillium citrinum,* in tRNA sequence analysis, LIX, 61
in tRNA digestion, LIX, 81–83, 85, 87, 94, 104
in tRNA labeling procedure LIX, 71
solution, stability, LIX, 61
Nuclease-T
semisynthetic, XXXIV, 632–635
thermolysin hydrolysis, XLVII, 188
Nuclease-T-(6-48), synthetic, XXXIV, 632
Nuclease-T-(49-149), native, XXXIV, 632
Nuclease-T-agarose, XXXIV, 635
Nucleic acid, XLVI, 88, *see also* specific type
antinucleoside and antinucleotide antibodies, XL, 302
coupling with histone complexes, XL, 192
denaturation studies, XLIX, 11
electron microscope cytochemistry, XL, 22
methods staining both nucleic acids, XL, 23
extraction procedure, LIX, 311, 312
interactions, XLVIII, 295–299
magnetic circular dichroism, XLIX, 164–166
photoaffinity labeling, XLVI, 644–649

polynucleotide matrices, XXXIV, 463–475
applications, XXXIV, 472–475
preparation, XXXIV, 463–471
precursor incorporation, substrate entry, XXXIX, 495
purification, XXXIV, 463–475, *see also* specific compounds
removal by ethanol fractionation, LI, 71, 75, 76
with nucleases, LI, 61, 124, 190, 311
by protamine sulfate, LI, 32, 183, 252, 439, 440, 449, 519
by streptomycin sulfate, LI, 7, 43, 53, 80, 87, 124, 137, 215, 230, 276, 287, 302, 303, 324, 339, 340, 347, 402, 410, 414, 415, 420, 433, 443, 554, 562
Nucleic acid-agarose, XXXIV, 464–468
Nucleic acid-associated protein, staining, XL, 21
Nucleic acid-cellulose matrice, XXXIV, 469, 470
Nucleic acid derivative, in media, LVIII, 67
Nucleic acid-protein complex, XLVI, 168–180
cross linking, XLVI, 168–180
Nucleoid, isolation from peroxisomes, XXXI, 368–374
Nucleomedium
composition, XXXII, 804
for human cell culture, XXXII, 803
Nucleophilic group
reaction, XLIV, 263
rate determination, XLIV, 14
Nucleophilic reagent, XLVII, 517, 540
Nucleopore filter, for use with environmental samples, LVII, 80
Nucleoprotein, *see also* specific type
autoradiographical and cytochemical localization, XL, 3–41
circular dichroism spectra, analysis, XL, 237
complexes, circular dichroism analysis, XL, 209
preparation, XL, 227
light-scattering, of solutions, XL, 225
Nucleoside, *see also* specific types
modified, separation, LIX, 181
phosphorolysis, determination methods, LI, 526, 532, 539, 540
phosphorylation, XLVI, 324, 325
product of rat liver acid nucleotidase, LI, 271
separation on polyethyleneimine cellulose, XXXVIII, 17, 18
thin-layer chromatography, XXXVIII, 28–30, 158
Nucleoside-agarose, XXXIV, 475–479
Nucleoside antibiotic, XLIII, 187
Nucleoside deoxyribosyltransferase, LI, 446–455
activity, LI, 446
assay, LI, 446–448
distribution, LI, 452
isozymes, LI, 452
kinetic properties, LI, 452–454
from *Lactobacillus helveticus*, LI, 446–455
molecular weight, LI, 450
pH optimum, LI, 452
stability, LI, 450
substrate specificity, LI, 450, 452
Nucleoside diphosphate
ATP-P_i exchange preparation, LV, 362
substrate of ribonucleoside diphosphate reductase, LI, 227
Nucleoside diphosphate kinase, LI, 371–386
activity, LI, 371, 376, 386; LVII, 86
affinity for calcium phosphate gel, LI, 584
assay, LI, 371–373, 376–378; LV, 287–289; LVII, 30
cation requirements, LI, 385
contaminant, in luciferase preparation, LVII, 35, 38, 62
determination of guanosine nucleotides, LVII, 86, 92
from human erythrocytes, LI, 376–386
interference by, in ATP assay, LVII, 29
isoelectric variation, LI, 378, 379
kinetic properties, LI, 386
molecular weight, LI, 375, 385
pH effects, LI, 386
purification, LI, 373–375; LX, 581
reaction mechanism, LI, 375, 384

from *Salmonella typhimurium*, LI, 371–375
stability, LI, 375
substrate specificity, LI, 375, 384
temperature dependence, LI, 385
thiol group activity, LI, 385, 386
Nucleoside diphosphokinase, *see* Nucleoside diphosphate kinase
Nucleoside monophosphate
electrophoretic mobility, LI, 355
product of nucleoside triphosphate pyrophosphohydrolase, LI, 275
Nucleoside monophosphate analog, XLVI, 299–302
Nucleoside phosphate, XXXIV, 479–491
adenosine phosphate ligands, XXXIV, 485–490
periodate oxidized ribonucleoside, XXXIV, 490, 491
synthesis of ligands, XXXIV, 481–485
Nucleoside phosphate-agarose, XXXIV, 479–491, *see also* Nucleoside phosphate
Nucleoside phosphotransferase, LI, 387–394
activity, LI, 387, 390, 391
amino acid analysis, LI, 390
assay, LI, 387, 388
from carrot, LI, 387
effect of cations and anions, LI, 393
hydrolase activity, characteristics, LI, 392
isoelectric point, LI, 390
kinetic properties, LI, 391
molecular weight, LI, 390
pH optimum, LI, 392
properties, LI, 390–394
purification, LI, 388–390
purity, LI, 390
stability, LI, 394
substrate specificity, LI, 391, 392
Nucleoside Q, LIX, 129, 222
Nucleoside triphosphatase, cyclase assays, XXXVIII, 116, 117
Nucleoside triphosphate, *see also* specific type
concentration and volume, nucleotide cyclase assays, XXXVIII, 115–117
degradation, reduction, XXXVIII, 119, 120
as effectors, XLVI, 322
labeled, impurities, XXXVIII, 118
substrate of nucleoside triphosphate pyrophosphohydrolase, LI, 275
synthesis, LI, 371
thin-layer chromatography, XXXVIII, 35, 36
Nucleosidetriphosphate pyrophosphatase, LI, 275–285
assay, LI, 275, 276
inhibitors, LI, 279
kinetic properties, LI, 285
molecular weight, LI, 279
pH optimum, LI, 279, 285
properties, LI, 279, 281, 282, 285
purification, LI, 276–284
stability, LI, 285
substrate specificity, LI, 279, 285
Nucleoside triphosphate pyrophosphohydrolase, *see* Nucleosidetriphosphate pyrophosphatase
Nucleoside X, *see* N^3-(3-L-Amino-3-carboxypropyl)uridine
Nucleotidase
in adenylate cyclase systems, XXXI, 111
comparative properties, LI, 290
3′-Nucleotidase, plasma membrane and, XXXI, 90
5′-Nucleotidase
agarose chromatography, XXXII, 372, 373
assay, XXXI, 91, 92; XXXII, 124–131, 368, 369
in brush borders, XXXI, 122
cGMP assay, XXXVIII, 88
as diagnostic enzyme, XXXI, 19, 24, 35–37, 93, 262, 729
in fat-cell plasma membrane, XXXI, 65, 67
in liver, XXXI, 97
as marker enzyme, XXXII, 124, 393
in microsomal fractions, LII, 88
in microsomal peroxide production system, LII, 344
oxidation
procedure, LII, 92
in reductase assay, LII, 91
perchloric acid, LV, 201
in phagolysosomes, XXXI, 344, 345

phosphodiesterase, assay, XXXVIII, 133, 205–208, 210, 218, 224, 225, 240, 249, 250, 263, 264
in plasma membranes, XXXI, 88, 91, 100, 101, 148; XXXII, 91
properties, XXXII, 126, 373, 374
purification of, as sphingomyelin complex, XXXII, 368–374
viscosity barrier centrifugation, LV, 133, 134
Nucleotide, *see also* specific type
agarose hydrazide, XXXIV, 477, 478, 491
analogs, R_F values, LVI, 648, 652
anhydride analogs, XLVI, 306, 307
antisera, XXXVI, 17
assay of bound, XXXIV, 235, 236
binding sites, XLVI, 312–321
chromatographic separation, LI, 373
complexes, electrolytic preparation, XLVI, 313, 314
cyclic, *see* Cyclic nucleotide
derivatives, XLVI, 150
extraction methods, LVII, 81, 88
guanylate cyclase, XXXVIII, 195, 199
immobilized, XXXIV, 242–253, 475–479
affinity chromatography of enzymes, XXXIV, 245–253
ion-exchange chromatography, XXXVIII, 126–128
methylation, effect on oligonucleotide properties, LIX, 90–92
modified
effect on oligonucleotide properties, LIX, 90–99
electrophoretic mobilities, LIX, 73
identification, LIX, 70–74, 87–99
separation, LIX, 181
M value, for 3′-terminal removal, LIX, 89
periodate oxidation, XXXIV, 478
rapid labeling probes, LV, 245, 246
receptors, XLVI, 88
relative electrophoretic mobilities, LIX, 64
separation, LV, 251, 252
substrate of rat liver acid nucleotidase, LI, 271
terminal phosphate, XLVI, 102
thin-layer chromatography, XXXVIII, 28–30, 158
cyclic nucleotides, XXXVIII, 30–35
tightly bound, rate and extent of labeling, LV, 248, 249
Nucleotide adipic hydrazide, XXXIV, 478
Nucleotide-agarose, XXXIV, 229, 475–479
Nucleotide diphosphate sugar, pools in animal cells, L, 178, 179
Nucleotide-hydrazide-agarose, XXXIV, 491
Nucleotide pyrophosphatase, purification of flavin peptides, LIII, 455
Nucleotide standard solutions, stability, LVII, 76
Nucleus
from brain cells, isolation, XXXI, 452–457
bulk yields, XXXI, 276–279
characterization, XXXI, 23, 24, 33, 35, 259–262
differential sedimentation, XXXI, 727
DNase digestion, XL, 87
electron microscopy, XXXI, 28, 269
high M.W. RNA, XXXI, 272, 273
isolation, XXXI, 722; XL, 84; LVIII, 222
from liver, XXXI, 10–15, 722
from oviduct, XXXI, 263–279
procedures with metaphase chromosomes and mitotic apparatus, XL, 75
stability, XL, 87
for steroid receptor studies, XXXVI, 295, 296, 491
from thymus, XXXI, 246–252
lipid, XXXI, 35
from liver, preservation, XXXI, 5, 6
of plant cells, isolation, XXXI, 501, 510, 544, 558–565
preparation, LV, 36
in analysis of chromosomal nonhistone proteins, XL, 161
receptor binding, XXXVI, 299–304
Nucleus counting, protocol, LVIII, 143, 144
Nucleus fusion, LVIII, 366
Nucleus transfer, LVIII, 359
Nude mouse
breeding, LVIII, 372, 373

cell culture, LVIII, 378, 379
cell fusion, LVIII, 348
characterized, LVIII, 370
maintenance, LVIII, 372, 373
mass propagation, LVIII, 371, 375
nontumorigenic cell line, LVIII, 378
source, LVIII, 372, 373
transfer from, LVIII, 378, 379
transformed cells, LVIII, 368
tumorigenicity testing, LVIII, 370–379
Number fluctuation spectroscopy, XLVIII, 421, 451, 452
Nupercaine, phospholipase, LV, 6, 32
Nutation angle
definition, LIV, 156
in long-pulse method, LIV, 168
computation, LIV, 185, 186
optimization, LIV, 163–165
Nutrient, LVIII, 52
low-molecular-weight, LVIII, 555
optimum concentration, LVIII, 76
Nutritional deficiency, XL, 321
Nylon, *see also* Polyaminoaryl-nylon; Polyisonitrile-nylon; specific types
acid chlorides, XXXIV, 203
alkylation, XLIV, 122–134
bridge formation, XLIV, 166
carbodiimide procedures, XXXIV, 202, 203
chemical properties, XLIV, 118
in continuous flow analyzer, XLIV, 639–642
activation, XLIV, 640, 641
bore size, XLIV, 641
storage, XLIV, 642
covalent binding, XLIV, 560, 561
derivatives, XXXIV, 203, 204, 214
leakage, XXXIV, 204
protein, XXXIV, 203
specificity, XXXIV, 214
stability, XXXIV, 204
derivatized fiber, XXXIV, 200, 201
DNP-albumin, XXXIV, 202
elastic support, XLIV, 560, 561
enzyme conjugates, storage, XLIV, 121, 133
enzyme immobilization, XLIV, 118–134
film, XXXIV, 221, 222
hydrazide-substituted, storage, XLIV, 124
imidate salts, XLIV, 122–127
mesh, large scale, XXXIV, 210
microencapsulation, XLIV, 207–209
for Mössbauer spectroscopy sample cells, LIV, 376
partial acid hydrolysis, XLIV, 120, 121, 131
peptide bond cleavage, XLIV, 121
physical properties, XLIV, 118
powder, partially hydrolyzed
preparation, XLIV, 131
storage, XLIV, 131
spacers, XXXIV, 203
Nylon 6, XLIV, 118, 126
N-alkylation, XLIV, 130, 131
O-alkylation, XLIV, 126, 127
Nylon 66, XLIV, 118
Nylon 610, XLIV, 118
Nylon cloth
gauze, XXXI, 116
mesh, for liver homogenate filtration, XXXI, 8
in microcolumn, LVII, 281
for plating, LVIII, 320
Nylon column filtration, LVIII, 489, 490
Nylon-concanavalin A, XXXIV, 201
Nylon-dinitrophenol, XXXIV, 201
Nylon tubing, immobilized glucose oxidase, LVI, 489, 490
Nystatin, XLIII, 135, 186
Donnan potential, LV, 584
effect on bilayers, XXXII, 500, 535
in media, LVIII, 112–114
phospholipid bilayers, LV, 766
resistance
heme-deficient mutants, LVI, 559, 560
sterol, LVI, 558
Nytex cloth, XXXVI, 287

O

O–antigen chain, in bacteria cell wall, XXXI, 643, 649, 650
Oat
protoplast isolation, XXXI, 580
root
ATPase isolation, XXXII, 393–405

× maize, LVIII, 365
Obelia, fluorescent accessory protein, LVII, 259
Obelin, as marker protein, LVII, 318
O'Brien foreign body spud, XXXIX, 175, 180
Observation unit, continuous-flow pH measurement, LV, 621, 622
Ochramycin, XLIII, 136
Ochromonas danica, aldolase, XLII, 234
cis,cis-9,12-Octadecadienylglyceryl-1-ether, intermediate in phosphatidylcholine synthesis, XXXV, 445, 446
Octadecanal dimethyl acetal, intermediate in plasmalogen synthesis, XXXV, 480
Octadecanoic acid, quantum yield, in bioassay, LVII, 192
1-*O*-*cis*-9′-Octadecenyl-2-octadecanoylglycerol, *see* 2-Stearoyl selachyl alcohol
Octadecylamine, planar black membranes, LV, 772
Octadecyl silyl column, LII, 284
Octadecyltrichlorosilane, immobilized mitochondria, LVI, 551
Octadiene, substrate of ω-hydroxylase, LIII, 360
Octaene, XLIII, 136
Octanal, quantum yield, in bioassay, LVII, 192
n-Octane, assay of ω-hydroxylase, LIII, 356
Octanol
- antifoam agent, XLVII, 470, 493
- in picosecond continuum generation, LIV, 12
- in purification, of orotate phosphoribosyltransferase, LI, 71

1-Octanol, electron spin resonance spectra, XLIX, 383
Octanol cell, LIV, 5
3-Octaprenyl-4-hydrobenzoic acid
- intermediate, in ubiquinone synthesis, LIII, 601
- isolation, LIII, 605
- radiolabeling and isolation, LIII, 608, 609
- spectral properties, LIII, 604
- structural formula, LIII, 601

2-Octaprenyl-4-hydroxybenzoate, accumulation by mutants, LVI, 110
2-Octaprenyl-6-methoxy-1,4-benzoquinone
- accumulation by mutants, LVI, 110
- intermediate, ubiquinone biosynthesis, LIII, 601
- isolation, LIII, 607
- spectral properties, LIII, 604
- structural formula, LIII, 601

2-Octaprenyl-6-methoxyphenol
- intermediate, ubiquinone biosynthesis, LIII, 601
- isolation, LIII, 606
- spectral properties, LIII, 604
- structural formula, LIII, 601

2-Octaprenyl-3-methyl-5-hydroxy-6-methoxy-1,4-benzoquinone
- accumulation by mutants, LVI, 110
- intermediate, ubiquinone biosynthesis, LIII, 601
- spectral properties, LIII, 604
- structural formula, LIII, 601

2-Octaprenyl-3-methyl-6-methoxy-1,4-benzoquinone
- accumulation by mutants, LVI, 110
- intermediate, ubiquinone biosynthesis, LIII, 601
- isolation, LIII, 607
- spectral properties, LIII, 604
- structural formula, LIII, 601

2-Octaprenylphenol
- accumulation by mutants, LVI, 110
- intermediate, ubiquinone biosynthesis, LIII, 601
- isolation, LIII, 606
- spectral properties, LIII, 604
- structural formula, LIII, 601

Octene 1,2-oxide, substrate of epoxide hydrase, LII, 199, 200
Octopamine, XXXIV, 105
Octopine dehydrogenase, stereospecificity determination, LIV, 231
n-Octylamine
- cytochrome binding, LII, 264, 267, 276, 277
- cytochrome *P*–450 difference spectra, LII, 266, 269, 270
- in dimethylaniline monooxygenase assay, LII, 143

Octylamine-substituted Sepharose
in cytochrome *P*–450 purification, LII, 139
synthesis, LII, 138
Octyl dinitrophenol, uncoupling activity, LV, 465
β-D-Octylglucoside
structure, LVI, 736
uses, LVI, 745
n-Octyl-β-D-glucoside, self-association, XLVIII, 110
Octylguanidine, LV, 516
N-Octylmaleimide, in studies of luciferase active site, LVII, 180
p-*tert*-Octylphenylpolyoxyethylene, structure, LVI, 736
O-Octyl Sepharose, in purification, of *Lactobacillus* deoxynucleoside kinases, LI, 351
ODS column, *see* Octadecyl silyl column
Oestradiol Benzoate Injection B.P., XXXI, 409
Offset circuit, LIV, 41–43
OIA, *see* Oxindolylalanine
Oil red O, indicator for lipids, XXXV, 405
Oilseed, enzyme studies, XXXI, 522
3β-ol dehydrogenase, staining, XXXIX, 258
Old Yellow enzyme, *see* NADPH dehydrogenase
Oleaic acid
deuterated, mass spectrum, XXXV, 345
mass spectrum, XXXV, 345
of triglycerides, XXXV, 351–353
thin-layer chromatography, XXXV, 400
Oleandomycin, LVI, 31
bioautography for quantitative estimation, XLIII, 186
mitochondria, LVI, 32
paper chromatography, XLIII, 131
spectropolarimetry, XLIII, 356–359
circular dichroism spectrum, XLIII, 357
optical rotatory dispersion spectrum, XLIII, 358
thin-layer chromatography, XLIII, 198, 211
Oleic acid
in biomembrane phospholipids, XXXII, 542
in cupric oleate synthesis, LII, 312
cytochromeless strains, LVI, 122
freeze-fracture, XXXII, 46, 47
heme-deficient mutants, LVI, 558, 559
inhibitor, of adenylate kinase, LI, 465
of myristic acid, bioassay, LVII, 192
in media, LVIII, 54, 68
uncoupling activity, LV, 465
Oleic acid-agarose, XXXIV, 158
1-Oleoyl-2-myristoyl-*sn*-glycerol-3-phosphoric acid, intermediate in phosphatidic acid synthesis, XXXV, 439
Oleoyl phosphate, ATPases, LV, 343
Oleylamine, synthesis of oleylpolymethacrylic acid resin, LIII, 214
12-Oleyl-CoA 12-hydroxylase, LII, 20
Oleyl coenzyme A
product and inhibitor of stearyl-CoA desaturase, XXXV, 260, 261
substrate for palmityl thioesterase I, XXXV, 106
1-Oleyl-2-palmitoyl-*sn*-glycerol-3-phosphoryl-*sn*-1′-glycerol
intermediate in phosphatidylglycerol synthesis, XXXV, 466, 467
sodium salt, XXXV, 467
Oleylpolymethacrylic acid resin
purification of cytochrome *b*, LIII, 213–216
of cytochrome oxidase, LIII, 67, 68
synthesis, LIII, 213–215
3β-ol-hydroxysteroid dehydrogenase, as marker enzyme, XXXIX, 291
Oligodendroglia
chemical composition, XXXV, 578
fixation for electron microscopy, XXXV, 577
fractionation, XXXV, 575
isolation, XXXV, 570–572, 576
morphology, XXXV, 576, 577
Oligodendroglial cell
enzyme marker, XXXII, 329
as myelin source, XXXII, 323

Oligo(deoxythymidylate)cellulose
of polyadenylate plus myosin messenger ribonucleic acid, LX, 544, 545
purification of globin messenger ribonucleic acid, LX, 143
of ribonuclear protein particles, LX, 542, 543
γ-Oligo-L-glutamate, XXXIV, 5
Oligomycin, XLIII, 137; LVI, 33
AE particles, LV, 386, 390
ATPase assays, LV, 329, 330, 333
ATPase inhibitor, LV, 407
ATP synthesis, LV, 118, 671, 672
biological properties, LV, 503, 504
brown adipose tissue mitochondria, LV, 71, 72
calcium uptake, LVI, 338
carnitine-acylcarnitine translocase, LVI, 375
complex V, LVI, 586
effects on fluorescence probes, XXXII, 242
electron flow, LV, 242
exchange reactions, LV, 285, 287
fluorescence, LVI, 499, 500
immobilized mitochondria, LVI, 557
inhibitor of ATP hydrolysis, LVII, 41
measurement of Ca^{2+}/site ratio, LV, 650
M. phlei membrane systems, LV, 185
N. crassa mitochondria, LV, 147
preparation, LV, 502, 503
proteolipid ionophore, LV, 419
resistance, LVI, 118, 140
stability, LV, 503
structure, LV, 501, 502
submitochondrial particles, LV, 106, 108, 110, 111
toxicology, LV, 503
Oligomycin A, physical properties, LV, 476
Oligomycin B, physical properties, LV, 476
Oligomycin C, physical properties, LV, 477
Oligomycin D, physical properties, LV, 477
Oligomycin-sensitive adenosinetriphosphatase, LV, 302, 303
composition, LV, 303
of rat liver mitochondria, LV, 328, 329
analytical procedures, LV, 329, 330
isolation procedures, LV, 330–332
properties, LV, 333
stability, LV, 333
reconstitution, LV, 703, 704
yeast, LV, 351
assay method, LV, 352
properties of F_1 ATPase, LV, 353, 354
of F_1 subunits, LV, 356
of oligomycin-sensitive ATPase, LV, 356–358
purification of F_1 ATPase, LV, 352, 353
of F_1 subunits, LV, 354, 355
of sensitive ATPase, LV, 356
subunits, LV, 358
Oligomycin sensitivity conferring protein, LV, 391, 392
antibodies, LVI, 227, 228
assay, LV, 392, 393
properties, LV, 381–383, 396, 397
removal, LV, 112
as stalk, LV, 397
Oligonucleotide, XXXIV, 645, *see also* specific types
binding sites, XLVI, 622–624
bromoacetylation, XLVI, 673–675
CACCA-[^{3}H]Leu-Ac, preparation, LIX, 818
chemically reactive, XLVI, 669–676
cross linking, XLVI, 675
5′-end-group analysis, LIX, 78–80, 100, 101
5′-end-group labeling, LIX, 74–76, 100
enzymic digestion, LIX, 79
inhibition of initiation complex formation, LX, 265
long, migration rate, LIX, 100
preparation from tRNA, LIX, 75, 99, 100
purification, XLVI, 672

separation, LIX, 76–78, 100, 120, 343, 344
synthesis of 5′-modified, XLVI, 670–672
synthetic mixtures, XXXIV, 645–649
Oligopeptidase, in brush border, XXXI, 130
Oligosaccharide, XXXIV, 329–331, *see also* specific types
acid hydrolysis, L, 52
β-(*p*-aminophenyl)ethylamine derivatives, L, 163–169
chemical synthesis, L, 93–121
hydroxyl group reactivity, L, 108
Koenigs–Knorr reaction, L, 96–100, 105, 108, 109, 113
α-linkages, L, 100, 101
methods, L, 113–121
orthoester method, L, 100
polymer supports, L, 111–113
protein complex formation, L, 155–171
with chromophore, XLIII, 126, 127
from glycolipid degradation, XXXII, 364, 365
from human milk
fractionation, L, 216–220
structures, L, 217
from human urine
blood group related, L, 226, 227
isolation, L, 234, 235
lactation, L, 228, 229
not blood group related, L, 228
pregnancy, L, 228, 229
secretor status and pregnancy, L, 230, 231
hydrazido-agarose derivatives, XXXIV, 85
Oligosaccharide hapten, L, 163
Oligosaccharide-protein conjugate, phenylisothiocyanate derivatives and, L, 169–171
Oligothymidylic acid-cellulose, XXXIV, 472, 474
Oligouridylic acid, LX, 432, 433, 435, 436
Olivomycin, XLIII, 127
aglycone, XLIII, 334
OMA, *see* Optical multichannel analyzer
Omega dot glass, LVII, 313
Omnifluor, in scintillation cocktail, LI, 220, 544
Omnifluor-toluene scintillation fluid, LIX, 191, 363, 565, 567
Omni-Mixer, properties, XXXI, 662
Omohyoid muscle, incision, XXXIX, 170, 171
OMPdecase, *see* Orotidylate decarboxylase
Oncogenesis, fuco-glycosphingolipids, L, 247
Onco-RNA virus, XXXIV, 337
One-photon spectroscopy, principles, XLIX, 68
One-step reaction, *see* Single-step reaction
Onion, spherosome isolation, XXXI, 578
Onion shoot meristem, ATPase isolation, XXXII, 395, 405
ONPG, *see o*-Nitrophenyl-β-galactoside
Oocyte, hormone effects, XXXIX, 424
Oomycete, D(−)-lactate dehydrogenase, XLI, 294, 296
Operating board, for rat brain surgery, XXXIX, 168, 169
Opiate-glass, XXXIV, 622, 623
Opiate receptor, XXXIV, 619, 622; XLVI, 85, 601–606
assay, XLVI, 603
Oplophorus
coelenteramide, emission wavelength, LVII, 341
luciferin
definition, lack, LVII, 344
reaction mechanism, LVII, 348
OPRTase, *see* Orotate phosphoribosyltransferase
Opsonin
deficiency
assessment of, using granulocyte chemiluminescence assay, LVII, 490
associated clinical disorders, LVII, 484
neonatal, LVII, 480, 484–487
phagocyte chemiluminescence, LVII, 477, 479, 480, 482–487
types, LVII, 482, 483
Opsonization process
pathways, LVII, 482–484

phagocyte chemiluminescence, LVII, 464, 465
Optical activity, theories, LIV, 254–257
μ-*M* mechanism, LIV, 254, 256, 277
one-electron theory, LIV, 254, 255
Optical comparator, XLIX, 46
Optical constant, XLVIII, 170
Optical density
optimum, of sample, derivation, LIV, 45, 46
path length, LIV, 39
signal-to-noise ratio, LIV, 37, 38
Optical-difference spectroscopy
areas of application, LII, 258
atypical peaks, factors contributing, LII, 273
categories of difference spectra, LII, 259
Michaelis constant, LII, 261, 262
spectrophotometer, LII, 265
spin-state, LII, 259–261
structure, LII, 260, 261
Optical-electron paramagnetic resonance titrator, schematic, LIV, 125
Optical filtering, XLIX, 39–63
applications, XLIX, 49–59
best image selection, XLIX, 49–52
general techniques, XLIX, 42–49
indexing of diffraction pattern, XLIX, 55, 56
principle, XLIX, 41
of superimposed images, XLIX, 55–58
of two-sided images, XLIX, 55–58
Optical focusing, zero angle, XLVIII, 456, 457
Optical multichannel analyzer, LIV, 5
Optical probe, *see also* Probe, optical
high-pressure techniques, XLIX, 16, 21–24
Optical rail, XLVIII, 187
Optical rotation, XL, 215
of glycosphingolipids, XXXII, 365
Optical rotatory dispersion, XLIII, 347–373, *see also* Spectropolarimetry
conformation studies, XLIV, 366, 367
equations, XLIII, 355
Optical spectroscopy, combination with low-temperature EPR spectroscopy, LIV, 131, 132
Optical system, for Fast Analyzer, XXXI, 794–799
Optic nerve, myelin membrane, X-ray studies, XXXII, 219
Oral lesion, human acatalasemia, XLIV, 687
Orange
oxidase, LII, 10
peel, carotenoid pigment binding to protein, XXXI, 531
Orbital atomic moment, magnetic susceptibility measurements, LIV, 383, 384
Orcinol, indicator for glycolipids, XXXV, 406, 412, 413
Orcinol hydroxylase, *see* Orcinol 2-monooxygenase
Orcinol 2-monooxygenase, LII, 17
Orcinol reaction, for glycolipids, XXXII, 357
ORD, *see* Optical rotatory dispersion
Order parameter, XLIX, 381
Organ, perfused, for transport studies, XXXIX, 506–510
Organ absorbance spectrophotometry, LII, 48
procedure, LII, 55, 56
Organ culture
in bone collagen metabolism studies, XL, 331
defined, LVIII, 263
of endocrine cell, XXXIX, 165
of pineal gland, XXXIX, 392–394, 398–403
of testis, XXXIX, 284–288
Organelle, *see also* specific types
fractionation, LVIII, 222
internal volumes, LV, 551
negative staining, XXXII, 20–35
preparation, LVIII, 221–229
subcellular, LVIII, 223
Organ fluorometry, LII, 48
procedure, LII, 56, 57
Organic acid, *see also* specific substance
luminescent bacteria, LVII, 165
production system, XLIV, 184, 831–844
resolution by α-chymotrypsin, XLIV, 833–835
Organic synthesis, use of enzymes, XLIV, 831–844

Organism, physiological state, mitochondria, LV, 3
Organomercurial, reactivity, LVI, 615
Organomercurial-agarose, XXXIV, 544–547
  capacity tests, XXXIV, 545, 546
  histones, XXXIV, 549–552
  NADP-protected cysteine, XXXIV, 552–554
  regeneration, XXXIV, 552
Organophosphate pesticide, detection, XLIV, 648, 657, 658
Organosilane, XXXIV, 63
Organotin
  availability, LV, 507
  relative inhibitory potencies, LV, 510
  structure, LV, 507
  toxicity, LV, 508
Organ perfusion, LII, 48, 58, 59
Orientation factor
  assigned value, XLVIII, 354, 369
  definition, XLVIII, 353
  for several cone axes orientations, XLVIII, 372
  theoretical value, XLVIII, 369
Ornithine
  internal standard, XLVII, 34
  in media, LVIII, 63
  peptide containing, diisothiocyanate coupling, XLVII, 284
  transport, LVI, 248, 249
L-Ornithine, XLIII, 445, 458
  activator of carbamoyl-phosphate synthetase, LI, 22, 27, 33, 34, 105, 112
Ornithine carbamoyltransferase, assay of carbamoyl-phosphate synthetase, LI, 105, 112
Ornithine decarboxylase, hormone induction of biorhythms, XXXVI, 480
Orosomucoid, XXXIV, 689
Orotate, *see also* Orotic acid
  assay of phosphoribosylpyrophosphate synthetase, LI, 5
  chromatographic separation, LI, 158, 159
  inhibitor of dihydroorotate dehydrogenase, LI, 62, 68
  product of dihydroorotate dehydrogenase, LI, 63
  substrate of OPRTase:OMPdecase complex, LI, 135, 144, 145, 155
    of orotate phosphoribosyltransferase, LI, 69, 70
Orotate phosphoribosyltransferase, *see also* Orotate phosphoribosyltransferase:orotidylate decarboxylase complex
  activators, LI, 74, 152
  animal sources, LI, 72, 73
  assay, LI, 69, 70, 135, 136, 141, 143–145, 156, 157
  equilibrium constant, LI, 164
  inhibitors, LI, 74, 153, 163, 166
  kinetic properties, LI, 74, 140, 152
  molecular weight, LI, 73
  pH optimum, LI, 73, 152
  purification, LI, 70–73
  purity, LI, 73
  pyrophosphorolysis, LI, 154
  stability, LI, 73, 153, 154
  substrate specificity, LI, 152
  from yeast, LI, 69–74
Orotate phosphoribosyltransferase:orotidylate decarboxylase complex, LI, 135–167
  assay, LI, 135, 136, 156
  dissociation, LI, 141
  from Ehrlich ascites cells, LI, 155–167
  from erythrocyte, LI, 143–154
  kinetic properties, LI, 140, 141, 166, 167
  molecular weight, LI, 137, 138, 154, 164–166
  purification, LI, 137, 138, 146–151, 160–164
  purity, LI, 167
  from *Serratia marcescens*, LI, 135–143
  stability, LI, 167
Orotic acid, *see also* Orotate
  protein incorporation, in liver explants, XXXIX, 40
  radioactive, microsome labeling, XXXI, 218
Orotidine, inhibitor of orotate phosphoribosyltransferase, LI, 166

Orotidylate
  assay of orotate phosphoribosyltransferase, LI, 144
  chromatographic separation, LI, 158, 159
  extinction coefficient, LI, 158
  inhibitor of orotate phosphoribosyltransferase, LI, 74
  product of orotate phosphoribosyltransferase, LI, 69
  substrate of OPRTase-OMPdecase complex, LI, 135, 155, 157
    of orotidylate decarboxylase, LI, 74
Orotidylate decarboxylase, LI, 74–79, *see also* Orotate phosphoribosyltransferase:orotidylate decarboxylase complex
  assay, LI, 74, 75, 136, 145
  inhibitors, LI, 79, 153, 163, 166
  kinetics, LI, 78, 79, 140
  molecular weight, LI, 78
  in orotate phosphoribosyltransferase assay, LI, 69
  pH optimum, LI, 78, 152
  properties, LI, 78, 79
  purification, LI, 75–78
  purity, LI, 78
  stability, LI, 78, 153, 154
  storage, LI, 70
  from yeast, LI, 74–79
Orsellinate decarboxylase, immobilized assay, XLIV, 348
Orsellinic acid decarboxylase, *see* Orsellinate decarboxylase
[^{32}P]Orthophosphoric acid, in preparation of [γ-^{32}P]ATP, LIX, 61
Orthophosphoric diester phosphorylase, *see* Phosphodiesterase I
Oscilloscope
  delayed sweep, LIV, 44
  fast, LIV, 23
  matching detector, LIV, 41, 42
  nonstorage, LIV, 40
  offset circuit, LIV, 42, 43
  storage, LIV, 39
OSCP, *see* Oligomycin-sensitivity-conferring protein
Osmium ammine, XL, 32
Osmium tetroxide, XXXI, 21, 22; XLIV, 714
  for double bond position determination, XXXV, 347
  effect on membrane fracture, XXXII, 37
  for pancreatic subcellular electron microscopy, XXXI, 53, 54
  in Parducz fixative, XXXII, 55
Osmium tetroxide fixation, XL, 5
Osmium tetroxide-potassium ferrocyanide stain, XXXIX, 143, 144
Osmolarity
  adjustments for lower vertebrates, LVIII, 469
  calculation, LVIII, 200, 201
  determination, LVIII, 200, 201
  effect on culture, LVIII, 73, 136, 137
  internal volume, LV, 550
Osmometer, *see also* specific types
  block-type, XLVIII, 137
  design, solution types, XLVIII, 76
  Hellfrizt, XLVIII, 137
  magnetic
    design, XLVIII, 62, 63, 65–68
    sensitivity, XLVIII, 62
    theory, XLVIII, 62–65
  membrane
    electronic, design, XLVIII, 73
    Hewlett-Packard, XLVIII, 133, 136, 145–147
    Knauer, XLVIII, 134–136
    Wescan
      calibration, XLVIII, 141–143
      design, XLVIII, 133–136
      measurement of osmotic pressure, XLVIII, 144, 145
      membrane installation, XLVIII, 139–141
  stirring, XLVIII, 76
  vapor pressure
    calibration, XLVIII, 152–154
    commercial models available, XLVIII, 79, 147, 148
    operating procedures, XLVIII, 148–152
    preparation, XLVIII, 149
Osmometry
  historical considerations, XLVIII, 70, 71

membrane, XLVIII, 69–154
applications, XLVIII, 72–76
high speed, XLVIII, 71
history, XLVIII, 71
range of usefulness for m.w. measurements, XLVIII, 70
theory, XLVIII, 73
multicomponent system, theory, XLVIII, 73
self-associations, XLVIII, 103–121
for solute
heterogeneous, nonassociating, XLVIII, 97–103
homogeneous, nonassociating, XLVIII, 77–96
thermal, XLVIII, 78
three-component, ionizable systems, XLVIII, 73, 89–96
two-component systems, XLVIII, 77–87
vapor pressure, XLVIII, 78–80, 87–89
m.w. range of usefulness, XLVIII, 70
Osmotic coefficient, definition, XLVIII, 86–87
Osmotic lysis, of mycoplasmas, XXXII, 460, 461
Osmotic pressure
counteracting during dialysis, XLVIII, 335
definition, XLVIII, 69, 70, 81, 97
determination, using magnetic suspensions, XLVIII, 29, 61–68
measurement, in Hewlett-Packard Membrane Osmometer, XLVIII, 145–147
in Knauer High Speed Membrane Osmometer, XLVIII, 145
in Wescan High Speed Membrane Osmometer, XLVIII, 144, 145
mixed association, ideal, theory, XLVIII, 73
molecular weight, XLVIII, 82–87
relation between vapor pressure lowering, XLVIII, 88, 89
self-association, ideal, theory, XLVIII, 73
Osmotic pressure cell, schematic, XLVIII, 69
Osmotic shock, vesicle cation loading, LVI, 403, 404
Osmotic shock fluid, preparation, LI, 558, 567
Osmotic swelling, uncouplers, LV, 463
Ossamine, structure, LV, 511
Ossamycin
availability and preparation, LV, 512
energy-conserving systems, LV, 512
structure, LV, 511
toxicology, LV, 512
Osterizer, for plant-cell extraction, XXXI, 541, 593
Ostracod crustacea
general anatomy, LVII, 332
taxonomy, LVII, 331, 332
O–Syl detergent, XXXIX, 37
OTE, *see* Electrode, optically transparent
Ouabain
binding by (Na^+ + K^+)-ATPase, XXXII, 287, 288
binding site, XLVI, 75, 89, 523–531
photochemically active derivatives, LVI, 660
resistance, LVIII, 309, 314
as selective agent, LVIII, 353
Ouchterlony double-diffusion plate
commercial source, LII, 244
use, LII, 245, 246, 248
Ouchterlony double immunodiffusion
ferritin-affinity antibody conjugates, LVI, 712
monospecificity of affinity antibody, LVI, 703–705
purity of affinity antibody, LVI, 701–703
Output ratio
distortion, LVI, 143
parental contributions, LVI, 142, 143
Ovalbumin, XXXIV, 717
antibody binding, L, 55
as contaminant in progesterone receptor preparations, XXXVI, 209
gel electrophoresis, XXXII, 101
HABA column, XXXIV, 266
hollow-fiber retention data, XLIV, 296
inert proteic matrix, XLIV, 268
molecular weight, LIII, 351, 353
oligosaccharide structure, L, 279

osmometric studies, XLVIII, 75, 76, 78
osmotic pressure, XLVIII, 73
radiolabeled, in estrogen receptor studies, XXXVI, 170, 171, 174
SDS complex, properties, XLVIII, 6
sequence analysis, XLVII, 251
Ovarian cell
chinese hamster, *see* Chinese hamster ovarian cell
differentiation *in vitro*, XXXIX, 183–230
rat, LVIII, 96
Ovarian teratoma, LVIII, 376
Ovary
dissection procedure, XXXIX, 231, 232
oxidase, LII, 13, 14
perfusion, *in vitro*, XXXIX, 230–237
preparation for collagen studies, XL, 311
receptors for LH and hCG, XXXVII, 167–193
solubilization, XXXVII, 192, 193
viability criteria, XXXIX, 235–237
Overhauser effect, XLIX, 294, 295, 335; LIV, 177, 178
Overhauser enhancement, in NMR, XXXII, 200
Overvoltage, electrodes, XLVIII, 489
Oviduct
chick, progesterone receptor, XXXVI, 187–211
nuclei isolation from cells, XXXI, 263–279
Ovine follicle-stimulating hormone, LVIII, 102
Ovine growth hormone, in media, LVIII, 103
Ovoinhibitor, XXXIV, 5, 7
Ovomucoid, XXXIV, 5
in agglutinin binding, XXXII, 615, 616
from egg white, reductive alkylation, XLVII, 472, 474
gel electrophoresis, XXXII, 101
as matrix, XXXIV, 336
molecular weight, by electrophoresis, XXXVI, 99
Ovomucoid-agarose, XXXIV, 418, 450
Ovulation, in animals, in studies on LH-RH and FSH-RH, XXXVII, 237, 238
Oxacillin, XLVI, 536
Oxalate ion, chemiluminescent reaction, LVII, 525
Oxalate oxidase, LII, 19; LVI, 474
Oxalic acid, XLVI, 20, 21, 23
inhibitor of D-lactate dehydrogenase, LIII, 525
of rat liver acid nucleotidase, LI, 274
Oxalic acid dihydrate, XLVII, 544
Oxalic acid dihydrazide, preparation, LIX, 179
Oxaloacetate decarboxylase, XLVI, 23
Oxaloacetic acid, XLIV, 181
adenylate cyclase, XXXVIII, 169
aspartate loading, LVI, 260
assay, LVII, 181–188
cycling method, LV, 207
of malate dehydrogenase, LVII, 188
radiometric method, LV, 208
coenzyme A assay, LV, 210
complex II, LVI, 586
content
of brain, LV, 218
of hepatocytes, LV, 214
of kidney cortex, LV, 220
of rat heart, LV, 216, 217
of rat liver, LV, 213
deactivated, succinate dehydrogenase, LIII, 471–473
determination, XXXVII, 290
guanylate cyclase, XXXVIII, 195
inhibitor of succinate dehydrogenase, LIII, 33
in liver compartments, XXXVII, 287
renal gluconeogenesis, XXXIX, 14
substrate, citrate synthase, XLIV, 458, 474
Oxalyl chloride, chemiluminescence, LVII, 456, 563
Oxamate-agarose, XXXIV, 599
Oxamic acid, XXXIV, 298, 599; XLVI, 20
inhibitor of D-lactate dehydrogenase, LIII, 525
1,2-Oxathiolane-2,2-dioxide, *see* 1,3-Propane sultone

Oxazolidinyl *N*-oxyl group, structure, XLIX, 434
Oxazolone, XLVI, 200
  peptide synthesis, XLVII, 505, 506
Oxenoid, mechanism, LII, 39
Oxidase, *see also* specific types
  activity, of *E. coli* mutants, LVI, 109
  classes, LII, 6
  definition, LII, 3
  electron transferring, LII, 6, 7
  hydrogen acceptor, LVI, 448
  immobilization, LVI, 449
  mechanisms, LII, 34–40
  mixed function, *see also* Monooxygenase
    hepatic, *see* Dimethylaniline monooxygenase
  oxygenated forms, spectra, LII, 36
  phylogenetic occurrence, LII, 9–21
  prosthetic groups, LII, 4–6
  reaction classes, active-site types, LII, 33
  self-pasteurization, LVI, 473
  superoxide anion, LVI, 460, 461
Oxidation, *see also* Dioxygenation
  coupling to phosphorylation, model reactions, LV, 536–540
  mixed-function, LII, 6, 7, 39
  role of molecular oxygen, LII, 3–40
β-Oxidation
  control studies, XXXVII, 292
  of fatty acid, XXXV, 126–151
    regulation, XXXV, 149, 150
Oxidation-reduction potential
  determination, LIV, 396–410
    absorbance-charge technique, LIV, 400, 401
Oxidation-reduction potentiometry, *see* Redox potentiometry
Oxidation state
  determination, from resonance Raman studies, LIV, 243
  of iron-sulfur proteins, LIII, 264, 265
Oxidation system, for chemiluminescent assays, LVII, 426, 427, 438, 439
Oxidative phosphorylation, *see* Phosphorylation, oxidative
Oxidoreductase, *see also* specific types
  assay, XXXII, 397–401
    by Fast Analyzer, XXXI, 816
  gel electrophoresis, XXXII, 91
  in progesterone metabolism, XXXVI, 490
Oxime, of steroids, preparation, XXXVI, 17, 18
Oxime reagent, LI, 51
2-Oximino-5α-androstan-17β-ol-3-one, synthesis, XXXVI, 415
Oxindole, chromophore, extinction coefficient, XLVII, 445
Oxindole-62 lysozyme, XXXIV, 643–645
Oxindolylalanine
  absorption spectra, XLVII, 463
  alternative names, XLVII, 443
  determination, XLVII, 446
  preparation, XLVII, 447
  stability, XLVII, 446, 447
Oxipurinol, inhibitor of orotate phosphoribosyltransferase, LI, 166
1-Oxipurinol-5′-phosphate, inhibitor of orotidylate decarboxylase, LI, 153
7-Oxipurinol-5′-phosphate, inhibitor of orotidylate decarboxylase, LI, 153
Oxiran-Acrylharzperlen, support, for penicillin acylase, XLIV, 764
Oxirane, XLIV, 32–34
  assay, by sodium thiosulfate titration, XLIV, 34
  stability, XLIV, 33
  storage, XLIV, 33
Ox muscle, 3-phosphoglycerate kinase, purification, XLII, 129
μ-Oxobishemin, esters
  preparation, LII, 427–430
  spectral properties, LII, 429–431
N^5-(4-Oxo-1,3-diazaspiro[4,4]non-2-ylidene)-L-ornithine, XLVII, 157
6-Oxoestradiol, XLVI, 89, 479
2-Oxoglutarate, XLIV, 865
2-Oxoglutarate dehydrogenase
  coenzyme A assay, LVI, 371
  toluene-treated mitochondria, LVI, 547
Oxoid Bacteriological Peptone L37, XXXI, 668, 669
Oxolinic acid, XLIII, 211, 212
2-Oxo-4-methylpentanoate, transport, LVI, 292
D-2-Oxo-3-methylvaleric acid, synthesis of *Cypridina* luciferin, LVII, 369
Oxonium salt, XLVI, 70

Oxonol
bleaching, LV, 574, 576
membrane potentials, LV, 559, 571, 573, 694, 695
3-Oxo-5,10-secosteroid, XLVI, 461–468
synthesis of acetylenic and allenic, XLVI, 462, 463
4-Oxo-2,2,6,6-tetramethylpiperidine, synthesis, XLIX, 435
Oxy-cellulose, XLIV, 46
n-Oxyfurazan, XLVI, 510
Oxygen
Δ*g*-type singlet, lipid peroxidation, LII, 304
acrylic polymerization inhibition, XLIV, 57
affinity of cytochrome oxidase, LVII, 223
antibody inhibition studies, LII, 249
ATP production, LV, 292
concentration changes, electrode calibration, LVI, 554, 555
concentration determination, LII, 57
consumption
assay procedure, LIV, 491–494
calcium uptake, LVI, 339, 340
computation, LIV, 494–498
cuvettes, LIV, 489, 490, 491
experimental parameters, LIV, 489, 490
oxygraphic estimation, LIV, 495, 496
spectrophotometric assay, LIV, 485–498
spectrophotometric estimation, LIV, 496
wavelength, choice, LIV, 489, 490
contamination, LVIII, 202
in *Cypridina* luciferin-luciferase reaction, LVII, 356–358, 365
determination
calibration curve, LIV, 501–503
effect of bacterial concentration, LIV, 503
of pH, LIV, 503
of salt concentration, LIV, 503
in gases, LIV, 119, 120
instrumentation, LIV, 500, 501
reaction flask, LVII, 224, 225
sensitivity, LVII, 225
using luminescent bacteria, LIV, 499–505
method limitations, LIV, 509
procedure, LIV, 501–504
Edman degradation, XLVII, 346, 347, 351–353
effect on activity, XLVI, 478
on culture, LVIII, 74
on radical ion chemiluminescent reactions, LVII, 498
energy-linked transhydrogenase, LV, 789, 790
fermentation, LV, 289, 290
flow dialysis experiments, LV, 686
fluorescence quenching agent, XLIX, 225, 233
freeze-quench technique, LIV, 92
glucose isomerase inactivation, XLIV, 816
hemoglobin association kinetics, XLVIII, 274–276
interference with difference spectra, LII, 222
luminescence, LVI, 543
measurement of Ca^{2+}/site ratio, LV, 649, 652
of proton translocation and, LV, 631, 633, 634
methods of addition, LIV, 105, 106
molecular
assay of luciferase, LIII, 565
detection with bacterial luciferase, LIII, 564
enzyme-catalyzed reaction mechanisms, LII, 34–40
role in biological oxidations, LII, 3–40
inactivator of clostridial hydrogenase, LIII, 287
reaction with *Pseudomonas* cytochrome oxidase, LIII, 657
substrate of complex IV, LIII, 40
of cytochrome oxidase, LIII, 54, 73
of flavoprotein oxidases, LIII, 398
of flavoprotein oxygenases, LIII, 398
of *p*-hydroxybenzoate hydroxylase, LIII, 544

of melilotate hydroxylase, LIII, 552
of phenylalanine hydroxylase, LIII, 278
of salicylate hydroxylase, LIII, 528
utilization, uncoupling, LIII, 536–540
M. phlei membrane fragments, LV, 196
nonrequirement for bioluminescence, LVII, 272, 274
permeability of plastics, LIV, 376
peroxo state, LII, 35–37
proton magnetic resonance studies, LIV, 199, 200
regeneration of aequorin, LVII, 289
removal, LIV, 52
from gases, LII, 221, 222
by oxygen scavenging system, LII, 222, 223
requirement for growth, LVIII, 136
role in control of bacterial bioluminescence, LVII, 129, 165
in rubber septa, LIV, 126
singlet, XLVI, 13
luminol chemiluminescence, LVII, 421, 449
role in phagocyte chemiluminescence, LVII, 462–464, 488
solubility, LV, 634, 644
at low temperatures, LIV, 106
spin-label experiments, XLIX, 401
stretching frequency, LIV, 311
substrate of bacterial luciferase, LVII, 126, 133, 165
of firefly luciferase, LVII, 3, 37, 58, 74
of *Renilla* luciferase, LVII, 238
superoxo state, LII, 35–37
tributylphosphine, XLVII, 114
triplet quenching, LIV, 48, 52
triplet state, definition, LII, 34
uptake
amino acid transport, LV, 180
by brain mitochondria, LV, 57
by brown adipose tissue mitochondria, LV, 75, 76
hepatoma mitochondria, LV, 81, 83
measurement, LII, 143
P/O ratio, LV, 226
smooth muscle mitochondria, LV, 64, 65
by yeast mitochondria, LV, 150
Oxygen-16, nucleus
coherent scattering length, LIX, 672
scattering cross section, LIX, 672
Oxygen-17, lock nuclei, XLIX, 348
Oxygen-18, exchange
measurement
ATP exchanges, LV, 258
illustrative calculations, LV, 258–261
medium $P_i \rightleftarrows$ HOH exchange, LV, 258
total $P_i \rightleftarrows$ HOH exchange, LV, 257, 258
probes, LV, 252–256
Oxygenating reservoir, for heart perfusion, XXXIX, 50
Oxygenation, apparatus, LII, 52, 53
Oxygenator
for testis infusion, XXXIX, 276, 278
for thyroid gland perfusion, XXXIX, 361
Oxygen electrode, assay of cytochrome *c*, LIII, 160, 161
Oxygen microtechnique, for hormone studies on endocrine cells, XXXIX, 403–425
Oxygen plasma device, for ashing, LIV, 481
Oxygen pulse procedure
measurement of H^+/site ratio, LV, 645, 646
calculations, LV, 645
principle, LV, 642, 643
procedure, LV, 643–645
reagents and test medium, LV, 645
oxidative phosphorylation, LV, 226, 227
proton translocation, LV, 750, 751
Oxygen transferase, *see* Dioxygenase
Oxyhemoglobin, XLIV, 544, 545
carboxypeptidase A digestion, LIV, 488, 489
CD spectrum, LIV, 292
concentration determination, LII, 449
extinction coefficient, LIV, 488

hydrogen peroxide, LIV, 486
MCD spectrum, LIV, 292
oxidation state determination, LIV, 243
millimolar extinction coefficient, LII, 449
in oxygen consumption measurements, LIV, 485
preparation, LII, 457, 458, 476, 477
purification, LIV, 488
spectral properties, LII, 36, 459–463
1-[(1-Oxy-2-hydroxy-3-*t*-butylamino)propane]-3-bromoacetamidoisoquinoline, XLVI, 595
1-(Oxy-2-hydroxy-3-isopropylamine)propane-3-bromoacetamidoisoquinoline, XLVI, 595, 596
N-Oxyloxazolidine, from ketones, XXXII, 191
1-Oxyl-2,2,6,6-tetramethyl-4-piperidinal
saturation-recovery time, XLIX, 501, 502
second harmonic absorption out-of-phase spectra, XLIX, 509
structure, XLIX, 501
N-(1-Oxyl-2,2,6,6-tetramethyl-4-piperidinyl)iodoacetamide, XLVII, 409
N-(1-Oxyl-2,2,5,5-tetramethyl-3-pyrrolidinyl)iodoacetamide, XLVII, 409
Oxyluciferin
firefly
product, of firefly luciferase, LVII, 3
structural formula, LVII, 4, 27
luminescence decay, LVI, 534
Pholas dactylus
amino acid analysis, LVII, 390
purification, LVII, 391, 392
Oxyluciferin diacetate, structural formula, LVII, 27
o-Oxylyldithiol, identification of iron-sulfur centers, LIII, 272
Oxyntic cell
electrical studies, XXXII, 715, 716
electron microscopy, XXXII, 712–714
enzymes, XXXII, 716, 717
identification, XXXII, 712–714
ionic composition, XXXII, 716
isolation, XXXII, 707–717
phase contrast, XXXII, 712
viability tests, XXXII, 715
Oxysolve T scintillation mix, LIX, 120
Oxysorb, LIV, 52
Oxytetracycline, XLIII, 199, 318, 319
in vivo administration, LVI, 36, 37
Oxytocin
cAMP production, LVIII, 559
as neurosecretory granule marker, XXXI, 406, 408
synthesis, XLVII, 587
Ozonolysis
determination of double bond position, XXXV, 345
glycosphingolipids, L, 137
of prostaglandins, XXXV, 372

P

P5, hollow-fiber device, XLIV, 293
hydraulic permeability, XLIV, 298
P8, hollow-fiber device, XLIV, 293
P10, hollow-fiber device, XLIV, 293, 296
hydraulic permeability, XLIV, 298
P30, hollow-fiber device, XLIV, 297
P100, hollow-fiber device, XLIV, 293
hydraulic permeability, XLIV, 298
P700, measurement, XXXII, 867, 868
PA–114 A antibiotic, XLIII, 338
PA–114 B antibiotic, XLIII, 338
PA–312 antibiotic, XLIII, 338
Packed cell volume, measurement, in fat-cell fractionation, XXXI, 65
Pactamycin
effect on protein synthesis, LIX, 854
protein inhibitor, XL, 287, 288
PAGE, *see* Polyacrylamide gel electrophoresis
PALA, *see* *N*-Phosphonacetyl-L-aspartate
Palay perfusion method, XXXIX, 134
Palladium
atomic emission studies, LIV, 455
catalytic hydrogenation, XLVII, 548
Palladium charcoal, in 5-fluoro-2′-deoxyuridine 5′-(*p*-aminophenyl phosphate) synthesis, LI, 99
Pallidin
affinity chromatography, L, 313, 314
biological significance, L, 315, 316
hemagglutination assay, L, 312, 313
lactose inhibition, L, 315
properties, L, 314, 315

Palmitic acid
in assay of palmityl-CoA synthetase, XXXV, 117, 118
freeze-fracture, XXXII, 46
mass spectrometry, XXXV, 343
deuterated, XXXV, 342–344
high resolution, XXXV, 348
of triglycerides, XXXV, 351–353, 355
in phospholipid bilayers, saturation-transfer studies, XLIX, 497, 498
product of fatty acid synthase, XXXV, 73, 81
Palmitic acid-agarose, XXXIV, 158
Palmitoleic acid
inhibitor, of myristic acid bioassay, LVII, 192
quantum yield, in bioassay, LVII, 192
Palmitoleyl coenzyme A, substrate for palmityl thioesterase I and II, XXXV, 106, 109
D-*erythro*-*N*-Palmitoyl-3-*O*-benzoylceramide, intermediate in sphingomyelin synthesis, XXXV, 501
erythro-*N*-Palmitoyl-3-*O*-benzoyl-D-sphingosine, intermediate in sphingomyelin synthesis, XXXV, 500
erythro-*N*-Palmitoyl-3-*O*-benzoyl-D-sphingosyl-1-(2′-aminoethyl) phosphonate, intermediate in synthesis of sphingomyelin analogues, XXXV, 527
Palmitoylcarnitine, brown adipose tissue mitochondria, LV, 74–76
Palmitoyl-CoA deacylase, in assay of long chain fatty acyl-CoA, XXXV, 273
Palmitoyl-CoA hydrolase
correction, LVI, 370–372
interference in assay of palmityl-CoA synthetase, XXXV, 118
Palmitoyl-CoA synthetase, XXXV, 117–122
activators, XXXV, 122
in fat cells, XXXI, 67, 68
lipid content, XXXV, 121
stability, XXXV, 120, 121
Palmitoyl coenzyme A
in assay of palmityl-CoA deacylase activity of fatty acid synthetase, XXXV, 51
of palmityl-CoA synthase, XXXV, 117, 118
detergent action, XXXV, 55, 56, 105
inhibitor
of acetyl-CoA:amine acetyltransferase, XXXV, 253
of acetyl-CoA carboxylase, XXXV, 9, 17
of fatty acid synthase, XXXV, 56, 88
synthesis, XXXV, 48
2-*O*-Palmitoyl-*sn*-glycerol-3-(2′-aminoethyl) hydrogen phosphate, intermediate in lysoglycerophospholipid synthesis, XXXV, 488–490
1-Palmitoyl-2-oleoyl-*sn*-glycerol-3-phosphorylethanolamine, intermediate in phosphatidylethanolamine synthesis, XXXV, 458
D-*erythro*-*N*-Palmitoylsphingomyelin, intermediate in sphingomyelin synthesis, XXXV, 501
erythro-*N*-Palmitoyl-D-sphingosyl-1-(2′-aminoethyl) phosphonate, intermediate in synthesis of sphingomyelin analogues, XXXV, 526–528
Palmitoyl thioesterase, in assay of long-chain fatty acyl-CoA, XXXV, 273
Palmitoyl thioesterase I, XXXV, 102–106
assay, XXXV, 102, 103
of fatty acyl thioesters of CoA and ACP, XXXV, 106
inhibitors, XXXV, 106
stability, XXXV, 105
substrate specificity, XXXV, 102, 106
Palmitoyl thioesterase II, XXXV, 107–109
assay, XXXV, 107
inhibitors, XXXV, 109
stability, XXXV, 109
substrate specificity, XXXV, 102, 109
Palmityl, *see* Palmitoyl
Pancreas
artificial, LVIII, 184

bovine, fractionation, XXXI, 49–59
deoxyribonuclease, XXXIV, 464, 468
kallikreins, *see* Kallikrein A; Kallikrein B
microsome isolation, XXXI, 199, 200
phospholipase A_2, XXXII, 131, 137, 138, 147–154
subcellular fractions, XXXI, 41–59
chemistry and identification, XXXI, 46, 47
electron microscopy, XXXI, 53, 54
isolation, XXXI, 43–46
superfusion, XXXIX, 373–376
trypsin inhibitor, XLV, 821, *see also* Trypsin inhibitor
in vitro perfusion, XXXIX, 364–372
Pancreatic cell
media, LVIII, 90
mouse, media, LVIII, 59
rat, media, LVIII, 59
Pancreatic hormone, immunochemical assay, XXXVII, 144
Pancreatic islet tissue, LVIII, 126
Pancreatin
in mucosal cell isolation, XXXII, 665
tissue culture, LVIII, 125
Panose
activated hydroxyl groups, L, 109, 110
synthesis, L, 101, 102
Pantetheine, XLIII, 586
D-Pantetheine
in assay of acetyl(malonyl)-CoA-pantetheine transacylase, XXXV, 49
separation of CoA and pantetheine esters, XXXV, 49
substrate for acetoacetyl-CoA thiolases, XXXV, 134
synthesis, XXXV, 49
P–125 antibiotic, XLIII, 168
Pantothenic acid, in media, LVIII, 54, 64
PAPA, *see also* α-Benzylsulfonyl-*p*-aminophenylalanine
Papain, XXXIV, 5, 105; XLVI, 22, 155, 206, 221
affinity labeling, XLVII, 427, 428
antipain, inhibited, XLV, 683
assay
for cysteine proteases, XLV, 1
with *p*-nitrophenol *N*-benzyloxycarbonyl glycinate, XLV, 6
capacity, XXXIV, 545, 546
comparison to clostripain, XLVII, 166
conformation studies, XLIX, 143–148
conjugate
activity, XLIV, 110–116, 156
optimal effective pH, XLIV, 429–431
storage, XLIV, 111
sulfhydryl group determination, XLIV, 392
covalent chromatography, XXXIV, 538–544
capacity, XXXIV, 540
L-cysteine, XXXIV, 541
properties, XXXIV, 543, 544
separation from L-cysteine, XXXIV, 542, 543
thiol content of gels, XXXIV, 539, 540
cyanylation, XLVII, 131
in hormone receptor studies, XXXVII, 212
hydrolysis, XLVII, 41
immobilization, XLVII, 44
with collagen, XLIV, 244, 255
in collodion membrane, XLIV, 904, 909
on controlled-pore glass, XLIV, 156
on hydroxyalkyl methacrylate gel derivative, XLIV, 80–82
with inert protein, XLIV, 925
with magnetic carrier, XLIV, 278, 279
on nitrated methacrylic acid/3-fluorostyrene, XLIV, 113
on nitrated methacrylic acid/methacrylic acid 3-fluoroanilide, XLIV, 112
in polyacrylamide gel, XLIV, 902
in porous matrix, XLIV, 270–272
kinetics, XLIV, 425, 437
leupeptin, inhibited, XLV, 679
p-nitrophenol ester binding, XLIV, 81, 82
oscillatory behavior studies, XLIV, 926, 927

pH profile, of proteolytic activity, XLIV, 82
protein digestion, concentration and incubation period, XL, 38
sequence analysis, from the carboxyl terminus, XLVII, 359
short-term memory studies, XLIV, 925, 926
specific activity, XLIV, 255
sulfhydryl group, protection, XLIV, 274
tissue culture, LVIII, 125
tryptophanyl phosphorescence, XLIX, 247
Papain-L-cysteine, XXXIV, 541
Papain-cysteine separation, XXXIV, 542, 543
Papaverine, phosphodiesterase, XXXVIII, 72
Paper, aminated, XLIV, 271
Paper chromatogram, *see* Paper chromatography
Paper chromatography, XLIII, 100–172, *see also* specific substance
adenylyliminodiphosphate, XXXVIII, 426
alicyclic antibiotics, XLIII, 156–159
amino acid antibiotics, XLIII, 144–152
7-aminocephalosporanic acid, XLIII, 116–118
aminoglycosidic antibiotics, XLIII, 119–122
6-aminopenicillanic acid, XLIII, 110–116
anthracyclines, XLIII, 141–144
anthracyclinones, XLIII, 141–144
aromatic antibiotics, XLIII, 159–162
of atractyloside and carboxyatractyloside, LV, 521
bioautography, XLIII, 105–110
calculation, XLIII, 104, 105
cAMP acyl derivatives, XXXVIII, 402
cephalosporin C family, XLIII, 116–118
cephamycins, XLIII, 116–118
chromatograms
pH, XLIII, 164–168
salting-out, XLIII, 162–164
summarized, XLIII, 168–171
classification of antibiotics by, XLIII, 162–171
cyclic nucleotides, XXXVIII, 417, 418
detection methods, XLIII, 104
everninomycin group, XLIII, 128
lincomycin group, XLIII, 128
lysine derivative identification, XLVII, 477
macrocyclic lactone antibiotics, XLIII, 129–137
nitrogen-containing heterocyclic antibiotics, XLIII, 152–154
oligosaccharides with chromophore, XLIII, 126, 127
penicillin, XLIII, 110–116
peptide antibiotics, XLIII, 144–152
polyene macrolide, XLIII, 134–136
of proline and hydroxyproline, XL, 374
purification of flavin peptides, LIII, 456, 459
quinone antibiotics, XLIII, 137–144
R_f values, XLIII, 168–170
R_m values, calculations, XLIII, 104, 105
samples, preparation of, XLIII, 102, 103
separation of flavins, LIII, 427
of nucleosides and nucleotides, LI, 300, 323
of radiolabeled bases, LI, 220
solvents, XXXVIII, 414; XLIII, 103, 104
streptothricin group, XLIII, 123
tetracycline, XLIII, 137–141
Paper electrophoresis, XLIII, 280–286
Barton reagent for reducing properties, XLIII, 286
bioautography, XLIII, 282
classification of antibiotics using, XLIII, 280
detection and results, XLIII, 282–286
high-voltage electrophoresis, XLIII, 281, 282
ninhydrin reagent for amines, XLIII, 286
relative mobilities of antibiotics to alanine, XLIII, 282–285
Rydon Smith reagent for amides, XLIII, 286

in studies of thyroxine-binding proteins, XXXVI, 129–131
ultraviolet light detection, XLIII, 283, 296
Paper wick, LIX, 62, 63, 72
Papovavirus, cell transformation, XXXII, 586, 587
PAPS, *see* 3′-Phosphoadenosine 5′-phosphosulfate
Paracatalytic reaction, XLVI, 48–53
modification, XLVI, 48–53
criteria, XLVI, 52
Paracoccus
crude extract preparation, LIII, 336
growth, LIII, 335
high-potential, iron-sulfur protein, LIII, 330–334, 339
Paracoccus denitrificans
ATPase inhibitor, LV, 407
aureovertin, LV, 483
cytochrome, LIII, 207, 209, 210
cytochrome *c*, LIII, 229–231
fluorescent probe studies, XXXII, 234
Nbf-Cl, LV, 490
nitrate reductase, LIII, 642, 643
nitrite reductase, LIII, 643
transport systems, XXXI, 704
venturicidin, LV, 507
Paraffin
in bead polymerization, XLIV, 102, 103
for microencapsulation, XLIV, 328–330
Paraffin oil, in assay of oxygen uptake, LIV, 491
Parafilm column float, LIX, 327
Paraformaldehyde, in Karnovsky fixative, XXXIX, 137
Paragloboside, isolation, XXXII, 355
Para hydrogen
concentration determination, LIII, 299
conversion
apparatus, LIII, 301, 302, 305
catalysis of, by hydrogenase, LIII, 298, 299
procedure, LIII, 306–309
in equilibrium mixture, table of percentages, LIII, 306
preparation, LIII, 306, 307
storage, LIII, 307
Parainfluenza virus, LVIII, 28
Parainfluenza virus type I, *see* Sendai virus
Paramagnetic center, effects on nuclear resonances, LIV, 193, 194
Paramagnetic ion, binding study, of transfer RNA, LIX, 44–47
Paramagnetic non-Kramers ion, Mössbauer spectroscopy, LIV, 352–355
Paramagnetic resonance spectra, in flexibility gradient studies, XXXII, 170
Paramecium
mitochondrial mutations, LVI, 144
mitochondria studies, XXXII, 33
Paramyosin, cyanylation, XLVII, 131
Paramyxovirus, interferon producing, LVIII, 293
Paraoxon
detection, XLIV, 658
titration of carboxylesterase, XXXV, 213
Parapriacanthus beryciformes, luciferin, LVII, 342
Pararosaniline, in enzyme detection, XXXII, 90
Parathyroid gland
location and excision, XXXVII, 347, 348
tissue
incubation, XXXVII, 348, 349
protein extraction, XXXVII, 352
pulse labeling, XXXVII, 349–352
Parathyroid hormone
adenylate cyclase, XXXVIII, 151–153
affinity chromatography, XXXVII, 46
anti–bovine antisera, XXXVII, 53–56
cleavage, XXXVII, 359
effect on perfused kidney, XXXIX, 9, 10
on renal tubules, XXXIX, 15, 20
exogenously administered, XXXVII, 61–63
heterogeneity, XXXVII, 41–44
immunoprecipitation, XXXVII, 356, 357
in man, assay, XXXVII, 56–61
bovine comparison, XXXVII, 63, 64

in media, LVIII, 99
precursor, *see* Proparathyroid hormone
preparation, dosage, route of administration, XL, 319
radioactive, preparation and isolation, XXXVII, 358, 359
radioimmunoassay, XXXVII, 34, 38–66, 356
antisera, XXXVII, 50–52
applications, XXXVII, 53–64
hormone preparation, XXXVII, 47, 48
iodination, XXXVII, 49, 50
sequence-specific or region-specific, XXXVII, 44–64
tryptic digestion, XXXVII, 359
Parducz procedure, for SEM fixation, XXXII, 55
Parenchymal cell, intact
isolation from liver, XXXII, 625–632
of mammary gland, XXXII, 693–706
metabolism, XXXII, 629–631, 703–706
morphology, XXXII, 629, 701–703
Pargyline, XLVI, 161
inactivator of flavoprotein enzyme, LIII, 441
Parke-Davis CI-628 inhibitor, in estrogen receptor studies, XXXVI, 270
Parlodion, LIV, 333
solution, XL, 7
Paromomine, XLIII, 126, 623
Paromomycin, LVI, 31, 33
gas-liquid chromatography, XLIII, 228
ion-exchange chromatography, XLIII, 269
kanamycin acetylation, XLIII, 623
in media, LVIII, 113, 115, 116
resistance, LVI, 140
genetic locus, LVI, 196, 197
structure, XLIII, 614
Parotid, subcellular fractionation, XXXI, 43
Parr cell disruption bomb, XXXI, 293, 294, 481, 502
Parr pressure hydrogenator, LI, 99
Parsa medium, LVIII, 90
Parsley, ferredoxin from, proton magnetic resonance studies, LIV, 204, 206, 207
Partial molal volume, XLVIII, 21
Partial specific volume
of apoprotein, XLVIII, 16
of component *i*, XLVIII, 165
definition, XLVIII, 12
density, XLVIII, 15
of detergent, XLVIII, 16, 17
determination
for lipid-associated proteins, XLVIII, 11–23
using magnetic balancing, XLVIII, 42–44
effective
definition, XLVIII, 12, 13
of protein-lipid-detergent complex, XLVIII, 16
of homologous amphiphiles, XLVIII, 17–23
of lipid, XLVIII, 16, 17
measurement with glass diver, XLVIII, 23–29
Particle
size
distribution, XXXII, 13
effect on diffusion control, XLIV, 136, 138
submitochondrial
ATP synthesis, LVI, 494, 495
fluorescence changes, LVI, 500, 501
Partisil-10 ODS column, LII, 284
Partition chromatography, XLIII, 174, 175
reversed phase, XLIII, 180
Partition coefficient, XLIV, 400–402
definition, XLIV, 400
Partitioning effect, of charged matrix, XLIV, 400–408
Parvalbumin, carp, phthalylation, XLVII, 151, 155
PAS, *see* Periodic acid-Schiff method
PAS medium, formula, LII, 170
Passage effect, XLIX, 486, 506
Pasteur effect
cofactors of phosphorylation system, LV, 290–292
efficiency, factors influencing accessment, LV, 295, 296

historical, LV, 289, 290
quantitation, LV, 292–295
Pasteur pipette
electron spin resonance sample cell, XLIX, 435–437
in electrophoretic elution technique, LIX, 69
Path length, optical density, LIV, 39
Patulin
biosynthesis, *see also* specific enzyme
m-hydroxybenzyl alcohol dehydrogenase, XLIII, 540–548
6-methylsalicylic acid decarboxylase, XLIII, 530–550
effects on membranes, XXXII, 882
paper chromatography, XLIII, 154, 156
PB, *see* 2,3-Dimethoxy-5-methyl-6-pentyl-1,4-benzoquinone
P1 basic protein
isolation from nerve tissue, XXXII, 343, 344
properties, XXXII, 345
P2 basic protein
in peripheral nervous tissue, XXXII, 340
isolation, XXXII, 343
properties, XXXII, 345
PBG, *see* Progesterone-binding protein
PBS, *see* Phosphate-buffered saline
5P8 buffer, for erythrocyte ghost isolation, XXXI, 173, 175
PCMB, *see* *p*-Chloromercuribenzoate
PCP, *see* Pentachlorophenol
PCS scintillation cocktail, LI, 366
PDB, *see* 2,3-Dimethoxy-5-methyl-6-pentadecyl-1,4-benzoquinone
PDP 11/05 minicomputer, LII, 223
pdTp, *see* 3′-(4-Bromoacetylamidophenylphosphoryl)-deoxythymidine 5′-phosphate
Pea, *see also* *Pisum sativum*
agglutinin isolation, XXXII, 615
aleurone grain, XXXI, 576
chloroplast isolation, XXXI, 600–606
enzyme studies, XXXI, 521
glucose-6-phosphate isomerase, assay, purification, and properties, XLI, 388–392
mesophyll cell, LVIII, 361
organelles, XXXI, 533
seed hemagglutinin, XXXIV, 181
Peak-detector circuit, circuit diagram, LVII, 537, 539
Peak distortion, frequency sweep rate, LIX, 25
Peak wavelength, relation to peak wavenumber, for photobiology emissions, LVII, 573–575
Pea mesophyll × soybean culture, LVIII, 365
Pea mesophyll × *Vicia* culture, LVIII, 365
Peanut
agglutinin, *see* Lectin, from *Arachis hypogaea*
ribosome isolation, XXXI, 585, 586
seed, source of phospholipase D, XXXV, 226
spherosome isolation, XXXI, 578
Pea Popper, for plant-cell rupture, XXXI, 502, 560–562
Pear, organelles studies, XXXI, 533, 541
PE buffer, *see* Sodium phosphate-ethylenediaminetetraacetic acid buffer
Peclet number, XLIV, 803
Pectin derivative, as plant extraction interferents, XXXI, 531
Pectinol Rio, in protoplast isolation, XXXI, 580
PED buffer, formulation, LI, 137
Pederine, LVI, 31
PEG, *see* Polyethylene glycol
PEI-cellulose sheet, *see* (Polyethyleneimine)cellulose, thin-layer plates
Peking duck, Rous sarcoma virus, LVIII, 395
Pelagia noctiluca, bioluminescence, LVII, 271, 275
Peliomycin, XLIII, 340
oligomycin B, LV, 501, 502
physical properties, LV, 477
Pelleting assay, of nectin, XXXII, 434, 435
Pellicle, from ultrafiltration, XXXII, 3, 6
Pellicon membrane, XXXII, 3
Pellicon membrane filter, LIX, 500
Peltier device, principle, LVII, 542

Penam antibiotics, XLIII, 410
PenCP-S-1, *see* Penicillocarboxypeptidase S-1
PenCP-S-2, *see* Penicillocarboxypeptidase S-2
Penetration model, for protein, XLIX, 35, 36
D-Penicillamine, XL, 321
Penicillic acid, XLIII, 154, 156
Penicillin, XLVII, 409; LII, 66
  acylase, *see* Penicillin acylase
  4-alanine carboxypeptidase, XXXIV, 398
  analysis, XLIV, 760, 761
  antisera, XXXVI, 17
  binding components, XXXIV, 532
  biosynthesis, *see also* specific enzyme
    acyl-CoA:6-aminopenicillanic acid acyltransferase, XLIII, 474–476
    δ-(α-aminoadipyl)cysteinylvaline synthetase, XLIII, 471–473
    phenacyl:coenzyme A ligase, XLIII, 476–481
    phenylacetyl:coenzyme A hydrolase, XLIII, 482–487
  biosynthetic, XLIII, 115
  differential pulse polarography, XLIII, 385–387
  gas liquid chromatography, XLIII, 248–251
  for haploid cell culture, XXXII, 799
  high-pressure liquid chromatography, XLIII, 309, 312, 313, 319
  β-lactamase, XLIII, 640
  natural, XLIII, 115
  paper chromatography, XLIII, 110–116
    6-aminopenicillanic acid separation, XLIII, 115, 116
    applications, XLIII, 113, 114
    bioautography, XLIII, 111
    biosynthetic penicillins, XLIII, 115
    chemical detection methods, XLIII, 111–113
    labeled penicillins, XLIII, 115
    mobilities compared, XLIII, 113
    natural penicillins, XLIII, 115
    sampling, XLIII, 110
    solvent systems, XLIII, 110, 111
  Penicillium chrysogenum media, XLIII, 16
  reaction with penicillin-binding proteins, XXXIV, 401–405, *see also* Penicillin-binding protein
  semisynthetic, XLIII, 207, 208
  solvent systems for countercurrent distribution, XLIII, 340
  spectrofluorimetry for quantitative estimations of, XLIII, 186
  thin-layer chromatography, XLIII, 204
  in tissue culture media, XXXIX, 111, 112
  use to weaken cell walls, XXXI, 681
Penicillin acylase, XLIII, 485, 698–705, *see also* Penicillin amidase
  from *Acromobacter*, XLIII, 719
  from *Alcaligenes faecalis*, XLIII, 718
  assay, XLIII, 699–705
    biochromatographic, XLIII, 703–705
    buffer, XLIII, 700
    chromatographic assay, XLIII, 700
    hydroxylamine assay, XLIII, 669, 700, 701, 702
    β-lactamase as contaminant, XLIII, 700, 701
  from *Bacillus megaterium*, XLIII, 711–721
    culture, XLIII, 711
    deacylation, XLIII, 715
    fermentation, XLIII, 711, 712
    immobilization of enzyme, XLIII, 717, 721
    inhibitors, XLIII, 713
    kinetic properties, XLIII, 713, 714
    molecular weight, XLIII, 712
    pH optimum, XLIII, 713
    properties, XLIII, 712–721
    purification, XLIII, 712, 713
    specific activity, XLIII, 712
    substrate profile, XLIII, 715, 716
  Bacillus subtilis preparation, XLIII, 705
  bacterial, XLIII, 705–725
  from *Erwinia aroideae*, XLIII, 718
  from *Escherichia coli*, XLIII, 705–721
    assay, XLIII, 699–705
    cell disruption, XLIII, 708
    cell harvest, XLIII, 708

culture, XLIII, 706
deacylation, XLIII, 715
fermentation, XLIII, 706, 707
immobilization of enzyme, XLIII, 717, 721
inhibitors, XLIII, 713
kinetic properties, XLIII, 713, 714
molecular weight, XLIII, 712
pH optimum, XLIII, 713
properties, XLIII, 712–721
purification, XLIII, 709–711
specific activity, XLIII, 712
substrate profile
hydrolytic direction, XLIII, 715, 716
synthetic direction, XLIII, 716, 717
fungal, XLIII, 721–728, *see also* *Penicillium chrysogenum* acylase; *Penicillium fusarium* acylase
assay, XLIII, 699–728
specificity, XLIII, 726–727
from *Micrococcus roseus,* XLIII, 719
from *Nocardia*, XLIII, 720
pH optimum, XLIII, 700
from *Pleurotus ostreatus*, XLIII, 721
from *Proteus rettgeri*, XLIII, 718
from *Pseudomonas melanogenum*, XLIII, 719
from *Streptomyces lavendulae*, XLIII, 720
substrate, XLIII, 703
transformation of antibiotics, XLIII, 206
Penicillin acyltransferase, XLIII, 476, 485
Penicillin-agarose, XXXIV, 401, 402
Penicillin amidase, *see also* Penicillin acylase
activity assay, XLIV, 237, 761
fiber entrapment, industrial application, XLIV, 241
immobilization
by bead polymeriation, XLIV, 194
on Sephadex G–200, XLIV, 764, 765
on Sepharose 4B, XLIV, 760
purification, XLIV, 760–763
supports, XLIV, 764
Penicillin amidohydrolase, XLIII, 206
Penicillinase, XXXIV, 5; XLVI, 531, *see also* β-Lactamase
assays, *see* β-Lactamase
in enzyme electrode, XLIV, 592, 597
immobilization, on glass beads, XLIV, 670
physiological efficiency, XLIII, 660
R factor-mediated, XLIII, 678
specific anti–β-lactamase sera for quantitative study, XLIII, 86–100, *see also* Immunological technique
staphylococcal, XLIII, 671
substrates, XLIV, 615
TEM-R factor, XLIII, 684
thin-layer chromatography, XLIII, 207
γ-Penicillinase, XLIII, 640
Penicillin-binding component, *see* Penicillin-binding protein
Penicillin-binding protein, XXXIV, 401–405
affinity chromatography yield, XXXIV, 404
membranes preparation, XXXIV, 402
spacers, XXXIV, 404
Penicillin deacylation, *see* Penicillin acylase
Penicillin electrode, XLIV, 585, 592, 593, 617
interferences, XLIV, 615
response time, XLIV, 608
Penicillin G
assay, by microcalorimetry, XLIV, 671, 672
in media, LVIII, 113, 114, 116
Penicillin G potassium, XLIII, 382
nitrosated, XLIII, 382
Penicillin isocyanate, XLVI, 531–537
Penicillin N, XLIII, 409, 701
Penicillin V, XLIII, 408, 409, 418–425
exogenous, XLIII, 114
in media, LVIII, 113, 114, 116
Penicillium
media for, XLIII, 15
as source of double-stranded ribonucleic acid, LX, 382, 386, 550

Penicillium chrysogenum
acylase, XLIII, 722, 723, *see also* *Penicillium chrysogenum* acylase
acyl-CoA:6-aminopenicillanic acid acyltransferase, XLIII, 475
δ-(α-aminoadipyl)cysteinylvaline, XLIII, 471
for carbon molecular resonance spectroscopy, XLIII, 414
media, XLIII, 16
phenacyl:coenzyme A ligase, XLIII, 476–478
phenylacetyl:coenzyme A hydrolase, XLIII, 482
strain development program, XLIII, 25
Penicillium chrysogenum acylase, XLIII, 722, 723
6-aminopenicillanic acid, XLIII, 722
culture, XLIII, 722, 723
extraction, XLIII, 723
specific activity, XLIII, 723
unit definition, XLIII, 723
Penicillium citreoviridi, citreoviridin preparation, LV, 486, 487
Penicillium cyclopium, β-cyclopiazonate oxidocyclase, LIII, 450
Penicillium duponti, glucose-6-phosphate dehydrogenase, assay, purification, and properties, XLI, 201–205
Penicillium fusarium acylase, XLIII, 723–725
activators, XLIII, 725
chemical properties, XLIII, 725
culture, *Fusarium semitectum*, XLIII, 724
extraction, XLIII, 724
fractionation, XLIII, 724
inhibitors, XLIII, 725
IRC–50 column chromatography, XLIII, 725
kinetic properties, XLIII, 726
molecular weight, XLIII, 725
phenoxymethylpenicillin hydrolysis rate, XLIII, 726
pH optimum, XLIII, 725, 726
properties, XLIII, 725–728
stability, XLIII, 725
Penicillium janthinellum
carboxypeptidases, XLV, 587
cultivation, XLV, 434
enzyme production, XLV, 434
sequence, XLV, 597, 598
Penicillium lilacinum, leucinostatin, LV, 513
Penicillium melinii, cell-wall hydrolyzing enzyme, XXXI, 610
Penicillium notatum
mutarotase, XLI, 472
substrate specificities, XLI, 487
Penicillium patulum
media, XLIII, 15
6-methylsalicylic acid decarboxylase, XLIII, 534, 535
6-methylsalicylic acid synthetase, XLIII, 523, 524
patulin, oxidase, LII, 15
Penicillium rubrum, oxidase, LII, 19
Penicillium urticae, *m*-hydroxybenzyl alcohol dehydrogenase, XLIII, 540, 542
Penicillocarboxypeptidase S-1
amino acid composition, XLV, 594
assay
in mixture, XLV, 592
with ninhydrin, XLV, 591
with 2,4,6-trinitrobenzenesulfonic acid, XLV, 591
endopeptidase activity, XLV, 589
inhibitors, XLV, 595
kinetic, XLV, 596
pH dependence, XLV, 597
properties, XLV, 588, 593, 595
purification, XLV, 587, 590
specificity, XLV, 596
stability, XLV, 593
unit definition, XLV, 592
Penicillocarboxypeptidase S-2
amino acid composition, XLV, 594
assay
from mixtures, XLV, 592
with ninhydrin, XLV, 591
with 2,4,6-trinitrobenzenesulfonic acid, XLV, 591
inhibitors, XLV, 595
kinetic, XLV, 596
pH dependence, XLV, 597
properties, XLV, 589–599
enzymic, XLV, 595, 596
molecular, XLV, 593
purification, XLV, 587, 590

specificity, XLV, 596
stability, XLV, 593
units, XLV, 592
Penicilloic acid, XLIII, 71
Penicillopepsin
activity, XLV, 451
amino acid composition, XLV, 445
sequence, XLV, 445
assay, XLV, 441
bovine serum albumin, XLV, 443
trypsinogen, XLV, 441
characteristics, XLV, 434
inhibitors, XLV, 450
pH optimum, XLV, 452
properties, XLV, 443–452
molecular, XLV, 444
purification, XLV, 436
purity, XLV, 439
specificity, XLV, 450
stability, XLV, 443
tertiary structure, XLV, 445
Penta-*O*-acetyl-α-L-idopyranose, synthesis, L, 143, 144
rac-2,3,4,5,6-Pentaacetyl-*O*-myoinositol, intermediate in phosphatidylinositol synthesis, XXXV, 473–475
Pentachlorophenol
complex V, LV, 315
M. phlei membrane systems, LV, 185
as uncoupler, LV, 470
uncoupling activity, LV, 465, 470
Pentadecanoic acid
quantum yield, in bioassay, LVII, 192
yeast fatty acids, LVI, 562
Pentafluorophenyl isothiocyanate, XLVII, 350
3,3′,4′,5,7-Pentahydroxyflavone, *see* Quercitin
Pentane, XLIV, 838
extraction of radiolabeled quinone intermediates, LIII, 608
of ubiquinone, LIII, 573, 574
removal, LIX, 794
2,4-Pentanedione, *see* Acetylacetone
Pentene, XLIII, 135
Pentitol
as inhibitor for D-arabinose isomerase, XLI, 465
for L-arabinose isomerase, XLI, 461
for D-xylose isomerase, XLI, 471
Pentobarbital, LII, 54
as rat anesthetic, XXXIX, 166–167
Pentobarbital hydroxylase, *Ah* locus, LII, 232
Pentose phosphate, preparation of mixtures, for assay, XLI, 37
Pentose phosphate pathway, in cell characterization, XXXI, 23
Peppermint
enzyme studies, XXXI, 537–539, 541
terpene studies, XXXI, 536
Pepsin, XXXIV, 5
canine
composition, XLV, 458
preparation, XLV, 456
carbodiimide binding, XLIV, 77
concentration dependence of proteolytic activity, XLIV, 77–79
conjugation with phenylenediamine-agarose, procedure, XLIV, 40
diethylpyrocarbonate treatment, XLVII, 433
digestion of proteins, XL, 35
fluorescence quenching, XLIX, 222
hollow-fiber retention data, XLIV, 296
hydrolysis, XLVII, 41
immobilization on activated inorganic support, XLIV, 141
on Spheron derivatives, XLIV, 77–80
immobilization procedure, XLVII, 42
immobilized, relative activity, XLIV, 40
pH profile
for proteolytic activity, XLIV, 79
for stability, XLIV, 79, 80
porcine, structural analysis with thermolysin, XLVII, 184
purification of flavin peptide, LIII, 453
reductive denaturation studies, XLIV, 517, 518
SDS complex, properties, XLVIII, 6
Pepsinogen
canine
activators, XLV, 458
assay, XLV, 36
composition, XLV, 458

properties, XLV, 457, 458
purification, XLV, 452
purity, XLV, 457
specificity, XLV, 458
stability, XLV, 457
conjugate, intramolecular activation, XLIV, 520
fluorescence quenching, XLIX, 222
immobilization, on Sepharose, XLIV, 519, 520
membrane osmometry, XLVIII, 75
Pepstatin, XXXIV, 5; XLVII, 87, 88
acid proteases, inhibition, XLV, 689
activity, XLV, 691
assay, by casein, XLV, 690
kinetics, XLV, 692
properties
biological, XLV, 691
physicochemical, XLV, 691
purification, XLV, 690
Peptidase
extraction, from plant cells, XXXI, 538
Leucostoma, XLV, 397–403
Peptide, *see also* Dipeptide
active site, XXXIV, 711
attachment
carboxyl-terminal, XLVII, 305, 306
to glass, XLVII, 322
lysine residues, XLVII, 283, 309
bond formation, XLVI, 181, 184
in situ activation procedure, XLVII, 561
in solid-phase peptide synthesis, XLVII, 595–598
using isolated active esters, XLVII, 560, 561
in vitro, kinetics, LIX, 370
chain elongation, XLVII, 503–508
from amino terminal, XLVII, 503, 507, 508
from carboxyl terminal, XLVII, 504, 507
by fragment condensation, XLVII, 504, 508
cleavage, by hydrogen bromide, XLVII, 459–469, 546, 547, 566, 567
at proline residue, XLVII, 394, 403
by sodium in liquid ammonia, XLVII, 572–575
solid-phase technique, XLVII, 276, 277
at tryptophan residue, XLVII, 459–469
derivatization, with sulfonated isothiocyanates, XLVII, 260–263
detection with differential refractometer, XLVII, 212
by fluorescence methods, XLVII, 201, 202, 236–243
with ninhydrin, XLVII, 60, 185, 223–225
in organic solvents, XLVII, 212
on paper, XLVII, 240, 241
by radiolabeling, XLVII, 203, 204, 212
spectrophotometrically, XLVII, 204, 205, 211, 225
on thin-layer plates, XLVII, 240, 241
using *o*-phthalaldehyde, XLVII, 241–243
electrophoretic mobility
cysteic acid residues, XLVII, 59–61
of dansylated derivatives, XLVII, 65
histidine residues, XLVII, 58, 59
logarithmic plots, XLVII, 51–69
ester, saponification, XLVII, 569
extraction, from thin-layer plate, XLVII, 202, 203
heme peptide of cytochrome *c*, XXXIV, 194
highly apolar, solvents, XLVII, 563
hydrochloride salt, synthesis, XLVII, 557
hydrogenolysis, procedure, XLVII, 547, 548
hydrolysis
alkaline, XLVII, 237
by carboxypeptidase C, XLVII, 79–84

by dipeptidyl aminopeptidase, XLVII, 394, 395
by immobilized enzymes, XLVII, 40–45
by pepsin, XLVII, 41
to preserve tryptophan residues, XLVII, 616
standardization factors, XLV, 380
hypothalamic, LVIII, 527
introduction into biomembranes, XXXII, 545
as ionophore, LV, 440–442
large
purity check, XLVII, 551, 552
support, XLVII, 321, 333
mapping
determination of derivative homogeneity, LIII, 154
of MS2 coat peptides, LIX, 301–309
thin-layer method
data recording, XLVII, 202
electrophoresis, XLVII, 200
sample application, XLVII, 200
sample preparation, XLVII, 197–199
of mitochondrial inner membrane, LVI, 12, 13
modified, XXXIV, 182–195
affinity labeled, XXXIV, 185–188
lysyl, XXXIV, 185
tyrosyl, XXXIV, 185
antibodies, XXXIV, 183, 184
antibody-agarose column, XXXIV, 191
anti–DNP antibodies, XXXIV, 183
arsanilazotyrosyl, XXXIV, 190
BADE labeled protein, XXXIV, 188
BADL labeled protein-315, XXXIV, 188
cysteine, XXXIV, 191, 192
deoxythymidine 3′,5′-diphosphate derivatives, XXXIV, 183
dinitrophenol-binding myeloma protein, XXXIV, 183
group specific isolation, XXXIV, 184
isolation, XXXIV, 182–195
methionine, XXXIV, 184, 192
nitrotyrosine, XXXIV, 188, 189
PUDP-peptide, XXXIV, 187
purification, XXXIV, 183
ribonuclease, XXXIV, 183
tryptophan, XXXIV, 192, 193
tyrosine, XXXIV, 188
purification, XXXIV, 707
tRNA binding, XLVI, 181
separation by ion-exchange chromatography, XLVII, 27–31, 204–210
on microbore columns, XLVII, 210–236
by micro procedures, automated, XLVII, 220–236
on polyacrylamide gel slabs, XLVII, 328–330
by thin-layer methods, two dimensional, XLVII, 195–204
small, sequence analysis, XLVII, 318–321
solubility, enzyme hydrolysis, XLVII, 41
substrate, for thrombin assay, XLV, 160
synthesis, XLVI, 101, 104
α-amino group blocking by benzyloxycarbonyl group, XLVII, 510–512
basic approaches, XLVII, 502, 503
by *tert*-butyloxycarbonyl group, XLVII, 512–517
carboxyl activation methods, XLVII, 552–562
carboxyl component, synthesis, XLVII, 510–520
deprotection, XLVII, 545–549, 571–576
electrophoretic mobilities, XLVII, 68, 69
by fragment condensation, XLVII, 562–571
active ester method, XLVII, 568–571
azide method, XLVII, 562–568
of practice tetrapeptide, XLVII, 601
problems, XLVII, 504–510
product separation, XLVII, 507–510
racemization, XLVII, 505, 506, 521, 540, 563, 577, 578, 582

solid-phase, XLVII, 578–617
carboxyl-terminal amino acid attachment, XLVII, 590–592
cleavage from support, XLVII, 601–604
difficult sequences, XLVII, 606–608
equipment, XLVII, 604–606
monitoring procedures, XLVII, 598–601
product purification, XLVII, 608
side reactions, XLVII, 606
synthetic cycle, XLVII, 592–598, 609
stepwise method, condensation step, XLVII, 550–552
supports, XLVII, 584–590
synthetic
assessment of homogeneity, XLVII, 577, 578
purification, XLVII, 576
Peptide aldehyde, XLVI, 220–225
synthesis, XLVI, 221, 224
Peptide antibiotic
chiroptical studies, XLIII, 352
heteromer, XLIII, 146
paper chromatography, XLIII, 144–152, *see also* specific antibiotic
thin-layer chromatography, XLIII, 187
Peptide antibiotic-antimycin A, XLIII, 253
Peptide antibiotic lactonase, XLIII, 767–773
agar diffusion bioassay, XLIII, 770, 771
assay, XLIII, 770, 771
calcium phosphate gel chromatography, XLIII, 771
DEAE-cellulose column chromatography, XLIII, 771, 772
inhibitors, XLIII, 772
kinetic properties, XLIII, 772
pH optimum, XLIII, 772
properties, XLIII, 772, 773
purification, XLIII, 771, 772
Sephadex G–200 column chromatography, XLIII, 772
specific activity, XLIII, 772, 773
specificity, XLIII, 772
Peptide chloromethyl ketone, XLVI, 200, 201
Peptide growth factor, LVIII, 79
Peptide hormone
cell-receptor assay, XXXVII, 66–81
iodinated
preparation, XXXVII, 224–233
properties, XXXVII, 223–233
labeling, XXXVII, 28, 29
radioimmunoassay, XXXVII, 22–38
Peptidoglutaminase, *Bacillus circulans*, XLV, 485–492
Peptidoglycan, XXXIV, 398; XLVI, 403
Peptidorhamnomannan, XXXIV, 337
Peptidyl carrier protein, XLIII, 595–602
8α-Peptidyl flavin, prosthetic group, LII, 4
Peptidyllysine, activity of enzymes, XL, 343
Peptidylproline, activity of enzymes, XL, 343, 346
Peptidyl-tRNA, XLVI, 186, 626–633
endogenous, removal from ribosomes, LIX, 404
homologous series, XLVI, 707–711
translocation, assay, LIX, 359, 360
Peptidyltransferase, XLVI, 184, 627–629, 638, 676, 703
activity
fragment reaction assay, LIX, 779, 780
mitochondrial ribosomes, LVI, 91
assay for ribosome dissociation factor, LX, 296, 297
ribosomal, XLVI, 707–711
Peptococcus aerogenes, fructose-diphosphate aldolase class I, assay, purification, and properties, XLII, 249–258
Peptone, in reactivation medium, XLIV, 190
Peptostreptococcus elsdenii
D-lactate dehydrogenase, LIII, 439
assay, purification, and properties, XLI, 309–312
rubredoxin, LIII, 344
Per-*O*-acetyl-di-*N*-acetyl-α-chitobiosyl phosphate, preparation, L, 134–136
Perborate unit, XLIV, 685

Perchloric acid
activation of succinate dehydrogenase, LIII, 33, 472
assay of GTP hydrolysis, LIX, 361
concentration for tissue extraction, LV, 201, 202
determination of deuterium content of RNA, LIX, 646
in ethanol determination, LII, 357, 358
extraction of adenine nucleotides, LVII, 81
of guanine nucleotides, LVII, 88
in iron analysis, LIV, 439
metal-free, preparation, LIV, 476, 477
precautions, LI, 487
purification of flavins, LIII, 420, 423
for reaction termination in enzyme assays, LI, 4, 12, 42, 85, 107, 156, 158, 213, 214, 228, 418, 424, 425, 460
in wet digestion procedure, LIV, 481
Perfluorooctanoic acid, testosterone derivative, EC detection, XXXVI, 59
Performic acid, XLVII, 409
Perfusate
for thyroid gland, XXXIX, 362, 363
for *in vitro* pancreas, XXXIX, 366
Perfusion
of adrenal gland
comparison of methods, XXXIX, 335, 336
equipment, XXXIX, 329, 330
analytical procedures, LII, 55–58
capillary culture, unit, LVIII, 181
of endocrine cells, XXXIX, 165
extracorporeal, XLIV, 679, 680
flow rate, LVIII, 182
fluid, LII, 50, 51
of livers, XXXI, 95
metabolite measurement, LVI, 202
method, for cells, XXXIX, 134, 135
of ovaries *in vitro*, XXXIX, 230–237
of rat kidney, XXXIX, 3–11
perfusate, XXXIX, 6
of rat liver, XXXIX, 25–36
technique, XXXIX, 29–33
system types, LII, 49, 50
Perfusion apparatus, LII, 51–54
for hemicorpus, XXXIX, 75–77
for hepatocyte isolation, LII, 62, 63
for kidney, XXXIX, 4
for ovaries, XXXIX, 234–235
for rat liver, XXXIX, 25, 28, 29
Perfusion medium
for liver, XXXII, 626
for multilayered tissue cultures, XXXII, 568–574
Perfusion pump, LVIII, 181
Perfusor Braun Melsungen, XXXIX, 248
Perhydrol solution, in microsomal peroxide formation system, LII, 344
Perimycin, XLIII, 136
Periodate, *see also* Periodic acid
activation, XXXIV, 476
with agarose, XXXIV, 80–85
with cellulose, XXXIV, 82
with dextrans, XXXIV, 82
reductive amination, XXXIV, 82
treatment of ligands, XXXIV, 476
Periodate oxidation
determination of *in vivo* aminoacylation, LIX, 268–270
of glycoenzyme, XLIV, 317–320
of tRNA, LIX, 172, 173, 178, 269, 270
Periodate-oxidized agarose, XXXIV, 84
Periodate-oxidized compound, XXXIV, 476
Periodate-oxidized nucleotide, XXXIV, 478
Periodate-oxidized ribonucleoside phosphate, XXXIV, 490, 491
Periodic acid, cleavage of cross-linked polypeptides, LVI, 639
Periodic acid-Schiff method, for glycoprotein determination, XXXII, 87, 88
Periodic acid Schiff-methyl blue-orange G stain, for cloned pituitary cells, 132
Periodic structure, enhancement by optical filtering, XLIX, 39–63
Peripheral blood, lymphocyte isolation, XXXII, 633–636
Peristaltic pump
for perfusion apparatus, XXXIX, 26
for *in vitro* thyroid perfusion, XXXIX, 367
Peritoneal exudate cell
collection, LVIII, 505
stimulation, LVIII, 505

Peritoneal macrophage, *see* Macrophage, peritoneal
Peritoneum
  exudate, PMN granule isolation, XXXI, 346
  phagolysosome isolation, XXXI, 342, 343
Peritubular cell, cultures, XXXIX, 291, 292
Periwinkle, crown-gall studies, XXXI, 554, 556, 557
Perkin-Elmer differential scanning calorimeter, XXXII, 266
Perkin-Elmer F–40 gas-liquid chromatograph, LII, 357
Perkin-Elmer Model 180 Infrared Spectrometer, LIV, 319
Perkin-Elmer MPF-4 fluorometer, trypsinlike enzymes, XLV, 20
Perkin-Elmer spectrophotometer, XXXII, 249
Perlon wool, XXXIV, 174
Permablend III, scintillator, LI, 509
Permafluor liquid scintillation fluid, LI, 300, 323
Permanganate
  in ashing procedure, LIV, 481
  fixation, LVI, 719, 720
Permeability, measurement, of hollow-fiber membrane, XLIV, 301, 302
Permeability barrier, inhibitors, LVI, 32
Permeation chromatography, XLIV, 85, 86
Peroxidase, XXXIV, 336, 338, *see also* Horseradish peroxidase
  activity, XLIV, 200, 201, 271, 336, 716, 912, 913
  amino acid oxidase, LVI, 485
  assay with luminol, LVII, 405
  assay reagent, XLIV, 460, 474, 476, 632
  catalyst, in luminol chemiluminescence, LVII, 449–451
  conjugation, with antibody, XLIV, 711
  electron microscopic tracer, XXXIX, 145, 146
  in enzyme electrode, XLIV, 589
  glucose electrode, LVI, 483
  horseradish, thermolysin hydrolysis products, XLVII, 176
  immobilization
    on inert protein, XLIV, 908, 912
    in liposomes, XLIV, 224
    in polyacrylamide gel, XLIV, 902, 908
  immobilized, in chemiluminescent assays, LVII, 456
  in immunoassay, XLIV, 714–716
  metal substitution, LII, 493
  microestimation, with *Pholas* luciferin, LVII, 405, 406
  of plants, XXXI, 500, 501
    isozyme patterns, LII, 520, 521
    purification by affinity chromatography, LII, 514–521
  stain for detection, XLIV, 713
  of wheat germ, binding properties, LII, 516, 521
Peroxidase-hydroxamic acid complex, apparent dissociation constants, LII, 516
Peroxidase-labeled antibody method, for hormones, XXXVII, 133–144
  controls, XXXVII, 143, 144
  endocrine tissue fixation, XXXVII, 136
  light microscopic tissue fixation, XXXVII, 136, 137
  peroxidase-immunoglobulin conjugation, XXXVII, 134–136
  in polyethylene glycol-embedded tissues, XXXVII, 141–143
  simultaneous antigen localization, XXXVII, 137, 138
  thick tissue sections, XXXVII, 138, 139
  ultrathin sections, XXXVII, 139–143
Peroxidation, of lipids, in microsome damage, XXXI, 195
Peroxide
  biological sources, XXXII, 108
  IR studies, LIV, 307
  removal, using sodium hydrosulfite, LII, 98
Peroxido sulfate, *see* Ammonium persulfate
Peroxisome, *see also* Microbodies
  alternative names, LII, 494
  characterization, XXXI, 23, 24, 39

common enzymes absent, LII, 502–530
diagnostic enzyme, XXXI, 20
differential sedimentation, XXXI, 727
electron microscopy, XXXI, 26, 34
electrophoretic purification, XXXI, 754
enzymes, XXXI, 496, 501, 509; LII, 495–502
function, LII, 505
isolation, XXXI, 506, 549, 552, 553, 566, 590; LII, 494, 495
from liver
isolation, XXXI, 356–368
purification, XXXI, 10
morphometry, XXXII, 15
nucleoid isolation, XXXI, 368–374
peroxide generation, XXXII, 108
properties, XXXI, 363–367
purification, XXXI, 40, 41
separation, XXXI, 20
Peroxydisulfate, cooxidant, in luminol chemiluminescence, LVII, 448
C(4a)-Peroxyflavin, reaction intermediate, LIII, 550, 558
4α-Peroxyflavin, LII, 39
Peroxylamine disulfonate, *g* value, XLIX, 523
Peroxyoxalate
assay of lactate dehydrogenase, LVII, 455
chemiluminescence, LVII, 456–462
activator concentration, LVII, 458, 459
disadvantages, LVII, 462
effect of pH on, LVII, 459
hydrogen peroxide detection limit, LVII, 459
solvent system, LVII, 457, 458
Persulfate
in electrophoresis reagent, XXXII, 85
oxidant of cytochrome c_1, LIII, 226
oxidizing agent, in luminol chemiluminescence reactions, LVII, 410
Perturbation method, XLVIII, 308
oscillating sinusoidal pressurization method, XLVIII, 313
pulsed pressure, XLVIII, 312, 313
signal-averaging method, XLVIII, 313
Perylene
activator, in chemiluminescent reaction, LVII, 520
oxidation potential and role, LVII, 522
fluorphor, in chemiluminescence system, LVII, 459, 460
PES, *see* *N*-Ethyl phenazonium ethosulfate
Pestle, Teflon, for tissue homogenizers, XXXI, 4, 7, 8, 14
Petite mutant, *see also* Mutant
cytoplasmic, LVI, 14
isolation, LVI, 157, 158
stability, LVI, 157
Petite-negative yeast, mitochondrial mutants, XXXII, 838–843
Petri dish, XXXIV, 223; LVIII, 13
Petroleum ether, XLIV, 103, 113, 230, 231
for extraction of styrene oxide, LII, 417
in linoleic acid hydroperoxide synthesis, LII, 412
for oxide removal, LII, 193, 194
preparation of 2-cyano-6-methoxybenzothiazole, LVII, 20
Petunia
from protoplast, LVIII, 367
protoplast fusion, LVIII, 367
Petunia × *Atropa*, LVIII, 365
Petunia × carrot, LVIII, 365
Petunia hybrida mesophyll × *P. parodii* mesophyll, LVIII, 365
Pevicon block electrophoresis, α_2-macroglobulin, XLV, 643
PFO, *see* Perfluorooctanoic acid
3PF technique, *see* Three-photon method
pH
activity profile
changes, XLIV, 53, 346, 352, 353
of collagen complexes, XLIV, 259
of esterolytic activity of chymotrypsin and chymotrypsin-Spheron 300, XLIV, 75, 76
of glucoamylase, XLIV, 180
of immobilized papain, XLIV, 111
partitioning effect, XLIV, 401–404

of proteolytic activity
of chymotrypsin and chymotrypsin-Spheron 300, XLIV, 75
of papain, XLIV, 82
adenylate cyclases, XXXVIII, 147, 159, 168, 173
ATP-P_i exchange, LV, 360, 363
brown adipose tissue mitochondria, LV, 70
carrier choice, XLIV, 154, 155
changes
calculation
acetate concentration gradient, LV, 687, 688
acetate uptake by vesicles, LV, 686, 687
free acetate in upper chamber, LV, 686
caused by energized vesicles, LV, 366, 367
cyanine dyes, LV, 693, 694
determination across membranes
by change in external concentration, LV, 563–565
internal pH indicator, LV, 565–567
ion-distribution techniques, LV, 561–563
measurement, LV, 749, 750
proteolipid ionophores, LV, 418, 419
choice of coupling technique, XLIV, 148
in chromosome isolation, XL, 65
cross-linking, LVI, 635
difference
across mitochondrial membrane, LVI, 219, 220
anion gradients, LVI, 254
effect on coupling, XXXIV, 88
on culture, LVIII, 72, 135
on elution, XXXIV, 298
on hydrogen exchange rate, XLIX, 27, 28, 33
on membrane thermal transitions, XLIX, 14
on neutral protease elution, XXXIV, 440
on plant-cell fractionation, XXXI, 523
control, XXXI, 532–534
of elution buffers, XXXIV, 139
eriochrome blue absorbance, LVI, 316
guanylate cyclase, XXXVIII, 195, 199
in histone fractionation, XL, 132
light-induced changes, LVI, 406
measurement, by continuous-flow method
construction scheme, LV, 619–622
of proton translocation associated with aerobic oxidation of mitochondrial respiratory carriers, LV, 623–627
in stirred suspensions, LV, 616–618
of media, LVIII, 199, 200
for poikilotherm vertebrate cells, LVIII, 470
mitochondrial matrix, LVI, 253
partitioning effect, XLIV, 401
phosphodiesterases, XXXVIII, 214, 215, 232, 244, 247, 248, 256
phosphorylation potential, LV, 236, 238
protein kinase, XXXVIII, 314, 322
transhydrogenase, LV, 270
Phage fd DNA, XLVI, 90
Phage lysate, preparation for transduction experiments, LVI, 111
Phage P22, in *Salmonella typhimurium*, transduction, XLIII, 47
Phage P1 kc, in *Escherichia coli*, transduction, XLIII, 45–47
Phage Qβ replicase, *see* Protein, ribosomal, S1
Phagocyte, *see also* Macrophage
chemiluminescence
data analysis, LVII, 480–482
instrumentation, LVII, 473–480
opsonophagocytic dysfunction, LVII, 482–490
principles, LVII, 462–466, 468, 469
mononuclear, isolation and cultivation, XXXII, 758–765
uptake by macrophage, LVIII, 498, 499
Phagolysosome, preparation, XXXI, 339–345

Phagosome, of amoeba, isolation, XXXI, 695–698
Phalloidin, cleavage, XLVII, 462
Pharmacologic agent, effect on phagocytic cells, LVII, 490–494
Phase angle, LIV, 162
Phase-contrast microscopy, monolayer cultures, LVIII, 139
Phase fraction analysis, LVIII, 241
Phaseolus aureus, oxidase, LII, 21
Phaseolus lunatus, lectin, XXXIV, 333
Phaseolus vulgaris, lectin, XXXIV, 333
Phase shift, LIV, 340
 determination, LIV, 343–345
 transferability, LIV, 343
Phase transition, spin-label measurements, XXXII, 172–174
PHB, *see* Poly-β-hydroxybutyrate
pH chromatogram, XLIII, 164–168
Phenacetin
 difference spectra, LII, 272
 as endogenous substrate, LII, 272
 N-hydroxylation, Ah^b allele, LII, 231
 type of binding reaction with cytochrome *P*–450, LII, 264
Phenacetin *O*-deethylase, Ah^b allele, LII, 231
Phenacylchymotrypsin, XLVI, 89
Phenacyl-CoA ligase, XLIII, 476–481
 assay, XLIII, 447, 478
 crude enzyme extract, XLIII, 479
 mycelium production, XLIII, 478, 479
 Penicillium chrysogenum preparation, XLIII, 477, 478
 pH optimum, XLIII, 481
 purification, XLIII, 479–481
 specificity, XLIII, 481
 stability, XLIII, 481
Phenanthrene, internal standard, XLVII, 34
Phenanthrene dihydrodiol, chromatographic properties, LII, 294–296
Phenanthrene 9,10-oxide, substrate of epoxide hydrase, LII, 199, 200
Phenanthrene phenol, chromatographic properties, LII, 293, 294
Phenanthroline, ATP-Co(III) complex, XLVI, 314, 315, 318–321
1,10-Phenanthroline, XLIII, 502; XLVII, 89, 394; XLIX, 177, 178
 effect on *Cypridina* luciferase, LVII, 352
 inactivator of NADH dehydrogenase, LIII, 20
 inhibitor of amidophosphoribosyltransferase, LI, 178
  of phenylalanine hydroxylase, LIII, 285
  of uridine nucleosidase, LI, 294
 M. phlei membrane systems, LV, 185
 phosphodiesterase, XXXVIII, 248, 256
Phenathiazine drug, enzymatic oxidation, LII, 151
Phenazine ethosulfate
 as mediator-titrant, LIV, 408
 redox titration, LIII, 491
Phenazine methosulfate, LI, 69; LIV, 144, 148, 407, 408
 assay of D-lactate dehydrogenase, LIII, 519
 carboxamides, LV, 461
 colorimetric test for tetrazolium derivatives, XLIV, 276, 905
 complex II, LVI, 586
 cytochrome *c*, LV, 110
 in flavoprotein assays, LIII, 406
 flow dialysis experiments, LV, 683, 685, 686
 isoenzyme staining, LVIII, 168, 169
 Keilin-Hartree preparation, LV, 126
 reagent, glucose oxidase staining, XLIV, 713, 905
 in redox titration, LIII, 491
 storage, LIII, 26, 467
 substrate of acyl-CoA dehydrogenase, LIII, 503
  of complex II, LIII, 22, 24
 of succinate dehydrogenase, LIII, 26, 27, 34, 466–468
Phenethylbiguanide, XLVI, 548
Phenidon developer, XL, 10
Phenobarbital
 activities induced, LII, 373
 detection in biological samples, LII, 336, 341, 342
 dose for injection, LII, 92, 118
  for oral administration, LII, 111

effect on cytochrome *P*-450-linked metabolism, LII, 65, 69
on difference spectra of reduced cytochromes, LII, 274, 275
on microsomal fractionation, LII, 73
on protein and lipid metabolism in microsomes, LII, 75
inducer of cytochrome b_5, LII, 109
of cytochrome *P*-450, LII, 203, 270, 328, 329
of epoxide hydrase activity, LII, 194–195
photochemically active derivative, LVI, 660
Phenol, XXXIV, 6, 107, 108
in ammonia determination, LI, 498
cation scavenger, XLVII, 538, 547
coupling, XXXIV, 102, 103
diazo groups destroyed, XLIV, 93
extraction
analysis of phosphorylated acidic chromatin proteins, XL, 179
of nucleic acids, LIX, 311
of oligonucleotides, LIX, 99
of radiolabeled RNA, LIX, 831, 832
of ribosomal RNA, LIX, 445, 446, 452, 470, 556, 648
of tRNA, LIX, 120, 192, 206, 210, 220, 221, 230, 269, 342, 343, 818
indefinite self-association, XLVIII, 115
as ligands, XXXIV, 105
test, XLIV, 276
tyrosine modification, XLVII, 461
Phenolacetic acid reagent, XXXII, 867
Phenol-acetic acid-urea, for depolymerization of cytochrome oxidase subunits, LIII, 79
Phenolacetic *N*-hydroxysuccinimide ester, in modification of chloroplastic tRNA, LIX, 213
Phenolase, *see* Laccase
Phenolase complex, reaction mechanism, LII, 39
Phenolate ion, spectral properties, LVII, 22–24
Phenol-chloroform, extraction of RNA, LIX, 102, 103, 298, 299
Phenol-2,4-disulfonyl chloride, XLIV, 263
Phenol-dodecyl sulfate, extraction of RNA, LIX, 586
Phenol hydroxylase, *see* Phenol 2-monooxygenase
Phenol *o*-hydroxylase, *see* Tyrosinase
Phenol 2-hydroxylase, *see* Phenol 2-monooxygenase
Phenolic benzo[*a*]pyrene, *see* 3-Hydroxybenzo[*a*]pyrene
Phenolic chromophore, extinction coefficient, XLVII, 446
Phenolic substance
in plant-tissue extraction, XXXI, 528–544
mitochondria effects, XXXI, 530, 531
protein reactions, XXXI, 529, 530
removal, XXXI, 530, 537
Phenol 2-monooxygenase, oxy form, spectrum, LII, 17, 36
Phenol oxidase
inhibitors, XXXI, 538–540
in plant tissues, XXXI, 528
studies, XXXI, 544
Phenolphthalein, XLIV, 111
as mixed-function oxidase substrate, XXXI, 233
Phenol-Pronase-sodium dodecyl sulfate, extraction of RNA, LIX, 428
Phenol reagent, in aniline hydroxylase assay, LII, 409
Phenol red, XLIX, 312
absorption spectra, LVI, 310, 311
hydrogen ion measurements, LVI, 310–313, 336–338
in media, LVIII, 70
pH indicator, LVIII, 121, 122
Phenolsaffranine, in redox titration, LIII, 491
Phenol-sucrose, gradient method, ribosomal RNA isolation, LIX, 556, 557
Phenol sulfokinase, *see* Aryl sulfotransferase
Phenomenological constant, *see* Adair constant
Phenosafranine, as mediator-titrant, LIV, 409
Phenothiazine, XLIV, 855

Phenotype
expression, time, LVIII, 313
stability, LVIII, 308
Phenoxyacetyl coenyzme A, XLIII, 476, 486
ω-Phenoxyalkoxybenzamidine, XLVI, 122–127
m-(ω-Phenoxyalkoxy)benzamidine, XLVI, 122, 123
Phenoxybenzamine, XLVI, 604
Phenoxymethylpenicillin, XLIII, 476
penicillin acylase assay, XLIII, 699
substrate for fungal penicillin acylase, XLIII, 722
m-(3-Phenoxypropoxy)benzamidine, XLVI, 127
m-(3-Phenoxypropoxy)benzamidine *p*-toluenesulfonate, XLVI, 127
2-[*m*-(3-Phenoxy)propoxyphenyl]-2-imidazoline hydrochloride hydrate, XLVI, 120
Phentolamine, in studies on epinephrine binding, XXXI, 70, 71
Phenylacetic acid, XLIII, 482–487, *see also* Phenylacetyl-CoA hydrolase
assay, XLIV, 760, 761
removal, XLIV, 767
Phenylacetyl-CoA hydrolase, XLIII, 482–487
ammonium sulfate precipitation, XLIII, 485
assay, XLIII, 482, 483
buffer, XLIII, 487
DEAE-cellulose column, XLIII, 485
molecular weight, XLIII, 486
Penicillium chrysogenum preparation, XLIII, 484
phenylacetic acid, activation, XLIII, 432, 487
pH optimum, XLIII, 486
properties, XLIII, 485, 486
purification, XLIII, 484, 485
Sephadex G–100 column, XLIII, 485
shake cultures, XLIII, 484
specificity, XLIII, 485
thiol effect on assay, XLIII, 486, 487
unit definition, XLIII, 484
Phenylacetyl coenzyme A, XLIII, 476, 486
[1-^{14}C]Phenylacetyl coenzyme A, XLIII, 482, *see also* Phenylacetyl-CoA hydrolase
L-Phenylalaninal diethylacetal, XLVI, 223, 224
Phenylalanine, XXXIV, 5
aminoacylation, isomeric specificity, LIX, 274, 275, 279, 280
fluorine labeling, XLIX, 274, 290
in media, LVIII, 53, 62
peptide synthesis, XLVII, 509
solid-phase, XLVII, 610
phosphorescence, XLIX, 242–244
photochemically active derivatives, LVI, 660
radiolabeled, in protein turnover studies, XXXVII, 239, 242–244
release, measurement, XLVII, 76, 77
residue, detection, XLVII, 217
substrate, of phenylalanine hydroxylase, LIII, 278
thermolysin hydrolysis, XLVII, 175, 176
transport, XLVI, 151, 608
[^{14}C]Phenylalanine
in aminoacylation assay, LIX, 183
in poly(Phe) synthesis, LIX, 459
DL-Phenylalanine, XLIV, 81, 82
[^{3}H]Phenylalanine, preparation of aminoacylated tRNA, LIX, 126, 127, 818
L-Phenylalanine
enzyme electrode, XLIV, 589, 617, 618
isolation, XLIV, 757
magnetic circular dichroism spectrum, XLIX, 160
Phenylalanine-agarose, XXXIV, 397
Phenylalanine ammonia-lyase, XLVI, 67
in glyoxysomes, LII, 500
L-Phenylalanine chloromethyl ketone, XLVI, 202, 609, 611–613
Phenylalanine decarboxylase
conjugate, activity, XLIV, 271
immobilization, on inert protein, XLIV, 909
Phenylalanine hydroxylase, *see* Phenylalanine 4-monooxygenase
Phenylalanine methyl ester, substrate for carboxylesterase, XXXV, 190

Phenylalanine 4-monooxygenase, LII, 12; LIII, 278–286
activity, LIII, 278
assay, LIII, 278–280
inhibitors, LIII, 285
iron, LIII, 286
isozymes, LIII, 286
kinetic properties, LIII, 285
molecular weight, LIII, 286
purification, LIII, 280–283
from rat liver, LIII, 278–286
substrate specificity, LIII, 283, 284
tetrahydropterin specificity, LIII, 284, 285
Phenylalanine-tRNA ligase, XXXIV, 5
L-Phenylalaninol, XLVI, 222
Phenylalanyllysylchloromethyl ketone, XLVI, 206
N-Phenylalanylpuromycin aminonucleoside, in photoaffinity labeling studies, LIX, 802, 804, 805
Phenylalanyl-$tRNA^{Phe}$, XLVI, 75, 638, 640
Phenylalanyl-tRNA synthetase, XXXIV, 167, 513–516; XLVI, 85; LIX, 178; LX, 618
aminoacylation assay, LIX, 183
misactivation studies, LIX, 291
preparation of aminoacylated tRNA, LIX, 126, 127
purification, XXXIV, 169, 170; LIX, 262, 266
subcellular distribution, LIX, 233, 234
Phenylalanyl-transfer ribonucleic acid, XLVI, 625, 627–629, 707
acetylation, LX, 231
binding to 70 S ribosome complexes, LX, 624–626
to 80 S ribosomes, LX, 587, 588, 661, 662, 664, 689, 690
^{14}C-labeled
in cell-free peptide synthesis, LIX, 388
in isolation of translating ribosomes, LIX, 391
complex with 30 S ribosomal subunits and polyuridylate, LX, 232–234, 428, 429
formation of polyphenylalanine, *see* Polyphenylalanine
preparation, LX, 231, 580, 597, 598, 620, 622, 623
with terminal 3′-amino-3′-deoxyadenosine
electrophoretic properties, LIX, 145
preparation, LIX, 142–144
m-(ω-Phenylalkoxy)cyanophenyl ether, XLVI, 123
ω-Phenylalkylbromide, XLVI, 123
Phenylazide, triplet nitrene, LVI, 658
m-(4-Phenylbutoxy)benzamidine *p*-toluenesulfonate, XLVI, 123, 127
4-Phenylbutylamine-agarose, XXXIV, 416–420
preparation, XXXIV, 419, 420
m-(4-Phenylbutyl)benzamide *p*-toluenesulfonate, XLVI, 124
4-Phenylbutylbromide, XLVI, 127
α-Phenyl-ω-(3-cyanophenyl)alkane, XLVI, 122
Phenyldiazoacetate, XLVI, 40
Phenyldicarbaundecaborane
as penetrating ion, LV, 598, 600
proteoliposomes, LV, 764, 768
4-Phenyl-3,5-dimethylpyrazole-1-carboxyamidine, XLVI, 549, 552
binding with ^{14}C-labeled, XLVI, 554
N,N′-(1,2-Phenylene)bismaleimide, XLIV, 263
2,2′-(1,4-Phenylene)bis(5-phenyloxazole), XLVIII, 336, 343
in scintillation fluid, XXXII, 110; LI, 30, 157, 292, 355, 409, 540
p-Phenylenediamine, XL, 10
agarose derivatization, XLIV, 39, 40
p-Phenylene diisothiocyanate, XLVII, 268, 283, 284, 293, 294, 303, 309
m-Phenylethylbenzamidine *p*-toluenesulfonate, XLVI, 127
Phenyl-β-D-fucopyranoside, XXXIV, 369
Phenyl-β-D-glucopyranoside, XXXIV, 334
α-Phenylglyceric acid, XLVI, 547
α-Phenylglycidate, XLVI, 543, 544
synthesis, XLVI, 544, 545
β-Phenylglycidate, XLVI, 543
Phenylglyoxal, XLVI, 66; LV, 518
modification of arginine, LIII, 143, 152
9-Phenylguanine, XXXIV, 524
Phenylhydrazine, XLIV, 117, 144

cis-2-Phenyl-4-hydroxymethyl-5-(1′-pentadecenyl)-2-oxazoline, intermediate in sphingomyelin synthesis, XXXV, 499
Phenyl α-L-iduronide, L, 141
α-L-iduronidase assay, L, 148–150
synthesis, L, 145, 146
Phenylimidazole, photochemically active derivative, LVI, 660
N-Phenylimidazole, inhibitor of *Rhizobium P*–450, LII, 157
Phenylisocyanate, XLVI, 534
Phenyl isothiocyanate, XLVII, 336
radioactive, sequence analysis, XLVII, 247, 248, 254, 255, 273
storage, in solution, XLVII, 337
for *N*-terminus studies, LI, 133
Phenylmercury(II) acetate, inhibitor, XLIV, 615
Phenylmethyl sulfonyl fluoride, XXXIV, 738; LIX, 190
acylating agent, XLIV, 562, 567
adenosinetriphosphatase isolation, LV, 148
assay of adenosine monophosphate nucleosidase, LI, 264, 265
disadvantage, in purification of tryptophanyl-tRNA synthetase, LIX, 242
HeLa cell mitochondrial proteins, LVI, 75, 78
inhibitor, of carboxypeptidase Y, XLVII, 89
of palmityl thioesterase I, XXXV, 106
of thrombin, XLV, 173
isolation of mitochondria, LV, 345
in microsome isolation, LII, 92
preparation of *B. subtilis* ribosomes, LIX, 374
of chromophore-labeled sarcoplasmic reticulum vesicles, LIV, 51
of ribosomal proteins, LIX, 484, 490
of submitochondrial particles, LIII, 222, 228
of yeast submitochondrial particles, LIII, 113
protease action, LIX, 267
protease inhibitor, LVIII, 273
purification of aminoacyl-tRNA synthetases, LIX, 259
of blue-fluorescent protein, LVII, 228
of CPSase-ATCase complex, LI, 114
of CTP(TP):tRNA nucleotidyltransferase, LIX, 184
of cytochrome bc_1 complex, LIII, 115
of flavocytochrome b_2, LIII, 241, 242, 246, 251
of D-lactate dehydrogenase, LIII, 520
of purine nucleoside phosphorylase, LI, 521
of pyrimidine nucleoside monophosphate kinase, LI, 324, 328
in SKE buffer, LIII, 114
in storage buffer of ribosomal protein S2, LIX, 498
yeast cell breakage, LVI, 44
yeast cell disruption, LV, 416
10-Phenylphenothiazine
chemiluminescent reaction, LVII, 517
structural formula, LVII, 498
Phenylphosphate, *see also* Sodium phenylphosphate
phosphate donor of nucleoside phosphotransferase, LI, 387, 391
Phenylpropanoid compound, in plants, XXXI, 529, 530
m-(3-Phenylpropionamidobenzene)benzamidine *p*-toluenesulfonate, XLVI, 123, 124
β-Phenylpropionate, pseudosubstrate, of melilotate hydroxylase, LIII, 557
β-Phenylpropionate hydroxylase, LII, 20
3-Phenylpropionic acid, substrate analog, of chymotrypsin, XLIV, 365
2-D-Phenylpropionyl chloride
3-hydroxy fatty acids, 2-D-phenylpropionate derivatives, XXXV, 328–331
preparation, XXXV, 327
β-Phenylpropionyl-L-phenylalanine, XLIX, 168
β-Phenylpropionylthiocholine iodine, substrate, factor XIII assay, XLV, 185

2-Phenyl-2-propyloxycarbonyl group, removal, XLVII, 594
Phenyl succinate
cell fractionation, LVI, 217
transport, LVI, 248, 292
Phenylthiocarbamyl-amino acid APAM-polystyrene resin-amide
conversion to PTMA-amino acid, XLVII, 377, 378
synthesis, XLVII, 383
N^{α}-Phenylthiocarbamyl-amino acid methylamide
identification, XLVII, 379–381
preparation, XLVII, 375–379
quantitation, XLVII, 381, 382
Phenylthiohydantoin, *see* Amino acid, phenylthiohydantoin derivatives
Phenylthiol diazoacetate, XLVI, 40
4-Phenyl-3-thiosemicarbazide, incorporation in tRNA, LIX, 118
Phenyltrimethylammonium-agarose, XXXIV, 573
Phenytoin, detection in biological samples, LII, 336
Pheophytin, LIV, 28, 32
pH gradient, imposed, reconstitution of ATP synthesis, LV, 746–748
Phialidium, bioluminescence, LVII, 275
Philips calcium-sensitive electrodes, LVI, 347
pH indicator, internal, LV, 565–567
Phleomycin, XLIII, 150
Phloretin, cyanine dyes, LV, 693
Phlorizin acceptor, in brush borders, XXXI, 122
Phlorizin hydrolase, in microvillous membrane, XXXI, 130
Phoenix-Chance spectrophotometer, XXXII, 416
Pholas, luciferase, LIII, 560
Pholas dactylus
bioluminescence, LVII, 271, 272, 354, 355, 383–406
mechanism, LVII, 563, 564
reaction intermediate, structure, LVII, 524, 565
luciferin-binding protein, LVII, 528
Phorbol myristate acetate, stimulator, of phagocyte chemiluminescence, LVII, 466
Phosgene, XLIV, 120
[^{14}C]Phosgene, synthesis of [^{14}C]ethyl-2-diazomalonylchloride, LIX, 811, 813
Phosphatase, XXXIV, 710; XLII, 341–426, *see also* specific types
active transport studies, XLIV, 921–924
assays, XXXII, 396, 397
binding to BrCN-activated cellulose, LX, 7
dephosphorylation of polyuridylate fragments at the 3′-end, LX, 748–751
gel electrophoresis, XXXII, 91
kinetics, in enzyme system, XLIV, 435
phosphoprotein
activity, LX, 530–534
assay, LX, 526–529
functional role, LX, 522
molecular weights, LX, 529
purification, LX, 528–530
Phosphate
assay, XXXI, 19; LV, 212, 215, 221, 422
ATPase assay, LV, 314
ATPase F_1, LV, 323
binding
to ATPases, LV, 302
by beef heart mitochondrial ATPase, LVI, 527–530
calcium transport, LVI, 239, 339, 340
cell incorporation, substrate entry, XXXIX, 495
content of kidney cortex, LV, 220
Crabtree effect, LV, 296, 297
energy charge, LV, 233, 234
esterified, estimation, LV, 171, 172
exchange, LVI, 253
extramitochondrial, LV, 244
fluorescence quenching agent, XLIX, 233
inorganic
activator of phosphoribosylpyrophosphate synthetase, LI, 11, 16
of succinate dehydrogenase, LIII, 33, 472
allosteric inhibitor of adenosine monophosphate nucleosidase, LI, 263, 270, 271
assay, XXXIV, 235, 236

determination, LI, 272, 275, 285, 286
effect on cytochrome *c* chromatographic mobility, LIII, 147
inhibitor of adenylate deaminase, LI, 496, 502
of aspartate carbamyltransferase, LI, 57
of orotate phosphoribosyltransferase, LI, 163
of orotidylate decarboxylase, LI, 153, 163
in nucleotidase assay, XXXII, 126
product of adenylosuccinate synthetase, LI, 207
of carbamoyl-phosphate synthetase, LI, 21, 29
of cytidine triphosphate synthetase, LI, 84
of deoxythymidylate phosphohydrolase, LI, 285
of FGAM synthetase, LI, 193
of phosphoribosylglycinamide synthetase, LI, 179
of rat liver acid nucleotidase, LI, 271
of succino-AICAR synthetase, LI, 186
substrate of glycine reductase, LIII, 373
of purine nucleoside phosphorylase, LI, 517, 538
of pyrimidine nucleoside phosphorylase, LI, 432
of thymidine phosphorylase, LI, 437, 442
of uridine phosphorylase, LI, 423
ubiquinone redox state, LIII, 583
interference
in copper analysis, LIV, 445
in iron analysis, LIV, 440
intramitochondrial, LV, 244
leakage from mitochondria, LVI, 289
measurement of proton translocation, LV, 636, 642, 644, 645
in media, LVIII, 68
mitochondrial incubation, LV, 7
mitochondrial protein synthesis, LVI, 21
organic, in plants, pH effects, XXXI, 534
oxygen-18 content, reactions contributing, LV, 256
oxygen exchange, LV, 253, 254, 257, 258
Pasteur effect, LV, 290, 291
ΔpH determination, LV, 229
^{32}P-labeled
cellular, estimation of specific activity, LV, 172
equilibration of starved cells, LV, 169
organic, determination of acid-labile fraction, LV, 172, 174
purification, LV, 311
in *in vivo* preparation of radiolabeled RNA, LIX, 831
radioactive, precautions in handling, XXXVIII, 413, 414
separation
procedure, LV, 251, 252
reagents and materials, LV, 250, 251
spot test, XXXIV, 482
transport
parathyroid hormone effects, XXXIX, 20
Pasteur effect, LV, 292
Phosphate acetyltransferase, XXXIV, 267–271
coenzyme A assay, LV, 210
Phosphate analysis, for bound protein determination, XLIV, 388
Phosphate anion, inhibitor of NADH-cytochrome b_5 reductase, LII, 244
Phosphate buffer, *see also* Potassium phosphate
in bacterial luciferase reagent, LVII, 200
electrophoretic analysis of ribosomal proteins, LIX, 507, 509
interference, in quinone extraction, LIII, 580
preparation, LVII, 143, 145
of modified tRNA, LIX, 171
sodium/potassium, preparation, LVII, 172

staphylococcal protease hydrolysis, XLVII, 191
for storage of ribosomal RNA, LIX, 586
for subcellular fraction study, XXXII, 4
for transmission electron microscopy, XXXIX, 135
Phosphate-buffered saline, XXXII, 741
composition, XXXII, 617, 618
definition, XLIV, 711
erythrocyte ghost preparation, XXXI, 173, 175, 177
tissue fixation, XLIV, 712, 713
for transmission electron microscopy, XXXIX, 135
Phosphate ion, stabilizer, of luciferase, LVII, 199
Phosphate/oxygen ratio
mitochondrial integrity, LV, 8, 9
of phosphorylating vesicles, LV, 167
representative measurements in intact cells, LV, 173, 174
uncoupler assay, LV, 464
yeast mitochondria, LV, 150
Phosphate transport, LVI, 247–249, 292
assay
comparison with other methods, LVI, 357, 358
net uptake method, LVI, 354, 355
$^{32}P_i$ exchange method, LVI, 355–357
principle, LVI, 353
Phosphate transporter, reconstituted, LV, 706, 707, 710
activity, LV, 705
Phosphatidalcholine, *trans* isomer, intermediate in plasmalogen synthesis, XXXV, 484
Phosphatidate phosphatase, inhibition, XXXV, 506
Phosphatide, R_f values, XXXVII, 267
Phosphatide acylhydrolase, *see* Phospholipase A_2
Phosphatidic acid, XLIV, 227, *see also* specific compounds
analogues
methods of synthesis
via diester phosphonate analogues, XXXV, 503–506
via diether phosphonate analogues, XXXV, 506, 507
intermediate in phosphatidylethanolamine synthesis, XXXV, 455, 456
in phosphatidylserine synthesis, XXXV, 461
methods of synthesis
via acylation of *sn*-glycerol-3-phosphoric acid, XXXV, 439–441
via α-iodoglycerol, XXXV, 431–441
phospholipase A_2 effects, XXXII, 148
in plant extracts, XXXI, 526, 527
plasmalogenic, intermediate in phosphatidic acid synthesis, XXXV, 441
preparation, XXXII, 142, 143
product of phospholipase D, XXXV, 226
spin-labeled, in microsomes, XXXII, 193
thin-layer chromatography, XXXV, 400
Phosphatidylcholine, XLIV, 218, *see also* Lecithin; Phosphoglyceride; Phospholipid; specific compounds
biosynthesis, in endoplasmic reticulum, XXXI, 23
[^{3}H]choline-labeled, substrate for phospholipase D, XXXV, 227
in complex I–III, LIII, 8
in cytochrome b_5 assay, LII, 110
in cytochrome b_5 reduction, LII, 107
hydrolysis with snake venom phospholipase, XXXV, 317, 318
lateral diffusion, XXXII, 181–184
in lipid bilayers, XXXII, 516
lipophilic Sephadex chromatography, XXXV, 389, 392
liposomes, NADH-cytochrome b_5 reductase binding, LII, 108
in membranes, XXXII, 138
methods of synthesis
via *sn*-glycerol-3-phosphorylcholine, XXXV, 442–453
via phosphatidic acid, XXXV, 441, 442

in mitochondria, XXXI, 302, 303
in mixed-function oxidase system, LII, 202, 204
phospholipase effect, XXXII, 138
phospholipid exchange, XXXII, 140
^{32}P-labeled assay of phospholipid exchange enzyme, XXXV, 262–264
preparation, LV, 718
reductase activity, LII, 96
single-walled vesicles, XXXII, 485–489
physical properties, XXXII, 488, 489
spin-label studies, XXXII, 185, 186
substrate for phospholipase D, XXXV, 227, 231
synthesis of choline-labeled, XXXV, 533–541
thin-layer chromatography, XXXV, 400
Phosphatidylcholine analogue, *see also* specific compounds
methods of synthesis
via choline alkylphosphonate derivatives, XXXV, 517–520
via glycerol derivatives, XXXV, 514–517
via phosphinate analogues, XXXV, 520–525
Phosphatidylcholine cholinephosphohydrolase, *see* Phospholipase C
Phosphatidylcholine derivative, *see also* Lecithin
partial specific volume, XLVIII, 22
L-α-Phosphatidylcholine dimyristoyl, bioassay of phospholipase, LVII, 193
Phosphatidyl-*N,N*-dimethylethanolamine
intermediates in synthesis of choline labeled phosphatidylcholines, XXXV, 533
purification, XXXV, 535
Phosphatidylethanolamine, XLIV, 218, *see also* Cephalin; Phosphoglyceride; Phospholipid; specific compounds
adenylate cyclase, XXXVIII, 176, 177, 180
amino acid transport, LV, 188
assay with trinitrobenzenesulfonic acid, XXXV, 339
bacterial, in liposomes, XXXII, 505
column chromatography, XXXV, 411
in complex I–III, LIII, 8
from *E. coli* auxotroph, fatty acids, XXXII, 864
in ω-hydroxylase, LIII, 360
isolation, XXXII, 450
in lipid bilayers, XXXII, 516, 531
lipophilic Sephadex chromatography, XXXV, 389
in membranes, XXXII, 138
methods of synthesis
via ether phosphatidylethanolamine, XXXV, 458–460
via glycerol-α-iodohydrin diester, XXXV, 457, 458
via phosphatidic acid, XXXV, 455, 456
in mitochondria, XXXI, 302, 303
partial specific volume, XLVIII, 22
phospholipid exchange, XXXII, 140, 146
preparation, LV, 718
of thermophilic bacterium PS3, LV, 370
molecular species, LV, 371
thin-layer chromatography, XXXV, 400, 407
Phosphatidylethanolamine analogue, *see also* specific compounds
methods of synthesis
via dihydroxypropylphosphonic acid derivatives, XXXV, 512, 513
via glycerol derivatives, XXXV, 508–513
Phosphatidyl exchange protein, XXXII, 140–146
assay, XXXII, 141, 142
from beef liver, XXXII, 140–146
properties, XXXII, 145, 146
purification, XXXII, 144, 145
unit of activity, XXXII, 143
Phosphatidylglucose, intermediate in phosphatidylinositol synthesis, XXXV, 477
Phosphatidylglycerol, *see also* Phosphoglyceride; Phospholipid; specific compounds
lipoteichoic acid carrier, L, 388

methods of synthesis, XXXV, 465–469
partial specific volume, XLVIII, 22
phospholipase A_2 effects, XXXII, 148
product of phospholipase D, XXXV, 227
of thermophilic bacterium PS3, LV, 370
molecular species, LV, 371
thin-layer chromatography, XXXV, 400
Phosphatidylglycerol analogue, methods of synthesis, XXXV, 469
Phosphatidylinositol
adenylate cyclase, XXXVIII, 149, 177, 179, 180
column chromatography, XXXV, 411
methods of synthesis, XXXV, 471–477
phospholipid exchange, XXXII, 140
thin-layer chromatograhy, XXXV, 400
Phosphatidylserine, XLIV, 218, *see also* Phospholipid; specific compounds
adenylate cyclase, XXXVIII, 149, 176–178, 180
analogue, synthesis, XXXV, 465
assay with trinitrobenzenesulfonic acid, XXXV, 339
column chromatography, XXXV, 411
in membranes, XXXII, 138
methods of synthesis
via phosphatidic acid, XXXV, 461, 462
via substituted glycerophosphorylserine, XXXV, 463–465
partial specific volume, XLVIII, 22
reconstitution studies, LV, 708
thin-layer chromatography, XXXV, 400
Phosphinate, analogue of phosphatidylcholine, intermediate in synthesis of phosphatidylcholine analogues, XXXV, 520–525
3′-Phosphoadenosine 5′-phosphosulfate
biosynthesis, LVII, 245
product, of luciferin sulfokinase, LVII, 237
3′-Phosphoadenylyl sulfate, *see* 3′-Phosphoadenosine 5′-phosphosulfate
Phosphocellobiase
assay, XLII, 495
chromatography, XLII, 496
molecular weight, XLII, 497
occurrence, XLII, 497
properties, XLII, 496–497
purification, XLII, 495–496
Phosphocellulose, XXXIV, 607, 634; XLIV, 46
cAMP-binding protein, XXXVIII, 370
chromatography, purification of flavin peptides, LIII, 454
column, capacity for tRNA synthetase, XXXIV, 167, 168
filtration
hydrogen-exchange, XLVIII, 327
procedure, XLVIII, 343
peptide separation, XLVII, 204–210
phosvitin kinase, XXXVIII, 328, 329
preparation, XXXVIII, 371
of ribosomal protein, LIX, 557–561
pretreatment, LX, 5, 17, 102, 103, 467, 468, 642
prewashing procedure, LI, 384, 499, 500
protein kinases, XXXVIII, 297
purification
acetyltransferase, LX, 536, 537, 539
of adenylate deaminase, LI, 500
of adenylosuccinate synthetase, LI, 209
casein kinases, LX, 501, 502
of CTP(ATP):tRNA nucleotidyltransferase, LIX, 125
of deoxycytidylate deaminase, LI, 415
elongation factor
EF-1, LX, 142, 582, 639, 642, 643, 645, 646
EF-2, LX, 142, 582, 639, 642, 643, 645, 646, 652, 653, 656, 707, 709
heme-regulated translational inhibitor, LX, 471, 472, 474, 475, 477, 478, 486, 495
initiation factor
EIF, LX, 40, 42–44

eIF, LX, 20–22, 24–28, 105–107, 129, 130, 132–134, 149–152, 154, 157, 158, 242, 244, 267, 407, 469, 470
IF, LX, 6, 9, 96, 99, 207, 227–229, 233
of nucleoside diphosphokinase, LI, 380, 382, 384
nucleotide diphosphate kinase, LX, 581
protein kinases, LX, 509
Qβ-ribonucleic acid replicase, LX, 634
ribonucleic acid-binding proteins, LX, 395–397
of tRNA methyltransferases, LIX, 195, 196, 200–202
of RNase III, LIX, 829
wheat germ initiation factors, LX, 194, 198–201
Phosphocellulose column, preparation, XL, 199
Phosphocreatine, assay of thymidine kinase, LI, 366
Phosphodeoxyribomutase, *see* Phosphopentomutase
Phosphodiesterase, XXXIV, 605–610
activator, XXXVIII, 262, 263
assay, XXXVIII, 263, 264
preparation, XXXVIII, 254, 255, 264–267
properties, XXXVIII, 267–273
requirement, XXXVIII, 255
activity stain, on gels
assay method, XXXVIII, 259, 260
electrophoresis procedure, XXXVIII, 260
staining procedure, XXXVIII, 260, 261
affinity procedure, XXXIV, 609, 610
assay, XXXIV, 607, 608
by continuous titrimetric technique
apparatus and experimental conditions, XXXVIII, 214
conclusions, XXXVIII, 217
dynamic aspects, XXXVIII, 216, 217
pH selection, XXXVIII, 214, 215
principle, XXXVIII, 213
reagents, XXXVIII, 213, 214
sensitivity, XXXVIII, 215, 216
with radioactive substrates
one-step procedure, XXXVIII, 210, 211
principle, XXXVIII, 205, 206
reagents and constituents, XXXVIII, 208
theory, XXXVIII, 206, 207
two-step procedure, XXXVIII, 209, 210
8-azido-cAMP, LVI, 646
bovine brain, XXXVIII, 223, 224
assay, XXXVIII, 132–134, 224, 225
properties, XXXVIII, 232–239
purification, XXXVIII, 225–232
regional and subcellular distribution, XXXVIII, 237, 238
bovine heart
assay, XXXVIII, 218, 219
properties, XXXVIII, 222, 223
purification, XXXVIII, 219–221
cAMP XXXI, 111
assay, XXXVIII, 62, 63
authenticity, XXXVIII, 26
inhibitor, XXXVIII, 275, 276, 282
cGMP assay, XXXVIII, 73, 78, 79, 83, 85, 88, 89, 106, 108, 110, 111
cyclase assays, XXXVIII, 120, 121, 124
cyclase preparations, XXXVIII, 143
Dictyostelium discoideum, XXXVIII, 244, 245
assay, XXXVIII, 245
properties, XXXVIII, 247, 248
purification, XXXVIII, 245–247
Escherichia coli
assay, XXXVIII, 249, 250
culture, XXXVIII, 250, 251
extract preparation, XXXVIII, 251
properties, XXXVIII, 255, 256
purification, XXXVIII, 251–255
kinetics, activator, XXXVIII, 272
liver, multiple forms
alternate preparations, XXXVIII, 259, 285
assay, XXXVIII, 257

chromatography, XXXVIII, 258
properties, XXXVIII, 259
tissue extract preparation, XXXVIII, 257
with phospholipase C, XXXII, 160
in plasma membranes, XXXI, 88
purification, XXXVIII, 79
rod outer segments, XXXVIII, 154, 155
skeletal muscle
assay, XXXVIII, 240, 241
properties, XXXVIII, 243, 244
purification, XXXVIII, 241–243
in testicular metabolism, XXXIX, 253, 260
Phosphodiesterase I
assay, XXXI, 91, 93
in digestion of tRNA, LIX, 59, 60, 79, 81, 83, 87, 89, 94, 103, 122, 126, 127, 132, 140, 175, 187
gel electrophoresis, XXXII, 83, 90
in liver, XXXI, 97
as marker enzyme, XXXI, 93
in plasma membrane, XXXI, 91, 100, 101
storage, as solution, LIX, 60
Phosphodiester bond, alkaline hydrolysis, L, 255
Phosphoenolpyruvate, *see also* Phosphoenolpyruvic acid
aminoacylation of tRNA, LIX, 298
assay, LV, 221
of adenylosuccinate synthetase, LI, 213
of ADP, LVII, 61
cAMP, XXXVIII, 62
cyclase, XXXVIII, 119, 120, 143
of cyclic AMP phosphodiesterase, LVII, 96, 97
of FGAM synthetase, LI, 196, 197
of GTP hydrolysis, LIX, 361
of guanylate kinase, LI, 475, 476, 483
of guanylate synthetase, LI, 219
of nucleoside diphosphokinase, LI, 377
of polypeptide chain elongation, LIX, 358
of pyrimidine nucleoside monophosphatekinase, LI, 323
of uridine-cytidine kinase, LI, 300, 309, 315, 316
cell-free protein synthesis system, LIX, 300, 358, 448, 459, 850
content of kidney cortex, LV, 220
conversion of GDP to GTP, LVII, 91
guanylate cyclase, XXXVIII, 195
hydrogen-isotopically labeled, preparation, XLI, 110–115
in isolated renal tubules, XXXIX, 16, 18, 19
preparation of 3′-amino-3′-deoxy-ATP, LIX, 135
of *E. coli* crude extract, LIX, 297
E-[3-^{2}H]Phosphoenolpyruvate, preparation, XLI, 112–114
Z-[3-^{3}H]Phosphoenolpyruvate, preparation, XLI, 110, 111
Phosphoenolpyruvate carboxylase, gluconeogenic catalytic activity, XXXVII, 281
Phosphoenolpyruvate kinase, *see* Pyruvate kinase
Phosphoenolpyruvate phosphotransferase
in bacterial transport, XXXI, 698, 706, 707
E. coli, XXXI, 707–709
Phosphoenolpyruvate-uridine-5′-diphospho-*N*-acetyl-2-amino-2-deoxyglucose 3-*O*-pyruvyltransferase, XLVI, 543
Phosphoenolpyruvic acid, *see also* Phosphoenolpyruvate
interatomic distances, XLIX, 352, 353, 358
transport, LVI, 248, 249, 292
Phosphoenzyme, in sarcoplasmic reticulum, XXXI, 241, 243
Phosphofructokinase, XXXIV, 5, 532; XLVI, 90
from *Aerobacter aerogenes*, XLII, 63–66
assay, XLII, 63, 64
chromatography, XLII, 65
molecular weight, XLII, 66
properties, XLII, 66
purification, XLII, 64–66
from *Bacteroides symbiosis*, XLII, 63
binding site, XLVI, 76
cAMP binding, XXXVII, 286
carbethoxylation, XLVII, 437, 439

from *Clostridium pasteurianum*, XLII, 86–91
assay, XLII, 86, 87
chromatography, XLII, 88
molecular weight, XLII, 89
properties, XLII, 89, 91
purification, XLII, 87–90
conjugate, activity, XLIV, 271
control studies, XXXVII, 293, 294
from *Escherichia coli*, XLII, 91–98
assay, XLII, 91–93
chromatography, XLII, 94, 95
electrophoresis, XLII, 96
fluorescent derivatives, XLII, 97
molecular weight, XLII, 97
properties, XLII, 84, 96–98
purification, XLII, 93–96
from human erythrocytes, XLII, 110–115
assay, XLII, 110, 111
chromatography, XLII, 112, 113
electrophoresis, XLII, 114
inhibition, XLII, 114
molecular weight, XLII, 114
properties, XLII, 114, 115
purification, XLII, 111–114
iterative labeling, XXXVIII, 396–398
from mammalian tissues, XLII, 99–110
assay, XLII, 99–101, 105
properties of, XLII, 105
purification, XLII, 102–104, 106–110
Pasteur effect, LV, 291, 293
from porcine liver and/or kidney, XLII, 99–105
assay, XLII, 99–101
electrophoresis, XLII, 105
molecular weight, XLII, 105
preparation, XLII, 71
properties, XLII, 105
protein determination, XLII, 101
purification, XLII, 102–104, 107
from rabbit liver, XLII, 67–71
assay, XLII, 67
chromatography, XLII, 68, 69
electrophoresis, XLII, 70
properties, XLII, 70, 71
purification, XLII, 67–70
from rabbit muscle, XLII, 71–77
assay, XLII, 71, 72
electrophoresis, XLII, 70
properties, XLII, 77
purification from fresh muscle, XLII, 72–75
from frozen muscle, XLII, 75–77
site-specific reagent, XLVII, 483
from swine kidney, XLII, 375, 382–385
assay, XLII, 382
inhibition, XLII, 385
molecular weight, XLII, 385
properties, XLII, 385
purification, XLII, 383–385
from yeast, XLII, 78–85
amino acid composition, XLII, 85
assay, XLII, 78, 79
chromatography, XLII, 81
molecular weight, XLII, 85
properties, XLII, 81, 83–85
purification, XLII, 79–82
Phosphofructokinase A, from rabbit muscle, XLII, 71
Phosphofructokinase B, from pig liver, XLII, 71
6-Phospho-β-D-galactosidase
amino acids, XLII, 494
assay, XLII, 491
chromatography, XLII, 492, 493
distribution, XLII, 493
molecular weight, XLII, 494
properties, XLII, 494
purification, XLII, 491–493
Phosphoglucomutase
AMP assay, LV, 209
associated with cell fractions, XL, 86
as diagnostic enzyme, XXXI, 24
inorganic phosphate assay, LV, 212
6-Phosphogluconate dehydrogenase
associated with cell fractions, XL, 86
from *Candida utilis*, XLI, 237–240
activators, XLI, 240
amino acids, XLI, 240
assay, XLI, 237, 238
chromatography, XLI, 238
molecular weight, XLI, 240
properties, XLI, 239, 240
purification, XLI, 238, 239

from human erythrocytes, XLI, 220–226
amino acids, XLI, 226
assay, XLI, 220–222
electrophoresis, XLI, 225
properties, XLI, 225, 226
purification, XLI, 222–224
from *Neurospora crassa*, XLI, 227–231
assay, XLI, 227, 228
chromatography, XLI, 229
genetics, XLI, 231
inhibitors, XLI, 231
properties, XLI, 230, 231
protein determination, XLI, 227
purification, XLI, 228–230
reduction of $NADP^+$ analogs, XLIV, 874–876
from sheep liver, XLI, 214–220
assay, XLI, 215
chromatography, XLI, 216, 217
inhibitors, XLI, 220
properties, XLI, 218, 219
purification, XLI, 215–217
from *Streptococcus faecalis*, XLI, 232–237
activators and inhibitors, XLI, 236
assay, XLI, 232
chromatography, XLI, 233, 234
electrophoresis, XLI, 234
properties, XLI, 235, 236
purification, XLI, 232–236
6-Phosphogluconic acid, assay, LV, 221
Phosphoglucose isomerase, XLVI, 381, 382, *see also* Glucosephosphate isomerase
3-Phosphoglyceraldehyde, in isolated renal tubule, XXXIX, 16
D-3-Phosphoglycerate dehydrogenase
from hog spinal cord, XLI, 282–285
assay, XLI, 282
chromatography, XLI, 283–285
distribution, XLI, 284
properties, XLI, 284
purification, XLI, 282–284
from *Pisum sativum* seedlings, XLI, 278–281
assay, XLI, 278, 279
inhibitors and activators, XLI, 281
properties, XLI, 280, 281
purification, XLI, 279–281
from wheat germ, XLI, 285–289
assay, XLI, 285, 286
chromatography, XLI, 287, 288
properties, XLI, 289
purification, XLI, 286–288
sources, XLI, 289
Phosphoglycerate kinase
activity on ATP analogs, XLIV, 876
assay in nonaqueous tissue fractions, LVI, 204
GTP assay, LV, 208
inorganic phosphate assay, LV, 212
localization, LVI, 205
yeast
phthalylation, XLVII, 155
in tRNA sequence analysis, LIX, 61
3-Phosphoglycerate kinase, XXXIV, 246
from baker's yeast, XLII, 134–138
activators and inhibitors, XLII, 138
amino acids, XLII, 138
assay, XLII, 134
chromatography, XLII, 136
molecular weight, XLII, 138
properties, XLII, 138
purification, XLII, 134–137
cGMP assay, XXXVIII, 106, 107
from *Escherichia coli*, XLII, 139–144
assay, XLII, 139–141
chromatography, XLII, 142, 143
molecular weight, XLII, 144
properties, XLII, 143, 144
purification, XLII, 141–143, 444
gluconeogenic catalytic activity, XXXVII, 281
from human erythrocytes, XLII, 144–148
amino acids, XLII, 148
assay, XLII, 144, 145
chromatography, XLII, 146, 147
molecular weight, XLII, 148
properties, XLII, 147, 148
purification, XLII, 145–147
from skeletal muscle, XLII, 127–134
assay, XLII, 127–129
chromatography, XLII, 130
electrophoresis, XLII, 133
molecular weight, XLII, 131, 133

properties, XLII, 133, 134
protein measurement, XLII, 128
purification, XLII, 129–133
Phosphoglycerate mutase
activity determination, XLVII, 497
affinity labeling, XLVII, 423, 496, 497
amino acids, XLII, 445
assay, XLII, 436–438
from chicken breast muscle
molecular weight, XLII, 446
preparation and purification, XLII, 435, 440–442
properties, XLII, 446–449
from *Escherichia coli*
molecular weight, XLII, 446
properties, XLII, 446
purification, XLII, 438, 444
from liver, molecular weight, XLII, 445
molecular weight, XLII, 446
from pig heart, purification, XLII, 438, 445
from pig kidney
chromatography, XLII, 443
molecular weight, XLII, 446
preparation and purification, XLII, 435, 442–444
properties, XLII, 446–449
properties, XLII, 445–450
from rabbit muscle
molecular weight, XLII, 446, 447
properties, XLII, 446–449
purification, XLII, 438, 445
from rice germ
molecular weight, XLII, 435
purification and properties, XLII, 431, 434, 435
from sheep muscle
molecular weight, XLII, 446
purification and properties, XLII, 438, 446–449
site-specific reagent, XLVII, 483
from wheat germ, XLII, 429–435
assay, XLII, 429–431
chromatography, XLII, 433
molecular weight, XLII, 435
purification, XLII, 431–434
from yeast
molecular weight, XLII, 446, 447
preparation and purification, XLII, 435, 438–440
properties, XLII, 446–449
3-Phosphoglycerate phosphatase
assay, XLII, 405, 406
chromatography, XLII, 408
properties, XLII, 408, 409
purification, XLII, 406–408
sources, XLII, 409
from sugarcane leaves, purification, XLII, 406, 407
Phosphoglycerate phosphomutase
from human erythrocytes, XLII, 450–454
assay, XLII, 450–452
chromatography, XLII, 452
molecular weight, XLII, 453
properties, XLII, 453, 454
purification, XLII, 452, 453
Phosphoglyceric acid, XLVI, 388, 390
3-Phosphoglyceric acid
assay, LV, 221
phosphorylation potential, LV, 244, 245
Phosphoglyceride, *see also* Phospholipid, specific compounds
analysis of positional distribution of fatty acids, XXXV, 317–319
deacylation
with pancreatic lipase, XXXV, 317
with snake venom phospholipase, XXXV, 317, 323, 325
thin-layer chromatography analysis, XXXV, 319, 325
synthesis from diglyceride, XXXV, 321
with diglyceride kinase, XXXV, 321
with phenyldichlorophosphate, XXXV, 324, 325
3-*sn*-Phosphoglyceride, phospholipase A_2 effects, XXXII, 148
Phosphoglyceromutase
from *Escherichia coli*, XLII, 139–144
assay, XLII, 139–141
chromatography, XLII, 142, 143
molecular weight, XLII, 144
properties, XLII, 143, 144
purification, XLII, 141–143

Phosphoglycolate phosphatase, as marker enzyme, XXXI, 735
Phosphoglycolic acid, XLVI, 388
conformation studies, XLIX, 358
2-Phosphoglycolic acid, XLVI, 18
2-Phosphoglycolohydroxamate, XLVI, 23, 25
1-Phosphoimidazole, formation, LV, 540
Phosphoinositide, in glycolipids, XXXII, 350
Phospho-2-keto-3-deoxygluconate aldolase
from *Pseudomonas putida*, XLII, 258–264
amino acids, XLII, 262
assay, XLII, 259, 260
molecular weight, XLII, 262
properties, XLII, 262–264
purification, XLII, 260–262
in sodium pyruvate synthesis in TOH or D_2O, XLI, 106–110
Phosphokinase, assay reagent, XLIV, 630
Phospholactic acid, conformation studies, XLIX, 358
Phospholipase, *see also* specific type
bioassay, LVII, 193, 194
calcium activation, XXXI, 591
effect on erythrocyte membranes, XXXII, 131–140
inhibitor, LV, 6, 32
microorganismal mitochondria, LV, 142
in protein inactivation studies, XXXVII, 210
from snake venom, XXXV, 317, 323, 325
Phospholipase A
activity determination, 497n
amino acid transport, LV, 188
isolation from cobra venom, XXXII, 379
microsome damage, XXXI, 195
partial purification, 497n
purification of monoamine oxidase, LIII, 497, 498
in release of D-β-hydroxybutyrate apodehydrogenase, XXXII, 375, 379, 380
large-scale, XXXII, 381, 382
in testis hormone receptor studies, XXXVII, 186, 187
Phospholipase A_1
in bacterial membrane, XXXI, 649
for dihydroorotate dehydrogenase solubilization, LI, 66
from *E. coli* membrane, XXXII, 91
snake venom, ubiquinone reductase activity, LIII, 13
Phospholipase A_2
assay of succinate dehydrogenase, LIII, 468
diethylpyrocarbonate treatment, XLVII, 433
for dihydroorotate dehydrogenase solubilization, LI, 63
from *Naja naja*, XXXII, 137
effect on erythrocytes, XXXII, 137, 138
from porcine pancreas, XXXII, 131, 147–154
unit of activity, XXXII, 149
zymogen form, XXXII, 147
Phospholipase B
enzymatic activity, XXXI, 522
inhibition, XXXI, 524
in plant extracts, XXXI, 527
Phospholipase C
from *Bacillus cereus*, XXXII, 131, 139, 161
purification, XXXII, 154–161
from *Clostridium perfringens,* use in preparing isolated fat cells, XXXV, 556
from *Clostridium welchii*, XXXII, 139, 159
effect on liposomes, XXXII, 512
properties, XXXII, 159–161
purification of monoamine oxidase, LIII, 497, 498
substrates, XXXII, 159, 160
Phospholipase D, XXXV, 226–232
activators, XXXV, 230, 231
assay, XXXV, 227, 228
enzymatic activity, XXXI, 521
inhibitors, XXXV, 231
pH optimum and K_m, XXXV, 232
in phosphatidic preparation, XXXII, 142, 143
in phospholipid spin labeling, XXXII, 192, 193

purification, XXXV, 228–230
specificity, XXXV, 231
stability, XXXV, 232
transphosphatidyl activity, XXXV, 232

Phospholipid, *see also* Phosphoglyceride, specific compounds
activation of acetyl-CoA carboxylase, XXXV, 17
of dihydroorotate dehydrogenase, LI, 62
of phenylalanine hydroxylase, LIII, 284
analogue, synthesis, XXXV, 502
analysis, XXXI, 19
assay of complex I, LIII, 14
of NADH-cytochrome *c* reductase, LIII, 9
ATP-P_i exchange, LVI, 600
bacteriorhodopsin planar membranes, LV, 752
BCF_0-BCF_1, LV, 195
bilayer membranes uncouplers, LV, 466
in biomembranes, XXXII, 542
bovine factor V, XLV, 110, 116
bovine factor VIII, XLV, 85
bovine factor IX, XLV, 76
column chromatography, XXXV, 409–411, 417, 418, 420
in complex I, LIII, 12
in complex II, LIII, 24
in complex III, LIII, 89, 96, 97
in complex I–III, LIII, 8
complex V, LV, 314, 315
composition of brown adipose tissue mitochondria, LV, 78
in cytochrome oxidase, LIII, 64, 76
cytochrome oxidase vesicles, LV, 749
DCCD, LV, 498
in *E. coli* 0111A cell membranes, XXXI, 641, 642
of *E. coli* auxotrophs, XXXII, 861
enzymatic deacylation, XXXI, 522
from erythrocytes, XXXII, 134, 135
exchange among, XXXII, 140
factor XI, XLV, 67
filter membranes, LV, 604, 605
in ω-hydroxylase, LIII, 360
indicators, on thin-layer chromatography, XXXV, 403–405, 408, 412, 413, 416, 417
of insect muscle mitochondria, LV, 23
IR spectroscopy, XXXII, 252, 253
in isolated MEOS fraction, LII, 362, 363
isolation, XXXII, 346
Keilin-Hartree preparation, LV, 123
in lipid bilayers, saturation-transfer studies, XLIX, 498, 499
lipophilic Sephadex chromatography, XXXV, 388, 389
liposome, XXXII, 48, 504, 505
formation, XLIV, 218–227
preparation, LII, 208; LV, 541, 542
in liver fractions, XXXI, 29, 35
in lung lamellar bodies, XXXI, 424
as markers, XXXII, 15
in membranes, XXXII, 131–271
X–ray studies, XXXII, 219
methanolysis, XXXV, 417, 418, 422, 423
in microsomal fraction, LII, 88
in mitochondria, XXXI, 302
mixed brain, preparation, XLV, 44
monolayer, support, for ribonuclease, XLIV, 905, 906
M. phlei ETP, LV, 185
oligomycin-sensitive ATPase, LV, 302, 332, 358, 713
in pancreas subcellular fractions, XXXI, 47, 48, 52
partial specific volume, XLVIII, 22
penetrating anions, LV, 601
photochemically active derivatives, LVI, 660
in planar lipid bilayer membranes, XXXII, 515, 516
in plasma membranes, XXXI, 87, 101, 148, 149, 165–167, 694
preparation, LV, 608, 609, 717, 718
radioactive, in microsome studies, XXXI, 219, 220
reconstituted alanine transport vesicles, LVI, 433–435
for reconstitution of mixed-function oxidase system, LII, 204
reconstitution procedures, LV, 370, 371, 701, 702, 704–709

spin-labeling studies, XXXII, 167, 192
suspension, preparation, LV, 311, 316, 364, 714
synthesis
hormone induction of biorhythms, XXXVI, 480
methods, XXXV, 430–502
TSH effects, XXXVII, 262–268
thin-layer chromatography, XXXV, 397, 398, 400, 403–405, 408, 412, 416, 417
transhydrogenase, LV, 283
treatment of filters, LV, 589
Phospholipid exchange enzyme, XXXV, 262–269
specificity, XXXV, 268, 269
Phospholipid:protein ratio
in microsomal subfractions, LII, 83
transhydrogenase reconstitution, LV, 813, 814
N-Phosphonacetyl-L-aspartate, in aspartate transcarbamylase active site titration, LI, 126
Phosphonate ester, removal in phosphatidic acid analogue synthesis, XXXV, 508
Phosphonate lipid synthesis, *see* Phospholipid, analogue
Phosphonocephalin, intermediate in synthesis of phosphatidylethanolamine analogues, XXXV, 512
Phosphonomycin, XLIII, 254; XLVI, 543, *see also* Fosfomycin
Phosphopantetheine
in acyl carrier protein, XXXV, 114
in fatty acid synthase, XXXV, 42, 43, 45, 46, 59, 73, 87
substrate for palmityl-CoA synthase, XXXV, 122
turnover in acyl carrier protein, XXXV, 101
4′-Phosphopantetheine, XLIII, 577
Phosphopentomutase, regulation, LI, 438, 517
Phosphoprotein, intermediate, LI, 394
Phosphoprotein phosphatase, plasma membrane, XXXI, 90
Phosphor, radioactive, quantum yield determination, LVII, 598–600
Phosphoramidon
activity, XLV, 695
assay, by casein, XLV, 693
kinetics, XLV, 695
properties, XLV, 694
purification, XLV, 694
thermolysin, inhibition, XLV, 693
Phosphorescence, XLIX, 236–249; LIV, 2, 48
aqueous snowed matrix, XLIX, 238
beam chopper, XLIX, 239
chromophore proximities, XLIX, 245–247
decay, XLIX, 241, 242
measurement error sources, XLIX, 241
polarization, XLIX, 242
principle, XLIX, 237
protein structure, XLIX, 236–249
quantum yield of emission, XLIX, 240, 241
sample cell, XLIX, 239
sample cooling, XLIX, 239
solvent medium, XLIX, 237–239
spectrum, measurement, XLIX, 240
Phosphoribokinase, from *Pseudomonas saccharophila*, XLII, 120–123
assay, XLII, 120, 121
inhibition, XLII, 123
properties, XLII, 123
purification, XLII, 121–123
Phosphoribosylamine
assay of amidophosphoribosyltransferase, LI, 172
product of amidophosphoribosyltransferase, LI, 171
substrate of phosphoribosylglycinamide synthetase, LI, 179
Phosphoribosylglycinamide
assay of amidophosphoribosyltransferase, LI, 172
product of phosphoribosylglycinamide synthetase, LI, 179
Phosphoribosylglycinamide synthetase, LI, 179–185
from *Aerobacter aerogenes*, LI, 179–185

amidophosphoribosyltransferase
assay, LI, 172
assay, LI, 179–182
inhibitors, LI, 185
kinetic properties, LI, 185
molecular weight, LI, 185
pH optimum, LI, 185
from pigeon liver, LI, 180
properties, LI, 185
purification, LI, 182–184
Phosphoribosyl pyrophosphate
activator of carbamoyl-phosphate synthetase, LI, 33, 34
of CPSase-ATCase-dihydroorotase complex, LI, 119
assay of orotate phosphoribosyltransferase, LI, 144, 156
preparation, LI, 70
product of phosphoribosylpyrophosphate synthetase, LI, 3, 12
purification, LI, 552, 560
separation, LI, 4, 13
substrate of amidophosphoribosyltransferase, LI, 171
of CPSase-ATCase complex, LI, 118
of guanine phosphoribosyltransferase, LI, 549
of hypoxanthine-guanine phosphoribosyltransferase, LI, 13
of hypoxanthine phosphoribosyltransferase, LI, 543, 548, 549
of OPRTase-OMPdecase complex, LI, 135
of orotate phosphoribosyltransferase, LI, 69, 155
Phosphoribosyl pyrophosphate amidotransferase, XLVI, 420, 424
purification, LI, 183
Phosphoribosyl pyrophosphate-ATP-ligase, XLVI, 294
Phosphoribosylpyrophosphate synthetase, LI, 3–17
activators, LI, 11, 16
activity, LI, 3, 12
assay, LI, 3–6, 12–14
inhibitors, LI, 11, 17
kinetics, LI, 10, 11, 17
molecular weight, LI, 11, 16
properties, LI, 9–12, 16, 17
purification, LI, 6–9, 14–16
from rat liver, LI, 12–17
from *Salmonella typhimurium*, LI, 3–11
stability, LI, 9, 10, 16
states of aggregation, LI, 11
storage, LI, 9, 10, 16
substrate specificity, LI, 10, 16
Phosphoribulokinase, from *Chromatium*, XLII, 115–119
assay, XLII, 115, 116
chromatography, XLII, 118
electrophoresis, XLII, 119
inhibition, XLII, 119
molecular weight, XLII, 119
properties, XLII, 119
purification, XLII, 116–119
Phosphoric acid, XLVII, 492
as lock nuclei source, LIV, 189
preparation of 6-amino-1-hexanol phosphate, LI, 253
of eosin isothiocyanate, LIV, 50
of regenerated cellulose, LVII, 145
of ribosomal protein, LIX, 649
of tributylammonium phosphate, LI, 254
in reverse-phase chromatography, LII, 285
for washing phosphocellulose, LI, 500
Phosphoric acid buffer, for phosphocellulose chromatography, XLVII, 207
Phosphoric-carboxylic anhydride, *see* Carboxylic-phosphoric anhydride
Phosphorus
deficiency, XL, 321
lock nuclei, LIV, 189
in nuclear envelope, XXXI, 289
resonance frequency, LIV, 189
Phosphorus-31
lock nuclei, XLIX, 348, 349
nucleus
coherent scattering length, LIX, 672

scattering cross section, LIX, 672
Phosphorus-32
in nucleotidase detection, XXXII, 126, 127
in steroid autoradiography, XXXVI, 153
in studies of phospholipid synthesis, XXXVII, 265–268
Phosphorus oxychloride
phosphorylation of ligand-adenosine conjugates, LVII, 119
preparation of luciferin, LVII, 19, 20
Phosphorus:oxygen ratio
immobilized mitochondria, LVI, 556
pet9 mitochondria, LVI, 127
Phosphorus pentasulfide, XLVII, 291
Phosphorus pentoxide, XLIV, 131; XLVII, 492
Phosphorus tribromide, XLVI, 39
Phosphorylase, *see also* specific types
cAMP derivatives, XXXVIII, 399, 400
control studies, XXXVII, 293
in fat cells, XXXI, 68
gonadotropin regulation, XXXIX, 237
inactive, cAMP inhibitor assay, XXXVIII, 279–281
from rabbit muscle, hydroxylamine cleavage, XLVII, 140
from sarcoplasmic reticulum, XXXII, 292
from swine kidney, XLII, 375, 389–394
assay, XLII, 389
molecular weight, XLII, 393
properties, XLII, 393, 394
purification, XLII, 390–393
Phosphorylase *a*
AMP assay, LV, 209
gel electrophoresis, XXXII, 101
inorganic phosphate assay, LV, 212
molecular weight, LIII, 350, 351
in perfused muscle, XXXIX, 73
Phosphorylase *b*, XXXIV, 127, 128; XLVI, 241, 293, 316–318
activation, XLVI, 293
site-specific reagent, XLVII, 481, 484
Phosphorylase kinase
in perfused muscle, XXXIX, 73
protein kinase inhibitor, XXXVIII, 350, 351
Phosphorylation
in assay methods, XLI, 79–81
by ETP of *M. phlei*, LV, 183, 184
initiation factor, LX, 505–511, 521, 522
eIF-2 subunit, LX, 287–290, 467, 468, 470, 476–490
procedure, LX, 505, 525, 526
protein kinases effecting, LX, 507–511
sites, LX, 505–507, 510
of nucleosides, XLVI, 324, 325
oxidative
assay, LV, 153, 155
in intact cells, LV, 169–174
in phosphorylating vesicles, LV, 166, 167
ATPase inhibitor, LV, 401, 404, 406
calcium ion accumulation, LVI, 533, 534
everted vesicles, LVI, 235
hepatoma mitochondria, LV, 82, 83
measurement, LV, 164, 175
cultivation of bacteria, LV, 165
in intact cells, LV, 168–175
in phosphorylating vesicles, LV, 165–168
principle, LV, 165
in mitochondria, XXXII, 671, 672
enzyme complexes I–IV, LIII, 3, 4, 6
measurement, LVII, 38–50
uncouplers, LVII, 38
by *N. crassa* mitochondria, LV, 147
phenol effects, XXXI, 531
rate measurements, LV, 225–227
reconstitution, LV, 701, 703, 708
activity, LV, 703, 705
site I
assay, LV, 723–726
properties after reconstitution, LV, 725
reconstitution, LV, 716, 722
site II
assay, LV, 734
properties, LV, 734, 735
reconstitution, LV, 732–734

site III
assay, LV, 731, 732
properties, LV, 732, 733
reconstitution, LV, 728–731
steady-state measurements, LV, 227–229
sterol or fatty acid depletion, LVI, 563, 566, 571, 572
submitochondrial particles, LV, 105
sonicating medium, LV, 106
substrates, transport, LVI, 247–249
system
complexes I-V
properties exhibited, LVI, 578, 579
uncertainties regarding, LVI, 577–579
enzymic activities, LVI, 585–587
prosthetic groups and characterized protein components
complex I, LVI, 580, 581, 586–589
complex II, LVI, 580, 581, 586, 589, 590
complex III, LVI, 580, 582, 583, 586, 590–592
complex IV, LVI, 580, 583, 584, 586, 592–596
complex V, LVI, 580, 584–586, 596–602
uncouplers, LV, 462, 463
assay, LV, 463, 465
types, LV, 466–472
uncoupling, XLVI, 83
of primary hydroxyl groups, XLI, 119
by protoplast ghosts, LV, 182
reconstitution in detergent-treated ETP, LV, 188, 193
regulation of protein synthesis, LX, 496
self, of heme-regulated translational inhibitor, LX, 490–494
spin-labeling, XLIX, 443
40 S ribosomal subunits, LX, 525
Phosphorylation potential
calculation, LV, 236
concept and definition, LV, 235–237
cytosolic, estimation, LVI, 214
determination, LV, 237
importance, LV, 239–243
in intact cells, LV, 243–245
measurement with mitochondrial suspensions, LV, 237–239
values, LV, 236
Phosphorylation ratio, LI, 388
Phosphorylation system, cofactors, Pasteur effect, LV, 290–292
Phosphoryl azide, XLVI, 78
Phosphorylazide diester, XLVI, 87
Phosphorylcholine, XXXIV, 106
α-Phosphoryl nitrene, XLVI, 78
Phosphotransacetylase, *see* Phosphate acetyltransferase
Phosphotungstic acid, XXXIX, 143; XL, 21, 27
in negative stains, XXXII, 22–24, 29, 34
osmolality, XXXII, 30
Phosvitin kinase, cytoplasmic
assay, XXXVIII, 328
properties, XXXVIII, 329
purification, XXXVIII, 328, 329
Photagogikon
comparative chemical structures, LVII, 341
definition, LVII, 343
Photinus pyralis, *see also* Luciferase, firefly; Luciferin, firefly
bioluminescence, LVII, 1–122
chemical substitution of luciferin, LVII, 568
pH, LVII, 563
Photoactivation, XLVI, 337, 478
in affinity labeling, LX, 720, 721, 723–725, 727, 728, 733–740, 742, 745
Photoaffinity labeling, XLVI, 11, 69–114
absorption characteristics, XLVI, 92, 93
activation, XLVI, 478
apparatus, XLVI, 105–107
competition experiments, LVI, 668–672
controls, XLVI, 110, 111
covalent incorporation, LIX, 797, 798
dark reaction, XLVI, 107, 111
design of reagents, LVI, 658, 659

direct, XLVI, 335–339
factors affecting, LVI, 646
hydrophobic probe, XLVI, 191
hypervariable region, XLVI, 86
intermediates, lifetime, XLVI, 92
mechanisms, XLVI, 79
mitochondrial uncoupler binding site, LVI, 673–683
peaks, broad, assignment of maximum, LVI, 676
photochemical considerations, LVI, 663–666
pseudo, XLVI, 92, 656
reaction time, XLVI, 183
reagents, *see also* specific substances
concentration, XLVI, 107, 108
criteria, XLVI, 71, 78
stability, XLVI, 97, 98
synthesis, XLVI, 91–105
receptors, XLVI, 112–114
of ribosomes, LIX, 796–815
site localization, LIX, 798–808
Photoaffinity reagent
advantages, LVI, 643
reactions, LVI, 655–658
Photoattachment, XLVI, 89, 90
Photobacterium
catabolite repression, LVII, 164
characteristics, LVII, 153, 157
luciferase, LIII, 560
Photobacterium belozerskii, *see Beneckea harveyi*
Photobacterium fischeri
aldehyde mutant, LVII, 189
alternative names, LVII, 157
base ratio, LVII, 161
diagnostic taxonomic characters, LVII, 161
growth and harvesting, LVII, 142, 224
growth range, temperature, LVII, 157
isolation, LVII, 158
luciferase, LVII, 133, 147, 148, 173, 201
assay, LVII, 139, 140
specific activity, LVII, 150
luminescent system, LVII, 129
malate dehydrogenase, LVII, 181
natural habitats, LVII, 157, 160
oxidase, LII, 17
PAP levels, LVII, 256
storage temperature, LVII, 156, 157
Photobacterium fisheri, culture, LIV, 499, 500
Photobacterium leiognathi
alternative names, LVII, 157
base ratio, LVII, 161
diagnostic taxonomic characters, LVII, 161
growth, LVII, 143
growth range, temperature, LVII, 157, 158
isolation, LVII, 158
luciferase, LVII, 132, 149
luminescent system, LVII, 129, 149
natural habitats, LVII, 157, 160
storage temperature, LVII, 156, 157
Photobacterium mandapamensis, *see Photobacterium leiognathi*
Photobacterium phosphoreum
alternative names, LVII, 157
base ratio, LVII, 161
blue-fluorescent protein, LVII, 226–234
culture, LIV, 499, 500
diagnostic taxonomic characters, LVII, 161
fluorescent proteins, LVII, 230
growth, LVII, 142, 143, 224, 227, 228
growth range, temperature, LVII, 157, 158
isolation, LVII, 158
luciferase, LVII, 133, 149
assay, LVII, 139
specific activity, LVII, 150
luminescent system, LVII, 129
natural habitats, LVII, 157, 160
oxidase, LII, 17
storage temperature, LVII, 156, 157
Photocathode, photoelectric efficiency, LVII, 584
Photochemical reactions, of ketone, XLVI, 470–479
addition of C-H bonds, XLVI, 471
of nucleophiles, XLVI, 470
apparatus, XLVI, 472–475
effect of oxygen, XLVI, 478
oxidation-reduction, XLVI, 472
Photo-cross linking, *see* Cross linking

Photocurrent, *see* Autocorrelation function; Power spectrum
  detection modes, XLVIII, 423
  intensity, XLVIII, 422
  relation to scattered light, XLVIII, 423–425
Photodeacylation, XLVI, 89
Photodensitometry, XLIII, 185, 186
Photodetector, optical probes, LV, 577
Photodiode, LIV, 19, 21, 23
  biplanar vacuum, LIV, 40
  for nanosecond absorbance spectroscopy, LIV, 40
Photodiode-amplifier combination, in flash photolysis system, LIV, 98
Photodiode array, as multichannel emission detector, LIV, 461
Photodissociation, ligand binding, LIV, 510–517
Photodynamic damage, cyanine dyes, LV, 693
Photoelectric current, magnitude, LIV, 34
Photoetching, of filter masks, XLIX, 47, 48
Photogen
  comparative chemical structures, LVII, 341
  definition, LVII, 343
Photogenic organ, of euphausiid shrimp, LVII, 331
Photographic film, *see also* specific type
  for display of picosecond spectrometric data, disadvantages, LIV, 14
Photography
  choice of film, LVIII, 338
  chromosome study, LVIII, 336–339
  electron microscopy, LVI, 721
  film development, LVIII, 338
  printing, LVIII, 339
Photoimmobilization, *see* Immobilization, photochemical
Photoinactivation, kinetics, XLVI, 467
Photoinhibition, by ATP analog, XLVI, 277
Photoirradiation
  assay of atractyloside-binding protein, LVI, 418
  preparation of apoflavoprotein, LIII, 435, 436
Photoluminescent indicator, calcium, LVI, 303
Photolysis, XLVI, 105–109, 526, 653–655
  duration, XLVI, 108, 109
Photometer, LVII, 38, 217, 425, 529–540
  aequorin assay, LVII, 279, 293–297, 327
  calibration, LVII, 300, 386, 543–545, 560–600
  cAMP assay, XXXVIII, 62
  commercial instruments
    comparative ATP concentration curves, LVII, 558
    comparative light standard concentration curves, LVII, 557
    evaluation methods, LVII, 556–558
    general characteristics, LVII, 551–555
    modification, for light-free injections, LVII, 556
  design principles, LVII, 550, 551
  with dual phototubes, LVII, 259, 260
  luminescence, LVI, 532
  multifunction, LVII, 535–540
    autoranging instrument amplifier, LVII, 535, 537, 538
    circuit diagram, LVII, 532
    current to voltage transducer, LVII, 531, 535
    digital integrator, LVII, 536, 538, 539
    high-voltage shutdown circuit, LVII, 537, 540
    peak-detector circuit, LVII, 537, 539
  simple, LVII, 532–535
    block diagram, LVII, 530
    circuit diagram, LVII, 532
    criteria for component selections, LVII, 534
    fabrication, LVII, 534
    housing, LVII, 532, 533
    parts list, LVII, 534
    postconstruction adjustments, LVII, 534, 535
  spectral efficiency, determination, LVII, 578–583
Photometric Fast Analyzer, principles and use, XXXI, 794–796

Photometric method, for dilution assay, XLIII, 63–65
dose-response line, XLIII, 64
equations, XLIII, 64
interference, XLIII, 63, 64
Photomultiplier
fatigue effects, LIV, 40
in flash photolysis apparatus, LIV, 54
for nanosecond absorbance spectroscopy, LIV, 40
Photomultiplier-amplifier combination, in flash photolysis system, LIV, 97, 98
Photomultiplier tube, XLVIII, 462–466
bialkali, in chemiluminescence measurements, LVII, 474
in commercial instruments, LVII, 551
current to voltage conversion, LVII, 531, 532
gain, XLVIII, 463, 464
for light detection
advantages, LVII, 530
selection criteria, LVII, 530, 531
noise, XLVIII, 464
photoproteins, for use with, LVII, 318, 319
sensitivity, LVII, 576, 577
shield, LVII, 534
Photon
dimensions, LVII, 572
standard light unit equivalent, LVII, 386
Photon absorption, time range, LIV, 1, 2
Photon irradiance, LVII, 572
noise equivalent at given wavelength, LVII, 577
Photooxidation, XL, 307
affinity-sensitized, XLVI, 12, 13
coupling, XLVI, 563, 564, 566
denaturation, XL, 306
dye-sensitized, XLVI, 88, 562–567
microscopy, XL, 307
modification of methionine, LIII, 139
of proteins, XLVI, 561–567
Photophosphorylation, ATPase inhibitor, LV, 404
Photophosphorylation assay, LVII, 50–56
instrumentation, LVII, 55
principle, LVII, 52–54
procedure, LVII, 55, 56
Photopotential
assay from bacteriorhodopsin-containing liposomes, LV, 608–611
rate of generation, LV, 592
Photoprotein bioluminescent systems, LVII, 272, 276, 277
Photorespiration, ribulose-1,5-diphosphate carboxylase role, XLII, 481
Photoselection, XLIX, 182–184, 187, 188, 200, 204–206
definition, XLVIII, 350
principle, LIV, 47, 48
Photosynthesis
aldolase, XLII, 234
in algae, measurements, XXXII, 867, 868
bisulfite effects, XXXI, 539
fructose-1,6-diphosphatase, XLII, 398
3-phosphoglycerate phosphatase, XLII, 409
picosecond spectroscopy, LIV, 23–32
ribulose-1,5-diphosphate carboxylase role, XLII, 481
time range, LIV, 2
Photosystem I
fluorescence lifetimes, LIV, 25
proteoliposomes, LV, 760, 761, 768
Phototransistor array, as multichannel emission detector, LIV, 461
Phototube calibration
absolute, LVII, 583–585
BEA technique, LVII, 585–587
step-by-step method, LVII, 583–585, 590, 591
using Bowen quantum counter, LVII, 591–593
pH-recording unit, assembly, LV, 619
Phrenosine, *see also* Glycosphingolipid; Glycosylceramide
thin-layer chromatography, XXXV, 416
o-Phthalaldehyde, XLVII, 8, 13
in fluorometry of indoles, XXXIX, 383, 384
peptide detection, XLVII, 236
preparation, XLVII, 17, 241, 242
Phthalaldehydric acid, reaction intermediate, in luminol chemiluminescence, LVII, 421

Phthalate ion, chemiluminescence, LVII, 412, 416, 417
Phthalic anhydride
  reaction with proteins, XLVII, 150–153
  in uric acid assay, XLIV, 632
Phthalic hydrazide, chemiluminescence, LVII, 416
2-Phthalimidoethylphosphonic acid, intermediate in synthesis of phosphatidylethanolamine analogues, XXXV, 510, 511
Phthalimidoethylphosphonyl monochloride, intermediate in synthesis of sphingomyelin analogues, XXXV, 528
2-Phthalimidoethylphosphoryl dichloride, intermediate in phosphatidylethanolamine synthesis, XXXV, 459
6-Phthalimidopenicillanic acid isocyanate, XLVI, 532, 534–536
Phthalocyanin, MCD studies, LIV, 296
Phthalylation, of amino groups, XLVII, 149–155
  product stability, XLVII, 153, 154
  unblocking, XLVII, 155
N-Phthalyl phosphonate lipid, intermediate in synthesis of phosphatidylethanolamine analogues, XXXV, 510
Physarum polycephalum
  disruption, LV, 137
  ribonuclease, in tRNA sequence analysis, LIX, 107
Phytanic acid, substrate of liver α-hydroxylase, LII, 317
Phytochrome, conformational transitions, XLIV, 527
Phytohemagglutinin, XXXIV, 331
  affinity electrophoresis, XXXIV, 178–181
  O-glycosyl polyacrylamide, XXXIV, 361–367, *see also* *O*-Glycosyl polyacrylamide
  lymphocyte reaction, XXXII, 636
  in lymphocytoid line studies, XXXII, 812
  as mitogen, LVIII, 487, 492
Phytomitogen, XXXIV, 7
Phytosphingosine, silica gel chromatography, XXXV, 529–533
Phywe flow cytometer, LVIII, 236
Picoline, XLVI, 291
2-Picolylchloride 1-oxide, reaction conditions, XLVII, 410
Picosecond clock, LIV, 13
Picosecond continuum, LIV, 12–14
  generation, LIV, 12
  parameters, LIV, 13
  photograph, LIV, 12
  variations, LIV, 13
Picosecond pulse
  characteristics, LIV, 4–11
  generation, LIV, 4–8
  photograph, LIV, 9, 11
  shape, LIV, 8–11
  variation, LIV, 8
  width, LIV, 8
Picosecond spectrometer
  double-beam system, advantages, LIV, 17
  output, LIV, 14–16
  partial reflector, LIV, 5
  schematic representation, LIV, 5
Picosecond spectroscopy
  characteristics, LIV, 4
  of hemoproteins, LIV, 3–32
Picric acid
  complex V, LV, 315
  as internal filter, LVI, 672
  ionophores, LV, 439, 444
Picric acid method, XLVII, 591, 599, 600
Picromycin, XLIII, 131
Picrylsulfonic acid, *see also* Trinitrobenzenesulfonic acid
  alkylamine glass beads, XXXVIII, 181–183
Piericidin A
  binding to NADH dehydrogenase, LIII, 415
  complex I, LVI, 586
  inhibitor of complex I, LIII, 13
    of complex I–III, LIII, 9
    of reconstituted electron-transport system, LIII, 50, 51
  isolation, LV, 457, 458
  NADH oxidation, LV, 456–458
Piezoelectric transducer, XLVIII, 313
Piezometer, mercury, XLIX, 17

Pig, *see also* Hog; Swine
adipose tissue, phosphofructokinase, purification, XLII, 107
brain, phosphofructokinase, purification, XLII, 107
brain mitochondria, L-glycerol-3-phosphate dehydrogenase, assay, purification, and properties, XLI, 254–259
costal cartilage, in somatomedin bioassay, XXXVII, 101–104
granulosa cell isolation, XXXIX, 199
heart
phosphofructokinase, purification, XLII, 107, 110
phosphoglycerate mutase, purification, XLII, 438, 445
kidney
glucosamine-6-phosphate deaminase, XLI, 497
phosphofructokinase, assay, purification, and properties, XLII, 99–105
phosphoglycerate mutase, preparation, purification, and properties, XLII, 435, 442–444, 446–449
liver
acyl-CoA dehydrogenases, LIII, 506–518
electron-transferring flavoprotein, LIII, 506–518
galactokinase, assay, purification, and properties, XLII, 43–47
glycolic acid oxidase, assay, purification, and properties, XLI, 337–343
mitochondria preparation, LIII, 509, 512
phosphofructokinase, assay, purification, and properties, XLII, 99–105
preparation, XLII, 71
as source of initiation factor eIF-2, LX, 240–242
muscle
phosphofructokinase, purification, XLII, 107
phosphoglycerate kinase, purification, XLII, 129, 131, 133
ovarian follicle isolation, XXXIX, 188–194
serum, for granulosa cell culture, XXXIX, 207, 208
small intestine, phosphofructokinase, purification, XLII, 107
spleen, phosphofructokinase, purification, XLII, 110
Pigeon liver, acetone powder extract, AIR synthetase, LI, 194
Pigment, membrane, as optical probes, LV, 573
Pigmentary hormone
bioassay, XXXVII, 121–130
from plasma, extraction, XXXVII, 122–124
from tissue, extraction, XXXVII, 124, 125
Pigmentation, genetic marker, LVIII, 565
Pimaricin, XLIII, 135
Pimelic acid, XLVI, 506
Pimelic acid dihydrazide, preparation, LIX, 179
Pineal gland
avaian, cell dissociation, LVIII, 130
biological activity, methods for study, XXXIX, 376–397
catecholamine and cyclic nucleotide metabolism, XXXIX, 395–397
enzymes, assay, XXXIX, 387–392
as neuroendocrine transducer, XXXIX, 379
nonindolic compounds, XXXIX, 397
organ culture
indole metabolism, XXXIX, 392–394
techniques, XXXIX, 398–403
rat, cell dissociation, LVIII, 129, 130
removal, effects, XXXIX, 382
Pineapple
peroxidase, LII, 514
stem, acidic cysteine protease inhibitors, *see* Bromelain inhibitor, pineapple stem
Pinner reaction, XLVI, 117, 123, 124, 127
Pinocytosis, by protoplasts, XXXI, 578
Piomycin, XLIII, 153
Pipecolate oxidase, LII, 18

Piperazine-N,N'-bis(2-ethanesulfonic acid), XXXII, 110, 111
assay of aequorin, LVII, 298
isolation buffer, XL, 77, 87
in protein binding buffer, LIX, 322, 328
Piperidine
electron spin resonance spectra, XLIX, 383
solvent, for spectral analysis of porphyrins, LII, 445
Piperidinol nitroxide, *see* 2,2,6,6-Tetramethylpiperidine-1-oxyl-4-ol
Piperidinyl N-oxyl group, structure, XLIX, 434
Piperidone nitroxide, *see* 2,2,6,6-Tetramethylpiperidine-1-oxyl-4-one
Piperonyl butoxide, metabolite binding to reduced cytochrome, LII, 274, 275
PIPES, *see* Piperazine-N,N'-bis(2-ethanesulfonic acid)
Pipetman, LIX, 194
Pipette
automatic, jet mixing in cuvettes, LII, 402
as column, XXXIV, 500
transfer, calibration procedure, LIV, 128
Pipetting error, in radioligand assay, XXXVII, 20, 21
Pipetting technique, LVIII, 38, 39
Piricularia oryzae, oxidase, LII, 10
Pisum sativum
lectin, XXXIV, 334
seedlings, 3-phosphoglycerate dehydrogenase, assay, purification, and properties, XLI, 278–281
PITC, *see* Phenyl isothiocyanate
Pitch, magnetic field standard, XLIX, 523
Pituitary
estrogen receptors, XXXVI, 168, 175
nuclear estrogen receptors, XXXVI, 286
steroid hormone determination, XXXVI, 150, 151
Pituitary cell
ACTH-producing, LVIII, 531
anterior
calf, microcarrier suspension culture, LVIII, 207, 208
cloning, XXXIX, 128–132
suspension culture, LVIII, 203
corticotropin-producing, LVIII, 531
β-endorphin-producing, LVIII, 531
enkephalin-producing, LVIII, 531
growth hormone formation, LVIII, 528
hormone-producing, LVIII, 528
mouse, characteristics, LVIII, 531
prolactin formation, LVIII, 528
rat, characteristics, LVIII, 529
for secretion assay for hypophysiotropic substances, XXXVII, 84–93
in somatic cell hybrids, LVIII, 534
source, LVIII, 528
Pituitary gland
anterior
perfusion, XXXIX, 179–182
secretory granule isolation, XXXI, 410–419
fractionation of whole tissue, XXXVII, 363–365
ovine and bovine, XXXVII, 363, 364
rat and rabbit, XXXVII, 364, 365
human, fractionation, XXXVII, 380, 381, 389, 390
neurosecretory gland isolation, XXXI, 403–410
removal for assay procedures, XXXVII, 83
Pituitary hormone
effect on corpus luteum, XXXIX, 241
human, purification, XXXVII, 380–389
immunochemical assay, XXXVII, 144
ion exchange separation, XXXVII, 382, 383
purification, XXXVII, 360–380
in secretion assay for hypophysiotropic substances, XXXVII, 82–93
standard, purified, XXXIX, 210
Pituitary Hormone Radioimmunoassay Kit, XXXVII, 83
Pituitary-hypothalamic axis, methods for study, XXXIX, 166–182

Pituitary receptor binding assay, of hypothalamic releasing factors, XXXVII, 213–219
Pituitary stalk
  blood collection, XXXIX, 175–179
  cannulation, XXXIX, 176
Pituitary tumor cell, LVIII, 377, 527–535
  hormone-producing, LVIII, 528
Pit viper venom, coagulant enzyme, *see* Batroxobin
Placenta
  circulation, XXXIX, 245
  human, adenylosuccinate synthetase, LI, 212
  lactogen, XXXIV, 731
  membrane, ATPase, XXXII, 306
  oxidase, LII, 13
  ruminant, aldose reductase, assay, purification, and properties, XLI, 165–170
  in vitro perfusion, XXXIX, 244–252
Plakalbumin, osmotic pressure, XLVIII, 73
Planck's equation, transition volume, XLIX, 17–19
Plant, *see also* specific type
  auxin receptor, XLVI, 84
  cell, *see* Plant cell
  chlorophyll, LIV, 23
  cold-sensitive, critical growth temperature, XXXII, 176
  ferredoxin, physical properties, LIII, 262
  hybrid fusion, LVIII, 364–366
  membrane-component isolation, XXXII, 391–422
  mitochondria, LV, 24, 25
  mutarotase from higher, assay, purification, and properties, XLI, 484–487
  oxidases, LII, 9–21
  proembryonic stock culture, XXXII, 732
  proteinase inhibitors, XLV, 695
  from protoplasts, LVIII, 367, *see also* Protoplast
  subcellular fraction isolation, XXXI, 487
  sucrose phosphatase, assay, purification, and properties, XLII, 341–347
  vacuole, *see* Tonoplast
Plant cell, LVIII, 478–486
  bacterial transformation, XXXI, 553–558
    interruption technique, XXXI, 556
  cloning, LVIII, 482–484
  differentiation, LVIII, 484
  enlargement, XXXI, 494
  explants, LVIII, 479
  free, isolation and culture, XXXII, 723–732
  plating, LVIII, 482–484
  suspension culture, LVIII, 481, 482
Plant tissue, *see also* specific types
  fractionation, XXXI, 501–519
    antioxidants, XXXI, 538–540
    artifactual organelles, XXXI, 544
    breaking techniques, XXXI, 502, 503
    density gradient centrifugation, XXXI, 505–519
    gradients
      equivolumetric, XXXI, 512
      hyperbolic, XXXI, 514
      inverse sample, XXXI, 513
      isokinetic, XXXI, 510
      isometric, XXXI, 511
      isosmotic, XXXI, 508, 509
    green tissue, XXXI, 544–553
      extraction methods, XXXI, 547, 548
    inert atmospheres, XXXI, 540
    isolation and purification media, XXXI, 545–547
    isolation techniques, XXXI, 532–542
    lipid-breakdown detection, XXXI, 525, 526
    lipid oxidation problem, XXXI, 520–528
    pH control, XXXI, 532–534
    phenolics and quinones, XXXI, 528–544
    plant-material choice, XXXI, 525, 531, 532
    polymer use, XXXI, 534–538
    protein determination, XXXI, 542–544
    rapid procedures, XXXI, 540–542
    storage problems, XXXI, 532

hydrophobic interactions, XXXI, 530
3-phosphoglycerate phosphatase, XLII, 409
Plasdone polymer, in plant-tissue fractionation, XXXI, 535
Plaskon 2300 CTFE powder, LIX, 221
Plasma
assay
for factor XIII activity, XLV, 178
for vitamin D metabolites, LII, 394
charcoal treatment, XXXVI, 91–97
clotting, XLV, 31
corticosteroid-binding globulin isolation, XXXVI, 104–109
cyclic nucleotide assay, XXXVIII, 105
defibrinated, and prothrombin assay, XLV, 125
drug isolation, LII, 333, 334
insulin-free, XXXIV, 747–750
pigmentary hormone extraction, XXXVII, 122–124
platelet-rich, for thyroid gland perfusion, XXXIX, 363
prostaglandin E extraction, XXXII, 117
thromboplastin antecedent, *see* Factor XI, bovine
Plasma-binding protein
in steroid hormone assays, XXXVI, 34–48
assay evaluation, XXXVI, 43–46
methodology, XXXVI, 41–43
sample preparation, XXXVI, 38, 39
separation of unbound material, XXXVI, 39, 40
solutions, XXXVI, 37, 38
Plasma cell tumor, LVIII, 349
Plasma-coated charcoal, *see* Charcoal, plasma-coated
Plasmacytoma, LVIII, 376
hybrid, LVIII, 350
mouse, cell fusion, LVIII, 349
Plasmalemma
in angiosperm cells, XXXII, 723
of pancreas, XXXI, 42
isolation, XXXI, 46
properties, XXXI, 578
Plasmalogen, *see also* specific compounds
analogue synthesis, XXXV, 484
methanolysis, XXXV, 423
methods of synthesis, XXXV, 477–484
via *sn*-glycerol-*cis*-1′-alkenyl ether, XXXV, 479–481
via *sn*-glycerol-2,3-cyclocarbonate, XXXV, 478, 479
thin-layer chromatography, XXXV, 423
Plasma membrane
from amoeba, isolation, XXXI, 686–698
characterization, XXXI, 693, 694
electron microscopy, XXXI, 692
antigens on, studies using hybrid antibodies, XXXII, 66–70
from ascites cells, XXXI, 144
assay of electrical potentials across, LV, 612
buoyant densities, XXXI, 84, 85
characterization, XXXI, 23–25, 31, 35
chemical composition, XXXI, 86, 87, 148
2′,3′-cyclic nucleotide-3′-phosphohydrolase, XXXII, 128
diagnostic enzymes, XXXI, 19, 20, 185, 186
differential sedimentation, XXXI, 727, 730–732
electron microscopy, XXXI, 26, 85, 86, 101, 140–142, 146; XXXII, 31
electrophoretic purification, XXXI, 755, 756
enzymes, XXXI, 88–102; XXXII, 86, 87, 91
of erythrocytes, XXXI, 168–172
from fat cells, XXXI, 63, 64
characterization, XXXI, 65, 67
in glucose transport studies, XXXI, 68, 69
from fibroblasts, XXXI, 144, 161–168
fine structures, XXXI, 85, 86
fluorescein mercuric acetate method, XXXI, 157–159
fragments
chemical analysis, XXXI, 101
isolation, XXXI, 90–102
functional status, test, LII, 64
globular knobs, XXXI, 85, 86, 90

glycosphingolipids, XXXII, 345, 347
gradient separation, XXXI, 96, 98, 99
from HeLa cells, XXXI, 144
from hepatomas, isolation, XXXI, 75–90
from intestinal brush borders, XXXI, 123–134, 144
IR spectroscopy, XXXII, 252
isolation, LV, 36
from kidney, XXXI, 144
large-scale isolation, XXXII, 119–121
from L cells, XXXI, 156–162
as lipid bilayer, XXXII, 513
lipids, XXXI, 35
from liver
isolation, XXXI, 75–102, 144, 180–191
preservation and purification, XXXI, 6, 7, 9–15
local specialization, XXXI, 85, 86
marker enzymes, XXXII, 347, 393
morphology, XXXI, 85, 86
from myeloma cells, XXXI, 144
negative staining, XXXII, 31
NMR studies, XXXII, 202
nucleic acids, XXXI, 86, 87
5′-nucleotidase isolation, XXXII, 374
preparation, LVIII, 224
properties, XXXI, 83–90
prostaglandin binding, XXXII, 110
proteins of, gel electrophoresis, XXXII, 92
purity, XXXI, 84, 85
from *S. faecalis* preparation, XXXII, 429, 430
of synaptosomes, isolation of, XXXI, 445–452
from thyroid, XXXI, 144–149
Tris method, XXXI, 159, 160
zinc method, XXXI, 160, 161
Plasmanate, XXXIX, 371
Plasma protein
effect on corpus luteum, XXXIX, 241
hormone binding, XXXVI, 3
support, for multistep enzyme system, XLIV, 917–918
Plasmid
resistance, XLIII, 41–55
bacterial species with, XLIII, 43
compatibility properties, XLIII, 52–55
tests for, XLIII, 54, 55
conjugation, XLIII, 43–45
conjugative, XLIII, 42
curing agents, XLIII, 49
determinants, XLIII, 43
DNA, preparation, XLIII, 48
fi character determination, XLIII, 54
genetic marker, XLIII, 44
genetic transfer, XLIII, 43–49
resistance characters associated with, XLIII, 42
resistance transfer factor, XLIII, 43
structure, XLIII, 43
transduction, XLIII, 45–48
with P1, XLIII, 46
with P22, XLIII, 47
transformation, XLIII, 48–50
frequency, XLIII, 49
kinetics of resistance phenotypes, XLIII, 50
transfer, LVI, 115, 116
unc mutants, LVI, 115
Plasmid NTPI, in promoter site mapping, LIX, 846, 847
Plasmin, XXXIV, 5; XLVI, 206
activation
from plasminogen, XLV, 272
of rabbit plasminogen, XLV, 282
active-center serine residue, XLV, 270
amino-acid composition, XLV, 267, 269
amino terminal sequences, XLV, 271
antithrombin-heparin cofactor neutralization, XLV, 668
assay, XLV, 257
chains, amino acid composition, XLV, 267, 269
histidine residues, XLV, 270
human
hydrodynamic properties, XLV, 265, 266
isoelectric focusing, XLV, 264
preparation, XLV, 262
inhibition, XLV, 263, 826
by leupeptin, XLV, 679

nomenclature forms, XLV, 258
preparation from plasminogen, XXXIV, 429, 430
properties, XLV, 256
rabbit
amino acid composition, XLV, 285
sequence, XLV, 283
inhibitors, XLV, 285
kinetics properties, XLV, 285
molecular weight, XLV, 284
preparation, XLV, 278
properties, XLV, 284, 285
purity, XLV, 284
storage, XLV, 284
radiolabeling, XLV, 263
specificity, XLV, 257
Plasminogen, XXXIV, 46, 424–432
activation, to plasmin, XLV, 272
activator, assay, XLV, 21
activator complex, action, XLV, 254, *see also* Urokinase
amino acid composition, XLV, 267, 268
amino terminal sequences, XLV, 271
assay, XLV, 257, 273
bovine, XLV, 259
butesin-agarose, XXXIV, 428–430
cat, XLV, 259
dog, XLV, 259
equimolar complex, preparation, XLV, 269
human
activation by streptokinase, XLV, 254
hydrodynamic properties, XLV, 265, 266
isoelectric focusing, XLV, 264
preparation, XLV, 259
from plasma fraction III, XLV, 260, 261
with urokinase, human, XLV, 263
lysine-agarose, XXXIV, 425–427
multiple forms, XXXIV, 430, 431
nomenclature, XLV, 258
plasmin, XXXIV, 429, 430
rabbit
activation to plasmin, mechanism, XLV, 282
amino acid composition, XLV, 282
sequence, XLV, 283
assay
colorimetric, XLV, 273, 274
potentiometric, XLV, 275
units, XLV, 276
ligand binding, XLV, 283
properties, XLV, 280
chemical, XLV, 281
physical, XLV, 280
purification, XLV, 276
purity, XLV, 280
separation, XLV, 259
specificity, XLV, 257, 276
storage, XLV, 280
Plasminogen-streptokinase complex, equimolar, preparation, XLV, 262
Plasmin-streptokinase complex, preparation, XLV, 269
Plasmodesmata
in angiosperm cells, XXXII, 723, 724
in plant cells, XXXI, 489
Plastic, *see also* specific type
chlorinated, interference, in Mössbauer spectroscopy, LIV, 375
derivatized, XXXIV, 221
enzyme degradation, LII, 146
hydrogen fluoride stable, XLVII, 605
methods for derivatization, XXXIV, 223, 224
oxygen permeability, LIV, 376
Plastic binder clip, LIX, 62, 63
Plastic tubing
in freeze-quench technique, LIV, 88
as microcolumn, LVII, 281
Plastic ware
fibroblast culture, LVIII, 449
iron transport studies, LVI, 390
liver cells, LVIII, 541
macrophage culture, LVIII, 499, 500
multiwell tray for cloning, LVIII, 321, 332
suppliers, LVIII, 16
Plastid
in algae, XXXII, 865
isolation, XXXI, 506
in plant cells, XXXI, 494, 495
Plastocyanin
from *Chlorella*, XLVII, 67
K-edge Cu transitions, LIV, 339, 340

Raman frequencies and assignments, XLIX, 144
X-ray absorption spectroscopy, LIV, 332
resonance Raman spectrum, XLIX, 142
Plastoquinone-3, instability, LIII, 594
Platelet
contamination, LVIII, 492, 493
disruption, XXXI, 153, 154
growth factor, LVIII, 79
guanylate cyclase, XXXVIII, 122
activation with time, XXXVIII, 201, 202
assay, XXXVIII, 199, 200
distribution among blood cells, XXXVIII, 200
divalent cations, XXXVIII, 201
homogenate preparation, XXXVIII, 200
subcellular distribution, XXXVIII, 201
isolation, XXXI, 149–155
optical probes, LV, 573
plasma-membrane isolation, XXXI, 149–155
properties, XXXI, 150
release action, XXXI, 149
removal from lymphocytes, LVIII, 490
Platelet factor V, human, role, XLV, 122
Plating
adrenal tumors, LVIII, 573
liver cells, LVIII, 538, 539
plant cell, LVIII, 482–484
Platinum
atomic emission studies, LIV, 455
as electrode, LVI, 454, 457
Platinum-195, lock nuclei, XLIX, 349; LIV, 189
Platinum agent, for oxygen removal, LIV, 117
Platinum-carbon electrode, for freeze-fracture, XXXII, 42–44
Platinum wire, in micropipette design, LVII, 315
Pleated sheet, vacuum ultraviolet circular dichroism, XLIX, 218
Plectonema, aldolase, XLII, 233
Plethoric mouse assay, of erythropoietin, XXXVII, 111, 112
Pleuropneumonia-like organism, in cell cultures, XXXII, 803
Plexiglas
steroid adsorption, XXXVI, 93, 94
uncouplers, LVI, 554
PLPase A, *see* Phospholipase A
Plug-flow loading, XLIV, 160, 161
Pluronic F–68 detergent, enzyme-activated electrodes, LVI, 473
Pluronic F–127 surfactant, dye solutions, LV, 576
Pluronic polyol F–68, as hepatic perfusate, XXXIX, 36
20-PM10, hollow-fiber device, XLIV, 293
hydraulic permeability, XLIV, 298
PMB, *see* *p*-Mercuribenzoate
PMR, *see* Proton magnetic resonance
PMR spectroscopy, *see* Nuclear magnetic resonance, hydrogen-1
PMS, *see* *N*-Methyl phenazonium methosulfate
PMSF, *see* Phenylmethylsulfonyl fluoride
PMSG, *see* Pregnant mare serum gonadotropin
Pneumonia, phagocyte chemiluminescence, LVII, 489
PNP, *see* *p*-Nitrophenol
pO_2, polarography, LVI, 450, 451
Pockels cell, LIV, 5, 20, 23, 288
disadvantage, LIV, 44
Poikilotherm vertebrate cell, *see* Vertebrate cell, poikilotherm
Pokeweed
chloroplast isolation, XXXI, 502
mitogen, lymphocyte reaction, XXXII, 636
Polarimetric assay, XLIV, 350
Polarizability
definition, XLIX, 72, 73
incident light frequency, XLIX, 81
for Raman band intensities, XLIX, 77, 78
resonance scattering, XLIX, 74, 75
tensor, depolarization, XLIX, 75–77
theory, XLIX, 77, 78
Polarization, anomalous, LIV, 238
Polarization rotator, LIV, 23
fixed-position, LIV, 5
translatable, LIV, 5

Polarizer, *see also* specific type
in TPF apparatus, LIV, 9
vertical, LIV, 5
Polarographic anode, enzyme-activated
alcohol, LVI, 467, 468
care, LVI, 472, 473
cholesterol, LVI, 468–472
galactose, LVI, 465–467
glucose, LVI, 461–465
superoxide anion, LVI, 460, 461
urate, LVI, 468
Polarographic electrode
calibration, LVI, 554, 555
immobilized mitochondria, LVI, 552
Polarography, *see also* specific type
catalase measurement, LVI, 459
determination of 3-keto sugars, XLI, 22–27, 158
general rationale, LVI, 448–450
limitations of method, LVI, 340, 341
medical and other applications, LVI, 450–452
method, LIV, 399
nonoxidase substrates, LVI, 476–478
other oxidases, LVI, 473–476
reference electrode, LVI, 456
in studies of membrane phase changes, XXXII, 258–262
tissue electrodes, LVI, 476
Polarometer, *see* Fluorescence polarometer
Poliodal-2, homopolymer, XLIV, 84
Poliodal-4, homopolymer, XLIV, 84
Pollen tube, nuclei extraction, XXXI, 564, 565
Poly(A), *see* Polyadenylate; Poly(alanine)
Poly[d(A-T)$_n$·d(A-T)$_n$], XLVI, 90
Poly 610, XLIV, 207, *see also* Nylon
Polyacetal carrier, *see* Acryloyl-*N*,*N*-bis(2,2-dimethoxyethyl)amine
Polyacrolein, XLIV, 110
reduction with sodium borohydride, XLIV, 116
Polyacrylamide, XXXIV, 30–58
amide groups, XXXIV, 34
attachment of ligands, XXXIV, 30–58
capacity, XXXIV, 34
commercial, XXXIV, 31
coupling of diazonium salts, XXXIV, 49
derivatives, XXXIV, 34–37
acyl hydrazide, XXXIV, 34
adipic acid dihydrazide, XXXIV, 45
aminoethyl, XXXIV, 35, 37
aminoethyldithio, XXXIV, 43, 44
aminoethylglutaramyl-hydrazide, XXXIV, 42, 43
analytical procedures, XXXIV, 56–58
azo, XXXIV, 49
azophenyl-β-lactoside, XXXIV, 719
bromoacetamidocaproyl-hydrazide, XXXIV, 40, 41
tert-butyl carbazate, XXXIV, 44, 45
carbodiimides, XXXIV, 34
carboxyl, XXXIV, 35, 36
commercially available, XXXIV, 52
deamidation, preparation of carboxyl derivative, XXXIV, 35, 36
dinitrophenylaminoethyl, XXXIV, 49
dry weight determination, XXXIV, 56, 57
ethylenediamine, XXXIV, 37
functional group densities, estimation of, XXXIV, 56–58
glutarylhydrazide, XXXIV, 38, 39
glycinamidocaproyl-hydrazide, XXXIV, 41, 42
hydrazide, XXXIV, 36, 37, 50
of β-galactosidase, XXXIV, 352
spacers, XXXIV, 37–45, *see also* specific compounds
succinylhydrazide, XXXIV, 38, 39
sulfhydryl, XXXIV, 43, 44
trinitrophenylaminoethyl, XXXIV, 49
entrapped enzyme, LVI, 489, 490
flow rate, XXXIV, 32, 33
functional group
primary derivatives, XXXIV, 34–37, *see also* specific derivatives
quantitation, XXXIV, 34
spacers, XXXIV, 37–45
functional group density, *see* Functional group density

glucose oxidase adduct, LVI, 486
handling and washing, XXXIV, 32–33
lectin coupling, XXXIV, 338, 340, 341
ligands, attachment of specific, XXXIV, 45–51, *see also* specific ligands
as matrix, XXXIV, 8
for collagenase, XXXIV, 423
preparation, XXXIV, 32, 33
spacers, XXXIV, 37–45, *see also* specific compounds
storage, XXXIV, 33
TNBS test, XXXIV, 33
trypsin, XXXIV, 441
Polyacrylamide-agarose gel, inclusion of aggregated enzyme, XLIV, 272, 273
Polyacrylamide gel, *see also* Polyacrylamide gel electrophoresis
commercially prepared, use, XLVIII, 7
derivatization, XLVII, 267, 268
disc, identification of acyl-CoA dehydrogenases, LIII, 508
ethylene diacrylate, XLVI, 526
peptide separation, XLVII, 328–330
in phosphorylation assay for given histone fraction, XL, 141
prewashing, XLVII, 329
removal of protein, LIX, 313
separating, preparation, XLVIII, 9
stacking, preparation, XLVIII, 9
Polyacrylamide gel electrophoresis, *see also* Electrophoresis; Gel electrophoresis
acid gels, XXXII, 75, 76
assay of purity
casein kinases, LX, 503
elongation factors, LX, 604, 605, 668, 671–673, 681, 682, 701, 708
heme-regulated translational inhibitor, LX, 488, 489
initiation factors, LX, 10, 11, 23, 30, 45, 46, 57, 80–84, 145, 146, 159–161, 243, 346
messenger ribonucleic acid-binding proteins, LX, 395–398, 424–425, 439, 450, 453, 455
characterization of messenger ribonucleic acid-protein complexes, LX, 398–400
of chromatin protein kinases, XL, 202
dissociation of OPRTase-OMPdecase complex, LI, 141
gel preparation, XXXII, 73, 74, 76, 77, 84–86
of histones, XL, 127
identification, initiation factors phosphorylated by casein kinase, LX, 507–510
of labeled parathyroid proteins, XXXVII, 353–355
of membrane proteins, XXXII, 70–81, 92–102, 105, 106
molecular weights, 11, 28, 29, 45 63, 159–161, 229, 245, 395–397, 488, 489, 503, 644, 647, 648, 671, 672, 681, 682, 694, 695
determination, XXXII, 77, 78
of membrane proteins, XXXII, 92–102
monitoring acetylation of translational components, LX, 539, 540
monitoring of dephosphorylation of translational components, LX, 527
phosphodiesterase, XXXVIII, 228, 229, 260
activator, XXXVIII, 265–267
preparation of gels, LX, 268, 271, 505, 506, 518, 519
preparative, XXXII, 91
gel column preparation, LI, 342, 343
purification of aspartate carbamyltransferase, LI, 54–56
of deoxycytidine kinase, LI, 341–344
of dihydroorotase dehydrogenase, LI, 67
of purine nucleoside phosphorylase, LI, 523, 528
of thymidine kinase, LI, 357, 358
of progesterone receptors, XXXVI, 208, 209

purification of *Pseudomonas* toxin A, LX, 785
sample preparation, XXXII, 74, 75
SDS-containing, XXXII, 76, 77, 92–102
separation
of complex III peptides, LIII, 80, 81
initiation factor activities, LX, 175–180, 265–275
ribosome fragment binding sites, LX, 329, 347, 348, 369
solutions, XXXII, 95, 96
stains, XXXII, 74, 75, 81, 87
study
elongation factor EF-1, LX, 644, 694, 695
elongation factor EF-2, LX, 653
phosphorylation by heme-regulated translational inhibitor, LX, 470, 476
specificity of heme-regulated translational inhibitor, LX, 479, 491
trypsinolysis products from labeled elongation factor EF-G, LX, 743–745
technique, XXXII, 73–76
Polyacrylamide P–30, purification of glycine reductase, LIII, 378
Polyacrylic acid, activation sequence, XLIV, 85
Polyacrylic acid amide, cross-linked, XLIV, 20
Polyacrylic copolymer, XLIV, 84–107
Polyacrylic hydrazide-agarose, XXXIV, 76
derivative, XXXIV, 74
Polyacrylic polymer support
advantages, XLIV, 894, 895
artificial membranes, XLIV, 902
as beads, conjugation with enzyme, XLIV, 445, 446
chemical properties, XLIV, 53, 85
effects of solvents on volume, XLIV, 22
hollow-fiber devices, XLIV, 293, 311, 312
hydrophobicity studies, XLIV, 63, 64
multienzyme immobilization, XLIV, 181, 182
physical properties, XLIV, 53, 85
polymerization
as beads, XLIV, 57–60
as granules by bulk-polymerization, XLIV, 57
inhibitor, XLIV, 57
second, XLIV, 180
whole cell entrapment, XLIV, 185, 740, 741, 745
porosity, factors determining, XLIV, 56, 57
protein-protein interaction
effect of gel concentration, XLIV, 573, 574
of gel deformation, XLIV, 574–576
steric exclusion, XLIV, 666, 667
suitability for entrapment, XLIV, 171, 172
water content, XLIV, 172, 174
Polyacrylonitrile, methyl imidoester-containing, XLIV, 324
Poly[acryloyl-*N,N*-bis(2,2-dimethoxyethyl)amine] synthesis, XLIV, 101–103
Poly(adenosine, guanosine, uridine), effect on initiation by native 40 S ribosomal subunits, LX, 573, 575, 587
Poly(adenylate), XXXIV, 5
assay for translational control ribonucleic acid, LX, 546, 547
coding for transfer ribonucleic acid-lysine, LX, 615
complex with 70 S ribosomes and transfer ribonucleic acid-lysine, LX, 624
Polyadenylate mitochondrial RNA, LVI, 9, 10
[^{3}H]Polyadenylic acid:polydeoxythymidylic acid, preparation, LIX, 826
Poly(A)-agarose, XXXIV, 468
Poly(I:C)-agarose, XXXIV, 468
Poly(alanine), dye binding studies, XLIX, 186
Poly(DL-alanine)
hydrogen exchange rate, XLIX, 27
outexchange kinetics, XLVIII, 323
as spacer, XXXIV, 94
Poly(L-alanine), circular dichroism, XXXII, 221

Poly(DL-alanine)-agarose derivative, XXXIV, 73
Poly(DL-alanyllysyl)-agarose derivative, XXXIV, 73, 74
Polyallomer tubing, steroid binding, XXXVI, 161, 207
Polyallyl alcohol, XLIV, 108, 116
Polyamide-6 strip, LII, 442
Polyamine, *see also* Edeine; Spermidine; Spermine
  assay by fluorimetry, LX, 557, 558
  extraction from *Artemia* ribosomes, LX, 557
  inhibitor of rat liver carbamoyl-phosphate synthase, LI, 119
  in media, LVIII, 55
  occurrence, LX, 559
  tRNA stability, LIX, 18
  stimulation of initiation, LX, 563
  thin-layer chromatographic separation of dansylated derivatives, LX, 558, 559
Polyamine oxidase, in microbodies, LII, 496
Polyamino acid, XLIII, 391
Polyaminoaryl-nylon
  diazotization capacity, XLIV, 133
  immobilization, XLIV, 134
  synthesis, XLIV, 129, 132, 133
Polyanalyst apparatus, XXXII, 93, 94
Polyanion
  effect on cytoplasmic estrogen receptors, XXXVI, 181, 182
    on nuclear estrogen receptors, XXXVI, 184–186
Polyarginine
  inhibitor, of clostripain, XLVII, 167
  protein kinase inhibitor, XXXVIII, 357
Poly-γ-benzyl-L-aspartate, Raman spectrum, XLIX, 113
Polycarbonate bottle, monolayer culture, LVIII, 137
Polycarbonate centrifuge tube, XXXI, 13
Polycarbonate tube, calcium concentration, LVII, 282
Polychaeta, oxidase, LII, 13, 16
Polychlorotrifluoroethylene, coated with trioctylmethylammonium bromide, preparation, LIX, 119
Polychlorotrifluoroethylene powder, in preparation of RPC resin, LIX, 221
Polyclar AT
  insoluble polyvinylpyrrolidone, XLIII, 543, 545
  in plant-tissue fractionation, XXXI, 534–539, 592
  rutin binding, XXXI, 533, 534
Polycyclic aromatic hydrocarbon, *see* Hydrocarbon, polycyclic aromatic
Polycytidylate, *see also* Polycytidylic acid
  purification of ribosome protein S1, LX, 456, 457
  template, LX, 629, 630
Polycytidylic acid, LIX, 147, 157
  modification with *N*-hydroxysuccinimide esters, LIX, 162
    with sodium bisulfite and carbohydrazide, LIX, 147
[^{3}H]Polycytidylic acid:polyinosinic acid, preparation, LIX, 826
Poly(deoxyadenylate-deoxythymidylate), XXXIV, 466
Poly[deoxy(4-thiothymidylate)], XLVI, 89
Poly(deoxythymidine), XXXIV, 5
Poly(deoxythymidylate)-cellulose, XXXIV, 473
Poly(dimethylsiloxane), XLIV, 178
Polyene
  effects on membranes, XXXII, 882
  proton magnetic resonance spectroscopy, XLIII, 391
  thin-layer chromatography, XLIII, 187
Polyene antibiotic, resistance, LVI, 560
Polyene macrolide, XLIII, 134–136
Polyester cloth, for plating, LVIII, 320, 321
Polyether, as ionophores, LV, 442–446
Polyetherin A, LV, 445
Polyethylene
  for EPR tube filling, LIV, 113
  infrared spectral properties, LIV, 316
  for Mössbauer spectroscopy sample cells, LIV, 376
  steroid adsorption, XXXVI, 93, 94, 161
Polyethylene glycol, XLIV, 328
  ATPase purification, LV, 318, 794, 802–804

cell fusion, LVIII, 345, 353, 356–359
concentration of proteins, LX, 206, 207, 243, 395, 397, 438
for countercurrent distribution, XXXI, 763, 764
in dimethylaniline monooxygenase purification, LII, 147, 148
as fusing agent, LVIII, 351
fusion size, LVIII, 363
in lysozome transfer, LVIII, 359
in nucleus transfer, LVIII, 359
osmometric studies, XLVIII, 79
as plant cell protectant, XXXI, 536
protoplast fusion, LVIII, 362, 363
purification of elongation factors, LX, 665, 666, 677, 690, 691, 699
of Qβ ribonucleic acid replicase, LX, 633
spectrophotometric media, XLIV, 389
toluene-treated mitochondria, LVI, 548

Polyethylene glycol 4000, purification of molybdoferredoxin, LIII, 317, 320, 321

Polyethylene glycol 6000
cell-free protein synthesis system, LIX, 300
dissociation of *E. coli* ribosomes, LIX, 445
for microsomal protein precipitation, LII, 111, 115, 121
for precipitation of 1,25-dihydroxyvitamin D_3, LII, 393
preparation of coliphage R17 RNA, LIX, 364
of MS2 RNA, LIX, 299, 300
of ribosomal subunits, LIX, 483
purification of adenosine deaminase, LI, 510
of monoamine oxidase, LIII, 498, 499
of nitrogenase, LIII, 322
removal, LIX, 484
in separation of ribosomal fragments, LIX, 474

Polyethylene glycol 20000
concentration of ribosomal protein solutions, LIX, 499
storage of phosphoribosylglycinamide synthetase, LI, 184

Polyethylene glycol A1000, tissue embedment, XXXVII, 141–144

(Polyethyleneime)cellulose, thin-layer plate, LI, 158, 308, 315, 333, 373, 419, 509

Polyethyleneimine, XLIV, 124

Poly(ethyleneimine)cellulose
adenylate kinase assay, LV, 285, 286
ATP-ADP exchange assay, LV, 285, 286
ATP-P_1 exchange assay, LV, 286, 287
chromatography, background and general principle, LV, 283, 284
cyclic nucleotide column chromatography
column preparation, XXXVIII, 16, 17
fractionation procedures, XXXVIII, 17–20
nucleoside diphosphokinase assay, LV, 287–289
nucleotide purity, LV, 379
separation of nucleotides, LX, 514
thin-layer chromatography, XXXVIII, 27
biological materials, XXXVIII, 37, 38
cyclic nucleotides, XXXVIII, 30–35
elutions, XXXVIII, 38
materials and methods, XXXVIII, 28
nucleoside triphosphates, XXXVIII, 35, 36
nucleotide separation, XXXVIII, 28–30
purine nucleotides, related, XXXVIII, 36, 37
pyrimidine nucleotides, XXXVIII, 35

Polyethyleneimine-cellulose plate, assay of AMP formation during aminoacylation, LIX, 286

Poly(ethyleneimine) paper, ATP assay, LV, 539

Polyethylene oxide, osmometric studies, XLVIII, 75

Poly(ethylenesulfonic acid), isolation of mitochondria, LV, 4

Polyethylene tube, phosphorescence, LVII, 217
chemiluminescence, assay, LVII, 473–494
general principles, LVII, 462–466
isolation, LVII, 467, 470
Polyethylene tubing, for cannula, XXXIX, 177
Poly(C)-fiberglass, XXXIV, 470, 471
Polyfungin, XLIII, 135
Polyglutamic acid, dye binding studies, XLIX, 186
Poly(L-glutamic acid), circular dichroism, XXXII, 221
Polyglutamic acid hydrazide, XXXIV, 74
Polyglutamic acid hydrazide-agarose derivative, XXXIV, 74
Polyglycerol phosphate, L, 387
Polyglycerol phosphate polymerase, L, 394
Poly(glycosyl)ceramide
acetolysis, L, 215, 216
composition, L, 211, 212
isolation, L, 211–216
purification, L, 213, 214
Polygram Cell 300, XXXII, 129
Polygram CEL 300 plate, LIX, 777
Polygram polyamide-6, XLVII, 363
Polyguanylate polymerase, LX, 629, 630
Poly(hydroxyalkylmethacrylate), XLIV, 56, 85, *see also* Spheron
Poly-β-hydroxybutyrate, in bacterial cell wall, XXXI, 654, 665
Polyhydroxy compound, detection on paper chromatograms, XLI, 97
Poly(2-hydroxyethyl methacrylate), XLIV, 172–175
entrapment, XLIV, 692
water content, XLIV, 174, 176
Poly IC, *see* Polyriboinosinic:polyribocytidylic acid
Polyinosinic:polycytidylic acid, LX, 553, 554
Polyisonitrile-nylon
immobilization, XLIV, 133, 134
storage, XLIV, 131
synthesis, XLIV, 128, 130–132
Polyisoprenoid intermediate, extraction, LVI, 109, 111, 112
Polykaryocytic index, of cells, definition, XXXII, 582
Polylactic acid, XLIV, 217
biodegradable microcapsule membrane, XLIV, 698
Poly(L-leucine) derivative, XXXIV, 712
Polylysine, XXXIV, 7; LVIII, 49, 78
complex III or IV, LVI, 586
inhibitor, of clostripain, XLVII, 167
surface treatment, LVIII, 577
vessel pretreatment, LVIII, 137
Poly(D-lysine), LVIII, 49
Poly(L-lysine), XXXIV, 5, 611, 722
CD spectra, LIV, 258
coupling to succinylated cyclic nucleotides, XXXVIII, 98
inhibitor of complex IV, LIII, 45
magnetic circular dichroism spectrum, XLIX, 159, 160
Polylysine-agarose, *see* Polylysyl-agarose
Polylysyl-agarose, XXXIV, 94, 95
for avidin, XXXIV, 265
derivatives, XXXIV, 73, 74
preparation, XXXIV, 74, 75
Poly(L-lysyl-DL-alanine), XXXIV, 611, 681
Poly(L-lysyl-DL-alanine)-agarose, XXXIV, 91, 94, 95, 658
N-hydroxysuccinimide derivative, XXXIV, 660
Poly(L-lysyl-DL-alanyl-*p*-aminobenzamido)-agarose, XXXIV, 658, 659
Poly(L-lysyl-DL-alanyl-*N*-carboxymethyl-hydrazido)agarose, XXXIV, 659, 660
Polymer
in media, LVIII, 70
mitochondrial ribosomes, LVI, 84, 85
use in plant-tissue fractionation, XXXI, 534–538
Polymerization, *see also* Bead polymerization
common ingredients, XLIV, 172
degree of, of helices, vibrational frequency, XLIX, 112
entrapment, XLIV, 171–177, 182, 183
general procedure, XLIV, 111
initiation
agents, choice, XLIV, 173
photochemical, XLIV, 173–174, 194

by redox method, XLIV, 173, 193, 194
photoactivated, instrumentation, XLIV, 174, 194, 891
second, and interpenetrating networks, XLIV, 180
solution, XLIV, 172–176
suspension, XLIV, 176, 177
monomer concentration effects, XLIV, 178
vinyl, characteristics, XLIV, 172
Polymetaphosphate, in bacterial cell wall, XXXI, 654
Polymethacrylic acid anhydride, XLIV, 85
Polymethacrylic acid resin, synthesis of oleylpolymethacrylic acid resin, LIII, 213, 214
Polymethacrylic copolymer, XLIV, 84–107
Polymethyl acrylate, XXXIV, 76
Poly-γ-methyl-L-glutamate
fibers, preparation, XLIV, 231
vacuum ultraviolet circular dichroism spectrum, XLIX, 218, 219
Polymethylmetaacrylic bead, XXXIV, 714
Polymethyl methacrylate, XLIV, 194, 693
magnetic circular dichroism, XLIX, 159
Polymin P, purification of aminoacyl-tRNA synthetases, LIX, 260, 261
Polymorphonuclear leukocyte, *see* Leukocyte, polymorphonuclear
Polymyxin
biosynthesis, XLIII, 579–584, *see also* Polymyxin synthetase
effects on membranes, XXXII, 882
media, XLIII, 17, 18
paper chromatography, XLIII, 147
solvent systems for countercurrent distribution, XLIII, 340
structure, XLIII, 580
Polymyxin B, XLIII, 622
in media, LVIII, 113, 115
Polymyxin B_1, structure, XLIII, 330
Polymyxin G, XLIII, 200
Polymyxin synthetase:L-2,4-diaminobutyrate activating enzyme, XLIII, 579–584
assay, XLIII, 579–581
Bacillus polymyxa preparation, XLIII, 580, 581
crude extract preparation, XLIII, 582, 583
molecular weight, XLIII, 583
properties, XLIII, 583, 584
purification, XLIII, 581, 583
stability, XLIII, 583
Polynucleotide, *see also* specific type
hydrogen exchange, XLIX, 30–33
of exocyclic amino group protons, XLIX, 31
measurement, XLIX, 33
of ring NH proton, XLIX, 30, 31
matrix, XXXIV, 463–475
nascent, mitochondrial ribosomes, LVI, 86
removal, XXXIV, 166
Polynucleotide-cellulose, XXXIV, 472–475
Polynucleotide kinase, XXXIV, 469
from T4, in tRNA labeling procedure, LIX, 58, 61, 71, 75, 102, 103
Polynucleotide ligase, XXXIV, 473–475
Polynucleotide phosphorylase, XXXIV, 51, 254; XLVI, 670, 671
primer-dependent, in oligonucleotide sequence analysis, LIX, 88
Polyol, protection, XXXI, 539, 540
Polyolefin, hollow-fiber device, XLIV, 293
Polyoma cell, plasma membrane isolation, XXXI, 161
Polyoma virus
cell lines for study, XXXII, 584, 585
cell transformation, XXXII, 586–589
transformation, LVIII, 370
Polyornithine
sequence analysis, XLVII, 257, 258
succinylation, XLVII, 248
Polyoxin, XLIII, 153
Poly[oxy(2,6-dimethyl-1,4-phenylene)], XLVIII, 336, 343
Polyoxyethene[10]nonyl phenol ether, *see* Renex 690
Polyoxyethylene alcohol, partial specific volume, XLVIII, 19

Polyoxyethylene alkylphenol, partial specific volume, XLVIII, 19
Polyoxyethylene detergent, partial specific volume, LVI, 737
Polyoxyethylene monoacyl sorbitan, partial specific volume, XLVIII, 19
Polyoxyethylene(20) sorbitan trioleate, XLIV, 102
Polyoxyethylene sorbitol ester, structure, LVI, 736
Polypeptide
  α-helix, LIV, 257
    characteristic CD spectra, LIV, 258, 266
    in hemoproteins, LIV, 268, 269
  antifungal, XLIII, 334
  β-pleated sheet, LIV, 258, 268, 269
  carbamylation, LVI, 613
  chain
    J, XXXIV, 707
    L, XXXIV, 705
  circular dichroism, XXXII, 220, 221, 226
  of complex I, LVI, 587–589
  of complex II, LVI, 589, 590
  of complex III, LVI, 590–592, 637
  of complex IV, LVI, 592–596
  of complex V, LVI, 596–602
  conformation, from IR spectroscopy, XXXII, 251, 252
  3_{10} helix, LIV, 259
  hydrophobic, inhibitor binding, LVI, 584
  lipid interactions with, NMR studies, XXXII, 211
  mitochondrial, nucleotide binding, LV, 73, 74
  polarography, LVI, 477
  Raman spectroscopy, XXXII, 256, 257
  random coil form, LIV, 258
  small, resolution, LVI, 603, 604
  synthesis, assay, LIX, 367
Polypeptide antibiotic, XLIII, 313–315, 335
Polypeptide hormone, XLIII, 335, 337
  inactivation at receptor sites, XXXVII, 198–211
  tritium labeling of amino acids, XXXVII, 313–321
    peptide hydrolysis, XXXVII, 318, 319
    peptide purification, XXXVII, 316, 317
Polypeptide-SDS complex, anomalous behavior, XLVIII, 10
Polypeptin, XLIII, 340
Polyphenol oxidase
  assay, XXXI, 534
  inhibitor, XXXI, 538
  peroxisomes, LII, 503
Polyphenylalanine synthesis, LIX, 387–389, 448, 449, 459, 622, 819
  assay for initiation factor
    EF–1, LX, 93, 142, 540, 680
    EF–2, LX, 142, 640, 650, 651, 662, 664, 704, 705
    EF–3, LX, 679
    EF–1β, LX, 651
    EF–G, LX, 597, 598, 605, 606
    EF–Ts, LX, 597, 598, 605, 606, 665
    EF–Tu, LX, 597, 598, 605, 606, 650, 664, 665
   for protein S1, LX, 433–436
  dependence on adenosine triphosphate and guanosine triphosphate, LX, 683, 684
  translation of cellulose-bound polyuridylate, LX, 754–760, 763–765
Polyphenylalanylpuromycin, assay of translocation within ribosome, LX, 762–764, 767, 768
Polyphosphate, detection, XXXVIII, 412, 413
Polypolyacrylamide, XXXIV, 716
Polyporus, oxidase, LII, 10, 17, 21
Polyporus obtusus, pyranose oxidase, assay, purification, and properties, XLI, 170–173
Polyprenol, as substrates for C_{55}-isoprenoid alcohol phosphokinase, XXXII, 439–446
Polyprenol phosphate, acid lability, L, 426
Poly-L-proline, vacuum ultraviolet circular dichroism spectrum, XLIX, 218, 219
Polyproline I, vacuum ultraviolet circular dichroism, XLIX, 218
Polyproline II, vacuum ultraviolet circular dichroism, XLIX, 218

Polypropylene
hormone adsorption, XXXVII, 32, 35
for metal-free water storage, LIV, 473
Polypropylene centrifuge tube, commercial source, LI, 338
for freezing, in enzyme extraction, LI, 7
Polypropylene tube, LII, 392
calcium concentration, LVII, 282
in tRNA sequence analysis, LIX, 62
Polyribitol phosphate, biosynthesis
assays, L, 396, 398–402
preparation of compound I, L, 397, 398
of compound II, L, 398
reactions, L, 387
Staphlococcus aureus H membrane preparation, L, 394–396
Polyribitol phosphate polymerase
assay, L, 389
properties, L, 392–394
purification, L, 390, 391
Polyriboinosinic:polyribocytidylic acid, interferon producing, LVIII, 294, 295
Polyribosome, *see also* Ribosome
in plant cells, XXXI, 491, 504, 513
isolation, XXXI, 586–589
Polysaccharide, *see also* specific types
in agar
acidic, LVIII, 160
sulfated, LVIII, 160
bacterial, L, 250–272
acidic capsular, L, 253
acidic oligosaccharide components, L, 261–268
antibody receptor, L, 250–253
β-elimination, L, 255
extracellular, isolation, L, 254
isolation and purification, L, 253, 254
partial acid hydrolysis, L, 254, 255
Smith degradation, L, 255
capsular, XXXIV, 715
carboxyl reduction, L, 14
^{13}C–nuclear magnetic resonance spectroscopy, sialic acid residues, L, 44–46
N-deacetylation, L, 15–17
deamination, L, 17–20
degradation, viral induction, L, 269
enzymic digests, colorimetric assay of hexuronic acids, XLI, 29
meningococcal, ^{13}C nuclear magnetic resonance spectroscopy, L, 46
methylation analysis
chromium trioxide oxidation, L, 20–24
general procedure, L, 3–7
partial acid, hydrolysis, L, 8–14
mycobacterial, XXXV, 84, 85, 90–95
assay, XXXV, 91
MGLP, composition, XXXV, 85, 91, 95
MMP, composition, XXXV, 84, 85, 91, 94, 95
purification, XXXV, 85, 91–94
Pneumococcus type 14, acid hydrolysis, L, 16
removal, XXXIV, 499
sequential degradation, L, 33–38
Shigella dysenteriae type 3, hydrazine *N*-deacetylation, L, 16
structure determination
^{13}C-nuclear magnetic resonance, L, 39–50
Polysaccharide-agarose, XXXIV, 712–714
Polysaccharide gel, XXXIV, 441
Polysaccharide immunoadsorbent, XXXIV, 712–716, *see also* specific immunoadsorbents
Polysome, XXXIV, 7, 299, 496; XLVI, 336, 338
assay of GTP hydrolysis, LIX, 361
for polypeptide chain elongation, LIX, 356
of polypeptide synthesis, LIX, 367
for translocation of peptidyl-tRNA, LIX, 359
endogenous, preparation, LIX, 353–356, 363–366
properties, LIX, 368–371
isolated, protein synthesis, LVI, 27–29
ligands of, antibiotic binding, LIX, 859, 860
mitochondrial, isolation, LIX, 426
preparation
for affinity chromatography, XL, 270
in vitro, LIX, 368

rat liver, in *in vitro* ribosomal protein synthesis, LIX, 527–532
preparation, LIX, 526, 527
structural models, LIX, 746–750
structural parameters, LIX, 747
synthesizing specific protein, enrichment with affinity chromatography, XL, 266, 271
Polysphondylium pallidum
agglutinin, *see* Pallidin
culture conditions, L, 313
fructose-1,6-diphosphatase, assay, purification, and properties, XLII, 360–363
D(−)-lactate dehydrogenase, XLI, 294
Polystictus versicolor, oxidase, LII, 18
Polystyrene, *see also* Aminopolystyrene; TETA polystyrene resin
N-(3-aminopropyl)aminomethylated, synthesis, XLVII, 383
attachment, XLVII, 301
benzhydrylamine resin, XLVII, 587, 588
chloromethylated, XLVII, 263, 266, 584
amino acid attachment, XLVII, 590, 591
peptide synthesis, XLVII, 584, 590
in continuous flow analyzer, XLIV, 639
degradative efficiency, XLVII, 275, 276
derivatization, XLVII, 265–267
1,1-dimethylpropyloxycarbonylhydrazide resin, XLVII, 588
hormone adsorption, XXXVII, 32, 35
hydroxymethylated, XLVII, 585–587
amino acid attachment, XLVII, 591
peptide synthesis, XLVII, 585–587, 589
stability, XLVII, 585, 586
synthesis, XLVII, 585
lysozyme immobilization, XLIV, 288–290
Pam-resin, stability, XLVII, 586
peptidyl-resin complex, hydrolysis, XLVII, 598, 599
for small peptides, XLVII, 321
swelling, XLVII, 272, 298
use in spinning-cup sequenator, XLVII, 306, 307, 313, 316
Polystyrene bead
for column sieving fractionation, XXXI, 551
in plasma membrane isolation, XXXI, 162
Polystyrene capacitor, LVII, 539
Polystyrene-*p*-divinylbenzene polymer, derivatization, XLVII, 360, 361
Polystyrene latex sphere, photocurrent spectrum, XLVIII, 429
Polystyrene sphere, freeze fracturing, XXXII, 40
Polysulfone, hollow-fiber membrane, XLIV, 293
Poly[(4-thiouridylate)], XLVI, 89, 622, 623, 625
Polytoma mirum, cytochrome oxidase, LIII, 66
Polytomella cacca pringsheim, disruption, LV, 138
Polytron apparatus, XXXVI, 190
Polytron generator, XXXI, 136
Polytron grinder, for tissue homogenization, LVII, 387, 388, 391
Polytron homogenizer, XXXII, 671; XXXIX, 101
Polytron mixer, XXXI, 273
for plant-cell rupture, XXXI, 502
Poly(U)-agarose, *see* Polyuridylate-agarose
Poly(U)-fiberglass, *see* Polyuridylate fiberglass
Polyuridine, XLVI, 89, 622, 623, 625, 636, 641, 643, 662, 682
Polyuridylate, *see also* Polyuridylic acid
assay for initiation factor IF-3, LX, 232
for ribosome protein S1, LX, 418, 419, 423, 424, 447, 448
binding to 30 S ribosomal subunits, LX, 432
cellulose-bound
binding of translating ribosomes, LX, 754–760
preparation, LX, 746–753
translation in columns, LX, 755–779
in suspension, LX, 753–756

coding for transfer ribonucleic acid-phenylalanine, LX, 121, 615, *see also* Polyphenylalanine, synthesis
complex with 80 S ribosomes and *E. coli* transfer ribonucleic acid, LX, 661, 662
fragments
chain length, efficiency of translation, LX, 756
coupling with (carboxymethyl)cellulose hydrazide, LX, 751–753
dephosphorylation, LX, 748–751
periodate oxidation, LX, 747, 751
production by acid hydrolysis, LX, 747–749
hybridization to globin ribonucleic acid, LX, 72, 85–87
mitochondrial ribosomes, LVI, 91
removal of protein S1 from 30 S ribosomal subunits, LX, 453
ribosomal protein synthesis, LVI, 26
Sepharose-bound, kinetics of translation, LX, 755, 756
Polyuridylate-agarose, XXXIV, 468
isolation of messenger RNA, XXXIV, 496–499, *see also* Messenger RNA
Polyuridylate-fiberglass, XXXIV, 470, 471
Polyuridylic acid
matrix bound
preparation, LIX, 383–387
translation, LIX, 387–389
protein inhibitors, XL, 287, 288
Polyvinyl acetate, for fixing charcoal on nitrocellulose membranes, LIX, 238
Polyvinyl alcohol, XLIV, 108, 110, 115, 116, 328
effects of solvents on volume, XLIV, 22
entrapment, XLIV, 178
magnetic circular dichroism, XLIX, 159
Polyvinyl ether, XLIV, 117
Poly(vinylpyrrolidone), XLIV, 175
insoluble, XLIII, 543, 545
osmometric studies, XLVIII, 75
in plant-cell fractionation, XXXI, 505, 509, 534, 535
as plant-cell protectant, XXXI, 533, 536, 537, 592, 599, 603
plant mitochondria, LV, 24
suspension stabilizer in Spheron, XLIV, 67
POPOP, *see* 2,2′-(1,4-Phenylene)bis(5-phenyloxazole)
Population doubling, in cell culture, XXXII, 562, 563
Porasil C, immobilized mitochondria, LVI, 550
Porcelain ball mill, LIII, 240
Porcine, *see* Pig
Pore diameter
optimum, XLIV, 156
support choice, XLIV, 155–158
Porfiromycin, XLIII, 152, 153
Porichthys porosissimus, luciferin, LVII, 343
Porod volume, LIX, 713, 714
Poropak Q, LIII, 644
Porphobilinogen, LII, 12
isolation, XLIV, 847, 848
mutants, LVI, 173
product, δ-aminolevulinic acid dehydratase, XLIV, 844–847
structural formula, XLIV, 845
Porphobilinogen synthase
immobilization, on Sepharose 4B, XLIV, 846, 847
properties, XLIV, 844
Porphyridin, XLVI, 50, 51
Porphyrin
intermediates, heme-deficient mutants, LVI, 558
stability in light, LII, 450
structure, LII, 450
spectroscopic properties, LII, 435, 436
Porphyrin cytochrome *c,* preparation, LIII, 176, 177
Porphyrin ring
conformation studies, LIV, 243–245
doming, Raman spectra, LIV, 244
Portal vessel, hypophyseal, cannulation, XXXIX, 179–182
Possum, muscle, 3-phosphoglycerate kinase, purification, XLII, 129
Postmitochondrial supernatant, from rat liver, preparation, LIX, 516, 517

Poststaining, autoradiography, LVIII, 286, 287
Potassium
effect on pituitary hormones, XXXVII, 87
in media, LVIII, 68
in mitochondrial matrix, LVI, 254
release by parotid adrenergic receptors, XXXIX, 464–466
synthesis of ethyl[2-hydroxy-3-methoxycarbonyl-5-(4-methoxyphenyl)]valerate potassium enolate, LVII, 435
transport, parathyroid hormone effects, XXXIX, 20
uptake in electron transport studies, XXXI, 701, 702
Potassium-42
in membrane studies, XXXII, 884–893
properties, XXXII, 884
Potassium acetate
assay of ADP, LVII, 61
of guanylate kinase, LI, 476
determination of aminoacylation isomeric specificity, LIX, 273
electrophoretic separation of ribosomal protein, LIX, 431
extraction of nucleic acids, LIX, 312
preparation of aminoacylated tRNA isoacceptors, LIX, 298
of hydrazine-substituted tRNA, LIX, 120
ribosomal fragments, LIX, 472
Potassium aspartate, isolation of mitochondria, LV, 4
Potassium benzylpenicillin, XLIII, 360–362
Potassium bicarbonate
in ammonium sulfate fractionation, LI, 402–403
assay of carbamoyl-phosphate synthetase, LI, 106, 112, 113, 122
preparation of rat liver crude extract, LIX, 403
of ribosomes, LIX, 504, 505
Potassium bichromate, optical filter, XLIX, 192
Potassium borohydride
preparation of radiolabeled ribosomal proteins, LIX, 777
purification of glycine reductase, LIII, 378, 379
reduction of quinones, LIII, 586, 587
Potassium bromide, dissociation of flavoproteins, LIII, 431, 432, 433
Potassium *tert*-butoxide, XLVII, 591
Potassium carbonate, XLIV, 836
synthesis of *Diplocardia* luciferin, LVII, 377
of luciferin, LVII, 21
Potassium chloride
aminoacylation of tRNA, LIX, 216, 288, 298
assay of adenylate deaminase, LI, 491, 498
of AMP formation during aminoacylation, LIX, 287
of carbamoyl-phosphate synthetase, LI, 30, 122
of FGAM synthetase, LI, 195, 197
of guanylate kinase, LI, 475, 476, 483
of guanylate synthetase, LI, 219
of hypoxanthine phosphoribosyltransferase, LI, 543
of nucleotide incorporation, LIX, 182, 188
of polypeptide synthesis, LIX, 367
of purine nucleoside phosphorylase, LI, 539
of pyrimidine nucleoside monophosphate kinase, LI, 323
of radiolabeled ribosomal particles, LIX, 779
of tRNA hydrolysis, LIX, 183
of uridine-cytidine kinase, LI, 300, 309, 316
brain mitochondria, LV, 17
brown adipose tissue mitochondria, LV, 69, 70
cell-free translation system, LIX, 387, 459
crystallization of transfer RNA, LIX, 6
cyclic nucleotide separation, XXXVIII, 30, 31

density measurements, XLVIII, 29
determination of aminoacylation, isomeric specificity, LIX, 272
of [^{3}H]puromycin specific activity, LIX, 360
digestion of tRNA, LIX, 99
elution, *see* (Carboxymethyl)cellulose; Phosphocellulose; Sephadex A–50; Sephadex C–50
in formulation of diluted solid sodium dithionite, LIV, 130
high-salt buffer, LIX, 838
initiation factor buffer, LIX, 838
isolation of chloroplastic ribosomes, LIX, 434, 435
of mitochondria, LV, 3, 4
junction, measurement of proton translocation, LV, 633
lung mitochondria, LV, 20
misactivation studies, LIX, 286
preparation of aminoacylated tRNA, LIX, 126, 127, 130, 133
of 3′-amino-3′-deoxy-ATP, LIX, 135
of beef liver mitochondria, LIII, 497
of Chinese hamster ovary crude extract, LIX, 219, 232
of cross-linked ribosomal subunits, LIX, 535
of *E. coli* crude extract, LIX, 297
of mitochondrial enzyme complexes, LIII, 3, 6
of mitochondrial ribosomes, LIX, 425
of pig liver mitochondria, LIII, 512
of polysomes, LIX, 363
of postmitochondrial supernatant, LIX, 515
of protein-depleted 30 S subunits, LIX, 621
of radiolabeled ribosomal protein, LIX, 783
of rat liver crude extract, LIX, 403
of rat liver polysomes, LIX, 526
of rat liver tRNA, LIX, 230
of reconstituted ribosomal subunits, LIX, 652
of reduced ubiquinone, LIII, 39
of ribosomal proteins, LIX, 484, 837
of ribosomal RNA, LIX, 446, 648
of ribosomal subunits, LIX, 411–413, 416, 440, 505, 752
of ribosome-DNA complexes, LIX, 842
of ribosomes, LIX, 483, 504
of samples for neutron scattering, LIX, 655
of *T. vulgare* ribosomes, LIX, 752
of yeast submitochondrial particles, LIII, 113
protein binding buffer, LIX, 322, 328
purification of adenylate deaminase, LI, 492, 495, 499
of adenylosuccinate synthetase, LI, 208
of complex III, LIII, 88
of CPSase-ATCase-dihydroorotase complex, LI, 124
of CTP(ATP):tRNA nucleotidyltransferase, LIX, 125, 184
of cytochrome *c* reductase complexes, LIII, 57
of cytochrome oxidase, LIII, 68, 74
of guanylate synthetase, LI, 215
of hypoxanthine phosphoribosyltransferase, LI, 545
of tRNA methyltransferases, LIX, 194, 195, 197
of rubredoxin, LIII, 342
of tryptophanyl-tRNA synthetase, LIX, 240
reconstitution of ribosomal subunits, LIX, 441, 442, 837
of tRNA, LIX, 140, 141
reverse salt gradient chromatography, LIX, 216
sedimentation equilibrium studies, XLVIII, 175
separation of ribosomal subunits, LIX, 398, 399
stabilizer of adenylate kinase, LI, 471, 472
studies of protein-RNA interactions, LIX, 562, 563, 586
viscosity measurements, XLVIII, 54
in vitro RNA synthesis, LIX, 847

Potassium cholate, *see also* Sodium cholate
purification of complex II, LIII, 22
of complex III peptides, LIII, 84, 86, 88
of complex IV, LIII, 43
of cytochrome *b*, LIII, 216, 217
of cytochrome bc_1 complex, LIII, 114, 116
of cytochrome oxidase, LIII, 57, 68
of dimeric cytochrome *b*, LIII, 106
recrystallization, LIII, 83
Potassium chromate, optical filter, XLIX, 192
Potassium cyanide, XLIV, 290, 347; LII, 190, 529
assay of dihydroorotate dehydrogenase, LI, 60, 64
of succinate dehydrogenase, LIII, 467, 470, 473
of superoxide dismutase, LIII, 384, 385
inhibitor of brain α-hydroxylase, LII, 317
interference, in Nash reaction, LII, 299
in lignoceric acid synthesis, LII, 311
mitochondrial ATP production, LVII, 44, 45
preparation of cyanomethemoglobin, LII, 459
purification of succinate dehydrogenase, LIII, 485
Potassium deoxycholate, XXXII, 477; *see also* Sodium deoxycholate
purification of complex II, LIII, 23
of complex IV, LIII, 43
of cytochrome *b*, LIII, 214, 216, 217
of cytochrome bc_1 complex, LIII, 116
of cytochrome oxidase, LIII, 62, 67, 68
of dimeric cytochrome *b*, LIII, 106
Potassium ferrate(III)
affinity labeling, XLVII, 481, 484
structure, XLVII, 484
Potassium ferricyanide, XLIV, 87, 132, 256, 257, 290
absorptivity, XLIX, 157
antibody inhibition studies, LII, 249
assay of complex I, LIII, 13
of flavocytochrome b_2, LIII, 239
of *Pseudomonas* cytochrome oxidase, LIII, 648
of succinate dehydrogenase, LIII, 469
cytochrome b_5 reductase assay, LII, 207
electron acceptor, LII, 90, 96, 249
extinction coefficient, LII, 92, 103; LIII, 239
as magnetic circular dichroism standard, LIV, 290
NADH-cytochrome b_5 reductase assay, LII, 103
preferred oxidant, in redox potentiometry, LIV, 419
preparation of methemoglobin, LII, 459
purification of cytochrome *c*, LIII, 129, 131
of cytochrome *c*-551, LIII, 657
of high-potential iron-sulfur proteins, LIII, 338
reduction in NADH-cytochrome b_5 reductase assay, LII, 103
substrate of acyl-CoA dehydrogenases, LIII, 503
for washing affinity gel, LIII, 101
Potassium ferro/ferricyanide, as mediator-titrant, LIV, 408, 423, 424, 429
Potassium fluoride
assay of amidophosphoribosyltransferase, LI, 172
inhibitor of deoxythymidylate phosphohydrolase, LI, 290
of rat liver acid nucleotidase, LI, 274
in thymidylate synthetase assay, LI, 91
Potassium glycerophosphate
magnetic circular dichroism, XLIX, 159
in MCD studies, LIV, 293
Potassium hexacyanoferrate(III)
affinity labeling, XLVII, 481, 484, 487
structure, XLVII, 484

Potassium hydroxide
  alkaline hydrolysis of tRNA, LIX, 102, 192
  electrophoretic analysis of ribosomal proteins, LIX, 507, 544
  Fieser's solution, LIII, 299
  grease removal, XLIV, 620
  long chain fatty acid saponification mixture, LII, 312
  metal-free, LIV, 477
  preparation of carbamoylthiocarbonylthioacetic acid, LVII, 17
  prewashing of ion-exchange resins, LIX, 558
  tritiated, preparation, LV, 530
Potassium iodide
  assay reagent, XLIV, 664
  iodometric assay of lipid hydroperoxides, LII, 307
  modification of tyrosine, LIII, 148
  reversible dissociation of flavoproteins, LIII, 435
  synthesis of ethyl-4-iodo-2-diazoacetoacetate, LIX, 811
Potassium ion
  activator of adenylate deaminase, LI, 495, 496
    of CPSase-ATCase-dihydroorotase complex, LI, 118
    of guanylate kinase, LI, 474, 481, 489, 501
    of ribonucleoside triphosphate reductase, LI, 258
  brain mitochondria, LV, 57, 59
  calcium-selective electrodes, LVI, 348
  calibration of cyanine dye fluorescence, LV, 691
  concentration, proton translocation, LV, 643, 647
  cytochrome *m* stability, LII, 187
  effect on *Cypridina* luciferase, LVII, 351
  electrochemical potential on two sides of membrane, LV, 228
  endogenous mitochondrial, valinomycin-dependent release, LV, 669, 670
  gradient
    ATP synthesis, LV, 666, 667
      cation fluxes mediated by carboxylic ionophores, LV, 674, 675
      dependence of energy transduction on the gradient, LV, 672, 673
      experimental procedure, LV, 667, 668
      general features, LV, 669, 670
      properties, LV, 670–672
      summary, LV, 675
      yield and reproducibility, LV, 673, 674
    endergonic processes, LV, 666, 667
    use to support transport by *E. coli* membrane vesicles, LV, 679, 680
      general principle, LV, 676, 677
      generation of membrane potential, LV, 678, 679
      potassium-loaded vesicles, LV, 677, 678
      preparation of Na^+ vesicles, LV, 677
  inhibitor, of luciferase reaction, LVII, 70
  isocitrate assay, LV, 206
  liposomes, LV, 419
  membrane potential, LV, 229
  mitochondrial protein synthesis, LVI, 21
  M. phlei ETP, LV, 186
  phosphorylation potential, LV, 237
  ribosome dissociation, LIX, 367
  transport, ionophores, LV, 450
Potassium laurate, osmometric studies, XLVIII, 78
Potassium nitrate
  assay of nitrate reductase, LIII, 348, 350
  purification of nitrate reductase, LIII, 348
Potassium nitrite, preparation of carboxymethyl cellulose azide, LIX, 385
Potassium osmate, in derivatization of transfer RNA, LIX, 14

Potassium permanganate, XLIV, 685
Pholas luciferin chemiluminescence, LVII, 401
Potassium persulfate, XLIV, 102, 173, 174, 600, 741, 750, 890, 891
Potassium phosphate
assay of GTP hydrolysis, LIX, 361
dibasic, osmometric studies, XLVIII, 78
gradient elution
elongation factor EF-1, LX, 643, 654, 655
initiation factor purification, LX, 42, 54, 142, 189, 501, 502
Hummel-Dreyer gel filtration, LIX, 326
monobasic, osmometric studies, XLVIII, 78
as negative stain interferent, XXXII, 27
preparation of aminoacyl tRNA synthetase, LIX, 314
of carboxymethyl cellulose azide, LIX, 385
of cellulose-bound dithiodiglycolic dihydrazide, LIX, 386
of *E. coli* ribosomes, LIX, 444, 445
of rat liver crude extract, LIX, 403
of ribosomal subunits, LIX, 483
of ribosome-DNA complexes, LIX, 842
purification of aminoacyl-tRNA synthetases, LIX, 259
of CTP(ATP):tRNA nucleotidyltransferase, LIX, 123–126, 184
of guanylate kinase, LI, 478
of tRNA methyltransferase, LIX, 195, 198, 200
reference standard, in inorganic phosphate determination, LI, 275
sedimentation equilibrium studies, XLVIII, 175
Potassium phosphate buffer, for dehydrogenase isolation, XXXII, 380
Potassium phosphoenolpyruvate, assay of polypeptide synthesis, LIX, 367
Potassium phthalimide
synthesis of 4-(*N*-ethyl-*N*-[2-hydroxy-3-(*N*-phthalimido)propyl]amino)-*N*-methylphthalimide, LVII, 431
of *N*-methyl-4-*N*-[2-hydroxy-3-(*N*-phthalimido)propyl]aminophthalimide, LVII, 430
Potassium tetrachloroplatinate, derivatization of transfer RNA, LIX, 14
Potassium tetracyanoplatinate, derivatization of transfer RNA, LIX, 14
Potassium thiocyanate
hydrogen peroxide assay, LII, 346, 347
removal, LI, 25
for reversible dissociation of glutamine-dependent carbamoyl-phosphate synthetase, LI, 23–25, 34
Potato, *see also Solanum tuberosum*
carboxypeptidase inhibitor, *see* Carboxypeptidase inhibitor, potato
chymotrypsin inhibitor I, *see* Chymotrypsin inhibitor I, potato
enzyme studies, XXXI, 522, 525, 527, 538, 541, 591
lectin, L, 340–345
mitochondria isolation, XXXI, 594–596
organelles, XXXI, 533, 535–536
proteinase inhibitor, *see* Chymotrypsin inhibitor I, potato
Potato tuber
glucose-6-phosphate 1-epimerase, molecular weights, XLI, 493
glucosephosphate isomerase, anomerase activity, XLI, 57
PAP level, LVII, 256
Potato tuber moth, media, LVIII, 462
Potential change, bacteriorhodopsin-mediated, light-dependent, LV, 774, 775
Potential difference, ion selective electrodes, LVI, 360, 361
Potentiometric redox titration, EPR measurements, LIII, 492, 493
Potentiometric-spectrophotometric titration, LIV, 399–406
coulometric, LIV, 400, 401

indicator electrodes, LIV, 405, 406
mediator-titrant, choice, LIV, 429
reference electrode, LIV, 405
titration cells, LIV, 402–404
Potentiostat, automatic, LIV, 131
Potentiostat-coulometer, circuit diagram, LIV, 407
Potter-Elvehjem homogenizer, LI, 116, 160; LII, 312, 370, 379, 391; LIII, 41, 59, 60
large-type, XXXII, 377
for liver preparation, XXXI, 4, 5, 7, 8, 14, 229
for plant-cell rupture, XXXI, 502
Power spectrum, XLVIII, 424
heterodyne, XLVIII, 428
for depolarized scattering from anisotropic molecules, XLVIII, 441
for moving charged particles, XLVIII, 432
for multicomponent biopolymer solution, XLVIII, 430
for uniform translational motion, XLVIII, 430
homodyne, XLVIII, 428
data analysis of spectrum analyzer data, XLVIII, 481, 482
for depolarized scattering from anisotropic molecules, XLVIII, 441
for multicomponent biopolymer solution, XLVIII, 430
for particle undergoing translation, XLVIII, 480
for rodlike molecules, XLVIII, 437
for two relaxation process system, XLVIII, 480
for uniform translational motion, XLVIII, 431
Power supply, for multifunction photometer, LVII, 534, 540
10-PP, *see* 10-Phenylphenothiazine
PPBE Medium, composition, XXXI, 643
PPD, *see* 2-5-Diphenyl-1,3,4-oxadiazole
PPG, *see* Propargylglycine
PP-ribose-P, *see* Phosphoribosyl pyrophosphate
Practolol, XLVI, 591
Prague strain, Rous sarcoma virus, LVIII, 380, 394
Praseodymium, magnetic circular dichroism, XLIX, 177
Praseodymium nitrate, derivatization of transfer RNA, LIX, 14
Prealbumin, XXXIV, 7; XLVI, 435–441
affinity labeling, XLVI, 439–441
Precipitin reaction, quantitative, XL, 245, 248
Prednisolone, molar absorption coefficient, XLIV, 186
Prednisone, effects, sodium transport, XXXVI, 444
Preelectrophoresis, of polymerized gels, XL, 131
Pregnancy, stage, by embryo crown-rump length, XXXIX, 239
5α-Pregnane-3β,20β-diol, radioimmunoassay, XXXVI, 21
5β-Pregnane-3α,20α-diol, radioimmunoassay, XXXVI, 21
Pregnane-3,20-dione, separation, XXXVI, 493
5α-Pregnane-3,20-dione
molecular bioassay, XXXVI, 465
PBP affinity, XXXVI, 125
radioimmunoassay, XXXVI, 20, 21
5β-Pregnane-3,20-dione
molecular bioassay, XXXVI, 465
PBP affinity, XXXVI, 125
Pregnant mare serum gonadotropin, in embryo preparation, XXXIX, 298
Pregna-1,4,6-triene-3,20-dione, synthesis, XXXVI, 425
Pregn-4-ene-3,20-dione, *see* Progesterone
Δ^5-Pregnene-3,20-dione, XLVI, 461
Pregnenolone
autoradiography, XXXVI, 153
bioassay, XXXVI, 458
chromatographic separation, XXXVI, 487
in cytochrome P–450$_{scc}$ assay, LII, 125
formation, LV, 10
metabolism, in adrenal gland, LII, 124
metabolites, gas chromatographic and mass spectral properties, LII, 387
oxime derivative, XXXVI, 18
PBP affinity, XXXVI, 125
radioimmunoassay, XXXVI, 20, 21

thin-layer chromatographic properties, LII, 126
Pregn-4-en-20β-ol-3-one, radioimmunoassay, XXXVI, 20
Δ^4-Pregnen-20α-ol-3-one, radioimmunoassay, XXXVI, 21
Δ^4-Pregnen-20β-ol-3-one, radioimmunoassay, XXXVI, 21
Δ^5-Pregnenolone, as testosterone precursor, XXXIX, 277
Pregnenolone-16-carbonitrile, inducer of microsomal cytochrome *P*–450, LII, 270
Prehormone, definition, XXXVI, 68
Prekallikrein, XXXIV, 432–435; XLV, 315
activation, changes, XLV, 322
assay, XLV, 315–319
activation, XLV, 315–317
arginine esterase, XLV, 317
immunochemical, XLV, 318, 319
variations in disease, XLV, 317, 319
properties, XLV, 321, 322
purification, XLV, 320, 321
stability, XLV, 321, 322
Prelumirhodopsin, LIV, 19–21, 24
Premelanosome, formation, XXXI, 389
Preparative acrylamide gel electrophoresis
purification of azoferredoxin, LIII, 320, 321, 326
of molybdoferredoxin, LIII, 320
Preparative electrophoresis, purification of acyl-CoA dehydrogenases, LIII, 517
Preputial gland
androgen metabolism and receptor activity, XXXIX, 454–460
androgen production and interconversion rates, XXXVI, 74
Preservative, for proteinaceous membrane, XLIV, 268
Pressman cell, ionophores, LV, 438
Pressure, effect on reaction rate, LVII, 363, 364
Pressure-drop, *see also* Reactor
in collagen supports, XLIV, 244
particle size, XLIV, 136, 138
Pressure filtration, for concentration of ribosomal protein solutions, LIX, 500
Pressure filtration device, transport studies, LVI, 296, 297
Pressure-jump technique, XLVIII, 308–320
bursting membrane technique, XLVIII, 311, 312
equipment for 90° light-scattering, XLVIII, 313–317
for pulsed pressure-perturbation, XLVIII, 312, 313
for transmitted light, XLVIII, 311, 312
historical, XLVIII, 310
optical applications, XLVIII, 311–313
capabilities, XLVIII, 319, 320
procedure, XLVIII, 316–319
Pressure regulator, LVII, 316
Pressure system, for microinjection apparatus, LVII, 316–318
Prestaining, autoradiography, LVIII, 286
Prethrombin 1
bovine, XLV, 141
human, XLV, 142
Prethrombin 2
bovine, XLV, 142–144
carbohydrate composition, XLV, 2, 148
human, XLV, 144, 145
PRG synthetase, *see* Phosphoribosylglycinamide synthetase
Priapulid, oxidase, LII, 13
P–rib–PP, *see* Phosphoribosyl pyrophosphate
Primary charge effect, XLVIII, 89
Priming, interferon production, LVIII, 296
Primuline, indicator for lipids, XXXV, 405, 412, 413, 415, 417
Primycin, XLIII, 120
Printed circuit card, disadvantage, LVII, 530
Prism
laser optics, XLVIII, 187
rotating, LIV, 44
Pristane, priming, LVIII, 350
Privet berry, free cells, XXXII, 724
PRL, *see* Prolactin

Proacrosin, XLV, 325
assay, XLV, 326, 327
properties, XLV, 329
purification, XLV, 327–329
Probe
for membrane potential determination, LV, 556, 557
extrinsic, LV, 558, 559
intrinsic, LV, 557, 558
optical
calibration, LV, 581–585
inner filter effect, LV, 576
selection, LV, 570–574
sources, LV, 574
redistribution across membranes, LV, 585
Procarboxypeptidase A, XXXIV, 418
cobalt-substituted, magnetic circular dichroism, XLIX, 168
Procion brilliant orange M.G.S., XLIV, 906
Procollagenase, mouse bone and skin, XL, 352
Proflavin
binding to tRNA, LIX, 111–114
fluorescence-detected circular dichroism, XLIX, 210
inhibitor, thrombin, XLV, 174
phosphorescence studies, XLIX, 246
saturation-transfer studies, XLIX, 497
Proflavinylsuccinic acid hydrazide, incorporation in tRNA, LIX, 114
Progesterone, XXXVI, 126; XLVI, 76
AAG binding, XXXVI, 95, 96
adsorption to various materials, XXXVI, 93, 94
antibodies to derivatives, XXXVI, 54
antihormone, XL, 291, 292
autoradiography, XXXVI, 150, 153
bromo derivatives, XXXVI, 378–383, 385–388
chromatographic separation, XXXVI, 488
in corpus luteum, XXXIX, 240
biogenesis, XXXIX, 241–244
L-cysteine conjugation, XXXVI, 387, 388
cytoplasmic receptors, XXXVI, 188–211
partial purification, XXXVI, 198
proteins contaminating preparations, XXXVI, 209–211
density gradient centrifugation, XXXVI, 159
derivatives, XXXVI, 18, 19–20
diazo derivatives, XXXVI, 413, 417
effect on testicular enzymes, XXXIX, 282
embryo studies, XXXIX, 301
HSA binding, XXXVI, 95, 96
levels, in perfused placenta, XXXIX, 252
logit-log assay, XXXVII, 9
in media, LVIII, 102, 107
metabolites
gas chromatographic and mass spectral properties, LII, 387
isolation, XXXVI, 489–498
TLC, XXXVI, 144
as mixed-function oxidase substrate, XXXI, 233
molecular bioassay, XXXVI, 456–465
in monkey plasma, XXXIX, 197
nuclear receptors
DNA binding, XXXVI, 307–313
interactions, XXXVI, 292–313
as 5α-oxidoreductase substrate, XXXVI, 473
PBG binding, XXXVI, 97, 101–102, 125, 126
in plasma, XXXVI, 39
protein-binding assay, XXXVI, 36
radioimmunoassay, XXXVI, 20, 21, 30; XXXIX, 217, 219
receptor protein, XXXVI, 411
assay, XXXVI, 49
reduction by 20β-hydroxysteroid dehydrogenase, XXXVI, 398
in serum, removal, XXXVI, 92
as testosterone precursor, XXXIX, 256
Progesterone-6β, radioimmunoassay, XXXVI, 20
Progesterone-binding globulin, *see* Progesterone-binding protein
Progesterone-binding protein, XXXVI, 91
amino acid composition, XXXVI, 124

carbohydrate composition, XXXVI, preparation, XXXI, 419; XXXVII, 124 378-380 characteristics, XXXVI, 125-126 production, LVIII, 183 secretory granule, use in study, distribution, XXXVI, 126 XXXI, 410, 418 homogeneity, XXXVI, 123-124 synthesis of fragment, XLVII, 584 isoelectric focusing, XXXVI, 101-102 thyrotropin-releasing factor effects, isoelectric point, procedure, XXXVI, XXXVII, 87, 89, 92 100 Prolidase, see Proline dipeptidase molecular weight, by electrophoresis, Proline XXXVI, 99 purification, XXXVI, 120-126 aminoacylation, isomeric specificity, LIX, 274, 275, 279, 280 sedimentation coefficients, XXXVI, analogs, XL, 322 103 steroid binding, XXXVI, 97 autoradiography, XXXVI, 150 cleavage rates, XLVII, 312-344 specificity, XXXVI, 125 in steroid hormone assay, XXXVI, 34, detection system, XLVII, 8 enzyme hydrolysis, XLVII, 43 Progesterone 11α -hydroxylase, see measurement, XL, 366 Progesterone 11α -monooxygenase in media, LVIII, 53, 63 Progesterone 11α -monooxygenase, LII, peptide synthesis, XLVII, 507, 509 problems, XLVII, 606, 607 Progestin solid-phase, XLVII, 610 corticosterone-binding globulin assay, release, by carboxypeptidase Y, XXXIX, 217-219 XLVII, 84, 85 in monkey plasma, XXXIX, 197 separation from hydroxyproline, in plasma, assay, XXXIX, 196 method, XL, 374 Programmed multiple development thermolysin hydrolysis, XLVII, 179 chromatography, XLIII, 180 transport Proinsulin colicin, LV, 536 enzymatic conversion, XXXVII, 334, by M. phlei cells, LV, 180 detergent-treated ETP, LV, 193 extraction, XXXVII, 332 DETP, LV, 188 immunoassay, XXXVII, 344 ETP, LV, 186, 187 as insulin precursor, XXXVII, 326 ghosts, LV, 187 plasma levels, XXXVII, 340-343 reconstitution in radioimmunoassay, XXXVII, 41 proteoliposomes, LV, Prolabo, silica, XLIV, 913 197-199 Prolactin reconstitution in detergent-treated chromatographic separation, XXXVII, ETP, LV, 197 393-396 L-[14C]Proline, XL, 322 effect on mammary epithelial cells, paper chromatography, XL, 374 XXXIX, 445, 446 Proline carrier protein, purification, LV, homogeneity, XXXVII, 400-402 196 human Proline dipeptidase assay, XXXVII, 390, 391 hydrolysis, XLVII, 41, 43 purification, XXXVII, 389-402 immobilization, XLVII, 44 Proline racemase, XLVI, 23 flow chart, XXXVII, 397 immunological assay, XXXVII, 139 Proline residue in pituitary glands, XXXVII, 390 conformational state, XLIX, 103, 104 pituitary tumor cell, LVIII, 527 effect on Raman spectra, XLIX, 116

protein conformation, XLIX, 99, 101
Prolyl hydroxylase, LII, 12
Prolyl-tRNA synthetase, XXXIV, 170
subcellular distribution, LIX, 233, 234
Promethazine, phagocytosis, LVII, 492
Promitochondria
isolation from *S. cerevisiae*, XXXI, 627–632
properties, XXXI, 632
of yeast, LV, 28
Promonocyte, origin, XXXII, 758
Promoter, mapping by electron microscopy, LIX, 846, 847
Promoter ligand, XXXIV, 599
Pronase, XLIV, 390; XLVI, 426
digestion of proteins, XL, 35
concentration and incubation period, XL, 37
hydrolysis, XLVII, 41, 43
immobilization, XLVII, 44
inhibition, assay, XLV, 861, 870
kinetics, of substrate diffusion, XLIV, 437
in oxyntic cell isolation, XXXII, 712
in phospholipase studies, XXXII, 137
potato lectin hemagglutination activity, L, 344
source, XXXII, 707
tissue culture, LVIII, 125, 126, 129
1,3-Propanediamine, XLVII, 383
1,2-Propanediol, in saponification mixture, LII, 312
1,3-Propane sultone
alkylation, XLVII, 116–122
reaction conditions, XLVII, 410
reaction with tributylphosphine, XLVII, 113
toxicity, XLVII, 119
Propanol
in dimethylaniline monooxygenase purification, LII, 147
substrate of reconstituted MEOS system, LII, 367
1-Propanol
alcohol oxidase, LVI, 467
in chromatographic separation of uridylate, LI, 388
electron spin resonance spectra, XLIX, 383
purification of *Cypridina* luciferin, LVII, 367
synthesis of dithiodiglycolic dihydrazide, LIX, 385
2-Propanol, XLIV, 131, 562; LIX, 158
alkylamine conversion, XLVII, 376
crystallization of transfer RNA, LIX, 7
electron spin resonance spectra, XLIX, 383, 470
inhibitor, of carboxypeptidase C, XLVII, 78
lysine derivative identification, XLVII, 477
protein solubility, XLVII, 113
purification of *N*-hydroxysuccinimide ester of bromoacetic acid, LIX, 159
of pyrimidine nucleoside monophosphate kinase, LI, 324, 328
removal of disulfide impurities, XLVII, 549
sequencing, XLVII, 337
solid-phase peptide synthesis, XLVII, 609
synthesis of adenosine derivatives, LVII, 118
Propanol-ammonia-water
separation of nucleotides, LIX, 170
of oligonucleotides, LIX, 343
Propanol-hydrochloric acid-water
separation of RNase T2 digestion products
of tRNA, LIX, 169, 170
Propapain, XXXIV, 532
Proparathyroid hormone, XXXVII, 326
cleavage and structural analysis, XXXVII, 357–360
gel electrophoresis, XXXVII, 355, 356
identification, XXXVII, 345–360
immunoprecipitation, XXXVII, 356, 357
preparation and isolation, XXXVII, 358, 359
radioactive, preparation and isolation, XXXVII, 358, 359
radioimmunoassay, XXXVII, 356
tryptic digestion, XXXVII, 359

Propargyl amine, XLVI, 161
inhibitor of monoamine oxidase, LIII, 501
Propargylglycine, XLVI, 163
inactivator of flavoprotein enzymes, LIII, 441, 442, 445, 446, 448
structural formula, LIII, 441
Properdin, XXXIV, 7, 731
pathway, LVII, 483
Propidium azide, XLVI, 648, 649
Propidium iodide
fluorescence spectrum, LVIII, 238
fluorescent dye, LVIII, 236–238, 242
preparation, LVIII, 247
β-Propiolactone, virus inactivation, XXXII, 579
Propionibacterium shermanii
growth, XXXV, 237–238
pyruvate orthophosphate dikinase, purification and properties, XLII, 209–212
source of succinyl-CoA:propionate CoA-transferase, XXXV, 235
Propionic acid
activator of aspartate carbamyltransferase, LI, 49
chromatographic separation of bases and nucleotides, LI, 551
flow dialysis experiments, LV, 683–685
measurement of ΔpH, LV, 562
substrate for succinyl-CoA:propionate CoA-transferase, XXXV, 235, 242
Propionyl coenzyme A
assay, LV, 211
conformation studies, XLIX, 354, 356
molecular weight determination by NMR, XLIX, 322
substrate for acetyl-CoA carboxylase, XXXV, 16
for fatty acid synthase, XXXV, 58, 65
Proplastid
isolation, XXXI, 570
marker enzymes, XXXI, 735, 744, 745
as plant cell precursors, XXXI, 495
Proportionality constant, XLVIII, 88
Propranolol, XLVI, 591, 594, 598–601
adenylate cyclases, XXXVIII, 147, 149, 173, 179, 180, 191
analogs, XLVI, 594–598
1,1′-Propylene 2,2′-bipyridylium dibromide, as mediator-titrant, LIV, 409
Propylene glycol
in cytochrome P–$450_{11\beta}$ assay, LII, 125
for microsomal protein precipitation, LII, 191
Propylene oxide, in subcellular fraction embedding, XXXII, 6, 7
1-*n*-Propyl-6-hydroxy-1,4,5,6-tetrahydronicotinamide, oxidation by O_2, LV, 537
Propyl lipoamide glass, XXXIV, 288–294, *see also* Lipoamide glass
Prostaglandin
activator of adenylate kinase, LI, 466
adenylate cyclases, XXXVIII, 147, 148, 173
analysis, quantitative by gas chromatography–mass spectrometry, XXXV, 374, 375
antibody specificity, XXXV, 290–298
antigen preparation, XXXV, 288, 289
binding affinity, XXXII, 109, 110
biological activity, XXXII, 109
chemical derivatives for gas chromatography–mass spectrometry, XXXV, 363–370, 375
acetyl, XXXV, 363–365, 370, 375
O-benzyl oxime, XXXV, 367
deuterium labeled, XXXV, 365, 367
methyl ester, XXXV, 363–365, 369, 375
O-methyl oxime, XXXV, 363–369, 375
preparation, XXXV, 364, 365
trimethylsilyl ether, XXXV, 363–369
chromatographic purification, XXXII, 118, 119
deuterated, XXXV, 374, 375
effect on estrogen formation, XXXIX, 251
extraction and purification, XXXII, 115–119
gas chromatography, XXXV, 363, 365–367, 377

gas chromatography–mass spectrometry, XXXV, 359–377
quantitative determination, XXXV, 374–377
structural determination, XXXV, 367, 369–374
hydrogenation, XXXV, 372
immunology, XXXV, 287–298
levels, XXXII, 109
lipophilic Sephadex chromatography, XXXV, 393
mass spectrometry, XXXV, 360–364, 367–377
metabolites
antibodies, XXXV, 296, 297
gas chromatography–mass spectrometry, XXXV, 359–377
molecular weight determination, by mass spectrometry, XXXV, 367
ozonolysis, XXXV, 372
radioimmunoassay, XXXV, 287–298
ring transformations, XXXV, 372
standard, XXXIX, 210
structural determination, by mass spectrometry, XXXV, 367, 369–374
tritium labeled, XXXV, 290
Prostaglandin E
assay, XXXII, 110–115
binding studies, XXXII, 112
extraction, XXXII, 116, 117
Prostaglandin E_1
affinity receptor, XXXII, 110
sodium transport, XXXVII, 256
Prostaglandin E_2
effect on thyroid-stimulating hormone, XXXVII, 86, 87
stock solution, XXXII, 111
in urine, XXXII, 110
Prostate gland
androgen production and interconversion rates, XXXVI, 74
androgen receptor complex, XXXVI, 374
preparation for collagen studies, XL, 311
Protactinomycin, XLIII, 340
Protactinomycin-like antibiotic, XLIII, 340
Protamine, LVIII, 49
in media, LVIII, 70
as plant-cell protectant, XXXI, 537, 538
protein kinase assay, XXXVIII, 309, 313, 327, 328, 340, 348, 349, 357
Protamine kinase, XXXIV, 261–264
Protamine-Sepharose chromatography, of cytoplasmic progesterone receptors, XXXVI, 205–206
Protamine sulfate
ATPase purification, LV, 326
batch standardization, LI, 183
calcium transport assay, LVI, 241
for microsomal protein precipitation, LII, 93
in microsomal protein solubilization, LII, 146
in 5′-nucleotidase purification, XXXII, 370–372
in phospholipase C purification, XXXII, 157
purification of carbamoyl-phosphate synthetase, LI, 32
of hydrogenase, LIII, 287
of melilotate hydroxylase, LIII, 554
of nucleoside deoxyribosyltransferase, LI, 449
of phosphoribosylglycinamide synthetase, LI, 183
of purine nucleoside phosphorylase, LI, 519
of ribonucleoside triphosphate reductase, LI, 252
of salicylate hydroxylase, LIII, 533
of thymidine phosphorylase, LI, 439, 440
Protease, XLVI, 197, *see also* specific type
acid
N–terminal sequences, XLV, 449
pepstatin inhibited, XLV, 689
action, protection against, LIX, 267
Aeromonas neutral, *see Aeromonas,* neutral protease
alkaline, XXXIV, 5
active site sequence, XLV, 425
Aspergillus oryzae, XLV, 724
insulin, XLV, 430, 432

assay, XLV, 12
with bacterial luciferase, LIII, 564; LVII, 199–201
collagen supports, XLIV, 244
concentration, effect on stability, XLV, 431
contamination, qualitative assay, LIX, 264
cysteine, *see* Cysteine proteinase
effect on photolabeled proteins, XLVI, 345, 346
immobilization
choice of pore diameter, XLIV, 160
conditions, XLVII, 44
coupling time and autolysis, XLIV, 144
immobilized, conformational transitions, XLIV, 528–538
inhibitors
antithrombin-heparin cofactor, XLV, 668
boar seminal plasma, XLV, 834
from cereal grains, XLV, 723
ciliated membranes, respiratory, XLV, 869
guinea pig seminal vesicles, XLV, 825
from legume seed, XLV, 697
in histone extraction XL, 109
microbial, XLV, 678
naturally occurring, XLV, 638
from plant sources, XLV, 695, 696
from potatoes, XLV, 728
submandibular glands, dog, XLV, 860
insulin-specific, XXXIV, 5
α-lytic, amino acid composition, XLV, 423
as marker enzyme, XXXI, 735
of membranes, inactivation, XXXII, 98
microorganismal mitochondria, LV, 142
N. crassa mitochondria, LV, 148
neutral, XXXIV, 5, 437, 438
from *Bacillus subtilis*
circular dichroism spectrum, XLIX, 161
tryptophan content, XLIX, 161, 162
neutralization, LVIII, 77
in NMR studies of membranes, XXXII, 211
in protein inactivation studies, XXXVII, 210
sea urchin eggs, *see* Sea urchin egg protease
snake venom
amino acid composition, XLV, 465
assay method, XLV, 459
biological activities, XLV, 467
calcium and sulfhydryl group, role, XLV, 466
characteristics, XLV, 459
distribution, XLV, 467, 468
homogeneity, XLV, 462
inactivators, XLV, 466
metal ion requirement, XLV, 466
pH, XLV, 465
properties, XLV, 463, 464
purification, XLV, 460
specific activity, XLV, 460
specificity, XLV, 466
stability, XLV, 464
units, XLV, 460
staphylococcal
cleavage at glutamic acid, XLVII, 189–191
contaminant, XLVII, 190
reaction conditions, XLVII, 189, 190
specificity, XLVII, 190
stability, XLVII, 191
Staphylococcus aureus, *see* *Staphylococcus aureus* protease
tissue homogenization, LV, 4
wheat, XXXIV, 5
Protection method
in determination of ribosomal protein interactions, LIX, 591–602
enzymes, LIX, 600, 601
principles, LIX, 591, 592
procedure, using chemical modification, LIX, 592–595
using proteolytic protection, LIX, 595–598
Protective compound, photoaffinity labeling, LVI, 644, 645

Protein, *see also* Dipeptide; Metalloprotein, nonheme; Peptide; specific type
  adhesion, *see* Adhesion protein
  alkylated, isolation, XLVII, 115, 120
  antichromosomal, XL, 191
  assay, XXXI, 19
    in lipid extracts, XXXV, 334–339
  B1, XLVI, 321, 325–327
  B2, XLVI, 321, 325–327
  balance, substrate entry, XXXIX, 495
  B1-B2 complex, XLVI, 325–327
  binding sites, XLVI, 561–567
  bound
    colorimetric assay, XLIV, 387
    determination methods, XLIV, 94, 386–390
  brain mitochondria, LV, 58
  Ca(II)-dependent, XXXIV, 592–594
  carbethoxylation, XLVII, 434–438
  cell content, LVIII, 145
  chromatin, XL, 101
    acidic phosphorylated
      binding to DNA, XL, 181
      comparison of procedures, XL, 190
      isolation and characterization, XL, 171, 177, 181
      measurement, XL, 187
      reagents, XL, 171
    membrane filter analysis, XL, 185
    sucrose gradient analysis, XL, 185
  chromosomal immunochemical characteristics, XL, 191
  circular dichroism, XXXII, 220, 221
  cleavage of asparaginyl-glycyl bonds, XLVII, 132–145
    of aspartyl-prolyl bonds, XLVII, 145–149
    by chemical methods, XLVII, 127–161
    at cysteine bonds, after cyanylation, XLVII, 129–132
    with enzymes, XLVII, 165–191
    with hydroxylamine, XLVII, 132–145
  concentrating, method, LVI, 530
  concentration
    activity, XLIV, 383
    by centrifugation, LII, 105
    cross-linking, LVI, 635
    with DEAE-cellulose, LII, 136
    determination
      comparative methods, LIX, 500, 501, 586
      by dry weight method, XLVIII, 155–162
      using radiolabeling, XLVIII, 327
      using ultraviolet absorbancy, XLVIII, 161, 162
    effect on spectral observations, LII, 213, 219, 220, 225, 226, 262
    by solid sucrose dialysis, LII, 104
    by ultrafiltration, LII, 94, 99, 100, 120, 121, 154, 156, 174, 178
    units, XLVIII, 161
  concentration methods, LIX, 499, 500
  conformation
    electron spin resonance spectra, XLIX, 420–427
    from IR spectroscopy, XXXII, 251, 252
    theory, XLIX, 98, 99
    thermolysin, XLVII, 186–188
  conformational changes, XLVI, 65, 460, 512, 513, 515
  content in microsomal fractions, LII, 88
  copolymerization, XLIV, 195–201
    definition, XLIV, 195
  cross-reacting, LVIII, 177
  dansylation, XL, 159
  decarbethoxylation, XLVII, 435, 438, 439
  degradation, XXXVII, 250
    measurement, XL, 253
    rate studies, XXXVII, 243–245
  denaturation studies, XLIX, 10, 11, 420
    detergents, LVI, 738, 739
  derivatives of glass, XXXIV, 67–72
  desalting, LVI, 530
  detergent binding, cmc, LVI, 737
  determination, LV, 330
    after ammonium sulfate fractionation, LI, 438
    by biuret method, LIII, 4, 41, 55, 280, 497
    Bücher method, LI, 438, 518

of deuterium content, LIX, 645, 646
dithiothreitol, LI, 160
Folin method, LI, 264
Hartree modification of Lowry method, LI, 250
hemoglobin analysis, LI, 469
in initiation factors, LX, 31, 235, 236
Layne method, LI, 214
Leggett Bailey method, LI, 447
Lowry method, LI, 6; LIII, 35, 497
microprecipitation modification of Lowry method, LI, 239
modified biuret test of Schacterle and Pollack, LI, 292
for plant organelle purity, XXXI, 542–544
protocol, LVIII, 144–146
by trichloroacetic acid precipitation, LIII, 365
UV absorbance, LI, 339, 426, 469, 484, 503; LIII, 280
Waddel method, LI, 518
Warburg and Christian method, LI, 317, 339
Zamenhof method, LI, 526
dissociation constants, measurement, XLVIII, 321–346
dissociation kinetics, XLVIII, 327–331
distribution in mitochondrial subfractions, LV, 93, 94, 103
effect on chemiluminescent reactions, LVII, 439, 440
electron microscope, autoradiographic localization, labeling conditions, XL, 15
elongation, relation to initiation, LX, 161–163
enhancer of aromatic hydrocarbon solubility, LII, 237
enzymatic digestion, XL, 33
concentration and incubation periods, XL, 37
enzyme, in synthesis and degradation, XL, 241
enzyme-binding to applications, XLIV, 709, 710
exposed, on membranes, lactoperoxidase in determination, XXXII, 103–109
extended topographical structure
definition, XLIX, 105, 106
hydrogen-bonding patterns, XLIX, 109, 110
vibrational frequencies, XLIX, 113–115
in fat cells, XXXI, 64, 65
in fractions, XXXI, 66
fluorescence quenching, by solute, XLIX, 222–236
fluorescent labeling, XLVIII, 358
folding, XLIX, 420
gel electrophoresis, XXXII, 70–81
glycosylation, dolichol intermediates, L, 402–435
gravimetric determination, XXXV, 221–225
α-helix, *see* α-Helix
heme
axial ligands, XLIX, 172, 173
magnetic circular dichroism, XLIX, 170–175
resonance Raman spectra, XLIX, 127–137
sensitivity to redox state, XLIX, 170–172
spin state, XLIX, 174, 175
high potential iron sulfur, electron paramagnetic resonance studies, XLIX, 514
hormone-binding, XXXVI, 1–88
hydration theory, XLVIII, 74
on hydrocarbon/water interface
bacteriorhodopsin-mediated, light dependent volta potential changes, LV, 774, 775
chemicals, LV, 773, 774
device, LV, 773
interpretation, LV, 775, 776
principle, LV, 773
hydrogen exchange, XLIX, 24–39
for free peptide hydrogens, XLIX, 26–30
relative rates, XLIX, 25
of slow hydrogens, XLIX, 34–39
hydrolysis by hydrochloric acid treatment, LIX, 313
hydrophobic, preparation, LV, 713, 714
immunologically reactive, determination, XL, 242

incorporation into filters, LV, 589–591
L2, XLVI, 187
L27, XLVI, 187
ligand interaction, XXXVI, 3–16
equilibrium models, XXXVI, 3–9
kinetic models, XXXVI, 12
multivalency factors, XXXVI, 9
parameter fitting, XXXVI, 13–15
Sips and Hill plot, XXXVI, 10–12
as ligands, XXXIV, 51–56
acylazide derivatives, XXXIV, 52
carbodiimide coupling, XXXIV, 52, 53
coupling, XXXIV, 51–56
aryl amine derivative, XXXIV, 54
functional group densities, XXXIV, 52
glutaraldehyde, XXXIV, 56
spacers, XXXIV, 52
thiourea links, XXXIV, 55
lipid-associated, partial specific volume determination, XLVIII, 11–23
in liposomes, *see* Liposome-protein complex
in liver fractions, XXXI, 29, 35
loosely bound, removal, LVI, 421, 423
lysine content, XXXV, 339
magnetic circular dichroism, XLIX, 159–164
measurement with Folin phenol reagent, XL, 375
in media, LVIII, 70, 88
of membranes, XXXII, 70–131
gel electrophoresis, XXXII, 70–81
molecular weight determination by gel electrophoresis, XXXII, 92–102
messenger ribonucleic acid-binding
completeness of reaction, LX, 398
eight-subunit, LX, 397, 398
purification, LX, 395–398
specificity, LX, 398, 399
three-subunit, LX, 395, 396
two subunit, LX, 396, 397
metabolism, in perfused liver, XXXIX, 34
methylated, altered tryptic digestion products, LIX, 789, 790
mitochondrial
labeling procedure, LVI, 52
molecular weights, LVI, 75, 77
resolution, LVI, 602
sequencing, LVI, 197
solubilization, LVI, 673, 674
modification, XXXIV, 182–195, *see also* Peptides, modified
affinity labeling, XXXIV, 182
antibodies, XXXIV, 183, 184
general modification, group specific reagent, XXXIV, 182
selective modification, XXXIV, 182
site-directed modification, XXXIV, 182
NMR spectrum, principle, LIV, 192–194
nonhistone chromosomal, *see* Nonhistone chromosomal protein
nonionic detergents, LVI, 740, 741
nonribosomal particles
isolation of translational control ribonucleic acid from, LX, 543–545
purification, LX, 542, 543
nucleic acid-associated, staining, XL, 21
nylon derivatives, XXXIV, 203
oligomycin sensitivity-conferring, complex V, LVI, 584, 597
oligonucleotide-labeled, XLVI, 675, 676
penetration model, XLIX, 35, 36
phosphodiesterase assay, XXXVIII, 217
phosphorylated
fractionation, LX, 516, 517
identification by polyacrylamide gel electrophoresis, LX, 517–521
photooxidation, XLVI, 561–567
in plasma membranes, XXXI, 87
polarography, LVI, 477
potential modifying, LIII, 237
preparation for radioactivity determination, LVI, 23

pulse-labeling by *in vivo* incorporation of leucine
homogeneous labeling of polypeptides, LVI, 52
labeling kinetics, LVI, 52–54
principle, LVI, 51
quinone and phenol reactions, XXXI, 529, 530
radiolabeling, XLVIII, 332, 333
Raman spectroscopy, XXXII, 256, 257
solubilization, XXXII, 71–73
rate of turnover, effect of isotope reutilization, XL, 255
reaction with glutaraldehyde, XLIV, 551–555
reactive residues, percent composition, XLIV, 13
recovery from polyacrylamide gels, LIX, 313
reductive alkylation of amino groups, XLVII, 469–478
procedure, XLVII, 469–472
properties, XLVII, 473–475
resolution in polyacrylic gels, XLIV, 56
ribosomal, *see also* Ribosomal protein
S1
antibodies, LX, 429, 440–442
assay
with polyuridylate, LX, 418, 419, 447, 448
of purity, LX, 423–425, 439
binding
of messenger ribonucleic acid, LX, 14, 15, 337
by 30 S ribosomal subunits, LX, 430, 434, 435, 452–454
denaturation of polyribonucleotides, LX, 451, 452
functional role
elongation, LX, 426, 427, 436
initiation, LX, 417, 418, 426, 430, 431, 436, 446, 447, 456
isolation from *E. coli* ribosomes, LX, 448–451, 456–459
from 30 S ribosomal subunits, LX, 420, 421, 437–441
from 70 S ribosomes, LX, 421–423
properties, LX, 418, 425, 451, 452
separation from 30 S subunit, LX, 429
from 30 S (−S1) subunits, LX, 428–430
S2, binding of messenger ribonucleic acid, LX, 14, 15
rotational diffusion in membranes, measurement, LIV, 47–61
R-type, XXXIV, 305, 310–312
50 S, XLVI, 187
salt-extraction, principles, LIX, 481, 482
procedure, LIX, 484–486
saturation-transfer studies, XLIX, 488–494
separation on polyacrylamide gels, LIX, 313
solubility, solvents, XLVII, 113
solubilization, LII, 141
essential reagents, LII, 202
using cholate, LII, 111, 115, 118–120, 122, 128, 129, 195
using deoxycholate, LII, 82, 83, 191, 202, 360
using Renex 690, LII, 90
using Triton X–100, LII, 98, 99
solution
anaerobic storage, LIII, 315
deionization procedure, XLVIII, 158
density approximation, XLVIII, 160
stains, XXXII, 99
structure
reconstruction, energy transfer, XLVIII, 375, 376
relationship with CD spectra, XL, 235
substrate, radioactive, preparation, XL, 342
support, for enzyme, XLIV, 903, 904

synthesis, XLVI, 181, 315, 501–504, 515
assay, LX, 465
8-azido-GTP, LVI, 646
in cell-free system, LIX, 300, 301
fractional incorporation rate, calculation, LIX, 317–321
control studies, XXXVII, 295
cross linking, XLVI, 503, 504
effect of polyamines, LX, 555, 556
hormone induction of biorhythms, XXXVI, 480
inhibition, LX, 475, 476, 478, 480–484, 547, 548, 574–577, 780
inhibitors, XL, 241, 286, 288
effects on algae greening, XXXII, 869
initiation mechanism, LX, 62
by mitochondria, LVI, 18–23
inhibitors, LVI, 29–32
isolated polysomes, LVI, 27–29
isolated ribosomes, LVI, 24–27
in vivo assay, LVI, 99–101
mutant, selection, LVIII, 317–319
by *N. crassa* mitochondria, LV, 147, 148
rate, LX, 66, 67, 466, 467, 475, 476, 478, 480–482
in reticulocyte lysate, LX, 464–475, 513–516
synthetic, XXXIV, 631–638
translation, XLVI, 181
triangulation, *see* Protein pair scattering
trinitrophenylated, for protein assay in lipid extracts, XXXV, 334–339
tryptophan content, determination, XLIX, 161–164
turnover
hormone effects on perfused organs, XXXVII, 238–250
rate-limiting reactions, XXXVII, 245–250
measurements, LIX, 321
unordered topographical structure
definition, XLIX, 106
states, XLIX, 111
using DNA as messenger, LIX, 850, 851
viscosity barrier centrifugation, LV, 133, 134
visualization on gels, LVI, 71, 72
in vitro radiolabeling, LIX, 586
Protein-315, affinity labeling, XXXIV, 185
β-Protein, 5α-dihydrotestosterone binding, XXXVI, 313–319
Protein A, LIII, 373
amino acid content, LIII, 379
chemical properties, LIII, 379, 380
γ-globulin, LVI, 50
midpoint redox potential, LIII, 380
molecular weight, LIII, 379
purification, LIII, 377
stability, LIII, 380
Proteinase, XXXIV, 546, *see also* specific types
streptococcal, structural analysis with thermolysin, XLVII, 184
Proteinase K
in extraction of radiolabeled RNA, LIX, 836
for ribonuclease inactivation, XXXI, 720
in studies of protein-RNA interactions, LIX, 576
from *Tritirachium album,* specificity, LIX, 600
Protein B, LIII, 373
assay, LIII, 381, 382
inhibitors, LIII, 381
purification, LIII, 380, 381
Protein B1, LI, 227
composition, LI, 234
extinction coefficient, LI, 234
magnesium requirement, LI, 234
molecular weight, LI, 234
purification, LI, 231, 232, 234
stabilization by dithiothreitol, LI, 228, 230
Protein B2, LI, 227
extinction coefficient, LI, 235
molecular weight, LI, 235
purification, LI, 231, 232, 234
Protein-binding reaction, LVII, 424–445
assay, of homogeneous competitive reaction, LVII, 441, 442
heterogeneous, definition, LVII, 442
homogeneous, definition, LVII, 441

luciferase-mediated monitoring, LVII, 113–122
monitoring, with chemiluminescence, LVII, 424–445
oxidation systems, LVII, 426, 427, 438, 439
principle, LVII, 424
reaction procedure, LVII, 426
Protein body
isolation, XXXI, 518
marker enzymes, XXXI, 735, 745
Protein crystal
insolubilization, XLIV, 548–556
insolubilized, X-ray analysis, XLIV, 555
properties, XLIV, 546–548
Protein J, XXXIV, 133
Protein kinase, XXXIV, 5; XLVI, 83, 344, *see also* Casein kinase
activation in intact cells, XXXVIII, 358, 359
hormonal regulation, XXXVIII, 363–365
materials, XXXVIII, 359
methodology for adipose tissue, XXXVIII, 359–363
for other tissues, XXXVIII, 365–367
assay, LX, 499, 507
8-azido-cAMP, LVI, 646
bovine heart
assay, XXXVIII, 308–310
properties, XXXVIII, 313–315
purification, XXXVIII, 310–313
cAMP assay, XXXI, 106
activation
method, XXXVIII, 66–70
modifications, XXXVIII, 72, 73
sensitivity, reproducibility and validation, XXXVIII, 70–72
cAMP dependent
assay, XXXVIII, 287–290
preparation of, XXXVIII, 66, 67
subunit, XXXVIII, 299
cAMP inhibitor, XXXVIII, 281
catalytic subunit preparation, XXXVIII, 306–308, 314, 315, 326, 387
cGMP assay
materials, XXXVIII, 90–92
principle, XXXVIII, 90
procedure, XXXVIII, 92–95
cGMP dependent, XXXVIII, 329, 330
characterization, XXXVIII, 335–349
preparation, XXXVIII, 91
purification, XXXVIII, 330–335
standard assay, XXXVIII, 330
subunit, XXXVIII, 344–346
classification, XXXVIII, 290, 291
criteria, XXXVIII, 291–293
generality, XXXVIII, 293
as contaminants, LX, 477, 478
cyclase preparations, XXXVIII, 143
in cytoplasmic progesterone receptor preparations, XXXVI, 210, 211
erythrocyte, cAMP derivatives, XXXVIII, 398
examples, XXXVIII, 297–299
procedure, XXXVIII, 293–296
follicle-stimulating hormone activation, XXXIX, 253
heme-regulated translational inhibitor, LX, 476, 484, 490, 491
inhibition of methionyl-transfer ribonucleic acid binding, LX, 287–290
inhibitor
for cyclic AMP assay, XXXI, 105; XXXVIII, 49, 52, 53
preparation, XXXVIII, 52
liver cytosol
assay, XXXVIII, 323, 324
mechanism of cAMP action, XXXVIII, 323
properties, XXXVIII, 327, 328
purification, XXXVIII, 324–327
modulator, cGMP-dependent kinase, XXXVIII, 346–349
nomenclature, LX, 507–508
phosphorylation
histone, LX, 524
initiation factors, LX, 507–511, 525, 532, 533
protein inhibitor, XXXVIII, 296, 350, 351, 360, 361
assay, XXXVIII, 351–353

interaction with enzyme, XXXVIII, 356, 357
with substrates and activators, XXXVIII, 357, 358
properties, XXXVIII, 355, 356
purification, XXXVIII, 353–355
rabbit reticulocyte, XXXVIII, 297, 315, 316, 322
assay, XXXVIII, 316, 317
crude preparation, XXXVIII, 317, 318
purification of kinase I, XXXVIII, 318, 319
of kinase II, XXXVIII, 319–321
rabbit skeletal muscle, XXXVIII, 299–301
catalytic subunit, XXXVIII, 306–308
holoenzyme, XXXVIII, 301–306
regulation by cyclic adenosine monophosphate, LX, 484, 496
regulatory subunit
inhibition, XXXVIII, 296
preparation, XXXVIII, 327
from swine kidney, XLII, 375, 394–397
assay, XLII, 394–396
molecular weight, XLII, 397
properties, XLII, 397
purification, XLII, 396, 397
utilization of nucleotide triphosphates, LX, 287–290, 503, 508
Protein L1
E. coli
purification, LIX, 492–496, 502
RNA binding, LIX, 553, 574, 575
storage, LIX, 498
rat liver, molecular weight, LIX, 512
Protein L2
Bacillus subtilis, isolation, LIX, 440, 441
E. coli
purification, LIX, 492, 493, 494, 495, 496, 502
puromycin binding, LIX, 809, 810
RNA binding, LIX, 553, 559
rat liver, molecular weight, LIX, 512
Protein L3
E. coli
purification, LIX, 492, 495, 496, 497, 502
puromycin binding, LIX, 809
RNA binding, LIX, 553
rat liver
loss, during analysis, LIX, 521
molecular weight, LIX, 512
Protein L4
E. coli
purification, LIX, 491, 492
RNA binding, LIX, 553
rat liver, molecular weight, LIX, 512
Protein L5
E. coli
purification, LIX, 492, 547, 562
RNA binding, LIX, 553, 563, 568, 570, 580
rat liver, molecular weight, LIX, 512
Protein L6
E. coli
purification, LIX, 492–496, 502, 547
RNA binding, LIX, 553
storage, LIX, 499
rat liver, molecular weight, LIX, 512
Protein L7
E. coli
molecular weight, LIX, 719, 720
structural model, LIX, 720, 725
structural parameters, LIX, 719, 720
rat liver, molecular weight, LIX, 512
Protein L7/12, *E. coli*
purification, LIX, 492–494, 497, 546, 818
ribosomal functions, LIX, 824
storage, LIX, 499
Protein L8, rat liver, molecular weight, LIX, 512
Protein L9
E. coli, purification, LIX, 492–496
rat liver
loss, during analysis, LIX, 521
molecular weight, LIX, 512
Protein L10
E. coli
molecular weight, LIX, 719, 720

purification, LIX, 492–494, 496, 497, 818
storage, LIX, 498, 499
structural model, LIX, 725
structural parameters, LIX, 719, 720
rat liver, molecular weight, LIX, 512
Protein L11
E. coli
molecular weight, LIX, 718, 719
peptidyltransferase activity, LIX, 816
purification, LIX, 492–494, 496, 502, 818
puromycin binding, LIX, 809, 810
ribosomal functions, LIX, 823, 824
storage, LIX, 499
structural model, LIX, 725
structural parameters, LIX, 718, 719
rat liver
molecular weight, LIX, 512
purification, LIX, 512
Protein L12, rat liver, molecular weight, LIX, 512
Protein L13
E. coli
purification, LIX, 492, 494, 495, 497
puromycin binding, LIX, 806
RNA binding, LIX, 553
rat liver, molecular weight, LIX, 512
Protein L14
E. coli
purification, LIX, 492
puromycin binding, LIX, 806
rat liver, molecular weight, LIX, 512
Protein L15
E. coli
purification, LIX, 492, 493, 495, 496
puromycin binding, LIX, 806
rat liver, molecular weight, LIX, 512
Protein L16
E. coli
purification, LIX, 492–494, 496, 497, 502
ribosomal functions, LIX, 823
RNA binding, LIX, 553
rat liver, molecular weight, LIX, 512
Protein L17
E. coli
purification, LIX, 492, 494, 496, 502, 547
puromycin binding, LIX, 806, 809
rat liver, molecular weight, LIX, 512
Protein L18
E. coli
molecular weight, LIX, 718, 719
purification, LIX, 492–494, 496, 562
puromycin binding, LIX, 806, 807, 809, 810
RNA binding, LIX, 553, 563, 568, 570, 578–580
structural model, LIX, 725
structural parameters, LIX, 718, 719
rat liver, molecular weight, LIX, 512
RNA complex, *E. coli,* structural parameters, from scattering data, LIX, 723
Protein L19
E. coli
purification, LIX, 492, 494, 495, 497
puromycin binding, LIX, 806, 807
rat liver, molecular weight, LIX, 512
Protein L20
E. coli
puromycin binding, LIX, 806
RNA binding, LIX, 553
rat liver, molecular weight, LIX, 512
Protein L21
E. coli
purification, LIX, 492, 502
puromycin binding, LIX, 806, 807
rat liver, molecular weight, LIX, 512
Protein L22
E. coli
purification, LIX, 492, 495, 496, 502
puromycin binding, LIX, 806, 807
rat liver
loss, during analysis, LIX, 521
molecular weight, LIX, 512
Protein L23
E. coli
purification, LIX, 492, 495–497, 502

puromycin binding, LIX, 800, 801, 803, 806, 807, 810
RNA binding, LIX, 553, 581
rat liver
loss, during analysis, LIX, 521
molecular weight, LIX, 512
purification, LIX, 512
Protein L24
E. coli
purification, LIX, 492, 494–496, 502
puromycin binding, LIX, 806
RNA binding, LIX, 553, 563, 580, 581
storage, LIX, 498
rat liver, molecular weight, LIX, 512
Protein L25
E. coli
molecular weight, LIX, 718, 719
purification, LIX, 492–494, 496, 547, 562
puromycin binding, LIX, 806, 807
ribosomal functions, LIX, 823
RNA binding, LIX, 553, 563, 568, 570, 580
structural model, LIX, 725
structural parameters, LIX, 718, 719
rat liver, molecular weight, LIX, 512
Protein L26, rat liver, molecular weight, LIX, 512
Protein L27
E. coli
purification, LIX, 492–494, 496, 497, 502
puromycin binding, LIX, 809, 810
ribosomal functions, LIX, 822, 823
rat liver, molecular weight, LIX, 512
Protein L28
E. coli, purification, LIX, 492–494, 496, 502
rat liver, molecular weight, LIX, 512
Protein L29
E. coli, purification, LIX, 492, 494, 496
rat liver, molecular weight, LIX, 512
Protein L30
E. coli
purification, LIX, 492, 493, 495, 496, 502
storage, LIX, 499
rat liver, molecular weight, LIX, 512
Protein L31, rat liver, molecular weight, LIX, 512
Protein L32
E. coli, purification, LIX, 492, 495, 547
rat liver, molecular weight, LIX, 512
Protein L33
E. coli, purification, LIX, 492, 494, 496
rat liver, molecular weight, LIX, 512
Protein L34
E. coli, purification, LIX, 492
rat liver, molecular weight, LIX, 512
Protein L35, rat liver, molecular weight, LIX, 512
Protein L36, rat liver, molecular weight, LIX, 512
Protein-nucleic acid complex, *see* Nucleic acid-protein complex
Proteinoplast, *see* Proteoplast
Protein pair scattering, LIX, 656–669, 702–706, *see also* Neutron scattering
collimation, LIX, 662–665
data collection, LIX, 656, 667, 668
data processing, LIX, 668, 669
experimental range, choice, LIX, 661, 662
experimental setup, LIX, 659, 660
flux measurement, LIX, 667
sample cell, LIX, 656, 657, 660
sample changer, LIX, 665, 666
sample compositions, LIX, 654, 655, 666, 667, 702, 703
sample preparation, LIX, 655, 657
sample thickness, LIX, 656, 657
scattering buffer, LIX, 655
transmission measurements, LIX, 667
Protein-protein interaction, XXXIV, 592–594
mechanochemistry, XLIV, 571–576
in nonradiative energy transfer, LVII, 263–267
of *Pholas* luciferin-luciferase system, LVII, 393–398
Protein-protein pair, studies, protein modification, LIX, 591–602
Protein:ribonucleic acid ratio, LIX, 419, 420

Protein-RNA complex
analysis, LIX, 565–570
formation, LIX, 562–565
precipitation, by trichloroacetic acid, LIX, 837
preparation, LIX, 586, 587
structural models, from scattering data, LIX, 723, 724
Protein-RNA interaction, *see also* tRNA-protein interaction
analysis, LIX, 587–591
in bacterial ribosome, LIX, 551–583
chemical modification, LIX, 576, 577
circular dichroism, LIX, 579
electron microscopic visualizations, LIX, 581, 582
fluorescence techniques, LIX, 578
kinetic parameters, LIX, 577, 578
photochemical cross-linking, LIX, 576
scattering techniques, LIX, 578, 579
specificity, LIX, 569, 570
thermal transitions, LIX, 580, 581
Protein S1
E. coli
binding sites on 30 S subunit, LIX, 400, 401
isolation, LIX, 398–401
molecular weight, LIX, 718, 719
protein-protein interactions, LIX, 596, 599
purification, LIX, 485, 487, 489–651
radiolabeled, LIX, 791, 795
structural parameters, LIX, 718, 719
rat liver
loss, during analysis, LIX, 521
molecular weight, LIX, 512
Protein S2
E. coli
protein-protein interactions, LIX, 596, 599
purification, LIX, 485, 487, 489–651
storage, LIX, 498
rat liver
loss, during analysis, LIX, 521
molecular weight, LIX, 512
Protein S3
E. coli
purification, LIX, 485, 487–489, 651, 652
puromycin binding, LIX, 809
rat liver, molecular weight, LIX, 512
Protein S4
E. coli
molecular weight, LIX, 718, 719
protein-protein interactions, LIX, 596, 597–599
purification, LIX, 485–489, 625, 651
puromycin binding, LIX, 809
RNA binding, LIX, 553, 563, 575–577, 579, 581, 582
30 S subunit structure, LIX, 622–629
structural parameters, LIX, 718, 719
structure, LIX, 609
rat liver
loss, during analysis, LIX, 521
molecular weight, LIX, 512
RNA complex, *E. coli,* structural model from scattering data, LIX, 723, 724
Protein S5
E. coli
degradation, LIX, 484
purification, LIX, 485, 487–651
puromycin binding, LIX, 809
rat liver, molecular weight, LIX, 512
Protein S6
E. coli, purification, LIX, 485–488, 502, 650–652
rat liver
loss during analysis, LIX, 521
molecular weight, LIX, 512
Protein S7
E. coli
protein-protein interactions, LIX, 594–597
purification, LIX, 485, 486, 488, 651
RNA binding, LIX, 553, 576, 579
30 S subunit structure, LIX, 622–625
rat liver, molecular weight, LIX, 512

Protein S8
  E. coli
    purification, LIX, 485, 487–489, 502, 650, 651
    RNA binding, LIX, 553, 559, 563, 576, 577, 579, 581, 582
    30 S subunit structure, LIX, 622–625
    storage, LIX, 498
  rat liver, molecular weight, LIX, 512
Protein S9
  E. coli, purification, LIX, 485, 486, 502, 651
  rat liver
    loss, during analysis, LIX, 521
    molecular weight, LIX, 512
Protein S10
  E. coli, purification, LIX, 485–487, 650, 651
  rat liver, molecular weight, LIX, 512
Protein S11
  E. coli, purification, LIX, 651
  rat liver, molecular weight, LIX, 512
Protein S12
  E. coli
    purification, LIX, 651, 652
    puromycin binding, LIX, 806, 807
  rat liver, molecular weight, LIX, 512
Protein S13
  E. coli
    purification, LIX, 485, 488–490, 651
    puromycin binding, LIX, 806, 807
  rat liver, molecular weight, LIX, 512
Protein S14
  E. coli
    chloramphenicol binding, LIX, 802, 803, 806, 807
    protein-protein interactions, LIX, 596
    purification, LIX, 485, 487, 489, 651, 652
    puromycin binding, LIX, 809, 810
  rat liver, molecular weight, LIX, 512
Protein S15
  E. coli
    purification, LIX, 485, 487–490, 651, 652
    RNA binding, LIX, 553, 577, 579
    30 S subunit structure, LIX, 622–625
  rat liver, molecular weight, LIX, 512
Protein S16
  E. coli, purification, LIX, 485, 487, 489, 652
  rat liver, molecular weight, LIX, 512
Protein S17
  E. coli
    purification, LIX, 485, 488, 489, 652
    RNA binding, LIX, 553
  rat liver, molecular weight, LIX, 512
Protein S18
  E. coli
    protein-protein interactions, LIX, 596, 599
    purification, LIX, 651, 652
    puromycin binding, LIX, 806, 807, 809, 810
  rat liver, molecular weight, LIX, 512
Protein S19
  E. coli
    protein-protein interactions, LIX, 594–597
    purification, LIX, 485, 488–490, 651
    puromycin binding, LIX, 806, 807
  rat liver, molecular weight, LIX, 512
Protein S20
  E. coli
    purification, LIX, 485, 487–489, 651, 652
    RNA binding, LIX, 553, 577
  rat liver
    loss, during analysis, LIX, 521
    molecular weight, LIX, 512
Protein S21
  E. coli
    protein-protein interactions, LIX, 596, 599
    purification, LIX, 485, 487, 489, 651
  rat liver
    loss during analysis, LIX, 521
    molecular weight, LIX, 512
Protein S22, rat liver, molecular weight, LIX, 512
Protein S23, rat liver, molecular weight, LIX, 512

Protein S24, rat liver, molecular weight, LIX, 512
Protein S25, rat liver, molecular weight, LIX, 512
Protein S26, rat liver, molecular weight, LIX, 512
Protein Sa1, rat liver, molecular weight, LIX, 512
Protein Sa2, rat liver, molecular weight, LIX, 512
Protein Sa3, rat liver, molecular weight, LIX, 512
Protein-SDS complex
  gel electrophoresis, XXXII, 92–102
  properties, XLVIII, 6
Proteoglycan, LVIII, 268
  differentiation, LVIII, 268, 269
  preparation, LVIII, 274
  Smith-degraded, XXXIV, 6
Proteolipid
  ATPase, LV, 350, 358
    synthesis, LV, 351
  of ATPase complex, carbodiimide, LVI, 171, 172
  dicyclohexylcarbodiimide-binding, LV, 383
  ionophore of yeast ATP synthetase, LV, 414, 415
    assay procedures, LV, 418–421
    isolation, LV, 415–418
  lipophilic Sephadex chromatography, XXXV, 389
  proton ionophore activity, LV, 418, 419
  from sarcoplasmic reticulum, isolation, XXXII, 291, 293
Proteoliposome
  associated with interface of biphasic decane plus phospholipid/water system, LV, 769, 772
    with planar phospholipid membranes of phospholipid-impregnated filters
      applicability, LV, 768
      general remarks, LV, 763, 766–768
      kinetics of association, LV, 768
  glucose carrier, LV, 707
  incorporation into filters, LV, 589–591
  planar membranes, LV, 756, 757
  preparation, LV, 197
  reconstitution
    bacteriochlorophyll plus bacteriorhodopsin, LV, 762
    bacteriochlorophyll reaction center complex, LV, 760
    bacteriorhodopsin, LV, 758–760
    cytochrome oxidase, LV, 761
      plus bacteriorhodopsin, LV, 762
    H^+–ATPase, LV, 761, 762
      plus bacteriorhodopsin, LV, 762
      plus cytochrome oxidase, LV, 762
    photosystem I, LV, 760, 761
    principle, LV, 757, 758
    reaction center and antenna complexes, LV, 760
  synthetic penetrating ions as probes for membrane potential, LV, 763
Proteolysis
  BKA-protein complex, LVI, 409
  yeast cell breakage, LVI, 44
Proteolytic activity
  acrosin, XLV, 331
  assay determination, casein substrates, XLV, 26, 27
    fluorometric, XLV, 24
  in factor V, bovine, XLV, 118
  in factor VII, XLV, 53
  in factor X, bovine, XLV, 96
Proteolytic enzyme
  catalytic constant, XLIV, 528, 529
  immobilization with diazonium compounds, XLIV, 18
  inhibitors, XLVI, 38
Proteoplast, in plant cells, XXXI, 495
Proteus mirabilis
  resistance plasmids, XLIII, 45
  transport systems, XXXI, 704
Proteus rettgeri
  penicillin acylase, XLIII, 718
  ubiquinone, LIII, 578
Proteus vulgaris
  enzyme A, XLIII, 735
  glucosamine-6-phosphate deaminase, XLI, 497
  5-keto-D-fructose reductase, XLI, 137
  oxygen-stable hydrogenase, LIII, 297, 310

Prothrombin, XLVII, 66
abnormal, congenitally, XLV, 135
activation, XLV, 154
amino terminal sequence, XLV, 149
components, XLV, 141–148
immunological properties, XLV, 145, 146
bovine
activation components, XLV, 136, 140, 146
amino acid composition, XLV, 137
assay, XLV, 124
first stage, XLV, 125, 126
second stage, XLV, 126
carbohydrate composition, XLV, 138
characteristics, XLV, 123
components, physical, XLV, 146
properties, XLV, 135
prothrombinase activation, XLV, 124
purification, XLV, 127, 128
synthesis, XLV, 123
calcium binding sites, XLV, 151, 153
canine, isolation, XLV, 132
carbohydrate composition, XLV, 138
derivatives, XLV, 134
human
activation component, XLV, 136, 140
amino acid composition, XLV, 137
fragmentation, XLV, 138, 139
properties, XLV, 135
purification, XLV, 132, 133
labeling
fluorescein isothiocyanate, XLV, 134
[^{3}H]sialyl, XLV, 134
reagent, factor V, XLV, 110
vitamin-K dependence, XLV, 151
Prothrombinase, activation of prothrombin, XLV, 124
Prothrombin fragment 1
amino acid composition, XLV, 147
chromatography, XLV, 142
properties, XLV, 141
purification, XLV, 142
Prothrombin fragment 2
amino acid composition, XLV, 148
carbohydrate composition, XLV, 2, 148
preparation, XLV, 144
production, XLV, 142
Protocatechuate 3,4-dioxygenase, LII, 11
oxy form, spectrum of, LII, 36
Protocatechuate 4,5-dioxygenase, LII, 11
3,4-Protocatechuate dioxygenase, Mössbauer studies, LIV, 379
Protoheme, *see also* Heme B
carbon monoxide binding
data evaluation, LIV, 512–515, 523
theoretical treatment, LIV, 507–511
definition, LIII, 203
resonance Raman spectroscopy, LIV, 240
spectral properties, LIII, 126
Protohemin dimethyl ester, preparation, LV, 538
Protomycin, XLIII, 158, 340
Proton, *see also* Hydrogen ion
anion transport, LVI, 253, 255
ejection, calcium-induced, LVI, 341, 342
excretion in urine, LV, 12
extramitochondrial changes, transport, LVI, 256, 257
mitochondrial permeability, LVI, 563, 566
permeability, oligomycin, LV, 504
product of complex III, LIII, 35
of NADH:cytochrome *c* oxidoreductase, LIII, 5
substrate of complex IV, LIII, 40
of cytochrome oxidase, LIII, 54, 73
of *p*-hydroxybenzoate hydroxylase, LIII, 544
of melilotate hydroxylase, LIII, 552
of NADH:ubiquinone oxidoreductase, LIII, 11
of salicylate hydroxylase, LIII, 528
translocation
complexes I, III, or IV, LVI, 585
by cytochrome oxidase
phospholipid vesicles, LV, 748
oxygen pulse experiments, LV, 750, 751
preparation of vesicles, LV, 749

proton translocation assay, LV, 749, 750
respiratory control, LV, 749
equipment, LV, 629–633
general principles, LV, 628–633
measurement, LV, 627, 628
calculation of → H^+/O values and essential controls, LV, 635–638
method, LV, 635
mitochondrial preparations and reagents, LV, 633, 634
scope of stoichiometric → H^+ measurements, LV, 638–640
stoichiometry linked to ATP synthesis, LV, 241, 242
uptake, phosphorylating vesicles, LV, 168
Proton-adenosinetriphosphatase, proteoliposomes, LV, 761, 762
Proton gradient
ATPase, LV, 302
light-dependent, assay, LV, 779, 780
uncouplers, LV, 466
Proton magnetic resonance, *see* Nuclear magnetic resonance, hydrogen-1
Proton-motive force
estimation in mitochondria, LV, 567–569
importance, LV, 547
Proton pump, reconstituted, LV, 710
activity, LV, 703, 705, 708
Proton site ratio, of mitochondrial electron transport, LV, 641, 642
calculations, LV, 645
comments, LV, 645, 646
modified oxygen pulse procedure, LV, 642, 643
procedure, LV, 643–645
with Ca^{++}, LV, 648, 649
with K^+ (plus valinomycin), LV, 647, 648
reagents and test medium, LV, 643
steady-state method, LV, 646, 647
Proton-transfer reaction
calculation of reaction time, LV, 622
rapid electrometric measurement, LV, 614, 615
instrumentation, LV, 615, 616
pH measurement, by continuous-flow method, LV, 618–627
in stirred suspensions, LV, 616, 618
Protoplasm, in plant cells, protection, XXXI, 529
Protoplast, *see also* Spheroplasts
formation, LVIII, 361
fusion, LVIII, 359–367
with polyethylene glycol, LVIII, 362, 363
protocol, LVIII, 362–364
selection, LVIII, 360, 366, 367
subsequent development, LVIII, 366, 367
yield, LVIII, 360
isolation, XXXI, 578–583
lysis, yeast cell breakage, LVI, 48
nuclei extraction, XXXI, 563, 564
from pea, LVIII, 361
plating, LVIII, 483
separation, LVIII, 361
stabilization, LVIII, 360, 361
tobacco leaf, LVIII, 362
from *Vicia*, LVIII, 361
wall regeneration, LVIII, 361
yeast, preparation, LV, 26, 27
Protoplast ghost
amino acid transport, comparison to ETP, LV, 186, 187
of *M. phlei*
preparation, LV, 180–182
properties, LV, 182
Protoporphyrin IX
heme-deficient mutants, LVI, 559
resonance Raman spectroscopy, LIV, 240
Protoporphyrin IX dimethyl ester
hydrolysis, LII, 452
preparation, LII, 452
Protosol, LIX, 567, 826, 832, 833
Prototheca zopfii, disruption, LV, 138
Prototroph, XLIII, 38, 39
Protozoa
disruption, LV, 135–137
radiolarian, bioluminescence, LVII, 274
Protuberance, of plant cells, as fractionation contaminant, XXXI, 545

Providencia, gentamicin acetyltransferase, XLIII, 624
Proximity effect, XLVI, 10
PRPP, *see* Phosphoribosyl pyrophosphate
PRPP synthetase, *see* Phosphoribosylpyrophosphate synthetase
Pseudodiploid cell line, LVIII, 323
Pseudo Hurler polydystrophy, *see* Mucolipidosis III
Pseudomonad
  coenzyme Q homologs, LIII, 593
  crude extract preparation, LIII, 554
  culture conditions, LIII, 529, 531, 532, 553, 554
  cytochrome *c*, LIII, 138
  D-4-deoxy-5-ketoglucarate hydro-lyase, XLII, 273
  hydrogenase reaction, LIII, 287
  salicylate hydroxylase, LIII, 528–543
  sarcosine dehydrogenase, LIII, 450
Pseudomonad MSU-1
  D-aldohexose dehydrogenase, assay, purification, and properties, XLI, 147–150
  L-*arabino*-aldose dehydrogenase, assay, purification, and properties, XLI, 150–153
  D-fuconate dehydratase, assay, purification, and properties, XLII, 305–308
  2-keto-3-deoxy-L-arabonate aldolase, assay, purification, and properties, XLII, 269–272
Pseudomonas
  4-aminobutanal dehydrogenase, LIV, 229
  cytochrome oxidase, LIII, 647–657
    activity, LIII, 647
    assay, LIII, 647–649
    composition, LIII, 654
    crystallization, LIII, 652, 653
    hemes, LIII, 647
    inhibitors, LIII, 657
    molecular weight, LIII, 655
    purification, LIII, 649–653
    purity, LIII, 654
    reaction with azurin, LIII, 656
      with cytochrome *c*-551, LIII, 656
      with molecular oxygen, LIII, 656, 657
      with nitrite, LIII, 656, 657
    redox potentials, LIII, 656
    spectral properties, LIII, 654, 655
    stability, LIII, 654
    subunits, LIII, 656
  melilotate hydroxylase, LIV, 229
  oxidases, LII, 11–21
Pseudomonas acidovorans, oxidase, LII, 17
Pseudomonas aeruginosa
  aldehyde dehydrogenase, assay, purification, and properties, XLI, 348–354
  crude extract preparation, LIII, 650
  culture and maintenance, LIII, 649, 650
  cytochrome, LIII, 647
  cytochrome *c*, CD spectrum, LIV, 276, 278
  cytochrome *c*-551, LIII, 210
  cytochrome c_{551}-azurin, LIV, 78
  cytochrome oxidase, LIII, 647–657
  3′,4′-dideoxykanamycin B, XLIII, 271, 272
  gentamicin acetyltransferase, XLIII, 624, 625
  IIIa β-lactamase, XLIII, 673
  kanamycin acetyltransferase, XLIII, 624
  neomycin phosphotransferase, XLIII, 626
  streptomycin phosphotransferase, XLIII, 627
  toxin A
    assay
      catalysis of adenosine diphosphate ribosylation, LX, 788–790
      hamster ovary cell toxicity, LX, 787, 788
      mouse lethality, LX, 787
      radioimmunoassay, LX, 789–793
    inhibition of protein synthesis, LX, 780, 781, 793
    production
      bacterial strains, LX, 781, 782
      culture conditions, LX, 782, 783

media, LX, 782
purification, LX, 783–787
immunoadsorption, LX, 785–787
ion-exchange methods, LX, 784
Pseudomonas arvilla, oxidase, LII, 11, 12
Pseudomonas B-16, transport system, XXXI, 704
Pseudomonas cocovenenans, bongkrekic acid, LV, 528
Pseudomonas denitrificans, coenzyme Q homolog, LIII, 593
Pseudomonas desmolytica, p-hydroxybenzoate hydroxylase, LIII, 544
Pseudomonas fluorescens
aspartate carbamyltransferase from, LI, 51–58
coenzyme Q homolog, LIII, 593
crude extract preparation, LIII, 546
culture, LIII, 545
glyoxylate reductase, assay, purification, and properties, XLI, 343–348
p-hydroxybenzoate hydroxylase, LIII, 544–552
oxidase, LII, 11, 15, 16
Pseudomonas hydrophila, XLIV, 811
Pseudomonas melanogenum, penicillin acylase, XLIII, 719
Pseudomonas oleovorans
aldehyde dehydrogenase, assay, purification, and properties, XLII, 313–315
crude extract preparation, LIII, 342
growth, LIII, 358
ω-hydroxylase, LIII, 356–360
oxidase, LII, 12
rubredoxin, LIII, 262, 340–346; LIV, 359
Pseudomonas ovalis, oxidase, LII, 11
Pseudomonas perfectomarinus, nitric oxide reductase, LIII, 644, 645
Pseudomonas putida, XLIV, 745
cell lysis, LII, 153
cytochrome *P*–450, LII, 152; LIII, 207, 211
cytochrome $P–450_{cam}$, LIV, 353–355, 375
ferredoxin, LIII, 262
freeze-thaw autolysis, LII, 172
growth conditions, LII, 151, 152, 169–172
p-hydroxybenzoate hydroxylase, LIII, 544
2-keto-3-deoxy-6-phosphogluconic aldolase, assay, purification, and properties, XLII, 258–264
oxidase, LII, 17
salicylate hydroxylase, LIII, 528
Pseudomonas saccharophila
L-arabinose metabolism, and L-2-keto-3-deoxyarabonate isolation, XLI, 102
L-2-keto-3-deoxyarabonate dehydratase, purification, XLII, 309
methylglyoxal synthetase, preparation and purification, XLI, 503–505
phosphoribokinase, assay, purification, and properties, XLII, 120–123
Pseudomonas solanaceum, bioautography, XLIII, 117
Pseudomonas testeroni, oxidase, LII, 11
Pseudouridine, XLVI, 185, 693
chromatographic mobility, LIX, 73
effect on base-pairing in tRNA, LIX, 47, 48
electrophoretic mobility, LIX, 91
identification, LIX, 92
Pseudouridine monophosphate, identification, LIX, 89
Psicofuranine, inhibitor of GMP synthetase, LI, 217, 224
Psychosine, *see also* Glycosphingolipid
enhancer of brain α-hydroxylase, LII, 317
thin layer chromatography, XXXV, 396, 400, 402
Psychosine sulfate, XXXIV, 4
Pteridine cofactor, in melatonin biosynthesis, XXXIX, 379
Pterin
definition, LIII, 284
substrate of phenylalanine hydroxylase, LIII, 284
Pteroic acid, XXXIV, 283
N^{α}-Pteroyl-L-lysine, XXXIV, 283

Pteroyllysine-agarose, purification of dihydrofolate reductase, XXXIV, 281–288, *see also* Dihydrofolate reductase pteroyl-lysine-agarose
Pteroyl oligo-γ-L-glutamyl endopeptidase, XXXIV, 5
PTH, *see* Parathyroid hormone
Ptilosarcus guerneyi, source, LVII, 248
Puck's balanced salt solution, LVIII, 120, 121, 142
Puck's medium, XXXII, 744
Puck's nutrient component, LVIII, 522
PUDP, *see* 5′-(4-Diazophenylphosphoryl)uridine 2′(3′)-phosphate
Pulegone reductase, extraction, XXXI, 537–539
Pullulanase, immobilized, XLIV, 470
Pulmonary artery, cannulation, XXXIX, 46, 47
Pulse elution, XXXIV, 238
Pulse-labeled cell, LVIII, 244
Pulse labeling
  of *C. reinhardi* precursors, XXXII, 868
  of parathyroid tissue slices, XXXVII, 349–352
  in vivo cell cycle analysis, XL, 59
Pulse radiolysis, method, LIV, 86
Pump
  for column chromatography, XLVII, 211, 227, 228
  gradient, LII, 174
  peristaltic, LII, 174
  for sequenators, XLVII, 313
  for thyroid gland perfusion, XXXIX, 361, 362
  tube sizes, XLVII, 224
Punch, hat shape, for freeze etching, XXXII, 37
Purex, replica cleaning, XXXII, 43
Purging, dry light, in circular dichroism, XL, 231
Purine, *see also* specific types
  auxotroph, LVIII, 309
  hydrogen exchange, XLIX, 25
  indefinite self-associations, XLVIII, 112, 113
  in media, LVIII, 54, 66, 67
Purine base
  elution, XXXVIII, 17, 18
  thin-layer chromatography, XXXVIII, 28–30, 158
Purine deaminase, XLVI, 22
Purine dinucleotide, XLVI, 315
Purine nucleoside, XLIII, 153, 154
Purine-nucleoside phosphorylase, LI, 517–543
  active site, LI, 529
  activity, LI, 517, 524, 530, 531, 542
  affinity for calcium phosphate gel, LI, 584
  amino acid composition, LI, 536, 538
  assay, LI, 518, 525, 526, 531, 532, 538–541
  from Chinese hamster, LI, 538–543
  crystallization, LI, 534, 535
  extinction coefficient, LI, 536
  frictional ratio, LI, 536
  from human erythrocytes, LI, 530–538
  inhibitors, LI, 530, 536, 537
  isoelectric point, LI, 529, 543
  isozymes, LI, 518, 536, 537
  kinetic properties, LI, 537, 543
  molecular weight, LI, 520, 530, 536, 543
  partial specific volume, LI, 536
  pH optimum, LI, 521, 529
  photooxidation, LI, 530
  purification, LI, 519–524, 527–529, 532–535, 541, 542
  purity, LI, 529
  from rabbit liver, LI, 524–530
  from rat liver, LI, 521–524
  reaction mechanism, LI, 537
  regulation, LI, 438, 517
  removal, with calcium phosphate gel, LI, 503, 504
  from *Salmonella typhimurium*, LI, 517–524
  secondary structure from CD data, LI, 536
  sedimentation coefficients, LI, 536, 543
  sources, LI, 524, 525
  stability, LI, 521, 530, 536
  Stokes radius, LI, 536
  substrate specificity, LI, 521, 529, 537
  subunit structure, LI, 520, 530, 543

translational diffusion coefficient, LI, 536
Purine nucleotide, brown adipose tissue mitochondria, LV, 70–74
Purine nucleotide thioether, XLVI, 289–295
6-(Purine 5′-ribonucleotide)-5-(2-nitrobenzoic acid) thioether, XLVI, 241
Purine trinucleotide, XLVI, 315
Purity, criteria of cell, XXXIV, 196
Purkinje cell
adenine nucleotide effect, XXXIX, 439
norepinephrine effects, XXXIX, 435
Puromycin, XLVI, 75, 90, 181, 190, 335, 629–631, 634–636, 682, 711–717
analogs, XLVI, 662–669
in antigen studies, XXXII, 67
assay
of GTP hydrolysis, LIX, 361, 362
for initiation factors, LX, 18, 19, 32, 33, 36, 38, 41, 51, 119–121, 169, 170, 208, 296, 297, 407, 571
of radiolabeled ribosomal particles, LIX, 779, 780
of ribosome dissociation factor, LX, 296, 297
binding model, LIX, 810
biosynthesis, *see also* specific enzyme
adenine, XLIII, 515
2-amino-2-deoxy-D-lyxose 5-phosphate, XLIII, 514
2-amino-2-deoxy-D-ribose 5-phosphate, XLIII, 514
3-aminopentose, XLIII, 514
O-demethylpuromycin, XLIII, 515
N^6,N^6-dimethyladenine, XLIII, 515
5′-phosphate esters, XLIII, 515
puromycin *S*-adenosylmethionine: *O*-demethylpuromycin *O*-methyltransferase, XLIII, 508–515
N^6,N^6,*O*-tridemethylpuromycin, XLIII, 514
competence, LIX, 396
effect on protein synthesis, LIX, 853, 854
incorporation into ribosomes, XLVI, 712, 713
inhibition, LVI, 31
inhibitor, of chain elongation, LIX, 370
measurement of translocation within ribosome, LX, 762–764
for photoaffinity labeling, LIX, 798, 800–802
preparation of ribosomal subunits, LIX, 404, 505
protein inhibitor, XL, 289
in release of nascent peptides, LIX, 859, 860
in ribosome release studies, LIX, 861
structural formula, LIX, 812
[^{3}H]Puromycin
assay for translocation of peptidyl-tRNA, LIX, 359
specific activity, determination, LIX, 360
Puromycin *S*-adenosylmethionine:*O*-demethylpuromycin *O*-methyltransferase, XLIII, 508–515
S-adenosylmethionine, XLIII, 509
assay, XLIII, 509–511
cell-free extract, XLIII, 512
DEAE-cellulose column chromatography, XLIII, 512
O-demethylpuromycin, XLIII, 509
pH optimum, XLIII, 514
properties, XLIII, 513, 514
purification, XLIII, 511–513
puromycin biosynthesis, scheme, XLIII, 508, *see also* Puromycin, biosynthesis
salt fractionation, XLIII, 512
Sephadex G–200 gel filtration, XLIII, 513
specific activity, XLIII, 510–513
specificity, XLIII, 514
stability, XLIII, 514
unit definition, XLIII, 510
Puromycin-potassium hydroxide, dissociation of mitochondrial ribosomes, LIX, 425, 426
Purple membrane, incorporation into vesicles capable of light-stimulated ATP synthesis, LV, 777–780

Putidaredoxin, LII, 175–185
colorimetric determination of ferrous 2,4,6-tripyridyl-*s*-triazene complex, LII, 176
in cytochrome P–450_{cam} activity assay, LII, 156
gel electrophoresis, LII, 175
isolation and purification
flow chart, LII, 174
procedure, LII, 178–180
reagents, LII, 175
magnetic circular dichroism, XLIX, 175
optical properties, LII, 184, 185
partial purification, LII, 154
purity criteria, LII, 184
stability, LII, 183, 184
Putidaredoxin reductase, LII, 173–185
in cytochrome P–450_{cam} activity assay, LII, 156
fluorometric measurement of flavin content, LII, 176
gel electrophoresis, LII, 175
isolation and purification, LII, 176–178
flow chart, LII, 173
procedure, LII, 176–178
reagents, LII, 175
optical properties, LII, 184, 185
partial purification, LII, 154, 155
purity criteria, LII, 184
stability, LII, 183, 184
Putrescine
inhibitor of rat liver carbamoyl-phosphate synthase, LI, 119
in media, LVIII, 55, 63
Putrescine oxidase, LII, 18
assay, with *Diplocardia* bioluminescence, LVII, 381
PVA, *see* Polyvinyl alcohol
PVC, *see* Packed cell volume
PVP, *see* Polyvinylpyrrolidone
Py, *see* Pyrene
Pycnometer, XLVIII, 15
applications, XLVIII, 18–22
Pyocyanine, as mediator-titrant, LIV, 408, 423, 429, 433
Pyocyanine perchlorate, as electron donor, XXXI, 701
Pyranose oxidase, LII, 21
from *Polyporus obtusus*, XLI, 170–173
assay, XLI, 170
production, XLI, 171
properties, XLI, 172
purification, XLI, 172, 173
Pyrazofurin-5′-phosphate, inhibitor of orotidylate decarboxylase, LI, 162
Pyrazole, inhibitor of alcohol dehydrogenase, LII, 355, 357
Pyrazolo[3,4-*d*]pyrimidine, XXXIV, 525, 526
Pyrene
activator, oxidation potential, LVII, 522
chemiluminescent reaction with TMPD, LVII, 502–504
structural formula, LVII, 498
Pyrene dimer, chemiluminescence, LVII, 568
Pyrene excimer, emission, LVII, 503
N-(1-Pyrene)maleimide, XLVII, 416, 417, 420, 421
N-(3-Pyrene)maleimide, XLVII, 416, 417
Pyrene sulfonate, as fluorescent probe, XXXII, 235, 239
Pyrex tube, calcium concentration, LVII, 282
Pyridine, *see also* specific types
in carbonyl group determination, XLIV, 385
cleavage, XLVII, 147
extraction, XLVII, 345, 346
in 5-fluoro-2′-deoxyuridine 5′-(*p*-nitrophenylphosphate) synthesis, LI, 98
in heme isolation, LII, 425
preparation of dehydroluciferin, LVII, 25
of luciferyl adenylate, LVII, 25, 27
purification, LII, 423
of flavin peptides, LIII, 454, 455
procedure, XLVII, 223, 337
redistillation, XLVII, 337
in silylation of steroids, LII, 380, 381
in siroheme extraction and characterization, LII, 440, 441
support swelling, XLVII, 272

synthesis of benzyl luciferyl disulfate, LVII, 252
of 6-*N*-[(3-biotinylamido)-2-hydroxypropyl]amino-2,3-dihydrophthalazine-1,4-dione, LVII, 430
of 6-(*N*-ethyl-*N*-[2-hydroxy-3-(thyroxinylamido)propyl])-amino-2,3-dihydrophthalazine-1,4-dione, LVII, 432
of 7-[*N*-ethyl-*N*-(4-thyroxinylamido)butyl]amino-naphthalene-1,2-dicarboxylic acid hydrazide, LVII, 437
in Triton X–45 anionic derivative preparation, LII, 145
Pyridine-acetic acid
for drug metabolite isolation, LII, 334
for silylation, LII, 332, 335
Pyridine-acetic acid buffer, preparation, XLVII, 222; LIII, 458
Pyridineazoresorcinol, XLIX, 177, 178
Pyridine-chloride, in porphyrin ester synthesis, LII, 432
Pyridine-chloroform, for extraction of heme A, LII, 423
Pyridine-chloroform-bicarbonate-isooctane, in heme isolation, LII, 424
Pyridine dinucleotide, synthesis of analogs, XLVI, 252, 253
2-Pyridine disulfide agarose gel, XLIV, 41, 42
use, XLIV, 42
Pyridine hemochrome, reduced, extinction coefficient, LII, 457
Pyridine hydrochloride, XLVII, 42
Bpoc group removal, XLVII, 594
imidazole blocking, XLVII, 540
in situ preparation, LVII, 20, 21
synthesis of benzyl luciferin, LVII, 251
Pyridine-*N*-methylmorpholinetrifluoroacetic acid buffer, XLVII, 361
Pyridine nucleotide, *see also* Nicotinamide
analogs, XLVI, 260
content of mitochondria, LV, 637
determination, LV, 263–267
deuterated, preparation, LIV, 226, 227
difference extinction coefficient, LIII, 591
distribution between cytosol and mitochondria in hepatocytes, LV, 215
ratio in mitochondria, LV, 239, 240
redox changes, measurement, LV, 668
redox state, transport studies, LVI, 255, 256
spin-labeled, XLVI, 260
substrate regenerating system, LV, 262, 270–272
Pyridine nucleotide dehydrogenase, localization, LVI, 230
Pyridine nucleotide-linked activity, enzymes, after electrophoresis, XLI, 66–73
Pyridine nucleotide-linked enzyme, *see* Enzyme, NAD-linked
Pyridine nucleotide transhydrogenase, *see* NAD(P) transhydrogenase
Pyridine-water, coupling yields, XLVII, 270
Pyridinium acetate
in fingerprinting procedure, LIX, 65
purification of flavin peptides, LIII, 454, 455
Pyridinium chlorochromate, synthesis of *Diplocardia* luciferin, LVII, 377
Pyridinium formate, in fingerprinting procedure, LIX, 66
Pyridinium ion, fluorescence quenching agent, XLIX, 225, 233
Pyridinium oxidase, LII, 20
Pyridoxal, in media, LVIII, 65
Pyridoxal enzyme, XLVI, 31, 37, *see also* specific enzymes
Pyridoxal kinase, XXXIV, 5
Pyridoxal 5′-monophosphate
in Hummel-Dreyer gel filtration, LIX, 326
in sucrose gradient buffers, LIX, 329
Pyridoxal phosphate, XLIII, 440, 442; LV, 517
affinity labeling, XLVII, 481, 483, 493
in δ-aminolevulinic acid synthetase assay, LII, 351, 352
modification of lysine, LIII, 141
structure, XLVII, 483
Pyridoxal 5-phosphate, in media, LVIII, 65

Pyridoxal 5′-phosphate, XXXIV, 4, 294; XLVI, 90, 163, 432
  immobilization, XLIV, 885, 886
Pyridoxal-5′-phosphate analog, XLVI, 441–447
Pyridoxal phosphate enzyme, XLVI, 163, 164, 427
Pyridoxamine enzyme, XLVI, 45–47
Pyridoxamine phosphate, fluorescent donor, XLVIII, 378
Pyridoxamine 5′-phosphate, XXXIV, 4, 294–300; XLVI, 33
Pyridoxamine phosphate-agarose, XXXIV, 294
Pyridoxamine-5′-phosphate analog, XLVI, 441–447
Pyridoxamine phosphate gels, preparation, XL, 269
Pyridoxamine-5′-phosphate hydrochloride, XLVI, 442
Pyridoxamine phosphate oxidase, LII, 18; LIII, 403; LVI, 474
Pyridoxamine pyruvate transaminase, XLVI, 21
5-Pyridoxate 2,3-dioxygenase, LII, 16
Pyridoxine, in media, LVIII, 54, 65
Pyridoxine 4-oxidase, LII, 18; LVI, 474
Pyridoxine 5′-phosphate oxidase, XXXIV, 302
3-(2-Pyridyl)-5,6-diphenyl-1,2,4-triazine, in colorimetric ultramicro assay for reducing sugars, XLI, 29
2,4-Pyridyl disulfide, XXXIV, 538
S-β-(4-Pyridylethyl)-L-cysteine, internal standard, XLVII, 34, 37, 38
S-β-(4-Pyridylethyl)-DL-penicillamine, internal standard, XLVII, 34, 38
Pyrimidine
  auxotroph, LVIII, 309
  cyclic nucleotides, separation, XXXVIII, 35
  derivatives, differential absorptivities, LI, 395
  in media, LVIII, 67
Pyrimidine deaminase, XLVI, 22
Pyrimidine nucleoside, XLIII, 152, 153, 155
Pyrimidine nucleoside monophosphate kinase, LI, 321–331, *see also* Cytidylate kinase
  activators, LI, 331
  activity, LI, 321, 322
  assay, LI, 322, 323
  cation requirements, LI, 329
  inhibitors, LI, 331
  kinetic properties, LI, 329, 330
  from Novikoff ascites cells, LI, 326–329, 330, 331
  purification, LI, 323–329
  from rat liver, LI, 323–326, 329, 330
  regulation, LI, 331
  substrate specificity, LI, 329, 330
  sulfhydryl reducing agent requirement, LI, 330, 331
Pyrimidine-nucleoside phosphorylase, LI, 432–437, *see also* Thymidine phosphorylase; Uridine phosphorylase
  activity, LI, 432, 437
  assay, LI, 432, 433
  from *Bacillus stearothermophilus*, LI, 437
  from *Haemophilus influenzae*, LI, 432–437
  kinetic properties, LI, 436, 437
  pH effects, LI, 435
  purification, LI, 433–435
  stability, LI, 435
  substrate specificity, LI, 435, 436
Pyrimidine nucleotide, XLVI, 90
Pyrimidine residue, cleavage sites, LIX, 106
Pyrimidine ribonucleoside kinase, *see* Uridine kinase
Pyrite, phase shift studies, LIV, 343
Pyrocatechase, LII, 6, 38
Pyrocypris, in luciferin-luciferase specificity studies, LVII, 336
Pyrogallol
  alkaline, for oxygen removal, LIV, 117
  autoxidation, as source of free radicals, LIII, 387
  inhibitor of *Pholas* luciferin oxidation, LVII, 399
  in polyphenol oxidase assay, XXXI, 534
  stimulator of *Pholas* light emissions, LVII, 397
  substrate of *Pholas* luciferase, LVII, 398

Pyroglutamate derivative, in TRF synthesis, XXXVII, 410, 411
Pyroglutamic acid, in phospholipase A_2, XXXII, 147
Pyronin, as optical probes, LV, 573
Pyronin B, in gel electrophoresis, LII, 325
Pyronine G, LIX, 507, 542, 544
Pyronine Y, LIX, 431
Pyrophorus, bioluminescence, LVII, 271, 354, 355
Pyrophosphatase
assay, XXXII, 397
inorganic
effect on bioluminescence, LVII, 29, 33
in firefly crude extract, LVII, 4, 62
in plasma membranes, XXXI, 88, 90
Pyrophosphate
adenylate cyclase, XXXVIII, 146, 147, 160
alumina column chromatography, XXXVIII, 44
in anaerobic continuous titration, LIV, 126
determination, LI, 275
inhibitor, of adenine phosphoribosyltransferase, LI, 566
of adenylate deaminase, LI, 502
of aspartate carbamyltransferase, LI, 57
of GMP synthetase, LI, 224
of orotate phosphoribosyltransferase, LI, 153, 163
of orotidylate decarboxylase, LI, 163
inorganic
effect on bioluminescence, LVII, 5, 29
inhibitor, of luciferase light reaction, LVII, 4
product, of firefly luciferase, LVII, 3, 37, 58, 74
stabilizer, of luciferase, LVII, 199
interference, in iron analysis, LIV, 440
in oxygen determination, LIV, 119
product of adenine phosphoribosyltransferase, LI, 558, 568
of amidophosphoribosyltransferase, LI, 171
of GMP synthetase, LI, 213, 219
of guanine phosphoribosyltransferase, LI, 549
of hypoxanthine-guanine phosphoribosyltransferase, LI, 13
of hypoxanthine phosphoribosyltransferase, LI, 543, 549
of nucleoside triphosphate pyrophosphohydrolase, LI, 275
of orotate phosphoribosyltransferase, LI, 69
R. rubrum chromatophores, LV, 596
[^{32}P]Pyrophosphate
assay of tryptophanyl-tRNA synthetase, LIX, 238
in misactivation studies, LIX, 286
Pyrophosphate buffer, for microsome isolation, LII, 111
Pyrophosphorolysis, orotate phosphoribosyltransferase, LI, 154
Pyrophosphoryltransferase, radiolabeled, preparation, LIX, 790, 791
Pyrosulfite, *see* Metabisulfite
Pyrrolidinyl *N*-oxyl group, structure, XLIX, 434
Pyrrolnitrin, XLIII, 409
Pyruvate, *see also* Pyruvic acid
adenylate cyclase, XXXVIII, 160, 161, 168, 169
effects on deoxycorticosterone hydroxylation, XXXVII, 302–304
oxidation, control studies, XXXVII, 292
Pyruvate analog, XXXIV, 598, 599
Pyruvate carboxylase
gluconeogenic catalytic activity, XXXVII, 281, 282
labeling, XLVI, 139
Pyruvate decarboxylase, XLVI, 50, 51
Pyruvate dehydrogenase
brain mitochondria, LV, 58

control studies, XXXVII, 293, 294
energy transfer studies, XLVIII, 378
white adipose tissue mitochondria, LV, 16
Pyruvate-ferredoxin oxidoreductase, from *Clostridium acidi-urici*, XLI, 334–337
assay, XLI, 334
properties, XLI, 336, 337
purification, XLI, 334–336
Pyruvate formate-lyase, *see* Formate acetyltransferase
Pyruvate kinase, XXXIV, 5, 165, 246, 248, 602; XLVI, 21, 241, 248, 299, 306, 307, 529, 548
activity, LVII, 74, 86, 96, 108
affinity for calcium phosphate gel, LI, 584
affinity labeling, XLVI, 550–553
assay of adenylosuccinate synthetase, LI, 213
of ADP, LVII, 74
of AMP, LV, 208
of ATPase LV, 313, 320, 329
of ATPase inhibitor LV, 408, 422
of cAMP, XXXVIII, 62, 63, 65
of cGMP, XXXVIII, 73, 74, 78, 82
of cyclase, XXXVIII, 119, 143
of cyclic nucleotide phosphodiesterases, LVII, 96, 97, 98
by Fast Analyzer, XXXI, 816
of FGAM synthetase, LI, 197
of GTP hydrolysis, LIX, 361
of guanosine nucleotides, LVII, 86, 91
of guanylate kinase, LI, 475, 476, 483
of guanylate synthetase, LI, 219
of modulator protein, LVII, 108, 109
of nucleoside diphosphokinase, LI, 377
of polypeptide chain elongation, LIX, 358
of polypeptide synthesis, LIX, 367
of pyrimidine nucleoside monophosphate kinase, LI, 323
of uridine-cytidine kinase, LI, 309, 315, 316
in ATP regeneration, XLIV, 698
from *Bacillus licheniformis*, XLII, 157–166
activators and inhibitors, XLII, 166
assay, XLII, 158–160, 165
kinetic characterization, XLII, 162–166
preparation, XLII, 157
purification, XLII, 160–162
cell-free protein synthesis, LIX, 448, 850
coimmobilization, XLIV, 472
conformation studies, XLIX, 356, 358
conjugate, activity, XLIV, 271
control studies, XXXVII, 293
detection, by ultraviolet absorbance, XLI, 66, 67
electrophoresis, XLI, 68–70, 72
enzymic detection, XLI, 70
from human erythrocytes, XLII, 182–186
activators and inhibitors, XLII, 186
assay, XLII, 182, 183
kinetics and thermostability, XLII, 186
molecular weight, XLII, 185
properties, XLII, 185, 186
purification, XLII, 183–185
immobilization on cellulose sheets, XLIV, 908
on filter paper, XLIV, 906
by microencapsulation, XLIV, 214
inhibition, reversible, XLVI, 550, 551
isozymes from grassfrog *Rana pipiens*, XLII, 166–175
activators and inhibitors, XLII, 175
assay, XLII, 167, 168
chromatography, XLII, 169–174
molecular weight, XLII, 175
properties, XLII, 174, 175
purification, XLII, 168–174
in D-mannose microassay, XLI, 3
mercuri-estradiol reaction, XXXVI, 389
metal site, XLVI, 549
in parenchyma cells, XXXII, 706
Pasteur effect, LV, 290, 291

poly(Phe) synthesis, LIX, 459
preparation of 3′-amino-3′-deoxy-ATP, LIX, 135, 136
site-specific reagent, XLVII, 481, 483, 485
stability, 109n
substrate specificity, LVII, 82
from yeast, XLII, 176–182
assay, XLII, 176
electrophoresis, XLII, 181
molecular weight, XLII, 181
properties, XLII, 181, 182
purification, XLII, 176–180
Pyruvate orthophosphate dikinase
from *Acetobacter xylinum*, XLII, 192–199
assay, XLII, 192–196
chromatography, XLII, 196
distribution, XLII, 199
inhibitors, XLII, 198
physiological function, XLII, 198
properties, XLII, 197–199
purification, XLII, 196, 197
from *Bacteroides symbiosus*, XLII, 187–191, 199–209
activators and inhibitors, XLII, 191
assay, XLII, 187, 188, 199, 200
chromatography, XLII, 205
inhibitors, XLII, 208
mechanism, XLII, 208
molecular weight, XLII, 191, 207
properties, XLII, 191, 206–209
protein determination, XLII, 206
purification, XLII, 188–191, 200–206
from leaves, XLII, 212–219
activation and inactivation, XLII, 217, 218
assay, XLII, 212–215
molecular weight, XLII, 219
properties, XLII, 217–219
purification, XLII, 216, 217
sources and extraction, XLII, 215
from *Propionibacterium shermanii*, XLII, 209–212
assay, XLII, 199, 200
chromatography, XLII, 210, 211
inhibitors, XLII, 211
mechanism, XLII, 212
molecular weight, XLII, 211
properties, XLII, 211, 212
protein determination, XLII, 210
purification, XLII, 209–211
Pyruvate oxidase, LII, 19
Pyruvate P_i dikinase, *see* Pyruvate orthophosphate dikinase
Pyruvic acid, XLVI, 20; *see also* Pyruvate
assay, LV, 221
concentration determination, with enzyme electrode, XLIV, 877–880
content of kidney cortex, LV, 220
exchange, LVI, 253
hemin requirement, LVI, 172, 174, 175
internal standard, XLVII, 491
labeled with isotopic hydrogen, sodium salt, enzymic synthesis, XLI, 106–110
labeling with isotopic hydrogen, XLI, 110, 114
measurement of H^+/site ratios, LV, 648
in media, LVIII, 55, 66
N. crassa mitochondria, LV, 147
polarography, LVI, 451, 452
product of D-lactate dehydrogenase, LIII, 519
rat brain mitochondria, LV, 57
renal gluconeogenesis, XXXIX, 14–16, 18
steroid hydroxylation, LV, 10
substrate, of lactate dehydrogenase, XLIV, 459
transport, LVI, 248, 249, 256, 292
white adipose tissue mitochondria, LV, 15, 16
yeast mitochondria, LV, 150
[3-^{3}H]Pyruvic acid, sodium salt, enzymic preparation of crystalline, XLI, 107–110

Q

Q, *see* Ubiquinone
Qβ ribonucleic acid replicase
assay by polyguanylate polymerase activity, LX, 629, 630

denaturation and renaturation, LX, 637, 638
functional role, LX, 628
properties, LX, 634, 635
purification, LX, 632–635
subunits, LX, 418, 628, 637
QAE, *see* Quaternary aminoethyl
Q-banding, *see* Quinacrine banding
QLS, *see* Quasi-elastic light scattering
Q2–3067 optical couplant, LVII, 319
QPD, *see* Nuclear magnetic resonance, quadrature detection
Quadrature detection, LIV, 157
for elimination of fold-over, LIV, 159, 160
Quadrifidin, XLIII, 340
Quadrilineatin, XLIII, 163
Quadrupole doublet, LIV, 351, 365, 366
Quadrupole splitting
definition, LIV, 348
dimensions, LIV, 348
for non-Kramers systems, LIV, 351
Quality control
cell line, LVIII, 439
media, LVIII, 453, 454
Quality factor, EPR measurements, LIII, 489
Quantitation
bound ligands, XXXIV, 349
ligand, XXXIV, 683, 698
Quantum counter
definition, LVII, 300
for photometer calibration, LVII, 300
Quantum yield, definition, LVII, 358, 412
Quarter-wave plate, LIV, 9
Quarter-wave retarder, vacuum ultraviolet circular dichroism, XLIX, 215, 216
Quartz diffusion cell, XLIV, 910
Quartz EPR tubes, specifications, LIII, 489
Quartz syringe, LVII, 141
Quasi-buffering, LIV, 71
Quasidiploid cell line, LVIII, 323
Quasi-elastic light scattering, XLVIII, 415–495
applicability, XLVIII, 419–421
density-gradient centrifugation, XLVIII, 492, 493
electrophoresis, XLVIII, 494
with nonspherical particles, XLVIII, 436–442
physical basis, XLVIII, 416–419
relation between scattered light and photocurrent, XLVIII, 422–425
theoretical principles, XLVIII, 421–452
Quasi-reference electrode
in cyclic voltammetry, LVII, 511
in electrochemiluminescence step experiments, LVII, 513
Quaternary aminoethyl-Sephadex
atractyloside-binding protein, LVI, 416
cyclic nucleotide separation
resin preparation, XXXVIII, 13, 14
separation techniques, XXXVIII, 14–16
enzyme decontamination, XXXVIII, 65
fractionation, of nonhistone proteins, XL, 164
protein kinase, XXXVIII, 319–320
purification of purine nucleoside phosphorylase, LI, 522
Quaternary aminoethyl-Sephadex A-25, developers, XLVII, 214
Quaternary aminoethyl–Sephadex A–50
purification of aequorin, LVII, 308, 309
of green-fluorescent protein, LVII, 309
Quaternary ammonium group, solid-phase peptide synthesis, XLVII, 590, 595, 600, 606
Quebrachitol, XLVI, 387
Quench
correction, XLVIII, 336
using spin-labels, XLIX, 425
Quench-flow apparatus, ubiquinone oxidation kinetics, LIII, 583, 584
Quenching
in liquid scintillation counting, XL, 295
methods of correction, XL, 297
magnetic susceptibility determinations, LIV, 383
of triplet state, LIV, 48
Quenching constant, LVII, 569, 570

Queracetin, cyanine dyes, LV, 693
Quercetin, XLVI, 89
ATPase, LVI, 181
availability, LV, 491
biochemical properties, LV, 491, 492
physical properties, LV, 476
structure, LV, 491
toxicology, LV, 491
Quercitinase, LII, 38
Quercitin 2,3-dioxygenase, LII, 9
Quick-freeze procedure, for liver, XXXI, 5
Quick-thaw procedure, for liver, XXXI, 5, 19
Quinacillin, XLVI, 536
Quinacrine
as curing agent, XLIII, 51
fluorescence quenching, everted vesicles, LVI, 237, 238
submitochondrial particles, LV, 112
Quinacrine banding, LVIII, 324, 325
Quinacrine stain, LVIII, 325
Quinine bisulfate, emission spectrum, XLVIII, 369
Quinine sulfate
as fluorescence standard, LII, 238–400, 402
standard, of fluorescent quantum yield, LVII, 595, 600
Quinocycline, XLIII, 340
Quinolinate, as enzyme inhibitor, XXXVII, 294
Quinoline, XLVII, 457
8-Quinolinol, hydroxylamine determination, LIII, 637
Quinol phosphate, oxidation by Br_2, LV, 537
Quinomycin, XLIII, 180, 203
Quinone, LIII, 13, *see also* Menaquinone; Ubiquinone
activation agent, XLIV, 35
cofactor for dihydroorotate dehydrogenase, LI, 63
in cytochrome bc_1 complex, LIII, 116
extraction, LVI, 109, 111, 112
in glucose determination, XLIV, 586, 617
glucose electrode, LVI, 483
intermediate
isolation, LIII, 605–609
radiolabeling, LIII, 608, 609
spectral identification, LIII, 603, 604
in ubiquinone biosynthesis, LIII, 600–609
M. phlei membrane systems, LV, 186
in plant-tissue extraction, XXXI, 528–544
protein reactions, XXXI, 529, 530, 537
redox behavior, as mediator-tirants, LIV, 402
Quinone antibiotic, XLIII, 137–144
Quinonoid dihydropteridine, product of phenylalanine hydroxylase, LIII, 278
QUSO, as hormone adsorbent, XXXVII, 34

R

R, *see* Rubrene
Rabbit
alveolar macrophage isolation, LVII, 472
antimouse antiserum, LVIII, 276, 277
aortic intimal cell, LVIII, 96, 97
bone marrow, ribonucleoside diphosphate reductase, LI, 239
brain, aldolases, XLII, 245–247
corneal cell line, LVIII, 414
embryo cell, media for, LVIII, 58
erythrocyte, nucleoside triphosphate pyrophosphohydrolase, LI, 276–279
immunization, LVI, 224, 225
injection sites, for antibody production, LII, 242
intimal cell, effect of hormone, LVIII, 106, 107
kidney
D-2-hydroxy acid dehydrogenase, assay, purification, properties, XLI, 325–329
D-α-hydroxyacid oxidase, LIII, 439
liver
aldolases, XLII, 227, 245
cytochrome *P*–450, LII, 111–115
fructose-1,6-diphosphatase, assay, purification, and properties, XLII, 354–359, 369–374

homogenate preparation, LI, 279, 280, 527
microsome isolation, XXXI, 199
nucleoside triphosphate pyrophosphohydrolase, LI, 279
phosphofructokinase, assay, purification, and properties, XLII, 67–71, 107, 110
purine nucleoside phosphorylase, LI, 524–530
triosephosphate isomerase, assay, purification, and properties, XLI, 438–442
muscle
fructose-diphosphate aldolase, XLII, 227, 239, 243, 244
glucosephosphate isomerase, anomerase activity, XLI, 37
glyceraldehyde-3-phosphate dehydrogenase, assay, purification, and properties, XLI, 264–267
phosphofructokinase, assay, purification, and properties, XLII, 71–77
electrophoresis, XLII, 70
phosphoglycerate kinase, purification, XLII, 129, 131, 133
phosphoglycerate mutase, purification and properties, XLII, 438, 445–449
triosephosphate isomerase assay, purification and properties, XLI, 447–453
oxyntic cell isolation, XXXII, 709
in situ muscle perfusion, XXXIX, 67–73
skeletal muscle, myofibrils, ATPase inhibitor, LV, 402, 405
spermatozoa, lactate dehydrogenase–X purification, XLI, 323
testis infusion, XXXIX, 277–282
tissue
acetone powder, preparation, LI, 208
adenylosuccinate synthetase, LI, 208–211
PAP levels, LVII, 256, 257
in vitro perfusion of ovaries, XXXIX, 230–237
Rabies virus, microcarrier culture, LVIII, 192
Racemization, mechanism, peptide synthesis, XLVII, 505, 506
Racemomycin, XLIII, 123
Radiation, *see also* specific types
filters, XLVI, 106
systems, XLVI, 105
Radical ion
chemiluminescence, LVII, 496, 497
experimental techniques, LVII, 510–520
generation, LVII, 510
Radioactive labeling, XLVI, 110
Radioactivity
determination, in acrylamide slab gels, LIX, 521
labeling, by reductive alkylation, XLVII, 476
low incorporation, XLVII, 332
peptide detection, XLVII, 217
Radioautography, *see* Autoradiography
Radio frequency attenuator, for long-pulse methods, LIV, 168
Radio frequency field, XLIX, 256, 257
Radio frequency pulse
field strength, choice, LIV, 156
frequency, choice, LIV, 157
generation, LIV, 154
recycle time, LIV, 163
optimization, LIV, 165, 166
strength, in long-pulse method, LIV, 168
variable frequency source, LIV, 187, 188
width
choice, LIV, 156
in long-pulse method, LIV, 168, 169
Radioimmunoassay, XXXIV, 708
antibody characterization, XXXVII, 26–28
antibody to edestin-oligosaccharide conjugates, L, 163–169
antisera development, XXXVI, 17–23
assay set-up, XXXVII, 32, 33
competitive, LVIII, 420, 421
cyclic nucleotides, XXXVIII, 90–105

hormone standards, XXXVII, 30
hormone-tracer labeling, XXXVII, 28–30
immunization, XXXVI, 22–23
of parathyroid hormone, XXXVII, 38–56
of peptide hormones, XXXVII, 22–38
procedures, XXXVI, 26–34
detailed, XXXVI, 30–34
of prostaglandin, XXXV, 287–298
preparation of antigen, XXXV, 288, 289
specificity, XXXV, 290–298
for *Pseudomonas* toxin A
labeling of antitoxin, LX, 790, 791
percent antibody binding to toxin, LX, 791
solid-phase toxin binding, LX, 791–793
radioactive hormone, XXXVI, 23–26
result interpretation, XXXVII, 35–38
separation of bound and free hormone, XXXVII, 33–35
specimen preparation, XXXVII, 31, 32
statistical analysis, XXXVII, 3–22
of steroid hormones, XXXVI, 16–34
viral structural proteins, LVIII, 417–421
Radioisotope
experiment, double label
design, XL, 293
instrument settings, XL, 299
mathematics, XL, 296
sample preparation, XL, 302
for hormone labeling, XXXVII, 145–147
choice, XXXVII, 147, 148
Radiolabeling
in vitro, of ribosomal protein, LIX, 777–779, 783–787, 790, 791, 794, 795
in vivo, of ribosomal RNA precursors, LIX, 830, 831
Radiolabeling procedure
with formaldehyde, XLVIII, 332, 333
with methyl iodide, XLVIII, 332
Radioligand assay
curve fitting, XXXVII, 4–6
empirical methods, XXXVII, 6–13
generalized logistic model, XXXVII, 11, 12
hyperbolas, XXXVII, 6
logit-log method, XXXVII, 7–11
orthogonal polynomials, XXXVII, 6, 7
empirical quality control, XXXVII, 16–21
of hormones, XXXVII, 3–22
of LH and hCG, XXXVII, 176–178
quality control chart, XXXVII, 20
response variables, XXXVII, 4
statistical analysis, XXXVII, 3–22
variance and weighting, XXXVII, 13, 14
Radiometer, calcium-sensitive electrodes, LVI, 347
Radioreceptor assay, of peptide hormones, XXXVII, 66–81
bioassays compared, XXXVII, 68, 69
HGH, XXXVII, 70–77
historical aspects, XXXVII, 67, 68
insulin, XXXVII, 78–81
Radish, organelles, XXXI, 491
Radium, as stable light source, for photometer calibration, LVII, 300, 556–558
Radius of gyration, LIX, 637, 638, 685, 690, 710, 713, 714
definition, LIX, 710
of *E. coli* ribosomes, LIX, 688
relation to molecular weight, LIX, 720, 721
of ribosomal subunits, function of protein distribution, LIX, 765, 768–770
with two bodies, LIX, 756, 759
Radulum, oxidase, LII, 17
Raffinose, XXXIV, 333
chromatographic constants, XLI, 17
hollow-fiber dialytic permeability, XLIV, 297
hollow-fiber retention data, XLIV, 296
optical rotation, XLIV, 350
polarography, LVI, 467, 477
substrate, for α-galactosidase, XLIV, 350
Raman hyperchromism, XLIX, 115
Raman hypochromism, XLIX, 115

Raman scattering, XXXII, 247; LIV, 12, 13
  coherent anti-Stokes, LIV, 249
Raman spectrophotometer, XXXII, 248, 253, 254
Raman spectroscopy, XXXII, 253–257; XLIX, 65–149, *see also* Anti-Stokes Raman spectroscopy, coherent; Resonance Raman spectroscopy
  artifacts, XLIX, 88, 95, 96
  band intensity
    Albrecht vibronic expansion approach, XLIX, 79–82
    dispersion theory, XLIX, 78, 79
    forbidden, XLIX, 79, 81
    parameters, XLIX, 83
    polarizability theory, XLIX, 75–78, 81
    quantum theory, XLIX, 77–82
  CH group, XLIX, 98
  conformationally sensitive backbone vibrational modes, XLIX, 99–102
  data processing, XLIX, 89
  depolarization of incident light, XLIX, 75
  differential type, XXXII, 256
  error from Rayleigh scattering, XLIX, 87, 88
    from thermal lens effect, XLIX, 91
  excitation frequency, XLIX, 85, 86
  experimental conditions, XLIX, 94, 95
  fluorescence, XXXII, 254
    interference, XLIX, 83
  grating ghosts, XLIX, 88, 95
  internal standard, XLIX, 90
  laser plasma lines, XLIX, 95
  monochromator, XLIX, 87, 88
  N–C^{α}–C′ unit, skeletal stretching modes, XLIX, 100, 101
  nonresonance scattering, XLIX, 69–71
  one-photon process, theory, XLIX, 68, 69
  parameters measured, XLIX, 83
  photoelectric detection, XLIX, 88, 89
  principles, XLIX, 67–72; LIV, 233, 234
  resonance enhancement, LIV, 134
  results and interpretation, XXXII, 256, 257
  sample cell, XLIX, 87, 91, 93
  sample preparation, XXXII, 254, 255; XLIX, 89–93
  satellite, source, XLIX, 88
  scattering geometry, XLIX, 86, 91
  self-absorption, XLIX, 91, 93
  solution concentration, XLIX, 91, 93
  spectra interpretation, XLIX, 97–127
  spectrometer, components, XLIX, 84–89
  two-photon process, theory, XLIX, 68, 69
  Tyndall scattering, XXXII, 254, 255
  vibrational modes of side-chains, XLIX, 116–118
    of topographical structures, XLIX, 111–116
  vibrational Raman effect, theory, XLIX, 72–74
  virtual state, XLIX, 71
Raman tensor, polarization, LIV, 237–239
RAMPRESA sampling apparatus, description, LVI, 296–301
RAMQUESA sampling apparatus, description, LVI, 298–301
Rana pipiens, isozymes of pyruvate kinase, assay, purification, and properties, XLII, 166–175
Raney nickel, effect on thymidylate synthetase, LI, 97
Rapeseed, from protoplasts, LVIII, 367
Rapeseed mesophyll × soybean culture, LVIII, 365
Rapid-freeze EPR technique, LIII, 404, 406, 414
Rapid-freezing apparatus, LIV, 86–91
Rapid mixing and quenching device, design, LVI, 298–301, 492–494
Rapid-scan correlation method, XLIX, 343, 344; LIV, 169–171, 187
Rare element, yeast growth media, LVI, 59
Rat
  alveolar macrophage isolation, LVII, 472
  ascites hepatoma cell, *see also* Ehrlich ascites cells; Novikoff ascites cell
    CPSase-ATCase-dihydroorotase enzyme complex, LI, 111–121
    growth, LI, 115

brain
  aldolase, electropherograms, XLI, 72
  hexokinase, assay, purification, and properties, XLII, 20–25
  metabolite content, LV, 218, 219
brown adipose tissue, LV, 66
costal cartilage, in somatomedin bioassay, XXXVII, 95–101
embryo cell, media, LVIII, 58
epithelial cell, media, LVIII, 57
fetus, liver explant, XXXIX, 36–40
follicular cell, effect of hormone, LVIII, 100, 101
glial C_6, effect of hormone, LVIII, 96, 103
glioma C_6, LVIII, 97
  effect of hormone, LVIII, 103–105
  media, LVIII, 57
granulosa cell isolation, XXXIX, 198, 199
heart
  isolation and perfusion, XXXIX, 43–60
  metabolite contents, LV, 216
  mitochondria, aureovertin, LV, 482
hemicorpus, perfusion method, XXXIX, 73–82
hepatoma, LVIII, 544
hypophysectomized
  commercial, XXXIX, 340
  preparation, XXXIX, 337–340
karyotype, LVIII, 342
kidney
  cortex, metabolite content, LV, 220
  glucosephosphate isomerase, anomerase activity, XLI, 57
  L-α-hydroxyacid oxidase, LIII, 439
  perfusion, XXXIX, 3–11
liver
  adenine phosphoribosyltransferase, LI, 574–580
  adenylate kinase, LI, 459–467
  aldolase, electropherograms, XLI, 73
  cells
    internal volume, LV, 551
      media, LVIII, 90
    perfusion, LVIII, 128, 129
    plating, LVIII, 538, 539
  CPSase-ATCase-dihydroorotase complex, LI, 111–121
  crude extract preparation, LI, 116, 118, 324, 426, 460, 461, 521, 575, 576; LIII, 281; LIX, 403
  cytoplasmic fraction, preparation, LI, 427
  deoxythymidine kinase, LI, 360–365
  dimethylglycine dehydrogenase, LIII, 450
  glucokinase, assay, purification, and properties, XLII, 31–39
  glucosephosphate isomerase, anomerase activity, XLI, 57
  guanylate kinase, LI, 485–490
  L-gulono-γ-lactone oxidase, LIII, 450
  homogenates prepared, XXXI, 7, 8
  lysosomal extract preparation, LI, 273
  lysosomes
    acid nucleotidase, LI, 271–275
    internal volume, LV, 551
  metabolite contents, LV, 213
  mitochondria
    ATPase inhibitor, LV, 402, 405, 408–414
    aureovertin, LV, 482
    electrogenicity of calcium translocation, LV, 639, 640
    internal volume, LV, 551
    isolation, LVI, 18, 19
      of oligomycin-sensitive ATPase, LV, 328–333
    kinetics of glutamate and aspartate efflux, LVI, 258–262
    magnesium efflux, LVI, 315, 316
    membrane potential, LV, 555
    preparation of ribosomes, LVI, 24–27
    purification of F_1 ATPase, LV, 320–328, 333–337
    reconstitution of ATP-dependent functions, LV, 736–741

perfusion, XXXIX, 25–36
phenylalanine hydroxylase, LIII, 278–286
phosphofructokinase, purification, XLII, 107
plasma membrane, XXXI, 75–102
postmitochondrial supernatant, preparation, LIX, 516, 517
preparation of homogenate, LV, 33–35
preservation, XXXI, 3–6
purine nucleoside phosphorylase, LI, 517, 518, 521–524
pyrimidine nucleoside monophosphate kinase, LI, 321–331
regenerating, preparation, LI, 362
ribosomal subunits from preparation, LIX, 402–410, 505, 506
ribosomes, preparation, LIX, 504, 505
tRNA, large-scale preparation, LIX, 230, 231
sarcosine dehydrogenase, LIII, 450
subcellular fractionation, XXXI, 6–41
submitochondrial particles, aureovertin, LV, 482
uridine phosphorylase, LI, 423–431
muscle, glucosephosphate isomerase, anomerase activity, XLI, 57
neuroblastoma, LVIII, 96
ovarian cell, LVIII, 96
pancreatic cell, media for, LVIII, 59
pineal gland, cell dissociation, LVIII, 129, 130
pituitary cell, LVIII, 529
serum, sympathetic neurons, LVIII, 580
skeletal muscle
adenylate deaminase, LI, 490–497
crude extract preparation, LI, 492
submaxillary gimmel factor, LVIII, 101
submaxillary gland, extract, LVIII, 104
testis
gonadotropic function studies, XXXIX, 252–271
in vitro studies, XXXIX, 252–271
Rate-zonal centrifugation, of plant cells, XXXI, 545, 549
Rattlesnake venom, *see* Crotalase
Rattus norwegicus, cytochrome oxidase from, LIII, 66
Rayleigh horn beam trap, XLIX, 201, 207
Rayleigh interference optics
absolute fringe displacement, XLVIII, 170, 173
laser light source, XLVIII, 185–191
plate reading requirements, XLVIII, 170
sedimentation equilibrium, XLVIII, 169–171
Rayleigh peak, definition, LIV, 234
Rayleigh scattering, XXXII, 247; XLIX, 69, 72, 73, 87, 88, 90
Raytheon sonic oscillator, LI, 80
10-kc, LIII, 234
properties, XXXI, 662
R–banding, *see* Banding, reverse
RBA virus, cells transformed by plasma membrane isolation, XXXI, 167, 168
RC-2 centrifuge, XXXI, 96
Rd, *see* Rubredoxin
Reaction age, LIV, 85–87
Reaction boundary, *see also* Boundary profile
diffusion, XLVIII, 195
Reaction equilibrium
equations, XLVIII, 385–387
measurement, using fluorescence methods, XLVIII, 380–415
Reaction kinetics
for association equations, XLVIII, 387–392
instrumentation, XLVIII, 393–415
measurement, using fluorescence methods, XLVIII, 380–415
quasi-elastic laser light scattering, XLVIII, 444–450
rate, measurement, using bioluminescence, LVII, 361–364
rate constant determination
by dilution jump technique, XLVIII, 390–391
from initial and integrated rate measurements, XLVIII, 388, 389

by relaxation technique, XLVIII, 391, 392
theoretical stoichiometric expressions, XLVIII, 382, 383
Reaction vessel, measurement of proton translocation, LV, 629–631
Reacti-vial, LII, 332
Reactor
for alanine production, XLIV, 880–882
for L-amino acid production, XLIV, 752–758
artificial enzyme membranes, XLIV, 910, 911
batch, XLIV, 251, 718, 722, 723
catalyst packing density, XLIV, 770
collagen-invertase, constructional details, XLIV, 252
column dimension, XLIV, 743, 754
column pressure drop, XLIV, 742, 743, 754
contact efficiency, XLIV, 771, 772
for continuous aminoacylase production, XLIV, 752–755
design equation, XLIV, 772
effect of temperature, XLIV, 735, 736, 738
enzyme, XLIV, 721, 722
enzyme loading factor, XLIV, 769, 770
external mass transport, XLIV, 771
flow patterns, XLIV, 723, 733–736, 754, 755
flow rate, XLIV, 734, 744, 753, 754, 756
fluidized bed, XLIV, 341, 722, 854, 855
for fructose production
design, XLIV, 817–821
effect of metal ions, XLIV, 815, 816
pressure drop, XLIV, 820, 821
process parameters, XLIV, 812–817
scale-up, XLIV, 820
for glucose isomerization, XLIV, 768–776
for glucose production, XLIV, 776–792
bacterial contamination, XLIV, 784, 785
cost analysis, XLIV, 788–792
flow pattern, XLIV, 782, 789
operation, XLIV, 783–792
process detail, XLIV, 783
supports, comparative data, XLIV, 779–783
immobilization methods, XLIV, 725–729
industrial, design parameters, XLIV, 260, 261, 742, 743, 769–772, 774
internal transport resistance, XLIV, 771
for lactose production, XLIV, 792–809
column back-flushing, XLIV, 804
cost analysis, XLIV, 805–809
enzymes, comparative data, XLIV, 794–805
half-life studies, XLIV, 799–801
mass transfer studies, XLIV, 800–804
pressure drop, XLIV, 805
scale-up, XLIV, 804, 805
supports, comparative data, XLIV, 794–805
temperature effects, XLIV, 794
for lactose reduction of milk, XLIV, 822–830
enzyme leakage, XLIV, 827, 828
enzyme source, XLIV, 827
flow pattern, XLIV, 824
microbial contamination, XLIV, 826, 829, 830
operation, XLIV, 825, 826, 828–830
operation, general characteristics, XLIV, 735–738, 756
operational stability, XLIV, 735–738, 744, 758, 771, 773
packed-bed, XLIV, 252, 338–340, 722–724
plug-flow, XLIV, 722–724
for porphobilinogen production, XLIV, 844–849
for production of amine oxides of tertiary amines, XLIV, 854, 855
recycle, XLIV, 253, 254
regeneration, XLIV, 758
spiral-wound, XLIV, 251–253, 724
stirred batch, XLIV, 722–724
assay using, XLIV, 336–338, 342–345

substrate diffusion, XLIV, 732, 733
supports for, choice, XLIV, 725–729, 751, 752
types, XLIV, 722–725, 729–735
Reading error, LIV, 46
Read-out device, LIV, 39, 40
Reagent, *see also* specific types
for electron microscopy, LVI, 719
for fluorescence studies, LVI, 498
photosensitive, LVI, 655–658
for preparation of yeast mitochondria, LVI, 130
Reaginic antibody, XXXIV, 709
Receptor, XXXIV, 171–177, *see also* specific receptors
affinity labeling, general methods, XLVI, 572–582
heterogeneous, XLVI, 186
identification, XLVI, 71, 573
multicomponent, XLVI, 183, 186
nucleotide, XLVI, 88
photoaffinity labeling, XLVI, 112–114
plant auxin, XLVI, 84
purification, XXXIV, 94, 661–668; XLVI, 573, 574
quantitation, XLVI, 573
for steroid hormones
estrogen, XXXVI, 331–365
purification, XXXVI, 329–426
structure-function studies, XLVI, 574–582
unreduced, XLVI, 582
Receptor binding assay, for hormone inactivation, XXXVII, 205–207
Receptor protein, isolation, XXXVI, 491–492
Receptor transforming factor, in EB-protein formation, XXXVI, 333, 334, 337
Reconstitution
of ribosomal subunits, LIX, 447, 448, 451, 455, 652–654, 779
of tRNA, LIX, 127
of 30 S pre-rRNA-protein complex, LIX, 837
Recorder
electrophysiological techniques, LV, 659–661
measurement of proton translocation, LV, 633
for microspectrophotometer, XXXIX, 419
rapid proton-transfer reactions, LV, 615, 616
Recording technique, for electrophysiological studies, XXXIX, 434–438
Recrystallization, temperature, of biological material, XXXII, 36
Recycling method, enhanced sensitivity of enzymatic assays, LV, 215, 217, 221
Red blood cell, *see also* Erythrocyte
external membrane potential probes, LV, 559
ionophores, LV, 439, 440, 447
optical probes, LV, 573, 582
resting potential, LV, 692
Red kidney bean agglutinin, iodination, L, 57, 58
Red No. 2, XLIV, 654
Redox mediator, *see also* Mediator-titrant
function, LIV, 422, 423
interference, in spectrophotometric titrations, LIV, 429
properties, LIV, 423–426
Redox molecular state, lifetime
computation, LIV, 180, 181
definition, LIV, 179
Redox poise, definition, LIV, 427
Redox potential
of medium, LVIII, 74
membrane vesicle isolation, LVI, 382
of ubiquinone and cytochrome *c*, LV, 240
Redox potentiometry, LIV, 411–435
data analysis, for multicomponent system, LIV, 434, 435
degassing procedure, LIV, 430
experimental controls, LIV, 426, 427
experimental procedures, LIV, 430–434
half-cell, definition, LIV, 411
instrumentation, LIV, 419–422
low-temperature analysis, LIV, 434
mediator properties, LIV, 422–426
poise and pulse approach, LIV, 427, 428
sample cell, anaerobic, LIV, 420

theoretical considerations, LIV, 411–418
Redox reaction
electron-transfer chemiluminescence, LVII, 499
free energy transfer, LIV, 412
transmembrane, model system, LV, 541–544
Redox system, distance measurements, LIV, 188, 190
Reducing sugar, *see* Sugar
Reductase, XLVI, 220
Reduction, on column, XXXIV, 293, 294
Reductive alkylation, *see* Alkylation
Reductive methylation, *see* Methylation, reductive
Reeve Angel glass fiber filters, LVII, 80
Reference beam, heterodyne spectroscopy, XLVIII, 483–485
Reference electrode, *see also* specific type
measurement of proton translocation, LV, 632, 633
ion selective electrodes, LVI, 363
Reflectance spectroscopy, XLIII, 181, 182
Refractive index, relation to concentration of solute, XLVIII, 169
Refractive index detector, XLVII, 212, 216, 217
Refractometer, clinical hand type, XXXII, 61
Refractometry, determination of protein concentration, LV, 304
Refrigeration, LVIII, 9
Refrigerator, liquid nitrogen type, XXXI, 4
Regeneration, of enzyme and carrier, XLIV, 164, 165
Regeneration system, reconstitution, LV, 710
Regulatory protein, XXXIV, 368–373, *see also araC* protein
Reichstein's compound S, XLIV, 187–190, 471
Reichstein's substance E, U, isolation, XXXVI, 501
Reinheitszahl, LII, 524
Reinickate salt, XLIV, 594
Reinzuchthefe, pure culture yeast, aminoacyl-tRNA synthetase, LIX, 259
Relaxation, conformational, ligand binding data evaluation, LIV, 524, 525
Relaxation amplitude, LIV, 65
evaluation in multiple-step reaction, LIV, 72
in single-step experiment, LIV, 67–69
Relaxation kinetics, XLVIII, 248–270
of heme proteins, LIV, 72–84
principles, LIV, 65–72
Relaxation method
for determination of rate constants, XLVIII, 391, 392
differential, LIV, 184
general principle, LIV, 3
Relaxation probe, principle, LIV, 193, 194
Relaxation spectrum, LIV, 65
numerical analysis by nonlinear least-squares method, LIV, 70
Relaxation time
evaluation in single-step experiments, LIV, 66–68, 72, 73
in two-step reaction, LIV, 70–72, 74, 75
longitudinal
for biological sample calculations, XLIX, 502
Carr-Purcell pulsed T_1 method, XLIX, 333–339
continuous-wave method, XLIX, 335
demagnetization recovery method, XLIX, 339, 340
evaluation, XLIX, 326–332, 423
Fourier-transform methods, XLIX, 333–342
inversion recovery method, XLIX, 333–339
long-pulse method, XLIX, 342, 343
progressive saturation method, XLIX, 342
rapid-scan correlation method, XLIX, 343, 344
rotational correlation time, XLIX, 422, 484
slow-tumbling, definition, XLIX, 481
steady-state method, XLIX, 340–342

theoretical considerations, XLIX, 360–365
very slow tumbling
definition, XLIX, 481
nitroxides, XLIX, 501, 502
of metallochromic indicators, LVI, 307
reciprocal, LIV, 72–75
spin-lattice, *see* Relaxation time, longitudinal
spin-spin, *see* Relaxation time, transverse
theoretical paramagnetic effect, XLIX, 326–332
transverse
evaluation, XLIX, 326–332
Fourier, transform methods, XLIX, 344, 345
linewidth method, XLIX, 344, 350
rotational correlation time evaluation, XLIX, 466, 467
theoretical consideration, XLIX, 365–368
Relaxing medium, for coelenterates, LVII, 248, 249
Release, action of antibiotics, LIX, 860–862
Renal, *see* Kidney
Renex 690
commercial source, LII, 110
in cytochrome b_5 assay, LII, 110
in cytochrome P–450 purification, LII, 111–115
for NADPH-cytochrome P–450 reductase solubilization, LII, 90
removal, LII, 203
Renilla
bioluminescent reaction, LIII, 559
luciferase, LIII, 560
Renilla kollikeri
luciferase-GFP interaction, specificity, LVII, 263, 264
source, LVII, 248
Renilla mülleri
luciferase-GFP interaction, specificity, LVII, 264
source, LVII, 248
Renilla reniformis
bioluminescent system, LVII, 237–244, 562
nonradiative energy transfer, LVII, 240–242, 259, 263, 264
in vivo emission, LVII, 240, 259
coelenteramide, LVII, 344
emission wavelength, LVII, 341
luciferase-GFP interaction, specificity in, LVII, 263, 264
luciferin, definition, lack, LVII, 344
oxidation mechanism, LVII, 347, 348
luciferin-binding protein, LVII, 258
relaxing procedure, LVII, 248
source, LVII, 248
Renin, XXXIV, 5; XLVI, 235–240
assay, XLVI, 237–240
production in isolated renal cells, XXXIX, 20–23
substrate, XLVI, 235
Renkin equation, XXXVI, 218
Rennin
crystal
immobilization, XLIV, 548
x-ray analysis, XLIV, 555, 556
immobilization, with collagen, XLIV, 244
Renograffin, as gradient material, XXXI, 509
Reorganization parameter, ESR, LVII, 509
Repeating Dispenser, XXXII, 112
Replicate plating
cloth, LVIII, 320
procedure, LVIII, 320, 321
Replication, XLVI, 644
in mitochondria, inhibitors, LVI, 31
Reproductive tissue, XXXIX, 107–425
Reptile
balanced salt solution, osmolarity adjustment, LVIII, 469
incubation temperature, LVIII, 471, 472
media, LVIII, 469
RER, *see* Endoplasmic reticulum, rough
Resazurin
in dental caries diagnosis, XLI, 56
in fluorometric assay of dehydrogenase activity, XLI, 53

as indicator of microbiological contamination of meat and milk, XLI, 56
Reservoir, for manual sequence analysis, XLVII, 353, 354
Resin, *see also* specific type
anion-exchange, XLVII, 28, 29
cation-exchange
size fractionation, XLVII, 6, 20
types, XLVII, 29–31
ion-exchange
amino acid analysis, XLVII, 23–27
applications, XLVII, 23–31
column packing, XLVII, 213, 227
conditioning, XLVII, 227
cross-linking, XLVII, 22
improved, amino acid analysis, XLVII, 19–31
mixed-bed, artifact, XLVII, 37, 38
particle size, XLVII, 20–22
peptide analysis, XLVII, 27–31
properties, XLVII, 20
in plant-tissue fractionation, XXXI, 535, 536
Resin acid, as plant extraction interferents, XXXI, 531
Resin standard, conversion to N^{α}-phenylthiocarbamyl methylamide derivative, XLVII, 377, 378
Resistance
membrane, measurement, LV, 662
multiple loci, LVIII, 315
Resistance determinant, XLIII, 43
Resistance plasmid, *see* Plasmid, resistance
Resistance transfer factor, XLIII, 43
Resistant strain, drug concentration, LVIII, 314, 315
Resolution, XXXIV, 250, 252, 253
effect of pH, XXXIV, 252
enhancement, LIX, 54–56
factors affecting, XXXIV, 250
test kit, XXXIV, 134–136
Resolving time, LIV, 33
relation to light intensity, LIV, 34
Resonance
line, definition, LIV, 448
line width, LIV, 179
Resonance denominator, XLIX, 79
Resonance fluorescence, XLIX, 71
Resonance Raman spectroscopy, XLIX, 65–149; LIV, 233–249, *see also* Raman spectroscopy
absorbing samples, techniques, LIV, 247–249
advantages, XLIX, 71, 72
chromophores, XLIX, 96, 97
circular polarization measurements, LIV, 239
excitation frequency, XLIX, 86, 132, 133, 135, 137
fluorescence, interference, LIV, 248, 249
laser power, XLIX, 94; LIV, 246, 247
mode assignments, LIV, 239–242
principles, XLIX, 67–72
Raman band polarizations, LIV, 237–239
scattering mechanisms, LIV, 235, 236
self-absorption, XLIX, 93
spectra interpretation, XLIX, 127–147
theoretical background, XLIX, 74
Resorcinol, XLVII, 603
indicator for gangliosides and hematosides (sialic acid), XXXV, 407, 412
reaction for glycolipids, XXXII, 357
Resorcinol hydroxylase, LII, 17
Resorufin
fluorescence spectra, LII, 375, 376
in fluorometric determination of dehydrogenase activity, XLI, 53–56
as mediator-titrant, LIV, 408
sodium salt, in 7-ethoxyresorufin synthesis, LII, 374
Respiration
cyanide-insensitive, in *N. crassa*, LV, 26
efrapeptin, LV, 495
initial rate of ATP synthesis, LVI, 496
organotins, LV, 508, 509
sterol-depleted mitochondria, LVI, 563
of temperature-sensitive mutants, LVI, 137
uncouplers, LV, 463
Respiratory burst, LVII, 488

Respiratory chain
activity, proton translocation, LV, 627–640
calcium-dependent activation, LVI, 339–341
carriers, calcium-dependent redox shift, LVI, 341, 342
characteristic → H^+/O values, LV, 637, 638
of *M. phlei* ETP, LV, 184
of *N. crassa* mitochondria, LV, 147
phosphorylation site, LV, 240
role of ubiquinone, LIII, 575–576
Respiratory control
brown adipose tissue mitochondria, LV, 70–72
of cytochrome oxidase vesicles, LV, 749
hepatoma mitochondria, LV, 81, 83
rate measurements, LV, 225–227
steady-state measurements, LV, 227–229
uncoupler assay, LV, 464
Respiratory control ratio
of brain mitochondria, LV, 59
of hepatoma mitochondria, LV, 82, 83
immobilized mitochondria, LVI, 556
microorganismal mitochondria, LV, 143
mitochondrial integrity, LV, 8
smooth muscle mitochondria, LV, 64, 65
of stored mitochondria, LV, 30
of submitochondrial particles, LV, 108
of yeast mitochondria, LV, 150
Respiratory enzyme, mitochondrial, assay, LVI, 101, 102
Respiratory rate, of malonate-activated ESP, LV, 108
Resting cell, human lymphocyte, LVIII, 486–494
Resting potential, determination, LV, 691, 692
Restriction enzyme
mitochondrial DNA, LVI, 5, 6, 186–188
petite mutant DNA, LVI, 156, 162, 163
Retardation coefficient, of macromolecules, XXXII, 93
Reticulocyte
countercurrent separation, XXXII, 637
differential centrifugation, XXXI, 721
electron microscopy, XXXII, 31, 32
membrane, XLVI, 86
protein kinases, XXXVIII, 297, 315, 316, 322
assay, XXXVIII, 316–317
crude preparation, XXXVIII, 317–318
purification of kinase I, XXXVIII, 318–319
of kinases IIa and IIb, XXXVIII, 319–321
Reticulocyte lysate, rabbit
fractionated vs. unfractionated, initiation factor function, LX, 70, 71
globin messenger ribonucleic acid preparation, LX, 69
initiation factor preparation, LX, 15–29, 67, 101–108, 124–135, 172–181
messenger ribonucleic acid distribution, LX, 85–87
phosphoprotein synthesis, LX, 511–516
preparation, LX, 65, 125, 250, 352, 353, 411, 463, 464, 528, 536, 581
ribosome content, LX, 67
source of elongation factors, LX, 639–641
of heme-regulated translational inhibitor, LX, 471, 485
of phosphoprotein phosphatase, LX, 528
translational activity, LX, 65–67, 464–467
use in assay for eIF-2, LX, 391, 392
Reticuloendothelial cell, *see also* Macrophage
of liver, XXXIX, 36
Retina
outer rod segments, X-ray studies, XXXII, 216, 219
pigment cell, clones, XXXII, 770
rhodopsin isolation, XXXII, 306–309
Retinal, *cis-trans* isomerism, XXXII, 309
Retinal-binding peptide, sequence analysis, XLVII, 331

Retinal-binding protein, XXXIV, 7
Retinoic acid, in media, LVIII, 103
Retinol monophosphate, L, 405
Retinyl ester hydrolase, in microvillous membrane, XXXI, 130
Retrovirus, avian, LVIII, 379
Retsch-mill KMI, LIX, 445
Reversed phase chromatography, *see also* High-pressure liquid chromatography
  buffers, LIX, 213
  purification of chloroplastic phenylalanine tRNA, LIX, 214
  separation of CHO tRNA isoacceptors, LIX, 221–228
    of transfer RNAs, LIX, 111, 207
  in studies of amino acid transfer among isoacceptors, LIX, 294
Reverse salt gradient chromatography
  column preparation, LIX, 217, 221
  procedure, LIX, 217, 221, 222
  of tRNA, LIX, 215–218
  sample preparation, LIX, 217
Reverse transcriptase, XXXIV, 728, 729, *see also* Transcriptase, reverse
Reversion
  precautions against, LVI, 119, 120
  resolution of *syn*$^-$ and *rho*$^-$ mutants, LVI, 104
Revertant
  as isogenic controls, LVI, 119
  of *ole* mutants, LVI, 575, 576
  temperature-sensitive mutations, LVI, 136, 137
Revilla, oxidase, LII, 20
Rexyn 1–300, LII, 476, 488
Reynolds number, measurement, LVI, 215, 216
R factor, XLIII, 41–55, *see also* Plasmid resistance
  genetic transfer, XLIII, 43–49
R factor-containing strain, pathogens, XLIII, 611, 612
Rhamnose, chromatographic constants, XLI, 16, 17, 20
L-Rhamnulose, preparation, XLI, 92
L-Rhamnulose 1-phosphate, preparation, and dicyclohexylammonium salt, XLI, 91–93
L-Rhamnulose-1-phosphate aldolase
  activators and inhibitors, XLII, 268, 269
  assay, XLII, 264, 265
  chromatography, XLII, 266
  from *Escherichia coli*, XLII, 265
  molecular weight, XLII, 269
  properties, XLII, 268, 269
  purification, XLII, 265–268
Rhein
  complex I, LVI, 586
  inhibitor, of complex I, LIII, 13
  isolation, LV, 459, 460
  NADH oxidation, LV, 458–460
Rhenium, atomic emission studies, LIV, 455
Rheomacrodex, in placental perfusion, XXXIX, 246, 249
Rhesus monkey, ovarian follicle isolation, XXXIX, 194–198
Rhipicephalus appendiculatus, media, LVIII, 463
Rhizobium, oxidase, LII, 16
Rhizobium japonicum
  cell disruption in French press, LII, 160
  cytochrome *P*–450, LII, 157–166
Rhizobium meliloti
  extracellular polysaccharide
    side chain sequence determination, L, 36–38
    Smith degradation, L, 27
Rhodacyanine, as optical probes, LV, 573
Rhodamine, as optical probes, LV, 573
Rhodamine B
  fluorescent quantum yield, LVII, 590
  indicator for lipids, XXXV, 406
  solution quantum counter, LVII, 591, 592
Rhodamine 6G, LIV, 8; LV, 515
  indicator for lipids, XXXV, 406
  tuning range, LIV, 246
Rhodamine 6G laser, wavelength, XLIX, 85
Rhodomicrobium vannielii, high-potential iron-sulfur protein, LIII, 330
Rhodomycin, XLIII, 340, 352
Rhodomycinone, XLIII, 146

Rhodopseudomonas capsulata
ATPase, LV, 341, 342
aureovertin, LV, 483
ubiquinone, LIII, 579
Rhodopseudomonas gelatinosa
growth, LIII, 335
high-potential iron-sulfur protein, LIII, 329, 332, 333, 339
Rhodopseudomonas sphaeroides, *see Rhodopseudomonas*
Rhodopseudomonas spheroides, XLIV, 848
aureovertin, LV, 483
chromatophores
electron spin resonance spectra, XLIX, 407
hydrogen transport, LVI, 312
D-gluconate dehydratase activity, XLII, 304
D-β-hydroxybutyrate dehydrogenase, LIV, 229
membranes, optical probes, LV, 573
membrane vesicles, anaerobic active transport, LVI, 386, 387
phototrophically grown
culture media and growth conditions, LVI, 381
membrane vesicle preparation, LVI, 382, 383
Rhodopsin, XXXIV, 339; LIV, 18
in biomembrane reconstitution, XXXII, 554
from bovine retina, isolation, XXXII, 306–309
difference spectrum, LIV, 24
free peptide hydrogen exchange, XLIX, 30
kinetics, LIV, 18–21, 24
photocalorimetry, LVII, 548, 549
pineal gland enzymes, XXXIX, 380
properties, XXXII, 308, 309
in protein rotation studies, LIV, 57
purification, XXXII, 307, 308
in retinal rod outer segment, XXXII, 219
Rhodospirillum, hydrogenase reaction, LIII, 287
Rhodospirillum molischianum, cytochrome c_2 from, CD spectrum, LIV, 276
Rhodospirillum rubrum
ATPase, LV, 341, 342
ATP formation, measurement, LVII, 51–54
bacteriochlorophyll reaction center, proteoliposomes, LV, 760
chromatophore membrane, fluorescent probe studies, XXXII, 234, 238, 242, 243
chromatophores
aureovertin, LV, 483
membrane potential, LV, 555, 559, 595–596
proton gradient, LV, 452
citreoviridin, LV, 487
crude extract preparation, LIII, 318, 320
cytochrome c_2, LIII, 209, 210
CD spectrum, LIV, 277, 278
cytochrome c', LIII, 207
enzyme extraction, XXXI, 772
growth, LIII, 318
membranes, optical probes, LV, 573
nitrogenase, LIII, 318–321
ribulose-1,5-diphosphate carboxylase
assay, purification, and properties, XLII, 457–461
purification and properties, XLII, 461, 468–472
transhydrogenase, LV, 272
ubiquinone, LIII, 578
Rhodospirillum tenue
growth, LIII, 335
high-potential, iron-sulfur protein, LIII, 329, 331–334, 339
Rhodotorula, aldehyde reductase assay, purification, and properties, XLI, 361–364
Rhodotorula gracilis
glucose-6-phosphate 1-epimerase, molecular weight, XLI, 493
glucosephosphate isomerase, anomerase activity of, XLI, 57
Rhozyme-41, facilitator of rapid filtration of biological fluids, LVII, 69, 72
Rhythm, biological, hormonal factors, XXXVI, 474–481
Ribi press
for bacterial cell rupture, XXXI, 659
cell disruption, LV, 803
for plant-cell rupture, XXXI, 502

properties, XXXI, 663
Ribitol, D-ribulose preparation, enzymic method, XLI, 104–106
Riboflavin, XLIII, 520
biosynthesis, XLIII, 515–520, *see also* Guanosine triphosphate-8-formylhydrolase
in electrophoresis reagent, XXXII, 85
extinction coefficient, LIII, 421
initiator photochemical polymerization, XLIV, 174, 194, 571, 572, 600, 889, 891
in media, LVIII, 54, 65
peracetylation, LIII, 458
as prosthetic group, LIII, 420
separation from flavin compounds, LIII, 425, 426
substrate of bacterial luciferase, LVII, 151
Riboflavin-cellulose, XXXIV, 300
Riboflavin 4′,5′-cyclic phosphate, LIII, 422
Riboflavin 4′-phosphate, LIII, 422
Riboflavin 5-phosphate, in media, LVIII, 65
Riboflavin 5′-phosphate
assay of nitric oxide reductase, LIII, 645
in complex I, LIII, 12
in complex I–III, LIII, 7
determination by enzymic analysis, LIII, 421, 422
by fluorometric analysis, LIII, 423, 425–429
extinction coefficient, LIII, 421
in flavodoxin, LIII, 627, 630
molar fluorescence, LIII, 428
in NADH dehydrogenase, LIII, 17
redox potential, LIII, 400
reduced
activator, of succinate dehydrogenase, LIII, 33
assay with bacterial luciferase, LIII, 504, 565
reduction, with hydrogen, LIII, 565
by light, LIII, 566
with sodium dithionite, LIII, 565, 566
removal by photoirradiation, LIII, 435, 436
9-β-D-Ribofuranosyl-6-methylthiopurine, XLVI, 353
9-β-D-Ribofuranosyl-6-thiopurine, XLVI, 354
9-(β-D-Ribofuranosyluronic acid)adenine, XLVI, 303
Ribokinase, interference, in phosphoribosylpyrophosphate synthetase assay, LI, 3
Ribonuclease, XXXIV, 469, 492; XLVI, 89, 90, *see also* specific types
affinity labeling, XXXIV, 64, 185
3-aminotyrosine, pancreatic, XXXIV, 5
from *Bacillus cereus*, LIX, 106, 107
bovine pancreatic
activity, modification, XLVII, 160
arginine blockage, XLVII, 159–160
carbethoxylation, XLVII, 431–433, 439–441
carboxymethylation, XLVII, 116, 126
disulfide cleavage, XLVII, 112
end-group analysis, XLVII, 82, 93
histidine content, XLVII, 431, 432
hydroxylamine cleavage, XLVII, 134, 140, 142
octa *S*-sulfo derivative, reduction, XLVII, 124–126
peptide detection, XLVII, 239
peptide separation, XLVII, 205, 206, 207–210
reaction with DMSO/HCl, XLVII, 445, 458
reduction, XLVII, 116, 121
reductive alkylation, XLVII, 472, 474, 475
sequence analysis, from the carboxyl terminus, XLVII, 359
staphylococcal protease hydrolysis, XLVII, 190
sulfopropylation, XLVII, 121
synthesis, XLVII, 584
cAMP-binding protein preparation, XXXVIII, 369
circular dichroic spectrum, XLIX, 161
conformation studies using circular dichroism, XLIV, 367
using electron spin resonance, XLIV, 368

conjugate, activity, XLIV, 36, 271
as contaminant, LX, 65, 87, 423, 439, 440, 442
in proteolytic enzyme preparations, LIX, 598
determination of polyuridylate bound to cellulose, LX, 750
digestion of unhybridized messenger ribonucleic acid, LX, 86, 385, 547, 548
of dihydroorotate dehydrogenase, LI, 61
E. coli, XXXIV, 5
effect on estrogen receptor studies, XXXVI, 173
electron-nuclear distance, XLIX, 423
as electrophoretic marker, XXXII, 80
entrapment in *N,N'*-methylenebisacrylamide, XLIV, 171
exclusion, methods, LIX, 464
fluorescence quenching, XLIX, 222
high-pressure transition studies, XLIX, 21
hollow-fiber retention data, XLIV, 296
immobilization, by adsorption at air-water interface, XLIV, 905
matrix-bound, LIX, 472, 474
membrane preparation, LV, 803
membrane vesicle preparation, LVI, 381, 383
in milk, XXXII, 109
mitochondrial ribosomes, LVI, 86
molecular weight, by gel electrophoresis, XXXVI, 99
osmotic pressure studies, XLVIII, 73, 75
pancreatic, XXXIV, 492–494; LIX, 190
for digestion of ribosomal RNA, LIX, 570–574
in MS2 peptide analysis, LIX, 301
preparation of ribosomal protein, LIX, 517
removal of 3'-terminal nucleoside, LIX, 188
of tRNA, LIX, 90, 99, 106
in tRNA sequence analysis, LIX, 60
Sephadex-bound, preparation, LIX, 463
of single-stranded DNA, LIX, 840
in pancreatic microsomes, XXXI, 52, 53, 57, 59
phosphodiesterase activator, XXXVIII, 254
from *Physarum polycephalum*, LIX, 107
in plant ribosome isolation, XXXI, 584–586, 589
in plasma membranes, XXXI, 88
protection assay, as assay of ribosomal activity, LIX, 460
protective agents, XXXI, 504, 505
protein digestion, concentration and incubation period, XL, 38
purification, XXXIV, 494
of CPSase-ATCase-dihydroorotase complex, LI, 124
of double-stranded ribonucleic acid, LX, 552
of ω-hydroxylase, LIII, 358
of nitrate reductase, LIII, 348
of nitrogenase, LIII, 322, 327
of rubredoxin, LIII, 342
single-stranded activity, assay, LIX, 826
spheroplasts, LVI, 175
of succino-AICAR synthetase, LI, 190
in sucrose, XXXI, 720
tissue culture, LVIII, 125
for trimming initiation complexes, LX, 311–314, 318, 319, 326–328, 345, 346, 367, 368
tryptophan content, XLIX, 161, 162
unfolding, XLIX, 4
of uridine-cytidine kinase, LI, 311
Ribonuclease A, XXXIV, 119, 120; XLVI, 362, 477, 641, 642, 671
in bacterial protein isolation, LII, 178
classification of tyrosyl residues, XLIX, 118, 119
conjugate, reductive denaturation studies, XLIV, 518, 519
digestion, with carboxypeptidase A-Sephadex, XLIV, 533, 537
in gel electrophoresis, XXXII, 75

immobilization on acrylamide-acrylic acid copolymer, XLIV, 61, 62
on acrylamide-2-hydroxyethylmethacrylate copolymer, XLIV, 61, 62
on polyacrylamide beads, XLIV, 60, 61
outexchange kinetics, XLVIII, 324
in pancreatic juice, XXXI, 57
phosphorescence studies, XLIX, 244
Raman spectra, XLIX, 119–123
reductive denaturation studies, XLIV, 516, 526
synthetic, XXXIV, 637, 638
viscosity, XLVIII, 53, 57
volume changes, XLVIII, 38–42, 44
Ribonuclease-agarose, XXXIV, 183, 186, 187
Ribonuclease H
assay, LIX, 826
purification, LIX, 829
Ribonuclease II, immobilization
by adsorption on phospholipid monolayer, XLIV, 905, 908
in polyacrylamide gel, XLIV, 902, 908
Ribonuclease III
assay, LIX, 825, 826
in cleavage of 30 S pre-rRNA, LIX, 832, 833, 837
in digestion of affinity-labeled RNA, LIX, 808
purification, LIX, 826–830
storage, LIX, 829, 833
Ribonuclease S, XXXIV, 635
data analysis, XLVIII, 344–346
dissociation, XLVIII, 342–346
semisynthetic, XXXIV, 635–637
Ribonuclease S–agarose, XXXIV, 637
Ribonuclease St, from *Streptomyces erythreus,* arginine blocking, XLVII, 160
Ribonuclease T1, XLVI, 177, 178, 671
for digestion of ribosomal RNA, LIX, 570, 571
of tRNA, LIX, 60, 75, 90, 99, 105, 120, 343
preparation of single-stranded DNA, LIX, 840
Ribonuclease T2, XLVI, 175; LIX, 190
from *Aspergillus oryzae*, LIX, 178
for digestion of tRNA, LIX, 60, 71, 79, 171, 175, 180, 188, 192, 344
Ribonuclease U_2, for digestion of tRNA, LIX, 105
Ribonucleic acid, XXXIV, 6; XLVI, 358, *see also* specific type
centrifugal separation, XXXI, 723–726
in chromatin, XXXI, 278, 279
degradation, agar supports, XLIV, 23
density gradient studies, XXXI, 510
determination of deuterium content, LIX, 646, 647
digestion, XXXIX, 151
DNA-dependent, synthesis in thyroid cells, XXXII, 758
double-stranded
assay by activation of protein kinase, LX, 287–290
binding to initiation factor eIF-2, LX, 381, 388, 389
bonded to cellulose, purification of initiation factor eIF-2, LX, 247–255
detection, LX, 553–555
functional role, LX, 549
inhibition of initiation, LX, 380, 392, 554
mitochondria, LVI, 7
preparation, LX, 386, 551, 552
radioiodination, LX, 387
elution from gels, by electrophoresis, LIX, 834
in endoplasmic reticulum, XXXI, 23
fluorescent labeling, XLVIII, 358
high M.W. from, isolation, XXXI, 272–274
hybridized, XXXIV, 471
isolation, LVIII, 568, 569
in liver cell fractions, XXXI, 208, 209
localization, labeling conditions, XL, 13
mammary synthesis, XXXIX, 454
as marker, XXXII, 15
messenger, *see* Messenger ribonucleic acid
in microsomal fractions, LII, 88
in microsomes, XXXI, 228
mitochondrial, deletion mapping, LVI, 162

in pituitary neurosecretory granules, XXXI, 407, 410
in plant cells, XXXI, 495, 559, 562
plant ribosomes, XXXI, 584, 585
in plasma membrane, XXXI, 65–66, 86–87, 101, 165, 166, 167
poly(A)-containing, XXXIV, 497
polysomal m, XXXIV, 496
purification, XXXIV, 501, 502
recovery from polyacrylamide, LIX, 69, 70
ribosomal, *see* Ribosomal ribonucleic acid
5 S, domains, XLIX, 11
23 S, XLVI, 89, 632, 637–644, 682
30 S, XLVI, 626
single-stranded, XXXIV, 468
in studies of estrogen effects on protein synthesis, XXXVI, 454–455
synthesis
inhibitors, XL, 275, 280, 282
rhythmicity, XXXVI, 480
somatomedin effects on, XXXVII, 94
steroid-receptor effects, XXXVI, 319–327
transfer, *see* Transfer ribonucleic acid
translational control
assay
hybridization with polyadenylate, LX, 546, 547
inhibition of protein synthesis *in vitro*, LX, 547, 548
functional role, LX, 541, 548, 549
hybridization with messenger ribonucleic acid, LX, 548
isolation, LX, 542–545
molecular weight, LX, 545
purification, LX, 546
from yeast, hydrolysate preparation, LIX, 63, 71
in zymogen granule, XXXI, 51, 52
Ribonucleic acid polymerase, *see* RNA polymerase
Ribonucleoprotein
centrifugal separation of, XXXI, 723–726
staining, in electron microscope cytochemistry of nucleic acids, XL, 24
Ribonucleoside
dialdehydes
preparation, LIX, 181
reaction with hydrazides, LIX, 181
separation, LIX, 181
protected, phosphorylation, XXXVIII, 410–411
Ribonucleoside 5′-diphosphate, XLVI, 321
Ribonucleoside diphosphate reductase, XLVI, 321–327; LI, 227–246
active site, LI, 237
allosteric regulation, LI, 235, 236
assay, LI, 227–229, 238, 239
catalytic properties, LI, 235–237
from *E. coli*, LI, 227–237
effector binding, LI, 236
iron requirement, LI, 237, 244
kinetic properties, enzyme concentration, LI, 243, 244
molecular weight, LI, 235
properties, LI, 243–246
purification, LI, 229–234, 239–243
purity, LI, 242, 243
from rabbit bone marrow, LI, 237–246
redox properties, LI, 236
specificity sites, LI, 236
storage, LI, 241, 242
substrate binding, LI, 236
substrate specificity, LI, 235, 236, 245
subunit
properties, LI, 234, 235
purification, LI, 231, 232, 234
purity, LI, 234
structure, LI, 227, 234, 235
Ribonucleoside triphosphate, LX, 322, 323, 328, *see also* specific type
substrate of ribonucleoside triphosphate reductase, LI, 246
Ribonucleoside-triphosphate reductase, LI, 246–259
absorption coefficient, LI, 250
activators, LI, 258
adenosylcobalamin requirement, LI, 246, 247, 258
assay, LI, 247–250

isoelectric point, LI, 257
kinetic properties, LI, 258
from *Lactobacillus leichmannii*, LI, 246–259
ligand binding, LI, 259
molecular weight, LI, 257
partial primary structure, LI, 257
purification, LI, 250–252
purity, LI, 257
reaction mechanism, LI, 259
reducing substrates, LI, 258
regulation, LI, 257, 258
stability, LI, 257
substrate specificity, LI, 257, 258
5′-Ribonucleotidase, activity and distribution, XXXII, 124
Ribonucleotide, dialdehydes, separation, LIX, 181
Ribonucleotide reductase, XXXIV, 254–261; XLVI, 321, *see also* Ribonucleoside diphosphate reductase
B1 subunit, XXXIV, 254, 257, 258
mammalian, XXXIV, 258–260
purification of, XXXIV, 253–261
Ribose
inhibitor, of uridine nucleosidase, LI, 294
in media, LVIII, 66
in nucleotide, hydrogen exchange, XLIX, 25
product of uridine nucleosidase, LI, 290
D-Ribose
chromatographic constants, XLI, 16, 17
D-ribulose preparation, XLI, 104
α-D-Ribose 1,5-diphosphate, preparation, XLI, 83, 84
Ribose-phosphate
inhibitor, of orotate phosphoribosyltransferase, LI, 163
of orotidylate decarboxylase, LI, 163
Ribose 1-phosphate
inhibitor of purine nucleoside phosphorylase, LI, 530
product of purine nucleoside phosphorylase, LI, 517, 530
of uridine phosphorylase, LI, 423
Ribose 5-phosphate
assay of phosphoribosylglycinamide synthetase, LI, 180
determination, LI, 264
inhibitor of aspartate carbamyltransferase, LI, 57
of uridine nucleosidase, LI, 294
phosphate donor of nucleoside phosphotransferase, LI, 391, 392
product of adenosine monophosphate nucleosidase, LI, 263
substrate of phosphoribosylpyrophosphate synthetase, LI, 3, 10, 12, 13
D-Ribose 5-phosphate, preparation and analysis of mixtures with D-ribulose 5-phosphate and D-xylulose 5-phosphate, XLI, 37–40
Ribosephosphate isomerase
from *Candida utilis*, XLI, 427–429
assay, XLI, 427
inhibitors, XLI, 429
properties, XLI, 429
purification, XLI, 428, 429
from skeletal muscle, XLI, 424–426
assay, XLI, 424
chromatography, XLI, 425
electrophoresis, XLI, 426
inhibitors, XLI, 426
properties, XLI, 426
purification, XLI, 425, 426
D-Ribose 5-phosphate ketol isomerase, *see* Ribosephosphate isomerase
Ribose-5-phosphate kinase, interference, in phosphoribosylpyrophosphate synthetase assay, LI, 3
Ribose-5-phosphate pyrophosphokinase, *see* Phosphoribosylpyrophosphate synthetase
Ribosomal couple
definition, LIX, 376
formation, cause, LIX, 382
isolation, LIX, 377–379
Ribosomal particle
general structural feature, LIX, 742
reconstituted peptidyltransferase activity, LIX, 779–782
Ribosomal protein, *see also* specific type
acetic acid extraction, *see* Acetic acid

chemical modification, with fluorescein isothiocyanate, LIX, 819
concentration, LIX, 499, 500
determinations, LIX, 500, 501
dissociation, from ribosomal RNA, LIX, 440, 441
E. coli
electrophoretic analysis, LIX, 458, 538–550
fingerprint patterns, LIX, 548, 599
intermolecular distances, LIX, 607, 705, 731
interprotein distances, LIX, 705
preparation, LIX, 453–455, 557–562, 649–652
elution from diagonal gels, LIX, 544, 545
enzyme digestion, LIX, 575, 576
from *Escherichia coli,* sedimentation equilibrium studies, XLVIII, 182
fractionation techniques, comparison, LIX, 562
hydrated volume, LIX, 605
intermolecular association studies, LIX, 591–602
purification under nondenaturing conditions, LIX, 481–502
radii of gyration, theoretical derivation, LIX, 637
radiolabeled
activity measurements, LIX, 779, 780
double-labeling method, LIX, 517
properties, LIX, 780–782
in situ, LIX, 792–794
in solution, LIX, 792–794
in vitro, LIX, 517, 776–782
in vivo, LIX, 517
rat liver
analysis, LIX, 502–515, 520–525
fingerprint patterns, LIX, 530
preparation, LIX, 506, 517–521
synthesis by cell-free systems, LIX, 515–534
reductive methylation, LIX, 777–779
from ribosomal fragments, preparation, LIX, 470–473
salt-extraction, principles, LIX, 481, 482
procedure, LIX, 484–486
specific release, LIX, 818
storage, LIX, 498, 499, 585
tertiary structure, LIX, 720, 721
three-dimensional models, LIX, 602–611
in vitro synthesis, LIX, 515–534
Ribosomal ribonucleic acid, XLVI, 89, 622, 640, 642, 643
analysis, by slab gel electrophoresis, LIX, 458
cleavage, with matrix-bound RNase, LIX, 465, 467, 472–474
concentration equivalents, LIX, 452
digestion, LIX, 570–574
from *E. coli,* thermal transition studies, XLIX, 11
isolation and analysis of components, LVI, 90, 91
of mitochondria, LVI, 7, 8
from *Neurospora* mitochondria, preparation, LIX, 427, 428
in photoaffinity labeling studies, LIX, 806, 808
precursors, purification, LIX, 830–832
preparation from *E. coli* ribosomes, LIX, 452, 453, 556, 557
from ribosomal fragments, LIX, 470, 476
in protein turnover, XXXVII, 249, 250
from rat liver, quasi-elastic laser light scattering, XLVIII, 493
5 S
isolation from *E. coli* ribosomal subunits, LIX, 492, 555
purity determination, LIX, 556, 567
structural models, LIX, 715, 722, 725
structural parameters, LIX, 722
X–ray scattering data, LIX, 715, 722
16 S, XLVI, 187
chemical compounds, LIX, 760
partial specific volume, LIX, 760
purification, LIX, 453, 625
purity determination, LIX, 556, 567
scattering parameters, LIX, 760
structural model, LIX, 722, 729, 730

structural parameters, LIX, 722, 723, 729
in studies of protein-RNA interactions, LIX, 586
X–ray diffraction data, LIX, 740
23 S, XLVI, 187, 194
antibiotic binding, LIX, 865
chemical composition, LIX, 759
partial specific volume, LIX, 760
preparation, LIX, 452, 453
purity determination, LIX, 556, 567
scattering parameters, LIX, 759, 760
scattering data, LIX, 721–723
30 S precursor
cleavage by RNase III, LIX, 832–836
isolation, LIX, 832
storage, LIX, 586
in vivo radiolabeling, LIX, 830, 831
from whole ribosomes, preparation, LIX, 452
Ribosomal subunit, *see also* Ribosome, specific subunit
analysis by sucrose gradient centrifugation, LIX, 413–416
antibiotic binding, LIX, 862–866
buoyant densities, LIX, 407, 410, 421
concentration determination, LIX, 655
CsCl gradient analysis, LIX, 416–421
derived, preparation, LIX, 403–405
properties, LIX, 405–407
E. coli, isolation, LIX, 399, 400, 443–461, 483, 554, 555
Ehrlich cell
CsCl equilibrium density gradient analysis, LIX, 410–421
preparation, LIX, 403, 404, 407, 408
properties, LIX, 408–410
gel electrophoretic separation, LIX, 397–402
magnesium precipitation, LIX, 415
protein S1 binding sites, LIX, 400, 401
rat liver, preparation, LIX, 402–410, 505, 506
reconstitution, LIX, 437–449, 654
30 S
buoyant densities, LIX, 753
electron microscopy, LIX, 618–621
molecular weight, LIX, 688, 727
preparation for electron microscopy, LIX, 618
proteins
chemical composition, LIX, 760
chromatographic fractionation, LIX, 558–560
content, LIX, 753
discontinuous SDS acrylamide slab gels, LIX, 497, 498
purification, LIX, 486–490
salt extraction, LIX, 484–486
storage, LIX, 498
protein-RNA interactions, LIX, 730, 731, 733
radius of gyration, LIX, 688, 691, 692, 726, 727
relation of proteins to structure, LIX, 775
RNA packing tightness, LIX, 761–763, 770–775
RNA and protein distribution, LIX, 763–770
scattering data, LIX, 688, 726, 728, 761–775
structural models, LIX, 724, 770
structural parameters, LIX, 726, 727, 732
X–ray diffraction data, LIX, 740
40 S
preparation, LIX, 505, 506
proteins, molecular weights, LIX, 512
50 S
Bacillus
preparation, LIX, 375–382
reconstitution, LIX, 437–443
E. coli
buoyant densities, LIX, 753
cross-linked, extraction of protein, LIX, 537, 538
preparation, LIX, 535–537
dissociation, LIX, 445–447
fine structure, LIX, 736–742
molecular weight, LIX, 688, 690

proteins
chemical composition, LIX, 760
chromatographic fractionation, LIX, 558–561
content, LIX, 753
discontinuous SDS acrylamide slab gels, LIX, 497, 500
purification, LIX, 491–495
salt extraction, LIX, 490, 491
storage, LIX, 498, 499
radius of gyration, LIX, 688, 692–695, 737, 738
reconstitution, LIX, 443, 447, 448, 455–458
activity assay, LIX, 448, 449, 458–460
protein analysis, LIX, 458
RNA analysis, LIX, 458
RNA packing tightness, LIX, 761–763, 770–775
RNA and protein distribution, LIX, 763–770
scattering data, LIX, 688, 738, 760–775
shape model from scattering data, LIX, 733, 734, 736, 737
using spherical harmonics, LIX, 700–703
structural models, LIX, 725, 727, 770
structural parameters, LIX, 732, 734, 735
X-ray diffraction data, LIX, 740, 741
separation, by zonal centrifugation, LIX, 375–382, 451, 452
storage, LIX, 852
Ribosome, XXXIV, 299, 471; XLVI, 75, 84, 85, 89, 336, *see also* Ribosomal couple; specific types; Vacant couple
affinity labeling, XLVI, 187
aggregation in perfused muscle, XXXIX, 81, 82
A_{260} units, LVI, 83
Bacillus subtilis, isolation, LIX, 371–382, 438
binding to single-stranded DNA, LIX, 837–851
binding site, XLVI, 621–637, 683–702
mRNA, XLVI, 621
oligonucleotides, XLVI, 622–624
peptidyl-tRNA, XLVI, 626–633
polynucleotides, XLVI, 625, 626
centrifugal separation, XXXI, 717, 727
cytochrome oxidase subunits, LVI, 596
cytoplasmic, mitochondrial proteins, LVI, 74–76
dissociation, XLVIII, 254–257; LIX, 375–382, 451, 452
effect of sodium ion, LIX, 367
electron microscopy, 612–629
from *Escherichia coli*
labeling procedure, XLVII, 476
quasi-elastic laser light scattering, XLVIII, 493
Euglena chloroplast, isolation, LIX, 434–437
eukaryotic
80 S
binding of edeine, LX, 561
of messenger ribonucleic acid, LX, 50, 52, 256–265, 358
complex with methionyl-transfer ribonucleic acid, LX, 573
with phenylalanyl-transfer ribonucleic acid, LX, 587–589
with transfer ribonucleic acid and polyuridylate, LX, 661, 662
dissociation, LX, 52, 62, 291, 294, 295
formation, LX, 62, 76, 356, 365, 366, 571
preparation, LX, 72–74, 307, 308, 560, 561, 580, 660, 661, 688
protection of messenger ribonucleic acid segments, LX, 361, 378–380
separation, LX, 72–74, 580

whole
binding of messenger ribonucleic acid, LX, 285–287, 351–361, 408–410, 415–417
fractionation, LX, 61–87
polyamines, LX, 556, 557, 559, 560
preparation from Krebs ascites cells, LX, 91
from reticulocyte lysate, LX, 113, 114, 125, 139, 282, 283, 438, 516, 517
from wheat germ, LX, 193
fragments, preparation, LIX, 465, 467, 472–474
separation, LIX, 467, 474–480
free, assay, LIX, 367
high-salt wash, LIX, 838, 839
incorporation of inactive proteins, LIX, 815–824
isolated, protein synthesis, LVI, 24–27
labeling
with chloramphenicol analogs, XLVI, 702–707
design, XLVI, 189
chloramphenicol binding, XLVI, 706, 707
mitochondrial, LVI, 10, 11, 40
analysis, LVI, 84–91; LIX, 428, 433
dissociation into subunits, LVI, 83, 84
functions, LVI, 91
immunological studies, LVI, 88–90
isolation, LVI, 80–83
from *N. crassa*, LV, 148
preparation, LIX, 425, 426
proteins, LVI, 91; LIX, 428–433
protein synthesis, LIX, 426, 427
RNA, LIX, 427–429
morphometry, XXXII, 15, 19
negative staining, XXXII, 29, 31
Neurospora crassa, properties of cytoplasmic and mitochondrial, LVI, 89
phosphorylating vesicles, LV, 168
photoaffinity labeling, XLVI, 711–717; LIX, 796–815
with aminoacyl-tRNA analogs, XLVI, 676–683
of 23 S RNA, XLVI, 637–644
of plant cells, XXXI, 489, 491, 495, 583
core particles, XXXI, 516
isolation, XXXI, 501, 504, 509, 510, 514, 533, 534, 585, 586
in posttranslocation state, preparation, LIX, 394–396
preparation, LIX, 363, 451, 483, 554, 647, 752, 817, 827, 828, 839
pressure-jump studies, subunit, XLVIII, 310
in pretranslocation state, preparation, LIX, 397
prokaryotic, 70 S
binding
aminoacyl-transfer ribonucleic acids, LX, 615, 616, 625, 626
elongation factor EF-G, LX, 719, 720
study by photoactivated affinity labeling, LX, 724, 725
formylmethionyl-transfer ribonucleic acid, LX, 5, 6, 9, 11, 205, 431
phenylalanyl-transfer ribonucleic acid, LX, 431–435, 626–628
complex with polyadenylate and transfer ribonucleic acid lysine, LX, 624
with guanosine nucleotides and elongation factor EF-G, LX, 722, 723, 726–731, 734–737, 740, 741
dissociation, LX, 13, 14, 212–215
equilibrium with other factors, LX, 11–15
formation, LX, 317, 325–328
initiation of protein synthesis, LX, 11–15, 216
messenger ribonucleic acid, LX, 313, 314
preparation, LX, 314, 315, 419, 596, 617

protection of messenger ribonucleic acid from ribonuclease, LX, 311–314, 317, 318
ribosome protein S1, LX, 421–423, 425, 439
S1-depleted, LX, 432–435
tight couples, LX, 12, 13
translation of cellulose-bound polyuridylate, LX, 754–760, 763–779
product elution with decreasing magnesium ion concentration, LX, 759, 760
requirement of gain and then loss of elongation factor EF-G, LX, 766–779
trimming with ribonuclease, LX, 318
protein content, LIX, 752, 753
purified, affinity chromatography, XL, 272
rat liver, isolation, LIX, 504, 505
reconstitution, LIX, 437–443
release
antibiotic action, LIX, 861, 862
assay, LIX, 861
release factor, LIX, 860–862
removal
from microsomal subfractions, using EDTA treatment, LII, 76
of peptidyl-tRNA, LIX, 404
by ultracentrifugation, LI, 311
ribonuclease as contaminant in studies, XXXII, 109
RNA, isolation and analysis of components, LVI, 90, 91
tRNA binding 181, XLVI, 182
30 S, XLVI, 89
assay for elongation factor EF-G, LX, 607–610
binding
to elongation factor EF-G, LX, 719, 720
study by photoactivated affinity labeling, LX, 724, 725
to formylmethionyl-transfer ribonucleic acid, LX, 209–211, 332, 333, 340–343
to phenylalanyl-transfer ribonucleic acid, LX, 431–434
to polyuridylate, LX, 433
complex with messenger ribonucleic acid, LX, 218–221
energy transfer studies, XLVIII, 378
equilibrium with other subunits, LX, 11–15
initiation of protein synthesis, LX, 11–13
interaction with initiation factors, LX, 13–15, 210, 211, 230–239, 336, 337
preparation, LX, 13, 231, 232, 338, 419, 438, 609
ribosome protein S1 preparation from, LX, 418–421, 425
S1-depleted
assay of purity, LX, 439, 453, 454
binding
to aminoacyl-transfer ribonucleic acid, LX, 431–433
of polyuridylate, LX, 433
preparation from 30 S ribosomal subunits, LX, 428–430, 434, 438, 439, 452, 453
reactivation with S1, LX, 430, 434, 435
separation from ordinary 30 S subunits, LX, 452, 453
from S1-depleted subunits, LX, 452, 453
structure, XLVIII, 376, 378
ternary complex with initiation factor IF-3 and MS2 ribonucleic acid, LX, 343–346
34 S, LX, 333–338, 341–343
40 S
acetylation, LX, 539–541

binding of initiation factors, LX, 67, 74, 75, 83
of methionyl-transfer ribonucleic acid, LX, 32, 33, 37–39, 41, 42, 49–52, 59, 60, 62, 72, 74, 246, 276, 332, 357, 358, 461, 562–566
inhibition, LX, 576
measurement
by filtration, LX, 256–265, 283, 570
by gradient centrifugation, LX, 276–280, 283–285, 570, 586–588
use in assay for initiation factor EIF-2A, LX, 116, 309, 310
of polyamines, LX, 556, 560
of transfer ribonucleic acid, LX, 408–410
dephosphorylation, LX, 522, 523, 527, 533
juncture with 60 S ribosomal subunit, LX, 32, 33, 38, 41, 42, 50, 62, 294, 295, 308, 309, 357
native
activity in initiation, LX, 573, 574
binding of methionyl-transfer ribonucleic acid, LX, 569–571, 574, 575
functional role, LX, 566, 567
joining of 60 S ribosomal subunit, LX, 571
preparation, LX, 568
types, LX, 567, 571–574
phosphorylation, LX, 530, 533
preparation, LX, 69, 72–75, 114, 115, 139, 186, 257, 308, 309
protection of messenger ribonucleic acid segments, LX, 360, 361, 365–368
43 S
binding of methionyl transfer ribonucleic acid, LX, 74
composition, LX, 82
46 S, LX, 333–337, 340–343
48 S
composition, LX, 79, 80, 84, 86, 87
formation, LX, 82
50 S, XLVI, 89, 90
binding to elongation factor EF-G, LX, 719, 720
study by photoactivated affinity labeling, LX, 724, 725
energy transfer studies, XLVIII, 377, 379
equilibrium with other subunits, LX, 11–15, 212–215
initiation of protein synthesis and, LX, 11–13, 15, 333–335, 342
60 S
acetylation, LX, 439–441
binding of edeine, LX, 560
junction with 40 S ribosome subunits, LX, 32, 33, 38, 41, 42, 50, 62, 74, 82, 294, 295, 357, 358, 565, 566, 571
preparation, LIX, 505, 506; LX, 69, 72–76, 114, 115, 139, 186, 257
proteins, molecular weights, LIX, 512
66 S, composition, LX, 80
70 S
binding, XLVI, 710
E. coli
buoyant densities, LIX, 753
molecular weight, LIX, 688, 744
partial specific volume, LIX, 744
protein content, LIX, 753
radius of gyration, LIX, 695–697, 743, 744
scattering data, LIX, 688
structural parameters, LIX, 743
urea-lithium chloride dissociation, LIX, 440
X-ray diffraction data, LIX, 741
puromycin incorporation, XLVI, 712, 713
80 S
beef pancreas, structural parameters, LIX, 743
rabbit reticulocytes, structural parameters, LIX, 743
rat liver
structural model, LIX, 746

structural parameters, LIX, 743–746
X–ray diffraction data, LIX, 741
Triticum vulgare
buoyant density, LIX, 753
neutron scattering data, LIX, 742
protein content, LIX, 753
X–ray diffraction data, LIX, 741
X–ray scattering data, LIX, 774
yeast, X–ray diffraction data, LIX, 741
species specificity, LIX, 371, 372
storage, LIX, 852
sucrose density gradient analysis, LIX, 375–381
translating capacity, LIX, 392
isolation, LIX, 382–397
Ribosome cycle, in protein turnover, XXXVII, 248–250
Ribosome dissociation factor
assay by peptidyltransferase reaction, LX, 296, 297
preparation from native ribosomal subunits of rat liver, LX, 52, 290–294
properties, LX, 52, 294
by sucrose gradient centrifugation, LX, 294, 295
Ribosome-protection assay, *see* Messenger ribonucleic acid, protected fragments, preparation
Ribosome-selection assay, for active messenger ribonucleic acid segments, LX, 367
Ribostamycin
in media, LVIII, 116
structure, XLIII, 613
1-Ribosyloxipurinol 5′-phosphate, substrate of GMP synthetase, LI, 224
3-*N*-Ribosyl purine-5′-phosphate, product of orotate phosphoribosyltransferase, LI, 143
Ribosylthymine, *see* 5-Methyluridine
3-*N*-Ribosylxanthine-5′-phosphate
inhibitor of orotate phosphoribosyltransferase, LI, 153
of orotidylate decarboxylase, LI, 153
Ribothymidine, *see also* 5-Methyluridine
substrate of nucleoside phosphotransferase, LI, 391
D-Ribulose, preparation, XLI, 103–106
Ribulosebisphosphate, XLVI, 388–398
Ribulosebisphosphate carboxylase, XXXI, 768; XLVI, 23, 142, 152, 388–398
characterization, XLVI, 389–392
from hydrogen bacteria, XLII, 461–468
assay, XLII, 462, 463
properties, XLII, 471, 472
purification, XLII, 463–468
labeling lysyl residues, XLVI, 395–397
from leaves, XLII, 481–484
assay, XLII, 482
extraction, XLII, 481
purification, XLII, 482–484
from *Rhodospirillum rubrum*, XLII, 457–461, 468–472
amino acids, XLII, 471
assay, XLII, 457, 458
chromatography, XLII, 459, 460, 469, 470
inhibitors, XLII, 461
molecular weight, XLII, 460, 472
properties, XLII, 460, 461, 471, 472
purification, XLII, 458–460, 468–471
from spinach leaves, XLII, 472–480
assay, XLII, 472–474
chromatography, XLII, 475–477
kinetic properties and inhibitors, XLII, 478–480
mechanism of action, XLII, 479, 480
molecular weight, XLII, 477
properties, XLII, 477–480
purification, XLII, 474–477
Ribulosebisphosphate carboxylase/oxygenase, XLVI, 389
4-Ribulose-1,5-bisphosphate oxygenase
assay, XLII, 484–486

properties, XLII, 487
purification, XLII, 486
Ribulose-1,5-diphosphate carboxylase, *see* Ribulosebisphosphate carboxylase
Ribulose diphosphate mixed-function oxidase, LII, 9
4-Ribulose-1,5-diphosphate oxygenase, *see* 4-Ribulose-1,5-bisphosphate oxygenase
Ribulose-5-phosphate, inhibitor of uridine nucleosidase, LI, 294
D-Ribulose 5-phosphate
preparation and analysis of mixtures with D-ribose 5-phosphate and D-xylulose 5-phosphate, XLI, 37–40
spectrophotometric determination, XLI, 40, 41
D-Ribulosephosphate 3-epimerase, assay, XLI, 37, 63–66
L-Ribulosephosphate 4-epimerase
from *Aerobacter aerogenes*, XLI, 412–419
assay, XLI, 412–415
chromatography, XLI, 416, 417
properties, XLI, 418, 419, 422, 423
purification, XLI, 415–418
from *Escherichia coli*, XLI, 419–423
assay, XLI, 419, 420
lyophilization, XLI, 422
molecular weight, XLI, 423
properties, XLI, 422, 423
purification, XLI, 420–423
from *Lactobacillus plantarum,* properties, XLI, 423
Ribulose-5-phosphate kinase, *see also* Phosphoribulokinase
interference, in phosphoribosylpyrophosphate synthetase assay, LI, 3
Rice
aleurone grain, XXXI, 576
germ, phosphoglycerate mutase, purification and properties, XLII, 431, 434, 435
peroxidase, LII, 515
polish, protein bodies, XXXI, 518
Ricin
activity, L, 330
assay, L, 330, 331
chain separation, L, 332–334
properties, L, 334, 335
purification, L, 332–334
resistance, LVIII, 309
Ricinus agglutinin, *see also* Lectin, from *Ricinus communis*
activity, L, 330
affinity chromatography, L, 334, 335
assay, L, 330, 331
chain separation, L, 334
properties, L, 334, 335
purification, L, 332–334
Ricinus communis
lectin, XXXIV, 333
oxidase, LII, 20
Ridox catalyst, preparation and use, LIV, 117, 118
Rieske protein, *see* Cytochrome bc_1, complex
Rifampicin, XLVI, 352
assay of amino acid transport, LV, 179
fluorescent acceptor, XLVIII, 378
in promoter site mapping studies, LIX, 847
as protein synthesis inhibitor, XXXII, 870
Rifampin, XLIII, 304, 305, 319
Rifampin quinone, XLIII, 305
Rifamycin, LVI, 31
biosynthesis, XLIII, 409
effects on membranes, XXXII, 882
paper chromatography, XLIII, 137, 138
resistance as genetic marker, XLIII, 44
solvent system for countercurrent distribution, XLIII, 340
thin-layer chromatography, XLIII, 200
Rimocidin, XLIII, 135
Rinaldini's solution, LVIII, 452
Ring current shift effects, LIX, 32–36
Ringer's solution, LVIII, 120, 121
composition, XXXII, 797
modified, XXXII, 707
for organ perfusion, LII, 50, 51
Ringing, frequency sweep rate, LIX, 25
Rise time, LIV, 33
of biplaner vacuum photodiodes, LIV, 40
photomultipliers, LIV, 40

Ristocetin, XLIII, 127
Ristomycin, XLIII, 127
Rivanol, purification of *Cypridina* luciferase, LVII, 350
RNA, *see* Ribonucleic acid
tRNA, *see* Transfer ribonucleic acid
tRNA^{Ala}
from *E. coli*, LIX, 28, 193
isoacceptors, identification, LIX, 226, 228
tRNA^{Arg}
isoacceptors, identification, LIX, 226, 228
NMR spectra, LIX, 35, 38
tRNA^{Asn}, isoacceptors, identification, LIX, 222, 223
tRNA^{Asp}
from *Euglena gracilis* chloroplast, isolation, LIX, 211, 212, 214
isoacceptors, LIX, 222, 223
from yeast, crystallization conditions, LIX, 6, 7
tRNA^{Cys}, isoacceptors, identification, LIX, 227, 228
tRNA^{fMet}, XLVI, 182, 686, 695
from *E. coli*
assay of CTP(ATP):tRNA nucleotidyltransferase, LIX, 125, 126
crystallization conditions, LIX, 6, 7
NMR studies, LIX, 40, 41, 56
from yeast, crystallization conditions, LIX, 6, 7
tRNA^{Gln}
from *E. coli*
crystallization conditions, LIX, 8, 9
NMR studies, LIX, 43, 44
isoacceptors, identification, LIX, 222–224
tRNA^{Glu}, isoacceptors, identification, LIX, 222, 223, 224
tRNA^{Gly}, isoacceptors, identification, LIX, 227, 228
tRNA^{His}
isoacceptors, identification, LIX, 222, 223
NMR spectra, LIX, 31, 47
tRNA^{Ile}, XLVI, 90, 172, 173, 178, 179
isoacceptors, identification, LIX, 226, 228
sequence and structure, LIX, 341
tritium mapping, LIX, 340, 342
two-dimensional chromatogram, LIX, 343, 344
tRNA^{Leu}
from *E. coli*
base pairing, LIX, 52
crystallization conditions, LIX, 6, 7
isoacceptors, identification, LIX, 227, 228
separation, LIX, 217, 218
tRNA^{Lys}
from *E. coli*
in assay of CTP(ATP):tRNA nucleotidyltransferase, LIX, 125
NMR spectra, LIX, 35, 37, 40, 45–49
thermal unfolding studies, LIX, 48–50
isoacceptors, identification, LIX, 222–224
tRNA^{Met}
from *E. coli,* NMR studies, LIX, 40, 47
isoacceptors
identification, LIX, 225, 228
kinetics of formation, LIX, 278, 279
tRNA$^{Met}_{f}$, *see* tRNA^{fMet}
tRNA^{Phe}, XLVI, 172, 185, 632, 686, 695
from Chinese hamster ovary cells, isoacceptors, identification, LIX, 224, 225
from *Escherichia coli,* crystallization conditions, LIX, 6, 7
2,4-dinitrobenzoylated, preparation, LIX, 170
paramagnetic, chromatographic behavior, LIX, 168
preparation, LIX, 171
photolabile, preparation, LIX, 171
thermal unfolding studies, LIX, 50
from *Euglena gracilis,* purification, LIX, 211–214

from *Neurospora crassa,* oligonucleotide sequence analysis, LIX, 100
preparation, LIX, 207
soaking conditions for derivatization, LIX, 14
source, LIX, 147
structural features, LIX, 15–19
with terminal 3′-amino-3′-deoxyadenosine
electrophoretic properties, LIX, 145
preparation, LIX, 142–144
from yeast, crystallization conditions, LIX, 6, 7
density measurement, LIX, 5
hydrazine incorporation into wybutine residues, LIX, 114–121
modification of 3′-terminus, LIX, 178–190
NMR studies, LIX, 30, 35, 37, 39–45, 47, 51, 52
nucleotide sequence in cloverleaf configuration, LIX, 16
number of Watson-Crick base pairs, LIX, 28
shortened species, LIX, 186, 187
tRNA^{Pro}, isoacceptors, identification, LIX, 227, 228
kinetics of formation, LIX, 278, 279
tRNA^{Ser}, isoacceptors, identification, LIX, 227, 228
tRNA^{Thr}, isoacceptors, identification, LIX, 226, 228
tRNA^{Trp}
assay of tryptophanyl-tRNA synthetase, LIX, 237
from *E. coli,* reconstitution, LIX, 133, 134
venom treatment, LIX, 132, 133
from yeast, isoacceptors, identification, LIX, 225, 228
tRNA^{Tyr}, XLVI, 90, 172
from *Escherichia coli,* crystallization conditions, LIX, 6, 7
from *Euglena gracilis* chloroplast, isolation, LIX, 214
NMR studies, LIX, 52
from yeast, isoacceptors, identification, LIX, 222, 223
tRNA^{Val}, XLVI, 90, 632, 686, 694
from *E. coli,* NMR spectra, LIX, 23, 24, 28, 33–36, 38, 40, 42, 43, 45, 46, 54, 55
number of Watson-Crick base pairs, LIX, 28
in test of deacylation of CHO tRNA, LIX, 220
from yeast, isoacceptors, identification, LIX, 227, 228
RNA-agarose, XXXIV, 468, 469
tRNA-agarose, XXXIV, 513
tRNA-aminoacyl-tRNA synthetase complex, XLVI, 170
tRNA–C–C–A($^1NH_2$), in misactivation studies, LIX, 285, 288–291
RNA-cellulose, XXXIV, 470
RNA-dependent DNA polymerase, *see* DNA polymerase, RNA-dependent
RNA-fiberglass, XXXIV, 470
tRNA-fMet, *see* tRNA^{fMet}
tRNA methyltransferase, bacterial, LIX, 190–203
assay, LIX, 190, 191
inhibitor, LIX, 202
product analysis, by fluorography, LIX, 192, 193
specificity, LIX, 202
RNA nucleotidyltransferase, *see* RNA polymerase
tRNA-nucleotidyltransferase, XXXIV, 254
RNA polymerase, XXXIV, 5, 464, 469; XLVI, 89, 90; LX, 323, *see also* RNA nucleotidyltransferase
assay, XXXVI, 322; XLVI, 350
associated with cell fractions, XL, 85, 86
atomic emission studies, LIV, 459
in chromatin, XXXI, 279
DNA-dependent, in nuclei, XXXI, 259, 261, 262
DNA-dependent, XLVI, 346–358
affinity labeling, XLVI, 355, 356
atomic fluorescence studies, LIV, 469
daily rhythms, XXXVI, 477, 479
DNA transcription assay, XXXVIII, 372
E. coli, in *in vitro* RNA synthesis, LIX, 847

energy transfer studies, XLVIII, 378
estradiol effects, XXXVI, 320
hormone induction of biorhythms, XXXVI, 480
inhibition, XLVI, 350
in nuclei, XXXI, 263, 722
assay, XXXI, 270–271
preservation, XXXI, 271–272
tRNA-polyuridylamide, XXXIV, 471
tRNA-protein complex, structure mapping, by tritium labeling method, LIX, 332–350
tRNA-protein interaction, detection and quantitation
by density gradient zone centrifugation, LIX, 329, 330
by gel filtration, LIX, 325–328
by nitrocellulose membrane filtration, LIX, 322–324
$tRNA^{Ile}$-$tRNA^{Ile}$ synthetase, XLVI, 173, 177, 179, 180
RNase, *see* Ribonuclease
RNase-agarose, *see* Ribonuclease-agarose
RNA synthetase, XXXIV, 506–516, *see also* specific enzyme
$tRNA^{Ile}$ synthetase, XLVI, 172, 173
RNA tumor virus
cell transformation, XXXII, 591
microcarrier culture, LVIII, 191
transformation, LVIII, 369
RNP, *see* Ribonucleoprotein
Ro 2–2985 antibiotic, LV, 445
Ro 7–9409 antibiotic, LV, 445
Ro 20–0006 antibiotic, LV, 445
Ro 21–6150 antibiotic, LV, 445
Rocker platform, for calcium influx studies, XXXIX, 529, 530
Rodent neoplasm, LVIII, 584–590
Rodent respirator, XXXIX, 170
Rod outer segment, cyclic nucleotide metabolizing enzymes, XXXVIII, 153–155
Rolitetracycline, XLIII, 385
Roller bottle, LVIII, 13, 14, 397
for cell culture, XXXII, 566, 567, 811
for fibroblasts, LVIII, 450
incubation, LVIII, 185
Melinex spiral, LVIII, 13
monolayer culture, LVIII, 137
Roller bottle perfusion apparatus, LVIII, 13
Roller flask, LVIII, 398
Roller tube, LVIII, 203
Root
ATPase bound to membranes, XXXII, 392–406
growth and harvest, XXXII, 394
Rose, protoplast isolation, XXXI, 579, 581, 582
Rose Bengal, as optical probe, LV, 573
Rose chamber, LVIII, 14
assembly, XXXIX, 293
seminiferous tubule isolation, XXXIX, 292–295
Roseothricin, XLIII, 280
Rosett cell
for sonication of bacteria, LIII, 533
of mitochondria, LIII, 512
Rosette, XXXIV, 721
assays, XXXIV, 215–218
Ross cell disintegrator, properties, XXXI, 662
Ross filter, for X-ray diffraction, XXXII, 214
Rotary shaker, LVIII, 14
Rotating ring-disk electrode, LVII, 515, 516
Rotational diffusion coefficient
for long rigid rod, XLVIII, 436
for lysozyme, XLVIII, 442
quasi-elastic laser light scattering, XLVIII, 421
Rotational motion, light scattering, XLVIII, 436–442
Rotatory strength
computation, LIV, 253
dipole-dipole coupling, LIV, 255, 256
μ-m mechanism, LIV, 256, 257
one-electron theory, LIV, 255
Rotaventin, XLIII, 340
Rotenoid
inhibitor of complex I, LIII, 13
of complex I–III, LIII, 8
Rotenone
ATP synthesis, LV, 452
calcium depletion, LVI, 351
calcium influx measurements, LVI, 348

carnitine-acylcarnitine translocase, LVI, 375
coenzyme Q, LV, 456
complex I, LVI, 586
in enzyme assay, XXXI, 20
ferricyanide reduction, LVI, 231
fluorescence changes, LVI, 501
immobilized mitochondria, LVI, 557
inhibitor of complex I, LIII, 13
of reconstituted electron-transport systems, LIII, 50, 51
measurement of Ca^{2+}/site ratio, LV, 650
in mitochondrial incubation medium, LVII, 38
NADH dehydrogenase solubilization studies, LIII, 415, 416
NADH oxidase, LV, 266, 270
NADH oxidation, LV, 455, 456
N. crassa mitochondria, LV, 147
phosphorylation potential, LV, 238, 239
transport studies, LVI, 256, 257, 259, 260
yeast mitochondria, LV, 27
Rothschild method, of microsome fractionation, XXXI, 228–230
Rotochem Fast Analyzer, XXXI, 800
Rotor, for centrifugation, XXXI, 716, 717
Roucaire thermostatic pump, XXXIX, 247
Roughton-Millikan continuous-flow system, LIV, 85
Rouse theory, XLVIII, 443
Rous sarcoma virus, LVIII, 379–393
assay, LVIII, 388–390
Bryan high-titer strain, LVIII, 380
cloning, LVIII, 390–392
focus-forming assay, XXXII, 591, 592
harvesting, LVIII, 399, 400
helper virus, LVIII, 380
large-scale growth, LVIII, 393–403
nontransforming, LVIII, 403
Prague strain, LVIII, 380, 394
quantitation, LVIII, 388–390
Schmidt-Ruppin strain, LVIII, 380, 381, 394
storage, LVIII, 397
transformation, LVIII, 369, 370
plasma membrane isolation, XXXI, 161
transformation-defective mutant, LVIII, 390
yield, LVIII, 400
Roussel-Jouan Model CD spectrophotometer, LIV, 251
RPC-5, *see* Reversed phase chromatography
RPMI 1640 medium, LVIII, 56, 89, 91, 155, 213
composition, LVIII, 62–70
RRDE, *see* Rotating ring-disk electrode
RRF, *see* Ribosome release factor
rRNA, *see* Ribosomal ribonucleic acid
rRS, *see* Resonance Raman spectroscopy
RS, *see* Raman spectroscopy
RST2 cell, human, interferon, LVIII, 293
RTC
fluorescent acceptor, XLVIII, 362, 363
spectral properties, XLVIII, 362
structural formula, XLVIII, 362
RT-factor, *see* Receptor transforming factor
R–type protein, XXXIV, 305, 310–312
Rubber, steroid adsorption, XXXVI, 93, 94
Rubber resuspending cone, LI, 340
Rubber septa, oxygen, LIV, 126
Rubber tubing, contamination, LIV, 119
Rubidin, XLIII, 340
Rubidium, uptake in electron transport studies, XXXI, 701, 702
Rubidium-86
distribution, determination, LV, 553, 554
in membrance studies, XXXII, 884–893
properties, XXXII, 884
Rubidium-87, lock nuclei, XLIX, 349; LIV, 189
Rubidium ion, activator of ribonucleoside triphosphate reductase, LI, 258
Rubidomycin, XLIII, 145
Rubiflavin, XLIII, 340
Rubradirin, XLIII, 340
Rubredoxin, LIII, 340–346, 625–627
amino acid composition, LIII, 627
amino acid sequence, LIII, 343–345

assay, LIII, 341, 617, 618
of ω-hydroxylase, LIII, 356, 357
biological role, LIII, 340, 346
classification of iron proteins, LIII, 260
clostridial, magnetic circular dichroism, XLIX, 175
from *Clostridium pasteurianum*, LIII, 340, 341
definition, LIII, 261, 340
from *Desulfovibrio gigas*, LIII, 617, 618, 625–626, 627, 631
in *Desulfovibrio* strains, LIII, 632
from *Desulfovibrio vulgaris,* chemical cleavage, XLVII, 148
EXAFS spectrum, LIV, 341, 342
Fe *K*-edge absorption spectrum, LIV, 335
iron-sulfur center, LIII, 344, 346; LIV, 203
Mössbauer studies, LIV, 359–362, 369
molar extinction coefficient, LIII, 618
molecular weight, LIII, 343, 627
phase shift studies, LIV, 343, 344
properties, LIII, 262, 343–346
prosthetic group, LIII, 627, 631
from *Pseudomonas oleovorans*, LIII, 340–346, 356
purification, LIII, 622, 625, 626
redox potentials, LIII, 631
spectral properties, LIII, 346
Rubredoxin:NAD oxidoreductase, *see* Rubredoxin-NAD$^+$ reductase
Rubredoxin-NAD$^+$ reductase
assay of rubredoxin, LIII, 617, 618
from *Desulfovibrio gigas*, LIII, 619–621, 631
purification, LIII, 619–621
strain specificity, LIII, 618, 633
Rubrene
activator, in chemiluminescent reaction, LVII, 520
oxidation potential, LVII, 522
electron-transfer chemiluminescence, LVII, 496, 497, 500–502
structural formula, LVII, 498
Rubromycin, XLIII, 340
Ruby laser, LIV, 8, 96
capabilities, LIV, 45
wavelength, XLIX, 85, 95
Rudinger method, of carboxyl group activation, XLVII, 565, 567, 568
Russell viper venom
activator factor X, bovine, XLV, 90, 100
assay, factor V, XLV, 110
coagulant protein, XLV, 191
inhibitors, *see* Venom inhibitors
phospholipid mixture, reagent, factor X, bovine, XLV, 90
Rutamycin, LV, 477
ATPase, LV, 358
complex V, LV, 315
oligomycin D, LV, 502
Ruthenium red stain, XXXIX, 144
calcium efflux, LVI, 350
calcium transport, LVI, 335, 336, 339, 342, 344, 346
in electron microscope cytochemisry of nucleic acids, XL, 32
Rutin, binding, XXXI, 533, 534
R_M value, *see* Chromatography
Rye, protoplast isolation, XXXI, 580

S

$S_{0.5}$, definition, XLIV, 3
S–13, *see* 5-Chloro-3-*tert*-butyl-2′-chloro-4-nitrosalicylanilide
S–ρ separation, of plant cell organelles, XXXI, 505–507
sA, *see* 2-Methylthio-N^6-(2-isopentenyl)adenosine
Sabouraud dextrose broth, LVIII, 20
Saccharide, *see also* specific types
acid hydrolyzates, preparation, XLI, 20
alkaline degradation during ion-exchange chromatography, XLI, 21
automated determination by ion-exchange chromatography of borate complexes, XLI, 10–21
constituents, L, 93
qualitative analysis, XLI, 18
transformation reactions, L, 96
Saccharide phosphate, from dolichol diphosphate derivatives, L, 432
Saccharo-1,4-lactone, XXXIV, 4
Saccharomyces, free cells, XXXII, 723

Saccharomyces carlsbergensis
disruption, LV, 141
spheroplasts, XXXI, 613
Saccharomyces cerevisiae, *see also* Yeast
agglutination by antiserum against edestin-β-(*p*-aminophenyl)ethylamine mannotetraose, L, 169
coenzyme Q homologs, LIII, 593
complex III, LIII, 113–121
cytochrome c_1, LIII, 221–231
yield, strain dependence, LIII, 224, 225
cytochrome oxidase, LIII, 73–79
cytoplasmic vesicles, XXXI, 626
disruption, LV, 140
double-stranded ribonucleic acid, LX, 552
elongation factors, LX, 678
mitochondria, ATPase inhibitor, LV, 402, 404, 405
mitochondrial mutants, XXXII, 838–843
protein yield, brand dependence, LIII, 114
spheroplast isolation, XXXI, 572, 573
succino-AICAR synthetase, LI, 190
viruslike particles, LX, 549, 550
Saccharomyces fragilis, spheroplasts, XXXI, 613
Saccharomyces pastorianus, XLIII, 110
Safety cabinet, ventilated, LVIII, 10, 40–42
Safety equipment, *see* specific type
Safety guidelines, LVIII, 42, 43
Safety technique, LVIII, 36–43
Safranine, membrane potentials, LV, 559, 573
Safranine T, as mediator-titrant, LIV, 409
Safranin O, XLIV, 654
Safrole, metabolite binding to reduced cytochrome, LII, 274
Sage Model 255-1 syringe pump, LIX, 373
Salicin
lectin, XXXIV, 334
substrate, of β-glucosidase, XLIV, 272, 433, 917
Salicylaldehyde, modification of lysine, LIII, 140
Salicylaldoxime, interaction with cytochrome *c* oxidase, LIII, 195
Salicylamide, inhibitor of cytochrome *P*-450-linked drug metabolism, LII, 67, 68
Salicylanilide, as uncouplers, LV, 467
Salicylate anion, slow hydrogen exchange, XLIX, 34
Salicylate hydroxylase, *see* Salicylate 1-monooxygenase
Salicylate 1-monooxygenase, LII, 17; LIII, 527–543
activity, LIII, 527, 528, 535
altered form from mutated strain, LIII, 542, 543
assay, LIII, 529, 530
as external monooxygenase, LIII, 527
flavin determination, LIII, 429, 528, 534
inhibition by salts, LIII, 541
kinetic properties, LIII, 535–539
molecular weight, LIII, 528, 535
oxy form, spectrum, LII, 36
oxygenated flavin intermediate, LIII, 540, 541
partial specific volume, LIII, 535
from *Pseudomonas*, LIII, 527–543
pseudosubstrates, LIII, 528, 537–539
purification, LIII, 532–534
purity, LIII, 532–534
reaction mechanism, LIII, 540, 541
reduced enzyme-substrate complex, reaction with oxygen, LIII, 540, 541
reduction studies, LIII, 541, 542
sedimentation coefficient, LIII, 535
substrate specificity, LIII, 535, 536, 538, 539
subunits, LIII, 528, 535
uncoupling effect, LIII, 536–540, 549
Salicylic acid
reversible dissociation of flavoproteins, LIII, 434
substrate of salicylate hydroxylase, LIII, 528, 529, 535, 538, 539
Saligenin, XLIV, 918
product, of β-glucosidase, XLIV, 272
Saline, phosphate-buffered, *see* Phosphate-buffered saline
Saline A solution, composition, XXXI, 308

Saline citrate solution, in isolation of histones, XL, 106
Salinomycin, LV, 445
as ionophore, LV, 442
Saliva
collection, from insect salivary gland, XXXIX, 469, 470
drug isolation, LII, 333
glycoprotein, XXXIV, 337
Salivary gland
epidermal growth factor isolation, XXXVII, 427
of insects, hormone effects, XXXIX, 466–476
Salmine
inhibitor of complex IV, LIII, 45
in media, LVIII, 70
Salmon cell, incubation temperature, LVIII, 471
Salmonella, membrane isolation, XXXI, 642–653
Salmonella tranaroa, isolation of lipopolysaccharide A O-antigen, L, 317
Salmonella typhimurium
N-acetylmannosamine kinase, assay, purification, and properties, XLII, 53–58
ATPase, *E. coli* stripped membranes, LV, 800
bacteriophage P22, XLIII, 45
carbamoyl-phosphate synthetase, LI, 29
CMP-dCMP kinase, LI, 332
crude extract preparation, LI, 32, 420, 439, 519, 553
deoxycytosine triphosphate deaminase, LI, 418
deoxyribose-5-phosphate aldolase, assay, purification, and properties, XLII, 276–279
FGAM synthetase, LI, 200
glycosyltransferase complex reconstitution, XXXII, 449–459
growth, LI, 6, 32, 373, 420, 439, 519, 553
guanine phosphoribosyltransferase, LI, 549–558
histidine transport mutants, XXXII, 849–856
hypoxanthine phosphoribosyltransferase, LI, 549–558
membrane isolation, XXXI, 698
penicillin bioautography, XLIII, 111
phosphoribosylpyrophosphate synthetase, LI, 6, 7
purine nucleoside phosphorylase, LI, 517–521
succino-AICAR synthetase, LI, 190
thymidine phosphorylase, LI, 437–442
for transduction, XLIII, 47
transport systems, XXXI, 704
UMP kinase, LI, 332
Salmonella typhimurium LT-2, ATP-phosphoribosyltransferase, in protein-tRNA interaction studies, LIX, 323, 328
Salmonidae, enolase, XLII, 334
Salt, *see also* specific types
concentration
column pH artifact, XLIX, 33
effect of flavin fluorescence measurements, LIII, 424
interference, in bioluminescent reaction, LVII, 122
Salt bridge, XLIV, 149
Salt extraction analysis, phosphorylated acidic chromatin proteins, XL, 177
Salt gradient, for centrifugation, XXXI, 716, 717
Salting out, interfacial, LIX, 263
Salting-out chromatograph, XLIII, 162–164
Salt solution, *see also* specific type
balanced, XXXII, 709; LVIII, 120–122
composition, XXXII, 707; LVIII, 121
plant, LVIII, 480
vertebrate, LVIII, 468, 469
stock, LVIII, 198
SAM, *see S*-Adenosylmethionine
Samarium, magnetic circular dichroism, XLIX, 177
Samarium acetate, in derivatization of transfer RNA, LIX, 14
Samia cynthia
cytochrome *c*, LIII, 138
media, LVIII, 457
Sample
degassing procedures, LIV, 120, 122

optical density, LIV, 38, 39
optimun, derivation, LIV, 45, 46
Sample cell
bubble elimination, LIV, 317
concentric, LIV, 155, 188
in flash photolysis, LIV, 53
glass, water-jacketed, LVII, 39
in IR spectroscopy, window material, LIV, 314–317
in magnetic circular dichroism spectroscopy, LIV, 291, 293
pathlength determination, LIV, 317, 318
in pulse Fourier-transform NMR spectroscopy, LIV, 154, 155, 188
quartz, with graded seal, LIV, 124
specifications, in nanosecond absorbance spectroscopy, LIV, 39, 45
Sample holder, XLIX, 392, 436, 471, 472, 478–480
Sample injection value, 8-part, LIX, 221
Sample pan, for differential scanning calorimeters, XXXII, 267
Sampling
automatic method, LVI, 297–301
in electron microscope autoradiography and cytochemistry, XL, 3
timing, LVI, 355, 356
Sanborn recorder, XXXIX, 56
Sand
disruption of *N. crassa* cells, LV, 145, 147
Keilin-Hartree preparation, LV, 120, 125, 126
for mitochondria isolation, XXXI, 300
Sandhoff's disease
enzymic diagnosis, L, 485–487
globoside, L, 484
β-hexosaminidase A, L, 484
β-hexosaminidase B, L, 484
Sandwich gradient
for pancreas subfractionation, XXXI, 45, 50
Sanfilippo A syndrome, heparin sulfamidase, L, 448, 449
Sanfilippo B syndrome, α-*N*-acetylglucosaminidase, L, 450
Sanger's reagent, XLIV, 284
Sangivamycin
biosynthesis, XLIII, 516
toyocamycin nitrile hydrolase, XLIII, 759–762
SANS, *see* Neutron scattering
Saponification, of long chain fatty acids, LII, 312, 313
Sapphire, XLIX, 22
infrared spectral properties, LIV, 316
Saprospira thermalis, aldolase, XLII, 233
Sarcina
cell-wall isolation, XXXI, 659
oxidase, LII, 18
Sarcina lutea
bioautography, XLIII, 117
echinomycin assays, XLIII, 771
efrapeptin, LV, 495
penicillin bioautography, XLIII, 111
Sarcolemma, adenylate cyclase, XXXVIII, 146
Sarcoma cell
oxidase, LII, 13
protein content, LVIII, 145
Sarcoma/leukemia virus, cell lines for study, XXXII, 584, 585
Sarcoplasmic reticulum, XLVI, 86
assay, for characterization, XXXI, 243
ATPase, isolation, XXXII, 291, 292
ATPase rotation, LIV, 58
calcium transport, LVI, 321, 322
description, XXXI, 238
disc gel electrophoresis, XXXII, 292
gel electrophoresis, XXXII, 79–81, 479
hydrogen and calcium ion transport, LVI, 338
ionophores, LV, 454
isolation, from skeletal muscle, XXXI, 238–246
membrane reconstitution, XXXII, 475–481
negative staining, XXXII, 29
NMR studies, XXXII, 206
phospholipids, LV, 701
preparation, XXXII, 294, 295
properties, XXXI, 241; XXXII, 480
proteins, isolation, XXXII, 291–302
proteolipid, isolation, XXXII, 291, 293

spin-label studies, XXXII, 171, 175, 194
vesicle, labeling with eosin derivatives, LIV, 51
Sarcosine dehydrogenase, LIII, 402
flavin linkage, LIII, 450
Sarcosine demethylase, XLVI, 163
Sarcosine oxidase, LII, 18; LVI, 474
Sarin, inhibitor, cholinesterase, XLIV, 651
Sarkosyl detergent, in 5′-nucleotidase purification, XXXII, 374
Saturation-transfer spectroscopy, XLIX, 396, 397, 480–511; LIV, 171–178
applications, XLIX, 488–501
in physical chemistry, XLIX, 500, 501
to proteins, XLIX, 488–494
binding tightness, XLIX, 489, 502, 503
dispersion and absorption, XLIX, 506, 507
FM noise, elimination, XLIX, 504
instrumentation, XLIX, 503–505
modulation frequency, XLIX, 506
passage effects, XLIX, 486
phase stability, XLIX, 505
pulse sequence, LIV‡ 172, 173
sensitivity, XLIX, 505, 506
spectrometer, LIV, 172
theoretical considerations, XLIX, 480–488
using chemical exchange, LIV, 172–177, 180, 181
using cross-relaxation, LIV, 177, 178
Savoye cabbage, phospholipase D, XXXII, 143
Sawbench technique, of macroautoradiography, XXXVI, 138
Saxitoxin, neurotoxic action, XXXII, 310
SB-2 anion exchange paper, LI, 338, 346
SBMV, *see* Southern bean mosaic virus
Scallop, octopine dehydrogenase, LIV, 229, 231
Scanning, of histone gels, XL, 129
Scanning calorimetry, XLIX, 3–14
Scanning electron microscopy, XXXII, 50–60; XXXIX, 154–156
drying, XXXII, 56–58
fixation, XXXII, 55, 56
microscopy procedure, XXXII, 58, 59
preparative methods, XXXII, 55–58
shape studies, XXXII, 53
Scatchard plot, LII, 265
binding site heterogeneity, XLVIII, 280–286
extrapolation of slope to Y axis, XLVIII, 285, 286
of fluorescence probe data, XXXII, 243
of hemoglobin-oxygen binding, XLVIII, 274–276
for ligand-protein binding, XLVIII, 299–307
simulations, XLVIII, 300, 301
meaning, XLVIII, 271, 272
negative cooperativity, XLVIII, 283–285
of progesterone receptors, XXXVI, 195, 196
of prostaglandin binding data, XXXII, 110, 112, 114, 121–123
in protein-ligand interaction, XXXVI, 3, 6, 7, 15
quantitative relationship to Hill plot, XLVIII, 278, 279
of steroid binding studies, XXXVI, 244
Scattering amplitude, XLVIII, 425
computation, LIX, 757
function of protein content, LIX, 758, 759
Scattering cross section
for biological nuclei, LIX, 672
definition, LIX, 671
Scattering density, excess, LIX, 674, 690, 700, 708, *see also* Scattering amplitude
during solvent exchange, LIX, 677
Scattering form factor, intramolecular, XLVIII, 426
for flexible-chain polymers, XLVIII, 443
for rodlike molecules, XLVIII, 436, 437
Scattering intensity, LIX, 677, 709, *see also* Cross term scattering function; Fluctuation scattering function; Shape scattering function; Vacuum scattering function
function of scattering angle, LIX, 763–767

small-angle, LIX, 700
Scattering vector, definition, XLVIII, 425, 426
Scavenger, photoaffinity labeling, LVI, 644, 645
SCE, *see* Calomel electrode, saturated
Scenedesmus, hydrogenase reaction in, LIII, 287
Scheie syndrome, α-L-iduronidase activity, L, 443
Schiff base, XLVI, 133, 137, 138, 162, 163, 353, 357, 391
formation, XLVI, 52, 432, 469
intermediates, XLVI, 8
reduction, XLVI, 357
Schiff stain, for gel electrophoresis, XXXII, 96, 97, 99–101
Schizosaccharomyces pombe
adenylosuccinate synthetase, LI, 212
disruption, LV, 141
mitochondrial mutants, XXXII, 838–843; LVI, 121
Schlieren optics, sedimentation equilibrium, XLVIII, 168, 169
Schmidt-Lantermann cleft, XXXI, 435
Schmidt-Ruppin strain, Rous sarcoma virus, LVIII, 380, 381, 394
Schneider medium, LVIII, 464–466
preparation, LVIII, 464–466
Schöniger combustion method, XLIV, 133
Schuster-Koo mechanism, LVII, 520
Schwann cell, XXXI, 435
2′,3′-cyclic nucleotide-3′-phosphohydrolase, XXXII, 128
enzyme marker, XXXII, 329
as myelin source, XXXII, 323
Sciatic nerve
basic proteins, XXXII, 342
myelin membrane, X-ray studies, XXXII, 219
NMR studies, XXXII, 206
Scintillation cocktail, XLVIII, 336; LI, 4, 13, 30, 52, 85, 91, 113, 157, 220, 292, 300, 338, 362
Bray, effect of pH, LII, 415
for calcium studies, XXXIX, 550
dioxane-based, LII, 393, 398
for fatty acids, LII, 312, 315, 321
in α-hydroxylase assay, LII, 312
for sarcoplasmic reticulum studies, XXXI, 244, 245
for steroids, LII, 126
toluene-based, LII, 126, 312, 315, 321, 396, 398, 404
for vitamin D metabolites, LII, 393, 396, 398
Scintillation counting, XL, 294
double-label technique, spillover correction, XLVIII, 337
liquid, XL, 295
mixture, LV, 252
nucleosides and nucleoside products, LI, 339
Scintillation vial
assay of malate dehydrogenase, LVII, 183
background reduction, LVII, 475, 480, 481
bioluminescent assays, LVII, 218
phosphorescence, LVII, 183, 217
reuse, LI, 157
Scopoletin, in hydrogen peroxide assay, LII, 347, 349
Scott unit, XLIV, 157
Screening
of carbodiimide-resistant candidates, LVI, 165, 166
reproducibility, LVI, 166
routine, LVI, 168
primary, *E. coli* mutants, LVI, 108
secondary, *E. coli* mutants, LVI, 108, 109
SDH, *see* Succinate dehydrogenase
SDS, *see* Sodium dodecyl sulfate
Sea anemone, *see also* specific types
inhibitors
amino acid composition, XLV, 887
assay, XLV, 882
kinetics, XLV, 887
molecular weight, XLV, 886
N–terminal residues, XLV, 886
properties, XLV, 885
purification, XLV, 882
specificity, XLV, 886
stability, XLV, 885
Sea snake, neurotoxins, XXXII, 312
Sea urchin
egg protease
activators, XLV, 352

assay, XLV, 343
bioassays, XLV, 343
chemical, XLV, 345
esterase activity, XLV, 345, 346
sperm receptor hydrolase, XLV, 344
vitelline delaminase, XLV, 343
in cortical granule, XLV, 343
exudate preparation, XLV, 347
distribution, XLV, 353
gametes, shedding, XLV, 346
inhibitors, XLV, 352
isoelectric focusing, XLV, 349
isoelectric precipitation, XLV, 347
kinetics, XLV, 353
preparation, XLV, 346
properties, XLV, 351
purification, XLV, 346
specificity, XLV, 353
stability, XLV, 351
hatching enzyme, XLV, 354, *see also* Hatching enzyme
casein, digestion of modified, XLV, 356
chromatography, XLV, 364
dialysis, XLV, 363
fertilization envelope dissolution, XLV, 355
gametes shedding, XLV, 360
inhibitors, XLV, 368
kinetics, XLV, 367
metals, dependence, XLV, 367
molecular weight, XLV, 367
pH, XLV, 367
preparation, XLV, 359
crude enzyme, XLV, 360–362
incubating embryos, XLV, 362
yield, XLV, 363
properties, XLV, 366
purification, XLV, 363
urea treatment, XLV, 363
purity, XLV, 367
specificity, XLV, 369
stability, XLV, 366
temperature, XLV, 367
Sea water
artificial, XXXII, 315; XXXVIII, 197
preparation, LVII, 154, 155, 190
in preparation of aequorin crude extract, LVII, 304
Seawater complete medium
formulation, LVII, 154
for growth of luminescent bacteria, LVII, 141
preparation, LVII, 155
Seawater minimal medium
formulation, LVII, 154
preparation, LVII, 155
Sebacic acid, XXXIV, 477; XLIV, 118
Sebacic acid dihydrazide, Sepharose activation, XLIV, 882
Sebacyl chloride, XLIV, 207, 208
Sebacylic acid dihydrazide, preparation, LIX, 179
(4*R*)-5,10-Secoestra-4,5-diene-3,10,17-trione, XLVI, 463, 467
(4*S*)-5,10-Secoestra-4,5-diene-3,10,17-trione, XLVI, 463, 467
5,10-Secoestr-5-yne-3,10,17-trione, XLVI, 463, 467, 469
Seconal
inhibitor of complex I, LIII, 13
of complex I–III, LIII, 8
(4*R*)-5,10-Seco-19-norpregna-4,5-diene-3,10,20-trione, XLVI, 463, 467
(4*S*)-5,10-Seco-19-norpregna-4,5-diene-3,10,20-trione, XLVI, 463, 467
5,10-Seco-19-norpregn-5-yne-3,10,20-trione, XLVI, 463, 467, 469
Secretin, fragemnt, synthesis, XLVII, 588
Secretory granule, isolation from anterior pituitary glands, XXXI, 410–419
Secretory inhibitor I, porcine pancreatic, hydrolysis, XLVII, 172, 173
Section-Scotch tape technique, of macroautoradiography, XXXVI, 138, 153
Sectorial dilution, in plant-cell fractionation, XXXI, 511
Sedimentation
asymptotic boundary profiles, XLVIII, 208–211
centrifugal, LVI, 282, 283
computer-simulated, XLVIII, 212–242
box size, XLVIII, 229–231
procedure outline, XLVIII, 233–235

fractionation experiment, XLVIII, 245, 269, 270
irreversible reactions, XLVIII, 249–252
partial specific volumes, XLVIII, 11–23
stopped-flow dilution, XLVIII, 270
theoretical considerations, XLVIII, 13, 14
for kinetically controlled association-dissociation, XLVIII, 249, 260
Sedimentation coefficient
average, XLVIII, 218
calculation, XLVIII, 219–224
definition, XLVIII, 218
brown adipose tissue mitochondria, LV, 69
determination during dimerization reaction, XLVIII, 246, 247
viscosity barrier centrifugation, LV, 129, 130
Sedimentation equilibrium
absorbance optics, XLVIII, 171–173
Archibald technique, XLVIII, 168, 169
centrifugation speeds, XLVIII, 171
data analysis, XLVIII, 173–184
with Rayleigh interference optics, XLVIII, 169–171, 173–178
with scanning absorbance optics, XLVIII, 178
data error, XLVIII, 183
equation of state
for heterogeneous ideal two-component system, XLVIII, 166, 167
for homogeneous ideal two-component system, XLVIII, 166
for nonideal multicomponent system, XLVIII, 164–166
high speed, disadvantage, XLVIII, 171
homogeneous ideal two-component system, XLVIII, 166
methods, types, XLVIII, 168–173
molecular weight measurements, XLVIII, 163–185
nonideal multicomponent system, XLVIII, 164–166
preferential solvation, XLVIII, 175
use, XLVIII, 18–22
using schlieren optics, theory, XLVIII, 168, 169
Sedimentation velocity
ligand-mediated protein interactions, XLVIII, 242–248
for molecular weight determinations, XLVIII, 6
profile for self-associating solutes, calculation, XLVIII, 212–242
use, XLVIII, 18–22
Sedoheptulose-bisphosphatase, from *Candida utilis*, XLII, 347–353
activators and inhibitors, XLII, 353
assay, XLII, 348
molecular weight, XLII, 353
properties, XLII, 353
purification, XLII, 350, 351
Sedoheptulose-1,7-diphosphatase, *see* Sedoheptulose-bisphosphatase
Sedoheptulose 1,7-diphosphate
in determination of sedoheptulose 7-phosphate, XLI, 34–36
preparation, XLI, 77–79
Sedoheptulose 7-phosphate, determination, XLI, 34–36
Seebeck effect, LVII, 576
Seedling, etiolated, for plant-cell extraction, XXXI, 532
Segregation
problem, LVI, 146, 147
rates, LVI, 153, 154
Segregation analysis, mitochondrial markers, LVI, 144–146
Selectacel, in glycolipid isolation, XXXII, 352
Selecta filter, LIX, 447
Selective exclusion, XXXIV, 132, 133
Selenite
in media, LVIII, 500
radiolabeled, in glutathione peroxidase assay, LII, 507
^{75}Se-labeled, for radiolabeling formate dehydrogenase, LIII, 369, 371
Selenium
deficiency, LVIII, 74
formate dehydrogenase, LIII, 371, 372
in glutathione peroxidase, LII, 512, 513

growth requirement for production of protein A, LIII, 375
in media, LVIII, 52, 69, 86, 89, 90
neuroblastoma, LVIII, 98
oxygen metabolism, LII, 8
volatization losses, in trace metal analysis, LIV, 466, 480, 481
Selenium-77, lock nuclei, XLIX, 349; LIV, 189
Selenocyanate ion
in IR spectroscopy, LIV, 304
stretching frequency, LIV, 311
Selenoprotein, *see* Protein A
Self-absorption, XLIX, 91, 93
in autoradiography, LVIII, 289, 290
Self-association, *see also* Association reaction; Binding; Cooperativity; Dimerization; Dissociation; Interaction; Ligand binding
analysis method of Derechin, XLVIII, 116, 117
of Kreuzer, XLVIII, 117
of Lewis and Knott, XLVIII, 117
of Teller, XLVIII, 117, 118
indefinite, XLVIII, 110–116
type I, XLVIII, 111–113
type II, XLVIII, 113
type III, XLVIII, 113–115
type IV, XLVIII, 115, 116
of monomer-*n*-mer, XLVIII, 109, 110
of monomer-*n*-mer-*m*-mer, XLVIII, 116
nonideal, XLVIII, 106
osmotic pressure, XLVIII, 103–121
simulated sedimentation velocity profiles, XLVIII, 212–242
theory, XLVIII, 73
Self-association equilibrium, equations of state, XLVIII, 214
SEM, *see* Scanning electron microscopy
Semen, phosphomonoesterase, assay of guanylate kinase, LI, 476
SE method, *see* Sephadex SE method
Semicarbazide, XLIV, 552
for acetaldehyde binding, LII, 359
in ethanol determination, LII, 359
for formaldehyde binding, LII, 302
interaction with cytochrome *c* oxidase, LIII, 194
for trapping oxaloacetate, LIII, 472, 474
Semiconductor, *n*-type, LIV, 405
Seminal plasma, 5′-nucleotidase, XXXII, 368
Seminal plasma inhibitor
boar
amino acid composition, XLV, 840
assay, XLV, 835
carbohydrate composition, XLV, 843
characteristics, XLV, 834
composition, XLV, 843, 844
kinetics, XLV, 844
properties, XLV, 842
purification, XLV, 835
from spermatozoa, XLV, 841
purity, XLV, 843
specificity, XLV, 844
stability, XLV, 842
units, XLV, 835
guinea pig vesicle inhibitors
amino acid composition and sequence, XLV, 831, 832
assay, XLV, 826
characteristics, XLV, 825
molecular weight, XLV, 831
properties, XLV, 831, 832
purification, XLV, 827
reactive site, XLV, 831
stability, XLV, 831
human
assay
acrosin inhibition, XLV, 849
chymotrypsin inhibition, XLV, 848
leukocytic proteinases, XLV, 850
trypsin inhibition, XLV, 848
biological function, XLV, 859
characteristics, XLV, 847
composition, XLV, 857
fractions, separation, XLV, 854
kinetics, XLV, 859
molecular weight, XLV, 857
N–terminal residues, XLV, 858
properties, XLV, 857
purification, XLV, 851
reactive site, XLV, 859
specificity, XLV, 858

stability, XLV, 857
Seminal vesicle, aldose reductase, assay, purification, and properties, XLI, 165–170
Seminiferous tubule
isolation, XXXIX, 291
in Rose chambers, XXXIX, 292–295
Seminolipid, arylsulfatase A, L, 474
Semiquinone, EPR signal properties, LIV, 134, 137, 138
Semiquinone analogue, LIV, 424
Sendai virus
cell fusion, LVIII, 353–356
effect on cell fusion, XXXII, 575, 580–582
inactivated, for cell fusion, XXXIX, 122–124
in nucleus transfer, LVIII, 359
preparation, XXXII, 578, 579
storage, LVIII, 354
Sennoside, extraction, LV, 459
Sephacryl 200, in *Rhizobium P*–450 purification, LII, 161
Sephacryl S–200, purification and molecular weight determination of phosphoprotein phosphatases, LX, 528, 529, 531
Sephadex, *see also* Dextran
blocking of chymotrypsin, XLIV, 24
for brain nuclei isolation, XXXI, 457
as carrier, XXXIV, 5
for column sieving fractionation, XXXI, 551
detergent removal, LV, 702
lack of fluorescent impurities, XLIV, 363, 364
lectin purification, XXXIV, 336
lipophilic, XXXV, 378–395
chromatography
bile acids, XXXV, 395
bile acid sulfates, XXXV, 393
ethanediol esters and ethers, XXXV, 389
fatty acids, XXXV, 388, 392
galactolipids, XXXV, 388
lipids, XXXV, 386, 388–391
phosphatidylcholines, XXXV, 389, 392
phosphatidylethanolamine, XXXV, 389
phospholipids, XXXV, 388, 389, 392
prostaglandins, XXXV, 393
proteolipids, XXXV, 389
steroids, XXXV, 389, 393, 394
steroid sulfates, XXXV, 389, 393
sterols, XXXV, 388, 389, 392, 394
steryl esters, XXXV, 388, 391, 392
terpenoid compounds, XXXV, 393, 395
tocopherols, XXXV, 395
triglycerides, XXXV, 383, 388–391
vitamin D, XXXV, 395
vitamin K, XXXV, 395
waxes, XXXV, 391, 392
isomerization of 1,2-diglycerides into 1,3-diglycerides, XXXV, 390
mechanism of chromatographic separation, XXXV, 386
noninterference with subsequent gas chromatography–mass spectrometry, XXXV, 383, 384
preparation, XXXV, 378–383
recycling chromatography, XXXV, 385
side chains
carboxymethoxypropyl, XXXV, 383
chlorohydroxypropyl, XXXV, 381, 382
dibutylaminohydroxypropoxypr- opyl, XXXV, 382, 388, 389, 392
diethylaminoethoxypropyl, XXXV, 383
hydroxyalkoxypropyl, XXXV, 380, 381, 383, 386–388, 390–395
hydroxypropyl, XXXV, 380, 387–390, 392–395
mercaptohydroxypropoxypropyl, XXXV, 382
methyl, XXXV, 379, 380, 388, 395
trimethylsilyl, XXXV, 381

solvent systems, XXXV, 387
stability, XXXV, 383
in thin-layer chromatography, XXXV, 385
as matrix, XXXIV, 19
multienzyme system, XLIV, 181, 456
submitochondrial particles, LV, 111, 112
Sephadex A–25
incorporation of terminal nucleotide analogs, LIX, 188
purification of elongation factor EF-1α, LX, 669
of shortened tRNAs, LIX, 186
Sephadex A–50, purification
elongation factor
eEF–Ts, LX, 602–604
eEF–Tu, LX, 600–604, 700, 701
EF–1, LX, 654–656, 678, 679
EF–2, LX, 667, 668
EF–1$\beta\gamma$, LX, 671
EF–G, LX, 600–602, 610–613
heme-regulated translational inhibitor, LX, 468, 471, 486, 487, 495
initiation factors, LX, 5, 7, 9, 17, 96, 99, 100, 187, 189
Qβ ribonucleic acid replicase, LX, 634, 635
Sephadex bead, charged, use for hydrogen out-exchange, XLIX, 33
Sephadex C–50, purification
elongation factors, LX, 652, 653, 656, 666, 667, 669, 674, 675, 697, 699
initiation factors, LX, 40, 54–56, 142, 187–189, 192
Sephadex CM–50, purification of specific aminoacyl-tRNA synthetases, LIX, 262, 264, 265
Sephadex CM–C50, carboxyl group activation, XLIV, 542
Sephadex column, high-pressure technique, XLIX, 33
Sephadex G–10, XLVII, 83, 208
in cytochrome P–450_{cam} isolation, LII, 155, 156
for desalting, XLIV, 866
determination of deuterium content of RNA, LIX, 646
iodinated cyclic nucleotide derivatives, XXXVIII, 100
purification of cytochrome c derivative, LIII, 150, 152, 171
of luciferin, LVII, 36
Sephadex G–15, XLVII, 115
in cytochrome m purification, LII, 183
preparation of rat liver polysomes, LIX, 526
purification of cytochrome c, LIII, 131
of cytochrome c derivatives, LIII, 152, 172
of dihydroorotate dehydrogenase, LI, 66
removal of endogenous amino acids, LIX, 517
Sephadex G–25, XLIV, 532, 711; XLVII, 83, 115, 116, 137, 152, 218
assay of carbamoyl-phosphate synthetase, LI, 31
ATPase desalting, LV, 806
ATPase inhibitor, LV, 410
for deoxycholate removal, LII, 210
for desalting, LII, 99, 100
fine, preparation of radiolabeled ribosomal protein, LIX, 777, 778
glutathione peroxidase assay, LII, 507
heterogeneous competitive protein-binding assay, LVII, 442
hydroxyapatite column, preparation, LIII, 316
in preparation of chromophore-labeled sarcoplasmic reticulum vesicles, LIV, 51
of F_1 and membrane factor, LV, 743
of hemoglobin, LIV, 488
of luciferyl adenylate, LVII, 27
of matrix-bound RNase, LIX, 463
of mitochondrial enzyme complexes, LIII, 3
of S100 extract, LIX, 357
purification
of adenosine monophosphate nucleosidase, LI, 266
aminoacylated transfer ribonucleic acid, LX, 622
of apoflavoproteins, LIII, 433, 435
of azurin, LIII, 659
of complex II, LIII, 23
of cytidine deaminase, LI, 410
of cytochrome c-551, LIII, 659

of cytochrome *c* derivatives, LIII, 45, 149, 170, 177
of dCTP deaminase, LI, 420, 421
of glutathione reductase, LII, 507
of high-potential iron-sulfur proteins, LIII, 338
initiation factors, LX, 96, 99, 100
of luciferase, LVII, 31
of luciferase oxygenated-flavin intermediate, LIII, 570
of luciferase reaction intermediate, LVII, 196
of nitrogenase, LIII, 320
nuclease-resistant messenger ribonucleic acid hybrids, LX, 547
of phenylalanine hydroxylase, LIII, 282
of reconstituted hemoglobin, LII, 455
reticulocyte lysate, LX, 391
of RNase III, LIX, 828, 830
of uridine-cytidine kinase, LI, 311
for reconcentration, XLVIII, 245
removal of EDTA from purified aequorin, LVII, 311
of sulfhydryl compounds, LIX, 387
retrieval of stored aequorin, LVII, 281
separation of amino acids and tRNA, LIX, 312
spin-labeled acyl-atractyloside,
for sucrose removal, LII, 209
urea removal, LV, 797
Sephadex G–50, XLIV, 456, 872; XLVII, 160, 218, 219, 465, 466
activation of succinate dehydrogenase, LIII, 475
ATPase inhibitor, LV, 412
cyanogen bromide activation, XLIV, 457
hydroxylated, XXXII, 443
lipid filtration through, XXXII, 49
hydroxypropyl, DCCD-binding protein, LV, 432
measurement of glutamate binding, LVI, 427
of phosphate binding, LVI, 527, 528
nucleotide-depleted ATPase F_1, LV, 378
reattachment of nucleotides, LV, 379
preparation of monomeric cytochrome *c*, LIII, 161
purification
of aeguorin and GFP, LVII, 307, 308, 309
of cytidine triphosphate snythetase, LI, 81
of cytochrome *c* derivatives, LIII, 176
of desulfovibrione electron carrier, LIII, 633, 634
elongation factor EF-G, LX, 742, 743
guanosine triphosphate bound to initiation factor EF-G, LX, 740
of flavin peptides, LIII, 455
of flavodoxin, LIII, 625
of NAD:rubredoxin oxidoreductase, LIII, 619
transfer ribonucleic acid, LX, 414
retrieval of stored aequorin, LVII, 311
Sephadex G–50M, as microcarrier, LVIII, 186
Sephadex G–75
ATPase inhibitor and, LV, 411
molecular weight determination of aequorin, LVII, 283
protein kinase inhibitor, XXXVIII, 355, 356
purification of adenine phosphoribosyltransferase, LI, 572
of adenylate kinase, LI, 461
of adrenodoxin LII, 136
of aequorin, LVII, 280, 283
of amidophosphoribosyltransferase, LI, 176
of blue-fluorescent protein, LVII, 229
of cytochrome b_5, LII, 100
of cytochrome *c* derivatives, LIII, 149
of desulfovibrione electron carriers, LIII, 633, 634

of electron-transferring flavoprotein, LIII, 517
of ferredoxin, LIII, 623
of flavodoxin, LIII, 625
of high-potential iron-sulfur protein, LIII, 336–338
of initiation factors, LX, 29, 105, 107, 151, 157, 562, 563
of phosphoribosylglycinamide synthetase, LI, 184
of *Rhizobium* cytochrome *P*–450, LII, 160, 161
of superoxide dismutase, LIII, 389
of UMP-CMP kinase, LI, 335
Superfine
preparation of monomeric cytochrome *c*, LIII, 161
purification of blue-fluorescent protein, LVII, 231, 232
Triton X–100 removal, LII, 209, 210
Sephadex G–75–40, coupling factor B, LV, 388
Sephadex G–100, XLIV, 532; LIX, 236
analysis of initiation factor complexes, LX, 58, 59
cAMP-binding protein, XXXVIII, 370
phosphodiesterase, XXXVIII, 254, 256
phosphodiesterase activator, XXXVIII, 255, 269
preparation of ribosomal proteins, LIX, 651
of ribosomal RNA, LIX, 446
of tRNA, LIX, 211, 312
proline carrier proteins, LV, 196
protein kinase, subunits, XXXVIII, 306, 307, 361, 362, 364, 365
purification
of adenine phosphoribosyltransferase, LI, 576, 577
of adenosine deaminase, LI, 511
of adenylate kinase, LI, 471
of aequorin, LVII, 280, 281
of carbon monoxide-binding heme protein *P*–460, LIII, 638
of commercial cytochrome *c*, LIII, 384
of cytidine deaminase, LI, 397, 398
of cytidine triphosphate synthetase, LI, 81
of cytochrome *b*, LIII, 214, 217
of cytochrome oxidase, LIII, 60
of cytochrome P–450_{cam}, LII, 153, 155
of cytosine deaminase, LI, 396
of dCTP deaminase, LI, 421
of CTP(ATP):tRNA nucleotidyltransferase, LIX, 125, 126
of denitrifier *c*-type cytochrome, LIII, 645
of deoxythymidylate phosphohydrolase, LI, 288
of dimeric cytochrome *b*, LIII, 105–106
elongation factor EF-2, LX, 639, 645, 646
of glutathione reductase, LII, 507, 508
of guanylate kinase, LI, 478, 486
of guanylate synthetase, LI, 222
of hydrogenase, LIII, 289
of *p*-hydroxybenzoate hydroxylase, LIII, 547
initiation factors, LX, 24, 25, 96, 99
of luciferin sulfokinase, LVII, 249
of nitrite reductase, LIII, 643
of nitrogenase, LIII, 318, 322, 324
of nucleoside diphosphokinase, LI, 381, 383
of nucleoside phosphotransferase, LI, 390
of nucleoside triphosphate pyrophosphohydrolase, LI, 277
of orotate phosphoribosyltransferase, LI, 72
of orotidylate decarboxylase, LI 76
of oxidoreductases, LVII, 203
of peptide constituents of cytochrome bc_1 complex, LIII, 107
of *Pholas* luciferase, LVII, 388
of *Pholas* luciferin, LVII, 390
of *Pseudomonas* cytochrome oxidase, LIII, 650–651, 653
of purine nucleoside phosphorylase, LI, 534, 542
of ribonucleoside triphosphate reductase, LI, 252

of 30 S ribosomal proteins, LIX, 487
of 50 S ribosomal proteins, LIX, 493, 494
of thymidine phosphorylase, LI, 444
transfer ribonucleic acid, LX, 111, 112
of tryptophanyl-tRNA synthetase, LIX, 241
of uridine nucleosidase, LI, 293
separation of elongation factors, EF-G and EF-Tu, LX, 599–601
Superfine
in Hummel-Dreyer gel filtration, LIX, 327
isolation of luciferase-GFP complex, LVII, 266, 267
Sephadex G–150
concentration of ribosomal proteins, LIX, 499
purification
of adenylosuccinate synthetase, LI, 210
of cytidine deaminase, LI, 410
elongation factor
EF-G, LX, 611, 612
EF-2, LX, 706, 707, 710
of glutathione reductase, LII, 507, 508
initiation factors, LX, 198–201
nucleoside diphosphate kinase, LX, 581
of nucleoside diphosphokinase, LI, 381
of purine nucleoside phosphorylase, LI, 528
ribosome protein S1, LX, 422, 423
of salicylate hydroxylase, LIII, 534
of 50 S ribosomal proteins, LIX, 494
of thymidine kinase, LI, 357
of thymidine phosphorylase, LI, 440
for separation of OPRTase-OMPdecase complex aggregated forms, LI, 151
Sephadex G–200
adenylate cyclase, XXXVIII, 166, 168
benzoquinone activation, XLIV, 35
CAT-protein complex, LVI, 412
cyanogen bromide activation, XLIV, 30, 31, 529, 764, 765
dissociation of carbamoyl-phosphate synthetase, LI, 23, 24, 34
fractionation of irradiated ternary complex of guanosine triphosphate analogs, LX, 733, 734, 736, 737
for gel filtration, XLIV, 712
guanylate cyclase, XXXVIII, 194, 195
phosphodiesterase
brain, XXXVIII, 234–236
muscle, XXXVIII, 242, 243
preparation of ribosomal proteins, LIX, 519, 520
preparative electrophoretic purification of acyl-CoA dehydrogenases, LIII, 517
protein kinase, XXXVIII, 319, 321
purification
of adenosine monophosphate nucleosidase, LI, 267, 269
of amidophosphoribosyltransferase, LI, 174
of aspartate carbamyltransferase, LI, 174
of bacterial luciferase, LVII, 147
of cytidine triphosphate synthetase, LI, 81
of cytochrome oxidase subunits, LIII, 78
of deflavoenzyme, LIII, 434
of deoxycytidylate deaminase, LI, 416
elongation factor EF-1$\alpha\beta\gamma$, LX, 674
of FGAM synthetase, LI, 199
initiation factors, LX, 25, 95, 105, 106, 227, 228
of melilotate hydroxylase, LIII, 555
of nitrate reductase, LIII, 642
of nitric oxide reductase, LIII, 645
of nitrogenase, LIII, 317, 322, 323, 327
of nucleoside diphosphokinase, LI, 374
of OPRTase-OMPdecase complex, LI, 137
of phenylalanine hydroxylase, LIII, 282, 283

Pseudomonas toxin A, LX, 784
of pyrimidine nucleoside phosphorylase, LI, 435
of ribonucleoside diphosphate reductase, LI, 240
serum containing antibodies against protein S1, LX, 441, 442
of uridine-cytidine kinase, LI, 302, 304
of uridine phosphorylase, LI, 428
separation of elongation factors EF-1β and EF-1γ, LX, 672, 673
for separation of OPRTase-OMPdecase complex aggregated forms, LI, 151
as support, XLIV, 529, 764, 765
Sephadex gel, *see also* specific types
microbore column packing, XLVII, 213
Sephadex ion exchanger, XXXII, 158
resolution, XLVII, 220
Sephadex LH–20
chromatographic separation of vitamin D metabolites, LII, 392, 394, 398
DCCD-binding protein, LV, 429
preparation of luciferyl sulfate analogs, LVII, 252
purification
of cytochrome *P*–450 purification, LII, 120
of 7-[*N*-ethyl-*N*-(4-thyroxinylamido)butyl]amino-naphthalene-1,2-dicarboxylic acid hydrazide, LVII, 437
of luciferase oxygenated-flavin intermediate, LIII, 570
of prostaglandin, XXXII, 115, 118
of siroheme, LII, 441
of steroid, XXXVI, 39
separation of hydrocarbon-nucleoside adducts, LII, 292
steroid separation, LII, 379
tryptophan derivative separation, XLVII, 447
of vitamin D_3 metabolites, XXXVI, 529–533
Sephadex SE method, of nonhistone chromosomal protein preparation, XL, 150
controls and reproducibility, XL, 153
Sephadex-sulfopropyl, purification
casein kinases, LX, 501, 502
protein kinases, LX, 509
Sepharose, *see also* Agarose
adrenodoxin-substituted, preparation, LII, 133
affinity purification of antibodies, L, 171–175
amino acid content, XLIV, 497
aniline-substituted, for cytochrome *P*-450 purification, LII, 128, 129
preparation, LII, 127
bisoxirane activation, XLIV, 32, 33
chromatography, using reverse salt gradients, LIX, 215–218, 298
cyanogen bromide activation, XLIV, 16, 17
deoxythymidine-3′(4-amino-phenylphosphate)carboxyhexyl-substituted, preparation, LI, 366
glutamine binding, procedure, LI, 31, 32
octylamine-substituted, synthesis, LII, 138
as support, XLIV, 340
use in affinity chromatography of flavoproteins, LII, 90, 92
Sepharose-ε-aminocaproyl-cyclic adenosine monophosphate, preparation, XXXVIII, 386, 387
Sepharose 4B, *see also* Agarose
5′-amino-5′-deoxyuridine-substituted, preparation, LI, 309–311
N^6-(6-aminohexyl)-ATP-substituted, LI, 9
5′-AMP-substituted, preparation, LI, 493
benzoquinone activation, XLIV, 35
Blue dextran-substituted, LI, 9, 349
bromoacetamidoethyl-substituted, preparation, LI, 310, 311
chromatography of messenger ribonucleic acid, LX, 403, 404
coupled with anti-*Pseudomonas* toxin A
preparation, LX, 786

purification of Toxin A, LX, 786, 787
with heparin, purification of elongation factor EF-1, LX, 692, 694
with anti-S1 IgG
preparation, LX, 429
purification of 30 S (−S1) ribosomal subunits, LX, 428–430, 434
coupling of cytochrome *c*, LIII, 101
cyanogen bromide activation, XLIV, 28–30, 454, 529, 930; LI, 310
derivatization, with 5-fluoro-2′-deoxyuridylate, LI, 100
dGTP-substituted, preparation, LI, 252–256
molecular sieve, for lipid separation, XXXII, 50
phosphodiesterase, XXXVIII, 227–230, 234
preparation of polysomes, LIX, 365, 368
purification
aminoacyl-transfer ribonucleic acid synthetase, LIX, 263; LX, 620
of CTP(ATP):tRNA nucleotidyltransferase, LIX, 184
of coupling factor, LV, 189, 190
S100 supernatant, LX, 442, 443
translational control ribonucleic acid, LX, 543–545
as support, XLIV, 336, 337, 454, 457, 458, 491–494, 529, 530, 765, 846, 847, 930, 931
synthesis of succinyl-atractyloside-amino-Sepharose, LV, 525
Sepharose 6B
CAT-protein complex, LVI, 412
coupled with heparin, preparation of initiation factors, LX, 124, 125, 127, 135, 167, 170–180, 196, 468
hexylation, XLIV, 44
preparation of $TF_0 \cdot F_1$, LV, 369
protein kinase, XXXVIII, 304
purification
of ATPase, LV, 791–794, 804
Artemia salina cyst ribosomes, LX, 307, 308
of CPSase-ATCase complex, LI, 116
of dihydroorotate dehydrogenase, LI, 66, 67
elongation factor EF-1, LX, 639, 642, 643, 647, 648, 654, 656
elongation factor EF-2, LX, 652, 653, 656
heme-regulated translational inhibitor, LX, 471–473
of *Pholas* luciferase, LVII, 388
S1 depleted 30 S ribosomal subunits, LX, 428, 429
40 S ribosomal subunit-transfer ribonucleic acid complex, LX, 570
of TF_1 LV, 783
for separation of liposomes, XLIV, 224
Sepharose CL–4B, cross-linked, enzyme immobilization, XLVII, 41, 42
Sepharose CL–6B, purification of deoxynucleoside kinases, LI, 348
Sepharose-concanavalin A, in gonadotropin purification, XXXVII, 170, 171
Sepharose-cytochrome *c*, preparation, LV, 113, 114
Sepiapterin, substrate of phenylalanine hydroxylase, LIII, 285
Septamycin, LV, 446
code number, LV, 445
Septumless loop injector, LII, 282
serine-glyoxylate aminotransferase, in microbodies, LII, 496
Sequenator, *see also* Spinning-cup sequenator
adaptations for high-sensitivity, XLVII, 323, 324
design for automatic converter of thiazolinones
liquid-phase, XLVII, 386, 389–391
solid-phase, XLVII, 386–389, 391
dual-purpose, XLVII, 311–316
models available, micromethods, XLVII, 323
Sequence analysis, XLVII, 245–404
of amino-blocked peptides, XLVII, 306, 311

automated
using radiolabeled protein, XLVII, 249–251, 256–260
using synthetic carrier, XLVII, 248, 249, 251–255, 257, 258
Braunitzer reagents, XLVII, 260–263
carboxypeptidase digestion, XLVII, 81–83, 90–93
cleavage
agent, XLVII, 339–341
kinetics, XLVII, 341–345
conversion, XLVII, 347
coupling reaction, *see* Coupling
extraction, XLVII, 345–351
group specific modified peptides, XXXIV, 184
manual, XLVII, 335–357
instrumentation, XLVII, 351–355
procedure, XLVII, 355–356
micromethod
mechanical adaptations, XLVII, 323, 324
sequencing program, XLVII, 323–326
solid-phase, XLVII, 321–335
in spinning cup sequenators, XLVII, 247–260
with [^{35}S]PITC, XLVII, 323–325
polystyrene resins, XLVII, 306, 307
precautions, XLVII, 336, 337
of proteins eluted from SDS-gels, XLVII, 330
radioactive phenylisothiocyanate, XLVII, 247, 248, 254, 255
solid-phase, XLVII, 258–320
from carboxyl terminus, XLVII, 357–369
program, XLVII, 362, 363
degradative efficiency, XLVII, 274–276
new supports, XLVII, 263–277
polystyrene resin, XLVII, 306, 307
in spinning-cup sequenators, XLVII, 299–320
strategies, XLVII, 287
thermolysin, XLVII, 180–182
by thioacetylating reagents, XLVII, 290–297
washing step, XLVII, 338, 339
SE 52 reagent, XXXII, 360, 362
Serial sectioning, electron microscopy, LVI, 720, 721
Serine
in active site of carboxypeptidase, XLVII, 85, 89
aminoacylation isomeric specificity and, LIX, 274, 275, 279, 280
benzyloxycarbonylation, XLVII, 512
chemical properties, XLIV, 12–18
fluorescent labeling, XLVIII, 358
hydrophobicity, XLIV, 14
hydroxyl group blocking, XLVII, 524, 527, 528, 543, 544, 613
in media, LVIII, 63
in membranes, XXXII, 823
misactivation, LIX, 289
mobility reference, XLVII, 64
phenylthiohydantoin
loss, XLVII, 339, 340
recovery, XLVII, 347
phosphate, characterization, XLVII, 65
recovery, XLVII, 371
thermolysin hydrolysis, XLVII, 175
thiazolinone, half-life, XLVII, 376
Serine hydroxymethyltransferase, inhibition, XXXII, 576
Serine protease, XLVI, 197, 205–216, 220–225
active site, XLVI, 198
L-Serine *O*-sulfate, XLVI, 37
Serine transhydroxymethylase, *see* Serine hydroxymethyltransferase
Serotonin, XXXIV, 105
effect on melanin, XXXVII, 129
fluorometric assay, XXXIX, 384
gas chromatography-mass spectrometry, XXXIX, 386, 387
as melatonin precursor, XXXIX, 379
Serotonin-*N*-acetyltransferase
assay, XXXIX, 389–391
in melatonin biosynthesis, XXXIX, 379, 380
Serratia marcescens
6-aminopenicillanic acid assay, XLIII, 700
growth, LI, 136
oxidase, LII, 18
Sertoli cell, LVIII, 103
clone cultures, XXXIX, 294, 295

in vitro studies, XXXIX, 283
Serum, *see also* specific type
certification, LVIII, 197
charcoal treatment, XXXVI, 91–97
cloning, LVIII, 163
commercial production, LVIII, 197
deprivation, synchrony, LVIII, 259–261
dialyzed, LVIII, 82, *see also* Serum protein
effect on growth, LVIII, 77–79, 94–109
on macrophage, LVIII, 501
for granulosa cell culture, XXXIX, 206–209
human, XXXIX, 207
minimal requirements, LVIII, 91–93
neuronal cell, LVIII, 587
pretreatment, LVIII, 95
as trypsin inhibitor, LVIII, 95, 128
weaning, LVIII, 91–93
Serum albumin, *see also* specific types
human
aminoethylation, XLVII, 116
antibody binding, L, 55
disulfide cleavage, XLVII, 112
fluorescence quenching, XLIX, 224
molecular heterogeneity, XLIX, 186
osmotic pressure studies, XLVIII, 74
partial specific volume, XLVIII, 29
reduction, XLVII, 116, 120, 121
sulfopropylation, XLVII, 120, 121
immobilized, on hydroxyalkyl methacrylate gel derivative, XLIV, 81
for liver perfusate, XXXIX, 27
in media, LVIII, 88
in plant-cell fractionation, XXXI, 504, 505
reticulated on glass fiber, LVI, 425
synthesis, in hepatic explants, XXXIX, 39
tryptophan modification, XLVII, 440
Serum-free medium, *see* Serumless medium
Serum glycoprotein, XXXIV, 338
Serumless medium, LVIII, 57
composition, LVIII, 62–70
embryonal carcinoma cells, LVIII, 105
Serum protein, LVIII, 82
minimalization, LVIII, 49–51
Serum stopper, LIII, 364
Serum transferrin, human, reductive methylation, XLVII, 475
Servall Omnimixer, LIII, 233, 234
Seryl-tRNA synthetase, XXXIV, 167, 170
subcellular distribution, LIX, 233, 234
Sesamum indicum, lectin, XXXIV, 334
SE–Sephadex, *see* Sulfoethyl-Sephadex
Sesquiterpene lactone
allergenicity, XXXI, 531
as plant extraction interferents, XXXI, 531
Sevin, detection, XLIV, 658
Sex hormone binding globulin, in radioimmunoassay, XXXVI, 32
Sex steroid, effect, pituitary hormones, XXXVII, 91
Sex steroid binding globulin, in steroid hormone assay, XXXVI, 34, 37, 38
S30 extract, *see Escherichia coli,* S30 extract
S100 extract, *see Escherichia coli,* S100 extract
SF6847, *see* 3,5-Di-*tert*-butyl-4-hydroxybenzylidine-malononitrile
SFNMR, *see* Nuclear magnetic resonance, stopped-flow
Shadowing, high-resolution, LIX, 612
determination of metal layer thickness, LIX, 615
instrumentation, LIX, 613–615
procedure, LIX, 617
Shandon-Southern gel electrophoresis unit, LI, 342, 523
Shape analysis, using neutron-scattering data, LIX, 700–702
Shape scattering function, LIX, 677, 700, 710
Sharples Cream separator, LI, 136
SHBG, *see* Sex hormone binding globulin
Shea's lanthanum stain, XXXIX, 144
Shearing, LVIII, 223
effect on cell, LVIII, 224
Shear stress, viscosity determinations, XLVIII, 53–55

Sheep
heart, phosphofructokinase, purification, XLII, 107
liver
L-fucose dehydrogenase, assay, purification, and properties, XLI, 173–177
phosphofructokinase, purification, XLII, 107, 110
6-phospho-D-gluconate dehydrogenase, assay, purification, and properties, XLI, 214–220
muscle
3-phosphoglycerate kinase, purification, XLII, 129
phosphoglycerate mutase, purification and properties, XLII, 438, 446–449
Shift probe, principle, LIV, 194
Shigella flexneri, resistance plasmids, XLIII, 45
Shockman shaker, XXXI, 658
properties, XXXI, 662
Shot noise, XLVIII, 423, 424, 461
definition, LIV, 33, 34
determination, XLVIII, 481, 482
in nanosecond absorbance spectroscopy, LIV, 37, 40
Shotten-Baumann reaction, XXXVI, 22
Showdomycin
biosynthesis, XLIII, 409
effect on guanosine triphosphate-8-formylhydrolase, XLIII, 517, 519
Shutter
electronic, LVII, 319
in flash photolysis system, LIV, 98
Sialic acid, L, 94
O-acyl group determination, L, 76, 77
biosynthesis, L, 374–386
characterization, L, 64–89
^{14}C–labeling of animal cells, L, 185
colorimetric assay, L, 70–75
diphenol assay, L, 70–72
enzymic assay, L, 75, 76
in erythrocytes, effect on phospholipase activity, XXXII, 137
fluorometric assay, L, 75
gas-liquid chromatography, L, 83–86
methanolysis, L, 83, 84
trimethylsilylation, L, 83, 84
glycoconjugates, L, 66
in glycolipids, XXXII, 360, 361
glycolyl group determination, L, 77, 78
in glycoprotein hormones, XXXVII, 321, 322
hydrolysis, XXXIV, 689
isolation and quantitation, L, 237
O-lactyl group determination, L, 78
mass spectrometry, L, 86–89
methanolysis, L, 68
nuclear magnetic resonance spectroscopy, L, 89
periodate oxidation, L, 73, 82, 83
in plasma membranes, XXXI, 85, 87, 101, 165–167
of erythrocytes, XXXI, 176, 177, 179
in progesterone-binding protein, XXXVI, 124
sources, L, 65–68
thin layer chromatography, L, 78–81
thiobarbituric acid assay, L, 72–75
tritium labeling, XXXVII, 147; L, 83
Sialidase reagent, XXXI, 176, 177
Sialoglycoconjugate
acid hydrolysis, L, 67
enzymic hydrolysis, L, 67, 68
Sialosylglycolipid, stain, XXXII, 357
Sialuria, L, 66
Sialylfucosyllacto-*N*-hexaose I, structure, L, 217
6′-Sialyllactosamine, human urine, L, 228
Sialyllactose, *p*-isothiocyanate-phenethylamine derivatives, L, 170, 171
3′-Sialyllactose
human urine, L, 228
structure, L, 225
6′-Sialyllactose
human urine, L, 228
structure, L, 225
Sialyllactose-bovine serum albumin, preparation, L, 170, 171
Sialyl oligosaccharide
from human milk
fractionation, L, 221–226
structures, L, 224, 225

protein conjugation, L, 169–171
Sialyltransferase
in Golgi apparatus, XXXI, 191
particulate *N*-acetylneuraminate monooxygenase, L, 380
in plasma membrane, XXXI, 102
Siamensis α-neurotoxin
chromatography, XXXII, 321
purification and properties, XXXII, 313, 314
receptors, XXXII, 314, 316, 319
Siccanin, XLIII, 185
Sickle-cell anemia, erythrocyte shape, XXXII, 52
Siderochrome, Mössbauer spectroscopy, LIV, 358
Sideromycin, XLIII, 146–149
Sieve filtration, transport studies, LVI, 286, 287
Sieving, dissociation procedure, of neurobiological cell fractionation, XXXII, 772–774
Sigma factor, XLVI, 352
Signal positioning, effect, LIV, 136
Signal-to-noise ratio
acquisition time, LIV, 158
bandwidth matching, LIV, 43
effect of photodiodes, LIV, 98
exponential filtering, LIV, 160, 161
number of measurements, LIV, 1
optical density, LIV, 37, 38
pulse recycle time, LIV, 166
quadrature detection, LIV, 160
time-averaging, LIV, 163, 171
for transfer NMR spectra, LIX, 25
Silane, *see also* specific compound
commercial source, XLIV, 139
coupled to glass, XXXIV, 63
coupling techniques, XLIV, 139, 140
removal, procedure, XLVII, 310
Silanization
aqueous, XXXIV, 64; XLIV, 139
of glass matrix, XXXIV, 63–72, *see also* Glass, silanization
organic, XXXIV, 64; XLIV, 140
Silanol group, ionic bonding, XLIV, 151
Silastic, XLIV, 210
medical adhesive, XXXIX, 361
tubing, XXXII, 742, 818
Silica, *see* Glass
Silica alumina pellet, porous support, for glucose oxidase/catalase, XLIV, 480
Silica gel
aureovertin chromatography, LV, 479, 480
4-azido-2-nitrophenylaminobutyryl-atractyloside, LV, 527
citreoviridin analysis, LV, 486, 487
drying agent, in simple photometer, LVII, 533, 534
efrapeptin purification, LV, 494
in electrophoretic separation of modified nucleotides, LIX, 181
proteolipid, LV, 417, 418
purification of 4-*N*-[3-chloro-2-hydroxypropyl)-*N*-ethyl]amino-*N*-methylphthalimide, LVII, 431
of dimethyl 7-*N*-(4-phthalimidobutyl)aminonaphthalene-1,2-dicarboxylate, LVII, 436
of *Diplocardia* luciferin, LVII, 377
thin layer chromatography plate
cGMP separation, XXXVIII, 77, 78, 91
Vaseline petroleum jelly impregnated, LIII, 639
venturicidin analysis, LV, 505
Silica gel 60
purification of 7-[*N*-(4-aminobutyl)-*N*-ethyl]aminonaphthalene-1,2-dicarboxylic acid hydrazide, LVII, 437
of dimethyl 7[*N*-ethyl 4-[*N*-phthalimidobutyl]aminonaphthalene-1,2-dicarboxylate, LVII, 437
of 6-[*N*-ethyl-*N*-(4-thyroxinylamido)butyl]amino-2,3-dihydrophthalazine-1,4-dione, LVII, 434
of 7-[*N*-ethyl-*N*-(4-thyroxinylamido)butyl]amino-naphthalene-1,2-dicarboxylic acid hydrazide, LVII, 437
Silica gel chromatography, XLIII, 291–296
adsorbent, XLIII, 291, 292
column, XLVII, 47, 50
eluotropic series, XLIII, 292, 293
elution column, XLIII, 295, 296
monitoring column, XLIII, 296

preparation of column, XLIII, 293–295
purification of coenzyme Q analogs, LIII, 595
of cytochrome c_3, LIII, 622
of desulfovibrione electron carriers, LIII, 633
of high molecular weight cytochrome c_3, LIII, 625
of rubredoxin, LIII, 626
sample application, XLIII, 295
solvent system, XLIII, 292–294
washing procedure, LIII, 615
Silica gel F_{254} plate, separation of quinone intermediates, LIII, 605–607, 609
Silica gel G plate, separation of quinone intermediates, LIII, 603, 605
Silica gel slide, preparation, XLVII, 348
Silica GF, bongkrekic acid, LV, 529
Silica sol, gradient properties, XXXI, 508, 509
Silicic acid, XLIV, 151
chromatographic separation of fatty acids, LII, 321
volatile, XXXV, 73
piericidin A purification, LV, 458
purification of ubiquinone, LIII, 639
Siliclad, XLVIII, 487; LI, 576
glassware treatment, XXXII, 688
Silicon
infrared spectral properties, LIV, 316
in media, LVIII, 69
Silicon-29, lock nuclei, XLIX, 348; LIV, 189
Silicon-58, from ^{28}Al, XXXII, 884
Silicon dioxide, pore properties, XLIV, 137
Silicone
glassware coating, XXXII, 688
tubing, XXXII, 742
Silicone Antifoam AF, Tween-containing cultures, LVI, 575
Silicone membrane, XLIV, 178
Silicone oil
cell fractionation, LVI, 209, 210, 216, 217
separation of cells from medium, LV, 203, 205, 552
of cytosol, LV, 243
of mitochondria, LV, 204, 228, 229, 239
transport studies, LVI, 259, 261, 283–285
Silicone rubber
microencapsulation, XLIV, 210
pad, preparation, XLIV, 619–621
steroid adsorption, XXXVI, 93, 94, 161
Silicon photodiode, LIV, 40
Silicon sheet, XLIV, 271
Silicon vidicon tube, LIV, 11, 25
advantages, LIV, 17
array, as multichannel emission detector, LIV, 461
Silicotungstic acid
particle, LV, 383
preparation, LV, 398
Silk, XLIV, 271
Silkworm, media, LVIII, 457
Silver
atomic emission studies, LIV, 455
in carbon coating, XXXIX, 155
effect on aequorin, LVII, 314, 315, 325
as reference electrode, LVI, 456
Silver(II), in porphyrin ring conformation studies, LIV, 245
Silver bromide, infrared spectral properties, LIV, 316
Silver chloride, infrared spectral properties, LIV, 316, 317
Silver grain yield, in steroid autoradiography, XXXVI, 151
Silver nitrate-sodium iodide, cAMP isolation, XXXVIII, 40
Silver/silver chloride electrode, LIV, 422
standard potentials, LIV, 423
Silylation
reagents, LII, 332
of steroid hydroxylates, LII, 380, 381
Simian virus 40
biological activity, LVIII, 412
cell lines for study, XXXII, 584, 585
cell transformation, XXXII, 586–590
lectin binding, XXXII, 623
cleavage maps, LVIII, 412
coat protein, antibody binding, L, 55, 56
culture, LVIII, 405

DNA, LVIII, 405
assay, LVIII, 411, 412
preparation, LVIII, 406–410
purified virions, LVIII, 409, 410
quantitation, LVIII, 411, 412
radiolabeling, LVIII, 410, 411
MEM, LVIII, 405, 406
plasma membrane isolation, XXXI, 161
preparation, LVIII, 404–412
T-protein, XXXIV, 731
transformation, LVIII, 370
Simon ligand
ionophore, neutral, LV, 443
structure, LV, 441
Sinclair's medium, LVIII, 58
Sindbis virus, microcarrier culture, LVIII, 192
Single-photon counting technique, LIV, 17
Single-step perturbation experiment, LIV, 66
Single-step reaction, evaluation of reciprocal relaxation time, LIV, 72–73
Singlet oxygen, *see* Oxygen, singlet
Singlet-singlet energy transfer, XLVIII, 347–379
Singlet state, time range, LIV, 2
Sintered-glass funnel, adapted, purification of cytochrome *c*, LIII, 129
Sinusoidal lining cell, *see* Macrophage
Sips equation, binding affinities, XLVIII, 386
Sips plot, XXXVI, 10–12, 14, 15, *see also* Hill plot
Sipunculans, oxidase, LII, 13
Siroheme, LII, 436–447
absorption spectrum, LII, 440, 441
CO complex, absorption spectrum, LII, 442, 443
demetallation, LII, 443
of *E. coli*, LII, 437, 439–441, 443
of *Neurospora*, LII, 440, 443
solubility in apolar organic solvents, LII, 437
structure, LII, 436–439
sulfite reductase, LVI, 560
thin-layer chromatography, LII, 442
Sirohydrochlorin
esterification, LII, 444, 445
fluorescence spectra, LII, 443, 444
preparation, LII, 443
Sirohydrochlorin octaethyl ester
spectral properties, LII, 446, 447
thin-layer chromatography, LII, 445, 446
Sisomicin, XLIII, 623
Site heterogeneity, qualitative effects, XLVIII, 276–278
SIT vidicon, *see* Silicon vidicon tube
Sium, cell cultures, XXXII, 729
Size distribution, in morphometry, XXXII, 13, 14
Skatole, chemiluminescence, LVII, 348
SKE buffer, formulation, LIII, 114
Skeletal muscle
dispersion, LVIII, 123
incubation, XXXIX, 92–94
perfusion methods, XXXIX, 65–106
rabbit muscle, XXXIX, 67–73
in situ, XXXIX, 67–73
3-phosphoglycerate kinase, assay, purification, and properties, XLII, 127–134
D-ribose-5-phosphate isomerase, assay, purification, and properties, XLI, 424–426
sarcoplasmic reticulum isolation, XXXI, 238–246
triosephosphate isomerase from human, XLI, 44
in vitro preparations, XXXIX, 82–94
Skeletal myoblast, *see* Myoblast, skeletal
Skellysolve B-chloroform, chromatographic purification of vitamin D metabolites, LII, 394
extraction, LII, 439–441
Skellysolve B-chloroform-methanol, chromatographic purification of vitamin D metabolites, LII, 392, 394, 395
SKF, *see* 2-Diethylaminoethyl-2,2-diphenylvalerate
SKF–525A
effect on hepatocyte metabolism, LII, 65
inhibitor of cytochrome *P*–450, LII, 96, 276, 277, 279

metabolite binding to reduced cytochrome, LII, 274
type of binding reaction with cytochrome *P*–450, LII, 264
Skin
biopsy
explants, LVIII, 446–448
method, LVIII, 445, 446
epithelium, differentiation, LVIII, 265
explant, LVIII, 446–448
of frog, in melatonin assay, XXXIX, 384, 385
human, effect on steroid binding studies, XXXVI, 96, 97
melatonin effects, XXXIX, 378
mouse, procollagenase, XL, 352
oxidase, LII, 12, 13
preparation for amino acid analysis, XL, 360
for collagen studies, XL, 311
for stress-strain testing, XL, 364
sesquiterpene lactone combination with proteins, XXXI, 531
Skull, in rat brain surgery, XXXIX, 169–175
Slab gel electrophoresis, *see also* Electrophoresis
in analysis of ribosomal RNAs, LIX, 458
apparatus, LIX, 65
radioactivity determination, LIX, 521
in second-dimensional separation of ribosomal protein, LIX, 506, 507
for separation of ribosomal proteins, LIX, 431
Slice incubation, of corpus luteum, XXXIX, 238–244
Slicing, for plant-cell rupture, XXXI, 502
Slide preparation, chromosomes, LVIII, 326–334
Slime mold, D(–)-lactate dehydrogenase, XLI, 294
Slit foci, LIV, 37, 38
S^+L^- test, LVIII, 423, 424
SMD medium, XXXIX, 205, 209, 211
Smear-mounting, in steroid autoradiography, XXXVI, 146, 147
Smectic mesophase, *see* Liposome
S medium, for granulosa cell culture, XXXIX, 205
S77 medium, LVIII, 546
SMI pipette, LVII, 221
Smith degradation
complex carbohydrates, L, 24–28
for glycolipid structure studies, XXXII, 363
interresidue hemiacetals, L, 25
polysaccharide cleavage, L, 255
SM–20 medium, LVIII, 59, 90
SM–201 medium, LVIII, 59
SMP, *see* Submitochondrial particle
SN–5949, *see* 2-Alkyl-3-hydroxyl-1,4-naphthoquinone
S100N, isoparaffin emulsifier, XLIV, 211, 328, 330
Snail
albumin gland inhibitor, *see* Albumin gland inhibitor, snail
cell, media, LVIII, 466
gut juice
in cell-wall hydrolysis, XXXI, 610, 611, 630, 631
preparation, XXXI, 612
recycling, XXXI, 632
in spheroplast isolation, XXXI, 614–622
yeast mitochondria preparation, LV, 158, 159
trypsin-kallikrein inhibitors, *see* Trypsin-kallikrein inhibitor, snail
Snake
toxins, preparation and properties, XXXII, 309–323
venom, *see also* specific type
5′-nucleotidase, XXXII, 368
oxidase, LII, 18
phosphodiesterase, *see* Phosphodiesterase I
phospholipase A_2, in venom, XXXII, 147, 148
SN–5949 inhibitor, complex III, LVI, 586
Snowberry fruit, free cells, XXXII, 724
Soap film, lipid bilayers, XXXII, 514
Sociocell culture, LVIII, 275
SOD, *see* Superoxide dismutase
Soda glass bead, bridge formation, XLIV, 166
Soda-lime glass, calcium concentration, LVII, 282
Sodamide, XLVII, 572

Sodium
effect on hypophysectomized rats, XXXIX, 315–317
in liquid ammonia, deprotection, XLVII, 572–575
in media, LVIII, 68
transport
ADH effects, XXXVII, 256
parathyroid hormone effects, XXXIX, 20
Sodium-22
in membrane studies, XXXII, 884–893
properties, XXXII, 884
Sodium-23
lock nuclei, XLIX, 348; LIV, 189
resonance frequency, LIV, 189
Sodium acetate, XLIV, 93, 749, 763; XLVII, 138
activator of pyrimidine nucleoside monophosphate kinase, LI, 331
assay of acid nucleotidase, LI, 272
of amidophosphoribosyltransferase, LI, 172
of quinones, LIII, 586, 587
of ribonucleoside triphosphate reductase, LI, 247, 248
in copper analysis, LIV, 444
coupling of cytochrome *c* to Sepharose 4B, LIII, 101
cyclic nucleotide separation, XXXVIII, 32
determination of *in vivo* levels of aminocylated tRNA, LIX, 271
extraction of nucleic acids, LIX, 311
hydrolysis of tRNA adducts, LIX, 164
in iron analysis, LIV, 439
modification reaction of tRNA, LIX, 160, 179
periodate oxidation of tRNA, LIX, 186, 269, 270
preparation of cellulose-bound dithiodiglycolic dihydrazide, LIX, 386
of initiator methionine tRNA, LIX, 207
modified tRNA, LIX, 168, 170, 171, 179, 180
ribosomal proteins, LIX, 484, 501, 518, 519, 557, 559
purification of nucleoside phosphotransferase, LI, 388, 389, 390
procedure, LIV, 437
of shortened tRNAs, LIX, 186
tRNA base replacement reaction, LIX, 120
RNA staining solution, LIX, 557, 567
separation of aminoacylated tRNA, LIX, 216
of tRNA isoacceptors, LIX, 222
termination of enzyme reaction, LIX, 221
T2 ribonuclease digestion, LIX, 188
Sodium ampicillin, XLIV, 592, 593
Sodium anthraquinone β-sulfonate, in Fieser's solution, LIII, 299
Sodium arsenate
assay of succino-AICAR synthetase, LI, 187
of thymidine phosphorylase, LI, 438
Sodium ascorbate
in assay of cytochrome *c*, LIV, 493
of formate dehydrogenase, LIII, 364
reductant, of cytochrome *c*, LIII, 45
Sodium aurous cyanide, derivatization of transfer RNA, LIX, 14
Sodium azide, XLVII, 107
assay of complex I, LIII, 13
of complex III, LIII, 39
of NADH-cytochrome *c* reductase, LIII, 9, 10
effect on hydrogen peroxide assay, LII, 344, 349, 350
inhibitor of brain α-hydroxylase, LII, 317
of catalase, LII, 344, 356, 357
of chloroperoxidase, LII, 529
of mitochondrial TMPD oxidase activity, LII, 411
preparation of chloroplastic tRNA, LIX, 210, 211
of mitochondrial tRNA, LIX, 207
preservative, XLIV, 269
purification of chloroplastic phenylalanine tRNA, LIX, 213
of ribonucleoside triphosphate reductase, LI, 251, 252

resistance as genetic marker, XLIII, 44
for storage of adenylate deaminase, LI, 499
of affinity gel, LI, 367, 493
of Sephadex column, LIX, 312
in storage solutions, LIV, 479
Sodium bathophenanthrolinesulfonate, effect on ribonucleoside diphosphate reductase activity, LI, 244
Sodium benzoate, inhibitor of NADPH-cytochrome *P*-450 reductase, LII, 96
Sodium bicarbonate
in (±)-L-methylenetetrahydrofolate preparation, LI, 91
in perfusion fluid, LII, 51
preparation of dialysis tubing, LVII, 10
of matrix-bound RNase, LIX, 463
synthesis of adenosine derivatives, LVII, 118
of dithioglycolic dihydrazide, LIX, 385
of ethyl-4-iodo-2-diazoacetoacetate, LIX, 811
Sodium bicarbonate-sodium hydrosulfite, assay of nitrate reductase, LIII, 348, 350
Sodium biphthalate, optical filter, XLIX, 192
Sodium bisulfite, XLIV, 256, 257, 473
in modification of tRNA, LIX, 147
Sodium borate
coupling of cytochrome *c* to Sepharose 4B, LIII, 101
preparation of radiolabeled ribosomal protein, LIX, 777, 783
substrate, catalase, XLIV, 685
Sodium borohydride, XLIV, 23, 44, 212; XLVII, 42
determination, XLVII, 472–473
in dihydrouracil substitution reaction, LIX, 112, 117, 119
labeling
elongation factor EF-G, LX, 721, 722
initiation factors, LX, 34, 35, 55, 67, 122
lysyl residue modification, XLVII, 493
preparation of deoxyhemoglobin, LII, 458
protein labeling, XLVIII, 332
radiolabeled
inhibition studies of tryptophanyl-tRNA borohydride, LIX, 254
preparation of radiolabeled ribosomal proteins, LIX, 586, 783, 786, 787, 789, 790, 791
quantitative determination of tRNA substitution, LIX, 120, 121
source, LIX, 236
reductive alkylation, XLVII, 469–478
removal, XLVII, 478, 491
steroid reduction, LII, 379
Sodium borosilicate glass, XXXIV, 59, 60
Sodium borotritide, XLVII, 491
Sodium bromide
OSCP preparation, LV, 394
preparation of membrane factor, LV, 744
Sodium 2-bromoethanesulfonate, reaction conditions, XLVII, 410
Sodium cacodylate, in tritium labeling studies, LIX, 342
Sodium carbonate
analysis of mitochondrial ribosomal proteins, LIX, 429
aniline hydroxylase assay, LII, 409
assay of adenosine monophosphate nucleosidase, LI, 264
of carbamoyl-phosphate synthetase, LI, 122
preparation of luciferin, LVII, 21
stock solution of chemiluminescent compounds, LVII, 425, 427
synthesis of chemiluminescent aminophthalhydrazides, LVII, 431, 436
Sodium carbonate-barium chloride, cAMP isolation, XXXVIII, 40
Sodium carbonate-cadmium chloride, cAMP isolation, XXXVIII, 40
Sodium carbonate-calcium chloride, cAMP isolation, XXXVIII, 40, 41
Sodium carboxymethylcellulose, osmotic pressure studies, XLVIII, 74
Sodium chloride
activator of pyrimidine nucleoside monophosphate kinase, LI, 331

assay of PAP, LVII, 253
of tRNA methyltransferase, LIX, 191
of tryptophanyl-tRNA synthetase, LIX, 237
in bacterial luciferase reagent, LVII, 200
benzoquinone activation, XLIV, 35
carbohydrazide modification of tRNA, LIX, 147
concentration, effect on bioluminescence, LVII, 166
crystallization of firefly luciferase, LVII, 10
of transfer RNA, LIX, 6, 8
determination of *in vivo* levels of aminoacylated tRNA, LIX, 271
in elution buffer, LI, 210
extraction of radiolabeled RNA, LIX, 836
fluorescein labeling of tRNA, LIX, 149
gradient elution, LX, 420, 501, 502, 528, 529, 537, 634, 784
hollow-fiber dialytic permeability, XLIV, 297
isolation of translating ribosomes, LIX, 388
as lock nuclei source, LIV, 189
minimum requirement, of luminous bacteria, LVII, 223
osmometric studies, XLVIII, 73, 74, 78
osmotic lysis of bacterial cells, LVII, 145
in oxygen determination, LIV, 501, 503
periodate oxidation of tRNA, LIX, 269, 270
preparation of *Bacillus* crude extract, LIX, 438
of chloroplastic tRNA, LIX, 210, 211
of hemoglobin, LIV, 488
of luciferyl adenylate, LVII, 27
of matrix-bound RNase, LIX, 463
of mitochondrial membranes, LIII, 102, 103
of mitochondrial tRNA, LIX, 206, 207
of modified tRNA, LIX, 170, 179, 180
of MS2 RNA, LIX, 299, 300
of pig liver mitochondria, LIII, 509
of rat liver tRNA, LIX, 230
of ribosomal proteins, LIX, 518, 559, 649
of ribosomal RNA, LIX, 452, 556, 648
of synthetic nucleotide polymers, LIX, 826
protein solution, LIX, 323, 330
purification of adenosine deaminase, LI, 504, 509, 510, 511
of aequorin, LVII, 281, 309, 310
of complex III, LIII, 103
of *Cypridina* luciferase, LVII, 370
of cytochrome *c*, LIII, 129, 130, 224
of cytochrome oxidase, LIII, 62
of *Diplocardia* luciferin, LVII, 376
of luciferin sulfokinase, LVII, 249
of modified tRNA, LIX, 188
of nitrate reductase, LIII, 349
of *Pholas* luciferase, LVII, 387
of *Pholas* luciferin, LVII, 388, 390
of reconstituted tRNA, LIX, 134
of $tRNA^{Phe}$, LIX, 213
of shortened tRNAs, LIX, 186
reverse salt gradient chromatography, LIX, 216
RNA elution, LIX, 69
separation of aminoacylated tRNA, LIX, 216
of tRNA isoacceptors, LIX, 221
storage of activated gels in, XLIV, 33
synthesis of *N*-trifluoroacetylthyroxine, LVII, 432
transfer RNA NMR spectra, LIX, 25
Sodium chloride complete medium, for growth of luminescent bacteria, LVII, 141
Sodium chloride-sodium citrate-EDTA buffer, LIX, 450, 648
Sodium cholate, *see also* Potassium cholate
chromatographic purification, 223n
cytochrome *P*–450 purification, LII, 127, 139, 140
mitochondrial disruption, LII, 139

preparation of cytochrome oxidase, LIII, 58, 59, 60, 63
protein solubilization, LII, 111, 115, 118–120, 122, 128, 129, 195
purification of cytochrome a_1, LIII, 640
of cytochrome c_1, LIII, 184, 186, 222, 223
Sodium citrate, XLIV, 92, 387, 476
determination of [^{3}H]puromycin specific activity, LIX, 360
electrophoretic separation of modified nucleotides, LIX, 181
preparation of total ribosomal RNA, LIX, 452, 648
tRNA digestion mixture, LIX, 107
RNA elution, LIX, 69
Sodium cyanide
inhibitor of uridine nucleosidase, LI, 294
preparation of cytochrome *c* derivative, LIII, 150
Sodium cyanoborohydride, XLVII, 478
Sodium deoxycholate, *see also* Potassium deoxycholate
cytochrome *P*–450 isolation, LII, 363, 364
cytochrome *P*–450LM$_2$ assay, LII, 205
in detergent mixtures, XXXI, 559
in gel electrophoresis, XXXII, 83, 85
isolation of chloroplastic ribosomes, LIX, 436
for lysis of Ehrlich ascites cells, LIX, 412
microsomal protein solubilization, LII, 82, 83, 191, 202, 360
for microsome isolation, XXXI, 216
preparation, LII, 189
of ribosomal subunits, LIX, 403
protein determination, LIII, 497
purification
of cytochrome b_5, LII, 99
of cytochrome c_1, LIII, 223
of hydrogenase, LIII, 311
of nitrate reductase, LIII, 642
of reductase, LII, 94
removal, by gel filtration, LII, 210
stearyl-CoA desaturase activity assay, LII, 189
Sodium dextran sulfate 500, preparation of coliphage R17 RNA, LIX, 364
Sodium dithionite, XLIV, 142, 862; XLVIII, 334, *see also* Dithionite
alkali removal, from gases, LII, 222
assay of hydrogenase, LIII, 301
commercial source, LII, 222
cytochrome b_5 assay, LII, 110
cytochrome *P*–450 assay, LII, 127, 138
for deoxygenation of oxyhemoglobin, LIV, 489, 491
difference spectra measurements, LII, 215, 216
enzyme reduction, LII, 152
in Fieser's solution, LIII, 299
hemin reduction, LII, 433
preferred reductant, in redox potentiometry, LIV, 418, 419
preparation of deoxyhemoglobin, LII, 458
of reduced ubiquinone-2, LIII, 39
purification of cytochrome c_1, LIII, 185
of high-potential iron-sulfur protein, LIII, 337, 338
of nitrogenase, LIII, 315, 316, 320–323, 326
as redox titrant, LIV, 409
reduction of *Cypridina* luciferin, LVII, 345
of cytochrome *c*-551, LIII, 658, 661
of flavin, LVII, 139
of ferricytochrome *c*, LIII, 383
of FMN, LIII, 565, 566
of *p*-hydroxybenzoate hydroxylase, LIII, 549
solid, diluted, preparation, LIV, 130
solution preparation, LIV, 126
solution stability, LIV, 127–419; LVII, 139
test for hemes, LII, 439
Sodium dodecyl sulfate, XXXIV, 656; XLIV, 370; LIX, 446, *see also* Polyacrylamide gel electrophoresis
analysis of mitochondrial ribosomal proteins, LIX, 429
ATPase, XXXII, 281, 282; XXXIV, 570
in circular dichroism studies, XXXII, 231, 232
as curing agent, XLIII, 51

in discontinuous gel electrophoresis of membrane proteins, XXXII, 92
for dissociation of uridine phosphorylase, LI, 429
electron spin resonance spectra, XLIX, 403, 404
in electrophoresis buffer, LII, 325, 330
electrophoretic analysis of *E. coli* ribosomal proteins, LIX, 496–500, 507, 509, 510, 538, 539
elution of protein from gels, LIX, 544, 545
extraction of nucleic acids, LIX, 311
of radiolabeled RNA, LIX, 831, 835
in gel electrophoresis, XXXII, 71, 76, 77, 79, 86
of enzymes, XXXII, 82, 83, 85, 86
inhibitor of adenylate kinase, LI, 465
of deoxythymidine kinase, LI, 365
of dimethylaniline monooxygenase, LII, 149
of glutathione peroxidase, LII, 510
micelles, ionic strength, LVI, 738
for microsome isolation, XXXI, 216, 220–223
partial specific volume, LVI, 737
peptide solubilization, XLVII, 332
preferential solvation, XLVIII, 175, 176
preparation of chloroplastic tRNA, LIX, 210
of mitochondrial tRNA, LIX, 206
of rat liver tRNA, LIX, 230
of ribosomal fragment RNA, LIX, 470
of ribosomal RNA, LIX, 445, 446, 452, 556, 557
protein denaturation, XLVIII, 5–7
for protein solubilization, XXXII, 71, 72
purification of complex III peptides, LIII, 84
of cytochrome oxidase subunits, LIII, 79
of peptide constituents of cytochrome bc_1 complex, LIII, 106–108
in reductase molecular weight determination, LII, 95
removal, LIX, 545
staphylococcal protease hydrolysis, XLVII, 191
structure, LVI, 735
temperature-concentration phase diagram, LVI, 739
for termination of enzyme reaction, LIX, 132, 833
use of urea, disadvantages, XLVIII, 7
Sodium dodecyl sulfate-diethylpyrocarbonate, extraction of RNA, LIX, 427, 428
Sodium dodecyl sulfate-phosphate gel, XL, 156
Sodium dodecyl sulfate-polyacrylamide gel electrophoresis, XL, 133, 146
of acidic chromatin protein, XL, 173
of chromatin nonhistone proteins, XL, 165
disrupted antigen-antibody complex, XL, 248
identification of specific glycoproteins and antigens, L, 54–64
of nonhistone chromosomal proteins, XL, 146
controls and reproducibility, XL, 148
of glycoproteins, XXXVI, 97, 98
reagents, XL, 172
two-dimensional, XL, 168
Sodium dodecyl sulfate-tris-glycine gel, XL, 156
Sodium fluorescein, XLIX, 210, 212
Sodium fluorescein laser, wavelength, XLIX, 85
Sodium fluoride
assay of glucose oxidase, LVII, 403
of phosphoribosylpyrophosphate synthetase, LI, 3
of thymidine kinase, LI, 355, 365
effect on nucleoside phosphotransferase, LI, 393
inhibitor of phosphoribosylglycinamide synthetase, LI, 185
of pyrimidine nucleoside monophosphate kinase, LI, 331
protein inhibitor, XL, 289
Sodium fluoride-hydrochloric acid, in preparation of apomyoglobin, LII, 478

Sodium formate
anaerobic TEA maleate buffer, LIII, 366
analysis of tRNA hydrazide derivatives, LIX, 180
assay of formate dehydrogenase, LIII, 364
of nucleoside diphosphokinase, LI, 373
chromatographic separation of nucleotides, LI, 373
modified, LIX, 181
cyclic nucleotide separation, XXXVIII, 32
Sepharose derivatization, LI, 32
Sodium glutamate, XLIV, 189
Sodium glycerol 2-phosphate, XLVII, 490
Sodium hydrogen carbonate, XLIV, 99
Sodium hydrogen sulfite, assay of deoxythymidylate phosphohydrolase, LI, 286
Sodium hydrosulfite, XLIV, 210, *see also* Sodium dithionite
in cytochrome b_5 assay, LII, 207
for heme reduction, LII, 426
isolation of luciferase reaction intermediate, LVII, 196
for peroxide removal, LII, 98
Sodium hydroxide
in acid nucleotidase assay, LI, 272
in ammonia determination, LI, 498
anilinothiazolinone hydrolysis, XLVII, 371, 372
in benzo[*a*]pyrene recrystallization, LII, 237
chemical cleavage of tRNA, LIX, 102
determination of acid-labile sulfide, LIII, 276
for extraction of phenolic metabolites, LII, 237
metal-free, LIV, 477
for oxygen removal from gases, LII, 221, 222
preparation of denatured DNA, LIX, 840
of water-soluble luminol salt, LVII, 477
purification of D-lactate dehydrogenase, LIII, 522
solvent, of 4-methoxythiooxanilamide, LVII, 19
washing of DEAE-cellulose, LIX, 194
Sodium hydroxide-ethanol, in benzo[*a*]-pyrene monooxygenase assay, LII, 414
Sodium *p*-hydroxybenzoate, assay of *p*-hydroxybenzoate hydroxylase, LIII, 545
Sodium hypochloride, in ammonia determination, LI, 498
Sodium hypochlorite
chemiluminescence assay, LVII, 427
in cleaning solution, XLIV, 310
in monochlorodimedone synthesis, LII, 523
Sodium iodide, XLIV, 749
Sodium iodoacetate, XLVII, 115, 116, 126
Sodium ion
activator of adenylate deaminase, LI, 501
of guanylate kinase, LI, 489
of ribonucleoside triphosphate reductase, LI, 258
amino acid transport, LV, 187, 197, 199
calcium efflux, LVI, 349, 350
calcium-selective electrodes, LVI, 348
concentration required for optimal growth of luminescent bacteria, LVII, 154
constant molarity, XLVII, 25, 27, 29
effect on *Cypridina* luciferase, LVII, 351
high concentration, effect on resin bed, XLVII, 11
inhibitor of adenine phosphoribosyltransferase, LI, 573
of luciferase reaction, LVII, 81
measurement of proton translocation, LV, 636
ribosome dissociation, LIX, 367
role in *Renilla* bioluminescence, LVII, 244
Sodium ion pump, reconstitution, LV, 701, 702, 704–707
activity, LV, 703, 705, 708, 709
Sodium isocitrate, in NADPH-generating system, LII, 274, 359, 378, 380

Sodium lactate, variability in commercial preparations, effects on bacterial growth, LIII, 614
Sodium laurate, ω and ω-1 hydroxylation procedure, LII, 320
Sodium lauryl sarcosinate, detergent, XXXII, 374
Sodium lauryl sulfate, XLIV, 290
  inhibitor of glutathione peroxidase, LII, 510
Sodium malonate, in acid nucleotidase assay, LI, 272
Sodium mersalyl, purification of complex III peptides, LIII, 83
Sodium metabisulfite, LIX, 147, 157
Sodium metaborate, XLIV, 212
Sodium metaperiodate, XLIV, 40
Sodium methanethiolsulfonate, synthesis, XLVII, 428
Sodium methoxide, XLVII, 599
  carboxyl group determination, XLIV, 131, 384
Sodium 2-(*N*-Morpholino)ethanesulfonate buffer, *see* 2-(*N*-Morpholino)ethanesulfonate sodium hydroxide buffer
Sodium nitrate
  as fusing agent, LVIII, 361, 362
  optical filter, XLIX, 192
Sodium nitrite, XLIV, 93, 95, 98, 103, 115–117, 133, 134, 143, 256, 311, 445, 446, 476, 933; XLVII, 566, 567
  assay of FGAM synthetase, LI, 195
    of nitrite reductase, LIII, 643
    of phosphoribosylglycinamide synthetase, LI, 180
  determination, XLVII, 138
    of tyrosine, LIII, 279
  preparation of immobilized enzyme rods, LVII, 204
  preservative, XLIV, 40
Sodium nitroprusside
  in ammonia determination, LI, 498
  inhibitor of hydrogenase, LIII, 314
Sodium oxamate, XLVII, 441
Sodium perborate
  assay by potassium permanganate titration, XLIV, 685
    by titanium sulfate colorimetric titration, XLIV, 685
Sodium perchlorate
  inhibitor of pyrimidine nucleoside monophosphate kinase, LI, 331
  purification of NADH dehydrogenase, LIII, 15
    of succinate dehydrogenase, LIII, 28
Sodium periodate, XLVII, 42
  in NADPH oxidation, LII, 92
  oxidation of tRNA, LIX, 186, 269
  oxygen donor, to cytochrome *P*–450, LII, 407
  viscose fiber activation, XLIV, 561
Sodium persulfate, XLIV, 273
  indole derivative chemiluminescence, LVII, 348
Sodium α-phenylglycidate, synthesis, XLVI, 545
Sodium phenylphosphate, assay of nucleoside phosphotransferase, LI, 387
Sodium phosphate, XLVIII, 9
  assay of photophosphorylation, LVII, 55
  crystallization of firefly luciferase, LVII, 10
  gradient elution, LX, 784
  NMR studies of tRNA, LIX, 25
  preparation of calcium phosphate gel, LIII, 182
    of modified tRNA, LIX, 167, 180
    of ribosomal proteins, LIX, 557
Sodium phosphate-ammonium sulfate-*n*-propanol, in tRNA sequence analysis, LIX, 62
Sodium phosphate buffer, chemiluminescence and, LVII, 438
Sodium phosphate-ethylenediaminetetraacetic acid buffer, formulation, LIII, 222, 223
Sodium phosphate-methylamine, in studies of protein-RNA interactions, LIX, 562
Sodium polyantholesulfonate, phagocytosis, LVII, 491
Sodium pyrophosphate
  assay of amidophosphoribosyltransferase, LI, 172, 173
    of tryptophanyl-tRNA synthetase, LIX, 238

purification of adenylate deaminase, LI, 500
of flavocytochrome b_2, LIII, 241
Sodium [32P]pyrophosphate, LIX, 235
Sodium pyruvate, enzymic synthesis of crystalline, labeled with isotopic hydrogen, XLI, 106–110
Sodium [3-3H]pyruvate, enzymic preparation of crystalline, XLI, 107–110
Sodium salicylate, assay of salicylate hydroxylase, LIII, 530
Sodium succinate
δ-aminolevulinic acid synthetase assay, LII, 352
assay of complex II, LIII, 26
of succinate dehydrogenase, LIII, 26, 27, 34, 467, 469, 470
dimethylaniline monooxygenase purification, LII, 147
isolation of mitochondria, LIII, 22
in mitochondrial incubation medium, LVII, 38
Sodium sulfanilate, purification of luciferin sulfokinase, LVII, 249
Sodium sulfate, LIX, 147, 157
activator of pyrimidine nucleoside monophosphate kinase, LI, 331
anhydrous, XLIV, 130, 132, 838
carbohydrazide modification of tRNA, LIX, 147
for drying organic phase, LII, 429
luciferase assay of bacterial count, LVII, 70, 72
precipitation, purification of nitrate reductase, LIII, 349
synthesis of dithiodiglycolic dihydrazide, LIX, 385
Sodium [35S]sulfate, *see also* [32S]Sulfate
for radiolabeling xylene cyanole FF dye, LIX, 70
Sodium sulfate-barium chloride, cAMP isolation, XXXVIII, 40
Sodium sulfide, XLIV, 163, 533; XLVII, 428
inhibitor of complex IV, LIII, 45
Sodium sulfite, XLVII, 408, 517
aequorin, LVII, 285
assay of deoxythymidylate phosphohydrolase, LI, 286
of ferredoxin, LIII, 617
of flavodoxin, LIII, 617
modification of cysteine, LIII, 144
Sodium taurocholate, XLIV, 447
purification of complex III peptides, LIII, 83, 84
Sodium tetraborate buffer, fluorescamine assay, LIX, 500
Sodium tetrathionate, XLVII, 408
in studies of luciferase active site, LVII, 180
Sodium thiocyanate
in cytochrome b_5 purification, LII, 99
inhibitor of pyrimidine nucleoside monophosphate kinase, LI, 331
Sodium thioglycolate
in anaerobic TEA-maleate buffer, LIII, 366
in electrophoresis buffer, LI, 342, 343
Sodium thiophenoxide, XLVII, 607
Sodium thiosulfate, XLIV, 41, 561
assay of cytochrome c_3, LIII, 616
crystallization of transfer RNA, LIX, 6, 8
in hydroxamate determination, XLVII, 138
preparation of tRNA for NMR studies, LIX, 25
synthesis of ethyl-4-iodo-2-diazoacetoacetate, LIX, 811
titration of oxirane groups, XLIV, 34
Sodium transport, in toad bladder, aldosterone effects, XXXVI, 439–444
Sörensen phosphate buffer, composition, XXXI, 399
Soft agar isolation method, LVII, 158
Solanesol, LIII, 592
Solanum, cell cultures, XXXII, 729
Solanum tuberosum, *see also* Potato
lectin, XXXIV, 333, *see also* Lectin, from potato
Solenoid, construction, XLVIII, 47, 48
Solenoid valve
for centrifuge, XXXII, 4
in pressure system, LVII, 316, 318
Teflon, LVII, 426
Soleus muscle, of rat, XXXIX, 84
in vitro studies, XXXIX, 91
Solid matrix, nucleic acid attached, XXXIV, 463–475, *see also* specific compounds

Solid phase synthesis, ribonuclease A, XXXIV, 637, 638
Solid state electrode, construction, LVI, 362
Solomon-Bloembergen equation, XLIX, 327, 352, 353; LIV, 188
  C values, XLIX, 331, 353
Soluene 350, LIX, 793
Solute, excluded volume, XLVIII, 85
Solute transport, MDCK cells, LVIII, 554, 555
Solution, *see also* specific types
  absorbance indicators, LVI, 308
  high concentration polyacrylamide step gels, LVI, 604, 605
  of optical probes, LV, 574, 576
  two-dimensional gels, LVI, 607, 608
  of uncouplers, LV, 463, 464
Solvent, *see also* specific types
  for ATP synthetase inhibitors, LV, 475
  choice, for mixed anhydride method, XLVII, 555
  effect on chemiluminescence rate constant, LVII, 522
  for extraction, XLVII, 345
  for ionophores, LV, 448
  for large apolar polypeptides, XLVII, 563
  for ligand binding studies, LIV, 528
  in media, LVIII, 70
  organic, effect on carboxypeptidase, XLVII, 78
  oxidant removal, XLVII, 347
  for peptide attachment, XLVII, 322
  polyethyleneimine cellulose chromatography, LV, 284, 288
  to prevent salt accumulation, XLVII, 339
  for sequence analysis, by thioacetylation, XLVII, 292
    of thiohydantoins, XLVII, 364, 365
  for thin-layer chromatography of PTH derivatives, XLVII, 328, 349
  for washing in sequence methods, XLVII, 338
Solvent exchange method, LIX, 675–678
Solvent mixture, for low-temperature spectroscopy, LIV, 105
Solvent saturation method, for elimination of water signal, LIV, 183
Solvent system, *see also* specific substance
  aglycosidic nonpolyene macrolides, XLIII, 136, 137
  7-aminocephalosporanic acid, XLIII, 116
  aminoglycosidic antibiotics, XLIII, 119, 120
  6-aminopenicillanic acid, XLIII, 110, 111
  cephalosporin, XLIII, 116
  cephamycin, XLIII, 116
  chloramphenicol, XLIII, 159
  countercurrent distribution, XLIII, 331, 341
  cycloheximide and other glutarimides, XLIII, 156, 157
  everninomicin, XLIII, 128
  griseofulvin, XLIII, 161
  for hydrophilic substances, XLIII, 104
  lincomycin, XLIII, 128
  mycetin-rhodomycin-cinerubin group, XLIII, 144
  nonpolyene glycosidic macrolide, XLIII, 129, 130
  novobiocin, XLIII, 160
  oligosaccharide, XLIII, 126
  penicillin, XLIII, 110, 111
  polyene macrolide, XLIII, 134
  purine nucleoside, XLIII, 153
  rifamycin, XLIII, 136, 137
  sideromycin, XLIII, 148
  silica gel chromatography, XLIII, 292–294
  steroid antibiotics, XLIII, 158, 159
  streptothricin group, XLIII, 123
  tetracycline, XLIII, 139, 140
  thin-layer chromatography, XLIII, 188–193, 196, 197, 200, 201
Soma, separation from neurites, LVIII, 302–307
Somatic cell, hybridization, XXXIX, 122–127
Somatomedin
  cartilage bioassay, XXXVII, 93–109
    buffers, XXXVII, 97
    with embryo chick cartilage, XXXVII, 104–106

with pig costal cartilage, XXXVII, 101–104
with rat costal cartilage, XXXVII, 95–101
in media, LVIII, 78, 99, 103
metabolic effects, XXXVII, 94
Somatostatin
commercial source, XXXVII, 415
effect on pituitary hormones, XXXVII, 91, 92
isolation, XXXVII, 404
Sonication
for bacterial cell lysis, LIX, 374
of bacterial cells, LI, 80, 92, 183, 215, 303, 311, 439, 553, 562
continuous, LII, 80–82
equipment, XXXII, 206, 207
liposome preparation, LII, 208
medium, submitochondrial particles, LV, 106
for microsome disruption, LII, 80–82, 118, 122, 360
of mast cells, XXXI, 402
for mitochondria disruption, LII, 128, 134, 139
multilamellar liposome formation, XLIV, 223
for nucleic acid shearing, LI, 287
for plant-cell rupture, XXXI, 503
reconstituted alanine transport vesicles, LVI, 433, 434
reconstitution studies
with freeze-thaw, LV, 707
without freeze thaw, LV, 704–707
reconstitution vesicles, LV, 372
Sonication buffer, XXXII, 873
Sonifier
Branson, LV, 107, 318, 325, 385, 720, 778
Branston, LV, 309
Bronson, LV, 418
Bronwill Biosonik, LV, 321, 737
Heat Systems-Ultrasonic, LV, 305
Raytheon, LV, 162, 183, 193, 194, 352, 719
for reconstitution studies, LV, 706
Tomy probe, LV, 371, 778
Sonifier Cell Disruptor, XXXI, 282
membrane preparation, LVI, 167
Sonifier Cell Disruptor Model W 140, LII, 360
Sophora japonica, lectin, XXXIV, 334
Sorbitan monooleate, XLIV, 211, 328, 330
Sorbitan sesquioleate, XLIV, 58, 177, 476, 662
Sorbitan trioleate, XLIV, 102, 204, 207
Sorbitol
autoradiography, XXXVI, 150
gradient properties, XXXI, 508
in isolation of chloroplastic tRNA, LIX, 208, 209
of chloroplasts, LIX, 434
of mitochondria, LV, 3
paper electrophoresis of tritiated, L, 51
radiolabeled, in protein turnover studies, XXXVII, 246
space, in perfused hemicorpus, XXXIX, 80
Sorbose, chromatographic constants, XLI, 16, 17
L-Sorbose, lectin, XXXIV, 334
L-Sorbose oxidase, LII, 21
Sorbyl coenzyme A, carnitine assay, LV, 211
Sorption, isotherm, XLIV, 153
Sorvall RC-3 centrifuge, XXXII, 113
Sorvall Ribi cell fractionator, membrane preparation, LVI, 112
Southern bean mosaic virus, as visual marker, XXXII, 63
Soxhlet extraction apparatus, LIII, 605
Soybean
aleurone grain, XXXI, 576
chymotrypsin inhibitors, XLV, 700
enzyme studies, XXXI, 522, 523, 532
ferredoxin, ^{1}H NMR studies, LIV, 204, 206
hemoglobin, CD spectrum, LIV, 275
nitrogenase, LIII, 328
organelles, XXXI, 518
trypsin inhibitors, XLV, 697, 700, 704
Soybean culture × barley mesophyll, LVIII, 365
Soybean culture × *Colchium* mesophyll, LVIII, 365
Soybean culture × corn mesophyll, LVIII, 365

Soybean culture × pea mesophyll, LVIII, 365
Soybean culture × rapeseed mesophyll, LVIII, 365
Soybean culture × tobacco mesophyll, LVIII, 365
Soybean culture × *Vicia* mesophyll, LVIII, 365
Soybean inhibitor, *see* Bowman-Birk soybean inhibitor; Kunitz soybean trypsin inhibitor
Soybean lecithin, *see also* Azolectin
  liposome preparation, LV, 418
  proteoliposome preparation, LV, 197
Soybean trypsin inhibitor, XXXIV, 3, 104, 448; XLIV, 568–570, 662; LVIII, 95; LIX, 598
  conjugate, acetylation, XLIV, 532
  in hamster cell harvesting, LI, 124
  in hormone receptor studies, XXXVII, 212
  immobilization, on Sepharose 4B, XLIV, 529
  steric exclusion, XLIV, 666, 667
Soybean trypsin inhibitor-agarose, XXXIV, 449
Spacer, XXXIV, 29–58, 72–76, 93–95, *see also* specific enzyme
  acetylcholinesterase, XXXIV, 591
  adsorption effect, XXXIV, 113, 114, 117, 118, 123
  anthranilate-agarose, XXXIV, 380, 381, 384
  araC protein, XXXIV, 373
  azo linkage, XXXIV, 103
  tert-butyl carbazate, XXXIV, 44, 45
  complement, XXXIV, 735
  cyanogen bromide, XXXIV, 79, 80
  β-galactoside, XXXIV, 356, 357
  glycogen-agarose, XXXIV, 126, 127
  hydrophilic, XXXIV, 117, 118, 123
  hydrophobic arm, XXXIV, 123
  N-hydroxysuccinimide esters, XXXIV, 86
  immunoadsorbent, XXXIV, 702
  insulin receptors, XXXIV, 663
  interference, XXXIV, 113, 114, 117, 118, 123
  leakage of ligands, XXXIV, 73
  macromolecular, XXXIV, 676, 681
  methionyl-tRNA synthetase, XXXIV, 508
  methotrexate, XXXIV, 273
  neutral protease, XXXIV, 438–440
  penicillin-binding proteins, XXXIV, 404
  polar, XXXIV, 117, 118
  polyacrylamide derivatives, XXXIV, 37–45
  polyfunctional, XXXIV, 72–75, 94
  polymeric, XXXIV, 611
  for proteins, XXXIV, 52
  role in retaining enzyme activity, XLIV, 38
  stable derivatives of agarose, XXXIV, 73
  thiol group, XXXIV, 535
  trypsin, XXXIV, 444
  types, XLIV, 70
  tyrosine, XXXIV, 395, 396
  use with fibers, XXXIV, 203
Space velocity
  in L-amino acid reactors, XLIV, 753
  definition, XLIV, 756
Span 80, *see* Sorbitan monooleate
Span 85, *see* Sorbitan trioleate
Spark source, LIV, 454–456
Sparsomycin, LX, 285, 351, 355, 356, 365, 410, 414–416
  effect on protein synthesis, LIX, 853, 854, 862
  inhibition, LVI, 31
    of polypeptide chain elongation, LIX, 359, 370
    of protein, XL, 287, 289
  in photoaffinity labeling studies, LIX, 804, 805
Spatial coherence, XLVIII, 461
Specificity site, of ribonucleoside diphosphate reductase, LI, 236
Speckle, reduction, XLVIII, 188
Specpure Graphite rod, XXXII, 42
Spectinomycin
  effect on protein synthesis, LIX, 855
  gas liquid chromatography, XLIII, 215–217
  as protein synthesis inhibitor, XXXII, 869
  proton magnetic resonance spectroscopy, XLIII, 390

spectropolarimetry, XLIII, 359, 360
structure, XLIII, 615
Spectral brightness, definition, LVII, 597
Spectral dissociation, correlation with Michaelis constant, LII, 261, 262
Spectral irradiance, LVII, 597
Spectral overlap integral, XLVIII, 353
Spectral width
choice, LIV, 158, 159
number of data points, LIV, 158
sampling rate, LIV, 157
Spectra-Physics dual-channel UV detector, Model 230, LIX, 188
Spectrapor dialysis tubing, LIII, 130, 521; LIX, 440
advantages, LIX, 482, 483
Spectrin, LIV, 58
isolation from erythrocyte membranes, XXXII, 275–277
SDS complex, properties, XLVIII, 6
Spectroelectrochemical cell, LIV, 402–404
Spectrofluorometer
for fluorescent probe studies, XXXII, 236
optical probes, LV, 577
Spectrograph
for atomic spectroscopy, LIV, 460
basic principle, LIV, 459
Spectrometer
for atomic spectroscopy, LIV, 460, 461
basic principle, LIV, 459
for circular dichroism, LIV, 251
interferometer type, LIV, 319
for magnetic circular dichroism spectroscopy, LIV, 288, 289
for pulse Fourier-transform NMR spectroscopy, LIV, 153, 154
Spectrometer-phototube detector, *see* Photometer
Spectrophotometer
double-beam laser, XLIV, 912, 919
for elution profiles, XLVII, 225
optical probes, LV, 576, 577
quadruple beam, LVI, 333
Spectrophotometric constant, pyridine nucleotide analogs, LV, 271
Spectrophotometry, *see also* Difference spectra; specific types
assay, XLIV, 336–345
determination of isoenzyme content in lactate dehydrogenase, XLI, 47, 49–52
of pyruvate kinase, XLI, 72
of D-xylulose 5-phosphate and D-ribulose 5-phosphate, XLI, 40
of hemoglobin, LVIII, 511
protein concentration, LII, 222
stopped-flow, LII, 221–226
data analysis, LII, 223, 224
titration, of active sites, XLIV, 391
Spectropolarimetry, XLIII, 347–373
atmosphere, XLIII, 370
biopolymer interaction with antibiotic, XLIII, 367–369
blank, XLIII, 372, 373
chromophores, XLIII, 350, 351
circular dichroism, XLIII, 347, 350
experimental parameters, XLIII, 369
gain settings, XLIII, 371
instrument calibration, XLIII, 370, 371
magnetic circular dichroism, XLIII, 347
molecular conformation, XLIII, 347, 348
optical rotary dispersion, XLIII, 347, 350
equations, XLIII, 355
phenomenology of, XLIII, 349, 350
reproducibility, XLIII, 373
sample concentration, XLIII, 371, 372
scanning speed, XLIII, 371
slit widths, XLIII, 371
solvent, XLIII, 369
survey measurements, XLIII, 372
temperature, XLIII, 369, 370
ultraviolet, XLIII, 350
Spectroscopy, *see* specific type
Spectrum
band width, fourth derivative analysis, LVI, 506, 507
convolution functions and higher derivatives, LVI, 509
experimental, fourth derivative analysis, LVI, 501–503
fourth derivative analysis
artifacts, LVI, 504, 505
band shape, LVI, 509–511

information retrieval, LVI, 508, 509
noise, LVI, 511–515
recording, importance of penline width, LVI, 515
simulated, fourth derivative analysis, LVI, 504–505
of spin probes, analysis, LVI, 519–522
Spectrum analyzer, XLVIII, 480–482
calibration, XLVIII, 482
truncation error, XLVIII, 480
Spegazzinine, LV, 515, 516
Sperm
guanylate cyclase, XXXVIII, 197
long-term storage, XXXI, 3
Sperm acrosin, *see* Acrosin, sperm
Spermatogenesis, *in vitro* studies, XXXIX, 283–296
Spermatozoa
electron microscopy, XXXII, 22, 33
high voltage electron microscopy, XXXIX, 156, 157
lactate dehydrogenase–X, purification, XLI, 322, 323
mitochondria, LV, 21, 22
Spermidine, XLIII, 560
assay of 5-methyluridine methyltransferase, LIX, 191
inhibitor of rat liver carbamoyl-phosphate synthetase, LI, 119
oxidation, LVI, 474
in liver peroxisomes, LII, 498
in plant-cell fractionation, XXXI, 505
stimulation of ribosomal protein synthesis, LX, 555, 556, *see also* Edeine
Spermidine oxidase, LII, 18
Spermidine trihydrochloride, isolation of chloroplastic ribosomes, LIX, 434, 435, 437
of chloroplasts, LIX, 434
Spermine, *see also* Polyamine
in crystallization of tRNA, LIX, 6
effector, of ammonia oxidase, LIII, 637
inhibitor of rat liver carbamoyl-phosphate synthetase, LI, 119
oxidation, in liver peroxisomes, LII, 498
regulation of eukaryotic protein synthesis, LX, 555–566
tRNA binding site, LIX, 18
stimulation of binding of messenger ribonucleic acid to 40 S ribosomal subunits, LX, 409, 410, 555, 556, 563, 564
cell, preparation, XXXII, 592
in measurement of termination point, XL, 51, 53
Spheron, XLIV, 56, 69, 74–82, 85
monomer components, XLIV, 67
pore distribution, XLIV, 67
Spheron 10, XLIV, 74
Spheron 100, XLIV, 74
Spheron 200, XLIV, 74
Spheron 500, XLIV, 74
Spheron 700, XLIV, 74
Spheron 1000 BTD, XLIV, 77–80
Spheroplast
isolation, from yeast and fungi, XXXI, 609–626
pet9 mitochondria, LVI, 127
preparation, LV, 135, 142; LVI, 380, 382
lysostaphin, LVI, 175
protoplast, XXXI, 610
vacuole isolation, XXXI, 572–574
yeast, LVI, 19
formation, LV, 158
preparation, LVI, 129
rupture, LVI, 129
separation of mitochondria, LVI, 130
Spherosome
in plant cells, XXXI, 489
isolation, XXXI, 544, 577, 578, 643, 644
4-Sphingenine, *see* Sphingosine
Sphingolipid
metabolic relationships, L, 457
stain, XXXII, 357
Sphingolipidose
comparison, L, 458, 459
definition, L, 456
enzyme sources for diagnosis, L, 457–461
enzymic diagnosis, L, 456–488
substrates for enzymic detection, L, 462

Sphingomyelin, *see also* Glycosphingolipids; specific compounds
adenylate cyclase, XXXVIII, 176, 177
analogue, *see also* specific compounds
methods of synthesis, XXXV, 525
biosynthesis, in endoplasmic reticulum, XXXI, 23
as cardiolipin substrate, XXXII, 160
in lipid bilayers, XXXII, 516
in liposomes, XXXII, 503, 505
in membranes, XXXII, 138
methods of synthesis, XXXV, 493–502
Nieman–Pick disease, L, 465
NMR studies, XXXII, 208, 209, 211
5′-nucleotidase complex with, purification, XXXII, 368–374
partial specific volume, XLVIII, 22
phospholipase effect, XXXII, 138
in plasma membranes, XXXI, 87
preparation of ^{14}C-labeled, L, 466
synthesis of choline-labeled, XXXV, 533–541
thin-layer chromatography, XXXV, 400, 406, 540
trans-*erythro*-D-Sphingomyelin, intermediate in sphingomyelin synthesis, XXXV, 493–501
Sphingomyelinase
assay, L, 465–467
deficiency, L, 465, *see also* Niemann–Pick disease
effect on erythrocytes, XXXII, 137, 138
magnesium dependent, L, 467, 468
Niemann–Pick disease, L, 465
Sphingosine
bases, in glycolipids, XXXII, 361, 362
enhancer, of brain α-hydroxylase, LII, 317
glycosphingolipids, L, 236
oxidative degradation of, XXXII, 364
silica gel chromatography, XXXV, 529–533
Spillover
correction, XLVIII, 337
of radioactivity counts, calculation, LV, 554
Spinach
chloroplast
ATPase inhibitor, LV, 402, 404, 405
cytochrome *f* isolation, XXXII, 412
ferredoxin-activated fructose-1,6-diphosphatase system, properties, XLII, 404, 405
fructose-1,6-diphosphatase phosphohydrolase, assay, purification and properties, XLII, 397–403
protein factor component, assay, XLII, 403
cytochrome *f*, XXXII, 91
enzyme studies, XXXI, 525, 528, 741
ferredoxin
characterization, LIII, 273
^{1}H NMR studies, LIV, 204, 206, 207
nomenclature, LIII, 265
purification, LIII, 267
fructose-bisphosphate aldolase, assay, purification, and properties, XLII, 234–239
leaf
PAP level, LVII, 256
ribulose-1,5-diphosphate carboxylase, assay, purification, and properties, XLII, 472–480
organelles, XXXI, 508, 519
oxidase, LII, 19
peroxidase, LII, 515
3-phosphoglycerate phosphatase, XLII, 409
protein, XXXI, 529
Spinal root, basic protein isolation, XXXII, 342–344
Spin conservation, electron-transfer chemiluminescence, LVII, 509, 510
Spinco rotor, characteristics, XXXI, 9
Spin count, XLIX, 441–460
definition, XLIX, 420, 451
Spin coupling, in heme proteins, LIV, 300–302
Spin diameter, definition, XLIV, 156
Spindle inhibitor, in cell synchronization studies, XXXII, 596
Spin filture culture, LVIII, 14

Spin Hamiltonian, XLIX, 426; LIV, 349–351
electron-electron exchange, XLIX, 385
for Kramers ions, LIV, 363
of nitroxide free radicals, XLIX, 372
in saturation-transfer experiment, XLIX, 482, 483
Spin label
classification, XLIX, 428
common reagents, XLIX, 369–371, 434, 443, 444
definition, XLIX, 427
Spin-labeling technique, *see* Electron spin resonance
Spin-label measurement
of flexibility-function relationships, XXXII, 176–181
of hydrocarbon-chain flexibility, XXXII, 166–172
of lateral diffusion, XXXII, 181–184
in membranes, XXXII, 161–198
of orientation, XXXII, 177–181
of phase transitions, XXXII, 172–174
preparation and availability of spin labels, XXXII, 186–189
sample cells, XXXII, 195–197
synthetic labels, XXXII, 185–193
for lipids, XXXII, 186–190
temperature control, XXXII, 195–197
theory, XXXII, 162–165
of transmembrane motion, XXXII, 184, 185
of yeast cells, XXXII, 828, 829–837
Spin-lattice relaxation time, for measurement of chemical exchange, LIV, 181
Spinner culture, LVIII, 14, 203
Spinneret, emulsion extrusion, XLIV, 228, 229
Spinner flask, LVIII, 46
for endocrine cell culture, XXXIX, 119
Spinning-cup sequenator
glass cup with coupled peptides, XLVII, 307–310
solid-phase program, XLVII, 306–311
glass beads, XLVII, 310, 311
glass cup, XLVII, 307–310
polystyrene, XLVII, 306, 307
Spin operator
electron, XLIX, 372, 385
electronic, LIV, 349
nuclear, XLIX, 372
Spin probe
CAT_n, methods of synthesis, LVI, 518, 519
definition, XLIX, 427
experimental applications, XLIX, 457–466
interpretation of data, LVI, 523–526
permanently charged cationic, binding, LVI, 517
sample preparation, LVI, 518, 519
spectra, aqueous phase, LVI, 519, 520
transmembrane mobility, measurement, LVI, 522, 523
use, principles, XLIX, 428, 429
Spin state
determination, from resonance Raman studies, LIV, 243–245
equilibria, MCD spectroscopy, LIV, 299, 300
Spiramycin, XLIII, 131, 198, 211; LVI, 31, 33
mitochondria, LVI, 32
resistance, LVI, 140
Spirillum lipoferum, nitrogenase, LIII, 328
Spirocheta aurantia, ferredoxin, LIII, 629
Spirolactone
as aldosterone antagonists, XXXVI, 444
antihormone, XL, 292
Spironolactone
antihormone, XL, 291, 292
effect on cytochrome P-$450_{11\beta}$, LII, 131, 132
Spleen
cAMP-receptorin, XXXVIII, 380
cell, XXXIV, 207
lymphocyte preparation, LVIII, 124
mitochondria, LV, 21
from mouse, in erythropoietin studies, XXXVII, 117
neonatal, cell fusion, LVIII, 357
oxidase, LII, 13
Spodoptera frugiperda, media, LVIII, 460

Spreading solution, LIX, 842
Spruce budworm, media, LVIII, 457
SP–Sephadex C–25, developers, XLVII, 214
Squalene, LII, 6
  heme-deficient mutants, LVI, 560
Squamous cell carcinoma, plasma membrane isolation, XXXI, 161
Squash, aleurone grain, XXXI, 576
Squid, giant axon, calcium transport, LVI, 325, 326
SQUID flux measurement device, LIV, 388–396
SRIF, *see* Somatostatin
SSBG, *see* Sex steroid binding globulin
SSWC, *see* Farghaley's minimal media
SSWM, *see* Farghaley's minimal media
Stability
  diffusion effects, XLIV, 183
  immobilization effects, XLIV, 3
Stachyose, chromatographic constants, XLI, 17
Stadie-Riggs microtome, XXXIX, 240; LII, 356
Stain, *see also* specific type
  fluorescent, XXXIV, 216
  for gel electrophoresis, XXXII, 74, 75, 81
    of glycoproteins, XXXII, 96
    of proteins, XXXII, 96, 99
  for negative staining technique, XXXII, 21–24
    comparison, XXXII, 29, 30
  for transmission electron microscopy, XXXIX, 141–145
Staining, *see also* specific types
  antibody, fluorescent, XL, 304
  autoradiography, LVIII, 285
  centromere region, LVIII, 325
  chondrocyte, LVIII, 563
  with DAPI
    microscopy, LVI, 731
    postvital, LVI, 730
    vital, LVI, 730, 731
  DNA, LVIII, 240, 246, 412
  fluorescent, LVIII, 233, 234, 236
    chromosome, LVIII, 325
  grid, XXXIX, 142, 143
  hemoglobin, LVIII, 510
  isoenzymes, LVIII, 168–170
  macrophage, LVIII, 502
  metaphase cell, LVIII, 172–174
  microcarrier culture, LVIII, 188, 189
  monolayer cultures, LVIII, 140
  negative
    of algae, XXXII, 873, 874
    applications, XXXII, 21
    grids, XXXII, 24–26
    image interpretation, XXXII, 31–35
    of membranes, XXXII, 20–35
    model, XXXII, 23
    ping-pong balls, XXXII, 21, 23, 24
    principles, XXXII, 21, 22
    procedure, XXXII, 27–29
    quantitative studies, XXXII, 34, 35
    stains for, XXXII, 23, 24, 29, 30
  nucleic acid-associated protein, XL, 21
  nucleolus organizing region, LVIII, 326
  solutions, LVIII, 246
  surface, extracellular, XXXIX, 143–145
Stainless-steel tubing
  in anaerobic procedures, LIV, 116
  in freeze-quench technique, LIV, 89
  in redox potentiometry apparatus, LIV, 421
Stains-all, LIX, 120, 398
Staley STAR–DRI, starch hydrolyzate, XLIV, 783, 785, 786
Standard absorber method, for sample cell pathlength determination, LIV, 317
Standard addition method, LIV, 484
Standard Lamp of Color Temperature, LVII, 578, 579, 582
Standard Lamp of Spectral Irradiance, LVII, 578, 579, 580, 582
Standard Lamp of Spectral Radiant Intensity, LVII, 578, 579, 582
Standard Lamp of Total Radiation, LVII, 581
Standard light unit, photon equivalent, LVII, 386
Stannic cytochrome *c*
  preparation, LIII, 178
  properties, LIII, 178–181

Stannous chloride, XLIV, 167–169
anilinothiazolinone hydrolysis, XLVII, 370, 371
for heme A reduction, LII, 427
for hemin reduction, LII, 433–435
nitropolystyrene reduction, XLVII, 265
reagent in water purification, XLVII, 14
synthesis of 4-amino-*N*-methylphthalimide, LVII, 427
Stannous octoate, XLIV, 178, 210
Stansted Cell Disrupter, LVIII, 223
Staphylococcal chloramphenicol acetyltransferase, *see* Chloramphenicol acetyltransferase, from *Staphylococcus*
Staphylococcal β-lactamase, *see* β-Lactamase, from *Staphylococcus aureus*
Staphylococcal nuclease, XXXIV, 495, 496, 592, 632; XLVI, 358–362
affinity labeled peptides, XXXIV, 187
azo linkage, XXXIV, 106
MOPC-315, XLVI, 155
purification, XXXIV, 50
reagents for affinity labeling, XLVI, 358–362
synthetic, XXXIV, 632–635
Staphylococcal nuclease-agarose, XXXIV, 186
Staphylococcal protein A
antibody reaction, L, 61
immunoglobulin binding, L, 55, 56
iodination, L, 59
Staphylococcus albus, in phagocytosis tests, XXXII, 652
Staphylococcus aureus
bioautography
for aminoglycosidic antibiotics, XLIII, 120
for cephalosporins, XLIII, 117
for penicillin, XLIII, 111
cell-wall isolation, XXXI, 653–655, 657, 698
efrapeptin, LV, 495
everninomycin and lincomycin groups, XLIII, 128
growth supplements, LVI, 174
heme-deficient mutants, LVI, 172, 173
growth characteristics, LVI, 174, 175
isolation, LVI, 174
media and supplements, LVI, 173, 174
preparation of respiratory-deficient cells and membranes, LVI, 175, 176
restoration of electron transport with hemin, LVI, 176, 177
stock cultures, LVI, 173
C_{55}-isoprenoid alcohol phosphokinase, XXXII, 439–446
β-lactamase, XLIII, 94, 664–672
neomycin phosphotransferase, XLIII, 627
phospho-β-galactosidase, XLII, 493
polyribitol phosphate polymerase, L, 387–402
protease
activators, XLV, 473
amino acid composition, XLV, 473, 474
assay, XLV, 471, 472
distribution, XLV, 474
esterase activity, XLV, 475
properties, XLV, 473
purification, XLV, 469
specificity, XLV, 473; LIX, 600
stability, XLV, 473
protein A, γ-globulins, LVI, 50
resistance plasmids, XLIII, 45
transport systems, XXXI, 702, 704
Staphylomycin, XLIII, 340
Staphylomycin S, XLIII, 767, 773
Starch
degradation, XLIV, 23
immobilization, XLIV, 178
as matrix, XXXIV, 329–331
pad, preparation, XLIV, 652–654
polarography, LVI, 477
purification of glycogen, XXXIV, 164
substrate of α-amylase, XLIV, 98
of β-amylase, XLIV, 40, 44
of glucoamylase, XLIV, 778, 783–786
support properties, XLIV, 20
Zulkowsky, XLIV, 44

Starch gel electrophoresis, in quantitative histone assay, XL, 124
Starch grain, in differential centrifugation, XXXI, 745, 746
Stark degradation, XLVII, 358, 359
Statham pressure transducer, XXXIX, 56
Statham strain gage, XXXIX, 61, 62
Steady state, chemostat culture, equation, LVI, 573, 574
Steady-state assumption, LIV, 71, 72
Steady-state method for T_1 evaluation, XLIX, 340–342
 principle, XLIX, 340
 procedure, XLIX, 341, 342
 spectrometer, XLIX, 340
Steady-state parameter, uncertainty, estimation, LVI, 273, 274
Steady-state rate method
 measurement
  of Ca^{2+}/site ratio
   medium and apparatus, LV, 652, 653
   procedure, LV, 653–655
  of H^+/site ratios
   apparatus, LV, 646
   procedure using Ca^{2+}, LV, 648, 649
    using K^+, LV, 647, 648
   reagents and medium, LV, 646, 647
Stearic acid
 as calorimetric calibrant, XXXII, 271
 freeze-fracture, XXXII, 46
 inhibitor of dimethylaniline monooxygenase, LII, 149
 in mass spectrometry of triglycerides, XXXV, 351, 352
rac-Stearoyl-2-*O*-benzylglycerol-3-iodohydrin, intermediate in lysoglycerophospholipid synthesis, XXXV, 485–488
α-Stearoylglycerol-L-α-iodohydrin, intermediate in phosphatidic acid synthesis, XXXV, 436, 437
DL-1-Stearoyl-3-lysolecithin, intermediate in lysoglycerophospholipid synthesis, XXXV, 485–488
2-Stearoyl selachyl alcohol, intermediate in phosphatidylcholine synthesis, XXXV, 448, 449
Stearylamine, XLIV, 227
Stearyl-CoA desaturase, XXXV, 253–262; LII, 46, 47
 assay, XXXV, 254–256; LII, 188–190
  in vesicle preparations, LII, 207, 208
 inhibitors, XXXV, 260, 261
 kinetics, LII, 211
 mechanism, XXXV, 261, 262
 molecular weight, LII, 192
 pH optimum and K_m values, XXXV, 261
 reactivity, LII, 101
 reconstitution, LII, 206–211
  into liposomes, LII, 206–211
 regeneration of activity, XXXV, 258, 259
 specificity, XXXV, 260
 stability, LII, 192
 substrate specificity, LII, 192, 193
 Triton X–100 removal, LII, 209, 210
Stearyl coenzyme A
 product of fatty acid synthase, XXXV, 84, 89
 in stearyl-CoA desaturase activity assay, LII, 189
 substrate for palmityl thioesterase I, XXXV, 106
  for stearyl-CoA desaturase, XXXV, 254
Steiner equation, XLVIII, 117, 128
Stellacyanin
 electronic absorption spectrum, XLIX, 141
 Raman frequencies and assignments, XLIX, 144
 resonance Raman spectrum, XLIX, 142
Stem bromelain, *see* Bromelain, stem
Stem-puncture method, for crown gall inoculation, XXXI, 555
Stendomycin
 Candida albicans, XLIII, 771
 peptide antibiotic lactonase, XLIII, 767
 solvent system for countercurrent distribution, XLIII, 340
 structure, XLIII, 768
Steramine H, antiseptic, XLIV, 826, 829

Stereospecificity, of hydride transfer, LIV, 223–232
data analysis, LIV, 229–231
experimental principles, LIV, 224–226, 231, 232
experimental procedure, LIV, 226–231
Sterility test, LVIII, 19–21
Sterilization, LVIII, 3–7, 10, 11, 75, 182
insect media, LVIII, 457
Stern equation, surface charge density, LVI, 524
Sternohyoid muscle, incision, XXXIX, 170, 171
Stern-Volmer equation, LVII, 569
deviations, XLIX, 223–225
plot, XLIX, 226–230
Stern-Volmer quenching constant, XLIX, 220, 222, 226
Steroid, XLVI, 75, 76, *see also* specific type
cell lines dependent, XXXII, 558, 559
effect on collagens, preparation, dosage, route of administration, XL, 319
on perfused liver, XXXIX, 34
in endocrine tissue and cells, production, XXXIX, 302–328
16,17-epoxidation, LII, 15
estrogenic, XL, 320
hydroxylations, LII, 337, 388
labeling, XLVI, 447–460
lipophilic Sephadex chromatography, XXXV, 389, 393, 394
receptors, XLVI, 76
standard, XXXIX, 210
thin-layer chromatography, XXXV, 405
transformation, XLIV, 184
trivial names, LII, 384
Steroid-agarose, XXXIV, 555–566, *see also* Steroid enzyme
Steroid antibiotic, XLIII, 158, 159
Steroid dehydrogenase, *see also* specific types
in progesterone metabolism, XXXVI, 490
α-Steroid dehydrogenase, occurrence, XXXIX, 215
Steroid 4-demethylase system, LII, 20
Steroid dioxygenase, LII, 11
Steroid enzyme, XXXIV, 557–566
estradiol-agarose, XXXIV, 562
hydroxysteroid dehydrogenase, XXXIV, 564–566
Δ^5-3-ketosteroid isomerase, XXXIV, 560–564
nortestosterone-agarose, XXXIV, 560–562, 563
Steroid hormone
adsorption to various materials, XXXVI, 93, 94
affinity-labeled, XXXVI, 374–410
applications, XXXVI, 388–410
antibody
in cells, assay, XXXVI, 52–58
quantification, XXXVI, 16–34
assays, XXXVI, 1–88
autoradiography, XXXVI, 135, 136
dry-mount type, XXXVI, 139–146
electron microscopic, XXXVI, 148, 149
light microscopic, XXXVI, 138–148
macroautoradiography, XXXVI, 137, 138
smear-mounting, XXXVI, 146, 147, 151
thaw-mounting, XXXVI, 146
touch-mounting, XXXVI, 146, 147, 151
biochemical processes affected, XXXVI, 427–481
biological rhythmicity, XXXVI, 474–481
19-carbon type, XXXVI, 37
21-carbon type, XXXVI, 36
cellular receptors, assay, XXXVI, 52–58
chromatographic separation of universal method, XXXVI, 485–489
cytoplasmic receptors, XXXVI, 133–264
diazo derivatives, XXXVI, 413–419
electron-capture detection, XXXVI, 58–67
enzyme binding, XXXVI, 390–405
extraction, XXXVI, 492, 493
isolation, XXXVI, 493–496
mercurated, XXXVI, 422–426
metabolites, XXXVI, 483–546

molecular bioassay, XXXVI, 456–465
nomenclature, XXXVI, 18
nuclear receptors, XXXVI, 265–327
plasma-binding protein assay, XXXVI, 34–48
general principles, XXXVI, 35, 36
protein-binding assay, XXXVI, 34–48
radioimmunoassay, XXXVI, 16–34
antisera development, XXXVI, 17–23
receptor, *see also* Cytoplasmic steroid hormone receptors; Nuclear steroid hormone receptors
affinity labels, XXXVI, 411–426
purification, XXXVI, 330–426
receptor protein assay, XXXVI, 49–58
serum-binding proteins, XXXVI, 89–132
sulfur-containing, synthesis, XXXVI, 419–422
tracer superfusion method for cell metabolism, XXXVI, 75–88
in vivo production and interconversion rates, XXXVI, 67–75
Steroid hydroxylase, *see* specific steroid monooxygenase
Steroid Δ-isomerase, XLVI, 23
Steroid ketone, photoexcited, XLVI, 469–479
Steroid 2α-monooxygenase, LII, 14
Steroid 2β-monooxygenase, LII, 14
Steroid 6α-monooxygenase, LII, 14
Steroid 6β-monooxygenase, LII, 14
Steroid 7α-monooxygenase, LII, 14
Steroid 11β-monooxygenase, LII, 14
Steroid 12α-monooxygenase, LII, 14
Steroid 15β-monooxygenase, LII, 14
Steroid 16α-monooxygenase, LII, 14
Steroid 16β-monooxygenase, LII, 14
Steroid 17α-monooxygenase, LII, 14
Steroid 18-monooxygenase, LII, 14
Steroid 19-monooxygenase, LII, 14
Steroid 20-monooxygenase, LII, 14
Steroid 20α,22*R*-monooxygenase, LII, 14
Steroid 21α-monooxygenase, LII, 14
Steroid 22-monooxygenase, LII, 14
Steroid 25-monooxygenase, LII, 14
Steroid 26-monooxygenase, LII, 14
Steroidogenesis
ovarian, in *in vitro* preparations, XXXIX, 237
in perfused placentas, XXXIX, 250–252
Y1 adrenal cell line, LVIII, 571
Steroid receptor, affinity labeling of, XXXVI, 374–410
Steroid sulfatase, multiple sulfatase deficiency, L, 454
Steroid sulfate, lipophilic Sephadex chromatography, XXXV, 389, 393
Sterol, *see also* specific compounds
in biomembranes, XXXII, 542–544
estimation, LVI, 562, 563
heme-deficient mutants, LVI, 558–560
lipophilic Sephadex chromatography, XXXV, 394
in liposomes, XXXII, 505
mitochondrial composition, function, LVI, 563–566
synthesis of 3β-hydroxysterols, rate, XXXV, 281, 282
Sterol glycoside, thin-layer chromatography, XXXV, 403
Sterox AJ 100 detergent, LVI, 741
Steryl ester, *see also* specific compounds
lipophilic Sephadex chromatography, XXXV, 391, 392
STI, *see* Soybean trypsin inhibitor
Stigmasterol, plant membranes, XXXII, 543
Stilbene sulfonate, XLVI, 167
Stirring effect, reduction, LVI, 457, 458
Stock culture, LVIII, 117–119
defined, LVIII, 117
Stock salt solution, LVIII, 198
Stoichiometric → H^+ measurement, scope, LV, 638–640
Stokes-Einstein equation, XLVIII, 419
Stokes' law, in gel filtration, XXXVI, 213
Stokes radius, XXXII, 93, 101
Stokes Raman scattering, XLIX, 69, 73
frequency shift, XLIX, 83
Stopcock, commercial source, LIV, 125, 126
Stopped-flow apparatus, assay of bacterial luciferase, LVII, 137

Stopped flow method, macromolecular interaction kinetics, XLVIII, 308, 309
Stopped-flow spectrophotometry, mechanistic studies of flavoproteins, LIII, 404
Stopping solution, formulation, LIX, 831
Storage, *see* Cell storage; specific cell type
by freezing, *see* Freezing
Storage disease, lysosomal, therapy, XLIV, 700–702
Storage facility, LVIII, 3
Storage medium, for liver preservation, XXXI, 4
Strain development, XLIII, 24–41, *see also* Mutations
Straus and Goldstein expression, inhibitor data, LV, 473, 474
Streak camera, LIV, 11
commercial sources, LIV, 32
principle, LIV, 25
Streptamine, XLIII, 444, 445
from dihydrostreptomycin, XLIII, 448, 449
monoamidinated, XLIII, 449
Streptamine derivative, XLIII, 121
Streptidine
biosynthesis, XLIII, 429
from dihydrostreptomycin, XLIII, 448, 449
Streptidine moiety, XLIII, 430, 444
Streptidine 6-phosphate, XLIII, 629, 631
Streptimidon, XLIII, 158
Streptococcal group-specific polysaccharide antibody, XXXIV, 713
Streptococcus
cell rupture, XXXI, 660–662, 666
D(−)-lactate dehydrogenase, XLI, 294
Streptococcus faecalis
aspartate carbamyltransferase, LI, 41–50
ATPase, LV, 298
ATPase isolation, XXXII, 428–439
cell-wall isolation, XXXI, 658
chloramphenicol acetyltransferase, XLIII, 738
Dio-9, LV, 515
growth, LI, 43
membranes, optical probes, LV, 573
methyl group source, LIX, 202
nectin, LV, 397
nectin isolation, XXXII, 428–439
6-phosphogluconate dehydrogenase, assay, purification, and properties, XLI, 232–237
pyruvate formate-lyase, XLI, 518
Streptococcus lactis
cyanine dyes, LV, 691
membranes, optical probes, LV, 573
phospho-β-galactosidase, XLII, 493
Streptokinase
amino acid composition, XLV, 254
assay, XLV, 245, 246
degradation, XLV, 255
function, XLV, 244
plasmin complexes, XLV, 256
plasminogen, activation, XLV, 254
properties, XLV, 252
chemical, XLV, 253
physical, XLV, 253
purification, XLV, 248
purity, XLV, 252
stability, XLV, 252
Streptolidine, XLIII, 123, 258, 262
Streptolidine sugar compound, XLIII, 261
N-guan-Streptolidylgulosamine, XLIII, 123
Streptolydigin, XLIII, 390
Streptomyces
cell wall hydrolyzing enzyme, XXXI, 610
DD-carboxypeptidase-transpeptidase, XLV, 614, 615
leupeptin, XLV, 679
venturicidin preparation, LV, 505
Streptomyces 3022a, chloramphenicol, XLIII, 735–737
Streptomyces alboniger, puromycin *S*-adenosylmethionine:*O*-demethylpuromycin *O*-methyltransferase, XLIII, 508, 511, 514
Streptomyces albus, XLIV, 60
Streptomyces aureofaciens
S-adenosylmethionine:dedimethylamino-4-aminoanhydrotetracycline *N*-methyltransferase, XLIII, 603, 605
chlortetracycline, media, XLIII, 19
NADP:tetracycline 5a(11a)-dehydrogenase, XLIII, 606

Streptomyces bikiniensis
L-alanine:1D-1-guanidino-1-deoxy-3-keto-*scyllo*-inositol aminotransferase assay, XLIII, 462–465
L-glutamine:keto-*scyllo*-inositol aminotransferase, XLIII, 442, 443
1-guanidino-1-deoxy-*scyllo*-inositol-4-phosphate phosphohydrolase assay, XLIII, 459
inosamine kinase, XLIII, 445, 446, 448
inosamine-phosphate amidinotransferase, XLIII, 452, 456
myo-inositol, XLIII, 438
streptomycin 6-kinase, XLIII, 631
streptomycin-6-phosphate phosphohydrolase assay, XLIII, 466, 468
Streptomyces clavuliger, β-lactam antibiotic, XLIII, 117
Streptomyces coelicolor Müller, clindamycin, XLIII, 756
Streptomyces diastatochromogenes, oligomycins, LV, 502
Streptomyces erythreus
erythromycin
A, B, and C, XLIII, 487, 496
media, XLIII, 20
fungal penicillin acylase, XLIII, 722
protease, XLVI, 229
inhibition, XLVI, 233
Streptomyces fradiae
neomycin, media, XLIII, 22
streptomycin-6-phosphate phosphohydrolase, XLIII, 469
Streptomyces galbus
L-arginine:inosamine-phosphate amidinotransferase, XLIII, 457
streptomycin-6-phosphate phosphohydrolase, XLIII, 459
Streptomyces glebosus, myo-inositol, XLIII, 438
Streptomyces griseocarneus
L-arginine:inosamine-phosphate amidinotransferase, XLIII, 457
streptomycin-6-phosphate phosphohydrolase, XLIII, 469
Streptomyces griseus
S-adenosylmethionine:indolepyruvate 3-methyltransferase, XLIII, 498
L-alanine:1D-1-guanidino-1-deoxy-3-keto-*scyllo*-inositol aminotransferase, XLIII, 464
L-arginine:inosamine-phosphate amidinotransferase, XLIII, 457
L-glutamine:keto-*scyllo*-inositol aminotransferase, XLIII, 442
mannosidostreptomycin, XLIII, 638
oxidase, LII, 16
protease B, XLVI, 206
streptomycin, media, XLIII, 23
streptomycin 3μ-kinase, XLIII, 633
streptomycin 6-kinase, XLIII, 631
streptomycin-6-phosphate phosphohydrolase, XLIII, 465, 469
Streptomyces humidus, L-arginine:inosamine-phosphate amidinotransferase, XLIII, 457
Streptomyces hydrogenans, 20β-hydroxysteroid dehydrogenase, XXXVI, 390
Streptomyces hygroscopicus, chymostatin purification, XLV, 686
Streptomyces hygroscopicus forma glebosus
L-arginine:inosamine-phosphate amidinotransferase assay, XLIII, 452
1-guanidino-1-deoxy-*scyllo*-inositol-4-phosphate phosphohydrolase assay, XLIII, 461
myo-inositol, XLIII, 433, 434
keto-*scyllo*-inositol, XLIII, 440
Streptomyces hygroscopicus var. *ossamyceticus,* ossamycin, LV, 512
Streptomyces kanamyceticus
aminoglycoside antibiotics, XLIII, 265
streptomycin-6-phosphate phosphohydrolase, XLIII, 469
Streptomyces β-lactamase, *see* β-Lactamase from *Streptomyces albus*
Streptomyces lavendulae, penicillin acylase, XLIII, 720, 721
Streptomyces luteogriseus, peliomycin, LV, 503
Streptomyces michigaensis, antipain purification, XLV, 684

Streptomyces mobaraensis, LV, 456
piericidin A isolation, LV, 457, 458
Streptomyces netropsis, fungal penicillin acylase, XLIII, 266
Streptomyces niveus, novobiocic acid synthetase, XLIII, 502
Streptomyces noursei, fungal penicillin acylase, XLIII, 722
Streptomyces ornatus
L-alanine:1D-1-guanidino-1-deoxy-3-keto-*scyllo*-inositol aminotransferase, XLIII, 464
L-arginine:inosamine-phosphate amidinotransferase, XLIII, 457
L-glutamine:keto-*scyllo*-inositol aminotransferase, XLIII, 442
Streptomyces rimosus
S-adenosylmethionine:dedimethylamino-4-aminoanhydrotetracycline *N*-methyltransferase, XLIII, 603, 604
guanosine triphosphate-8-formylhydrolase, XLIII, 515, 516
toyocamycin, XLIII, 760
Streptomyces roseus, leupeptin purification, XLV, 681
Streptomyces rutgersensis, rutamycin, LV, 503
Streptomyces testaceus, pepstatin purification, XLV, 690
Streptomyces venezuelae
chloramphenicol, media, XLIII, 18
glucose isomerase, XLIV, 161, 250, 770
Streptomyces violascens, oxidase, LII, 18
Streptomycin, XLVI, 634, 635; LVI, 31; LIX, 371
biosynthesis, *see also* specific enzyme
aminoglycoside modifying enzymes, XLIII, 616–618
dihydrostreptomycin-6-phosphate 3′,α-kinase, XLIII, 634–637
L-glutamine:keto-*scyllo*-inositol aminotransferase, XLIII, 439–443
guanidinated inositol moieties of streptomycin, XLIII, 429–433
1-guanidino-1-deoxy-*scyllo*-inositol-4-phosphate phosphohydrolase, XLIII, 459–461
inosamine kinase, XLIII, 444–451
myo-inositol:NAD$^+$-2-oxidoreductase, XLIII, 433–439
mannosidostreptomycin hydrolase, XLIII, 637–640
streptidine 6-phosphate, XLIII, 629
streptomycin 3′-kinase, XLIII, 632–634
streptomycin 6-kinase, XLIII, 628–632
streptomycin-6-phosphate phosphohydrolase, XLIII, 465–470
bromoacetamidophenylhydrazone, XLVI, 667
bromoacetylated, XLVI, 664
synthesis, XLVI, 664–667
countercurrent distribution, XLIII, 333
derivatives, chromatographic behavior, XLIII, 635
effect on membranes, XXXII, 882
on protein synthesis, LIX, 854, 855, 860, 862
haloacylated, XLVI, 662–669
in lactose reduction reactor, XLIV, 826
in media, LVIII, 113, 115
paper chromatographic data, XLIII, 121
photoactivated analog, XLVI, 660–662
reactions, XLVI, 663, 664
relative mobilities, XLIII, 122
resistance, nutrition, LVI, 172
solvent systems for countercurrent distribution, XLIII, 342
streptidine moiety, XLIII, 430, 444
Streptomyces griseus, media, XLIII, 23
structure, XLIII, 614
thin-layer chromatography in pharmacokinetic studies, XLIII, 211
in tissue culture media, XXXIX, 111, 112
use in crown gall studies, XXXI, 557
Streptomycin adenylyltransferase, XLIII, 618, 619

Streptomycin hydrochloride, synthesis of *N*-(ethyl-2-diazomalonyl)streptomycyl hydrazone, LIX, 815
Streptomycin 3″-kinase, XLIII, 632–634
  assay, XLIII, 633, 634
  biological distribution, XLIII, 631
  reaction scheme, XLIII, 632
  specificity, XLIII, 634
  stability, XLIII, 633
Streptomycin 6-kinase, XLIII, 628–632
  assay, XLIII, 629–631, 632
  biological distribution, XLIII, 631
  properties, XLIII, 631, 632
  reaction scheme, XLIII, 628
  specificity, XLIII, 631
  stability, XLIII, 631
Streptomycin-6-phosphatase, XLIII, 465, 476
  assay
    alternative, XLIII, 470
    method I, [3′,α-^{3}H]dihydrostreptomycin 6-phosphate, XLIII, 466, 467
    method II, streptomycin-6-phosphate phosphatase, XLIII, 467, 468
    method III, partially purified streptomycin-6-phosphate phosphatase, XLIII, 468
  biological distribution, XLIII, 469
  [3′,α-^{3}H]dihydrostreptomycin 6-phosphate, XLIII, 466
  inhibitors, XLIII, 470
  purification, XLIII, 460
  reaction scheme, XLIII, 465
  specificity, XLIII, 469
  streptomycin-6-phosphate phosphatase, XLIII, 468
Streptomycin-6-phosphate phosphatase, XLIII, 460, *see also* Streptomycin-6-phosphatase
Streptomycin-6-phosphate phosphohydrolase, *see* Streptomycin-6-phosphatase
Streptomycin phosphotransferase, XLIII, 616
  assay, XLIII, 618, 619
  from *Escherichia coli*, XLIII, 627
  from *Pseudomonas aeruginosa*, XLIII, 627
Streptomycin-spectinomycin adenylyltransferase, XLIII, 616, 625, 626
Streptomycin sulfate, XLIII, 382–384
  phosphodiesterase, XXXVIII, 252
  preparation of S100 extract, LIX, 357
  purification of adenine phosphoribosyltransferase, LI, 562
    of aspartate carbamyltransferase, LI, 43, 53
    of CPSase-ATCase-dihydroorotase complex, LI, 124
    of cytidine deaminase, LI, 402, 410
    of cytidine triphosphate synthetase, LI, 80, 87
    of dCTP deaminase, LI, 420
    of deoxycytidine kinase, LI, 339, 340
    of deoxycytidylate deaminase, LI, 414, 415
    of deoxythymidylate phosphohydrolase, LI, 287
    of guanine phosphoribosyltransferase, LI, 554
    of guanylate synthetase, LI, 215
    of ω-hydroxylase, LIII, 358
    of hypoxanthine phosphoribosyltransferase, LI, 554
    of *Lactobacillus* deoxynucleoside kinases, LI, 347
    of nucleoside triphosphate pyrophosphohydrolase, LI, 276
    of OPRTase-OMPdecase complex, LI, 137
    of phosphoribosylpyrophosphate synthetase, LI, 7
    of pyrimidine nucleoside monophosphate kinase, LI, 324
    of pyrimidine nucleoside phosphorylase, LI, 433
    of ribonucleoside diphosphate reductase, LI, 230
    of thymidine kinase, LI, 357, 367
    of thymidine phosphorylase, LI, 443
    of tryptophanyl-tRNA synthetase, LIX, 240

of uridine-cytidine kinase, LI, 302, 303
Streptonigrin, XLIII, 342
Streptothricin
chiroptical studies, XLIII, 352
detection, XLIII, 123
paper chromatography, XLIII, 123
structure, XLIII, 257
Streptothricin-like antibiotic
ion-exchange chromatography, XLIII, 256–263
structure, XLIII, 257
Streptovaricin, XLIII, 137, 390, 409
Streptovitacin, XLIII, 158, 342
Streptozotocin, XLIII, 342
Strepzyme method, for yeast spheroplast formation, XXXI, 613, 614
Stress-plate modulator, LIV, 288
Stress-strain testing of skin, preparation, XL, 364
Stress value, LIX, 609
Strip-chart recorder, for recording of light signals, LVII, 378
Strittmatter method, reversible dissociation of flavoproteins, LIII, 431, 432
Stroma, as fractionation contaminant, XXXI, 545
Stromal cell, LVIII, 266
primary culture, LVIII, 278
Strongylocentrotus franciscanus, hatching enzymes, XLV, 372, *see also* Sea urchin, hatching enzyme
Strongylocentrotus purpuratus, *see also* Sea urchin, egg protease, hatching enzyme
guanylate cyclase
assay, XXXVIII, 196–198
properties, XXXVIII, 198, 199
Strontium, atomic emission studies, LIV, 455
Strontium-90
in membrane studies, XXXII, 884–893
properties, XXXII, 884
Strontium ion
activator of deoxycytidine kinase, LI, 345
antipyrylazo III, LVI, 326, 328
eriochrome blue, LVI, 315
inhibitor of deoxythymidine kinase, LI, 365
Strophanthidin, XLVI, 523
Strophanthidin 3,5-bis-4-benzoylbenzoate, XLVI, 89
Structure refinement, LIX, 14, 15
Stuart factor, *see* Factor X
Stuart-Prower factor, *see* Factor X
STV medium, composition, XXXI, 309
Stylatula elongata
luciferase-GFP interaction specificity, LVII, 264
source, LVII, 248
Styrene, XLIV, 289
Styrene-divinylbenzene copolymer bead, in bacterial cell rupture, XXXI, 661
Styrene glycol, product of epoxide hydrase, LII, 193
Styrene oxide, XLVI, 543; XLVII, 424, 428
reaction conditions, XLVII, 424
substrate of epoxide hydrase, LII, 193, 199
Styrene oxide hydrase, *see also* Epoxide hydrolase
alternate activity, LII, 416
Styrene/4-vinylbenzoic acid copolymer, synthesis, XLIV, 289
Styryl, as optical probes, LV, 573
m-Styrylbenzamidine *p*-toluenesulfonic acid, XLVI, 127
SU, *see* Sumner units
s^4U, *see* 4-Thiouridine
Subcell culture, defined, LVIII, 263
Subcellular fraction
morphometry, XXXII, 3–20
negative staining, XXXII, 21
Subcellular fractionation, viscosity barrier, LV, 127, 128
practical procedures, LV, 131–135
principle, LV, 128–131
Subculture, LVIII, 95
GH cells, LVIII, 532
invertebrate cell line, LVIII, 451, 452
poikilotherm vertebrate cells, LVIII, 476, 477
Y1 cell line, LVIII, 572, 573
Suberic acid, XLVI, 506
Suberic acid dihydrazide, preparation, LIX, 179

Submandibular gland, proteinase inhibitor, dog
assay, XLV, 860
composition, XLV, 865
molecular weight, XLV, 865, 866
properties, XLV, 865
purification, XLV, 861
reactive site, XLV, 866
specificity, XLV, 865, 867
structural homology, XLV, 868
Submaxillary gland, oxidase, LII, 12
Submicrosomal vesicle, preparation, LII, 81, 82
Submitochondrial fraction, procedures for pressing and separating, LV, 92, 93
Submitochondrial particle, XXXIV, 568
affinity chromatography, LV, 114
depletion of ATPase, inhibitor, LV, 422
effects of alcohols, LIV, 149, 150
electrophoretic purification, XXXI, 760
fractionation, LV, 277
internal volumes, LV, 551
measurement of ΔpH, LV, 563
of $P_i \rightleftarrows$ HOH exchanges, LV, 257
of proton translocation, LV, 637
preparation, LV, 114, 393, 394
extraction, LV, 719, 720
from heavy, beef-heart mitochondria, LIII, 574
from yeast, LIII, 113, 114, 222
rate of ATP labeling, LV, 249, 250
separation of right-side-out and inside-out, LV, 112, 113
principle, LV, 113
procedure, LV, 113, 114
tightly coupled of beef heart, LV, 105, 106
activation of succinate dehydrogenase, LV, 107, 108
ATPase depleted preparation, LV, 110–112
cytochrome *c* enriched preparations, LV, 109, 110
fractionation by density gradient centrifugation, LV, 108, 109
preparation of EDTA particles, LV, 106, 107
Submitochondrial vesicle, *see* Vesicle, submitochondrial
Substance P, analog, synthesis, XLVII, 588
Substitution, degree, definition, XLIV, 46
Substitution reaction, with agarose, XXXIV, 77–93, *see also* Agarose, activation method
Substrate, *see also* specific types
of adrenal mitochondria, LV, 10
binding, carrier isolation, LVI, 420
of brain mitochondria, LV, 57
of brown adipose tissue mitochondria, LV, 16, 17, 75, 77
concentration
energy charge, LV, 234
membrane kinetics, XLIV, 914, 915
concentration determination
fluorometric silicone rubber pad method, XLIV, 642
rate method, XLIV, 624
conformation, distances, XLIX, 353–359
diffusion, XLIV, 176, 179, 182, 183, 415–426
of *E. coli* ATPase, LV, 795
endogenous
depletion by defatted bovine serum albumin, LII, 272
by low-temperature solvent extraction, LII, 271
by NADPH, LII, 271
entry into cells, XXXIX, 495–513
enzyme-bound, geometry, XLIX, 322–359
of heart mitochondria, LV, 14
for inner membrane vesicles, LVI, 359
of insect muscle mitochondria, LV, 23, 24
ion fluxes, XXXIX, 493–573
of liver mitochondria, LV, 20, 21
of lung mitochondria, LV, 20
measurement of proton translocation, LV, 636
membrane potential, pH difference and phosphorylation potential, LV, 228
mitochondrial incubation, LV, 7, 8
of *N. crassa* mitochondria, LV, 26

nonoxidase, measurement, LVI, 476–478
of plant mitochondria, LV, 25
of renal mitochondria, LV, 11
respiratory mutants, LVI, 172
size
effect on coupling, XLIV, 27
restrictions, XLIV, 580
of smooth muscle mitochondria, LV, 15, 64
of spermatozoan mitochondria, LV, 22
of spleen mitochondria, LV, 21
of striated muscle mitochondria, LV, 13
of thyroid mitochondria, LV, 18
of white adipose tissue mitochondria, LV, 15, 16
of yeast mitochondria, LV, 27
Substrate analog, XLVI, 8, 9
Substrate permease, of mitochondria, LV, 9
Subtilin, XLIII, 146, 330, 342
from *Bacillus subtilis,* thermolysin hydrolysis, XLVII, 185
Subtilisin, XXXIV, 417; XLIV, 936; XLVI, 22, 27, 75, 206, 207
amino acid composition, XLV, 423
conformation studies, using circularly polarized luminescence, XLIX, 187
effect on luciferase activity, LVII, 201
inhibition assay, XLV, 861
spin-label studies, XLIX, 445, 446, 453, 454
tryptophan content, XLIX, 162
Subtilisin BPN′, XLVI, 215
Subtilisin VII, in cytochrome *P*–450 isolation, LII, 362
Subtilopeptidase A, kinetics, of substrate diffusion, XLIV, 437
Subunit exchange chromatography, kinetics, XLIV, 543–545
Succinate, *see also* Sodium succinate; Succinic acid
perdeuterated, preparation of deuterated bacteria, LIX, 640, 641
Succinate alternative oxidase, LII, 21
Succinate anion, inhibitor of adenine phosphoribosyltransferase, LI, 573
Succinate-CoQ reductase, components, LVI, 580, 581
Succinate-cytochrome *c* oxidoreductase, in plasma membrane, XXXI, 92, 185, 186
Succinate-cytochrome *c* reductase
assay, LV, 102
in brush borders, XXXI, 121
as diagnostic enzyme, XXXI, 19, 20, 24, 31, 36, 37, 38
effect of DABS labeling, LVI, 621
iron-sulfur center, redox titration, LIII, 490
Keilin-Hartree preparation, LV, 126
in liver, XXXI, 97
plasma membrane, XXXI, 90, 100
preparation, LIII, 183
purification of complex III, LIII, 87
reconstitution, LIII, 22, 25, 32, 49, 50
resonance Raman studies, LIV, 245
storage, LIII, 183
toluene-treated mitochondria, LVI, 547
Succinate dehydrogenase, LIII, 397, 405–409, *see also* Complex II
absence of cytochromes, LVI, 568
activation, in submitochondrial particles, LV, 107, 108
activation-deactivation, LIII, 470–475
activators, LIII, 33
activity, LIII, 27
amino acid analysis of flavin peptide, LIII, 457
assay, XXXI, 742; LIII, 27, 34, 35, 406, 466–470
in electron-transferring flavoprotein, LIII, 466, 471, 474–476
inhibitors, LVI, 585, 586
in bacterial membrane, XXXI, 649
in brain homogenates, XXXI, 464, 466
catalytic site, structure, LIII, 479, 482, 483
comparison of preparations, LIII, 475–479, 480, 481, 484, 485
in complex II, LIII, 21, 27–35
complex III, LVI, 583
component stoichiometries, LIV, 145
composition, LIII, 30, 476, 484, 485

contaminant, purification of cytochrome bc_1 complex, LIII, 115
deactivation by oxaloacetate, LIII, 470–472
effect of DABS or PMPS labeling, LVI, 621
electron-transfer components, thermodynamic parameters, LIII, 490–493
electron-transferring flavoprotein, activity studies, LIII, 410, 411, 414, 416, 418, 466, 471, 474–476
EPR characteristics, LIII, 265, 486–495
EPR spectrum, LIV, 148–150
effect of alcohols, LIV, 150
error in protein determination, LIII, 4, 35
extraction, using chaotropic ions, XXXI, 772–777
flavin linkage, LIII, 449, 450
flavin site, LIII, 479, 482
inhibitors, LIII, 33, 34
iron-sulfur center, LIII, 31
EPR characteristics, LIII, 486–495; LIV, 138–141, 143
isolation of flavin peptide, LIII, 554
Keilin-Hartree preparation, LV, 124–126
kinetic properties, LIII, 31, 32, 406
localization, LVI, 230
mammalian, LIII, 466–483
as marker enzyme, XXXI, 735, 740, 742, 760
membrane-bound and soluble, comparative studies, LIII, 410–413
in mitochondria, XXXI, 65, 165, 166, 735
in mitoplasts, XXXI, 323
molecular weight, LIII, 30, 54, 485
purification, LIII, 27–30
comparison of methods, LIII, 476–479, 484, 485
by cyanide extraction, LIII, 485, 486
from electron-transferring flavoprotein, LIII, 477
partial, LIII, 88
purity, LIII, 30
reconstitution, LV, 699, 700
reconstitutively active, properties, LIII, 485
from *Rhodospirillum rubrum*, LIII, 54
sedimentation coefficient, LIII, 30
soluble preparations, LIII, 475
spectral properties, LIII, 31
substrates, LVI, 586
substrate site, LIII, 482, 483
subunits, LVI, 590
subunit structure, LIII, 30, 31, 485
turnover number, LIII, 32, 411, 412, 477, 478
ubiquinone extraction, LIII, 576
Succinate oxidase
effect of DABS or PMPS labeling, LVI, 621
irradiation, LV, 185
Keilin-Hartree preparation, LV, 121, 124, 125
in mitochondria, XXXI, 393
submitochondrial particles, LV, 106
Succinate oxidase ubiquinone extraction, LIII, 573, 575–578
Succinate-PMS-DPIP activity, LIII, 115
Succinate-PMS reductase activity, LIII, 117
Succinate thiokinase, *see also* Succinyl–CoA synthetase
cGMP assay, XXXVIII, 73, 74, 78, 81, 82
Succinate:ubiquinone oxidoreductase, *see* Succinate-ubiquinone reductase
Succinate-ubiquinone reductase, LIII, 21–27, *see also* Complex II
assay, LIII, 26, 27
EPR characteristics, LIII, 486, 488
iron-sulfur centers, redox titration, LIII, 490, 491
reconstitution, LIII, 53
Succinic acid, LIX, 157
activation of succinate dehydrogenase, LIII, 33, 35, 472, 473
amino acid transport, LV, 186, 187
brain mitochondria, LV, 57
carbodiimide-resistant mutants, LVI, 164
content
of brain, LV, 218

of rat liver, LV, 213
of mycobacterial polysaccharide, XXXV, 91
cytochrome *c* reduction, LV, 110
di-*N*-hydroxysuccinimide ester, *see* Disuccinimidyl succinate
failure of mutants to grow, LVI, 109, 112
ferricyanide, LVI, 232, 233
fluorescence, LVI, 499, 500
growth of *E. coli*, LV, 174
hepatoma mitochondria, LV, 81–83, 87
in hydroxybiphenyl assay, LII, 400–402
measurement of Ca^{2+}/site ratios, LV, 652–654
of H^+/site ratios, LV, 643, 645, 647–649
M. phlei ETP, LV, 184
M. phlei membrane fragments, LV, 196
N. crassa mitochondria, LV, 147
oxidation
characteristic → H^+/O values, LV, 638
inhibitors, LV, 460, 461
phosphorylating vesicles, LV, 166
preparation of disuccinimidyl succinate, LIX, 159
renal gluconeogenesis, XXXIX, 14
renal tubule permeability, XXXIX, 19
substrate
for CoA-transferase, XXXV, 235, 242, 299
of complex II, LIII, 21
of succinate dehydrogenase, LIII, 27
transport, LVI, 248, 292
pet9 mitochondria, LVI, 128
ubiquinone redox state, LIII, 583
yeast mitochondria, LV, 150
Succinic acid dihydrazide, preparation, LIX, 179
Succinic anhydride, XLIV, 144, 506, 932
in agarose derivatization, LI, 100
in hemisuccinate steroid derivative preparation, XXXVI, 18, 19
luciferase binding studies, LVII, 132
modification of bacterial luciferase subunits, LVII, 175, 176
of lysine, LIII, 139
synthesis of 4-(4-methoxyphenyl)-4-oxobutyric acid, LVII, 434
in Triton X–45 anionic derivative, preparation, LII, 145
Succinic dehydrogenase
in *E. coli* auxotrophs, XXXII, 864
in mitochondria, XXXII, 712
in renal tubule gluconeogenesis studies, XXXIX, 19
Succinic dihydrazide, XXXIV, 83, 84
Succinic semialdehyde dehydrogenase, brain mitochondria, LV, 58
Succinic thiokinase, *see also* Succinyl-CoA synthetase
in succinyl-CoA generating system, LII, 352
Succinimide ester, XLVI, 96
Succinimidyl bromoacetate
properties, LIX, 158, 159
radiolabeled, LIX, 166
reaction with modified RNA, LIX, 160
Succinimycin, XLIII, 342
Succino-AICAR, *see* **N-(5-Amino-1-ribosyl-4-imidazolylcarbonyl) L-aspartic acid 5′-phosphate**
Succino-AICAR synthetase, *see* *N*-(5-Amino-1-ribosyl-4-imidazolylcarbonyl) L-aspartic acid 5′-phosphate synthetase
Succinoxidase, LIII, 22, 25, 407, *see also* specific types
in granulosa cells, XXXIX, 423
Succinylamidohexamethyliminoagarose, XXXIV, 379, 385
Succinylation, XXXIV, 446
Succinyl-atractyloside, synthesis, LV, 524, 525
Succinyl-atractyloside-amino-Sepharose, synthesis and structure, LV, 525
Succinyl-CoA:propionate CoA-transferase, XXXV, 235–242, 299
in acetate assay, XXXV, 299
specificity, XXXV, 242
stability, XXXV, 241
in succinyl-CoA assay, XXXV, 235, 236, 299

Succinyl-CoA synthetase, XLIV, 882
from *Escherichia coli,* affinity labeling, XLVII, 427
succinyl CoA assay, LV, 210
Succinyl coenzyme A
assay, XXXV, 235, 236, 299; LV, 210, 211
content of rat heart, LV, 216
generating system, LII, 352
preparation from succinic anhydride, XXXV, 235, 299
substrate for CoA-transferase, XXXV, 235, 242, 299
2′-*O*-Succinyl-cyclic adenosine monophosphate, synthesis, XXXVIII, 97, 105
2′-*O*-Succinyl-cyclic guanosine monophosphate, synthesis, XXXVIII, 97, 98
2′-*O*-Succinyl-cyclic nucleotide, *see also* specific types
synthesis, XXXVIII, 96, 97
tyrosine methyl ester derivatives, XXXVIII, 99
iodination, XXXVIII, 99, 100
Succinyldiaminodipropylaminoagarose, XXXIV, 401
6-Succinyl-7,8-[^{3}H]dihydromorphine, XXXIV, 620
Succinyl glass, synthesis, XXXVIII, 183
Succinylhydrazide derivative, XXXIV, 38, 39
6-Succinylmorphine, XXXIV, 620
N^{α}-Succinyl-L-phenylalanine *p*-nitroanilide, as substrate
bdellin assay, XLV, 798
bronchial mucus assay, XLV, 870
seminal plasma inhibitor assay, XLV, 848
submandibular gland inhibitors, XLV, 860
Sucrase
in brush borders, XXXI, 130
in microvillous membrane, XXXI, 130
Sucrase-isomaltase complex, XLVI, 377, 379
Sucrose, LIX, 65
adrenodoxin reductase purification, LII, 133, 134
analysis of mitochondrial ribosomal proteins, LIX, 429
assay of deoxythymidine kinase, LI, 362
chromatographic constants, XLI, 17
crude extract preparation, LI, 273, 301, 324, 328, 426, 427, 460
cytochrome isolation, LII, 118
as density gradient, XXXI, 507, 508, 719, 720
analysis of mitochondrial ribosomes, LVI, 84–86
properties, XXXI, 508
dissociation of *E. coli* ribosomes, LIX, 445
electrophoresis buffer, LII, 325
electrophoretic analysis of protein-RNA interactions, LIX, 567
of ribosomal proteins, LIX, 506, 507
of RNA fragments, LIX, 574
gradient, purification of ATPase, LV, 791–793
harvesting of CHO cells, LIX, 219, 232
in histone determination, XL, 111
hollow-fiber retention data, XLIV, 296
impermeable space, intact mitochondria, LV, 143
inhibitor of brain α-hydroxylase, LII, 317
isolation of chloroplastic ribosomes, LIX, 435
of chloroplastic tRNA, LIX, 208, 209
of chloroplasts, LIX, 434
of mitochondria, LV, 3, 6
isotonic, in isolation of thymus nuclei, XXXI, 246–252
labeled, extramatrix space, LVI, 259, 344, 345
magnetic circular dichroism studies, XLIX, 159
for microsome isolation, LII, 98, 144, 190, 195, 201, 359
in mitochondrial incubation medium, LVII, 38
as negative-stain interferent, XXXII, 27, 28
osmolality, XXXII, 30
osmometric studies, XLVIII, 78

partial specific volume determinations, XLVIII, 17
phosphorescence studies, XLIX, 238
in plant-cell fractionation, XXXI, 504
polarography, LVI, 477
preparation
of chromophore-labeled sarcoplasmic reticulum vesicles, LIV, 51
from commercial cane sugar, LIX, 555
of *E. coli* endogenous polysomes, LIX, 354, 355
of mitochondrial enzyme complexes, LIII, 3, 5, 42, 57, 102, 103, 497
of rat liver crude extract, LIX, 403, 413
of rat liver polysomes, LIX, 526, 527
of rat liver postmitochondrial supernatant, LIX, 515
of rat liver tRNA, LIX, 230
of ribosomes, LIX, 422, 425, 438, 504, 554, 752
product, of α-galactosidase, XLIV, 350
purification of carbon monoxide-binding heme protein P–460, LIII, 638
of complex II, LIII, 22, 23
of complex III, LIII, 103
of complex III peptides, LIII, 83, 84
of CPSase-ATCase complex, LI, 116
of cytochrome oxidase, LIII, 74, 75
of deoxythymidine kinase, LI, 362, 363
of dihydroorotate dehydrogenase, LI, 66
of glycine reductase, LIII, 376, 377
of luciferase, LVII, 62
of mitochondrial tRNA, LIX, 205
of nitrogenase, LIII, 320, 326
of purine nucleoside phosphorylase, LI, 523
of ribosomal RNA species, LIX, 453
of thymidine phosphorylase, LI, 441, 442
quinone extraction, LIII, 580
radiolabeled, in protein turnover studies, XXXVII, 246
in reactivation medium, XLIV, 190
removal by gel filtration, LII, 209
tRNA sequence analysis, LIX, 67
separation of ribosomal subunits, LIX, 451, 452
in SKE buffer, LIII, 114
solid, for solution concentration, LII, 104
space
calculation, LVI, 261, 262
measurement, LVI, 210–212, 287, 288
Special Enzyme Grade, XXXII, 119
substrate, in invertase assay, XLIV, 232, 238, 239
Ultrapure grade, XXXII, 376
in whole cell storage, XLIV, 189
Sucrose buffer, for liver fractionation, XXXI, 10, 14
Sucrose density gradient centrifugation
adenylate cyclase, XXXVIII, 150, 151
analysis of tRNA-protein complexes, LIX, 329, 330, 565, 837
analytical, instrumentation, LIX, 373
assay of distribution of initiation factors, LX, 50, 70–75, 79, 80
of messenger ribonucleic acid binding by protein, LX, 398, 408–449
of free ribosomes, LIX, 367
of ribosome dissociation factor, LX, 294, 295
in cytidine triphosphate synthetase purification, LI, 88
fractionation of irradiated ternary complex of guanosine triphosphate analogs, LX, 733
hyperbolic gradient, preparation of ribosomal subunits, LIX, 445, 451, 555
isolation of mitochondria, LIX, 206
molecular weight, LX, 45
monitoring 40 S ribosomal subunit-methionyl-transfer ribonucleic acid complex formation, LX, 570, 573, 574, 587
of phosphorylated acidic chromatin proteins, XL, 185

preparation of reconstituted ribosomal subunits, LIX, 654
of ribosomal RNA, LIX, 556
of ribosomal RNA precursors, LIX, 832
of ribosomal subunits, LIX, 408, 413–416, 505, 506, 554, 752
of ribosomes, LIX, 404, 405, 554
of S1 depleted 30 S ribosomal subunits, LX, 438, 452–455
preparative, instrumentation, LIX, 373
purification
of complex III, LIII, 103, 105
of *E. coli* endogenous polysomes, LIX, 355
initiation factors, LX, 13, 21, 22, 33, 38, 105–107, 127, 130, 141–144, 150, 151, 153, 155, 182
messenger ribonucleic acid, LX, 394, 395, 403, 405
messenger ribonucleic acid-binding protein, LX, 397
ribonucleic protein particles, LX, 542
ribosomal acetyltransferase, LX, 537
ribosomal subunits, LX, 559, 560, 569, 571–573
viruslike particles from yeast, LX, 551
ribosomal binding sites, LX, 318, 319, 326, 356–358, 415, 416
ribosomal subunit complexes, LX, 122, 123, 259, 277–279, 284, 294, 295, 356–358
Sucrose gradient
cation-containing, for microsome isolation, XXXI, 200
for liver fractionation, XXXI, 40, 41
for microsome isolation, XXXI, 198, 199
preparation, XXXI, 716, 735, 736
for thyroid plasma membranes, XXXI, 145, 146
Sucrose gradient electrophoresis, bacterial membrane separation, XXXI, 650–653
Sucrose gradient ultracentrifugation, of cytoplasmic progesterone receptors, XXXVI, 206, 207
Sucrose hydrolase, immobilization, by bead polymerization, XLIV, 195
Sucrose pad preparation of CHO cell tRNA, LIX, 220
Sucrose-phosphatase, from plants, XLII, 341–347
activators and inhibitors, XLII, 346
assay, XLII, 341–344
properties, XLII, 346
purification, XLII, 344–346
Sucrose phosphate, purification, XLII, 342
Sucrose-phosphate phosphohydrolase, *see* Sucrose-phosphatase
Sucrose reagent
for bacterial cell wall separation, XXXI, 645–647
for liver preservation, XXXI, 4, 8
for nuclei isolation, XXXI, 256, 257
for plasma membrane isolation, XXXI, 77, 94, 135
Sucrose solution, saturated, as cryogenic solvent system in MCD studies, LIV, 293
Sucrose-TKM reagent, for nuclei isolation, XXXI, 257
Sudan Black B, as lysosome stain, XXXI, 473
Sugar, XXXIV, 7, *see also* Saccharide; specific type
O-acetylated, XXXIV, 364
analysis, XLI, 3–10
cane, preparation of sucrose, LIX, 555
keto, colorimetric assay, XLI, 29–31
3-keto, polarographic determination, XLI, 22–27
microdetermination
N-acetylation, L, 52, 54
fucose loss during, L, 53
internal standard, L, 52
neutral, L, 52
radioisotope labeling, L, 50–54
reducing
assay, LI, 263, 264
colorimetric ultramicro assay, XLI, 27–29
transport, K^+ gradient, LV, 679
Sugar-agarose, *see also* Monosaccharide-agarose
Sugarbeet, leaf, 3-phosphoglycerate phosphatase, XLII, 409

Sugarcane
leaf, 3-phosphoglycerate phosphatase, purification, XLII, 406, 407, 409
stem, sucrose phosphatase, purification, XLII, 344, 345
Sugar derivative, thin layer chromatography, XXXIV, 322, 323
Sugar nucleotide, XXXIV, 479
Sugar phosphatase, partial purification, LI, 273
Sugar phosphate phosphohydrolase, *see* Sugar phosphatase
Sugar-starch adsorbent, XXXIV, 329–331
Suicide reagent, XLVI, 9, 10
Suicide substrate, definition, LIII, 437, 438
Sulfadimethoxine, XLIII, 186
Sulfamic acid
preparation of immobilized enzyme rods, LVII, 204
synthesis of benzyl luciferyl disulfate, LVII, 252
Sulfanilamide, XXXIV, 4
nitrite determination, LIII, 638
Sulfanilic acid, XLIV, 445; XLVII, 138
assay of nitrate reductase, LIII, 348, 350
Sulfatase, placental defect, XXXIX, 251
Sulfate, *see also* Sulfuric acid
activator of phosphoribosylpyrophosphate synthetase, LI, 16
of succinate dehydrogenase, LIII, 33
incorporation, biological mechanism, LVII, 244, 245
inhibitor of adenine phosphoribosyltransferase, LI, 573
internal standard, XLIX, 90
in media, LVIII, 68
radioactive, incorporation into mitochondrial protein, LVI, 60
transport, LVI, 292
[^{35}S]Sulfate, LIX, 536, *see also* Sodium [^{35}S]sulfate for radiolabeling *E. coli* ribosomal subunits
Sulfate adenylyltransferase, activity, LVII, 245
Sulfate binding protein, from *Salmonella typhimurium,* thermolysin hydrolysis products, XLVII, 176
Sulfate deficiency, XL, 321
Sulfate ion
inhibitor of luciferase reaction, LVII, 81
of UDP-glucose pyrophosphorylase, LVII, 93
stabilizer of luciferase, LVII, 199
Sulfatide, *see also* Glycosphingolipid
arylsulfatase A, L, 472
column chromatography, XXXV, 416, 417, 421
thin-layer chromatography, XXXV, 396, 400, 402, 403, 406, 407, 416
Sulfation, factor, cell culture, XXXII, 561
Sulfatoglycosphingolipid
characterization, L, 236
myelin formation, L, 247
structures, L, 242, 243
Sulfhydrylagarose, XXXIV, 658, 682
Sulfhydryl compound
activator of formate dehydrogenase, LIII, 371
isolation of mitochondria, LV, 4
oxidation, LII, 45
Sulfhydryl-containing reagent, in histone determination, XL, 111
Sulfhydryl derivative, XXXIV, 43, 44
Sulfhydryl group
concentration determination, XLIV, 385, 392, 522
determination, LIII, 275
quinone binding, XXXI, 537
reaction with carbodiimides, XLIV, 18
in ribosome protein S1, LX, 424
Sulfhydryl-inorganic carriers, preparation, LVI, 490, 491
Sulfhydryl oxidase, XLIV, 525
Sulfhydryl reagent, XLVII, 407–430, *see also* Cysteine, modification; Methionine, modification
affinity labeling groups, XLVII, 409, 422
blocking and labeling, XLVII, 407–412
cross-linking, XLVII, 408, 420–422
in plant-cell extraction, XXXI, 524

in protein inactivation studies, XXXVII, 210
reporter groups, XLVII, 408, 409, 413–420
simple oxidizing agent, XLVII, 409
transport, LVI, 248, 249
Sulfhydryl reducing agent
activators of deoxycytidine kinase, LI, 345
inhibitors of cytidine deaminase, LI, 411
of nucleoside triphosphate pyrophosphohydrolase, LI, 279
Sulfide
acid-labile
in complex I, LIII, 12
in complex II, LIII, 24
in complex III, LIII, 96, 97
in hydrogenase, LIII, 292, 314
in NADH dehydrogenase, LIII, 17
quantitative determination, LIII, 275–277
in succinate dehydrogenase, LIII, 30, 476, 484
complex IV, LVI, 586
complex V, LV, 315
interaction with cytochrome *c* oxidase, LIII, 194
M. phlei membrane systems, LV, 185
Sulfite
adduct, of flavoprotein oxidases, LIII, 398
as antioxidant, XXXI, 538
autoxidation, as source of free radicals, LIII, 387
inhibitor of *Pholas* luciferin chemiluminescence, LVII, 400
Sulfite oxidase, LII, 15; LVI, 475
toluene-treated mitochondria, LVI, 547
Sulfite reductase
cytochromeless strains, LVI, 122
EPR characteristics, LIII, 265
of *Escherichia coli,* siroheme in, LII, 437, 439
flavin determination, LIII, 429
heme-deficient mutants, LVI, 560
reversible dissociation, LIII, 436
(Sulfoethyl)cellulose chromatography, XLV, 731
Sulfoethyl-Sephadex, XLIV, 532, 765
S-Sulfo group
reductive cleavage, with tributylphosphine, XLVII, 123–126
stability, XLVII, 123
Sulfolipase, enzymatic activity, XXXI, 522
Sulfolipid, thin layer-chromatography, XXXV, 403
Sulfonamide, spin probes for carbonic anhydrase, XLIX, 464
Sulfonyl azide, XLVI, 78
Sulfonyl chloride, fluorescent labeling, XLVIII, 358
Sulfonyl fluoride
chemical fates, XLIX, 450–453
structures, XLIX, 444
α-Sulfonyl nitrene, XLVI, 78
4-Sulfophenylisothiocyanate, XLVII, 67, 261
S-3-Sulfopropylcysteine, amino acid analysis and, XLVII, 122
Sulfopropyl-Sephadex, LX, 501, 502, 509
atractyloside-binding protein, LVI, 416, 417
Sulfoquinovosyl diglyceride, enzymatic hydrolysis, XXXI, 522
Sulfosalicylic acid
as membrane stain, XXXII, 878
substrate of salicylate hydroxylase, LIII, 538
Sulfotransferase, activity, LVII, 245
Sulfoxide, XLVI, 135
Sulfur
analysis, for bound protein determination, XLIV, 388
labeling with rhombic, XLVI, 354, 358
radioactive, assay of PAPS, LVII, 245
synthesis, of 7-methoxynaphthalene-1,2-dicarboxylic anhydride, LVII, 436
Sulfur-32, nucleus
coherent scattering length, LIX, 672
scattering cross section, LIX, 672
Sulfur-33, lock nuclei, XLIX, 348
Sulfur-35
advantages, XLVII, 326
in steroid autoradiography, XXXVI, 153

Sulfur-containing steroid, synthesis, XXXVI, 419–422
Sulfur dioxide, synthesis of dimethyl 7-aminonaphthalene-1,2-dicarboxylate, LVII, 436
Sulfur dioxygenase, LII, 11
Sulfuric acid
  in ashing procedure, LIV, 480, 481
  assay of amidophosphoribosyltransferase, LI, 172, 173
    of deoxythymidylate phosphohydrolase, LI, 286
  diphenylamine reagent, LI, 248
  for drying hydrogen chloride, LVII, 20
  esterification of iron porphyrins, LII, 429
  extraction of adenine nucleotides, LVII, 81
    of guanine nucleotides, LVII, 88
  in iron analysis, LIV, 439, 443
  metal-free, preparation, LIV, 476
  sulfoxide stability, XLVII, 455
  synthesis of 3,4-dihydronaphthalene-1,2-dicarboxylic acid anhydride, LVII, 435
    of dimethyl 7-aminonaphthalene-1,2-dicarboxylate, LVII, 436
Sulfuric acid-silica gel plate, isolation of quinone intermediates, LIII, 607
Sulfur reagent, XXXIV, 607
Sulfur trioxide-trimethylamine complex, XLVII, 488
Sumner unit, XLIV, 167
Sum rule, LIV, 255, 264
Sunflower, crown-gall studies, XXXI, 554
Superbrite glass bead, in cell-wall isolation, XXXI, 658, 659
Superconducting magnet, LIV, 288
Superconducting magnetic flux device, LIV, 388, 392
Superfusion technique, for endocrine tissue and cells, XXXIX, 165, 302–328
  apparatus, XXXIX, 323–326
Superinduction, interferon, LVIII, 295
Supernatant factor, ribosomal protein synthesis, LVI, 26, 27
Supernatant reverse transcriptase assay, LVIII, 416
Supernatant 100S, LX, 427, 436, 442, 443, 619, 620
Superoxide anion, XLVI, 321
  oxidases, LVI, 460, 461, *see also* Superoxide dismutase
  product, of flavoenzymes, LIII, 402
  sources, LIII, 387
Superoxide dismutase, LI, 69; LIII, 382–393; LIV, 93; LVI, 475
  activity, LIII, 382
  assay, LIII, 383–388
    with luminol, LVII, 404, 405
  biological role, LIII, 382
  cobalt-substituted, magnetic circular dichroism, XLIX, 168, 175
  contaminant, of commercial cytochrome *c* preparations, LIII, 384
  cuprozinc protein, from bovine erythrocytes, LIII, 383, 388, 389
  electron paramagnetic resonance studies, XLIX, 514
  inhibitor of phagocyte chemiluminescence, LVII, 492
    of *Pholas* luciferin-luciferase light emission, LVII, 396, 397, 399, 401
  kinetic properties, LIII, 386, 387
  lipid peroxidation, LII, 304
  magnetic circular dichroism, XLIX, 175
  mangano protein, from human liver, LIII, 389–393
  metal content, LIII, 383, 389, 393
  molecular weight, LIII, 384, 389
  partial specific volume, LIII, 393
  peroxisomes, LII, 503
  pH effects, LIII, 386, 387
  purification, LIII, 388–392
  sequence analysis, XLVII, 331
  spectral properties, LIII, 389, 393
  stability, LIII, 389, 393
  subunits, LIII, 389
  types, tests to distinguish, LIII, 386
Superoxide ion
  luminol chemiluminescence, LVII, 420
  phagocyte chemiluminescence, LVII, 463–465, 488, 489
  precursor, of singlet molecular oxygen, LVII, 499

Superoxide radical, *Pholas* luciferin chemiluminescence, LVII, 399–401
Support, *see also* specific type
activation
by direct coupling, XLIV, 26–37, *see also* Covalent binding
photochemical, XLIV, 280, 281
polymeranalogous transformation, XLIV, 70
anionic, XLIV, 107, 108
binding ability, parameters influencing, XLIV, 108
choice for large peptides, XLVII, 321
for peptide synthesis, XLVII, 584–590
for short peptides, XLVII, 321, 333
choice parameters, XLIV, 19, 20, 134–138
for coenzymes, XLIV, 155, 869–873
comparative data
surface area, XLIV, 727–729
surface charge, XLIV, 726
using aminoacylase, XLIV, 751
using penicillin acylase, XLIV, 764, 765
concentration of reactive groups, XLIV, 383–386
density of particles, XLIV, 721
effect of charged groups, XLIV, 400–408, 726
excess amino groups, blocking, XLVII, 286, 287
for gel permeation chromatography
selection, XLVII, 98–100
swelling, XLVII, 100, 101
for hydrolytic enzymes, XLVII, 41, 42
hydrophilic, XLIV, 270
effect of solvents, XLIV, 22
synthetic, XLIV, 20
hydrophobic, XLIV, 270
inorganic, XLIV, 134–148
activation with γ-mercaptopropyltrimethoxysilane, XLIV, 146
alkylamine derivatives, XLIV, 139–142
alkylhalide silane derivatives, XLIV, 146
arylamine derivatives, XLIV, 142–144
hydration and cleaning, XLIV, 139
hydrogen ion concentration operating range, XLIV, 163
load, analysis of carboxyl-terminal residue, XLVII, 591, 592
lysine-rich inert protein, XLIV, 264, 265
neutral, XLIV, 108
oxidation, procedure, XLVII, 42
polyacrylamide, pH sensitivity, XLVII, 98
polysaccharide
oxygen sensitivity, XLVII, 98, 104
pH sensitivity, XLVII, 98
types, XLIV, 20
porous
enzyme aggregation, XLIV, 265, 270–272
selection factors, XLIV, 272
properties, XLIV, 19, 20, 107
recovery after use, XLVII, 107
regeneration, XLIV, 165
for solid-phase sequencing, XLVII, 263–277
types, XLIV, 18–20
Support film, for negative staining, XXXII, 24, 25
Surface, XXXIV, 195–225, *see also* Cell, surface; Fiber, fractionation
Surface antigen, XXXIV, 195–225, *see also* Cell, surface; Fiber, fractionation
Surface area, determination, XXXII, 10–13
Surface binding microautoradiography, of protein hormones, XXXVII, 164–167
Surface charge, adsorption rate, XLIV, 151
Surface chemistry, monolayers, bilayer electrochemistry, XXXII, 554
Surface-connected system, plasma-membrane, XXXI, 155
Surface potential, methods of measurement, LVI, 516, 517
Surfonic N–95 detergent, LVI, 741
partial specific volume, XLVIII, 19, 20
Surgical kit, for human cell biopsy, XXXII, 806
Surgical wire, XXXII, 99

Suspension, preparation for mutagenic treatment, XLIII, 26, 27
Suspension culture, LVIII, 14, 46–49, 81
cell lines, LVIII, 202–206
chondrocyte, LVIII, 563
of endocrine cell lines, XXXIX, 118–120
HTC cells, LVIII, 550, 551
human adenovirus, LVIII, 426
invertebrate cell, LVIII, 454, 455
media, LVIII, 56, 58, 89
operations, LVIII, 205
plant cell, LVIII, 481, 482
primary cell, LVIII, 206–210
scaling-up, LVIII, 211–221
Suspension polymerization, *see* Bead polymerization
Sutan, herbicide, detection, XLIV, 658
SVD, *see* Phosphodiesterase I
Svedberg unit, XXXVI, 157
SV3T3 cell, adhesion studies, XXXII, 609
SV40 virus, *see* Simian virus 40
SWC, *see* Seawater complete medium
Sweet potato
β-amylase, XLIV, 98
peroxidase, binding properties, LII, 521
Sweet protein, free peptide hydrogen exchange, XLIX, 30
Swelling capacity
effect of cross-linking, XLIV, 108
pH dependence, XLIV, 108
Swine, *see also* Hog; Pig
kidney
fructose-1,6-diphosphatase, assay, purification, and properties, XLII, 375, 385–389
glycogen synthetase, assay, purification, and properties, XLII, 375–381
phosphofructokinase, assay, purification, and properties, XLII, 375, 382–385
phosphorylase, assay, purification, and properties, XLII, 375, 389–394
protein kinase, assay, purification, and properties, XLII, 375, 394–397
Swinny syringe filter holder, XXXIX, 182
Swiss 3T3 cell, LVIII, 96, 97
SWMH, *see* Seawater minimal medium
Sylgard encapsulating resin, XXXIX, 93
preparation of stable light source, LVII, 300
Sympathetic ganglion-like PC12 cell, LVIII, 585
Sympathetic neuron
cell preparation, LVIII, 575, 576
cholinergic activity, LVIII, 575
co-culture, LVIII, 575, 581
conditioned medium, LVIII, 581, 582
differentiation, LVIII, 574
long-term culture, LVIII, 574–584
media, LVIII, 578–580
microculture, LVIII, 582, 583
phenotypic characteristics, LVIII, 583, 584
use of Methocel, LVIII, 580
Symphoricarpos, free cells, XXXII, 724
Synaptogenesis, electrolectin activity, L, 300
Synaptosome
differential sedimentation, XXXI, 727
electrophoretic purification, XXXI, 755
isolation, XXXI, 446; LV, 55
mitochondria, LV, 18
optical probes, LV, 573
plasma membrane fractions, isolation, XXXI, 445–452
preparation of two populations of mitochondria, LV, 56
Syncatalytic inactivation, XLVI, 11, 41–47
Syncatalytic modification, XLVI, 41–47
Synchronous growth, LVIII, 218
Synchrony, LVIII, 248–262
calcium deprivation, LVIII, 261, 262
criteria, LVIII, 249
degree, LVIII, 249–257, 259
double-thymidine block, LVIII, 261
evaluation, LVIII, 249–252
Ficoll gradients, LVIII, 257, 258
gradient centrifugation, LVIII, 256–258
hydroxyurea, LVIII, 259–261

isoleucine deprivation, LVIII, 258, 259
mitotic detachment, LVIII, 252–256
myoblast, LVIII, 511, 518–520
physical methods, LVIII, 256, 257
serum deprivation, LVIII, 259–261
Synchrotron radiation, LIV, 324, 325
research facility, LIV, 326, 327
Synechococcus lividus, ferredoxin, ^{1}H NMR studies, LIV, 208
Synpore filter, membrane potential, LV, 589
Syrian hamster karyotype, LVIII, 342
Syringe
collection, continuous-flow pH measurement, LV, 622
micrometer, for potentiometric-spectrophotometric titrations, LIV, 404
rapid mixing devices, LV, 245, 246
reactant
continuous-flow pH measurement, LV, 621
measurement of proton translocation, LV, 631
Syringe adapter, LVIII, 9
Systox, pesticide, detection, XLIV, 658
Szent-Györgyi-Blum continuous flow apparatus, XXXII, 276

T

T, *see* 5-Methyluridine
T_1, *see* Relaxation time, longitudinal
T_2, *see* Relaxation time, transverse
T_3, *see* 3,5,3′-Triiodothyronine
T_4, *see* Thyroxine
14T27, as yeast spin label, XXXII, 829, 832
Tadpole, collagenase, zymogen, XL, 351
D-Tagatose, cultured cell growth, L, 190
Taka-amylase, *see Aspergillus oryzae,* carboxyl proteinase
Taka-amylase A, reductive denaturation studies, XLIV, 517
Talc, as hormone adsorbent, XXXVII, 34
3-*O*-α-D-Talopyranosyl-D-glucose synthesis, L, 121
TAME, *see* N^{α}-Toluenesulfonyl-L-arginine methyl ester
Tanford equation, for solvent chemical potential, XLVIII, 84
Tank buffer, for gel electrophoresis, XXXII, 74, 76
Tannase, specific activity, XLIV, 255
Tannin
in plant cells, XXXI, 501, 529, 532, 544
precipitation of penicillin acylase, XLIV, 762
Tanning, collagen membrane, XLIV, 247, 248
Tanol, *see* 1-Oxyl-2,2,6,6-tetramethyl-4-piperidinal
Tantalum, cathode, LIV, 456
T antigen, in transformed cells, XXXII, 590
Tap water, Keilin-Hartree preparation, LV, 119, 120, 124, 125
Tartaric acid
inhibitor, of lactate dehydrogenase, XLIV, 428
of rat liver acid nucleotidase, LI, 274
Tartaric acid dihydrazide, XLIV, 88, 91, 103
Tartaryl diazide, cross-linking, LVI, 631, 632
D-Tartronate semialdehyde 2-phosphate, preparation of, and complex formation with enolase, XLI, 121–124
Tartryl diazide, LIX, 550
Taurine, XLVII, 34
internal standard, XLVII, 34, 39
in media, LVIII, 63
Taurochenodeoxycholate, sodium salt, partial specific volume, XLVIII, 21
Taurocholate
micelles, ionic strength and pH effects, LVI, 747
partial specific volume, XLVIII, 21
Taurodeoxycholate
micelles, ionic strength and pH effects, LVI, 746
sodium salt, partial specific volume, XLVIII, 21
structure, LVI, 735
usefulness, LVI, 749
Taurodeoxycholate 7α-hydroxylase, LII, 14

Tay–Sachs disease
cell ganglioside levels, L, 247
enzymic diagnosis, L, 484–487
G_{M2}-ganglioside, L, 484
β-hexosaminidase A, L, 484
Tay-Sachs ganglioside, *see* *N*-Acetylgalactosaminylgalactosylglucosylceramide; *N*-Acetylneuraminosylgalactosylglucosylceramide
Tay–Sachs trisaccharide synthesis, L, 109
$TBABF_4$, *see* Tetra-*n*-butylammonium fluoroborate
TBG, *see* Thyroxine-binding globulin; Thyroxine-binding protein
TBS, *see* Tris-saline buffer
TCA, *see* Trichloroacetic acid
T cell, XXXIV, 216, 719, 721; LVIII, 178, 493
3T3 cell, LVIII, 376
monolayer, LVIII, 133
mouse, DNA, content, LVIII, 148
plasma membrane isolation, XXXI, 161
3T3 cell line, properties and uses, XXXII, 584, 589, 590
TCH, *see* Triethylenediaminecadmium hydroxide
TCPO, *see* Bis-2,4,6-trichlorophenyloxalate
TDH, *see* Tartaric acid dihydrazide
TdR kinase:ATP-thymidine 5′-phosphotransferase, *see* Deoxythymidine kinase
Tea, enzyme studies, XXXI, 534, 544
TEAB, *see* Triethylammonium bicarbonate
TEA buffer, *see* Triethanolamine buffer
TEAE, *see* Triethylaminoethyl
TEAE-cellulose, *see* (Triethylaminoethyl)cellulose
TEA-maleate buffer, *see* Triethanolamine-maleate buffer
TeBG, *see* Testosterone-estradiol-binding globulin
Technicon AutoAnalyzer
in kidney perfusion studies, XXXIX, 4, 8
L-methionine assay, XLIV, 747
Tec-Mar homogenizer, LI, 324
Teepol wetting agent, XXXI, 580
Teflon, steroid adsorption, XXXVI, 93, 94, 161, 162
Teflon film
for lipid monolayers, XXXII, 546, 547
membrane potential, LV, 589
Teflon insulated socket, LVII, 529, 531
Teflon tubing
in chemiluminescent reaction apparatus, LVII, 426
for freeze-quench techniques, LIV, 88
in hollow-fiber device, XLIV, 295
Teichoic acid, XXXIV, 7, 337
bacterial polysaccharides, L, 253
biosynthesis, L, 387–402
α-glycosylated, XXXIV, 338
Tektronix AM502 differential amplifier, LIV, 54
Tektronix graphic computer Model 4051, LIV, 319
Tektronix memory oscillator, XXXIX, 62
Tektronix oscilloscope, LIV, 39, 40, 54
Tektronix transient digitizer, R7912, LIV, 39
Teleost
balanced salt solution osmolarity adjustment, LVIII, 469
incubation temperature, LVIII, 471
media, LVIII, 469
Telethermometer, LIII, 531
for tissue temperature monitoring, XXXIX, 368
Teletypewriter, in Fast Analyzer, XXXI, 801
TEM, *see* Transmission electron microscopy
TEMED, *see* *N,N,N′,N′*-Tetramethylethylenediamine
Temik, pesticide, detection, XLIV, 658
Temp-Blok Module Heater, XXXII, 619
TEMPENE, *see* 2,2,6,6-Tetramethyl-4-ene-*N*-oxylpiperidine
Temperature
colicin action, LV, 535, 536
density measurements, XLVIII, 15
effect, XXXIV, 116, 117
on bioluminescence assay, LVII, 217
on culture, LVIII, 55, 135, 450
on reaction rate, LVII, 363, 364

of elution, XXXIV, 252
EPR data, LIII, 488–490
high pressure ion-exchange chromatography, XXXVIII, 21
isopycnic, determination, XLVIII, 26
low, cytochrome kinetics, LIV, 102–111
micelle formation, LVI, 740, 744
phosphorylation potential, LV, 235, 236
for sensitive mutants, LVI, 132
for storage of mitochondria, LV, 31
Temperature control
electron paramagnetic resonance spectra and, XLIX, 520, 521
saturation-transfer spectroscopy, XLIX, 504
Temperature-jump method of Czerlinski and Eigen, XLVIII, 308, 311
Temperature jump technique, LIV, 66–69
TEMPO, *see* 2,2,6,6-Tetramethylpiperidine-1-oxyl
Tenax wax plate, XXXII, 654, 656
Tendon, isolation of cells, XL, 313
TENK buffer, composition, XXXVI, 338
Tentoxin
availability and preparation, LV, 492
biological properties, LV, 493
physical properties, LV, 476
structure, LV, 492
Terasaki plate, XXXII, 734, 736
Teratoma, ovarian, LVIII, 376
Terbium
circularly polarized luminescence, XLIX, 187
magnetic circular dichroism, XLIX, 177
Terbium acetate, derivatization of transfer RNA, LIX, 14
Terephthalaldehyde, XLIV, 115
Tergitol, in yeast mutant isolation, XXXII, 824, 830
Tergitol NP27 detergent, LVI, 741
Ternary complex, abortive, XXXIV, 239
Ternary complex dissociation, *see* Initiation factor EIF-2
Terpene
metabolism, in plants, XXXI, 536
as plant-extraction interferents, XXXI, 531
Terpenoid compound, lipophilic Sephadex chromatography, XXXV, 393, 395
Terramycin, XLIII, 334
Terumo microsyringe, LIX, 864
TES buffer, *see* *N*-Tris(hydroxymethyl)methyl-2-aminoethanesulfonic acid
Tesla, equivalents, LIV, 289
Tesla conversion factor, XLIX, 156
Test grid, for morphometry, XXXII, 11
Testicular cell
effect of hormone, LVIII, 103
mouse TM4, LVIII, 96, 97
succinoxidase activity, XXXIX, 424
Testicular feminization, in rats and mice, hormone studies, XXXIX, 454–460
Testis
cell culture, XXXIX, 288–296
gonadotropin binding, XXXIX, 268–271
histology, XXXIX, 286, 287
infusion, XXXIX, 272–282
interstitial cell function, *in vitro*, XXXIX, 252–271
lactate dehydrogenase–X from mouse, rat, and bull, XLI, 318–322
mammalian, organ culture, XXXIX, 284–288
mitochondria, LV, 11
oxidase, LII, 13, 14
preparation, XXXIX, 256–261
protein kinase, XXXVIII, 365, 366
radioautography, XXXIX, 287, 288
receptor for LH and hCG, XXXVII, 167–193
physical parameters, XXXVII, 191, 192
properties, XXXVII, 175, 176, 185–192
solubilization, XXXVII, 181, 182
transplants, XXXIX, 288
Testis-specific protein, sequence analysis, with thermolysin, XLVII, 184
Testosterone, XLVI, 476
adsorption to various materials, XXXVI, 93, 94
antibodies to derivatives, XXXVI, 54, 56
antiglucocorticoid, XL, 292
antiserum, XXXIX, 262, 263

assay, in interstitial cells, XXXIX, 261, 262
autoradiography, XXXVI, 139, 150, 151, 155
biosynthesis, in interstitial cells, XXXIX, 256
3-*O*-carboxymethyl oxime of, preparation, XXXIX, 263, 264
cellular receptors, assay, XXXVI, 56–58
chromatographic separation, XXXVI, 487
circulatory enrichment, XXXIX, 272
competitive inhibitor of cytochrome P-$450_{11\beta}$, LII, 132
competitive protein-binding assay, XXXIX, 264, 265
cytochrome binding, LII, 276, 277
  reaction type, LII, 267
derivatives, XXXVI, 18–21
effect on ductus deferens tumor growth, XXXIX, 115
  on levator ani muscle, XXXIX, 98
  $(Na^+ + K^+)$-ATPase, XXXVI, 435
electron-capture detection, XXXVI, 59–64
estrogen production, XXXIX, 251
in incubation media direct assay, XXXIX, 265, 266
logit-log assay, XXXVII, 9
in media, LVIII, 102, 103
metabolism, in preputial glands and kidneys, XXXIX, 454–460
metabolites, gas chromatographic and mass spectral properties, LII, 385
molecular bioassay, XXXVI, 465
nuclear receptor binding, XXXVI, 318, 319
as 5α-oxidoreductase substrate, XXXVI, 466–468, 472, 473
PBP affinity, XXXVI, 125
plasma levels, in stress, XXXIX, 279
protein-binding assay, XXXVI, 37
radioimmunoassay, XXXVI, 16, 20, 21, 24; XXXIX, 262–264
receptors for purification, XXXVI, 366–374
5α-reduction, XXXIX, 455–457
stabilizer of purified cytochrome P-$450_{11\beta}$, LII, 131
substrate of cytochrome P–$450_{11\beta}$, LII, 132
  of cytochrome P–450LM forms, LII, 117
sulfur-containing derivatives, XXXVI, 419
synthesis, LV, 11
in testicular lymph, XXXIX, 276
testicular production, XXXIX, 262–266
tritiation, XXXIX, 263
in urine, EC determination, XXXVI, 63, 64
in vivo production and interconversion, XXXVI, 68–75
Testosterone diazoacetate, synthesis, XXXVI, 419
Testosterone-estradiol-binding globulin
  analysis, XXXVI, 117
  isolation, XXXVI, 109–120
  summary, XXXVI, 119
Testosterone ethyl diazomalonate, synthesis, XXXVI, 417, 418
Testosterone glucuronide, *in vivo* production, XXXVI, 68, 69
Testosterone 6β-hydroxylase, Testosterone 6β-monooxygenase
Testosterone 7α-hydroxylase, *see* Testosterone 7α-monooxygenase
Testosterone 16α-hydroxylase, *see* Testosterone 16α-monooxygenase
Testosterone 6β-monooxygenase, *Ah* locus, LII, 232
Testosterone 7α-monooxygenase, *Ah* locus, LII, 232
Testosterone 16α-monooxygenase, *Ah* locus, LII, 232
Test tube, calcium concentration, LVII, 282, 301
Tetanus toxin, XXXIV, 610
Tetanus toxin receptor, gangliosides, L, 250
TETA polystyrene resin, *see* Triethylenetetramine polystyrene resin
2,3,4,6-Tetra-*O*-acetyl-β-D-glucopyranose, L, 125
2,3,4,6-Tetra-*O*-acetyl β-D-glucopyranosyl phosphate preparation, L, 128, 129

3,4,5,6-Tetra-*O*-acetyl-*myo*-inositol, synthesis, XLVI, 372
3,4,5,6-Tetra-*O*-acetyl-*myo*-inositol-1,2-thionocarbonate, synthesis, XLVI, 372, 374
2,3,4,6-Tetra-*O*-acetyl-β-D-mannopyranose, L, 125
2,3,4,6-Tetra-*O*-acetyl-β-D-mannopyranosyl phosphate purification, L, 126
synthesis, L, 124–126
Tetraacetylriboflavin, monobromination, LIII, 458
2,4,5,7-Tetrabromofluorescein, *see* Eosin
Tetra-*n*-butylammonium fluoroborate
electron-transfer chemiluminescence, LVII, 511, 512
structural formula, LVII, 498
Tetracarcinoma, LVIII, 376
2,4,4′,5-Tetrachlorodiphenylsulfone, LV, 515
Tetrachloroethane, solvent, synthesis of chemiluminescent aminophthalhydrazides, LVII, 434
3,3′,4′,5-Tetrachlorosalicylanilide, proton ionophores, LV, 420
4,5,6,7-Tetrachloro-2′-trifluoromethylbenzimidazole
synthesis, LV, 468
uncoupling activity, LV, 465
Tetracosanoyl coenzyme A, product of fatty acid synthase, XXXV, 84, 89
Tetracycline, XLVI, 182; LVI, 31
bioautography, XLIII, 45, 140
biosynthesis, *see also* specific enzyme
S-adenosylmethionine:dedimethylamino-4-aminoanhydrotetracycline *N*-methyltransferase, XLIII, 603–606
chemical detection, XLIII, 141
chiroptical studies, XLIII, 352
effect on membranes, XXXII, 882
on protein synthesis, LIX, 853, 854, 862
gas liquid chromatography, XLIII, 251–253
high-pressure liquid chromatography, XLIII, 315–319
calculation, XLIII, 316, 317
inhibition of initiation complex formation, LX, 343
of *in vitro* protein synthesis, LIX, 370
in media, LVIII, 113, 115, 116, 565
NADP:tetracycline 5a(11a)-dehydrogenase, XLIII, 606, 607
paper chromatography, XLIII, 137–141, 142
bioautography, XLIII, 140
detection, XLIII, 140, 141
solvent system, XLIII, 139, 140
for photoaffinity labeling, LIX, 798, 802, 805
photodensitometry, XLIII, 185
solution spectrophotometry with thin-layer chromatography, XLIII, 186
spectropolarimetry, XLIII, 363, 364
thin-layer chromatography, XLIII, 199, 204, 211
Tetracycline hydrochloride, XLIII, 135
circular dichroism spectrum, XLIII, 364
differential pulse polarography, XLIII, 382, 384, 385
optical rotatory dispersion spectrum, XLIII, 364
Tetradecanal
assay of luciferase, LIII, 565
bioassay, LVII, 191
quantum yield, in bioassay, LVII, 192
Tetradecanoic acid
assay with bacterial luciferase, LIII, 564
bioassay, LVII, 191–193
Tetradecyltrimethylammonium chloride, properties of protein complexes, XLVIII, 7
Tetradifon, LV, 515
Tetraene, XLIII, 135
Tetraethylammonium, photochemically active derivative, LVI, 660
4,5,6,7-Tetrafluorotryptophan, inhibitor of tryptophanyl-tRNA synthetase, LIX, 252
Tetrahexosylceramide structure, L, 273
3α,5β-Tetrahydroaldosterone
isolation and measurement, XXXVI, 505–507
isomers, XXXVI, 507, 508, 511, 512

synthesis, XXXVI, 507
Tetrahydrobiopterin, LII, 6
Tetrahydrocannabinol, in microsomal fractions, LII, 88
Tetrahydroenopterin, substrate of phenylalanine hydroxylase, LIII, 285
(±)-L-Tetrahydrofolate, in (±)-L-methylenetetrahydrofolate preparation, LI, 91
Tetrahydrofolate dehydrogenase, XXXIV, 281; XLVI, 68
  copurification, with thymidylate synthetase, LI, 94
Tetrahydrofolic acid
  inhibitor of guanine deaminase, LI, 516
  substrate of phenylalanine hydroxylase, LIII, 284
Tetrahydrofuran, XLVII, 488, 529, 550, 555
  in diazo steroid synthesis, XXXVI, 413
  radical ion chemiluminescence reactions, LVII, 498, 504, 505, 507
  synthesis of chemiluminescent aminophthalhydrazides, LVII, 433
    of *N*-(ethyl-2-diazomalonyl)streptomycyl hydrazone, LIX, 815
Tetrahydropteridine, substrate of phenylalanine hydroxylase, LIII, 278
3,4,5,6-Tetrahydrouridine, inhibitor of cytidine deaminase, LI, 405, 407
3,4,5,6-Tetrahydrouridine 5′-phosphate, XLVI, 25
(3,5/4,6)-Tetrahydroxycyclohex-1-ene, *see* Conduritol B
Tetrahymena
  glyoxylate cycle enzymes in, XXXI, 566
  mitochondria studies, XXXII, 33
  oxidase, LII, 21
Tetrahymena pyriformis
  disruption, LV, 136
  oxidase, LII, 18
  uridylate-cytidylate kinase from, LI, 331–337
Tetralin, XLVII, 546, 547
Tetramethylammonium, cyclic nucleotide levels, XXXVIII, 94
Tetramethylammonium chloride, in hydride transfer stereospecificity experiments, LIV, 228
3,3′,5,5′-Tetramethylbenzidine, for staining of cytochrome *P*–450 in gels, LII, 331
Tetramethyl-*p*-benzoquinone, *see* Duroquinone
2,2,5,5-Tetramethyl-3-carbhydrazinoylpyrrolin-1-oxyl, preparation, LIX, 179
2,2,5,5-Tetramethyl-3-(carboxy-*N*-hydroxysuccinimide ester)pyrrolin-1-oxyl, preparation of modified tRNA, LIX, 171
2,2,5,5-Tetramethyl-3-carboxy-1-oxyl pyrrolidine-*p*-nitrophenyl ester
  preparation, XLIX, 437
  substrate of α-chymotrypsin, XLIX, 438–442
2,2,5,5-Tetramethyl-3-carboxypyrrolidine-1-oxyl, structure, XLIX, 437
Tetramethyl-1,3-cyclobutanedione, XLIX, 377
2,2,6,6-Tetramethyl-4-ene-*N*-oxylpiperidine, as yeast spin label, XXXII, 828, 831
Tetramethylenephenylenediamine, as electron mediator, LV, 668
N,N,N′,N′-Tetramethylethylenediamine, XLIV, 57, 58, 177, 185, 187, 188, 193, 194, 273, 455, 458, 476, 662; XLVIII, 9, 10; LIX, 431, 507, 509, 539, 544
  for chromaffin granule isolation, XXXI, 383
  for gel electrophoresis, XXXII, 85, 94
  in gel preparation, LI, 342, 343; LII, 325
2,3,5,6-Tetra-*O*-methyl-D-galacitol, acetylation, L, 4
3,7,11,15-Tetramethylhexadecanoic acid, *see* Phytanic acid
N-4-(2,2,6,6-Tetramethyl-1-oxylpiperidinyl)bromacetamide, XLIV, 368
N-(2,2,6,6-Tetramethyl-1-oxypiperid-4-yl)-*N′*-cyclohexylcarbodiimide, *see* Carbodiimide
Tetramethylphenylenediamine, hydrogen donor, of cytochrome *P*–450 peroxidase activity, LII, 411

N,N,N′,N′-Tetramethylphenylenediamine
assay of cytochrome *c* oxidase, LIII, 46
of succinate dehydrogenase, LIII, 466, 470
purification, LIII, 470
reductant, of cytochrome *c*, LIII, 159, 160
N,N,N′,N′-Tetramethyl-*p*-phenylenediamine
chromatophore planar membranes, LV, 756
as electron donor, XXXI, 701
electron-transfer chemiluminescence, LVII, 497, 499, 501–504
as mediator-titrant, LIV, 408, 423
model redox reactions, LV, 542
reaction with 9,10-dimethylanthracene, LVII, 504
energy balance, LVII, 501
with pyrene excimer, LVII, 502–504
with rubrene, LVII, 497, 501
structural formula, LVII, 498
transport studies, LVI, 257, 260
2,3,5,6-Tetramethylphenylenediamine, as redox mediator, LIV, 423, 431
2,2,6,6-Tetramethyl-4-phosphopiperidine-1-oxyl, synthesis, XLIX, 458
2,2,6,6-Tetramethylpiperidine-1-oxyl
electron spin resonance spectra, XXXII, 174–176; XLIX, 378, 384, 385
as yeast spin label, XXXII, 828, 831
2,2,6,6-Tetramethylpiperidine-1-oxyl-adenosine triphosphate, XLIX, 355
2,2,6,6-Tetramethylpiperidine-1-oxyl-4-ol, electron spin resonance spectra, XLIX, 398, 402, 403
2,2,6,6-Tetramethylpiperidine-1-oxyl-4-one, electron spin resonance spectra, XLIX, 395, 399–402, 413
2,2,6,6-Tetramethylpiperidinol-1-oxyl, electron spin resonance spectra, XLIX, 378
2,2,6,6-Tetramethyl-4-piperidone, CAT_n spin probe synthesis, LVI, 518
N-(2,2,5,5-Tetramethyl-3-pyrolidinyl-1-oxyl)bromoacetamide, modification of cysteine, LIII, 143
N-(2,2,5,5-Tetramethyl-3-pyrolidinyl-1-oxyl)iodoacetamide, modification of methionine, LIII, 139
Tetramethylsilane, XLIX, 256, 260
Tetramycin, *see* Tetracycline hydrochloride
Tetranitromethane, XLIII, 662; XLVI, 49–52, 64; XLVII, 67, 409; LV, 517
assay of superoxide dismutase, LIII, 387
inhibitor of crotalase, XLV, 235
modification of tyrosine, LIII, 172
in studies of luciferase active site, LVII, 180
Tetrapentylammonium, as penetrating ion, LV, 599, 600
Tetraphenylborate
calcium-sensitive electrodes, LVI, 347
ionophores, LV, 439
as penetrating ion, LV, 598, 599
Tetraphenylboron
as chelator, LVIII, 126
membrane potential, LV, 678
transhydrogenase assay, LV, 273, 274, 811, 814
transport, LVI, 407
uptake, LV, 168
Tetraphenylphosphonium, proteoliposomes, LV, 765
Tetrasaccharide, qualitative analysis, XLI, 18
Tetrasodium iminodiphosphate, preparation, XXXVIII, 421, 422
Tetra-*p*-sulfanatophenylporphinylterbium(III), XLIX, 269
Tetrathionate, carnitine assay, LV, 211, 212
Tetrazolium chloride, for petite negative yeast mutant enrichment, XXXII, 839, 840
Tetrazolium dye reduction, by aldolase in its detection, XLI, 67
Tetrazolium overlay
pet9 strains, LVI, 126
rho colonies, LVI, 120, 125
Tetrazolium test, XLIV, 276
Tetrin, XLIII, 135
Tetrodotoxin, XLVI, 260, 288
neurotoxic action, XXXII, 310
Tetrose bisphosphate, XLVI, 20
TF-VII, *see* Tissue factor VII

TGA, *see* Thermogravimetric analysis
Thalassemia, erythrocyte shape, XXXII, 52
Thallium, atomic emission studies, LIV, 455
Thallium-203, lock nuclei, XLIX, 348, 349; LIV, 189
Thallium-205
  lock nuclei, XLIX, 348, 349; LIV, 189
  resonance frequency, LIV, 189
Thallium acetate, as lock nuclei source, LIV, 189
Thallium ethylate, XL, 24, 29
THAM, *see* Tris(hydroxymethyl)aminoethane
Thawing medium, for liver preservation, XXXI, 4
Thaw-mounting, of autoradiographic specimens, XXXVI, 146
Theca cell, hormone effects, XXXIX, 424
Thenoyltrifluoroacetone
  complex II, LVI, 586
  inhibitor of dihydroorotate dehydrogenase, LI, 68, 69
    of reconstituted electron-transport system, LIII, 50, 51
    of succinate dehydrogenase, LIII, 410
    of succinate-ubiquinone reductase, LIII, 26
  succinate oxidation, LV, 460, 461
Theobromine, phosphodiesterase, XXXVIII, 244
Theophylline
  cAMP isolation, XXXVIII, 41
  cyclase assays, XXXVIII, 121, 136, 143, 171, 177
  effect on adenylate cyclase, XXXIX, 253
    on water flow, XXXVII, 256
  guanylate cyclase, XXXVIII, 196
  in interstitial cell studies, XXXIX, 259
  phosphodiesterase, XXXVIII, 244, 256, 272
Thermal conductivity, cell, in para hydrogen conversion technique, LIII, 307, 308
Thermal lens effect, XLIX, 91
Thermionic cooling device, XXXII, 216
Thermistemp temperature control system, XXXIX, 25
Thermistor, XLIV, 598, 667–676
  low-temperature, for Mössbauer spectroscopy, LIV, 377
Thermoanalysis, assay, XLIV, 659–676
Thermoanalytical method, XXXII, 263
Thermobacterium acidophilus, nucleoside deoxyribosyltransferase, LI, 454
Thermogram, principle, LVII, 542
Thermogravimetric analysis, XXXII, 263
Thermolysin, XXXIV, 435–440, 592
  acyl azide binding, XLIV, 95
  amino acid composition, XLV, 423
  assay, XXXIV, 436
    by spectroscopy, XLIV, 89–91
  autodigestion, XLVII, 188
  from *Bacillus thermoproteolyticus*
    hydroxylamine cleavage, XLVII, 140, 142
    specificity, LIX, 601
  cobalt-substituted, magnetic circular dichroism, XLIX, 168
  conjugates
    activity assay, XLIV, 94
    bound protein determination, XLIV, 94
    kinetic behavior, XLIV, 95
    thermal stability, XLIV, 97, 98
  diazo binding, XLIV, 93
  histidine modification, XLVII, 433, 438, 441
  immobilization, XLVII, 44
    on aminoaryl derivatized copolyacrylamide, XLIV, 93, 94
    on Enzacryl AH, XLIV, 95
  neodymium-substituted, magnetic circular dichroism, XLIX, 178, 179
  phosphoramidon, inhibited, XLV, 693
  primary cleavage agent, XLVII, 184
  properties, XLVII, 175
  protein cleavage, XLVII, 175–189
    applications, XLVII, 180–188
    cleavage sites, XLVII, 176–178
    specificity, XLVII, 175
  purification, XXXIV, 437
  reaction conditions, XLVII, 183

source, XLIV, 89
spacer, XXXIV, 438–440
thermostability, XLVII, 181
Thermometer, Hewlett-Packard electronic, LII, 223
Thermomycolin
active site sequence, XLV, 424
amino acid composition, XLV, 423
assay, XLV, 418
characteristics, XLV, 415
glucagon, digestion, XLV, 428, 429
inhibitors, XLV, 424
insulin digestion, XLV, 428, 430
properties, XLV, 421, 422
protease type, XLV, 421
purification, XLV, 419
source, XLV, 416
specificity, XLV, 424
stability, XLV, 431
Thermophilic bacterium, culture, LV, 782
Thermopile
blackened, in measurement of light, LVII, 575, 576
photometer calibration, LVII, 581, 582
principle, LVII, 542
Thermoviridin, XLIII, 342
Therm-O-Watch electronic controller, XXXII, 748
Theta antigen, XXXIV, 216
ganglioside, L, 250
on mouse thymocytes, XXXII, 67
THF, *see* Tetrahydrofuran
Thiamin, XXXIV, 303
in media, LVIII, 54, 65
Thiamin-binding protein, XXXIV, 303, 304
Thiamin dehydrogenase
flavin linkage, LIII, 450
flavin peptide
amino acid analysis LIII, 457
fluorescence, LIII, 462
interactions, LIII, 465
Thiamin monophosphate, XXXIV, 303
in media, LVIII, 65
Thiamin monophosphate-agarose, XXXIV, 304
Thiamin pyrophosphate, XXXIV, 303, 304
in media, LVIII, 65
Thiamin pyrophosphate-agarose, XXXIV, 303–304
Thiamphenicol, XLIII, 254
administration, LVI, 35, 37, 39
Thianthrene
chemiluminescent reaction, LVII, 519, 525, 526
in electrochemiluminescence step experiment, LVII, 512
electron-transfer chemiluminescence, LVII, 497
structural formula, LVII, 498
Thiazine, XLVI, 150
Thick-layer chromatography, purification of *Cypridina* luciferin, LVII, 368
Thiele modulus, XLIV, 421
7β-Thienylacetamidocephalosporadesic acid, *see* Diacetylcephalothin
β-2-Thienylalanine, substrate of phenylalanine hydroxylase, LIII, 284
β-(2-Thienyl)-DL-alanine, XLVII, 32
internal standard, XLVII, 32, 34, 36
β-(2-Thienyl)-DL-serine, XLVII, 34, 40
Thin-layer chromatography, XLIII, 172–213, *see also* specific substance
actinomycin, XLIII, 187
adsorbents, XLIII, 176
adsorption chromatography, XLIII, 173, 174
adsorption coefficient calculation, XLIII, 174
alumina, XLIII, 176, 187, 189
amidase, XLIII, 207
of amino acid thiohydantoins, XLVII, 363–365
antibiotics resistant to β-lactamase or amidase, XLIII, 208
antifoam agents, XLIII, 204, 205
basic water-soluble antibiotics, XLIII, 187, 200
bioautography, XLIII, 179, 186, 194, 195, 198
biosynthesis of antibiotics, XLIII, 201–213, *see also* Biotransformation

biotransformation of antibiotics, XLIII, 201–213
by single-enzyme, XLIII, 206–209
in vivo, XLIII, 209–213
biphenyl metabolites, LII, 403, 404
buffers, XLIII, 179
cAMP inhibitor, XXXVIII, 277
cellulose, cAMP separation, XXXVIII, 156–158
on cellulose 300 (MN), XLIII, 200
cGMP separation, XXXVIII, 77–78, 91
chambers, XLIII, 178
chromatoplate, XLIII, 175, 176, 179
circular, XLIII, 180
for classification of antibiotics, XLIII, 186–201
for conjugation of antibiotics, XLIII, 210
dichlorofluorescein indicator, XXXV, 319, 324, 325
drum, XLIII, 178
Eastman chromatogram sheets, XLIII, 191
7-ethoxycoumarin, LII, 374
7-ethoxyresorufin, LII, 374
evaporation of mobile phase, XLIII, 178
extraction process, XLIII, 210
fermentation, XLIII, 201–206
antifoam agents, XLIII, 204, 205
chloroform, XLIII, 205
hexane as solvent, XLIII, 205
intracellular antibiotics, XLIII, 205
pH, XLIII, 205
sample preparation, XLIII, 204
tetracycline, XLIII, 204
water-soluble antibiotics, XLIII, 205, 206
fluorescent indicator, XLIII, 178
of glycerides and phosphoglycerides, XXXV, 319, 321, 324, 325
of glycolipids, XXXII, 356–358
of glycosphingolipids, XXXV, 396–425
gradient, XLIII, 179, 180
of 2-hydroxy and 3-hydroxy fatty acids, XXXV, 327, 328
hydroxylated steroids, LII, 380
identification of coenzyme Q analogs, LIII, 596
of quinone intermediates, LIII, 602, 603
of ubiquinones, LIII, 639
indicators for lipids, XXXV, 403–408, 412, 413, 415–417
iodine vapor, indicator for glycerides, XXXV, 321
ion-exchange, XLIII, 180, 181, *see also* Ion-exchange thin-layer chromatography
Kieselguhr G, XLIII, 176, 196, 197
of lipids, XXXV, 396–425
with lipophilic Sephadex, XXXV, 385
macrolide antibiotics, XLIII, 187
mobile phase, XLIII, 177, 178
nucleoside antibiotics, XLIII, 187
organic supports, XLIII, 176, 177
partition chromatography, XLIII, 174, 175
peptide antibiotics, XLIII, 187
in pharmacokinetics, XLIII, 209–213
of N^{α}-phenylthiocarbamyl methylamide-amino acids, XLVII, 379–382
of phenylthiohydantoins, XLVII, 348–351
of phospholipids, XXXV, 397, 398, 400, 403–405, 408, 412, 416, 417
of plant lipids, XXXI, 526, 527
plate cleaning, XXXVI, 33
polyene antibiotics, XLIII, 187
(polyethyleneimine)cellulose, XXXVIII, 27
biological material, XXXVIII, 37, 38
cyclic nucleotide separations, XXXVIII, 30–35
elutions, XXXVIII, 38
materials and general methods, XXXVIII, 28
nucleoside triphosphates, XXXVIII, 35, 36
nucleotide separation, XXXVIII, 28–30
pyrimidine nucleotide separation, XXXVIII, 35
related purine nucleotides, XXXVIII, 36, 37
preparative, XLIII, 184, 185

programmed multiple development chromatography, XLIII, 180
protoporphyrin IX dimethyl ester, LII, 452
purification of coenzyme Q_2 analog, LIII, 595
of flavin peptides, LIII, 456
of quinone intermediates, LIII, 605–609
quantitative, XLIII, 185, 186
bioautography, XLIII, 186
photodensitometry, XLIII, 185
with solution spectrophotometry, XLIII, 186
spectrofluorimeter, XLIII, 185
of radiolabeled peptides, XLVII, 256
in radioligand assays, XXXVII, 10
rapid-column chromatography, XLIII, 182
reflectance spectroscopy, XLIII, 181, 182
reversed-phase partition chromatography, XLIII, 180
R_f values, XLIII, 192, 193
reproducibility, XLIII, 182, 183
R_M values, XLIII, 183, 184
samples
preparation for spotting, XLIII, 178
uniform application to plate, XLIII, 177
screening antibiotics for antitumor activity, XLIII, 203
separation of antibiotics, XLIII, 186–201
of bases, LI, 239
of bases and nucleosides, LI, 550, 551
of flavin compounds, LIII, 427
of nucleosides and nucleotides, LI, 308, 315, 316
of orotate, orotidylate and urididylate, LI, 158, 159
of uridine and uracil, LI, 291, 292
on Sephadex G–15, XLIII, 198, 199
silica gel, XLIII, 176, 180, 188
siroheme, LII, 442
sirohydrochlorin octamethyl ester, LII, 445, 446
solid support, XLIII, 176
solvent equilibrium, XLIII, 182
solvent systems, XLIII, 188, 189–191
source of plates, XLVII, 327
spotting, preparation of sample, XLIII, 178
spray reagent, XLIII, 178
steroid purification, XXXVI, 38, 39, 47
steroids, LII, 126
two-dimensional
materials, LIX, 62, 63
peptide mapping, XLVII, 195–204
procedure with labeled tRNA hydrolysate, LIX, 71–74
water-soluble basic antibiotics, XLIII, 187, 200
Thin-layer electrophoresis, XLIII, 286, 287
Thioacetamide, peptide deprotection, XLVII, 517, 549
Thioacetylmorpholide, synthesis, XLVII, 290, 291
Thioacetylpiperidine, XLVII, 296
Thioacetylthioglycolic acid, sequence analysis, XLVII, 296–298
4′-Thioadenosine, substrate of adenosine deaminase, LI, 507
Thioamide, substrate of amine oxidase, LII, 151
Thiobacillus thiooxidans, oxidase, LII, 11
Thiobarbituric acid
assay
for determination of lipid peroxidation, LII, 306
interference with, LII, 309
determination of nucleoside phosphorolysis, LI, 526
in estimation of 3-deoxy-2-ketonaldonic acids, XLI, 32
Thiocapsa pfennigii
growth, LIII, 335
high-potential iron-sulfur protein, LIII, 329, 332, 333, 339
Thiocarbonyl diimidazole, XLVI, 165
2-Thiocarboxamido-6-methoxybenzothiazole, preparation, LVII, 25
Thiocholine ester, as esterase substrates, XXXII, 88
Thioctic acid, *see* α-Lipoic acid
Thiocyanate
calcium-sensitive electrodes, LVI, 347

inhibitor of firefly luciferase, LVII, 15
of salicylate hydroxylase, LIII, 541
ionophores, LV, 439
in IR spectroscopy, LIV, 304
measurement of proton translocation, LV, 635, 636
permeability, LVI, 252
stretching frequency, LIV, 311
Thiocyanate ion, fluorescence quenching agent, XLIX, 225, 231
Thiocyanogen
chemical properties, XLVII, 412
S-trityl group removal, XLVII, 538
4-Thiodeoxythymidylate, inhibitor of deoxythymidylate phosphohydrolase, LI, 289
Thiodigalactoside-binding lectin, *see* Electrolectin
Thiodiglycol, purity check, XLVII, 15
Thioester, XLVI, 158
cleavage or formation, assay of carnitine palmitoyltransferase, LVI, 369, 370
Thioester group, XLIV, 12–14
stability, XLIV, 14
Thio ether, XLVI, 289–295
oxidation by I_2, LV, 537
β-Thiogalactoside-binding protein, XXXIV, 7
Thioglycolate broth, LVIII, 20
Thioglycolic acid, XLIV, 524
activator of formate dehydrogenase, LIII, 371
in iron analysis, LIV, 439, 443
peptide hydrolysis, XLVII, 616
stability, in solution, LIV, 436
6-Thioguanine
hybrid selection, LVIII, 352
resistant strains, LVIII, 314
substrate of guanine deaminase, LI, 516
6-Thioguanosine, inhibitor of adenosine deaminase, LI, 507
6-Thioguanosine 5′-monophosphate, XLVI, 302
Thiohemiacetal, XLVI, 25
Thiohydantoin, XLVI, 167
6-Thioinosine, inhibitor of adenosine deaminase, LI, 507
6-Thioinosine 5′-monophosphate, XLVI, 302
Thioketone, XLVI, 88
Thiol, determination, XXXIV, 537
Thiolase, *see also* Acetyl-CoA acetyltransferase
in glyoxysomes, XXXI, 569
Thiol-containing peptide, XXXIV, 546
Thiol-disulfide interchange, XXXIV, 532–544
glutathione-agarose conjugate, XXXIV, 536–538
glutathione-2-pyridyl disulfide-agarose, XXXIV, 537, 538
mixed-disulfide matrix, XXXIV, 533–536
thiol determinations, XXXIV, 537
thiol-polymer conjugate, XXXIV, 535
Thiol ester, factor XIII assay, XLV, 183
Thiolesterase, interference, in assay of 3-ketoacyl-CoA thiolase, XXXV, 130
Thiol group, oxidative phosphorylation system, LVI, 585
7-Thiolincomycin, synthesis of *S*-(3-carbethoxy-3-diazo)acetonyl-7-thiolincomycin, LIX, 814
Thiol inhibitor, complex II
Thiolnitrobenzoate anion, extinction coefficient, XLIV, 522
Thiol-polymer conjugate, XXXIV, 535
Thiol reagent
identification of iron-sulfur centers, LIII, 272
inhibitor of complex II, LIII, 26
of succinate dehydrogenase, LIII, 33
p-Thiomethylphenyl ester moiety, peptide synthesis, XLVII, 522
Thionicotinamide adenine dinucleotide, transhydrogenase assay, LV, 270–272
Thionine, as mediator-titrant, LIV, 408
5-Thio-2-nitrobenzoate, XLIII, 742; XLVII, 411, 434
Thionyl chloride, XLIV, 145; XLVI, 236, 237; XLVII, 268, 523
synthesis of oleylpolymethacrylic acid resin, LIII, 213, 214
Thiooxidase, LII, 10
Thiopeptin, XLIII, 147
Thiophosgene, XXXIV, 55; XLIV, 86, 99, 141; XLVI, 166
2-Thiopyridone, XLIV, 385

Thioredoxin
assay, LI, 231
reducing substrate of pyrimidine nucleoside monophosphate kinase, LI, 330
of ribonucleoside diphosphate reductase, LI, 227
of ribonucleoside triphosphate reductase, LI, 247, 258
reduction, LI, 227
spectrophotometric determination, LI, 249
Thioredoxin reductase
assay, LI, 231
of ribonucleoside diphosphate reductase, LI, 227
flavoprotein classification, LIII, 397
for reduction of thioredoxin, LI, 247
Thiosemicarbazide
glycogen staining, XXXIX, 152
incorporation in tRNA, LIX, 118
incubation of native collagen, XL, 363
Thiostrepton, LVI, 31
chromatographic data, XLIII, 146, 147
effect on protein synthesis, LIX, 853, 854, 862
inhibition of protein chain elongation, LX, 327–329, 724, 725, 741
inhibitor of polypeptide chain elongation, LIX, 359
peptide antibiotic lactonase, XLIII, 767–773
Sarcina lutea, XLIII, 771
structure, XLIII, 768
Thiosulfate, transport, LVI, 292
Thiosulfate oxidase, LII, 21
Thiourea, XLVI, 165
derivatives, XLVI, 167
linkage to glass, XXXIV, 68
to protein, XXXIV, 55
in MEOS assay, LII, 366
substrate of amine oxidase, LII, 151
Thioureylene, substrate of amine oxidase, LII, 151
6-Thiouric acid, molar extinction coefficient, LI, 532
4-Thiouridine, XLVI, 185, 632, 684, 685, 692, 693
base, XLVI, 184
chromatographic behavior, LIX, 169
electrophoretic mobility, LIX, 91
NMR spectrum, LIX, 39
4-Thiouridine monophosphate, XLVI, 89
6-Thioxanthylate, substrate of guanylate synthetase, LI, 224
Thomas Model C Teflon-glass homogenizer, LI, 521
Thomas-Sheaff object marker, XXXII, 794
Thorium, colloidal, as stain, XXXIX, 152, 153
Thorotrast tracer, for electron microscopy, XXXIX, 148, 152
Three-photon method, principle, LIV, 8–11
Threonine, XLIV, 12–15
aminoacylation isomer specificity, LIX, 274, 275, 279, 280
esterification, XLVII, 524
fluorescent labeling, XLVIII, 358
hydrogen bromide cleavage, XLVII, 603
hydrophobicity, XLIV, 14
hydroxyl group blocking, XLVII, 527, 528, 543, 544, 613
in media, LVIII, 53, 62
misactivation, LIX, 289
phenylthiocarbamyl-*n*-propylamide derivative, identification, XLVII, 377
phenylthiohydantoin derivative
coeluting artifact, XLVII, 273
loss, XLVII, 339, 340
recovery, XLVII, 347
radiolabeled, in pulse labeling of parathyroid tissue, XXXVII, 349
recovery, XLVII, 371
thermolysin hydrolysis, XLVII, 175
thiazolinone, half-life, XLVII, 376
Threonine deaminase, XXXIV, 5
L-Threonine deaminase-Leu-tRNA^{Leu}, protein-tRNA interaction studies, LIX, 324, 326, 330
L-Threonine dehydratase, XLVI, 64
Threonyl-tRNA synthetase, XXXIV, 170
purification, LIX, 262, 263, 267
subcellular distribution, LIX, 233, 234
L-Threo-1-*p*-phenylphenyl-2-dichloroacetamido-1,3-propanediol, XLIII, 366
D-Threose 2,4-bisphosphate, XLVI, 21

Thrombin, XXXIV, 5, 445–448; XLVI, 128, 206, 207
amino acid composition, XLV, 232
assay, XXXIV, 447, 448; XLV, 126, 156–160
clotting time, XLV, 159
hirudin, XLV, 677
substrates, XLV, 160
bovine
factor V, XLV, 110, 117
multiple forms, XLV, 170
preparation, XLV, 161
structure, XLV, 175
cleavage sites, XLV, 168
equine, purification, XLV, 166
fibrinogen clotting, XLV, 35, 36
human
active site titrations, XLV, 166
esterase activity, XLV, 167
multiple forms, XLV, 172
purification, XLV, 165, 166
inhibitors, XLV, 173
neutralization, XLV, 656
antithrombin-heparin cofactor specificity, XLV, 667
porcine, purification, XLV, 166
purification, XXXIV, 447; XLV, 175
role, XLV, 156
specificity, XLV, 168
succinylation, XXXIV, 446
yeast electron microscopy, LVI, 720
α-Thrombin
bovine
structure, XLV, 175
substrates, action, XLV, 102
in thrombin purification, human, XLV, 166
human, structure, XLV, 176
β-Thrombin
bovine
action, on substrates, XLV, 172
structure, XLV, 175
human, structure, XLV, 176
γ-Thrombin, bovine, substrates, action, XLV, 172
Thromboplastin, tissue, XLV, 109
Thunberg optical cell, LIII, 269
Thylakoid, in algae, XXXII, 865
Thymic antigen, XXXIV, 717
Thymidine
DNA incorporation, in liver explants, XXXIX, 40
DNA inhibitor, XL, 276
^{3}H-labeled
localization of DNA, XL, 13
in vivo cell cycle analysis, XL, 59
incorporation, LVIII, 251, 252
inhibitor of cytosine deaminase, LI, 400
in media, LVIII, 67
optical spectra, XLIX, 165
2-oxoglutarate dioxygenase, LII, 12
product of deoxythymidylate phosphohydrolase, LI, 285
substrate of nucleoside phosphotransferase, LI, 391
of pyrimidine nucleoside phosphorylase, LI, 432, 436
of thymidine kinase, LI, 365
of thymidine phosphorylase, LI, 437, 441, 442
of uridine phosphorylase, LI, 424, 429
synchrony, LVIII, 261
tritiated, as mutagen, LVIII, 316, 317
Thymidine 3′,5′-di(*p*-nitrophenyl phosphate), XXXIV, 495
Thymidine 2′-hydroxylase, *see* Thymidine 2-oxoglutarate dioxygenase
Thymidine kinase, XXXIV, 5; XLVI, 89; LI, 354–360, 365–371, *see also* Deoxythymidine kinase
activity, LI, 365
assay, LI, 354–356, 365–367
from blast cells of myelocytic leukemia, LI, 365
cation requirement, LI, 358, 370
deficiency, in cell hybridization, XXXII, 576, 577
from *E. coli*, LI, 354–360
inhibitor, LI, 371
in intestinal mucosa, XXXII, 666
kinetic properties, LI, 359, 360, 370
lack of, in hybrid cells, XXXIX, 123, 124
molecular weight, LI, 359, 369, 370
pH optimum, LI, 358
properties, LI, 358–360, 369–371
purification, LI, 356–358, 362–364

reaction mechanism, LI, 371
regulation, LI, 354
stability, LI, 358, 369
substrate specificity, LI, 358, 370
temperature sensitivity, LI, 359
ultraviolet sensitivity, LI, 359
Thymidine 5′-phosphate 3′-(*p*-nitrophenol phosphate), XXXIV, 495
Thymidine phosphorylase, LI, 437–445, *see also* Pyrimidine nucleoside phosphorylase
activity, LI, 437, 438, 442
assay, LI, 438, 442, 443
from *E. coli*, LI, 442–445
from *Haemophilus influenzae*, LI, 432
kinetic properties, LI, 441, 445
molecular weight, LI, 441, 445
pH optimum, LI, 442, 445
regulation, LI, 438, 517
from *Salmonella typhimurium*, LI, 437–442
separation from uridine phosphorylase, LI, 427, 429
stability, LI, 442, 445
substrate specificity, LI, 441, 444
subunit structure, LI, 441, 445
transferase activity, LI, 431
Thymidylate
inhibitor of orotate phosphoribosyltransferase, LI, 163
of orotidylate decarboxylase, LI, 163
Thymidylate synthase, LI, 90–101
assay, LI, 90–92, 101
inhibitors, LI, 96
isoelectric point, LI, 96
kinetic properties, LI, 96
from *Lactobacillus casei*, LI, 90–97
molecular weight, LI, 95
pH optimum, LI, 96
properties, LI, 95–97
purification, LI, 92–95, 98–104
spectral properties, LI, 96
stability, LI, 96
stabilization by thiols, LI, 96, 97
subunit structure, LI, 96
ternary complex formation, LI, 97
Thymidylate synthetase, XXXIV, 520–523; XLVI, 164, 307–312, *see also* Thymidylate synthase
inhibition, XXXII, 576
quantitation, XXXIV, 522, 523
Thymine
determination, spectrophotometric, LI, 433, 438
extinction coefficient, LI, 433, 438
in media, LVIII, 67
2-oxoglutarate dioxygenase, LII, 12
product of cytosine deaminase, LI, 395
of thymidine phosphorylase, LI, 437, 442
substrate of orotate phosphoribosyltransferase, LI, 152
Thymine 7-hydroxylase, *see* Thymine, 2-oxoglutarate dioxygenase
Thymocyte, XXXIV, 206, 211, 713, 750–755
prostaglandin binding, XXXII, 109
species-specific antigens, XXXII, 67–69
Thymol blue, XLIV, 131
Thymus
cell, canine, LVIII, 414
lymphocyte preparation, LVIII, 124
mitochondria, LV, 19
nuclei isolation, XXXI, 246–252
protein kinases, XXXVIII, 298
Thyristor, in flash photolysis apparatus, LIV, 96
Thyroglobulin, XXXIV, 610, 611
in thyroid cells, XXXII, 745
Thyroglobulin-agarose, XXXIV, 611
Thyroid
cell
iodide transport, XXXII, 756–758
isolation and culture, XXXII, 745–758
morphology, XXXII, 749–752
separation, XXXIX, 229
histology, XXXI, 144, 145
mitochondria, LV, 18
pineal effects, XXXIX, 382
plasma membrane, XXXIV, 694
isolation, XXXI, 144–149
prostaglandin binding, XXXII, 110

stimulator, long-acting, effect on adenylate cyclase, XXXI, 147
Thyroid gland, perfusion, XXXIX, 359–364
Thyroid hormone
effect on collagens, preparation, dosage, route of administration, XL, 320
metabolites, XXXVI, 483–546
in serum, measurement, XXXVI, 126–132
serum-binding proteins, XXXVI, 89–132
Thyroid stimulating hormone, *see also* Thyrotropin receptor
cation-exchange chromatography, XXXVII, 371–373
effects on glucose oxidation, XXXVII, 262–268
on thyroidal iodide transport, XXXVII, 256–262
high purification, XXXVII, 374, 375
human purification, XXXVII, 382, 388, 389
LH activity separation, XXXVII, 373, 374
inactivation, XXXVII, 375
preparation, XXXVII, 370–375
sialic acid content, XXXVII, 323
thyroid cell response, XXXII, 745–746, 753, 755, 757, 758
thyrotropin releasing factor effects, XXXVII, 87
tumor secretion, XXXVII, 92
unit of activity, XXXVII, 362
Thyroid-stimulating hormone receptor, gangliosides, L, 250
Thyroid-stimulating hormone-releasing hormone, autoradiography, XXXVI, 150, 155
Thyrotropic hormone, *see* Thyrotropin receptor
Thyrotropic hormone-agarose, XXXIV, 692, 693
Thyrotropin, *see also* Thyroid-stimulating hormone
plasma-membrane receptor sites, XXXI, 145
secretory granule use in study, XXXI, 410, 418
Thyrotropin receptor, XXXIV, 692–695
cultured thyroid cells, XXXIV, 694
lithium diiodosalicylate solubilized, XXXIV, 694
trypsin-released, XXXIV, 695
TSH-agarose, XXXIV, 694
Thyrotropin releasing factor
antisera, XXXVI, 17
commercial source, XXXVII, 415
effects on TSH and prolactin, XXXVII, 87, 92
in media, LVIII, 99
physical constants, XXXVII, 415
pituitary receptor binding assay, XXXVII, 214–219
purification, XXXVII, 414, 415
radiolabeled, XXXVII, 213
solvent extraction, XXXVII, 402–407
synthesis, XXXVII, 408–415
classic, XXXVII, 408, 409
solid phase, XXXVII, 409, 411–415
Thyroxine, XXXIV, 385–389
adenylate cyclases, XXXVIII, 148, 149, 173, 176
antibody against preparation, LVII, 425, 426
autoradiography, XXXVI, 150
conversion to L-triiodothyronine determination, XXXVI, 537–546
in humans, XXXVI, 541–546
effect on pituitary hormones, XXXVII, 91, 92
free, in serum, XXXVI, 127, 128
labeled, purification, XXXVI, 128
in media, LVIII, 63
M. phlei membrane systems, LV, 186
plasma protein interaction, XXXVI, 3
protein-binding assay, XXXVI, 35
synthesis of *N*-trifluoroacetylthyroxine, LVII, 432
TBG binding, XXXVI, 126, 129–132
in thyroid cells, XXXII, 745
L-Thyroxine, analog, XLVI, 435
Thyroxine-agarose, XXXIV, 386, 387
storage, XXXIV, 389
Thyroxine-binding globulin, human, XXXIV, 385–389
preparation, XXXIV, 387, 388

properties, XXXIV, 388
Thyroxine-binding pre-albumin, XXXVI, 126
capacity determination, XXXVI, 129–132
measurement, XXXVI, 132
Thyroxine-binding protein, in serum, measurement, XXXVI, 126–132
Thyroxyl-*N*-carbonylmethylprealbumin, XLVI, 436
Tick cell line, media, LVIII, 463
Ticosan, XXXIX, 419
Time averaging, electron spin resonance spectra, XLIX, 406, 407
Time compression procedure, XLVIII, 481
Time-release plot, XLVII, 75, 81, 82
Time-sharing equipment, type, LIV, 104
multiwavelength, LIV, 108–111
Tin
atomic emission studies, LIV, 455
in media, LVIII, 69
radioactive, counting, LV, 508
volatilization losses, in trace metal analysis, LIV, 481
Tin(IV), in porphyrin ring conformation studies, LIV, 245
Tin-117, lock nuclei, XLIX, 349; LIV, 189
Tin-119, lock nuclei, XLIX, 349; LIV, 189
Tin oxide, in semiconductor electrodes, LIV, 405
TiP, *see* Tumor-inducing principle
Tipping angle, resonance, XLIX, 282, 290
Tissue, *see also* specific types
androgen production and interconversion rates, XXXVI, 74, 75
concentration of substances, LVI, 449
cyclic nucleotide levels, XXXVIII, 93, 94, 104
extraction techniques, XXXVIII, 7–9, 14, 24, 25, 68, 75
fragment, bone, in collagen metabolic studies, XL, 326
metabolism, polarography, LVI, 478, 479
metabolite contents, LV, 222
mitochondria, LVI, 3
perchloric acid extraction, LV, 201, 202
preparation
for cAMP assay, XXXVIII, 54, 55, 100
in study of hormone effect on collagen, XL, 310
preservation of ultrastructure, XLIV, 553
purification of extracts, cGMP assay, XXXVIII, 87, 88, 100
rapid fixation, XXXVIII, 3–6, 75
storage, XXXVIII, 7
rapid freeze procedure, LIX, 311
soft, collagen metabolism studies, XL, 324
defatting, drying and demineralization, XL, 312
Tissue culture, LVIII, 119–131
cell line, in chromosome isolation, XL, 76
in collagen degradation studies, XL, 349
defined, LVIII, 263
of endocrine cells, XXXIX, 165
of granulosa cells, XXXIX, 199–210
incubator, XXXIX, 37
long-term cell, preservation, XXXI, 3
medium, XXXIX, 191, 199
multilayered, XXXII, 568–574
plasma membrane isolation from L cells, XXXI, 156–162
primary culture isolation, XXXIX, 112–116
Tissue dispersion, LVIII, 119–131
mechanical, LVIII, 130, 131
shearing, LVIII, 130, 131
solutions, LVIII, 124
Tissue disruption, LVIII, 119–131
Tissue dissociation, *see* Tissue dispersion
Tissue electrode, oxidases, LVI, 476
Tissue extract, guanosine nucleotide determinations, LVII, 85–94
Tissue factor, assay, XLV, 47
Tissue factor apoprotein
assay, XLV, 43, 44
carbohydrate composition, XLV, 41
characterization, XLV, 37, 40
preparation, XLV, 39, 43
purification, XLV, 37–39

relipidation, XLV, 44
Tissue factor VII, preparation, XLV, 104
Tissue grinder, XXXII, 120
Bolab, LV, 80
Tissue homogenate, drug isolation, LII, 333
Tissue powder, delipidated, preparation, XLV, 38
Tissue preparation, LVIII, 122, 123
mincing, LVIII, 124
Tissue press, XXXI, 95
Tissue retractor, for rat brain surgery, XXXIX, 168, 169
Titania, *see* Titanium oxide
Titanic chloride, XLIV, 48, 166, 167
Titanium, atomic emission studies, LIV, 455
Titanium oxide, XLIV, 153
available preparations, XLIV, 158
controlled-pore, urease immobilization, XLIV, 167–169
durability, XLIV, 135, 136
hydrogen ion concentration operating range, XLIV, 163
pore properties, XLIV, 137
Titanium sulfate, in catalase determination, LII, 497
Titanous chloride, XLIV, 91, 92, 166
Titrant flask, design, LIV, 126
Titration
continuous, apparatus, LIV, 124–126
oxidoreductive
anaerobic, LIV, 123–131
continuous, LIV, 124–130
Titrator
automatic, XLIV, 111
glass cap, commercial source, LIV, 126
Titrimetry, assay, XLIV, 346
TK, *see* Thymidine kinase
TKE buffer, XXXVI, 171
TKM reagent, for nuclei isolation, XXXI, 268
TLC, *see* Thin-layer chromatography
TLME, *see* *p*-Toluenesulfonyl-L-lysine methyl ester
T lymphocyte, XXXIV, 716, 721, 750–755
separation, XXXII, 636
TME, *see* Tyrosine methyl ester
TMK buffer, LIX, 296
TMP, *see* Thiamin monophosphate
TMP synthetase, *see* Thymidylate synthase
TMS, *see* Tetramethylsilane
TMV, *see* Tobacco mosaic virus
TNS, *see* 2-*p*-Toluidinylnaphthalene-6-sulfonic acid
Toad bladder, sodium transport in, aldosterone effects, XXXVI, 439–444
Tobacco
cell, callus culture, LVIII, 479
chloroplast isolation, XXXI, 602
crown-gall studies, XXXI, 554
enzyme studies, XXXI, 538
leaf, protoplast, LVIII, 302
organelles, XXXI, 493, 499
from protoplast, LVIII, 367
protoplast isolation, XXXI, 579
spherosome isolation, XXXI, 577
Tobacco hornworm, media, LVIII, 457
Tobacco mesophyll × chicken red cell, LVIII, 365
Tobacco mesophyll × HeLa culture, LVIII, 365
Tobacco mesophyll × soybean culture, LVIII, 365
Tobacco mosaic virus, XXXIV, 192
assembly, XLVIII, 309
light scattering properties, XLVIII, 437–439
membrane osmometry, XLVIII, 72, 74
as visual marker, XXXII, 63
Tobacco mosaic virus protein
chemical cleavage, XLVII, 148
thermolysin cleavage sites, XLVII, 176, 177
Tobramycin, XLIII, 612
Tobramycin acetyltransferase, XLIII, 623
Tocopherol
in lipid bilayer formation, XXXII, 491, 515
lipophilic Sephadex chromatography, XXXV, 395
α-Tocopherol, lipid peroxidation and, LII, 304
α-Tocopherol phosphate, *see* Vitamin E
Tocopherylquinone, reduction, by complex I, LIII, 13

Togavirus, interferon producing, LVIII, 293
Toggle valve, LVII, 316
Toluene, XLIV, 50, 58, 140, 146, 177, 230, 231, 480, 662; XLVIII, 336
cell permeability, LVI, 544
preservative in regenerated cellulose, LVII, 145
purification of 4-nitrophthalic anhydride, LVII, 427
of orotate phosphoribosyltransferase, LI, 70
as red blood cell hemolysing agent, LII, 448, 456
in scintillation cocktail, LI, 4, 30, 91, 157, 220, 292, 300, 323, 338, 355, 362, 409, 419, 509, 540, 544, 551; LII, 126, 312, 315, 321, 396, 398, 404; LIX, 183, 191, 232, 236, 323, 358, 363, 388, 396, 459, 565, 567, 793, 825, 832, 833
solvent, for dimethyldichlorosilane, LIX, 4
synthesis of 4-(4-methoxyphenyl)butyric acid, LVII, 434
treatment of mitochondria, LVI, 546
for yeast plasmolysis, LI, 292, 396
Toluene 2,4-diisocyanate, in immune conjugate preparation, XXXII, 61, 62
p-Toluenesulfonic acid, XLVI, 118; XLVII, 446, 449, 452, 455, 458, 466, 524
choline salt as intermediate in synthesis of phosphatidylcholine analogues, XXXV, 525
synthesis of adenosine derivative, LVII, 118
of dithiodiglycolic dihydrazide, LIX, 384
p-Toluenesulfonic acid monohydrate, XLVII, 544
p-Toluenesulfonyl, *see* Tosyl
N^{α}-Toluenesulfonyl-L-arginine methyl ester
assay, isotopic, XLV, 21
kallikrein assay, XLV, 305, 313
α_2-macroglobulin assay, XLV, 641
in plasminogen assay, XLV, 273
prekallikrein, XLV, 316
synthesis, XLV, 22
for thrombin assay, XLV, 160
trypsin inhibitor assay, XLV, 815
venom inhibitor assay, XLV, 875
p-Toluenesulfonyl chloride, XLIV, 130; XLVI, 231
p-Toluenesulfonyl-L-lysine methyl ester, kallikrein, XLV, 313
Toluidine blue
chondrocyte staining, LVIII, 563
for pellicle cutting, XXXII, 7
Toluidinonaphthalenesulfonic acid, inhibitor, of bioluminescence, LVII, 49
2,6-Toluidinonaphthalenesulfonic acid, luminescence, LVI, 543
2-*p*-Toluidinylnaphthalene-6-sulfonic acid, XLIV, 535
conformation studies, of chymotrypsin, XLIV, 365
Toluylene blue, as mediator-titrant, LIV, 408
Tolypocladium inflatum, growth conditions, LV, 494
Tolypomycin, XLIII, 137
Tomato
chloroplast isolation, XXXI, 602
mitochondria isolation, XXXI, 593
protoplasts, XXXI, 578, 579
ripening, XXXI, 495
vacuole isolation, XXXI, 575
Tonofilament, electron microscopy, XXXII, 32
Tonoplast, marker enzyme, XXXII, 393
Topoculture, XXXIV, 221, 224
Topographical state
definition, XLIX, 106
vibrational frequencies, XLIX, 111–116
Topographical structure, *see also* Protein
classes, XLIX, 104
definition, XLIX, 104
Torenia baillonii × *T. fournieri* petal, LVIII, 365
Torpedo
electric tissue, current output, XXXII, 310
nicotinic receptors, XXXII, 315–319, 322, 323
TORSCA multidimensional scaling program, LIX, 609–611

Tosyl, as blocking group, XLVI, 199
L-(1-Tosylamido-2-phenyl)ethyl chloromethyl ketone, XLVII, 168, 172
in studies of enzyme active sites, XXXVI, 390
L-1-Tosylamido-2-phenylethyl diazomethyl ketone, LVI, 660
N^{ω}-Tosyl-L-arginine, synthesis, XLVII, 534
N^{α}-Tosyl-L-arginine chloromethyl ketone, XLVI, 206, 229
synthesis, XLVI, 231, 232
Tosyl-L-arginine methyl ester, substrate, trypsin, XLIV, 530, 531, 561, 562, 573
Tosylation, XLIV, 32
Tosyl chloride, XLVII, 533, 536, 541
Tosyl group, removal, XLVII, 572, 573, 611, 612
N^{im}-Tosyl group, stability, XLVII, 540
Tosylhomoarginine methyl ester, inhibitor, of clostripain, XLVII, 168
Tosylhydrazone, XLVI, 94
N^{ε}-Tosyl-L-lysine, synthesis, XLVII, 535, 536
Tosyllysine chloromethyl ketone, XLVI, 206
active site label, at clostripain, XLVII, 168
Tosylnitroarginine, XLVI, 232
Tosylnitroarginine chloromethyl ketone, XLVI, 232
Tosyl-nylon, XXXIV, 212
N-Tosyl-L-phenylalanine, XLVI, 76, 130, 197
Tosylphenylchloromethyl ketone, XLVI, 206
Total power, definition, LVII, 597
Touch-mounting, in steroid autoradiography, XXXVI, 146
Toxicity, media, LVIII, 117, 453, 454
Toxin inhibitor, of protein synthesis, LX, 780, 781, *see also Pseudomonas aeruginosa* toxin A
Toyocamycin
biosynthesis, XLIII, 515, 516, *see also* specific enzyme
toyocamycin nitrile hydrolase, XLIII, 759–762
paper chromatography, XLIII, 154
RNA inhibitor, XL, 283
Toyocamycin nitrile hydrolase, XLIII, 759–762
activators, XLIII, 762
ammonium sulfate fractionation, XLIII, 760
assay, XLIII, 759, 760
heat effect, XLIII, 762
hydroxyapatite adsorption, XLIII, 762
inhibitors, XLIII, 762
metal ions effect, XLIII, 762
Michaelis constant, XLIII, 762
pH optimum, XLIII, 762
protamine sulfate treatment, XLIII, 760
purification, XLIII, 760–762
specific activity, XLIII, 760
specificity, XLIII, 762
Streptomyces rimosus preparation, XLIII, 760
unit definition, XLIII, 760
TpψpCpGp
inhibition of phenylalanyl-transfer ribonucleic acid binding to 70 S ribosomes, LX, 627, 628
preparation, LX, 618
TPA, *see* Triphenylamine
TPB, *see* Tetraphenylboron
TPCK, *see* N-Tosyl-L-phenylalanine
TPCK trypsin, in digestion of MS2 peptides, LIX, 301
TPF, *see* Two-photon fluorescent technique
T4 phage, XXXIV, 7
TPN, *see* Nicotinamide adenine dinucleotide phosphate
TPNH, *see* Nicotinamide adenine dinucleotide phosphate, reduced
TPNH-cytochrome *c* reductase, *see* NADPH-cytochrome *c* reductase
TPNH-dehydrogenase, *see* NADPH dehydrogenase
TPTA, *see* Tri-*p*-tolylamine
Trace element, in media, LVIII, 52, 69, 95
Trace metal, analysis
airborne contamination, LIV, 479
atomic spectroscopic methods, LIV, 446–471
blanks, preparation, LIV, 483

of enzyme preparations, procedure, LIV, 471–483
method verification, LIV, 483, 484
recovery studies, LIV, 481, 482
sample preparation, LIV, 479–482
standard additions method, LIV, 484
standards, LIV, 482, 483
Tracer iodination, of peptide hormones, XXXVII, 228, 229
Tracer superfusion method, for steroid dynamic studies, XXXVI, 75–88
Trachea
cell, embryonic bovine, LVIII, 29
incision, XXXIX, 170, 171
Tradescantia
free cells, XXXII, 724
nuclei extraction, XXXI, 564, 565
Tramestes, oxidase, LII, 21
Tranexamic acid, internal standard, XLVII, 34
Tranfection-infectivity assay, LVIII, 434, 435
Transaldolase, XLVI, 50, 51
activity determination, XLVII, 495
analytical chromatography, XLII, 292
from *Candida utilis*, XLII, 290–297
assay, XLII, 291–293
denaturation studies, XLIV, 498–500
immobilization on Sepharose 4B, XLIV, 494
purification, and isoenzymes, XLII, 291
in sedoheptulose 1,7-biphosphate preparation, XLI, 77–79
site-specific reagent, XLVII, 484
Transaldolase isoenzyme I
chromatography, XLII, 293, 294
molecular weight, XLII, 297
properties, XLII, 297
purification, XLII, 293–296
Transaldolase isoenzyme II, XLII, 291
Transaldolase isoenzyme III
molecular weight, XLII, 297
properties, XLII, 297
purification, XLII, 295, 296
Transaminase, XXXIV, 295–299, *see also* Aminotransferase; specific enzymes
in liver compartments, XXXVII, 288
Transamination, XLVI, 428
Transcarboxylase, *see* Methylmalonyl-CoA carboxyltransferase
Transcobalamin II, XXXIV, 305
Transcriptase, reverse, XXXIV, 728, 729
atomic emission studies, LIV, 455
Transcription, XLVI, 644
in mitochondria, inhibitors, LVI, 31
mitochondrial and extramitochondrial, differential inhibition, LVI, 33–35
Trans-*N*-deoxyribosylase, *see* Nucleoside deoxyribosyltransferase
Transduction
oxidative phosphorylation deficient mutants, LVI, 112
with phage P1 kc in *Escherichia coli*, XLIII, 45–47
with phage P22 in *Salmonella typhimurium*, XLIII, 47
preparation of phage lysates, LVI, 111
technique, LVI, 111, 112
Transferase, assay by Fast Analyzer, XXXI, 816
Transfer efficiency
correction of acceptor stoichiometry, XLVIII, 368, 369
distance
for random labeling case, XLVIII, 373–375
for specific-site labeling case, XLVIII, 368–373
theoretical considerations, XLVIII, 359, 360
evaluation
from emission data, XLVIII, 365–368
theoretical, XLVIII, 354–356
Transfer factor, of blood steroids, XXXVI, 69–71
Transfer factor G, labeled GDP preparation, XXXVIII, 87
Transfer ribonucleic acid, XLVI, 102, 170, 335, *see also* Aminoacylation; Aminoacyl-transfer ribonucleic acid; specific tRNA species
activity of probe-modified, XLVI, 697
aliphatic proton, LIX, 22
aliphatic resonance, high-field, LIX, 57

aminoacylation, LIX, 130–132, 216, 217, 298, 818
as test of terminal structure, LIX, 126, 127, 132, 183
A site, XLVI, 697, 698, 700, 701
chemical cleavage at 7-methylguanosine, LIX, 100–102
from CHO cells, preparation, LIX, 220
class 1, definition, LIX, 27
class 3, base pairing, LIX, 52, 53
commercial source, LIX, 194
conformation studies using fluorescence-detected circular dichroism, XLIX, 212–214
coupling, XLVI, 692–697
with protein affinity-labeling reagents, LIX, 156–166
principle, LIX, 157
procedure, LIX, 158–162
criteria of labeling, XLVI, 699
crystallization, conditions, LIX, 6, 7
deacylation, LIX, 211, 213, 214
denaturation studies using scanning calorimetry, XLIX, 11, 12
dephosphorylation procedure, LIX, 75
dialdehyde derivatives, XXXIV, 50
digestion with alkali, LIX, 102, 192
with alkaline phosphatase, LIX, 175, 186, 343
with *B. cereus* ribonuclease, LIX, 106
with nuclease P1, LIX, 81–83, 85, 87, 94, 104
with pancreatic ribonuclease, LIX, 60, 90, 99, 106, 188
with phosphodiesterase I, LIX, 59, 60, 79, 81, 83, 87, 89, 94, 103, 122, 126, 127, 132, 140, 175, 187
with ribonuclease T1, LIX, 60, 75, 90, 99, 105, 120, 343
with ribonuclease T2, LIX, 60, 71, 79, 171, 175, 180, 188, 192, 344
with ribonuclease U_2, LIX, 105
direct sequence analysis, LIX, 102–109
dye compounds, dye determination, spectrophotometric, LIX, 112
3′-end nucleoside, determination of, LIX, 188, 189
eukaryotic
aminoacyl, *see* Aminoacyl-transfer ribonucleic acid
preparation from *Artemia salina* cysts, LX, 305, 306
from reticulocyte lysate, LX, 111, 112, 139, 141–143
from wheat germ, LX, 413, 414
fluorescein-labeled, amino acid acceptor activity, LIX, 152–155
determination of moles of dye bound, LIX, 151, 152
optical properties, LIX, 152, 153
preparation, LIX, 147–150
principle, LIX, 146
helix-coil exchange rates, determination, LIX, 50–52
high-field aliphatic resonance, LIX, 57
^{3}H-labeled, XLVI, 177
hydrazide derivatives, LIX, 175–180
preparation, LIX, 179, 180
N-hydroxysuccinimide ester adducts, stability, LIX, 163–165
importation by mitochondria, LVI, 8, 9
incorporation of amines, LIX, 110–121
of fluorescent probes, LIX, 146–156
of hydrazines, LIX, 110–121
irradiation, XLVI, 698
I site, XLVI, 697, 698
isoacceptors
in cultured Chinese hamster ovary cells, LIX, 218–229
identification, LIX, 222–229
separation, LIX, 215–218, 221, 222, 298
in studies of codon-anticodon recognition, LIX, 292–309
ligand interaction, LIX, 17–19
magnesium effects, LIX, 53, 54
mitochondrial, LVI, 8, 9
isolation, LIX, 206, 207

modification with *N*-hydroxysuccinimide esters, LIX, 170
with sodium bisulfite and carbohydrazide, LIX, 147
modified, protein cross-linking, LIX, 165, 166
modified nucleosides, LIX, 21
modified nucleotides
analysis of, LIX, 59
identification, LIX, 70–74
NMR spectra, analysis, LIX, 27–36
NMR studies, resolution enhancement, LIX, 54–56
ring-current shift effects, LIX, 32–36
in solution, relation to crystal structure, LIX, 30
nonradioactive, sequence analysis, *in vitro* ^{32}P labeling, LIX, 58–109
overall structure, LIX, 16, 30
periodate-oxidized, LIX, 187
reaction with [^{14}C]isonicotinic acid hydrazide, LIX, 175, 181
stability, LIX, 174, 175, 180, 181
phenylalanine, from yeast, nuclear magnetic resonance spectrum, XLIX, 266
photochemically active derivatives, LVI, 656
precipitation by cetyltrimethylammonium bromide, LIX, 236
prokaryotic
binding to 70 S ribosomes, LX, 626, 627
complex with 70 S ribosomes and polyadenylate, LX, 624
with 80 S ribosomes and polyuridylate, LX, 661, 662
in protein turnover, XXXVII, 241, 245, 247
P site, XLVI, 697, 698, 701
from rat liver, large-scale preparation, LIX, 230, 231
reconstitution, LIX, 127–129, 140
regulatory functions, LIX, 268, 322
removal, XXXIV, 168, 169
from ribosomes, LIX, 555
ribosomal binding sites, XLVI, 181, 182, 683–702
sample preparation for NMR studies, LIX, 25
secondary resonance assignments, LIX, 30–36
secondary structure, LIX, 16, 17
separation from aminoacyl-tRNA synthetase, LIX, 342, 343
specificity, LIX, 20
storage, LIX, 186
structural stability, LIX, 20
structure determination
by crystallographic method, LIX, 3–21
in solution, LIX, 21–57
using nuclear magnetic resonance, LIX, 21–57
structure-function relationships, LIX, 19–21
substitution, quantitative determination, LIX, 120, 121
with terminal 2′-amino-2′-deoxyadenosine, preparation, LIX, 140–142
with terminal 3′-amino-3′-deoxyadenosine, preparation, LIX, 140–142
with terminal 2′- and 3′-deoxyadenosine, preparation, LIX, 121–134
3′-terminus
modification with periodate, LIX, 172–181, 188
nucleotide incorporation, assay, LIX, 182, 183
tertiary resonance assignments, LIX, 37–47
tertiary structural features, LIX, 17–19
thermal flexibility, LIX, 20, 21
thermal unfolding sequence, LIX, 47–50
valine, from *E. coli,* nuclear magnetic resonance correlation spectrum, XLIX, 263
from yeast, preparation, LIX, 235
Transfer ribonucleic acid-like moiety identification, LIX, 202
Transferrin
in fetal serum, XXXII, 560

gel electrophoresis, XXXII, 101
magnetic circular dichroism, XLIX, 175
in media, LVIII, 95, 99–107, 109, 500
molecular weight, LIII, 351, 353
solution densities, XLVIII, 28, 29
Transfer RNA, *see* Transfer ribonucleic acid
Transformation, XLIII, 48–50, *see also* Cell, transformed; Cell transformation; Plasmid, resistance
malignant, LVIII, 296, 368
neoplastic, LVIII, 296–302
soft agar assay, LVIII, 297, 299, 300
Transform Technology 220 Fourier transform system, LIV, 228
Transformylase, absence, in *E. coli* mutant, LI, 214
Transglutaminase, *see* Glutaminyl-peptide γ-glutamyltransferase
Trans-*N*-glycosidase, *see* Nucleoside deoxyribosyltransferase
Transglycosylation, XLVI, 404
Transhydrogenase, *see also* Nicotinamide, Pyridine
ATPase inhibitor, LV, 407, 412, 413
ATP-dependent, assay, in mutants, LVI, 113, 114
brown adipose tissue mitochondria, LV, 77
DCCD, LV, 499
energy-dependent, assay, LV, 789, 790
mitochondrial, rhein, LV, 459
phospholipids, LV, 701
steroid hydroxylation, LV, 10
substrate analogs, LV, 262, 270–272
venturicidin, LV, 506
Transhydrogenation
everted vesicles, LVI, 237
inhibition by arylazido-β-alanine ATP, XLVI, 278, 279
Transhydrogenation activity
of complex I, LIII, 12
of complex I–III, LIII, 8
of NADH dehydrogenase, LIII, 17
Transient dichroism, data analysis model, LIV, 56
Transition
differential heat-capacity calorimetry, XLIX, 3–14
forbidden, circularly polarized luminescence, XLIX, 187
vibrational, Raman spectroscopy, XLIX, 68–149
virtual, definition, XLIX, 70, 71
Transition state
analogs, XLVI, 9, 221
as affinity labeling reagents, XLVI, 15–28
criteria, XLVI, 17–20
kinetics, XLVI, 15–17
mechanisms, XLVI, 15–17
Transketolase, XLVI, 50, 51
in sedoheptulose 1,7-biphosphate preparation, XLI, 77–79
site-specific reagent, XLVII, 484
Translation, XLVI, 644
of initiation complex, antibiotic action, LIX, 857–859
in mitochondria, inhibitors, LVI, 31
mitochondrial and extramitochondrial, differential inhibition, LVI, 33–35
Translocase, of brain mitochondria, LV, 17
Translocation
inhibition, by cooling, LIX, 359
of peptidyl-tRNA, assay, LIX, 359, 360
Translocator, *see also* Carrier
kinetics, general rate equation, LVI, 262–264
types, LVI, 253–255
Transmembrane motion, spin-label studies, XXXII, 184, 185
Transmission electron microscopy, XXXIX, 133–154
buffers, XXXIX, 135, 136
cytochemistry, XXXIX, 149–154
electron-opaque tracers, XXXIX, 145–149
of endocrine tissue, XXXIX, 133–154
enzymatic tracers, XXXIX, 145–148
fixation techniques, XXXIX, 133–138
postfixation, dehydration, and embedding, XXXIX, 138, 139
stains, XXXIX, 141–145

Transplantation antigen, mouse, sequence analysis, XLVII, 256, 257
Transport
in bacterial membrane vesicles, XXXI, 698–709
carrier mediated, criteria, LVI, 246, 247
growth of cells, LVI, 399–401
light-induced, LVI, 398, 399
amino acid transport experiments, LVI, 404, 405
linked to membrane potential, LV, 656, 657
culture media, LV, 657, 658
electrophysiological techniques, LV, 658, 662
flux techniques, LV, 663–666
measurement of light-induced changes in pH and membrane potential, LVI, 406, 407
passive carrier type, in substrate entry, XXXIX, 498–502
preparation of envelope vesicles, LVI, 401, 402
properties and handling of envelope vesicles, LVI, 402–404
Transport coefficient, average, XLVIII, 218
Transport mechanism, hormone effects, XXXVI, 429–433
Transport selectivity, of ionophores, factors affecting, LV, 440
Transport study
flow dialysis, LVI, 387, 388
kinetic measurements
automatic sampling methods, LVI, 297–301
inhibitor stop method, LVI, 290–296
pressure filtration device, LVI, 296, 297
methods
direct, LVI, 257, 258
indirect, LVI, 255–257
preparations
incubation conditions, LVI, 281, 282
labeling of intramitochondrial metabolites, LVI, 281
loading of mitochondria with metabolites, LVI, 280, 281
separation methods
centrifugal filtration, LVI, 283–285
centrifugal layer filtration, LVI, 285, 286
centrifugal sedimentation, LVI, 282, 283
sieve filtration, LVI, 286, 287
steady-state distribution measurements
efflux, back exchange, LVI, 289, 290
intramitochondrial volume, LVI, 287, 288
uptake, LVI, 288, 289
Transport system
amino acid, XLVI, 151, 607–613
biotin, XLVI, 608, 613–617
criteria for labeling, XLVI, 613
glucose, XLVI, 89, 167, 608
lactose, XLVI, 86, 608
mitochondrial, overview, LVI, 245–252
phenylalanine, XLVI, 151, 608
regulation, LVI, 250
tyrosine, XLVI, 608
Tranylcypromine, XLVI, 38
Trasylol
in pancreas superfusion, XXXIX, 374
plasminogen, activation, XLV, 263
as trypsin inhibitor, XXXVII, 31
Traube's rule, applications, XLVIII, 18–22
Trehalase, in brush borders, XXXI, 122, 129, 130, 132
Trehalosamine, XLIII, 272
Trehalose, polarography, LVI, 477
α,α-Trehalose, XXXIV, 173
Tremella mesenterica, glucan ^{13}C-nuclear magnetic resonance spectrum, L, 47
TRF, *see* Thyrotropin releasing factor
Triacanthine
chromatographic mobility, LIX, 73
electrophoretic mobility, LIX, 91
identification, LIX, 83, 89, 92–94
NMR studies, LIX, 57
Triacetic acid lactone, XXXV, 45, 83
Triacetoneamine, *see* 4-Oxo-2,2,6,6-tetramethylpiperidine

1α,3β,25-Triacetoxycholesta-5,7-diene, synthesis, XXXVI, 528
Tri-*N*-acetylchitotriose-starch adsorbent, XXXIV, 330, 331
Tri-*O*-acetyl-1,2-epoxy-α-D-glucopyranose, L, 104, 105
Tri(*N*-acetylglucosamine)-agarose, XXXIV, 639–645
Triacylglycerol lipase
  immobilization
    by microencapsulation, XLIV, 214
    on polyacrylamide, XLIV, 442
  in microbodies, LII, 496
  modification, effect on activity, XLIV, 447–450
Trialkyl oxonium group, XLVI, 579
Triangulation method, LIX, 702–706, *see also* Protein pair scattering
Triarylethylene derivative, XLVI, 90
Triatoma infestans, media, LVIII, 463
Triazine, coupling, XLIV, 47
  enzyme immobilization, LVI, 491
  to inorganic supports, XLIV, 141, 142
Triazine-agarose, XXXIV, 286
2,4,6-Tribromo-4-methylcyclohexadienone, XLVII, 443, 456, 459
  disulfide bond cleavage, XLVII, 111–116, 120–122
    mechanism, XLVII, 112
  oxygen sensitivity, XLVII, 114
  S-sulfo group reductive cleavage and, XLVII, 123–126
  titration, to determine reducing power, XLVII, 114
  toxicity, XLVII, 114, 125
Tributylamine
  in tributylammonium phosphate preparation, LI, 254
  in *N*-trifluoroacetyl 6-amino-1-hexanol phosphate preparation, LI, 254
Tributylammonium phosphate, XXXIV, 489
  preparation, LI, 254
  in *N*-trifluoroacetyl 6-aminohexanol 1-pyrophosphate preparation, LI, 254
Tributylammonium pyrophosphate, synthesis of 8-azido-ATP, LVI, 650, 651
Tri-*n*-butyl phosphate, as antifoam agent, LI, 43
Tributyrin, substrate, of lipase, XLIV, 447
Tricalcium citrate-cellulose chromatography, factor VIII, bovine, XLV, 87
Tricaprylylmonomethylammonium chloride, preparation of RPC resin, LIX, 221
Tricarboxylic acid cycle, *see* Krebs cycle
Tricene, in dimethylaniline monooxygenase assay, LII, 143
Trichloroacetamide, preparation of luciferin, LVII, 17, 19
Trichloroacetic acid, XLIV, 90, 533, 632, 685, 690, 845, 852
  apoadrenodoxin preparation, LII, 136
  assay of carbamoyl-phosphate synthetase, LI, 30, 31
    of FGAM synthetase, LI, 195
    of phosphoribosylglycinamide synthetase, LI, 180, 181
    of succino-AICAR synthetase, LI, 188
  ATPase inhibitor, LV, 410, 424
  determination of radioactivity in slab gels, LIX, 521
  extraction of adenine nucleotides, LVII, 81
    of guanine nucleotides, LVII, 88
  formaldehyde assay, LII, 299, 345
  in gel staining solution, LIX, 545
  hydrogen peroxide determination, LII, 345
  hydroxylamine determination, LIII, 637
  α-hydroxylase assay, LII, 315
  in preparation of carboxypeptidase-treated oxyhemoglobin, LIV, 488
  protein determination, LI, 88; LIII, 365
  purification of flavins, LIII, 420, 423, 453, 456
    of D-lactate dehydrogenase, LIII, 522
  pyrophosphate determination, LI, 275
  in quantitative histone determination, XL, 113

for reaction termination, LIX, 180, 182, 191, 231, 238, 301, 315, 358, 360, 367, 449, 459, 825
in aniline hydroxylase assay, LII, 409
removal, with diethyl ether, LIX, 497
from gels, LIX, 546
for ribosomal protein precipitation, LIX, 470, 497, 518, 538, 565, 837
for tRNA precipitation, LIX, 271, 273, 287, 288
termination
of assay of phenylalanine hydroxylase, LIII, 279
of bacterial metabolism, LIX, 269
thiobarbituric acid assay, LII, 306
Trichloroacetonitrile, drying, XXXVIII, 412
Trichloroethanol, fluorescence quenching, XLIX, 224
Trichloromethanesulfonyl chloride, XLVII, 456
3,4′,5-Trichlorosalicylanilide
synthesis, LV, 467
uncoupling activity, LV, 465
Trichoderma viride, XLIV, 166
Trichomonas, rat infection, XXXII, 667, 668
Trichophyton mentagrophyta, fungal penicillin acylase, XLIII, 721
Trichoplusia ni cell, LVIII, 450
media, LVIII, 457, 460
storage, LVIII, 453
suspension culture, LVIII, 455
Trichosporon, oxidase, LII, 17
Tricine buffer, *see* *N*-Tris(hydroxymethyl)methylglycine
Tricine-KOH, *see* *N*-Tris(hydroxymethyl)methylglycine-potassium hydroxide
Tricosyl bromide, lignoceric acid synthesis, LII, 311
Tridecanoic acid, quantum yield, in bioassay, LVII, 192
2-Tridecanone oxidative lactonase, LII, 20
Triene, XLIII, 135
Triethanolamine
assay of adenosine monophosphate nucleosidase, LI, 265
of adenylate kinase, LI, 460
of carbamoyl-phosphate synthetase, LI, 30
of hydrogen peroxide, LII, 348
of phosphoribosylpyrophosphate synthetase, LI, 3, 4
preparation of *T. vulgare* ribosomes, LIX, 752
purification of CPSase-ATCase-dihydroorotase complex, LI, 124
of monoamine oxidase, LIII, 497–199
of nucleoside diphosphokinase, LI, 372
of purine nucleoside phosphorylase, LI, 519
of thymidine phosphorylase, LI, 439
quinone extraction, LIII, 580
in TEA-maleate buffer, LIII, 364
Triethanolamine·HCl, *see* Triethanolamine-hydrochloric acid
Triethanolamine-hydrochloric acid, XLVII, 493
preparation of cross-linked ribosomal subunits, LIX, 535
of ribosomal subunits, LIX, 411, 412, 419
Triethanolamine-maleate buffer, anaerobic
formulation, LIII, 364
preparation, LIII, 366
Triethanolamine-Tris-chloride buffer, *see* Triethanolamine-tris(hydroxymethyl)-aminomethane-chloride buffer
Triethanolamine-tris(hydroxymethyl)aminomethane-chloride buffer, in purification of nitrogenase, LIII, 320
Triethylamine, XLIV, 132, 141, 384; XLVI, 124
activator, in peroxyoxalate chemiluminescence reaction, LVII, 457, 459, 461
amino group neutralization, XLVII, 595, 596
in condensation reaction, XLVII, 550
in derivatization of glass, XLIV, 932
drying, XXXVIII, 412
fluorescence quencher of 3-hydroxybenzopyrene, LII, 410
peptide synthesis, disadvantages, XLVII, 563, 564

preparation of dehydroluciferin, LVII, 25
of triethylammonium bicarbonate, LIX, 61
spray, XLVII, 196
synthesis of 6-*N*-[(3-biotinylamido)-2-hydroxypropyl]amino-2,3-dihydrophthalazine-1,4-dione, LVII, 430
of oleylpolymethacrylic acid resin, LIII, 214
of *N*-trifluoroacetylthyroxinylethyl carbonic anhydride, LVII, 432
(Triethylaminoethyl)cellulose, XLIV, 46
adrenodoxin reductase purification, LII, 134
peptide separation, XLVII, 204–210
purification of rubredoxin:NAD oxidoreductase, LIII, 621
washing procedure, LIII, 615
(Triethylaminoethyl)cellulose chromatography, ancrod, XLV, 207
Triethylammonium acetate, XLVII, 292, 294
amino acid analysis of flavin peptides, LIII, 456
Triethylammonium acetate-acetic acid buffer, purification of flavin peptide, LIII, 454
Triethylammonium bicarbonate
in P^3-(6-aminohex-1-yl)dGTP preparation, LI, 255–256
digestion of tRNA, LIX, 171
preparation, LIX, 61
in tRNA sequence analysis, LIX, 63, 66, 70, 102
synthesis of 7-[*N*-ethyl-*N*-(4-throxinylamido)butyl]aminonaphthalene-1,2-dicarboxylic acid hydrazide, LVII, 437
Triethylammonium chloride, removal, LVII, 430, 432
Triethylammonium hydrogen carbonate, *see* Triethylammonium bicarbonate
Triethylenediaminecadmium hydroxide, XLVIII, 74
Triethylenetetramine polystyrene resin, XLVII, 264, 301
capacity, XLVII, 269
coupling reaction, choice, XLVII, 271
peptide coupling, XLVII, 280, 282
stability, XLVII, 266, 273
synthesis, XLVII, 266
Triethylorthoformate, XLVI, 223
Triethyloxonium tetrafluoroborate, alkylating agent, XLIV, 123, 124, 126, 640
Triethyl phosphate, XXXIV, 229
phosphorylation of ligand-adenosine conjugates, LVII, 119
Triethyltin
complex V, LV, 315
purification, LV, 507, 508
Triethyltin chloride, physical properties, LV, 477
Triethyltin sulfate, complex V, LVI, 586
Triethyoxysilane-glass, XXXIV, 63
N^8-(6-Trifluoroacetamidohexyl)-8-aminoadenosine 5′-diphosphate, XXXIV, 488, 489
N^8-(6-Trifluoroacetamidohexyl)-8-aminoadenosine 5′-phosphate, XXXIV, 488
Trifluoroacetic acid, XLVI, 200, 201
background spots, XLVII, 332
cleavage, XLVII, 342, 604
coupling yields, XLVII, 270
homoserine attachment, XLVII, 283
hydrogen bromide cleavage, XLVII, 546, 547, 603
nuclear magnetic resonance reference, XLIX, 288
prevention of porphyrin aggregation, LII, 436
removal
of *tert*-butyl-type protectors, XLVII, 545
by N_2, XLVII, 376
for removal of isopropylidene residue, LVII, 119
salt accumulation, XLVII, 339
support swelling, XLVII, 272
synthesis of *N*-trifluoroacetylthyroxine, LVII, 432
in *N*-trifluoroacetyl 6-amino-1-hexanol phosphate preparation, LI, 253, 254
wash, function, XLVII, 269
d-Trifluoroacetic acid
determination of deuterium content of protein, LIX, 645, 646
of RNA, LIX, 646, 647

Trifluoroacetic acid-methylene chloride, peptide cleavage from support, XLVII, 604
Trifluoroacetic anhydride, synthesis of *N*-trifluoroacetylthyroxine, LVII, 432
N-Trifluoroacetylalaninal, XLVI, 225
N-Trifluoroacetyl-ε-aminocaproic acid, XXXIV, 325
N-(*N*-Trifluoroacetyl-ε-aminocaproyl)-2-amino-2-deoxy-D-glucopyranose, XXXIV, 325, 326
N-Trifluoroacetyl 6-amino-1-hexanol phosphate, XXXIV, 483
  preparation, LI, 253, 254
N-Trifluoroacetyl 6-amino-1-hexanol-pyrophosphate
  chromatographic behavior, LI, 255
  preparation, LI, 254, 255
N-Trifluoroacetyl-L-phenylalaninal, XLVI, 222, 223
N-Trifluoroacetyl-L-phenylalaninol, XLVI, 222
N-Trifluoroacetyl-*O*-phosphoryl-6-amino-1-hexanol imidazolide, XXXIV, 483, 484
N-Trifluoroacetylthyroxine
  synthesis, LVII, 432
    of 6-[*N*-ethyl-*N*-(4-thyroxinylamido)butyl]amino-2,3-dihydrophthalazine-1,4-dione, LVII, 433
    of 7-[*N*-ethyl-*N*-(4-thyroxinylamido)butyl]amino-naphthalene-1,2-dicarboxylic acid hydrazide, LVII, 437
    of *N*-trifluoroacetylthyroxinylethyl carbonic anhydride, LVII, 432
N-Trifluoroacetylthyroxinyl ethyl carbonic anhydride
  synthesis, LVII, 432
    of 6-(*N*-ethyl-*N*-[2-hydroxy-3-(thyroxinylamido)propyl])-amino-2,3-dihydrophthalazine-1,4-dione, LVII, 432
N-Trifluoroacetyluridine diphosphate-hexanolamine, XXXIV, 484
3,3,3-Trifluoro-2-diazopropionyl chloride, XLVI, 95
2,2,2-Trifluoroethanol, solvent, synthesis of chemiluminescent aminophthalhydrazides, LVII, 430, 431, 436
Trifluoromethane, fluorine labeling, XLIX, 290
p-Trifluoromethoxycarbonylcyanide phenylhydrazone
  calcium transport, LVI, 239, 325–327
  fluorescence, LVI, 500
  immobilized mitochondria, LVI, 556
  transport studies, LVI, 256
5-Trifluoromethyl deoxyuridine, substrate of deoxythymidine kinase, LI, 364
Trifluoromethyldiazoacetyl compound, XLVI, 74, 95
4,4,4-Trifluoro-1-thien-2-ylbutane-1,3-dione, LV, 517
Trigger mechanism, LIX, 248
Triglyceride
  analysis of mixtures by mass spectrometry, XXXV, 355–359
    of positional distribution of fatty acids, XXXV, 320–325
  biosynthesis, in endoplasmic reticulum, XXXI, 23
  deacylation with Grignard reagent, XXXV, 320, 322–324
    with pancreatic lipase, XXXV, 320–322, 324
    thin-layer chromatography analysis, XXXV, 320, 321, 324, 325
  enzymes attacking, XXXI, 522
  gas chromatography–mass spectrometry, XXXV, 358, 359
  gas liquid chromatography, XXXV, 331
  hydrolysis by lipoprotein lipase, XXXV, 181
  lipophilic Sephadex chromatography, XXXV, 383, 388–391
  mass spectrometry, XXXV, 348–359
    after gas chromatography, combined use, XXXV, 358, 359
    analysis of triglyceride mixtures, XXXV, 355–359
    applications, XXXV, 358
    double bond position determination, XXXV, 352
    molecular weight distribution, determination, XXXV, 357, 358

order of fatty acids, determination, XXXV, 353
thin-layer chromatography, XXXV, 397, 404, 405
in plasma membranes, XXXI, 87
in spherosomes, XXXI, 577
synthesis
control studies, XXXVII, 292, 295
substrate entry, XXXIX, 495
Triglyceride lipase, in fat cells, XXXI, 67, 68
Trihexosylceramide
Fabry's disease, L, 480
isolation, XXXII, 350
3α,18,21-Trihydroxy-5β-pregnane-11,20-dione
isolation and measurement, XXXVI, 505–507
as major aldosterone metabolite, XXXVI, 504
synthesis, XXXVI, 509–511
11β,17α,21-Trihydroxypregn-4-ene-3,20-dione, *see* Cortisol
2,3,5-Trihydroxytoluene 1,2-dioxygenase, LII, 19
1,24(*R*),25-Trihydroxyvitamin D_3, vitamin D_3 metabolism, LII, 389
Triiodothyronine, XXXIV, 388
adenylate cyclases, XXXVIII, 148, 173, 176
effect on melanin, XXXVII, 129
in media, LVIII, 99, 101
in thyroid cells, XXXII, 745, 756
3,5,3′-Triiodo-L-thyronine
autoradiography, XXXVI, 150
TBG binding, XXXVI, 126
2,4,6-Triisopropylbenzenesulfonyl chloride, XXXIV, 646
intermediate in phosphatidylethanolamine synthesis, XXXV, 456
Trimethylamine, solvent, spectral analysis of porphyrins, LII, 445
Trimethylamine dehydrogenase, iron-sulfur center, identification, LIII, 273
Trimethylamine oxidase, LII, 21
Trimethylaminoacylcarnitine, transport, LVI, 248
p-(Trimethylammonium)benzenediazonium fluoroborate, XLVI, 583
Trimethylammonium bromide, partial specific volume, XLVIII, 18
Trimethylammonium chloride, partial specific volume, XLVIII, 18
Trimethylammonium methyleneboronic acid, XLVI, 25
2,4,6-Trimethylbenzyl ester moiety, peptide synthesis, XLVII, 522
Trimethylborate, as lock nuclei source, LIV, 189
Trimethylchlorosilane, silylation of hydroxylated steroids, LII, 380
[1-^{2}H]2,4,6-Tri-*O*-methyl-1,5-di-*O*-trideuteriomethyl-D-galacitol, L, 33
2,4,6-Tri-*O*-methyl-D-galactose, L, 28
2,4,6-Tri-*O*-methyl-D-glucose, L, 28
N^ε,N^ε,N^ε-Trimethyllysine, identification, XLVII, 477
3,4,6-Tri-*O*-methyl-D-mannose, L, 9
Trimethyloxonium fluoroborate, XLVI, 579, 583
Trimethylphenylphosphonium bromide, transport, LVI, 407
2,4,6-Trimethylpyrimidine, in ninhydrin solution, LIX, 777
Trimethylsilylating reagent, XXXII, 360, 362
Trimethylsilylation
fatty acid unsaturation position determination, XXXV, 347
^{18}O measurements, LV, 252
procedure, XLVII, 395
of prostaglandins, XXXV, 364, 365
of Sephadex, XXXV, 381
Trimethylsilylimidazole, silylating reagent, LII, 332
3-(Trimethylsilyl)propane sulfonate, XLIX, 256
Trimethylsilylsodium [U-^{2}H]propionate in proton magnetic resonance experiments, LIV, 229
Trimethyltin chloride, physical properties, LV, 477
Trimyrestin, bioassay of lipase, LVII, 193
2,4,6-Trinitrobenzenesulfonic acid, XXXIV, 33, 48, 107, 515; XLVII, 76, 77, 151, 472, 477
amino group determination and, XLVII, 302

measurement of free amines, LVI, 624
modification of lysine, LIII, 142
phosphatidylserine and phosphatidylethanolamine assay, XXXV, 339
for protein assay, in lipid extracts, XXXV, 334–339
in studies of luciferase active site, LVII, 180
test, XXXIV, 54
Trinitrobenzenesulfonic acid reagent, XLIV, 103
Trinitro compound fixative, XXXIX, 137, 138
Trinitrophenylaminoethyl derivative, XXXIV, 49
N′-Trinitrophenylhydrazide, XXXIV, 49
Tri-*n*-octylamine, XLIX, 458, 459
Tri-*N*-octylammonium hydroxide, in P^3-(6-aminohex-1-yl)dGTP preparation, LI, 255
Triolein
^{14}C-labeled, marker for liposomes, XXXV, 263, 264
emulsion, preparation of, XXXV, 183
Triose isomerase, in plants, XXXI, 740
Triosephosphate dehydrogenase, XLVI, 20, 21
from honey bee, XLI, 273–278
assay, XLI, 274
chromatography, XLI, 275
inhibition, XLI, 277
molecular weight, XLI, 277
properties, XLI, 276–278
purification, XLI, 275, 276
Triose-3-phosphate dehydrogenase, in plasma membrane, XXXI, 89
Triosephosphate isomerase, XLIV, 337; XLVI, 18, 23, 140, 141, 143, 381, 382, 432; XLVII, 66
affinity labeling procedure, XLVII, 497, 498
assay, XLI, 37
assay reagent, XLIV, 495, 496
conjugate, activity, XLIV, 271
from human erythrocytes, XLI, 442–447
assay, XLI, 442–444
electrophoresis, XLI, 445
isolation, XLI, 444–446
properties, XLI, 446, 447
protein determination, XLI, 444
from human and horse liver, XLI, 430–434
assay, XLI, 430
chromatography, XLI, 431
isozymes, XLI, 433
molecular weight, XLI, 133
properties, XLI, 433, 434
purification, XLI, 430, 433
from human skeletal and cardiac muscle, XLI, 446
immobilization, on inert protein, XLIV, 909
inorganic phosphate assay, LV, 212
as marker enzyme, XXXI, 735, 744
assay, XXXI, 745
molecular weight and separation from 3-phosphoglycerate kinase, XLII, 131
from rabbit liver, XLI, 438–442
assay, XLI, 438
chromatography, XLI, 439, 440
inhibitors, XLI, 441
properties, XLI, 440–442
purification, XLI, 439, 440
from rabbit muscle, XLI, 447–453
assay, XLI, 447–449
chromatography, XLI, 451
inhibitors, XLI, 452
properties, XLI, 451–453
purification, XLI, 449–451
site-specific reagent, XLVII, 483
from yeast, XLI, 434–438
assay, XLI, 434, 435
chromatography, XLI, 436
inhibitors, XLI, 438
properties, XLI, 437, 438
purification, XLI, 435–437
Triphenylamine, activator in chemiluminescent reaction, LVII, 522, 523
Triphenylmethane dye, as optical probes, LV, 573
Triphenylmethyl group, *see* Trityl group
Triphenylsulfonium chloride, LV, 516
Triphenyltin chloride, physical properties, LV, 477

4-*N*-[(5′-Triphospho)-1′-ribosylamino]-2,5-diamino-6-hydroxypyrimidine, XLIII, 515, 520
Triplet exciton, interaction, LVII, 518, 519
Triplet interception, LVII, 516, 517
Triplet probe
preparation, LIV, 49, 50
for rotational diffusion measurements, principle, LIV, 47–49
Triple-trapping method, LIV, 105, 108
Triplet state, time range, LIV, 2
Tri-*n*-propyltin chloride, physical properties, LV, 477
2,4,6-Tripyridyl-*s*-triazine, in colorimetric ultramicro assay for reducing sugars, XLI, 27
Tris, *see* Tris(hydroxymethyl)aminomethane
Trisaccharide, qualitative analysis, XLI, 18
Tris-acetate buffer, *see* Tris(hydroxymethyl)aminomethane-acetate buffer
Tris base, *see* Tris(hydroxymethyl)aminomethane base
Trisbipyridylruthenium(II), electron-transfer chemiluminescence, LVII, 499
Tris-borate, *see* Tris(hydroxymethyl)aminomethane-borate
Tris buffer, *see* Tris(hydroxymethyl)aminomethane
Tris-cacodylate buffer, *see* Tris(hydroxymethyl)aminomethane-cacodylate buffer
Tris-2-carboxyethylphosphine, XLVII, 112
Tris-chloride buffer, *see* Tris(hydroxymethyl)aminomethane-hydrochloric acid buffer
Tris-EDTA buffer, *see* Tris(hydroxymethyl)aminomethane-ethylenediaminetetraacetic acid buffer
2′,3′,5′-Tris-*p*-fluorosulfonylbenzoyladenosine, XLVI, 243, 244
Tris-glycine buffer, *see* Tris(hydroxymethyl)aminomethane-glycine buffer
Tris·HCl buffer, *see* Tris(hydroxymethyl)aminomethane-hydrochloric acid buffer
Tris·HCl-NaCl buffer, *see* Tris(hydroxymethyl)aminomethane-hydrochloric acid-sodium chloride buffer
Tris-hydrochloride, *see* Tris(hydroxymethyl)aminomethane hydrochloride
Tris(hydroxymethyl)aminomethane, LIX, 235
aggregation inhibitor, XLIV, 266
assay of aminoacyl-tRNA synthetases, LIX, 231
of modulator protein, LVII, 109
of tRNA methyltransferase, LIX, 191
boiling, in extraction of bacterial ATP, LVII, 68, 81
in buffer for transmission electron microscopy, XXXIX, 136
cell-free protein synthesis system, LIX, 850
chemiluminescent assay of lactate dehydrogenase, LVII, 461
for chloroplast isolation, XXXI, 602, 603
crystallization of transfer RNA, LIX, 6
electrophoretic analysis of ribosomal protein, LIX, 506
for erythrocyte ghost isolation, XXXI, 168, 169
for gel electrophoresis, XXXII, 84, 85
as inhibitor for D-arabinose and L-arabinose isomerase, XLI, 465
in iron analysis, advantages, LIV, 440
in luciferase reagent, LVII, 70
for mitochondria isolation, XXXI, 597
mitochondrial protein synthesis, LVI, 21
neutral salt, in purification of luciferin sulfokinase, LVII, 249
for nuclear envelope isolation, XXXI, 280
preparation of denatured DNA, LIX, 840
of mitochondrial tRNAs, LIX, 206, 207
of MS2 RNA, LIX, 299

purification of aminoacyl-tRNA synthetases, LIX, 258
of luciferin sulfokinase, LVII, 247
of tRNA methyltransferase, LIX, 198
resuspension, for platelets, XXXI, 150
reverse salt gradient chromatography, LIX, 216
tRNA sequence analysis, LIX, 67, 69
saline, XXXII, 507
in seawater complete medium, LVII, 155
use in conversion to Tris salt, XLIV, 878
Tris(hydroxymethyl)aminomethane-acetate buffer
assay of ADP, LVII, 61
digestion of tRNA, LIX, 126, 132
electrophoretic analysis of protein-RNA complexes, LIX, 567
elution of proteins from gels, LIX, 544, 545
extraction of radiolabeled RNA, LIX, 831, 834, 836
preparation of ribosomal fragments, LIX, 472
of ribosomal protein, LIX, 453
Tris(hydroxymethyl)aminomethane base
in formulation of diluted solid sodium dithionite, LIV, 130
in relaxing medium, LVII, 248
in tRNA-protein interaction studies, LIX, 323
solid, for pH adjustment, LVII, 146
in sucrose gradient buffers, LIX, 329
Tris(hydroxymethyl)aminomethane-borate
in electrophoretic elution technique, LIX, 69
two-dimensional gel electrophoresis, LIX, 68
Tris(hydroxymethyl)aminomethane-cacodylate buffer, for determination of ferrocytochrome *c* oxidation, LIII, 45
Tris(hydroxymethyl)aminomethane-ethylenediaminetetraacetic acid buffer
preparation, LVII, 32
purification of luciferin sulfokinase, LVII, 32, 36
Tris(hydroxymethyl)aminomethane-glycine buffer, in purification of *Cypridina* luciferase, LVII, 370
Tris(hydroxymethyl)aminomethane-hydrochloric acid buffer
assay of AMP formation during aminoacylation, LIX, 287
of CTP(ATP):tRNA nucleotidyltransferase, LIX, 123
of GTP hydrolysis, LIX, 361
of malate, LVII, 187
of nucleotide incorporation, LIX, 182
of polypeptide chain elongation, LIX, 357
of polypeptide synthesis, LIX, 367
of radiolabeled ribosomal particles, LIX, 779
of RNase III, LIX, 825
of tRNA terminal hydrolysis, LIX, 183
for translocation, LIX, 359
of tryptophanyl-tRNA synthetase, LIX, 237, 238
of *in vivo* energy transfer, LVII, 260
cell-free translation system, LIX, 387, 448, 459
chemiluminescence, LVII, 438, 439
determination of aminoacylation isomeric specificity, LIX, 273
of tRNA-ester modification, LIX, 161
diluent, for microperoxidase, LVII, 425
E. coli lysis solution, LIX, 354
electrophoretic analysis of *E. coli* ribosomal proteins, LIX, 538, 539
of RNA fragments, LIX, 574
filter binding assay, LIX, 848
harvesting of CHO cells, LIX, 219, 232
homogeneous competitive protein-binding assay, LVII, 441
inhibitor, in hydrogen peroxide assay, LVII, 380
initiation factor buffer, LIX, 838
in in vitro RNA synthesis, LIX, 847

interference with membrane oxidation of D-lactate, LIII, 524
ligand binding studies, LVII, 120
misactivation studies, LIX, 286
preparation of aminoacyl tRNA synthetase, LIX, 314
of *Bacillus* crude extract, LIX, 438
of cellulose-bound dithioglycolic dihydrazide, LIX, 386
of chloroplastic tRNA, LIX, 210, 211
of cross-linked ribosomal subunits, LIX, 535
of *E. coli* crude extract, LIX, 194, 297
of hydrazine-substituted tRNA, LIX, 120
of IF-free polysomes, LIX, 363
of inactivated ribosomal proteins, LIX, 819
of matrix-bound RNase, LIX, 463
of postmitochondrial supernatant, LIX, 515
of protein-depleted 30 S subunits, LIX, 621
of radiolabeled ribosomal protein, LIX, 777, 778, 783
of rat liver tRNA, LIX, 230
of ribosomal proteins, LIX, 431, 484, 837
of ribosomal RNA, LIX, 446, 556, 557, 648
of ribosomal subunits, LIX, 398, 399, 403, 404, 440, 444, 505, 554, 752
of ribosome, LIX, 374, 425, 426, 434, 435, 438, 445, 451, 483, 504, 554, 752, 817, 818, 838
of ribosome-DNA complexes, LIX, 841
of samples for neutron scattering, LIX, 655
of S100 extract, LIX, 357
of synthetic nucleotide polymers, LIX, 826
purification of aequorin, LVII, 281
of chloroplastic $tRNA^{Phe}$, LIX, 213
of CTP(ATP):tRNA nucleotidyltransferase, LIX, 123
of cyclic GMP, LVII, 99
of *E. coli* endogenous polysomes, LIX, 354, 355
of luciferin sulfokinase, LVII, 249
of *Pholas* luciferase, LVII, 387
of tRNA methyltransferase, LIX, 194, 195, 197, 201
of RNase III, LIX, 826, 827
of tryptophanyl-tRNA synthetase, LIX, 241, 242
reconstitution of ribosomal subunits, LIX, 441, 447, 451, 652, 837, 864
of tRNA, LIX, 127, 128, 133, 134
removal of protein from polyacrylamide gels, LIX, 313
in RNA aminoacylation, LIX, 216, 221, 288, 298, 818
deacylation, LIX, 211, 270
digestion, LIX, 186, 188, 833, 834
labeling, LIX, 71, 99, 102, 103, 149, 150
modificaton, LIX, 119, 120, 147, 160
solvent, for phosphodiesterase I, LIX, 60
studies of protein-RNA interactions, LIX, 562
for transmission electron microscopy, XXXIX, 136
in washing of DEAE-cellulose, LIX, 194
Tris(hydroxymethyl)aminomethane-hydrochloric acid-sodium chloride buffer, in separation of creatine kinase isoenzymes, LVII, 62
Tris(hydroxymethyl)aminomethane-maleate buffer
composition, XXXIX, 481
in preparation of 3′-amino-3′-deoxy-ATP, LIX, 135
Tris(hydroxymethyl)aminomethane-sucrose-histidine buffer
in preparation of mitochondrial enzyme complexes, LIII, 3, 41
in purification of complex III, LIII, 86, 87
Tris(hydroxymethyl)aminomethane-sulfate buffer
in assay of succinate dehydroxygenase, LIII, 469, 470
in electrophoretic separation of ribosomal proteins, LIX, 431

in purification of cytochrome oxidase, LIII, 62, 63
N-Tris(hydroxymethyl)methyl-2-aminoethanesulfonic acid
in chloroplast isolation, XXXI, 602
interference, in OPRTase-OMPdecase complex assay, LI, 167
N-Tris(hydroxymethyl)methylglycine
for chloroplast isolation, XXXI, 602, 603
preparation of ribosomal RNA, LIX, 557
for storage of ribosomal RNA, LIX, 586
in studies of protein-RNA interactions, LIX, 586
N-Tris(hydroxymethyl)methylglycine-potassium hydroxide, in preparation of mitochondrial ribosomes, LIX, 422
Trishydroxymethylphosphine, XLVII, 112
Trisialoganglioside
isolation, XXXII, 354, 355
structure, L, 273
Tris-maleate, *see* Tris(hydroxymethyl)aminomethane-maleate buffer
Tris method, of plasma membrane isolation, XXXI, 156–162
Tris-sucrose-histidine buffer, *see* Tris(hydroxymethyl)aminomethane-sucrose-histidine buffer
Tris-sulfate buffer, *see* Tris(hydroxymethyl)aminomethane-sulfate buffer
Triticum vulgare
lectin, XXXIV, 333, 335
ribosomes, preparation, LIX, 752
Tritium
assay of hydrogenase, LIII, 298
of radioactivity, vacuum system, LIII, 303
as hormone label, XXXVII, 145–147
labeling of amino acids in polypeptide hormones, XXXVII, 313–321
of phosphoenolpyruvate and pyruvate, XLI, 110–115
of pyruvic acid, XLI, 106–110
of sialylated glycoprotein hormones, XXXVII, 321–326
labeling parameter
for adenine nucleotides, LIX, 339
definition, LIX, 337
for nucleic acids, LIX, 337, 338
lock nucleus, LIV, 189
Tritium exchange, *see also* Hydrogen exchange
assay, LIII, 304–306
calculation of $H_{rem/}$[Fe], XLVIII, 336–338
color quenching, XLVIII, 336
dilution procedure, XLVIII, 334
first separation, XLVIII, 334
inexchange procedure, XLVIII, 333, 342
for mapping protein-tRNA complexes, LIX, 332–350
outexchange, XLVIII, 334, 335
protein dissociation, XLVIII, 321–346
protocol, XLVIII, 325, 326
sampling, XLVIII, 335, 336, 343
scintillation counting, XLVIII, 336
theory, XLVIII, 327
Tri-*p*-tolylamine
chemiluminescent reaction, LVII, 505–507
emission spectrum, temperature dependence, LVII, 506
structural formula, LVII, 498
Triton gel system, phosphorylation of histone fractions, XL, 142
Triton N–101, partial specific volume, XLVIII, 19, 20
Triton WR–1339, XXXI, 324; LII, 504
in peroxisome isolation, XXXI, 357
Triton X–45, in dimethylaniline monooxygenase purification, LII, 145
Triton X–100, XXXIV, 567, 570, 688, 690; XLIV, 225, 226; XLVIII, 343; LVI, 741
absorption maxima, LII, 105
activator, of dihydroorotate dehydrogenase, LI, 62
of palmityl-CoA synthase, XXXV, 122
adenylate cyclase, XXXVIII, 140, 149, 155, 167, 168
alanine carrier isolation, LVI, 432
antibody studies with nitrate reductase, LIII, 352, 353

assay of adenylate kinase, LI, 468
of dihydroorotate dehydrogenase, LI, 60, 64
of long-chain aldehydes, LVII, 190, 191
of monoamine oxidase, LIII, 496
assay reagent, XLIV, 460, 474
ATPase solubilization, LV, 346, 356
BKA-protein complex, LVI, 413, 414
CAT-protein complex, LVI, 409, 411
for cell lysis, LII, 64
cell membrane solubilization, XLVII, 250
cross-linking, LVI, 633
in cytochrome b_5 reduction, LII, 107
in detergent mixtures, XXXI, 559, 560
in dimethylaniline monooxygenase purification, LII, 145
effect on enzyme gel electrophoresis, XXXII, 91
as eluent, XXXIV, 666
facilitator, of rapid filtration of biological fluids, LVII, 72
function, XLIV, 476
in gel electrophoresis, XXXII, 83, 85, 86
guanylate cyclase, XXXVIII, 198, 199, 201
insulin receptors, XXXIV, 656
in liposome rupture, XLIV, 708
for lysis of Ehrlich ascites cells, LIX, 412
of mitochondria, LVI, 25, 27
of non-bacterial cells, LVII, 68, 69, 71
in lysosome microscopy, XXXI, 34
for microsomal protein solubilization, LII, 98, 99, 104
for microsome isolation, XXXI, 216, 225
mitochondrial proteins isolated, LIII, 98
mitochondrial ribosome isolation, LVI, 80
M. phlei membrane fragments, LV, 195
in organic hydroperoxide solubilization, LII, 510
partial specific volume, XLVIII, 19, 20
preparation of rat liver polysomes, LIX, 526
of $TF_0{\cdot}F_1$, LV, 368, 369
purification of adenylate kinase, LI, 470, 471
of complex III, LIII, 88, 92–95, 103
of cytochrome a_1, LIII, 639
of cytochrome bc_1 complex, LIII, 116
of cytochrome oxidase, LIII, 75
of dihydroorotate dehydrogenase, LI, 61, 62, 65
of α-glycerophosphate dehydrogenase, LIII, 409
of D-lactate dehydrogenase, LIII, 521, 522
of *Lactobacillus* deoxynucleoside kinases, LI, 351
of mitochondrial membranes, LIII, 103
of monoamine oxidase, LIII, 499
of NADH dehydrogenase, LIII, 418
of rat liver acid nucleotidase, LI, 273
removal, XXXIV, 569; LII, 99, 105; LVI, 412
in scintillation cocktail, LI, 30, 220, 338, 540, 544; LII, 315, 321, 398, 404; LIX, 396
solubilization of BCF_0-BCF_1, LV, 194
solvent, of aldehyde, LVII, 138
for stabilizing NADH-cytochrome b_5 reductase, LII, 107
in stearyl-CoA desaturase activity assay, LII, 189
in stearyl-CoA desaturase solubilization, LII, 191
storage, to prevent peroxidation, LII, 145, 146
termination of complex III assay, LIII, 39
treatment of ETP, LV, 193
use in preparation of tissue extracts, XXXV, 131
yeast atractyloside-binding protein, LVI, 416
Triton X–102, LVI, 741
in dimethylaniline monooxygenase purification, LII, 145

Triton X–114, LVI, 741
purification of cytochrome oxidase, LIII, 56, 57, 58
Triton X–165, LVI, 741
Triton X–45-succinate, preparation, LII, 145
Triton X–100-sucrose solution, for peroxisome nucleoid isolation, XXXI, 368–374
Triton X–xylene solution, disadvantages, XLVIII, 336
Tritosome, LI, 271
electrophoretic purification, XXXI, 753, 754
separation, XXXI, 40
N-Trityl, XXXIV, 508
2-Tritylaminoethyl iodide, intermediate in lysoglycerophospholipid synthesis, XXXV, 491, 492
Tritylation, hydroxyl group activation, L, 109, 110
(Trityl)cellulose, XXXIV, 647
N-Trityl-*O*-ethanolamine, intermediate in phosphatidylethanolamine synthesis, XXXV, 456, 460
Trityl group
peptide synthesis, XLVII, 520
removal, XLVII, 538, 539
Trityl-oligonucleotide, XXXIV, 645, 646
Trizma, purification of firefly luciferase, LVII, 7
tRNA, *see* Transfer ribonucleic acid
Tronac 450 Isoperibol calorimeter, LVII, 542
light-detection signal, calibration, LVII, 543–545
modification for second sensor, LVII, 543, 544
Tronac 550 Isothermal calorimeter
light-detection signal, calibration, LVII, 543–545
modification for second sensor, LVII, 541, 543
Tropane diester, hydrolysis, XLIV, 835, 836
Troponin C, frog, phthalylation, XLVII, 155
Troponin I, cyanylation, XLVII, 132
Tropylium ion
identification by mass spectroscopy, LIII, 596
structural formula, LIII, 596
Trough, for monolayer formation, XXXII, 539, 540
Trout cell, incubation temperature, LVIII, 471
Trowell T8 medium, XXXIX, 445
Trypan blue
cell viability, LVIII, 151, 152
dye exclusion test, LII, 64
in cell adhesion studies, XXXII, 601
in cell isolation test, XXXII, 700
in liver cell studies, XXXII, 628–630
in oxyntic cell studies, XXXII, 715
for viable cell detection, XXXIX, 120
use in assessing hepatocyte membrane integrity, XXXV, 584
Trypanosome antigen, sequence analysis, XLVII, 334
Trypsin, XXXIV, 5, 417, 418, 440–445; XLVI, 96, 128, 155, 156, 197, 206, 207, 215, 229, 233
activity resembling, in beta granules, XXXI, 53, 378
in adrenal cortex cell isolation, XXXII, 674–676, 689, 690, 692
at alkaline pH
hydrolysis, XLVII, 170–174
stability, XLVII, 172
amino acid analog substrates, XLVII, 430
amino acid sequences, XLV, 296
assay, XXXIV, 445
by Fast Analyzer, XXXI, 816
ATPase, LV, 795
autodigestion, XLIX, 454–457
autolysis, rate determination, XLIV, 572, 573
azo linkage, XXXIV, 104, 105
BCF_0-BCF_1, LV, 195
bovine, reductive methylation, XLVII, 475
from bovine pancrease, specificity, LIX, 601
coentrapment, XLIV, 475–478
comparison to clostripain, XLVII, 165, 166
conformation studies, XLIV, 363, 364, 533–535

conjugate
active site titration, XLIV, 391, 392
activity assay, XLIV, 198, 200, 267, 271, 346, 530, 561
change in digestion products, XLIV, 437, 438
effect of support stretching, XLIV, 562, 563, 565
with polycarboxylic acid, XLIV, 179
reductive denaturation studies, XLIV, 518, 519
stability, XLIV, 537
contamination, LVIII, 125
coupling factor, LV, 189, 191, 192
in cytochrome *P*–450 conversion to cytochrome *P*–420, LII, 411
digestion
procedure, XLVII, 199
of proteins, XL, 35
concentration and incubation period, XL, 38
effect on luciferase activity, LVII, 201
on purified NADH-cytochrome b_5 reductase, LII, 108
energy transfer studies, XLVIII, 378
fibroblast cultures, L, 440
gel electrophoresis, XXXII, 101
in hormone receptor studies, XXXVII, 212
immobilization, XLVII, 44
on (aminoethyl)cellulose, XLIV, 48, 52
in cellophane membrane, XLIV, 905, 909
by direct covalent binding, XLIV, 909
on hydrazide derivative of hydroxyalkyl methacrylate gel, XLIV, 82, 83
on imidoester-containing polyacrylonitrile, XLIV, 323, 324
by microencapsulation, XLIV, 214
on nylon tube, XLIV, 120
in polyacrylamide beads, XLIV, 60–62, 171, 175, 176, 662, 663
in polyacrylamide gel, XLIV, 902, 909
on porous glass, XLIV, 153
by protein copolymerization, XLIV, 197, 198, 200
on Sephadex G–200, XLIV, 529
using *N*-carboxy-L-tyrosine, XLIV, 327
immobilized peptide cleavage, XLVII, 277
inhibition, by alkylated protein, XLVII, 474, 475
inhibition assay, XLV, 870
inter-α, inhibitor, *see* Inter-α-trypsin inhibitor
inhibitors
isolation, XLIV, 277
pancreatic, XLIV, 568–570
kinetics
of inhibition, XLIV, 428
perturbed by immobilization, XLIV, 403–405
of substrate diffusion, XLIV, 436–438
ligands, XXXIV, 441, 442
matrix, XXXIV, 443
in media, LVIII, 139
microcalorimetric assay, XLIV, 664, 665
microsome damage, XXXI, 195
modification, XLIV, 311, 312
in mucosal cell isolation, XXXII, 665
NADH dehydrogenase, LVI, 588
in neurobiological cell fractionation, XXXII, 771, 772
neutralization, LVIII, 77
pH-activity profiles, XLIV, 403
phosphodiesterase, XXXVIII, 235, 236, 272
phosphodiesterase activator, XXXVIII, 268
in phospholipase studies, XXXII, 137
porcine, hydroxylamine cleavage, XLVII, 140, 142
protein kinase inhibitor, XXXVIII, 351, 353
purification, XXXIV, 444, 445
of flavin peptides, LIII, 453
reductive denaturation studies, XLIV, 517–519
sequence analysis, XLVII, 180
source, XLIV, 196

soybean, membrane osmometry, XLVIII, 75
spacer, XXXIV, 444
spin-label studies
after phosphonylation, XLIX, 445, 446, 455–457
after sulfonylation, XLIX, 447–450
suitable carrier, XLIV, 63
in testis hormone receptor studies, XXXVII, 186, 187
in thyroid-cell isolation, XXXII, 746–748
treatment of ETP, LV, 188
tryptophan content, XLIX, 162
use in isolating neurons and astrocytes, XXXV, 573, 574
wash solution, for cell cultures, XXXII, 799, 805
α-Trypsin, XXXIV, 450
phosphorescence studies, XLIX, 247
titration, concentration incubation method, XLV, 17
β-Trypsin, XXXIV, 448–451
acetylation, XLIV, 532
autolysis, XLIV, 531, 532
inhibition, XLVI, 233
titration, XLV, 15, 16
Trypsin-EDTA solution, metaphase preparation, LVIII, 171
Trypsin inhibitor, LVIII, 95, 128, *see also* Bowman-Birk soybean inhibitor; Kunitz soybean trypsin inhibitor
antipain, XLV, 683
black-eyed pea, energy transfer studies, XLVIII, 378
cereal grains, XLV, 723
chick pea, XLV, 699
colostrum, XLV, 806
assay method, XLV, 807
composition, XLV, 811
kinetics, XLV, 811
properties, XLV, 810
purification, XLV, 807
purity, XLV, 810
specificity, XLV, 811
stability, XLV, 810
garden bean
amino acid composition, XLV, 715, 716
sequence, XLV, 697, 698
properties, XLV, 714
purification, XLV, 711
purity, XLV, 714
units, XLV, 710
groundnut
amino acid composition, XLV, 721
assay, XLV, 716
esterolysis inhibition, XLV, 717
potentiometric, XLV, 718
properties, XLV, 721
purification, XLV, 718
purity, XLV, 721
specificity, XLV, 721
spectrophotometric, XLV, 717
stability, XLV, 721
units, XLV, 717, 718
yield and potency, XLV, 719
legume seeds, XLV, 697
reactive sites, XLV, 698
leupeptin, XLV, 679
lima bean
amino acid sequence, XLV, 697, 698, 708
kinetics, XLV, 708
properties, XLV, 707
reactive sites, XLV, 709
pancreatic
amino acid composition and sequence, XLV, 822
assay method, XLV, 815
characteristics, XLV, 813
chromatographic forms, XLV, 821
homogeneity, XLV, 821
preparation, XLV, 815
properties, XLV, 821
soybeans, XLV, 700
of thrombin, XLV, 174
specific activity, XLV, 823
soybean
osmometric studies, XLVIII, 104, 105
self-association, XLVIII, 117
thermolysin hydrolysis, XLVII, 186, 188
Trypsinization, LVIII, 50
adenovirus, LVIII, 433, 434
BHK cells, LVIII, 298
in cloning, LVIII, 156
enzymes, LVIII, 124–128
myoblasts, LVIII, 517

poikilotherm vertebrate cell, LVIII, 475, 476
Trypsinization flask, XXXIX, 114; LVIII, 127
Trypsin-kallikrein inhibitor, XXXIV, 5, 6
bovine specificity, XLV, 867
cuttlefish
amino acid composition, XLV, 795, 797
assay, XLV, 792
isoinhibitor isolation, XLV, 794
kinetics, XLV, 796
properties, XLV, 795
purification, XLV, 793
units, XLV, 793
snail, XLV, 772
amino acid composition, XLV, 784, 785
assay, XLV, 774
distribution, XLV, 784, 785
isoinhibitor
purification, XLV, 778
separation, XLV, 776
kinetic, XLV, 781
properties, XLV, 779
purification, XLV, 774
purity, XLV, 779
specificity, XLV, 780
stability, XLV, 779
structure, XLV, 785
Trypsinlike enzyme
activity assay, XLVII, 78
assay, XLV, 21–24
fluorometric, of proteolytic activity, XLV, 24
isotropic, XLV, 21
sensitive, XLV, 12
hydrolysis rates, XLV, 24
rate assays, XLV, 18–20
Trypsinogen, XXXIV, 448, 451, 592
activation, XXXIV, 448, 450
assay, penicillopepsin, XLV, 441
Trypsinolysis, limited, of elongation EF-G, LX, 743–745
Trypsin plasma inhibitor, *see* Bdellin
Trypsin resin, preparation, XLV, 775
Trypsin-Sepharose, XLVII, 42
Trypsin-trypsin inhibitor reaction, mixed association, XLVIII, 122
Tryptamine
inhibitor of tryptophanyl-tRNA synthetase, LIX, 252
in melatonin biosynthesis, XXXIX, 379
Tryptamine hydrochloride, XLVII, 455
Tryptazan, substrate of tryptophanyl-tRNA synthetase, LIX, 249
Trypticase soy broth, LVIII, 20
Tryptone, preparation of MS2 RNA, LIX, 299
Tryptone glucose yeast extract, XLIII, 5
Tryptophan, XXXIV, 4, 103
aminoacylation isomeric specificity, LIX, 274, 275, 279, 280
analogs, tryptophanyl-tRNA synthetase, LIX, 251–254
assay, for bound enzyme determination, XLIV, 255, 256, 389
N-carbethoxylation, XLVII, 440
chemical properties, XLIV, 12–18
chloromethyl ketone analog, inhibitor, of tryptophanyl-tRNA synthetase, LIX, 253, 254
conformation studies, with circularly polarized luminescence, XLIX, 185
derivatives, separation, XLVII, 447
detection, XLVII, 217
determination, XLVII, 446
as enzyme inhibitor, XXXVII, 294
fluorescence, XLIV, 364
quenching, XLIX, 224, 425
fluorescent donor, XLVIII, 362, 363
fluorine labeling, XLIX, 273, 274, 290
hydrogen bromide cleavage, XLVII, 547, 603, 604
in media, LVIII, 53, 62
as melatonin precursor, XXXIX, 379
in membranes, fluorescent studies, XXXII, 236, 245, 246
modification, XXXV, 128; XLVII, 442–453, 457
NH proton, NMR spectroscopy, LIV, 186
peptide bond cleavage, XLVII, 459–469
mechanism, XLVII, 461, 462
peptide synthesis, XLVII, 509
phosphorescence, XLIX, 242–247

protein rotational diffusion, LIV, 57
in protein-turnover studies, XXXVII, 246, 247
quantitative measurement, XLIX, 157, 161, 162
reaction with *o*-nitrophenylsulfenyl chloride, XLVII, 548
recovery, XLVII, 371
removal, LIX, 242
secondary function blocking, XLVII, 615, 616
spectral properties, XLVIII, 362
structural formula, XLVIII, 362
substrate of phenylalanine hydroxylase, LIII, 284
of tryptophanyl-tRNA synthetase, LIX, 247
thermolysin hydrolysis, XLVII, 177
type of binding reaction with cytochrome *P*–450, LII, 264
D-Tryptophan, XXXIV, 4
inhibitor of tryptophanyl-tRNA synthetase, LIX, 252
DL-Tryptophan, base-line fluorophore, XLIX, 210
DL-[^{14}C]Tryptophan, LIX, 235
assay of tryptophanyl-tRNA synthetase, LIX, 238
[^{3}H]Tryptophan, preparation of aminoacylated tRNA, LIX, 132
L-Tryptophan, LIX, 235
in affinity labeling of steroids, XXXVI, 386, 387
assay of tryptophanyl-tRNA synthetase, LIX, 238
isolation, XLIV, 757
magnetic circular dichroism spectrum, XLIX, 160–164
purification of tryptophanyl-tRNA synthetase, LIX, 239
L-[^{14}C]Tryptophan, LIX, 235
assay of tryptophanyl-tRNA synthetase, LIX, 237
Tryptophan-agarose, XXXIV, 389–394, *see also* Anthranilate synthetase complex
matrix, XXXIV, 390, 391
Tryptophanase, XLVI, 445
coimmobilization, XLIV, 471
Tryptophan 2,3-dioxygenase, LII, 13
hormone induction of biorhythms, XXXVI, 480
oxy form, spectrum, LII, 36
Tryptophan 5-hydroxylase, *see* Tryptophan 5-monooxygenase
D-Tryptophan methyl ester, XXXIV, 514
Tryptophan 5-monooxygenase, LII, 12
assay, XXXIX, 387, 388
Tryptophan oxidative decarboxylase, LII, 20
Tryptophan peptides, XXXIV, 192, 193
from cytochrome *c*, XXXIV, 193
from DNPS albumin, XXXIV, 193
Tryptophan pyrrolase, *see also* Tryptophan 2,3-dioxygenase
as marker enzyme, XXXII, 733
Tryptophan pyrrolooxygenase, LII, 11
Tryptophan residue
modification, LIII, 167–171
using *N*-bromosuccinamide, LIII, 171
using formic acid-hydrochloric acid, LIII, 167, 170
using 2-hydroxy-5-nitrobenzylbromide, LIII, 170
Tryptophan synthase, XLVI, 31, 32
carbethoxylation, XLVII, 434, 435
from *Escherichia coli,* staphylococcal protease hydrolysis, XLVII, 191
fiber entrapment
activity, XLIV, 237
in cellulose triacetate, XLIV, 235
industrial application, XLIV, 241
Tryptophan synthetase, *see* Tryptophan synthase
Tryptophanyl adenylate, substrate of tryptophanyl-tRNA synthetase, LIX, 247
DL-Tryptophanylhydroxamate, inhibitor of tryptophanyl-tRNA synthetase, LIX, 251
Tryptophanyl-tRNA synthetase
in aminoacylation test of tRNA hydrolysis, LIX, 132, 133
assay, LIX, 236–239
from beef pancreas, LIX, 234–257
with covalently bound tryptophan, LIX, 246, 247
diffusion constant, LIX, 244
extinction coefficient, LIX, 239

immunochemical properties, LIX, 248
inhibitors, LIX, 249–257
kinetic properties, LIX, 247
molecular weight, LIX, 244
multiple forms, LIX, 245
paracrystals, LIX, 246
pH optimum, LIX, 247
purity, LIX, 244
sedimentation coefficient, LIX, 244
stability, LIX, 244
subcellular distribution, LIX, 233, 234
substrate binding, LIX, 247, 248
substrate specificity, LIX, 249
subunit structure, LIX, 244, 245
tertiary structure, LIX, 245
TSH, *see* Thyroid-stimulating hormone; Thyrotropin receptors
TSH buffer, *see* Tris-sucrose-histidine buffer
TSH-releasing hormone, *see* Thyroid-stimulating hormone-releasing hormone
TSIM, *see* Trimethylsilylimidazole
TSP, *see* Trimethylsilylsodium [U-^{2}H]-propionate
Tsuchihashi fractionation, purification of superoxide dismutase, LIII, 388
TTB, *M. phlei* membrane systems, LV, 185
TTC, *see* Tetrazolium chloride
TTF, *see* Thenoyltrifluoroacetone
TTFB, *see* 4,5,6,7-Tetrachloro-2′-trifluoromethylbenzimidazole
TTY, *see* Teletypewriter
Tube, culture, disposable plastic, source, LIX, 194
Tubercidin, XLIII, 154, 156, 516
Tubercidin cyclic 3′,5′-phosphate, DNA transcription, XXXVIII, 375
Tubercidin 5′-phosphate, AMP nucleosidase binding studies, LI, 270
Tubercidylcobalamin, effector of ribonucleoside triphosphate reductase, LI, 258
Tubing, *see also* specific type
for freeze-quench techniques, LIV, 88, 89
for low-temperature EPR spectroscopy, LIV, 116, 119
d-Tubocurarine, XXXIV, 106
as receptor antagonist, XXXII, 309
Tubule, renal, isolation, XXXII, 658–664
Tubulin, XXXIV, 623–627; XLVI, 567–571
in brain tissue, XLVI, 570, 571
computer-simulated sedimentation, XLVIII, 236
dimerization, XLVIII, 243, 244, 246, 247
optical filtering, XLIX, 50, 56, 60–63
polymerization, 8-azido-GTP, LVI, 646
Tuftsin, XLVII, 170
Tumor
ACTH, XXXVII, 35
cells
agglutinins for, isolation, XXXII, 615, 621
2′,3′-cyclic nucleotide-3′-phosphohydrolase, XXXII, 128
indefinite population growth, XXXII, 562
thyrotropic, thyrotropin releasing factor binding, XXXVII, 217, 218
transformed cells compared, XXXII, 583
estrogen receptors, XXXVI, 248–254
hormone effects, XXXIX, 115
mammary, *see* Mammary tumors
of mammary gland, mitochondria isolation, XXXI, 307, 308
ovarian, hormone-dependent, culture, XXXII, 557
pituitary, in neurobiological studies, XXXII, 769
of plant cells, XXXI, 553–558
plasma membrane isolation, XXXI, 75–90
substrate-dependent, treatment, XLIV, 689–694, 703, 704
tissue cultures, in insulin precursor studies, XXXVII, 328
transplantable, LVIII, 276, 277
virus-induced, plasma-membrane isolation, XXXI, 163
Tumorigenic cell line, *see* Cell line, tumorigenic

Tumorigenicity
human adenovirus, LVIII, 425
testing
nude mouse, LVIII, 370–379
protocol, LVIII, 373–379
threshold, LVIII, 374
Tumor-inducing principle, crown gall bacterium, XXXI, 554
Tumor-specific cell surface antigen, *see* Antigen, tumor-specific-cell surface antigen
Tungstate, osmolality, XXXII, 30
[^{185}W]Tungstate, radiolabeling formate dehydrogenase, LIII, 369, 371
Tungsten
atomic emission studies, LIV, 455
filament, in flash photolysis system, LIV, 98
formate dehydrogenase, LIII, 371
for high-resolution shadowing, LIX, 612
as reference electrode, LVI, 456
Tungsten filament lamp, defined color temperature, LVII, 578
Tungsten-halide lamp, in flash photolysis, LIV, 53
Tungsten-rhenium, for high-resolution shadowing, LIX, 612
Tungsten-tantalum, for high-resolution shadowing, LIX, 612
Tungstoborate, as negative stain, XXXII, 22
Tungstosilicate
as negative stain, XXXII, 22, 24
osmolality, XXXII, 30
Tunicamycin
glycoprotein biosynthesis inhibition, L, 188
polyribitol phosphate biosynthesis, L, 396
Tunneling effect, ligand binding and, LIV, 508, 509, 526–529
Tunneling rate, evaluation, LIV, 509, 528
Turanose, degradation on heating in boric acid/2,3-butanediol or in borax, XLI, 21
Turbidimetry, procedure, in quantitative histone determination, XL, 113
Turbidity, of dicyclohexylcarbodiimide, compensation, LVI, 166
Turbidity method, for plant protein determination, XXXI, 543
Turbidometric assay, *see* Photometric method, for dilution assay
Turner fluorometer, XXXII, 678; LI, 468
in hormone determination, XXXIX, 308
Turner Spectro 210, spectrophotofluorometer, LIII, 428
Turnip, root, peroxidase in, LII, 514
Turnip yellow mosaic virus, molecular weight, LIII, 350, 351
Tween, XXXIV, 709
ATPase, XXXIV, 570
in yeast mutant isolation, XXXII, 824
Tween 20, XLIV, 204, 208, 210; LVI, 741
carnitine palmitoyltransferase extraction, LVI, 373
in cytochrome P–$450_{11\beta}$ assay, LII, 126
in cytochrome P–450 purification, LII, 127, 129
in media, LVIII, 70
partial specific volume, XLVIII, 19, 20
purification of cytochrome oxidase, LIII, 56, 58
Tween 21, LVI, 741
Tween 40, LVI, 741
in detergent mixtures, XXXI, 559
Tween 60, LVI, 741
Tween 80, XLIV, 185, 187, 190; LVI, 741
cytochromeless strains, LVI, 122
effect on enzyme electrophoresis, XXXII, 91
on epoxide hydrase activity, LII, 417
in epoxide hydrase assay, LII, 194
fatty acid composition, LVI, 571
growth of *M. phlei*, LV, 179
heme-deficient mutants, LVI, 559–561
interference, in fluorometric assays, LII, 402
in media, LVIII, 70
M. phlei membrane systems, LV, 185
partial specific volume, XLVIII, 19, 20
purification of cytochrome oxidase, LIII, 56, 58, 68, 69
solvent, for biphenyls, LII, 400
ultraviolet absorption, LVI, 744

Tween 81, LVI, 741
Tween 85, *see* Polyoxyethylene(20) sorbitan trioleate
Two-photon fluorescence technique, general principle, LIV, 8, 9
Tygon, steroid adsorption, XXXVI, 93, 94, 161
Tygon tubing
  disadvantage, in peroxyoxalate chemiluminescence assays, LVII, 460
  in microcolumn, LVII, 281
  for perfusion, XXXIX, 5
Tylosin, XLIII, 131, 495
  in media, LVIII, 113, 115, 116
  membranes, LVI, 32
Tyndall scattering, in Raman spectroscopy, XXXII, 254, 255
Type C viral group specific antigen, XXXIV, 729, 730
Tyramine, XXXIV, 105
  as spacer, XXXIV, 72
Tyramine oxidase, LII, 18; LVI, 474
Tyrocidine
  biosynthesis, XLIII, 560, *see also* specific enzyme
    tyrocidine synthetase, XLIII, 585
  compared with gramicidin S, XLIII, 579
    with polymycin, XLIII, 584
  countercurrent distribution, XLIII, 324, 325
    partition isotherms, XLIII, 330
    solvent systems for, XLIII, 342
    for structural studies, XLIII, 33
  effect on membranes, XXXII, 882
  structure, XLIII, 586
Tyrocidine B, tryptophanyl phosphorescence, XLIX, 248
Tyrocidine C, tryptophanyl phosphorescence, XLIX, 248
Tyrocidine-synthesizing system, XLIII, 586–595, *see also* Tyrocidine synthetase
Tyrocidine synthetase, XLIII, 585–602
  amino acid activating subunits, XLIII, 598
  assay, XLIII, 589, 590
  ATP-PP_i exchange, XLIII, 588, 589, 596
  Bacillus brevis preparation, XLIII, 587, 588
  carrier protein
    isolation from purified polyenzyme, XLIII, 600, 601
    purification from crude extracts, XLIII, 598–600
    sodium dodecyl sulfate gel electrophoresis, XLIII, 599, 601, 602
  dissociation of polyenzymes, XLIII, 595–602
    in crude extracts, XLIII, 596, 597
    of purified polyenzymes, XLIII, 597
  enzymatic synthesis, XLIII, 588, 589
  extract preparation, XLIII, 590
  heavy enzyme
    dissociation of, XLIII, 597, 598
    molecular weight, XLIII, 595
    purification, XLIII, 592–595
  intermediate enzyme
    molecular weight, XLIII, 595
    purification, XLIII, 592, 594, 595
  light enzyme, purification, XLIII, 591, 592
  ornithine-dependent PP_i-ATP exchange, XLIII, 588, 589, 596
  polyenzymes, dissociation products, XLIII, 595–602
  purification, XLIII, 590–595
  Sephadex G–200 filtration, XLIII, 590, 591
Tyrode's balanced salt solution, LVIII, 120, 121, 452
  composition, XXXIX, 479
  medium, XXXIX, 106
  for placental perfusion, XXXIX, 245
Tyrode's cacodylate buffer, composition, XXXIX, 481
Tyrosinase, XXXIV, 6, *see also* Tyrosine phenol-lyase
  assay, LVIII, 568
  melanocytes, LVIII, 564
  synthesis of, in melanin formation, XXXI, 389, 394
  toxicity, LVIII, 567
β-Tyrosinase, *see* Monophenol monooxygenase
Tyrosinate, phosphorescence, XLIX, 243

Tyrosine, XXXIV, 394–398; XLIV, 49
  aminoacylation, isomeric specificity, LIX, 274, 275, 279, 280
  aromatic ring alkylation, XLVII, 542
  O-carbethoxylation, XLVII, 438, 439
  chemical properties, XLIV, 12–18
  content, of histones, in direct quantitative determination, XL, 114
  detection, XLVII, 217, 259
  determination, LIII, 279, 280
  enzyme aggregation, XLIV, 263
  fluorescence, XLIV, 364
    in direct quantitative histone determination, XL, 114
  fluorine labeling, XLIX, 273, 274, 290
  glucose electrode, LVI, 483
  hydrogen bromide cleavage, XLVII, 547, 603
  hydroxyl group blocking, XLVII, 527, 528, 541–543, 615
  inhibitor, of trypsin, XLV, 174
  in media, LVIII, 53, 62
  misactivation, LIX, 291
  modification, XLVII, 484
  paper chromatography, XXXII, 788
  peptide synthesis, XLVII, 509
  phosphorescence, XLIX, 242–245, 248
  product of phenylalanine hydroxylase, LIII, 278
  protein ^{19}F-labeling, XLIX, 275
  in proteins, iodination, XXXII, 103, 106
  residue
    of ATPases, LV, 299
    4-chloro-7-nitrobenzofurazan, LV, 490
    enzyme immobilization, XLIV, 930
    modification, LIII, 137, 148, 149
      with lactoperoxidase, LIII, 137, 148
      using lactoperoxidase-potassium iodide-hydrogen peroxide, LIII, 171–173
      using tetranitromethane, LIII, 171, 172
  as spacer, XXXIV, 395, 396
  thermolysin hydrolysis, XLVII, 175
  topographical state in insulin, XLIX, 127
    in ribonuclease A, XLIX, 121
  transport, XLVI, 608
  vibrational modes of side chains, XLIX, 118
L-Tyrosine
  enzyme electrode, XLIV, 585, 589
  magnetic circular dichroism spectrum, XLIX, 160–164
L[U-^{14}C]-Tyrosine, radiolabeling of quinone intermediates, LIII, 608
m-Tyrosine, substrate of phenylalanine hydroxylase, LIII, 284
L-Tyrosine-agarose, XXXIV, 395
Tyrosine aminotransferase, XXXIV, 295–299; XLVI, 445
  daily rhythms, XXXVI, 477, 479
    hormone factors, XXXVI, 480
  in hepatocyte cell lines, XXXII, 740
  hormone induction, XXXII, 733
Tyrosine chloromethyl ketone, XLVI, 609, 611, 612
Tyrosine decarboxylase, XLVI, 445
  conjugate, activity, XLIV, 271
  in enzyme electrode, XLIV, 589
  immobilization, on inert protein, XLIV, 909
Tyrosine hydroxylase, *see* Tyrosine 3-monooxygenase
Tyrosine *o*-hydroxylase, LII, 12
Tyrosine α-ketoglutarate transaminase, *see* Tyrosine aminotransferase
Tyrosine methyl ester
  in steroid radioimmunoassay, XXXVI, 25, 31, 33
  succinyl cyclic nucleotide derivatives, XXXVIII, 99
Tyrosine 3-monooxygenase, XXXIV, 6
  assay, XXXII, 785–788
  in cultured cells, XXXII, 766
  in neurobiology, XXXII, 765
Tyrosine peptide, XXXIV, 188
  from carboxypeptidase, XXXIV, 189
Tyrosine phenollyase, XLIV, 886; LII, 6, 9
  histidine modification, XLVII, 433
  oxy form, spectrum, LII, 36
  prosthetic group, LII, 4
  reaction mechanism, LII, 39
Tyrosine thiohydantoin, marker, XLVII, 364
Tyrosine transaminase, *see* Tyrosine aminotransferase

Tyrosyl-tRNA synthetase, XXXIV, 170, 503–506; XLVI, 90
phosphocellulose capacity for, XXXIV, 167
purification, LIX, 262
subcellular distribution, LIX, 233, 234
L-Tyrosyl-D-tryptophan, XXXIV, 4

U

U–12,241 antibiotic, rhodomycin-like, XLIII, 342
U–12,898 antibiotic, XLIII, 342
U–13,714 antibiotic, XLIII, 342
Ubiquinol, oxidation, proton uptake, LV, 627
Ubiquinol-cytochrome *c* reductase, reconstituted, activity, LV, 709, 710
Ubiquinone, *see also* Coenzyme Q; Menaquinone; Quinone intermediates
analogs, model redox reactions, LV, 543
assay of oxidative phosphorylation at site I, LV, 724
biosynthesis pathway in *E. coli*, LIII, 601
biosynthetic intermediates, detection, LVI, 109–111
brown adipose tissue mitochondria, LV, 75
cofactor for dihydroorotate dehydrogenase, LI, 62, 64
complex I, LVI, 580, 586
complex II, LVI, 586
complex III, LIII, 89, 96, 97; LVI, 580, 582, 586
in complex I–III, LIII, 8
difference extinction coefficient, LIII, 586, 588
enzyme activity in bacterial membranes, LIII, 578, 579
in mitochondria, LIII, 575–578
EPR characteristics, LIV, 134, 137, 138
extinction coefficient, LIII, 574, 586
extraction from submitochondrial particles, LIII, 574
genes for biosynthesis, map location, LVI, 110
intermediates in synthesis, identification, LIII, 603, 604
Keilin-Hartree preparation, LV, 123
in kinetic studies of bacteriochlorophyll, LIV, 28, 30
lack, in promitochondria, XXXI, 627
from *Nitrosomonas*, LIII, 639
oxidation, before extraction, LIII, 580
phosphorylation site, LV, 240
product, of complex III, LIII, 35
redox reaction
determination, LIII, 585–591
kinetics, LIII, 582–584
redox state, determination, LIII, 581, 582
reincorporation, LIII, 574, 575
role in respiratory chain, LIII, 575–579
specificity of complexes, LVI, 586
spectral properties, LIII, 604
structural formula, LIII, 601
substrate of complex II, LIII, 21
of NADH:ubiquinone oxidoreductase, LIII, 11, 12
thin-layer chromatography, LVI, 109, 110
Ubiquinone–1
solubilization, LIII, 13
substrate of NADH dehydrogenase, LIII, 18, 20
Ubiquinone–2
reduced
assay of complex, LIII, 3, 38, 39
effect of nonpolar solvents/water, LIII, 13
of organic solvents, LIII, 13
preparation, LIII, 39
substrate, of complex II, LIII, 24, 26
Ubiquinone–10, in complex I, LIII, 12
Ubiquinone–30, *see* Coenzyme Q_6
Ubiquinone–45, *see* Coenzyme Q_9
Ubiquinone–50, *see* Coenzyme Q_{10}
Ubiquinone isoprenolog
solution preparation, LIII, 14
substrates of NADH dehydrogenase, LIII, 18
Ubiquinone Q_1, assay of succinate dehydrogenase, LIII, 466
Ubiquinone Q_2, assay of succinate dehydrogenase, LIII, 466

Ubisemiquinone, EPR spectrum, LIV, 146, 147
UDP, *see* Uridine diphosphate
UDPgalactose:*N*-acetylglucosamine galactosyltransferase, in plasma membranes, XXXI, 180, 184
UDPgalactose hydrolase, in plasma membrane, XXXI, 102
UDPgalactose-lipopolysaccharide galactosyltransferase
  purification, XXXII, 453
  reconstitution of activity, XXXII, 449–459
UDPglucose hydrolase, in bacterial membranes, XXXI, 649
UDPglucose-lipopolysaccharide glucosyltransferase
  purification, XXXII, 453
  reconstitution of activity, XXXII, 449–459
UDPglucose pyrophosphorylase, *see* Glucose-1-phosphate uridylyltransferase
UDPglucuronosyltransferase
  Ah^b allele, LII, 231
  in microsomal fractions, LII, 88
UDPglucuronyltransferase, *see* UDP-glucuronosyltransferase
Ulex europaeus, lectin, XXXIV, 332, 334
Ulex galli, lectin, XXXIV, 334
Ulex mamus, lectin, XXXIV, 334
Ulex parviflorus, lectin, XXXIV, 332
3-Ulose
  determination, XLI, 158
    polarographic, XLI, 22–27
    qualitative, XLI, 23
    quantitative, XLI, 24–27
Ultracentrifuge, continuous laser optics, XLVIII, 185–191
Ultrafiltration
  factors producing, XLIV, 298
  in hollow-fiber membrane devices, XLIV, 298–300
  preparation of ribosomal proteins, LIX, 650
  of ribosomal RNA, LIX, 648
  virus, LVIII, 228
Ultrafiltration cell, XXXII, 380
Ultramicro assay, colorimetric, for reducing sugars, XLI, 27–29
Ultrasectioning, of filtration pellicles, XXXII, 7
Ultra Turrax homogenizer, XXXI, 259; XXXII, 150; XXXVII, 290; LII, 378; LV, 4
Ultraviolet absorption
  1,N^6-ethenoadenosine 3′,5′-monophosphate, XXXVIII, 430
  ribosome protein S1 spectrum, LX, 424, 425
Ultraviolet excitation, optics, XLIX, 86, 89
Ultraviolet light
  collagen membrane immobilization, XLIV, 248
  detection with paper electrophoresis, XLIII, 283, 286
  high pressure ion-exchange chromatography, XXXVIII, 22
  irradiation, mutation induction, XLIII, 29–31
    caffeine as mutagen, XLIII, 36
    ethyleneimine as mutagen, XLIII, 36
    8-methoxypsoralen as mutagen, XLIII, 36
  as mutagen, LVIII, 312
  polyacrylamide gel polymerization, XLIV, 92, 93
  RNA-cellulose, XXXIV, 470
Ultraviolet spectrophotometry
  for peptide detection, XLVII, 204–206, 214, 215
  N^α-phenylthiocarbamyl methylamide-amino acid identification, XLVII, 381
  in plant-cell extraction, XXXI, 544
  sensitivity, XLVII, 215
  of steroid hormone receptors, XXXVI, 412
Ultrogel AcA 34
  chromatography of elongation factors, LX, 683–691, 697
  column
    ATPase, LV, 341
    BKA-protein complex, LVI, 414
    CAT-protein complex, LVI, 412
  purification of formate dehydrogenase, LIII, 367
    of initiation factor IF-2, LX, 207

of purine nucleoside phosphorylase, LI, 523
of ribonucleoside diphosphate reductase, LI, 229, 232, 234
Ultrogel AcA 44, chromatography of elongation factors, LX, 697–699, 701
UM–10, semipermeable membrane, XLIV, 300
Umbelliferone, *see also* 7-Hydroxycoumarin
hydrogen ions, LVI, 303
Umbelliferone phosphate, XLVI, 62
UMP, *see* Uridine 5′-monophosphate
UMP-CMP kinase, *see* Uridylate-cytidylate kinase
Unamycin, XLIII, 135
Uncoupler, *see also* specific types
activity, comparison, LV, 464, 465
bacteriorhodopsin-mediated potential changes, LV, 774, 775
binding site, photoaffinity labeling, LVI, 673–683
calcium uptake, LVI, 338
classic, LV, 466, 467
benzimidazoles, LV, 468
benzylidinemalononitriles, LV, 468, 469
carbonylcyanide phenylhydrazones, LV, 469, 470
dicoumarol, LV, 471
fluoroalcohols, LV, 470
halophenols, LV, 470
nitro phenols, LV, 470
salicylanilides, LV, 467
complex V, LV, 315; LVI, 586
Crabtree effect, LV, 296, 297
definition, LV, 462, 463
diffusion potentials, LV, 584
hepatoma mitochondria, LV, 84, 85
immobilized mitochondria, LVI, 556
luminescence, LVI, 543
M. phlei membrane systems, LV, 185, 186
nonclassic
deaspidin, LV, 472
ionophores, LV, 471, 472
long-chain fatty acids, LV, 472
macrocyclic antibiotics, LV, 471, 472
of oxidative phosphorylation, LV, 462, 463
assay, LV, 463–465
types, LV, 466–472
Pasteur effect, LV, 290, 293
pet9 mitochondria, LVI, 127, 128
phosphate efflux, LVI, 354
Plexiglas, LVI, 554
reconstituted membranes, LV, 226
transhydrogenase, LV, 815
Uncoupling effect, hydroxylases, LIII, 536–540, 549
Undecanal, quantum yield, bioassay, LVII, 192
Undecaprenol phosphate, polyribitol phosphate synthesis, L, 387
Underwater decoupling, XLIX, 262
Unicellular organism, XXXIX, 483–492
membrane-component isolation, XXXII, 422–445
preparation, XXXI, 607–609
Union Carbide A1100, *see* γ-Aminopropyltriethoxysilane
Union Carbide silicone antifoam, LII, 170
Unisil gel, in α-hydroxylase assay, LII, 312, 314
p-*UpGpA, XLVI, 623
UpUpUpU*, XLVI, 623
UpUpUpU**, XLVI, 623
Uracil
assay of orotate phosphoribosyltransferase, LI, 144, 145
chromatographic separation, LI, 316, 409
derivatives, osmometric studies, XLVIII, 80
extinction coefficient, LI, 433
hemin requirement, LVI, 172, 174, 175
indefinite self-association, XLVIII, 113
inhibitor of orotate phosphoribosyltransferase, LI, 166
of uridine phosphorylase, LI, 429
in media, LVIII, 67
menaquinone deficient mutants, LVI, 108

product of cytosine deaminase, LI, 395
of uridine nucleosidase, LI, 290
of uridine phosphorylase, LI, 423
repressor of carbamoyl-phosphate synthetase synthesis, LI, 21
in starvation medium, LI, 32
uptake, mycoplasmal, LVIII, 26, 27
Uracil arabinoside, substrate of pyrimidine nucleoside phosphorylase, LI, 436
Uranium, as immunochemical marker, XXXVII, 133
Uranyl acetate
grid stain, XXXIX, 142
as negative stain, XXXII, 22, 24, 29, 34
osmolality, XXXII, 30
for staining DNA-ribosome complexes, LIX, 842, 843
Uranyl acetate-collidine buffer block stain, XXXIX, 142
Uranyl acetate-maleate buffer block stain, XXXIX, 141
Uranyl acetate reagent, XXXI, 21
Uranyl formate, as negative stain, XXXII, 22
Uranyl magnesium acetate, as stain, XXXI, 23
Uranyl salt, in DNA localization, XL, 28
Urate oxidase, LII, 10, 13
activity, effect of xanthine concentration, XLIV, 918, 919
assay by Fast Analyzer, XXXI, 820, 821
conjugate, activity, XLIV, 200, 271
as diagnostic enzyme, XXXI, 20, 24, 37, 39, 326, 327, 746
in enzyme electrode, XLIV, 591
in enzyme thermistor, XLIV, 674
in glyoxysomes, XXXI, 565, 569
immobilization
in cellophane membrane, XLIV, 905, 908
on controlled-poreglass, LVI, 491
by entrapment, XLIV, 176
by microencapsulation, XLIV, 214
in multistep enzyme system, XLIV, 918
on nylon tube, XLIV, 120
in microbodies, LII, 496, 500
in peroxisomes, XXXI, 364, 367, 370, 374
sources, LVI, 468
therapeutic use, XLIV, 695, 696, 704
Urea
in adrenodoxin chromophore reconstitution, LII, 137
alanine carrier, LVI, 435
assay
in continuous flow analyzer, XLIV, 638
fluorometric, XLIV, 630
by microcalorimetry, XLIV, 671–673
buffer for gel electrophoresis, LX, 10, 29, 30, 268–270, 272–274, 369, 517, 518, 604
carbamylation of polypeptides, LVI, 613
colorimetric determination, LI, 105, 106
cycle
in hepatocyte cell lines, XXXII, 740
liver mitochondria, LV, 21
in cytochrome *c* denaturation studies, LIV, 279
denaturation studies with chymotrypsinogen A, XLIV, 522, 524
with glutamate dehydrogenase, XLIV, 510–512
with monomers and oligomers, XLIV, 498–500
with trypsin derivatives, XLIV, 518, 519
determination, by Fast Analyzer, XXXI, 822
dissociation of bacterial luciferase, LIII, 567, 569
for dissociation of uridine phosphorylase, LI, 429
effect on aspartate carbamyltransferase, LI, 48
on carboxypeptidase A crystals, XLIV, 549
on electron spin resonance spectra, XLIX, 438, 439
on estrogen receptor studies, XXXVI, 174
electrode, *see* Electrode, urea

electrophoretic analysis of protein from cross-linked ribosomal subunits, LIX, 542, 544
of ribosomal proteins, LIX, 431, 506
RNA fragments, LIX, 574
as eluent, XXXIV, 664
elution of protein from gels, LIX, 544, 545
enzyme deactivation, XLIV, 364, 879
enzyme desorption, XLIV, 140, 141, 147, 152
enzyme electrode, XLIV, 585, 587–589
fingerprinting procedure, LIX, 65
in histone fractionation, XL, 132
hollow-fiber dialytic permeability, XLIV, 297
hollow-fiber retention data, XLIV, 296
inhibitor of deoxythymidine kinase, LI, 365
mobility reference, XLVII, 57
particle, preparation, LV, 738, 741
peptide hydrolysis, XLVII, 41
peptide separation, XLVII, 207
in preparation of bacterial luciferase subunits, LVII, 172, 178
of protein from cross-linked ribosomal subunits, LIX, 537
of ribosomal protein, LIX, 440, 447, 470, 518, 557, 559, 649
of ribosomal RNA, LIX, 557, 648
of RNA hydrolysate, LIX, 63
protein solubility, XLVII, 113, 150
purification of cytochrome oxidase subunits, LIII, 78
of initiation factors, LX, 8–10
replacement of guanidine, LV, 784
tRNA digestion mixture, LIX, 107
tRNA sequence analysis, LIX, 68
slab gels, LIX, 431
staphylococcal protease stability, XLVII, 191
storage buffer of ribosomal proteins, LIX, 585
stripped membrane particles, LV, 796, 797
studies of protein-RNA interactions, LIX, 563, 564
submitochondrial particles, LV, 111, 112
substrate, of urease, XLIV, 104, 348
thermolysin hydrolysis, XLVII, 183
ultrapure, source, LIX, 65
use with sodium dodecyl sulfate, disadvantages, XLVIII, 7
in washing solution, XLIV, 77

Urease, XXXIV, 105
aldehydrol coupling, XLIV, 105
assay, spectrophotometric, XLIV, 104
attachment to mercapto-agarose gel, XLIV, 42
conjugate
activity assay, XLIV, 105, 271, 347, 348
kinetic behavior, XLIV, 105, 106
optimal effective pH, XLIV, 430, 431
storage stability, XLIV, 106, 168, 169
in enzyme electrode, XLIV, 587–589, 597, 600
fiber entrapment, biomedical application, XLIV, 242
in fluorometric urea assay, XLIV, 630
HABA column, XXXIV, 266, 267
immobilization
by bridge formation, XLIV, 167–169
in cellophane membrane, XLIV, 905
on collagen, XLIV, 255, 260, 902, 906, 909
on controlled-pore ceramic, XLIV, 167–169
by entrapment, XLIV, 176, 903, 909
in gel, XLIV, 909
on glass beads, XLIV, 670, 671
on inert protein, XLIV, 909
by microencapsulation, XLIV, 214, 330
on nylon tube, XLIV, 120
in polyacrylamide-agarose gel, XLIV, 272, 273
jack bean
absorbance at 280 and 278.5 nm, XXXV, 224, 225
osmometric studies, XLVIII, 75

membrane potential studies, XLIV, 924, 925
purification, XLIV, 389
source, XLIV, 104
specific activity, XLIV, 255
spontaneous structuration studies and, XLIV, 927, 928
therapeutic application, XLIV, 679–684, 695, 696
by extracorporeal perfusion, XLIV, 683, 684
by intraperitoneal injection, XLIV, 681–683
L-3-Ureido-2-aminopropionate, *see* L-Albizziin
Ureidosuccinate, colorimetric determination, LI, 105, 107
Ureogenesis
localization, LVI, 207
in perfused liver, XXXIX, 34
Urethane, as anesthetic for kidney removal, XXXIX, 13
Uric acid
assay
in continuous flow analyzer, XLIV, 638
fluorometric, XLIV, 631–633
of orotate phosphoribosyltransferase, LI, 144, 145
elution, XXXVIII, 17, 18
enzyme electrode, XLIV, 585, 591
extinction coefficient, LI, 432, 447
glucose electrode, LVI, 483
gout, XLIV, 704
hollow-fiber dialytic permeability, XLIV, 297
polarography, LVI, 452, 468
thin-layer chromatography, XXXVIII, 28–30
Uricase, *see* Urate oxidase
Uridine, XLVI, 323
chromatographic mobility, LIX, 73, 188
chromatographic separation, LI, 300, 308, 315, 316, 409
determination, LI, 408, 409
extinction coefficient, XLIX, 460
inhibitor of cytidine deaminase, LI, 405
of nucleoside phosphotransferase hydrolase activity, LI, 392
of orotate phosphoribosyltransferase, LI, 166
in media, LVIII, 67
NMR spectrum, LIX, 35, 39
optical spectra, XLIX, 165
product of cytidine deaminase, LI, 395, 401, 408, 411
residue, modification, by carbohydrazide and bisulfite, LIX, 155
substrate of nucleoside phosphotransferase, LI, 391
of pyrimidine nucleoside phosphorylase, LI, 432, 436
of uridine-cytidine kinase, LI, 299, 305, 308, 313, 314
of uridine nucleosidase, LI, 290, 294, 295
of uridine phosphorylase, LI, 423, 429
5-[^{3}H]Uridine, in *in vivo* preparation of radiolabeled RNA, LIX, 830, 831, 834
Uridine 2′,3′-cyclic monophosphate, *see* Cyclic uridine monophosphate
Uridine-2′,3′-cyclic-phosphate hydrolase, hydrolysis rate, XXXII, 128
Uridine-cytidine kinase, *see* Uridine kinase
Uridine diphosphate
chromatographic separation, LI, 315, 316, 333
inhibitor of carbamoyl-phosphate synthetase, LI, 21, 119
of deoxyctyidine kinase, LI, 345
of orotate phosphoribosyltransferase, LI, 163
of orotidylate decarboxylase, LI, 163
product of pyrimidine nucleoside monophosphate kinase, LI, 321
of UMP-CMP kinase, LI, 331
substrate of nucleoside diphosphokinase, LI, 375
of ribonucleoside diphosphate reductase, LI, 235, 236, 245

Uridine 5′-diphosphate, substrate, of pyruvate kinase, LVII, 82
Uridine diphosphate *N*-acetylglucosamine
animal cell pools, L, 178
dolichol intermediates, L, 416
Uridine diphosphate *N*-acetylglucosamine polyribitol phosphate glycosyltransferase, L, 394
Uridine 2′(3′),5′-diphosphate-agarose, XXXIV, 119, 360
Uridine diphosphate galactose, animal cell pools, L, 178
Uridine diphosphate glucose, XL, 353
animal cell pools, L, 178
assay, LV, 221
dolichol intermediates, L, 408–416
inhibitor of carbamoyl-phosphate synthetase, LI, 119
in sugar transfer, XXXI, 23
Uridine 5′-diphosphate-glucuronic acid, LII, 69
Uridine diphosphate hexanolamine-agarose, XXXIV, 485
Uridine diphosphate muramyl pentapeptide, preparation of radioactive, L, 399, 400
Uridine 5′-diphosphate-4-*O*-(2,2,6,6)-tetramethyl-4-piperidinyl-1-oxyl
extinction coefficient, XLIX, 460
spectra, XLIX, 461
structure, XLIX, 458
synthesis, XLIX, 458–460
Uridine kinase
activity, LI, 299, 308, 314
assay, LI, 299–301, 308, 309, 314, 315
from *Bacillus stearothermophilus*, LI, 299–307
cation requirements, LI, 305, 319
from *E. coli*, LI, 308–314
kinetic properties, LI, 306, 314, 320, 321
molecular weight, LI, 314, 320
from murine neoplasm, LI, 314–321
from Novikoff ascites cells, LI, 299–307
pH optimum, LI, 314
purification, LI, 301–305, 311–313, 317–319
regulation, LI, 307, 314, 320
from *Salmonella typhimurium*, LI, 314, 320
stability, LI, 303, 305, 312, 314, 318, 319
substrate specificity, LI, 305, 306, 313, 314, 319, 320
temperature effects, LI, 306, 307
Uridine 5′-monophosphate
electrophoretic mobility, LIX, 64
M value range, LIX, 89
Uridine monophosphate-agarose, XXXIV, 477
Uridine 5′-monophosphate-morpholidate, XLIX, 458
Uridine nucleosidase, LI, 290–296
activity, LI, 290
amino acid composition, LI, 295
assay, LI, 291, 292
inhibitors, LI, 294, 295
isoelectric point, LI, 294
metal content, LI, 294
molecular weight, LI, 294
pH optimum, LI, 294
purification, LI, 292, 293
purity, LI, 294, 295
stability, LI, 293, 294
substrate specificity, LI, 294, 295
from yeast, LI, 290–296
Uridine nucleotide, effect on orotidylate decarboxylase, LI, 79
Uridine 5-oxyacetic acid methyl ester methyltransferase
assay, LIX, 191–193
inhibition, LIX, 203
purification, LIX, 200, 201
Uridine 2′(3′)-phosphate, electrophoretic separation, LIX, 177
Uridine 5′-phosphate, *see* Uridine 5′-monophosphate
Uridine phosphorylase, LI, 423–431, *see also* Pyrimidine nucleoside phosphorylase
activity, LI, 423
amino acid analysis, LI, 429
assay, LI, 424–426; LVIII, 25, 26
from *Haemophilus influenzae*, LI, 432
inhibitors, LI, 429
kinetic properties, LI, 430
molecular weight, LI, 429
pentosyl transfer, LI, 430, 431

pH optima, LI, 429
purification, LI, 426–428
purity, LI, 428
from rat liver, LI, 423–431
reaction mechanism, LI, 431
reverse direction assay, LI, 425
sources, LI, 423
stability, LI, 428, 429
substrate specificity, LI, 429
subunit structure, LI, 429
tissue distribution, LI, 431
Uridine triphosphate
chromatographic separation, LI, 315, 316, 333
extinction coefficient, LI, 79
inhibitor of aspartate carbamyltransferase, LI, 57, 58
of carbamoyl-phosphate synthetase, LI, 21, 111, 119
of cold inactivation of carbamoyl-phosphate synthetase activity, LI, 105, 107
of cytidine deaminase, LI, 400
of orotate phosphoribosyltransferase, LI, 163
of orotidylate decarboxylase, LI, 163
of uridine-cytidine kinase, LI, 307, 320
in media, LVIII, 67
substrate of cytidine triphosphate synthetase, LI, 79, 84
of nucleoside triphosphate pyrophosphohydrolase, LI, 279
of phosphoribosylpyrophosphate synthetase, LI, 10
Uridine 5′-triphosphate
ATP-P_i exchange preparation, LV, 361, 362
complex V, LV, 314
in GTP assay procedure, LVII, 89
separation, LV, 287, 288
Uridine 5′-triphosphate-agarose, XXXIV, 477
Uridylate
chromatographic separation, LI, 158, 159, 315, 316, 333
inhibitor of carbamoyl-phosphate synthetase, LI, 21, 33, 34, 111
of orotate phosphoribosyltransferase, LI, 153, 163
of orotidylate decarboxylase, LI, 153, 155, 163
product of orotidylate decarboxylase, LI, 74, 135
of uridine-cytidine kinase, LI, 299, 308
substrate of pyrimidine nucleoside monophosphate kinase, LI, 321, 329
of UMP-CMP kinase, LI, 331, 336
Uridylate-cytidylate kinase, LI, 331–337, *see also* Uridylate kinase
activity, LI, 331, 332
assay, LI, 332–334
kinetic properties, LI, 337
pH optimum, LI, 336
purification, LI, 334–336
reaction mechanism, LI, 337
stability, LI, 336
substrate specificity, LI, 336
from *Tetrahymena pyriformis*, LI, 331–337
Uridylate kinase, in bacterial extracts, LI, 332
Uridylyltransferase, from yeast, antibody binding, L, 55
Urine
cloning, LVIII, 163
cortisol-metabolite isolation, XXXVI, 499–503
drug isolation, LII, 333, 334
estradiol determination, XXXVI, 64–67
estrone determination, XXXVI, 64–67
gas chromatographic analysis, LII, 336
melatonin isolation, XXXIX, 396, 397
preparation for assay, XXXVIII, 76
prostaglandins, XXXII, 110
extraction, XXXII, 116, 117
sample, determination of bacterial count, LVII, 71, 72
testosterone determination, XXXVI, 63, 64
Urine A pentasaccharide structure, L, 227

Urine A trisaccharide structure, L, 228
Urine B pentasaccharide structure, L, 227
Urine B trisaccharide structure, L, 228
Urine O(H) disaccharide, *see* 2-Fucosylgalactose
Urine O(H) tetrasaccharide, *see* Lactodifucotetraose
Urobilinogen, autoradiography, XXXVI, 150
Urocanic acid, continuous production, XLIV, 745
Urografin, in buoyant density measurements, XXXI, 84, 85
Urokinase, XXXIV, 451–459
  activators, XLV, 242
  affinity material
    characteristics, XXXIV, 455–459
    synthesis, XXXIV, 453–455
  assay, XXXIV, 453; XLV, 239, 240
  composition, XLV, 243
  distribution, XLV, 243
  eluting agents, XXXIV, 455
  inhibitors, XLV, 242
  isoelectric focusing, XLV, 249
  plasminogen, activation, XLV, 263
  preparation, of activator-free plasmin, XLV, 278, 279
  properties, XLV, 242
    physical, XLV, 243
  purification, XLV, 241
  specific activity, XLV, 247
  specificity, XLV, 242
  units, XLV, 241, 247
Uronic acid
  colorimetric assay with *m*-hydroxydiphenyl, XLI, 31
  methylation of polysaccharides, L, 14, 28, 29
Usnic acid, XLIII, 154
  M. phlei membrane systems, LV, 186
Ussing chamber, LVIII, 555–557
Uterus
  artificial chamber, XXXIX, 248–252
  cytosol preparation, XXXVI, 49, 337
  estrogen receptor studies, XXXVI, 175, 350
  incubation media, XXXIX, 103, 104
  nuclear steroid receptor complex, XXXVI, 357–362
    for estrogens, XXXVI, 286
  nuclei isolation from cells, XXXI, 259, 260
  nuclei preparation, XXXVI, 278–280
  preparation, for collagen studies, XL, 311
  protein synthesis
    analysis, XXXVI, 449–454
    estrogen effects, XXXVI, 444–455
  removal, for *in vitro* studies, XXXIX, 102
  in vitro responses, XXXIX, 101–106
UTP, *see* Uridine 5′-triphosphate
UV, *see* Ultraviolet light

V

V_{max}, rate constant, apparent, definition, XLIV, 3
Vacant couple
  definition, LIX, 376
  dissociation, LIX, 379–381
  isolation, LIX, 378
cis-Vaccenyl coenzyme A, substrate for palmityl thioesterase I, XXXV, 106
Vaccine, microencapsulation, XLIV, 215
Vacuole
  acid, in plant-tissue fractionation, XXXI, 532, 533
  isolation, XXXI, 572–576
  meristematic, isolation, XXXI, 574, 575
Vacuum evaporator, LIX, 613, 614
Vacuum filtration, effect on nucleotide concentrations, LVII, 80, 88
Vacuum manifold, glass, LIII, 302
Vacuum scattering function, LIX, 710
δ-Valerolactam, XLVI, 218
Validamycin, XLIII, 254–256, 272, 273
Valine, XLIII, 411, 413
  aminoacylation, isomeric specificity, LIX, 274, 275, 279, 280
  esterification, XLVII, 522
  in media, LVIII, 53, 62
  peptide synthesis, XLVII, 509
  radiolabeled, in protein turnover studies, XXXVII, 239, 240, 244

solid-phase peptide synthesis, XLVII, 610
thermolysin hydrolysis, XLVII, 175, 176
[^{14}C]Valine
assay of initiation factor activity, LIX, 368
of polypeptide chain elongation, LIX, 358
of polypeptide synthesis, LIX, 367
L-Valine, isolation, XLIV, 757
L-Valine:tRNA-ligase, *see* Valyl-tRNA synthetase
Valine-tRNA synthetase, XXXIV, 170
Valinomycin, XXXII, 883
anion transport, LV, 453
ATP synthesis, LV, 452
chemiosmotic hypothesis, LV, 451
complex V, LV, 315
diffusion potentials, LV, 584
effect on bilayers, XXXII, 500, 535, 552
on membrane, XXXII, 882
endogenous K^+ release, LV, 669, 670
expulsion of protons, LV, 367
flow dialysis experiments, LV, 683, 685
as ionophore, LV, 442, 444
with K^+ ions uncoupling activity, LV, 465
measurement of proton translocation, LV, 628, 635, 637, 642, 643, 647, 648
model redox reactions, LV, 542, 543
M. phlei ETP, LV, 186
phosphate efflux, LVI, 354
potassium diffusion, LVI, 431, 434
proteolipid ionophores, LV, 419
regenerating systems, LV, 710
solutions, LV, 448
structure, LV, 436
transport by *E. coli* membrane vesicles, LV, 676, 679
use with cyanine dyes, LV, 691
Valonia ventricosa, free cells, XXXII, 723
Valve, cardiac, for perfused hearts, XXXIX, 55, 60
Valyl-tRNA, XLVI, 624
[^{14}C]Valyl-tRNA, assay of polypeptide chain elongation, LIX, 358
Valyl-tRNA synthetase, XLVI, 90, 152
in misactivation studies, LIX, 289–291
purification, LIX, 262–264, 266, 267
subcellular distribution, LIX, 233, 234
Vanacryl P, vanillin methacrylate/allylic alcohol copolymer, XLIV, 84
Vanadium
atomic emission studies, LIV, 455
in media, LVIII, 52, 69
Vanadium-51, lock nuclei, LIV, 189
Vanadium-54, lock nuclei, XLIX, 348
Vanadous sulfate, for oxygen removal, LIV, 117
Vancomycin, XLIII, 127, 204, 211
van der Waals forces, in lipid monolayers, XXXII, 544
Vanillin methacrylate, XLIV, 84
Vanillin methacrylate/allylic alcohol copolymer, XLIV, 84
Vannas capsulotomy scissors, XXXIX, 175, 177
Vapor-diffusion method, LIX, 4, 5
Vapor pressure, lowering, relation between osmotic pressure, XLVIII, 88, 89
Vapor pressure osmometry, XLVIII, 69–154
Vargula harveyi, LVII, 336
Vargula hilgendorfi, LVII, 336
Vargula tsujii, LVII, 336
Varian HR–220 NMR spectrometer, LIV, 228
Varian Spin Decoupler, LIV, 178
Varian Variable Temperature Unit, XXXII, 195
Varian XL–100-FT system, LIV, 178, 185, 187, 190
Varidase, streptokinase, purification, XLV, 251
Varsol, XLVII, 197
Vascular tissue, cellular activity studies, XXXIX, 477–482
Vaseline, petroleum jelly impregnated-silica gel TLC plates, *see* Silica gel thin layer chromatography plates
Vasopressin
adenylate cyclase, XXXVIII, 151, 152, 173
antisera, XXXVI, 17

cAMP production, LVIII, 559
effect on perfused kidneys, XXXIX, 10
iodinated, properties, XXXVII, 231
[Lys^8], XXXIV, 7
disulfide cleavage, XLVII, 112
as neurosecretory granule marker, XXXI, 406, 408
Veillonella, hydrogenase reaction, LIII, 287
Vein, isolation of mitochondria, LV, 60, 61
Velocity sedimentation, of germinal cells, XXXIX, 290, 291
Velveteen cloth, for plating, LVIII, 320
Venom, *see* specific type
Venom exonuclease, *see* Phosphodiesterase I
Venom inhibitor
amino acid composition and sequence, XLV, 879
assay, XLV, 875
characteristics, XLV, 874
distribution, various species, XLV, 881
kinetics, XLV, 878
properties, XLV, 877
purification, XLV, 876
purity, XLV, 877
specificity, XLV, 878
stability, XLV, 878
units, XLV, 876
Ventilation, laminar air, LVIII, 10
Venturicidin
analysis, LV, 505
ATPase assay, LV, 329, 330, 333
availability, LV, 505
complex V, LV, 315; LVI, 586
effects on energy-conserving systems
bacterial, LV, 507
chloroplast, LV, 506
mitochondrial, LV, 505, 506
N. crassa mitochondria, LV, 147
preparation, LV, 505
resistance, LVI, 118, 140
structure, LV, 504, 505
toxicology, LV, 505
Venturicidin A, physical properties, LV, 477
Venturicidin B, physical properties, LV, 477
Veratridine, cyanine dyes, LV, 693
Vernamycin B
peptide antibiotic lactonase, XLIII, 767–773
structure, XLIII, 768
VERO cell line, properties and uses, XXXII, 585
Veronal-saline buffer, XXXII, 507
Versene, purification of uridine-cytidine kinase, LI, 302, 303
Vertebrate cell, LVIII, 466–477
lower, LVIII, 467
osmolarity adjustments, LVIII, 469
poikilotherm, LVIII, 466–477
media, LVIII, 470
monolayer culture, LVIII, 474–476
pH, LVIII, 470
primary culture, LVIII, 474–476
subculture, LVIII, 476, 477
temperature requirements, LVIII, 470–472
Vesicle
ATP synthetase, reconstitution, LV, 715
energized
measurement of fluorescence, LV, 366
pH changes caused, LV, 366, 367
formation by cholate-dialysis technique, LV, 722, 723, 729, 730, 733, 734
by cholate-dilution technique, LV, 730, 731, 734
inner membrane, preparation, LV, 737, 738
inside-out, electrophoretic purification, XXXI, 759, 760
internal volumes, LV, 551
negative staining, XXXII, 29
phospholipid, containing complex I and ATP synthetase, LV, 716–722
phosphorylating
preparation, LV, 165, 166
properties, LV, 167, 168
reconstituted
by dialysis, LV, 371, 372
phospholipids, LV, 370, 371
properties, LV, 372

by sonic oscillation, LV, 372
of sonicated microsomes, LII, 81, 82
submitochondrial
preparation, XXXI, 292–299
properties, XXXI, 297
Vesicular stomatitis virus, microcarrier culture, LVIII, 192
Vexar, polymeric netting, XLIV, 251, 252
Viability, *see* Cell viability
Viability test, LII, 64, 65
Vibrating-sample device, LIV, 386–388, 391, 392
Vibrio, luminescent, *see Photobacterium*
Vibrio cholerae, neuraminidase, L, 67
Vibrio fischeri, luciferase, LVII, 218
Vibrio percolans, bioautography, XLIII, 117
Vibrio succinogenes, succinate dehydrogenase, LIII, 450
Vibromixer, XXXII, 668
Vibronic coupling operator, XLIX, 80
Vibronic expansion theory of Albrecht, XLIX, 79–82
Vicia, protoplast, LVIII, 361
Vicia cracca, lectin, XXXIV, 333, 334
Vicia culture × pea mesophyll, LVIII, 365
Vicia faba lectin, hemagglutinating activity, L, 335
Vicia mesophyll × soybean culture, LVIII, 365
Vicinal hydroxyl
in agar polysaccharides, XLIV, 27
oxirane groups, XLIV, 32
Victoreen formula, in X-ray absorption spectroscopy data analysis, LIV, 335, 336
Vidicon, *see* Silicon vidicon tube
Vinblastine, XXXIV, 623; XLVIII, 243, 246
effects on water flow, XXXVII, 256
inhibitor of chromosome movement, XL, 279, 284
microtubular disruptive, XL, 322
mitotic inhibitor, LVIII, 256
Vinblastine-agarose, XXXIV, 623
Vinca, oxidase, LII, 14
Vincristine, XXXIV, 623
4-Vinylbenzoic acid, XLIV, 289
Vinyl chloride, as inhibitors, XLVI, 37, 38
Vinyl coupling, XLIII, 399, 400
5-Vinyldeoxyuridine, substrate of thymidine kinase, LI, 370
Vinyl ether, XLIV, 108
Vinylethyl ether, XLIV, 117
Vinyl glycine, XLVI, 33
inactivator of flavoprotein enzymes, LIII, 442, 443
synthesis, XLVI, 39, 40
L-Vinylglycine, structural formula, LIII, 443
Vinyl glycolate, inactivator of flavoprotein enzymes, LIII, 443, 445, 446
Vinylmethyl ether, XLIV, 179
Vinyl monomer, relative reactivities, XLIV, 181
2-Vinyloxyethyl-4-nitrobenzoate, XLIV, 117
2-Vinylpyridine, reaction with tributylphosphine, XLVII, 113
4-Vinylpyridine, reaction with tributylphosphine, XLVII, 113
N-Vinylpyrrolidone, XLIV, 108, 114
N-Vinylpyrrolidone/2-hydroxyethylmethacrylate copolymer, polymerization procedure, XLIV, 175
N-Vinylpyrrolidone/maleic acid anhydride copolymer
papain immobilization, XLIV, 115
polymerization, XLIV, 114, 115
Vinyl sulfonylethylene ether, order of reactivity to nucleophilic groups, XLIV, 34
Viocin, XLIII, 200
Viokase, XXXII, 741, 742
in media, LVIII, 532, 572, 573
neuronal cell culture, LVIII, 587
Violarin, XLIII, 168
Violet iron alum, for Parducz fixation, XXXII, 55
Viomycin, XLIII, 120, 211
in media, LVIII, 113, 115
Viral antigenic determinant, LVIII, 418
Viral contamination, *see* Contamination, viral
Virginiamycin, XLIII, 305, 319

Virial coefficient, XLVIII, 98
first, sedimentation equilibrium, XLVIII, 174
second
derivation, XLVIII, 84
elimination, XLVIII, 108
evaluation, XLVIII, 109, 120
for indefinite self-association, XLVIII, 111
for nonideal mixed associations, XLVIII, 128, 129
Virus, *see also* specific type
acquisition, LVIII, 226, 227
banding, isopycnic, LVIII, 415
cell-fusion induction, XXXII, 575, 576
cell transformation, XXXII, 583, 586
plasma membrane transformed, XXXI, 163, 166–168
detection, cell fusion, LVIII, 348, 349
DNA, assay, LVIII, 434, 435
envelope glycoprotein, XXXIV, 338
harvesting, LVIII, 399, 400
infectious, biohazard, LVIII, 37, 42, 43
isolation, XXXI, 509; LVIII, 222, 227
in media, LVIII, 225–228
microcarrier culture, LVIII, 191, 192
molecular weight determination, XLVIII, 68
negative staining, XXXII, 21
nontransforming, large-scale growth, LVIII, 403
preparation, LVIII, 222
production
effect of serum, LVIII, 226, 227
mechanization, LVIII, 400, 401
ultrafiltration, LVIII, 228
purification, LVIII, 401–403, 414–416
ribonuclease as contamination in studies, XXXII, 109
ribonucleic acid, shear stress, XLVIII, 56, 57
testing, LVIII, 28, 29
xenotropic, LVIII, 413, 414
Virus capsule protein, precursors, XXXVII, 326
Viruslike particle
extraction of double-stranded ribonucleic acid, LX, 551, 552
isolation from mycelia, LX, 550
from yeasts, LX, 549, 550
purification, LX, 551
Viscometer-densimeter
magnetic
design, XLVIII, 50–52, 57
sample size, XLVIII, 52
sensitivity, XLVIII, 52, 53
speed, XLVIII, 52, 53
shear stress, XLVIII, 53, 54
Viscose fiber
elastic support, XLIV, 560
enzyme binding procedure, XLIV, 561
Viscosity
barrier, subcellular fractionation, LV, 127, 128
practical procedures, LV, 131–135
principle, LV, 128–131
changes, effect on fluorescence, XLVIII, 349, 351, 352
coefficient, calculation, LVI, 216
determination using magnetic suspensions, XLVIII, 29
intrinsic, molecular weight determination, XLVIII, 6
Vision, ultrafast intermediates, LIV, 19
Visking dialysis tubing, LII, 104; LIII, 43
disadvantage of, ribosomal proteins, LIX, 482
Vitamin, *see also* specific types
fat-soluble, LVIII, 54, 55
in media, LVIII, 54, 64, 65
yeast growth media, LVI, 59
Vitamin A, in media, LVIII, 65
Vitamin B, in media, LVIII, 54, 65
Vitamin B_{12}, XXXIV, 305–314, *see also* Vitamin B_{12}-binding proteins
carboxylate derivatives, XXXIV, 306–308
hollow-fiber dialytic permeability, XLIV, 297
hollow-fiber retention data, XLIV, 296
logit-log assay, XXXVII, 9
structure, XXXIV, 306
Vitamin B_{12}-binding hormone, molecular weight, by electrophoresis, XXXVI, 99

Vitamin B_{12}-binding protein, XXXIV, 305–314
elution, XXXIV, 309, 310
intrinsic factor, XXXIV, 312
preparation of samples, XXXIV, 309
purification
hog gastric, XXXIV, 312–314
human, XXXIV, 310–312
separation, XXXIV, 312–313
transcobalamin II, XXXIV, 312
R-type proteins, XXXIV, 310
Vitamin B_{12}-diaminodipropylamine-agarose, XXXIV, 37
Vitamin C, deficiency, XL, 322
Vitamin D, LII, 388, 398
deficiency, XL, 321
effects on calcium influx, XXXIX, 523, 524
lipophilic Sephadex chromatography, XXXV, 395
in media, LVIII, 65
metabolic pathways, LII, 388, 389
Vitamin D_3
competitive protein binding assay, XXXVI, 534–536
metabolites
preparation and biological evaluation, XXXVI, 516–536
structures, XXXVI, 513
Vitamin D_3 25-hydroxylase, LII, 20
Vitamin E, in media, LVIII, 65
Vitamin K
biosynthesis factor X, bovine, XLV, 93
lipophilic Sephadex chromatography, XXXV, 395
in media, LVIII, 65
prothrombin dependence, XLV, 151
Vitamin K_1, reduction, by complex I, LIII, 13
Vitamin K_3
reduction, by complex I, LIII, 13
substrate, of NADH dehydrogenase, LIII, 18
Vitelline delaminase, assay of sea urchin egg protease, XLV, 343
Vitreoscilla, o-type cytochrome, LIII, 209, 211
V–605 medium, LVIII, 56
Voltage-to-frequency converter, LVII, 538
Voltammetry, cyclic, LVII, 511, 512
Volume
particle, from neutron scattering data, LIX, 686, 713–714
total, determination, XXXII, 10–13
Volume change
determination, XLVIII, 32–43
of mixing, XLVIII, 37, 38
of transport of perturbant, XLVIII, 39–42
of protein, XLVIII, 38, 39
VUCD, *see* Circular dichroism, vacuum ultraviolet
Vycor syringe, LVII, 141
Vycor tube, calcium concentration in, LVII, 282
Vyon disk, in small-column chromatography, LVII, 442

W

Walden inversion, Koenigs–Knorr reaction, L, 99
Waldenström macroglobulin, affinity column purification, L, 175
Walker carcinosarcoma 256 cell, LVIII, 86
Walking ghost artifact, XLIX, 88
Wall effect, in plant-cell fractionation, XXXI, 505
Walnut, organelles, XXXI, 535, 536
Warburg apparatus, assay, XLIV, 348
Warburg respirometric technique, LIII, 616, 617
Waring blender, LI, 6
for plant-cell rupture, XXXI, 502
yeast cell breakage, LVI, 47
Washing, Keilin-Hartree preparation, LV, 119, 120
Washing buffer, for platelets, XXXI, 150
Wash medium
for frog haploid cell culture, XXXII, 799
for human mutant cell culture, XXXII, 805
Watasenia, coelenterazine, LVII, 344
Water
chemical shift, temperature sensitivity, LIV, 200

Continental treatment system, LIV, 473, 474
deuterated, XXXV, 279–287, *see also* Deuterium oxide
distillation, LVIII, 195
drinking, inhibitor administration, LVI, 35
effect on radical ion chemiluminescent reactions, LVII, 497, 498
electron spin resonance spectra, XLIX, 383
enzyme-activated electrodes, LVI, 473
infrared spectrum, LIV, 303, 309
internal, determination of volume, LV, 549–552
as lock nuclei source, LIV, 189
metal-free preparation, LIV, 472–474
NMR signal, elimination of interference, LIV, 182–187
oxygen exchange, LV, 253, 254
in picosecond continuum generation, LIV, 12
preparation, LVIII, 94
cost, LVIII, 195
product, of complex IV, LIII, 40
of cytochrome oxidase, LIII, 54, 73
of phenylalanine hydroxylase, LIII, 278
of salicylate hydroxylase, LIII, 528
purification, LVII, 425
for HPLC, LII, 282
purification procedure, XLVII, 13, 14
quality, LVIII, 8
solvent, in luminol chemiluminescence reactions, LVII, 410
system, suppliers, LVIII, 17
treatment, LVIII, 8
tritiated
fatty acid synthesis rate measurement, XXXV, 279–283, 287
3β-hydroxysterol synthesis rate measurement, XXXV, 281, 282
intramitochondrial volume, LVI, 287, 288
water space measurement, LVI, 260
Water elimination Fourier-transform, XLIX, 258
method, *see* Relaxation method differential
Water flow, antidiuretic hormone effects, XXXVII, 251–256
Watermelon, free cells, XXXII, 724
Water monitoring, XLIV, 647–658, *see also* Air monitoring
Waters Associates HPLC column, LII, 392, 395, 406
Water solubility, antibiotic, XLIII, 187, 200
Wavelength
choice, optical probes, LV, 578
cyanine dyes, LV, 690
selection, absorption indicators, LVI, 308
Wavenumber, definition, XLIX, 83
Wax, lipophilic Sephadex chromatography, XXXV, 391, 392
Waymouth's medium MB 752/1, XXXIX, 445; LII, 66; LVIII, 57, 88, 90
composition, LVIII, 62–70
W126 cell, plasma membrane isolation, XXXI, 161
W138 cell, plasma membrane isolation, XXXI, 161
Weber-Teale solution scatter, LVII, 593–595
Wedge-shaped gradient, in plant-cell fractionation, XXXI, 513, 514
Weight fraction
apparent, XLVIII, 108
evaluation, XLVIII, 107, 108
for indefinite self associations, XLVIII, 112
of monomer, definition, XLVIII, 109, 110
for monomer-*n*-mer-*m*-mer self-associations, XLVIII, 116
for type II indefinite self-associations, XLVIII, 113
for type III indefinite self-associations, XLVIII, 114, 115
Wendt oscillating sinusoidal pressurization method, XLVIII, 313
Werkman medium, *K. pneumoniae* culture, LI, 182
Westphalia desludging centrifuge, LIII, 327

Whatman cellulose powder CF–11, flow rate maintenance, LIII, 183
Whatman DE 81 paper, LI, 339; LIX, 65, 67
Whatman GF/C glass fiber filter, LVII, 80
Whatman 3 MM paper wick, LIX, 62, 63, 72
Whatman No. 50 filter paper, purification of cytochrome *c*, LIII, 130
Whatman 3 paper, XLIV, 271
Whatman PC 1520 column, XLIV, 913
Wheat, *see also Triticum vulgare*
  aleurone isolation, XXXI, 576
  organelle studies, XXXI, 542
  protoplast isolation, XXXI, 580
  ribosome isolation, XXXI, 585, 586, 589
Wheat germ
  adenylosuccinate synthetase, LI, 212
  agglutinin isolation, XXXII, 611–615
  isomeric specificity of aminoacylation in, LIX, 279–281
  peroxidase, LII, 515, 516, 521
  D-3-phosphoglycerate dehydrogenase, assay, purification, and properties, XLI, 285–289
  phosphoglycerate mutase, assay, purification, and properties, XLII, 429–435
  as source of elongation factor EF-2, LX, 706
    of initiation factors, LX, 181–204, 411
Wheat germ agglutinin, XXXIV, 341–346
  iodination, L, 57, 58
Wheat germ agglutinin-agarose, XXXIV, 669, 670
Whey
  commercial use, XLIV, 792–809
  source, XLIV, 792
White's Basal Media, for cell culture, XXXII, 730
White's medium, LVIII, 56
  composition, XXXI, 557, 558
White noise, XLVIII, 424
WI–38 cell
  cloning, LVIII, 155
  DNA content
  population doubling, XXXII, 561
  protein content, LVIII, 145
  subcultivation, XXXII, 564
Wicksell's transformation, for particle size distribution, XXXII, 13, 19
Wide-Range Cryostat, XXXVI, 142, 143
Wiener-Khintchine theorem, XLVIII, 424, 430, 437
Wiener skewing, XLVIII, 191
Wilhelmy surface balance, XXXII, 541
Willems Polytron PT20st homogenizer, LVII, 249
Willems Polytron tissue homogenizer, XXXVII, 290
Williams' medium E, LVIII, 59, 90
Williams' medium G, LVIII, 59, 90
  composition, LVIII, 62–70
Wilson purified hormone, preparation, dosage, route of administration, XL, 319
Wilzbach method of tritium labeling, XXXVII, 313
Wisconsin 38 cell, callus culture, LVIII, 479
WISH cell, protein content, LVIII, 145
W medium, for granulosa cell culture, XXXIX, 205
Wolff rearrangement, XLVI, 72, 78
Wollenberger medium, LVIII, 9
Wombat, muscle, 3-phosphoglycerate kinase, purification, XLII, 129, 133
WO_5 medium, LVIII, 57
Wood shaving, effect on liver microsomes of laboratory animals, LII, 272
Woods polarization anomaly, XLIX, 88
Woodward Reagent K, *see N*-Ethyl-5-phenylisooxazolium 3′-sulfonate
Wool
  disulfide cleavage, XLVII, 111, 112
  low sulfur protein, osmometric studies, XLVIII, 75
Word length
  of analog-to-digital converter, aqueous solutions, LIV, 183
  of computer, water signal, LIV, 183
Working culture, defined, LVIII, 118
Wright's stain, for thyroid cells, XXXII, 749
Wrist shaker, LVIII, 203
Wulzen's cone, in pituitary gland, XXXI, 403

Wurster's blue
flavoprotein assays, LIII, 406
free radical, LII, 411
Keilin-Hartree preparation, LV, 126
as mediator-titrant, LIV, 408, 410
Wybutine
electrophoretic mobility, LIX, 91
identification, LIX, 92
substitution for, LIX, 110–121
Wybutine 5′-phosphate, identification, LIX, 89

X

X50, hollow-fiber device, XLIV, 293–298
hydraulic permeability, XLIV, 298
permeability changes, XLIV, 307
X–206, as ionophore, LV, 443
X–464 antibiotic, LV, 445
X–537 A antibiotic, XLIII, 352, 409; LV, 445
X–14547A antibiotic, structure, LV, 445
Xanthine, XXXIV, 524
assay of orotate phosphoribosyltransferase, LI, 144, 145
of superoxide dismutase, LIII, 384
chromatographic separation, LI, 220
competitive inhibitor, urate oxidase, XLIV, 918
product of guanine deaminase, LI, 512
substrate of guanine phosphoribosyltransferase, LI, 557
of nucleoside deoxyribosyltransferase, LI, 452
Xanthine dehydrogenase
flavin determination, LIII, 428
in microbodies, LII, 496, 500
reversible dissociation, LIII, 435
Xanthine oxidase, XXXIV, 6, 527, 528; LII, 6, 17; LIII, 397, 401, 404; LIV, 93; LVI, 474
assay of nucleoside deoxyribosyltransferase, LI, 447
of purine nucleoside phosphorylase, LI, 518, 531
of superoxide dismutase, LIII, 384
conjugate, activity, XLIV, 271
electron paramagnetic resonance studies, XLIX, 514
flavin determination, LIII, 428
immobilization
on inert protein, XLIV, 908
in multistep enzyme system, XLIV, 918
magnetic circular dichroism, XLIX, 175
Pholas luciferin chemiluminescence, LVII, 401
reductive titration, LIV, 124
reversible dissociation, LIII, 434
SDS complex, properties, XLVIII, 6
Xanthine oxidase/oxygen/hypoxanthine, in assay of superoxide dismutase, LVII, 405
Xanthine 5′-phosphate:ammonia ligase (AMP), *see* Guanosine 5′-phosphate synthetase
Xanthine-5′-phosphate:L-glutamine amidoligase (AMP), *see* Guanosine 5′-phosphate synthetase
Xanthomycin, XLIII, 342
Xanthophyll, in chromoplasts, XXXI, 495
Xanthosine, absorption maximum, LI, 525
Xanthosine diphosphate, substrate of nucleoside diphosphokinase, LI, 375
Xanthosine triphosphate, substrate of nucleoside triphosphate pyrophosphohydrolase, LI, 279
Xanthydrol, modification of lysine, LIII, 140
Xanthylate
spectrophotometric determination, LI, 214
substrate of GMP synthetase, LI, 213, 219, 224
X chromosome, hybridization studies on, XXXIX, 127
XC test, LVIII, 421–423
7X detergent, XXXII, 742
Xenon arc lamp, LIV, 35
relative brightness, LIV, 95
Xenon gas laser, wavelength, XLIX, 85
Xenopus muelleri, cytochrome oxidase, LIII, 66
Xenotropic virus, *see* Virus, xenotropic

Xi, XLVIII, 110, 112–115
derivation, XLVIII, 108
20–XM50, hollow fiber device, XLIV, 293, 313, 320
45–XM50, hollow fiber device, XLIV, 293, 314
XMP aminase, *see* Guanylate synthetase
X–press, XXXI, 660
X–ray
cell transformation, XXXII, 586
data analysis, Q-function, XXXII, 216, 217, 219, 220
leaks, human hazard, XXXII, 215
X–ray absorption fine structure spectrum, extended, LIV, 324
analysis, LIV, 340–345
data collection, LIV, 336, 337
sample parameters, LIV, 331–333
X–ray absorption spectroscopy, LIV, 323–345
applicability, LIV, 323, 324
data analysis of edge spectra, LIV, 337–340
of EXAFS spectra, LIV, 340–345
initial reduction, LIV, 334–337
data collection, LIV, 334
detectors, LIV, 330, 331
experimental parameters, LIV, 333
fluorescence detection, LIV, 330, 331, 334, 335
instrumentation, LIV, 324–331
monochromator, LIV, 328, 329
sample parameters, LIV, 331, 332
sample preparation, LIV, 332, 333
sensitivity, LIV, 334
X–ray source, LIV, 324–328
X–ray crystallography
for determination of tRNA structure, LIX, 3–21
lasalocid-barium complex, LV, 446
phase of diffracted beam, determination, LIX, 11, 12
principles, LIX, 10–15
X–ray diffraction
of F_1 ATPase crystals, LV, 334–337, 343
small-angle method, XXXII, 211–220
data collection, XXXII, 215
equipment, XXXII, 212–215
model building, XXXII, 218, 219
of nonordered membranes, XXXII, 216, 217
Q-function in data analysis, XXXII, 219, 220
sample preparation, XXXII, 215, 216
of stacked membranes, XXXII, 217–220
swelling transform, XXXII, 218
X–ray film, LI, 372
flashed, in fluorography, LIX, 192
X–ray image intensifying system, XXXII, 214
X–ray scattering
distance of scattering centers, LIX, 714
instrumentation, LIX, 752–754
in studies of protein-RNA interactions, LIX, 578–579
theory, LIX, 756–759
Xylene, XLVIII, 343
assay of deuterium oxide in water samples, LIX, 644–645
m-Xylene, in density gradient, LIX, 5
Xylene cyanole blue, LIX, 64, 66, 83, 107
Xylene Cyanole FF dye, LIX, 67
mobility reference, XLVII, 57
radiolabeling procedure, LIX, 70
D-Xylopyranose oxidase, LVI, 474
Xylose, XXXIV, 318
chromatographic constants, XLI, 16, 17
D-Xylose, oxidation, LVI, 474
D-Xylose isomerase, XLIV, 811
from *Lactobacillus brevis*, XLI, 466–471
assay, XLI, 466, 467
chromatography, XLI, 468
electrophoresis, XLI, 470
inhibitors, XLI, 471
molecular weight, XLI, 470
properties, XLI, 470, 471
purification, XLI, 467–470
Xylosidase, XLVI, 86
Xylosyl adenine, substrate of adenosine deaminase, LI, 507
3-Xylosylglucose, human urine, L, 228
Xylosylserine
human urine, L, 229
structure, L, 229

Xylosyltransferase, XXXIV, 6
D-Xylulose 5-phosphate
preparation and analysis of mixtures with D-ribose 5-phosphate and D-ribulose 5-phosphate, XLI, 37–40
spectrophotometric determination, XLI, 40, 41
D-Xylulose-5-phosphate phosphoketolase, from *Leuconostoc mesenteroides,* purification, XLI, 413, 414
o-Xylyl-α,α'-dithiol, identification of iron-sulfur clusters, LIII, 269, 272, 274

Y

Y, *see* Wybutine
Y1 adrenal cell line, LVIII, 570–574
effect of ACTH, LVIII, 571
initiation of stock, LVIII, 572
karyotype, LVIII, 571
media, LVIII, 571
monolayer growth, LVIII, 572
stability, LVIII, 570
steroidogenesis, LVIII, 571
storage, LVIII, 573
subculture, LVIII, 572, 573
Yamane's medium, LVIII, 57
composition, LVIII, 670
YC, *see* Yellow compound
Y chromosome, LVIII, 174
YDG medium, XXXII, 838, 839
YD medium, XXXII, 838, 839
Yeast, XLVI, 50, *see also Saccharomyces cerevisiae*; specific type
acetyl-CoA:long-chain base acetyltransferase, XXXV, 242
aldehyde dehydrogenase assay, purification, and properties, of baker's, XLI, 354–360
anaerobic growth, XXXI, 629, 630
anomerase activity of glucosephosphate isomerase from baker's, XLI, 57–61
aurovertin-resistant, LVI, 182
autolysis, carboxypeptidase Y activity, XLV, 575
breakage of cells, LVI, 44, 46–48
cell, for oxygen depletion, LVII, 225
commercial source, LIX, 259–260
countercurrent distribution, XXXI, 768
crude extract preparation, LIX, 123, 260
culture conditions, LV, 151, 152, 155–157
cytidine deaminase, LI, 394–401
cytochrome b_2, LIII, 439
CD spectrum, LIV, 282, 283
cytochrome deficient
cultivation for determination of spectra, LVI, 124
fermentor cultures, LVI, 124, 125
germination of spores, LVI, 124
maintenance cultures, LVI, 124
nutritional requirements, LVI, 122
cytochrome oxidase, components, LVI, 41
cytosine deaminase, LI, 394–401
disruption, LV, 135, 140, 141
electron microscopy, LVI, 719–721
fatty acid composition, LVI, 570
of Tween 80 supplemented mutant, LVI, 572, 576
β-D-fructofuranoside fructohydrolase, assay, purification, and properties, XLII, 504–511
glucokinase, assay, purification, and properties, XLII, 25–30
glucose-6-phosphate 1-epimerase from baker's, XLI, 488–493
glyoxylate cycle enzymes, XXXI, 566
growth, LVI, 43
harvest, LV, 415, 416
labeling media, LVI, 59
heme-deficient mutants, investigation of mitochondrial function and biogenesis, LVI, 558–568
hexokinases, assay, purification, and properties, XLII, 6–20
3-hydroxy-3-methylglutaryl-CoA synthase, XXXV, 173–177
inorganic pyrophosphatase, in pyrophosphate determination, LI, 275
invertase, assay, purification, and properties of external and internal, XLII, 504–511
5-keto-D-fructose reductase, assay, purification, and properties, XLI, 132–138

labeling of cells, LVI, 44, 45
lipid mutant, isolation and culture, XXXII, 819–843
materials, petite clone isolation, LVI, 158
mechanical breakage, LVI, 19
mit mutants, further characterization, LVI, 106
mitochondria, XLVI, 84; LV, 149–151; LVI, 3, 4
ATPase isolation by chloroform method, LV, 341, 342
atractyloside-binding protein, LVI, 414–418
complex III
EPR signal properties, LIV, 144
reductive titration, LIV, 124
DNA base composition, LVI, 6
DNA mapping, LVI, 182–185, 194, 195
enzymatic preparation, LV, 157–159
genome map, LVI, 15
information content, LVI, 5, 6
isolation, LV, 26, 27; LVI, 19, 20
large scale preparation, LV, 161, 162
mechanical preparation, LV, 151–157
modification by diet, LVI, 568–577
preparation of ribosomes, LVI, 24
properties, LV, 27, 28
ribosomal RNA, LVI, 7, 8
ribosomes, LVI, 10, 11
size of DNA, LVI, 4, 5
small scale preparation, LV, 162, 163
mitochondria isolation, XXXII, 825–829
mitochondrial marker segregation, LVI, 139, 140
estimating parameters for a random segregation model, LVI, 150, 151
gene conversion, LVI, 152, 153
mating, LVI, 141, 142
mitochondrial mutants, LVI, 140, 141
pedigree studies, LVI, 147–150
random diploid analysis, LVI, 142–144
segregation analysis, LVI, 144–146
segregation problem, LVI, 146, 147
segregation rate, LVI, 153, 154
uniparental inheritance, LVI, 151, 152
zygote clone analysis, LVI, 144
mutant
defective in mitochondrial function, LVI, 95–97
media and reagents, LVI, 97, 98
procedures, LVI, 98–106
growth and media, LVI, 123, 124
stability and storage, LVI, 105, 106
oligomycin-sensitive ATPase, LV, 303, 351
assay method, LV, 351
properties of F_1 ATPase, LV, 353–354
of F_1 subunits, LV, 356
of sensitive ATPase, LV, 356–358
purification of F_1 ATPase, LV, 352, 353
of F_1 subunits, LV, 354, 355
of sensitive ATPase, LV, 356
op mutation, LVI, 126
organelles, XXXI, 510, 517
orotate phosphoribosyltransferase, LI, 70
orotidylate decarboxylase, LI, 73, 75
OSCP, LV, 397
oxidase, LII, 13, 19
petite mutants for DNA gene amplification, background, LVI, 154–156
petite-negative, mitochondrial mutants, XXXII, 838–843
pet9 strains, LVI, 126
growth of fermentor cultures, LVI, 128
mitochondrial properties, LVI, 127, 128
preparation of mitochondria, LVI, 128–130

properties, LVI, 126, 127
rho$^-$ mutations, LVI, 127
phosphofructokinase, assay, purification, and properties, XLII, 78–85
3-phosphoglycerate kinase from baker's, assay, purification, and properties, XLII, 134–138
phosphoglycerate mutase, preparation, purification, and properties, XLII, 435, 438–440, 446
plasmolysis, LI, 292, 396
promitochondria, LV, 28
isolation, XXXI, 628
proteolipid ionophore from mitochondrial ATP synthetase, LV, 414–421
pyruvate kinase, assay, purification, and properties, XLII, 176–182
random segregation model, estimating parameters, LVI, 150, 150
spheroplast and membrane vesicles, XXXI, 609–626
spin labeling studies on, XXXII, 828–837
staining with DAPI, LVI, 728–733
strains
cytochrome *c* oxidase and cytochrome *c*, LVI, 42, 43
variability, LVI, 119
succinate dehydrogenase, LIII, 450
temperature-sensitive mutations, mitochondrial function, LVI, 131, 132
transaldolase from baker's and brewer's, purification, XLII, 291
triosephosphate isomerase, assay, purification, and properties, XLI, 434–438
unsaturated fatty acid auxotroph, characteristics, LVI, 570
uridine nucleosidase, LI, 290–296
in vivo labeling of mitochondrial translation products, LVI, 59, 60
Yellow compound
function, LVII, 284
product, of aequorin reaction, LVII, 284
structural formula, LVII, 285
Yellow jacket, glycerol-3-phosphate dehydrogenase, XLI, 244
Yellow Springs Instruments glucose electrode, LVI, 462–465
YEPD medium, XXXII, 823, 824
YG medium, XXXII, 838, 839
Yield efficiency, XLVII, 305
Ylid mechanism, LIX, 335, 348
Y medium, XXXIX, 224
for granulosa cell culture, XXXIX, 205, 209, 211
Yphantis technique, XLVIII, 171, 173
Yphantis-Waugh partition cell, XLVIII, 245, 261
Yttrium, atomic emission studies, LIV, 455
Yunker, Vaughn and Cory medium, LVIII, 461
YWye, *see* Wybutine

Z

Zamboni's fluid, in endocrine tissue fixation, XXXVII, 136
β-Zeacarotene hydroxylase, LII, 20
Zeaxanthine epoxidase, LII, 20
Zeeman effect, LIV, 285, 286
Zeeman interaction, electronic, LIV, 349, 350
Zeeman term, XLIX, 151, 372
nuclear, LIV, 350, 351
Zeiss microspectrophotometric equipment, XXXIX, 417
Zeiss particle analyzer, XXXII, 13, 19
Zeitgebers, for biological clocks, XXXVI, 476, 481
Zero-field splitting, XLIX, 513, 516–518, 527; LIV, 349
magnetic susceptibility measurements, LIV, 383, 385
ZFS, *see* Zero-field splitting
Ziegler needle knife, XXXIX, 175, 180
Zinc
activated, preparation, XXXV, 513
cleavage, reductive, identification of amino acid bound to flavin ring, LIII, 464
detection by atomic emission spectroscopy, LIV, 458
by atomic fluorescence spectroscopy, LIV, 469

identification of cysteinyl flavins, LIII, 464
in media, LVIII, 52, 69, 89
phospholipase C requirement, XXXII, 160
in superoxide dismutase, LIII, 389
volatilization losses, in trace metal analysis, LIV, 480
Zinc(II), electron paramagnetic resonance studies, XLIX, 514
Zinc-65
in membrane studies, XXXII, 884–893
properties, XXXII, 884
Zinc acetate
in aspartate transcarbamoylase dissociation, LI, 36
determination of acid-labile sulfide, LI, 68; LIII, 276
Zinc amalgam, preparation, LVII, 434
Zinc-barium solution, for cyclic AMP assay, XXXI, 106, 112
Zinc carbonate
cAMP isolation, XXXVIII, 40, 41
cGMP isolation, XXXVIII, 197
Zinc chloride, XLIV, 529, 533
Zinc dust, in AICAR synthesis, LI, 189
Zinc ion
activator of adenine phosphoribosyltranferase, LI, 573
of adenylate deaminase, LI, 495
of deoxycytidine kinase, LI, 345
of thymidine kinase, LI, 370
of uridine-cytidine kinase, LI, 319
cyclase, assays, XXXVIII, 123
effect on bioluminescent spectra of luciferase/luciferin, LVII, 5
inhibitor of adenylosuccinate synthetase, LI, 212
of *Cypridina* luminescence, LVII, 349, 350, 351
of deoxythymidine kinase, LI, 365
of guanine phosphoribosyltransferase, LI, 557
of hypoxanthine phosphoribosyltransferase, LI, 557
of phosphoribosylpyrophosphate synthetase, LI, 11
of uridine nucleosidase, LI, 294
method of plasma membrane isolation, XXXI, 156–162, 163
microsome aggregation, LII, 87
in reconstituted hemoglobin, LII, 491
Zinc octaethylporphyrin, XLIX, 159
Zincon, LV, 517
Zinc protein, cobalt-substituted, electron paramagnetic resonance studies, XLIX, 514
Zinc-substituted cytochrome *c*
preparation, LIII, 177
properties, LIII, 178–181
Zinc sulfate
in formaldehyde assay, LII, 299
purification of cyclic AMP, LVII, 98
Zinc sulfate-barium chloride, cAMP isolation, XXXVIII, 40
Zinc sulfate-barium hydroxide, cAMP isolation, XXXVIII, 39–41
Zinc sulfate-calcium chloride, cAMP isolation, XXXVIII, 40
Zinc sulfate-sodium carbonate, cAMP isolation, XXXVIII, 40, 41
Zirconium, atomic emission studies, LIV, 455
Zirconium oxide
durability, XLIV, 135
pore properties, XLIV, 137
Zirconium phosphate, XLIV, 216
Zirconium-titanium-nickel alloy, for oxygen removal, LIV, 117
Zizanin, XLIII, 342
Zona fasciculata-reticularis, steroid production, XXXIX, 314
Zonal centrifugation
assay of free ribosomes, LIX, 367
definition, XXXI, 715
duration of run, XXXI, 738, 739
of *E. coli* cell extracts, XXXI, 640, 641
in glyoxysome isolation, XXXI, 571
of D-β-hydroxybutyrate apodehydrogenase, XXXII, 382
isolation of chloroplasts, LIX, 209–210
of mitochondria, LIX, 206
of nuclei, XXXI, 259
of liver fractions, XXXI, 729, 730
of $(Na^+ + K^+)$-ATPase, XXXII, 283, 284, 289

preparation of ribosomal subunits, LIX, 404–405, 408, 647–648, 654, 752
procedural manual, LIX, 405
ribosomal release assay, LIX, 861
rotors, XXXI, 717–719
separation of ribosomal fragments, LIX, 467–468, 474
study of tRNA-protein interaction, LIX, 329–330, 331
Zonal rotor
for density gradient centrifugation, XXXI, 505, 506
types, XXXI, 717–719
Zone electrophoresis, aspartate transcarbamoylase subunit dissociation studies, LI, 36–37
Zone gel filtration, LIX, 328–329, 331
Zonula adhaerans, of plasma membranes, XXXI, 85
Zorbax-octadecylsilane column, XLVII, 46, 47
Zoxazolamine, paralysis test, *Ah* b allele, LII, 232, 233
Zoxazolamine 6-hydroxylase, *Ah* b allele, LII, 231
Zygote
bud position, genetic purity, LVI, 147, 148
clone analysis, LVI, 144
cytoplasmic mixing, LVI, 148, 150
pedigree experiments, LVI, 145, 146
yield, LVI, 141
Zymase, complex, yeast, immobilization, by microencapsulation, XLIV, 214
Zymogen, XXXIV, 593
blood coagulation factors, XLV, 34
factor X, XLV, 96
granule, of pancreas, XXXI, 41
chemistry, XXXI, 47, 51–53
electron microscopy, XXXI, 5
isolation, XXXI, 3–46, 49, 50
for phospholipase A_2, XXXII, 147, 150, 151
large-scale activation, XXXII, 153, 154
of plasmin, XXXIV, 424
synthesis, XLV, 35
of tadpole collagenase, XL, 351
Zymosan
opsinization, LVII, 486
phagocyte chemiluminescence, LVII, 471, 476–478, 480
suspension, in assay of leukocyte oxygen uptake, LIV, 494
Zymosterol, heme-deficient mutants, LVI, 558, 559